ABR9559

WITHDRAWN

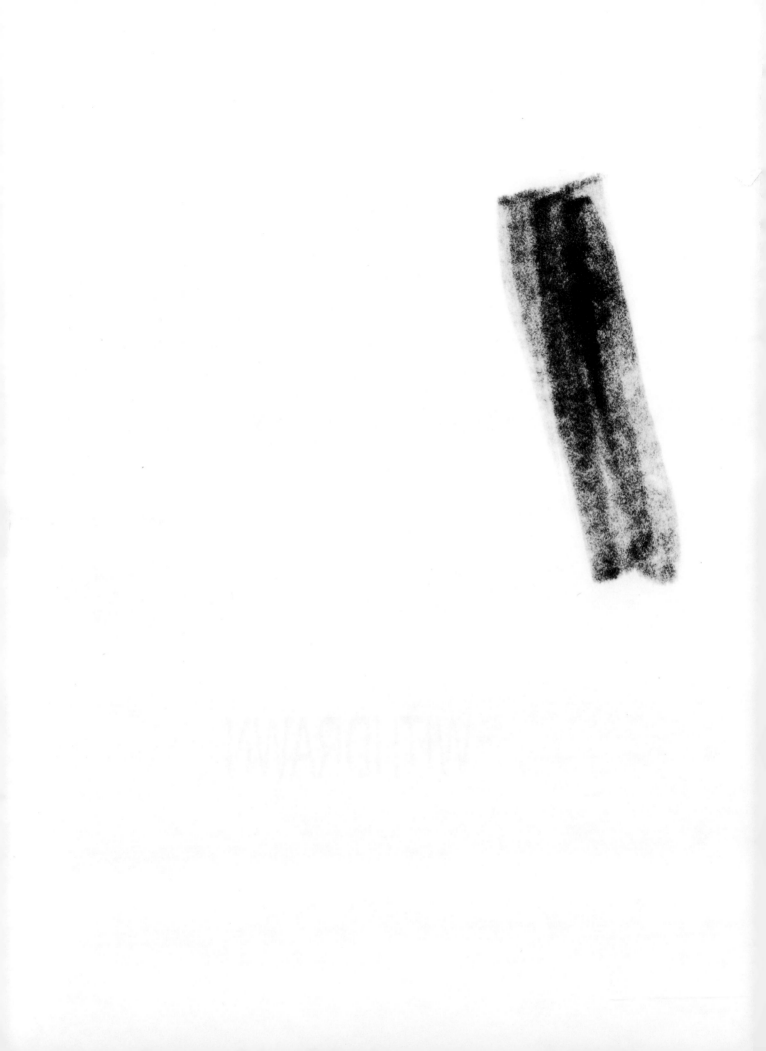

Biostratigraphy of Jamaica

Frontispiece. In commenting on this illustration by Henry De la Beche of a large gastropod, Peter Jung (1987, p. 90) remarked:

> Exactly 160 years ago giant gastropods obviously belonging to some species of *Campanile* have been recorded for the first time from the Caribbean Eocene. De la Beche (1827, pp. 169–171, Pl. 21) had collected them from limestones in the Chapelton area of Jamaica. This finding led De la Beche to believe that the Jamaican limestones were of the same age as the Calcaire Grossier of the Paris Basin.
>
> Jung, P., 1987, Giant gastropods of the genus *Campanile* from the Caribbean Eocene: Ecologae geologicae Helvetiae, v. 80, p. 889–896.

As far as we know, this was the first correlation and age determination suggested for Jamaican rocks, based on fossil evidence.

Photograph provided by Eric Robinson, President, Geologists' Association. For the use of the figure on the frontispiece from De la Beche's 1827 paper, permission was kindly provided by the Geological Society of London.

Geological Society of America
Memoir 182

Biostratigraphy of Jamaica

Edited by

Raymond M. Wright
Petroleum Corporation of Jamaica
36 Trafalgar Road
Kingston 10, Jamaica, West Indies

and

Edward Robinson
Department of Geology
University of the West Indies
Mona, Kingston 7, Jamaica, West Indies

1993

Copyright © 1993, The Geological Society of America, Inc. (GSA). All rights reserved. GSA grants permission to individual scientists to make unlimited photocopies of one or more items from this volume for noncommercial purposes advancing science or education, including classroom use. Permission is granted to individuals to make photocopies of any item in this volume for other noncommercial, nonprofit purposes provided that the appropriate fee ($0.25 per page) is paid directly to the Copyright Clearance Center, 27 Congress Street, Salem, Massachusetts 01970, phone (508) 744-3350 (include title and ISBN when paying). Written permission is required from GSA for all other forms of capture or reproduction of any item in the volume including, but not limited to, all types of electronic or digital scanning or other digital or manual transformation of articles or any portion thereof, such as abstracts, into computer-readable and/or transmittable form for personal or corporate use, either noncommercial or commercial, for-profit or otherwise. Send permission requests to GSA Copyrights.

Copyright is not claimed on any material prepared wholly by government employees within the scope of their employment.

Published by The Geological Society of America, Inc.
3300 Penrose Place, P.O. Box 9140, Boulder, Colorado 80301

Printed in U.S.A.

GSA Books Science Editor Richard A. Hoppin

Library of Congress Cataloging-in-Publication Data

Biostratigraphy of Jamaica / edited by Raymond M. Wright and Edward Robinson.
    p.   cm. — (Memoir / Geological Society of America ; 182)
    Includes bibliographical references and index.
    ISBN 0-8137-1182-7
    1. Animals, Fossil—Jamaica. 2. Paleontology, Stratigraphic.
3. Paleontology—Jamaica. I. Wright, Raymond M., 1940–
II. Robinson, Edward, 1934–   . III. Series: Memoir (Geological Society of America) ; 182.
QE750.J25B56   1993
560.9'7292—dc20   93-29179
                                                  CIP

10 9 8 7 6 5 4 3 2 1

Dedication

Norman Frederick Sohl
(1924–1993)

This memoir is dedicated to the memory of Norman F. Sohl, leading authority on Cretaceous biostratigraphy, and gastropods in particular, who died on April 14, 1993.

The foremost expert on the ecology and evolutionary history of Cretaceous molluscs in the Caribbean, he produced important studies on the Upper Cretaceous molluscs in Jamaica. He would have written a foreword to this volume were it not for his death.

Contents

Preface ix

Acknowledgments xi

1. *Late Cretaceous Calcareous Nannoplankton Zonation of Jamaica* 1
   J. G. Verdenius

2. *Campanian Calcareous Nannofossils in the Sunderland Inlier,
   Western Jamaica* 19
   M. M. Jiang

3. *Review of Upper Cretaceous Orbitoidal Larger Foraminifera from
   Jamaica, West Indies, and Their Connection with Rudist Assemblages* 29
   J. P. Krijnen, H. J. Mac Gillavry, and H. van Dommelen

4. *Ostracode Biostratigraphy of the* Titanosarcolites-*Bearing Limestones
   and Related Sequences of Jamaica* 65
   J. E. Hazel and T. Kamiya

5. *Upper Cretaceous Ammonites from Jamaica and Their Stratigraphic
   and Paleogeographic Implications* 77
   J. Wiedmann and W. Schmidt

6. *Jamaican Cretaceous Echinoidea* 93
   S. K. Donovan

7. *Cretaceous and Cenozoic Brachiopoda of Jamaica* 105
   D.A.T. Harper

8. *The Fossil Arthropods of Jamaica* 115
   S. F. Morris

9. *Crinoids, Asteroids, and Ophiuroids in the Jamaican Fossil Record* 125
   S. K. Donovan, C. M. Gordon, C. J. Veltkamp, and A. D. Scott

10. *Calcareous Nannofossil Stratigraphy of the Neogene Formations
    of Eastern Jamaica* 131
    M.-P. Aubry

11. **Neogene Planktonic Foraminiferal Biostratigraphy of Eastern Jamaica** 179
    W. A. Berggren

12. **Miocene-Pliocene Bathyal Benthic Foraminifera and the Uplift
    of Buff Bay, Jamaica** 219
    M. E. Katz and K. G. Miller

13. **Taxonomy, Biostratigraphy, and Paleoecologic Significance of Calcareous-
    Siliceous Facies of the Neogene Montpelier Formation,
    Northeastern Jamaica** 255
    F. J-M. R. Maurrasse

14. **Jamaican Paleogene Larger Foraminifera** 283
    E. Robinson and R. M. Wright

15. **Tertiary Cephalopods from Jamaica** 347
    W. Schmidt and P. Jung

16. **The Fossil Record of Terrestrial Mollusks in Jamaica** 353
    G. A. Goodfriend

17. **The Freshwater Mollusca of Jamaica** 363
    C.R.C. Paul, P. Hales, R. A. Perrott, and F. A. Street-Perrott

18. **Jamaican Cenozoic Echinoidea** 371
    S. K. Donovan

19. **Jamaican Tertiary Marine Vertebrata** 413
    D. P. Domning and J. M. Clark

20. **Quaternary Land Vertebrates of Jamaica** 417
    G. S. Morgan

21. **Contribution Toward a Tertiary Palynostratigraphy for Jamaica: The Status
    of Tertiary Paleobotanical Studies in Northern Latin America and Preliminary
    Analysis of the Guys Hill Member (Chapleton Formation, Middle Eocene)
    of Jamaica** 443
    A. Graham

Index 463

Preface

Toward the end of the 1970s and lasting until the mid-1980s the Petroleum Corporation of Jamaica conducted a major oil and gas exploration program onshore and offshore Jamaica. Although no commercial quantities of oil or gas were found in this phase of exploration activity, a significant body of information on the geology of Jamaica in general, and biostratigraphy in particular, was generated. A major purpose of this volume is to present the biostratigraphic results of the work of that period, as well as important studies conducted independently of oil and gas matters by many workers. The literature on Jamaican biostratigraphy, beginning with such early workers as Henry Thomas De la Beche, James Sawkins, and Robert T. Hill, is mature, but diffused in a large variety of publications that are often difficult to access, particularly for students and nonspecialists in biostratigraphy. Jamaica's long history of geological research began with Sir Henry De la Beche, who produced the initial geological map of eastern Jamaica in 1827; this was probably the first geological map produced in the Western Hemisphere. De la Beche was later instrumental in founding the Geological Survey of Great Britain, the world's first organized geological survey.

The composite mixture of papers in this volume, covering varying fossil groups, will enable the student, the nonspecialist, as well as the specialist, to find a significant body of information in a single source. Further, the volume accentuates the description and photographic illustration of the fossil groups, so that the more common and index species from the Jamaican fossil record can be identified readily by reference to the pertinent papers. We have tried to have the contents of each chapter presented in a lucid, readable style, emphasizing clarity without eschewing scholarship and authenticity.

In allocating papers for this volume, the editors have been guided by the stratigraphic usefulness of some groups of fossils and the paucity of previously published information on other groups. We have attempted to include as complete a spectrum as possible of fossil types known in Jamaica. Unfortunately, there are some prominent omissions, such as the corals, the inoceramids, rudists and other Cretaceous molluscs, as well as the bryozoa and algae. Let it be noted that this was not our original intention. Invitations were sent to workers competent to address these missing fossil groups as a result of their research on Jamaica, but geologists are busy people; they are always surveying, teaching, researching, administering, and writing. Some, alas, in high hopes, did undertake to try and fill some of the gaps, but because of overcommitment were unable to complete the work. It is our hope that at least some of this work will appear in a future publication.

In addition to the standard microfossil groups important to biostratigraphy, such as foraminifers, calcareous nannofossils, and palynomorphs, this volume brings together the current state of knowledge of previously less studied groups in the Jamaican fossil record such as the echinoids, crinoids, ammonites, and vertebrates. Some of these papers might not have been written but for the requirement of this memoir.

The near-compendium of papers on Jamaican biostratigraphy is truly an international effort. The 21 papers are written or cowritten by 34 persons from 12 countries, most of whom have spent long periods of time doing fieldwork in Jamaica. This shows clearly the interest that Jamaican biostratigraphy has generated in researchers from around the world. Indeed, Jamaica is probably the Caribbean country most studied with respect to paleontology.

Raymond M. Wright
Edward Robinson

Acknowledgments

The will to prepare this book and the concepts behind its final execution were sprung in the first quarter of 1988. With a span of nearly six years from idea to production, we have accumulated many debts, and we wish to acknowledge persons who have helped in several ways over these years.

First and foremost, our gratitude goes to Faydene Gillings-Grant who brought great commitment to the project in corresponding with authors, caring for manuscripts, and in willingly taking on extra responsibilities to ensure that the book would be completed. Her work went far beyond what might have been reasonably expected, and she was always forgiving. We offer special thanks to Sheila Budram and Doreen Morgan for secretarial support and assistance with correspondence and communications.

For his wise advice and careful readings, we are grateful to Stephen Donovan, who began this project as a coeditor, but found it difficult to continue due to work pressure. Richard Hoppin, GSA books science editor was extremely generous with his insights, experience, and time.

In addition, we would like to express appreciation to the following people for their reviews, readings, comments, kind help, and dialogues: Elsie Aarons, John Aarons, Geoffrey Adams, Locksley Allen, Gale Bishop, Daniel Blake, Hans Bolli, D. Brunton, David Bukry, Jackie Burnett, Burchard Carter, William Cobban, J. Collins, Geoff Eaton, Rodney Feldmann, Norman Frederiksen, Stefan Gartner, Robert Goll, Glenn Goodfriend, David Haig, Rex Harland, Leo Hickey, David Jarzen, Peter Jung, Wann Langston, Jr., Daphne Lee, David Lewis, A. Logan, Rosalie Maddocks, Michael McKinney, Peter McLaughlin, Jr., David Meyer, Robert Olsson, Hugh Owen, Christopher Paul, Katharina Perch-Nielsen, Clayton Ray, William Riedel, Gary Rosenberg, Annika Sanfilippo, John Saunders, Robert Savage, Ortwin Schultz, Andrew Smith, Norman Sohl, Ellen Thomas, Leonard Tjalsma, Jan van Hinte, Joost Verbeek, Edith Vincent, David Webb, Charles Woods, John Wrenn, C. Wright, and Keith Young.

We are sure that we have omitted some persons whom we should acknowledge, and we ask their forebearance and express our appreciation. As we look back over the names herein, we can see how many so graciously helped and how large are the obligations we have acquired. This publication would not have been possible without the contributions of all the above, together with the authors who worked hard and cooperated with enthusiasm.

The Petroleum Corporation of Jamaica gave ample financial and administrative support, directly and indirectly, which covered the costs involved in the coordination and initial preparation of this work.

Late Cretaceous calcareous nannoplankton zonation of Jamaica

Jacob G. Verdenius
IKU Petroleum Research, N-7034 Trondheim, Norway

ABSTRACT

Six biozones can be recognized in the Turonian to the Lower Maastrichtian of Jamaica. In descending stratigraphic order, the boundaries separating these six zones are coincident with the last occurrence of *Calculites obscurus, Quadrum trifidum, Eiffellithus eximius, Eiffellithus gorkae, Lithastrinus septenarius,* and *Lithastrinus moratus,* respectively. This scheme provides a consistent zonation that has been recognized in three wells and one surface section. The ages of the last occurrences of the zonal markers, as well as of the last and first occurrences of other relevant taxa, have been derived from a comparison of several syntheses of Cretaceous nannofossil biostratigraphy. Additional Jamaican sections and spot samples bearing Late Cretaceous nannoplankton are correlated to this regional zonation. Changes in nannofossil abundance in the four major stratigraphic sections indicate environmental shifts and intervals of nondeposition on Jamaica during the Late Cretaceous. One *Eiffellithus* species that apparently has not been published previously is described in open nomenclature.

INTRODUCTION

The Cretaceous rocks of Jamaica, which are partly igneous and partly of sedimentary and organic-carbonate origin, are exposed in restricted areas, commonly at the bottom of topographic depressions that are surrounded by Tertiary deposits (Zans et al., 1962). These "Cretaceous inliers" (McFarlane, 1977) are in fact erosional windows through a several-layered shield of Early Tertiary carbonate rocks. There is considerable lithostratigraphic variation in the Cretaceous from one inlier to the next, and our knowledge of the Cretaceous subcrop below the Early Tertiary shield is restricted. It is generally assumed, however, that Cretaceous rocks occur below the exposed Lower Tertiary limestones.

The geological study of Jamaica has a long history. Fossils have been used extensively in the stratigraphic breakdown of the sediments of the island and in the recognition of map units. Much of this work has been done with the help of endemic fossils, e.g., rudist molluscs and foraminifera. Stratigraphy and zonations utilizing different fossil groups will facilitate the global correlation of the sediments of Jamaica. This justifies the effort to apply a zonation to Late Cretaceous of Jamaica on the basis of calcareous nannoplankton.

The nannoplankton zonation represented herein was developed by the following steps:

1. Recording of calcareous nannofossils in well samples, surface sections, and isolated field exposures. Raw data and biostratigraphic results of these studies were presented to the Petroleum Corporation of Jamaica as confidential reports from 1984 to 1991. For the present chapter, these results have been edited to a uniform standard.

2. Compilation of a standard calcareous nannofossil range chart from literature.

3. Choice of nannofossil index taxa and zones for the Late Cretaceous of Jamaica.

4. Regional review of all available Late Cretaceous nannofossil evidence from Jamaica.

BIOSTRATIGRAPHIC FRAMEWORK

Range chart

A nannofossil range chart covering the Late Cretaceous, with which to correlate our biostratigraphic results from Jamaica, has been compiled from Sissingh (1977), Verbeek (1977), Doeven (1983), and Perch-Nielsen (1985). A numerical age scale was used for this range chart, and the chronostratigraphy has been plotted according to the scale in Haq et al. (1987). Instead of the nannofossil zonation of Roth (1978) that is referred to by Haq et al. (1987), I prefer to use the nannofossil biozonation of Sissingh (1977) because most literature data, as well as our previous biostratigraphic results, are correlated to Sissingh's CC zonation. A more recent discussion of this zonation can be found in Perch-Nielsen (1985).

Verdenius, J. G., 1993, Late Cretaceous calcareous nannoplankton zonation of Jamaica, *in* Wright, R. M., and Robinson, E., eds., Biostratigraphy of Jamaica: Boulder, Colorado, Geological Society of America Memoir 182.

In the correlation of the zonations erected by the abovementioned authors, as well as in the correlation of zones to stages, I have followed the synthesis of Perch-Nielsen (1985). The numerical ages for CC zones that coincide to a chronostratigraphic boundary can be found directly in the Haq et al. (1987) scale. The remaining zonal boundaries have been interpolated at their stratigraphic position between the nearest numerical ages of the Haq et al. (1987) scale. Our numerical ages for the CC zonation are therefore of variable reliability and must be regarded as tentative. The zonal boundaries with their numerical ages and stratigraphic designation have been listed in Table 1.

Calcareous nannoplankton zonation of Jamaica

After the initial study of the available well and surface samples, it became apparent that a regional Late Cretaceous nannofossil zonation for Jamaica was required in order to ensure the best results from the available material. This proposed zonation is tentative, because the available data base has several obvious defects. The number of stratigraphic sections is low (three wells and one surface section), and in the three exploration wells—Hertford 1, Windsor 1, and Retrieve 1—that were made available by the Petroleum Corporation of Jamaica the samples are irregularly spaced with extensive unsampled intervals. With few exceptions, the material consists of cuttings samples. Nannofossil recovery is quite variable, and caving probably has affected the deepest occurrence of several taxa. In the surface material from previous surveys, a few samples with an adequate nannofossil assemblage for zonation were identified, but most samples were barren or yielded only a poor nannofossil assemblage of low diversity. It was assumed that the poor sample quality was caused by near-surface diagenesis and leaching, and when the surface section in the St. Ann's Great River was sampled in 1989 for the Petroleum Corporation of Jamaica, an effort was made to obtain unweathered samples. In comparison to other surface material, these samples from the St. Ann's Great River section are markedly richer and more diverse, but even the best of them are only of moderate quality and not suitable as a biostratigraphic yardstick for the Late Cretaceous of Jamaica.

The biostratigraphy of the three exploration wells is the main source of information for our nannofossil zonation of the Late Cretaceous of Jamaica. For this reason, a zonation was developed on last stratigraphical occurrences in the three wells. In the fossiliferous well samples, the nannofossil diversity and abundance are markedly better than in most of the field sections studied. Therefore I assume that the present zonation, with its admittedly low stratigraphic resolution, is the best that can be obtained and is of sufficient detail to correlate sections and samples on Jamaica that contain nannofossils. Single samples of exceptional quality may obtain a more accurate time label via a correlation to the global standard nannofossil zonation. However, on the assumption that the studied material is representative, it seems unrealistic to expect a consistently higher level of accuracy than can be obtained with the proposed zonation of this study.

CALCAREOUS NANNOFOSSIL ZONES

All biostratigraphically useful calcareous nannofossils that occur in the Late Cretaceous of Jamaica have been recorded in Figure 1. Their ranges have been derived from Verbeek (1977), Doeven (1983), and Perch-Nielsen (1985). The stratigraphic information from these sources can be correlated to the CC zonation of Sissingh (1977). Based on the stratigraphically useful calcareous nannofossil taxa that are reasonably common in the four reference sections, six biozones have been erected for the Late Cretaceous. The zones are discussed in stratigraphically descending order in the following paragraphs.

Calculites obscurus Partial Range Zone

Definition. From the last occurrence of *Calculites obscurus* to the last occurrence of *Quadrum trifidum*.
Age. Early Maastrichtian.
Possible correlation. CC 24; NC 21 (part); middle part of the Arkhangelskiella cymbiformis Zone in Doeven (1983).
Occurrence. Retrieve 1 well.
Remarks. The Calculites obscurus Zone has been recognized in the Retrieve 1 well only (refer to Fig. 4), and its correlation potential is unknown. Doeven (1983) recorded a simultaneous last occurrence of all *Calculites* species within his

TABLE 1. AGES OF LATE CRETACEOUS EVENTS

Level	Stratigraphic Designation	Approximate Numerical Age (Ma)
Top CC 26	Tertiary-Cretaceous boundary	66.5
Base CC 26	Identical to base NC 23	67.3
Base CC 25	Late-Early Maastrichtian boundary	71.0
Base CC 24		73.0
Base CC 23	Zone CC 23 spans the Maastrichtian-Campanian boundary (74 Ma)	74.5
Base CC 22	Identical to base NC 20	75.5
Base CC 21	Late-Early Campanian boundary	78.3
Base CC 20	Identical to base NC 19	80.0
Base CC 19		81.5
Base CC 18	Near the Campanian-Santonian boundary	83.5
Base CC 17	Zone CC 17 spans the Campanian-Santonian boundary (84 Ma)	84.5
Base CC 16		86.0
Base CC 15	Identical to base NC 17, which is slightly over Santonian-Conician boundary at 88 Ma	87.5
Base CC 14	Slightly below the Santonian-Coniacian boundary	88.2
Base CC 13	Identical to base NC 15, slightly above the base of the Conician	88.7
Base CC 12	Identical to base NC 14	90.5
Base CC 11	Turonian-Cenomanian boundary	92.0
Base CC 10	Identical to base NC 10	98.5

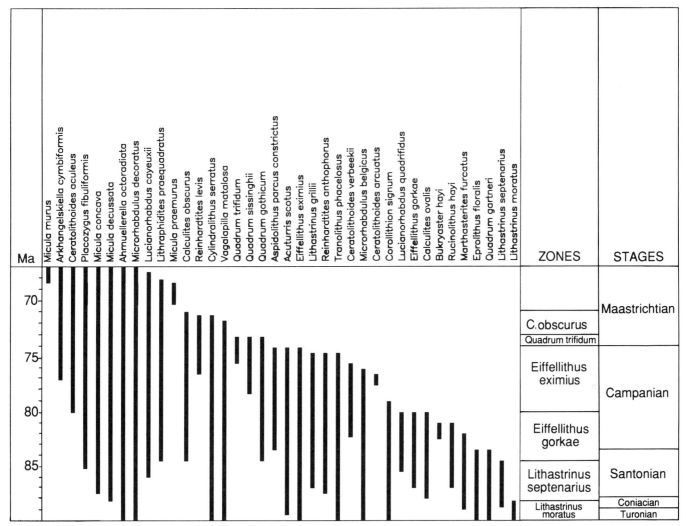

Figure 1. Nannofossil range chart and Jamaican nannofossil zones in the Late Cretaceous (latest Turonian to Maastrichtian).

Arkhangelskiella cymbiformis Zone. According to Perch-Nielsen (1985), *Calculites obscurus* is common up to the CC 24/25 boundary and rare in CC 25. It is possible that the top of my Calculites obscurus Zone, i.e., the last occurrence of the index taxon, is stratigraphically higher than the end of CC 24.

Quadrum trifidum Partial Range Zone

Definition. From the last occurrence of *Quadrum trifidum* to the last occurrence of *Eiffellithus eximius*.
Age. Earliest Maastrichtian.
Probable correlation. CC 23; NC 20-21 in part; later part of Quadrum trifidum Zone in Doeven (1983); within the earlier part of the Quadrum trifidum Zone of Verbeek (1977).
Occurrence. Hertford 1 and Retrieve 1 wells.
Remarks. Quadrum trifidum occurs intermittently in the above-mentioned wells and is a common taxon in many field samples with an otherwise low-diversity and low-abundance nannoplankton assemblage. It is also a common reworked element in Lower Tertiary sediments of the Rio Grande area, eastern Jamaica.

In the Hertford 1 well (refer to Fig. 3), *Quadrum trifidum* occurs in the highest sample with a Cretaceous nannoflora. In Retrieve 1 (refer to Fig. 4), the top of *Quadrum trifidum* is in the highest sample with a rich calcareous nannoflora. Consequently it is possible that the highest occurrence of *Quadrum trifidum* in this material is environmentally controlled.

Eiffellithus eximius Partial Range Zone

Definition. From the last occurrence of *Eiffellithus eximius* to the last occurrence of *Eiffellithus gorkae*.
Age. Late Campanian.
Possible correlation. CC 23 (part), CC 22-20; later part of the Quadrum trifidum Zone to Cerotolithoides aculeus Zone

of Verbeek (1977); Quadrum trifidum Zone to Ceratolithoides aculeus Zone of Doeven (1983).

Occurrence. Hertford 1, Windsor 1, and Retrieve 1 wells.

Remarks. Eiffellithus eximius occurs in all three wells and appears to be a useful regional marker. According to Verbeek (1977), the last occurrence of this taxon is near the Campanian-Maastrichtian boundary. The last occurrences of *Reinhardtites anthophorus* and the *Aspidolithus parcus* group at the top of this zone also indicate an approximation to the Maastrichtian-Campanian boundary, and I assume that, with the resolution of the present study, the two events coincide. However, Doeven (1983) recorded a long interval with intermittent occurrences of *Eiffellithus eximius* in the Late Campanian and the Maastrichtian.

Within the Eiffellithus eximius Zone, the earliest occurrence of *Quadrum trifidum* and the total range of *Ceratolithoides arcuatus* may allow for a further subdivision, but since these index forms are rare in the four stratigraphic sections, the regional stratigraphic resolution will not be improved by defining zones or subzones on the base of their presence.

Eiffellithus gorkae Partial Range Zone

Definition. From the last occurrence of *Eiffellithus gorkae* to the last occurrence of *Lithastrinus septenarius*.

Age. Latest Santonian–Early Campanian.

Possible correlation. CC 19-17; NC 18; Broinsonia parca Zone and later part of Zygodiscus spiralis Zone in Verbeek (1977) (*Zygodiscus spiralis* = *Placozygus fibuliformis*); Broinsonia parca Zone and later part of Rucinolithus hayi Zone in Doeven (1983).

Occurrence. Hertford 1, Windsor 1, and Retrieve 1 wells.

Remarks. Perch-Nielsen (1985) recorded *Eiffellithus gorkae* from two intervals: a late occurrence in the Maastrichtian and an earlier tentative occurrence in the Early Campanian and Santonian. Our zonal marker is considered to coincide to the earlier occurrence of this taxon, and its range has been drawn accordingly. We did not find *Eiffellithus gorkae* in the Maastrichtian Calculites obscurus or Quadrum trifidum Zones.

In Hertford 1 (refer to Fig. 3), the stratigraphically highest occurrence of *Eiffellithus gorkae* lies in a relatively well sampled interval with comparatively abundant and diverse nannofossil assemblages. This event coincides with the stratigraphically highest occurrence of *Bukryaster hayi*, which supports an Early Campanian age. In the Windsor 1 well, *Marthasterites furcatus* is consistently present from within the Eiffellithus gorkae Zone downward. However, the occurrences of these two taxa are too rare in our material for useful correlation.

Lithastrinus septenarius Partial Range Zone

Definition. From the last occurrence of *Lithastrinus septenarius* to the last occurrence of *Lithastrinus moratus*.

Age. Latest Coniacian–Santonian.

Possible correlation. CC 16-14; earlier part of the Zygodiscus spiralis Zone and the entire interval incorporating the Rucinolithus hayi to Broinsonia lacunosa zones of Verbeek (1977); earlier part of the Rucinolithus hayi Zone and later part of the Marthasterites furcatus Zone of Doeven (1983).

Occurrence. Windsor 1 and Retrieve 1 wells.

Remarks. Lithastrinus septenarius is present in the Windsor 1 (refer to Fig. 5) and Retrieve 1 (refer to Fig. 4) wells as part of nannoplankton assemblages without other age-significant taxa that are markedly less rich in taxa end specimens than the assemblages immediately above and below. The last occurrence of *Eprolithus floralis* coincides with the last occurrence of *Lithastrinus septenarius,* and the last occurrence of *Quadrum gartneri* slightly precedes this level (Fig. 1). These events can be used as secondary markers for the top of this zone. It is assumed that stratigraphically higher occurrences of these secondary markers in the Retrieve 1 well (refer to Fig. 4) and the St. Ann's Great River section (refer to Fig. 6) are due to reworking. In the Hertford 1 well (refer to Fig. 3), where *Lithastrinus septenarius* was not recorded, the top of the zone has been drawn at the stratigraphically highest occurrence of *Quadrum gartneri.*

Lithastrinus moratus Partial Range Zone

Definition. From the last occurrence of *Lithastrinus moratus*. As the nannofossil assemblages deteriorate downward in this zone, its lower boundary could not be defined with a biostratigraphical marker.

Age. Coniacian (and earlier).

Possible correlation. CC 13 and earlier zones; Marthasterites furcatus and older zones of Verbeek (1977); earlier part of the Marthasterites furcatus Zone and older zones of Doeven (1983).

Occurrence. Hertford 1, Windsor 1, and Retrieve 1 wells.

Remarks. In the Hertford 1 (refer to Fig. 3) and Windsor 1 (refer to Fig. 5) wells, the highest stratigraphical occurrence of *Lithastrinus moratus* is immediately below the lowest stratigraphical occurrence of *Micula decussata*, which confirms the position of the Lithastrinus moratus Zone immediately before Zone CC 14. In Hertford 1 (refer to Fig. 3), *Marthasterites furcatus* and *Liliasterites angularis* are both present in the deepest productive sample, which supports a stratigraphic assignment very near to the Coniacian-Turonian boundary.

Evidence of older nannoplankton assemblages. A number of older nannofossils occur at several levels in the localities examined here. *Corollithion kennedyi* (refer to Fig. 6) is restricted to the Cenomanian, whereas *Nannoconus donnatensis* and *Nannoconus fragilis* (refer to Fig. 6) are known from the Albian only. *Conusphaera mexicana* (refer to Figs. 3 and 4) and several *Nannoconus* taxa (refer to Figs. 3 and 4) are known to occur in Albian and older strata, and *Nannoconus quadriangulus* (refer to Fig. 6) is restricted to the Late Aptian. To our knowledge, no open marine sediments of these ages are exposed on Jamaica. Early Cretaceous sediments with abundant *Nannoconus* have been described from Cuba (Brönnimann, 1955). Erosion of similar deposits somewhere in the Caribbean area during the Late Cretaceous may have reworked Early Cretaceous nannofloras into the Late Cretaceous sediments of Jamaica.

STRATIGRAPHY AND CORRELATION

In addition to the previously mentioned three exploration wells, the proposed nannoplankton biostratigraphy of the Late Cretaceous deposits of Jamaica was applied to one corehole well, one extensive field section, and a number of small field sections and spot samples. These stratigraphic results are displayed on Figure 2 from west (left) to east (right). In addition, the results of a previous study by Jiang and Robinson (1987) are included in this figure. The formation names on Figure 2 were derived from the Ministry of Mining geological map of Jamaica 1:250,000 (McFarland, 1977) and from Jiang and Robinson (1987). For the sake of convenience, geographic areas or inliers are represented by one column that summarizes the data from several separate outcrops. Where the age assignment is of apparently good quality, sediments are indicated by a full column. Sedimentary breaks are represented by an interrupted column. Where the data are of little diagnostic value, a narrowly hachured column is drawn over the interval.

Many of the field samples in this study were taken without special consideration of their possible nannofossil richness. It appears that if these samples are not barren, their nannofossil assemblage most often is restricted to one to three taxa that are reputed to be resistant to diagenesis, and that most specimens are corroded. In the St. Ann's Great River section, however, special care was taken to sample clayey and marly sediments below the water table. In most fossiliferous samples from this section, fragile taxa occur commonly, and the number of taxa per sample is consistently much higher than in the other field samples, where no special attention was given to lithology and exposure. Therefore it is assumed that the poor nannofossil recovery from these field samples it due to leaching or other processes at the surface. The nannofossil diversity of the St. Ann's Great River section and the three exploration wells is probably a primary feature that can be used for a paleo-environment interpretation. I consider the relatively high nannofossil diversities of 15 or more taxa to represent an open marine depositional environment; less diverse associations may represent restricted marine conditions, and assemblages of few commonly occurring taxa may be evidence of marginal marine conditions. The abrupt upper contacts of some of the high-diversity associations are interpreted as evidence of a sea-level fall, with erosion and/or a period of nondeposition. In Figure 2, the above-mentioned criteria have been applied to the St. Ann's Great River section and the three exploration wells. The paleo-environment is recorded to the left of the chronostratigraphic column by a full line where precise ages are available and by a broken line where the biostratigraphy is less accurate. Truncation is marked by a horizontal wavy line. In the following paragraphs the age and depositional environments of the localities are discussed, and a regional synthesis of these results is attempted at the end of this chapter.

Jerusalem Mountain

The Jerusalem Mountain Formation consists of hard limestone that is not favorable for nannofossil study. Most samples from this area are barren or contain only sparse nannofossils. Corrosion and a marked predominance of robust species is evident, which suggests postdepositional dissolution as an additional cause of poor nannofossil recovery. The samples examined in the Jerusalem Mountain area range over the Eiffellithus eximius to the Quadrum trifidum zones. *Quadrum trifidum* is present in several of the better-preserved samples and restricts their age to latest Campanian and earliest Maastrichtian.

Jiang and Robinson (1987) reached essentially similar conclusions from their study of this same area. They attributed a probable latest Campanian age to samples just below the Jerusalem Limestone Member.

Lucea Inlier

Exposures are rare in this area and their relative stratigraphic relationship is often uncertain. Two samples from the Hanover Formation at Rock Spring can be attributed to the later part of the Eiffellithus eximius Zone.

Jiang and Robinson (1987) attributed a Santonian age to samples from the Jericho, Harvey River, and Mount Peace formations in the Lucea Inlier. These authors regarded the absence of *Broinsonia parca* as biostratigraphically significant. In the zonation presented herein, these formations are attributed to the Lithastrinus septenarius Zone and the earliest part of the Eiffellithus gorkae Zone, because of the co-occurrence of *Rucinolithus* sp. (probably *Rucinolithus hayi*) and *Lithastrinus grillii,* in the absence of *Broinsonia parca.* Other nannofossil assemblages from the Lucea Inlier probably can be attributed to the CC 18 to CC 22 zonal interval, according to Jiang and Robinson (1987). From these results, it can be assumed that the composite sedimentary section of this area reaches from the Late Santonian to the Late Campanian.

Hertford 1

Sampling of this exploration well was intermittent, and there are considerable unsampled intervals. The available nannofossil-productive samples (Fig. 3) range from the Lithastrinus moratus Zone to the Quadrum trifidum Zone.

A rich nannoplankton assemblage is present in the deepest sample of this well, with both *Marthasterites furcatus* and *Liliasterites angularis* being present, giving evidence of a Turonian age. *Conusphaera mexicana* may indicate reworking of earliest Albian or earlier strata. Immediately above the deepest sample, the nannoplankton assemblage is reduced to an impoverished flora composed of only a few commonly occurring taxa. This sharp shift may be evidence of a marginal marine environment, possibly preceded by subaerial exposure and nondeposition or erosion. In the Lithastrinus septenarius Zone, the nannoflora recovers gradu-

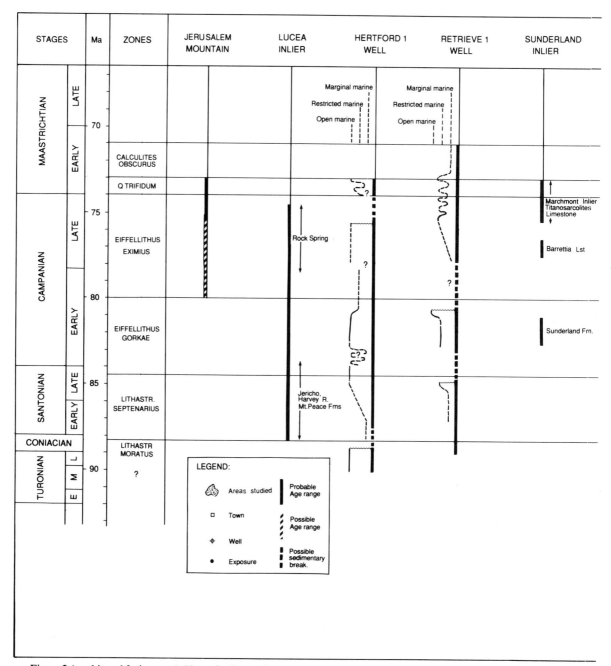

Figure 2 (on this and facing page). Nannofossil zonation and ages of Late Cretaceous outcrops, Jamaica.

ally, and an open marine environment probably existed during the earliest part of the Eiffellithus gorkae Zone. Several shifts in nannoplankton diversity in the upper part of this zone imply a variable environment, again followed by a homogeneous, open marine interval. An assumed shift to a more stable restricted marine environment, near to the top of the Eiffellithus gorkae Zone, is stratigraphically constrained by the latest occurrence of *Bukryaster hayi*. This restricted marine environment continued into the overlying Eiffellithus eximius Zone, which is represented by few samples with variable nannoplankton assemblages. The marked nannoflora impoverishment immediately above the top of the Eiffellithus eximius Zone in this well suggests truncation. There is no evidence for the length of a possible sedimentary break. The overlying Quadrum trifidum Zone has a low-diversity nannoflora with a single, rich sample at the top. From this evidence, two open marine intervals can be assumed.

Retrieve 1

Below 7,500 ft (2,273 m) the well is virtually barren of calcareous nannoplankton. The samples are rather regularly

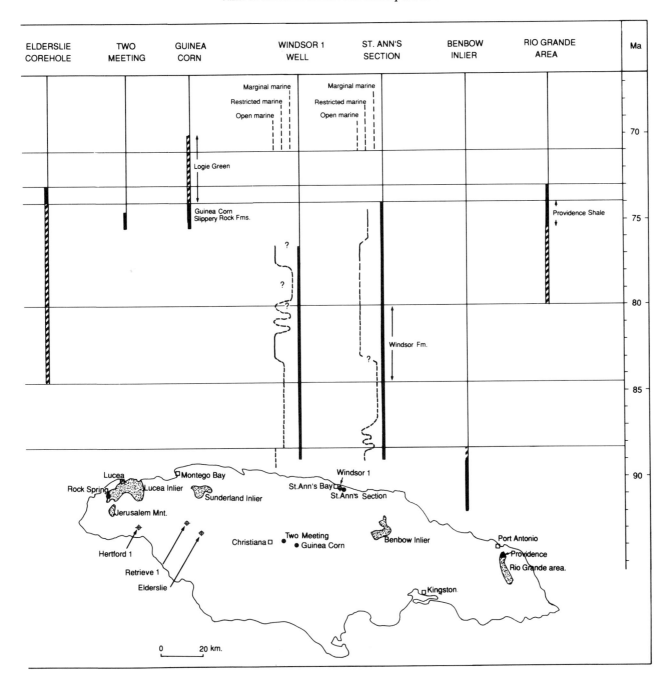

spaced in the fossiliferous interval, except in the lowermost part of the Lithastrinus moratus Zone and Lithastrinus septenarius Zone (Fig. 4). All of the nannofossil zones up to the Calculites obscurus Zone are represented. The uppermost Cretaceous is overlain by a thin Paleogene unit.

Only one sample can be attributed to the Lithastrinus moratus Zone. *Conusphaera mexicana* and *Eprolithus antiquus* in the same sample give evidence of reworking of Early Cretaceous marine deposits. The overlying Lithastrinus septenarius Zone is also found in only one sample. The Lithastrinus moratus and Lithastrinus septenarius zones represent two short marine incursions that were followed by a depositional break and sedimentation in a marginal marine environment. In the Eiffellithus gorkae Zone, a more open marine environment developed that can be dated more accurately by the range of *Ceratolithoides verbeekii*. The rich, open marine assemblage slightly higher in the well has a sharp upper contact, which is assumed to represent truncation by a marginal marine event. The nannoflora is of low diversity in the lower part of the Eiffellithus eximius Zone. The occurrence of *Quadrum gartneri* may indicate reworking. The next rich nanno-

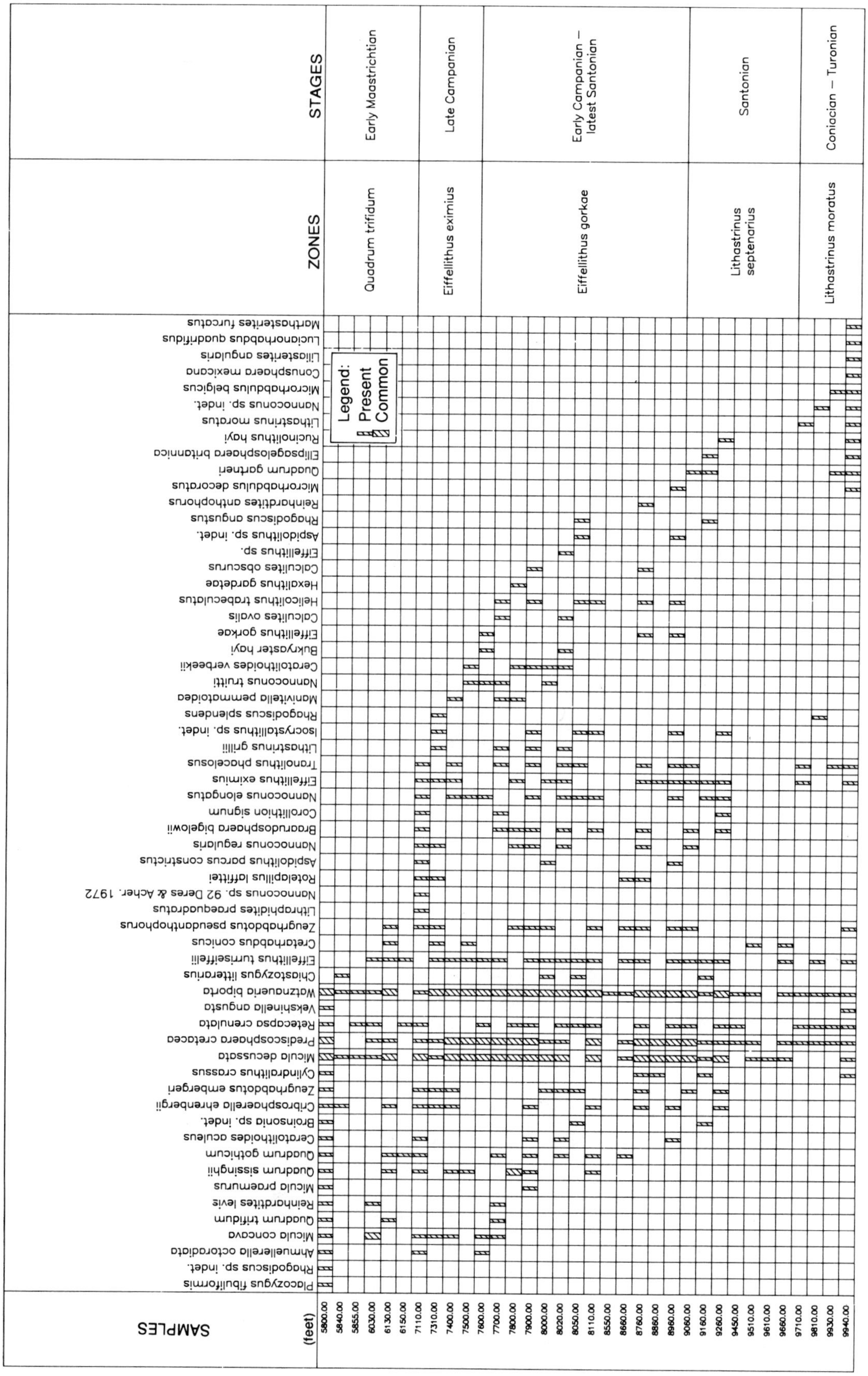

Figure 3. Hertford 1 well: Nannofossil distribution chart.

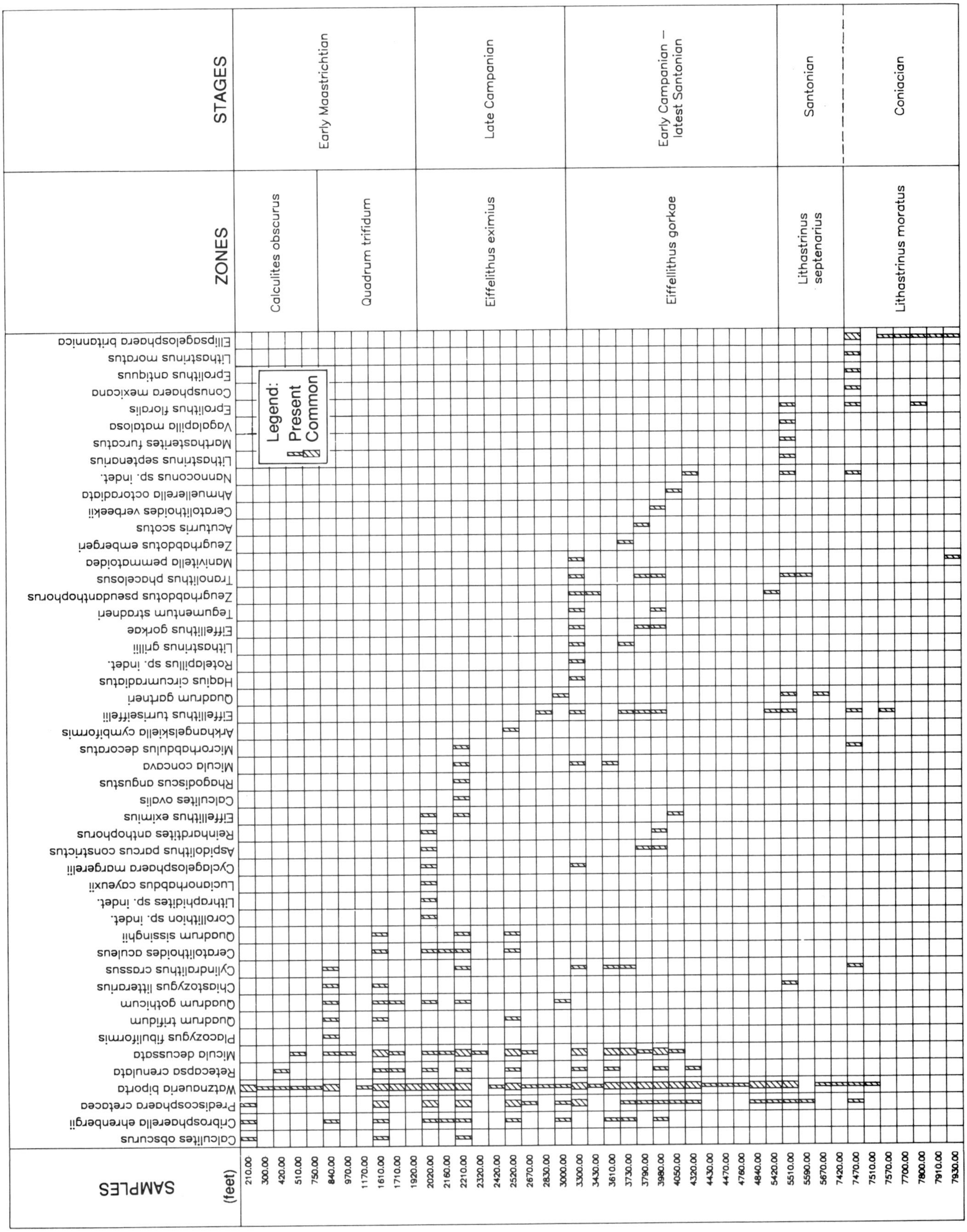

Figure 4. Retrieve 1 well: Nannofossil chart.

fossil assemblage develops above the lowest occurrence of *Quadrum trifidum*. This suggests that only the uppermost part of the Eiffellithus eximius Zone is present at this locality and that the lower part of this zone has been truncated. The pattern of alternating nannofossil-poor and nannofossil-rich samples continues in the Quadrum trifidum Zone. The uppermost interval of this well has a very poor nannofossil assemblage of the Calculites obscurus Zone.

Sunderland Inlier and surroundings

An unfavorable depositional environment and weathering of surface samples may be the cause of the poor nannoplankton yield from this material. Jiang and Robinson (1987) obtained better results from the same area, however. They assigned an Early Campanian age to samples from the Sunderland Formation, which can be attributed to a short interval in the Eiffellithus gorkae Zone because of the presence of *Bukryaster hayi*. The lithostratigraphic suite from the Newmans Hall Formation to the lower Shepherds Hall Formation can be attributed to the middle part of the Eiffellithus eximius Zone, which is characterized by *Ceratolithoides arcuatus*. Because this suite includes the Stapleton Formation with the rudist mollusc *Barrettia gigas*, a Late Campanian nannofossil age is obtained for the *Barrettia* Limestone in this area.

Quadrum trifidum is found in samples from the Marchmont Inlier, which also contain the *Titanosarculites* rudist fauna (Jiang and Robinson, 1987). Consequently these samples can be attributed to the interval from the uppermost Eiffellithus eximius Zone to the Quadrum trifidum Zone (latest Campanian to earliest Maastrichtian in age).

Elderslie corehole

Few samples are fossiliferous, and most specimens are poorly preserved. A single specimen of *Quadrum gartneri* was found as the only nannofossil of stratigraphic value. This specimen may mark the same horizon of reworking that is suspected in the Retrieve 1 well. Rare occurrences of *Calculites obscurus, Calculites ovalis,* and *Quadrum gothicum* rarely occur higher in the corehole. Assignment to the interval from the *Eiffellithus gorkae* Zone to the *Quadrum trifidum* Zone is possible on this evidence. The nannofossil evidence is weak, however, because of the extremely poor assemblages without diagnostic index taxa. A previous biostratigraphic analysis of the same interval in this corehole concluded a Maastrichtian age on the basis of palynomorphs and foraminifera (Petroleum Corporation of Jamaica confidential report).

Two Meeting section

This exposure displays a marly facies of the Guinea Corn Formation. Its accurate stratigraphic relation to the Guinea Corn type section is not known, however. *Quadrum trifidum* occurs in the lowermost sample, and *Lithastrinus grillii* occurs in the uppermost sample, which constrains this exposure to the top of the Eiffellithus eximius Zone, of latest Campanian age. Considering the stratigraphic resolution of this nannofossil biostratigraphy, the Two Meeting exposure and the Guinea Corn type section are of similar age.

Guinea Corn type section

Adequate nannofossil associations were found in one sample from the Slippery Rock Formation and two samples from the overlying Guinea Corn Formation. The co-occurrence of *Eiffellithus eximius, Aspidolithus parcus constrictus,* and *Quadrum trifidum* is evidence of the latest part of the Eiffellithus eximius Zone, of latest Campanian age. The faunal evidence of Jiang and Robinson (1987) in the Guinea Corn Formation is quite similar. In the Logie Green section, Jiang and Robinson obtained an Early Maastrichtian age for a sample that is overlying the Guinea Corn Formation.

Windsor 1

The Lithastrinus moratus to Eiffellithus eximius zones are represented in this well (Fig. 5). The number of fossiliferous samples is low, however, and all conclusions have a relatively high degree of uncertainty.

The rich assemblages of the stratigraphically higher samples in the Lithastrinus moratus Zone probably indicate open marine conditions. The presence of *Marthasterites furcatus* is evidence of a Coniacian, possibly Turonian, age. Samples below the interval attributed to the Lithastrinus moratus Zone are very poor and without consistent age-significant nannoplankton. The presence of *Assipetra infracretacea* in one sample may indicate reworking of Albian or earlier strata.

The Lithastrinus septenarius Zone is found in only two samples, both of only moderate diversity. An interval with a rich nannoflora in the Eiffellithus gorkae Zone can be accurately dated, based on the last occurrence of *Marthasterites furcatus*. Presumably a restricted marine environment is recorded in the Lithastrinus septenarius Zone and the earlier part of the Eiffellithus gorkae Zone. Higher in this well, several changes in nannoplankton assemblage composition may indicate separate open marine incursions. The lowest fossiliferous sample in the overlying Eiffellithus eximius Zone includes *Ceratolithoides arcuatus*. Consequently there is no nannofossil record from the later part of the Eiffellithus gorkae Zone up to the *Ceratolithoides arcuatus* level in the Eiffellithus eximius Zone.

St. Ann's Great River field section

The St. Ann's Great River section is in the vicinity of the Windsor 1 exploration well and ranges over approximately the same stratigraphic interval, from the Lithastrinus moratus Zone to the Eiffellithus eximius Zone (Fig. 6). The nannoflora of this section is richer in taxa and specimens than the Windsor 1 well.

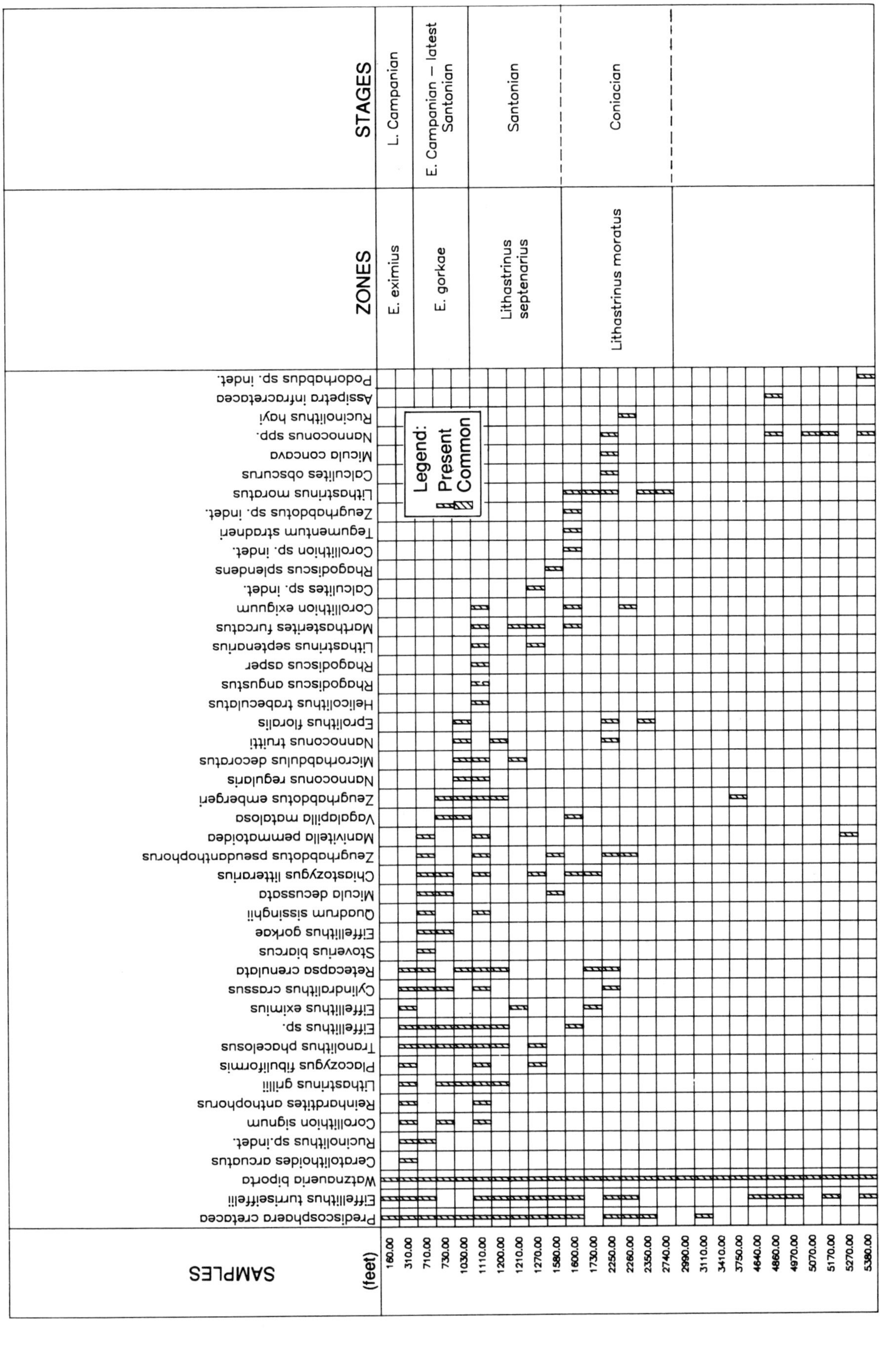

Figure 5. Windsor 1 well: Nannofossil distribution chart.

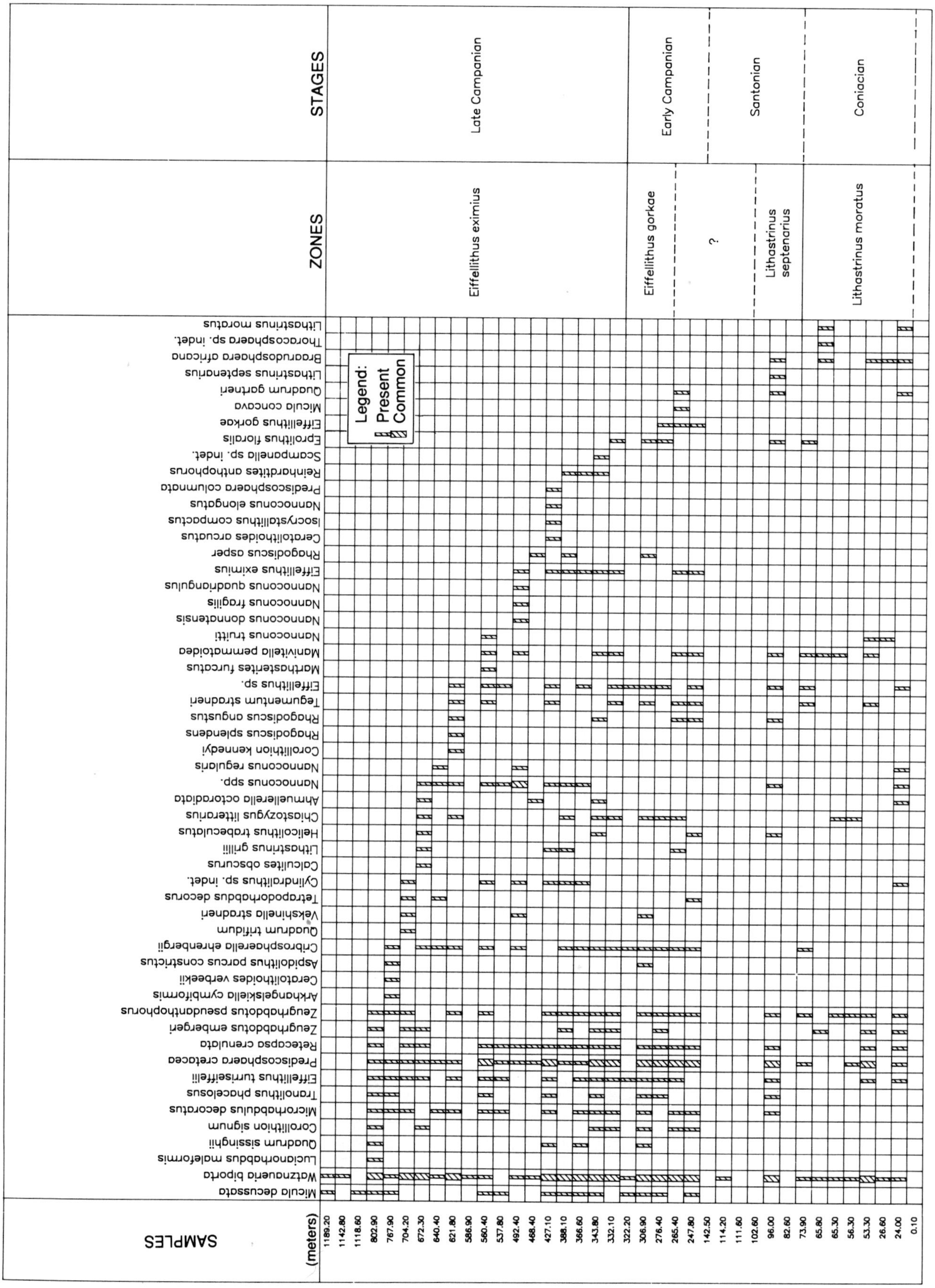

Figure 6. St. Ann's Great River section: Nannofossil distribution chart.

The base of the section shows a variable but fairly open marine environment in the Lithastrinus moratus Zone and the lower part of the Lithastrinus septenarius Zone. The overlying thick, barren interval is in turn overlain by an interval with a rich nannoflora characterized by the highest occurrence of *Eiffellithus gorkae* (i.e., the top of this zone). The boundary between the Lithastrinus septenarius and Eiffellithus gorkae zones is probably in the barren interval, which is assumed to represent marginal marine conditions. The entire section above the Eiffellithus gorkae Zone is attributed to the Eiffellithus eximius Zone. Single occurrences of *Ceratolithoides arcuatus* and *Quadrum trifidum* suggest that this zone is essentially complete in the St. Ann's Great River section. The occurrence of *Quadrum trifidum* could be interpreted as evidence of the presence of the Quadrum trifidum Zone. However, the presence of *Tranolithus phacelosus, Aspidolithus parcus constrictus,* and *Lithastrinus grillii,* which do not occur above the Eiffellithus eximius Zone, make this improbable.

The *Barrettia gigas* Limestone occurs in the uppermost part of the section, above the lowest occurrence of *Quadrum trifidum.* Therefore, this limestone cannot be older than latest Campanian, which is slightly younger than the *Barrettia* Limestone in the Sunderland Inlier that on the nannofossil evidence from Jiang and Robinson (1987) correlates to the middle part of the Eiffellithus eximius Zone, of Late Campanian age.

The stratigraphy of the St. Ann's Great River has been reviewed by Meyerhoff and Krieg (1977a), who assigned it an Early Coniacian to Late Campanian age on the basis of planktonic foraminifera (Esker, 1969) and ammonites (Sohl, 1976). My results confirm this conclusion. The correlation of the samples of the present study to the stratigraphic subdivision in Meyerhoff and Krieg (1977a) is not satisfactory. A further discussion of this issue will necessitate the correlation of my field data to those in the sources of Meyerhoff and Krieg (1977a), which is outside the scope of this chapter.

Samples studied by Jiang and Robinson (1987) from the Windsor Formation in the St. Ann's Great River can be attributed to the Eiffellithus gorkae Zone, whereas the overlying St. Ann's Great River Formation is not older than the Eiffellithus eximius Zone.

Benbow Inlier

Jiang and Robinson (1987) recorded a Turonian nannoplankton assemblage in one sample from the Rio Nuevo Formation. In the lower part of the same formation, a Middle Albian rudist fauna had previously been collected (Sohl, 1976). Samples from two localities in the overlying Tiber River Formation yielded an Early and Late Turonian nannoflora, respectively. *Lithastrinus moratus* is quoted as evidence of a Turonian age. However, this nannofossil (the index taxon of our Lithastrinus moratus Zone) ranges into the Coniacian, which implies that some of these samples may be younger. Nevertheless, the Early Turonian age of at least one sample from the Tiber River Formation is sufficiently confirmed by the presence of *Lithraphidites acutus.* These are the oldest Jamaican in-place nannofossils on record.

Rio Grande area

Because of Late Cretaceous and Tertiary tectonic activity, this area is characterized by a complex structure that obscures the stratigraphic relation of isolated exposures (Krijnen and Lee Chin, 1978; Wadge and Draper, 1978). Most of the fossil material is from scattered exposures of several formations that have been studied in an effort to identify the Cretaceous-Tertiary boundary. All Cretaceous samples can be attributed to the Eiffellithus eximius–Quadrum trifidum zones. In some of the richer samples, the presence of *Quadrum trifidum* further restricts their position to the later part of the Eiffellithus eximius Zone and the Quadrum trifidum Zone. In other samples, the additional presence of *Reinhardtites anthophorus, Tranolithus phacelosus,* or *Lithastrinus grillii* requires an assignment to the *Eiffellithus eximius* Zone. The samples with the richest consistent nannoplankton associations correlate with the latest part of the Eiffellithus eximius Zone and the Quadrum trifidum Zone, of latest Campanian and earliest Maastrichtian age.

Jiang and Robinson (1987) recorded *Eiffellithus eximius* and *Quadrum trifidum* in the Providence Shale at Breastworks. Consequently, this locality also can be attributed to the latest part of the Eiffellithus eximius Zone, of latest Campanian age.

In a few samples, mutually exclusive taxa are found that together span all six nannoplankton biozones of the Late Cretaceous of Jamaica. Since Cretaceous nannofossils can be abundant in Early Tertiary samples from the Rio Grande area, it seems reasonable to suggest that these mutually exclusive nannofossils may be reworked and mask the rare evidence of a Paleocene age.

DEPOSITION AND ENVIRONMENT

The earliest nannofossil-productive strata are of Turonian age (Hertford 1 and Benbow Inlier). In all three wells and in the St. Ann's Great River section, marine Coniacian sediments occur. In the Early Santonian, a change to a marginal marine environment in the St. Ann's region may correlate to a sedimentary break in the more westward Hertford 1 and Retrieve 1 wells. Open marine conditions were restored during the Late Santonian (85 Ma) in West Jamaica, whereas a restricted marine environment was maintained in the St. Ann's region. An Early Campanian (83 Ma) marine event is recognized over the entire area. In West Jamaica, this event was preceded by environmental shifts and possibly a depositional break. A Late Campanian (77.5 Ma) marine event in the St. Ann's area can probably be correlated to West Jamaica. During the latest Campanian and Early Maastrichtian, marked environmental shifts occurred in the Hertfort 1 and Retrieve 1 wells, whereas the majority of the contemporaneous surface sections of West and Central Jamaica display shallow marine limestones. The virtually contemporaneous Guinea Corn section and Two Meeting exposure, of rudist limestone and marine marl, respectively, give evidence of an environmental shift over a short distance. In East Jamaica, the Providence Shale represents a more open marine environment during the latest Campanian.

CONCLUSIONS

The biostratigraphic use of Late Cretaceous calcareous nannofloras from Jamaica is hampered by their poor preservation. Near-surface dissolution in prominent topographical exposures probably destroyed the less robust nannofossils, particularly in limestones. Exposures near the permanent water table and well samples are more productive, both in total nannofossil yield and number of taxa, but even in the best sections nannofossil presence is intermittent. Consequently the biostratigraphic boundaries in the sections cannot be drawn with certainty everywhere.

In many samples Sissingh's (1977) cosmopolitan zonation scheme could not be applied, and the proposed low-resolution Late Cretaceous nannoplankton biozonation of Jamaica is based on index taxa that occur in most nannofossil-productive lithologies. Apart from their stratigraphic significance, changes in the stratigraphic or regional assemblage patterns give additional information on the paleogeographic framework of Jamaica during the Late Cretaceous.

ACKNOWLEDGMENTS

I am indebted to the Petroleum Corporation of Jamaica for the opportunity to execute the Jamaica nannofossil studies and to complete this chapter. Thanks go in particular to Dr. Raymond M. Wright for his continuous interest in the progress of the project. Thanks are due to Messrs. Winston Scott and Locksley Allen for their indispensable help and cheerful company during the fieldwork on Jamaica. Assistance from the PCJ staff in matters of regional geology, planning, and logistics is gratefully acknowledged. Special thanks are due to Dr. Edward Robinson and Prof. Jan E. van Hinte for their advice on stratigraphic issues. Dr. Hans Van Der Eem and Hubert Jansen (formerly Center for International Geohistory Analysis Research and Services, Amsterdam) cooperated in an early synthesis of Jamaican well results. Sample preparation, drafting, and typing during the Jamaica project and the completion of this paper were done at IKU Petroleum Research, Trondheim, Norway. Colleague geologists at IKU assisted with various advice. Jorunn Os Vigran and Robert Goll checked the text for stratigraphic and linguistic inconsistencies. The paper was reviewed by Dr. Jackie A. Burnett (University College, London) and Dr. Joost W. Verbeek (Geological Survey of the Netherlands). It is a pleasure to thank them all.

APPENDIX 1: ALPHABETICAL LIST OF SPECIES RECORDED IN THE DISTRIBUTION CHARTS, FIGURES 3–6

Acuturris scotus (Risatti, 1973) Wind and Wise, *in* Wise and Wind, 1977.
Ahmuellerella octoradiata (Górka, 1957) Reinhardt, 1964.
Arkhangelskiella cymbiformis Vekshina, 1959.
Aspidolithus parcus constrictus (Hattner et al., 1980) Perch-Nielsen, 1984.
Assipetra infracretacea (Thierstein, 1973) Roth, 1973.
Braarudosphaera africana Stradner, 1961.
Braarudosphaera bigelowii (Gran and Braarud, 1935) Deflandre, 1947.
Bukryaster hayi (Bukry, 1969) Prins and Sissingh, *in* Sissingh, 1977.
Calculites obscurus (Deflandre, 1959) Prins and Sissingh, *in* Sissingh, 1977.
Calculites ovalis (Stradner, 1963) Prins and Sissingh, *in* Sissingh, 1977.
Ceratolithoides aculeus (Stradner, 1961) Prins and Sissingh, *in* Sissingh, 1977.
Ceratolithoides arcuatus Prins and Sissingh, *in* Sissingh, 1977.
Ceratolithoides verbeekii Perch-Nielsen, 1979.
Chiastozygus litterarius (Górka, 1957) Manivit, 1971.
Conusphaera mexicana Trejo, 1969.
Corollithion exiguum Stradner, 1961.
Corollithion kennedyi Crux, 1981.
Corollithion signum Stradner, 1963.
Cretarhabdus conicus Bramlette and Martini, 1964.
Cribrosphaerella ehrenbergii (Arkhangelsky, 1912) Deflandre, *in* Piveteau, 1952.
Cyclagelosphaera margerelii Noël, 1965.
Cylindralithus crassus Stover, 1966.
Cylindralithus serratus Bramlette and Martini, 1964.
Eiffellithus eximius (Stover, 1966) Perch-Nielsen, 1968.
Eiffellithus gorkae Reinhardt, 1965.

Eiffellithus turriseiffelii (Deflandre, *in* Deflandre and Fert, 1954) Reinhardt, 1965.
Ellipsagelosphaera britannica (Stradner, 1963) Perch-Nielsen, 1968.
Eprolithus antiquus Perch-Nielsen, 1979.
Eprolithus floralis (Stradner, 1962) Stover, 1966.
Haqius circumradiatus (Stover, 1966) Roth, 1978.
Helicolithus trabeculatus (Górka, 1957) Verbeek, 1977.
Hexalithus gardetae Bukry, 1969.
Isocrystallithus compactus Verbeek, 1976.
Liliasterites angularis Svabénická and Stradner *in* Stradner and Steinmetz, 1984.
Lithastrinus grillii Stradner, 1962.
Lithastrinus moratus Stover, 1966.
Lithastrinus septenarius Forchheimer, 1972.
Lithraphidites praequadratus Roth, 1978.
Lucianorhabdus cayeuxii Deflandre, 1959.
Lucianorhabdus maleformis Reinhardt, 1966.
Lucianorhabdus quadrifidus Forchheimer, 1972.
Manivitella pemmatoidea (Deflandre *in* Manivit, 1965) Thierstein, 1971.
Marthasterites furcatus (Deflandre *in* Defrandre and Fert, 1954) Deflandre, 1959.
Microrhabdulus belgicus Hay and Towe, 1963.
Microrhabdulus decoratus Deflandre, 1959.
Micula concava (Stradner *in* Martini and Stradner, 1960) Verbeek, 1976.
Micula decussata Vekshina, 1959.
Micula murus (Martini, 1961) Bukry, 1973.
Micula praemurus (Bukry, 1973) Stradner and Steinmetz, 1984.
Nannoconus donnatensis Deres and Achéritéguy, 1980.
Nannoconus elongatus Brönnimann, 1955.
Nannoconus fragilis Deres and Achéritéguy, 1980.
Nannoconus quadriangulus Deflandre and Deflandre, 1967.
Nannoconus regularis Deres and Achéritéguy, 1980.
Nannoconus truitti Brönnimann, 1955.
Nannoconus sp. 92 *in* Deres and Achéritéguy, 1972.
Placozygus fibuliformis (Reinhardt, 1964) Hoffman, 1970.
Prediscosphaera columnata (Stover, 1966) Perch-Nielsen, 1984.
Prediscosphaera cretacea (Arkhangelsky, 1912) Gartner, 1968.
Quadrum gartneri Prins and Perch-Nielsen *in* Manivit et al., 1977.
Quadrum gothicum (Deflandre, 1959) Prins and Perch-Nielsen *in* Manivit et al., 1977.
Quadrum sissinghii Perch-Nielsen, 1984.
Quadrum trifidum (Stradner *in* Stradner and Papp, 1961) Prins and Perch-Nielsen *in* Manivit et al., 1977.
Reinhardtites anthophorus (Deflandre, 1959) Perch-Nielsen, 1968.
Reinhardtites levis Prins and Sissingh *in* Sissingh, 1977.
Retecapsa crenulata (Bramlette and Martini, 1964) Grün *in* Grün and Allemann 1975.
Rhagodiscus angustus (Stradner, 1963) Reinhardt, 1971.
Rhagodiscus asper (Stradner, 1963) Reinhardt, 1967.
Rhagodiscus splendens (Deflandre, 1953) Verbeek, 1977.
Rotelapillus laffittei (Noël, 1957) Noël, 1973.
Rucinolithus hayi Stover, 1966.
Stoverius biarcus (Bukry, 1969) Perch-Nielsen, 1984.
Tegumentum stradneri Thierstein *in* Roth and Thierstein, 1972.
Tetrapodorhabdus decorus (Deflandre *in* Deflandre and Fert, 1954) Wind and Wise *in* Wise and Wind, 1977.
Tranolithus phacelosus Stover, 1966.
Vagalapilla matalosa (Stover, 1966) Thierstein, 1973.
Vekshinella angusta (Stover, 1966) Verbeek, 1977.
Vekshinella stradneri Rood et al., 1971.
Watznaueria biporta Bukry, 1969.
Zeugrhabdotus embergeri (Noël, 1959) Perch-Nielsen, 1984.
Zeugrhabdotus pseudanthophorus (Bramlette and Martini, 1964) Perch-Nielsen, 1984.

APPENDIX 2: DESCRIPTION OF A PREVIOUSLY NOT IDENTIFIED TAXON

Eiffellithus Reinhardt (1965)

Eiffellithus sp.
Figures 7.1, 7.2

Diagnosis. A species of *Eiffellithus* that is characterized by a small central cross with elongate diamond-shaped arms, approximately aligned in the long and short axis of the coccolith.
Description. The elliptical coccolith has a narrow rim of oblique elements and a closed or nearly closed central area, filled by 8 to 10 plates with radial sutures. The four arms of the central cross are elongate diamond-shaped and approximately coincide to the short and long axis of the coccolith. The longitudinal arms are longest and reach from the center of the coccolith at least halfway to the rim.
Remarks. The following differences of *Eiffellithus* sp. from some previously described *Eiffellithus* taxa have been noted:
In *Eiffellithus gorkae,* the central cross is of similar form but of different orientation, i.e., in between the long and short axes. In *Eiffellithus eximius,* the central cross has the same orientation as in *Eiffellithus* sp., but the arms of the crossbars are markedly longer and of equal width over most of their length, bifurcating nearest to the rim. *Eiffellithus primus* Applegate and Bergen 1988 is smaller in overall size and, in polarized light, displays a shift in the orientation of the central cross from axial to slightly off-axial, when the direction of polarity is changed from parallel to an angle of 45 relative to the longitudinal axis of the coccolith.
Occurrences. In the Hertford 1 and Windsor 1 wells, the St. Ann's Great River section and at Rock Spring (West Jamaica).
Stratigraphic range. Coniacian to Late Campanian.
Size range: length 7 to 11 μm, width 6 to 8 μm.
Figured specimens. Figure 7.1: Windsor 1 well, 1,100 to 1,110 ft (335 to 338 m) depth interval; negative 91-7-35. Figure 7.2: St. Ann's Great River Section, 96.0 m; negative 91-4-13. Both negatives are filed in the nannofossil film archive, IKU, Trondheim, Norway.

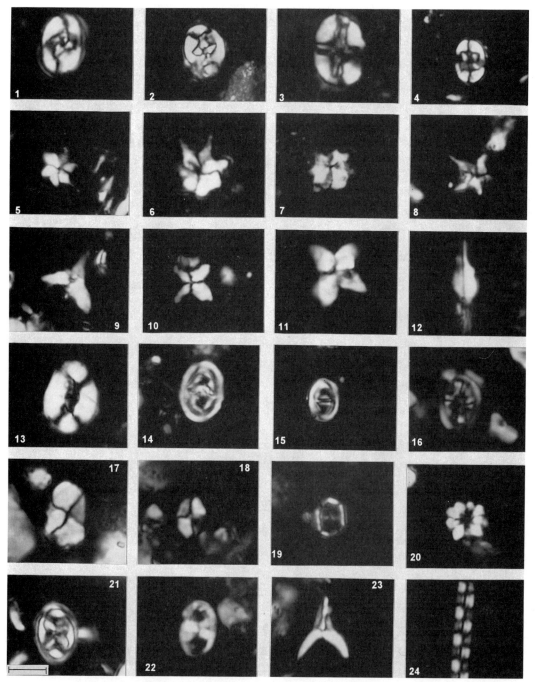

Figure 7. Length of scale bar is 5 μm. 1. *Eiffellithus* sp. Windsor 1, 1,110 ft. 2. *Eifellithus* sp. St. Ann's Great River section, 96.0 m. 3. *Eiffellithus eximius* (Stover) Perch-Nielsen. St. Ann's Great River section, 427.1 m. 4. *Eiffellithus gorkae* Reinhardt. St. Ann's Great River section, 276.4 m. 5. *Lithastrinus grillii* Stradner. Hertford 1, 8,020 ft. 6. *Lithastrinus septenarius* Forchheimer. Windsor 1, 1,270 ft. 7. *Lithastrinus moratus* Stover. St. Ann's Great River section, 65.8 m. 8. *Bukryaster hayi* (Bukry) Prins and Sissingh. Hertford 1, 8,020 ft. 9. *Quadrum trifidum* (Stradner) Prins and Perch-Nielsen. Rio Grande area. 10. *Quadrum sissinghii* Perch-Nielsen. Rio Grande area. 11. *Quadrum gothicum* (Deflandre) Prins and Perch-Nielsen. Hertford 1, 7,110 ft. 12. *Lithraphidites praequadratus* Roth. Hertford 1, 7,110 ft. 13. *Aspidolithus parcus constrictus* (Hattner et al.) Perch-Nielsen. Hertford 1, 8,000 ft. 14. *Reinhardtites anthophorus* (Deflandre) Perch-Nielsen. Hertford 1, 8,760 ft. 15. *Zeugrhabdotus pseudanthophorus* (Bramlette and Martini) Perch-Nielsen. Hertford 1, 8,000 ft. 16. *Ahmuellerella octoradiata* (Górka) Reinhardt. Hertford 1, 7,600 ft. 17. *Calculites obscurus* (Deflandre) Prins and Sissingh. Hertford 1, 8,760 ft. 18. *Calculites ovalis* (Stradner) Prins and Sissingh. Hertford 1, 8,020 ft. 19. *Corollithion signum* Stradner. Windsor 1, 1,110 ft. 20. *Eprolithus floralis* (Stradner) Stover. St. Ann's Great River section, 73.9 m. 21. *Helicolithus trabeculatus* (Górka) Verbeek. Windsor 1, 1,110 ft. 22. *Tranolithus phacelosus* Stover. St. Ann's Great River section, 24 m. 23. *Ceratolithoides aculeus* (Stradner) Prins and Sissingh. Hertford 1, 7,110 ft. 24. *Microrhabdulus decoratus* Deflandre. St. Ann's Great River section, 306.9 m.

REFERENCES CITED

Applegate, J. L., and Bergen, J. A., 1988, Cretaceous calcareous nannofossil biostratigraphy of sediments recorded from the Galicia margin, *in* Boillot, G. and others, eds., Proceedings of the Ocean Drilling Program: Scientific results: v. 103, p. 293–348.

Brönnimann, P., 1955, Microfossils *incertae sedis* from the Upper Jurassic and Lower Cretaceous of Cuba: Micropaleontology, v. 1, p. 28–51.

Doeven, P. H., 1983, Cretaceous nannofossil stratigraphy and paleoecology of the Canadian Atlantic margin: Geological Survey of Canada Bulletin 356, p. 1–69.

Easker, G. C., 1969, Planktonic foraminifera from the St. Ann's Great River Valley, Jamaica: Micropaleontology, v. 15, p. 210–220.

Haq, B. U., Hardenbol, J., and Vail, P. R., 1987, Chronology of fluctuating sea levels since the Triassic: Science, v. 235, p. 1156–1167.

Jiang, M.-J., and Robinson, E., 1987, Calcareous nannofossils and larger foraminifera in Jamaica rocks of Cretaceous to early Eocene age, *in* Ahmad, R., ed., Proceedings of a Workshop on the Status of Jamaica Geology, Kingston, March 1984: Kingston, Geological Society of Jamaica, Special Issue, p. 24–51.

Krijnen, J. P., and Lee Chin, A. C., 1978, Geology of the northern, central and south-eastern Blue Mountains, Jamaica, with a provisional compilation map of the entire inlier: Geologie en Mijnbouw, v. 57, p. 243–250.

McFarlane, N., 1977, compiler, Jamaica—Geology 1:250,000, Geological data prepared by Mines and Geology Division: Kingston, Ministry of Mining and Natural Resources.

Meyerhoff, A. A., and Krieg, E. A., 1977, Petroleum potential of Jamaica: Kingston, Ministry of Mining and Natural Resources Special Report, p. 1–131.

Perch-Nielsen, K., 1985, Mesozoic calcareous nannofossils, *in* Bolli, H. M., Saunders, J. B., and Perch-Nielsen, K., eds., Plankton stratigraphy: Cambridge, Cambridge University Press, p. 329–426.

Roth, P. H., 1978, Cretaceous nannoplankton biostratigraphy and oceanography of the northwestern Atlantic Ocean: Initial Report of Deep Sea Drilling Project 44, Washington, U.S. Government Printing Office, p. 731–759.

Sissingh, W., 1977, Biostratigraphy of Cretaceous calcareous nannoplankton: Geologie en Mijnbouw, v. 56, p. 37–65.

Sohl, N. F., 1976, Cretaceous geology of the Jamaica inliers [Unpublished report]: Kingston, Mines and Geology Division, p. 1–23.

Verbeek, J. W., 1977, Calcareous nannoplankton biostratigraphy of Middle and Upper Cretaceous deposits in Tunisia, southern Spain and France: Utrecht Micropaleontological Bulletins, v. 16, p. 1–157.

Wadge, G., and Draper, G., 1978, Structural geology of the southeastern Blue Mountains, Jamaica: Geologie en Mijnbouw, v. 57, p. 347–352.

Zans, V. A., Chubb, L. J., Versey, H. R., Williams, J. B., Robinson, E., and Cooke, D. L., 1962, Synopsis of the geology of Jamaica: Kingston, Geological Survey Department, p. 1–72.

MANUSCRIPT ACCEPTED BY THE SOCIETY NOVEMBER 11, 1992

Campanian calcareous nannofossils in the Sunderland Inlier, western Jamaica

Mark M. Jiang*
ARCO Oil and Gas Company, Plano, Texas 75075

ABSTRACT

Calcareous nannofossil assemblages in Campanian rocks in the Sunderland Inlier, western Jamaica, are documented and assigned to Sissingh's (1977, emended 1978) zones 19, 21, and 22, his zone 20 being missing due either to a local fault or to an unconformity. Of the 107 calcareous nannofossil species identified, five new combinations are proposed: *Chiastozygus dennisonii*, n. comb., *Cretarhabdus sinuosus*, n. comb., *Tranolithus bitraversus*, n. comb., *Uniplanarius sissinghii*, n. comb., and *Zygodiscus pontilithus*, n. comb.

INTRODUCTION

Cretaceous formations in Jamaica are exposed at the surface in a number of inliers (Fig. 1). Studies concerned with dating and correlation of these rocks have been primarily based on molluscs and foraminifera. Calcareous nannofossils have so far been reported only by Jiang and Robinson (1987, 1989). Samples examined in these studies are numerous and include suites from the Lucea, Jerusalem Mountain, Marchmont, Maldon, Sunderland, Central, Benbow, and Blue Mountains inliers (Fig. 1). Discus-

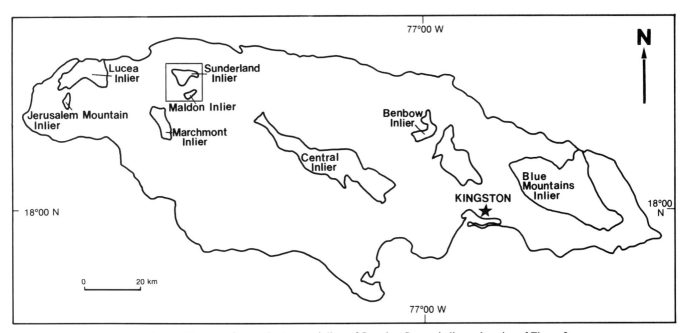

Figure 1. Locality map of some Cretaceous inliers of Jamaica. Square indicates location of Figure 2.

---

*Present address: ARCO Oil and Gas Company, 15375 Memorial Drive, Houston, Texas 77079.

Jiang, M. M., 1993, Campanian calcareous nannofossils in the Sunderland Inlier, western Jamaica, *in* Wright, R. M., and Robinson, E., eds., Biostratigraphy of Jamaica: Boulder, Colorado, Geological Society of America Memoir 182.

sions in these studies, however, primarily emphasized age interpretations, and detailed aspects of biostratigraphic significance were not demonstrated. The present study elaborates part of the previous work and documents in detail the floral associations and succession in the Campanian rocks in the Sunderland Inlier (Fig. 1). Based on this, a revised biostratigraphic interpretation for the section is also made. The Sunderland section was selected because sample collection is more complete there than elsewhere.

STRATIGRAPHIC SETTING AND SAMPLES

Figure 2 shows the Sunderland Inlier general geology and sample transect. Cretaceous formations in the inlier are bounded to the east, south, and west by the Yellow Limestone Group of Paleogene age. Chubb (1958) established the sequence, from top to bottom, as follows: Shepherd's Hall Formation—variegated mudstones, sandstones, and conglomerates; Stapleton Forma-

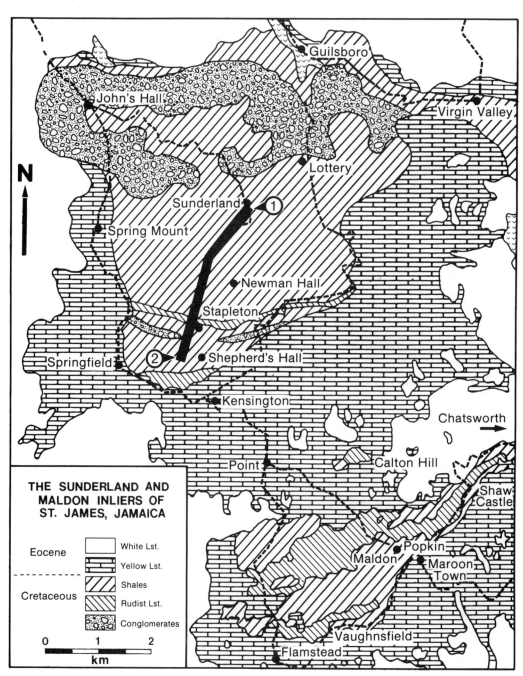

Figure 2. Locality map showing line of section along which samples were collected through the Sunderland Inlier. Heavy dashed lines in the map indicate roads. Ornament with V's represents alluvium. Geology from Zans et al., 1963. (1) Lowest sample; (2) highest sample.

tion—marls and rudist (*Barrettia*) limestones; Newman Hall Formation—gray shales with calcareous concretions; Sunderland Formation—brown-weathering shales and sandstones; and John's Hall Formation—conglomerates.

Figure 3 shows the vertical extent of the Cretaceous formations as measured by Krijnen and van Dommelen (Krijnen, 1972, p. 9, Fig. 2; van Dommelen, 1972, p. 56, Text—Fig. 15).

Seventy-seven samples (SU-23 through SU-100, less SU-26) collected from the Sunderland Inlier were used in this study. They were made available through the courtesy of the Petroleum Corporation of Jamaica. They are in a stratigraphic order, but intervals between samples are in most cases unclear due to local covered geologic structures. Of the 77 samples, SU-23 (the oldest) through SU-59 are from the Sunderland Formation, some of these being from the same Orange River locality as those sampled by Chubb (1958) and Krijnen (1972) (see Fig. 3). SU-60 through SU-64 are the Newman Hall Formation. SU-65 through SU-100 (the youngest) represent the Shepherd's Hall Formation. The Stapleton and John's Hall formations were not examined because shallow water carbonate and conglomerate lithologies are usually not suitable for a calcareous nannofossil study.

TECHNIQUES AND METHODS

This study was carried out using a high power (×1500) optical microscope. Slide preparations were made following a standard procedure: A small piece of fresh sample was pulverized with an applicator (first with a mortar and pestle, if needed) in a small glass vial. A suspension was made by adding clean water. After stirring, the suspension was allowed to settle for about 5 to 10 seconds. A portion of the suspension in the upper two-thirds of the suspension column was pipetted onto a cover glass on a slide-warming plate. When dry, the cover glass, then evenly spread with a thin film of sample material, was mounted onto a labeled glass slide with piccolyte on a hot plate. This technique was routinely used because it provided a way to eliminate the coarse fraction with little biasing of the original floral assemblage.

Relative abundances of calcareous nannofossil species were derived from direct counts of specimens on the slides, which were carried out in two steps. During the initial examination of each slide, numbers of individual species were continuously recorded until the total reached 300. These counts were subsequently used for manipulation for abundances, assuming each single count of 300 represented 0.33% of the flora. The second step consisted of searching the same slide for at least 30 minutes. Rare species not observed in the first step of counting, but found during the second step of searching, were assigned an abundance of less than 0.33%.

Preservation of the floral assemblages is generally good, but fragments of coccoliths were noted. If these broken specimens could be identified to a species level, they were counted. If not, they were ignored. Species of *Thoracosphaera* were not counted as they are no longer considered calcareous nannofossils.

CALCAREOUS NANNOFOSSILS

One hundred and five calcareous nannofossil species belonging to 50 genera and two indeterminate forms were observed. Figure 4 shows the distribution of these species. Species are listed in alphabetical order of generic and species names, and relative abundances of individual species are presented in a hierarchy of seven classes as follows: (1) Very rare (<0.33%): present but not encountered in the first 300 specimens; (2) rare (0.33 to 1%): 1 or 2 specimens in 300; (3) a few (1 to 2%): 3 through 5 specimens in 300; (4) common (2 to 5%): 6 through 14 specimens in 300; (5) very common (5 to 10%): 15 through 29 specimens in 300; (6) abundant (10 to 20%): 30 through 59 specimens in 300; and (7) very abundant (>20%): 60 or more specimens in 300. Rela-

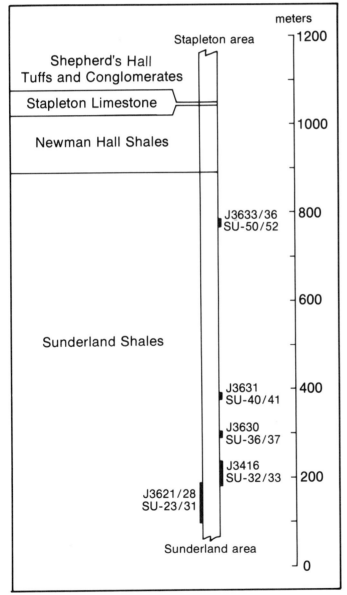

Figure 3. Stratigraphy of the Campanian formations in part of the Sunderland Inlier, modified from Krijnen (1972). J = samples studied by Krijnen (1972); SU = samples examined in this study.

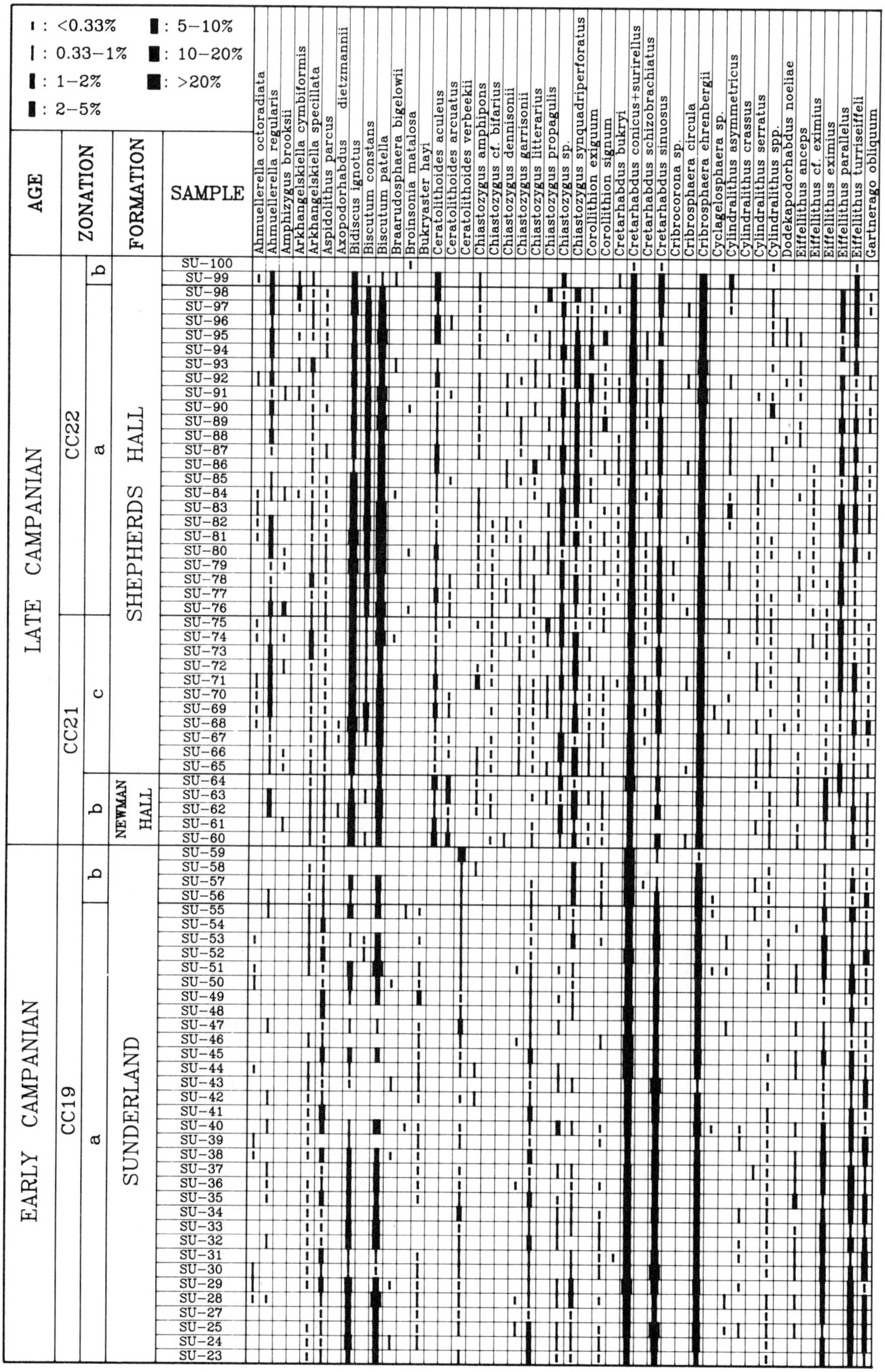
: <0.33%
: 0.33–1%
: 1–2%
: 2–5%
: 5–10%
: 10–20%
: >20%
AGE
ZONATION
FORMATION
SAMPLE
LATE CAMPANIAN
EARLY CAMPANIAN
CC22
CC21
CC19
a
b
c
SHEPHERDS HALL
NEWMAN HALL
SUNDERLAND
Ahmuellerella octoradiata
Ahmuellerella regularis
Amphizygus brooksii
Arkhangelskiella cymbiformis
Arkhangelskiella specillata
Aspidolithus parcus
Axopodorhabdus dietzmannii
Bidiscus ignotus
Biscutum constans
Biscutum patella
Braarudosphaera bigelowii
Broinsonia matalosa
Bukryaster hayi
Ceratolithoides aculeus
Ceratolithoides arcuatus
Ceratolithoides verbeekii
Chiastozygus amphipons
Chiastozygus cf. bifarius
Chiastozygus dennisonii
Chiastozygus garrisonii
Chiastozygus litterarius
Chiastozygus propagulis
Chiastozygus sp.
Chiastozygus synquadriperforatus
Corollithion exiguum
Corollithion signum
Cretarhabdus bukryi
Cretarhabdus conicus+surirellus
Cretarhabdus schizobrachiatus
Cretarhabdus sinuosus
Cribrocorona sp.
Cribrosphaera circula
Cribrosphaera ehrenbergii
Cyclagelosphaera sp.
Cylindralithus asymmetricus
Cylindralithus crassus
Cylindralithus serratus
Cylindralithus spp.
Dodekapodorhabdus noeliae
Eiffellithus anceps
Eiffellithus cf. eximius
Eiffellithus eximius
Eiffellithus parallelus
Eiffellithus turriseiffeli
Gartnerago obliquum
SU-100
SU-99
SU-98
SU-97
SU-96
SU-95
SU-94
SU-93
SU-92
SU-91
SU-90
SU-89
SU-88
SU-87
SU-86
SU-85
SU-84
SU-83
SU-82
SU-81
SU-80
SU-79
SU-78
SU-77
SU-76
SU-75
SU-74
SU-73
SU-72
SU-71
SU-70
SU-69
SU-68
SU-67
SU-66
SU-65
SU-64
SU-63
SU-62
SU-61
SU-60
SU-59
SU-58
SU-57
SU-56
SU-55
SU-54
SU-53
SU-52
SU-51
SU-50
SU-49
SU-48
SU-47
SU-46
SU-45
SU-44
SU-43
SU-42
SU-41
SU-40
SU-39
SU-38
SU-37
SU-36
SU-35
SU-34
SU-33
SU-32
SU-31
SU-30
SU-29
SU-28
SU-27
SU-25
SU-24
SU-23

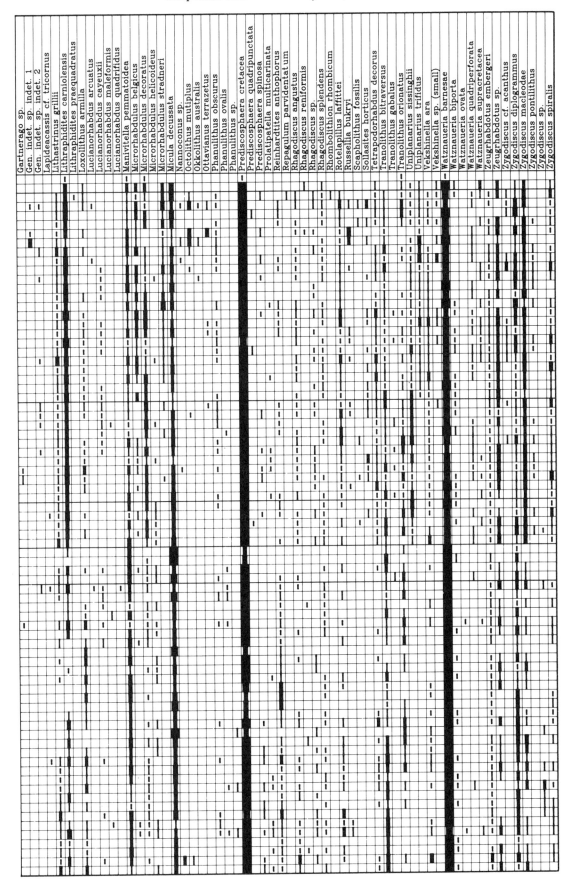

Figure 4. Distribution of calcareous nannofossils in the Sunderland Inlier, western Jamaica.

tive abundances of species in the topmost sample, SU-100, were not evaluated because this sample yielded only extremely rare specimens of coccoliths. In the checklist chart (Fig. 4), occurrences of species in this sample are shown using the same symbol as that for the abundance category of "very rare."

The dominant calcareous nannofossil species in the samples are *Prediscosphaera cretacea* and *Watznaueria barnesae*. Consistently abundant taxa include *Micula decussata*, *Cribrosphaerella ehrenbergii*, *Cretarhabdus conicus* plus *surirellus*, and *Cretarhabdus sinuosus*. Species that are intermittently abundant are *Bidiscus ignotus*, *Biscutum constans*, *Biscutum patella*, and *Lithraphidites carniolensis*. Other species vary greatly in abundance from very common to very rare (Fig. 4). Taxonomic notes of these species are given in the appendix.

The Sunderland Formation

Age-diagnostic species recorded in the Sunderland Formation include *Aspidolithus parcus*, *Eiffellithus eximius*, *Ceratolithoides verbeekii*, and *Bukryaster hayi*. *Aspidolithus parcus* occurs as low as SU-23, the lowest sample examined in this study. This species is considered a reliable marker for the Campanian stage. *Ceratolithoides verbeekii* is in most cases rare but fairly persistent in the Sunderland Formation. Perch-Nielsen (1979, 1985) considered it to be essentially restricted to the late Early Campanian. *Bukryaster hayi* is also rare but is present in most of the Sunderland samples; its highest occurrence is at SU-55.

The Newman Hall Formation

In addition to *Aspidolithus parcus* and *Eiffellithus eximius*, age-diagnostic species recorded in the Newman Hall Formation include *Eiffellithus parallelus*, *Ceratolithoides aculeus*, *Ceratolithoides arcuatus*, and *Uniplanarius sissinghii*. The last three species first appear simultaneously in SU-60, the lowest sample examined for the Newman Hall Formation. The association indicates an early Late Campanian age. An overlap in the range of *Ceratolithoides aculeus* and *Ceratolithoides verbeekii*, the forerunner of *Ceratolithoides aculeus*, was not observed. The acme of *Ceratolithoides arcuatus* seems restricted to the Newman Hall Formation, although rare specimens of the species may be found in samples from higher horizons. *Eiffellithus parallelus* first occurs in SU-63. *Eiffellithus eximius* is common in the Sunderland and Newman Hall formations; it is present, but is rare, in the overlying Shepherd's Formation.

The Shepherd's Hall Formation

Age-diagnostic species recorded in the Shepherd's Hall Formation are *Aspidolithus parcus*, *Arkhangelskiella cymbiformis*, *Ceratolithoides aculeus*, *Reinhardtites anthophorus*, *Lithastrinus grillii*, *Eiffellithus eximius*, *Eiffellithus parallelus*, *Uniplanarius sissinghii*, *Uniplanarius trifidus*, and *Lithraphidites praequadratus*. *Uniplanarius trifidus* first occurs in SU-76; this datum is used to indicate the lower limit of the late Late Campanian. Although rare, *Eiffellithus eximius* persists as high as sample SU-78. A form closely resembling *Eiffellithus eximius*, herein documented as *Eiffellithus* sp. cf. *E. eximius*, continues to SU-86. *Arkhangelskiella cymbiformis* is present in the Shepherd's Hall Formation, but its lowest occurrence is difficult to determine because the species is in many cases difficult to separate from *Arkhangelskiella specillata*, a species that occurs much farther down. The highest occurrence of *Lithastrinus grillii* is at SU-98. *Lithraphidites praequadratus* is present only in sample SU-98.

CALCAREOUS NANNOFOSSIL ZONATION AND CORRELATION

Figure 5 shows the calcareous nannofossil zonation of Sissingh (1977, 1978), emended by Perch-Nielsen (1985) for the Campanian and adjacent intervals. This zonation is discussed because, with only minor modifications, it is most compatible with the floral succession recorded in the Campanian formations in the Sunderland Inlier. Listed below are Sissingh's Campanian zones observed in the Sunderland Inlier, starting from the oldest zone.

The Phanulithus ovalis Zone (zone 19)

Definition. Last occurrence (LO) of *Marthasterites furcatus* to the first occurrence (FO) of *Ceratolithoides aculeus*.

Age. Late Early Campanian.

Occurrence. All samples examined from the Sunderland Formation (SU-23 to SU-59).

Remarks. The *Phanulithus ovalis* Zone is characterized by the presence of *Ceratolithoides verbeekii*; *Phanulithus ovalis* is present but extremely rare in these samples. Sissingh (1977, 1978) subdivided the zone into a lower subzone 19a and an upper subzone 19b, based on the last occurrence of *Bukryaster hayi* within the zone. This marker species is also present in the samples and persists up to SU-59. Accordingly, most of the Sunderland Formation (SU-23 to SU-55) can be further restricted to subzone 19a, and only the topmost four Sunderland samples (SU-56 to SU-59) are assigned to subzone 19b.

The Ceratolithoides aculeus Zone (zone 20)

Definition. FO of *Ceratolithoides aculeus* to the FO of *Uniplanarius sissinghii*.

Age. Late Early Campanian.

Occurrence. This zone was not observed in the studied samples.

Remarks. This zone and the superjacent subzone 21a were not recorded, which indicates a prominent hiatus in the section. This hiatus closely coincides with the Sunderland–Newman Hall formational contact, but it is unclear whether the hiatus is related to a local fault or to an unconformity.

The *Uniplanarius sissinghii* Zone (zone 21)

Definition. FO of *Uniplanarius sissinghii* to the FO of *Uniplanarius trifidus*.

Age. Early Late Campanian.

Occurrence. All the samples from the Newman Hall Formation (SU-60 to SU-64) and those from the lower part of the Shepherd's Hall Formation (SU-65 to SU-75).

Remarks. Sissingh (1977, 1978) thought the total range of *Ceratolithoides arcuatus* to be confined to the *Uniplanarius sissinghii* Zone (zone 21), and based on that he subdivided the zone into the three subzones 21a, 21b, and 21c, subzone 21b corresponding to the total range of *Ceratolithoides arcuatus*. In the Jamaica section the acme of *Ceratolithoides arcuatus* is restricted to zone 21, but the total range of this marker species actually extends beyond the top of the zone. Hence the top of subzone 21b is placed at the top of the acme of *Ceratolithoides arcuatus*, which in Jamaica also closely corresponds to the top of persistent *Eiffellithus eximius* (Fig. 4).

According to this revised definition, the five samples examined from the Newman Hall Formation (SU-60 to SU-64) are assigned to subzone 21b because they yielded persistent *Ceratolithoides arcuatus* and *Eiffellithus eximius*. The 11 samples from the lower part of the Shepherd's Hall Formation (SU-65 to SU-75) contained only sporadic *Ceratolithoides arcuatus* and *Eiffellithus eximius* and are assigned to subzone 21c. On stratigraphic position, the Stapleton (*Barrettia*) Limestone is restricted to zone 21, but whether it belongs to subzone 21b or 21c is uncertain.

Eiffellithus parallellus is another reliable marker species. In Jamaica it first occurs in subzone 21b just below the top of "persistent" *Eiffellithus eximius*.

The *Uniplanarius trifidus* Zone (zone 22)

Definition. FO of *Uniplanarius trifidus* to the LO of *Reinhardtites anthophorus*.

Age. Late Late Campanian.

Occurrence. Samples from the middle and upper part of the Shepherd's Hall Formation (SU-76 to SU-100).

Remarks. Sissingh (1977, 1978) subdivided the *Uniplanarius trifidus* Zone (zone 22) into subzones 22a and 22b, based on the FO of *Reinhardtites levis* within the zone. Perch-Nielsen (1979, 1985) showed that the LO of *Lithastrinus grillii* is closely associated with the FO of *Reinhardtites levis*; hence, empirically, subzone 22a can be recognized by the association of *Reinhardtites anthophorus*, *Uniplanarius trifidus*, and *Lithastrinus grillii*, without *Reinhardtites levis*, and subzone 22b by that of *Reinhardtites anthophorus*, *Uniplanarius trifidus*, and *Reinhardtites levis*, without *Lithastrinus grillii*. Both authors assigned subzones 22a and 22b to the Late Campanian.

Because most samples from the middle and upper part of the Shepherd's Hall Formation (SU-76 to SU-98) are characterized by the presence of *Reinhardtites anthophorus*, *Uniplanarius trifidus*, and *Lithastrinus grillii*, without *Reinhardtites levis*, they are assigned to subzone 22a. On stratigraphic position, and assuming continuous deposition, the overlying two samples SU-99 and SU-100 seem assignable to subzone 22b, as *Lithastrinus grillii* was not seen. However, neither were *Aspidolithus parcus* and

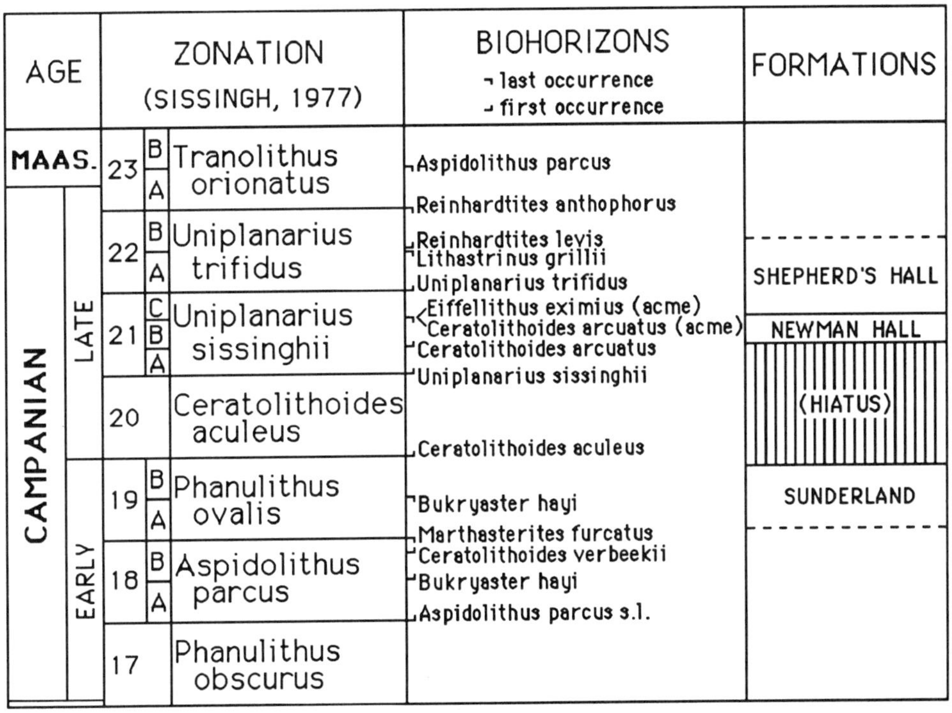

Figure 5. Correlation of Sunderland section with Campanian calcareous nannofossil zonation.

Reinhardtites anthophorus found in these two samples. Thus the two topmost samples from the Shepherd's Hall Formation may as well be assigned to the *Tranolithus orionatus* Zone (zone 23), a zone that straddles the Campanian-Maastrichtian boundary.

The highest occurrence of *Eiffellithus eximius* is within subzone 22a in Jamaica, but this marker species gradually becomes atypical and is extremely rare toward its extinction (see also taxonomic notes for details). As Doeven et al. (1982) have concluded, the "true" top of the species is actually not as reliable as its "persistent" top for biostratigraphic correlation.

Roth (1978) suggested *Lithraphidites praequadratus* should first occur somewhere near the Campanian-Maastrichtian boundary. In this study the species was found to occur in subzone 22a in the Late Campanian.

SUMMARY

Seventy-seven samples (SU-23 through SU-100, less SU-26) from the Sunderland Inlier, western Jamaica, were analyzed for their calcareous nannofossil content. Important datums recorded in the section are, from bottom to top, the LO of *Bukryaster hayi* at SU-55; the LO of *Ceratolithoides verbeekii* at SU-59; the FO of *Ceratolithoides aculeus, Ceratolithoides arcuatus,* and *Uniplanarius sissinghii* at SU-60; the FO of *Eiffellithus parallelus* at SU-63; the last persistent occurrence of *Eiffellithus eximius* at SU-64; the FO of *Uniplanarius trifidus* at SU-76; and the LO of *Lithastrinus grillii* at SU-98. This succession of datums is compatible with Sissingh's (1977, emended 1978) zonation framework for the Campanian Stage.

In terms of Sissingh's zonation, the Sunderland Formation in the section is assigned to zone 19 of late Early Campanian age. The Newman Hall Formation belongs to zone 21b of the early Late Campanian. The Shepherd's Hall Formation straddles zones 21c and 22 and is mainly placed in the late Late Campanian. A hiatus was proven at the contact of the Sunderland Formation and the Newman Hall Formation, which may indicate either a local fault or an unconformity.

ACKNOWLEDGMENTS

I am deeply grateful to Dr. Edward Robinson of the Department of Geology, Florida International University, and Dr. Stefan Gartner of the Department of Oceanography, Texas A&M University. Both provided guidance, offered helpful suggestions, and critically reviewed this text. My appreciation is also due the Petroleum Corporation of Jamaica for permission to use the 77 outcrop samples from the Sunderland Inlier. The calcareous nannofossil checklist chart (Fig. 4) was prepared by Nancy S. Jiang using VAX with a plotter at the Geochemical and Environmental Research Group, Department of Oceanography, Texas A&M University.

APPENDIX: CALCAREOUS NANNOFOSSIL SPECIES RECORDED

Listed below are calcareous nannofossil species recorded during the course of this study. They are mostly well documented in the literature and need no discussion. Hence, in the interests of brevity, only the simplest synonymy list has been prepared. For a more complete discussion and references of each species the reader is referred to Perch-Nielsen (1985) and Jiang (1989).

Ahmuellerella octoradiata (Gorka, 1957) Reinhardt, 1966.
Ahmuellerella regularis (Gorka, 1957) Reinhardt and Gorka, 1967.
Amphizygus brooksii Bukry, 1969.
Remarks. *Amphizygus brooksii brooksii* Bukry (1969) and *Amphizygus brooksii nanus* Bukry (1969) were not separated in this study.
Arkhangelskiella cymbiformis Vekshina, 1959.
Remarks. *Arkhangelskiella cymbiformis* differs from *Arkhangelskiella specillata* by having a relatively smaller central structure and a slightly broader rim.
Arkhangelskiella specillata Vekshina, 1959.
Aspidolithus parcus (Stradner, 1963) Noel, 1969.
Remarks. Hattner et al. (1980) subdivided *Broinsonia parca* (= *Aspidolithus parcus*) into two subspecies: primitive *Broinsonia parca parca* and advanced *Broinsonia parca constricta.* In this study both forms were observed, but the advanced *Aspidolithus parcus constrictus* is numerically much more significant than primitive *Aspidolithus parcus parcus.*
Axopodorhabdus dietzmannii (Reinhardt, 1965) Wind and Wise *in* Wise and Wind, 1977.
Bidiscus ignotus (Gorka, 1957) Hoffmann, 1970.
Biscutum constans (Gorka, 1957) Black, 1967.
Biscutum patella Risatti, 1973.
Remarks. This species has been documented as *Biscutum coronum* Wind and Wise *in* Wise and Wind (1977).
Braarudosphaera bigelowii (Gran and Braarud, 1935) Deflandre, 1947.
Broinsonia matalosa (Stover, 1966) Doeven, 1983.
Bukryaster hayi (Bukry, 1969) Prins, 1971.
Ceratolithoides aculeus (Stradner, 1961) Prins and Sissingh *in* Sissingh, 1977.
Ceratolithoides arcuatus Prins and Sissingh *in* Sissingh, 1977.
Ceratolithoides verbeekii Perch-Nielsen, 1979.
Chiastozygus amphipons (Bramlette and Martini, 1964) Gartner, 1968.
Chiastozygus sp. cf. *C. bifarius* Bukry, 1969.
Chiastozygus dennisonii (Worsley, 1971) Jiang n. comb.
=*Eiffellithus dennisoni* Worsley, 1971, p. 1307, Plate 1, Figs. 11–13.
=*Chiastozygus mediaquadratus* Risatti, 1973, p. 22, Plate 6, Figs. 10–11.
=*Chiastozygus dennisonii* (Worsley) Jiang, 1989 (unpublished dissertation), p. 285 and 286, Plate 14, Figs. 15–16.
Chiastozygus garrisonii Bukry, 1969.
Chiastozygus litterarius (Gorka, 1957) Manivit, 1971.
Chiastozygus propagulis Bukry, 1969.
Chiastozygus synquadriperforatus Bukry, 1969.
Chiastozygus sp.
=*Chiastozygus* sp., Jiang, 1989 (unpublished dissertation), p. 289, Plate 9, Figs. 19–20.
Corollithion exiguum Stradner, 1961.
Corollithion signum Stradner, 1963.
Cretarhabdus bukryi Black, 1973.
=*Cretarhabdus striatus bukryi* Black, 1973, p. 55.
Cretarhabdus conicus Bramlette and Martini, 1964.
Cretarhabdus schizobrachiatus (Gartner, 1968) Bukry, 1969.
Cretarhabdus sinuosus (Noel, 1970) Jiang n. comb.
=*Heterorhabdus sinuosus* Noel, 1970, p. 48–49, Text-Fig. 9; Plate 13, Figs. 1a–c, 2, 3, 4, 6.
=*Cretarhabdus sinuosus* (Noel) Jiang, 1989 (unpublished dissertation), p. 295 and 296, Plate 11, Fig. 2b.

Cretarhabdus surirellus (Deflandre *in* Deflandre and Fert, 1954) Reinhardt, 1970.
Cribrocorona sp.
Remarks: This bright circular form resembles *Cribrocorona gallica* (Stradner, 1963) Perch-Nielsen (1973), but its central structure is much smaller and not discernible in the light microscope.
Cribrosphaerella circula (Risatti, 1973) Verbeek, 1977.
Cribrosphaerella ehrenbergii (Arkhengelsky, 1912) Deflandre, 1952.
Cyclagelosphaera sp.
Remarks. This small (3.5 µm to 4.5 µm) bright circular form probably has been called *"margarelli"* in the literature.
Cylindralithus asymmetricus Bukry, 1969.
Cylindralithus crassus Stover, 1966.
Cylindralithus serratus Bramlette and Martini, 1964.
Cylindralithus spp.
Remarks. This group of species may include *Cylindralithus nudus* Bukry (1969), *Cylindralithus sculptus* Bukry (1969), and *Cylindralithus oweinae* Perch-Nielsen (1973).
Dodekapodorhabdus noeliae Perch-Nielsen, 1968.
Eiffellithus anceps (Gorka, 1957) Reinhardt and Gorka, 1967.
Eiffellithus sp. cf. *E. eximius* (Stover, 1966) Perch-Nielsen, 1968.
Remarks. This form appears almost identical to *Eiffellithus eximius*. The only difference is that its bifurcated crossbars orient at a slightly larger angle (15 to 25°) to the major and minor axes of the coccolith. In this respect, it also appears somewhat similar to *Eiffellithus parallelus*, whose bifurcated crossbars orient at an angle close to 45°.
Eiffellithus eximius (Stover, 1966) Perch-Nielsen, 1968.
Eiffellithus parallelus Perch-Nielsen, 1973.
Eiffellithus turriseiffelii (Deflandre *in* Deflandre and Fert, 1954) Reinhardt, 1965.
Gartnerago obliquum (Stradner, 1963) Reinhardt, 1970.
Gartnerago sp.
=*Gartnerago* sp., Jiang, 1989 (unpublished dissertation), p. 309 and 310, Plate 11, Figs. 13–16.
Lapideacassis sp. cf. *L. tricornus* Wind and Wise *in* Wise and Wind, 1977.
Lithastrinus grillii Stradner, 1962.
Lithraphidites carniolensis Deflandre, 1963.
Lithraphidites praequadratus Roth, 1978.
Loxolithus armilla (Black *in* Black and Barnes, 1959) Noel, 1965.
Lucianorhabdus arcuatus Forchheimer, 1972.
Lucianorhabdus cayeuxii Deflandre, 1959.
Lucianorhabdus maleformis Reinhardt, 1966.
Lucianorhabdus quadrifidus Forchheimer, 1972.
Manivitella pemmatoidea (Deflandre *in* Manivit, 1965) Thierstein, 1971.
Microrhabdulus belgicus Hay and Towe, 1963.
Microrhabdulus decoratus Deflandre, 1959.
Microrhabdulus helicoideus Deflandre, 1959.
Microrhabdulus stradneri Bramlette and Martini, 1964.
Micula decussata Vekshina, 1959.
Nannoconus sp.
Remarks. This *Nannoconus* sp. appears similar to *Nannoconus farinacciae* Bukry (1969) but is smaller and very rare.
Octolithus multiplus (Perch-Nielsen, 1973) Romein, 1979.
Okkolithus australis Wind and Wise *in* Wise and Wind, 1977.
Ottavianus terrazetus Risatti, 1973.
Phanulithus obscurus (Deflandre, 1959) Wind and Wise *in* Wise and Wind, 1977.
Phanulithus ovalis (Stradner, 1963) Wind and Wise *in* Wise and Wind, 1977.
Phanulithus sp.
=*Phanulithus* sp., Jiang, 1989 (unpublished dissertation), p. 341, Plate 8, Fig. 11a.
Prediscosphaera cretacea (Arkhangelsky, 1912) Gartner, 1968.
Remarks. *Prediscosphaera grandis* Perch-Nielsen (1979) was not separated from *Prediscosphaera cretacea* in this study.
Prediscosphaera quadripunctata (Gorka, 1957) Reinhardt, 1970.
Prediscosphaera spinosa (Bramlette and Martini, 1964) Gartner, 1968.
Prolatipatella multicarinata Gartner, 1968.
Reinhardtites anthophorus (Deflandre, 1959) Perch-Nielsen, 1968.
Repagulum parvidentatum (Deflandre *in* Deflandre and Fert, 1954) Forchheimer, 1972.
Rhagodiscus angustus (Stradner, 1963) Reinhardt, 1971.
Rhagodiscus reniformis Perch-Nielsen, 1973.
Rhagodiscus splendens (Deflandre, 1953) Verbeek, 1977.
Rhagodiscus sp.
=*Rhagodiscus* sp., Jiang, 1989 (unpublished dissertation), p. 354, Plate 15, Fig. 16.
Rhombolithion rhombicum (Stradner and Adamiker, 1966) Black, 1973.
Rotelapillus laffittei (Noel, 1957) Perch-Nielsen, 1985.
Russellia bukryi Risatti, 1973.
Scapholithus fossilis Deflandre *in* Deflandre and Fert, 1954.
Sollasites horticus (Stradner, Adamiker, and Maresch *in* Stradner and Adamiker, 1966) Black, 1968.
Tetrapodorhabdus decorus (Deflandre *in* Deflander and Fert, 1954) Wind and Wise *in* Wise and Wind, 1977.
Tranolithus bitraversus (Stover, 1966) Jiang n. comb.
=*Parhabdolithus? bitraversus* Stover, 1966, p. 145, Plate 6, Figs. 20a–b, 21a–b, 22a–b; Plate 9, Fig. 19.
=*Tranolithus bitraversus* (Stover) Jiang, 1989 (unpublished dissertation), p. 361 and 362, Plate 13, Figs. 2–4.
Tranolithus gabalus Stover, 1966.
Tranolithus orionatus (Reinhardt, 1966) Reinhardt, 1966.
Uniplanarius sissinghii (Perch-Nielsen, 1986) Jiang n. comb.
=*Uniplanarius gothicus* (Deflandre, 1959) Hattner and Wise, 1980, p. 68, Plate 32, Fig. 4; Plate 42, Figs. 4–5.
=*Quadrum sissinghii* Perch-Nielsen, 1986, p. 838–839, Plate 3, Figs. 3–5.
Uniplanarius trifidus (Stradner *in* Stradner and Papp, 1961) Hattner and Wise, 1980.
Vekshinella ara Gartner, 1968.
Vekshinella sp.
Remarks. This small (4.5 µm to 6 µm) elliptical form probably has been frequently called *"crux"* in the literature.
Watznaueria barnesae (Black *in* Black and Barnes, 1959) Perch-Nielsen, 1968.
Watznaueria biporta Bukry, 1969.
Watznaueria ovata Bukry, 1969.
Watznaueria quadriperforata (Bukry, 1969) Reinhardt, 1971.
Watznaueria supracretacea (Reinhardt, 1965) Wind and Wise *in* Wise and Wind, 1977.
Zeugrhabdotus embergeri (Noel, 1959) Perch-Nielsen, 1984.
Zeugrhabdotus sp.
Remarks. This small (5 µm to 7 µm) elliptical form probably has been called *"erectus"* or *"noeli"* in the literature.
Zygodiscus sp. cf. *Z. acanthus* (Reinhardt, 1965) Reinhardt, 1966.
Zygodiscus diplogrammus (Deflandre *in* Deflandre and Fert, 1954) Gartner, 1968.
Zygodiscus macleodae Bukry, 1969.
Remarks. *Zygodiscus macleodae* Bukry (1969) is a senior synonym of *Zygodiscus minimus* Bukry (1969) and *Zygodiscus tarboulensis* Shafik and Stradner (1971).
Zygodiscus pontilithus (Bukry, 1969) Jiang n. comb.
=*Percivalia pontilitha* Bukry, 1969, p. 54, Plate 30, Figs. 11–12; Plate 31, Fig. 1.
=*Zygodiscus pontilithus* (Bukry) Jiang, 1989 (unpublished dissertation), p. 377, Plate 14, Figs. 17–19.
Zygodiscus spiralis Bramlette and Martini, 1964.
Zygodiscus sp.
=*Zygodiscus* sp., Jiang, 1989 (unpublished dissertation), p. 378,

Plate 15, Figs. 7–8.

Remarks. Zygodiscus sp. appears somewhat similar to *Reinhardtites anthophorus,* but it is smaller and the elliptical rim structure is narrower. The spiral central stem of the species is also much reduced and less prominent. This species probably has been called *"elegans"* in the literature.

Gen. indet. sp. indet. 1

=Gen. indet. sp. indet. 1, Jiang, 1989 (unpublished dissertation), p. 378 and 379, Plate 17, Figs. 1–3.

Remarks. This form is polygonal or roughly circular in outline and consists of seven segments. It superficially resembles *Lithastrinus septenarius* Forchheimer (1972), but the seven segments are wedge shaped and the suture lines that separate the segments are in many cases not clearly shown in cross-polarized light. This form may be a short morphotype of *Nannoconus* preferentially settled in plane view.

Gen. indet. sp. indet. 2

=Gen. indet. sp. indet. 5, Jiang, 1989 (unpublished dissertation), p. 380, Plate 7, Figs. 16, 20.

Remarks. This form is flask shaped and resembles *Laguncula* Black (1971). It appears bright and is distinctive in cross-polarized light.

REFERENCES CITED

Chubb, L. J., 1958, The Cretaceous rocks of central St. James: Geonotes, v. 1, p. 3–11.

Doeven, P. H., Gradstein, F. M., Jackson, A., Agterberg, F. P., and Nel, L. D., 1982, A quantitative nannofossil range chart: Micropaleontology, v. 28, p. 85–92.

Hattner, J. G., Wind, F. H., and Wise, S. W., 1980, The Santonian/Campanian boundary: Comparison of nearshore-offshore calcareous nannofossil assemblages: Cahiers de Micropaleontologie, v. 3, p. 9–26.

Jiang, M.-J., 1989, Biostratigraphy and geochronology of the Eagle Ford shale, Austin chalk, and lower Taylor marl in Texas, based on calcareous nannofossils [Ph.D. dissertation]: College Station, Texas A&M University, 496 p.

Jiang, M.-J., and Robinson, E., 1987, Calcareous nannofossils and larger foraminifera in Jamaican rocks of Cretaceous and early Eocene age; *in* Ahmad, R., ed., Proceedings, Workshop on the Status of Jamaican Geology, Kingston, March 1984: Journal of the Geological Society of Jamaica, Special Issue, p. 24–51.

Jiang, M.-J., and Robinson, E., 1989, Zonation of some Jamaican Cretaceous rocks using calcareous nannofossils: Transactions, Caribbean Geological Conference, 10th, Cartagena, Colombia, August 14–22, 1983: p. 243–249.

Krijnen, J. P., 1972, Morphology and phylogeny of pseudorbitoid foraminifera from Jamaica and Curacao; a revisional study: Scripta Geologica, v. 8, 133 p.

Perch-Nielsen, K., 1979, Calcareous nannofossils from the Cretaceous between the North Sea and the Mediterranean: Aspekte der Kreide Europas, IUGS Ser. A, no. 6, p. 223–272.

Perch-Nielsen, K., 1985, Mesozoic calcareous nannofossils, *in* Bolli, H. M., Saunders, J. B., and Perch-Nielsen, K., eds., Plankton stratigraphy: Cambridge, Cambridge University Press, p. 329–426.

Roth, P. H., 1978, Cretaceous nannoplankton biostratigraphy and oceanography of the north-western Atlantic Ocean: Initial Reports of the Deep Sea Drilling Project, v. 44, p. 731–759.

Sissingh, W., 1977, Biostratigraphy of Cretaceous calcareous nannoplankton (with Appendix by Prins, B., and Sissingh, W.): Geologie en Mijnbouw, v. 56, p. 37–65.

Sissingh, W., 1978, Microfossil biostratigraphy and stage-stratotypes of the Cretaceous: Geologie en Mijnbouw, v. 57, p. 433–440.

van Dommelen, H., 1972, Ontogenesis, phylogenetic, and taxonomic studies of the American species of *Pseudovaccinites* and of *Torreites* and the multiple-fold Hippuritids [Ph.D. dissertation]: Amsterdam, University of Amsterdam, 125 p.

Zans, V. A., Chubb, L. A., Versey, H. R., Williams, J. B., Robinson, E., and Cooke, D. L., 1963, Synopsis of the geology of Jamaica: An explanation of the 1958 provisional geological map of Jamaica. Kingston, Geological Survey Department, Bulletin 4, 72 p.

Manuscript Accepted by the Society November 11, 1992

Review of upper Cretaceous orbitoidal larger foraminifera from Jamaica, West Indies, and their connection with rudist assemblages

Jan P. Krijnen
Geological Survey of New South Wales, 29-57 Christie Street, St. Leonards, NSW 2065, Australia
H. J. Mac Gillavry
Rubenslaan 1, Flat 226, BM Bilthoven, The Netherlands
H. van Dommelen
Organization of Pure Scientific Research (Z.W.O.), Koningin Sophie Straat, The Hague, The Netherlands

ABSTRACT

Cretaceous orbitoid larger foraminifera of the families Pseudorbitoididae, Orbitoididae, and Orbitocyclinidae are known from the Green Island and Grange inliers (western Jamaica), from the Sunderland and Central inliers (west-central Jamaica), and from the Blue Mountain Inlier (eastern Jamaica). They occur in prominent limestone units that are regarded as ranging in age from Campanian to Early Maastrichtian on the basis of nannofossils and accompanying rudist assemblages. Cretaceous pseudorbitoid, orbitoid, and orbitocyclinid foraminifera are useful in interpreting fluctuations of the depositional environment and for time-stratigraphic correlation of Late Cretaceous limestones across Jamaica.

INTRODUCTION

Cretaceous orbitoidal larger foraminifera and rudists were sampled from several horizons in five Cretaceous inliers in Jamaica during three periods of extensive fieldwork. Subsequent studies of the collected material include an analysis of a selected group of rudists (van Dommelen, 1971) and four papers on pseudorbitoidal foraminifera (Krijnen, 1971, 1972, 1978, 1993). These studies were aimed at providing a better understanding not only of the morphological characteristics and evolutionary development of these fossils but also of their stratigraphic significance in the Caribbean region. For the same purpose, ample material collected from the Central Inlier is presently being processed for a forthcoming statistical study of the pseudorbitoid genus *Vaughanina*, which occurs abundantly in several horizons together with *Orbitoides media megaloformis*.

The present study essentially intends to focus attention on describing as well as comparing the morphology of the various forms encountered in Jamaica. It aims at providing an aid to recognition of these forms and may assist those engaged in future fieldwork. We hope it may also result in obtaining additional samples needed for follow-up evolutionary studies of several groups of these foraminifera.

Past fieldwork was carried out during a three-month trip across Jamaica led by Professor H. J. Mac Gillavry (from February to May 1967) and during two assignments of the first author to the Geological Survey Department in Jamaica (from 1973 to 1976 and from 1978 to 1981).

The project, under the supervision of Professor Mac Gillavry, consisted of sampling of rudists and larger foraminifera from representative road and river traverses within the major Cretaceous inliers. All the samples, presently housed in the National Museum of Geology and Mineralogy (Leyden, The Netherlands), were thin sectioned or washed, and the microfauna subsequently analyzed by Professor Mac Gillavry. The material from the Green Island and Sunderland inliers was found to be rich in foraminifera and rudists. Several traverses in the Central Inlier proved to be equally rich in foraminifera and rudists at a number of horizons. Its association of faunas, therefore, was also found to be of significant stratigraphic value. To date only a few of these sections have been incorporated in published theses (van Dommelen, 1971; Krijnen, 1972). The bulk of Mac Gillavry's analyses of the sections from the Central Inlier, however, has never been published, and we are glad to have the opportunity to do so now.

During the field campaign, Professor Mac Gillavry was assisted by the two coauthors, who were still postgraduate stu-

Krijnen, J. P., Mac Gillavry, H. J., and van Dommelen, H., 1993, Review of upper Cretaceous orbitoidal larger foraminifera from Jamaica, West Indies, and their connection with rudist assemblages, *in* Wright, R. M., and Robinson, E., eds., Biostratigraphy of Jamaica: Boulder, Colorado, Geological Society of America Memoir 182.

dents at the time. Mrs. Mac Gillavry joined in most of the traverses, and her presence on many occasions in the field, with friends in Kingston, and while staying in unforgettable places like Green Park House, Katie Heird's Guest House, and a number of other idyllic guest houses across the island, contributed to a pleasant familylike atmosphere during our stay in Jamaica.

The first author's second and third periods of fieldwork in Jamaica were arranged through a Technical Assistance Agreement between the governments of The Netherlands and Jamaica and related to assessing the feasibility of a hydroelectric project in the Blue Mountain range. For a regional geological study of this inlier, he was seconded to the Geological Survey Department of Jamaica, initially for a three-year period from 1973 to 1976; that was followed by another term of three years from 1978 to 1981. During these years the drainage systems of Spanish River, Swift River, and Rio Grande, with its major tributary systems of Guava River (with Corn Husk River and Mocho Toms River) and Back Rio Grande (with Stony River), were systematically mapped. Compilations of the last five eastern 1:50,000 geological sheets were completed, based on the results of numerous smaller and larger geological mapping programs conducted by both the Geological Survey of Jamaica and the Geology Department of the University of the West Indies.

OCCURRENCES AND REGIONAL GEOLOGICAL SETTING

The Cretaceous sequences of the western, west-central, and central inliers in Jamaica comprise a variety of rock types that include waterlaid tuffs and conglomerates, tuffaceous shales, and limestones. In the western and central part of Jamaica, these deposits were laid down on an unstable shelf. Chubb (*in* Zans et al., 1962) considered periodic subsidence the cause of rhythmic deposition of at least twelve cycles of sedimentation, each normally starting with a conglomerate, being succeeded by shales, and ending with a limestone. However, only three major limestone horizons were encountered in the western, west-central, and central inliers.

In the eastern part of Jamaica, the Cretaceous succession, which includes substantial amounts of volcanic and volcaniclastic units, limestones, and shales, is in excess of 1,000 m and is intruded by late-phase granodioritic plutons and plugs. The depositional regime of the sequence is associated with a developing volcanic island arc. The presence of at least two Late Cretaceous bioclastic limestone units interbedded in the predominantly volcanic sequence may suggest an overall shallow marine environment of deposition. This seems to be confirmed not only by subaerial volcanic eruption centers found in volcanic sequences of the Bellevue Formation near Coopers Hill (H. McQueen, personal communication, 1976), but also by the presence of limestone debris and hardly eroded specimens of orbitoidal larger foraminifera in the matrix of unsorted immature volcaniclastic sequences of the Blue Mountain Formation.

The limestones, in particular the Rio Grande Limestone, vary from reef limestone to rubble beds within marly to tuffaceous sediments. In a number of cases the proximity to near-coast reefs can be demonstrated by the presence of interbedded clusters and fragments of bioherms. Generally these deposits contain a wealth of fossils, of which rudists and orbitoidal larger foraminifera form a major constituent. Rudists have to a large extent been used for biostratigraphic purposes, but the value of many, if not most, species as index fossils may be restricted because of adaptation to extreme environmental conditions. An optimal environment for advanced forms of rudists toward the end of the Cretaceous may have led to an explosive generation of species, each of which adapted to special conditions associated with available niches in order to survive vigorous competition. As a result, such species may have become less widespread and as such less useful for regional-scale correlation. The foraminifera, on the other hand, appear to be more suitable in that respect. The various species seem to occur abundantly and consistently across the Caribbean and parts of the Pacific Tethys, where they represent valuable markers for subdividing the Campanian-Maastrichtian interval.

The Western and West-Central Inliers

In the western and west-central Cretaceous inliers—that is, the Green Island Inlier in the Parish of Hanover, the Grange Inlier in the Parish of Westmoreland, and the Sunderland Inlier in the Parish of St. James—the rock sequence is dominated by thick-bedded biostromal to bioclastic calcarenites and rudist rubble beds, underlain by a predominant shaly succession and overlain by sandy sequences with intercalated conglomerate horizons.

The age of these successions has been regarded as Campanian, based on the presence of *Barrettia gigas* (Chubb *in* Zans et al., 1962). On the basis of nannoplankton, the age of the *Barrettia* limestone unit is presently considered to be late Middle Campanian and the age range of the entire shale, limestone, and conglomerate sequence from middle Early Campanian to possibly early Late Campanian (Jiang and Robinson, 1987).

The sequences containing orbitoidal larger foraminifera and rudists in the western and west-central Cretaceous inliers are generally confined to the Green Island Formation in the Green Island and Grange inliers and to the Sunderland, Newman's Hall, Stapleton, and Shepherd's Hall formations in the Sunderland Inlier.

The Green Island and Grange sections. The succession in the Green Island Inlier comprises several tens of meters of shale, tuffaceous sediments, conglomerates, and an interbedded dense to rubbly thick-bedded "*Barrettia* Limestone" (Zans et al., 1962) of the Green Island Formation (McFarLane, 1977). Detailed mapping of this unit in the Green Island Inlier by the first author, in 1975, showed the regional extent of the limestone unit (see Fig. 1). The limestone and underlying shale are rich in foraminifera and rudists, and a summary of successive fauna associations is shown in Figure 2. The Green Island inlier is the type locality of the orbitoid larger foraminifer *Pseudorbitoides trechmanni* H. Douvillé, 1922. Abundant free specimens of microspheric and

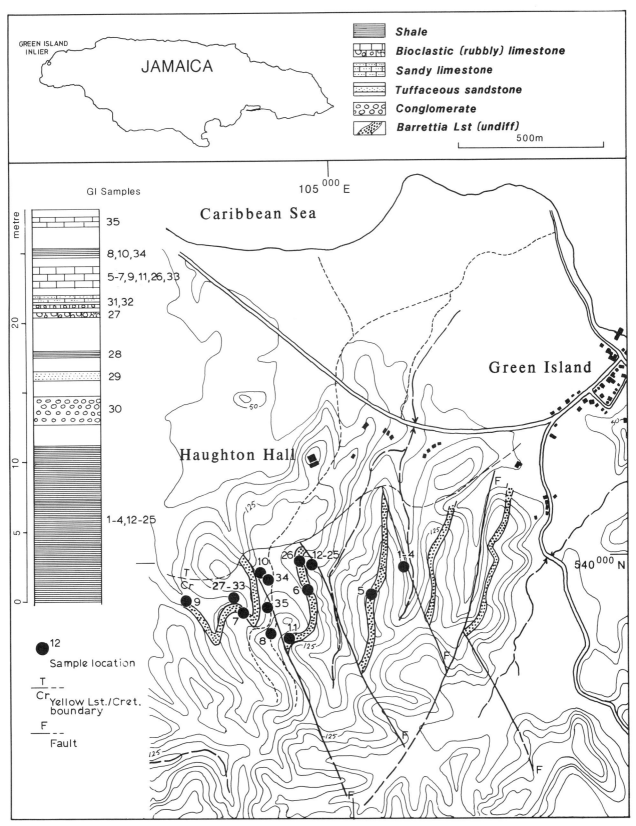

Figure 1. Stratigraphy of the Green Island Formation, Green Island Inlier, with outcrop pattern of the *Barrettia* Limestone member and distribution of sample locations.

megalospheric forms were sampled from the shale unit at the base, from the sandy to rubbly limestone beds below the limestone member, and from the surrounding cultivation soil. The dense *Barrettia* Limestone Member itself also contains an abundance of *P. trechmanni* but in addition contains, in sample GI6 only, several specimens and fragments of the related form *Radiorbitoides pectinatus* (Krijnen, 1993).

Rudists are represented by an abundance of forms and specimens in several horizons in the succession. The *Barrettia* Limestone is at places mainly a *Barrettia*-rubble bed. Weathering of the unit has released many perfect, or almost perfect, specimens of *Barrettia gigas, B. multilirata,* and *Parastroma trechmanni* (see van Dommelen, 1971) and fragments of *"Durania" nicholsasi.* Mac Gillavry noted that the species *nicholasi* Whitfield is not a *Radiolites* nor a *Sphaerulites, Lapeirousia,* or *Durania* and may represent a new genus. It is mentioned as *Bournonia, new section,* in Mac Gillavry, 1937, p. 41, and as *"Durania" nicholasi* in Mac Gillavry (1977), the quotation marks denoting that this generic designation is believed to be incorrect but that it is often so written. The best thing to do would probably be to write *Nov. gen. nicholasi* (Whitfield, 1897). In our notes several specimens have been identified as *Durania nicholasi* or *Radiolites nicholasi.* Following Mac Gillavry (1977), these names, with generic specification in quotation marks, have been retained here (refer to Figs. 3, 7, 8, 10, and 11).

The fossils are often partly or wholly buried in the soil

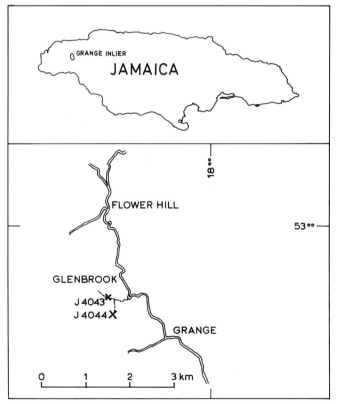

Figure 2. Foraminiferal and rudist assemblages of the Green Island and Grange Inliers. *Parastroma trechmanni* was exclusively found in the Grange Inlier.

on the east-sloping hillside, about 500 m due south of Haughton Hall. An almost perfect specimen of *"Durania" nicholasi* was found as float in the gully about 300 m southeast of Haughton Hall. The sandy to rubbly limestone underlying the *Barrettia*-rubble yielded small specimens of *Barrettia, ?Vaccinites,* and *?Sauvagesia.* Quite a number of specimens of *Antillocaprina* and *Mitrocaprina* were found in the cultivation soil above the *Barrettia* Limestone level (H. J. Mac Gillavry, personal communication, 1975).

The succession in the Grange Inlier in north-central Westmoreland at Glenbrook, between Flower Hill to the north and Grange to the southeast, includes extremely fossiliferous beds. Two horizons were extensively sampled for rudists and foraminifera (see Fig. 3 for sample locations), which are predominantly made up of fragmental limestone embedded in a clayey matrix. From the matrix of one of the samples (J4044) some 10 megalospheric specimens of *Pseudorbitoides trechmanni* were recovered. Horizontal sections across the equatorial layer revealed a biserial to quadriserial nepionic chamber arrangement similar to that of the majority of the specimens from Green Island. Since the configuration of nepionic chambers is regarded to be subject to an evolutionary trend in the majority of orbitoid larger foraminifera (Mac Gillavry, 1963), an evolutionary level similar to the fauna from Green Island is suggested (Krijnen, 1971, 1972).

Many of the limestone fragments are right valves of *Parastroma trechmanni,* except for one juvenile specimen of *Barrettia multilirata* and one specimen of *Antillocaprina* sp. (van Dommelen, 1971).

The age of the units in the Green Island and Grange inliers has been revised on the basis of nannoplankton and is regarded now as late Middle Campanian (Jiang and Robinson, 1987).

The Sunderland section. In the Sunderland Inlier, the succession containing larger foraminifera comprises in ascending stratigraphic order the Sunderland, the Newman's Hall, the Stapleton, and the Shepherd's Hall formations. The units cover the southern part of the inlier, and the combined thickness of the

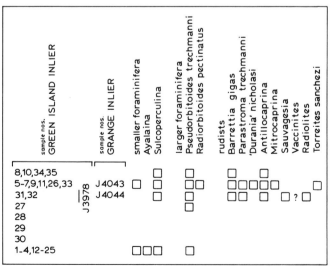

Figure 3. Sample locations in the Grange Inlier.

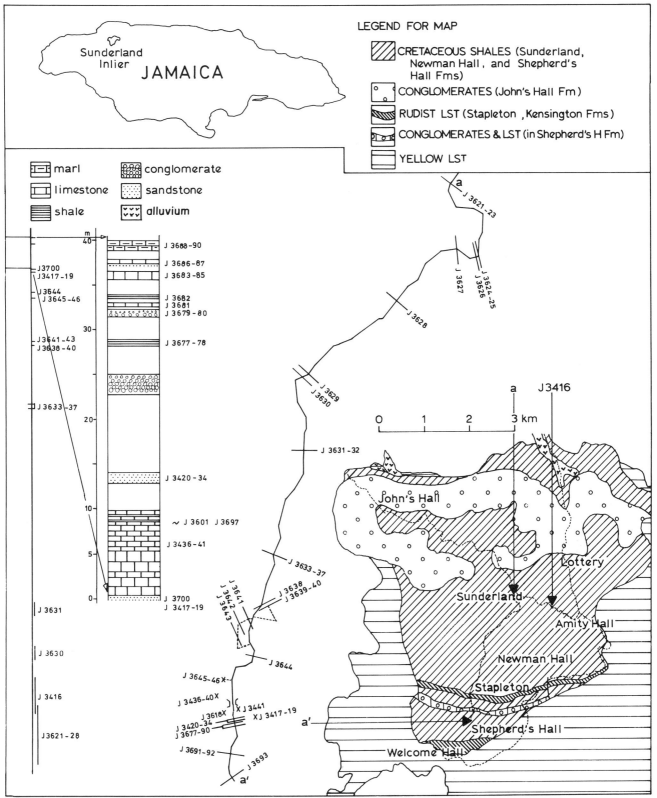

Figure 4. Stratigraphy, geology, and sample locations of the Sunderland section (a-a'), in the Sunderland Inlier, west-central Jamaica. The geology is slightly modified after Chubb, *in* Zans et al. (1962).

sequence is approximately 1,500 m (see Fig. 4). A number of horizons contain excellent marker fossils, which can be grouped into characteristic fauna associations (see Fig. 5).

The shale of the Sunderland Formation was found to be predominantly rich in planktonic foraminifera. Those identified by H. Bolli included the species *Globotruncana stuarti, G. fornicata,* and *G. lapparenti ss.* (Chubb, 1958, p. 7). Meinster and Kuhry of the Microplankton Section of the University of Amsterdam identified a slightly different association that included the forms *G. elevata, G. fornicata, G. rosetta, G. linneiana, G. falsostuarti, G. arca, G. coronata, G. bulloides,* and *G. ventricosa.* Between Sunderland and Amity Hall, sample J3416—taken from a sandy-silty level—proved to contain the primitive species *Pseudorbitoides* cf *P. (?) chubbi.* The overlying Newman's Hall Formation was found to be generally devoid of orbitoidal larger foraminifera, except for a 10-cm-thick marly horizon, a few centimeters below the base of the Stapleton Limestone. This horizon is the earliest occurrence of *Radiorbitoides pectinatus* in association with *Sulcoperculina* and fragments of rudists. The occurrence of calcareous-shaly concretions and marly horizons in this unit and the absence of planktonic foraminifera may indicate a shallowing transitional period between full-marine upper to lower neritic environment of deposition of the Sunderland Formation and littoral to upper neritic fore-reef slope conditions during deposition of the Stapleton Limestone member.

The Stapleton Formation, described as a more-than-7-m thick limestone unit conformably overlying the Newman's Hall Formation, is a typical fore-reef deposit, with numerous rudists and rudist fragments deposited in a rubble deposit. Not a single specimen was found to be in growth position. Identified rudists include *Barrettia gigas* in association with *Torreites sanchezi.* The larger foraminifer *Pseudorbitoides trechmanni* occurs abundantly, particularly near the top of the unit. A detailed study of the nepiont of this form may show a slightly more advanced evolutionary stage than those of the faunas from the western inliers. It suggests that the Stapleton Limestone may be slightly younger than the *Barrettia* Limestone in the Green Island and Grange inliers.

The Shepherd's Hall Formation conformably overlies the Stapleton Formation and is described as an approximately 900-m-thick unit consisting of tuffs, marls, mudstone, limestone, and conglomerate horizons (Zans et al., 1962). The outcrop of conglomerates along the roadside, about 100 to 150 m due south of the bridge of Stapleton, is the designated type locality of *Radiorbitoides pectinatus.* It comprises three foraminiferal horizons that occur approximately 20 to 30 m stratigraphically above the top of the Stapleton Limestone Member (Fig. 4).

The pseudorbitoidal forms of samples J3700, J3680, J3686, and J3688 were originally assigned as a subspecies of *Pseudorbitoides trechmanni* Douvillé, 1922 (Krijnen, 1972). On the basis of a reassessment of the diagnostic value of the radially arranged equatorial chambers in the adult equatorial layer, however, the taxonomic position was raised to genus level and the form renamed *Radiorbitoides pectinatus* (Krijnen, 1993). This form, presently the only species representing a possibly new lineage, has been regarded as an off-shoot from the *Pseudorbitoides* lineage. The nepionic chamber arrangement of *R. pectinatus* from the Newman's Hall Formation (sample J3700) and from the Shepherd's Hall Formation (samples J3680, J3686, and J3688) showed a slight difference in the number of primary spiral chambers and the size of the protoconch in the successive samples. This difference, however, is only marginal and considered too small to justify the separation of a second species of *Radiorbitoides.* Other than the usual rudist fragments, small forms of *?Antillocaprina* sp. were identified.

Age of the various formations in the Sunderland Inlier. Based on nannofossils from the Sunderland Inlier (Jiang and Robinson, 1987), *Pseudorbitoides chubbi* from the Sunderland Formation is now regarded as of approximately middle Early Campanian age (Nannofossil Zone 19a) and *P. trechmanni* of the upper Newman's Hall Formation and the Stapleton Limestone of late Middle Campanian age (Nannofossil Zone 21b-21c), whereas *Radiorbitoides pectinatus* of the Shepherd's Hall Formation

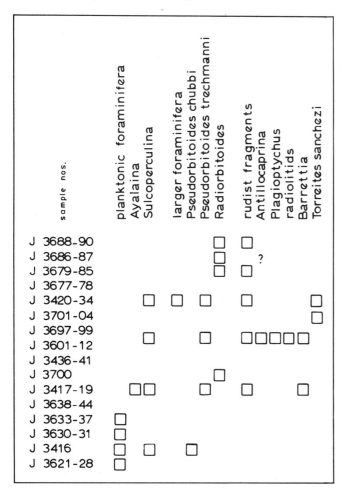

Figure 5. Foraminiferal and rudist assemblages encountered along the Sunderland–Welcome Hall traverse, Sunderland Inlier. Samples J3700-3704 and J3601-3612 were taken along the road from Amity Hall to Welcome Hall. Sample J3416 is located along the road from Amity Hall to Sunderland.

may range from late Middle Campanian to possibly early Late Campanian (Nannofossil Zone 21b-22a). *P. trechmanni* from the Stapleton area shows a slightly more evolutionarily advanced nepiont in comparison to the forms from Green Island. Although the Stapleton Limestone is considered to be of a similar age to the *Barrettia* Limestone member in the Green Island Inlier (Krijnen, 1972; Jiang and Robinson, 1987), a slightly younger age of this horizon should not be ruled out.

The Central Inlier

In the northwestern Central Inlier the rudist-foraminifer-bearing interval is predominantly a calcarenitic to rubbly limestone, marl, and shale of the Guinea Corn Formation (Zans et al., 1962; Wright, 1974). The sequence is folded into open synclines and anticlines and faulted particularly in the limb zones, with northeast-southwest direction of faulting broadly parallel to the fold axes. Faulting may be related to a phase of step-faulting during, or immediately after, an east-west compression phase, with progressive stepping-up of fault blocks to the northwest.

The fossiliferous foraminiferal-rudist zone is exposed along several road traverses. Generally the sequence is repeated several times as a result of a combination of shallow dips, variable dip directions, changing road directions, and the presence of faults. Outcrops and exposures were sampled in detail along road traverses from White Shop to Grantham (the Logie Green section), from White Shop to Tweedside (the Sanguinetti section), and along the Borobridge-Christiana and Borobridge-Alston-Baillieston road (the Alston sections).

The Logie Green section. This approximately 100-m-thick sequence was measured and sampled along the road from Sanguinetti to Grantham (see Fig. 6). It comprises more or less massive, thin- to medium-bedded limestones at the base, a middle section consisting of a claystone horizon followed by thin-bedded marls and shales, and an extremely rich foraminiferal horizon associated with thin-bedded marls and limestones near the top (see Fig. 7-II). The eastern part of the traverse appears to consist of pronounced medium- to thick-bedded limestone horizons (Fig. 7-I). A change into a more calcareous environment of deposition to the east suggests eastward-shallowing water depths. The sequence dips to the north and forms the southern limb of a west-southwest–plunging faulted syncline. Various horizons are rich to extremely rich in larger foraminifera and found to be associated with rudists (see Fig. 7). The foraminiferal assemblages throughout the sequence generally contain abundant *Kathina*, whereas *Chubbina cardenasensis, Sulcoperculina,* and the orbitoid larger foraminifera *Orbitoides media megaloformis* and *Vaughanina cubensis* are mainly confined to the upper half of the sequence. Throughout the sections the rudists *Titanosarcolites, Antillocaprina, Biradiolites,* and radiolitids are common, as to a lesser extent are *Chiapasella, Thyrastylon, "Durania" nicholasi,* and *Sauvagesia*.

The Sanguinetti section. This section is exposed along the road from White Stop to Tweedside (see Fig. 6). It comprises an approximately 100-m-thick sequence of medium- to thick-bedded limestone at the base, succeeded by an alternation of claystone, marls, and limestones, almost similar to the western part of the Logie Green section (see Fig. 6, traverse b to b' and b''; Fig. 8-I). About 1 km southwest of Tweedside the sequence comprises less massive limestones and relatively more claystone and marls (Fig. 6, traverse c to c'; Fig. 8-II), which seems to confirm the suggestion that the environment of deposition shallows to the east or southeast and deepens to the west to northwest. The sequence southwest of Tweedside dips to the east, and it represents the southwestern continuation of the syncline at Logie Green. The stratigraphic distribution of foraminifer and rudist assemblages is shown in Figure 8. As in the Logie Green section, the foraminifer *Kathina* occurs throughout the entire sequence, and *Sulcoperculina, Orbitoides media megaloformis, Vaughanina cubensis,* sporadically *Asterorbis* cf *A. havanensis,* and *Chubbina cardenasensis* are mainly restricted to the upper to extreme upper levels of the sequence. Throughout the section the identified rudists include radiolitids, *Antillocaprina, Titanosarcolites, Biradiolites,* and *Sauvagesia*. Rare *Prebarrettia sparcilirata* appears to be restricted to the upper levels. *Plagioptychus* was recorded from lower horizons in the sequence.

MacGillavry (1977) considered the Guinea Corn Formation in the area of Figure 6, with occurrence of *Kathina* and general lack of other larger foraminifera, to represent a sheltered (?back-reef) environment, with the layer of abundant *Orbitoides* (see also p. 29) as a temporary open-sea influx. It is in this layer that *Prebarrettia sparcilirata* is found.

The Alston sections. These sections are repeatedly exposed along the road from Borobridge to Baillieston via Alston and from Borobridge to Christiana (see Fig. 9). They comprise an approximately 100-m-thick succession of limestones, marls, limestone rubble beds, sandstones, and claystones. Along the road from Borobridge to Christiana (Fig. 9, traverse a-b; see also Fig. 10-I), thick-bedded limestone seems to occur particularly toward the top of the sequence. The basal part is distinctly finer grained and consists of claystones, sandstones, and thin-bedded limestones.

Along the Borobridge-Alston traverse, the basal part of the sequence comprises limestone and limestone rubble beds, grading upward into limestone conglomerate horizons, with predominant finer-grained shale and sandstone beds toward the top (see Fig. 10-II). The southern Alston section, along the Alston-Baillieston road, is characterized by a dominant sandstone unit in the middle portion of the sequence (see Fig. 11). A deeper-water environment of deposition may be envisaged for the basal portion of the sequence, characterized by mass dumping of limestone debris from shallow near-reef conditions in a possibly upper to middle neritic environment. In the upper portion of the northern Alston sequences, where limestone and sandstone horizons frequently alternate with claystone, marl, and limestone rubble beds, conditions of sedimentation may have fluctuated between a marine upper neritic to transitional littoral environment of deposition.

The sequence extends as an approximately 300- to 400-m-

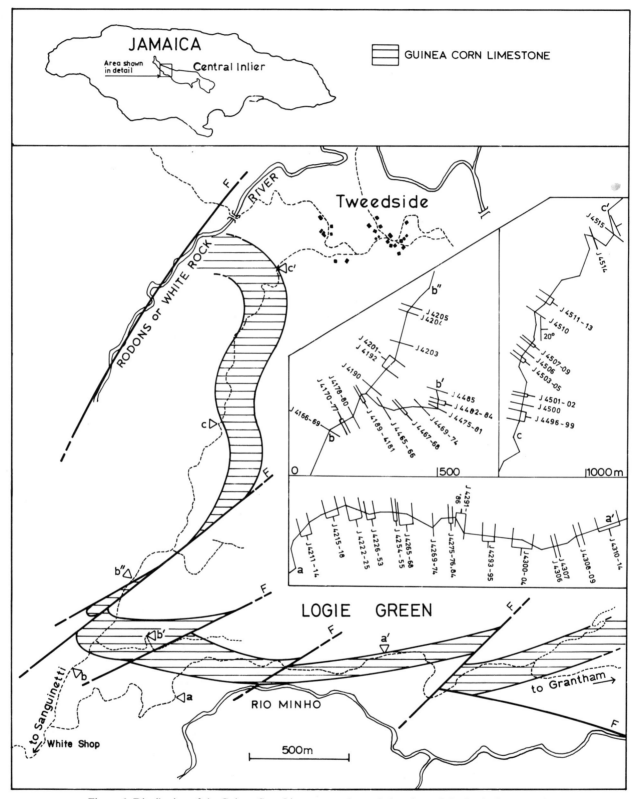

Figure 6. Distribution of the Guinea Corn Limestone and sample locations of the Logie Green (a-a') and Sanguinetti (b-b'-b" and c-c') road traverses, Central Inlier.

wide belt to the east, where it is exposed along the Sanguinetti-Tweedside road and described as the Sanguinetti section (see Fig. 6). To the south of Alston the unit dips to the north and northwest and represents the northwestern limb of a fault-bounded anticline that has its axis immediately to the east Baillieston and approximately along the Rodons or White Rock River (see Fig. 6). This anticline is truncated by a west-southwest- to east-northeast-trending fault that extends southwest of Alston approximately along the Cave River (see Fig. 9). The westward extension of the limestone belt is shifted to the north and is compressed into a north-south-trending syncline and anticline (northwest of Alston).

Foraminifera and rudist assemblages are shown in Figures 10 and 11. Among the smaller benthonic foraminifera, *Sulcoperculina* and *Kathina* are by far the dominant genera, with *Sulcoperculina* occurring throughout the entire sequence and *Kathina* possibly more confined to the upper half of the unit. *Chubbina* and *Ayalaina* are rare and only recorded near the top of the sequence. The orbitoid larger foraminifera *Orbitoides media megaloformis*, *Asterorbis* cf *A. havanensis*, and *Vaughanina cu-*

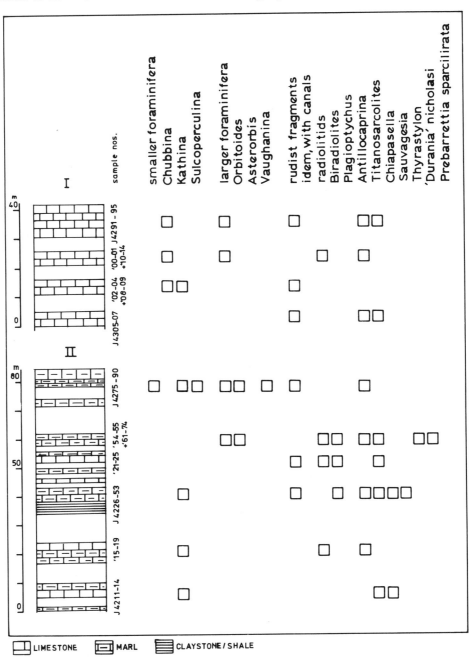

Figure 7. Stratigraphy and successive foraminiferal and rudist assemblages of the Logie Green traverse. I, road traverse White Shop–Grantham (eastern part of a-a' on Fig. 6); II, road traverse White Shop–Grantham (western part of a-a' on Fig. 6).

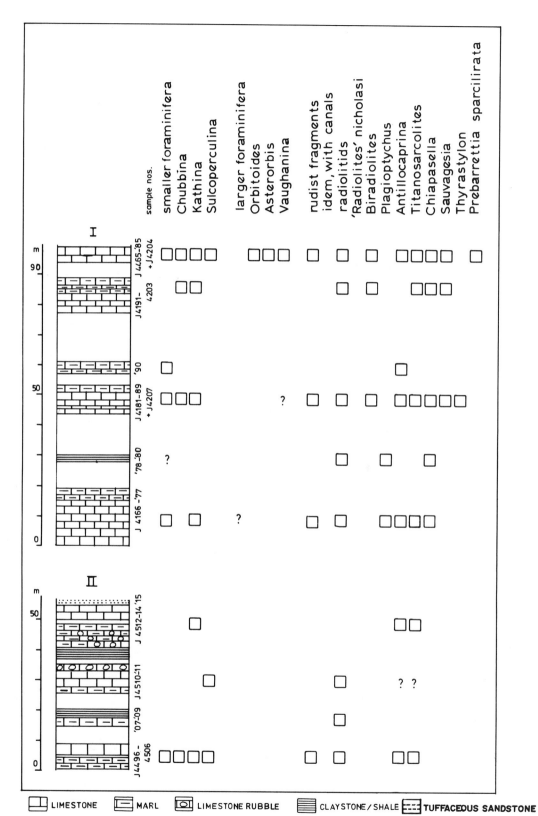

Figure 8. Stratigraphy and successive foraminiferal and rudist assemblages of the Sanguinetti traverses. I, White Shop–Tweedside road traverse (a-b'-b" on Fig. 6); II, White Shop–Tweedside road traverse (c-c' on Fig. 6).

bensis are apparently not confined to the upper part of the section as in the Logie Green and Sanguinetti sections. The most frequent occurrences are recorded from the Alston-Borobridge and Alston-Baillieston traverses, and if the lithological successions are interpreted correctly, a more or less uniform vertical distribution of these forms may be assumed.

Associated rudists are radiolitids fragments of rudists with canals (occurring in numerous horizons), *Antillocaprina, Titanosarcolites,* and an occasional *Biradiolites, Hippurites, Plagioptychus,* and *Prebarrettia sparcilirata.*

The age of the sequences. A Maastrichtian age for the sequence was based on the occurrence of *Titanosarcolites* (Chubb

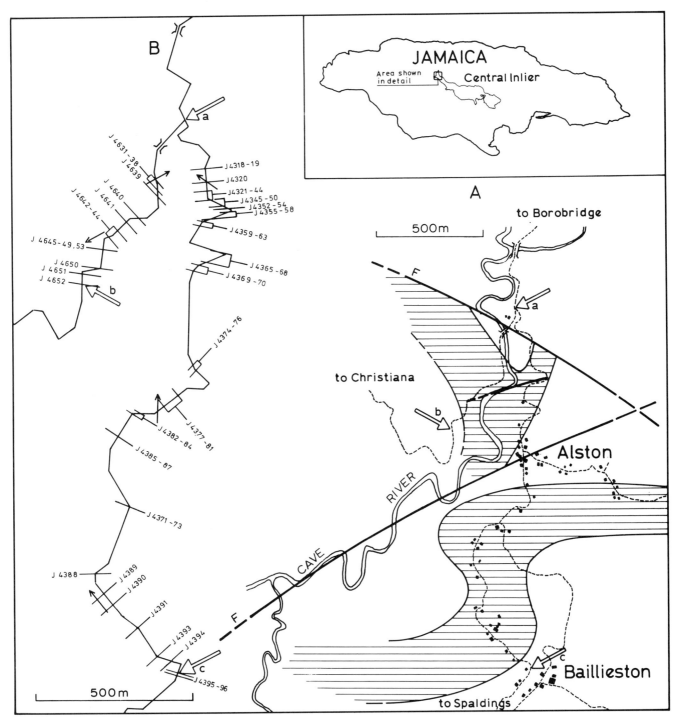

Figure 9. Distribution of the Guinea Corn Limestone (hatched), sample locations along the Borobridge-Christiana road (traverse a-b), and sample locations along the Borobridge-Alston-Baillieston road (traverse a-c), Central Inlier.

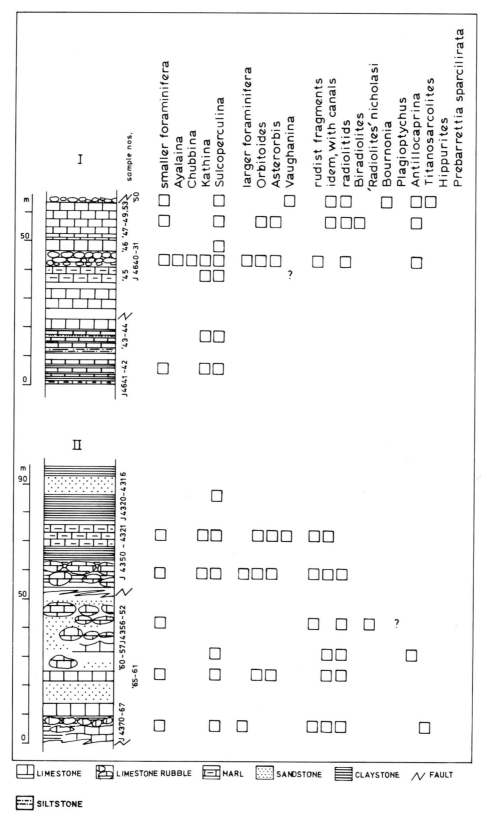

Figure 10. Stratigraphy and successive foraminiferal and rudist assemblages of the Alston traverses. I, Borobridge–Christiana road traverse (a–b on Fig. 9; II, Borobridge–Alston road traverse (northern a–c on Fig. 9).

in Zans et al., 1962), but a late Late Campanian to Early Maastrichtian age, based on nannoplankton, is regarded as more precise (Jiang and Robinson, 1987).

The Blue Mountain Inlier

The majority of the deposits in the Blue Mountain Inlier consists of volcanic extrusives and volcaniclastic sediments intruded by late-stage granodiorites. Limestones are only a minor constituent. So far two Cretaceous limestone horizons have been recognized: the Back Rio Grande Limestone member and the Rio Grande Limestone. They were mapped throughout the Blue Mountain Inlier and may have been associated with reef buildup along a developing volcanic island chain. The numerous occurrences of these limestones in the Blue Mountain Inlier are largely due to the effect of folding, followed by step-faulting, which may have resulted in a grabenlike structure in the central part of the inlier.

The Back Rio Grande Formation. The type section of the limestone member of this formation crops out in the Back Rio Grande, near the Cathalina River confluence (Fig. 12c-1). It consists almost entirely of an approximately 7-m-thick sequence of reef-rubble beds, with sandy tuffaceous layers at the top. Equivalent occurrences have been found in the Corn Husk River (Fig. 12c-3), along the road to Bath Fountain (Fig. 12d-1), and in the streambed of the Devils River (Fig. 12d-2).

The association of foraminiferal and rudist faunas is shown in Figure 13. Noteworthy is the presence of abundant *Pseudorbitoides trechmanni* and *Sulcoperculina* in friable tuffaceous horizons at the very top of the sequence in the Back Rio Grande near the Cathalina River confluence. This locality is the type locality of *Barrettia monilifera,* which occurs predominantly in rubble horizons stratigraphically lower in the exposed sequence. The equivalent limestone units in Corn Husk River, along the road to Bath Fountain, and in the Devils River all contain the extremely rich *P. trechmanni* fauna.

The age of the sequences. The pseudorbitoids from the Back Rio Grande show a configuration of nepionic equatorial chambers identical to that of *P. trechmanni* from Green Island (Krijnen, 1978). Accordingly a similar evolutionary stage of the juvenarium is assumed, and an equivalent late Middle Campanian age for the Back Rio Grande Limestone has been suggested (Jiang and Robinson, 1987).

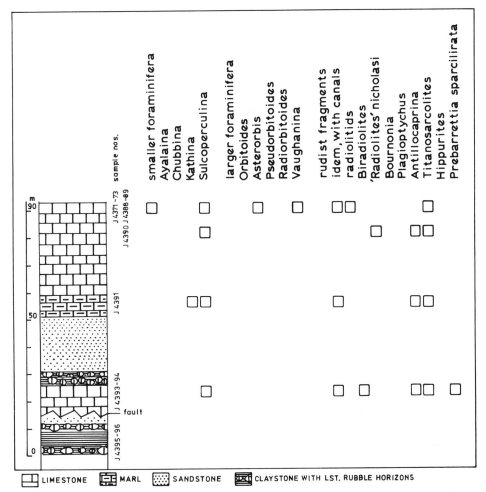

Figure 11. Stratigraphy and successive foraminiferal and rudist assemblages of the Alston-Baillieston road traverse (southern a-c on Fig. 9).

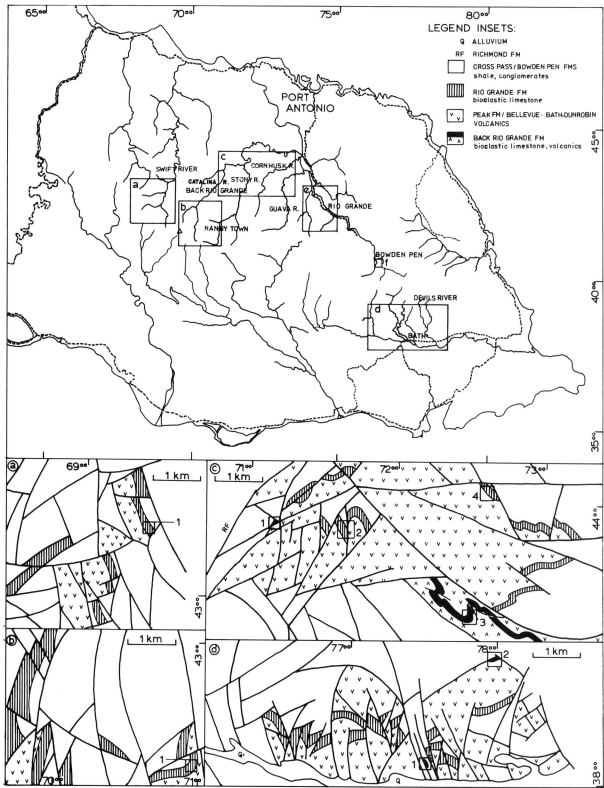

Figure 12. Limestone occurrences in the Blue Mountain Inlier and sample locations. a-1, Swift River (see Fig. 14); b-1, Nanny Town area (see Fig. 14); c-1, Back Rio Grande (see Fig. 15); c-2, Back Rio Grande– Stony River confluence (see Fig. 14); c-3, Corn Husk River (see Fig. 15); c-4, Mocho Toms River (see Fig. 14); d-1, road from Bath to Bath Fountain (see Fig. 15); d-2, Devils River (see Fig. 15; no detailed map); e, Rio Grande section (see Fig. 14; for detailed map, see Fig. 13); f, Bowden Pen (see Fig. 14; no detailed map).

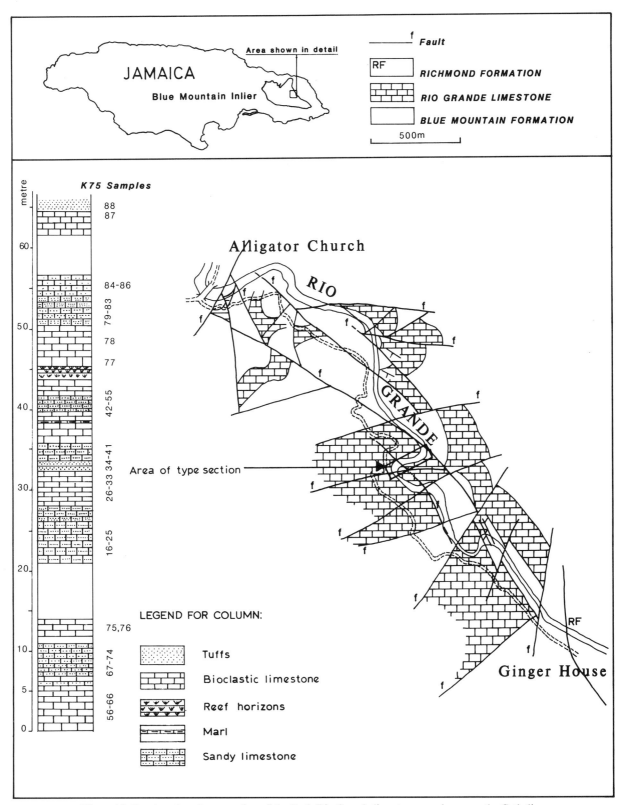

Figure 13. Stratigraphy of type section of the Back Rio Grande limestone member, near the Cathalina River–Back Rio Grande confluence, and foraminiferal and rudist assemblages from various other occurrences of the limestone member. See also Figure 12.

The Rio Grande Limestone. The type section of this unit is alongside the banks of the Rio Grande, immediately southeast of Alligator Church (see Fig. 14). The sequence predominantly comprises thick- to medium-bedded sandy calcarenitic to rubbly bioclastic limestones with a total thickness of approximately 80 m. The sequence attains its maximum thickness in the Rio Grande valley. Other localities of Rio Grande Limestone include those on high ridges above Nanny Town (Fig. 12b-1) and along the ridge toward the Blue Mountain Peak area (the western limestone outcrops shown in inset Fig. 12b), several outcrops in the Swift River valley (Fig. 12a), several outcrops to the northeast and east of the type section of the Back Rio Grande Limestone Member in the Back Rio Grande streambed (just west of the Back Rio Grande–Stony River confluence; see Fig. 12c-2), above the streambed of the Mocho Toms River (Fig. 12c-4), and in quite a number of localities to the west and east of Bath (Fig. 12d). Fragments of Rio Grande Limestone occur frequently at least in one horizon in conglomerates of the Bowden Pen Formation in the upper reaches of the Rio Grande (Fig. 12f).

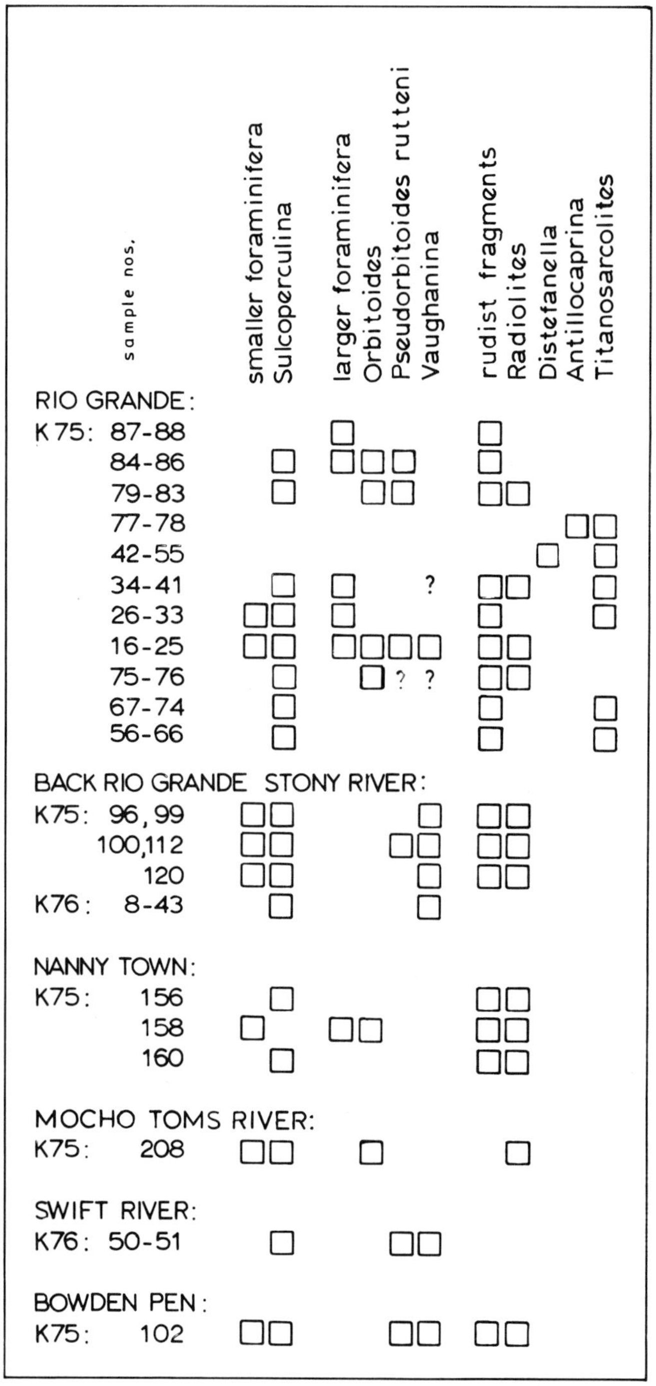

The faunal association found at a number of localities is summarized in Figure 15. In the type locality of the Rio Grande Limestone, at least eight horizons with larger foraminifera have been encountered, which generally consist of an assemblage of abundant *Sulcoperculina, Orbitoides,* and *Vaughanina* and to a lesser extent *Pseudorbitoides* cf *P. rutteni* (see Krijnen, 1978). *Orbitoides* and *Vaughanina* are the most diagnostic larger foraminifera in the succession, and these forms have been found in almost all the occurrences of Rio Grande Limestone in the inlier. The limestone outcrops near Bath are an exception. A tentative correlation of this limestone unit with the Rio Grande Limestone is based exclusively on field evidence.

Spectacular reefs of *Distefanella mooretownensis,* various species (a relatively small one in particular) of *Titanosarcolites,* and radiolitids occur intermittently in the succession in the Rio Grande valley.

The age of the sequences. Consistent with the Early Maastrichtian age of the upper part of the Guinea Corn Formation in the Central Inlier, the presence of an abundance of *Orbitoides* and *Vaughanina* in the basal part of the limestone sequence in the Rio Grande valley suggests that the Rio Grande Limestone is also of Early Maastrichtian age. However, the presence of possible *Pseudorbitoides* cf *P. rutteni* may indicate that the Rio Grande Limestone reaches into younger levels, that is, ranges from early to late Early Maastrichtian.

Conclusions

Based on ample material collected from five major Cretaceous inliers across Jamaica, the following subdivisions of the Campanian to Early Maastrichtian interval are distinguished on the basis of characteristic fauna associations (see also Fig. 16).

Early Campanian. The number of samples and faunas from this interval is limited (and exclusively from the Sunderland Formation, Sunderland Inlier). No rudists found in these samples; the pseudorbitoid foraminifer *Pseudorbitoides chubbi* occurs in one sample only.

The environment during deposition of the Sunderland Formation is regarded as generally being middle to lower neritic on the basis of the presence of rich faunas of planktonic foraminifera. Occasional sandier horizons with pseudorbitoids may have represented brief regressive interludes of shallower-marine inner sublittoral conditions.

Figure 14. Stratigraphy and distribution of the Rio Grande Limestone between Alligator Church and Ginger House.

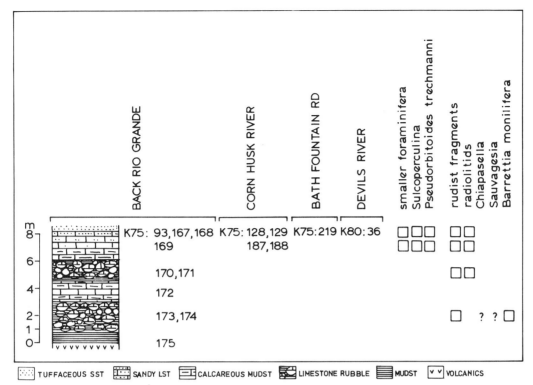

Figure 15. Foraminiferal and rudist assemblages from various occurrences of the Rio Grande Limestone. See also Figures 12 and 13.

Middle Campanian. Samples with abundant faunas of larger foraminifera and rudists were collected from the Green Island, Newman's Hall, Stapleton, and Shepherd's Hall formations (in the Green Island, Grange, and Sunderland inliers) and from the Back Rio Grande Limestone (in the Blue Mountain Inlier). Characteristic larger foraminifera are *P. trechmanni* and *Radiorbitoides pectinatus,* the latter possibly ranging into the Late Campanian. Characteristic rudists for this interval are *Barrettia gias, B. multilirata, B. monilifera, Parastroma trechmanni, Vaccinites, Mitrocaprina,* and *Torreites sanchezi.* "*Durania*" *nicholasi* (= ? "*Radiolites*" *nicholasi*) and species of *Antillocaprina, Sauvagesia, Chiapasella,* and *Radiolites* have also been found in younger stratigraphic levels and are thus regarded as noncharacteristic for this time interval.

The limestone and sandy-conglomeratic units in the Green Island Inlier, the Sunderland Inlier (Stapleton and Shepherd's Hall formations), and the Blue Mountain Inlier (Bath Fountain Road, Devils River, Corn Husk River) are all regarded as deposited in a fore-reef, upper to middle neritic depositional regime. However, the *Barrettia*-rubble beds of the Green Island and Grange inliers as well as the *Barrettia monilifera*-bearing rubbly limestone sequence in the Back Rio Grande (Blue Mountain Inlier) may have been deposited in somewhat deeper-water middle neritic conditions.

Late Campanian–early Early Maastrichtian. Numerous samples from this interval have been collected from the Guinea Corn Limestone, Central Inlier. Characteristic larger foraminifera include *Orbitoides media megaloformis, Vaughanina cubensis,* and *Asterorbis* cf *A. havanensis,* in association with the rudists *Titanosarcolites, Prebarrettia sparcilirata, Bournonia, Biradiolites, Plagioptychus,* and *Thyrastylon.*

The environment of deposition may have deepened from the Logie Green area to the Alston area in the northwest, from sheltered (?back-reef) conditions in the Logie Green east-traverse— via a shallow (?back-reef) regime interrupted by a marine orbitoid-rich transgressive interval in the Logie Green west and Sanguinetti traverses—to full-marine conditions in the Alston traverses.

Early Maastrichtian. Samples from this interval from the Rio Grande Limestone, Blue Mountain Inlier, yielded the *Orbitoides media megaloformis-Vaughanina cubensis* assemblage, here without *Asterorbis* but accompanied with *Pseudorbitoides* cf. *P. rutteni. P. rutteni* is supposed to be an Early Maastrichtian species described from Cuba, and the introduction of an equivalent form in several samples from the Rio Grande Limestone may point to a younger age interval, for example, Early Maastrichtian, than that of the Guinea Corn Limestone. Characteristic rudists in this interval include an abundance of a small species of *Titanosarcolites* and *Distefanella mooretownensis.*

The presence of orbitoid larger foraminifera throughout the dominant clastic limestone unit suggests that the Rio Grande Limestone, for a major part, consists of foreslope debris deposits

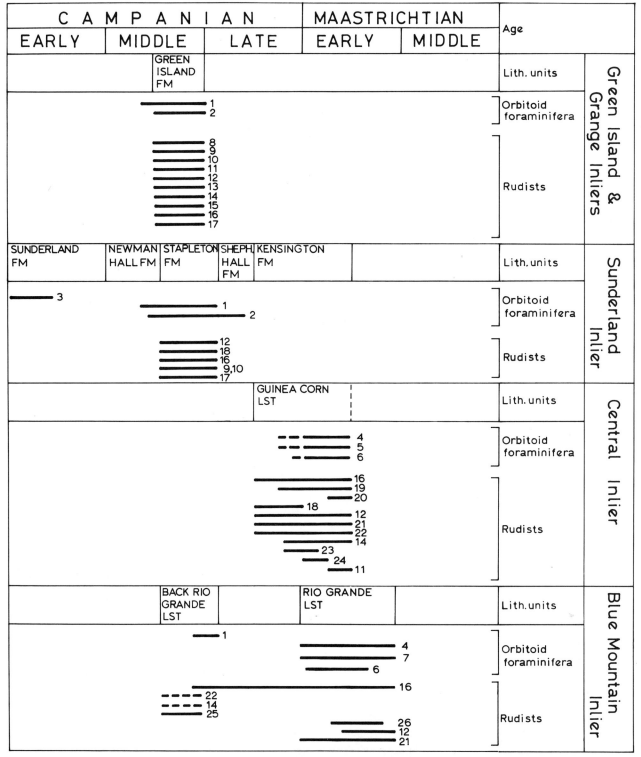

Figure 16. Foraminiferal and rudist assemblages in various Late Cretaceous formations of five Cretaceous inliers in Jamaica. **Foraminifera:** 1. *Pseudorbitoides trechmanni*; 2. *Radiorbitoides pectinatus*; 3. *Pseudorbitoides chubbi*; 4. *Orbitoides media megaloformis*; 5. *Asterorbis* cf *A. havanensis*; 6. *Vaughanina cubensis*; 7. *Pseudorbitoides* cf *P. rutteni*. **Rudists:** 8. *Parastroma trechmanni*; 9. *Barrettia multilirata*; 10. *Barrettia gigas*; 11. *"Durania" nicholasi*; 12. *Antillocaprina*; 13. *Mitrocaprina*; 14. *Sauvagesia*; 15. *Vaccinites*; 16. *Radiolites*; 17. *Torreites sanchezi*; 18. *Plagioptychus*; 19. *Bournonia*; 20. *Biradiolites*; 21. *Titanosarcolites*; 22. *Chiapasella*; 23. *Thyrastylon*; 24. *Prebarrettia sparcilirata*; 25. *Barrettia monilifera*; 26. *Distefanella mooretownensis*.

accumulated in an upper neritic environment. Intercalated horizons of organic buildup point to temporary extreme shallow littoral conditions.

PALEONTOLOGY

Introduction

The Jamaican orbitoidal larger foraminifera comprise a number of benthonic forms with an essentially similar morphology: The comparatively large test is generally lens shaped and characterized by the presence of a median layer (the so-called equatorial layer) and the formation of layers of lateral chambers on both sides of this median layer.

There is some variation in the development of the equatorial layer, as, for instance, the actinate or stellate equatorial layer in some forms in populations of *Vaughanina* and *Pseudorbitoides*. In these specimens the equatorial layer is interrupted by sectors (so-called interradii) filled by lateral chambers. Another deviation of the typical orbitoidal test is the change of the equatorial layer from a single- to a multilayer structure during individual growth. Such an equatorial layer has been noted in *Orbitoides*. In the radially thickened "rays" of the equatorial layer in *Asterorbis*, the equatorial layer remains single layered, but the equatorial chambers attain a size many times bigger than those in between such rays.

The equatorial layer is generally composed of an initial stage (or juvenarium) and a subsequent adult stage in which chamber formation is cyclical. The chamber arrangement of the juvenarium differs in the various genera and comprises: (1) a normally thick-walled embryo, consisting of two to four embryonic chambers, surrounded by either (2) periembryonic chambers, or (3) short spirals of postembryonic primary chambers. In the cyclical adult equatorial layer, the chambers are usually arranged in a single layer. Cyclicity in growth is generally established by rings of arcuate to ogival or spatulate chambers of equal budding stage number, so that successive rings form a system of multiple intersecting oblique tiers of chambers fanning out from the juvenarium toward the periphery. In *Orbitoides* and *Asterorbis* the equatorial chambers are simple in shape and interconnected by means of openings at the base of the chamber walls (basal stolons in a four-stolon system) and of annular and diagonal openings (at and near the base of the walls in a six-stolon system). Where the equatorial layer becomes multilayered in *Orbitoides*, rows of stolons are visible in vertical sections across the equatorial chamber wall. These may be the result of several superposed sets of basal and radial stolons in each equatorial chamber (Van Gorsel, 1978). The equatorial chambers in the adult cyclical stage in *Vaughanina* are elongately annular in shape. These appear to be traversed by radial rods attached against the roof and floor of the equatorial chambers (Brönnimann, 1954b). The equatorial chambers in *Pseudorbitoides* and related forms are arc shaped and orbitoidlike but divided vertically by the pseudorbitoid layer of radial elements (Mac Gillavry, 1963; Krijnen, 1972).

Lateral chambers are present in all forms of Cretaceous orbitoidal foraminifera, but there are differences in size, thickness of chamber walls, and number of layers between the various species. The more robust tests of *Orbitoides* show larger lateral chambers with clearly thicker walls than, for instance, the smaller and more delicate forms of *Vaughanina* and the relatively flat *Asterorbis*. In the typical orbitoidal form in *Orbitoides*, the layers of lateral chambers completely cover both sides of the equatorial layer, even in juvenile individuals. This, however, is not always the case in *Vaughanina*, *Radiorbitoides*, and some forms (probably juvenile forms of the megalospheric generation) of *Pseudorbitoides trechmanni*. Often the periphery of these specimens shows a flange consisting of a bare equatorial layer. In this area only prominent radial elements appear to be present, but scrutiny of vertical sections across such flanges shows that either roof and floor (in *Vaughanina*) or top and bottom "halves" of the equatorial chambers (in *Radiorbitoides* and *Pseudorbitoides*) are draped over the ridges of the radial elements and slightly depressed in the spaces in between the radial elements. The formation of lateral chambers seems to lag behind the growth of the equatorial layer.

In aktinate forms of *Vaughanina* and *Pseudorbitoides*, where the equatorial layer is interrupted by interradii filled with lateral chambers, it is argued that the formation of lateral chambers may already have been present during growth of the juvenarium and may have hampered the formation of equatorial chambers by blocking the formation of cyclical equatorial chambers at one or a number of places at the periphery of the juvenile test (Krijnen, 1972, Fig. 8, Plate 27).

The structure of lateral chambers in *Orbitoides* may differ on either side of the equatorial layer, as in forms collected from hardground horizons within the Maastrichtian stratotype. The asymmetry of these tests is caused by the presence of a thin top (apical) layer of open lateral chamberlets and by a thick, compact bottom (abapical) lateral layer. Such a structure may be related to the adaptation of individuals to a higher-energy environment, to the extent that the thick and massive abapical layer may have had an anchorage function (see Baumfalk and Willemsen, 1986).

Orbitoidal foraminifera are considered to form micro- and megalospheric forms, which represent two alternating stages in the reproduction process of larger foraminifera. The microspheric generation is generally characterized by larger tests with a more primitive configuration of small juvenile chambers; the megalospheric forms, on the other hand, are smaller in size, with a more advanced juvenarium comprising bigger chambers. Based on the reproduction process of modern foraminifera, it is assumed that with the orbitoidal foraminifera, the microspheric generation asexually produced the haploid megalospheric generation by division of the nucleus. The megalospheric generation, in turn, produced gametes that, after amalgamation, produced the diploid microspheric generation. Megalospheric and microspheric forms were found, or are known to occur, in *Pseudorbitoides*, *Radiorbitoides*, and *Orbitoides*. In *Pseudorbitoides* and *Orbidoites* the microspheric forms are distinctly larger in their adult stages, in contrast to those in the populations of *Radiorbitoides*. In soil samples in the vicinity of the rudist limestones from Green Island and Stapleton, the number of microspheric forms is quite large and possibly in the order of 10% of the total number of micro- and megalospheric forms. Dimorphism was not observed in the samples with *Vaughanina* and *Asterorbis*.

THE PSEUDORBITOIDES FAMILY

All members of this family show the typical orbitoidal subdivision of the test in equatorial layer and adjacent layers of lateral chambers. Characteristic for the *Pseudorbitoides* and allied forms is the presence of radial elements that appear to traverse, or divide, the equatorial chambers throughout the entire adult (neanic) equatorial layer or are only present in a peripheral zone of the equatorial layer. The presence of a sulcoperculinoid juvenarium, particularly visible in the long-spiraled less-advanced forms of *Sulcorbitoides*, *Pseudorbitoides*, and *Vaughanina*, points to *Sulvoperculina* as being the likely ancestral form of the pseudorbitoids.

The genus *Sulcorbitoides* is regarded as the first pseudorbitoidal genus, developed out of *Sulcoperculina* (Brönnimann, 1954a; Mac Gillavry, 1963). It is characterized by the formation of secondary chambers generated by the spiral of sulcoperculinoid chambers and by the lengthening of the rudimentary radial elements from the sulcus of the sulcoperculinoid chambers into these developing secondary chambers.

The genus *Pseudorbitoides* is regarded to have been developed from *Sulcorbitoides*, merely by increase in the growth of the radial elements, by increase in the extent of the equatorial layer of arcuate secondary equatorial chambers, and by the development of layers of lateral chambers.

The name *Radiorbitoides* has recently (Krijnen, 1993) been assigned to populations of pseudorbitoids from below and above the Stapleton Limestone (Sunderland Inlier), which were earlier identified as *Pseudorbitoides trechmanni pectinata* Krijnen, 1972. *Radiorbitoides* was also found present, apparently mixed with *Pseudorbitoides trechmanni*, in the Green Island Inlier, so that the regional distribution of this form was considered to be more widespread than originally assumed. *Radiorbitoides* is characterized by a radial arrangement of the secondary equatorial chambers and, unlike both *Pseudorbitoides* and *Vaughanina*, has developed radial plates that tend to border these radial rows of equatorial chambers (Krijnen, 1993). The form is regarded as an offshoot from the *Pseudorbitoides* lineage.

The genus *Vaughanina* has been described in detail by Brönnimann (1954b), who recognized the presence of a true equatorial layer composed of equatorial chambers with roofs and floors. As interpreted by Mac Gillavry (1963) and Krijnen (1972), these roofs and floors may be regarded as the equivalent of the equatorial chamber "halves" on either side of the pseudorbitoid layer in *Pseudorbitoides* but fused into a roof and a floor plate bordering the pseudorbitoidal layer of radial elements. The equatorial chambers have annular walls that extend from floor to roof. These chambers are traversed by two sets of alternating vertical "hanging" and "standing" radial plates, which are attached to the roof and floor and extend into the equatorial layer for about half its height, leaving a median gap between the two systems of plates.

The following summarized key to the genera is to a large degree based on Brönnimann's studies (1954–1958) but emended according to a reinterpretation of particularly the equatorial layer by Mac Gillavry (1963) and Krijnen (1967, 1972) and to a downgrading of the diagnostic value of the "historbitoid" and "rhabdorbitoid" structure of radial elements.

Family PSEUDORBITOIDIDAE Rutten, 1935

1. Equatorial chambers subdivided, vertically, into chamber "halves" by pseudorbitoid layer of radial elements ....
.... Subfamily PSEUDORBITOIDINAE Rutten, 1935
    (a) Equatorial chambers arcuate; radial elements are two sets of rods ....
.... *Sulcorbitoides* Brönnimann, 1954a
    (b) Equatorial chambers arcuate; radial elements are essentially a single set of plates but may become complex rods within incipient radii ....
.... *Pseudorbitoides* H. Douvillé, 1922
    (c) Equatorial chambers initially arcuate, later truncated and bordered by radial plates ....
.... *Radiorbitoides* Krijnen, 1993
2. Alternating "hanging" and "standing" set of radial plates attached to roofs and floors of the equatorial chambers, with median gap between the two sets of radial plates; adult equatorial chambers concentrically elongated, with chamber walls perforated by radial stolons ....
.... Subfamily VAUGHANININAE Rutten, 1935
    (a) Circular and actinate forms ....
.... *Vaughanina* D. K. Palmer, 1934
    (b) Conical forms *Conorbitoides* Brönnimann, 1958b and *Ctenorbitoides* Brönnimann, 1958b (not found in Jamaica).

Genus *Pseudorbitoides* H. Douvillé, 1922

Type species. *Pseudorbitoides trechmanni* H. Douvillé, 1922.
Diagnosis. Species of *Pseudorbitoides* are characterized by the presence of a pseudorbitoid layer of radial elements that is formed in the adult stage of the equatorial layer. It initially divides the equatorial layer into two layers of equatorial chamber "halves" but may increase in thickness to the extent that it almost fully traverses the equatorial layer from roof to floor. *Pseudorbitoides* lacks the median gap between two sets of radial plates of *Vaughanina*. *Pseudorbitoides* also differs from *Vaughanina* in the shape of the equatorial chambers, which are arcuate in *Pseudorbitoides* and annular in *Vaughanina*.

The formation of radial plates in *Pseudorbitoides* does not affect the cyclical growth pattern of the equatorial chambers in the adult equatorial layer, which remains a pattern of multiple intersecting oblique tiers up the periphery. In this respect *Pseudorbitoides* differs from *Radiorbitoides*, in which the growth of the radial plates appears to induce a radial arrangement of the equatorial chambers.

Pseudorbitoides differs from *Sulcorbitoides* by its larger adult equatorial layer.

Description. Dimorphism is present. Micro- and megalospheric specimens are generally lenticular in shape; the test circular, subcircular to occasional dented, or stellate in outline and with a papillate to pustulate surface. Microspheric specimens, particularly young individuals, may show heavy pillars in the umbonal region. The structure of both micro- and megalospheric specimens is orbitoidal, and the lateral chambers usually cover the equatorial layer completely.

The equatorial layer is composed of true equatorial chambers. The juvenile portion consists of primary and juvenile secondary chamber spirals and the surrounding adult equatorial layer of cyclically arranged secondary chambers. These latter chambers are arcuate to truncated arcuate in shape and arranged in multiple intersecting oblique tiers up to the periphery. In the more-primitive forms with uniserial juvenarium, the pseudorbitoid layer appears to already divide the first-formed secondary equatorial chambers. In the more-advanced forms from Green Island and Stapleton, a concentric zone of undivided secondary equatorial chambers is formed prior to the introduction of the pseudorbitoid layer. Mac Gillavry (1963) proposed that the onset of the formation of radial elements may have been subject to a proterogenetic shift and may have been delayed to an ever-later stage of the equatorial layer in successive populations.

***Pseudorbitoides chubbi* Brönnimann, 1958a**
(Fig. 17: 1–6)

1958a *Pseudorbitoides(?) chubbi* sp. nov.- Brönnimann, pp. 422–436, Plate 1, Figs. 1–3; p. 425, tf. 2a–c; p. 426, tf. 3a–d; p. 427, tf. 4b, f–i.
1972 *Pseudorbitoides(?) chubbi* Brönnimann-Krijnen, p. 45, Plates 2–4.

Figure 17. Pseudorbitoids from the Sunderland Inlier (after Krijnen, 1972). 1–6: *Pseudorbitoides chubbi* Brönnimann. (1, 2) Exterior view of aktinate forms—1: J 3416-1, 41×; 2: J 3416-16, 36×. (3) Centered vertical section, J 3416-22, 64×. (4) Horizontal section of a microspheric form, J 3416-67, 85×. (5, 6) Horizontal sections of megalospheric forms—5: J 3416-1, 55×; 6: J 3416-8, 43×. 7–14: *Radiorbitoides pectinatus* (Krijnen). (7, 12) Exterior view of the megalospheric forms—7: J 3680-61, 31×; 12: side view with protruding single set of radial plates, J 3680-52, 45×. (8, 10) Horizontal sections of megalospheric specimens—8: J 3700-58, 31×; 10: type specimen J 3680-61, 35×. (9) Horizontal section of microspheric specimen, J 3700-8, 23×. (11, 13) Centered vertical section of megalospheric specimen, J 3680-27—11: 34×; 13: 112×. (14) Horizontal section across megalospheric juvenarium with relatively large quadriserial nepionic chambers and absence of radial elements in the secondary chambers immediately surrounding the nepiont, J 3700-206, 130×.

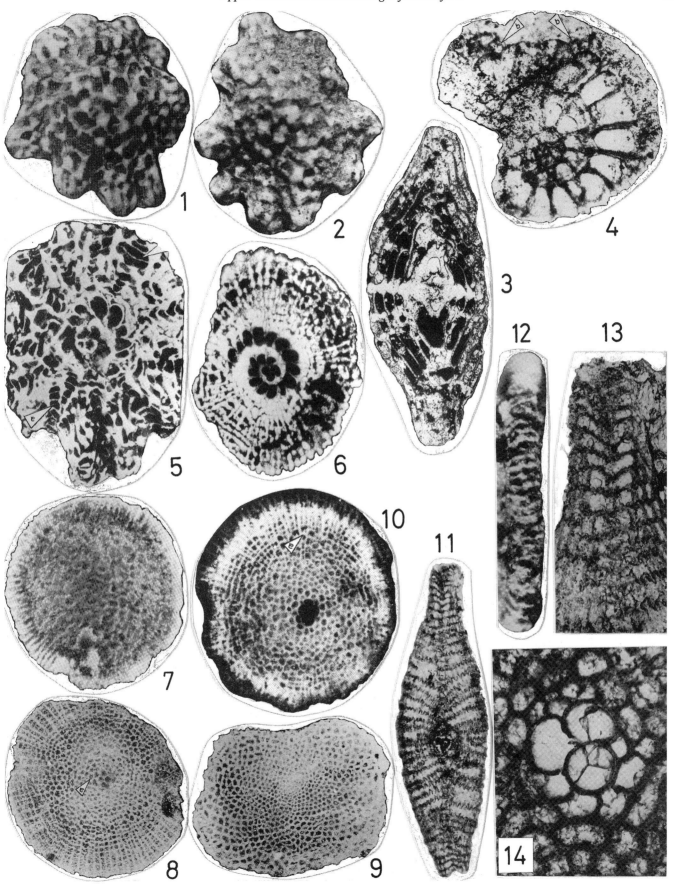

Type locality. Elm Creek, Kinney County, Texas.
Jamaican locality (sample J 3416). Roadside gully along the road from Sunderland to Amity Hall, approximately 800 m east-southeast of Sunderland, Sunderland Inlier (Fig. 4).
Material. 30 measured free specimens.
Description. Type material characterized by thin forms that are less than 1 mm in diameter. The juvenarium consists of a uniserial spiral of 15 to 20 spiral chambers. The diameter of the protoconch ranges from 50 to 130 microns. The Jamaican material consists of regularly notched and stellate forms with a typical orbitoid structure, comprising a true equatorial layer, with layers of lateral chambers on either side (Fig. 17-1, 2, 5). Diameter and thickness of 30 specimens measured vary from respectively .40 to 1.40 mm and .20 to .75 mm. After grinding horizontal sections, the bulk of the specimens turned out to be megalospheric forms; only one test proved to be a juvenile microspheric specimen (Fig. 17-4) measuring .75 mm in diameter and .20 mm in thickness. The juvenarium of the megalospheric forms is a uniserial spiral consisting of 16 to 25 primary equatorial chambers, including the proto and deuteroconch (Fig. 17-5, 6). The diameter of the protoconch varies from 48 to 76 microns and that of the deuteroconch from 35 to 64 microns. The juvenarium of the microspheric form consists of a uniserial spiral of a total of 37 primary equatorial chambers (Fig. 17-4), a protoconch of 23 microns, and a deuteroconch of 17 microns.

The cyclical phase of the megalospheric forms is comparatively large and shows a cyclical arrangement of arcuate to truncated secondary equatorial chambers divided by the pseudorbitoid layer (Fig. 17-6). The radial elements within the pseudorbitoid layer are of *Sulcorbitoides*-type of radial rods and *Pseudorbitoides*-type of radial plates. In *Historbitoides*-like incipient radii, the radial elements are irregularly interconnected by horizontal bars. Such radii are generally present. In some specimens there may be as many as nine radii (Fig. 17-5). Interradii are generally composed of lateral chambers.

Pseudorbitoides trechmanni H. Douvillé, 1922
(Fig. 18-1–4; Fig. 19-1–7)

1922 *Pseudorbitoides Trechmanni* g. nov., sp. nov.-Douvillé, p. 204, Fig. 1.
1955a *Pseudorbitoides trechmanni* Douvillé-Brönnimann, p. 58, Plates 9, 10, text Figs. 1, 2–7.
1972 *Pseudorbitoides trechmanni trechmanni* Douvillé-Krijnen, p. 46, Plates 5–9, 14–20.

Type locality. Green Island Inlier, Jamaica.
Material. Abundant free specimens and random sections of megalospheric and microspheric forms from the type area (samples J3978, GI-19, and GI-33); from the Sunderland Inlier, central Jamaica (samples J3601, J3697, and J3419); and from the Blue Mountain Inlier (sample K75-219, K75-93, K75-128, K75-129, and K80-36).
Description of megalospheric forms. Tests generally lenticular and circular to subcircular in outline (Fig. 19-3), in rare cases with one or more notches that mark the presence of interradii. The external diameter ranges from .30 to 2.55 mm, the thickness from .15 to 1.00 mm. The juvenarium is characterized by a large protoconch measuring from 40 to 243 microns and an equally large deuteroconch measuring from 36 to 246 microns. These chambers are surrounded by one to four spirals of periembryonic chambers, depending on the configuration of chambers being uniserial, biserial, or quadriserial. The primary chamber spiral is usually considered to be the longer of the two spirals embracing the protoconch. The number of primary chambers (which include both proto- and deuteroconch) varies from five to six (see Fig. 18-1, 2). The cyclically arranged equatorial chambers are arcuate to truncated arcuate (Fig. 18-4). They are vertically divided by the pseudorbitoid layer of radial plates, which often starts at a short distance distally from the juvenarium (Fig. 18-4). The equatorial chambers are arranged in intersecting oblique curves or spires fanning out from the center of the equatorial layer (Fig. 18-3c). Growth of the equatorial chambers does not seem to be affected by the formation of radial elements. On the contrary, the growth of the radial elements tends to be perpendicular to the distal wall of the equatorial chambers. As a result, growth may cause radial elements to converge and, in extreme cases, to run into one another to form historbitoid radii, where kinks in the generally annular chamber pattern show that during the cyclical growth the concentrical expansion of the equatorial layer at one or more places along the edge has lagged behind (Krijnen, 1972).

Description of microspheric forms. Tests also lenticular with generally circular to subcircular outline (Fig. 18-3c). The external diameter varies from .50 to 7.35 mm and the thickness from .20 to 2.50 mm. The smaller specimens are juvenile forms that are fairly common among the bulk of megalospheric forms in the soil in the vicinity of the limestone outcrops. The juvenarium is much smaller than that of the megalospheric forms, is invariably uniserial, and is composed of 14 to 22 primary chambers (proto- and deuteroconch included) (Fig. 18-3b). The size of the protoconch is much smaller than that of megalospheric forms (17 to 36 microns). The secondary equatorial chambers are distinctly arcuate to truncated arcuate, arranged in generally oblique mutually intersecting tiers fanning out from the center (Fig. 18-3c). Between the juvenarium and the first appearance of the radial elements, the equatorial chambers do not seem to be vertically divided by the pseudorbitoid layer: The ontogenetic introduction of this layer appears to be delayed, possibly as a result of an acceleration in the production of secondary chambers after completion of the juvenarium. The general concept that the radial elements are linked with the formation of secondary equatorial chambers apparently does not apply all the time. The introduction of these structural elements may, however, be linked with a specific number of completed budding stages (4 to 5 in megalospheric forms; 18 to 31 in microspheric forms) and may be related to the commencement of cyclicity in the growth of the equatorial layer. As in the megalospheric forms, the development of historbitoid incipient radii is occasionally recorded. The presence of these structures may be correlated with disharmonic growth in the intended concentrical development of the equatorial layer (Krijnen, 1972).

Figure 18. *Pseudorbitoides trechmanni* Douvillé from the Green Island Inlier. (1, 2) Horizontal sections across the juvenarium of megalospheric specimens—1a: asymmetric quadriserial nepiont with a primary spiral of five chambers, GI19-9, 180×; 1b: interpreted chamber configuration; 2a: asymmetric quadriserial nepiont with a primary spiral of six chambers, GI19-2, 180×; 2b: interpreted chamber configuration. (3) Horizontal section of microspheric specimen GI19-5—3c: overview of equatorial layer with faint radial elements developed in a later stage of the development of the equatorial layer, 45×; 3a: detail of the uniserial juvenile portion of the equatorial layer, 180×; 3b: interpreted chamber configuration with 17 primary chambers. (4) Horizontal section showing the presence of radial elements in chamber walls immediately surrounding the nepiont (compare Fig. 19-1), GI19-7, 180×.

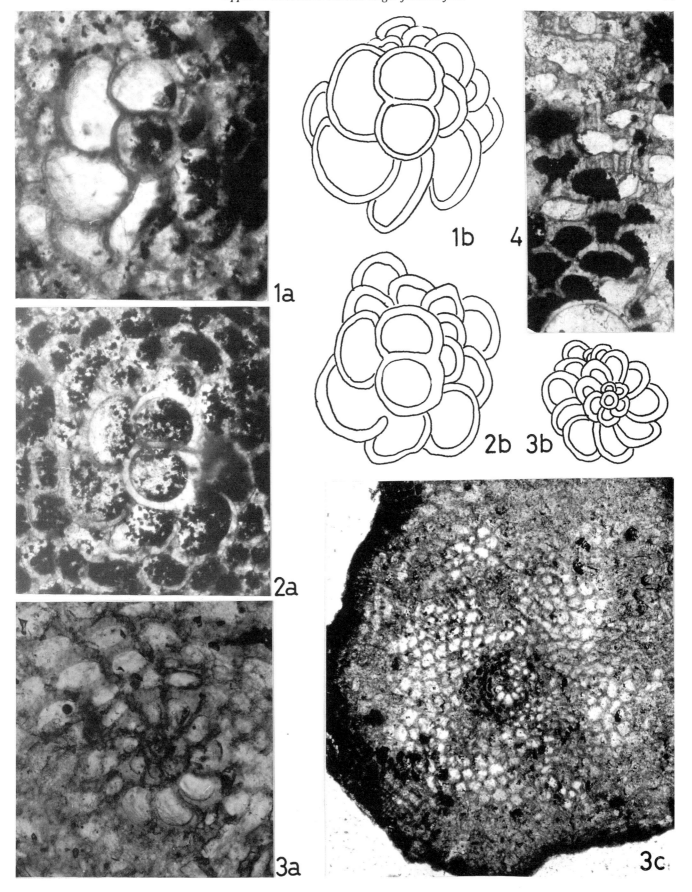

Pseudorbitoides cf *P. rutteni* Brönnimann, 1955a
(Fig. 19-8–11)

1955a *Pseudorbitoides rutteni* sp. nov.-Brönnimann, pp. 68–75, Plates 11, 12, text Figs. 8–17.
1963 *Pseudorbitoides rutteni* Brönnimann. Seiglie and Ayala-Castañares, pp. 45, 46, Plate 10, Figs. 2, 3, Plates 11, 12, 13.

Type locality. Road between Camajuani and Santa Clarita, Las Villas Province, Cuba.
Jamaican material. Numerous random sections in limestones from the Blue Mountain Inlier (samples K75-79, K75-100, K75-137, K75-112, and K75-102).
Description. Test lenticular, orbitoid in structure, with layers of lateral chambers on both sides of the equatorial layer (Fig. 19-8–11). Dimorphism may be present, but no vertical section shows a microspheric juvenarium with certainty. Several vertical and somewhat oblique sections of specimens measure a diameter that varies from 0.6 to 3.5 mm and a thickness of 0.2 to 1.0 mm. In several cases the vertical diameter of a megalospheric protoconch could be measured; it was found to vary from 96 to 115 microns. Thickness of the juvenarium was 136 to 158 microns. No horizontal section across the nepiont is available, and a biserial to quadriserial configuration of nepionic chambers was tentatively interpreted on the basis of central vertical sections. The adult equatorial layer is very distinct and gradually widens from approximately 60 microns near the center to 217 microns near the periphery. The pseudorbitoid layer is often complex *Historiboides*like and points to a stage in the development of radial elements similar to that in some microspheric forms of *P. trechmanni* from the Stapleton Formation, Sunderland Inlier (Krijnen, 1972, Plate 20, Figs. 2, 3, 6).

Genus *Radiorbitoides* Krijnen, 1993

Type species. *Pseudorbitoides trechmanni pectinata* Krijnen, 1972 (Plate 21, Fig. 3; Plate 22, Fig. 2).
Diagnosis. *Radiorbitoides* is characterized by the presence of a *Pseudorbitoides*-type layer of radial plates that initially divides the equatorial layer vertically but ultimately traverses the equatorial layer from bottom to top. The radial elements gradually form strong plates, which develop into partitions between radial rows of equatorial chambers. This feature differs from *Pseudorbitoides,* in which the layer of radial plates vertically divides the equatorial layer up to the periphery, without affecting the growth pattern of the equatorial chambers. *Radiorbitoides* differs from *Vaughanina* by not having an equatorial layer composed of annular equatorial chambers with distinct roofs and floors and by radial elements not being attached to the roof and floor of the equatorial chambers. The nepionic arrangement is more advanced than that in the known species of *Vaughanina.*

Radiorbitoides pectinatus (Krijnen, 1972)
(Fig. 17-7–14)

1972 *Pseudorbitoides trechmanni pectinata* Krijnen, pp. 49–51, Plates 10–13, 21–26, Text Figs. 10–16, Tables 4, 7–18.
1993 *Radiorbitoides pectinatus* Krijnen, Figs. 2.1–5; 3.1–5.

Holotype. Megalospheric specimen J3680-61 (Krijnen, 1972, Plate 21, Fig. 3, Plate 22, Fig. 2).

Type locality. J3680, Shepherd's Hall Formation, approximately 100 m south of the bridge of Stapleton, Sunderland Inlier.
Paratypes. Megalospheric and microspheric specimens from localities J3700, J3680, J3686, and J3688 from the Sunderland Inlier (Krijnen, 1972, Plates 11–13, 22–26).
Description. Both micro- and megalospheric forms are lenticular in shape, with a narrow flange and an inflated peripheral margin (see Fig. 17-7, 11, 12). The microspheric generation is distinctly larger in size, ranging from .75 to 4.45 mm in diameter and from .25 to .80 mm in thickness, compared with the megalospheric generation, which measures .30 to 2.25 mm in diameter and .10 to .75 mm in thickness. The general structure of the test is orbitoid, e.g., composed of a median equatorial layer and layers of lateral chambers on either side of the equatorial layer (Fig. 17-11). The flange is devoid of lateral chambers (see Fig. 17-7). In this area the radial plates of the pseudorbitoid layer slightly protrude beyond the peripheral margin, causing the pectinate outline of the flange (see Fig. 17-12). Pillars are generally absent in material from the Sunderland Inlier, but small pustules may be seen at the corners of the polygonal-shaped lateral chambers covering the umbonal regions of the test.

Horizontal thin sections of micro- and megalospheric forms show an equatorial layer with arcuate secondary chambers surrounding a uniserial nepionic spiral in microspheric forms (Fig. 17-9) and a biserial to quadriserial nepiont in megalospheric specimens (Fig. 17-8, 10, 14). The protoconch in microspheric specimens could only be measured twice, the diameter in both cases being 27 microns. The diameter of proto- and deuteroconch in megalospheric specimens ranges from 62 to 148 microns and from 59 to 188 microns. In a later ontogenetic stage, the equatorial chambers become truncated and line up in radial rows, bordered by radial plates (Fig. 17-10). Radial stolons in the equatorial chamber walls are regularly visible, and growth of the equatorial layer may well have taken place by means of these stolons alone. Equal growth rate of the equatorial layer in all directions may have caused the formation of concentric "rings" of equatorial chambers (Fig. 17-8, 9).

In vertical sections all specimens show the distinct orbitoid division of the test in equatorial layer and layers of lateral chambers (Fig. 17-11). The pseudorbitoid layer occupies a substantial portion of the equatorial layer, which widens rapidly from the nepiont toward the periphery (Fig. 17-11, 13). Where the radial plates are obliquely cut, the equatorial chamber "halves" are V-shaped, particularly toward the peripheral margin (Fig. 17:13). At the extreme margin, the thickened equatorial layer is not covered by lateral chambers.

Figure 19. *Pseudorbitoides* spp. (1, 3) *P. trechmanni* Douvillé from the type area near Green Island—1: horizontal section of megalospheric specimen showing the development of radial elements soon after completion of the nepionic stage (compare Fig. 18-4), GI19-7, 46×; 3: exterior view of specimen with flange not covered with lateral chambers, GI17, 30×. (2, 4–7) Centered to near-centered vertical section of *P. trechmanni* Douvillé from the Back Rio Grande, all 48×, sample K75-93. (8–11) Centered and tangential vertical section of *P.* cf *P. rutteni* Brönnimann from the Blue Mountain Inlier, all 45×—8, 10: sample K75-79 Rio Grande; 9: sample K75-21 Rio Grande; 11: sample K75-102A Bowden Pen.

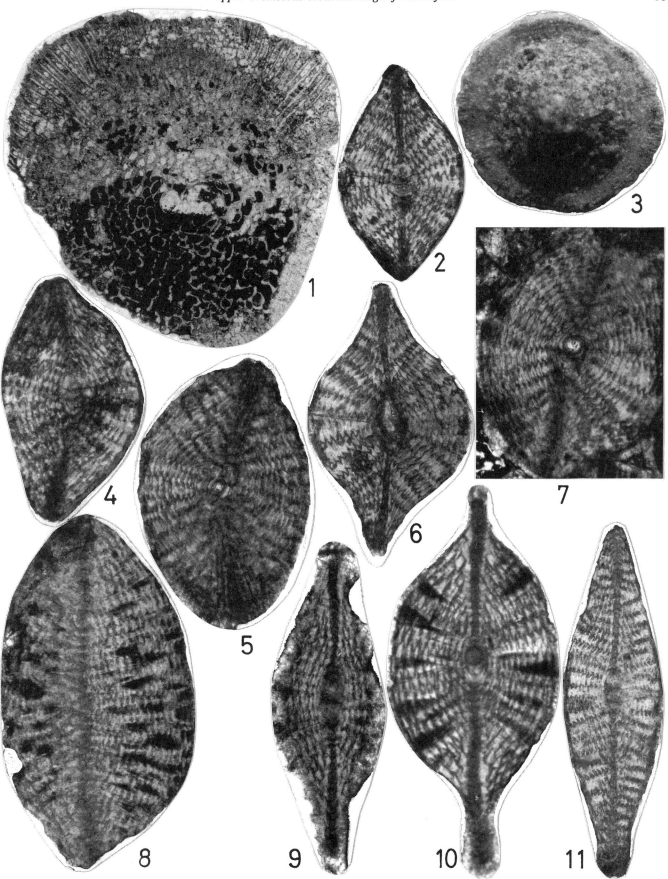

The description of the forms found in the Green Island Inlier is exclusively based on two thin sections. Most of the specimens present are megalospheric forms, lenticular in shape, and ranging from 1.17 up to 4.21 mm in diameter and from 0.29 up to 1.01 mm in thickness of the test (Krijnen, 1993, Fig. 2.1-2, Fig. 3.1-4). One microspheric form is significantly larger than those from the faunas from the Sunderland Inlier, measuring 6.78 mm in diameter and 2.18 mm in thickness (Krijnen, 1993, Fig. 3.3).

One horizontal section is more or less cut through the center of the test (Krijnen, 1993, Fig. 2.1, Fig. 3.1). The arrangement of initial chambers is unfortunately partly obscured and may be uniserial or biserial with rather small embryonic chambers. The equatorial chambers surrounding the initial embryonic or nepionic chambers are clearly arcuate and are formed according to the general pattern of oblique spirals of chambers in orbitoidal foraminifera. Continuous radial plates are introduced at a later stage in ontogenetic development, but rather than vertically dividing the equatorial chambers as in *Pseudorbitoides,* they tend to border them. Radial rows of equatorial chambers have thus been formed, each row separated from the adjacent row by a radial plate. Where the plates diverge to the extent that they exceed the average mutual distance, intercalating plates have developed, and the radial row of equatorial chambers is split in two. The thickness of the plates is in the order of 10 microns, and the distance between them at the peripheral margin measures up to 45 microns. The chambers in each radial row communicate by means of radial stolons (Krijnen, 1993, Figs. 2.1, 2.4).

Ample specimens from the Sunderland Inlier show that a slight protrusion of radial plates beyond the peripheral margin causes the pectinate outline of the foraminiferal test (Fig. 17-12). The radial symmetry of the equatorial layer and the pectinate outline of the test are typical for this species.

Some fairly centered and various near-centered vertical sections are available from material from Stapleton, showing the distinct orbitoidal subdivision of equatorial layer and lateral chamber layers. The thickness of the equatorial layer increases more rapidly from the center of the test outward as compared with that in *Pseudorbitoides trechmanni.* In megalospheric forms an increase in layer thickness from 66 microns near the center to 275 microns near the periphery has been measured, whereas in the microspheric forms the thickness of the equatorial layer near the periphery amounts to 443 microns.

Pillars have been found in micro- as well as in megalospheric forms. In the microspheric form shown in Krijnen, 1993, Fig. 3.3, at least 30 pillars are partly embedded in the layers of lateral chambers, slightly protruding from the surface of the test. The length of these pillars may range up to 340 microns, the thickness up to 100 microns.

Genus *Vaughanina* D. K. Palmer, 1934

Type species. *Vaughanina cubensis* Palmer, 1934.
Diagnosis. The adult portion of the equatorial layer is composed of a layer of chambers developed between a common roof and floor plate. The equatorial chambers are initially arcuate, but during individual growth these rapidly lengthen in a tangential direction, eventually becoming annular. The radial elements of the pseudorbitoid layer form two systems of plates, each half the height of the equatorial chambers. The plates of one system are attached to the roof plate and alternate with the plates of the other system, which are attached to the floor plate. A permanent median gap exists due to the fact that the plates of both systems reach only half way across the height of the equatorial chambers.

Vaughanina cubensis Palmer, 1934
(Fig. 20-1-13; Fig. 21-1-4)

Type locality. Palmer Station 1214, Havana Province, Cuba.
Jamaican localities. Alston, Sanguinetti, and Logie Green traverses, western Central Inlier; several localities of outcropping Rio Grande Limestone in the Blue Mountain Inlier.
Material. Abundant free specimens and random sections.
Description. Specimens from the Central Inlier and Blue Mountain Inlier show a generally lenticular test and are typically orbitoid, with a true equatorial layer and layers of lateral chambers flanking it (Fig. 20-2, 4, 5, 7). Aberrant forms with aktinorbitoid interradii are generally present. These forms show a clearly notched outline where such an interradius occurs (Fig. 20-8, 10). Much rarer are specimens with an additional "equatorial layer" developed at right angles to the main median plane, on one side of the test (see Fig. 20-12). The formation of lateral chambers may be slower than the growth of the equatorial layer, with the result that an equatorial flange may have developed devoid of lateral chambers (Fig. 20-11). Measurements on free specimens from the Central Inlier show the diameter of the test to vary from 0.66 to 2.68 mm and its thickness from 0.29 to 1.02 mm. Random (vertical) sections from the Blue Mountain Inlier do not differ greatly in this respect, with values of diameter and thickness of 0.393 to 1.913 mm and 0.267 to 0.935 mm, respectively. The equatorial layer can be subdivided into a juvenarial stage, consisting of a uniserial spiral of primary chambers, and a number of secondary chambers, surrounding the spiral chambers. The adult portion of the equatorial layer comprises annular equatorial chambers that are fully traversed by two alternating sets of radial elements (Fig. 20-2, 6; Fig. 21-1a–4a): one set attached to the roof, the other to the floor plate. Communication between the equatorial chambers exists via radial stolons. From free specimens from the Central Inlier, the diameter of the protoconch was found to vary from 30 to 47 microns and the width of the deuteroconch from 35 to 69 microns. The number of primary chambers proved to be relatively constant, varying from 9 to 13 (Fig. 21-1bc–4bc). Measurements on several vertical sections of specimens from both Central and Blue Mountain inliers showed that the internal height of the protoconch and deuteroconch varies from 20 to 58 microns and from 23 to 55 microns, respectively. The thickness of the juvenarium was also measured in various centered sections: It varies from 49 to 146 microns.

Figure 20. *Vaughanina cubensis* Palmer (1–7) Tangential and near-centered vertical sections from the Blue Mountain Inlier, all approximately 45×—1, 6: K75-100 Stony River; 2, 4, 7: K75-102A Bowden Pen; 3, 5: K75-96, Stony River. (8, 9, 12, 13) Exterior of specimens from the Central Inlier, J4285 Logie Green, all approximately 45×—8: form with one notch = aktinorbitoid interradius; 9: form with vermicular pattern of ridges over the surface; 12: aberrant form with a bifurcating equatorial layer; 13: normal circular specimen. (10, 11) Exterior of specimens from sample J4276—10: form with aktinorbitoid notch, 45×; 11. circular form with flange not covered by lateral chambers, 175×.

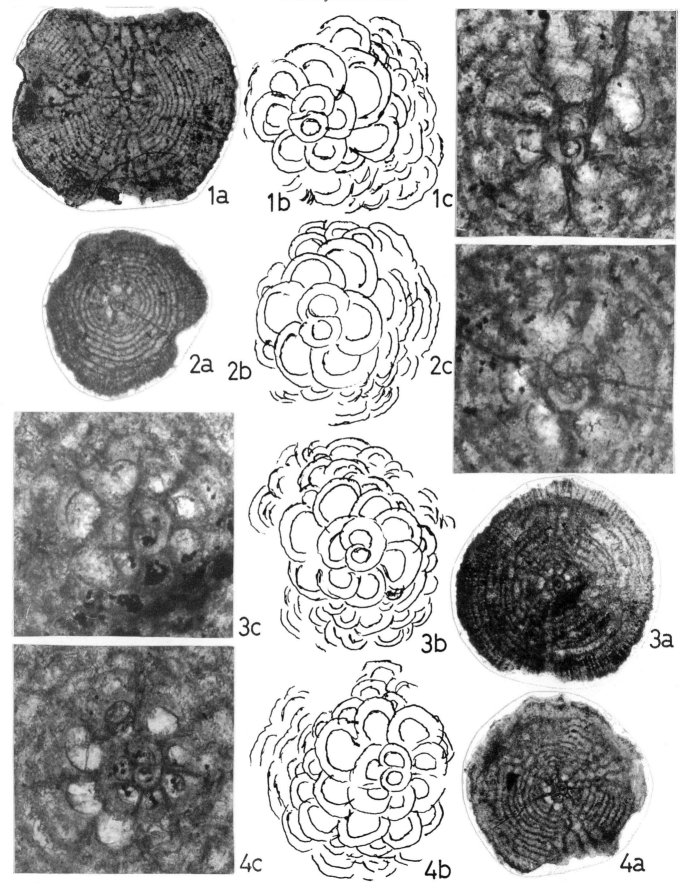

Figure 21. *Vaughanina cubensis* Palmer from the Central Inlier, sample J4480, Logie Green–Sanguinetti traverse, showing a consistent primary spiral of about 11 to 14 chambers, all "a" figures approximately 45×, all "b" and "c" figures 140×—1: specimen 143; 2: specimen 169; 3: specimen 155; 4: specimen 180.

The Orbitoides Family

Orbitoides not only is the most important genus of the family Orbitoididae but also is regarded as the most widespread genus of the Cretaceous orbitoids (Van Gorsel, 1978). Modern studies on members of this family include those of Neumann (1958, 1972a, b), Van Hinte (1966a, b), Baumfalk and Van Hinte (1985), and Baumfalk and Willemsen (1986), all of whom regarded *Orbitoides* as a single evolutionary series, possibly evolved from a biserial *Guembelina*-like planktonic ancestor (Mac Gillavry, 1963; Van Hinte, 1968).

A general characteristic is, except for early members of the evolutionary series, the lenticular shape of the test, which may be more or less asymmetric in shape and ornamentation. Internally, advanced species of the evolutionary species show a somewhat wobbly equatorial layer (see Fig. 22-7, 8) with distinct arcuate equatorial chambers (Fig. 22-6) and the large thick-walled bi- to quadrilocular embryo (Fig. 22-2, 6; Fig. 23-1–6).

Genus *Orbitoides* d'Orbigny, 1848

Type species. *Orbitolites media* d'Archiac, 1837.
Diagnosis. External features of the test varying from flat to lenticular, more or less circular in outline and in advanced forms slightly to pronounced asymmetric. In planoconvex individuals the surface of the convex side may show a central knob and an irregular vermicular pattern of radial ridges and depressions (Fig. 22-1). The planar side is often smooth (Fig. 22-3).

Dimorphism is present, with the megalospheric generation generally far more abundant than the microspheric generation. Early megalospheric members of the lineage are characterized by a relatively long series of initial uniapertural chambers, arranged in a highly variable pattern with varying directions of coiling (Van Hinte, 1966b).

Advanced megalospheric forms are internally characterized by a large thick-walled bi- to quadrilocular embryo, immediately surrounded by periembryonic chambers issued by stolons developed in the embryonic wall. These periembryonic chambers are all comparable to the so-called progressive chamber *sensu* Van Hinte (1966a) and represent the start of cyclical growth of the equatorial layer. The equatorial chambers are relatively large. These generally form a somewhat wobbly equatorial layer.

The microspheric generation consists of forms with a small uniserial spiral of primary chambers or with a biserial configuration of chambers reminiscent of that of the planktonic foraminifer *Guembelina*. For this reason it is assumed that *Orbitoides* might have evolved from a biserial planktonic *Guembelina*-like ancestor.

Orbitoides media megaloformis Papp and Küpper, 1953
(Fig. 22-1–8; Fig. 23-1–6)

1953 *Orbitoides media megaloformis* Papp and Küpper, Plate 1, Figs. 8a, b, 9.

Type locality. Between Dreistetten and Bad Fischau, Niederösterreich, Austria.
Jamaican material. *Orbitoides media megaloformis* appears to be the only species of the *Orbitoides* family found in Jamaica. It occurs abundantly as free specimens in one specific interval in the Guinea Corn Formation of the Sanguinetti and Logie Green road traverses in the western Central Inlier. *Prebarrettia sparcilirata* is found in this horizon. In the Alston area *Orbitoides* is less restricted and occurs in several horizons. This agrees with the assumed deeper-water environment in this area. *Orbitoides media megaloformis* is also found in several random thin sections from various localities of the Rio Grande Limestone in the Blue Mountain Inlier.
Description of megalospheric forms. Test asymmetric lenticular (Fig. 22-7, 8), the convex side of the test ornamented with vermicular radiating ridges and depressions (Fig. 22-1), and an umbonal pustule in juvenile specimens (Fig. 22-4, 5). Maximum diameter and thickness measured on free specimens and more or less oriented vertical sections vary from 0.85 to 8.58 mm and from 0.25 to 1.96 mm, respectively. In four horizontal sections obtained so far, the embryon is quadrilocular (Fig. 23-1–6), with external dimensions varying from 387 to 632 microns in width and 330 to 493 microns in height. From the same sections, the number of epiauxiliary chambers ranges from six to nine. The majority of the specimens show asymmetry of the layers of lateral chambers on either side of the equatorial layer, with more massive bands of lateral chambers one one side (Fig. 22-7, 8). In at least one Jamaican specimen these massive bands of lateral chambers occur on the apical side in contrast with *Orbitoides apiculatus* from Maastricht, The Netherlands (see Baumfalk and Willemsen, 1986).

Microspheric specimens have not been encountered with certainty as yet, and so details of the nepionic structure are not yet available.
Remarks. The size of embryo and the number of epiauxiliary chambers of the Jamaican specimens exceed the species limits proposed by Van Hinte (1966a, 1976) for *Orbitoides media* (the average sum of the external width and height of the embryo between 550 and 700 microns; the number of periauxiliary chambers between 4 and 5.5). The Jamaican figures partly fit the limits set by Van Hinte (1976) for *O. media megaloformis* (the average sum of the external width and height of the embryo between 700 and 750 microns; the number of periauxiliary chambers from 5.5 to 10). Of the four specimens measured, at least two may exceed the limits of the size of the embryo of *O. media megaloformis*. We, therefore, may deal here with a species that may be slightly more advanced than the fauna from the type Dordonian (Aubeterre). Although the name *O. media megaloformis* has been retained, the fauna from Jamaica may be characteristic for the very Early Maastrichtian.

58   J. P. Krijnen and others

Figure 22. *Orbitoides media megaloformis* Papp and Küpper. (1, 3–5) Exterior view of individuals from the Central Inlier, sample J4480 Logie Green–Sanguinetti traverse, showing in figure 1 the vermicular pattern of fused pustules on the usually (but not always, as shown in 8) convex abapical side of the adult test, 12×. (2) Centered vertical section of fairly juvenile specimen, K75-21 Blue Mountain Inlier, Rio Grande, 48×. (6) Horizontal section of megalospheric specimen, J4522-1, Logie Green–Sanguinetti traverse, 55×. (7, 8) Nearly centered vertical sections, showing the different structure of the apical open lateral chambers and the more massive abapical ones—7: J4476 Logie Green–Sanguinetti traverse, 48×; 8: J4361 Alston, Borobridge–Alston traverse, 48×.

The Orbitocyclina Group

The taxonomic position of *Asterorbis* is uncertain, but a close resemblance of this genus to *Orbitocyclina* suggests that it may be ranged under the group of American Orbitocyclinidae (Van Gorsel, 1978). Van Gorsel (1978) mentioned the possibility that *Asterorbis* is only a variant of *Orbitocyclina,* the stellate pattern being produced only under certain ecological conditions. Nevertheless, he is inclined to consider *Asterorbis* a distinct genus on the basis of illustrative evidence provided by Rutten (1935), Hanzawa (1962), and Seiglie and Ayala-Castañares (1963), which demonstrates that this genus, unlike *Orbitocyclina,* has a more advanced quadriserial nepiont showing a symmetric embryon with two fairly large auxiliary chambers.

Genus *Asterorbis* Vaughan and Cole, 1932

Type species. *Asterorbis rooki* Vaughan and Cole, 1932.
Diagnosis. Test orbitoid and stellate in outline, with four to eight rays. The nepionic initial part of the equatorial layer is quadriserial, with two equal-size relatively large principal auxiliary chambers. The genus resembles *Orbitocyclina* apart from the stellate equatorial layer.

Asterorbis cf. *A. havanensis* Palmer, 1934
Fig. 24-1–10

1934 *?Asterorbis havanensis* Palmer, p. 252, text Figs. 13–14.

Type locality. Palmer Station 784, Central Havana Province, Cuba.
Jamaican material. Random sections and a few free specimens from the Alston traverse, Central Inlier.

Description. The test is orbitoid with equatorial layers and lateral chambers but stellate in outline (Fig. 24-1, 2). Several free specimens show the presence of four radii, which are raised because of an increase in size of the equatorial chambers from the center to the periphery, within the radii (Fig. 24-3, 7, 8). The maximum diameter of the test of a number of free specimens and the diameter of the majority of centered or near-centered vertical sections range from 1.60 to 4.04 mm; the maximum thickness varies from 0.30 to 1.21 mm. Numerous vertical sections more or less along a radius show the distinct increase of equatorial chambers during growth. A few horizontal sections show the generally very delicate nature of the small equatorial chambers. Preservation of the material is also very poor, and no clear picture of nepionic chambers could be obtained so far. The proto- and deuteroconch seem to be equal in size, and based on two horizontal sections their diameters vary from 41 to 53 and from 51 to 53 microns, respectively. In one vertical section the height of the protoconch measures 77 microns. Generally in random vertical sections the equatorial layer is shown to be relatively thin. Near the center its thickness (height of the equatorial chambers) varies from 35 to 80 microns. A rapid increase in chamber height is shown where the section cuts along a radius (Fig. 24-7, 8). The chamber height in such radii may attain 310 microns.

ACKNOWLEDGMENTS

Acknowledgments are due to Dr. Raymond Wright of the Petroleum Corporation of Jamaica, who kindly provided the information necessary for key figures for the chapter. We are indebted to both Dr. David W. Haig from the Department of Geology, University of Western Australia, Nedlands, and Dr. Norman F. Sohl from the U.S. Geological Survey, Reston, Virginia, who critically read the first draft of this manuscript and suggested a number of useful changes. We also wish to thank two successive directors of the Geological Survey of New South Wales, Dr. Neville Markham and Mr. John Cramsie, for their approval for the use of departmental analytical and laboratory equipment; Dr. John Pickett, who allowed the first author to share SEM time; and Otti Mueller and David Barnes for doing most of the photographic work.

Dr. C. Winkler Prins, acting on behalf of the Board of Directors of the National Museum of Geology and Mineralogy (Leyden, The Netherlands), is thanked for approving the loan of most of the sediment thin sections of the Jamaica collection and the same microscope with which the thesis material was analyzed back in 1971.

The first author wishes to thank Alice Warren, for her encouragement at times of faltering enthusiasm and periods of stagnation due to increased departmental and other activities, while preparing text, line drawings, and photographs for this chapter.

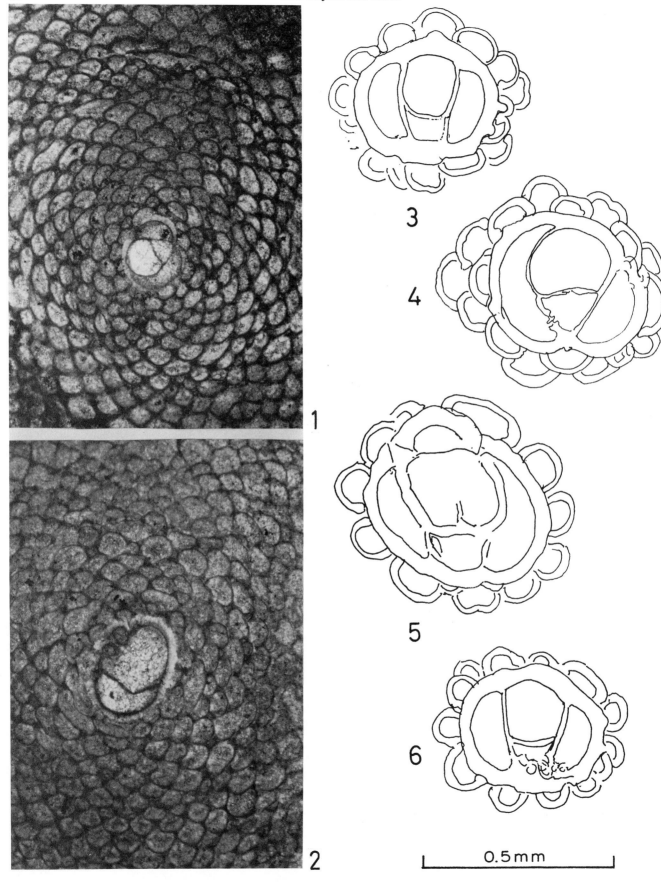

Figure 23. *Orbitoides media megaloformis* Papp and Küpper, from the Central Inlier, Logie Green. (1, 2) Horizontal sections of megalospheric specimens—1: J4480-2, 55×; 2: J4480-3, 55×. (3-6) Interpretation of nepionic chamber configuration in megalospheric specimens, all circa 110×—3: J4522-1; 4: J4480-2; 5: J4480-1; 6: J4285.

REFERENCES CITED

d'Archiac, A., 1837, Mémoire sur la formation crétacé du sud-ouest de la France: Mémoires de la Société géologique de France, ser. 1, v. 2, p. 157–192.

Baumfalk, Y. A., and Van Hinte, J. E., 1985, *Orbitoides media* (d'Archiac) in the Campanian deposits of the A10 Motorway at Mirambeau (Charente Maritime): Cretaceous Research, v. 6, p. 181–189.

Baumfalk, Y. A., and Willesen, F., 1986, Ecophenotypic variation of the larger foraminifer *Orbitoides apiculata* from the Maastrichtian stratotype: Geologie en Mijnbouw, v. 65, p. 23–34.

Brönnimann, P., 1954a, Upper Cretaceous Orbitoidal foraminifera from Cuba, Part I: *Sulcorbitoides* nov. gen: Contributions from the Cushman Foundation for Foraminiferal Research, v. 5, p. 55–61.

Brönnimann, P., 1954b, Upper Cretaceous Orbitoidal foraminifera from Cuba, Part II: *Vaughanina* Palmer, 1934: Contributions from the Cushman Foundation for Foraminiferal Research, v. 5, part 3, p. 91–105.

Brönnimann, P., 1955a, Upper Cretaceous Orbitoidal foraminifera from Cuba, Part III: *Pseudorbitoides* H. Douvillé, 1922: Contributions from the Cushman Foundation for Foraminiferal Research, v. 6, p. 57–76.

Brönnimann, P., 1955b, Upper Cretaceous Orbitoidal foraminifera from Cuba, Part IV: *Rhabdorbitoides* nov. gen: Contributions from the Cushman Foundation for Foraminiferal Research, v. 6, p. 97–104.

Brönnimann, P., 1956, Upper Cretaceous Orbotoidal foraminifera from Cuba, Part V: *Historbitoides* nov.gen: Contributions from the Cushman Foundation for Foraminiferal Research, v. 7, part 2, p. 60–66.

Brönnimann, P., 1958a, New pseudorbitoids from the Upper Cretaceous of Guatemala, Texas and Florida: Eclogae Geologica Helvetica, v. 51, p. 422–437.

Brönnimann, P., 1958b, New Pseudorbitoididae from the Upper Cretaceous of Cuba, with remarks on encrusting foraminifera: Micropaleontology, v. 4, p. 165–186.

Chubb, L. J., 1958, The Cretaceous rocks on central St. James: Geological Society of Jamaica, Geonotes, 1, p. 3–11.

van Dommelen, H., 1971, Ontogenetic, phylogenetic and taxonomic studies on the American species of *Pseudovaccinites* and of *Torreites,* and the multiple-fold hippuritids [Ph.D. thesis]: University of Amsterdam, 125 p.

Douvillé, H., 1922, Orbitoïdés de la Jamaïque: *Pseudorbitoides Trechmanni,* nov.gen., nov.sp: Comptes Rendus de la Société géologique de France, v. 17, p. 203–204.

van Gorsel, J. T., 1978, Late Cretaceous Orbitoidal foraminifera, *in* Hedley, R. H., and Adams, C. G., eds., Foraminifera: London, Academic Press, v. 3, p. 1–120.

Hanzawa, S., 1962, Upper Cretaceous and Tertiary three-layered foraminifera and their allied forms: Micropaleontology, v. 8, p. 129–186.

van Hinte, J. E., 1966a, *Orbitoides* from the Campanian type section: Koninklijke Nederlandse Akademie van Wetenschappen, Proceedings, Serie B, v. 69, p. 79–110.

van Hinte, J. E., 1966b, *Orbitoides hottingeri* n.sp. from Northern Spain: Koninklijke Nederlandse Akademie van Wetenschappen, Proceedings, Ser. B, v. 69, p. 388–402.

van Hinte, J. E., 1968, The Late Cretaceous larger foraminifer *Orbitoides douvillei* (Silvestri) at its type locality Belves, S.W. France: Koninklijke Nederlandse Akademie van Wetenschappen, Proceedings, Ser. B, v. 71, p. 359–372.

van Hinte, J. E., 1976, A Cretaceous time scale: American Association of Petroleum Geologists Bulletin, v. 60, p. 269–287.

Jiang, M. J., and Robinson, E., 1987, Calcareous nannofossils and larger foraminifera in Jamaican rocks of Cretaceous to Early Eocene age: Proceedings, Workshop Status of Jamaican Geology, 1984: Geological Society of Jamaica, Special Issue, p. 24–51.

Krijnen, J. P., 1967, Pseudorbitoid foraminifera from Curaçao: Koninklyke Nederlandse Akademie van Wetenschappen, Proceedings, Ser. B, 70, 2, p. 144–164.

Krijnen, J. P., 1971, Analysis of pseudorbitoids from Glenbrook, Jamaica, and discussion of a study by Cole & Applin (1970): Scripta Geologica, v. 4, p. 1–15, 2 pls.

Krijnen, J. P., 1972, Morphology and phylogeny of pseudorbitoidal foraminifera from Jamaica and Curaçao, a revisional study: Scripta Geologica, v. 8, p. 1–77, 27 pls.

Krijnen, J. P., 1978, Pseudorbitoids from the Parguera Limestone, Puerto Rico, and from the Back Rio Grande Limestone, Jamaica, with remarks on the pseudorbitoidal evolutionary pattern: Geologie en Mijnbouw, v. 57, p. 233–242, 1 pl.

Krijnen, J. P., 1993, A new genus and lineage of pseudorbitoid foraminifera in Jamaica: The Journal of the Geological Society of Jamaica, v. 29 (in press).

Mac Gillavry, H. J., 1937, Geology of the province of Camaguey, Cuba, with revisional studies in rudist paleontology (mainly based upon collections from Cuba) [Ph.D. thesis]: University of Amsterdam, 168 p.

Mac Gillavry, H. J., 1963, Phylomorphogenesis and evolutionary trends of Cretaceous orbitoidal foraminifera, *in* von Koeningswald, G.H.R., et al., eds., Evolutionary trends in Foraminifera: Amsterdam, Elsevier, p. 139–197.

Mac Gillavry, H. J., 1977, Senonian rudists from Curaçao and Bonaire: A typical Antillean fauna: Abstracts 8th Caribbean Geological Conference, Curaçao 9–24 July 1977, p. 101–102.

McFarlane, N., 1977, Jamaica Geology 1:250,000: Mines and Geology Division, Ministry of Mining and Natural Resources, Jamaica.

Neumann, M., 1958, Révision des orbitoïdidés du Crétacé et de l'Eocène en Aquitaine occidentale: Mémoires de la Société géologique de France, n. ser., v. 37, Mémoire 83, p. 1–174.

Neumann, M., 1972a, A propos des orbitoïdés du Crétacé supérieur et de leur signification stratigraphique. I. Genre *Orbitoides* d'Orbigny (1847): Revue

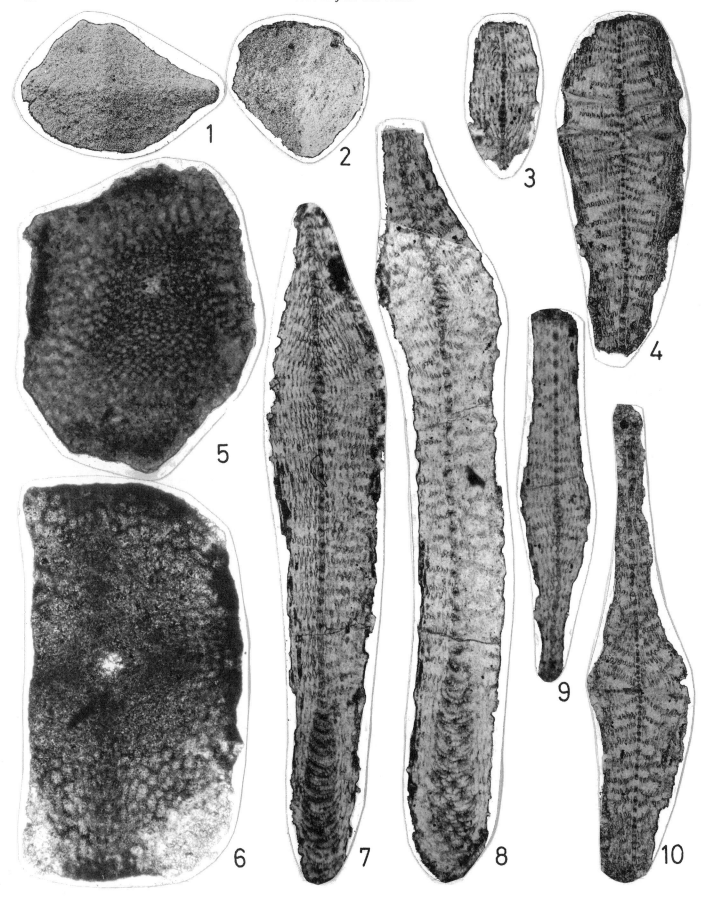

Figure 24. *Asterorbis* cf. *A. havanensis* Palmer from the Central Inlier, Alston traverses. (1, 2) Exterior view showing four elevated rays, J4333, Borobridge-Alston traverse, 11×. (3, 4, 7–10) Vertical sections showing a strongly thickening equatorial layer where a ray is intersected, all circa 48×—3, 4: J4634, 8: J4387, Borobridge-Christiana traverse; 7: J4387, 9: J4385 Alston-Spaldings traverse; 10: J4348 Borobridge-Alston traverse. (5, 6) Centered horizontal sections showing the stellate configuration of the equatorial layer—5: J4333-4, 48×, 6: J4333-1, 48×, Borobridge-Alston traverse.

◄───

de Micropaléontologique, v. 14, p. 197–226.

Neumann, M., 1972b, Sur les orbitoïdidés du Crétacé supérieur et du Tertiaire. II. Structure and classification: Revue de Micropaléontologique, v. 15, p. 163–189.

d'Orbigny, A. D., 1848, *in* Lyell, C., On the relative age and position of the so-called Nummulite Limestone of Alabama: Quarterly Journal of the Geological Society of London, v. 4, p. 10–16.

Palmer, D. K., 1934, Some large fossil foraminifera from Cuba: Sociedad Cubana Historia Natural, Memoria, v. 8, p. 235–264, 5 pls.

Papp, A., and Küpper, K., 1953, Die foraminiferenfauna von Guttaring und Klein St Paul. II. Orbitoiden aus Sandsteinen von Pemberger bei Klein St Paul: Sitzungsberichte der Osterreichische Akademie der Wissenschaften, Mathematisch-naturwissenschaftliche Klasse, Abt I, v. 162, p. 65–82.

Rutten, M. G., 1935, Larger foraminifera of Northern Santa Clara Province, Cuba: Journal of Paleontology, v. 9, p. 527–545.

Seiglie, G. A., and Ayala-Castañares, A., 1963, Sistematica y bioestratigrafia de los foraminiferos grandes del Cretacico superior (Campaníano y Maastrichtiano) de Cuba: Paleontologia Mexicana, v. 13, p. 1–56.

Vaughan, T. W., and Cole, W. S., 1932, A restudy of the foraminiferal genera *Pseudorbitoides* and *Vaughanina:* Journal of Paleontology, v. 17, p. 97–100.

Whitfield, R. P., 1897, Observations on the genus *Barrettia* Woodward with description of two new species: Bulletin of the American Museum of Natural History, v. 19, p. 233–246.

Zans, V. A., Chubb, L. J., Versey, H. R., Williams, J. B., Robinson, E., and Cooke, D. L., 1962, Synopsis of the geology of Jamaica: Jamaica, Geological Survey Department Bulletin, v. 4, 72 p.

MANUSCRIPT ACCEPTED BY THE SOCIETY NOVEMBER 11, 1992

Geological Society of America
Memoir 182
1993

Ostracode biostratigraphy of the Titanosarcolites-*bearing limestones and related sequences of Jamaica*

Joseph E. Hazel
Department of Geology and Geophysics, Louisiana State University, Baton Rouge, Louisiana 70803
Takahiro Kamiya
Department of Earth Sciences, Kanazawa University, Kanazawa, Ishikawa, 920 Japan

ABSTRACT

The Cretaceous limestone and limestone rubble beds of Jamaica that contain the rudist *Titanosarcolites* with associated macroinvertebrates, and the shales between them, contain a diverse ostracode assemblage. The same is true of the shales associated with the so-called oyster limestones. Samples from the Central, Maldon, Marchmont, and Jerusalem Mountain inliers have yielded more than 123 species, only two of which can be confidently assigned to known species. Many of these taxa are rare or are still under study, but 38 common forms allow the division of the studied interval into three very distinct zones, the oldest of which can be divided into three subzones. The *Titanosarcolites* and associated shales of the Central, Maldon, and Marchmont inliers can all be placed in the oldest zone delineated. Using the subzones of this zone, lithic units can be correlated from inlier to inlier. The small Jerusalem Mountain Inlier has two different ostracode assemblages. The *Titanosarcolites* beds of this inlier are placed in one of these, and the younger oyster limestone beds and associated shales fall in the other.

The presence of the genera *Schizoptocythere, Buntonia,* and *Ovocytheridea* indicates a tropical Tethyan affinity for the Jamaican ostracode fauna. It is distinctly unlike the richly fossiliferous deposits of the North American Coastal Plain.

A review of the chronostratigraphic position of the Campanian-Maastrichtian boundary and an alternate interpretation of recently published calcareous nannofossil data for the Jamaica deposits support the conclusion of macrofossil workers that the *Titanosarcolites* beds are Maastrichtian and not partly Campanian in age.

INTRODUCTION

The study of Cretaceous ostracodes in the Caribbean region did not begin until 1946, when van den Bold (1946) recorded eight species from the Upper Cretaceous of Cuba and five from British Honduras. In 1950, he added two localities and four additional species to the list from Cuba (van den Bold, 1950). In 1974, Lyubimova and Sanchez-Arango described 25 species and one new genus from the Campanian and Maastrichtian of Cuba (Lyubimova and Sanchez-Arango, 1974). Compared to other fossil groups, the Cretaceous ostracodes of the Caribbean are very poorly known.

The explanation for this lack of information lies partly in the fact that many of the facies that have been sampled for microfossils in the Caribbean region are of deeper-water origin and represent high sedimentation rates, neither of which is conducive to high ostracode density or diversity. However, the principal reason for the paucity of literature, at least as concerns the shoal-water facies, is lack of investigation.

In 1966 a U.S. Geological Survey–Smithsonian Institution–University of the West Indies field party collected numerous samples from stratigraphic intervals in the Cretaceous inliers of central and western Jamaica (Fig. 1) that are generally rich in ostracodes. In an abstract, Hazel (1971) called attention to a diverse ostracode fauna in Jamaica. These samples form the basis of the present report. Although this collection is 24 years old, no

---

Hazel, J. E., and Kamiya, T., 1993, Ostracode biostratigraphy of the *Titanosarcolites*-bearing limestones and related sequences of Jamaica, *in* Wright, R. M., and Robinson, E., eds., Biostratigraphy of Jamaica: Boulder, Colorado, Geological Society of America Memoir 182.

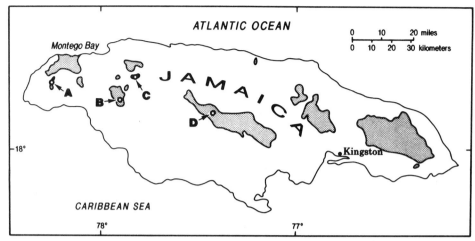

Figure 1. Index map showing the Cretaceous inliers of Jamaica. Letters indicate collecting areas in the following inliers: A = Jerusalem Mountain; B = Marchmont; C = Maldon; D = Central.

other work on Jamaican Cretaceous ostracodes has been published, and only the work of Lyubimova and Sanchez-Arango (1974) is pertinent to the understanding of the Jamaican fauna. This chapter is a report on the ostracode biostratigraphy of that part of the Jamaican sequence that contains the distinctive rudist *Titanosarcolites* and associated forms (see Kauffman and Sohl, 1974). It was hoped that the study of the Ostracoda would result in a zonation that would act as a check on the correlations based mainly on the rudists found in the limestone and limestone rubble beds and, at the same time, provide biostratigraphic control for the intervening shales. As documented below, it is believed that these goals have been obtained.

STAGE PLACEMENT OF THE *TITANOSARCOLITES* AND YOUNGER BEDS

Before the ostracode data are addressed, it is necessary to consider assignment of the *Titanosarcolites*-bearing beds to the universally used stages based on European type sections. Traditionally, these beds have been considered to be of Maastrichtian age (Chubb, 1962, 1971). Sohl and Kollman (1985) consider *Titanosarcolites* and associated actaeonellid gastropods to indicate a middle Maastrichtian and younger age. However, neither the rudists, the actaeonellids, nor the ostracodes, for that matter, give a direct tie to the European type area.

Recently, Jiang and Robinson (1987, 1989), in a study of the calcareous nannoplankton from the *Titanosarcolites*-bearing limestones and associated deposits, have questioned the Maastrichtian age assignment. They concluded that the *Titanosarcolites* beds and even the deposits that occur above them in the Jerusalem Mountain Inlier of Westmoreland Parish (Fig. 1) are of Campanian age. This is largely because of the presence of the species *Quadrum trifidum* (Stradner) in most of the productive samples. The extinction of this taxon has been used by several nannoplankton workers as a zonal boundary definer and as an indicator of the Campanian-Maastrichtian boundary (see Perch-Nielsen, 1985). However, it is not at all clear what nannoplankton events mark the lower Maastrichtian or the upper Campanian in the type areas of these stages. Most planktic foraminifer workers (e.g., Caron, 1985) use the extinction of *Globotruncanita calcarata* (Cushman) as indicating the end of the Campanian, but this taxon has not been reported from western Europe.

In a chronostratigraphic framework model for the Upper Cretaceous and Cenozoic developed using the graphic correlation technique, which is tied to a time-scale (Hazel and Brouwers, 1982), there is a difference in time of about 3.7 m.y. between the extinction of *G. calcarata* at about 75.6 Ma and that of *Quadrum trifidum* at about 71.9 Ma. (It should be pointed out that the late Campanian and Maastrichtian part of the framework model has undergone major revision since the publication of Hazel and Brouwers [1982].)

On the basis of benthic foraminifers, Reiss et al. (1985) suggested that the Campanian-Maastrichtian boundary of western Europe correlates with the basal Ghareb Formation in the Negev Desert of Israel. This is just above what is known as the Phosphorite Unit, which contains the ammonite *Anaklinoceras reflexum* Stephenson (Gvirtzman et al., 1989). The last authors (p. 125) observed the first appearance of *Gansserina wiedenmayeri* (Gandolfi) at the same level as the diagnostic benthic foraminifers and ". . . recommend, for all practical purposes and for the time being, the FA (first appearance) of . . . *G. falsostuarti* and/or *G. wiedenmayeri* as the best and the closest planktic guide for the base of the Maastrichtian." However, *Globotruncana falsostuarti* has been reported elsewhere from chronostratigraphic positions that would seem slightly older than this. Based on the chronostratigraphic framework model mentioned above, the entry level of *G. falsostuarti* is in the chronozone of the *Didymoceras nebrascense* Zone of the American Western Interior.

The assertion by Gvirtzman et al. (1989) that the entry level of *Gansserina wiedenmayeri* is a marker for the Campanian-Maastrichtian boundary has considerable merit. Based on the data presented by Gvirtzman et al. (1989), it can be shown that the first appearance of *G. wiedenmayeri* is chronostratigraphically very close to the first evolutionary appearance of *Baculites reesidei* Elias of the Western Interior of the United States. In Obradovich and Cobban (1975), W. A. Cobban suggested that the Campanian-Maastrichtian boundary could fall as low as the base of the chronozone of the *Baculites reesidei* Zone. In the North American Gulf Coastal Plain this event probably correlates with the lower part of the *Nostoceras stantoni* Zone of Young (1986).

Based on the aforementioned chronostratigraphic framework model, the extinction of *Quadrum trifidum* is still about 1.4 m.y. younger than the entry level of *Gansserina wiedenmayeri*. There is nothing in the nannofossil data presented in Jiang and Robinson (1987, 1989) that indicates that any of the *Titanosarcolites* localities are Campanian in age. In fact, the extinction of *Quadrum trifidum* postdates the entry level of *Gansserina gansseri* (Bolli) by about 0.5 m.y. *Gansserina gansseri* is a widely used marker for the base of the middle Maastrichtian. Thus, as Sohl and Kollman (1985) indicate, the first evolutionary appearance of *Titanosarcolites* may be well above the base of the Maastrichtian. The presence of *Quadrum trifidum* with the youngest ostracode assemblage found (see below) in the Jerusalem Mountain Inlier (Fig. 1) does suggest that there may be no late or even later middle Maastrichtian in the inliers studied, unless it is represented by what Jiang and Robinson (1987) refer to as the Masemure Formation above the oyster-bearing beds.

COLLECTING AREAS

General

Beds containing such rudists as *Titanosarcolites, Antillocaprina, Biradiolites,* and *Thyrastylon* crop out in several of the Jamaican Cretaceous inliers—the Jerusalem Mountain Inlier in Westmoreland Parish; the Sunderland, Calton Hill, and Maldon inliers in St. James Parish; and the Central Inlier in Clarendon Parish—as well as on the north flank of the Blue Mountains near Port Antonio in Portland Parish. Our samples were collected in the Central, Maldon, Marchmont, and Jerusalem Mountain inliers (Fig. 1). Sample location data can be found below under "Collections."

Central Inlier

About 150 m (500 ft) of limestone, the Guinea Corn Limestone, containing rudists and interbedded silty shales with rudists and other fossils, crops out in the valley of the Rio Minho between Grantham and Guinea Corn in Clarendon Parish (Coates, 1968). Several samples containing ostracodes were collected from the lower 90 m (290 ft) of the exposure. The upper part of the Guinea Corn consists mostly of hard limestones and was not sampled.

Maldon Inlier

Over 490 m (1,600 ft) of shale and limestone is exposed in the Maldon Inlier (Chubb, 1958, 1962; Coates, 1977). The basal unit of the inlier consists of about 90 m (300 ft) of reddish shale, the lower part of which is badly weathered. This unit is referred to as the Woodland Shale. Parts of the upper Woodland contain abundant mollusks, ostracodes, and the foraminifer *Ayalaina.* The Woodland is overlain by the Maldon Limestone, about 150 m (500 ft) of rudist limestones and interbedded calcareous shales. The shales and limestone rubble beds have yielded abundant ostracodes. The Popkin Formation overlies the Maldon and is approximately 180 m (600 ft) of brown and gray shale, with some quartz pebble conglomerate in the lower part. Ostracodes are abundant in the brown and gray shales of the upper Popkin; the lower part is poorly exposed and could not be sampled. Another mainly limestone unit with abundant rudists, the Vaughnsfield Limestone, overlies the Popkin and consists of about 80 m (260 ft) of limestone with interbedded shales. The lower 15 m (50 ft) consists mainly of massive limestone, but a clay lens in the lower few meters yielded abundant ostracodes. The upper Vaughnsfield is poorly exposed and weathered and was not collected for microfossils. Volcanic shales, conglomerates, and tuffs of the Garlands Formation overlie the Vaughnsfield.

Marchmont Inlier

The classic localities along the Cambridge-to-Catadupa railroad studied by Hill (1899) and by others are found in this inlier. In addition, very fossiliferous shales and limestones are present in and adjacent to Ducketts Land Settlement, just southwest of the railroad localities (Fig. 2). Ostracodes are diverse and abundant in our samples from these areas. However, faults and covered

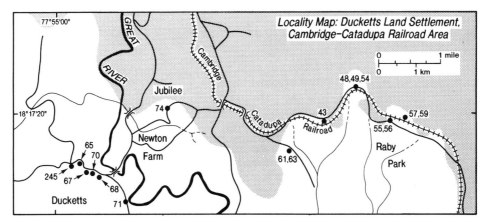

Figure 2. Index map showing collecting localities in the Marchmont Inlier. Numbers are sample numbers. Pattern indicates Tertiary outcrop; white is Cretaceous.

intervals are common, and we were not able to establish a stratigraphic succession of lithic units either along the railroad or at Ducketts. One of the objects of the ostracode study was to see if the ostracodes could be used to suggest a succession that could then be field tested by mapping.

There is no formal or informal nomenclature for the Maastrichtian rocks of the Marchmont Inlier. Chubb (1962) reported "three rudist limestones, all yielding the *Titanosarcolites* fauna, with intervening shales." These are underlain by shales containing the bivalve *Veniella.* Chubb speculated that the limestones might be equivalent to the limestones of the Maldon Inlier and that the *Veniella* shale was equivalent to the Woodland Shale.

Jerusalem Mountain Inlier

As pointed out by Trechmann (1924), there are two main types of Cretaceous limestone in the Jerusalem Mountain section. In the upper part of the section there are massive gray limestones containing oysters and no rudists, underlain by at least 15 m (50 ft) of somewhat weathered, reddish, calcareous shale. In the lower approximately 30 m (100 ft) of the exposed section are limestone and limestone rubble beds containing *Titanosarcolites* and other rudists, interbedded and intercalated with calcareous shale.

Technically, there is no formal nomenclature for the Jerusalem Mountain Inlier. Jiang and Robinson (1987) introduced terms originally used by the Jamaica Stanolind Oil Company in the 1950s. They refer to the calcareous beds as the Jerusalem Mountain Formation, using Jerusalem Limestone for the oyster-bearing limestones and referring to the lower calcareous shales and limestone rubble beds as the Thicket River Limestone Member.

Ostracodes occur in considerable abundance in all the samples processed that were taken from the Jerusalem Mountain Formation. The preservation of the microfauna here, however, is poorer than in most samples from the other inliers. Nonetheless, it is fair in some samples, and the common species are easily separated. Mollusks other than rudists, solitary corals, and bryozoans are common in the shales below and between the oyster-bearing limestones. The floating crinoid *Saccocoma,* the worm *Hamulus,* and the foraminifer *Kathina* are also found in the shales. Trechmann (1924, p. 387) considered the oyster-bearing limestone to be the stratigraphically highest Cretaceous limestone in Jamaica. Chubb (1962) suggested a possible correlation of the *Titanosarcolites* limestone at Jerusalem Mountain with the Vaughnsfield Formation of the Maldon Inlier. Jiang and Robinson (1987) indicated that there are evenly stratified Cretaceous sandstones and shales, which they refer to as the Masemure Formation, above the oyster-bearing limestone. These were not observed when the ostracode samples were taken.

THE OSTRACODE FAUNA

There are more than 123 species of ostracodes in the 50 samples examined in detail to date. The Cytheracea, with at least 103 species, are the most diverse and abundant. At least 15 cytherellid species are present. The species of some groups (bairdiids, Krithinae, xestoleberids, marine cyprids) have not been fully studied as yet. No myodocopids or cladocopids were found. Of the species studied so far, only two can be assigned to previously described taxa with some confidence. These are *Cytherella tuberculifera* Alexander, which is known to occur in rocks of late Campanian and Maastrichtian age in the North American Coastal Plain (Crane, 1965), and *Schizoptocythere segurai* (Lyubimova and Sanchez-Arango),which was found in rocks said to be of Maastrichtian age in Cuba (Lyubimova and Sanchez-Arango, 1974). Many of the taxa are represented by few specimens. Thirty-eight of the more common species are particularly valuable in developing a zonation that is at least applicable in Jamaica. This zonation is discussed below. For the purpose of this chapter, the undescribed forms are left in open nomenclature.

Preservation of the ostracodes, except in the Jerusalem Mountain Inlier, is generally good, particularly in the shales of the Maldon Inlier and the shales and limestone rubble beds of the Marchmont Inlier. Recrystallization is common, but in most cases it is not coarse, and finer ornamental features of the carapace are easily distinguished. However, very few single valves are present in any samples for any species, except platycopids. The carapaces are filled with crystalline calcite, and it has proved futile to try to pull the valves apart. Therefore, most species are known only from their exteriors. This does not seriously hamper delineation at the species level, because for most groups there are, in general, quite sufficient external characters. However, ignorance of muscle scars, hingement, and marginal areas has, in a few cases, made generic and subfamilial identifications difficult.

In the rocks studied, ostracodes are the most abundant, diverse, and consistently occurring microfossils in the size fraction greater than 177 microns. In nearly all samples examined they greatly outnumber benthonic small foraminifers; planktonic foraminifers are virtually absent. Large foraminifers occur abundantly in some samples; *Kathina* is scattered throughout the sampled interval, and *Chubbina* and *Ayalaina* are present in several samples. Cellariform cheilostome bryozoans are not uncommon; membraniporiform cheilostomes and a few cyclostomes also occur. The small floating crinoid *Saccocoma* was also found in some samples, as were species of dasycladacean algae.

BIOGEOGRAPHY

The Jamaican ostracode assemblage is quite distinct in generic composition from the richly fossiliferous deposits of the Coastal Plain of the United States. A fauna of similar age has been described from western Cuba by Lyubimova and Sanchez-Arango (1974). It has some of the distinct elements of the Jamaican assemblages, such as *Schizoptocythere,* but the Cuban fauna also has elements that are characteristic of the Coastal Plain of the United States, such as *Ascetoleberis.*

Seven species of *Schizoptocythere* have been found in Jamaica. This genus was proposed by Siddiqui and Al-Furaih (1981) and based on Paleocene and Eocene species from India, Pakistan, and Saudi Arabia. Specimens that can be assigned to this genus have been illustrated from Paleocene deposits from West Africa (Reyment, 1963; Khan, 1970). *Cythereis segurai* Lyubimova and Sanchez-Arango from the Maastrichtian of Cuba, is a *Schizoptocythere* that also seems to occur in Jamaica. Another species that can be assigned to *Schizoptocythere, S. compressa* (Hazel and Paulson), is known from the Campanian of the American Coastal Plain (see Hazel and Brouwers, 1982).

The genus *Ovocytheridea* is very common in the more calcareous facies in Jamaica. This taxon is common in western and northern Africa, where it occurs in deposits ranging in age from at least Turonian to Paleocene (Reyment, 1960, 1963; van den Bold, 1964; Apostolescu, 1961, 1963). The genus *Buntonia* occurs commonly in the Jamaican Maastrichtian. It is known from the Cretaceous and lower Tertiary of Africa and has been observed in early Danian deposits from Puerto Rico (unpublished U.S. Geological Survey data), but it does not appear in the American Coastal Plain until the late Paleocene and has not been reported at all from the lower Tertiary or Cretaceous of Europe. *Schizoptocythere, Ovocytheridea,* and *Buntonia* suggest a tropical Tethyan affinity for the Jamaican ostracode assemblage. However, except for some data from Cuba, we know very little about the Cretaceous Caribbean ostracodes. Whether the Jamaican assemblage is typical of other areas in the region remains to be seen.

BIOSTRATIGRAPHY

The results of the biostratigraphic study were positive and somewhat surprising. Two distinctly different ostracode assemblages are associated with the *Titanosarcolites* beds and associated shales. The younger of these is found only in the Jerusalem Mountain Inlier. A third distinct assemblage occurs above the *Titanosarcolites* facies and is associated with the oyster limestones of the Jerusalem Mountain Inlier. The older of the ostracode assemblages is widespread among the inliers, except Jerusalem Mountain, and can be divided into three parts. The sections studied are placed in three zones. These are Oppel zones in the sense of the North American Commission on Stratigraphic Nomenclature (1983), in that each zone is characterized by more than two taxa and the boundaries of the zones are based on two or more first and/or last occurrences of the included characterizing taxa. How widely applicable these are in the Caribbean is unknown. As more information is compiled for the Caribbean Cretaceous, it is hoped that an interval zonation of widespread usage within the region can be developed. The distribution of the biostratigraphically useful ostracodes in the samples is given in Table 1. These taxa are illustrated in Figures 3, 4, and 5.

Ostracode Zone 1 (OZ1)

This zone is present in all the inliers investigated except Jerusalem Mountain. Many species are restricted to it (Table 1).

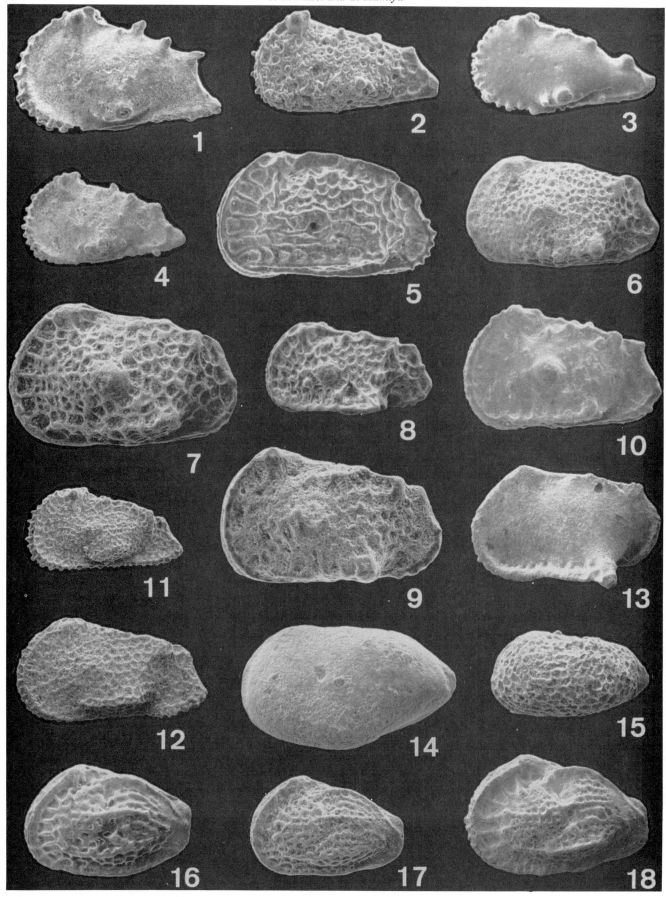

Figure 3. Ostracodes from the Maastrichtian of Jamaica. Numbers 1–10, 14, 15 ×80; numbers 11–13, 16–18 ×60. All photomicrographs are left lateral views of carapaces. 1—*Schizoptocythere* sp. 3, female, sample J70, OZ1b; 2—*Schizoptocythere segurai* (Lyubimova and Sanchez-Arango), female, sample J91, OZ1c; 3—*Schizoptocythere* sp. 1, female, sample J91, OZ1c; 4—*Schizoptocythere* sp. 4, female, sample J40, OZ3; 5—*Limburgina* sp. 3, female, sample J70, OZ1b; 6—*Limburgina* sp. 5, female, sample J15, OZ2; 7—*Limburgina* sp. 4, female, sample J15, OZ2; 8—*Limburgina* sp. 2, male, sample J91, OZ1c; 9—*Limburgina* sp. 1, female, sample J95, OZ1a; 10—*Spinoleberis*? sp. 1, female, sample J56, OZ1b; 11—*Platycosta*? sp. 1, female, sample J70, OZ1b; 12—*Cletocythereis* sp. 1, female, sample J55, OZ1b; 13—*Alatacythere* sp. 1, female, sample J48, OZ1a; 14—*Eocytheropteron* sp. 2, male, sample J15, OZ2; 15—*Eocytheropteron* sp. 1, male, sample J240, OZ1a; 16—*Buntonia* sp. 1, female, sample J67, OZ1b; 17—*Buntonia* sp. 3, female, sample J15, OZ2; 18—*Buntonia* sp. 2, female, sample J33, OZ2.

Of the approximately 60 rare and common species found in deposits assigned to OZ1, only an *Amphicytherura*, which was found in all three zones, and "*Cytherura*" sp. 1 occur in the younger zones. The distribution of the ostracodes at localities where superposition of the samples is not a problem (the Maldon Inlier and the Rio Minho section of the Central Inlier) indicates that OZ1 can be divided into three biostratigraphically useful subzones: OZ1a, OZ1b, and OZ1c.

The lowest subzone, OC1a, is an Oppel-type unit that includes samples J95 and 96 from the upper part of the Woodland Shale of the Maldon Inlier. In the Marchmont Inlier, the shales along or near the Cambridge-Catadupa Railroad (Fig. 2) (samples J48, 49, 50, 54, 57, 59, 61, and 63) and the reddish shales near the edge of the inlier at Newton Farm (J74) also have an OZ1a assemblage. The lowest sample of the Guinea Corn Formation (J240) is also placed here. *Aequacytheridea* sp. 1, *Eocytheropteron* sp. 1, *Brachycythere* sp. 1, and *Alatacythere* sp. 1 have only been found in these samples.

The Maldon Limestone of the Maldon Inlier, the Guinea Corn samples above J240, the faulted limestones and shales along the main road through Ducketts Land Settlement (Fig. 2), the limestones with intervening shale on the Lambs River road south of Ducketts, and limestone outcrops along the Cambridge-Catadupa Railroad are placed in a second Oppel-type subzone, OZ1b. Several species are only found in these samples (Table 1). The fauna of the upper Popkin and the Vaughnsfield is included in a third subzone, OZ1c. Nearly all of the common species found in the upper Popkin and lower Vaughnsfield also occur in the other subzones of OZ1. However, these samples contain some rare elements not seen elsewhere and the common and distinctive *Brachycythere* sp. 1.

Ostracode Zone 2 (OZ2)

This zone includes the *Titanosarcolites* limestone and associated shales (Thicket River Member of the Jerusalem Mountain Formation) in the Jerusalem Mountain Inlier. The fauna is quite distinct from that of OZ1 (Table 1). The generic composition is similar, but the species for the most part are different. In addition, there are a number of new species that are still under study.

As mentioned above, Chubb (1962) stated that the *Titanosarcolites* limestone at Jerusalem Mountain may be equivalent to the Vaughnsfield Formation of the Maldon Inlier. However, the ostracode assemblages are very different. Even though there were no productive samples from the upper part of the Vaughnsfield, a more likely scenario is that the volcaniclastics below the Jerusalem Mountain Formation are equivalent at least in part to the volcanics above the Vaughnsfield. The fauna of the lower Jerusalem Mountain Formation probably represents a return to conditions similar to those that existed during deposition of the Maldon and Vaughnsfield formations following a volcanic episode that eliminated the OZ1c fauna.

Ostracode Zone 3 (OZ3)

The shales just below and between the oyster limestones at Jerusalem Mountain contain the third major assemblage. Particularly distinctive is the abundance of small Cytherideinae, mainly in the form of *Aequacytheridea*.

The biostratigraphic zonation is summarized in Table 2.

Although the ostracodes are poorly known from the Caribbean region, it is obvious that they have great potential for establishing a microfossil biostratigraphy in the shoal-water lithofacies that includes not only the rudist-dominated frameworks but also the intervening shales.

COLLECTIONS

Central Inlier

Guinea Corn Formation (lower part): Interbedded limestone and shale in bluffs on the right side of the Rio Minho, about

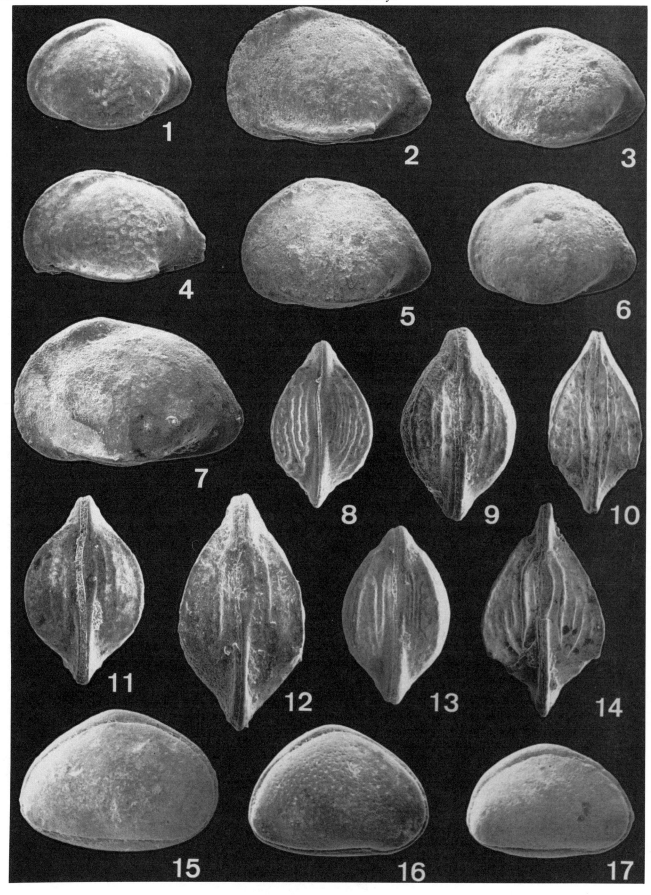

Figure 4. Ostracodes from the Maastrichtian of Jamaica. All specimens ×60. All specimens are carapaces. 1, 8—*Brachycythere* sp. 1:1 is left lateral view female, sample J91, OZ1c; 8 is ventral view female, sample J91; 2, 14—*Brachycythere* sp. 2:2 is left lateral view female, sample J70, OZ1b; 14 is ventral view female, sample J70; 3, 11—*Brachycythere* sp. 3:3 is left lateral view female, sample J70, OZ1b; 11 is ventral view female, sample J245, OZ1b; 4, 10—*Brachycythere* sp. 4:4 is left lateral view male?, sample J70, OZ1b; 10 is ventral view female, sample J70; 5, 9—*Brachycythere* sp. 7:5 is left lateral view female, sample J34, OZ3; 9 is ventral view female, sample J34; 6, 13—*Brachycythere* sp. 6:6 is left lateral view female, sample J48, OZ1a; 13 is ventral view female, sample J54, OZ1a; 7, 12—*Brachycythere* sp. 5:7 is left lateral view female, sample J15, OZ2; 12 is ventral view female, sample J15; 15—*Ovocytheridea* sp. 2, right lateral view female, sample J71, OZ1b; 16—*Ovocytheridea* sp. 1, right lateral view female, sample J70, OZ1b; 17—*Ovocytheridea* sp. 3, right lateral view female, sample J15, OZ2.

2.25 km (1.40 mi) N54°W of the bridge at Frankfield, Clarendon Parish. Five samples collected from shale interbeds. J240 taken about 18 m (60 ft) above the base of the exposed Guinea Corn; 241 is 23 m (75 ft) above 240; 242 is 12 m (38 ft) above 241; 243 is 23 m (75 ft) above 242; 244 is 12 m (38 ft) above 243. J242 is the same as locality 64 of Sohl and Kollmann (1985), and 244 is the same as their locality 71.

Maldon Inlier

(1) Exposures in fields north of Flamstead to Maroon Town road in St. James Parish about 1.77 km (1.10 mi) southwest of Maroon Town. This is virtually the same as locality 117 of Sohl and Kollmann (1985).

Vaughnsfield Limestone. About 15 m (50 ft) of dense rudist-bearing limestone. In the lower few meters is a gray clay lens. J88 collected from the clay.

Popkin Formation. About 15 m (50 ft) of brown to gray clay underlying the Vaughnsfield. J90 from the upper 1.5 m (5 ft) of the Popkin; 89 is 3 m (10 ft) below 90; 91 about 8.5 m (28 ft) below 90.

(2) Road cut 0.83 km (0.51 mi) north of Flamstead on the road to Tangle River. Several meters of gray to brown blocky clay of the Popkin are exposed. J94 taken here. J94 is in the upper Popkin stratigraphically below 91, but its exact position is unknown.

(3) Road cut near Chatsworth School at Shaw Castle, 1.45 km (0.90 mi) northwest of Maroontown. J99 from thin, very fossiliferous clay between rudist-bearing limestones. The locality is in the lower part of the Maldon Limestone and is the same as locality 119 of Sohl and Kollman (1985).

(4) Exposures of *Titanosarcolites*-rich limestone and interbedded clays of the Maldon Limestone in a road cut and in the school yard of the Maldon School, 0.21 km (0.13 mi), northwest of Maroon Town. J92 taken at road cut near the entrance to the school. J93 from clays intercalated with a large *Titanosarcolites* in the schoolyard. J93 is about 6 m (20 ft) below 92. The Maldon School locality is 118 of Sohl and Kollmann (1985).

(5) About 2.7 m (9 ft) of blocky reddish clay of the uppermost Woodland Shale exposed just north of Tangle River on the road from Flamstead to Tangle River. J95 taken from the lower part of the exposure and 96 from the upper part.

Marchmont Inlier

In Westmoreland Parish an undetermined thickness of unnamed, faulted rudist-bearing limestones and intervening and intercalated calcareous clays are exposed at Ducketts Land Settlement (locality 92 of Sohl and Kollmann (1985) and in road cuts and railroad cuts east of Ducketts for 6.45 km (4.0 mi) (Fig. 2). The stratigraphic relationships between the exposures are not obvious, but they can be ordered biostratigraphically (see above). Samples J43, 48, 49, 54, 56, 57, 61, and 63 were taken along or near the Cambridge-Catapuda Railroad. J74 is at Newton Farm. J65, 67, 68, 70, 71, and 245 are in the Ducketts area.

Jerusalem Mountain Inlier

In Westmoreland Parish, limestones and shales of the Jerusalem Mountain Formation are exposed along the Jerusalem Mountain to Belle Isle road at about 18°19′29″N, 78°13′14″W. See locality 110 of Sohl and Kollmann (1985).

Jerusalem Limestone Member. At least two massive, 1- to 1.5-m- (3- to 5-ft-) thick limestones with *Lopha arizpensis,* but no rudists, separated by calcareous shales occur at or near the top of the Cretaceous section. Sample J40 is from a shale just above the lowest limestone.

Unnamed Shale. About 12 m (40 ft) of reddish brown calcareous shale occurs below the Jerusalem Limestone. Samples J21, 25, 26, 27, 28, and 29 are from this unit. Exact stratigraphic separation of the samples is unknown because of unknown dip. J28 and 29 are near the base of the unit; 21 is near the top; 25, 26, and 27 are from the middle part.

Thicket River Member. About 30.5 m (100 ft) of limestones and rudist rubble beds with *Titanosarcolites* and other macrofossils, separated by calcareous clays, are exposed in road cuts. Calcareous clays are also intercalated with the rudist specimens. J9, 15, 19, 32, and 33 are from the clays associated with the rudist rubble beds. J9, 15, and 19 are from the lower part of the sequence, and 32 and 33 are from the upper beds.

Figure 5. Ostracodes from the Maastrichtian of Jamaica. Numbers 1–11, 14, 16–18 ×91; numbers 12, 13, 15 ×68. All specimens are carapaces except number 14. 1, 8—*Aequacytheridea* sp. 2:1 is right lateral view female, sample J91, OZ1c; 8 is dorsal view female, sample J91; 2, 10—*Aequacytheridea* sp. 3:2 is right lateral view female, sample J28, OZ3; 10 is dorsal view female, sample J28; 3, 6, 7—*Aequacytheridea* sp. 4:3 is right lateral view male, sample J25, OZ3; 6 is right lateral view female, sample J40, OZ3; 7 is dorsal view female, sample J25; 4, 11—*Aequacytheridea* sp. 1:4 is right lateral view female, sample J59, OZ1a; 11 is dorsal view female, sample J59; 5, 9—*Aequacytheridea* sp. 5:5 is right lateral view female, sample J74, OZ1a; 9 is dorsal view female, sample J74; 12, 13— *Asciocythere* sp. 1:12 is right lateral view male, sample J88, OZ1c; 13 is right lateral view female, sample J91, OZ1c; 14—*Cytherelloidea* sp. 1, right valve female, sample J91, OZ1c; 15—*Cytherella tuberculifera* Alexander, left lateral view female, sample J242, OZ1b; 16—*Bythoceratina* sp. 1, left lateral view female, sample J91, OZ1c; 17, 18—"*Cytherura*" sp. 1:17 is left lateral view female, sample J70, OZ1b; 18 is left lateral view male, sample J70.

ACKNOWLEDGMENTS

The authors are grateful to Dr. N. F. Sohl, U.S. Geological Survey, for many helpful discussions of Jamaican geology and paleontology and for a review of the manuscript. Drs. S. K. Donovan, University of the West Indies; E. Robinson, Florida International University; and R. F. Maddocks, University of Houston, also provided helpful reviews. Dr. Sohl; Dr. E. G. Kauffman, University of Colorado; and Dr. A. G. Coates, George Washington University, accompanied Hazel in the field in Jamaica. The U.S. Geological Survey made the fieldwork possible. Kamiya was able to work on the project during a stay at Louisiana State University as a visiting assistant professor.

TABLE 1. LIST OF OSTRACODE SPECIES USED IN THE BIOSTRATIGRAPHIC ANALYSIS

Species	Zone	Sample Occurrences	Figure
Aequacytheridea sp. 1	OZ1a	59, 95, 96, 240	5-4, 11
Aequacytheridea sp. 2	OZ1a,b,c	48, 54, 55, 56, 59, 61, 63, 70, 71, 88, 91, 92, 242, 243, 245	5-1, 8
Aequacytheridea sp. 3	OZ3	21, 25, 26, 27, 28, 40	5-2, 10
Aequacytheridea sp. 4	OZ3	21, 25, 26, 27, 28, 34, 40	5-3, 6, 7
Aequacytheridea sp. 5	OZ1a	74	5-5, 9
Alatacythere sp. 1	OZ1a	48, 50, 59	3-13
Asciocythere sp. 1	OZ1a,b,c	48, 63, 65, 71, 74, 88, 89, 90, 91, 95, 96, 245	5-12, 13
Brachycythere sp. 1	OZ1c	88, 89, 90, 91, 94	4-1, 8
Brachycythere sp. 2	OZ1a,b,c,	55, 63, 70, ?71, 91, 99	4-2, 14
Brachycythere sp. 3	OZ1b	55, 56, 65, 70, 71, 242, 243, 245	4-3, 11
Brachycythere sp. 4	OZ1b,c	55, 56, 70, 71, 88, 99, 241, 242, 245	4-4, 10
Brachycythere sp. 5	OZ2	15, 19, 33	4-7, 12
Brachycythere sp. 6	OZ1a	48, 50, 54, ?59	4-6, 13
Brachycythere sp. 7	OZ3	21, 25, 26, 27, 28, 34, 35, 40	4-5, 9
Buntonia sp. 1	OZ1a,b,c,	55, 56, 63, 65, 67, 68, 70, 71, 88, 89, 90, 93, 95, 240, 241, 242, 243, 244	3-16
Buntonia sp. 2	OZ2	15, 32, 33	3-18
Buntonia sp. 3	OZ2	?9, 15, 19, 33	3-17
Bythoceratina sp. 1	OZ1a,c	48, 59, 91, 95	5-16
Cletocythereis sp. 1	OZ1b	55, 56, 67, 68, 70, 71, 93, 99, 243, 244	3-12
Cytherella tuberculifera	OZ1b,c	55, 56, 65, 67, 70, 71, 90, 93, 99, 241, 242 245	5-15
Cytherelloidea sp. 1	OZ1a,c	54, 63, 88, 89, 90, 91, 95, 96	5-14
"*Cytherura*" sp. 1	OZ1b,c; OZ2	15, 19, 33, 55, 56, 67, 68, 70, 71, 88, 89, 90, 92, 99, 242, 244	5-17, 18
Eocytheropteron sp. 1	OZ1a	57, 240	3-15
Eocytheropteron sp. 2	OZ2	15, 32, 33	3-14
Limburgina sp. 1	OZ1,a,b,c	48, 50, 54, 61, 63, 70, 88, 90, 93, 95, 99	3-9
Limburgina sp. 2	OZ1a,b,c	55, 63, 70, 88, 90, 91, 93, 95	3-8
Limburgina sp. 3	OZ1b	55, 56, 70, 71, 92, 243	3-5
Limburgina sp. 4	OZ2	9, 15, 19, 33	3-7
Limburgina sp. 5	OZ2	15, 19, 33	3-6
Ovocytheridea sp. 1	OZ1a,b,c	55, 56, 63, 65, 67, 68, 70, 71, 89, ?90, 93, 99, 240, 241, 242, 243, 245	4-16
Ovocytheridea sp. 2	OZ1b	55, 67, 70, 71, 243	4-15
Ovocytheridea sp. 3	OZ2	15, 19, 33	4-17
Platycosta? sp. 1	OZ1b,c	70, 88, 91	3-11
Schizoptocythere segurai	OZ1a,c	54, 91	3-2
Schizoptocythere sp. 1	OZ1a,b,c	48, 50, 54, 59, 61, 63, 74, 88, 89, 90, 91, 95, 96, 242	3-3
Schizoptocythere sp. 3	OZ1b,c	56, 70, 71, 99, 245	3-1
Schizoptocythere sp. 4	OZ3	21, ?25, 26, 27, 28, 40	3-4
Spinoleberis? sp. 1	OZ1b,c	55, 56, 71, 91, 93, 99, 241, 244	3-10

TABLE 2. STRATA ASSIGNED TO THE FIVE OSTRACODE BIOSTRATIGRAPHIC UNITS

Ostracode Zone 3		Shales associated with the Jerusalem Limestone (the oyster limestone) of the Jerusalem Mountain Formation, Westmoreland Parish.
Ostracode Zone 2		Shales interbedded and intercalated with rudist limestones of the Thicket River Limestone of the Jerusalem Mountain Formation, Westmoreland Parish
Ostracode Zone 1	Subzone 1c	Shales of the Popkin Formation and clays interbedded with the lower part of the overlying Vaughnsfield Limestone in the Maldon Inlier, St. James Parish.
	Subzone 1b	Faulted limestones and shales along the main road through Ducketts Land Settlement; limestones and interbedded shales along the Lambs River road south of Ducketts; limestones along the Cambridge-Catadupa railroad—all in the Marchmont Inlier, St. James Parish; the Maldon Limestone of the Maldon Inlier; middle part of the Guinea Corn Formation along the Rio Minho, Clarendon Parish.
	Subzone 1a	Woodland Shale in the Maldon Inlier in St. James Parish; shales along the Cambridge-Catadupa railroad and reddish shales at Newton Farm in the Marchmont Inlier, St. James Parish; lower Guinea Corn Formation along the Rio Minho, Clarendon Parish.

REFERENCES CITED

Apostolescu, V., 1961, Contribution a l'etude paleontologique (ostracodes) et stratigraphique des bassins Cretaces et Tertiaires de L'AFrique Occidentale: Revue de L'Institut Francais du Petrole, v. 16, p. 779–867.

Apostolescu, V., 1963, Essai de zonation par les ostracodes dans le Cretace du bassin du Senegal: Revue de L'Institut Francais du Petrole, v. 18, p. 1675–1694.

Bold, W. A. van den, 1946, Contribution to the study of Ostracoda with special reference to the Tertiary and Cretaceous microfauna of the Caribbean region: Amsterdam, DeBussy, 167 p.

Bold, W. A. van den, 1950, A checklist of Cuban Ostracoda: Journal of Paleontology, v. 24, p. 107–109.

Bold, W. A. van den, 1964, Ostracoden aus der Oberkreide von Abu Rawash, Agypten: Palaeontographica, v. 123, p. 111–136.

Caron, M., 1985, Cretaceous planktic Foraminifera, in Bolli, H. M., Saunders, J. B., and Perch-Nielsen, K., eds., Plankton stratigraphy: Cambridge, England, Cambridge University Press, p. 17–86.

Chubb, L. J., 1958, The Cretaceous rocks on central St. James: Geonotes, v. 1, p. 3–11.

Chubb, L. J., 1962, Cretaceous formations, in Zans, V. A., Chubb, L. J., Versey, H. R., Williams, J. B., Robinson, E., and Cooke, D. L., Synopsis of the geology of Jamaica: An explanation of the 1958 provisional map of Jamaica: Jamaica, Geological Survey Department Bulletin 4, 72 p.

Chubb, L. J., 1971, Rudists of Jamaica: Paleontographica Americana, v. 7, p. 160–257.

Coates, A. G., 1968, The geology of the Cretaceous Central Inlier around Arthurs' Seat, Clarendon, Jamaica: Transactions 4th Caribbean Geological Conference, p. 309–315.

Coates, A. G., 1977, Jamaican coral-rudist frameworks and their geologic setting: American Association of Petroleum Geologists, Studies in Geology 4, p. 83–91.

Crane, M. J., 1965, Upper Cretaceous ostracodes of the Gulf Coast area: Micropaleontology, v. 11, p. 191–254.

Gvirtzman, G., Almogi-Labin, A., Moshkovitz, S., and Lewy Z., 1989, Upper Cretaceous high-resolution multiple stratigraphy, northern margin of the Arabian platform, central Israel: Cretaceous Research, v. 10, p. 107–135.

Hazel, J. E., 1971, Ostracode biostratigraphy in the Maestrichtian rudist sequences of Jamaica: Transactions 5th Caribbean Geological Conference, p. 120.

Hazel, J. E., and Brouwers, E. M., 1982, Biostratigraphic and chronostratigraphic distribution of ostracodes in the Coniacian-Maastrichtian (Austinian-Navarroan) in the Atlantic and Gulf Coastal Province, in Maddocks, R. F., ed., Texas Ostracoda: Guidebook of Excursions and Related Papers for the 8th International Symposium on Ostracoda: Houston, Texas, Department of Geosciences, University of Houston, p. 166–198.

Hill, R. T., 1899, The geology and physical geography of Jamaica; a study of a type of Antillean development: Harvard Museum Comparative Zoology Bulletin, v. 34, p. 1–226.

Jiang, M.-J., and Robinson, E., 1987, Calcareous nannofossils and larger foraminifera in Jamaican rocks of Cretaceous to early Eocene age, in Ahmad, R., ed., Proceedings of a Workshop on the Status of Jamaican Geology, March 1984: Journal Geological Society Jamaica, Special Issue, p. 24–51.

Jiang, M.-J., and Robinson, E., 1989, Zonation of some Jamaican Cretaceous rocks using calcareous nannofossils: Transactions 10th Caribbean Geological Conference, 1983, p. 243–249.

Kauffman, E. G., and Sohl, N. F., 1974, Structure and evolution of Antillean Cretaceous rudist frameworks: Naturforschende Gesellschaf Basel Verhandlungen, v. 84, p. 399–467.

Khan, M. H., 1970, Cretaceous and Tertiary rocks of Ghana: Ghana Geological Survey Bulletin 40, 43 p.

Lyubimova, P. S., and Sanchez-Arango, J. R., 1974, Los Ostracodos del Cretacico Superior y del Terciario de Cuba: Havana, Instituto Cubano del Libro, 171 p.

North American Commission on Stratigraphic Nomenclature, 1983, North American Stratigraphic Code: American Association Petroleum Geologists Bulletin, v. 67, p. 841–875.

Obradovich, J. D., and Cobban, W. A., 1975, A time-scale for the Late Cretaceous of the Western Interior of North America: Geological Association of Canada Special Paper 13, p. 31–54.

Perch-Nielsen, K., 1985, Mesozoic calcareous nannofossils, in Bolli, H. M., Saunders, J. B., and Perch-Nielsen, K., eds., Plankton stratigraphy: Cambridge, Cambridge University Press, p. 329–426.

Reiss, Z., and six others, 1985, Late Cretaceous multiple stratigraphic framework of Israel: Israel Journal of Earth-Sciences, v. 34, p. 147–166.

Reyment, R. A., 1960, Studies on Nigerian Upper Cretaceous and Lower Tertiary Ostracoda. Part I: Senonian and Maestrichtian Ostracoda: Stockholm Contributions in Geology, v. 7, p. 1–258.

Reyment, R. A., 1963, Studies on Nigerian Upper Cretaceous and Lower Tertiary Ostracoda. Part II: Danian, Paleocene, and Eocene Ostracoda: Stockholm Contributions in Geology, v. 10, p. 1–286.

Siddiqui, Q. A., and Al-Furaih, A.A.F., 1981, A new trachyleberid ostracod genus from the early Tertiary of western Asia: Palaeontology, v. 24, p. 877–890.

Sohl, N. F., and Kollman, H. A., 1985, Cretaceous Actaeonellid gastropods from the Western Hemisphere: U.S. Geological Survey Professional Paper 1304, 104 p.

Trechmann, C. T., 1924, The Cretaceous limestones of Jamaica and their molluscs: Geological Magazine, v. 61, p. 385–410.

Young, K., 1986, Cretaceous marine inundations of the San Marcos Platform, Texas: Cretaceous Research, v. 7, p. 117–140.

MANUSCRIPT ACCEPTED BY THE SOCIETY NOVEMBER 11, 1992

Upper Cretaceous ammonites from Jamaica and their stratigraphic and paleogeographic implications

Jost Wiedmann
Geologisch-Paläontologisches Institut, Universität, Sigwartstrasse 10, D-7400, Tübingen, Germany
Winfried Schmidt
Fundaçao Estadual de Proteçao Ambiental (FEPAM), Avenida A. J. Renner 10, 90250 Porto Alegre, RS, Brazil

ABSTRACT

Nine ammonite species from the Upper Cretaceous of Jamaica are revised and described briefly, including one new species, *Pachydiscus* (P.) *jamaicensis* n., of Lower Campanian age.

The Cretaceous ammonite fauna is classified into four different levels: Level 1 with *Peroniceras* cf. *moureti* Grossouvre, Lower Coniacian; Level 2 with *Nowakites lemarchandi* (Grossouvre), late Santonian; Level 3 with *Pachydiscus* (P.) *koeneni* Grossouvre and *Glyptoxoceras retrorsum* (Schlueter), early Campanian; and Level 4 with *Pachydiscus* (P.) *oldhami* (Sharpe), late Campanian. No older pelagic facies have been recognized. These ammonite levels correlate perfectly with inoceramid "zones" proposed by Kauffman (1966). The biostratigraphy established in this paper facilitates correlation between the different sedimentary facies of the scattered and isolated Upper Cretaceous outcrops in Jamaica.

Paleobiogeographic relations can be established with Europe (Level 4), southern India (Level 3), Madagascar (Level 2), and Japan (Level 1).

INTRODUCTION

Problems of dating and correlation of the Jamaican Cretaceous are related to the scattered distribution of the outcrops (Fig. 1), the diversity of lithofacies (see Figs. 2, 3, and 4; Table 1), and the scarcity of convincing microfaunas in these mainly clastic sediments. The more calcareous layers of the Jamaican Cretaceous have yielded rare foraminifera (Pessagno, 1978), inoceramids, and rudists (Trechmann, 1924; Kauffman, 1966; Chubb, 1971) and have allowed the establishment of a rough dating scheme.

Ammonites could play a convincing and most important role for Cretaceous biozonation and correlation, but unfortunately they are rather rare throughout the Caribbean. Jamaican Cretaceous ammonites were reported and partly described by Spath (1925), Trechmann (1927), Sohl (1978), and Schmidt (1986). The purpose of this chapter is a revision of that material, as far as it is available, in order to improve or correct previous identifications and to reinterpret the ages and paleogeographic significance of these faunas. All specimens, except for one newly described ammonite, are contained in the collections of the Natural History Museum, London (BMNH).

The rediscovery of Spath's and Techmann's fossil localities and sections of over 50 years ago has proved to be difficult. The localities along the Port Antonio–Fellowship road could not be found exactly, but detailed stratigraphic sections of two fossil horizons in the Providence Shales are described, based on several field trips to this area by one of the authors (W. S.). These localities are believed to be the same as those of Spath and Trechmann, since numerous smaller fragments of ammonites were found there, whereas other beds along the Port Antonio–Fellowship road yielded no cephalopods.

REGIONAL STRATIGRAPHY

The occurrence of ammonites is restricted to four areas: (1) Lucea Inlier, in western Jamaica (Fig. 2); (2) St. Ann Inlier, in the central North (Fig. 3); (3) Providence Shale, near Port Antonio in eastern Jamaica (Fig. 4B, 4C); and, (4) Blue Mountain Inlier, in the central East (Fig. 4A).

Wiedmann, J., and Schmidt, W., 1993, Upper Cretaceous ammonites from Jamaica and their stratigraphic and paleogeographic implications, *in* Wright, R. M., and Robinson, E., eds., Biostratigraphy of Jamaica: Boulder, Colorado, Geological Society of America Memoir 182.

In some of these sections, ammonites are associated with inoceramids. However, the stratigraphic significance of this latter important mollusc group is even more difficult to establish with precision. A series of inoceramid zonations has been proposed (Chubb, 1955, 1958, 1960, 1962; Kauffman, 1966, 1978), but none of the specimens upon which the schemes were based have ever been illustrated.

Lucea Inlier

Previous studies by Chubb (1962) and Grippi (1980) established a stratigraphical framework for the Lucea Inlier, western Jamaica. This was supplemented by Schmidt (1986). The Hanover Group of the Lucea Inlier (Fig. 2) exposes a 5,000-m-thick sequence of predominantly clastic and volcani-

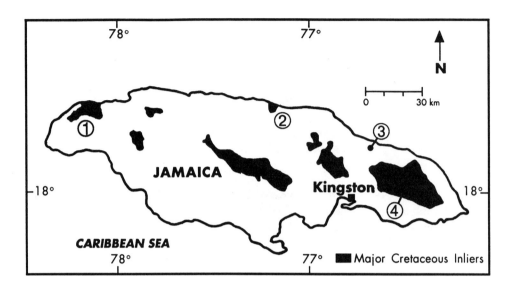

① Lucea Inlier ② St. Ann Inlier ③ Providence Shale ④ Blue Mtn. Inlier

Figure 1. Major Cretaceous inliers and ammonite locations in Jamaica.

TABLE 1. CORRELATION TABLE OF LITHOSTRATIGRAPHIC UNITS, JAMAICAN CRETACEOUS

Western Jamaica Lucea Inlier	Central North St. Ann Inlier	Eastern Jamaica Blue Mountain Area		
		Bowden Pen Formation		Maastrichtian
	Black Shale	Upper Providence Shale	Upper	Campanian
Barrettia Limestone	Barrettia Limestone		Middle	
Hanover Shale Clifton Limestone		Lower Providence Shale	Lower	
	Inoceramus Shale		Upper	Santonian
			Lower	
			Upper	Coniacian
	Windsor Shale		Lower	

clastic rocks. These consist, in order of abundance, of siltstones, shales, volcaniclastic sandstones, polymict conglomerates, mudstones, bioclastic limestones, and mafic intrusive rocks. The thick pile of sediments represents a slope-to-basin sequence of volcaniclastics transported by gravity-flow mechanics and is partially deposited in submarine fans.

Biohermal deposits, called the Clifton Limestone (Fig. 2), occur midway up the section. They yield rudist bivalves, both in life position and detached; echinoids; and assorted shell-debris layers intercalated with calcareous shales and calcarenite beds. A lithological change to a high-energy shoal environment occurs toward the top of the Clifton Limestone, where a 5-m-thick bed of oolithic grainstone contains numerous smaller echinoids.

At the Maryland Waterfall, 0.5 km southwest of Maryland Road Junction, Hanover Parish, a well-preserved ammonite was found by one of the authors (W. S.) in the oolithic grainstone member of the Clifton Limestone (Fig. 2). It was attributed to *Pachydiscus cayeuxi* Grossouvre of a late Santonian age

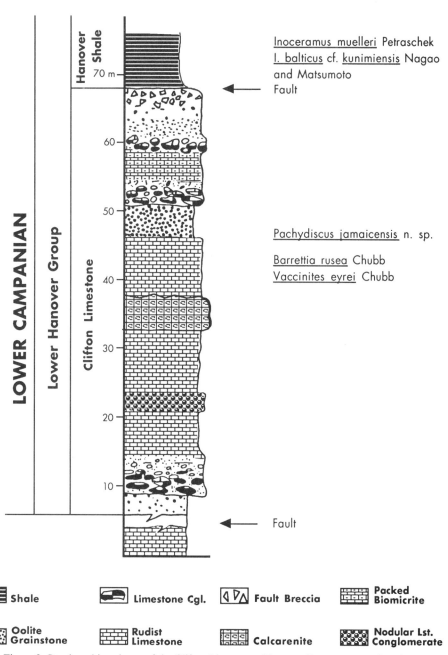

Figure 2. Stratigraphic column of the Clifton Limestone, Hanover Group, western Jamaica.

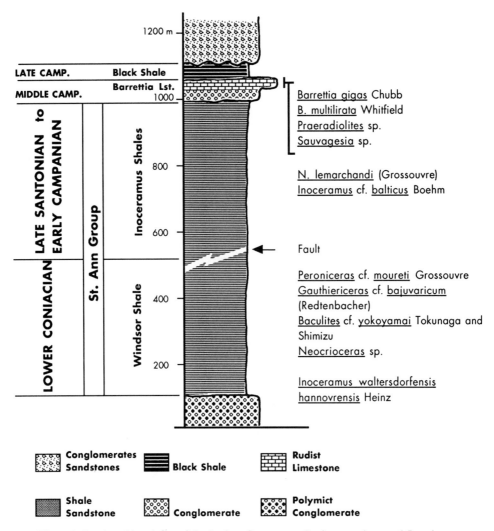

Figure 3. Stratigraphic column of the St. Ann Group, near St. Ann, north-central Jamaica.

(Schmidt, 1986). The specimen is here described and figured (see Fig. 5.3) to belong to a new species, *Pachydiscus jamaicensis* n. sp. It may connect the Santonian *P. cayeuxi* with *P. spissus* Collignon of Middle Campanian age. Accordingly, a Lower Campanian age can be inferred for the Clifton Limestone. This age is roughly compatible with the Campanian age assigned to this unit by Chubb (1962, table 2), and it correlates even better with the "very early Campanian" age suggested for the Clifton Limestone by Kauffman (1966:39). His dating was based on collections of:
Inoceramus muelleri Petraschek and
I. balticus cf. *kunimiensis* Nagao and Matsumoto.
Both are from the Hanover Shale on Harvey's River Road, Hanover Parish, from levels "probably above the Clifton Limestone." The co-occurrence of these two inoceramids would support a Lower Campanian age also for a large portion of the Lower Hanover Shale, which overlies the Clifton Limestone.

In addition, the rudists *Barrettia rusea* Chubb and *Vaccinites eyrei* Chubb associated with *Pachydiscus jamaicensis* n. sp. were attributed to a Lower Campanian age (Sohl and Kollmann, 1985). Jiang and Robinson (1987), however, considered the lower part of the Hanover Group, including the Clifton Limestone, to belong to the Santonian, according to the calcareous nannofossils *Eiffellithus eximius* (Stover), *Micula decussata* Veksh., *Rucinolithus* sp., and *Lithastrinus grillii* Stradner.

St. Ann Inlier

A small outcrop of Cretaceous strata of approximately 4 km² is located in the valley of St. Ann's Great River on the northern coast of Jamaica. The Upper Cretaceous sequence consists mainly of shales (St. Ann Group) with a thickness of about 900 m (Fig. 3). They are subdivided into a lower, Windsor Shale and an upper, *Inoceramus* Shales. Toward the top of these shales, conglomerates and coarse sandstones are intercalated. These are overlain by the *Barrettia* Limestone of Middle Campanian age.

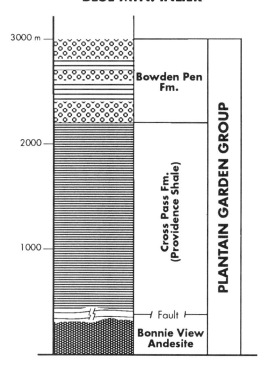
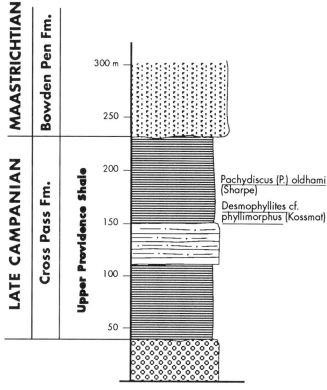
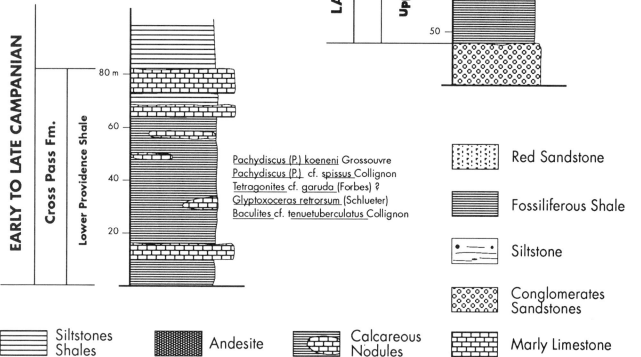

Figure 4. (A) General stratigraphic column of the Providence Shale and (B, C) detailed sections from the Port Antonio–Fellowship Road, eastern Jamaica.

Despite the fact that the history of the stratigraphic interpretation of this section has been rather chaotic, it may become one of the most important late Cretaceous sequences of the Antilles.

Misdating the fossil record, Chubb (1958) was led to the assumption that a major fault dissects the shale unit and places the northern down-dropping limb of an anticline against an apparently older south-dipping structure. He assigned a Campanian age to the shales of the northern limb and proposed a Cenomanian to Coniacian age to those of the southern limb. In reality, however, the succession seems to be continuous, with ages decreasing southward (Trechmann, 1924; Sohl, 1978). The central fault may exist, but its throw is considerably less pronounced than formerly thought and may only separate Lower Coniacian from Lower Santonian sediments. A hiatus may also exist between the Windsor Shale and the *Inoceramus* Shales.

Windsor Shale. The shale unit was considered by Chubb (1962) to represent the youngest, Campanian part of the sequence. Age dating was based on planktonic foraminifera (*Contusotruncana contusa* [Cushman], *Contusotruncana fornicata* [Plummer], *Marginotruncana linneiana* [D'Orbigny]), determined by Brönnimann (in Chubb, 1962). All attempts to locate these samples or to duplicate the sampling have failed. In contrast to Chubb's dating, Sohl (1978) discovered in the middle part of this sequence an interesting mollusc assemblage that includes:
Peroniceras cf. *moureti* Grossouvre
Gauthiericeras cf. *bajuvaricum* (Redtenbacher)
Neocrioceras sp.
Baculites cf. *yokoyamai* Tokunaga and Shimizu
and *Inoceramus waltersdorfensis hannovrensis* Heinz.
This assemblage provides a precise Lower Coniacian age. Unfortunately, we are unable to include these important fossils in this chapter.

Inoceramus Shales. The history of the dating of the southern and consequently younger shale unit is more complicated. Trechmann (1936) discovered an ammonite at a level 250 m below the *Barrettia* Limestone. It was determined as "*Nowakites* aff. *paillettei* (D'Orbigny)" by Spath (in Trechmann, 1936) and thought to represent the Upper Coniacian or Lower Santonian.

As early as 1927, Trechmann described a rich mollusc assemblage from another level, 270 to 300 m below the *Barrettia* Limestone, correctly identifying *Inoceramus* cf. *balticus* Boehm, and considered the assemblage to be of "?Middle Senonian" age (Trechmann, 1927:33). Chubb, however, on several occasions (1955, 1958, 1960) insisted on a Coniacian, or even older, age for the *Inoceramus* Shales. In 1955 (p. 184, 190 ff.) he cited the Cenomanian bivalve *Inoceramus crippsi* Mantell from Trechmann's mollusc level, citing at the same time (1955) *Inoceramus inconstans* Woods of Coniacian age from a river cobble. In 1958 (p. 149) and 1960 (p. 91) Chubb added another Coniacian species, *Inoceramus deformis* Meek, again from the mollusc and *Nowakites* level, using its occurrence to further support the Coniacian age of the assemblage. These erroneous ages for the *Inoceramus* Shales were confirmed by two citations of planktonic foraminifera. Brönnimann (in Zans et al., 1962) reported the Middle Turonian *Praeglobotruncana helvetica* (Bolli), and Esker (1969) listed a rich association of planktonics from a level immediately above the fault that mixed Turonian, Coniacian, and Santonian species. Nevertheless, he considered this association to represent the late Coniacian. It is worth mentioning that Esker was unable by resampling to confirm Brönnimann's Turonian fauna.

Pessagno (1978), in revising Esker's "Coniacian" association, confirmed the presence of *Dicarinella concavata* (Brotzen), thus indicating a Lower Santonian age for this shale unit.

Interestingly, Kauffman (1966) came to a conclusion similar to that arrived at here when he tried to revise Chubb's inoceramid data. None of the inoceramid specimens Chubb had described and assigned to the three above-mentioned species has been located in the Geological Survey of Jamaica collections or rediscovered in the sections except for one, which Chubb identified as *I.* cf. *deformis*. This specimen from a stream cobble, however, was found to be a rather late Santonian–Campanian *Endocostea* sp. (Kauffman, 1966). Another specimen from the *Inoceramus* Shales contained in the collections of the Geological Survey of Jamaica is indeed an *Inoceramus balticus* Boehm of a similar age (Kauffman 1966), fully confirming Trechmann's (1927) dating. No trace of the alleged Cenomanian *I. crippsi* has ever been discovered.

The specimen of *Nowakites,* now figured for the first time (Fig. 5.1), agrees with the age assignment given by the inoceramids. It is considered here to represent *Nowakites lemarchandi* (Grossouvre), a species ranging in age from the Coniacian to the Santonian. The species has been described by Bilotte (1984, Plate 32, Fig. 9) from the Upper Santonian of the Pyrenees. The co-occurrence of *Nowakites lemarchandi* with *Inoceramus balticus* infers a late Santonian age for the middle part of the *Inoceramus* Shales, but given the long ranges of the species, even an early Campanian age cannot be excluded. This would agree better with a presumably mid-Campanian age for the overlying *Barrettia* fauna (including *Barrettia gigas* Chubb, *B. multilirata* Whitfield,

Figure 5. (All specimens figured in natural size). 1, *Nowakites lemarchandi* (Grossouvre) BMNH C. 38120, late Santonian *Inoceramus* Shales, "800 feet below *Barrettia* Limestone," St. Ann Group, St. Ann's Great River, near St. Ann, northern Jamaica—A: lateral view; B: ventral view. 2, *Pachydiscus (P.) koeneni* Grossouvre BMNH C. 25256, lower Providence Shale, Roadside section, 1.5 km south of Port Antonio, Port Antonio–Fellowship Road. Early Campanian—A: lateral view; B: ventral view. 3, *Pachydiscus (P.) jamaicensis* n. sp. Holotype, BMNH C. 90957, Clifton Limestone, Maryland Waterfall, 0.5 km southwest of Maryland road Junction, Hanover, western Jamaica. Lower Campanian—A: lateral view; B: front view; C: ventral view. 4, *Pachydiscus (P.)* cf. *spissus* Collignon BMNH C. 25257, same locality as Figure 5.2. Early Campanian. Lateral view. 5, *Pachydiscus (P.) oldhami* (Sharpe) BMNH C. 74026, upper Providence Shale, Roadside section, 3.2 km south of Port Antonio, Port Antonio–Fellowship Road. Upper Campanian. A, B: lateral views; C: ventral view.

and *Praeradiolites* sp.) and a late Campanian age for the capping "*Diozoptyxis* shales." Kauffman (1966) mentioned the occurrence of inoceramids in these higher gray clay-shales, exposed along the road 2.5 km north of the outcropping *Barrettia* Limestone at New Ground, St. Ann's Great River. They were tentatively determined as *Inoceramus proximus subcircularis* Meek and Hayden inferring a late Campanian age. But there must be some confusion, either with this locality or with the sample, since we are at the indicated place, 2.5 km north of the outcropping *Barrettia* Limestone at New Ground, in the Coniacian (N. Sohl, personal communication).

The existence of a late Campanian–Maastrichtian inoceramid association in the Newman Hall Shale of the St. James Inlier, halfway between the localities in western Jamaica and the St. Ann Inlier in north-central Jamaica, is supported by the presence of the "late Campanian lineages" of *Inoceramus (Endocostea) typicus* Whitfield and *I. barabini* Morton from three localities near its type locality (Kauffman, 1966).

Providence Shale near Port Antonio

The Providence Shale is the equivalent of and is included in the Cross Pass Formation (Fig. 4) in the new edition of the Geological Map of Jamaica (Balkissoon, 1987). Lithologically the unit consists of gray and reddish shales, siltstones, and sandstones. Occasionally, conglomerate beds and coarse volcaniclastic sandstones occur. Intercalations of thin, gray limestone beds and marly limestones are present throughout the section. Thickness is estimated to be 700 to 900 m.

The Providence Shale–Cross Pass Formation is underlain by the Bonnie View Andesite, a massive and fairly homogenous porphyry. The igneous rocks have been interpreted as an intrusive sill (Krijnen and Lee, 1977). South of Port Antonio the Bonnie View Andesite is fault bounded against the fossiliferous section containing ammonites. The overlying Bowden Pen Formation (?Maastrichtian) is not observed in the Port Antonio area.

Two ammonite localities were described by Trechmann (1927) from the Providence Shale. In Figure 4 an attempt is made to relate these two ammonite layers to precise sections. The lower portion of the Providence Shale is exposed 1.5 km S of Port Antonio (Fig. 4B). The ammonite fauna of the "Baculite Bed" was described by Spath (1925) and referred to the Campanian. After revision, the following species are recognized:
Pachydiscus (P.) koeneni Grossouvre
Pachydiscus (P.) cf. *spissus* Collignon
Tetragonites cf. *garuda?* (Forbes)
Glyptoxoceras retrorsum (Schlueter) and
Baculites cf. *tenuetuberculatus* Collignon.
This might also be the same level of rich inoceramid occurrence recorded by Kauffman (1966) that included in particular:
Inoceramus balticus balticus Boehm
I. balticus kunimiensis Nagao and Matsumoto and
I. balticus toyajoanus Nagao and Matsumoto.
The complete association points to an early Campanian age for this horizon. Additional collections referring to this inoceramid association were mentioned as occurring on the "track leading to Peter's Hill from Crawle River, and in the bed of Pindars River, about ¾ mile north of Arthur's Seat," and "from the Blue Mountain Shale 500 feet below the main peak" (Kauffman, 1966:38).

The second ammonite locality, 3.2 km south of Port Antonio, belongs to the upper portion of the Providence Shale (Fig. 4C). The ammonites from this locality are identified here as:
Pachydiscus (P.) oldhami (Sharpe) and
Desmophyllites cf. *phyllimorphus* (Kossmat).
Both species support a late Campanian age for the upper part of the Providence Shale.

In conclusion, therefore, the entire Providence Shale fauna indicates a more or less complete Campanian sequence in eastern Jamaica. The overlying Bowden Pen Formation is Maastrichtian in age.

STRATIGRAPHIC RESULTS

Summarizing the biostratigraphic ammonite record in the Upper Cretaceous of Jamaica, only four different ammonite levels can be recognized. These are (Table 2) from base to top: Level 1: the Lower Coniacian level with *Peroniceras* cf. *moureti* Grossouvre; Level 2: the late Santonian (eventually Santonian-Campanian boundary) level with *Nowakites lemarchandi* (Grossouvre); Level 3: the early Campanian level with *Pachydiscus (P.) koeneni* Grossouvre and *Glyptoxoceras retrorsum* (Schlueter); and Level 4: the late Campanian level with *Pachydiscus (P.) oldhami* (Sharpe).

Level 1

The oldest known ammonite level in Jamaica occurs in the Windsor Shale of the St. Ann Group. It was discovered by Sohl (1978) and permits an exact dating of the commencement of pelagic sedimentation. This event is Lower Coniacian in age, based on the presence of *Peroniceras* cf. *moureti* and *Gauthiericeras* cf. *bajuvaricum*. The co-occurrence of *Inoceramus waltersdorfensis hannovrensis* confirms the dating.

Level 2

As mentioned above, the long-ranging index species of the second level, *Nowakites lemarchandi*, is known to range upward from Coniacian to late Santonian strata, but its persistence into the early Campanian cannot be denied. In Jamaica, it is found in the *Inoceramus* Shales of the St. Ann Group, which had been dated erroneously as Cenomanian to Coniacian by Chubb (1955, 1958, 1960). Neither his foraminiferal data nor his inoceramid ages have been confirmed by subsequent work. Pessagno's revision (1978) of Esker's (1969) planktonic fauna revealed, however, a Lower Santonian age for the base of this unit, based on the occurrence of *Dicarinella concavata*. Kauffman (1966) revised the inoceramids co-occurring with *Nowakites lemarchandi* in the middle part of these shales. He recognized a first appearance of

TABLE 2. PROVISIONAL CORRELATION OF JAMAICAN LATE CRETACEOUS PELAGIC FACIES SHOWING INOCERAMID "ZONES" AND AMMONITE LEVELS

Inoceramid "zones"†	Ammonite Levels§		
Inoceramus proximus subcircularis M. & H.	?		Maastrichtian
	(4) *Pachydiscus (P.) oldhami* (Sharpe)	Upper	Campanian
	?	Middle	
Inoceramus balticus spp.	(3) *Pachydiscus (P.) koeneni* Gross. *Glyptoxoceras retrorsum* (Schluet.)	Lower	
Inoceramus muelleri Petr. *Inoceramus balticus* Boehm *Endocostea* sp.	(2) *Nowakites lemarchandi* (Gross.)	Upper	Santonian
	?	Lower	
	?	Upper	Coniacian
Inoceramus waltersdorfensis hannovrensis Heinz	(1) *Peroniceras cf. moureti* Gross.	Lower	

†Kauffman, 1966, 1978.
§This chapter.

Inoceramus balticus together with *Endocostea* sp., supporting a late Santonian to early Campanian age for this part of the *Inoceramus* Shales. Kauffman (1966) stated that his "Zone of *Inoceramus balticus*" is among the richest and most widespread of Caribbean inoceramid zones, occurring at the same time in a variety of different sedimentary facies. In Puerto Rico a division into two subzones, of late Santonian and early Campanian age, is feasible. In Jamaica, according to Kauffman (1966), only the upper subzone may be represented. This would, however, imply the synchroneity of the *Nowakites* level with the later, early Campanian, subzone. Therefore, a correlation of the *Inoceramus* Shales with the lower part of the Providence Shale and the Hanover Group has to be considered. However, for the time being, the level with *Nowakites lemarchandi* is better placed at the Santonian-Campanian boundary.

Level 3

This level, containing *Pachydiscus (P.) koeneni,* is the most fossiliferous and the most widespread of those with an ammonite-inoceramid association. It is most typically represented by the lower portion of the Providence Shale that, on the faunal evidence, is found to be the equivalent of the Clifton Limestone of the Hanover Group. Both sections yield early Campanian ammonites in coeval existence with the inoceramid association *Inoceramus muelleri* Petraschek and *I. balticus* spp. From the inoceramid data, the Hanover Shale overlying the Clifton Limestone can also be included in the early Campanian.

Level 4

A late Campanian age is assigned to the upper part of the Providence Shale yielding *Pachydiscus (P.) oldhami.* Here again, inoceramids confirm the dating. The "late Campanian lineages" of *I. barabini* and *I. (Endocostea) typicus* were recorded from the Newman Hall Shale, St. James Inlier, Central Jamaica, and considered to represent a "late Campanian–Maastrichtian inoceramid association" (Kauffman, 1966:39).

Age of the Barrettia Limestone

Due to the revision of the late Cretaceous mollusc assemblages of Jamaica, the controversial previous dating of the *Barrettia* Limestone can now be corrected, and this prominent limestone can be included in the Middle Campanian.

It was again Trechmann (1924) who came to a similar conclusion when he compared the different rudistid limestones of Jamaica. He considered them to be vertically restricted and all of Campanian to Maastrichtian age.

PALEOBIOGEOGRAPHY

The paleogeographic significance of the Upper Cretaceous ammonites from Jamaica coincides with that of the inoceramids. Both faunas are distinctly cosmopolitan. The only exception is the endemic Lower Campanian *Pachydiscus (P.) jamaicensis* n. sp.; but even this new species can easily be related to, or is transitional to, other cosmopolitan species. From Figure 6, the Jamaican ammonite faunas can easily be traced to Europe and to a lesser

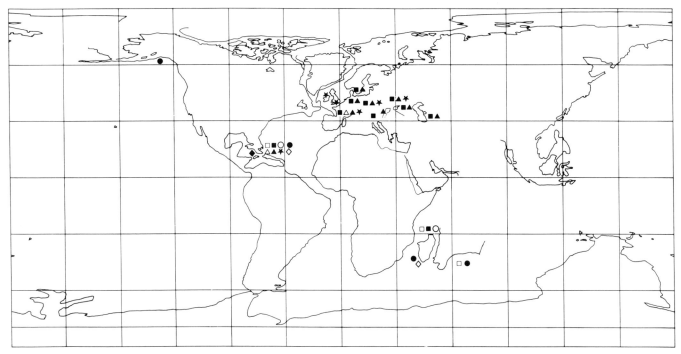

Figure 6. Paleobiogeography of Campanian (and Santonian) Jamaican ammonite species. Paleogeography of the Santonian adapted from Smith et al. (1981).

degree to southern India and Madagascar. The expected connections with the North American faunas are, however, much reduced. The inoceramids exhibit a similar distribution pattern but with somewhat more pronounced North American and Japanese relationships. The east-west–oriented ammonite distribution may be related to the efficacy of a North Equatorial current during the late Cretaceous. The cosmopolitan distribution of the Campanian faunas, moreover, might be connected with a general global sea-level rise during the Campanian (Sliter, 1976; Vail et al., 1977; Haq et al., 1987).

SYSTEMATIC DESCRIPTIONS OF AMMONITES

Suborder Lytoceratina Hyatt, 1889
Superfamily Tetragonitaceae Hyatt, 1900
Family Tetragonitidae Hyatt, 1900

Genus *Tetragonites* Kossmat, 1895
incl. *Epigoniceras* Spath, 1925

***Tetragonites* cf. *garuda* (Forbes)**
Fig. 7.4

1925 *Epigoniceras* sp. ind.-Spath, 29, Plate 1, Fig. 2.
cf. 1846 *Ammonites Garuda* Forbes, 102, Plate 7, Fig. 1.

The unique specimen BMNH C. 25258 is a tetragonitid inner whorl that is chambered throughout, the suture line being visible on the cast. Spath (1925) placed this poorly preserved and crushed specimen in the vicinity of *T. epigonus* Kossmat, proposing at the same time a separation of the group of *Tetragonites epigonus* Kossmat as *Epigoniceras* n. g. from *Tetragonites* s. str. This separation was, however, based on a misinterpretation of Kossmat's (1895, Plate 17) suture line drawings of tetragonitids, and Howarth (1958), Matsumoto (1959), and Wiedmann (1962) have included *Epigoniceras* again in *Tetragonites*.

In contrast to Kossmat's (1895) assumption that *Ammonites garuda* Forbes may form the inner whorls of *Pseudophyllites indra* (Forbes), it is considered here to be distinct (see also Collignon, 1956:88) and referred to *Tetragonites*. The Jamaican specimen agrees perfectly with those specimens figured by Forbes (1846), Stoliczka (1865), and Collignon (1956) in having rounded whorl section, rapid increase in whorl width, weakness of constrictions, and a relatively small umbilicus.

T. cf. *garuda* is found together with *Pachydiscus (P.) koeneni* Grossouvre in the lower Providence Shale (early Campanian) near Port Antonio. *T. garuda* is known from the Campanian Valudayur Beds of southern India and the Upper Santonian–Lower Campanian of Madagascar.

Suborder Ancyloceratina Wiedmann, 1966
Superfamily Turrilitaceae Gill, 1871
Family Turrilitidae Gill, 1871
Subfamily Diplomoceratinae Spath, 1926
Genus *Glyptoxoceras* Spath, 1925
incl. *Neoglyptoxoceras*, *Epiglyptoxoceras* Collignon, 1969

***Glyptoxoceras retrorsum* (Schlueter)**
Figs. 7.1, 7.2

1872 *Ancyloceras retrorsum* Schlueter, 97, Plate 30, Figs. 5–10.
1925 *Glyptoxoceras* cf. *rugatum* (Forbes).-Spath, 31, Plate 1, Fig. 4.

Figure 7. 1, *Glyptoxoceras retrorsum* (Schlueter) BMNH C. 25261, lower Providence Shale, Roadside section, 1.5 km south of Port Antonio, Port Antonio–Fellowship Road. 30 m above *P. (P.) koeneni*. Early Campanian. 1/1. 2, same species, BMNH C. 25259, same locality, same age. 1/1. 3, *Baculites* cf. *tenuetuberculatus* Collignon BMNH C.25260, locality and age as above. 1/1. 4, *Tetragonites* cf. *garuda* Forbes BMNH C. 25258, lower Providence Shale. Same locality as above, level with *P. (P.) koeneni*—A: lateral view; B: ventral view. 2/1. 5, *Desmophyllites* cf. *phyllimorphus* (Kossmat) BMNH C. 90948, upper Providence Shale. West Town River Bank, 300 m south of road junction, bottom road leading to Bermuda. Late Campanian. A: lateral view; B: ventral view. 1/1.

1986 *Neoglyptoxoceras*(?) *retrorsum* (Schlueter).-Kennedy, 106, Fig. 38; Plate 16, Figs. 1–4, 6, 7; Plate 17, Figs. 1, 2 (and synonymy).

Only two of the three specimens figured by Spath (1925, Plate 1, Fig. 4) are preserved (BMNH C.25261, C.25259), and these are refigured here in Figures 7.1 and 7.2. The fragments were compared by Spath with *Gl. rugatum* (Forbes) but possess rather straighter and denser ribbing, constrictions, and a circular whorl section. There is nearly complete agreement, especially of the larger fragment (Fig. 7.1), with *Gl. retrorsum* (Schlueter) and above all with the lectotype recently refigured by Kennedy (1986, Fig. 38A–C). Both specimens correspond perfectly in the anisoceratid torsion of the preserved whorl, in compressed oval whorl sections, and in the laterally rursiradiate ribbing with a rib index of 5. *Gl. retrorsum* is of Campanian age, and the Jamaican occurrence confirms the dating of the lower Providence Shale.

We do not accept here Collignon's (1969) proposal to separate the giant Lower Campanian criocones from Madagascar, on the basis of size and suture line, from the Lower Maastrichtian group of *Gl. rugatum*. This is not the place to revise the Madagascan faunas, but the type of Collignon's new genus *Neoglyptoxoceras*, *N. magnificum* Collignon, is found to be identical with *Gl. retrorsum*.

Gl. retrorsum is a cosmopolitan Campanian species. The type material is from the Middle Campanian of northwestern Germany. It is also known from the Lower Campanian of southern France, central Poland, and Madagascar (see above) and from the Middle and Upper Campanian of Austria, western and southern Russia, and Sweden. The Jamaican specimens come from a level 30 m above *Pachydiscus (P.) koeneni* Grossouvre in the early Campanian lower Providence Shale near Port Antonio.

Family Baculitidae Gill, 1871

Genus *Baculites* Lamarck, 1799

***Baculites* cf. *tenuetuberculatus* Collignon**
Fig. 7.3

1925 *Baculites* sp. ind.-Spath, 31, Plate 1, Fig. 1.
cf. 1966 *Baculites capensis* var. *tenuetuberculata* Collignon, 6, 22, Plate 457, Figs. 1863, 1864; Plate 463, Figs. 1894, 1895.

Three fragmentary and crushed specimens are preserved on a single slab (BMNH C. 25260). Spath (1925) assigned these specimens to the group of *B. incurvatus* Dujardin, which can be characterized by periodic crescentic bullae on the lateral sides. As can be seen on the largest of the Jamaican specimens, the bullae are cone shaped near the ventral margin. This type of ornamentation is specific among the group of *B. capensis* Woods of Santonian age. Concerning the characteristics of these bullae, the Jamaican specimens correspond nearly perfectly to Collignon's "var. *tenuetuberculata*" and especially with the younger cotypes from the Middle Santonian of Madagascar (Collignon, 1966, Plate 463, Figs. 1894, 1895). In contrast to the Madagascar specimens, the present form has traces of shorter intermediate ventral ribs. Moreover, the exact whorl section cannot be determined because of the crushed state of the specimens. Therefore, we refrain from including the Jamaican specimens in the Madagascan species.

As mentioned above, the Madagascar *B. tenuetuberculatus* is of Lower and Middle Santonian age. The Jamaican specimens are from a level 30 m above that of *Pachydiscus (P.) koeneni* Grossouvre in the lower Providence Shale near Port Antonio, of early Campanian age.

Suborder Ammonitina Hyatt, 1889
Superfamily Desmocerataceae Zittel, 1895
Family Desmoceratidae Zittel, 1895

Genus *Desmophyllites* Spath, 1929

***Desmophyllites* cf. *phyllimorphus* (Kossmat)**
Fig. 7.5

cf. 1897 *Desmoceras phyllimorphum* Kossmat, 175, Plate 25, Fig. 10.

A single poorly preserved specimen (BMNH C. 90948), which exhibits the main features of the genus *Desmophyllites*—namely, smooth lateral sides with strongly projected constrictions and flares and a small umbilicus. It shares its main characteristics with *D. phyllimorphus* (Kossmat), i.e., the rounded whorl section with maximum width above mid-whorl height and a broad venter. The measurements are:

	Dm	WH	WW	UW
BMNH C. 90948				
	ca. 78.0 mm	43 mm (0.53 WH/Dm)	31 mm (0.40 WW/Dm)	---
Kossmat's holotype				
	81.5 mm	43.5 mm 0.53 (WH/Dm)	30.5 mm (0.37 WW/Dm)	---

Only the umbilical width of the present specimen seems to be larger than in *D. phyllimorphum* and any other *Desmophyllites*. Therefore, it can only doubtfully be referred to that species.

D. cf. *phyllimorphus* is from the late Campanian upper Providence Shale, West Town River bank. Kossmat's species is known from the Ariyaloor Beds of southern India, the Upper Campanian to Lower Maastrichtian of southern Alaska, and the Lower Maastrichtian of Madagascar.

Family Pachydiscidae Spath, 1922

Genus *Nowakites* Spath, 1922

***Nowakites lemarchandi* (Grossouvre)**
Fig. 5.1

1894 *Puzosia Le Marchandi* Grossouvre, 173, Plate 22, Fig. 5.
1936 *Nowakites* aff. *paillettei* (D'Orbigny).-Trechmann, 253.
1984 *Nowakites le marchandi* (Grossouvre).-Bilotte, Plate 32, Fig. 9.

The single specimen (BMNH C. 38120) is figured here for the first time. It is easy to see from the figure that the Jamaican specimen is quite different from the Santonian *N. pailletteanus* (D'Orbigny) to which it was referred by Spath in Trechmann (1936). Although in *N. pailletteanus* the ribbing is uniform and constrictions are rather weak, the present specimen has strong principal ribs accompanied by sharp constrictions. About 10 principal ribs occur per whorl. They bear distinct umbilical tubercles. In between these principal ribs, between three and six secondary ribs are intercalated, of differing length and strength. Ribs and constrictions are sigmoidal on the flanks; when they pass over the venter, they are of nearly equal strength and slightly curved forward. The whorl section is broadly rounded; the umbilicus is rather small.

All these features are those of *N. lemarchandi* to which the Jamaican specimen is best referred. Moreover, it matches with the holotype in its measurements:

Dm	WH	WW	UW
BMNH C. 38120			
ca. 42 mm	22 mm (0.52 WH/Dm)	21 mm (0.5 WW/Dm)	ca. 11 mm (0.26 UW/Dm)
Grossouvre's holotype			
65.0 mm	30 mm 0.46 (WH/Dm)	?	16.4 mm (0.25 UW/Dm)

The holotype has 10 principal ribs per whorl and about 50 secondaries.

The holotype of *N. lemarchandi* is of Lower Coniacian age; the specimen figured by Bilotte (1984) is from the late Santonian; both are from southern France. The Jamaican specimen was found in the *Inoceramus* Shales at St. Ann's Great River. A late Santonian age for these *Inoceramus* Shales can be assumed, but also an early Campanian age cannot be totally denied.

Genus *Pachydiscus* Zittel, 1884
Subgenus *Pachydiscus* Zittel, 1884

Pachydiscus (P.) koeneni Grossouvre
Fig. 5.2

1872 *Amm. Galicianus* Favre.-Schlueter, 63, Plate 19, Figs. 3–5; Plate 20, Fig. 9.
1885 *Amm. Oldhami* Sharpe.-Moberg, 23, Plate 3, Fig. 1.
1894 *Pachydiscus Koeneni* Grossouvre, 178.
pars *Pachydiscus Oldhami* Sharpe.-Grossouvre, 182, Plate 22, Fig. 1.
1913 *Pachydiscus Oldhami* Sharpe.-Nowak, 362, Plate 41, Fig. 16; Plate 43, Fig. 31; Plate 45, Fig. 43.
1925 *Parapachydiscus* aff. *stallauensis* (Imkeller).-Spath, 29, Plate 1, Fig. 3.
1951 *Pachydiscus* cf. *Koeneni* Grossouvre.-Mikhailov, 60, Plate 10, Fig. 4.
1959 *Pachydiscus koeneni* Grossouvre.-Naidin and Shimansky, 185, Plate 9, Fig. 1.
1964 *Pachydiscus koeneni* Grossouvre.-Giers, 263, Fig. 5, Plate 5, Fig. 1, ?2.
pars 1974 *Pachydiscus koeneni* Grossouvre.-Naidin, 186, only Plate 65, Fig. 2.
1980 *Pachydiscus koeneni* Grossouvre.-Blaszkiewicz, 42, Plate 26, Figs. 1, 2; Plate 27, Figs. 1–4; Plate 28, Figs. 1–4; Plate 34, Figs. 1, 4.
pars 1986 *Pachydiscus haldemsis* (Schlueter).-Kennedy, 45, Fig. 17, Plate 5, Figs. 7–14.

The Jamaican specimen (BMNH C. 25256) was considered by Spath (1925) to have an affinity with *P. stallauensis* Imkeller (1901, 57, Plate 3, Fig. 5). This Campanian, Gosau species from the Northern Alps is, however, rather similar to or even identical with *P. haldemsis* (Schlueter). We do not follow Kennedy (1986) in synonymizing *P. haldemsis* and *P. koeneni* Grossouvre. We fully agree with Blaszkiewicz's (1980) interpretation of the latter species, which can easily be separated from *P. haldemsis* by its irregular ribbing and the strength of the principal ribs provided with umbilical tubercles. Whereas in *P. koeneni* the part on the lower whorl flank of these principal ribs remains strengthened even with age (Schlueter, 1872, Plate 19, Fig. 3; Kennedy, 1986, Fig. 17), in *P. haldemsis* the strength of the ribbing is more uniform throughout. The principal ribs are more pronounced on the upper flank toward the venter, with tubercles on the ventro-lateral shoulders of the inner whorls, and show a reduction of the strength of the ribs on the lower whorl flank with age.

The specific identity is supported by comparing the measurements:

Dm	WH	WW	UW
BMNH C. 25256			
51 mm	22 mm (0.43 WH/Dm)	>12 mm (>0.23 WW/Dm)	12 mm (0.23 UW/Dm)
Holotype of *P. koeneni*			
123 mm	52 mm 0.42 (WH/Dm)	>28 mm (>0.23 WW/Dm)	33.5 mm (0.27 UW/Dm)

Both specimens are laterally crushed. The holotype (in Schlueter, 1872; Kennedy, 1986) bears 22 principal and 28 shorter secondary ribs per whorl; the Jamaican specimen 17 and 27, respectively.

P. koeneni is a long-ranging Campanian species, known from northwestern Germany, Poland, southeastern Russia, Bulgaria, Sweden, and southwestern France. The specimen figured is from the lower Providence Shale near Port Antonio and is considered to indicate the early Campanian.

Pachydiscus (P.) cf. *spissus* Collignon
Fig. 5.4

1925 *Parapachydiscus* cf. *gollevillensis* (D'Orbigny).-Spath, 30, Plate 1, Fig. 3
cf. 1955 *Pachydiscus spissus* Collignon, 72, Plate 26, Fig. 1.

The unique specimen (BMNH C. 25257) was erroneously attributed to the Maastrichtian *P. gollevillensis* (D'Orbigny) by Spath (1925). It is, however, much closer to the Middle Campanian *P. spissus* Collignon from Madagascar and especially to the specimen figured from Ankilizato (Collignon, 1970, 38, Pl. 623, Fig. 2311). All these specimens agree in the density of ribbing, with about 15 radial to slightly forward projected main ribs, bearing umbilical tubercles, and an irregular number of shorter secondaries (two to five). A typical feature of *P. spissus* is the strengthening of the secondaries on the external lateral sides.

There is a great similarity with *P. jamaicensis* n. sp. described next. *P. spissus* is known only from the Middle Campanian of Madagascar. The Jamaican specimen was found together with *P. koeneni* in the lower Providence Shale (early Campanian) near Port Antonio.

Pachydiscus (P.) jamaicensis n. sp.
Fig. 5.3

Holotype. Specimen BMNH C. 90957 from the Clifton Limestone, Maryland Waterfall, Lucea Inlier.
Diagnosis. Involute pachydiscid with slightly projected trifurcating ribs, main ribs with umbilical tubercles, and periodic constrictions. Rounded whorl section broader than high.
Description. The holotype of *P. (P.) jamaicensis* n. sp. is a phragmocone. The umbilicus is relatively small, and the whorl section is broadly rounded, somewhat broader than high. Maximum whorl width is near the steep umbilical border. The venter is broadly rounded. Most of the

11 stronger main ribs with umbilical tubercles are trifurcating at the umbilical tubercle. Shorter secondaries are intercalated to a total of 46 ventral ribs. All ribs are slightly forward projected and cross the venter. About six constrictions can be discerned on the last whorl. The measurements of the holotype are:

Dm	WH	WW	UW	WW/WH
BMNH C. 90957				
65 mm	30 mm	33 mm	14 mm	1.10 mm
	(0.46 WH/Dm)	(0.51 WW/Dm)	(0.22 UW/Dm)	

Comparisons. *P. (P.) jamaicensis* n. sp. has a type of ribbing similar to *P. spissus* Collignon from the Middle Campanian of Madagascar, in which 15 main ribs rise at the umbilical border, dividing to produce 60 ribs across the venter. No constrictions can be observed. The oval whorl is higher than wide (WW/WH = 0.91). The Upper Santonian *P. cayeuxi* Grossouvre has a rib index of 9:42, a WW/WH ratio of 0.97, and a wider umbilicus, and the type of ribbing is more irregular.

Occurrence. *P. (P.) jamaicensis* n. sp. was found in the Clifton Limestone at Maryland Waterfall, Hanover. It is considered to be Lower Campanian in age.

Pachydiscus (P.) oldhami (Sharpe)
Fig. 5.5

1855 *Amm. Oldhami* Sharpe, 32, Plate 14, Fig. 2.
1986 *Pachydiscus (P.) oldhami* (Sharpe).-Kennedy, 40, Figs. 4A, 15, 16, 18; Plate 3, Figs. 1–3; Plate 4, Figs. 4, 5; Plate 5, Figs. 1–3 (and synonymy).

A single specimen (BMNH C. 74026) is figured here for the first time. *P. oldhami* Sharpe belongs obviously to the same group of pachydiscids discussed above. It has, however, a larger umbilicus, a more rounded whorl section, and a typical effacing of ribs on the lateral sides of the inner whorls. In this respect, the Jamaican specimen corresponds perfectly with the juvenile specimens figured by Kennedy (1986, Plate 5, Figs. 1–6).

The measurements also agree:

Dm	WH	WW	UW
BMNH C. 74026			
49 mm	19.5 mm	17.5 mm	15 mm
	(0.4 WH/Dm)	(0.36 WW/Dm)	(0.31 UW/Dm)
Sharpe's holotype			
99 mm	40 mm	34 mm	32 mm
	(0.4 WH/Dm)	(0.34 WW/Dm)	(0.32 UW/Dm)

P. oldhami is an Upper Campanian species. This age again fits perfectly with the presumed age of the upper Providence Shale in which *P. oldhami* was found 3.2 km south of Port Antonio. The species is reported from England, northern Ireland, southwestern France, Poland, and Russia.

ACKNOWLEDGMENTS

The authors appreciate critical remarks and photographs of the BMNH ammonites by H. Owen and a critical review by N. Sohl.

REFERENCES CITED

Balkissoon, I., compiler, 1987, Geological map of Jamaica—Port Antonio: Geological Survey Division, sheet 28, scale 1:50,000.

Bilotte, M., 1984, Le Crétacé Supérieur des plates-formes est-pyrénéennes. Atlas: Strata, Série 2, Mémoires, v. 1, 45 plates.

Blaszkiewicz, A., 1980, Campanian and Maastrichtian ammonites of the middle Vistula River valley, Poland: A stratigraphic-paleontological study: Prace Instytut Geologiczny), v. 92, 63 p.

Chubb, L. J., 1955, The Cretaceous succession in Jamaica: Geological Magazine, v. 92, p. 177–195.

Chubb, L. J., 1958, Cretaceous inlier of St. Ann's Great River: Geonotes, v. 1, p. 148–152.

Chubb, L. J., 1960, Correlation of the Jamaican Cretaceous: Geonotes, v. 3, p. 85–97.

Chubb, L. J., 1962, Recent work on the Cretaceous of Jamaica [abs.]: Abstracts of the 3rd Caribbean Geological Conference Kingston, p. 1–42.

Chubb, L. J., 1971, Rudists of Jamaica: Palaeontographica Americana, v. 7, p. 161–257.

Collignon, M., 1955, Ammonites néocréatacées du Menabe (Madagascar). 2. Les Pachydiscidae: Service des Mines, Annales Géologiques, v. 21, 98 p.

Collignon, M., 1956, Ammonites néocrétacées du Menabe (Madagascar). 4. Phylloceratidae. 5. Les Gaudryceratidae. 6. Tetragonitidae: Service des Mines, Annales Géologiques, v. 23, 106 p.

Collignon, M., 1966, Atlas des fossiles caractéristiques de Madagascar (Ammonites). 14. (Santonien): Service Géologique Republique Malgache, 134 p.

Collignon, M., 1969, Atlas des fossiles caractéristiques de Madagascar (Ammonites). (Campanien inférieur): Service Géologique Republicque Malgache, 216 p.

Collignon, M., 1970, Atlas des fossiles caractéristiques de Madagascar (Ammonites). 16. (Campanien moyen. Campanien supérieur): Service Géologique Republic Malgache, 82 p.

Esker, G., 1969, Planktonic foraminifera from the St. Ann's Great River valley, Jamaica: Micropaleontology, v. 15, p. 210–220.

Forbes, E., 1846, Report on the fossil invertebrata from southern India, collected by Mr. Kaye and Mr. Cunliffe: Transactions of the Geological Society of London, ser. 2, v. 7, p. 97–174.

Giers, R., 1964, Die Grossfauna der Mukronatenkreide (unteres Obercampan) im östlichen Münsterland: Fortschritte in der Geologie von Rheinland und Westfalen, v. 7, p. 213–294.

Grippi, J., 1980, Geology of the Lucea Inlier, western Jamaica: Journal of the Geological Society of Jamaica, v. 19, p. 1–24.

Grossouvre, A. de, 1894, Recherches sur la Craie supérieure. 2e partie—Paléontologie. Les ammonites de la Craie supérieure: Mémoires pour servir á l'explication de la carte géologique détaillée de la France, 264 p.

Haq, B. U., Hardenbohl, J., and Vail, P., 1987, Chronology of fluctuating sea-level since the Triassic: Science, v. 235, p. 1156–1167.

Howarth, M. K., 1958, Upper Jurassic and Cretaceous ammonite faunas of Alexander Land and Graham Land: Falkland Islands Dependencies Survey,

Scientific Reports, v. 21, p. 1–16.
Imkeller, H., 1901, Die Kreidebildungen und ihre Fauna am Stallauer Eck und Enzenauer Kopf bei Tölz: Palaeontographica, v. 48, p. 1–64.
Jiang, M.-J., and Robinson, E., 1987, Calcareous nannofossils and larger foraminifera in Jamaican rocks of Cretaceous to Early Eocene age, in Ahmad, R., ed., Proceedings of a workshop on the status of Jamaican Geology, Kingston, 1984: Kingston, Geological Society of Jamaica, p. 24–51.
Kauffman, E. G., 1966, Notes on Cretaceous Inoceramidae (Bivalvia) of Jamaica: Journal of the Geological Society of Jamaica, v. 8, p. 32–40.
Kauffman, E. G., 1978, Middle Cretaceous bivalve zones and stage implications in the Antillean subprovince, Caribbean Province, in Evénements de la partie moyenne du Crétacé; rapports sur la biostratigraphie des regions cles (Anonymous): Annales du Muséum d'Histoire Naturelle de Nice, v. 4, p. xxx.1–xxx-11.
Kennedy, W. J., 1986, Campanian and Maastrichtian ammonites from northern Aquitaine, France: Special Papers in Palaeontology, v. 36, 145 p.
Kossmat, F., 1895 (Part 1), 1896 (Part 2), 1897 (Part 3), Untersuchungen über die südindische Kreideformation: Beiträge zur Paläontologie und Geologie Österreich-Ungarns und des Orients, v. 9(3–4), v. 11(1, 3), 217 p.
Krijnen, J. P., and Lee Chin, A. C., 1977, Stratigraphy of the northern, central and southeastern Blue Mountains, Jamaica: Kingston, n.p., 3 p.
Matsumoto, T., 1959, Cretaceous ammonites from the upper Chitina valley, Alaska: Kyushu University Faculty of Science Memoirs, ser. D, v. 8, pt. 3, p. 49–90.
Mikhailov, N. P., 1951, Ammonites of the Upper Cretaceous from southern European USSR and their significance to zonal stratigraphy (in Russian): Trudy Instytut geologitshesky Academiya Nauk, SSSR, ser. Geol., v. 50, 143 p.
Moberg, J. C., 1885, Cephalopoderna i Sveriges Kritsystem, 2. (Artbeskrifning): Sveriges Geologiska Undersökning, Series C, no. 73, 64 p.
Naidin, D. P., 1974, Ammonoidea (in Russian), in Krymgolts, G. Ya., ed., Atlas of Upper Cretaceous faunas of the Don basin: Moscow, Nedra, p. 158–195.
Naidin, D. P., and Shimansky, V. N., 1959, Cephalopods (in Russian), in Moskvin, M. M., ed., Atlas of Upper Cretaceous fauna from northern Caucasus and Crimea: Moscow, Gostoptekhizdat, p. 9–23.
Nowak, J., 1913, Untersuchungen über die Cephalopoden der oberen Kreide in Polen: Bulletin de l'Acámie des Sciences de Cracovie, Classe des Sciences Mathematiques et Naturelles, ser. B, p. 335–415.
Pessagno, E. A., Jr., 1978, Middle Cretaceous planktonic foraminiferal biostratigraphy of the Antillean-Caribbean region and eastern Mexico. Notes on the ammonites of eastern Mexico, in Evénements de la partie moyenne du Crétacé; rapports sur la biostratigraphie des regions cles (Anonymous): Annales du Muséum d'Histoire Naturelle de Nice, v. 4, p. xxviii.1–xxviii-10.
Schlueter, C., 1871/72, Cephalopoden der oberen deutschen Kreide. Part 1: Palaeontographica, v. 21, parts 1-5, 120 p.
Schmidt, W., 1986, Stratigraphy and depositional environment of the Lucea and Grange Inliers, western Jamaica: Abstracts of the 11th Caribbean Geological Conference, Bridgetown, Barbados, p. 97–98.
Sharpe, D., 1853–1857, Description of the fossil remains of Mollusca found in the Chalk of England: Transactions of Palaeontographical Society of London, pts. 1–3, 68 p.
Sliter, W. V., 1976, Cretaceous foraminifers from the south-western Atlantic Ocean, Leg 36, Deep Sea Drilling Project, in Barker, P. F., Dalziel, I.W.D., and others, Initial Reports of the Deep Sea Drilling Project: Washington, D.C., U.S. Government Printing Office, v. 36, p. 519–573.
Smith, A. G., Hurley, A. M., and Briden, J. C., 1981, Phanerozoic paleocontinental world maps: Cambridge, Cambridge University Press, 102 p.
Sohl, N. F., 1978, Notes on middle Cretaceous macrofossils from the Greater Antilles, in Evénements de la partie moyenne du Crétacé; rapports sur la biostratigraphie des regions cles (Anonymous): Annales du Muséum d'Histoire Naturelle de Nice, v. 4, p. xxxi.1–xxxi.6.
Sohl, N. F., and Kollmann, H. A., 1985, Cretaceous actaeonellid gastropods from the western hemisphere: U.S. Geological Survey Professional Paper, v. 1304, 95 p.
Spath, L. F., 1925, On Senonian ammonites from Jamaica: Geological Magazine, v. 62, p. 28–32.
Stoliczka, F., 1863–1866, The fossil cephalopoda of the Cretaceous rocks of southern India, Ammonitidae, with revision of the Nautilidae: Memoir of the Geological Survey of India, Palaeontologia indica, series 3, v. 1, p. 41–216.
Trechmann, C. T., 1924, The Cretaceous limestones of Jamaica and their mollusca: Geological Magazine, v. 61, p. 385–410.
Trechmann, C. T., 1927, The Cretaceous shales of Jamaica: Geological Magazine, v. 64, p. 49–51.
Trechmann, C. T., 1936, Basal complex question in Jamaica: Geological Magazine, v. 73, p. 257–267.
Vail, P. R., Mitchum, R. M., Jr., and Thompson, S., III, 1977, Global cycles of relative changes of sea level. (Seismic stratigraphy and global changes of sea level, pt. 4), in Payton, C. E., ed., Seismic stratigraphy—Applications to hydrocarbon exploration: American Association of Petroleum Geologist Memoir, v. 26, p. 83–97.
Wiedmann, J., 1962, Ammoniten aus der vascogotischen Kreide (Nordspanien). 1. Phylloceratina, Lytoceratina: Palaeontographica, Abt. A, v. 118, pts. 4–6, p. 119–237.
Zans, V. A., Chubb, L. J., Versey, H. R., Williams, J. B., Robinson, E. and Cooke, D. L., 1962, Synopsis of geology of Jamaica: Bulletin of the Geological Survey of Jamaica, v. 4, p. 1–72.

MANUSCRIPT ACCEPTED BY THE SOCIETY NOVEMBER 11, 1992

Geological Society of America
Memoir 182
1993

Jamaican Cretaceous Echinoidea

Stephen K. Donovan
Department of Geology, University of the West Indies, Mona, Kingston 7, Jamaica, West Indies

ABSTRACT

Twelve species of fossil echinoid have been recognized from the Jamaican Cretaceous. Only two taxa, the cidaroid *Phyllacanthus leoni* (Lambert and Sánchez Roig) and the arbacioid *Goniopygus supremus* Hawkins, are known from the Cretaceous of both Jamaica and Cuba. They are also the two most common echinoids in the Jamaican Cretaceous. Two other cidaroid species, including *Temnocidaris*? (*Stereocidaris*) sp., have been identified from disarticulated radioles only. Two other species of regular echinoid, the acrosaleniid *Heterosalenia occidentalis* Hawkins and the temnopleuroid *Scoliechinus axiologus* Arnold and Clark, have been described. The late Cretaceous age of *S. axiologus* is uncertain. Irregular echinoids are less common. Only *Hemiaster*? sp. (or spp.) is abundant, although it is usually poorly preserved. Other irregular echinoid species include the unique holectypoid *Metholectypus trechmanni* Hawkins, the cassiduloid *Pygopistes*? *rudistarum* (Hawkins), and a cassiduloid sp. indet. Two taxa, *Orthopsis* sp. and a nucleolitid sp. nov., are awaiting description. Most of these taxa are of Campanian and/or Maastrichtian age. Jamaican Cretaceous echinoids are best known from the Marchmont and Central inliers.

INTRODUCTION

The Jamaican Cretaceous echinoid fauna comprises 12 known species belonging to eight orders. Two of these species are awaiting description, and many of the other taxa were originally described on the basis of unique specimens. However, this is the most diverse Cretaceous echinoid fauna known from the islands of the Caribbean Plate and is thus of particular interest. The most common echinoid fossils in the Cretaceous of Jamaica, and therefore potentially the most useful biostratigraphic markers, are radioles derived from the cidaroid *Phyllacanthus leoni* (Lambert and Sánchez Roig) and probably from the arbacioid *Goniopygus supremus* Hawkins. These form distinctive, monospecific radiole "bands" in, for example, the Guinea Corn Formation (Upper Campanian to Middle Maastrichtian; Jiang and Robinson, 1987) of the Central Inlier (Q in Fig. 1) and (*P. leoni* only) the Maastrichtian of the Jerusalem Mountain Inlier (C in Fig. 1). It is significant to note that the tests of these two species are more common than those of most of the other Jamaican Cretaceous echinoids and that both taxa are also known from Cuba.

The first report of an echinoid from the Jamaican Cretaceous was by Hill (1899, p. 118), who mentioned *Salenia* from the Blue Mountain Inlier (Z in Fig. 1). However, this genus has not subsequently been recognized from the island. Trechmann (1929, p. 485) reported "indeterminate echinoderm spines" from the same inlier, to which Chubb (1961) added *Hemiaster*? sp. and *Phymosoma*? or *Pseudodiadema*? sp.

Hawkins (1923) described the first three named species from the Jamaican Cretaceous: *Metholectypus trechmanni*, *Heterosalenia occidentalis*, and *Botriopygus rudistarum*. The holotype of *M. trechmanni* is the only described specimen within this genus, and *B. rudistarum* is also rare. In the same paper Hawkins described a "diadematoid" radiole (probably *Goniopygus supremus* Hawkins), an interambulacral plate of *Leiocidaris* sp. indet. (= *Phyllacanthus leoni* (Lambert and Sánchez Roig)), and a collection of spatangoids referred to *Hemiaster* sp. or spp. Another test of *Hemiaster* sp. and the holotype of *Goniopygus supremus* were also described by Hawkins (1924). The last new Cretaceous(?) echinoid to be named from the island was *Scoliechinus axiologus* Arnold and Clark, 1927. Cutress (1980) recognized *P. leoni* and two other species of cidaroid from the Cretaceous of Jamaica.

The first stratigraphic scheme of the Jamaican fossil echinoids (Donovan, 1988, Table 1) recognized 11 species from the Cretaceous. This was subsequently increased to 13 (Donovan and Bowen, 1989, text-fig. 2), but it has since been argued (Donovan, 1990) that *Trochalosoma chondra* (Arnold and Clark) is probably Eocene in age (*Scoliechinus axiologus* may also be Eocene;

Donovan, S. K., 1993, Jamaican Cretaceous Echinoidea, *in* Wright, R. M., and Robinson, E., eds., Biostratigraphy of Jamaica: Boulder, Colorado, Geological Society of America Memoir 182.

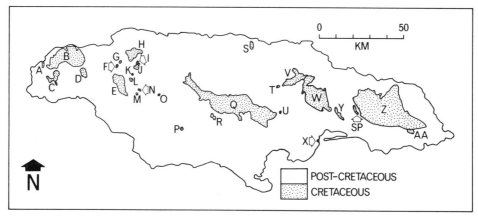

Figure 1. The distribution of Cretaceous inliers in Jamaica (redrawn after Porter et al., 1982, Fig. 8.3). Key to inliers: A = Green Island, B = Lucea, C = Jerusalem Mountain, D = Grange, E = Marchmont, F = Seven Rivers, G = Mafoota, H = Sunderland, I = Calton Hill, J = Maldon, K = Garlands, L = Mocho-Sweetwater, M = Barracks River, N = Elderslie, O = Aberdeen, P = Nottingham, Q = Central, R = Banana Ground, S = St. Ann, T = Mount Diablo, U = Giblatore, V = Benbow–Guys Hill, W = Above Rocks, X = Lazaretto, Y = Jacks Hill, Z = Blue Mountain, AA = Sunning Hill. SP is the so-called St Peter's Inlier, which is probably a suite of deformed Tertiary rocks (Jackson, 1986).

see below). A revision of the Jamaican Cretaceous echinoids has so far considered *Pygopistes*? (= *Botriopygus*) *rudistarum* (Donovan and Bowen, 1989), *G. supremus* and *H. occidentalis* (Donovan, 1990), and *S. axiologus* and an indeterminate cassiduloid (Donovan, 1992).

The stratigraphic (Fig. 2) and geographic (Table 1) distributions documented herein are revised after Donovan and Bowen (1989). The morphological terminology used in the systematic descriptions follows Melville and Durham (1966) and Smith (1984). The classification follows Moore (1966) and Smith (1984) unless otherwise stated. The following abbreviations are used in the text: BMNH, Natural History Museum, London; USNM, U.S. National Museum, Smithsonian Institution, Washington, D.C., MCZ, Museum of Comparative Zoology, Harvard.

SYSTEMATIC DESCRIPTIONS

Class ECHINOIDEA Leske
Order CIDAROIDA Claus

Remark. The cidaroid classification used herein follows Smith and Wright (1989). The following key differentiates between the radioles of the three species of cidaroid identified from the Jamaican Upper Cretaceous:

1. Radioles slender, elongate, with numerous columns of spinules 2
    Radioles broader, with fewer columns of coarser spinules .... indeterminate genus B
2. Columns of spinules continuous .... *Phyllacanthus leoni*
    Columns of spinules offset .... *Temnocidaris* (*Stereocidaris*) sp. B

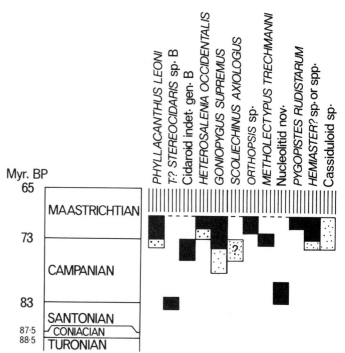

Figure 2. Known stratigraphic distribution of fossil echinoids in the Cretaceous of Jamaica (revised after Donovan and Bowen, 1989, text-fig. 2). Radiometric dates of boundaries after Harland et al. (1982). Solid black areas = probable ranges; stippled areas = possible ranges. The vertical ruling in the Upper Maastrichtian indicates that this part of the succession is rarely (if ever) present and fossiliferous in Jamaica. Stratigraphic data derived largely from Kauffman (1966, 1968), Chubb (1971), McFarlane (1977), Cutress (1980), Green (n.d.) and Krijnen (1993).

TABLE 1. KNOWN GEOGRAPHIC DISTRIBUTION OF THE CRETACEOUS ECHINOIDS OF JAMAICA*

	Phyllacanthus leoni	T.? Stereocidaris sp. B	Cidaroid indet. sp. B	Heterosalenia occidentalis	Goniopygus supremus	Goniopygus sp.	Scoliechinus axiologus	Orthopsis sp.	Metholectypus trechmanni	Nucleolitid nov.	Pygopistes rudistarum	Hemiaster? sp. or spp.	Cassiduloid sp.
A†									♦				
B										♦			
C	♦										♦		
D													
E	♦			♦	♦			♦			♦	♦	
F													
G													
H						?				?			
I													
J													
K													
L													
M													
N													
O													
P													
Q	♦				♦				♦		♦	♦	
R													
S		♦	♦	♦									
T													
U													
V										♦			
W													
X													
Y													
Z												♦	
AA													
Cuba	♦		?	♦									

*Revised after Donovan and Bowen, 1989, Figure 3.
†Localities A to AA refer to the inliers in Figure 1. ♦ = definitely present; ? = presence uncertain.

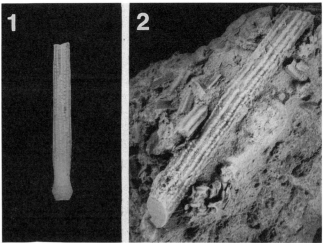

Figure 3. Cidaroid radioles. 1: *Temnocidaris*? (*Stereocidaris*) sp. B *in* Cutress, 1980, USNM 232519, ×2.1 2: *Phyllacanthus leoni* (Lambert and Sánchez Roig *in* Sánchez Roig, 1926), USNM 232518, ×2.25.

Family CIDARIDAE Gray
Subfamily CIDARINAE Gray
Tribe STEREOCIDARINI Mortensen

Genus *Temnocidaris* Cotteau
Subgenus *Stereocidaris* Pomel
Temnocidaris? (*Stereocidaris*) sp. B *in* Cutress, 1980
(Figure 3.1)

Description. A species of cidaroid known only from disarticulated radioles. The base is low and tapered, with a broad acetabulum, about one and a half times wider distally than proximally and about one third as high as the maximum width. The milled ring is low and slightly wider than the collar. The collar tapers gently, and the neck is slightly constricted. The shaft is long and cylindrical, with a sculpture of ridges formed from the coalescence of elongate spinules. Ridges are offset about halfway along the figured specimen (Fig. 3.1). The distal tip is not known.

Material. Known only from two radiole fragments, USNM 232519.

Occurrence. These radioles came from the *Inoceramus*-bearing shales of the St. Ann Inlier (Cutress, 1980, p. 19, 46; S in Fig. 1 herein). This probably represents the interval Upper Santonian to Lower Campanian (Chubb, 1955, p. 178; Kauffman, 1966, 1968) and belongs to either the Windsor or Cascade Formation (McFarlane, 1977).

Remarks. These Jamaican radioles compare poorly with the Cuban taxon described from numerous radioles and rare interambulacral plates by Cutress (1980, p. 44–47, Plate 1, Figs. 10–13). The Jamaican specimens probably represent a different species but cannot be identified further on the basis of the known specimens.

Tribe CIDARINI Gray

Genus *Phyllacanthus* Brandt
Phyllacanthus leoni (Lambert and Sáchez Roig
in Sánchez Roig, 1926)
(Figures 3.2, 4)

Description. (See also Cutress, 1980, p. 122–127, Plate 12, Figs. 5–11.) Test globular and high.

Apical system unknown, but smaller than the peristome and pentagonal in outline (Fig. 4.1).

Ambulacra are broad, weakly sinuous, and composed of numerous plates. The poriferous zone is sunken. Pore pairs are conjugate and broadly spaced. The interporiferous zone is narrower than the combined width of the adjacent poriferous zones. Each plate bears a primary tubercle and one or more secondary tubercles.

There are eight to nine plates per interambulacral column. Primary tubercles are large and perforate, either smooth or crenulate adapically and smooth adorally. Scrobicular tubercles are closely packed around the primary tubercles, elliptical in outline, and separated by ridges. Tertiary tubercles are closely packed over the remainder of the plate.

The peristome is broad and has a rounded pentagonal outline.

Radioles are broad, cylindrical and elongate. The periacetabulum is

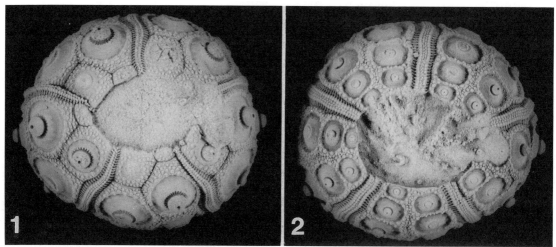

Figure 4. *Phyllacanthus leoni* (Lambert and Sánchez Roig *in* Sánchez Roig, 1926), USNM 232518, test. 1: apical view. 2: oral view. Both ×2.15. Test slightly crushed laterally.

crenulate or smooth beneath a gently tapering base that is wider than high. The milled ring is low and ribbed. The collar is as high as or higher than the base. The neck is low. The shaft is sculptured by numerous fine ridges formed from coalesced, rounded spinules (Fig. 3.2).

Material. Jamaican specimens of *P. leoni* include USNM 232517 (7 tests, 66 radioles), 232518 (1 test, 1 radiole; Figs. 3.2, 4 herein), BMNH E82395a-c (3 radioles), and E83296a-c (3 radioles).

Occurrence. This species is locally common in the Maastrichtian of Jamaica. It has only been collected as "complete" tests from the Jerusalem Mountain Inlier (C in Fig. 1) and Cuba (Cutress, 1980, p. 127). Tests from the Jerusalem Mountain Inlier occur with disarticulated test plates and radioles. Disarticulated radioles also occur in the Marchmont and Central inliers (E and Q in Fig. 1).

Remarks. Cutress (1980, p. 120–122) made this taxon the type species of *Prophyllacanthus* Cutress. However, Smith and Wright (1989, p. 12) considered this genus to be a junior synonym of *Phyllacanthus* Brandt. *P. leoni* is locally very common as disarticulated plates (particularly radioles) in the Maastrichtian of Jamaica. The only other species that is so frequently encountered is *Goniopygus supremus* Hawkins, whose radioles have smooth shafts and are thus easily distinguished. *P. leoni* is unusual in being found in both limestone and siliciclastic sequences.

Incertae familiae

Cidaroid indeterminate genus B *in* Cutress, 1980
(Figure 5)

Description. (See also Cutress, 1980, p. 141–142, Plate 14, Fig. 12.) A species of cidaroid only known from disarticulated radioles. Collar gently tapering and unsculptured. Shaft cylindrical and tapering distally. The shaft is sculptured by 11 longitudinal ribs with large rounded spinules in close association. The shaft is granular between ribs.

Material. Seven fragments of radioles, USNM 243089.

Occurrence. These specimens come from the St. Ann Inlier (S in Fig. 1) from Upper Campanian strata (Dr. N. F. Sohl, written communication).

Remarks. The spinulation of this taxon is much coarser than that of the two other species of Jamaican Cretaceous cidaroid.

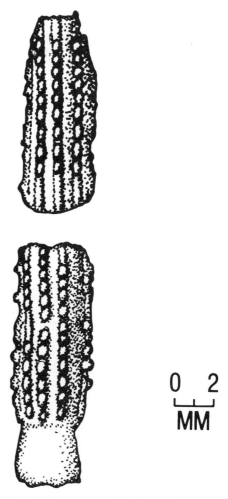

Figure 5. Cidaroid indet. gen. sp. B *in* Cutress, 1980, USNM 243089, radiole fragments. Drawn after Cutress (1980, Plate 14, Fig. 12).

Order CALYCINA Gregory
Family ACROSALENIIDAE Gregory

Genus *Heterosalenia* Cotteau
Heterosalenia occidentalis Hawkins, 1923
(Figure 6)

Description. (See also Hawkins, 1923, p. 206–213, Plate 9, Figs. 4–6; Donovan, 1990.) The test is circular in ambital outline, flattened adorally and adapically, with a strong lateral convexity (Fig. 6.3).

The periproct is eccentric, being displaced toward ocular I (Fig. 6.1, 6.2) and elongated perpendicular to I-3. The periproctal outline varies from being unevenly elliptical (Fig. 6.1) to triangular with convex sides (Fig. 6.2). The apical system is dicyclic with ocular I insert and has a central, hexagonal suranal plate. The five genital plates are large, hexagonal, and in contact, except where they are separated by the periproct adjacent to ocular I. Genital pores are present near the adoral margins of all genital plates. The madreporite is slightly larger than the other genital plates. The five ocular plates are either trapezoid (ocular I only), triangular, or kidney shaped in outline.

The ambulacra are narrow and sinuous adapically, becoming broader and straighter adorally (Fig. 6). The ambulacral plating is simple aborally, becoming compound bigeminate to trigeminate (acrosaleniid compounding) adorally. The pore pairs are depressed. The poriferous series expands to form narrow, multiserial, triangular phyllodes adjacent to the peristome. The tubercles of the interporiferous zone are small, imperforate, and noncrenulate. The interporiferous zone widens away from the apical system, becoming widest in a swollen area close to the peristome, from which the interporiferous zone narrows adorally.

The interambulacra are widest at the ambitus, narrowing adorally and adapically. The large primary tubercles are perforate and crenulate, with broad, conical bosses, and are surrounded by densely packed secondary and tertiary tubercles.

The peristome is large, with a rounded, pentagonal outline and triangular buccal notches.

The only fragment of radiole is circular in section, tapering, and smooth.

A bivariate analysis of this species is given in Donovan (1990, text-fig. 5).

Material. Holotype, BMNH E16643 (figured Hawkins, 1923, Plate 9, Figs. 4, 5); paratype, BMNH E16644 (figured Hawkins, 1923, Plate 9, Fig. 6). Five other specimens, BMNH E75846, E83303, and USNM 442324 to 442326 (figured Donovan, 1990, Plate 3).

Occurrence. This species is known only from the Maastrichtian (Fig. 2). The type locality is the *Titanosarcolites* Limestone of the Marchmont Inlier (E in Fig. 1) in the railway cutting between Cambridge and Catadupa, parish of St. James. With one exception, all of the known specimens come from this inlier and probably from this horizon (Donovan, 1990). The exception comes from near locality 4 of Robinson (1988), in the type area of the Guinea Corn Formation (Upper Campanian to Middle Maastrichtian; Jiang and Robinson, 1987) in the Central Inlier (Q in Fig. 1).

Remarks. *H. occidentalis* is the youngest *Heterosalenia* (Mortensen, 1935) and one of the last of the acrosaleniids, which are not known from the Cenozoic (Fell and Pawson, 1966, p. U375).

Order ARBACIOIDA Gregrory
Family ARBACIIDAE Gray

Genus *Goniopygus* L. Agassiz
Goniopygus supremus Hawkins, 1924
(Figure 7)

Description. (See also Hawkins, 1924, p. 313–316, Plate 18, Figs. 1, 2; Donovan, 1990.) The test is low and circular in outline, with a domed adapical surface. Adorally the test is flattened and has a broad peristome.

The apical system is large and dicyclic (Fig. 7.1). The five large, elongate heptagonal genital plates are widest close to the periproct. The size and shape of the madreporite are similar to those of the other genital plates. The genital pores are located just adorally to the genital plates. The five ocular plates are broad and kidney shaped in outline. The periproct is trilobate and elongate in the direction I-3 (Fig. 7.1), sometimes with a chamfered margin and the development of elongate processes in the angles.

The ambulacra are straight and widest at the ambitus, tapering slightly adorally and more adapically. The arbacioid compounding of the ambulacral plates is usually trigeminate but sometimes quadrigeminate (Donovan, 1990). The arrangement of pore pairs in a plate is straight or gently arcuate. Pore zones are polyporous adjacent to the peristome. Primary ambulacral tubercles are prominent, noncrenulate, and imperforate.

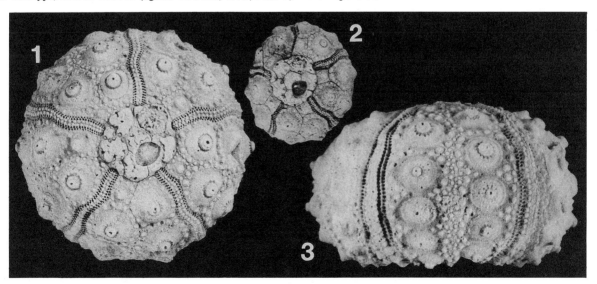

Figure 6. *Heterosalenia occidentalis* Hawkins, 1923 (after Donovan, 1990, Plate 3). 1: USNM 442325, apical view, ×2.2. 2: USNM 442324, apical surface of small test, ×2.0. 3: USNM 442326, lateral view, interambulacrum 3 central, ×2.3.

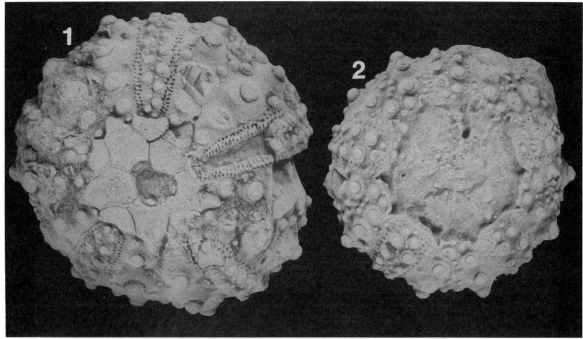

Figure 7. *Goniopygus supremus* Hawkins, 1924. 1: USNM 442319, apical view, ×2.8 (after Donovan, 1990, Plate 1, Fig. 1). 2: USNM 442322, oral view, ×2.6 (after Donovan, 1990, Plate 1, Fig. 2).

Primary tubercles of the interambulacral plates are large, noncrenulate, and imperforate. Secondary tubercles are absent adapically, becoming plentiful adorally (Fig. 7).

The peristome is broad, with a rounded pentagonal outline and well-developed, triangular buccal notches.

Radioles are smooth, circular in section and tapering distally, where the shaft becomes polygonal with a rounded tip.

A bivariate analysis of this species appears in Donovan (1990, text-fig. 2).

Material. Holotype, BMNH E17205 (figured Hawkins, 1924, Plate 18, Figs. 1, 2). Fourteen other tests from Jamaica, BMNH E75836–E75838, E75840–E75844; USNM 442318–442323 (442318, 442319, and 442322 figured Donovan, 1990, Plate 1, Figs. 1–3; see also Fig. 7 herein). One hundred ninety-six radioles attributed to *G. supremus* (Donovan, 1990), BMNH E83297a-c, E83298a-c, E83299a-c, E83300a-c, E83301, and E83302.

Occurrence. Most specimens of this species in Jamaica are Campanian? or Maastrichtian (Fig. 2). The holotype comes from the *Titanosarcolites* Limestone of the Marchmont Inlier (E in Fig. 1) in the Great River valley, near Catadupa, parish of St. James. Many of the other specimens come from the type section or type area. USNM 442318 is from the Guinea Corn Formation (Upper Campanian to Middle Maastrichtian; Jiang and Robinson, 1987) of the Central Inlier (Q in Fig. 1). BMNH E75836 and E75837 were collected from "1000 feet below *Barrettia* Limestone, St. Ann's Great River" (label), St. Ann Inlier (S in Fig. 1) of probable Campanian age or older (McFarlane, 1977; Donovan, 1990). This species is also known from Cuba (Weisbord, 1934, p. 16–18, Plate 1, Figs. 1–3; Sánchez Roig, 1949, p. 50, 51).

Remarks. The original description of this species by Hawkins (1924) was based on an aberrant specimen, as was recognized by Arnold and Clark (1927; but see below for a discussion of Arnold and Clark's specimens) and Weisbord (1934). This has been emphasized in a recent biometric analysis by Donovan (1990). As originally diagnosed, this species was typified by having quadrigeminate ambulacral plating, but other specimens are mainly or entirely trigeminate. In consequence, the subgenus *Goniopygus (Tetragoniopygus)*, type species *G. supremus*, diagnosed on the basis of quadrigeminate ambulacral plating (Fell and Pawson, 1966, p. U412), is invalid.

Radioles attributed to *G. supremus* are common at some horizons and are distinguished from *Phyllacanthus leoni* in being smooth (Donovan, 1990, Plate 2) rather than having a coarse spinulation.

Goniopygus sp.

Description. See Arnold and Clark (1927, p. 12, 13).
Material. MCZ 3384 (two tests) and 3383.
Occurrence. These specimens came from the west bank of the Rio Nuevo, parish of St. Mary, close to Pembroke Hall (Arnold and Clark, 1927, p. 13). This is part of the Benbow–Guys Hill Inlier (V in Fig. 1). The rocks of this inlier are Albian to Upper Cretaceous (McFarlane, 1977; E. Robinson, written communication).
Remarks. Although these three specimens were originally attributed to *Goniopygus supremus* by Arnold and Clark (1927), they are poorly preserved and highly waterworn. Until well-preserved specimens are found, it is considered judicious to classify them with caution.

Order TEMNOPLEUROIDA Mortensen
Family TOXOPNEUSTIDAE Troschel

Genus *Scholiechinus* Arnold and Clark
Scholiechinus axiologus Arnold and Clark, 1927
(Figure 8)

Description. (See also Arnold and Clark, 1927, p. 23, 24, Plate 2, Figs. 7, 8; Donovan, 1992.) The test is rounded pentagonal in outline and has slightly convex sides. The ambitus is low, at about 25% of the height of the test. The corona curves under the oral surface from the ambitus and also curves gently to the apex.

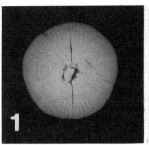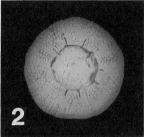

Figure 8. *Scoliechinus axiologus* Arnold and Clark, 1927, holotype, MCZ 3268. 1: apical view. 2: oral view. Both ×1.

The apical disc is not preserved, but it is much smaller than the peristome and polygonal in outline (Fig. 8.1).

The ambulacra are wide, about 26% of the test diameter at the ambitus, and straight. Pore pairs are arranged in triads (Donovan, 1992, Fig. 2C, D). Primary ambulacral tubercles are smooth, imperforate, and noncrenulate and about as large as interambulacral secondary tubercles. The poriferous zones also include secondary tubercles, which are slightly smaller than primaries, and small tertiary tubercles. Apically, smooth perradial areas are flanked by tubercles. Tubercles become abundant about 50% of the height of the test below the apex and are present up to the margin of the peristome. Ambulacra are widest at the ambitus, where the interporiferous zone is as wide as the combined widths of the poriferous zones. The interporiferous zone tapers to the peristome.

Each interambulacrum is naked interradially adapically to the ambitus, bordered by densely packed tubercles. From the ambitus adorally the tubercles are densely packed. Four sizes of tubercle are apparent, all smooth, imperforate, and noncrenulate. Primary tubercles are offset in double columns adradially. Large (primary and secondary) tubercles are most densely packed adorally.

The peristome is large, central, rounded pentagonal in outline and depressed. Elongate buccal notches are present.

Material. Holotype, MCZ 3268, plus two other specimens, MCZ 3498 and BMNH EE2802.

Occurrence. The type locality is in the "Leyden region, St. James Parish" (Arnold and Clark, 1927, p. 23). Previous authorities have suggested the age of this species to be Cretaceous (Donovan, 1988, 1992; Donovan and Bowen, 1989), questionably Cretaceous (Mortensen, 1943; Fell and Pawson, 1966), or Eocene (Kier and Lawson, 1978). Leyden is within the outcrop area of the Cretaceous rocks of the Sunderland Inlier (H in Fig. 1). Donovan (1992) concluded that *S. axiologus* probably comes from the Newmans Hall and Sunderland shales, of late Campanian age (Green, n.d.). However, the BMNH specimen (Donovan and Lewis, 1993) is from Spring Mount, parish of St. James, from the Eocene Chapelton Formation. The age of the type specimen is therefore now considered to be uncertain. Further collecting at Spring Mount and around Leyden will be necessary in order to determine the true stratigraphic range of this taxon. The locality of MCZ 3498 is unknown.

Order HOLECTYPOIDA Duncan
Family HOLECTYPIDAE Lambert

Genus *Metholectypus* Hawkins
Metholectypus trechmanni Hawkins, 1923
(Figure 9)

Description. (See also Hawkins, 1923, p. 201–205, Plate 9, Figs. 1–3.) The test is small, pentagonal in outline, high and globular, with a convex apex and only slightly flattened adorally.

The apical disc is small, with four genital pores (genital 5 is imperforate). The madreporite is approximately central.

The periproct is inframarginal and very close to the peristome. The periproct is elongated III-5, broad and longer than the peristome is wide.

Ambulacra are straight, narrow, and slightly inflated. Poriferous zones are narrow. Pore pairs are uniserial and isoporous. Ambulacra are composed of simple primary plates throughout.

Interambulacral plates are numerous, broad, and low. Tubercles of ambulacra and interambulacra are numerous and small, with an increase in size and density adorally.

The peristome is central, small, and an irregularly rounded pentagon in outline.

Material. Holotype, BMNH E16642.

Occurrence. The type locality is the Green Island Inlier (A in Fig. 1), where the holotype was collected from the *Barrettia* Limestone of late Campanian age (Krijnen, 1993).

Remarks. Rose and Olver (1985, p. 86, 87) discussed the affinities of this monotypic genus in detail.

Order CASSIDULOIDA Claus
Family NUCLEOLITIDAE L. Agassiz and Desor

Genus *Pygopistes* Pomel
Pygopistes? rudistarum (Hawkins, 1923).
(Figure 10)

Description. (See also Hawkins, 1923, p. 213, 214, Plate 9, Figs. 7–9; Donovan and Bowen, 1989.) The test is inflated, oval in outline, with a well-rounded anterior and a more angular posterior. The test is widest at about 75% of the distance from the anterior to the posterior.

The apical system is slightly anterior of center, with the four genital pores arrayed subtetragonally, although the plating is indistinct. The madreporite is apparently cruciform.

The periproct is vertically elongated and occurs on the posterior margin, where the test is angled adorally inward.

The ambulacral petals are moderately wide, open, and slightly raised above the surface of the test. All petals extend to near the ambitus. The interporiferous zone is wider than the combined width of the poriferous zones of the petal and is widest at the ambitus. Poriferous zones are narrow and taper adorally. Pores are conjugate within the petals. Inner pores are circular and outer pores slightly elongate. Near the ambitus, pores occur in shallow depressions and are circular in outline. Pores are smallest adorally. Phyllodes are double pored, with up to about 20 pore pairs in each column.

Interambulacra are broad, with a dense covering of primary and secondary tubercles. Tubercles show no variation in size over the entire test. Sutures between ambulacral and interambulacral plates are generally indistinct. Bourrelets are very weakly developed.

The peristome is slightly anterior of center, not depressed, and is an elongate, round-angled pentagon in outline. The peristome is oblique, with the long axis aligned III-5.

Material. Holotype, BMNH E16645. One other specimen, BMNH E83008.

Occurrence. The type section is the *Titanosarcolites* Limestone (Maastrichtian: McFarlane, 1977) in the Great River valley below Catadupa, on the border of the parishes of Westmoreland and St. James (Marchmont Inlier: E in Fig. 1). E83008 is from near the top of the type section of the Guinea Corn Formation (Robinson, 1988, locality 3) in the Central Inlier (Q in Fig. 1) and is also Maastrichtian (Donovan and Bowen, 1989).

Remarks. Hawkins originally placed this species in the genus *Botriopygus* d'Orbigny, a junior synonym of *Pygorhynchus* L. Agassiz (Kier, 1966, p. U506). However, a recent reassessment of *B. rudistarum* by Donovan and Bowen (1989) suggested that it is closest to *Pygopistes* Pomel, while possibly belonging to a new genus.

Figure 9. *Metholectypus trechmanni* Hawkins, 1923, holotype, BMNH E16642. 1: apical view. 2: lateral view (anterior to right). 3: oral view. All ×2.

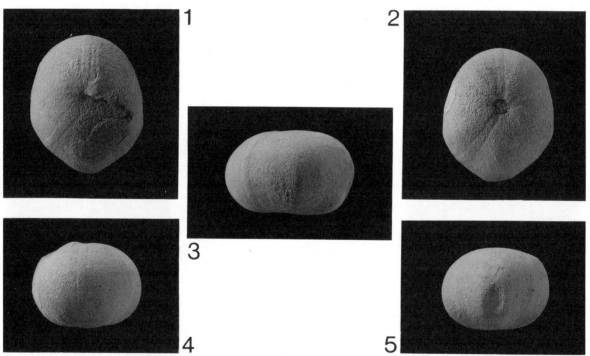

Figure 10. *Pygopistes? rudistarum* (Hawkins, 1923), holotype, BMNH E16645 (after Donovan and Bowen, 1989, Plate 1). 1: apical view. 2: oral view. 3: lateral view (periproct visible extreme right). 4: anterior view. 5: posterior view. All ×1.

Incertae familiae

Cassiduloid gen. et sp. indet.
(Figure 11)

Description. (See also Donovan, 1992.) Only the apical (Fig. 11) and part of the lateral surfaces are exposed. Ambulacra petaloid, not sunken. Petals closed, with the ambulacra expanding adorally of the petals. The petals have conjugate anisopores, occurring adjacent to the adoral suture of each plate. Pores adoral of the petals are isoporous. The interambulacral plates are arcuate and increase in height and length away from the apical system. Primary tubercles are closely spaced, depressed into the test, and perforate.

Material. A single specimen, BMNH E83009.
Occurrence. This test came from the Guinea Corn Formation (Upper Campanian to Middle Maastrichtian; Jiang and Robinson, 1987) of the Central Inlier (Q in Fig. 1).
Remarks. This is the largest irregular echinoid from the Jamaican Cretaceous. Unfortunately, the preservation is too poor to permit a more precise identification.

Figure 11. Cassiduloid gen. et sp. indet., BMNH E83009, apical view, ×1.

Order SPATANGOIDA Claus
Family HEMIASTEROIDAE H. L. Clark

Genus *Hemiaster* L. Agassiz
Hemiaster? sp. or spp.
(Figure 12)

Description. (See also Hawkins, 1923, p. 206; 1924, p. 316, Plate 18, Fig. 3.) The test is thin, moderately high, broad, and elongate (up to 40+ mm in length; Fig. 12), with a broadly elliptical ambital outline that is blunt posteriorly and has a marked anterior sulcus.

The apical disc varies in position from anterior of center to posterior of center.

The periproct is high on the posterior test, vertically elongate, with the vertical axis about twice the horizontal axis.

The four petals and the anterior sulcus are weakly to strongly depressed. The posterior petals are short, about one half to two thirds the length of the anterior petals. The anterior sulcus is broad. Ambulacral pores are elliptical in outline in the petaloid zone. Adorally the ambulacra are flush with the surface of the test. Pores are very small adambitally, increasing in size adorally, with single pores in standard positions around the peristome. Tubercles are tiny and situated along the adapical horizontal margin of each plate.

Adapically the interambulacra are composed of large plates, about twice as high as long, bearing small, regularly spaced tubercles. Adorally the plastron is mesamphisternous, with the labrum overhanging the periproct. Primary tubercles are slightly inset into the test adorally and adapically, with secondary tubercles scattered over the remaining surface. Fascioles have not been observed as a result of poor preservation.

The peristome is kidney shaped and overhung posteriorly by the labrum.

Material. Twelve tests in the collection of the BMNH, E16646–E16656 and E17206.

Occurrence. This is the most common irregular echinoid in the Jamaican

Figure 12. *Hemiaster*? sp., BMNH E17206. 1: apical view. 2: lateral view (anterior to left). 3: oral view. All ×2.

Cretaceous. Unlike other irregular echinoids from the island, it is found usually (but not invariably) within siliciclastic sequences, rather than in rudist limestones. The BMNH specimens come from the Blue Mountain Inlier and the Maastrichtian *Veniella* Shales of the Marchmont Inlier (Z and E, respectively, in Fig. 1). Specimens have also been collected from the Jerusalem Mountain and Central inliers (C and Q in Fig. 1). The spatangoid echinoids that Chubb (1958, p. 8) recognized from shales within the Stapleton Limestone Formation of the Sunderland Inlier (H in Fig. 1) were presumably *Hemiaster*. All of these occurrences are Maastrichtian, possibly extending back into the Upper Campanian.

Remarks. This taxon, or taxa, is almost invariably poorly preserved as a result of the thin test's being crushed and obscured by sediment. At least two species appear to be present, based on the relative position of the apical system, but this may be an artifact of preservation. Until the distribution of the fascioles can be discerned, it is suggested that these specimens are referred to *Hemiaster* with caution.

OTHER TAXA

Two further, but as yet undescribed, echinoid species are known from the Jamaican Cretaceous.

(1) About 30 specimens of a nucleolitid *sensu lato* have been collected from the oolitic facies of the Lower Campanian Clifton Limestone, Lucea Inlier (Schmidt, 1988, Fig. 9; B in Fig. 1 herein). No complete tests of this echinoid are known, and most specimens are badly waterworn, often appearing as sections through the corona. It differs from other cassiduloids in the Jamaican Cretaceous in having the periproct situated in a posterior groove on the aboral surface and in having well-developed bourrelets.

(2) Donovan and Lewis (1993) have recorded two tests, BMNH EE2785 and EE2786, that belong to the regular echinoid genus *Orthopsis*. EE2785 comes from the Marchmont Inlier, Catadupa district, from the Maastrichtian *Titanosarcolites* Limestone. EE2786 is from the Maastrichtian Guinea Corn Formation of the Central Inlier. A further test in the author's collection, from the Maastrichtian of the Marchmont Inlier, may be conspecific.

ACKNOWLEDGMENTS

I am most grateful to David N. Lewis and Andrew B. Smith (BMNH), Felicita d'Escrivan (MCZ), and Frederick J. Collier and Jan Thomas (USNM) for loaning and allowing me access to specimens in their care. Specimens in Figures 9–12 were photographed by the Photographic Unit of the BMNH. Other photographs were taken by the author during the periods of Smithsonian Short Term Visitor Grants in 1987 and 1989, using equipment kindly made available by Dr. David L. Pawson (Department of Invertebrate Zoology). This paper was improved following constructive reviews by David Lewis, Andrew Smith, Burchard Carter, and Ted Robinson. I particularly thank Dave Lewis for his help in describing *Hemiaster*.

REFERENCES CITED

Arnold, B. W., and Clark, H. L., 1927, Jamaican fossil echini: Memoirs of the Museum of Comparative Zoology, Harvard, v. 50, p. 1–75.

Chubb, L. J., 1955, The Cretaceous succession in Jamaica: Geological Magazine, v. 92, p. 177–195.

Chubb, L. J., 1958, The Cretaceous rocks of central St. James: Geonotes, v. 1, p. 3–11.

Chubb, L. J., 1961, Blue Mountain Shale: Geonotes, v. 4, p. 1–7.

Chubb, L. J., 1971, Rudists of Jamaica: Palaeontographica Americana, v. 7, p. 161–257.

Cutress, B. M., 1980, Cretaceous and Tertiary Cidaroida (Echinodermata: Echinoidea) of the Caribbean area: Bulletins of America Paleontology, v. 77, no. 309, 221 p.

Donovan, S. K., 1988, A preliminary biostratigraphy of the Jamaican fossil Echinoidea, *in* Burke, R. D., Mladenov, P. V., Lambert, P., and Parsley, R. L., eds., Echinoderm biology: Proceedings, Sixth International Echinoderm Conference, Victoria, British Columbia, 23–28 August 1987: Rotterdam, A. A. Balkema, p. 125–131.

Donovan, S. K., 1990, Jamaican Cretaceous Echinoidea. 2. *Goniopygus supremus* Hawkins, 1924, *Heterosalenia occidentalis* Hawkins, 1923 and a comment on *Trochalosoma chondra* (Arnold and Clark, 1927): Mesozoic Research, v. 2, p. 205–217.

Donovan, S. K., 1992, Jamaican Cretaceous Echinoidea (Echinodermata). 3. *Scoliechinus axiologus* Arnold and Clark, 1927, and an indeterminate cassiduloid: Proceedings of the Biological Society of Washington, v. 105, p. 23–31.

Donovan, S. K., and Bowen, J. F., 1989, Jamaican Cretaceous Echinoidea. 1. Introduction and reassessment of ?*Pygopistes rudistarum* (Hawkins, 1923) n. comb.: Mesozoic Research, v. 2, p. 57–65.

Donovan, S. K. and Lewis, D. N., 1993, The H. L. Hawkins collection of Caribbean fossil echinoids: An annotated catalog of rediscovered specimens from the University of Reading, England: Caribbean Journal of Science, v. 29 (in press).

Fell, H. B., and Pawson, D. L., 1966, Echinacea, *in* Moore, R. C., ed., Treatise on invertebrate paleontology, Part U, Echinodermata 3(2): Geological Society of America and University of Kansas Press, New York and Lawrence, p. U367–U440.

Green, G. W., n.d., Jamaica 1:50,000 Geological Sheet 5. Queen of Spains Valley: Kingston, Ministry of Mining and Natural Resources.

Harland, W. B., Cox, A. V., Llewellyn, P. G., Pickton, C.A.G., Smith, A. G., and Walters, R., 1982, A geologic time scale: Cambridge, Cambridge University Press, 131 p.

Hawkins, H. L., 1923, Some Cretaceous Echinoidea from Jamaica: Geological Magazine, v. 60, p. 199–216.

Hawkins, H. L., 1924, Notes on a new collection of fossil Echinoidea from Jamaica: Geological Magazine, v. 61, p. 312–324.

Hill, R. T., 1899, The geology and physical geography of Jamaica: Study of a type of Antillean development: Bulletin of the Museum of Comparative Zoology, Harvard, v. 34, 256 p.

Jackson, T. A., 1986, St. Peter's Inlier—fact or fiction: Journal of the Geological Society of Jamaica, v. 23 (for 1985), p. 44–49.

Jiang, M.-J., and Robinson, E., 1987, Calcareous nannofossils and larger foraminifera in Jamaican rocks of Cretaceous to early Eocene age, *in* Ahmad, R., ed., Proceedings, Workshop on the Status of Jamaican Geology, Kingston, 14–16 March 1984: Special Issue of the Journal of the Geological Society of Jamaica, p. 24–51.

Kauffman, E. G., 1966, Notes on Cretaceous Inoceramidae (Bivalvia) of Jamaica: Journal of the Geological Society of Jamaica, v. 8, p. 32–40.

Kauffman, E. G., 1968, The Upper Cretaceous *Inoceramus* of Puerto Rico, *in* Saunders, J. B., ed., Transactions of the Fourth Caribbean Geological Conference, Port-of-Spain, Trinidad, 28 March–12 April 1965, p. 203–218.

Kier, P. M., 1966, Cassiduloids, *in* Moore, R. C., ed., Treatise on invertebrate paleontology, Part U, Echinodermata 3(2): Geological Society of America and University of Kansas Press, New York and Lawrence, p. U492–U523.

Kier, D. M., and Lawson, M. H., 1978, Index of living and fossil echinoids

1924–1970: Smithsonian Contributions to Paleobiology, v. 34, 182 p.

Krijnen, J. P., 1993, A new genus and lineage of pseudorbitoid foraminifera in Jamaica: Journal of the Geological Society of Jamaica (in press).

McFarlane, N., 1977, Jamaica 1:250,000 Geological Sheet: Kingston, Ministry of Mining and Natural Resources.

Melville, R. V., and Durham, J. W., 1966, Skeletal morphology, *in* Moore, R. C., ed., Treatise on invertebrate paleontology, Part U, Echinodermata 3(1): Geological Society of America and University of Kansas Press, New York and Lawrence, p. U220–U251.

Moore, R. C., ed., 1966, Treatise on invertebrate paleontology, Part U, Echinodermata 3 (in 2 volumes): Geological Society of America and University of Kansas Press, New York and Lawrence, v. 1, p. xxx + U1–U366a; v. 2, p. U367–U695.

Mortensen, T., 1935, A monograph of the Echinoidea. II. Bothriocidaroida, Melonechinoida, Lepidocentroida and Stirodonta: Copenhagen, Reitzel, 647 p.

Mortensen, T., 1943, A monograph of the Echinoidea. III(2). Camarodonta I: Copenhagen, Reitzel, 533 p.

Porter, A.R.D., Jackson, T. A., and Robinson, E., 1982, Minerals and rocks of Jamaica: Kingston, Jamaica Publishing House, 174 p.

Robinson, E., 1988, Late Cretaceous and early Tertiary sedimentary rocks of the Central Inlier, Jamaica: Journal of the Geological Society of Jamaica, v. 24 (for 1987), p. 49–67.

Rose, E.P.F., and Olver, J.B.S., 1985, Slow evolution in the Holectypidae, a family of primitive irregular echinoids, *in* Keegan, B. F., and O'Connor, B.D.S., eds., Echinodermata: Proceedings, Fifth International Echinoderm Conference, Galway, 24–29 September 1984: Rotterdam, A. A. Balkema, p. 81–89.

Sánchez Roig, M., 1926, Contribucion a la paleontologi Cubana: Los equinodermos fósiles de Cuba: Boletin de Minas, v. 10, 179 p. [Not seen.]

Sánchez Roig, M., 1949, Los equinodermos fosiles de Cuba: Paleontologia Cubana, v. 1, p. 1–302.

Schmidt, W., 1988, Stratigraphy and depositional environment of the Lucea Inlier, western Jamaica: Journal of the Geological Society of Jamaica, v. 24 (for 1987), p. 15–35.

Smith, A. B., 1984, Echinoid palaeobiology: London, George Allen and Unwin, 191 p.

Smith, A. B., and Wright, C. W., 1989, British Cretaceous echinoids. Part 1, general introduction and Cidaroida: Monograph of the Palaeontographical Society, London, v. 141 (for 1987), p. 1–101.

Trechmann, C. T., 1929, Fossils from the Blue Mountains of Jamaica: Geological Magazine, v. 66, p. 481–491.

Weisbord, N. E., 1934, Some Cretaceous and Tertiary echinoids from Cuba: Bulletins of American Paleontology, v. 20, p. 165–270.

Manuscript Accepted by the Society November 11, 1992

Cretaceous and Cenozoic Brachiopoda of Jamaica

David A.T. Harper
Department of Geology, University College, Galway, Ireland

ABSTRACT

Thirteen species of fossil articulate brachiopod are documented from the Cretaceous and Cenozoic rocks of Jamaica. *Dyscritothyris* sp. cf. *D. cubensis* and *Terebratulina*? sp. are described from the upper Cretaceous; the Cenozoic fauna comprises the rhynchonellide *Probolarina*? sp., the thecideid *Lacazella* sp. cf. *L. caribbeanensis* and the terebratulides *Gryphus*? sp., *Tichosina* sp., *Argyrotheca* sp. cf. *A. barrettiana*, *A.* sp. cf. *A. magnicostata*, *A.* sp. 1 and *A.* sp. 2, *Terebratulina* sp. cf. *T. palmeri*, *Terebratulina* sp., and *Hercothyris* sp. cf. *H. semiradiata*.

INTRODUCTION

Fossil brachiopods are relatively sparse but widely dispersed throughout the Cretaceous and Cenozoic rocks of Jamaica. Although to date fewer than 60 specimens are known from the entire island, the material is currently assigned to 13 separate taxa, indicating a relatively high diversity and a considerable potential for further research. The limited previous work on the phylum has recently been reviewed by Harper and Donovan (1990), where the early contributions of de la Beche (1827), Sawkins and Brown (1869), Cockerell (1894), and Hill (1899) are noted. In this chapter 13 brachiopod species are described and illustrated. The fauna is similar to those described from the adjacent Dominican Republic (Logan, 1987) and elsewhere in the Caribbean as a whole (Cooper, 1979).

STRATIGRAPHIC DISTRIBUTION

The stratigraphic distribution of the phylum in Jamaica is illustrated in Figure 1. Three broad horizons have to date yielded fossil brachiopods: the upper Campanian–lower Maastrichtian, the lower-middle Eocene, and the Pliocene-Pleistocene. The group occupied a relatively wide range of environments, and some genera, for example, *Argyrotheca*, are represented by several species throughout parts of the succession.

The Guinea Corn Formation (upper Campanian–middle Maastrichtian) contains a rich fauna of rudist bivalves together with other molluscs, corals, echinoids, foraminifers, and ostracods. Brachiopods are extremely rare, represented by the small, pedunculate *Dyscritothyris*, which may have occupied a cryptic reef environment (Harper and Donovan, 1990). The broken and abraded *Terebratulina*? from the deeper water and near contemporary environments of the Providence Shale (lower Maastrichtian) are almost certainly allochthonous. Small, thin-shelled species of *Argyrotheca* are known from deeper-water facies within the lower Eocene of the island, whereas the middle Eocene Swanswick Formation has yielded species of *Hercothyris* and *Probolarina*? from abundantly fossiliferous bioclastic limestones in shallower-water facies.

The Manchioneal Formation (basal Pleistocene) has yielded, to date, the most diverse of the Jamaican brachiopod faunas. Species of *Gryphus*?, *Tichosina*, *Argyrotheca*, *Terebratulina*, and *Lacazella* have been documented from bioclastic limestone facies. Higher in the succession the reef facies of the Falmouth Formation has produced specimens of *Terebratulina*. This preliminary review has demonstrated that although brachiopods are rare members of the Jamaican fossil benthos, the fauna as a whole is relatively diverse. Sustained investigations will almost certainly increase this diversity. Brachiopods have thus a considerable potential in future environmental and stratigraphic studies of the Cretaceous and Cenozoic rocks of Jamaica and those elsewhere in the Caribbean.

SYSTEMATIC PALEONTOLOGY

Despite the recent efforts to expand the recorded fauna of fossil brachiopods from Jamaica (Harper and Donovan, 1990), sample sizes of the 13 known taxa remain small. Ontogenetic and size-independent intraspecific variation is thus presently difficult to define and assess. Moreover, irrespective of the small size of samples, little information is as yet available regarding the internal features of the majority of the described species; a number of generic assignments are thus provisional. Nevertheless the following descriptions, based largely on external features—for example, the outlines, profiles, and ornament of the valves—provide an initial data base for the Jamaican fauna. Where possible, species descriptions are supplemented by measurements and some statistical informa-

---

Harper, D.A.T., 1993, Cretaceous and Cenozoic Brachiopoda of Jamaica, *in* Wright, R. M., and Robinson, E., eds., Biostratigraphy of Jamaica: Boulder, Colorado, Geological Society of America Memoir 182.

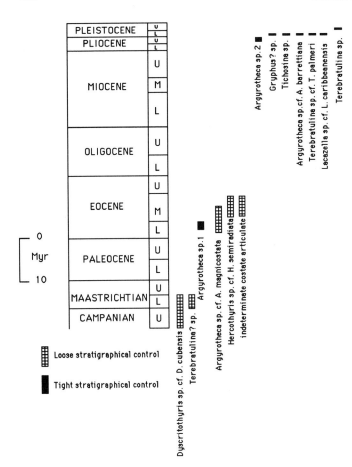

Figure 1. Stratigraphic distribution of all documented fossil Brachiopoda from Jamaica (modified after Harper and Donovan 1990, Fig. 2); absolute time scale from Harland et al. 1982.

tion; some interspecific comparisons have been effected by Principle Component Analysis (PCA). The following abbreviations have been used for measurements (in mm): slpv—sagittal length of pedicle valve, slbv—sagittal length of brachial valve, mwi—maximum width, hwi—hinge width, pmwi—position of maximum width, dpv—maximum depth of pedicle valve, and dbv—maximum depth of brachial valve. All the material is housed in the British Museum (Natural History), London.

Order RHYNCHONELLIDA Kuhn, 1949
Superfamily RHYNCHONELLACEA Gray, 1848
Family BASILIOLIDAE Cooper, 1959

Genus *Probolarina* Cooper, 1959

Type species. By original designation, *Rhynchonella holmesii* Dall, 1903, from the Castle Hayne Formation (Eocene), North Carolina, U.S.A.

Probolarina? sp.
Fig. 4.15, 4.16

1989 Ribbed brachiopod *in* Donovan et al., p. 6.
1990 Ribbed brachiopod nov.? *in* Harper and Donovan, p. 28.

Material, horizon, and locality. This single, poorly preserved conjoined pair is from the Swanswick Formation (late Middle Eocene), Beecher Town, parish of St. Ann (GR 510 532).
Description. Small, unequally biconvex valves of subcircular outline with maximum width anterior to midvalve length; hinge line, with obtuse, rounded cardinal extremities, about three-fifths maximum width. Anterior commissure uniplicate; apical angle approximately rectangular. Pedicle valve about as long as wide and about one-third as deep as long. Anterior profile with low, rounded median fold developing anteriorly; flanks weakly convex. Lateral profile evenly convex with subdued umbo. Pedicle foramen small, oval, and hypothyridid; deltidial plates present but incomplete.

Brachial valve about as long as wide and about one-third as deep as long. Anterior profile with shallow sulcus pronounced on anterior half of valve surface; flanks weakly convex, flattened laterally. Lateral profile flatly convex. At least six, rounded costae developed on anterior halves of both valves; elsewhere valve surface essentially smooth.

Measurements.

	slpv	slbv	mwi	hwi	pmwi	dpv	dbv
BC1003	6.0	3.8	6.3	4.0	4.0	2.0	2.0

Remarks. The paucity and poor preservation of the Swanswick material of this species does not permit a confident generic assignment. The shell, however, is similar in shape and style of ornament to *Probolarina transversa* Cooper, 1988 from the Eocene Santee Formation, South Carolina. That species is characterized by a transverse outline and relatively few costae (Cooper, 1988, p. 7). Despite a lack of detailed morphological information, the Swanswick species is quite different from all other taxa described to date from the Caribbean area. In the absence of more material and knowledge of the valve interiors, the valves are tentatively assigned to Cooper's genus.

Order SPIRIFERIDA Waagen, 1883

Remarks. The recent assignment of the thecideidine brachiopods to the spiriferides (Baker, 1990) is followed herein.

Suborder THECIDEIDINA Elliott, 1958
Superfamily THECIDEACEA Gray, 1840
Family THECIDEIDAE Gray, 1840
Subfamily LACAZELLINAE Backhaus, 1959

Genus *Lacazella* Munier-Chalmas, 1881

Type species. By original designation, *Thecidea mediterranea* Risso, 1826; a Recent species from the Mediterranean Sea.

Lacazella sp. cf. *L. caribbeanensis* Cooper, 1977
Fig. 2.17

cf. 1864 *Thecidium mediterraneum* (Risso); Davidson, p. 21, Plate 2, Fig. 5.
cf. 1887 *Thecidium mediterraneum* (Risso); Davidson, p. 158.
cf. 1971 *Lacazella mediterranea* (Risso); Meile and Pajaud, p. 470, Plate 1, Figs. 1–4.
cf. 1977 *Lacazella caribbeanensis* Cooper, p. 132, Plate 4, Figs. 12–19.
cf. 1979 *Lacazella caribbeanensis* Cooper; Logan, p. 75, Plate 10, Figs. 9–14.
cf. 1979 *Lacazella caribbeanensis* Cooper; Cooper, p. 28, Plate 1, Figs. 2–5.
cf. 1987 *Lacazella caribbeanensis* Cooper; Logan, p. 51, Plate 12, Figs. 42–49.
1990 *Lacazella* sp.; Harper and Donovan, p. 29, Fig. 3.

Material, horizon, and locality. This single brachial valve (B 21998) from the "newer Pliocene" was located in the Lucas Barrett collection in the British Museum (Natural History), London; precise details of the locality are not known.

Figure 2. Fossil *Argyrotheca* and *Lacazella* from Jamaica. 1, 2, *Argyrotheca* sp. cf. *A. magnicostata* Cooper, 1979, dorsal and ventral views of conjoined pair, BC1002, from the Yellow Limestone Group, Port Maria (GR 582 540), ×15. 3, 4, *Argyrotheca* sp. 2, ventral and dorsal views of conjoined pair, BC1014, from the upper Pliocene, Innes Bay (GR 8018 4008), both ×15. 6–8, 10, *Argyrotheca* sp. 1–6, 10: external and internal views of brachial valve, BC1015, both ×15; 7, 8: external and ventral views of brachial valve, BC1016, both ×15, both from the Lower Eocene, Rio Sambre. 5, 9, 11–16, *Argyrotheca* sp. cf. *A. barrettiana* (Davidson, 1866)—5: pedicle valve exterior, BC1015, from the Manchioneal Formation, Manchioneal, ×8, 9: brachial valve exterior, BC1016, from the Manchioneal Formation, Folly Point, ×8; 11, 12: ventral and dorsal views of conjoined pair, B41835, from the "newer Pliocene" of Jamaica, Lucas Barrett collection, both ×3; 13: dorsal view of conjoined pair, B41835, from the "newer Pliocene" of Jamaica, Lucas Barrett collection, ×3; 14, 15: dorsal and ventral views of conjoined pair, B21998, from the "Tertiary" of Jamaica, both ×3; 16: dorsal view of conjoined pair, B21998, from the "Tertiary" of Jamaica, ×8. 17, *Lacazella* sp. cf. *L. caribbeanensis* Cooper, 1977, brachial valve interior, B21998, from the "Tertiary" of Jamaica, Lucas Barrett collection, ×12.

Description. Small brachial valve of rounded, transversely quadrate outline, about five-sixths as long as wide, roughly flat, and with maximum width at about mid-valve length; anterior commissure apparently rectimarginate. Hinge line with obtuse, rounded cardinal extremities; hinge width about two-thirds maximum width. Ornament of variably accentuated concentric growth lines. Dorsal interior with square, robust cardinal process and relatively deep visceral cavity. Brachial lobes arcuate, each relatively narrow, anteriorly convergent, and separated by wide sinus, approximately one-third valve width; lophophore groove deep and narrow. Thin jugum flanked by pair of septa extending anteriorly to near one-half valve length. Elsewhere internal surface papillose, particularly anterolaterally. Exterior smooth with variably accentuated growth lines.

Measurements.

	slbv	mwi	hwi	pmwi	dbv
B21998	3.0	3.8	2.0	1.7	c.flat

Remarks. Cooper's species, *L. caribbeanensis,* was first described from living specimens in waters off Jamaica and the Dominican Republic

(Cooper, 1977). Subsequent fossil records from rocks ranging in age from Eocene to Miocene were noted by Cooper (1979) from various localities on Cuba; moreover, he figured the species from the Miocene of Matanzas Province, Cuba. More recently Logan (1987) has documented the species from the Gurabo Formation (Miocene-Pliocene), Rio Gurabo section, in the northern Dominican Republic. Logan (1987) discussed the three described species of *Lacazella* and noted that *L. caribbeanensis* was characterized by its smaller size, a hemispondylium with two separate plates, together with smoother margins to the "ascending apparatus" and intervening median ridge in the brachial valve; furthermore, the internal markings in the Caribbean species are more subdued. The sparse Jamaican material is, on the basis of its dorsal features, accordingly compared with *L. caribbeanensis*.

Order TEREBRATULIDA Waagen, 1883
Suborder TEREBRATULIDINA Waagen, 1883
Superfamily TEREBRATULACEA Gray, 1840
Family TEREBRATULIDAE Gray, 1840
Subfamily TEREBRATULINAE Gray, 1840

Genus *Gryphus* Mergerle von Muhlfeldt, 1811

Type species. By original designation, *Anomia vitrea* Born, 1778; a Recent species from the Mediterranean.

Gryphus? sp.
Fig. 3.1, 3.2, 3.16, 3.17

1990 *Gryphus* sp.; Harper and Donovan, p. 26.

Material, horizon, and locality. A conjoined pair from the basal Pleistocene Manchioneal Formation at the type section (Robinson, 1967; 1969, pp. 12–14, Fig. 4, GR 7995 4125), Manchioneal, parish of Portland. Several conjoined valves from the Barrett collection from "the Tertiary and newer Pliocene" deposited in the British Museum (Natural History), London, are tentatively assigned here; precise localities of these are not known.

Description. Moderately large, subequally biconvex valves of elongately oval outline. Maximum width about twice hinge width, occurring at mid-valve length; cardinal extremities obtuse and rounded. Anterior commissure essentially rectimarginate; lateral commissures straight. Pedicle valve about five-sixths as wide as long and about one-fourth as deep as long; anterior and lateral profiles evenly curved with maximum convexity medianly; umbo small and subdued. Relatively small, elongately oval pedicle foramen, apparently mesothyridid. Brachial valve about nine-tenths as wide as long and about one-fourth as deep as long; anterior and lateral profiles evenly curved, slightly less convex than those of pedicle valve, with small subdued umbo. External ornament smooth but with some variably accentuated growth lines.

Measurements.

	slpv	slbv	mwi	hwi	pmwi	dpv	dbv
BC1008	22.0	19.8	17.7	11.0	12.7	7.0	4.0

Remarks. The paucity of available material and lack of information regarding the internal features of this species obviate a definitive generic assignment and a formal comparison with possible congeneric forms. Nevertheless, the outline and profiles of the conjoined shells are more similar to those of the Recent species *G. vitreus* (Born) than to those of the fossil species of the genus previously described from the Caribbean (Cooper, 1979); moreover, the near planar anterior commissure separates it from the coeval, uniplicate *Tichosina*. The new Jamaican material is possibly the first post-Oligocene record of the genus in the Caribbean area (Cooper, 1977).

Genus *Tichosina* Cooper, 1977

Type species. By original designation, *Terebratula floridensis* Cooper; a Recent species from the Caribbean and the Gulf of Mexico.

Tichosina sp.
Figs. 3.3-15, 3.18, 3.19

1930 *Terebratula* cf. *cubensis* Pourtales; Trechmann, p. 214, Plate 12, Fig. 1.
1930 *Terebratula* cf. *bartletti* Dall; Trechmann, p. 214.
1930 *Terebratula* sp.; Trechmann, p. 214, Plate 12, Fig. 2.
1990 *Tichosina* sp. Harper and Donovan, p. 22, Fig. 3.

Material, horizon, and localities. Four conjoined pairs from the type section of the Manchioneal Formation (basal Pleistocene); two conjoined pairs from the Navy Island Member of the Manchioneal Formation (basal Pleistocene), San San Bay (see Robinson 1967 and 1969 for further details of stratigraphy and correlation). Additionally eight specimens were located in the Lucas Barrett collection in the British Museum (Natural History), London, from the "Tertiary and newer Pliocene" of Jamaica.

Description. Large ventribiconvex valves of elongately oval to teardrop shape outline. Maximum width over twice hinge width, occurring at about two-thirds valve length; cardinal extremities obtuse and rounded. Anterior commissure uniplicate; lateral commissures curved. Pedicle valve almost nine-tenths as wide as long and about one-third as deep as long. Anterior profile uniformly convex; lateral profile with maximum curvature at umbo; elsewhere surface evenly curved except over anterior third, where valve slopes steeply toward anterior commissure. Pedicle foramen oval, permesothyridid. Brachial valve almost as wide as long and about one-fifth as deep as long. Anterior profile flatly convex with shallow, broad sulcus developing anteriorly from near mid-valve length; lateral profile convex over posterior half of valve; elsewhere surface slopes anteriorly. Ornament of variably accentuated, but subdued, concentric growth lines.

Measurements.

	slpv	slbv	mwi	pmwi	dpv	dbv
B21999	13.0	12.0	11.1	7.5	4.0	2.5
B21997	23.1	20.4	20.0	17.7	9.0	5.3
B41851	25.0	21.4	21.5	14.2	8.8	5.6

Remarks. The Jamaican material is most similar to *T?* *lecta* (Guppy), first described from the Eocene of Trinidad (Guppy, 1866) and subsequently redescribed from the type area by Cooper (1979) together with the Miocene and Pliocene of the Dominican Republic (Logan, 1987). A pooled sample of all three forms was investigated by PCA (data from Cooper, 1979; Logan, 1987) for the following variates: slpv, slbv, mwi, and thickness (= dpv + dbv). Differentiation was achieved with respect to scores on the third eigenvector. The type specimens from Trinidad have higher scores (0.20, 0.42) on this eigenvector than the specimens from Jamaica and the Dominican Republic. The direction cosines of the third eigenvector (−0.69, −0.20, 0.60, 0.36) indicate the type *T?* *lecta* to have relatively shorter, wider, and more globose shells than those of the other two forms. Larger samples of all three are required, however, to confirm these apparent differences.

Figure 3. Fossil *Gryphus*?, *Tichosina,* and *Dyscritothyris* from Jamaica. 1, 2, 16, 17, *Gryphus*? sp—1, 2: dorsal and ventral views of conjoined pair, BC1008, from the Manchioneal Formation, both ×2; 16, 17: ventral and dorsal views of conjoined pair, B21999, from the "Tertiary" of Jamaica, Lucas Barrett collection, both ×2. 3–15, 18, 19, *Tichosina* sp.—3: dorsal view of conjoined pair, B41834, from the "newer Pliocene" Jamaica, Lucas Barrett collection, ×3; 4, 5, 6, 7, 12: dorsal, ventral, anterior, posterior, and lateral views of conjoined pair, B21997, from the "Tertiary" of Jamaica, Lucas Barrett collection, all ×1; 8, 9, 14, 15: ventral, dorsal, posterior, and lateral views of conjoined pair, B41851, from the "newer Pliocene" of Jamaica, Lucas Barrett collection, all ×1; 10: dorsal view of conjoined pair, B8656, ×1; 11: dorsal view of conjoined pair, B8659, ×1; 13: dorsal view of conjoined pair, B41851, from the "newer Pliocene" of Jamaica, Lucas Barrett collection, ×1; 18, 19: dorsal and lateral views of conjoined pair, BB8658, both ×1 (figured Trechmann, 1930, Plate 12, Fig. 1). 20–23, *Dyscritothyris* sp. cf. *D. cubensis* Cooper, 1979, anterior, posterior, dorsal, and ventral views of conjoined pair, BC1000, from the Guinea Corn Formation, all ×8.

Genus *Dyscritothyris* Cooper, 1979

Type species. By original designation, *Dyscritothyris cubensis* Cooper, 1979; from the Upper Cretaceous of Cuba.

Dyscritothyris sp. cf. *D. cubensis* Cooper
Fig. 3.20-23.

cf. 1979 *Dyscritothyris cubensis* Cooper, p. 16, Fig. 1, Plate 7, Figs. 3-8.
1990 *Dyscritothyris* sp. cf. *D. cubensis* Cooper; Harper and Donovan, p. 25.

Material, horizon, and locality. Three conjoined pairs from the type section of the Guinea Corn Formation (Upper Campanian–mid-Maastrichtian), Rio Minho west of Guinea Corn (see Jiang and Robinson, 1987), on the B4 Frankfield to Spaldings road, parish of Clarendon (GR 419 455); the exposure is on the opposite side of the river from the road.

Description. Small, ventribiconvex valves of rounded subpentagonal outline; anterior commissure rectimarginate to slightly uniplicate. Maximum width at or near mid-valve length; hinge width about three-fifths maximum width. Pedicle valve about nine-tenths as wide as long and about one-fourth as deep as long. Anterior profile curved medianly with convex, laterally sloping flanks; lateral profile with subdued umbo and maximum convexity at mid-valve length. Pedicle foramen small, oval, and submesothyridid. Brachial valve about nine-tenths as long as wide and about one-fifth as deep as long. Anterior profile flat with slight concavity medianly; flanks convex and slope laterally. Lateral profile evenly convex with subdued umbo. Ornament of fine concentric growth lines.

Measurements.

	slpv	slbv	mwi	hwi	pmwi	dpv	dbv
BC1000	5.2	4.5	4.8	3.5	3.0	1.4	1.0
BC1019	6.7	5.9	5.8	3.0	4.0	1.8	1.5

Remarks. Although detailed information on the internal structures of the Guinea Corn species is lacking, the valve outline, profiles, and development of the pedicle foramen are in close agreement with those of the type and only known species of *Dyscritothyris* from the Upper Cretaceous of Cuba. The small sample from Jamaica (N = 3) was compared by PCA with that from the Upper Cretaceous of Cuba (N = 3; data from Cooper, 1979, p. 15), based on four variates: slpv, slbv, mwi, and thickness (dpv + dbv). No discrimination was possible on the 2nd, 3rd, and 4th (size-independent) eigenvectors; both are thus similar in overall shape, but the larger Cuban specimens have higher scores on the first eigenvector.

Superfamily TEREBRATELLACEA King, 1850
Family MEGATHYRIDIDAE Dall, 1870

Genus *Argyrotheca* Dall, 1900

Type species. By original designation, *Terebratula cuneata* Risso, 1826; a Recent species from the Mediterranean Sea.

Argyrotheca sp. cf. *A. magnicostata* Cooper, 1979
Fig. 2.1, 2.2.

cf. 1979 *Argyrotheca magnicosta* [sic] Cooper, p. 22, Plate 6, Figs. 26-30.
1990 *Argyrotheca* sp. cf. *A. magnicostata* Cooper; Harper and Donovan, p. 26.

Material, horizon, and locality. A conjoined pair from the Yellow Limestone Group (upper Lower to lower Mid Eocene), 1.5 km NW of Port Maria, parish of St. Mary (GR 582 540).

Description. Minute, ventribiconvex valves of elongately oval outline with maximum width at or near hinge line; cardinal extremities slightly obtuse and rounded. Pedicle valve about as wide as long and about one-half as deep as long. Anterior profile with maximum convexity medianly; here and elsewhere curvature modified, locally, by costae; lateral profile uniformly convex with prominent umbo. Ventral interarea slightly curved, apsacline, and about one-fifth valve length; apparently large and open delthyrium partly obscured. Brachial valve about four-fifths as long as wide; anterior and lateral profiles weakly convex. Ornament of broad costae, subdued over posterior third of valve surface, with rounded, semicircular profiles and relatively deep interspaces; two strong median ribs developed on both pedicle and brachial valve exteriors, thickening anteriorly and each flanked by at least two less-prominent ribs. Valve surface coarsely pitted.

Measurements.

	slpv	slbv	mwi	hwi	pmwi	dpv	dbv
BC1002	1.9	1.4	1.7	1.5	1.0	0.5	flat

Remarks. This small specimen is compared with *A. magnicostata* Cooper, 1979 from the Eocene of Cuba. Although Cooper, in his description of the species (1979, p. 22), cites it as "*Argyrotheca magnicosta*," the new species is first introduced as "*Argyrotheca magnicostata*" (1979, p. iv); the latter citation is accepted as correct. The Jamaican specimen is small and has correspondingly few ribs; nonetheless, the ribs are relatively broad and the interspaces narrow, similar to those of the Cuban species.

Argyrotheca sp. 1
Fig. 2.6-8, 2.10

Material, horizon, and locality. Three brachial valves from the upper Lower Eocene (*Globorotalia pentacamerata* Zone), Rio Sambre.

Description. Exterior. Minute, weakly convex valves of transverse, subquadrate outline. Hinge line about four-fifths maximum width, itself located at about one-half valve length; cardinal extremities obtuse, rounded. Anterior commissure essentially rectimarginate. Valve about three-fourths as long as wide. Anterior profile feebly convex, modified medianly by faint, narrow sulcus, anteriorly persistent; lateral profile with maximum curvature posteriorly; valve surface flattens anteriorly. Ornament of variably accentuated concentric lines delimiting well-defined growth stages. Elsewhere valve surface coarsely pitted; faint, incipient costae sporadically developed.

Dorsal interior. Relatively thick, short socket ridges, concave posterolaterally and together with hinge line define deep, conical sockets. Socket ridges flank low notothyrial platform with poorly defined, oval cardinal process situated centrally; strong median septum, arising anterior to notothyrial platform, differentially thickened with high crest at mid-valve length. Large elongately oval muscle scars feebly impressed, laterally adjacent to median septum. Lateral partitions incipient and subdued.

Measurements and statistics.

	slbv	mwi	hwi	pmwi	dbv
BC1015	1.9	2.3	2.0	1.0	0.3
BC1016	1.6	2.1	1.8	0.5	0.3

Remarks. These minute and distinctive shells are most similar to those of *A. anomala* Cooper, 1979 from the middle Oligocene of Cuba. That species possesses obsolescent costae and adventitious lateral partitions within the brachial valve. The available material is considered inadequate for a specific assignment.

Argyrotheca sp. 2
Fig. 2.3, 2.4

1990 *Argyrotheca* sp.; Harper and Donovan, p. 21.

Material, horizon, and locality. A conjoined pair from the Innes Bay section (upper Pliocene), on the A4 E coast road, about 3 km S of Manchioneal and 1.5 km N of Happy Grove (see Robinson and Lamb, 1970), parish of Portland (GR 8018 4008).

Description. Minute, ventribiconvex valves of subpentagonal outline with maximum width at mid-valve length; hinge line about four-fifths maximum width with obtuse, rounded cardinal extremities. Pedicle valve over four-fifths as long as wide and about one-half as deep as long. Anterior profile with maximum convexity medianly, modified by profiles of costae; lateral profile with convex umbo; elsewhere valve surface slopes anteriorly. Ventral interarea slightly curved, apsacline, and about two-fifths as long as valve; delthyrium large and open. Brachial valve about four-fifths as long as wide with roughly flat anterior and lateral profiles interrupted by costae. Ornament of broad costae with four on each valve expanding anteriorly and separated by relatively narrow, deep interspaces. Valve surface coarsely pitted.

Measurements.

	slpv	slbv	mwi	hwi	pmwi	dpv	dbv
BC1014	1.4	1.1	1.6	1.0	1.2	0.3	0.3

Remarks. This minute species is characterized by four relatively broad costae on each valve, separated by narrow interspaces. It is thus similar to *A. magnicostata* Cooper, 1979 and *A.* sp. cf. *A. magnicostata*, herein, but differs in having a more teardrop shaped outline and a longer ventral interarea.

Argyrotheca sp. cf. *A. barrettiana* (Davidson, 1866)
Fig. 2.5, 2.9, 2.11–16.

1866 *Argiope barrettiana* Davidson, p. 103, Plate 12, Fig. 3.
1866 *Argiope antillarum* Crosse and Fischer, p. 270, Plate 8, Fig. 7.
1887 *Cistella barrettiana* Davidson, p. 145, Plate 22, Figs. 1, 2.
1920 *Argyrotheca barrettiana* (Davidson); Dall, p. 329.
1930 *Terebratella* (?) sp.; Trechmann, p. 214, Plate 12, Fig. 3.
1977 *Argyrotheca barrettiana* (Davidson); Cooper, p. 107, Plate 22, Figs. 9–21, Plate 23, Figs. 6, 7, Plate 32, Figs. 22–32.
1990 *Argyrotheca* sp.; Harper and Donovan, p. 21–22.

Material, horizon, and locality. A conjoined pair from the Navy Island Member (basal Pleistocene), Manchioneal Formation (see Robinson, 1969), east side of Folly Point, Port Antonio, parish of Portland (GR 743 467), the rest from Manchioneal.

Description. Medium-sized, ventribiconvex valves of transverse to semicircular outline with maximum width either at or near hinge line or mid-valve length; cardinal extremities rounded, acute, or perpendicular. Pedicle valve about four-fifths as long as wide and about two-fifths as deep as long. Anterior profile with maximum convexity medianly where valve surface slightly carinate; flanks flat or weakly concave, particularly near lateral margin. Lateral profile uniformly convex; umbo subdued. Ventral interarea flat to slightly curved, apsacline, and about one-third valve length. Delthyrium large and open.

Brachial valve over four-fifths as long as wide and about one-fourth as deep as long. Anterior profile with narrow sulcus originating near umbo; flanks slightly convex adjacent to sulcus but flatten laterally. Dorsal interarea short and anacline. Ornament of strong costae and more rarely costellae; median sulcus with pair of costellae arising by internal branching within 5-mm growth stage. Eleven to 17 ribs present on 2,1,1,1,1,1, and 1 valves; median costae of average thickness 0.8 mm at 5-mm growth stage. Accentuated growth lamellae numbering about 3 to 6 per mm at 5-mm growth stage more marked on larger valves. Valve surfaces densely punctate.

*Measurements and statistics.**

	slpv	slbv	mwi	hwi	pmwi	dpv	dpv
Means							
	6.20	5.36	7.56	6.43	3.29	2.40	1.36
Variance-covariance matrix							
	3.40	3.48	3.94	3.32	1.88	0.81	1.16
		3.11	3.99	3.13	1.62	0.88	1.08
			5.39	5.00	1.81	1.47	1.04
				5.55	1.06	1.49	0.31
					1.72	0.71	0.72
						0.63	0.17
							0.75

**Conjoined pedicle and brachial valves (N = 9).*

Remarks. The outline and profile of this species together with the style of radial ornament are most similar to those of *A. barrettiana* (Davidson), a Recent species, described from the Caribbean.

Superfamily CANCELLOTHYRIDACEA Thomson, 1926
Family CANCELLOTHYRIDIDAE Thomson, 1926
Subfamily CANCELLOTHYRIDINAE Thomson, 1926

Genus *Terebratulina* d'Orbigny, 1847

Type species. By original designation, *Anomia retusa* Linné, 1767; a Recent species from Norwegian waters.

Terebratulina? sp.
Fig. 4.18, 4.19

1927 *Terebratulina* (?) sp.; Trechmann, p. 63, Plate 3, Fig. 15.
1990 *Terebratulina* sp.; Harper and Donovan, p. 26.

Material, horizon, and locality. Trechmann (1927) described and figured a probable member of the genus from the Providence Shales, at Providence, near Port Antonio, parish of Portland. Chubb (1963) considered this unit to be of Maastrichtian age. A few further specimens from Providence, broken and poorly preserved (Donovan and Harper collection), are also tentatively assigned to *Terebratulina*, probably from near the same locality (GR 743 455), probably within the upper part of Balkissoon's (1989) Cross Pass Formation.

Remarks. The conjoined pair, described and illustrated by Trechmann (1927), is relatively large, unequally biconvex, and of elongate oval outline; moreover, the shells are partly exfoliated. The anterior and lateral profiles of the brachial valve are both uniformly convex, whereas the anterior profile of the pedicle valve is modified by a faint sulcus; the lateral profile is evenly convex. The radial ornament is differentiated; about five ribs are present per 2 mm, medianly, at the 10-mm growth stage. Although some new material with a similar ornament was collected by Harper and Donovan (1990) from the Providence Shales, the available information remains inadequate to confirm the generic assignment of this species.

Measurements.

	slpv	slbv	hwi	mwi	pmwi	dpv	dpv
BB8660	27.8	23.9	13.8	21.3	21.9	10.0	5.5

Figure 4. Fossil *Terebratulina, Hercothyris,* and *Probolarina*? from Jamaica. 1–10, 17, *Terebratulina* sp. cf. *T. palmeri* Cooper, 1979—1–4: ventral, dorsal, anterior, and posterior views of conjoined pair, B22000, from the "newer Pliocene" of Jamaica, Lucas Barrett collection, all ×3; 5, 6: dorsal and ventral views of conjoined pair, B41834, from the "newer Pliocene" of Jamaica, Lucas Barrett collection, both ×3; 7, 8: ventral and dorsal views of conjoined pair, B22000, from the "newer Pliocene" of Jamaica, Lucas Barrett collection, both ×3; 9, 10: dorsal and ventral views of conjoined pair, B22000, from the "newer Pliocene" of Jamaica, Lucas Barrett collection, both ×3; 17: dorsal view of conjoined pair, B22000, ×3. 11, 12, 13, *Hercothyris* sp. cf. *H. semiradiata* Cooper, 1979, dorsal, ventral, and anterior views of conjoined pair, BC1004, from the Swanswick Formation, Beecher Town, ×2, ×2, and ×4, respectively. 14, *Terebratulina* sp., pedicle valve exterior, BC1009, from the Falmouth Formation, ×8. 15, 16, *Probolarina*? sp., ventral and dorsal views of conjoined pair, BC1003, from the Swanswick Formation, Beecher Town, both ×4. 18, 19, *Terebratulina*? sp., dorsal and ventral views of conjoined pair, BB8660, ×1, from the Providence Shales, Providence (figured Trechmann, 1927, Plate 3, Fig. 15).

Terebratulina sp. cf. *T. palmeri* Cooper, 1979
Fig. 4.1–10, 4.17

1930 *Terebratulina caput-serpentis* Linn.; Trechmann, p. 214, Plate 12, Fig. 4.
1979 *Terebratulina? palmeri* Cooper, p. 6, Plate 1, Figs. 6–23, Plate 7, Figs. 9–20.
1990 *Terebratulina* sp.; Harper and Donovan, p. 21–22.

Material, horizon, and locality. About 10 valves from the Manchioneal Formation at the type section (Robinson, 1967; 1969, p. 12–14, Fig. 4. GR7995 4125), Manchioneal, parish of Portland, and from the Navy Island Member of the Manchioneal Formation, east side of Folly Point, Port Antonio, parish of Portland (GR 743 467); eight conjoined valves are present in the Lucas Barrett collection in the British Museum (Natural History), London, from the "newer Pliocene" of Jamaica.

Description. Small, dorsibiconvex valves of elongately oval to subtrigonal outline. Maximum width about twice hinge width, occurring at or near mid-valve length; cardinal extremities obtuse and rounded. Anterior commissure uniplicate. Pedicle valve about nine-tenths as wide as long and about one-fifth as deep as long. Anterior profile with narrow, relatively deep median sulcus originating near prominent umbo and developing anteriorly; flanks convex medianly with steep lateral slopes. Lateral profile convex with maximum curvature at mid-valve length. Large suboval pedicle foramen perforates apsacline interarea. Brachial valve about as long as wide and about one-fourth as deep as long. Anterior profile subcarinate with prominent median fold and sloping, slightly concave flanks; lateral profile uniformly convex with marked umbo. Ornament of variably accentuated costae and costellae numbering 5 to 8 per 2 mm at 5-mm growth stage on 1,3,2, and 1 valves.

*Measurements and statistics.**

slpv	slbv	mwi	hwi	pmwi	dpv	dbv
Means						
9.47	8.24	8.30	3.53	5.54	1.76	2.47
Variance-covariance matrix						
7.64	7.17	6.89	2.08	3.95	1.98	2.67
	6.76	6.51	1.99	3.72	1.85	2.48
		6.28	1.97	3.58	1.77	2.34
			0.92	0.95	0.52	0.68
				2.22	0.96	1.26
					0.57	0.73
						1.20

*Conjoined pedicle and brachial valves (N = 7).

Remarks. The distinctive outline, development of the dorsal fold, and ventral sulcus together with the style and density of the radial ornament suggest a close comparison with *T? palmeri* Cooper from the Miocene of Cuba. The new material obtained from the Manchioneal Formation during this study was compared by PCA ordination with specimens of *Terebratulina* from the "new Pliocene" within the Lucas Barrett collection; no separation was achieved along any of the eigenvectors, and thus the two samples are considered conspecific.

Terebratulina sp.
Fig. 4.14

1990 *Terebratulina* sp.; Harper and Donovan, p. 22.

Material, horizon, and locality. Four valves from a Pleistocene raised reef that is possibly part of the Falmouth Formation (last interglacial); the road cutting on the north coast highway A1, east side of Round Hill, Hanover (GR 204 564).

Remarks. The Pleistocene *Terebratulina*, although similar to *T.* sp. cf. *T. palmeri*, differs in having a less well developed fold and sulcus and apparently a finer, less-accentuated radial ornament. Larger samples of the Round Hill species are required, however, to confirm these apparent differences.

Measurements.

	slpv	mwi	pmwi	dpv
BC1009	5.8	4.0	3.1	flat

Superfamily DALLINACEA Beecher, 1893
Family HERCOTHYRIDIDAE Cooper, 1979

Genus *Hercothyris* Cooper, 1979

Type species. By original designation, *Hercothyris borroi* Cooper, 1979; from the Eocene of Cuba.

Hercothyris sp. cf. *H. semiradiata* Cooper, 1979
Fig. 4, 11–13

cf. 1979 *Hercothyris semiradiata* Cooper, p. 27, Plate 5, Figs. 24–29.
1990 *Hercothyris* sp. cf. *H. semiradiata* Cooper, 1979; Harper and Donovan, p. 21.

Material, horizon, and locality. One slightly damaged, conjoined pair from from the Swanswick Formation (middle Eocene), Beecher Town, St. Ann (Donovan et al., 1989).

Description. Medium-sized, ventribiconvex valves of elongately oval outline. Hinge width about three-fourths maximum width, which occurs at about mid-valve length; cardinal extremities rounded and obtuse. Anterior commissure uniplicate. Pedicle valve about nine-tenths as wide as long and about one-third as deep as long. Anterior profile with narrow, pronounced median sulcus, developing at 5-mm growth stage, and concave flanks that flatten laterally. Lateral profile convex with subdued umbo. Large, subcircular pedicle foramen, submesothyridid. Brachial valve as long as wide and about one-fourth as deep as long. Anterior profile carinate with narrow fold arising at 5-mm growth stage; flanks concave. Lateral profile uniformly convex with subdued umbo. Feeble costae developed on early growth stages, particularly posterolaterally, where abraded. Elsewhere smooth valve surface interrupted by faint concentric growth lines.

Measurements.

	slpv	slbv	mwi	hwi	pmwi	dpv	dbv
BC1004	12.6	11.3	11.3	6.5	7.0	c4.0	c3.0

Remarks. To date two species of *Hercothyris* have been described. The type species *H. borroi* Cooper, 1979 is from the Eocene of the Camaguey and Pinar del Rio provinces of Cuba and differs from *H. semiradiata* Cooper, 1979 from the Eocene of the Camaguey Province, Cuba, in having a transverse outline and a smooth exterior. The Jamaican specimen is elongate and possesses faint ribs, somewhat abraded, on the posterolateral valve exteriors; accordingly the form is assigned, tentatively, to *H. semiradiata* pending acquisition of further material.

ACKNOWLEDGMENTS

I am very grateful to Dr. S. K. Donovan for his constant help and encouragement at all stages of this project. A Short Term Visitor fellowship from the Smithsonian Institution, Washington, D.C., permitted examination of G. A. Cooper's type and figured material from the Caribbean; I thank the staff of the Department of Paleobiology for access to these collections in their care. Dr. C.H.C. Brunton allowed the study and loan of specimens from the Lucas Barrett and C. T. Trechmann collections in the British Museum (Natural History), London. Professor E. Robinson donated specimens from his own collections as did Stephen Donovan. The manuscript was improved by constructive reviews from C.H.C. Brunton, D. E. Lee, and A. Logan; Daphne Lee's advice on a generic assignment is greatly appreciated.

REFERENCES CITED

Backhaus, E., 1959, Monographie der cretacischen Thecideidae (Brachiopoda): Hamburg Geologisches Staatsinstitut Mitteilungen, v. 28, p. 5–90, Plates 1–7.

Baker, P. G., 1990, The classification, origin, and phylogeny of the thecideidine brachiopods: Palaeontology, v. 33, p. 175–191.

Balkissoon, I. G., 1989, Compilation and suggestions for a standardized nomenclature for eastern Jamaica: Journal of the Geological Society of Jamaica, v. 25 (for 1988), p. 27–37.

Beecher, C. E., 1893, The development of *Terebratalia obsoleta*: Transactions of the Connecticut Academy of Arts and Sciences, v. 1, p. 141–144, Plates 22–23.

Born, I. E., 1778, Index rerum naturalium Musei Caeserei Vindobonensis, pars. 1: Testacea [Verzeichnis der naturlichen seltheiten des K.K. vol. 1 Schaltiere]: Naturalien Cabinet zu Wien, 458 p., 1 plate.

Chubb, L. J, 1963, Cretaceous formations, *in* Zans, V. A., Chubb, L. J., Versey, H. R., Williams, J. B., Robinson, E., and Cooke, D. L., Synopsis of the geology of Jamaica: Bulletin of the Geological Survey Department, Jamaica, v. 4 (for 1962), p. 6–20.

Cockerell, T.D.A., 1894, A list of the Brachiopoda, Pelecypoda, Pteropoda, and Nudibrachiata of Jamaica, living and fossil: Nautilus, v. 7, p. 103–107, 113–118.

Cooper, G. A., 1959, Genera of Tertiary and Recent rhynchonelloid brachiopods: Smithsonian Miscellaneous Collections, v. 139, p. 1–90, 22 plates.

Cooper, G. A., 1977, Brachiopods from the Caribbean Sea and adjacent waters: University of Miami Studies in Tropical Oceanography 14, 140 p., 35 plates.

Cooper, G. A., 1979, Tertiary and Cretaceous brachiopods from Cuba and the Caribbean: Smithsonian Contributions to Paleobiology, no. 37, 45 p., 7 plates.

Cooper, G. A., 1988, Some Tertiary brachiopods of the east coast of the United States: Smithsonian Contributions to Paleobiology, no. 64, 45 p., 9 plates.

Crosse, H., and Fischer, P., 1866, Note sur la distribution geographique des Brachiopodes aux Antilles: Journal de Conchiliologie, ser. 3, v. 6, p. 265–273, 10 plates.

Dall, W. H., 1870, A revision of the Terebratulidae and Lingulidea with remarks on some descriptions of some Recent forms: American Journal of Conchology, v. 6, p. 88–168, plates 6–8.

Dall, W. H., 1900, Some names which must be discarded: Nautilus, v. 14, p. 44–45.

Dall, W. H., 1903, Contributions to the Tertiary fauna of Florida: Bulletin of the Wagner Free Institute of Science of Philadelphia, v. 3, p. 1219–1620, plates 48–60.

Dall, W. H., 1920, Annotated list of the Recent Brachiopoda in the collections of the United States National Museum, with descriptions of thirty-three new forms: Proceedings of the United States National Museum, v. 57, p. 262–377.

Davidson, T., 1864, On the Recent and Tertiary species of the genus *Thecidium*: Geological Magazine, v. 1, p. 12–23, plates 1–2.

Davidson, T., 1866, Notes on some Recent Brachiopoda dredged by the late Lucas Barrett off the north-east coast of Jamaica, and now forming part of the collection of Mr. R. MacAndrew: Proceedings of the Zoological Society of London, v. 34, p. 102–104, plate 12.

Davidson, T., 1886–1888, A monograph of Recent Brachiopoda: Transactions of the Linnean Society of London (2d ser.), Zoology, v. 4, p. 1–248, plates 1–30.

de la Beche, H. T., Sir, 1827, Remarks on the geology of Jamaica: Transactions of the Geological Society of London (2d ser.), v. 2, p. 173, 179.

Donovan, S. K., Gordon, C. M., Schickler, W. F., and Dixon, H. L., 1989, An Eocene age for an outcrop of the "Montpelier Formation" at Beecher Town, St. Ann, Jamaica, using echinoids for correlation: Journal of the Geological Society of Jamaica, v. 26, p. 5–9.

Elliott, G. E., 1958, Classification of thecidean brachiopods: Journal of Paleontology, v. 32, p. 373.

Gray, J. E., 1840, Brachiopoda: London, Synopsis of the contents of the British Museum (42nd Edition), 370 p.

Guppy, R.J.L., 1866, On Tertiary Brachiopoda from Trinidad: Quarterly Journal of the Geological Society of London, v. 22, p. 295–297, plate 19.

Harland, W. B., Cox, A. V., Llewellyn, P. G., Pickton, C.A.G., Smith, A. G., and Walters, R., 1982, A geologic time scale: Cambridge, Cambridge University Press, 131 p.

Harper, D.A.T., and Donovan, S. K., 1990, Fossil brachiopods of Jamaica: Journal of the Geological Society of Jamaica, v. 27, p. 25–30.

Hill, R. T., 1899, The geology and physical geography of Jamaica: Study of a type of Antillean development: Bulletin of the Museum of Comparative Zoology, Harvard University, v. 34, p. 1–226.

Jiang, M-J., and Robinson, E., 1987, Calcareous nannofossils and larger Foraminifera in Jamaican rocks of Cretaceous to early Eocene age, *in* Ahmad, R., ed., Proceedings, Workshop on the Status of Jamaican geology: Special Issue of the Journal of the Geological Society of Jamaica, p. 24–51.

King, W., 1850, A monograph of the Permian fossils of England: Palaeontographical Society Monograph, no. 3, 258 p., 29 plates.

Linné, C., von, 1767, Systema naturae, 12th edition: Stockholm, 1154 p.

Logan, A., 1979, The Recent Brachiopoda of the Mediterranean Sea: Bulletin de l'Institute Oceanographie Monaco, v. 72, p. 1–112, plates 1–10.

Logan, A., 1987, Neogene paleontology in the northern Dominican Republic. 6. The Phylum Brachiopoda: Bulletins of American Paleontology, v. 93, p. 44–55, Plate 12.

Megerle von Muhlfeldt, J. K., 1811, Entwurf eines neuen system's der Schalthiergehause: Gesellschaft Naturforschende Freunde Magazin, v. 5, p. 38–72, plate 3.

Meile, B., and Pajaud, D., 1971, Presence de Brachiopodes dans le grand Banc des Bahamas: Comptes Rendus hebdomadaires des Seances de l'Academie Sciences (ser. D), v. 273, p. 469–472, plate 1.

Munier-Chalmas, M., 1881, Note sommaire sur les genre de la famille des Thecideidae: Bulletin de Societe Geologique de France (3d ser.), v. 8, p. 278–279.

d'Orbigny, A., 1847, Sur les Brachiopodes ou Palliobranches: Comptes Rendus Academie de Science, v. 25, p. 266–269.

Risso, A., 1826, Histoire naturelle des principales productions de l'Europe meriodionale et particulierement de celles des environs de Nice et des Alpes maritimes, part 4: Paris, v. 7, p. 1–439, plates 1–12.

Robinson, E., 1967, Biostratigraphic position of the late Cenozoic rocks of Jamaica: Journal of the Geological Society of Jamaica, v. 9, p. 32–41.

Robinson, E., 1969, Geological field guide to Neogene sections in Jamaica, West Indies: Journal of the Geological Society of Jamaica, v. 10, p. 1–24.

Robinson, E., and Lamb, J. L., 1970, Preliminary palaeomagnetic data from the Plio-Pleistocene of Jamaica: Nature, v. 227, p. 1236–1237.

Sawkins, J. G., and Brown, C. B., 1869, Reports on the geology of Jamaica, with an appendix on the palaeontology of the Caribbean by R. Etheridge: Memoir of the Geological Survey of Great Britain, 339 p.

Thomson, J. A., 1926, A revision of the subfamilies of the Terebratulidae (Brachiopoda): Annals and Magazine of Natural History (9th ser.), v. 18, p. 523–530.

Trechmann, C. T., 1927, The Cretaceous shales of Jamaica: Geological Magazine, v. 64, p. 27–42, plate 3.

Trechmann, C. T., 1930, The Manchioneal beds of Jamaica: Geological Magazine, v. 67, p. 199–218, plates 1–2.

Waagen, W. H., 1883, Brachiopoda, *in* Salt Range fossils, part 4 (2): Palaeontologica Indica Memoir (13th ser.), v. 1, p. 547–610, plates 50–57.

Manuscript Accepted by the Society November 11, 1992

The fossil arthropods of Jamaica

Samuel F. Morris
British Museum (Natural History), Cromwell Road, London SW7 5BD, United Kingdom

ABSTRACT

The Cretaceous of Jamaica has so far yielded only eleven specimens of decapods belonging to four genera; these are all recorded. One of them, *Carcineretes woolacotti*, is considered to be a back burrower and not a swimmer as previously thought. The flattened fifth leg of burrowers such as *C. woolacotti* is considered to be a preadaptation to swimming as seen in the more-advanced Portunoidea.

All previously published Cenozoic decapods and cirripedes are recorded. The decapods belong in five genera, and there are three genera of cirripedes. Fourteen decapod taxa are recorded for the first time from the Pleistocene raised-reef terrace of Rio Bueno, mostly in open nomenclature.

INTRODUCTION

The fossil record of the arthropods of Jamaica is extremely fragmentary in both time and space, as it is everywhere dependent on favorable taphonomy and diagenesis without any later undue pressures or stresses. The cuticles of crustaceans contain a significant proportion of organic material, which soon degenerates due to microbiological attack. Even the calcified exoskeleton becomes very friable within a few days in tropical climates. Most decapods live in the favored high-energy environments of the sublittoral littoral zones. It is in these environments that remains are most quickly destroyed. The best fossil remains, often found as calcareous or phosphatic nodules, are preserved within the areas of quiet sedimentation of low-energy environments. In the high-energy environments only the more resistant parts, usually only the fingers of decapods, survive the transport, reworking, and microbiological attack. Evidence of burrowing seldom survives in these inshore environments.

Most of the potentially fossiliferous rock of Jamaica is composed of sediments deposited in shallow-water conditions, and therefore little of the arthropod fauna is likely to be preserved. Consequently a biostratigraphy of the arthropods of Jamaica is not a meaningful concept. It is only possible to record the isolated stratigraphic and geographic occurrences and occasionally to interpret the life-style of the group (or species) and the environment in which it lived. The interpretation is usually dependent on analogy with similar groups known from the Recent. Apart from the Paleozoic trilobites (not found in Jamaica), arthropods have seldom been used as time indices, although cirripedes have been used as time indicators in the Upper Cretaceous Chalk of England, Holland, and Bohemia (Withers, 1935) and in the Eastern Gulf Region of the United States (Mellen, 1973). It is curious that Jamaica has a poor fossil record of cirripedes, especially of the scalpellids, whose disarticulated valves are reasonably resistant. The balanid cirripedes also have a poor record (Littlewood and Donovan, 1988), being preserved only as the encrusting basal plates on the mollusc *Crassostrea virginica* or inhabiting corals (Newman and Ladd, 1974).

Bishop (1972, 1981) used decapod-dominated assemblages for broad stratigraphic correlation across North America and for unraveling the biogeography of that region. Feldmann (1986) discussed the problems of producing a paleobiogeography of decapods with only sparse data. He thought that conclusions could only be drawn and tested on decapods by comparison with groups such as the molluscs that are more frequently preserved and collected.

Stock (1986), in his discussion on Caribbean biogeography, proposed a possible method for dating each Caribbean island by use of species/area curves for groundwater Malacostraca. However, all the work is based on the distribution of Recent specimens without any reference to fossil material. So far no fossil terrestrial or freshwater arthropods have been recorded from Jamaica.

The survey of the decapods of the Carolinas by Williams (1965) seems to indicate that Cape Hatteras has been an absolute barrier to northward migration of shallow-water forms for much of the upper Cenozoic, but at the present time Cape Lookout is a great barrier, surprisingly, to the northward extension of Antillean species. Williams is also useful in providing a taxonomic revision of Rathbun (various dates) for American species.

Morris, S. F., 1993, The fossil arthropods of Jamaica, *in* Wright, R. M., and Robinson, E., eds., Biostratigraphy of Jamaica: Boulder, Colorado, Geological Society of America Memoir 182.

CRETACEOUS ARTHROPODS

The only arthropod group (except ostracodes; see Hazel, this volume), so far described or recorded from the Cretaceous of Jamaica is the crabs. And of these only eleven specimens are known, belonging to four genera (Fig. 1). They all appear to be back-burrowing forms. This crab fauna is probably confined to the Maastrichtian (*Veniella* Shales of the Marchmont Inlier and Guinea Corn Formation of the Central Inlier) and is composed of *Notopocorystes* (*Cretacoranina*) *trechmanni* (Withers, 1927) and *Carcineretes woolacotti* Withers, 1922. Two other specimens, ?*Paranecrocarcinus* sp. and *Necrocarcinus* sp. indet., from the Cretaceous of Jamaica, without any details of horizon or locality, are preserved in the Lucas Barrett Collection in The Natural History Museum, London. All these genera are characteristically back-burrowers, and it is probably due to this mode of life that they are preserved at all. *Carcineretes woolacotti* was previously considered to be a swimming crab because the dactylus and propodus of the fifth pereiopod are flattened, similar to the flattened dactylus and propodus of the fifth pereiopods of the Recent swimming crabs of the family Portunidae. However, it is clear that the earliest development of the flattened fifth pereiopod in dromiaceans was to aid digging backward into the substrate and not primarily to aid swimming. The other obvious characteristic of back-burrowing forms is that they are narrowest at the posterior end of the carapace and widen forward. Special provision has to be made for the inhalant and exhalant respiratory currents so that the passageways are not blocked with sediment during burial in the substrate. In addition, the eyes are frequently carried on longer eyestalks to allow them to be kept clear of the sediment.

Back-burrowing crabs seldom produce a permanent burrow; the burrow is only used as a place of concealment during the crab's inactive period or as a place from which to catch unsuspecting passing prey. The main requirement for back-burrowers is a flexible substrate such as sand, sand and broken shells, or muds. Comparison of *Carcineretes* with other groups of Portunoidea is thought to be supported by the presence of long orbits such as are found in the Podophthalminae of the Portunidae, but this is a character more likely to be associated with their burrowing mode. Nevertheless, the Carcineretidae may well be early members of the portunid lineage in which adaptations to burrowing have been converted to swimming.

Subsequently, the change of swimming habits from a forward motion to one in which the animal moves sideways has imposed a clear morphologic change in which the widest part of the carapace is in the mid-line and the lateral margins have extended processes adding to the streamlining effect. In addition, the carapace of the Recent swimming forms has become thinner, reducing their weight. The Carcineretidae, as at present defined, are confined to the Upper Cretaceous, and it is in the Maastrichtian of the North American Gulf Coast that the nearest comparable form to *Carcineretes woolacotti*, in *Ophthalmoplax stephensoni* Rathbun, 1935, is found.

The raninid *Notocorystes* and the calappids ?*Paranecrocarcinus* and *Necrocarcinus* are also back-burrowers, and one can only suppose that they continued the primitive burrowing habit from the Dromiacea, an ancestral line. In evolutionary terms these two genera are widely separated; the raninid, primitively, still bears female and male genital orifices on the coxae of the third and fifth legs respectively, but in the more advanced calappids the female orifice has transferred to the sixth sternite.

Notopocorystes (*Cretacoranina*) has previously been recorded only from rocks of Albian to Santonian age, but the nominal subgenus is known from as late as early Maastrichtian of Tennessee, Mississippi, and New Jersey in the United States (Bishop, 1986). Similarly, *Paranecrocarcinus* occurs elsewhere in older rocks of Hauterivian to early Campanian age. *Necrocarcinus* has a much longer age range, from Aptian to Eocene; in any case the Jamaican specimen is too poorly preserved to give a precise name to it. The *Necrocarcinus* and ?*Paranecrocarcinus* are preserved in a light chocolate sandstone that contains the foraminifer *Lenticulina* sp., a genus ranging from Jurassic to Recent. *Lenticulina* is usually confined to deeper water. It is possible that these two specimens may have been collected from rocks older than the Guinea Corn Formation, that is, pre-Maastrichtian, possibly Campanian. Both genera are known from widespread localities: India, the Middle East, France, and England as well as the United States.

CENOZOIC ARTHROPODS

On the front of the marine transgression that characterized the Middle Eocene it would be expected that a reasonable amount of decapod material would be discovered, but it is not until high in the Yellow Limestone Group, Chapelton Formation, that arthopods (decapods) are found. They are associated with the large gastropod *Campanile giganteum* at Spring Mount, 10

Figure 1. *Carcineretes woolacotti* Withers, 1: dorsal view of nearly complete male specimen. Holotype British Museum (BM) In.20708 from Guinea Corn Formation of bed of Rio Minho a little to the west of Trout Hall, Clarendon Parish. ×1; 2, 3: dorsal and ventral views of male specimen. BM In.23000 from Guinea Corn Formation of Logie Green, Clarendon Parish. ×1. *Notopocorystes* (*Cretacoranina*) *trechmanni* (Withers), 4, dorsal view of carapace. BM In.26011. *Veniella* Shales from between Catadupa and Cambridge, St. James Parish. ×1. *Necrocarcinus* sp., 5, dorsal view of internal mould of carapace. BM In.23975. Upper Cretaceous of Jamaica. ×1. *Paranecrocarcinus*? sp., 6, dorsal view of internal mould of carapace. BM In. 23988. Upper Cretaceous of Jamaica. ×1. *Xanthilites rathbunae* Withers, 7: dorsal view of incomplete carapace. BM In.26007 from Chapelton Formation of Spring Mount, Hanover Parish. ×1; 8: dorsal view of anterior portion of carapace. Holotype BM In.23017 from Chapelton Formation of Glasgow, Hanover Parish. ×1; 9: outer lateral view of right propodus. Paratype BM In.23018 from Spring Mount, Hanover Parish. ×1; 10: outer view of finger. Paratype BM In.23021 from Chapelton Formation of Spring Mount, Hanover Parish. ×2. *Varuna*? sp., 11: outer view of merus. BM 23023. Chapelton Formation of Spring Mount, Hanover Parish. ×2.

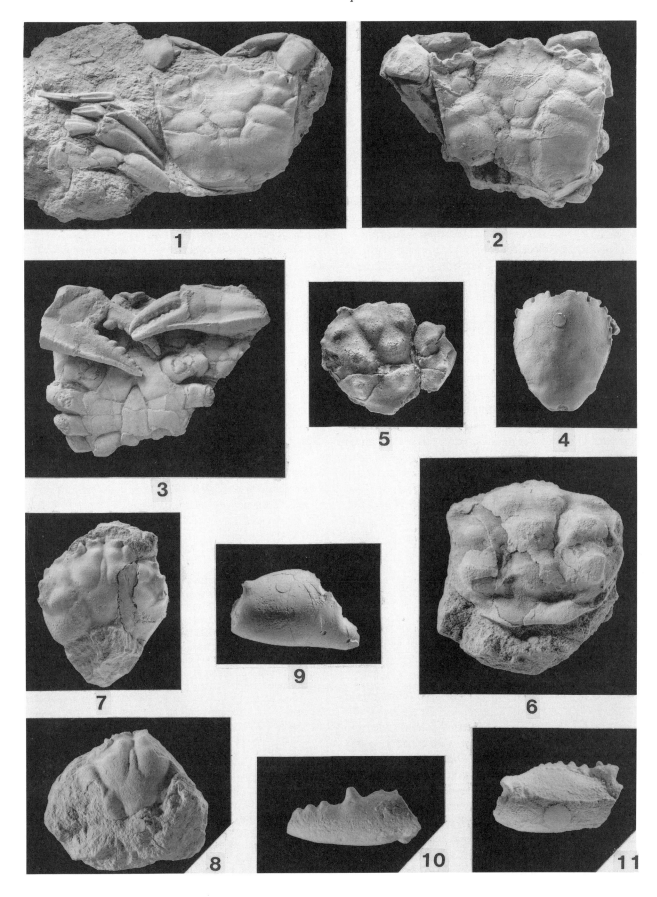

km southeast of Montego Bay, St. James Parish (Fig. 2). The decapods include *Callianassa subplana* Withers, 1924; *Callianassa trechmanni* Withers, 1924; *Xanthilites rathbunae* Withers, 1924 (? = *Xanthopsis* sp. of Rathbun, 1923 from the Middle Eocene, Plaisance Formation of Haiti; and ?*Varuna* sp.

Callianassa spp. are members of a group of burrowing shrimps that spend almost their entire lives within their burrows in sand or mud at or just below low tide and presumably owe their preservation to this fact. Of their exoskeleton, only the chelae are fully calcified, and in consequence usually only the chelae are ever found as fossils. The nominal taxon has become a "form genus" for chelae of this generalized type, having over 200 species assigned to the genus. The only characters available for analysis are the length/width ratios and the surface ornament.

It is only in the Holocene (7000 BP) of northern Australia and adjacent islands that subfossil, weakly calcareous carapaces are preserved within their burrows, allowing a thorough investigation of their taxonomic position within the superfamily Thalassinoidea.

The other two species recorded from the Chapelton Formation are both shallow-water forms: *Xanthilites? rathbunae* from the lower shore and ?*Varuna* sp. from inshore localities. Elsewhere—at Glasgow, 13 km south of Lucea, Hanover Parish—*Callianassa* and *Xanthilites* are associated with *Callinectes jamaicensis* Withers, 1924, a shallow-water predator, and with *Hepatiscus bartholomaeensis* (Rathbun, 1919), probably a back-burrower. One specimen of *Callianassa* sp. is recorded from Ulster Spring, Trelawny Parish. Four occurrences are known of indeterminate crab fragments from the Yellow Limestone Group: from Freemans Hall, Trelawny Parish; Mt. Friendship, St. Catherine Parish; Manchester, Manchester Parish; and Pimento Hill, St. Ann Parish.

In the overlying White Limestone Supergroup only a single decapod specimen is recorded, *Callianassa* sp. from St. Ann's Bay, St. Ann Parish, but at least three species of the coral-inhabiting cirripedes are known (Newman and Ladd, 1974)— one from the six-plated Archaeobalanidae, *Eoceratoconcha* cf. *E. renzi* Newman and Ladd, 1974 from the Miocene, Newport Formation of Santa Cruz, St. Elizabeth Parish, and a possible second species from the same locality and horizon, referred to ?*Eoceratoconcha* sp. by Newman and Ladd (1974). Also from Santa Cruz, Newman and Ladd described a species of the four-plated Ceratoconchinae *Ceratoconcha creusioides* Newman and Ladd, 1974.

From the Montpelier Formation *sensu lata* of Discovery Bay, parish of St. Ann, the same authors described a second species of the four-plated form *Ceratoconcha jungi* Newman and Ladd, 1974. Newman and Ladd discussed the various strategies that may have been involved in the evolution of the coral-inhabiting cirripedes, which are probably a polyphyletic group and came to the conclusion that they developed from a species resembling *Armatobalanus duvergieri* (de Alessandri) and *Armatobalanus* (*Hexacreusia*) *durhami* (Zullo).

Between the White Limestone Group and the younger Pleistocene Falmouth Formation, only a handful of crustacean remains are known (Fig. 3). The most significant of these are from the Round Hill Beds, August Town Formation, where Littlewood and Donovan (1988) recorded specimens of the barnacles *Balanus eburneus* Gould and *Balanus improvisus assimilis* Darwin encrusting *Crassostrea virginica*. They also found a single specimen of an indeterminate crab limb. Other occurrences of material from the August Town Formation are a crab merus, possibly a carpiliid crab, from Hope River Gorge, Portland Parish, and a crab fragment from Harbour View, Kingston. Dr. S. K. Donovan has recently collected a few specimens of *Ceratoconcha* aff. *C. barbadensis* (Withers, 1926) from the Pleistocene raised reef (probably Falmouth Formation) at Round Hill Bluff, Hanover Parish, inhabiting the coral *Agaricia* sp.

Dr. E. Robinson collected a few indeterminate crab fragments from the earliest Pliocene "San San Clays" at San San Bay.

From the Pleistocene (Calabrian) Manchioneal Formation, material has been collected suggesting that the specimens come from shallow, reef-associated deposits. At Manchioneal itself a galatheid (anomuran) fragment was recovered with indeterminate crab fragments, and at Alligator Pond, Manchester Parish, a specimen of a *Callianassa* sp. was collected.

But it is from the Pleistocene Falmouth Formation that the majority of Jamaican crustaceans have been recovered. Intensive collecting and sieving by Dr. S. K. Donovan and his students from the University of the West Indies have recently sorted thousands of crab and anomuran fragments from the patch reef, *Porites* zone raised-reef terrace of Rio Bueno (Liddell et al., 1984; Donovan et al., 1989).

Rathbun (1897, 1918, 1925, 1930, 1937) recorded approximately 120 genera of decapods from the Recent of Jamaica.

Figure 2. *Callinectes jamaicensis* Withers, 1, 2, outer and inner lateral views of left propodus. Holotype British Museum (BM) In.23016 from Chapelton Formation of Glasgow, Hanover Parish. ×2. *Hepatiscus bartholomaeensis* Rathbun, 3, dorsal view of carapace. BM In.26006 from Chapelton Formation of Glasgow, Hanover Parish. ×1. *Callianassa subplana* Withers, 4, 5, outer and lateral inner views of left manus. Holotype BM In.23013 from Chapelton Formation of Spring Mount, Hanover Parish. ×1. *Callianassa gigantea* Withers, 6, 7, outer and inner lateral views of right manus. Holotype BM In.23015 from Chapelton Formation of Glasgow, Hanover Parish. ×1. *Callianassa trechmanni* Withers, 8, 9, outer and inner lateral views of left propodus. BM In.26008 from Chapelton Formation of Spring Mount, Hanover Parish. ×1. *Uhlias* cf. *U. limbatus* Stimpson, 10, dorsal view of carapace. BM In.63676 from Falmouth Formation of Rio Bueno, ×6. *Mithraculus* cf. *M. forceps* A. Milne Edwards, 11, dorsal view of carapace. BM In.63769 from Falmouth Formation of Rio Bueno. ×4. *Uca* sp., 12, outer view of right dactylus. BM In.63776 from Falmouth Formation of Rio Bueno. ×6.5. *Pachygrapsus* sp., 13, dorsal view of fragment of carapace. BM In.63775 from Falmouth Formation of Rio Bueno. ×4. *Albunea* sp., 14, dactylus of one of the walking legs. BM In.63764 from Falmouth Formation of Rio Bueno. ×10. *Mithrax M.* cf. *caribbaeus* Rathbun, 15, outer view of left propodus. BM In.63768 from Falmouth Formation of Rio Bueno. ×4.5.

Present work by the writer, as yet unfinished, records 14 genera from the Jamaican Pleistocene of Rio Bueno, all assignable to Recent genera. Six of the crab genera contain species comparable to Recent species: *Mithrax* cf. *M. caribbaeus* Rathbun; *Phymodius* cf. *P. maculatus* Stimpson; *Calappa* cf. *C. gallus* (Herbst); *Micropanope* cf. *M. polita* Rathbun; *Micropanope* cf. *M. spinipes* A. Milne Edwards. In addition, three species are assigned with some degree of certainty to contemporary species *Mithrax spinosissimus* (Lamarck); *Eurypanopeus abbreviatus* (Stimpson), and *Panopeus herbstii* H. Milne Edwards.

Most of the material from the Pleistocene is fragmentary, being the result of sieving, and even though the average decapod fauna associated with corals is small (in the 2-mm to 1-cm size range), only dissociated fixed fingers and occasional meri are found. The exceptions are a few specimens of carapaces of *Mithraculus* cf. *M. forceps* and *Uhlias* cf. *U. limbatus*. Fragments of the pagurid *Petrochirus* and the porcellanid *Petrolisthes* are also found and easily sorted by their distinctive ornaments. Large numbers of xanthoid fingers are also present in the sample. These are easily distinguished by their dark fingers. Also present are fragments of limbs of the larger portunids. Other relatively easily sortable material is referred to *Uca, Albunea,* and *Callianassa* and—based on the characteristic terrace ridges of the carapace— fragments of *Pachygrapsus*.

Mithrax spp. and *Mithraculus* have been described from the coral-associated reefs of very similar age of Barbados (Collins and Morris, 1976). *Panopeus herbstii* is a predator upon the barnacle *Balanus eburneus,* and although the barnacle has not been recovered from the sample, it must have been present. The portunid crabs are well-known swimming opportunistic predators, probably not directly associated with the reefs, being rather larger than the average fauna (i.e., >1 mm to 1 cm). This suggests that some of the sample may have been washed in. This is also true for the burrowing shrimp *Callianassa,* which—in contrast to other Recent faunas—spends most of its life within the burrows and is only exposed following postmortality erosion. Some of the more robust xanthid fixed fingers also show signs of abrasion, due to having been reworked.

INVENTORY OF JAMAICAN ANTHROPODS

Cretaceous

Order Decapoda Latreille
Infraorder Brachyura Latreille
Superfamily Raninoidea de Haan
Family Raninidae de Haan

Genus *Notopocorystes* M'Coy
Subgenus *Cretacoranina* Mertin

N. (C) trechmanni (Withers, 1927), p. 177, Plate 7, Figs. 1, 2. Railway line between Catadupa and Cambridge, St. James Parish; *Veniella* Shales, Maastrichtian (1 specimen).

***Paranecrocarcinus* van Straelen**

?*P.* sp. Lucas Barrett Collection. Upper Cretaceous (?Maastrichtian) of Jamaica unlocalized (1 specimen).

***Necrocarcinus* Bell**

N. sp. Lucas Barrett Collection. Upper Cretaceous (?Maastrichtian) of Jamaica unlocalized (1 specimen).

Superfamily Portunoidea Rafinesque
Family Carcineretidae Beurlen

***Carcineretes* Withers**

C. woolacotti Withers, 1922, p. 535, Plate 16, Fig. 1, Plate 17, Figs. 2–6; Withers, 1924, p. 91, Plate 4, Figs. 1–4. Bed of Rio Minho, a little to the west of Trout Hall; Logie Green, 8 km to the west of Trout Hall, Chapelton, Clarendon Parish; river valley below Catadupa, St. James Parish; all Guinea Corn Formation, Maastrichtian (8 specimens).

Cenozoic

Order Decapoda Latreille
Infraorder Anomura H. Milne Edwards
Superfamily Thalassinoidea Latreille
Family Callianassidae Dana

***Callianassa* Leach**

C. gigantea Withers, 1924, p. 86, Plate 2, Figs. 13, 14. Glasgow, about 13 km south of Lucea, Hanover Parish; Chapelton Formation, Middle Eocene (1 specimen).
C. subplana Withers, 1924, p. 85, Plate 2, Figs. 9–12. Spring Mount, 9

Figure 3. *Phymodius* cf. *P. maculatus* (Stimpson), 1, 2, Falmouth Formation of Rio Bueno—1: outer view of left cheliped British Museum (BM) In.63771. ×4.5; 2: inner view of left propodus and merus of BM In.63772. ×4.5. *Micropanope* cf. *M. spinipes* A. Milne Edwards, 3, lateral view of right propodus and merus. BM In.63773 from Falmouth Formation of Rio Bueno. ×6. *Calappa* cf. *C. gallus* (Herbst), 4, outer view of right cheliped. BM In.63760. ×4.5. *Petrolisthes* sp., 5, 6, Falmouth Formation of Rio Bueno—5: outer view of right propodus. BM In.63765; 6: outer view of merus of left cheliped. BM In.63770. ×6. *Callianassa* sp. A, 7, outer view of right propodus. BM In.63778 from Falmouth of Rio Bueno. ×6. *Callianassa* sp. B, 8, inner view of right propodus. BM In.63779 from Falmouth Formation of Rio Bueno. ×6. Portunid genus indet., 9, inner view of incomplete propodus. Bm In.63780 from Falmouth Formation of Rio Bueno. ×4.5. *Petrochirus* sp., 10, merus of right cheliped. BM In.63781 from Falmouth Formation Rio Bueno. ×6. *Eurypanopeus abbreviatus* (Stimpson), 11, dactylus of right cheliped. BM In.63774 from Falmouth Formation of Rio Bueno. ×6. *Ceratoconcha* aff. *C. barbadensis* (Withers), 12, dorsal view of specimens on the coral *Agaricia* sp. BM In.63777 from the Pleistocene raised reef at Round Hill Bluff, Montego Bay. ×6. *Balanus eburneus* Gould, 13, specimens on the bivalve *Crassostrea virginica*. BM In.63663 from Round Hill Beds of Farquhar's Beach, Round Hill, Clarendon Parish. ×2. *Balanus improvisus assimilis* Darwin, 14, specimens on the bivalve *Crassostrea virginica*. BM In.63666 from Round Hill Beds of Farquhar's Beach, Clarendon Parish. ×2.

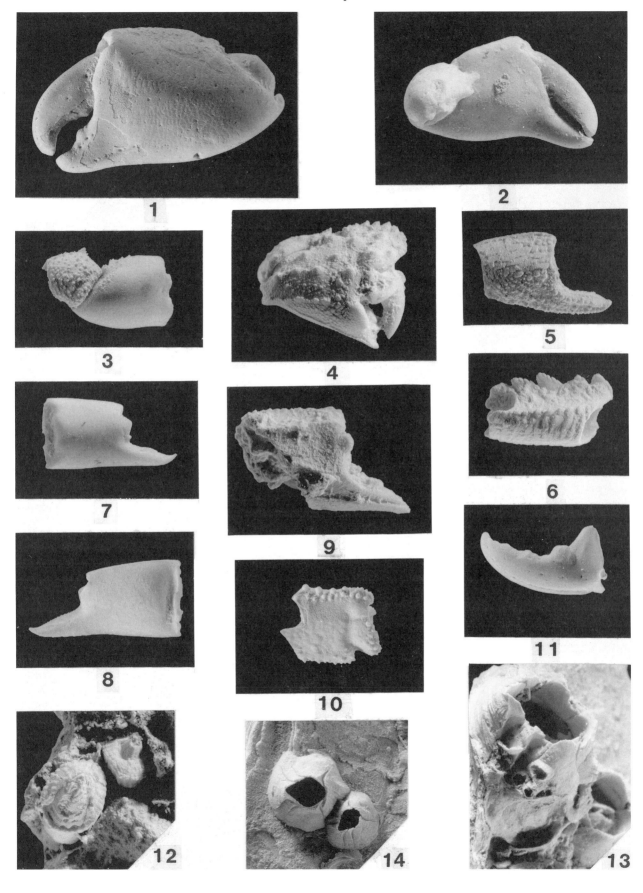

km southeast of Montego Bay; Chapelton Formation, Middle Eocene (2 specimens).
C. trechmanni Withers, 1924, p. 83, Plate 2, Figs. 1–8. Spring Mount, 9 km southeast of Montego Bay; Chapelton Formation, Middle Eocene (11 specimens).
Calianassa spp have been collected from the following localities:
(i) Ulster Spring, Trelawny Parish; Chapelton Formation, Middle Eocene.
(ii) New Ground, St. Ann's Bay, St. Ann Parish; White Limestone Formation, Middle Eocene–lower Middle Oligocene.
(iii) Alligator Pond on coast south of Mandeville; Quaternary (2 specimens).
(iv) Rio Bueno Harbour east shore, St. Ann Parish; Falmouth Formation, Pleistocene (many specimens).

Superfamily Paguroidea Latreille
Family Paguridae Latreille
Petrochirus Stimpson

P. sp. Rio Bueno Harbour east shore, St. Ann Parish; Falmouth Formation, Pleistocene.

Superfamily Galatheoidea Samouelle
Family Porcellanidae Haworth

Petrolisthes Stimpson

P. sp. Rio Bueno east shore, St. Ann Parish; Falmouth Formation, Pleistocene.

Superfamily Hippoidea Latreille
Family Albuneidae Stimpson

Albunea Weber

A. sp. Rio Bueno east shore, St. Ann Parish; Falmouth Formation, Pleistocene.

Infraorder Brachyura Latreille
Superfamily Calappoidea de Haan
Family Calappidae de Haan

Calappa Weber

C. cf. *C. gallus* (Herbst, 1803). Rio Bueno Harbour east shore, St. Ann Parish; Falmouth Formation, Pleistocene (1 specimen).

Hepatiscus Bittner
H. bartholomaeensis (Rathbun, 1919), p. 176, Plate 8, Fig. 3. Glasgow, 13 km south of Lucea, Hanover Parish; Chapelton Formation, Middle Eocene (1 specimen).

Family Leucosiidae Samouelle

Uhlias Stimpson

U. cf. *U. limbatus* Stimpson, 1871. Rio Bueno Harbour east shore, St. Ann Parish; Falmouth Formation, Pleistocene (2 carapaces).

Superfamily Majoidea Samouelle
Family Majidae Samouelle
Mithraculus White

M. cf. *M. forceps* A. Milne Edwards, 1875. Rio Bueno Harbour east shore, St. Ann Parish; Falmouth Formation, Pleistocene (5 carapaces).

Mithrax Latreille

M. cf. *M. caribbaeus* Rathbun, 1920. Rio Bueno Harbour east shore, St. Ann Parish; Falmouth Formation, Pleistocene.
M. spinosissimus Lamarck, 1818. Rio Bueno Harbour east shore, St. Ann Parish; Falmouth Formation, Pleistocene.

Superfamily Grapsoidea Macleay
Family Grapsidae Macleay

Pachygrapsus Randall

P. sp. Rio Bueno Harbour east shore, St. Ann Parish; Falmouth Formation, Pleistocene.

Varuna H. Milne Edwards

V? sp. Spring Mount, 9 km southeast of Montego Bay; Chapelton Formation, Middle Eocene (1 specimen).

Superfamily Ocypodoidea Rafinesque
Family Ocypodidae Rafinesque

Uca Leach

U. sp. Rio Bueno Harbour east shore, St. Ann Parish; Falmouth Formation, Pleistocene.

Superfamily Portunoidea Rafinesque
Family Portunidae, Rafinesque

Callinectes Stimpson

C. jamaicensis Withers, 1924, p. 87, Plate 3, Figs. 7, 8. Glasgow, 13 km south of Lucea, Hanover Parish; Chapelton Formation, Middle Eocene (1 specimen).

Superfamily Xanthoidea Dana
Family Xanthidae Dana

Xanthilites Bell

X? *rathbunae* Withers, 1924, p. 89, Plate 4, Figs. 5, 6. Glasgow, 13 km south of Lucea, Hanover Parish; Spring Mount, 9 km southeast of Montego Bay; Chapelton Formation, Middle Eocene (7 specimens).

Eurypanopeus A. Milne Edwards

E. abbreviatus (Stimpson, 1860). Rio Bueno Harbour east shore, St. Ann Parish; Falmouth Formation, Pleistocene.

Micropanope Stimpson

M. cf. *M. polita* Rathbun, 1893. Rio Bueno Harbour east shore, St. Ann Parish; Falmouth Formation, Pleistocene.

M. cf. *M. spinipes* A. Milne Edwards, 1880. Rio Bueno Harbour east shore, St. Ann Parish; Falmouth Formation, Pleistocene.

Panopeus H. Milne Edwards

P. herbstii H. Milne Edwards, 1834. Rio Bueno Harbour east shore, St. Ann Parish; Falmouth Formation, Pleistocene.

Phymodius A. Milne Edwards

P. cf. *P. maculatus* (Stimpson, 1860). Rio Bueno Harbour east shore, St. Ann Parish; Falmouth Formation, Pleistocene.

Indeterminate crab fragments have been collected from the following localities and horizons.
 (a) Yellow Limestone Gropu (Middle Eocene)
 (i) Dump B5 road at Manchester, National Grid Reference (NGR) 3760 4725–3745 4750; Dump Limestone, Guy's Hill Member of the Chapelton Formation.
 (ii) Margin of Windsor Forest Road, c. 1 km from Mt. Friendship, St. Catherine Parish.
 (iii) Freeman's Hall, exposure on west side of road from Ulster Spring to Lichfield, Trelawny Parish, NGR 3650 5045.
Swanswick Formation (late Middle Eocene)
Pimento Hill, Beecher Town near Ocho Rios, St. Ann Parish, NGR 510 532.
 (b) Coastal Group
 (i) Just north of Manchioneal Police Station, NGR M 7900 4125; Manchioneal Formation (includes a galatheid fragment).
 (ii) Farquhars Beach, Round Hill, Clarendon Parish, NGR H 413 347; August Town Formation.
 (iii) Harbour View, Kingston, NGR 5063 8180; August Town Formation, *Pecten* Bed.
 (iv) Hope River Gorge, Folly Peninsula, Port Antonio, Portland Parish; August Town Formation (1 carpiliid? crab fragment).
 (v) San San Bay, Portland Parish (locality ER143/21B); "San San Clays," Buff Bay Formation, earliest Pliocene.

Class Cirripedia Burmeister
Suborder Balanomorpha Pilsbry
Family Balanidae Leach

Balanus da Costa

B. eburneus Gould, 1841. Farquhar's Beach, Round Hill, Clarendon Parish, NGR H 413 347; Round Hill Beds, August Town Formation (on *Crassostrea virginica*). *B. improvisus* Darwin, 1854 *assimilis* Darwin, 1854. Farquhar's Beach, Round Hill, Clarendon Parish, NGR H 413 347; Round Hill Beds, August Town Formation (on *Crassostrea virginica*).

Family Archaeobalanidae Newman and Ross

Eoceratoconcha Newman and Ladd

E. cf. *E. renzi* Newman and Ladd, 1974. 6 km south-southeast of Santa Cruz, St. Elizabeth Parish; White Limestone Group, Newport Formation, Lower Miocene.
E? sp. Same horizon and locality.

Family Pyrgomatidae Gray
Subfamily Ceratoconchinae Newman and Ross

Ceratoconcha Kramberger-Gorjanovic

C. creusioides Newman and Ladd, 1974. Southampton, Santa Cruz, St. Elizabeth Parish; White Limestone Group, Newport Formation, Lower Miocene.
C. jungi Newman and Ladd, 1974. 1.5 km southwest of Discovery Bay, Trelawny Parish; White Limestone Group, Montpelier Formation, Lower Miocene.
C. aff. *barbadensis* (Withers, 1926). North coast road at Round Hill Bluff west of Montego Bay, Hanover Parish; Pleistocene raised reef.

ACKNOWLEDGMENTS

I must particularly thank Dr. S. K. Donovan for encouragement, provision of specimens, and references and his students at the University of the West Indies, especially Carla Gordon, who so assiduously sieved anad sorted the material from the Rio Bueno patch reefs. To Mark Donovan I extend my grateful thanks for the provision of further material. Mrs. J. Thomas, curator of the Department of Geology, University of the West Indies, kindly loaned me material in her care.

REFERENCES CITED

Bishop, G. A., 1972, Preservation of fossil crabs from the *Dakoticancer* Assemblage, upper Cretaceous Pierre Shale, South Dakota: Bulletin of the Georgia Academy of Science, v. 30, 80 p.

Bishop, G. A., 1981, Occurrence and fossilization of the *Dakoticancer* Assemblage, Pierre Shale, South Dakota, in Gray, J., et al., eds., Communities of the past: Stroudsberg, Pennsylvania, Hutchinson Ross, Publishing Company, p. 383–413.

Bishop, G. A., 1986, Occurrence, preservation and biogeography of the Cretaceous crabs of North America: Crustacean Issues, v. 4, p. 111–142.

Collins, J.S.H., and Morris, S. F., 1976, Tertiary and Pleistocene crabs from Barbados and Trinidad: Palaeontology, v. 19, p. 107–131.

Darwin, C., 1854, A monograph of the subclass Cirripedia with figures of all the species: The Balanidae . . . etc.: London, Ray Society, 648 p.

Donovon, S. K., Gordon, C. M., Littlewood, T. J., and Morris, S. F., 1989, Vagile benthos of a late Pleistocene patch reef, Falmouth Formation, Jamaica: Abstracts, Palaeontological Association Conference, v. 1989, p. 6.

Feldmann, R. M., 1986, Paleobiogeography of two decapod crustacean taxa in the Southern Hemisphere—Global conclusions with sparse data: Crustacean Issues, v. 4, p. 5–19.

Gould, A. A., 1841, A report on the Invertebrata of Massachusetts, comprising Mollusca, Crustacea, Annelida and Radiata: Cambridge, Massachusetts, State of Massachusetts Zoological and Botanical Survey, 373 p.

Herbst, J.F.W., 1803, Versuch einer naturgeschichte der krabben und krebse nebst einer systematischen beschreibung ihrer verschiedenen arten: Berlin and Stralsund, n. 1., v. 3 (3), p. 54.

Lamarck, J.B.P.A. DeM. De., 1818, Histoire naturelle des animaux sans vertebres: Paris, J. B. Baillière, v. 5, 612 p.

Liddell, W. D., Ohlhorst, S. L., and Coates, A. G., 1984, Modern and ancient carbonate environments of Jamaica: Sedimenta, v. 10, p. vii + 98 p.

Littlewood, D.T.J., and Donovan, S. K., 1988, Variation of Recent and fossil *Crassostrea* in Jamaica: Palaeontology, v. 31, p. 1013–1028.

Mellen, F. F., 1973, in Collins, J.S.H., and Mellen, F. F., eds., Cirripedes from the upper Cretaceous of Alabama and Mississippi, East Gulf Region, U.S.A.: Bulletin British Museum (Natural History), Geology, v. 23, p. 351–388.

Milne Edwards, A., 1875, Etudes sur les Xiphosures et les Crustaces de la region Mexicaine, in Mission scientifique au Mexique et dans l'Amerique centrale: Recherches Zoologiques pour servir a l'historie de la faune de l'Amerique centrale et du Mexique: Paris, Ministère de l'Instruction Publique, v. 5, p. 57–120.

Milne Edwards, A., 1880, *Ibid.*, Paris, Ministère de l'Instruction Publique, v. 5, p. 313–365.

Milne Edwards, H., 1834, Histoire naturelle des Crustaces: Paris, Librairie encyclopédique de Roret, v. 1, p. 468.

Newman, W. A., and Ladd, H. S., 1974, Origin of coral-inhibiting Balanids (Cirripedia, Thoracica), in Contributions to the Geology and Paleobiology of the Caribbean and Adjacent Areas: Verhandlungen der Naturforschenden Gesellschaft in Basel, v. 84, p. 381–396.

Rathbun, M. J., 1893, Descriptions of new genera and species of crabs from the West Coast of America and the Sandwich Islands: Proceedings of the United States National Museum, v. 16, p. 223–260.

Rathbun, M. J., 1897, List of the decapod Crustacea of Jamaica: Annals of the Institute of Jamaica, v. 1, p. 1–46.

Rathbun, M. J., 1918, The Grapsoid crabs of America: United States National Museum Bulletin, v. 97, p. 1–461.

Rathbun, M. J., 1919, West Indian Tertiary decapod Crustaceans: Publications, Carnegie Institution of Washington, v. 291, p. 159–184.

Rathbun, M. J., 1920, New species of spider crabs from the Straits of Florida and Caribbean Sea: Proceedings of the Biological Society of Washington, v. 33, p. 23–24.

Rathbun, M. J., 1923, Fossil crabs from the Republic of Haiti: Proceedings of the United States Museum, v. 2, p. 160.

Rathbun, M. J., 1925, The spider crabs of America: United States National Museum Bulletin, v. 129, p. 1–613.

Rathbun, M. J., 1930, The Cancroid crabs of America of the Families Euryalidae, Portunidae, Atelecyclidae, Cancridae and Xanthidae: United States National Museum Bulletin, v. 152, p. 1–609.

Rathbun, M. J., 1935, Fossil Crustacea of the Atlantic and Gulf Coastal Plain: Special Papers of the Geological Society of America, v. 2, p. 160.

Rathbun, M. J., 1937, The oxystomatous and allied crabs of America: United States National Museum Bulletin, v. 166, p. 1–278.

Stimpson, W., 1860, Notes on North American Crustacea in the Museum of the Smithsonian Institution, no. 2: Annals of the Lyceum of Natural History of New York, v. 7, p. 176–246.

Stimpson, W., 1871, Notes on North American Crustacea in the Museum of the Smithsonian Institution, no. 3: Annals of the Lyceum of Natural History of New York, v. 10, p. 92–136.

Stock, J. H., 1986, Caribbean biogeography and a biological calender for geological events: Crustcean Issues, v. 4, p. 195–203.

Williams, A. B., 1965, Marine decapod crustaceans of the Carolinas: Fish and Wildlife Service, United States Department of the Interior, Fishery Bulletin, v. 65, p. 1–298.

Withers, T. H., 1922, On a new Brachyurous crustacean from the upper Cretaceous of Jamaica: Annals and Magazine of Natural History, ser. 9, v. 10, p. 534–541.

Withers, T. H., 1924, Some Cretaceous and Tertiary decapod crustaceans from Jamaica: Annals and Magazine of Natural History, ser. 9, v. 13, p. 81–93.

Withers, T. H., 1926, Barnacles of the *Creusia-Pyrgoma* type from the Pleistocene of Barbados: Annals and Magazine of Natural History, ser. 9, v. 17, p. 1–6.

Withers, T. H., 1927, *Ranina trechmanni*, a new Cretaceous crab from Jamaica: Geological Magazine, v. 64, p. 176–180.

Withers, T. H., 1935, Catalogue of fossil Cirripedia in the Department of Geology, British Museum (Natural History), Cretaceous: London, British Museum (Natural History), p. 1–534.

MANUSCRIPT ACCEPTED BY THE SOCIETY NOVEMBER 11, 1992

Crinoid, asteroids, and ophiuroids in the Jamaican fossil record

Stephen K. Donovan and Carla M. Gordon
Department of Geology, University of the West Indies, Mona, Kingston 7, Jamaica, West Indies
Cornelis J. Veltkamp
Department of Environmental and Evolutionary Biology, University of Liverpool, P.O. Box 147, Liverpool L69 3BX, United Kingdom
A. Dawn Scott
Content District, Hope Bay, Portland, Jamaica, West Indies

ABSTRACT

Nonechinoid echinoderms are poorly known from the fossil record of Jamaica and only occur as disarticulated plates. Only crinoid columnals have hitherto been reported. However, asteroid marginal ossicles and ophiuroid vertebral ossicles are also locally common. Three species of isocrinid are recognized: cf. *Cenocrinus asterius* (Linné) from the Lower Pleistocene, *Diplocrinus* sp. from the Miocene, and an isocrinid sp. indet. from the Eocene. The microcrinoid *Applinocrinus cretacea* (Bather) and an indeterminate comatulid are both known from the Upper Cretaceous. Asteroid marginal plates (astropectinids and/or goniasterids) have been recovered from two Maastrichtian, three Eocene, and one Oligocene localities. Ophiuroid vertebral ossicles have been recovered from a patch reef in the back-reef lagoonal facies of the Upper Pleistocene (last interglacial) Falmouth Formation.

INTRODUCTION

Complete specimens of fossil echinoderms, apart from echinoids, are unknown from the fossil record of Jamaica, although disarticulated ossicles are locally common at some stratigraphic horizons and are thus potentially useful biostratigraphic markers. However, only crinoid columnals (Donovan, 1989) have hitherto been recorded. Although echinoderm biostratigraphy usually relies on the use of complete skeletons (see examples in Broadhead, 1980), disarticulated ossicles, such as crinoid columnals, have been used in both local and international correlation (for example, see Donovan, 1984a, 1985), so the potential exists to use the taxa described below as marker fossils. In particular, the recognition of isocrinid columnals from three horizons (albeit only from rare specimens so far) suggests that they may be of use in zoning deeper-water limestones in Jamaica (compare with the British Upper Cretaceous; Rasmussen, 1961, p. 409). Of the extant echinoderm classes, only the holothurians are so far unknown from the Jamaican fossil record, and the search for their disarticulated schlerites must be left to the micropaleontologists.

The geographic and stratigraphic distributions of the remains described herein are summarized in Figure 1 and Table 1, respectively. See Donovan (Chapter 6, Fig. 1) for a locality map of the Cretaceous inliers of Jamaica. Specimens described herein are deposited in the Natural History Museum, London (BMNH); National Museum of Natural History, Smithsonian Institution (USNM); U.S. Geological Survey collections in the National Museum of Natural History (USGS); and the Florida Museum of Natural History (UF).

SYSTEMATIC PALEONTOLOGY

Class CRINOIDEA J. S. Miller
Subclass ARTICULATA von Zittel
Order ISOCRINIDA Sieverts-Doreck
Suborder ISOCRININA Gislén
Family ISOCRINIDAE Gislén

Genus *Cenocrinus* Wyville Thomson

cf. *Cenocrinus asterius* (Linné, 1767)
(Figures 2, 3.8)

Description of material. Columnal outline pentagonal with rounded angles (Figs. 2.1, 3.8). Axial canal obscured by sediment. Articulation symplectial, arranged around five petaloid, depressed, diamond- to pear-shaped areola pits that correspond to the columnal angles. Areola pits closed and separate. Crenulae arrayed from perpendicular to subparallel to the circumference of the areola pits. Crenulae short and unbranched but sometimes fused between adjacent petals (Figs. 2.1, 3.8). Latus planar and unsculptured (Fig. 2.2).

Donovan, S. K., Gordon, C. M., Veltkamp, C. J., and Scott, A. D., 1993, Crinoids, asteroids, and ophiuroids in the Jamaican fossil record, *in* Wright, R. M., and Robinson, E., eds., Biostratigraphy of Jamaica: Boulder, Colorado, Geological Society of America Memoir 182.

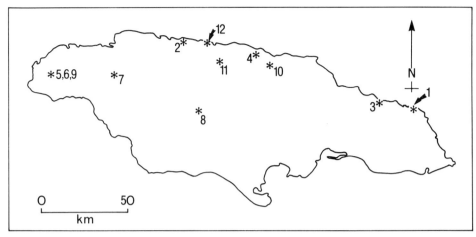

Figure 1. Outline map of Jamaica showing the relative positions of localities mentioned in the text. 1: San San Bay, parish of Portland, Lower Pleistocene—cf. *Cenocrinus asterius* (Linné). 2: Duncans Quarry, parish of Trelawny, Miocene—*Diplocrinus* sp. 3: Content, parish of Portland, Eocene—isocrinid sp. indet. 4: St. Anns Inlier, parish of St. Ann, Upper Cretaceous—comatulids sp. or spp. 5: Jerusalem Mountain Inlier, parish of Westmoreland, Maastrichtian—*Applinocrinus cretacea* (Bather). 6–11: asteroid marginal plate localities. 6: Jerusalem Mountain Inlier, parish of Hanover, Maastrichtian; 7: Marchmont Inlier, parish of Westmoreland, Maastrichtian; 8: north of Christiana, parish of Manchester, Eocene; 9: near Rock Spring, parish of Hanover, Eocene; 10: Beecher Town, parish of St. Ann, Eocene; 11: Browns Town, parish of St. Ann, Oligocene. 12: Rio Bueno Harbour, parish of St. Ann, Upper Pleistocene—ophiuroid vertebral ossicles.

TABLE 1. STRATIGRAPHIC DISTRIBUTION OF CRINOIDS, ASTEROIDS, AND OPHIUROIDS IN THE JAMICAN FOSSIL RECORD

Upper Pleistocene
    Ophiuroid vertebral ossicles indet.
Lower Pleistocene
    Cf. *Cenocrinus asterius* (Linné)
Miocene
    Diplocrinus sp.
Oligocene
    Asteroid marginal ossicles indet.
Eocene
    Asteroid marginal ossicles indet.
    Isocrinid sp. indet.
Maastrichtian
    Applinocrinus cretacea (Bather)
    Asteroid marginal ossicles indet.
Upper Cretaceous
    Comatulid brachial ossicles indet.

Dimensions. Columnal diameter = 4.8 mm; columnal height = 1.25 mm.
Material, locality, and horizon. The original locality of Robinson (1969, 1971; Fig. 1, locality 1 herein) is now obscured by a retaining wall, and his collection of crinoid columnals is lost (Donovan, 1989). A single internodal columnal, BMNH E70448, collected from a loose block in a banana plantation about localities ER143/21 to 24 (Robinson, 1969, Fig. 2, p. 7–9), is the only specimen to have been described (Donovan, 1989). San San Bay, parish of Portland, northeast Jamaica, GR 588663. Navy Island Member, Manchioneal Formation, Upper Coastal Group. Foraminiferan Biozone N22 (Banner and Blow, 1965), Lower Pleistocene (Calabrian).
Remarks. This specimen is the only crinoid ossicle to have hitherto been described from Jamaica and was discussed in detail by Donovan (1989). Recent *C. asterius* is known from about 200 to 400 m on the deep fore reef at Discovery Bay, parish of St. Ann, and Meyer et al. (1978, p. 424) recorded a total depth range for this species in the Caribbean of 183 to 585 m. A distal columnal from a Recent specimen of *C. asterius* is illustrated (Fig. 3.7).

Genus *Diplocrinus* Döderlein
***Diplocrinus* sp.**
(Figure 4)

Description of material. Columnal outline weakly pentastellate with strongly rounded angles (Fig. 4). Axial canal obscured by sediment. Articulation symplectial, arranged about five elongate, slender, lensoid areola pits that correspond to the columnal angles. Areola pits closed and separate. Crenulae perpendicular to subperpendicular to circumference of areola pits. Crenulae short and unbranched, with crenulae of adjacent petals tending to be fused close to the center of the columnal (Fig. 4). Triangular naked zones occur adjacent to circumference and between petaloid zones. Latus convex and unsculptured.
Dimensions. Columnal diameter = 5.7 mm; columnal height = 1.6 mm.
Material, locality, and horizon. A unique internodal columnal, UF

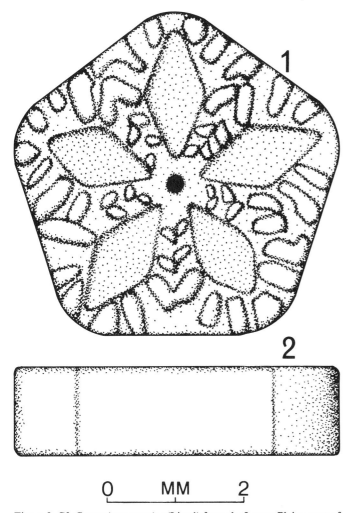

Figure 2. Cf. *Cenocrinus asterius* (Linné) from the Lower Pleistocene of San San Bay, parish of Portland. Restoration of the articular facet (1) and latus (2) (redrawn after Donovan, 1989, Fig. 1B, C).

38939, from a quarry about 5 km west of Duncans police station, on the south side of highway A1, parish of Trelawny, approximate GR 349570 (Fig. 1, locality 2; site 16 of Liddell et al., 1984). Sign Formation(?), Montpelier Group, White Limestone Supergroup, Miocene.
Remarks. The distinctive features shown by this columnal include the narrow, closed areola petals and the fusion of crenulae close to the position of the lumen. By comparison with the illustrated key of Roux (1977), it is apparent the Duncans Quarry specimen is closest to *Diplocrinus*. Neither of the species of *Diplocrinus* illustrated by Roux has closed petals (also see Macurda and Roux, 1981). However, Donovan (1984b) has noted open petals in *Neocrinus decorus* Wyville Thomson that are illustrated as closed in Roux (1977), and the reverse may be true for *Diplocrinus. D. maclearanus* (Wyville Thomson) is an extant Caribbean species (Meyer et al., 1978), but its columnals do not closely resemble those of the Duncans specimen (Macurda and Roux, 1981). Indeed, the Jamaican columnal is almost certainly a new species, but further specimens are required before it can be adequately diagnosed. The only Miocene isocrinid that has previously been described from the Antillean islands is *Balanocrinus haitiensis* Springer, 1924, whose columnals are markedly different from *Diplocrinus* sp.

Isocrinid sp. indet.
(Figure 5.1)

Dimensions. Columnal diameter = 3.2 mm; columnal height = 1.7 mm.
Material, locality, and horizon. A single internodal columnal, BMNH EE2182 (Fig. 5.1), collected by ADS from Content, near Hope Bay, parish of Portland, at about GR 70454640 (Fig. 1, locality 3). This specimen is late Middle Eocene in age and comes from an undescribed lithostratigraphic unit that conformably overlies the Font Hill Formation (E. Robinson, personal communication, 1990).
Remarks. Although undoubtedly an isocrinid columnal, this specimen is strongly pitted, making any attempt at further identification futile. Echinoids from the same locality show a similar style of preservation (Donovan et al., 1991).

Suborder COMATULIDINA A. H. Clark
Infraorder COMATULIDIA A. H. Clark
Incertae familiae

Comatulid sp. or spp. indet.
(Figures 5.2–5.6)

Material, localities, and horizons. Hundreds of disarticulated brachials have been collected from the Upper Cretaceous of the St. Ann Inlier, parish of St. Ann (Fig. 1, locality 4). Samples have the following USGS field numbers: J-66-55 (including USNM 458892–458896; Fig. 5.2–5.6), J-71-46, 71-45 (catalog number 30064), J-71-46, and J-71-48 (catalog number 30067).
Remarks. The disarticulated brachial plates referred to as comatulids herein could conceivably be derived from an isocrinid. However, the absence of any isocrinid columnals, which are generally more prominent fossils than brachials, suggests otherwise. Selective winnowing of isocrinid columnals might be invoked to explain their scarcity (compare with the different patterns of transport shown by the radioles of regular and irregular echinoids; Schäfer, 1972), but it is more probable that their absence is due to each comatulid having only one centrodorsal ossicle, but numerous brachials, producing horizons enriched in the latter. These specimens are awaiting detailed analysis (S.K.D. and C.R.C. Paul, research in progress).

Order ROVEACRINIDA Sieverts-Doreck

Remarks. See Simms (1988) for a discussion of the possible polyphyly of the order Roveacrinida.

Family SACCOCOMIDAE d'Orbigny

Genus *Applinocrinus* Peck

***Applinocrinus cretacea* (Bather, 1924)**

Locality and horizon. In the shales immediately below the oyster limestones (ostracode zone 3 of Hazel and Kamiya [this volume]) of the Upper Cretaceous (Maastrichtian) Jerusalem Mountain Formation, Jerusalem Mountain Inlier, parish of Westmoreland (Hazel and Kamiya, this volume; Fig. 1, locality 5).
Remarks. Hazel and Hamiya (this volume) report the presence of the planktic microcrinoid *Saccocoma* in association with other microfossils (particularly ostracodes) from the Upper Cretaceous of eastern Jamaica. Hazel sent all available specimens to the late R. E. Peck, who identified them as *Saccocoma cretacea* Bather (letter to Hazel, 29 December 1967). Peck also noted that "a few specimens appear to be brachials of another crinoid." Peck (1973) later made this species the type of the new genus *Applinocrinus*. The holotype theca and a primaxillary (1Br$_2$) of this species, both from the English chalk, were figured by Rasmussen

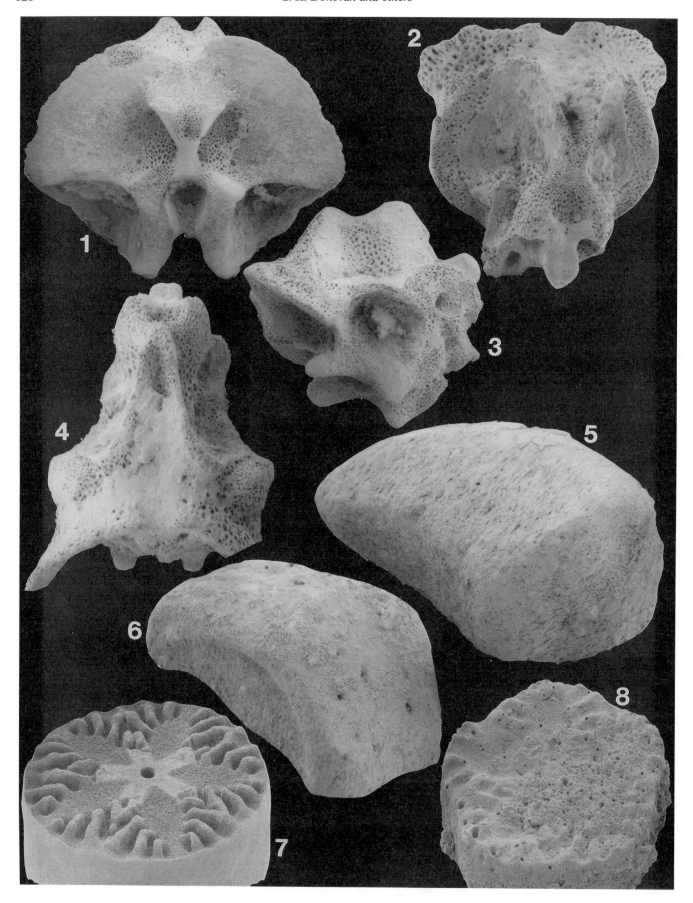

Figure 3. 1–4, ophiuroid vertebral ossicles indet., last interglacial, Falmouth Formation, parish of St. Ann. 1: BMNH E54438h, proximal facet, ×25. 2, 4, BMNH E54436g, terminal vertebra—2: distal view, ×70; 4: aboral view, ×70. 3: BMNH E54438a, lateral view, ×35. 5, 6, asteroid marginal ossicles indet., Eocene, Yellow Limestone Group, parish of Hanover—5: BMNH E54450a, wedge-shaped ossicle, ×35: 6: BMNH E54447a, oblique lateral view, ×16. 7, *Cenocrinus asterius* (Linné), Recent, fore reef, Discovery Bay, parish of St. Ann, BMNH E70449a, oblique view of articular facet, ×15. 8, cf. *Cenocrinus asterius* (Linné), Lower Pleistocene, Manchioneal Formation, parish of Portland, BMNH E70448, oblique view of articular facet, ×12. All figures are scanning electron micrographs of specimens coated with 60% gold-palladium.

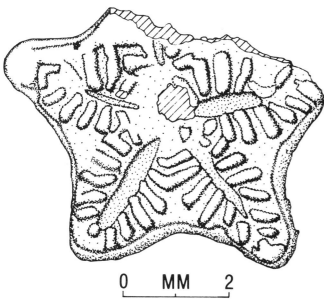

Figure 4. *Diplocrinus* sp., UF 38939, articular facet. Camera lucida drawing.

(1961, plate 57, Figs. 9, 10), and North American specimens were illustrated by Peck (1973).

Class ASTEROIDEA de Blainville
Subclass NEOASTEROIDEA Gale
Family ASTROPECTINIDAE Gray or
Family GONIASTERIDAE Forbes

Asteroid marginal ossicles indet.
(Figure 3.5, 3.6)

Material, localities, and horizons. Asteroid marginal ossicles have so far been recovered from the following localities in Jamaica.

(A) Maastrichtian *Titanosarcolites* Limestone, Jerusalem Mountain Inlier, parish of Hanover, GR 123522 (Fig. 1, locality 6). A solitary ossicle, BMNH E54441.

(B) Ducketts crossroads (USGS 30026, field number J-71-19; seven ossicles) and Ducketts (USGS 30030, field number J-71-22; at least 14 ossicles), both within the Marchmont Inlier, parish of Westmoreland (Fig. 1, locality 7). Maastrichtian.

(C) Middle Eocene, Albert Town Member, Chapelton Formation, Yellow Limestone Group, on Lorrimers road, north of Christiana, parish of Manchester, GR 376467 (Fig. 1, locality 8; locality 6 of Robinson, 1988). Eleven ossicles mounted on five scanning electron microscope (SEM) stubs; BMNH E54442a–c, E54443, E54444, E54445a, b, and E54446a–d.

(D) Lower or Middle Eocene, Yellow Limestone Group, near Rock Spring, parish of Hanover, GR 123528 (Fig. 1, locality 9). Thirteen ossicles mounted on four SEM stubs; BMNH E54447a–c, E54448a–d, E54449a–d and E54450a, b (Fig. 3.5, 3.6).

(E) Middle Eocene, Swanswick Formation, White Limestone Supergroup, Pimento Hill, Beecher Town, parish of St. Ann, GR 510532 (Fig. 1, locality 10). Locality 2 of Donovan and Gordon (1989; Donovan et al., 1989). A collection of 17 ossicles, BMNH E54451.

(F) Harold L. Dixon (research in progress) has collected large and distinctive marginal ossicles in association with *Clypeaster* sp. from the type area of the Oligocene Browns Town Formation, White Limestone Supergroup, Browns Town, parish of St. Ann (Fig. 1, locality 11).

Remarks. Dr. A. S. Gale (written communication) has suggested that the available Jamaican material was derived from astropectinids and/or

Figure 5. 1, isocrinid sp. indet., BMNH EE2182, strongly pitted articular facet, ×17. 2–6, brachials from comatulid crinoids, USGS field number J-66-55—2: USNM 458892, ×14. 3: USNM 458893, ×14; 4: USNM 458894, ×13; 5: USNM 458895, ×14; 6: USNM 458896, ×14. All figures are scanning electron micrographs of specimens coated with 60% gold-palladium (1) or gold (2–6).

goniasterids. However, until a much larger collection from each locality becomes available, precise taxonomic assignment is problematic.

Class OPHIUROIDEA Gray
Incerti ordinis
Ophiuroid vertebral ossicles indet.
(Figure 3.1–3.4)

Material, locality, and horizon. Thirty-five vertebral ossicles mounted on five SEM stubs: BMNH E54435a–g, E54436a–g, E54437a–g, E54438a–h, and E54439a–f (Fig. 3.1–3.4); numerous further, but unmounted, ossicles, BMNH E54440 (unfortunately, E54436g was lost in transit to the BMNH). These ossicles were picked by CMG from bulk samples from a locality on the east side of Rio Bueno Harbour (Fig. 1, locality 12; site 7 of Liddell et al., 1984; locality 3 of Donovan and Gordon, 1989), parish of St. Ann, GR 394572. Late Pleistocene Falmouth Formation (= reef terrace 1 of Cant, 1972; last interglacial in age); molluscan biomicritic wackestone facies of Larson (1983). This locality is interpreted as a patch reef in a back-reef lagoonal environment (Larson, 1983).

Remarks. Ophiuroid taxonomy is mainly based on the features of the disk and the gross anatomy of complete specimens. Although Murakami (1963) made a synoptic study of ophiuroid oral and dental plates, no similar reference allows identification of disarticulated vertebral ossicles. Thus, it is not even certain how many species are represented by the Rio Bueno material. Molluscs and echinoids from this locality belong to extant Caribbean, shallow-water species; it is probable that the ophiuroids are similarly referable.

ACKNOWLEDGMENTS

We are very grateful to Drs. Gordon Hendler (Natural History Museum of Los Angeles County) and Andrew Gale (Natural History Museum, London) for their comments on our ophiuroid and asteroid ossicles, respectively. Particular thanks to Joe Hazel (Louisiana State University) for providing a copy of Peck's letter on Jamaican *Saccocoma*, to Roger Portell for both telling us about and loaning the Montpelier Group isocrinid columnal, and to Chris Paul for providing a copy of Peck's *Applinocrinus* paper. Heather Davidson kindly took the SE micrographs of comatulid brachial plates. This paper is Discovery Bay Marine Laboratory, University of the West Indies, contribution number 513.

REFERENCES CITED

Banner, F. T., and Blow, W. H., 1965, Progress in the planktonic foraminiferal biostratigraphy of the Neogene: Nature, v. 208, p. 1164–1166.

Bather, F. A., 1924, *Saccocoma cretacea* n. sp., a Senonian crinoid: Proceedings of the Geologists' Association, v. 35, p. 111–121.

Broadhead, T. W., 1980, Biostratigraphic potential, *in* Broadhead, T. W., and Waters, J. A., eds., Echinoderms: Notes for a short course: University of Tennessee Studies in Geology, v. 3, p. 40–58.

Cant, R. V., 1972, Jamaica's Pleistocene reef terraces: Journal of the Geological Society of Jamaica, v. 12, p. 13–17.

Donovan, S. K., 1984a, A crinoid columnal morphospecies from the Upper Ordovician of Girvan and Kazakhstan: Scottish Journal of Geology, v. 20, p. 135–142.

Donovan, S. K., 1984b, Stem morphology of the Recent crinoid *Chaldocrinus* (*Neocrinus*) *decorus*: Palaeontology, v. 27, p. 825–841.

Donovan, S. K., 1985, Biostratigraphy and evolution of crinoid columnals from the Ordovician of Britain, *in* Keegan, B. F., and O'Connor, B.D.S., eds., Echinodermata: Proceedings, Fifth International Echinoderm Conference, Galway, 24–29 September 1984: Rotterdam, A. A. Balkema, p. 19–24.

Donovan, S. K., 1989, An isocrinid (Echinodermata: Crinoidea) from the Lower Pleistocene of Portland, eastern Jamaica: Journal of the Geological Society of Jamaica, v. 25 (for 1988), p. 3–7.

Donovan, S. K., and Gordon, C. M., 1989, Report of a field meeting to selected localities in St. Andrew and St. Ann, 25 February 1989: Journal of the Geological Society of Jamaica, v. 26, p. 51–54.

Donovan, S. K., Gordon, C. M., Schickler, W. F., and Dixon, H. L., 1989, An Eocene age for the "Montpelier Formation" at Beecher Town, St. Ann, Jamaica, using echinoids for correlation: Journal of the Geological Society of Jamaica, v. 26, p. 5–9.

Donovan, S. K., Scott, A. D., and Veltkamp, C. J., 1991, A late Middle Eocene echinoid fauna from Portland, northeastern Jamaica: Journal of the Geological Society of Jamaica, v. 28, p. 1–8.

Larson, D. C., 1983, Depositional facies and diagenetic fabrics in the Late Pleistocene Falmouth Formation of Jamaica [M.S. thesis]: Norman, University of Oklahoma, 228 p.

Liddell, W. D., Ohlhorst, S. L., and Coates, A. G., 1984, Modern and ancient carbonate environments of Jamaica: University of Miami, Sedimenta, v. 10, 98 p.

Linné, C., 1767, Systema naturae, v. 1 (2): Laurentius Salarius, Holmia, p. 533–1327. (Not seen.)

Macurda, D. B., Jr., and Roux, M., 1981, The skeletal morphology of the isocrinid crinoids *Annacrinus wyvillethomsoni* and *Diplocrinus maclearanus*: Contributions from the Museum of Paleontology, University of Michigan, v. 25, p. 169–219.

Meyer, D. L., Messing, C. G., and Macurda, D. B., Jr., 1978, Zoogeography of tropical western Atlantic Crinoidea (Echinodermata): Bulletin of Marine Science, v. 28, p. 412–441.

Murakami, S., 1963, The dental and oral plates of Ophiuroidea: Transactions of the Royal Society of New Zealand, Zoology, v. 4, p. 1–48.

Peck, R. E., 1973, *Applinocrinus*, a new genus of Cretaceous microcrinoids and its distribution in North America: Journal of Paleontology, v. 47, p. 94–100.

Rasmussen, H. W., 1961, A monograph on the Cretaceous Crinoidea: Biologiske Skrifter udgivet af Det Kongelige Danske Videnskabernes Selskab, v. 12, p. 1–428.

Robinson, E., 1969, Geological field guide to Neogene sections in Jamaica West Indies: Journal of the Geological Society of Jamaica, v. 10, p. 1–24.

Robinson, E., 1971, Late Tertiary erosion surfaces and Pleistocene sea levels in Jamaica, *in* Mattson, P. H., ed., Transactions of the Fifth Caribbean Geological Conference, St. Thomas, Virgin Islands, 1–5 July 1968: New York, Queens College Press, p. 213–221.

Robinson, E., 1988, Late Cretaceous and early Tertiary sedimentary rocks of the Central Inlier, Jamaica: Journal of the Geological Society of Jamaica, v. 24 (for 1987), p. 49–67.

Roux, M., 1977, The stalk-joints of Recent Isocrinidae (Crinoidea): Bulletin of the British Museum (Natural History), Zoology, v. 32, p. 45–64.

Schäfer, W., 1972, Ecology and palaeoecology of marine environments (translated I. Oertel, edited G. Y. Craig): Chicago, University of Chicago Press, 568 p.

Simms, M. J., 1988, The phylogeny of post-Palaeozoic crinoids, *in* Paul, C.R.C., and Smith, A. B., eds., Echinoderm phylogeny and evolutionary biology: Oxford, Clarendon Press, p. 269–284.

Springer, F., 1924, A Tertiary crinoid from the West Indies: Proceedings of the United States National Museum, v. 65, p. 1–8.

MANUSCRIPT ACCEPTED BY THE SOCIETY NOVEMBER 11, 1992

Calcareous nannofossil stratigraphy of the Neogene formations of eastern Jamaica

Marie-Pierre Aubry*
Centre de Paléontologie stratigraphique et Paléoécologie, Université Claude Bernard, 27-43 Bd du 11 Novembre, 69622 Villeurbanne Cedex, France, and Woods Hole Oceanographic Institution, Woods Hole, Massachusetts 02543

ABSTRACT

While numerous studies have been devoted to the planktonic and benthic foraminifera in the Neogene of Jamaica, little attention has been paid to the calcareous nannofossils. This contribution is thus a documentation of the calcareous nannofossils in the lower Miocene to upper Pliocene deposits of Jamaica. The Buff Bay section, the most extensive Neogene section in eastern Jamaica, is studied in great detail, and eight other sections, including the San San Bay and the Bowden type sections, are discussed as well. Current correlations between the zonal schemes established from calcareous nannofossils and planktonic foraminifera are discussed, based upon direct correlation between calcareous nannofossil and planktonic foraminiferal zones in these sections. Integration between calcareous planktonic microfossil stratigraphies, and magnetostratigraphy when available, leads to the delineation of regional unconformities and to the interpretation of the Neogene stratigraphic record of eastern Jamaica in terms of sequence stratigraphy. It is shown that the Buff Bay Formation, the San San Clay, and the Bowden Formation correspond to separate unconformable stratigraphic sequences. As a result, it is suggested that the San San Clay be regarded as a distinct formation rather than part of the Buff Bay or the Bowden formations as currently accepted. Several of the "holotype" and "paratype" localities that Blow (1969) designated for his Neogene planktonic foraminiferal zones (N-Zones) are located in Jamaica. Direct correlation between these and the calcareous nannofossil schemes of Martini (1971) and Okada and Bukry (1980) are established and the implications are discussed.

INTRODUCTION

Numerous studies have been devoted to the taxonomic description and stratigraphic and paleobathymetric distributions of calcareous microfossils (planktonic and benthic foraminifera and ostracodes) in the Neogene formations of Jamaica (see Berggren, this volume; Katz and Miller, this volume; van den Bold, this volume). In addition to the fact that the Upper White Limestone and Coastal groups have been tied to Bolli's planktonic foraminiferal zonal scheme (Robinson, 1969), the Montpelier, Buff Bay, and Bowden formations yield the "holotype" and/or "paratype" localities of half of the middle Miocene to Pleistocene planktonic foraminiferal Neogene (N) zones defined by Blow (1969, 1979).

In contrast, there have been no comprehensive studies of the calcareous nannofossils in the Neogene formations of Jamaica. There are only scarce and brief citations in the literature relative to their occurrence (Bramlette and Riedel, 1954; Gartner, 1967a, b; Hay et al., 1967; Black, 1967, 1971, 1972). *Amaurolithus tricorniculatus, Cyclococcolithus leptoporus, C. aequiscutum,* and *Discoaster perclarus* were reported from the San San Clay (the first three taxa by Gartner, 1967a, b; the latter by Hay in Hay et al., 1967); *Discoaster challengeri* was reported questionably from the Bowden Formation (Bramlette and Riedel, 1954; and *D. obtusus* was cited in the lower Miocene (undifferentiated; Black, 1972). Also, the holotypes of *Discolithina millepunctata* Gartner 1967b, *Dictyococcites antillarum* Black 1967, and *Helicosphaera burkei* Black 1971 were described, respectively, from the San San Clay (*D. millepunctata*) and the Upper White Limestone. These reports yield little stratigraphic information except that of *A. tricorniculatus* in the San San Clay, which indicates an upper

---

*Present address: Laboratoire de Géologie du Quaternaire, CNRS-Luminy, Case 907, 13288 Marseille, Cedex 9, France.

Aubry, M.-P., 1993, Calcareous nannofossil stratigraphy of the Neogene formations of eastern Jamaica, *in* Wright, R. M., and Robinson, E., eds., Biostratigraphy of Jamaica: Boulder, Colorado, Geological Society of America Memoir 182.

upper Miocene to lower Pliocene zonal assignment. While the planktonic foraminiferal zones of Bolli have been directly correlated to calcareous nannofossil schemes, based on the description of the calcareous nannofossil content of the type localities of these zones (Bramlette and Wilcoxon, 1967; Hay et al., 1967; Martini, 1971), Blow's zones have been only indirectly correlated to a calcareous nannofossil scheme (Gartner, 1969). As a result, the only information available concerning calcareous nannofossil stratigraphy of the Neogene of Jamaica is restricted to the report of the *Discoaster quinqueramus* Zone of Gartner (equivalent to the upper part of Zone NN11 and to Subzone CN9b), at the base of the San San Clay at San San Bay (Gartner, 1969).

The contribution of this chapter is fourfold. First, it documents the calcareous nannofossil assemblages in the Neogene formations developed in eastern Jamaica. The study is mainly centered on the biozonal succession in the Buff Bay section, but I also report on the age of samples collected from other sections, such as that of San San Bay. Second, it reconsiders the lithostratigraphic framework currently applied to the Neogene of eastern Jamaica in light of integrated calcareous nannofossil and planktonic foraminiferal stratigraphy. Third, it establishes the calcareous nannofossil zonal ages of the type localities chosen in Jamaica by Blow (1969, 1979) for the planktonic foraminiferal zones he defined (N-Zones), using the calcareous nannofossil zonations of Martini (1971) and Bukry (1973, 1975). Finally, as part of an integrated magnetobiostratigraphic effort, it provides new information pertinent to the late middle Miocene part of the time-scale. Although the last contribution will be only briefly discussed here, this chapter serves as a basis for further discussion.

REGIONAL SETTING AND LITHOSTRATIGRAPHY

The first comprehensive study of the geology of Jamaica is that of Hill (1899), who clarified the stratigraphic relationships suggested by earlier authors and established a coherent lithostratigraphic framework that remains essentially valid, although little used in subsequent works. Hill gave a thorough account of the regional extent and paleontological content of each lithostratigraphic unit and described and illustrated many sections, among them the ones discussed in this chapter.

More recent reviews of the geology of Jamaica can be found in Zans et al. (1962), Robinson (1969), Wright (1974), Steineck (1974, 1981), and Emery and Uchupi (1984).

The Neogene formations studied in this paper are those assigned to the Upper White Limestone (in part) and Coastal Group by Robinson (1969); they outcrop along the east coast of Jamaica between the villages of Buff Bay and Manchioneal, respectively west and southeast of Port Antonio (Fig. 1). They include the Buff Bay, Bowden, and Manchioneal formations (as part of the Coastal Group) and the Spring Garden Member of the Montpelier Formation.

The Montpelier Formation (Robinson, 1969) has been regarded as the youngest (Miocene) part of the White Limestone, a widespread unit that covers two-thirds of Jamaica, extends from middle Eocene to middle Miocene, and has long been thought of as a neritic deposit (Hose and Versey, 1956; Zans et al., 1962; Wright, 1968; Robinson, 1969). Steineck (1974) showed that, in fact, from middle Eocene to middle Miocene, two domains of carbonate sedimentation were juxtaposed, separated by the Duanvale-Wagwater escarpment (Fig. 2). While a shallow carbonate platform, the Cornwall-Middlesex Platform, extended south of the fault system, deep water carbonates were accumulating to the north in the Cayman trough. Appropriately, Steineck (1974, 1981) subdivided the genetically artificial White Limestone into two groups, both extending from middle Eocene to middle Miocene but of a different genetic origin (Table 1). The Clarendon Group includes the platform deposits, whereas the Montpelier Group comprises the deep water limestones that the Montpelier Formation (Hill, 1899) was originally meant to designate. For the sake of uniformity with other chapters in this volume (Berggren, this volume; Katz and Miller, this volume), a conservative, although not necessarily suitable, terminology is used in this chapter. The Montpelier Formation of Robinson (1969) should be regarded as the upper part of the Montpelier Group of Steineck (1974, 1981), and following Steineck (1974), the Spring Garden Member sensu Robinson (1969) may best be referred to as a formation.

The "Coastal series" of Hill (1899), originally denominated to group all rocks younger than the "Oceanic series" (i.e., younger than the Montpelier Group), has been designated the "Coastal Group" by Robinson (1969), who formalized the Neogene succession of Jamaica. The Coastal Group is restricted to the eastern part of Jamaica and is exposed in discontinuous sections in the Buff Bay, San San Bay, and Port Antonio areas and in the Manchioneal and Bowden districts. These sections have been described in detail by Robinson (1969). Of the seven formations of the Coastal Group (Table 1), only the Buff Bay, Dowden, and Manchioneal formations are discussed in this paper. Special attention is given to the San San Clay, regarded as a lateral equivalent of the Bowden Formation (Robinson, 1969).

Table 1 summarizes the middle Eocene to Pleistocene stratigraphic succession in Jamaica, with emphasis on the Neogene sequence. Correlations to the Paleogene chronostratigraphic scheme are made in broad terms because many uncertainties remain. For instance, although it is generally believed that there is a stratigraphic gap across the Oligocene/Miocene boundary (e.g., Zans et al., 1962; Robinson, 1969; Emery and Uchupi, 1984) there is growing evidence that the Montpelier Group may be continuous across this boundary (Steineck, 1981). Correlations to the Neogene chronostratigraphic scheme are discussed below.

CALCAREOUS NANNOFOSSIL STRATIGRAPHY OF THE BUFF BAY SECTION

Methods

As a result of the combined effects of such factors as faulting and folding since the late Pliocene, dense vegetation, intensive

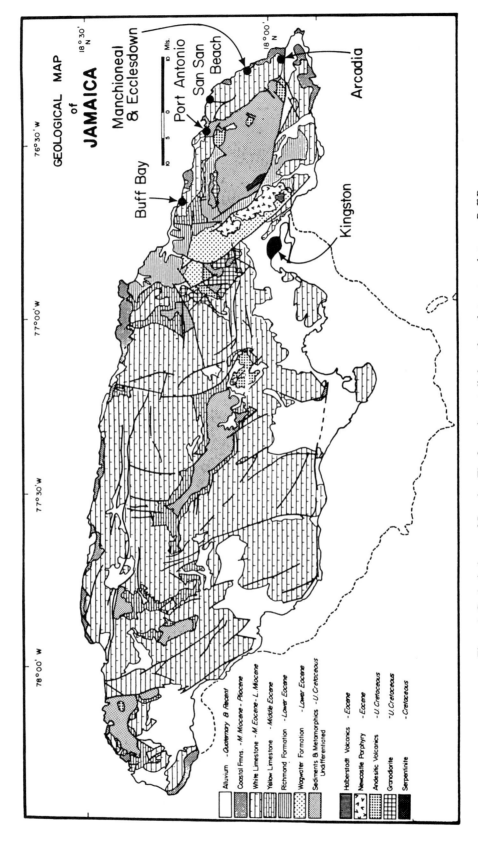

Figure 1. Geological map of Jamaica. The formations studied are those that outcrop between Buff Bay and Manchioneal (from Wright, 1974).

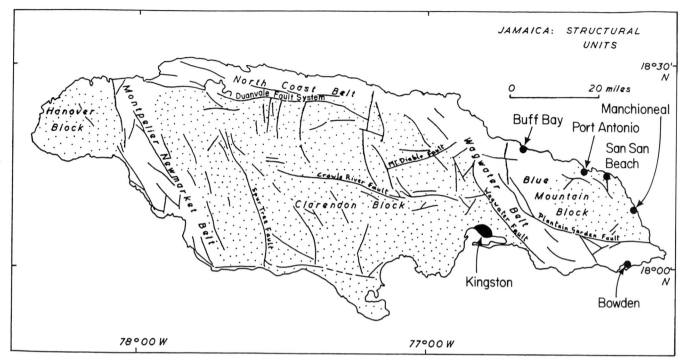

Figure 2. Structural map of Jamaica (from Wright, 1974).

erosion, and massive landslides, there are few good exposures in Jamaica. The Buff Bay section is one exception and exposes almost continuously a 320-m-thick Neogene section. The section, which is marked by intensive faulting at its base, comprises the Spring Garden Member of the Montpelier Formation and the Buff Bay Formation as they were defined by Robinson (1969) and an upper unit that has been correlated with the Bowden Formation by the latter author (1969). While the Buff Bay Formation overlies conformably the Spring Garden Member, the Bowden Formation correlative is unconformable with the Buff Bay Formation (Robinson, 1969). A collection of samples was originally made by E. Robinson (1969) and distributed to specialists around the world. These samples (ER 146/21 to ER 146/44), provided by Barry Kohl, served for a preliminary analysis of the calcareous nannofossil content of these formations, as discussed below (Figs. 3 and 4).

The section presently consists of six outcrops along Highway 4, designated here as Lower (Pots and Pans), Middle, Main, Slide, Dead Goat Gully, and West (Fig. 5). The Spring Garden Member is exposed in the Lower and Middle roadcuts, the Buff Bay Formation in the Main, Slide, and Dead Goat Gully roadcuts, and the Bowden equivalent in the West Roadcut.

The section was sampled in April 1987 under the guidance of E. Robinson (Florida International University) by a joint field party from the Lamont-Doherty Geological Observatory (M. van Fossen, D. V. Kent, K. G. Miller), the University of Lyon (M-P. Aubry), UNOCAL (G. Jones, C. Stuart, R. C. Tjalsma), and the Woods Hole Oceanographic Institution (W. A. Berggren). The section was measured and described by C. Stuart (Stuart et al., 1987). The exposed lithologies were similar to those described by Hill (1899) and Robinson (1969). The boundary between the Spring Garden Member and the Buff Bay Formation was moved slightly upward (Fig. 5) compared to its placement by Robinson (1969) (following C. Stuart, personal communication, 1987; see discussion in Katz and Miller, this volume). The location of the original samples ER 146/21 to 44 relative to the present road cut, which was retraced in 1974, was estimated by E. Robinson during the 1987 fieldwork.

Paleontological samples were taken from all outcrops; oriented samples for magnetostratigraphy were also collected, except from the Dead Goat Gully and West roadcuts. Samples from the Lower and Middle roadcuts are numbered (downsection) SG1 to SG32. Samples from the Main and Slide roadcuts are numbered (upsection), respectively, 165 to 222B and 223 to 240. Those from the Dead Goat Gully and West roadcuts are numbered, respectively, 241 to 243 and 244 to 252. Standard smear slides for calcareous nannofossil studies were prepared from all samples.

Correlation between planktonic foraminiferal and calcareous nannofossil biostratigraphies are discussed here and in Berggren (this volume). Integration of the biostratigraphic results with the magnetostratigraphic data (D. V. Kent and M. van Fossen, unpublished data, 1990) will be discussed elsewhere.

Biozonal scheme

The zonal scheme of Martini (1971) and the subzonal scheme of Bukry (1973, 1975; codified in Okada and Bukry, 1980) are jointly used in this chapter (Fig. 6). On the one hand,

TABLE 1. LITHOSTRATIGRAPHIC SUBDIVISIONS OF THE MIDDLE EOCENE TO RECENT FORMATIONS OF JAMAICA ACCORDING TO DIFFERENT AUTHORS

Series	Zans in Zans et al. (1962) Robinson (1969)		Steineck (1974, 1981)				
			South	Central			NE Coastal Jamaica
Neogene	Falmouth Formation						
	Port Morant Formation						
	Harbour View Formation	Coastal Group				Coastal Group	
	Manchioneal Formation*						
	Bowden Formation		August Town Formation				
	Buff Bay Formation* (including the San San Clays*)						
	August Town Formation			Subaerial erosion and karstification	Clarendon Group		Buff Bay Formation
	Montpelier Formation (including the Garden Spring Member*)					Duanvale-Wagwater Fault System	Spring Garden Formation*
	Newport Formation		Newport Limestone				Sign Beds
Middle Eocene-Oligocene	Walderston and Brown's Town limestones	White Limestone Group	Walderston Formation	Brown's Town Formation		Montpelier Group	Bonny Gate Member
	Somerset and Gibraltar limestones			Gibraltar Formation			
	Claremont Limestone		Claremont Limestone				Lloyds Member
				Swanswick Formation			
	Troy, Ipswich, and Swanswick limestones		Troy Formation				
	Yellow Limestone		Chapelton Formation			Font Hill Formation	

*Units studied in this chapter.

the subzonal scheme of Bukry allows greater stratigraphic refinement for the upper middle Miocene to lower Pliocene interval than does Martini's scheme. On the other hand, the common use by Bukry of alternate markers and of acmes to define zonal boundaries leads to ambiguities. In this study, priority has been given to the markers used by Martini (1971) to delineate zonal boundaries.

It was not possible to distinguish between Subzones CN11a and CN11b based on the acme of *Discoaster asymmetricus*, because the abundance of this species is highly variable. While Rio et al. (1990) encourage the use of the lowest occurrence (LO) of *Discoaster tamalis* as a good lower Pliocene marker for the Mediterranean area (LO slightly above the beginning of the acme of *D. asymmetricus*), this species was observed as low as in Subzone CN9b in Jamaica, and this taxon was found at all levels from this subzone to Subzone CN12a. Similar occurrence of *D. tamalis* in upper Miocene deposits from Italy was mentioned by Raffi and Rio (1979).

Calcareous nannofossil content of samples ER 146/21 to ER 146/44

Although the analysis of these samples was meant to be only preliminary, it appeared that it would be useful to thoroughly describe their nannofossil content and to delineate the biozonal framework of this study. In addition, while thir nannofossil content is critical for the interpretation of the whole section, the levels at which samples ER 146/35 to 37 were collected are now inaccessible (covered by a wall following a landslide). Finally, some of these samples serve as type-material for some of Blow's

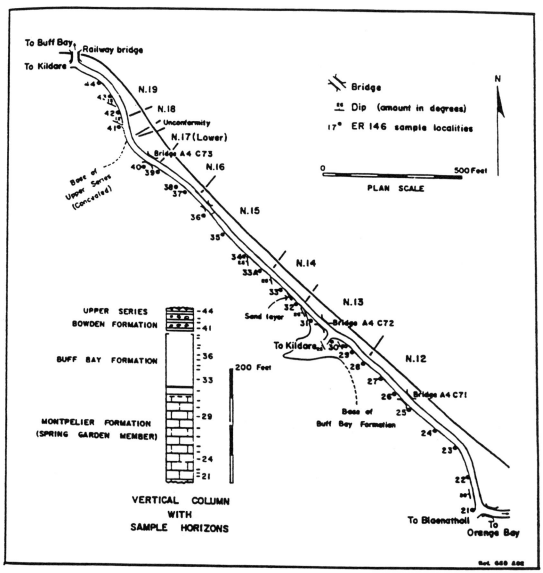

Figure 3. Plan of the coastal road (retraced in 1974 and renamed Highway 4) east of Buff Bay, showing the location of the ER samples and the approximate planktonic foraminiferal zonal boundaries (Banner and Blow's zones, 1965) (from Robinson, 1969).

zones, which warrants a detailed study of their calcareous nannofossil assemblages.

The ER sample collection was completed with two samples (WHA 81-3 and -11) collected in 1981 in the lower part of the Spring Garden Member and two other samples (TE 616 and 4 TE 181) taken in 1974 (all provided by B. Kohl), thought to be equivalent to the ER samples.

This suite of samples contains abundant, highly diversified assemblages, characteristic of warm waters, with abundant discoasters, lopadoliths, and cribriliths. Preservation is good in the upper Miocene (ER 146/33 to 40) and the Pliocene (ER 146/41 and 43) but poor in the middle Miocene (WHA 81-11 to TR 616) where discoasters are considerably overgrown. In the upper Miocene and Pliocene, dissolution has affected the assemblages to some degree, so that the tips of the arms of numerous discoasters are broken, preventing confident specific determination. This handicap has little stratigraphic relevance but precluded the satisfactory estimation of the abundance of various species at different levels. Abundance estimates given in Figure 4 were therefore broadly made. "Very abundant" (double thickness line) refers to predominant species in an assemblage. In the material studied, predominance is linked to poor preservation; the abundance of *Sphenolithus heteromorphus* in samples WHA 81-11 and -3 results from its resistance to diagenetic effects, as indicated by the thickly overgrown discoasters. "Rare" (thin line) refers to species of which only a few specimens were observed or could be confi-

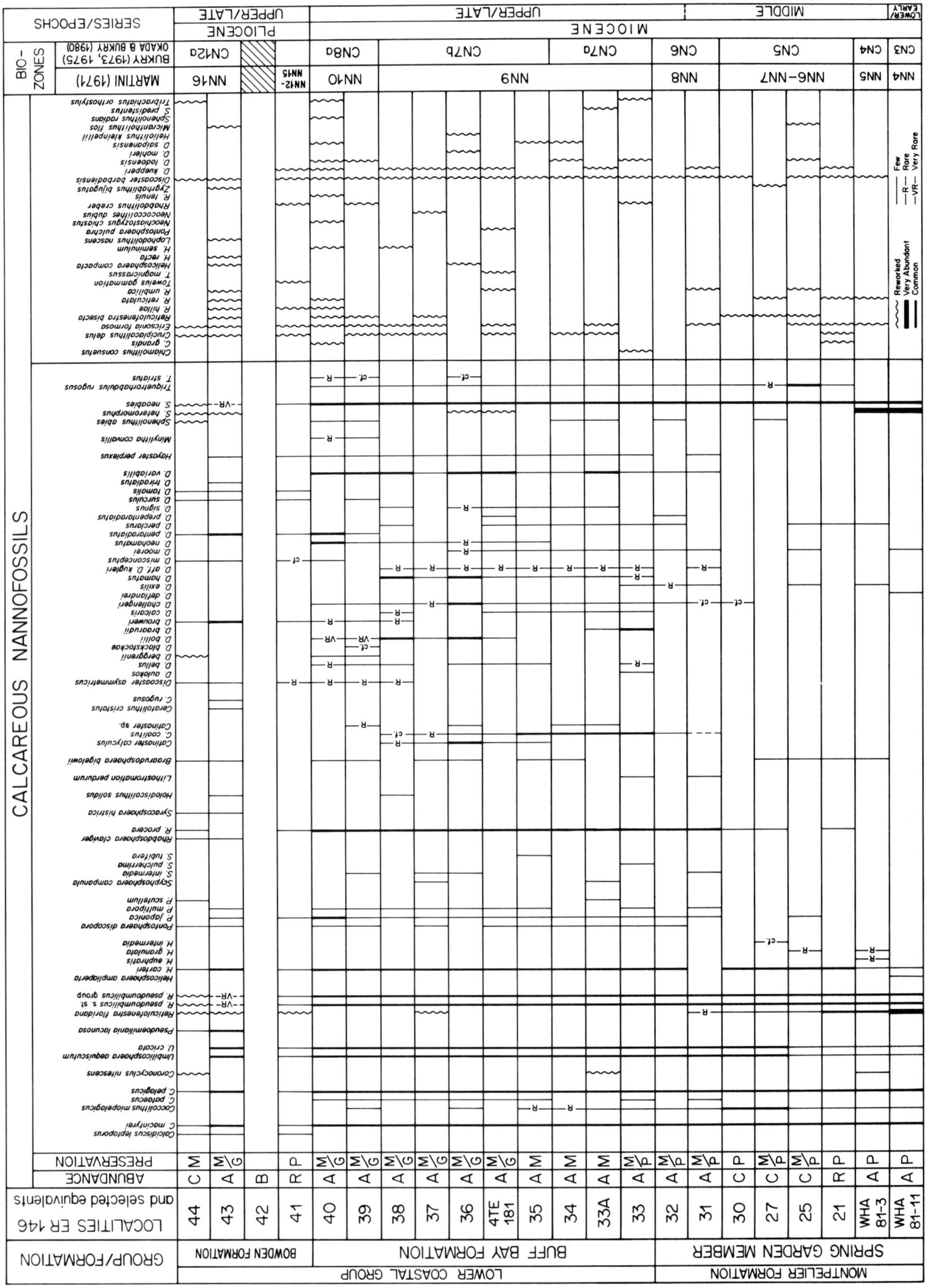

Figure 4. Distribution of calcareous nannofossil species in the ER samples (Robinson, 1969). Small taxa (<4 μ), such as *Scapholithus fossilis*, *Syracosphaera* sp., and others, are not recorded.

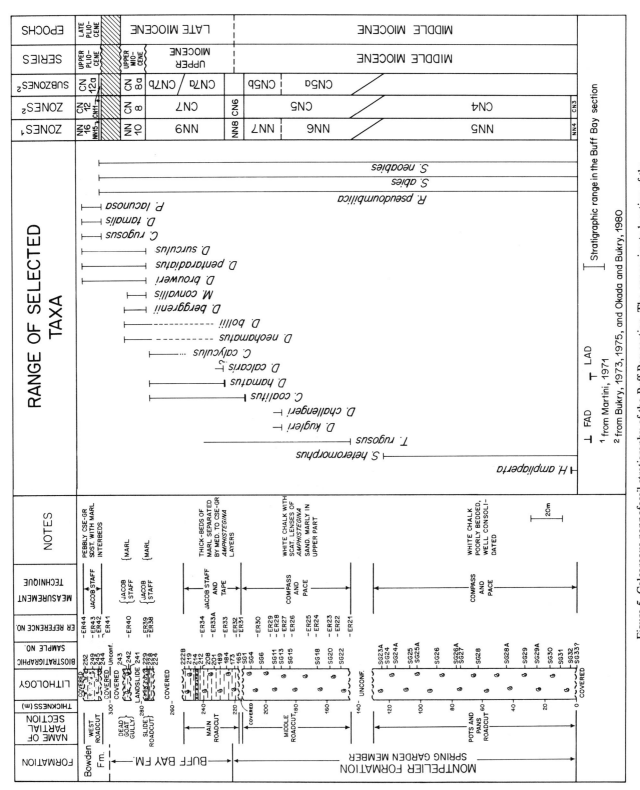

Figure 5. Calcareous nannofossil stratigraphy of the Buff Bay section. The approximate location of the ER samples with respect to the 1987 sampling is shown. Reference to these samples should be ER 146/21 to ER 146/44 rather than ER 21 to ER 44.

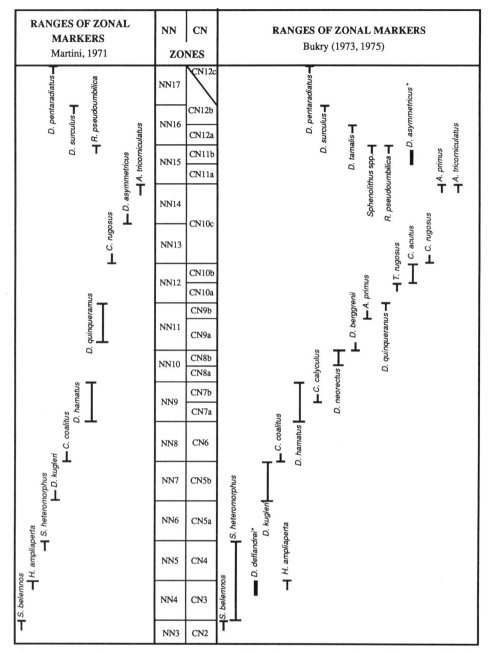

Figure 6. Correlation between the zonal schemes of Martini (1971) and Bukry (1973, 1975; codified in Okada and Bukry, 1980). *: Acme.

dently identified during the analysis. "Very rare" (dashed lines) refers to species of which only one or two specimens were encountered. All other species were regarded as "common" (thick line). The specific inventory has been as thorough as possible, although a number of taxa are not included in the range chart (Fig. 4). Lopadoliths are quite diversified in the assemblages, but most types are not common enough to allow specific determination based on more than one specimen. Only four species of the genus *Scyphosphaera* are reported here, although specific diversity is undoubtedly larger. A number of minute forms undescribed or poorly known, probably representative of the genus *Syracosphaera*, were encountered. They are not discussed here. Also, species of little or no stratigraphic significance (e.g., *Scapholithus fossilis, Helicosphaera burkei, H. stalis*), encountered in a few samples, are not included in the range chart but are illustrated.

Reworking is intensive in most samples; it seems to increase in the Pliocene samples, although this may be related to weaker diagenetic effects. Reworked species have been noted, although no special effort was made to inventory them. Their significance is discussed below.

From a stratigraphic point of view, this suite of samples

indicates that the Buff Bay section exposes sediments that extend from the uppermost lower Miocene (Zone NN4) to the lower upper Pliocene (Zone NN16) and includes two important stratigraphic gaps, one in the middle upper Miocene, the other spanning the upper upper Miocene and lower Pliocene (Figs. 4, 5).

Helicosphaera ampliaperta is common in sample WHA-81-11, assigned to Zone NN4 (=Zone CN3). It is absent in sample WHA-81-3, characterized by *Sphenolithus heteromorphus* and assigned to Zone NN5 (=Zone CN4). Because of poor preservation, no stratigraphic marker could be confidently identified in the next (upward) four samples (ER 146/21 to ER 146/30). They are assumed to represent a continuous section through the NN6-NN7 (=CN5) zonal interval. If not reworked, the highest occurrence of *Reticulofenestra floridana* may indicate a level within Zone NN7 (=Subzone CN5b) (Roth and Thierstein, 1972; Miller et al., 1985). However, without a confident determination of *Discoaster challengeri*, this correlation remains tentative. Sample ER 146/31 yields poorly preserved calcareous nannofossils, among them a few recrystallized globular forms that represent *Catinaster coalitus*. This species is well represented in sample ER 146/32, and both samples (ER 146/31 and 32) are assigned to Zone NN8 (=Zone CN6). *Catinaster coalitus* is very common in sample ER 146/33. Extremely rare, symmetrical, 5-rayed discoasters, poorly preserved but assignable to *Discoaster hamatus*, also occur, suggesting that sample ER 146/33 is located close to the NN8/NN9 zonal boundary. Assignment of this sample to Subzone CN7a is supported by the relatively common occurrence of *Discoaster calcaris* (Bukry, 1973). Samples ER 146/33A and 34 also belong to subzone CN7a. The co-occurrence of *Discoaster hamatus* and *Catinaster calyculus* in samples ER 146/35 to 38 characterizes Subzone CN7b (= the upper part of Zone NN9). Samples ER 146/39 and 40 yield *Discoaster bollii, D. neohamatus, D. pentaradiatus,* and *Minylitha convallis* (rare). These occurrences and the absence of *Catinaster coalitus* and *C. calyculus* indicate the mid to upper part of Subzone CN8a correlative with part of Zone NN10. The zonal position of sample ER 146/41 is difficult to determine within the interval NN12–NN15. Calcareous nannofossils are scarce, although discoasters are relatively common, and ceratoliths are absent (unless some broken or overgrown forms represent *Ceratolithus acutus*; despite a very long search, however, the presence of this species could not be established confidently). Extremely rare specimens of *R. pseudoumbilicus* and *Sphenolithus neoabies* occur in samples ER 146/43 and 44, where they are considered reworked. These samples are therefore assigned to Zone NN16. Alternatively (if *R. pseudoumbilicus* and *S. neoabies* are not reworked), they belong to the uppermost part of Zone NN15. With reference to Bukry's zonal scheme, samples ER 146/43 and 44 are assigned to Subzone CN12a, based on the occurrence of *Discoaster tamalis* and the absence of *R. pseudoumbilicus, Sphenolithus abies,* and *S. neoabies.*

Although the spacing between the ER samples in the lower part of the Buff Bay section is large, samples ER 146/35 to 44 were taken at closer intervals. Thus the juxtaposition of Subzone CN7b (Sample ER 146/38) and the mid to upper part of Subzone CN8a (Sample ER 146/39) suggests an unconformity between the two samples. The well-known unconformity between the Buff Bay Formation (Sample ER 146/40) and the Bowden Formation equivalent (Sample ER 146/41) (Robinson, 1969) is reflected by the juxtaposition of Zones NN 10 (Subzone CN8a) and NN15. These unconformities will be discussed below.

It is important to indicate here that some of the ER samples were taken close to zonal boundaries (Fig. 5), with the result that duplicate collection of samples may not yield the same stratigraphic information. Sample WHA 81-3 was thought to be equivalent to ER 146/21. While the former yields *Sphenolithus heteromorphus* (Zone NN5), the latter belongs to the undifferentiated NN6–NN7 zonal interval. Likewise, sample TE 616 was thought to be equivalent to sample ER 146/33, which is located close to the NN8/NN9 zonal boundary. Sample TE 616 is assigned to Zone NN8, and sample ER 146/33 is assigned to basal Zone NN9. It is possible that the assignment of sample TE 616 to Zone NN8 rather than to Zone NN9 results from poor preservation. It should be emphasized that the composition of the calcareous nannofossil assemblages in the section is greatly affected by preservation. Strong differences in preservation were observed in samples taken at the same level but in different years, demonstrating the profound effect of weathering. Samples collected in 1969 generally yield better-preserved calcareous nannofossils than equivalent samples taken in 1974, but most obvious is the general poor preservation in equivalent samples collected by us in 1987.

Calcareous nannofossil stratigraphy of the Buff Bay section

Because of very poor preservation the specific composition of the assemblages, particularly in the lower part of the section (Pots and Pans and Middle Roadcut outcrops), could not be established satisfactorily, and some biozonal boundaries were very difficult to draw. Also, for this reason and because of significant reworking and stratigraphic gaps, this section provides limited information as to the total range of stratigraphically important Neogene species.

The Spring Garden Member. The Spring Garden Member of the Montpelier Formation essentially spans the entire middle Miocene (Fig. 5). Its base lies within Zone NN4 very close to the NN4/NN5 zonal boundary, located between samples SG32 and SG31. The remainder of the Pots and Pans Roadcut belongs to Zone NN5. The NN5/NN6 zonal boundary probably occurs within the covered interval between the Pots and Pans and Middle roadcuts and possibly very close to the level of sample ER 146/21 (see above). Calcareous nannofossils are abundant but generally poorly preserved in this interval. Discoasters are heavily overgrown (particularly those of the *Discoaster deflandrei* group), which hampers their specific determination. There seems to be an upward increase in the abundance of *Discoaster exilis.* It is not clear, however, whether this enrichment is related to a parallel improvement in preservation.

The beds of the Spring Garden Member exposed in the Middle Roadcut comprise Zones NN6 and NN7. The uppermost beds may lie in Zone NN8 (see below). The abundance and the preservation of the calcareous nannofossils in these beds vary greatly, but preservation is mostly poor, and discoasters are heavily overgrown at most levels. Levels with better preservation were sampled in the middle part of the section between ~175 m and ~192 m (samples SG17 to SG13), but only Sample SG13 (~192 m) yielded a well-preserved assemblage. This is the only level in the outcrop where typical (common) specimens of *Discoaster kugleri* were encountered. The NN6/NN7 zonal boundary is therefore placed at this level. This placement should be regarded as tentative because this is also the level where *Discoaster challengeri* first occurs. These apparently simultaneous first occurrences may result from poor preservation below or may reflect an unconformity, the lower part of Zone NN7 being missing. According to Martini (1971) and Perch-Nielsen (1985), the lowest occurrence of *D. challengeri* is located in the upper part of Zone NN7. *Discoaster exilis* is dominant among the discoasters in all assemblages, although diversity increases in Zone NN7. *Discoaster sanmiguelensis*, very common in correlative levels from the Gulf of Mexico (Aubry, unpublished data) is extremely rare and was encountered only in Sample SG19.

The upper part of the Spring Garden Member (Stuart et al., 1987) is exposed in the Main Roadcut. Since its contact with the Buff Bay Formation is transitional, the boundary between both formations was conveniently drawn at a level delineated by the growth of lichens (see discussion in Katz and Miller, this volume). This places the top of the Spring Garden Member slightly above Sample 173, itself estimated to be slightly above Sample ER 146/32. With the boundary between the two formations placed at this level rather than between the Middle and Main Roadcut outcrops (Robinson, 1969), the Spring Garden Member extends well into Zone NN8. The NN7/NN8 zonal boundary is very difficult to locate due to poor preservation. *Catinaster coalitus* is common in Sample 167. However, very rare recrystallized globular forms that show features characteristic of *C. coalitus* (in particular the outline in side view and the alveolar pattern) occur down to Sample SG1. In Sample ER 146/31, at the base of the Main Roadcut section, such forms can confidently be assigned to *Catinaster coalitus*. In Sample SG1, they are very rare and their identification is more doubtful. Thus the NN7/NN8 boundary may be drawn either between the Middle and Main Roadcut sections or between samples SG1 and SG2 as it is tentatively drawn in Figure 5.

The Buff Bay Formation. Calcareous nannofossil preservation is generally better in the Buff Bay Formation than in the underlying unit, although the specific identification of the discoasters was often hampered by overgrowth and broken tips, particularly in the lower part of the formation. Preservation improves considerably upward from Sample 193. The part of the Buff Bay Formation exposed in the Main Roadcut is assigned to Zones NN8 and NN9 (Fig. 5). The NN8/NN9 zonal boundary is difficult to delineate. In the interval between samples 180 and 189, scarce, poorly preserved, symmetrical five-rayed discoasters occur. Extremely rare in Sample 180, they progressively increase in number upward. The NN8/NN9 zonal boundary is drawn arbitrarily at a level where the first *Discoaster hamatus* can be identified with certainty, that is, in Sample 184.

The *Catinaster calyculus* Subzone (Bukry, 1975; =CN7b) was defined as the stratigraphic interval between the first occurrence of *C. calyculus* and the last occurrence of *D. hamatus*. While Bukry (1973, 1975) considered that *C. calyculus* first occurs in the upper part of the *Discoaster hamatus* Zone (=Subzone CN7b =upper part Zone NN9), Martini (1980) indicated that its lowest occurrence is in Zone NN8 and commented that "*C. calyculus* developed from *C. coalitus* by extending the six rays of the distal side beyond their former bifurcation point at the rim of *C. coalitus*" (Martini, 1980, p. 558). The evolution of *C. calyculus* from *C. coalitus* as described by Martini (1980) was observed in the material studied here, but no specimen of *C. calyculus* was seen in the part of the section assigned to Zone NN8. The lowest transitional form was observed in Sample 184, which is the lowest sample assigned to Zone NN9 (as noted above, the NN8/NN9 zonal boundary may be as low as between samples 179 and 180). Transitional forms are, however, sporadic up to sample 203 and remain scarce up to sample 217. Between samples 218 and 222B, transitional forms become more common and show longer arms. Rare specimens in samples 218, 219, and 221 to 222B could be assigned to *C. calyculus*. Yet, typical specimens of *C. calyculus* with long and slender arms as described and illustrated by Martini and Bramlette (1963) are not common below sample ER 146/35, collected in an interval no longer accessible between the Main and Slide Roadcut outcrops. In these circumstances, the placement of the CN7a/CN7b subzonal boundary is dependent upon the taxonomic concept of *C. calyculus*. As for many other species of the calcareous nannofossils, little information regarding intraspecific variability was included in the definition of *C. calyculus*. Martini and Bramlette (1963) indicated only that in its upper range the rays are longer and curved. If a broad concept is given to the species (following Martini, 1980), the CN7a/CN7b boundary occurs low in the Buff Bay formation. If a narrow concept is used, the boundary is drawn either between samples 217 and 218 or between samples 222B and ER 146/35. The latter choice is preferred here, mainly to comply with the original definition of *C. calyculus* but also for convenience, the transitional forms being scarce below sample 218.

The upper part of the Buff Bay Formation is exposed in the Slide and in the Dead Goat Gully Roadcut outcrops (Fig. 5). The Slide and Main roadcuts are separated by an ~20-m-thick interval disrupted by a landslide. Analysis of samples ER 146/35 to ER 146/37 collected in this interval prior to the landslide reveals stratigraphic continuity between the Main and Slide Roadcut outcrops. Subzone CN7b extends up to the level of Sample 228, collected just below a 2-m-thick debris flow bed. There is a clear change in the nature of the calcareous nannofossil assemblages preserved below, in, and above the debris flow bed. This change is enhanced by a striking improvement in preservation in Sample 229 compared to Sample 228. Preservation above Sample 229 is moderate to good. Sample 228 yields an assemblage of discoasters dominated by *Discoaster hamatus* and *D. bollii* (compact), with rare *D. neohamatus. Catinaster coalitus* and *C. calyculus* are very rare. Sample 229 yields an assemblage of discoasters dominated by *Discoaster pentaradiatus, D. bollii* (less compact), *D. neohamatus,* and *D. brouweri.* The few specimens of *D. hamatus* and *C. coalitus* that also occur are regarded as reworked in this assemblage, which is assigned to Subzone CN8a. The absence of *Catinaster calyculus* and the presence of *D. bollii, D. brouweri, D. pentaradiatus, D. surculus,* and *Mynilitha convallis* warrant an assignment to the upper part of Subzone CN8a, correlative with the upper part of Zone NN10. The remaining part of the Buff Bay Formation above Sample 229 in the Slide and Dead Goat Gully roadcuts yields assemblages mostly similar in composition to that in Sample 229. The frequency of *D. bollii* decreases upward, whereas the frequency of *D. surculus* increases.

The Bowden Formation correlative. The West Roadcut exposes ~12 m of a mottled brownish-yellow, very coarse grained sandstone interbedded with gray marl. Robinson (1969) regarded this unit as correlative with the Bowden Formation, which has its type section in the Bowden District in eastern Jamaica. It is inferred to as the upper Buff Bay by Lamb and Beard (1972).

Calcareous nannofossils are rare to common in the marls and of good to moderate preservation. *Discoaster asymmetricus, D. brouweri, D. misconceptus, D. pentaradiatus,* and *D. tamalis* are common, and *D. surculus* and *Ceratolithus rugosus* are rare at all levels. *Reticulofenestra pseudoumbilicus* is rare in Sample 244, which is tentatively assigned to Zone NN15. The FAD (first appearance datum) of *Pseudoemiliania lacunosa* was thought to have occurred slightly after the FAD of *R. pseudoumbilicus* (see references in Berggren et al., 1985), but co-occurrence of both species has been reported from ODP (Ocean Drilling Program) Site 653 (Rio et al., 1990) and is observed in the San San Bay section (see below). As a result, the occurrence of this species in sample 244 may not be used as evidence for the reworking of *R. pseudoumbilicus* in it. No typical specimens of this latter species were observed above this level, and the NN15/NN16 zonal boundary is placed between samples 244 and 245. *Sphenolithus neoabies* is common in Sample 245, but only a few specimens of this species were encountered above it. It is possible that the LAD (last appearance datum) of *S. neoabies* occurs in Sample 245 and that the species is reworked in younger levels. Based on the occurrence of *D. tamalis,* the section above sample 244 is assigned to Subzone CN12a.

In summary, most of the exposed part of the Bowden Formation correlative belongs to Zone NN16 (Subzone CN12a). Its lower beds lie in Zone NN15.

Depositional history of the formations exposed in the Buff Bay section

Magnetobiostratigraphic correlations derived from the study of the Buff Bay section differ notably from those currently proposed in the upper middle Miocene (Miller et al., 1985). Although these new data may lead to a basic revision of the middle and late Miocene time scale, it is useful to situate the stratigraphy of the Neogene of Jamaica in a global chronological context. Age estimates of a number of calcareous nannofossil datums have been revised since the publication of the work of Berggren et al. (1985) (e.g., Backman and Pestiaux, 1987; Chepstow-Lusty et al., 1989; Channell et al., 1990; Backman et al., 1990). These revisions, which do not integrate planktonic foraminiferal and calcareous nannofossil biochronologies, are not included in this study for the sake of consistency between age estimates on calcareous nannofossil and planktonic foraminiferal datums.

The stratigraphy of the Neogene formations exposed in the Buff Bay section, eastern Jamaica, described by Robinson (1969), can be summarized as follows (Figs. 5, 7): (1) A lower ~275-m-thick unit (Unit 1) apparently continuous (but see below) from the base of the middle Miocene to the lower upper Miocene, that is, from a level very close to the NN4/NN5 (CN3/CN4) zonal boundary to a level in the upper part of Zone NN9 (in Subzone CN7b). It includes the Spring Garden Member and the lower 50 m of the Buff Bay Formation. The Spring Garden Member of the Montpelier Formation essentially represents the whole middle Miocene and encompasses the upper part of Zone NN4, Zone NN5 to NN7, and part of Zone NN8. The lower 50 m of the Buff Bay Formation comprise part of Zone NN8 and part of Zone NN9, that is, part of Subzone CN6, Subzone CN7a, and part of Subzone CN7b. (2) A medial unit (Unit 2), ~60 m thick (assuming stratigraphic continuity), which corresponds to the upper part of the Buff Bay Formation. The lower 50 m of this unit belong to Zone NN10 and to the upper part of Subzone CN8a. The uppermost beds of the Buff Bay Formation (~10 m) are not exposed. (3) An upper unit (Unit 3), ~16 m thick, regarded as correlative with the Bowden Formation. Its lower part is not exposed. The exposed upper 8 m encompass the NN15/NN16 zonal (=CN15/CN12a subzonal) boundary. (4) The stratigraphic gap between the Buff Bay and the Bowden equivalent formations encompasses the upper upper Miocene and lower Pliocene, from Zone NN11 to Zone NN14, and includes parts of Zones NN10 and NN15. (5) A stratigraphic gap within the Buff Bay Formation (between units 1 and 2 as described above) is delineated on the basis of calcareous nannofossil stratigraphy. It includes the upper part of Zone NN9 and part of Zone NN10, that is, the upper part of Subzone CN7b and the lower part of Subzone CN8a.

Although the calcareous nannofossil (and the planktonic foraminiferal, see Berggren, this volume) zonal succession in the Buff Bay section suggests that the Spring Garden Member–Buff Bay Formation represents an essentially continuous middle to upper Miocene sequence, the sedimentation rate curve (Fig. 8) established from the planktonic microfossils suggests that several lithologic breaks may be present in addition to that which occurs at the base of the Boulder Bed in the Slide Roadcut. The age estimates of the datums (Tables 2 and 3) used to establish the curve are from Berggren et al. (1985). The calcareous nannofossil datums were discussed above; the planktonic foraminiferal datums are discussed in Berggren (this volume).

There is a good agreement between the calcareous nannofossil and the planktonic foraminiferal datums in the lower and upper parts of the section, and the depositional histories of the lower part of the Spring Garden Member and of the Bowden Formation equivalent are straightforward. The part of the Spring Garden Member exposed in the Pots and Pans Roadcut was sedimented at a steady high rate (~7 cm/10^3 yrs). The shallower Bowden equivalent unit was deposited at a low rate (~1.5 cm/10^3 yrs).

It is more difficult to interpret the depositional history of the upper part of the Spring Garden Member and of the Buff Bay Formation, mostly because of strong discrepancies between the calcareous nannofossil and the planktonic foraminiferal datums but also because of discrepancies within each group (e.g., compare the FADS of *Discoaster hamatus* and *Catinaster calyculus*, the FAD of *Globoturborotalita nepenthes* and the LAD of *Globorotalia foshi robusta*, the LAD of *Neogloboquadrina* and the FAD of *N. acostaensis*). Several models can be drawn, but only

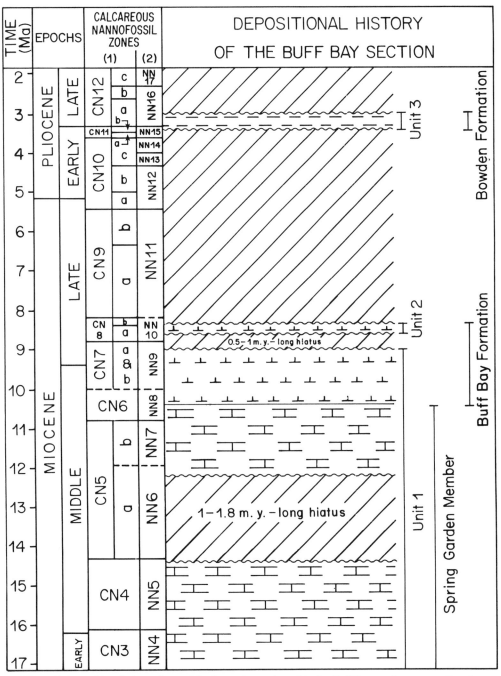

Figure 7. Depositional chronology of the formations exposed in the Buff Bay section, eastern Jamaica: (1) is from Okada and Bukry, 1980, modified; (2) is from Martini, 1971, modified. Current age estimates of the dashed zonal boundaries are being revised.

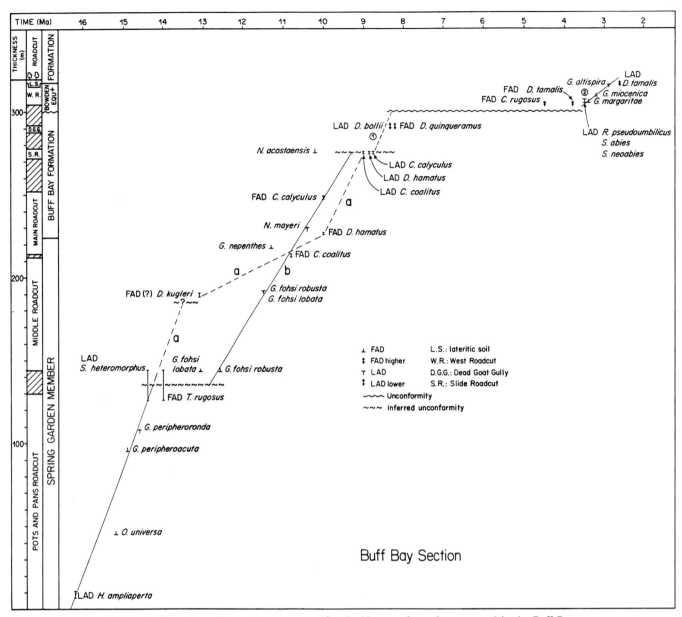

Figure 8. Tentative sedimentation rate curve for the Neogene formations exposed in the Buff Bay section, eastern Jamaica. Ages and locations of the datums are given in Table 2. (1): Minimal sedimentation rate for the stratigraphic interval between the FAD and the highest occurrence of *Discoaster hamatus* (Buff Bay formation, samples 184 to 228). (2) Minimal sedimentation rate for the interval between the intra–upper Miocene and the Miocene-Pliocene unconformities (upper part of the Buff Bay Formation, samples 229 to 243). See text for further explanation.

two appear to be reasonable. They are shown as models a and b in Figure 8 and are compared here.

In model a, the curve is drawn based only on the calcareous nannofossil datums. It suggests sharp changes in sedimentation rates and/or unconformities. In model b, the curve is drawn so as to reconcile as many datums as possible. This curve suggests continuous sedimentation from the base of the Middle Roadcut to the unconformity below the Boulder Bed in the Slide Roadcut.

However, model b implies that an unconformity occurs between the Pots and Pans and the Middle roadcuts. Indeed, this is supported by the juxtaposition of four datums, the LAD of *Sphenolithus heteromorphus* and the FADs of *Triquetrorhaldulus rugosus*, *Globorotalia foshi lobata*, and *G. foshi robusta*. The duration of the hiatus is somewhat difficult to determine because the unconformity occurs in a 15-m-thick covered interval. It is at least 1.4 m.y. (FAD of *T. rugosus* at 14 Ma; FAD of *G. foshi*

TABLE 2. CALCAREOUS NANNOFOSSIL DATUMS* USED TO ESTABLISH THE SEDIMENTATION RATE CURVE FOR THE NEOGENE FORMATIONS EXPOSED IN THE BUFF BAY SECTION, EASTERN JAMAICA

Datum Event	Sample	Level Height	Average	Age (Ma)
LAD *D. tamalis*	Above 252	>315.90	....	2.6
LAD *S. abies*	245/246	305.18–306.64	305.90	3.47
LAD *S. neoabies*	245/246	305.18–306.64	305.90	3.47
LAD *R. pseudoumbilicus*	244/245	305.18–306.64	305.90	3.5
FAD *D. tamalis*†	Below 244	<305.18	....	3.8
FAD *P. lacunosa*	Below 244	<305.18	....	3.4
FAD *C. rugosus*	Below 244	<305.18	....	4.5
FAD *D. quinqueramus*	Above 243	>290	....	8.2
LAD *D. bollii*	Above 243	>290	....	8.3
LAD *C. calyculus*	228/229	274.80–275.60	275.20	8.75
LAD *D. hamaltus*	228/229	274.80–275.60	275.20	8.85
LAD *C. coalitus*	228/229	274.80–275.60	275.20	9.0
FAD *C. calyculus*§	217/218 or 222B/ER 146/35	247.60–248.50 or above 251.70	248.05	10.0
FAD *D. hamatus*	183/184	226.05–226.45	226.25	10.0
FAD *C. coalitus*	SG2/SG1	211.93–213.78	212.85	10.8
FAD *D. kugleri***	SG14/SG13	187.45–190.65	189.05	13.1
FAD *T. nugosus*	SG23A/ER 146/21	125.73–144.72	135.22	14.0
LAD *S. heteromorphus*	SG23A/ER 146/21	125.73–144.72	135.22	14.4
LAD *H. ampliaperta*	SG32/SG31	10.50–6.68	8.59	16.2

*Ages are those estimated in Berggren et al., 1985.
†Unreliable datum since *D. tamalis* occurs in the uppermost Miocene of Jamaica.
§Magnetobiostratigraphic correlations in the Buff Bay section suggest that the age of this datum is younger than the FAD of *Discoaster hamatus*, contrary to what was observed in Miller et al., 1985.
**Dubious datum since this is the only level in the Montpelier Spring Garden Formation below the FAD of *Catinaster coalitus* where *Discoaster kugleri* occurs.

robusta at 12.6 Ma). However, since the NN5/NN6 zonal boundary is thought to occur close to the level of sample ER 146/21 (see discussion above), the hiatus may be over 1.8 m.y. long (LAD of *S. heteromorphus* at 14.4 Ma and FAD of *G. foshi robusta* at 12.6 Ma). Model b accounts for the anomalous range of four datums, and this is the first reason why it is preferred to model a. The second reason relates to the relationship between the FAD of *D. hamatus* and that of *C. calyculus*. Following model b, the FAD of *D. hamatus* appears anomalous with respect to the FAD of *C. calyculus,* and following model a the FAD of *C. calyculus* appears anomalous with respect to the FAD of *D. hamatus*. These two species first occur in a sequential fashion in the Buff Bay Formation, but they were observed to appear at the same level in DSDP (Deep Sea Drilling Project) Sites 558 and 563 (Miller et al., 1985), where apparently reliable magnetobiostratigraphic correlations served to establish the chronology of early late Miocene calcareous microfossil datums. Comparison between the stratigraphy of the Buff Bay Formation and of DSDP sites 558 and 563 suggests that an unrecognized unconformity occurs at the two DSDP sites just below the level where *D. hamatus* and *C. calyculus* first occur, leading to a notably younger age estimate for the FAD of *D. hamatus*. The anomalous FAD of *Discoaster kugleri* is also problematic. As noted above,

TABLE 3. PLANKTONIC FORAMINIFERAL DATUMS* USED TO ESTABLISH THE SEDIMENTATION RATE CURVE FOR THE NEOGENE FORMATIONS EXPOSED IN THE BUFF BAY SECTION, EASTERN JAMAICA

Datum Event	Sample	Height (m)	Age (Ma)
LAD *G. Altispira*	251	315	2.9
FAD *G. miocenica*	247	309	3.2
LAD *G. margaritae*	246	305.90	3.4
FAD *N. acostaensis*	228	275.20	10.2
LAD *N. mayeri*	189	229.60	10.4
FAD *G. nepenthes*	168	216.9	11.3
LAD *G. foshi robusta*	SG20	192	11.5
LAD *G. foshi lobata*	SG20	192	11.5
FAD *G. foshi robusta*	SG23	142.9	12.6
FAD *G. foshi lobata*	SG23	142.9	13.1
LAD *G. peripheroronda*	SG25	107.7	14.6
FAD *G. peripheroacuta*	SG26	90.0	14.9
FAD *O. universa*	SG28A	45.5	15.2

*Ages are those estimated in Berggren et al., 1985.

D. kugleri was encountered at a single level (in sample SG13) below the lowest occurrence of *Catinaster coalitus,* and this single occurrence may or may not correspond to the FAD of the species. In addition, the age of this FAD was derived from magnetobiostratigraphic correlations established at DSDP Sites 558 and 563 and was based on a broad concept of the species. Applying a strict concept to the taxon would lead to a much younger age estimate for its FAD (see Miller et al., 1985, Fig. 1), thus bringing the LO of *D. kugleri* in the Spring Garden Member in closer agreement with other datums. Lastly, model b is superior to model a in that it integrates datums from the planktonic foraminifera that otherwise are not accounted for. However, in model b the FAD of *G. nepenthes* and that of *N. acostaensis* appear anomalous, suggesting some diachrony. With regard to this, it should be noted that no satisfactory sedimentation rate curve could be drawn from the planktonic foraminiferal datums alone.

Blow (1979) drew an unconformity at the base of the Main Roadcut (between samples ER 146/31 and 32). Calcareous nannofossil stratigraphy offers no evidence to support this occurrence, and there are no indications from model a or b that this is the case.

As drawn, the sedimentation rate curve suggests high sedimentation rates for the upper part of the Spring Garden Member and for the Buff Bay Formation.

In summary, with reference to the time scale of Berggren et al. (1985), the following chronology can be estimated for the Buff Bay section (Figs. 7, 8): (1) Deposition of the lower part of the Spring Garden Member from at least ~16.5 Ma (older than the estimated age [16.4 Ma] of the LAD of *H. ampliaperta*) to ~14.4 Ma (close to the LAD of *S. heteromorphus*); (2) development of an unconformity present in a covered interval between 130 and 145 m. The duration of the associated hiatus is at least 1.4 m.y. and probably over 1.8 m.y. (LAD of *S. heteromorphus* at 14.4 Ma, FAD of *T. rugosus* at 14 Ma, FAD of *G. foshi robusta* at 12.6 Ma); (3) Deposition of the upper part of the Spring Garden Member and the lower part of the Buff Bay Formation from <12.6 Ma to >9 Ma (between the estimated ages of, respectively, the FAD of *G. foshi robusta* and the LAD of *C. coalitus*); (4) Development of an unconformity at 274.95 m (at the base of the Boulder Bed in the Slide Roadcut; most likely by slumping), leading to a hiatus of a few hiatus of a few hundred thousand years to possibly over 1 m.y. The lower surface of the unconformity is 9 Ma (estimated age of the LAD of *C. coalitus*) or older. Its upper surface is older than the LAD of *D. bollii,* estimated at 8.3 Ma (Berggren et al., 1985); (5) Deposition of Unit 2 within a few hundred thousand years, between the FADs of *Discoaster pentaradia*tus, *Discoaster surculus,* and *Minylitha convallis* and the LAD of *Discoaster bollii* (8.3 Ma); (6) Development of an unconformity at ~300 m with a hiatus of ~4.5 m.y. (before the LAD of *D. bollii* [8.3 Ma] on the one hand and between the LAD of *Amaurolithus* spp. [3.7 Ma] and the LAD of *Reticulofenestra pseudoumbilicus* [3.47 Ma] on the other hand); (7) Deposition of Unit 3 between ~3.5 Ma and >2.6 Ma (estimated age of the LAD of *D. tamalis*).

Reworking

Reworked calcareous nannofossils occur throughout the section. Their higher frequency in the Pliocene unit suggests intensified tectonic activity during the Pliocene. Calcareous nannofossils are mostly reworked from the Paleocene (Fig. 4 and Table 4), in particular from the lower Eocene as indicated by, for example, *Tribrachiatus orthostylus, Discoaster kuepperi,* and *D. lodoensis.* The species *Rhabdosphaera inflata* and *Discoaster sublodoensis* indicate reworking of lower middle Eocene strata, whereas *Discoaster saipanensis, Helicosphaera compacta,* and *Reticulofenestra reticulata* indicate reworking of middle middle to upper Eocene deposits. *Helicosphaera recta* and *Sphenolithus predistentus* are indicative of the reworking of lower upper and upper Oligocene strata. Species such as *Reticulofenestra bisecta, R. hillae, R. umbilicus, H. compacta,* and *Isthmolithus recurvus* may be reworked from Oligocene as well as from Eocene deposits. In addition, four species—*Heliolithus kleinpellii, Chiasmolithus consuetus, Fasciculithus tympaniformis,* and *Discoaster mohleri*—suggest the reworking of upper Paleocene sediments in these Neogene formations. While the origin of the Eocene and Oligocene calcareous nannofossil taxa may have been the Richmond Group, the Montpelier Group (sensu Steineck, 1974), and the more marine deposits of the Clarendon Group, the source of the Paleocene taxa may be the Nonsuch Limestone from which *Heliolithus kleinpellii* and *Discoaster mohleri* were reported (Jiang and Robinson, 1987). The diversity (and the good preservation) of the reworked Paleogene calcareous nannofossils encountered in the Neogene deposits from eastern Jamaica suggests very abundant and diversified lower Eocene to upper Eocene–Oligocene assemblages in the Paleogene deposits of Jamaica. Intra-Neogene reworking also occurs, as indicated by abnormal occurrences of *Sphenolithus heteromorphus, Coccolithus miopelagicus,* and *Reticulofenestra floridana* (which may also be reworked from the Paleogene), but seems to be more limited than the reworking from Paleogene deposits. Few specimens of *Sphenolithus belemnos* and *Triquetrorhabdulus carinatus* were encountered, which suggests that Miocene deposits older (belonging to Zone NN2 and perhaps NN3) than those described herein at the base of the Spring Garden Member were deposited in Jamaica.

Correlations between the planktonic foraminiferal and calcareous nannofossil zones in the Buff Bay section

Figure 9 shows how the calcareous nannofossil zones (Martini, 1971) and subzones (Okada and Bukry, 1980) correlate with the planktonic foraminiferal scheme of Blow (1969). This framework of direct correlations reveals several discrepancies with the correlation framework of Berggren et al. (1985) for the middle and upper Miocene.

1. The N10/N11 zonal boundary lies within Zone NN5 in the Spring Garden Member, and whereas the NN5/NN6 zonal boundary is shown to lie within Zone N10 in Berggren et al. (1985).

TABLE 4. PALEOGENE AND EARLY NEOGENE SPECIES REWORKED IN THE NEOGENE FORMATIONS EXPOSED IN THE BUFF BAY SECTION, EASTERN JAMAICA

SPECIES SUGGESTIVE OF UPPER PALEOCENE STRATA
Chiasmolithus consuetus
*Discoaster mohleri**
D. multiradiatus
*Fasciculithus tympaniformis**
*Heliolithus kleinpellii**
Neochiastozygus chiastus
N. concinnus

SPECIES SUGGESTIVE OF LOWER EOCENE STRATA
*Discoaster diastypus**
*D. kuepperi**
*D. lodoensis**
Helicosphaera seminulum
Lophodolithus nascens
Pontosphaera pulchra
Sphenolithus radians
Toweius magnicrassus
*Tribrachiatus orthostylus**

SPECIES SUGGESTIVE OF MIDDLE EOCENE STRATA
*Cruciplacolithus staurion**
*Discoaster sublodoensis**
*D. bifax**
*Helicosphaera heezenii**
H. lophota
Micrantholithus flos
Rhabdosphaera crebra
*R. inflata**
R. morionum
*Sphenolithus furcatolithoides**
Toweius gammation

SPECIES SUGGESTIVE OF OLIGOCENE STRATA
*Helicosphaera recta**
*Sphenolithus predistentus**

SPECIES SUGGESTIVE OF LOWER TO MIDDLE EOCENE STRATA
*Chiasmolithus solitus**
*C. grandis**
*Cruciplacolithus delus**
Neococcolithes dubius

SPECIES SUGGESTIVE OF UPPER EOCENE TO LOWER OLIGOCENE STRATA
*Isthmoliths recurvus**
Rhabdosphaera tenuis

OTHER PALEOGENE SPECIES
Discoaster barbadiensis (lower to upper Eocene)
Discoaster saipanensis (middle to upper Eocene)
Ericsonia formosa (lower Eocene to lower Oligocene)
Helicosphaera compacta (middle Eocene to upper Oligocene)
Reticulofenestra bisecta (upper middle Eocene to upper Oligocene)
R. floridana (middle Eocene to upper middle Miocene)
R. hillae (middle Eocene to lower Oligocene)
R. reticulata (middle and upper Eocene)
R. umbilicus (middle Eocene to lower Oligocene)
Zygrhablithus bijugatus (upper Paleocene to upper Oligocene)

NEOGENE TAXA OBVIOUSLY REWORKED IN THE FORMATIONS EXPOSED IN THE BUFF BAY SECTION
Coccolithus miopelagicus (NN2-NN8)
Reticulofenestra floridana (NP16-NN7)
Sphenolithus belemnos (NN2-NN3)
S. heteromorphus (NN4-NN5)
Triquetrorhabdulus carinatus (NP24 to NN2)

*Designates species restricted to this stratigraphic interval.

2. The position of the NN6/NN7 zonal boundary with respect to the N-zones is difficult to evaluate: first, because neither the N12/N13 nor the NN6/NN7 zonal boundaries could be firmly delineated in the Buff Bay section; second, because the base of Zone NN7 in Berggren et al. (1985) was drawn following a broad concept of *D. kugleri* (as applied in Miller et al., 1985).

3. The NN7/NN8 zonal boundary lies within the upper part of Zone N13 in the Buff Bay section, whereas it lies within Zone N14 in Berggren et al. (1985).

4. The NN8/NN9 zonal boundary correlates with the N14/N15 zonal boundary in the Buff Bay section. It lies within the lower part of Zone N16 in Berggren et al. (1985). This discrepancy may reflect a miscorrelation between calcareous nannofossil and planktonic foraminiferal zones at DSDP sites 558 and 563 (in Miller et al., 1985), as the result of an unrecognized unconformity but also because the FAD of *N. acostaensis* is strongly diachronous (see Berggren, this volume).

5. Zone NN10 correlates (at least in part) with Zone N17. This is a first-order correlation (between the "holotype" locality of Zone N17, see below, and calcareous nannofossil stratigraphy). Earlier correlations indicated that the N16/N17 zonal boundary lies in Zone NN11 (e.g., Brönnimann and Resig, 1971; Berggren and van Couvering, 1974).

6. The PL2/PL3 zonal boundary coincides with the NN15/NN16 zonal boundary in the Bowden Formation in agreement with Berggren et al. (1985).

Discrepancies between calcareous nannofossil and planktonic foraminiferal stratigraphic correlations as drawn in Berggren et al. (1985) and as directly delineated in the Buff Bay section result from unrecognized unconformities in deep sea sections but also from diachronous datum levels, particularly among the planktonic foraminifera (see discussion above) and from variable taxonomic concepts (see Berggren, this volume). The correlations established between planktonic foraminiferal and calcareous nannofossil stratigraphies in the Buff Bay section are closest to the correlation framework presented in Barron et al. (1985). However, there is no agreement between this latter correlation framework and the record of correlations established in the Buff Bay section for the upper Miocene. None of the three commonly used time scales (e.g., Barron et al., 1985; Berggren et al., 1985; Haq et al., 1987) offer correlations between planktonic foraminiferal and calcareous nannofossil zones that are supported by the correlation framework established in Jamaica.

CALCAREOUS NANNOFOSSIL STRATIGRAPHY OF OTHER DEPOSITS OF THE LOWER COASTAL GROUP

The San San Bay section and the type section of the Bowden Formation at Bowden are important with reference to Neogene biostratigraphy, for they yield some of the "holotype"/"paratype" localities of Blow's N Zones (1969, 1979). Both sections were visited during the field party of 1987 (Stuart et al., 1987), but sampling was restricted because of poor exposure. The discussions that follow about the calcareous nannofossil stratigraphy in each section are mainly based on analysis of the original ER samples collected in 1969 by J. B. Dunlap and B. Kohl with E. Robinson.

Six other sections were also collected during the field party of 1987. They are described in Stuart et al. (1987). I report below on the results of a reconnaissance study of this material.

San San Bay section

The ~175-m-thick section at San San Bay (Fig. 10) is currently interpreted as comprising three lithostratigraphic units (Robinson in Zans et al., 1962; Blow, 1969; Robinson, 1969). The Spring Garden Member of the Montpelier Formation, ~20 m thick, consists of chalky limestones and marls. It underlies

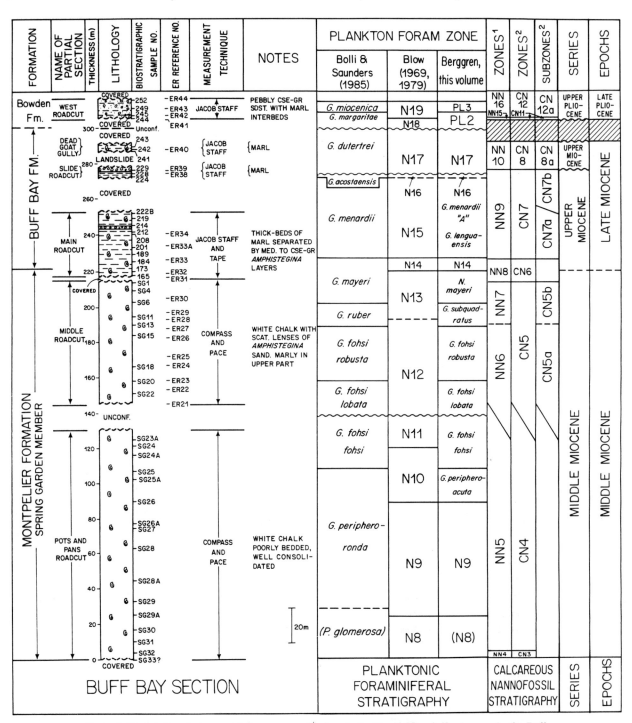

Figure 9. Correlations between calcareous nannofossil and planktonic foraminiferal zones in the Buff Bay section. (1): from Martini (1971); (2): from Okada and Bukry (1980).

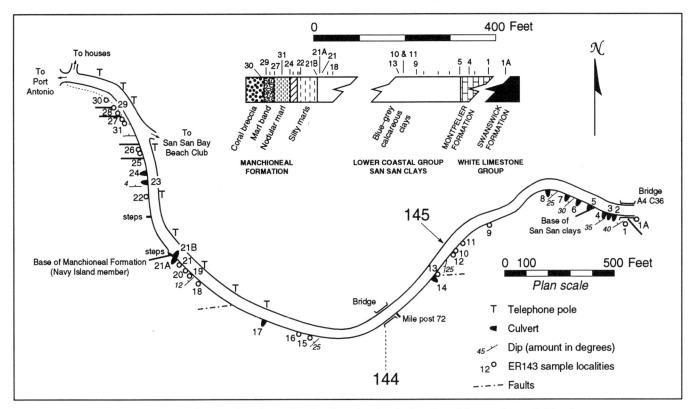

Figure 10. Plan of the road section at San San Bay, showing the location of the samples studied herein (from Robinson, 1969). Bold face designates the samples examined in this study.

conformably (Robinson, *in* Zans et al., 1962) the San San Clay, a uniform unit of soft brown calcareous clay and marl, almost 100 m thick, first described by Robinson (*in* Zans et al., 1962). The upper 50 m of the San San Bay section are formed by the Navy Island Member of the Manchioneal Formation, which apparently overlies conformably the San San Clay. The section at San San Bay has been mostly inaccessible since May 1967 because of the construction of a retaining wall. It is divided into a northern and southern block by a fault that runs east-northeast–west-southwest and intersects the coastal road twice (Fig. 10). The northern block comprises only the San San Clay.

Three samples were collected during the field party of 1987. Sample 144 was taken at milepost 72 (i.e., between samples ER 143/14 and 15; see Fig. 2 in Robinson, 1969). Sample 145 is probably equivalent to sample ER 143/10 or 11, and sample I corresponds to sample ER 143/17. In addition, samples ER 143/1, ER 143/14, and ER 143/23 and samples ER 143/6A, 6B, 7; ER 143/21A, 21B, 24; and ER 143/2 to 8 were provided by E. Robinson, J. Hardenbol, and B. Kohl, respectively. Also examined was a suite of samples, ER 1087 to ER 1097, collected by E. Robinson in 1969, provided by Jan Hardenbol (Exxon).

All samples examined yield abundant calcareous nannofossils. Their assemblages are described below according to their normal succession along the road regardless of their position with respect to the fault, because the normal sequential occurrence of events recorded in the San San Bay section implies that the fault has little bearing on the stratigraphic succession in the section. Preservation is generally much better than in the Buff Bay section. This is particularly noticeable for the Spring Garden Member. As in the Buff Bay section, reworking of Paleogene taxa is intensive at all levels.

Samples ER 143/1 to 4 yield typical middle Miocene discoaster assemblages with *Discoaster aulakos, D. challengeri* (common), *D. braarudii, D. exilis* (rare), *D. moorei* (common), *D. prepentaradiatus* (rare), and *D. signus* (rare). *Triquetrorhabdulus rugosus* is common. Although no typical specimen of *D. kugleri* was encountered, the occurrence of *D. challengeri* suggests an assignment to Zone NN7 (=Subzone CN5b).

Sample ER 143/5 yields a typical late Miocene assemblage with common *Discoaster brouweri, D. challengeri, D. pentaradiatus, D. quinqueramus,* and *D. surculus.* Ceratoliths are rare, represented by *Amaurolithus delicatus* and *A. primus.* As indicated by the co-occurrence of these latter taxa and *D. quinqueramus,* sample ER 143/5 belongs to Subzone CN9b, correlative with the upper part of Zone NN11. Similar assemblages occur up to the level of sample ER143/14. The frequency of *D. quinqueramus* varies greatly in this interval, which, based on its co-occurrence with *Amaurolithus delicatus, A. ninae, A. primus,* and *A. tricorniculatus,* may be assigned to Subzone CN11b.

Sample 144, taken at mile post 72 (Fig. 10), yields a well-preserved but unusual assemblage in which *Discoaster berggrenii* and *D. quinqueramus* predominate. Other discoaster species that occurred in the underlying samples are also present but are rare. The assemblage is further characterized by exceedingly rare *Amaurolithus primus* and (typical) *Ceratolithus acutus*. The FAD of this latter species defines the base of Subzone CN10b (correlative with the upper part of Zone NN12; see Fig. 6), to which it is mainly restricted.

Discoaster brouweri, D. challengeri, D. misconceptus, D. pentaradiatus, D. surculus, and *D. variabilis* are abundant and *D. asymmetricus* and *D. tamalis* are common in sample I (equivalent to ER 143/17). *Ceratolithus rugosus* is very common, whereas species of *Amaurolithus* (*A. delicatus* and *A. tricorniculatus*) are extremely rare. Rare specimens of *D. hamatus* and a few well-preserved *D. quinqueramus* were encountered. It is difficult to decide whether *A. delicatus* and *A. tricorniculatus* are also reworked at this level. Since ceratoliths of the genus *Amaurolithus* are very rare in the underlying levels, it is unlikely that they are reworked. They are considered indigeneous in this assemblage and indicative of a NN13-NN14 zonal interval.

Samples ER 143/21 and 21A yield abundant *Ceratolithus rugosus, Discoaster brouweri, D. challengeri, D. misconceptus, D. pentaradiatus, D. surculus, D. variabilis, Sphenolithus abies, S. neoabies,* common *Reticulofenestra pseudoumbilicus,* and few *Pseudoemiliana lacunosa.* They are assigned to Zone NN15.

Previous interpretations of the San San Bay section have been contradictory. Blow (1969, p. 296, Fig. 28) regarded the San San Clay as being younger than the Buff Bay Formation, a stratigraphic gap separating the two formations, but as overlapping largely with the Bowden Formation exposed in the Buff Bay section and in a more restricted manner with the Bowden Formation in the Drivers River and Bowden sections. In contrast, Robinson (*in* Lamb and Beard, 1972, p. 33, Fig. 16) interpreted the San San Clay as an intermediate unit between the Buff Bay and the Bowden formations, allowing for no overlap between it and either formation. While this author admitted the presence of a stratigraphic gap between the San San Clay and the Bowden but suspected continuity from the Buff Bay Formation and the San San Clay through "strata not accounted for" (Robinson, *in* Lamb and Beard, 1972), which justified his inclusion of the San San Clay in the Buff Bay Formation (Robinson, 1969).

Calcareous nannofossil stratigraphy suggests that there is stratigraphic continuity between sample ER 143/5 to sample ER 143/17, that is, from upper Zone NN11 to Zone NN13. The succession of calcareous nannofossil events in this interval is the same as in deep sea sections. In addition, it is in agreement with the record in sample ER 143/15 (Berggren, this volume) of a distinct association of forms transitional between *Globorotalia cibaoensis* and *G. puncticulata* at the boundary between Zones PL1a/PL1b (Berggren, 1977), correlative with the boundary between Zones NN12 and NN13 (see Berggren et al., 1985). Samples ER 143/15 and 16 were not available for examination, but it was established above that the NN12/NN13 zonal boundary occurs between samples 144 and ER 143/17. By correlation with planktonic foraminiferal stratigraphy it occurs at or near the level of sample ER 143/15.

Integrated calcareous nannofossil and planktonic foraminiferal stratigraphy further suggests a stratigraphic gap in the section, below the level of samples ER 143/21. Samples ER 143/18 to 20 were not available for calcareous nannofossil stratigraphy, but samples ER 143/21 and 21A were examined. They yield an assemblage indicative of Zone NN15 (see above). Stratigraphic distribution of the planktonic foraminifera in samples ER 143/5 to 8, 12, and 18 to 21A was given by Robinson (*in* Lamb and Beard, 1972, p. 33, Fig. 16), and Berggren (this volume) reexamined samples ER 143/5 to 7, 12, 14 to 17, 19, 21, and 21A. The data pertinent to our discussion are given in Table 5. The sedimentation rate curve (Fig. 11) based on these data must be seen as tentative: First, because the distance between samples can be established only approximately from Figure 2 in Robinson (1969) (Fig. 10 herein); second, because this distance is known only for a restricted number of samples. Yet, the absence of species of the genus *Amaurolithus* (LAD estimated at 3.7 Ma) and the simultaneous lowest occurrences of *Globorotalia crassaformis* (FAD estimated at 4.1 Ma) and *Pseudoemiliania lacunosa* (FAD estimated at 3.4 Ma) in sample ER 143/21 indicate the presence of an unconformity below this level. Robinson (*in* Lamb and Beard, 1972) recorded the presence of *G. crassaformis* in sample ER 143/19, but Berggren (this volume) did not. It should be noted that either interpretation supports the presence of an unconformity in the section. Sequential occurrence of *Globorotalis multicamerata* in sample ER 143/18 and *G. crassaformis s.l.* in sample ER 143/19 as recorded by Robinson (*in* Lamb and Beard, 1972) suggests that the unconformity is above this latter level. However, precise location of the inferred unconformity in the section requires establishing the highest occurrence of *Amaurolithus primus* and/or the lowest occurrence of *P. lacunosa*. The stratigraphic gap corresponds to Zone NN13 (partim), Zone

TABLE 5. CALCAREOUS NANNOFOSSIL AND PLANKTONIC FORAMINIFERAL DATUMS* USED TO ESTABLISH THE SEDIMENTATION RATE CURVE FOR THE SAN SAN CLAY EXPOSED IN THE SAN SAN BAY SECTION, EASTERN JAMAICA

Datum Event	Age (Ma)	Location†	Approximate Level
LAD *G. margaritae*	3.4	Above ER 143-21	Above 4.2
LAD *R. pseudoumbilicus*	3.5	Above ER 143-21	Above 4.2
LAD *A. primus*	3.7	ER 143-17/ER143-21	?-4.2
FAD *G. crassaformis*	4.1 (4.3)	ER 143-21 or ER 143-19	4-4.2
FAD *G. multicamerata*	4.3	ER 143-18/ER 143-19	?-4
PL1a/PL1b boundary	4.4	ER 143-15	?
FAD *C. acusus*	5	144/ER 143-14	2.4
LAD *G. dehiscens*	5.3	ER 143-7	1.8
FAD *A. primus*	6	Below ER 143-5	0.8

*Ages are those estimated in Berggren et al., 1985.
†Measurements given for distance between samples based on an arbitrary scale.

NN14, and Zone NN15 (partim). The immediate implication of this interpretation is that the upper part of the San San Clay correlates with the Bowden Formation in the Buff Bay section, as will be discussed below.

According to calcareous nannofossil stratigraphy, the interval between samples ER 143/5 and 14 belongs to Subzone CN9b. No levels representative of Subzone CN10a (upper part of Zone NN12) were recognized, suggesting that this subzone may be missing. Planktonic foraminiferal stratigraphy (Berggren, this volume) indicates, however, that the interval between samples ER 143/7 and ER 143/15 belongs to Zone PL1a, which is correlative with Subzones CN10a (upper part) and CN10b. This raises several questions regarding the occurrence of *D. quinqueramus* in an interval corresponding to Zone PL1a. That the discoaster is reworked can be argued on the grounds that it exhibits broad frequency variations in this interval and that the assem-

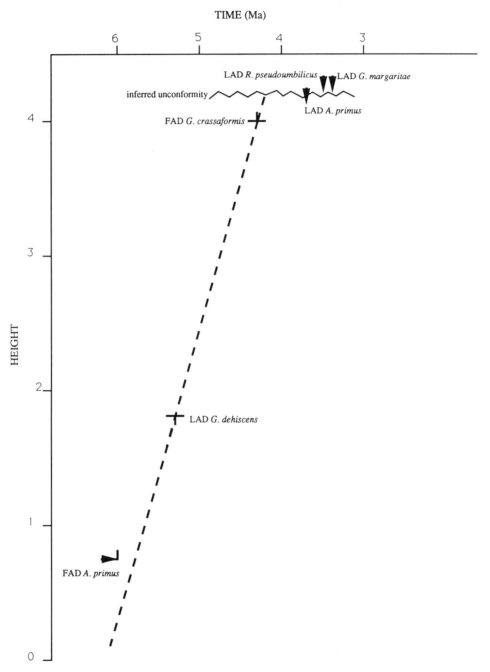

Figure 11. Provisional sedimentation rate curve for the San San Clay (San San Bay section). As it is constrained by a very limited number of tie-points, this curve is only tentative. However, it helps to visualize the level of the inferred unconformity in the upper part of the section (see text for further explanation). An arbitrary vertical scale based on Figure 10 is used.

blages include taxa reworked from the middle and lower upper Miocene, for example, *Coccolithus miopelagicus, Discoaster hamatus, D. neohamatus,* and *Sphenolithus heteromorphus.* Large fluctuations in abundance may not be a sufficient argument to support reworking, however, since Neogene discoasters exhibit large fluctuations in abundance throughout their range (see records in Backman and Shackleton, 1983; Chepstow-Lusty et al., 1989; Rio et al., 1990). It should be pointed out that reworking of *D. quinqueramus* in Subzone CN10a cannot be recognized through calcareous nannofossil stratigraphy alone because Subzone CN10a is an interval between the last occurrence of a taxon (*D. quinqueramus*) and the first occurrence of another taxon (*C. acutus*). Since in other sections similar correlation between the upper part of Zone NN11 and Zone PL1a can be established (Ceara Rise, Aubry, unpublished data), revision of the correlations between the calcareous nannofossil and the planktonic foraminiferal zones in the uppermost Miocene–lowermost Pliocene may be necessary. Until these correlations are better established, the interval between sample ER 143/5 and ER 143/14 is assigned to Subzones CN11b-CN10a undifferentiated. The CN10a/CN10b zonal boundary is extremely difficult to delineate because of the great scarcity of the ceratoliths, but it occurs at or below sample 144.

In summary, the stratigraphy of the San San Bay section can be outlined as follows:

1. A part of the Spring Garden Member, assignable to Zone NN7, is present in the San San Bay section. Its base is probably younger than ~12 Ma and its top older than ~10.8 Ma (estimated age of the FAD *C. coalitus*).

2. The main part of the section is of latest Miocene and earliest Pliocene age and spans the zonal interval CN11b-NN13, or PL1a to PL1c (lower part). The base of the San San Clay is younger than 6.5 Ma (estimated age of the FAD *A. primus*) and probably older than 6 Ma (estimated age of the FAD *A. tricorniculatus*). The age of sample ER 143/7 is older than 5.3 Ma (estimated age of the LAD *G. dehiscens*). The youngest lower Pliocene levels (probably near sample ER 143/19) are younger than 4.5 Ma (estimated age of the FAD *G. multicamerata*) or 4.1 Ma (estimated age of the FAD *G. crassaformis*) and older than 3.7 Ma (estimated age of the LAD *A. tricorniculatus*).

3. The upper part of the section, which is a few meters thick and underlies the Navy Island Member, is of early late Pliocene age. It is assigned to Zone NN15. Its base is younger than 3.7 Ma (estimated age of the LAD *A. tricorniculatus*); its top older than 3.5 Ma (estimated age of the LAD *R. pseudoumbilicus*).

4. The hiatus between the Spring Garden Member and the San San Clay is at least 4.5 m.y.

5. The hiatus that occurs in the upper part of the San San Clay is 0.5 to 1 m.y.

The San San Clay as described by Robinson (*in* Zan, et al., 1962; Robinson, 1969) extends from the upper Miocene to the lower upper Pliocene (between Subzone CN9b and Zone CN11, that is, from upper Zone NN11 to Zone NN15), and spans the Miocene/Pliocene boundary. It includes a stratigraphic gap corresponding to the upper lower Pliocene. As noted by Robinson (1969), the San San Clay is younger than the Buff Bay Formation. As noted by Blow (1969), its upper part (above the unconformity) correlates with the Bowden Formation.

As will be established below, there is no stratigraphic continuity between the Buff Bay Formation and the San San Clay; that is, no deposits representing Subzone CN9a occur in eastern Jamaica. There is therefore no reason to regard the San San Clay as part of the Buff Bay Formation. As will also be shown, there is no stratigraphic continuity between the Bowden Formation and the main (upper Miocene and lower Pliocene) part of the San San Clay; that is, there are no upper lower Pliocene deposits in eastern Jamaica. There is therefore no reason to regard the main part of the San San Clay as part of the Bowden Formation. I suggest restricting the San San Clay to the upper Miocene and lower Pliocene beds that occur in the San San Bay section. This leads to recognition of it as a formation distinct from the Buff Bay and the Bowden formations and, like these latter, as a separate sequence between two regional unconformities (see below).

Type section of the Bowden Formation

The type section of the Bowden Formation is located in the Bowden district in southeastern Jamaica. It is discussed by Robinson (1969). The base of the formation is formed by the Bowden Shell Bed described by Woodring (1925). Three samples were taken in this section during the field party of 1987. Sample 161 was taken at the level of sample ER 140. Sample 162 was taken 3 m above sample 161 in the same outcrop. Sample 163 was collected in a cliff behind abandoned sugar warehouses. In addition, sample ER 156, provided by B. Kohl, was examined.

In addition to the fact that calcareous nannofossils are scarce, diversity is low and preservation is poor in samples 161 to 163. No discoasters or ceratoliths were encountered, and no zonal assignment is possible.

Calcareous nannofossils are also rare in sample ER 156. Reworked Paleogene and middle Miocene taxa (e.g., *Sphenolithus heteromorphus*) are common, and it is likely that the rare specimens of *Reticulofenestra pseudoumbilicus,* and *Sphenolithus neoabies* are reworked as well. *Helicosphaera sellii, Oolithothus fragilis,* and possibly *Pseudoemiliania lacunosa* were encountered. Because of the absence of discoasters and ceratoliths, it is difficult to assign sample ER 156 to a calcareous nannofossil zone. It probably lies in the interval between Zones NN16 and NN18. It is noteworthy that, except for the absence of discoasters, sample ER 156 yields an assemblage similar to those that occur in the Bowden Formation at Folly Point.

The Bowden formation at other localities

Section at Folly Point. This section on the coast just West of Port Antonio exposes about 30 m of massively bedded marls of the Bowden Formation overlain by the Navy island Member (10 m) of the Manchioneal Formation.

Five samples were collected from the Bowden beds. Sample FP2 was collected in an erosional cliff facing the bay, high in the section. Sample FP3 was taken ~4 m below it. Samples FP4 to 6 were collected in an exposure adjacent to a driving school. Sample FP4 was taken at the base of the section and samples FP5 and FP6 respectively ~10 m and ~5 m above it.

Calcareous nannofossils are common in all samples, but preservation is rather poor, and discoasters are overgrown. *Calcidiscus macintyrei, Ceratolithus rugosus, Discoaster brouweri, D. misconceptus, D. pentaradiatus, Helicosphaera sellii, Oolithothus fragilis,* and *Pseudoemiliania lacunosa* characterize the assemblage. In addition, *Discoaster asymmetricus* (rare), *D. surculus* (very rare in samples FP2 and 3), and *D. tamalis* (exceedingly rare in sample FP 2, where it is believed to be reworked; common in samples PF4 to 6) were encountered.

Determination of the age of the Bowden Formation exposed in the section at Folly Point is very difficult because of reworking. Few specimens of *D. berggrenii* and *D. quinqueramus* were encountered. Rare, overgrown sphenoliths of *S. abies* and poorly preserved placoliths of *R. pseudoumbilicus* are believed to be reworked. The occurrence of *D. asymmetricus, D. surculus,* and *D. tamalis* may also be interpreted as resulting from reworking. On the other hand, the extreme rarity of *D. surculus* may reflect poor preservation or may be genuine. Semiquantitative analysis reveals large fluctuations in the abundance of *D. surculus* throughout its range (see Rio et al., 1990, p. 519, Fig. 3) and more particularly in the late Pliocene, even at tropical latitudes (see Chepstow-Lusty et al., 1988, p. 135, Fig. 15). Obvious reworking is restricted to a few form particularly resistant to dissolution. In addition, there is a change in composition in the assemblage, *D. tamalis* being common in samples FP4 to 6. It seems more reasonable to interpret this in terms of the natural sequence of stratigraphic events (see Backman and Shackleton, 1983; Chepstow-Lusty et al., 1989) rather than as a product of reworking. Thus, the Bowden Formation exposed at Folly Point is assigned to Zone NN16. Its upper levels (samples FP2 and FP3) are assigned to Subzone CN12b, its lower levels to Subzone CN12a based on the relatively common occurrence of *D. tamalis.*

The common occurrence of *Calcidiscus macintyrei* with *Pseudoemiliania lacunosa* in sample FP1 indicates that the Navy Island Member of the Manchioneal Formation belongs to the *Cyclococcolithina macintyrei* Zone of Gartner (1977).

Drivers River section. This section, located in the Manchioneal District along the road on the right bank of the Drivers River, exposes the same stratigraphic and paleontologic sequence as the section at Folly Point (Robinson, 1969). Five samples (130 at the base to 133A at the top) were collected.

Although the quality of preservation of the calcareous nannofossils (and accordingly the diversity of the assemblages) varies among them, all samples yield assemblages of similar composition. Based on the occurrence of *Discoaster tamalis* and the absence of *Sphenolithus abies, S. neoabies,* and *R. pseudoumbilicus,* the whole section is assigned to Subzone CN12a.

Samples 130 and 132, taken at about the same stratigraphic level at the base of the section, yield abundant, moderately preserved calcareous nannofossils, among them *Ceratolithus rugosus, Discoaster asymmetricus, D. brouweri, D. misconceptus, D. pentaradiatus, D. tamalis, D. surculus, Hayaster perplexus, Oolithothus fragilis,* and *Pseudoemiliania lacunosa. Sphenolithus abies* and *S. neoabies,* both very rare, are considered reworked, as are a few placoliths of *R. pseudoumbilicus.*

The section at Innes Bay. An exposure that covers the same stratigraphic interval of the Bowden Formation as the Drivers River section occurs at milepost 54 on the coast road on the north side of Innes Bay (Robinson, 1969). Four samples (151 at the top to 153 at the base) were collected.

Calcareous nannofossils are abundant and moderately to poorly preserved. The assemblages are similar to those in the Bowden Formation exposed in the Drivers River section. Rare *S. abies, S. neoabies,* and *R. pseudoumbilicus* are considered reworked. Based on the co-occurrence of *D. surculus* (common) and *D. tamalis,* the whole exposure is assigned to Zone NN16 and Subzone CN12a.

Road to Arcadia section. Although the base of the Bowden Formation was defined as the base of the Bowden Shell Bed (Woodring, 1925), marls that underlly it have been included in the Bowden Formation (Blow, 1969; Robinson, 1969). However, while Blow (1969, Fig. 29, p. 297) includes all marls in the Bowden Formation, Robinson includes only part of them in it and correlates the lower levels with the Buff Bay Formation.

Nine samples (154 to 160) were collected during the field party of 1987 in the Road to Arcadia section. In addition, samples ER 307, 305, and 300 were examined. Sample 160 is equivalent to ER 305, 158 to ER 304, 156 to ER 302, 155 to ER 301, and 154 to ER 300. Examination of samples taken from the same levels but years apart proves to be useful in controlling the effect of weathering on the preservation of the calcareous nannofossil assemblages. Despite efforts to collect the freshest material during the field party of 1987, preservation is much poorer in it than in the ER samples collected in 1969.

Samples 160 and 305 yield an extremely impoverished late Miocene calcareous nannofossil assemblage. *Discoaster quinqueramus,* overgrown, is the only discoaster common in the assemblage. It co-occurs with *Amaurolithus primus* (extremely rare), which indicates a biostratigraphic location in Subzone CN9b, possibly close to the CN9a/CN9b subzonal boundary. *Discoaster quinqueramus* was also unambiguously identified in sample ER 307 at the base of lithological "Unit i" (see Blow, 1969, p. 297, Fig. 29), which yields excessively scarce and poorly preserved calcareous nannofossils.

Sample 159, taken down road from sample 158, yields rare poorly preserved calcareous nannofossils. *Discoaster quinqueramus* was confidently identified, and this level is also assigned to Subzone CN9b.

Sample 158, equivalent to ER 304, is barren.

Although calcareous nannofossils are common in sample 157C, discoasters are rare and no ceratoliths were found. The discoasters are represented by *D. asymmetricus, D. brouweri, D.*

challengeri, D. misconceptus, D. pentaradiatus, D. surculus, and *D. variabilis.* Despite long search, *D. quinqueramus* was not found. Sample 157C therefore belongs to a level younger than Zone NN11, but in the absence of ceratoliths, it is not possible to locate it within the NN12–NN15 zonal interval. The rarity of discoasters (and the absence of ceratoliths) in this sample compared to their higher frequency in sample 160 probably reflects regional shallowing.

Discoasters are rare in sample 157B, 157A, 156 (= to sample ER 302), and 155 (= to sample ER 301) than in simple 157C and, except for *D. pentaradiatus* and *D. brouweri,* are mainly indeterminate because of overgrowth.

Preservation is much better in sample ER 300 than in sample 154 collected in 1987, attesting to the strong biasing effects of weathering on the nannopaleontologic record. Sample ER-300 yields the same discoaster assemblage as sample 157C. In addition, *Ceratolithus rugosus* is rare, and *Amaurolithus tricorniculatus* is extremely rare. Sample ER-300 belongs to Subzone CN10c (= zonal interval NN13-NN14).

The Road to Arcadia section thus extends from upper Miocene to lower Pliocene. The base of lithological "Unit i" in Blow (1969, Fig. 29) lies in Zone NN11, its top in Subzone CN10c. Precise location of sample ER 307 within Zone NN11 is somewhat uncertain because no ceratoliths were found at this level. Their absence may not be significant, however, if we consider (1) the utmost rarity of calcareous nannofossils in this sample, their poor preservation, and low diversity; (2) the scarcity of ceratoliths in younger levels of "Unit i" (e.g., none were found in sample 157; and (3) the general scarcity of ceratoliths in the upper Neogene deposits of eastern Jamaica. Unless it can be shown that there is a stratigraphic gap between the levels of samples ER 307 and 305, and considering the thinness of the interval between these two samples (less than 4 m; see Blow, 1969, p. 297, Fig. 29), the base of "Unit i" can be assigned to either Subzone CN9b or to the very top of Subzone CN9a. Because there is a strong suggestion that the absence of ceratoliths in sample ER 307 results from diagenetic processes, the base of "Unit i" is drawn within Subzone CN9b.

Blow (1969) indicated that the base of "Unit i" belongs to Zone N16 and is of Tortonian age. While the Tortonian/Messinian boundary may occur in the Road to Arcadia section (between samples ER 307 and 305, providing that the absence of species of *Amaurolithus* is genuine in sample ER 307), assignment to Zone N16 is in contradiction with calcareous nannofossil stratigraphy, which allows a definitive assignment to Zone NN11. It has been shown above that the N16/N17 zonal boundary in the Buff Bay section lies within the upper part of Zone NN10 (Fig. 9). In addition, the "holotype" locality of Zone N17 belongs to Zone NN10 (see discussion below). Assignment of the base of "Unit i" to Zone N16 by Blow (1969) may reflect diagenetic effects on the composition of planktonic foraminiferal assemblages. The age of the base of "Unit i" is thus well constrained on the basis of calcareous nannofossil biostratigraphy; with reference to the time scale of Berggren et al. (1985), it is younger than 6.4 Ma and probably older than 6 Ma (estimated age of the FAD of *A. tricorniculatus*) (if the base of "Unit i" lies in Subzone CN9a, it is only slightly older than 6.4 Ma, for the reasons given above). On the other hand, the age of its younger levels is difficult to estimate based on calcareous nannofossil biostratigraphy alone. Zonal assignment to the NN13-NN14 interval indicates that they are younger than 4.5 Ma (estimated age of the FAD of *C. rugosus*) and older than 3.7 Ma (estimated age of the LAD of *A. tricorniculatus*).

There is no biostratigraphic indication of whether the Road to Arcadia section is continuous; since *C. rugosus* was found only in sample ER 300 while no ceratoliths were found between the top of Zone NN11 and this level, it is difficult to consider its only occurrence as representing its FAD. Yet, *C. rugosus* is very common in Pliocene deposits from other Jamaican localities (e.g., in the San San Bay section). In addition, the Road to Arcadia section represents a shallowing upward sequence (Stuart et al., 1987), the level of sample ER 300 being the shallowest. *Ceratolithus rugosus* would be expected to occur below this sample if the underlying (i.e., deeper water) sediments were deposited within the range of the species. Thus, there are some hints that the occurrence of *C. rugosus* in sample ER 300 may indeed correspond to its FAD. Although not definitive evidence of continuity in sedimentation, planktonic foraminiferal stratigraphy (Berggren, this volume) brings additional support for interpreting the occurrence of *C. rugosus* in sample ER 300 as its FAD. A sinistral to dextral coiling change in the *Globorotalia limbata-pseudomiocenica* complex occurs between samples 155 (equivalent to sample ER 301) and 156 (equivalent to sample ER 302). In continuous sections, this coiling change occurs in the Thvera Subchron of the Gilbert Chron, which correlates with the upper part of Zone NN12 and is estimated to have occurred at ~4.7 Ma (W. A. Berggren, unpublished data). Magnetobiostratigraphic correlations have shown that *Ceratolithus rugosus* first occurs just abov the Thvera event, at ~4.5 Ma (Berggren et al., 1985). Thus, there is an apparently normal succession of events, and there is no reason to suspect that an unconformity occurs between samples ER 301 and 300. It is therefore estimated that the youngest levels of "Unit i" of Blow (1969) are slightly younger than 4.5 Ma.

Calcareous nannofossil stratigraphy of lithological "Unit i" of Blow (i.e., of the Road to Arcadia section) can be summarized as follows: (1) The lower part of the unit belongs to the upper part of Zone NN11 (i.e., to Subzone CN9b); (2) the NN11/NN12 zonal boundary occurs between samples 159 and 157C (i.e., probably close to the level of sample ER 304); (3) the main part of "Unit i" (i.e., from about the level of sample ER 304 to the level of sample ER 301) belongs to Zone NN12; and (4) the uppermost part of "Unit i" belongs to Zone NN13, and the NN12/NN13 zonal boundary is drawn between samples ER 301 and ER 300.

Road to Ecclesdown section. Seven samples (134 to 137C) were taken from this more than 150-m-thick section. Calcareous nannofossils are common to abundant, but preservation is generally poor and considerable mixing (*D. hamatus, D. neohamatus, D. quinqueramus* occurring together at the same levels) hampers biozonal determination. For these reasons, but also because of the scarcity of the zonal markers (in particular of the ceratoliths), biostratigraphic subdivision of this section is obscure. Also, the samples were taken far apart, which precludes delineation of a sequential occurrence of datums. The lowest level sampled (sample 137) belongs to the upper Miocene, possibly to Zone NN10. The next two samples (136 and 134), taken respectively 26 m and 64 m above sample 137, yield *Discoaster quinqueramus, Amaurolithus delicatus A. primus,* and *A. tricorniculatus,* which suggests a latest Miocene age for this interval and an assignment to Subzone CN9b. *Ceratolithus acutus* and *A. delicatus* occur in sample 137A, taken 117 m above sample 137. The latter species also occurs in sample 137B but was not found in sample 137C. *Ceratolithus rugosus* was not found. Although calcareous nannofossil biostratigraphy suggests that the section extends from upper Miocene to lower Pliocene, planktonic foraminiferal stratigraphy indicates that it extends well up into the upper Pliocene. The discrepancy between the planktonic foraminiferal and the calcareous nannofossil stratigraphic interpretations is probably a result of mixing, and the calcareous nannofossil stratigraphic interpretation should be regarded as very preliminary.

Stony Hill Road section. This section, in Port Antonio, exposes an upper Miocene interval. Ten samples were taken along the roadside exposures, but their stratigraphical relationships are unclear. In addition sample ER 767 was examined.

It is difficult to determine the age of the stratigraphically lower part of the section, because of very poor preservation and the possibility of mixing (as seen in the planktonic foraminifera; see Berggren, this volume). The lower part of the section probably belongs to Zone NN9 or NN10. Most of the section, however, can be assigned to upper Zone NN11 (Subzone CN9b), with abundant *D. quinqueramus* and rare *A. primus* and *A. delicatus.* Sample ER 767 yields an assemblage characteristic of this subzone. I suspect that an unconformity occurs in the section below Subzone CN9b. The youngest levels of the section may be lowermost Pliocene (see Berggren, this volume), although there was no strong evidence from calcareous nannofossils that levels younger than NN11 occur (*D. quinqueramus* being possibly reworked).

A suite of samples (ER 1080 to ER 1086) collected in 1969 by E. Robinson, provided by Jan Hardenbol (Exxon), was also studied. Preservation is excellent in these samples (illustrating once more the profound negative effect of weathering). All yield abundant, well-preserved, highly diversified calcareous nannofossil assemblages indicative of Subzone CN9b, with abundant *Discoaster quinqueramus* and relatively common ceratoliths, *Amaurolithus primus, A. delicatus,* and *A. tricorniculatus.* No levels younger or older than Subzone CN9b are represented by this suite of samples.

THE NEOGENE DEPOSITS OF EASTERN JAMAICA: IMPLICATIONS OF INTEGRATED LITHOSTRATIGRAPHY AND BIOSTRATIGRAPHY

The Neogene of eastern Jamaica is currently divided into four lithostratigraphic units: Spring Garden, Buff Bay, Bowden, and Manchioneal. The August Town Formation is now regarded as a lateral equivalent of the Bowden Formation and possibly of the Buff Bay Formation also (E. Robinson, personal communication, October 1990). No calcareous nannofossil data are available relative to this formation, and a sample from the locality of Garcia Pond, provided by E. Robinson, proved to be essentially barren (with only few, poorly preserved placoliths of *Reticulofenestra*). The Manchioneal, little discussed in this report, is essentially a Pleistocene unit unconformable upon the Bowden (see, for instance, Blow, 1969, p. 296, Fig. 28). The Spring Garden is a thick middle Miocene unit that likely extends into the lower Miocene. It is conformable with the Buff Bay Formation in the Buff Bay section but unconformable with the San San Clay in the San San Bay section. The lithologic characteristics of the Spring Garden may help in differentiating it consistently from members of the Lower Coastal Group. This latter comprises the Buff Bay and the Bowden formations. Presently, these lithostratigraphic terms are implicitly used so as to conform with biostratigraphic subdivisions delineated on the basis of planktonic foraminifera. The N18/N19 boundary essentially serves as the boundary between the Buff Bay and the Bowden formations. As a result, the beds exposed in the Road to Arcadia section, referred to as "lithologic Unit i" by Blow (1969, p. 297, Fig. 29) and placed by him in the Bowden Formation, are subdivided by Robinson (1969) so that the stratigraphically younger beds (ER 300 to ER 302, Zone N19 [Blow, 1969]) are placed in the Bowden Formation while the stratigraphically older beds (N17 and N18, Blow [1969]) are assigned to the Buff Bay Formation. Similarly, although younger than the Buff Bay Formation in its type locality, the San San Clay is regarded as part of the Buff Bay Formation (Robinson, 1969; Lamb and Beard, 1972). This practice is not satisfactory for a number of reasons. First, calcareous nannofossil stratigraphy indicates that there is no overlap between the Buff Bay Formation, the San San Clay, and the Bowden Formation. Second, it reveals an uneven development of the upper Miocene and Pliocene zones and subzones, which suggests the development of regional unconformities between the three formations. Third, integrated calcareous nannofossil and planktonic foraminiferal stratigraphy indicates that the Buff Bay Formation, the San San Clay, and the Bowden Formation represent three distinct stratigraphic sequences (in the sense of Van Wagoner et al., 1988).

Evidence for the delineation of sequences in the Lower Coastal Group (upper Miocene–Pliocene) of Jamaica

The three lines of evidence that indicate that the Lower Coastal Group can be subdivided into three sequences (Fig. 12) are examined in turn.

1. Calcareous nannofossil stratigraphy. The Buff Bay Formation. Calcareous nannofossil stratigraphy of the Buff Bay Formation in the Buff Bay section has been discussed in detail above. In its type locality, the Buff Bay Formation extends from Zone NN8 to Zone NN10. It correlates essentially with the lower part of the Tortonian as delineated in Berggren et al. (1985). No deposits correlative with the Buff Bay Formation occur in the sections studied herein, except in the Stony Hill Road section.

The San San Clay. The San San Clay as redefined above extends from the upper Miocene Zone NN11 (Subzone CN9b) to the lower Pliocene Zone NN13, that is, from upper Messinian to lower Zanclean. Deposits equivalent to the San San Clay occur in the Stony Hill Road, Road to Ecclesdown, and Road to Arcadia sections. Lithologic "Unit i" of Blow (1969) is almost equivalent to the San San Clay in its type section.

The Bowden Formation. Calcareous nannofossil stratigraphy does not allow zonal assignment of the type Bowden Beds. Through correlation with equivalent deposits such as those in the Innes Bay, Drivers River, and Folly Point sections, it can be shown that this is a stratigraphically restricted unit in the lower upper Pliocene. It belongs mostly to the lower part of Zone NN16, that is, to Subzone CN12a, but it extends into the upper part of Zone NN16 (Folly Point section) and into Zone NN15 (Buff Bay section).

There is perfect agreement between the identification of the Kaena and Mammoth subchrons of the Gauss Chron in the Bowden Formation exposed in the section at Innes Bay (Robinson and Lamb, 1970) and calcareous nannofossil assignment to Subzone CN12a (cf., Berggren et al., 1985). It is worth mentioning in passing that identification of the Olduvai Subchron of the Matuyama in the Navy Island Member of the Manchioneal Formation in the section at Innes Bay, in the classic Bowden locality, and in the type Navy Island Member (on Navy Island) (Robinson and Lamb, 1970) is also in agreement with the assignment of the correlative beds in the section at Folly Point to the *Calcidiscus macintyrei* Subzone of Gartner (1977).

2. Uneven development of upper Miocene and lower Pliocene biozones. While Subzone CN9b is broadly developed in eastern Jamaica, no deposits representative of Subzone CN9a were identified. This suggests that while the Messinian and the early Tortonian are well represented in eastern Jamaica, the late Tortonian corresponds to a stratigraphic gap. It further indicates that the Buff Bay and the San San Clay are unconformable, implying that the San San Clay cannot be regarded as part of the Buff Bay Formation.

Likewise, while stratigraphic continuity was observed from upper Miocene to lower Pliocene (from Subzone CN9b to Zone NN13), there is no strong evidence for stratigraphic continuity from Zone NN13 to Zone NN15, which suggests that the interval corresponding to Zone NN14 may be missing. This suggests, in turn, that the San San Clay and the Bowden Formation may be unconformable, which implies, in agreement with Robinson (1969) but in disagreement with Blow (1969), that the San San Clay cannot be included in the Bowden Formation.

Finally, there is a stratigraphic gap between the Bowden Formation and the Navy Island Member, as deduced from the absence in eastern Jamaica of deposits representing Zones NN17 and NN18.

3. Integration of calcareous nannofossil and planktonic foraminiferal stratigraphies. Although there are some discrepancies with regard to the relative ages of the formations of the Lower Coastal Group as deduced from calcareous nannofossil and from planktonic foraminifera stratigraphy (Berggren, this volume), these are minor, and planktonic foraminiferal stratigraphy supports the stratigraphic gaps delineated through calcareous nannofossil stratigraphy. Unless indicated, the planktonic foraminiferal ages discussed below are from Berggren (this volume).

Buff Bay section. Detailed study of the Buff Bay section leads to revision of the correlations between calcareous nannofossil and planktonic foraminiferal zonations, at least for the Miocene (Fig. 9). However, identification of Zone PL2 and PL3 in the Bowden Formation equivalent (West Roadcut) correlates well with identification of Zones NN15 and NN16, and both groups indicate the absence of (exposed) lower Pliocene deposits (NN14 in particular is not represented at the base of the Bowden Formation). In addition, calcareous nannofossil stratigraphy reveals the absence of upper Miocene Zone NN11.

San San Bay section. It has been shown above how integrated calcareous nannofossil and planktonic foraminiferal stratigraphies complement each other in the interpretation of the San San Bay section. It has been established that the lower part of Zone NN11 (Subzone CN9a) is not represented and that no deposits equivalent to Zone NN14 occur in the section.

Road to Arcadia Section. There is agreement between calcareous nannofossil and planktonic foraminiferal subdivisions of this section ("Unit i" of Blow, 1969). The calcareous nannofossil upper NN11-NN12-NN13 zonal succession is paralleled by the planktonic foraminiferal N17-PL1a-PL1b zonal succession. Both groups precisely constrain the age of the base of the section, ~6 Ma, and of its top, slightly younger than 4.5 Ma. There are no deposits representing Zones NN14 and NN15 between this unit and the type Bowden Formation, which is correlative with Zone NN16 (see discussion above) and Zone N20 (Blow, 1969, p. 297, Fig. 29).

Stony Hill Road section. Planktonic foraminiferal and calcareous nannofossil stratigraphies concur in locating the base of this section in the lower upper Miocene and in assigning it to Zone N16 and to an undetermined interval within Zones NN9-NN10, respectively. While the remainder of the section is assigned to Subzone CN9b, planktonic foraminiferal N17-PL1a-PL1b zonal succession (equivalent to the interval upper Zone

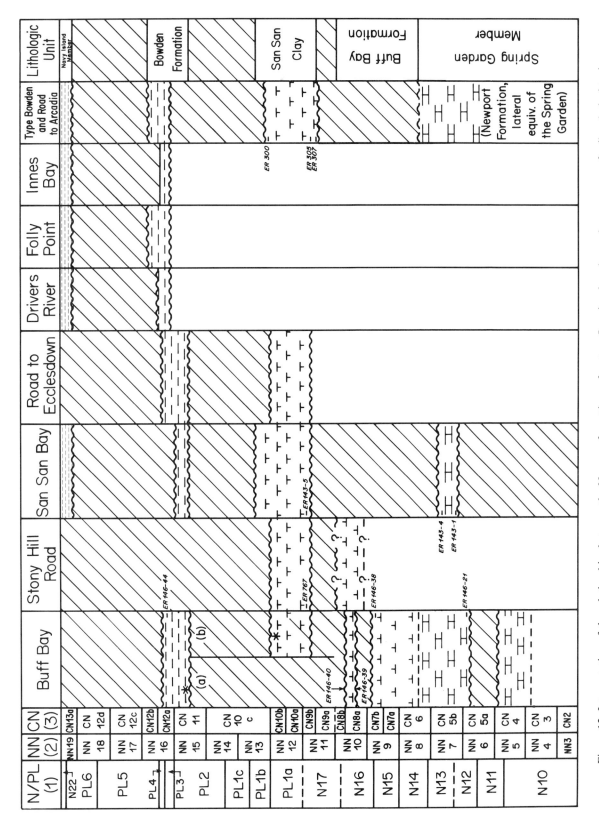

Figure 12. Interpretation of the relationships between the Neogene formations of eastern Jamaica, based on calcareous nannofossil and planktonic foraminiferal stratigraphy. (1): N-Zones from Blow (1969); PL-Zones from Berggren in Berggren et al. (1983); (2): NN Zones from Martini (1971); (3): CN Zones and Subzones from Bukry (1973, 1975) and Okada and Bukry (1980); (a): Interpretation of the Buff Bay section in which no beds younger than Subzone NN10 and older than Zone NN15 occur in the ~18-m-thick covered interval between the Buff Bay and the Bowden formations; (b) Interpretation of the Buff Bay section in which beds equivalent to the San San Clay occur within the ~18-m-thick covered interval between the Buff Bay and the Bowden formations; and *: position of the "holotype" locality of Zone N18 following interpretation (a) and (b).

NN11 to NN13) was recognized. Whatever the reasons for this discrepancy, the fact remains that (1) there is a stratigraphic gap corresponding to the lower part of Zone NN11 (Subzone CN9a), and (2) there are no deposits representative of Zone NN14 (the PL1b/PL1c zonal boundary lies within Zone NN13).

Road to Ecclesdown section. There is little agreement between the calcareous nannofossil and planktonic foraminiferal stratigraphies in this section. The latter group indicates a PL1b-PL1c-PL2-PL3 zonal succession (correlative with Zones NN13 to NN16), whereas the former group suggests an upper NN11-NN12-NN13 zonal succession. As discussed above, these results should be regarded as very preliminary. Similarly, the interpretation of this section given in Figure 12, which relies mainly on planktonic foraminiferal stratigraphy, should be regarded as very tentative.

Interpretative of the Lower Coastal Group

Based on the evidence discussed above, it appears reasonable to interpret the Buff Bay Formation, the San San Clay, and the Bowden Formation as three distinctive sequences (Fig. 12). Considering the maximum extent of these units in the different sections studied, these sequences can be interpreted as described below.

The Buff Bay Formation is the lowest sequence of the Lower Coastal Group and is of early late Miocene age (early Tortonian). It extends from Zone NN8 (upper part) to Zone NN10 (upper part) in terms of calcareous nannofossil stratigraphy and from Zone N14 to Zone N17 (lower part) in terms of planktonic foraminiferal stratigraphy. Following the age estimates of Berggren et al. (1985), it was deposited between ~10.4 and ~8.5 Ma (ongoing revision of the middle Miocene part of the time scale will lead to an older age estimate for the base of the Buff Bay, W.A. Berggren, personal communication, 1993).

As redefined in this work, the San San Clay, the medial sequence, is of late Miocene and early Pliocene age. It corresponds essentially to the Messinian and the early Zanclean. It extends from Zone NN11 (upper part; Subzone CN9b) to Zone NN13 in terms of calcareous nannofossil stratigraphy and from Zone N17 (upper part) to Zone PL1b in terms of planktonic foraminiferal stratigraphy. Following the age estimates of Berggren et al. (1985), it was deposited between ~6 and 4.5 Ma.

The Bowden Formation is the highest (youngest) sequence and is of early late Pliocene age (early late Piacenzian). It extends from Zone NN15 to Zone NN16 in terms of calcareous nannofossil stratigraphy and from Zone PL2 to Zone PL5 in terms of planktonic foraminiferal stratigraphy. Following the age estimates of Berggren et al. (1985), it was deposited between ~3.4 and ~2.5 Ma.

With reference to the maximum stratigraphic development of the formations of the Lower Coastal Group, the duration of the hiatuses associated with the unconformities that separate them can be estimated as discussed below.

The Buff Bay Formation is separated from the San San Clay by a stratigraphic gap that corresponds to the uppermost part of Zone NN10, all of Subzone CN9a, and the lower part of Subzone CN9b in terms of calcareous nannofossil stratigraphy and to the middle part of Zone N17 in terms of planktonic foraminiferal stratigraphy. The hiatus is at least 1.8 m.y. The lower surface of the unconformity is older than 8.2 Ma (estimated age of the FAD of *Discoaster quinqueramus*), and its upper surface is younger than 6.4 Ma (estimated age of the FAD of *Amaurolithus primus*).

The San San Clay is separated from the Bowden Formation by a stratigraphic gap that corresponds to the upper part of Zone NN13, Zone NN14, and the lower part of Zone NN15 in terms of calcareous nannofossil stratigraphy and to the upper part of Zone PL1c and PL2 in terms of planktonic foraminiferal stratigraphy. The hiatus is at least 0.4 m.y. The older surface of the unconformity is older than 4.1 Ma (estimated age of the acme of *D. asymmetricus*) and younger than 4.5 Ma (estimated age of the FAD of *Ceratolithus rugosus*); its younger surface is younger than 3.7 Ma (estimated age of the LAD of *Amaurolithus tricorniculatus*) and older than 3.5 Ma (estimated age of the LAD of *Reticulofenestra pseudoumbilicus*).

In addition, the Bowden Formation is separated from the Manchioneal Formation (Upper Coastal Group) by a stratigraphic gap that corresponds to the upper part of Zone NN16, Zone NN17, Zone NN18, and the lowermost part of Zone NN19 in terms of calcareous nannofossil stratigraphy and to the upper part of Zone PL5 and the lower part of Zone PL6 in terms of planktonic foraminiferal stratigraphy. Although generally regarded as a Pleistocene unit, the Navy Island Member of the Manchioneal Formation is of latest Pliocene age. As discussed above, it belongs to the *Calcidiscus macintyrei* Zone of Gartner (1977) and to Zone N22, and it was deposited during the Olduvai Subchron of the Matuyama. The hiatus that separates the two formations is ~0.5 m.y.; the older surface of the unconformity is older than 2.4 Ma (estimated age of the LAD of *D. surculus*), and the younger surface is younger than 1.9 Ma (the estimated age of the LAD of *D. brouweri*).

Discussion

It may appear hazardous at first to attempt to delineate sequences in a tectonically active region. From the distribution of the different formations in eastern Jamaica (Fig. 12), two sedimentary domains, probably tectonically derived and driven, can be distinguished. In the northeastern domain (around Buff Bay), sedimentation was more continuous during the Miocene than in the eastern domain (around Bowden). Pertinent to this, it should be noted that the Buff Bay Formation can be regarded as a sequence only relative to the San San Clay, for there is no stratigraphic gap between it and the Spring Garden Member. The Buff Bay and the Spring Garden formations may be deposits that are too deep (1,400 to 2,000 m, Katz and Miller, this volume) to yield a record of sequences. The unconformities between the Spring Garden and the San San Clay in the San San Bay section and between the Newport Formation (lateral equivalent of the Spring Garden Member) and "Unit i" of Blow (1969) (lateral equivalent of the San San Clay) in the Road to Arcadia section are likely to be derived from active local tectonics related to uplift. Development of the unconformity at ~300 m between the Buff Bay and the Bowden formations in the Buff Bay section may also be related to regional tectonics leading to uplift-induced shallowing from lower bathyal to upper bathyal (200 to 600 m, Katz and Miller, this volume). However, superimposed on this local tectonic regime, a pattern of sequences can be delineated in the upper upper Miocene to Pliocene stratigraphic record of eastern Jamaica (the Pleistocene stratigraphic record is not considered in this study) (Fig. 12). There are local variations as to the extent of the sequences. For instance, the basal beds of the Bowden formations are slightly older in the northeastern domain (Buff Bay and the San San Bay sections) than in the eastern one. The Bowden beds (as reinterpreted) are very thin in the San San Bay section compared to the local extent of this formation. The basal and the terminal beds of the San San Clay may be slightly older or younger in different sections. Yet, in all sections, a similar stratigraphic interval is missing. It includes at least Subzone CN9a between the San San Clay and the Buff Bay Formation, Zone NN14 between the former unit and the Bowdon formation, and Zones NN17 and NN18 between this latter and the Navy Island Member. It is beyond the scope of this contribution to discuss here the significance of this pattern of sequences, but a few remarks are appropriate.

On both sides of the unconformity that occurs in the Buff Bay section between the Buff Bay and the Bowden formations there is a covered stratigraphic interval, so that ~14 m of section are not exposed. It may be that the Buff Bay is directly unconformable with the overlying Bowden Formation (interpretation [a] in Figure 12), but it cannot be ruled out that a thin interval corresponding to the San San Clay may occur in this covered interval (interpretation [b] in Figure 12). The proximity of the Stony Hill Road section in which the Buff Bay Formation is unconformable with the San San Clay makes interpretation (b) in Figure 12 more likely.

There is a widespread unconformity in the North and South Atlantic between the Tortonian and the Messinian, so that Subzone CN9a is missing or restricted (Hodell and Kennett, 1986; Aubry, unpublished data). This unconformity may correlate (or not) with the unconformities associated with hiatus NH6 suggested by Keller and Barron (1987). It has no clear equivalent in the onlap/offlap record of Haq et al. (1988).

While there seems to be a good correlation between the unconformity in the upper lower Pliocene of eastern Jamaica with the 3.8-Ma sequence boundary of Haq et al. (1988), it seems more difficult to reconcile deposition of the shallow Bowden Formation with the correlative 3.0-Ma sequence boundary inferred from seismic stratigraphy by Haq et al. (1988), thought to reflect the onset of glaciation in the Northern Hemisphere. In addition, it is not clear whether the 2.4-Ma glacial event is reflected in the stratigraphy of eastern Jamaica.

In summary, sedimentary sequences have been (somewhat unexpectedly) delineated in the Neogene of Jamaica, a region strongly affected by regional tectonics. Comparison of this record with the global pattern of sequences delineated by Haq et al. (1988) indicates that, with one exception, there is no clear relation between the depositional pattern in Jamaica and the global sedimentary regime inferred from seismic stratigraphy. This suggests either that the age (not only the estimated age but the relative age) of the sequence boundaries needs to be revised or that the mechanisms that led to the formation of stratigraphic sequences even during glacial regimes remain to be established, as well as their significance. Unless it can be shown that regional tectonics in Jamaica compensated for glacioeustasy during the Pliocene, it is not apparent that glaciations had any influence on the Pliocene depositional/erosional patterns in Jamaica.

CALCAREOUS NANNOFOSSIL AGE OF THE JAMAICAN TYPE LOCALITIES OF BLOW'S PLANKTONIC FORAMINIFERAL ZONES (N ZONES, 1969)

Gartner (1969) was the first to develop a comprehensive calcareous nannofossil zonal scheme for the upper Miocene to the Recent, and the standard calcareous nannofossil zonation of Martini (1971) draws largely upon Gartner's work. The merit of this work was to develop a zonal scheme in relation to the broadly used planktonic foraminiferal zonation of Blow (1969), and the correlation pattern that Gartner established has been mostly followed ever since (e.g., Martini, 1971; Berggren et al., 1985). Relations of the ranges of calcareous nannofossil species to the N Zones were established by Gartner through the study of four cores, but only Core CAP 38 BP offered a complete succession from Zone N17 to Zone N23 and had been studied by Blow (1969, p. 38, Fig. 38). Gartner (1969), however, did not study the calcareous nannofossil content of sections that yield the type-localities of the N-Zones, several of which were located in Jamaica (Table 6). In this sense, the present study provides the first opportunity to draw direct correlations between Blow's zonal scheme and the currently used calcareous nannofossil zones. All uppermost Miocene to Pleistocene N zones, except Zone N21, have their "holotype" locality in Jamaica as well as one or two "paratype" localities. The calcareous nannofossil ages of the "holotype" and Jamaican "paratype" localities for Zones N17 to N20 are discussed below. Blow (1969) also designated in Jamaica a "paratype" locality for Zones N13 and N16. They are discussed in turn.

This study reveals large age differences for the "holotype" and "paratype" localities designated by Blow (1969) in Jamaica (Fig. 13 and Tables 6 and 7) and strong discrepancies as to the

TABLE 6. CALCAREOUS NANNOFISSIL ZONAL/SUBZONAL ASSIGNMENT
OF THE JAMAICAN "HOLOTYPE" AND "PARATYPE" LOCALITIES OF THE N-ZONES OF BLOW*

N Zone*	"Holotype" Locality*	"Paratype" Locality*	Paratype Locality	NN/CN Zone/Subzone†	
N20	Here proposed, at the locality and level of Sample ER.156, within the Bowden Formation, ca. 10 ft above the Bowden Shell Bed (within the "Bowden Beds" auctt. including Hill, 1899; within the "Pteropod Marl") of Sawkins, 1869, in part; from the Bowden section, Jamaica, W.I.	At the type and locality of sample ER.193, within beds referred to the Bowden Formation (Robinson, 1965) (within the "Pteropod Marl" of Sawkins, 1869, upper part); from the Driver's River section, Jamaica	At the type and locality of sample ER.538, within beds referred to the Bowden Formation (Robinson, 1965) (within the upper part of the probable equivalents of the "Pteropod Marl" of Sawkins, 1869); from the Folly Point section, Jamaica.	Sample ER.156:	?NN16
				Sample ER.163:	NN16, CN12a
				Sample ER.538	NN16
N19	Here proposed at the locality and level of sample ER.146/44 (the same locality and level as for the equivalent samples WHB.181A and Stainforth-sample.	At the type and locality of sample ER.300, within the Bowden Formation below the Bowden Shell Bed (lower Bowden Formation of Robinson, 1965) (within the "Bowden Beds" auctt. including Hill, 1899, and the Pteropod Marl [part] of Sawkins, 1869); from the Bowden section, Jamaica		Sample ER.146/44:	NN16, CN12a
				Sample ER.300	NN13
N18	At the locality and level of Sample ER.146/41, within the Bowden Formation (from the Buff Bay Beds of Cushman and Jarvis, 1930, non Hill, 1899, which were placed as the supposed equivalents of the type Manchioneal Formation [of Trechmann, 1930] by Trechmann, 1930); from the Buff Bay section, Jamaica, W.I.	At the locality and level of Sample WHB.181B, within the Bowden Formation (Robinson, 1965; from the "Buff Bay Beds" of Cushman and Jarvis, 1930, non Hill, 1899, referred by Trechmann [1920] to the supposed equivalents of the type Manchioneal Formation); from the Buff Bay section, Jamaica, W.I.	At the locality and level of sample ER.143/7 (= sample WHB.176B), within the San San Member (Robinson, 1965; possibly synonym of the "Pteropod Marl" of Sawkins, 1869, partim), Bowden Formation; from the San San Bay section, Jamaica, W.I.	Sample ER.146/41:	NN12-15
				Sample WHB.181B:	?
				Sample ER.143/7:	NN11/CN9b NN12/CN10a
N17	Here proposed, at the locality and level of Sample ER.146/40 within the type Buff Bay Beds of Hill, 1899 (non Cushman and Jarvis, 1930), emend. Robinson, 1965; from the Buff Bay section, Jamaica, W.I.	At the locality and level of Sample ER.305, within the brown clays and marls of the Bowden Formation (Robinson, 1965), the Bowden Beds of Hill (1899) et al.; from the Bowden section, Jamaica, W.I.		Sample ER.146/40	NN10, CN8a
				Sample ER.305:	NN11 CN9b
N16		At the locality and level of Sample ER.146/37, within the type Buff Bay Formation (Hill, 1899; Robinson, 1965; non "Buff Bay Beds," Cushman and Jarvis, 1930); from the Buff Bay section, Jamaica, W.I.		Sample ER.146/37:	upper NN9 CN7b
N13		At the locality and level of Sample ER.143/4, within the upper part of the Montpelier Formation (previously called the "Pelleu Island Formation", Robinson, 1962); from the San San Bay section, Jamaica, W.I.		Sample ER.143/4	NN7 (NN6)

*From Blow, 1979. Descriptions of the "holotype" and "paratype" localities are cited verbatim.
†From Martini (1971) and Bukry (1973, 1975; codified in Okada and Burky, 1980).

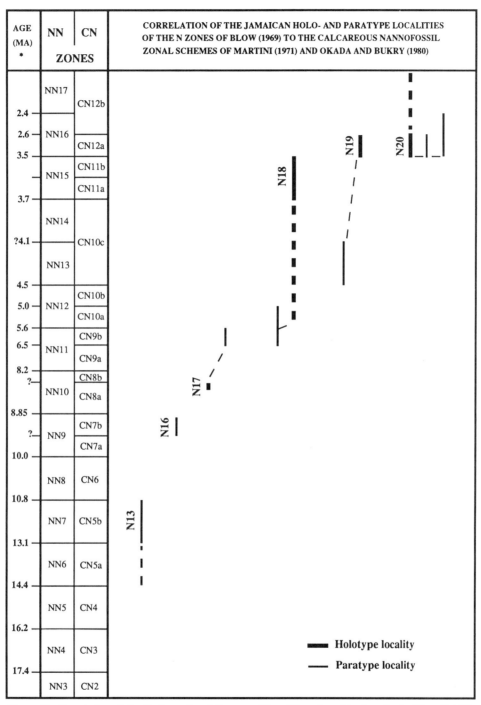

Figure 13. Correlation of the "holotype" and "paratype" localities selected in Jamaica by Blow (1969) in the definition of his so-called N-Zones. Thin oblique and horizontal dashed lines connect "holotype" and "paratype" localities of the same N-Zone. * from Berggren et al., 1985; planktonic foraminiferal zonal assignments from Berggren (this volume); NN zones from Martin, 1971; CN zones from Okada and Bukry, 1980.

TABLE 7. CALCAREOUS NANNOFOSSIL CONTENT OF THE "HOLOTYPE" AND "PARATYPE" LOCALITIES SELECTED IN EASTERN JAMAICA BY BLOW* IN THE DEFINITION OF HIS N-ZONES

Zone	Sample	Calcareous Nannofossil Content
N20	ER 156	Yields scarce poorly preserved calcareous nannofossils. Reworked taxa frequent. *S. abies* and *S. neoabies* are considered reworked. Biostratigraphically insignificant assemblage with *C. rugosus* and *D. misconceptus* (both extremely rare), and questionable *P. lacunosa*.
	ER 193	Yields common but poorly preserved calcareous nannofossils. *Discoaster pentaradiatus* and *P. lacunosa* are frequent. The assemblage is further characterized by *C. rugosus*, *D. brouweri*, *D. surculus*, *D. tamalis*, *H. sellii*, and *O. fragilis*. Located in Subzone CN12a, i.e., in the lower part of Zone NN16.
	ER 538	Yields common but poorly preserved calcareous nannofossils. Assemblage characterized by *D. brouweri*, *D. misconceptus*, *D. surculus* (rare), *D. tamalis*, *C. rugosus*, and *P. lacunosa*. Located within Zone NN16 and Subzone CN12a.
N19	ER 146/44	Yields common, moderately well preserved, calcareous nannofossils, but scarce discoasters. Assemblages characterized by *D. bouweri*, *D. misconceptus*, *D. surculus* (frequent), *D. tamalis*, *C. rugosus*, and *P. lacunosa*. Located within Zone NN16 and Subzone CN12a.
	ER 300	Yields common calcareous nannofossils but ceratoliths and discoasters are rare, the latter being often overgown or broken. With *R. pseudoumbilica*, *S. abies*, and *S. neoabies* (common), *D. asymmetricus* (rare), *D. brouweri*, *D. misconceptus*, *D. pentaradiatus*, *D. surculus*, *D. tamalis* (very rare), *A. tricorniculatus* (very rare), and *C. rugosus*. Assemblage indicative of an NN13-NN14 zonal interval, but level likely located in basal Zone NN13 (see text for discussion).
N18	ER 146/41	Yields scarce calcareous nannofossils except for relatively common discoasters, among which *D. asymmetricus* and *D. tamalis* are the most frequent. Other species include *D. brouweri*, *D. pentaradiatus*, *D. surculus*. Zonal assignment uncertain within an NN12-NN15 zonal interval because of the absence of ceratoliths, which may be a result of poor preservation or shallow water deposition or be stratigraphically genuine. The absence of *C. rugosus* may be interpreted as indicative of Zone NN12.
	ER 143/7	Yields abundant but highly fragmented calcareous nannofossils. Assemblage characterized by *D. berggrenii*, *D. brouweri*, *D. challengeri*, *D. pentaradiatus*, *D. quinqueramus*, *D. surculus*, *A. delicatus*, and *A. primus*. Located in the upper part of Zone NN11 and in Subzone CN9b, or in Zone NN12 (lower part) and Subzone CN10a (definitive age cannot be established because of possible reworking).
N17	ER 146/40	Yields abundant, well-preserved calcareous nannofossils although tips of discoaster arms are often broken. Assemblage characterized by *D. neohamatus* (common), *D. challengeri*, *D. pentaradiatus*, *D. berggrenii* (rare) and *M. convallis* (rare). Located in Zone NN10 upper part and in the Subzone CN8a.
	ER 305	Abundant but poorly preserved calcareous nannofossils. Discoasters strongly overgrown. Assemblages considerably improverished through diagenesis, dominated by *S. abies* and *S. neobies*, characterized by *D. quinqueramus* (common) and *A. primus* (extremely rare). Located in the upper part of Zone NN11, and in Subzone CN9b.
N16	ER 146/37	Yields abundant but fragmented and/or overgrown calcareous nannofossils. Assemblage characterized by *D. hamatus* (common), *C. calyculus* (few), *C. coalitus* (rare), *D. neohamatus* (few), *D. kugleri* (few), and *D. bollii* (few). Located in Zone NN9 and Subzone CN7b.
N13	ER 143/4	Abundant well-preserved calcareous nannofossils although discoasters are somewhat overgrown. Typical middle Miocene assemblages with *C. miopelagicus*, *D. challengeri* (common), *D. exilis* (rare), *D. moorei*, *T. rugosus*, indicative of an NN6-NN7 zonal interval. Likely located in Zone NN7 based on the (common) occurrence of *D. challengeri*.

*Blow, 1969.

correlations between the calcareous nannofossil zones and the N zones. It is not possible to review here the different correlation frameworks recognized in different deep sea and land sections, but early works that were decisive for the establishment of the correlations between calcareous nannofossil and N zones are considered as well as some recent time-scales. For further discussion on the validity of the N-zonal scheme, the reader is referred to Berggren (this volume), with whom I concur in discouraging the use of Blow's zonal scheme for the biostratigraphy of the Pliocene.

Zone N17

The Buff Bay Formation in the Buff Bay section yields the "holotype" level (sample ER 146/40) of Zone N17. In this section, Zone N17 extends from the level of sample ER 146/39 to the unconformity between the Buff Bay and the Bowden formations (Blow, 1969, p. 296, Fig. 28). Correlated to the stratigraphic log of Stuart et al. (1987), Zone N17, as delineated by Blow (1969) and also by Berggren (this volume), extends from the base of the debris flow in the Slide Roadcut to the unconformity at 300 m and includes the marls exposed in the Dead Goat Gully Roadcut (see Fig. 5). It has been established above that this interval belongs to the upper part of Zone NN10 (or upper part of Subzone CN8a). Thus the "holotype" level of Zone N17 lies well within this zone.

The Road to Arcadia section yields one (sample ER 305) of the two "paratype" localities of Zone N17 (the other being in Papua). Blow (1969) improperly indicated that this "paratype" locality is in the Bowden section, since there is no field continuity between the type Bowden and lithological "Unit i" to which Blow refers, which is exposed in the Road to Arcadia section (see Robinson, 1969). The level of sample ER 305 is located in the lower part of lithological "Unit i," close to the N16/N17 zonal boundary as drawn by Blow (1969, p. 297, Fig. 29). As described above, this level is within the upper part of Zone NN11 (Subzone CN9b), characterized by the co-occurrence of *Discoaster quinqueramus* and species of the genus *Amaurolithus*.

Following Blow's selection of its type localities, Zone N17 should correlate at least with the upper part of Zone NN10 and most of Zone NN11. Current biostratigraphic frameworks (e.g., Berggren et al., 1985; Haq et al., 1987) indicate, however, that the base of Zone N17 lies well within Zone NN11 (i.e., high in Subzone CN9a). This is in obvious contradiction with the direct correlation here established between the "holotype" level of Zone N17 and Zone NN10.

Correlation of Zone N17 to the calcareous nannofossil zonation constitutes an intricate problem beyond the scope of this contribution, and discussion will be restricted to a few remarks. The study by Gartner (1969) shows that the top of Zone N17 lies within Zone NN12 but provides no information with regard to the location of the base of Zone N17 with respect to the range of calcareous nannofossil species. It should be mentioned in passing that the *Discoaster quinqueramus* Zone of Martini (Zone NN11) differs from the *Discoaster quinqueramus* Zone of Gartner (1969), although Martini did not proceed to an emendation. Zone NN11 is a total range zone and corresponds to the range of the nominate taxon. The *Discoaster quinqueramus* Zone of Gartner (1969) is a concurrent range zone defined by the overlap of *Amaurolithus tricorniculatus* and *D. quinqueramus*. As defined by Gartner, it is equal to Subzone CN9b. DSDP Site 62.1 is the first deep sea site from which a continuous biostratigraphic succession from middle Miocene to Pleistocene was recorded, allowing direct correlations between planktonic foraminifera (Brönnimann and Resig, 1971) and calcareous nannofossils (Martini and Worsley, 1971). Brönnimann et al. (1971, p. 1741, Table 2) indicated that the N16/N17 zonal boundary lies with Zone NN11, probably based on the record obtained from that site, but failed to recognize that the location of the N16/N17 boundary varied between sites drilled during DSDP Leg 7. At Site 62.1, the N16/N17 boundary lies within the upper part of Zone NN11 (Subzone CN9b), but at sites 63 and 64 (with a discontinuous upper Miocene record) Zone N17 correlates with the lower part of Zone NN11 (Subzone CN9a). There is thus some ambiguity in determining the N16/N17 boundary, as discussed in Brönnimann and Resig (1971) and Berggren (this volume). It should be noted that Barron et al. (1985) tentatively correlate the N16/N17 and CN8a/CN8b zonal boundaries, which implies that the N16/N17 zonal boundary should lie low in Zone NN10. Finally, it should be pointed out that the exact correlations between the upper Miocene planktonic foraminiferal and calcareous nannofossil zones may be obscured by widespread but often unrecognized intra upper Miocene unconformities in the North and South Atlantic basin (Hodell and Kennett, 1986; Aubry, unpublished data).

Zone N18

The "holotype" locality (sample ER 146/41) and two (samples WHB.181B and ER 134/7) of the four "paratype" localities of Zone N18 are from Jamaica (the other "paratype" localities are from Papua and Fiji).

Only samples ER 146/41 ("holotype" locality) and ER 143/7 were available for examination. However, the calcareous nannofossil zonal age of sample WHB.181B must be very close (slightly older according to stratigraphic position; see Blow, 1969, p. 296, Fig. 28) to that of sample ER 146/41, both samples being from the basal part of the Bowden Formation in the Buff Bay section. Sample ER 143/7 is from the San San Clay in the San San Bay section.

Determination of the calcareous nannofossil zonal age of the "holotype" level of Zone N18 is extremely difficult because of the absence of ceratoliths (in the sample of locality ER 146/41 provided by Barry Kohl). It may lie at any level between NN12 and NN15. During the field party of 1987, E. Robinson estimated the level of sample ER 146/41 to be equivalent to the level of sample 244. If that is correct, sample ER 146/41 yields an early late Pliocene age and is located within Zone NN15 (or NN16). Planktonic foraminiferal stratigraphy (Berggren, this volume) supports this calcareous nannofossil age assignment. As remarked above, sample WHB.181B is likely to have the same age. It should be mentioned that a sample provided by Jan Hardenbol of locality ER 146/41 yields a rich calcareous nannofossil assemblage exactly similar to that in sample 244 collected by us in 1987.

Examined isolated, sample ER 143/7 would be assigned to Subzone CN9b. Examined in the context of the whole section at San San Bay, it would be assigned to Subzones CN9b–CN10a undifferentiated, based on the correlations established in the deep sea between calcareous nannofossil and planktonic foraminiferal zones (Berggren et al., 1985), the occurrence of *Discoaster quinqueramus* at this level possibly resulting from reworking. This implies that one of the "paratype" localities of Zone N18 may overlap with one of the "paratype" localities for Zone N17 or be stratigraphically very close to it. This leads to a question of the reliability of the N-zonal system, at least of part of it.

The "holotype" locality of Zone N18 and one of its "paratype" localities apparently belong to the lower upper Pliocene, and another "paratype" locality belongs to the upper Miocene, suggesting that Zone N18 is a zone of long duration. In fact, it is one of the Neogene planktonic foraminiferal zones with the shortest duration (see Berggren et al., 1985).

Blow (1969, p. 296, Fig. 28) draws the N17/N18 boundary between the levels of samples ER 143/5 and 6 in the San San Bay section, which implies that the N17/N18 boundary should lie within Subzone CN9b. However, he (1969, p. 297, Fig. 29) tentatively draws the same boundary at the level of sample ER 303 in the Road to Arcadia section, which then lies within Zone NN12. This latter correlation (N17/N18 zonal boundary within Zone NN12) was established by subsequent works (e.g., Gartner, 1969; Brönnimann et al., 1971) and is followed in Berggren et al. (1985) and Barron et al. (1985). However, Keller and Barron (1987) align the N17/N18 zonal boundary within the CN9b/CN10a subzonal (= NN11/NN12 zonal) boundary, and Haq et al. (1987) draw the N17/N18 zonal boundary in the uppermost part of Zone NN11 (which is in closest agreement with Blow's correlations in the San San Bay section). Since the reasons for the latter set of correlations are not well documented, it is not possible to discuss their validity.

More problematic is the position of the "holotype" locality of Zone N18 relative to the generally accepted position of the N18/N19 zonal boundary with respect to the calcareous nannofossil zonation. Direct correlations in Core CAP 38 BP between the planktonic foraminiferal and the calcareous nannofossil stratigraphic subdivisions, respectively by Blow (1969, p. 38, Fig. 38) and Gartner (1969, p. 587, Fig. 1), indicated that the N18/N19 zonal boundary lies within Zone NN13. Similarly, direct correlations at DSDP Site 25 (Blow, 1970; Hay, 1970) and at DSDP Site 62.1 (Brönnimann and Resig, 1971; Martini and Worsley, 1971; Brönnimann et al., 1971) (the NN12/NN13 zonal boundary in Brönnimann et al., 1971, being taken at the FAD of *Ceratolithus rugosus*, i.e., between sections 4 and 5 in Core 62.1-13 rather than between sections 2 and 5 of Core 62.1-12 as drawn in Martini and Worsley, 1971, Table 3) showed that the N18/N19 zonal boundary lies within the *Ceratolithus rugosus* Zone. However, there seems to be a general agreement to draw it within Zone NN12 (Berggren and van Couvering, 1974; Berggren et al., 1985; Barron et al., 1985; Haq et al., 1987). This is a minor point, however, if we consider that the type level of Zone N18 may lie in Zone NN15. This discrepancy is extremely difficult to account for. The only satisfactory, possibly far-fetched, explanation would be if the covered part of the Bowden Formation correlative in the Buff Bay section was of different age than the exposed beds and separated from those by an unconformity. The unconformity would occur very close to, but above, the level of the sample examined by Blow of locality ER 146/41 and would presently be in the covered part of the section. This suggestion is compatible with the interpretation of the Neogene stratigraphic record as discussed above.

Zone N19

The type locality (ER 146/44) of Zone N19 is from the Bowden Formation exposed in the Buff Bay section. One (ER 300) of its two "paratype" localities (the other is in Fiji) is from lithological "Unit i" (Blow, 1969, p. 297, Fig. 29) in the Road to Arcadia section (not from the Bowden section as indicated by Blow; see discussion above).

Sample ER 146/44 is from the uppermost levels of the Bowden Formation and yields a calcareous nannofossil assemblage characteristic of the lower part of Zone NN16 (= Subzone CN12a). Sample ER 300 corresponds to a level high in lithological "Unit i," assigned to Zone NN13.

Zone N19 should therefore span a stratigraphic interval corresponding at least to Zones NN13 to NN16 (lower part). This is in agreement with generally accepted correlations according to which the N18/N19 boundary lies within Zone NN12 and the upper part of Zone N19 correlates with the lower part of Zone NN16 (although in disagreement with Gartner (1969).

Zone N20

The Bowden Formation yields the "holotype" and two of the "paratype" localities of Zone N20. The "holotype" level (ER 156) is from the Bowden type section, slightly above the basal Bowden Shell Bed. The "paratype" locality ER 193 is from the Drivers River section and the "paratype" locality ER 538 from the Folly Point section.

Sample ER 156 yields an assemblage too poor to be stratigraphically significant. However, I have indicated above that despite the absence of discoasters and ceratoliths (see discussion relative to the type section of the Bowden Formation), its assemblage is quite comparable to the assemblages that occur in the Bowden Formation exposed at Folly Point and referrable to Zone NN16.

Sample ER 193 yields poorly preserved calcareous nannofossils, but its assemblage is similar to those that occur in samples collected during the field party of 1987 from the Drivers River section (see discussion above relative to the stratigraphy of this section), and like those, it is assigned to Zone NN16 (Subzone CN12a) based on the co-occurrence of *Discoaster surculus* and *D. tamalis* and the absence of *Reticulofenestra pseudoumbilicus*.

Sample ER 538 yields common but poorly preserved calcareous nannofossils. Based on the co-occurrence of *D. asymmetricus, D. surculus,* and *D. tamalis,* it is assigned to Zone NN16 and Subzone CN12a.

The "holotype" and "paratype" localities for Zone N20 are probably restricted to a thin stratigraphic interval within Zone NN16, although this cannot be established firmly for the "holotype" locality. It should be noted that at least one of the "paratype" localities (ER 193) overlaps biostratigraphically (or is in exact continuity with) the "holotype" locality of Zone N19. As the use of Zone N20 has been mostly abandoned, it would be useless to discuss this zone further with respect to calcareous nannofossil stratigraphy.

Zone N16

One of the two "paratype" localities of Zone N16 is within the Buff Bay Formation, in the Buff Bay section. The level of sample ER 146/37 is within the upper part of Zone NN9 and belongs to Subzone CN7b.

Zone N13

The "paratype" locality (ER 143/4) of Zone N13 is from the upper beds of the Montpelier Formation exposed in the San San Bay section. Although *Discoaster kugleri* was not encountered, assignment to Zone NN7 is warranted by the occurrence (common) of *D. challengeri.* Alternatively, this locality correlates with an undifferentiated NN6-NN7 zonal interval.

TAXONOMIC REMARKS

Many of the species encountered in the Neocene formations of eastern Jamaica are illustrated in Figures 14 to 19. Taxonomic remarks on a few species are made in the section below.

Helicosphaera burkei Black 1971

The holotype of *H. burkei* was described by Black (1971) from the Cobre Member of the upper White Limestone. Hill (1899, p. 78) assigned to the Cobre Formation "certain problematic beds of white limestone" exposed in the canyon cut by the Rio Cobre through a white limestone plateau developed between the towns of Bog Walk and Spanish Town, in southeastern Jamaica (west of Kingston). Considering its planktonic foraminiferal age (Zone N11, in Black, 1971), the Cobre member is a lateral equivalent of part of the Spring Garden Member.

The description that Black (1971, p. 618, 619) gave of this species is insufficient to characterize it. First, size is not a valid specific criteria for calcareous nannofossils. Second, the "rugged and untidy construction" and the "uneven outline of the central shield" (i.e., the basal plate) are effects of preservation rather than genuine characters. Wise (1973) remarked the presence of two holes in the basal plate offset with respect to the suture line, and this has become appropriately the distinguishing character of this taxon, which was more recently described as *Helicosphaera paleocarteri* (Theodoridis, 1984; see Aubry, 1990). While Wise noted the presence of two holes in the basal plate, Theodoridis indicated the presence of two shallow pits that do not penetrate the basal plate (implying that they are distal features). From the electromicrograph given by Black (1971, Plate 45.3, Fig. 23), it is clear that these are proximal features, but it is unclear whether they are holes or simply proximal pits.

Helicosphaera burkei is common at some levels in the Spring Garden and Buff Bay formations, but no attempts at delineating its stratigraphic distribution or variations in its abundance were made.

Discolithina millepunctata Gartner 1967b

This taxon was described from the upper Miocene beds of the San San Clay, and it is well established that it is a synonym of *Pontosphaera japonica* (Takayama 1967 (see Aubry, 1990). *Pontosphaera japonica* is common in the Neogene formations studied.

Dictyococcites antillarum Black 1967

The holotype of this form was described and illustrated from the *Globorotalia foshi robusta* Zone in the Upper White Limestone, without reference to a type-locality or a section. Although Black (1967) refers to the type level as lower Miocene (Burdigalian), this is a middle Miocene taxon, probably described from the Spring Garden Member or a lateral equivalent.

The description of this species is insufficient, and morphologic features that are effects of preservation, such as the "irregular outlines," "hummocky surfaces of the shields," and "crookedness of the rays," were regarded of specific value. In this work, this form has been placed in the *Reticulofenestra pseudoumbilicus* group.

Cyclococcolithus aequiscutum Gartner 1967b and *C. cricotus* Gartner 1967b

These two forms were described from the upper Pliocene (at 200-m depth; see Lamb and Beard, 1972) of Core 64-A-9-5E taken on the Sigsbee Knolls in the Gulf of Mexico, and the former was also reported from the San San Clay (Gartner, 1967b). Both taxa were observed in the Neogene of eastern Jamaica.

The extinction patterns of these placoliths are the same as those produced by placoliths in the genus *Umbilicosphaera,* to which both forms are reassigned. This leads to reinterpretation of the illustrations given by Gartner (1967b) as proximal (Plate 7, Fig. 5) and distal (Plate 7, Figs. 2 and 6) views of the placoliths. Cohen and Reinhardt (1968) transferred *Cyclocollithus cricotus* to the genus *Umbilicosphaera,* but as they gave to this genus a concept different than generally admitted (in which the distal shield is smaller than the proximal one), the validity of this transfer was questionable. In addition, Cohen and Reinhardt (1968) mistook *C. cricotus* for *Pseudoemiliania lacunosa,* as is apparent from the description (p. 296) and the illustrations (Plate 19, Figs. 1 and 5; Plate 21, Fig. 3) that they give of it.

Variations in abundance of both forms in the Neogene of Jamaica seem to occur in parallel, and it is likely that both coccoliths, which differ essentially in size, represent a case of dimorphism. Dimorphism may not be a rule in the genus *Umbilicosphaera,* but it was shown to occur in *U. hulburtiana* by Gaarder (1970), who pointed to the large variation in size of the coccoliths on the coccopheres. Further quantitative studies of these taxa (Aubry, in progress) will allow us to establish the validity of this interpretation.

Umbilicosphaera aequiscutum (Gartner, 1967b) n.c.

Basionym: *Cyclococcolithus aequiscutum* Gartner, 1967b, The University of Kansas Paleontological Contributions, paper 29, p. 4, Plate 7, Figs. 1–4.

Umbilicosphaera cricota (Gartner, 1967b) n.c.

Basionym: *Cyclococcolithus cricotus* Gartner, 1967b, The University of Kansas Paleontological Contributions, paper 29, p. 5, Plate 7, Figs. 5–7; non *Umbilicosphaera cricota* (Gartner) Cohen and Reinhardt, 1968, Neues Jahrbuch für Geologie und Paläontologie Abhandlungen, 131, p. 296, Plate 19, Figs. 1, 5, Plate 21, Fig. 3.

Umbilicosphaera cricota (Gartner, 1967b) may be a junior synonym of *U. aequiscutum* (Gartner, 1967b) and correspond to the microliths of this latter.

Figure 14. 1, 2: Sample SG13; Spring Garden Member (Zone NN7), Buff Bay section. 3–12: Sample ER146/36; upper part of the Buff Bay Formation (Zone NN9, Subzone CN7b), Buff Bay section. 1–5, 10, 11: ×2000. 6–9, 12: ×1250. 1, 2: *Discoaster kugleri.* 3, 9: *Discoaster bollii.* 3–9: compact morphotypes with a sturdy knob. 4–8: delicate morphotypes with a narrow stellate knob. 10: *Discoaster moorei.* 11: *Discoaster perclarus.* 12: *Discoaster challengeri* (slightly overgrown).

Discoaster bollii Martini and Bramlette, 1963

I have referred to the morphology of *D. bollii* as being more or less compact in relation to the stratigraphic levels where this species occurs in the Buff Bay Bay Formation. This is a very simplistic way to describe a morphologic change from predominantly forms similar to those illustrated in Martini and Bramlette (1963, Plate 105, Figs. 3, 4) to predominantly forms similar to that illustrated in Bramlette and Wilcoxon (1967, Plate 8, Fig. 11). Because of very high sedimentation rates, the Buff Bay Formation provides the opportunity to follow a progressive change in the outline and in the development of the knobs in *D. bollii.*

Discoaster tamalis Kamptner, 1967

A form similar to *Discoaster tamalis* with four rays intersecting at right angles but with delicate bifurcations at the tips of the rays was reported from the Lesser Antilles forearc region (ODP sites 671, 672 and 676; Clark, 1990). Such forms are also recorded in the Neogene from Jamaica, where they occur together with typical forms of *D. tamalis*. In this work they have been regarded as a variant of this species. Forms that are very similar to *D. asymmetricus* except for short, delicate bifurcations also occur in the Neogene of Jamaica and of other regions (e.g., Morocco, Rio Grande Rise, Aubry, unpublished data). Backman and Pestiaux (1987) suggested that *D. asymmetricus* and *D. tamalis* may be very closely related phylogenetically, based on patterns of covariation in abundance. The occurrence in both species of variants with bifurcation may reinforce this view.

FINAL REMARKS

This study is essentially a biostratigraphic study of the Neogene deposits of Jamaica. Although the Buff Bay section is studied in greater detail, considerable attention was given to all other sections. The difficulties encountered in delineating zonal boundaries, or more simply, in assigning a level to the proper biozone should not be underestimated. These difficulties are linked to several factors, among which are preservation, rarity of zonal markers such as ceratoliths, unusual potentially misleading stratigraphic ranges such as that of *Discoaster tamalis,* intense reworking that prevented the use of secondary markers such as *Reticulofenestra floridana,* and stratigraphically short sections, partly covered and/or interrupted by virtual unconformities. The stratigraphy of the lower Pliocene interval is particularly difficult because of the rarity of the ceratoliths and the potential reworking linked to intensified tectonic uplift during the latest Miocene and earliest Pliocene (Zans et al., 1962; Katz and Miller this volume). I have described the sections and their calcareous nannofossil contents in detail and with as much objectivity as possible. Integration of the calcareous nannofossil data with those obtained from the planktonic foraminifera for each section has helped me in controlling the consistency of the interpretations and in inferring previously unrecognized unconformities. A litho-biostratigraphic framework is attempted, which suggests a pattern of generalized depositional sequences alternating with widespread unconformities. This framework will be refined as new sections are examined and as some of the sections studied here are sampled in greater detail. Only then will it be possible to decipher satisfactorily the role of tectonic uplift and of glacio-eustasy in the shaping of the Neogene stratigraphic record of eastern Jamaica.

Figure 15. 1, 4, 6, 13, 14: Sample ER146/40; upper part of the Buff Bay formation (Zone NN10, Subzone CN8a), Buff Bay section. 2, 5: Sample ER146/36; upper part of the Buff Bay Formation (Zone NN9, Subzone CN7b), Buff Bay section. 3: Sample SG13; Spring Garden Member (Zone NN7), Buff Bay section. 7–12: Sample ER143/5; San San Clay (Zone NN11, Subzone CN9b), San San Bay section. Magnification ×2000 for all figures except 8 and 12–14, which are ×1250. 1: *Discoaster variabilis.* 2: *Discoaster subsurculus.* 3: *Discoaster exilis.* 4: *Discoaster calcaris.* 5: *Discoaster hamatus.* 6, 11: *Discoaster pentaradiatus.* 7, 10: *Discoaster quinqueramus.* 8, 12: *Discoaster surculus.* 9: *Discoaster brouweri.* 13, 14: *Discoaster sp.* cf. *Discoaster berggrenii.*

Taxonomic index

For complete citations of the species cited in this work, the reader is referred to the taxonomic indexes by Loeblich and Tappan (1966, 1968; 1969; 1970a, b; 1971, 1973), van Heck (1979a, b; 1980a, b; 1981a, b; 1982a, b), Steinmetz (1983a, b; 1984a, b; 1985a, b; 1986a, b; 1987a, b; 1988a, b; 1989a, b; 1990), and Siesser (1990) and to Aubry (1984, 1988, 1989, 1990).

ACKNOWLEDGMENTS

I gratefully acknowledge the organization of the field expedition of 1987 to Jamaica by W. A. Berggren, the kind cooperation of the members of the field party (W. A. Berggren, G. Jones, D. V. Kent, K. G. Miller, E. Robinson, C. Stuart, R. C. Tjalsma, M. Van Fossen), and the financial support of a consortium of oil companies (BP, Chevron, Marathon, Texaco, and Unocal), without which the fieldwork would not have been possible. Particular thanks are extended to D. V. Kent, K. G. Miller, and M. Van Fossen for discussion on the interpretation of the stratigraphy of the Buff Bay section; to W. A. Berggren for discussing the problems relative to the stratigraphy of the upper Miocene and lower Pliocene interval; and to D. Bukry, S. Gartner, K. Perch-Nielsen, E. Robinson, and R. C. Tjalsma for reviewing the manuscript. I am very thankful to B. Dale (University of Oslo), J. Hardenbol (Exxon), B. Kohl (Chevron), and E. Robinson for providing me with additional samples, particularly with those from the ER localities. The assistance of J. Cook for the artwork is gratefully acknowledged. This is Woods Hole Oceanographic Institution Contribution No. 7797.

Figure 16. All figures: Sample ER146/36; upper part of the Buff Bay Formation (Zone NN9, Subzone CN7b), Buff Bay section. 14: cross-polarized light. All other figures: Transmitted light. All figures: ×2000. 1–3: *Catinaster* sp. Convex side seen at three different levels from base (Fig. 1) to equatorial level (Fig. 3). 4–6: *Catinaster* sp. Concave side seen at three different levels from base (Fig. 4) to equatorial level (Fig. 6). 7–10: *Catinaster calyculus*. 7: convex side. 8–10: Concave side. 11–14: *Catinaster calyculus*. 11: Concave side. 12, 14: Side view. 13: convex side.

REFERENCES CITED

Aubry, M.-P., 1984, Handbook of Cenozoic calcareous nannoplankton, Book 1: New York, Micropaleontology Press, 266 p.

Aubry, M.-P., 1988, Handbook of Cenozoic calcareous nannoplankton, Book 2: New York, Micropaleontology Press, 279 p.

Aubry, M.-P., 1989, Handbook of Cenozoic calcareous nannoplankton, Book 3: New York, Micropaleontology Press, 279 p.

Aubry, M.-P., 1990, Handbook of Cenozoic calcareous nannoplankton, Book 4: New York, Micropaleontology Press, 381 p.

Backman, J., and Pestiaux, P., 1987, Pliocene discoaster abundance variations at DSDP Site 606, biochronology and paleoenvironmental implications, *in* Ruddiman, W. S., and Kidd, R. B., eds., Initial reports of the Deep Sea Drilling Project, Volume 7: Washington, D.C., U.S. Government Printing Office, p. 903–910.

Backman, J., and Shackleton, N. J., 1983, Quantitative biochronology of Pliocene and early Pleistocene calcareous nannofossils from the Atlantic, Indian and Pacific Oceans: Marine Micropaleontology, v. 8, p. 141–170.

Backman, J., Schneider, D. A., Rio, D., and Okada, H., 1990, Neogene low-latitude magnetostratigraphy from site 710 and revised age estimates of Miocene nannofossil datum events, *in* Duncan, R. A., and Backman, J., eds., Proceedings of the Ocean Drilling Program, Science Results, v. 115: College Station, Texas, Ocean Drilling Program, p. 271–276.

Banner, F. T., and Blow, W. H., 1965, Progress in the planktonic foraminiferal biostratigraphy of the Neogene: Nature, v. 208, p. 1164–1166.

Barren, J. A., Keller, G., and Dunn, D. A., 1985, A multiple microfossil biochro-

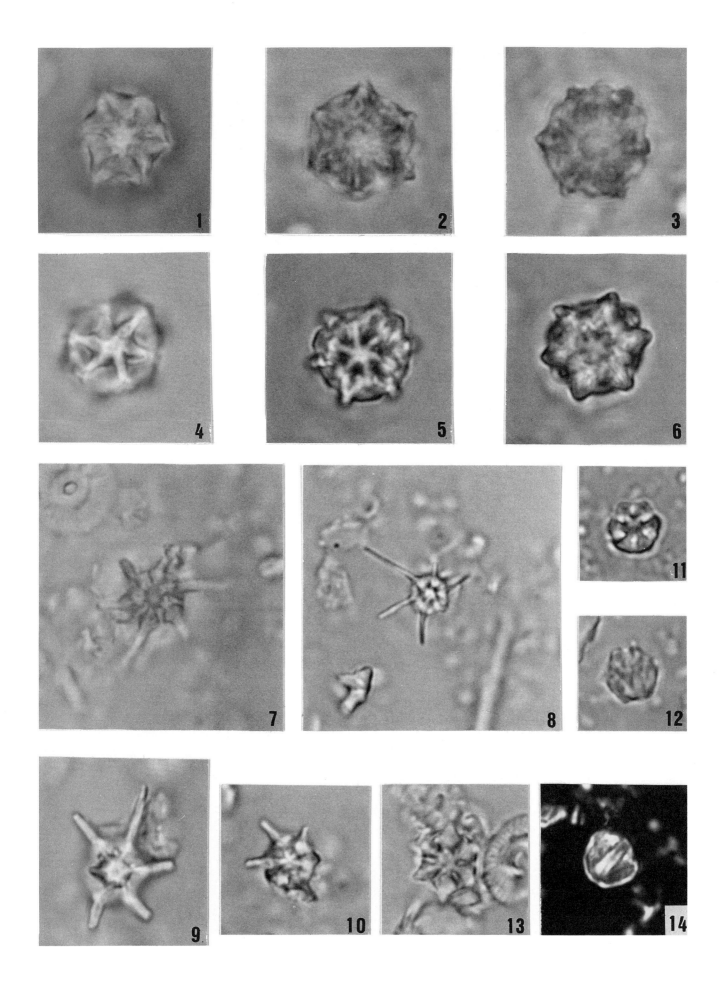

nology for the Miocene, *in* Kennett, J. P., ed., The Miocene ocean: Paleoceanography and biogeography: Geological Society of America Memoir 163, p. 21–36.

Berggren, W. A., 1977, Late Neogene planktonic foraminiferal biostratigraphy of the Rio Grande Rise (South Atlantic): Marine Micropaleontology, v. 2, p. 265–313.

Berggren, W. A., and van Couvering, J. A., 1974, The late Neogene: Biostratigraphy, geochronology and paleoclimatology of the last 15 million years in marine and continental sequences, *in* Developments in paleontology and stratigraphy, vol. 2: Amsterdam, Elsevier, 216 p.

Berggren, W. A., Aubry, M.-P., and Hamilton, N., 1983, Neogene magnetostratigraphy of Deep Sea Drilling Project 516 (Rio Grande Rise, South Atlantic), *in* Barker, P. F., and Carlos, R. L., eds., Initial reports of the Deep Sea Drilling Project, Volume 72: Washington, D.C., U.S. Government Printing Office, p. 675–713.

Berggren, W. A., Kent, D. V., and van Couvering, J. A., 1985, Neogene geochronology and chronostratigraphy, *in* Snelling, N. J., ed., The chronology of the geological record: Geological Society of London Memoir 10, p. 211–260.

Black, M., 1967, New names for some coccolith taxa: Proceedings of the Geological Society of London, no. 1640, p. 139–145.

Black, M., 1971, The systematics of coccoliths in relation to the paleontological record, *in* Funnel, B. M., and Riedel, W. R., eds., The micropaleontology of oceans: Cambridge, Cambridge University Press, p. 611–624.

Black, M., 1972, Crystal development Discoasteraceae and Braarudosphaeraceae (planktonic algae): Paleontology, v. 15 (part 3), p. 476–489.

Blow, W. H., 1969, Late middle Eocene to Recent planktonic foraminiferal biostratigraphy, *in* Brönnimann, P. R., and Renz, H. H., eds., Proceedings, First International Conference on Planktonic Microfossils, Geneva, 1967, Volume 1: Leiden, E. J. Brill, p. 199–421.

Blow, W. H., 1970, Deep Sea Drilling Project, Leg 4 foraminifera from selected samples, in Bader, R. G., eds., Initial reports of the Deep Sea Drilling Project, Volume 4: Washington, D.C., U.S. Government Printing Office, p. 383–400.

Blow, W. H., 1979, The Cainozoic Globigerinda, Vols. 1 and 2: Leiden, E. J. Brill, 1413 p.

Bolli, H. M., and Saunders, J. B., 1985, Oligocene to Holocene low latitude planktonic foraminifers, *in* Bolli, M. M., Saunders, J. B., and Perch-Nielsen, K., eds., Plankton stratigraphy: Cambridge, Cambridge University Press, p. 155–162.

Bramlette, M. N., and Riedel, W. R., 1954, Stratigraphic value of discoasters and some other microfossils related to Recent coccolithospheres: Journal of Palaeontology, v. 28, p. 385–403.

Bramlette, M. N., and Wilcoxon, J. A., 1967, Middle Tertiary calcareous nannoplankton of the Cipero section, Trinidad, W. I.: Tulane Studies Geology 5, p. 93–131.

Brönnimann, P., and Resig, J., 1971, A Neogene globigerinacean biochronologic time-scale of the southwestern Pacific, *in* Winterer, E. L., ed., Initial reports of the Deep Sea Drilling Project, Volume 7: Washington, D.C., U.S. Government Printing Office, p. 1235–1469.

Brönnimann, P., Martini, E., Resig, J., Riedel, W. R., Sanfilippo, A., and Worsley, T., 1971, Biostratigraphic synthesis: Late Oligocene and Neogene of the western tropical Pacific, *in* Winterer, E. L., ed., Initial reports of the Deep Sea Drilling Project, Volume 7: Washington, D.C., U.S. Government Printing Office, p. 1723–1745.

Bukry, D., 1973, Low-latitude coccolith biostratigraphic zonation, *in* Edgar, N. T., and Saunders, J. B., eds., Initial reports of the Deep Sea Drilling Project, Volume 32: Washington, D.C., U.S. Government Printing Office, p. 677–701.

Bukry, D., 1975, Coccolith and Silicoflagellate stratigraphy, northwestern Pacific Ocean, Deep Sea Drilling Project, Leg 32, *in* Larson, R. L., and Moberly, R., eds., Initial reports of the Deep Sea Drilling Project, Volume 15: Washington, D.C., U.S. Government Printing Office, p. 685–703.

Channell, J.E.T., Rio, D., Sprovieri, R., and Glacon, G., 1990, Biomagnetostratigraphic correlations from leg 107 in the Tyrrhenian Sea, *in* Kasten, K. A.,

Figure 17. All figures: ×2000. 1, 4: *Pontosphaera discopora.* Sample ER146/36; upper part of the Buff Bay Formation (Zone NN9, Subzone CN7b), Buff Bay section. 2, 5: *Coccolithus miopelagicus.* Sample SG32; Spring garden Member (Zone NN8), Buff Bay section. 3, 6: *Scyphosphaera* sp. Sample ER146/27; lower part of the Buff Bay Formation (Zone NN9, Subzone CN7a), Buff Bay section. 7, 8: *Calcidiscus macintyrei.* Sample SG13; Spring garden Member (Zone NN7); Buff Bay section. 9: *Scyphosphaera intermedia.* Sample ER146/36; upper part of the Buff Bay Formation (Zone NN9, Subzone CN7b), Buff Bay section. 10: *Amaurolithus primus.* Sample ER143/5; San San Clay (Zone NN11, Subzone CN9b), San San Bay section. 11, 12: ?*Rhabdosphaera* sp., Sample ER146/40; upper part of the Buff Bay Formation (Zone NN10, Subzone CN8a), Buff Bay section.

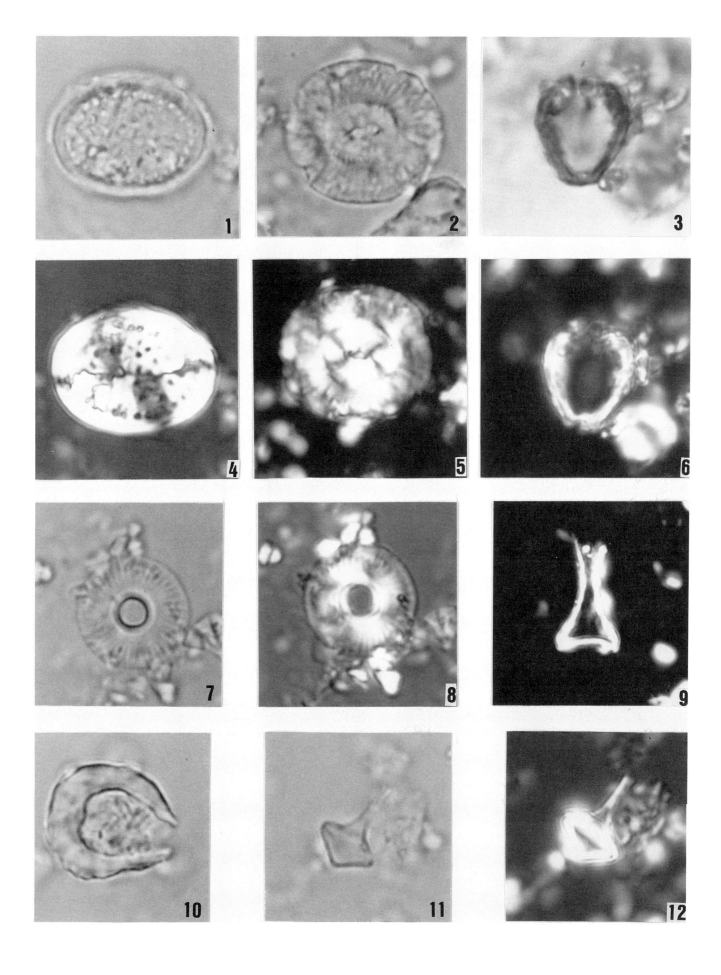

and Mascle, J., eds., Proceedings of the Ocean Drilling Program, Science Results, v. 107: College Station, Texas, Ocean Drilling Program, p. 699–682.

Chepstow-Lusty, A., Backman, J., and Shackleton, N. J., 1989, Comparison of upper Pliocene Discoaster abundance variations from North Atlantic sites 552, 607, 658 and 662: Further evidence for marine plankton responding to orbital forcing, *in* Ruddiman, W., and Sarthein, M., eds.: Proceedings of the Ocean Drilling Program, Science Results, v. 108: College Station, Texas, Ocean Drilling Program, p. 121–141.

Clark, M. W., 1990, Cenozoic calcareous nannofossils from the Lesser Antilles forearc region and biostratigraphic summary of leg 110, *in* Moore, J. C., and Mascle, A., eds., Proceedings of the Ocean Drilling Program, Science Results, v. 110: College Station, Texas, Ocean Drilling Program, p. 129–140.

Cohen, C.L.D., and Reinhardt, P., 1968, Coccolithophorids from the Pleistocene Caribbean deep sea core CP-28: Neues Jahrbuch für Geologie und Paläontologie, v. 131, p. 289–304.

Emery, K. O., and Uchupi, A., 1984, The Geology of the Atlantic Ocean: New York, Springer-Verlag, 1050 p.

Gaarder, K. R., 1970, Three new taxa of Coccolihineae: Nytt Magasin for Botanikk, Oslo, v. 17, p. 113–126.

Gartner, S., Jr., 1967a, Nannofossil species related to *Cyclococcolithus leptoporus* (Murray and Blackman): The University of Kansas Paleontological Contributions: Paper 28, p. 1–4.

Gartner, S., Jr., 1967b, Calcareous nannofossils from Neogene of Trinidad, Jamaica and Gulf of Mexico: The University of Kansas Paleontological Contributions, Paper 29, p. 1–7.

Gartner, S., Jr., 1969, Correlation of Neogene planktonic foraminifer and calcareous nannofossil zones: Transactions of the Gulf Coast Association of Geological Societies, v. 19, p. 585–599.

Gartner, S., Jr., 1977, Calcareous nannofossil biostratigraphy and revised zonation of the Pleistocene: Marine Micropaleontology, v. 2, p. 1–25.

Haq, B. U., Hardenbol, J., and Vail, P., 1987, The chronology of fluctuating sea level since the Triassic: Science, v. 235, p. 1156–1167.

Haq, B. U., Hardenbol, J., and Vail, P., 1988, Mesozoic and Cenozoic chronostratigraphic and cycles of sea-level change, *in* Sea-level changes: An integrated approach: Society of Economic Paleontologists and Mineralogists, Special Publication 42, p. 71–108.

Hay, W. W., 1970, Calcareous nannofossils from cores recovered on Leg 4, *in* Bader, R. G., ed., Initial reports of the Deep Sea Drilling Project, Volume 4: Washington, D.C., U.S. Government Printing Office, p. 455–501.

Hay, W. W., Mohler, H. P., Roth, P. H., Schmidt, R. R., and Boudreaux, J. E., 1967, Calcareous nannoplankton zonation of the Cenozoic of the Gulf Coast and Caribbean-Antillian area and transoceanic correlation: Transactions of the Gulf Coast Association of Geological Societies, v. 17, p. 428–459.

Hill, R. T., 1899, The geology and physical geography of Jamaica: A study of a type of Antillian development: Bulletin of the Museum of Comparative Zoologie (Harvard), v. 34, p. 1–226.

Hodell, D., and Kennett, J. P., 1986, Late Miocene and early Pliocene stratigraphy and paleoceanography of the south Atlantic and southwest Pacific Oceans: A synthesis: Paleoceanography, v. 1, p. 285–311.

Hose, H. R., and Versey, H. R., 1956, Paleontological and lithological divisions of the lower Tertiary limestones of Jamaica: Colonial Geology and Mineral Resources, v. 6, p. 19–39.

Jiang, M.-J., and Robinson, E., 1987, Calcareous nannofossils and larger foraminifera in Jamaican rocks of Cretaceous to early Eocene age, *in* Ahmad, R., ed., Proceedings of a Workshop on the Status of Jamaican Geology (1984): Geological Society of Jamaica, Special Issue, p. 24–51.

Keller, G., and Barron, J. A., 1987, Paleodepth distribution of Neogene deep sea hiatuses: Paleoceanography, v. 2, p. 697–713.

Lamb, J. L., and Beard, J. H., 1972, Late Neogene planktonic foraminifers in the Caribbean, Gulf of Mexico, and Italian stratotypes: University of Kansas Palaeontological Contributions, Article 57, 67 p.

Loeblich, A. R., and Tappan, H., 1966, Annotated index and bibliography of the calcareous nannoplankton: Phycologia, v. 5, p. 81–216.

Figure 18. 1, 9: Sample ER146/40; upper part of the Buff Bay Formation (Zone NN10, Subzone CN8a), Buff Bay section. 2: Sample SG13; Spring Garden Member (Zone NN7), Buff Bay section. 3: Sample 87-J-I; San San Clay (Subzone NN13); San San Bay section. 4–8, 10–14: Sample ER146/36; upper part of the Buff Bay Formation (Zone NN9, Subzone CN7b), Buff Bay section. 1b, 2b, 3b, 5, 7b, 9b, 10b, 12, 14: Cross-polarized light. All other figures: transmitted light. All figures: ×2000. 1–6: *Umbilicosphaera cricota*. Note the broad variations in the diameter of the central opening and in the width of the margin. 7: *Umbilicosphaera cricota* (left) and *Calcidiscus macintyrei* (right). 8: *Hayaster perplexus*. 9–11: *Ellipsodiscoaster lidzii*?. 12: *Reticulofenestra pseudoumbilicus*. 13, 14: *Umbilicosphaera aequiscutum*.

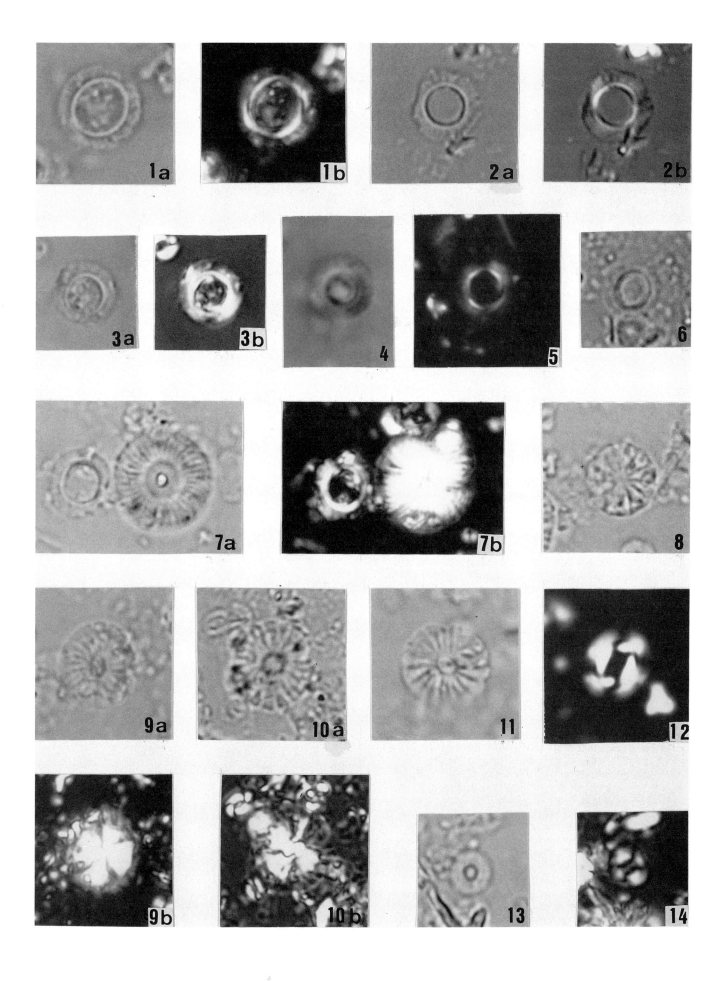

Loeblich, A. R., and Tappan, H., 1968, Annotated index and bibliography of the calcareous nannoplankton II: Journal of Paleontology, v. 42, p. 584–598.

Loeblich, A. R., and Tappan, H., 1969, Annotated index and bibliography of the calcareous nannoplankton III: Journal of Paleontology, v. 43, p. 568–588.

Loeblich, A. R., and Tappan, H., 1970a, Annotated index and bibliography of the calcareous nannoplankton IV: Journal of Paleontology, v. 44, p. 558–574.

Loeblich, A. R., and Tappan, H., 1970b, Annotated index and bibliography of the calcareous nannoplankton V: Physcologia, v. 9, p. 157–174.

Loeblich, A. R., and Tappan, H., 1971, Annotated index and bibliography of the calcareous nannoplankton VI: Phycologia, v. 10, p. 315–339.

Loeblich, A. R., and Tappan, H., 1973, Annotated index and bibliography of the calcareous nannoplankton VII: Journal of Paleontology, v. 47, p. 715–759.

Martini, E., 1971, Standard Tertiary and Quaternary calcareous nannoplankton zonation, in Farinacci, A., ed., Proceedings, Second Conference on Planktonic Microfossils, Rome, 1970, Volume 2: Rome, Tecnoscienza editzione, p. 739–785.

Martini, E., 1980, Oligocene to Recent calcareous nannoplankton from the Philippine Sea, Deep Sea Drilling Project Sea, Deep Sea Drilling Project, Leg 59, in Kroenke, L., and Scott, R., eds., Initial reports of the Deep Sea Drilling Project, Volume 59: Washington, D.C., U.S. Government Printing Office, p. 547–565.

Martini, E., and Bramlette, M. N., 1963, Calcareous nannoplankton from the experimental Mohole drilling: Journal of Paleontology, v. 37, p. 845–856.

Martini, E., and Worsley, T., 1971, Tertiary calcareous nannoplankton from the western equatorial Pacific, in Winterer, E. L., ed., Initial reports of the Deep Sea Drilling Project, Volume 7: Washington, D.C., U.S. Government Printing Office, p. 1471–1528.

Miller, K. G., Kahn, M. J., Aubry, M.-P., Berggren, W. A., Kent, D. V., and Melillo, A., 1985, Oligocene to Miocene biomagneto, and isotope stratigraphy of the western North Atlantic: Geology, v. 13, p. 257–261.

Okada, H., and Bukry, D., 1980, Supplementary modification and introduction of code numbers to the "low-latitude coccolith biostratigraphic zonation" (Bukry, 1973, 1975): Marine Micropaleontology, v. 5, p. 321–325.

Perch-Nielsen, K., 1985, Cenozoic calcareous nannofossils, in Bolli, H. M., Saunders, J. B., and Perch-Neilsen, K., eds., Plankton stratigraphy: Cambridge, Cambridge University Press, p. 427–554.

Raffi, I., and Rio, D., 1979, Calcareous nannofossil biostratigraphy of DSDP Site 132-Leg 13 (Tyrrhenian Sea–western Mediterranean): Rivista Italiana di Paleontologia e Stratigrafia, v. 85, p. 127–172.

Rio, D., Raffi, I., and Villa, G., 1990, Pliocene-Pleistocene calcareous nannofossil distribution patterns in the western Mediterranean, in Kasten, K. A., and Mascle, J., eds., Proceedings of the Ocean Drilling Program, Science Results, v. 107: College Station, Texas, Ocean Drilling Program, p. 513–533.

Robinson, E., 1969, Geological field guide to Neogene sections in Jamaica, West Indies: Journal of the Geological Society of Jamaica, v. 10, p. 1–24.

Robinson, E., and Lamb, J. L., 1970, Preliminary data from the Plio-Pleistocene of Jamaica: Nature, v. 227, p. 1236–1237.

Roth, P. H., and Thierstein, H., 1972, Calcareous nannoplankton; Leg 14 of the Deep Sea Drilling Project, in Hays, D. E., and Pimm, A. C., eds., Initial reports of the Deep Sea Drilling Project, Volume 14: Washington, D.C., U.S. Government Printing Office, p. 421–485.

Siesser, W. G., 1990, Bibliography and taxa of calcareous nannoplankton: International Nannoplankton Association Newsletter, v. 12, p. 23–33.

Steineck, P. L., 1974, Foraminiferal paleoecology of the Montpelier and lower coastal groups (Eocene-Miocene), Jamaica, West Indies: Paleogeography, Paleoclimatology, and Paleoecology, v. 16, p. 217–242.

Steineck, P. L., 1981, Upper Eocene to middle Miocene ostracode faunas and paleo-oceanography of the north coastal belt, Jamaica, West Indies: Marine Micropaleontology, v. 6, p. 339–366.

Steinmetz, J. C., 1983a, Bibliography and taxa of calcareous nannoplankton: International Nannoplankton Association Newsletter, v. 5, p. 4–13.

Steinmetz, J. C., 1983b, Bibliography and taxa of calcareous nannoplankton: International Nannoplankton, Association Newsletter, v. 5, p. 29–47.

Steinmetz, J. C., 1984a, Bibliography and taxa of calcareous nannoplankton:

Figure 19. 1–3: Sample WAH-81-11; basal part of the Spring Garden Member, Buff Bay section. 6, 12, 22, 25: Sample 87-J-I; San San Clay (Zone NN13); San San Bay section. 7, 8: Sample ER 143/5; San San Clay (Zone NN11, Subzone CN9b); San San Bay section. 11: Sample SG13; Spring Garden Member (Zone NN7), Buff Bay section. 14–16, 24, 27: Sample ER 146/40; upper part of the Buff Bay Formation (Zone NN10, Subzone CN8a), Buff Bay section. All other figures: Sample ER 146/36; upper part of the Buff Bay Formation (Zone NN9, Subzone CN7b), Buff Bay section. 1, 4, 6, 7, 9, 11, 13, 14, 16, 22, 28: Transmitted light. All other figures: Cross-polarized light. All figures: ×2000, except 7, 8: ×2500. 1–3: *Sphenolithus heteromorphus*. 4, 5: *Rhabdolithus* sp. 6, 12, 14, 15: *Rhabdosphaera procera*. 7, 8: *Sphenolithus abies*. 9, 10: *Scapholithus fossilis*. 11: *Holodiscolithus solidus*. 13, 17: *Calcidiscus macintyrei*. 16: *Mynilitha convallis*. 18, 22–25: *Helicosphaera burkei*. 19–21, 26, 27: *Helicosphaera vedderi*. 28–30: *Syracosphaera* sp.

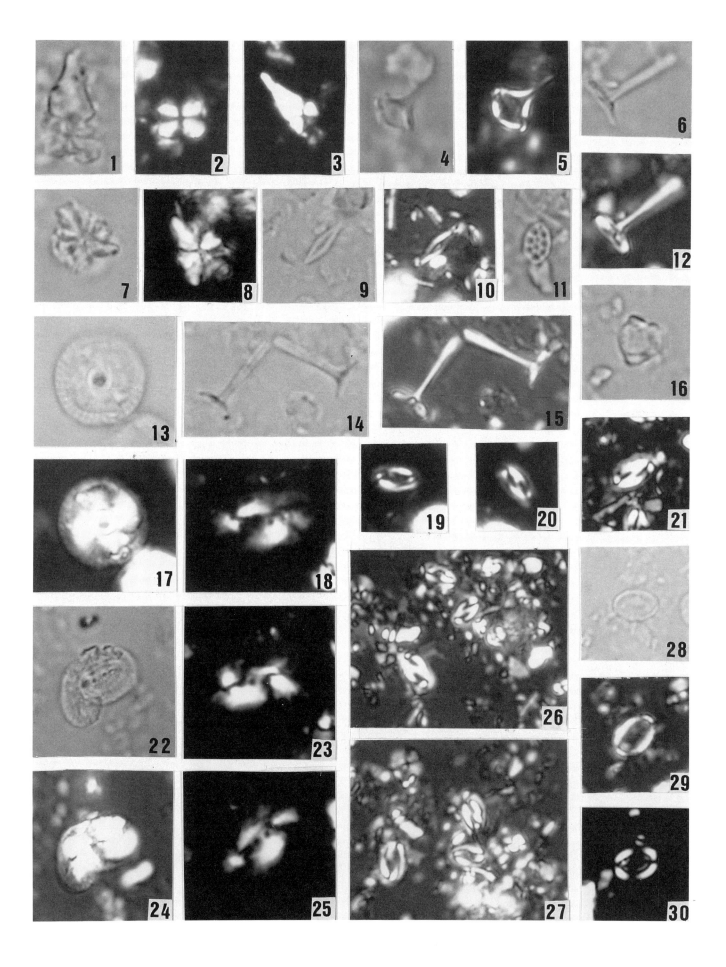

International Nannoplankton Association Newsletter, v. 6, p. 6–37.
Steinmetz, J. C., 1984b, Bibliography and taxa of calcareous nannoplankton: International Nannoplankton Association Newsletter, v. 6, p. 55–81.
Steinmetz, J. C., 1985a, Bibliography and taxa of calcareous nannoplankton: International Nannoplankton Association Newsletter, v. 7, p. 5–28.
Steinmetz, J. C., 1985b, Bibliography and taxa of calcareous nannoplankton: International Nannoplankton Association Newsletter, v. 7, p. 122–145.
Steinmetz, J. C., 1986a, Bibliography and taxa of calcareous nannoplankton: International Nannoplankton Association Newsletter, v. 8, p. 12–32.
Steinmetz, J. C., 1986b, Bibliography and taxa of calcareous nannoplankton: International Nannoplankton Association Newsletter, v. 8, p. 66–87.
Steinmetz, J. C., 1987a, Bibliography and taxa of calcareous nannoplankton: International Nannoplankton Association Newsletter, v. 9, p. 8–29.
Steinmetz, J. C., 1987b, Bibliography and taxa of calcareous nannoplankton: International Nannoplankton Association Newsletter, v. 9, p. 81–107.
Steinmetz, J. C., 1988a, Bibliography and taxa of calcareous nannoplankton: International Nannoplankton Association Newsletter, v. 10, p. 7–29.
Steinmetz, J. C., 1988b, Bibliography and taxa of calcareous nannoplankton: International Nannoplankton Association Newsletter, v. 10, p. 60–88.
Steinmetz, J. C., 1989a, Bibliography and taxa of calcareous nannoplankton: International Nannoplankton Association Newsletter, v. 11, p. 6–23.
Steinmetz, J. C., 1989b, Bibliography and taxa of calcareous nannoplankton: International Nannoplankton Association Newsletter, v. 11, p. 6–23.
Steinmetz, J. C., 1990, Bibliography and taxa of calcareous nannoplankton: International Nannoplankton Association Newsletter, v. 12, p. 21–57.
Stuart, C., Tjalsma, R. C., and Jones, G., 1987, Jamaica field excursion: Brea, California, Unocal Internal Report, 15 p. (unpublished.)
Takayama, T., 1967, First report on nannoplankton of the upper Tertiary and Quaternary of the southern Kwanto Region, Japan: Austria, Geologische Bundesanstalt, Jahrbuch 110, p. 169–198.
Theodoridis, S., 1984, Calcareous nannofossil biozonation of the Miocene and revision of the helicoliths and discoasters: Utrecht Micropaleontological Bulletin, v. 32, 271 p.
van Heck, S., 1979a, Bibliography and taxa of calcareous nannoplankton: International Nannoplankton Association Newsletter, v. 1, p. AB I–B27.
van Heck, S., 1979b, Bibliography and taxa of calcareous nannoplankton: International Nannoplankton Association Newsletter, v. 1, p. AB VI–B42.
van Heck, S., 1980a, Bibliography and taxa of calcareous nannoplankton: International Nannoplankton Association Newsletter, v. 2, p. 5–34.
van Heck, S., 1980b, Bibliography and taxa of calcareous nannoplankton: International Nannoplankton Association Newsletter, v. 2, p. 43–81.
van Heck, S., 1981, Bibliography and taxa of calcareous nannoplankton: International Nannoplankton: International Nannoplankton Association Newsletter, v. 3, p. 34–44.
van Heck, S., 1981b, Bibliography and taxa of calcareous nannoplankton: International Nannoplankton Association Newsletter, v. 3, p. 51–86.
van Heck, S., 1982a, Bibliography and taxa of calcareous nannoplankton: International Nannoplankton Association Newsletter, v. 4, p. 7–50.
van Heck, S., 1982b, Bibliography and taxa of calcareous nannoplankton: International Nannoplankton Association Newsletter, v. 4, p. 65–96.
van Wagoner, J. C., and six others, 1988, An overview of the fundamentals of sequence stratigraphy and key definitions, *in* Sea-level changes: An integrated approach: Society of Economic Paleontologists and Mineralogists Special Publication 42, p. 39–45.
Wise, S. W., 1973, Calcareous nannofossils from cores recovered during Leg 18, Deep Sea Drilling Project—Biostratigraphy and observations of diagenesis, *in* Kulm, L. D., and von Huene, R., eds., 1973, Initial reports of the Deep Sea Drilling Project, Volume 18: Washington, D.C., U.S. Government Printing Office, p. 569–615.
Woodring, W. P., 1925, Miocene mollusks from Bowden, Jamaica: Carnegie Institute of Washington Publication 336.
Wright, R. M., 1968, Tertiary biostratigraphy of central Jamaica: Tectonic and environmental implications: fifth Caribbean Geological Conference, St. Thomas, Virgin Islands, p. 1–21.
Wright, R. M., ed., 1974, Field guide to selected Jamaican geological localities: Ministry of Mining and Natural Resources, Mines Geological Division Special Publication 1, 57 p.
Zans, V. A., Chubb, L. J., Versey, H. R., Williams, J. B., Robinson, E., and Cooke, D. L., 1962, Synopsis of the geology of Jamaica: Jamaica, Bulletin of the Geological Survey Department, v. 4, 72 p.

MANUSCRIPT ACCEPTED BY THE SOCIETY NOVEMBER 11, 1992

Neogene planktonic foraminiferal biostratigraphy of eastern Jamaica

W. A. Berggren
Department of Geology and Geophysics, Woods Hole Oceanographic Institution, Woods Hole, Massachusetts 02543, and Department of Geology, Brown University, Providence, Rhode Island 02912

ABSTRACT

The Neogene deep water carbonate section exposed at Buff Bay, north coast of Jamaica, is about 300 m thick and spans the interval from early middle Miocene to mid-Pliocene (ca. 16 to 3 Ma). It has come to serve as a standard reference section for upper Neogene biostratigraphy because several of the N-zones of Blow (1969) are typified here. A diverse tropical planktonic foraminiferal fauna characterizes this section and enables a detailed biostratigraphic subdivision. One obvious hiatus in the upper part of the Bowden Formation spans about 4 m.y. of late Miocene–early Pliocene time; a shorter more subtle, ca. 1-m.y. intra-late Miocene hiatus is thought to exist separating beds of Zone N17 from Zone N16/15.

Faunal analysis reveals general agreement with mid-late Neogene biostratigraphy of Bolli, Blow, and Robinson. A distinct stratigraphic separation between the last occurrence of *Paragloborotalia mayeri* and the first occurrence of *Neogloboquadrina acostaensis* suggests that "restoration" of Zone N15 to the planktonic foraminiferal hagiography may be required and that its apparent absence at North Atlantic Deep Sea Drilling Project (DSDP) Sites 558 and 563 may be due to a widespread early late Miocene hiatus. A new type level in the Buff Bay Formation is chosen for the *Globorotalia menardii* Zone (Bolli, 1966) = *Globorotalia (Turborotalia) continuosa* Consecutive Range-Zone (Blow, 1969) because the type sample in Trinidad for these zones (which is the same for both zones) has been found to be correlative with a level within the *Glborotalia (Turborotalia) acostaensis–G. (T.) merotumida* (N16) Zone of Blow (1969).

Analysis of type level samples used to denote some classic late Neogene zones has shown that the type level of Blow's (1969) Zone N17 is in calcareous nannoplankton Zone NN10 (rather than NN11 as previously thought), the type level of Zone N18 is biostratigraphically equivalent to Zone PL 2 (~3.7 Ma, lower Pliocene; i.e., within Zone N19), and the type level of Zone N19 is equivalent to Zone PL3–PL4 (ca. 3+ Ma). Use of Blow's Pliocene N18 to N21 zones is discouraged.

The taxonomy and stratigraphy of over 40 taxa are discussed and most are illustrated by scanning electron microscope (SEM) micrographs.

REGIONAL SETTING

Studies on the geologic history of Jamaica date back only to the turn of this century. The earliest description of the stratigraphic section exposed near Buff Bay was made by Hill (1899) who made the first systematic survey of the island geology. Hill (1899) differentiated the relatively indurated marls of the Buff Bay Bed (above) from the underlying white chalky marl that grades downward into the Montpelier Beds and correlates the (younger) Bowden Beds (Pliocene) with the Buff Bay Beds.

A formal lithostratigraphic subdivision of Jamaican Cenozoic rocks was made by Robinson (1965, 1967, 1969b) in a series of studies. In doing so he modified some of Hill's (1899) original designations, including for example, the chalky white marl below

---

Berggren, W. A., 1993, Neogene planktonic foraminiferal biostratigraphy of eastern Jamaica, *in* Wright, R. M., and Robinson, E., eds., Biostratigraphy of Jamaica: Boulder, Colorado, Geological Society of America Memoir 182.

the Buff Bay Beds as an upper member of the Montpelier Formation. Steineck (1974) subsequently elevated the Spring Garden unit to formational rank, while noting that it is probably lithologically gradational with both the underlying Sign Beds and overlying Buff Bay Formations. The Buff Bay Formation is unconformably overlain, in turn, by the Bowden Formation. The latter unit consists of calcareous clays interbedded with bioclastic limestones containing a very diverse molluscan fauna at its type section on the southern coast of Jamaica (Robinson, 1969b). At Buff Bay deep fore-reef corals dominate (a deep reef talus deposit containing blocks of possibly? older resedimented material). The benthic foraminiferal fauna attest to a marked shallowing relative to the subjacent Buff Bay Formation (see Katz and Miller, this volume) and represent the antepenultimate stage of mid-late Neogene uplift of Jamaica. The age relationships of these Neogene units were determined by Robinson (1969a) using the recently developed planktonic foraminiferal zonation of Bolli (1957, 1966), itself based primarily on studies in the Caribbean region and studies in progress by Banner and Blow (1965b).

The interested reader is referred to Steineck (1974, 1981) and Aubry (this volume) for a more comprehensive review of Neogene lithostratigraphy of the North Coastal Belt of Jamaica (Fig. 1).

CARIBBEAN PLANKTONIC FORAMINIFERAL STUDIES

Planktonic foraminifera were described from Jamaican outcrops, usually as incidental items in a study devoted to benthic foraminifera—as early as 1919 with the description of *Sphaeroidina immatura* (Cushman) [= *Sphaeroidinellopsis immatura* (Cushman)]. Nearly a dozen late Neogene planktonic taxa were originally described from Jamaica (Table 1).

The systematic studies of Neogene planktonic foraminifera of the Caribbean by Bolli (1957, 1966, 1970), Bolli and Bermúdez (1965), and Blow (1959, 1969, 1979) form the basis upon which our understanding of the tropical biostratigraphic history of this group is based. A (now somewhat outdated) review and critique of these studies is provided by Blow (1969, p. 82–90) and some additional aspects of applying these zonal biostratigraphics are discussed below in this paper.

Jamaican Neogene stratigraphic relationships were placed in perspective by Robinson (1969b) who applied Bolli's (1957) zonal scheme. He subsequently distributed a suite of samples from Jamaican outcrops to specialists around the world and in particular samples from the Buff Bay section (ER 146-21 to ER146-44); these samples have since played an important role in the formulation of Neogene planktonic foraminiferal biostratigraphies. For example, the roadcut exposed near Buff Bay that forms the basis of this study was used by Blow (1969, 1979) in the development of his mid-late Neogene zonal scheme (see discussion below) as well as for the description of several new taxa (Table 1). These samples were also examined by Bolli (1970) and Liska (1985; Table 2).

I would draw attention to the fact that the wide sample spacing, common downslope transport and reworking, and absence of independent stratigraphic control (calcareous nannoplankton, magnetostratigraphy, e.g.) renders difficult the evaluation of previous work. These problems are addressed here and by other members of our field party (Katz and Miller, this volume; Aubry, this volume).

This study focuses upon the biostratigraphic development of planktonic foraminifera in the Spring Garden Member (Montpelier Formation), Buff Bay, and Bowden Formations exposed along the north coast of Jamaica, east of the village of Buff Bay. Comparison with material of other outcrops/sections is made where relevant.

METHODS

A joint field party from the Woods Hole Oceanographic Institution (W. A. Berggren, M. -P. Aubry), Lamont-Doherty Geological Observatory (K. G. Miller, D. V. Kent, M. van Fossen), and UNOCAL (G. Jones, R. C. Tjalsma, C. Stuart) sampled outcrop sections near Buff Bay, Jamaica, in April 1987, with the guidance of E. R. Robinson (Florida International University). Robinson estimated the location of his original (1969b) samples relative to the present roadcut, which was altered in 1974. There are six sections exposed along the roadcut designated here as: Lower (Pots and Pans), Middle, Main, Slide, Dead Goat Gully, and (above the main unconformity), West (Fig. 1). The Spring Garden Member is exposed in the lower (Pots and Pans) section; the Middle Section contains the upper part of the Spring Garden unit (of Robinson, 1969b). The Buff Bay Formation (Robinson, 1969b) is exposed in the Main, Slide and Dead Goat Gully Roadcuts. The Bowden Formation (as identified by Robinson, 1969b) is exposed in a narrow outcrop band several meters thick in the West Roadcut.

Paleontological and oriented paleomagnetic samples were collected from all sections, although Dead Goat Gully and West Roadcuts were unsuitable for paleomagnetic analysis. Samples from the Pots and Pans and Middle Roadcuts were numbered (downsection) SG 1 through SG 32. Samples from the Main and Slide Roadcuts were numbered (upsection) 165 to 230 (UNOCAL) = (BB [= Buff Bay] 1 to 66; WAB), from Dead Goat Gully (241 to 243) and West Roadcut (244 to 252). The UNOCAL numbering system for the main and Slide Roadcuts are used here and by Aubry (this volume) and Miller and Katz (this volume). The entire outcrop section was measured and described by C. Stuart (Fig. 2).

Lithologies are similar to those described previously (Hill, 1899; Robinson, 1969b; Steineck, 1974, 1984). A moderate (upward) adjustment of the Spring Garden/Buff Bay Formation boundary in the lower third of the Main section was suggested by C. Stuart based on presence of black lichen growth on the indurated white chalk below (Spring Garden Member) and their absence on sandier, less indurated chalks above (Buff Bay Formation).

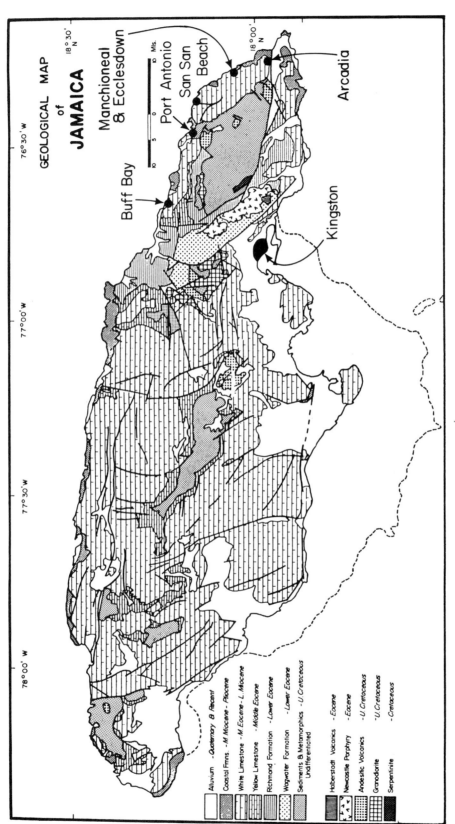

Figure 1. Geologic map of Jamaica showing location of Buff Bay section (northeast coast) and other localities mentioned in the text. Formations studies outcrop between Buff Bay and Bowden (from Wright, 1974).

TABLE 1. LATE NEOGENE PLANKTONIC FORAMINIFERAL TAXA ORIGINALLY DESCRIBED FROM JAMAICA*

	Taxon	Type Locality
1	*Globorotalia menardii* var. *multicamerata* Cushman and Jarvis, 1930	Probably from about horizon of sample ER 146-42 to 43 (1/2 mile east of Buff Bay)
2	*Globorotalia (Globorotalia) crassaformis viola* Blow, 1969	Sample WHB 181B, "Buff Bay Beds" of Cushman and Jarvis (? = ER 146-44)
3	*Globorotalia (Globorotalia) crassaformis ronda* Blow, 1969	Sample WHB 181, "Buff Bay Beds" of Cushman and Jarvis (? near locality ER 146-42)
4	*Globigerina altispira* Cushman and Jarvis, 1936	Recorded as having been collected at mile post 71, east of Port Antonio, Jamaica. The section of San San Bay, a few km to the east, is the more likely type locality (fide Robinson, 1969b, p. 9)
5	*Globorotalia (Globorotalia) truncatulinoides* Blow, 1969	Sample 187A (presumably close to level of sample ER 143-24), Navy Island Member, Manchioneal Formation, San San Bay (now inaccessible)
6	*Sphaeroidinellopsis subdehiscens paenedehiscens* Blow, 1969	ER 195, basal strata of Bowden Formation exposed along coast road just south of Drivers River
7	*Globorotalia cultrata exilis* Blow, 1969	ER 156 (= holotype sample of Zone N20 of Blow, 1969). Bowden Formation (type section), Bowden, Jamaica (See Bolli, 1970, p. 596, for further discussion on nature and stratigraphic position of this sample).
8	*Globorotalia menardii miocenica* Palmer, 1945	Bowden Shell Bed locality (ER 140), type locality, Bowden Formation, Bowden
9	*Sphaeroidina immatura* Cushman, 1919	Figured specimen from "Bowden Marl," Bowden
10	*Candeina nitida praenitida* Blow, 1969	Sample locality ER 307 along Arcadia Road, north of Bowden (correlates with Buff Formation)
11	*Globigerinoides parkerae* Bermúdez	0.8 miles east of Buff Bay, Jamaica, probably from level of sample ER 146-36

*Data from Robinson, 1969b.

TABLE 2. COMPARISON OF PLANKTONIC FORAMINIFERAL TAXA RECORDED BY DIFFERENT AUTHORS FROM ER SAMPLES

ER	Blow (1969)	Robinson§	Bolli (1970)	Liska (1985)
40	N17*	N	N cf.P A	
39		N L[1]		N L[1] A P R
38		N L		
37	N16†	N L	L	N M L R
36		N L		N M L R
35		N L		N M L R
34		N L[2]	L	
33A		N		N M R
33	N			N M R
32		N M[3]	N M[3]	
31		M	N M	

*Holotype locality.
†Paratype locality.
§ = Robinson, in Lamb and Beard (1972).
[1] = LO = last occurrence of lenguaensis; [2] = FO = first occurrence of lenguaensis; [3] = LO = last occurrence of mayeri; N = nepenthes; M = mayeri; A = acostaensis; L = lenguaensis; P = plesiotumida; R = ruber (= subquadratus).

Thin calcarenitic lenses are common in the Main and Slide sections. The coarse fraction contains inter alia, *Amphistegina* spp., *Asterigerina* spp., and coralline (and other reefal) debris. Reworked planktonic foraminifera have been found in these samples, usually infrequently, but nevertheless in a distinctive fashion (see discussion below).

In addition I have examined samples from several other outcrops along the north and east coast of Jamaica and results are discussed below.

PLANKTONIC FORAMINIFERAL BIOSTRATIGRAPHY OF THE BUFF BAY SECTION

The stratigraphic section exposed at Buff Bay spans an interval from middle Miocene to mid-Pliocene (ca. 16 to 3 Ma; Fig. 2). There is an obvious hiatus in the upper part of the Buff Bay Formation spanning about 4 m.y. of late Miocene–early Pliocene time (between samples 243, Buff Bay Formation, and sample 244, Bowden Formation). A shorter, more subtle intra–late Miocene hiatus is thought to exist at the base of a 2-m-thick cobble to boulder marl breccia debris-flow bed in the Buff Bay Formation at the 275 to 277 m level (sample 229) in the Slide Roadcut section (see paper by Aubry, this volume). The duration of this hiatus is presumably relatively short, on the order of 1 m.y. or less inasmuch as the unconformity is located within Zones NN9 and NN10 and within the range of N16/lower N17 (the first occurrence (FO) of *N. acostaensis* is in a marl, sample 228, just below the debris flow). Another short hiatus (~1.5 m.y. duration) occurs in the early middle Miocene within Zones N11 and N12.

The lower part of the Buff Bay section (Pots and Pans Roadcut) is about 130 m thick and consists of (generally) indu-

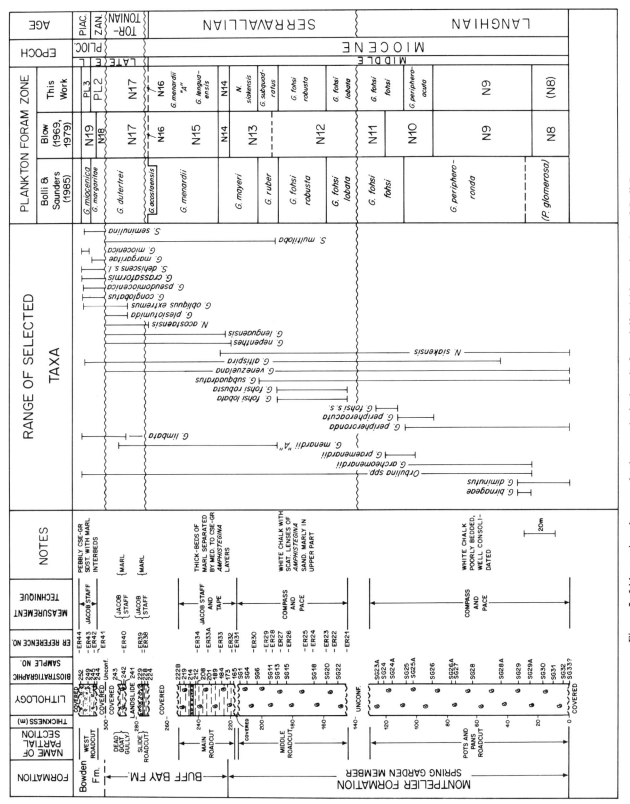

Figure 2. Lithostratigraphy, sample data, and planktonic foraminiferal biostratigraphy of Buff Bay section, Jamaica. The biostratigraphic scheme adopted in this work is an informal, essentially hybrid combination of the zonal stratigraphy of Bolli (1966) and Blow (1969, 1979). See text for further explanation.

rated chalks (in which the foraminifera exhibit variable preservation) of the Spirng Garden Member of the Montpelier Formation. Faunas are generally dominated by globoquadrinids (*venezuelana*), *Globigerinoides* (*sacculifer, subquadratus*), and globorotaliids (*peripheroronda*). The presence of *Globorotalia birnageae* (samples SG24, 29A) and *Globigerinoides diminutus* (samples SG33 to SG29) suggests that the base of the section is of earliest middle Miocene (Zone N8) age.

The first occurrence (FO) of *Orbulina suturalis* occurs in sample SG29A, but the evolutionary progression between *Praeorbulina* to *Orbulina* has not been observed in this section probably owing to the relatively poor preservation of planktonic foraminifera in samples below sample level SG26.

A generally good succession of the *Globorotalia fohsi* group is seen between samples SG26 and SG23A (Pots and Pans Roadcut) and samples SG22 and SG12 (Middle Roadcut). The last occurrence (LO) of *G. peripheroronda* (sample SG25), the FO and LO of *G. peripheroacuta* (SG26 and SG24, respectively), the LO of *G. fohsi fohsi* (SG23A), and the LO of *G. fohsi lobata* and *G. fohsi robusta* (SG12) aid in the delineation of Zones N10 to 12 in the Spring Garden Member. The simultaneous FO of *G. fohsi lobata* and *G. fohsi robusta* in sample SG 22 (at the base of the Middle Roadcut section) together with the simultaneous FO of *Triquetrorhabdulus rugosus* and LO of *Sphenolithus heteromorphus* indicate the presence of an unconformity between Zones N11 and N12, and with a hiatus of ~2 m.y., within the covered interval between the Pots and Pans and the Middle Roadcut sections (see Aubry, this volume). Rare individuals of *G. fohsi lobata* occur in samples SG3 and SG1 at the top of the Middle Roadcut in association with *Globorotalia menardii* "A," *G. praemenardii*, *Paragloborotalia Mayeri-siakensis* group, *Globigerina druryi*, and *Sphaeroidinellopsis multiloba*, about 7 to 10 m above the (apparent) LO of *Globigerinoides subquadratus* (sample SG6) and about 5 m below the (apparent) FO of *Globigerina nepenthes* (sample 168) in the lower part of the Main Roadcut. These rare individuals of *G. fohsi lobata* are interpreted as being reworked owing to the frequent occurrence of displaced amphisteginids, molluscs, and other neritic elements throughout the stratigraphic succession exposed at Buff Bay.

The FO of rare but distinct and typical individuals of *Globoturborotalita nepenthes* occurs in sample 168, about 3 m above the base of the Main Roadcut, in the upper part of the Spring Garden Member. Sample ER 146-31 was projected by Robinson into the base of the Main Roadcut approximately equivalent to our sample 165. Bolli (1970) recorded *G. nepenthes* from sample ER 146-31, while Robinson (1969b) did not. Despite diligent searching we have been unable to find any typical individuals of *G. nepenthes* in samples 165 to 167. This anomaly may be the result of a slight offset in the projection of (presumed) sample level ER 146-31 into the present roadcut or to somewhat different taxonomic concepts with regard to *G. nepenthes*.

The FO of (rare) *Globorotalia lenguaensis* is in sample 173 (about 3 m above the FO of *G. nepenthes* in sample 168); from sample 175 (<2 m higher) upwards, *G. lenguaensis* is a distinct and common component of middle and late Miocene faunas.

Faunas in the lower part of the Buff Bay Formation in the Main Roadcut are characterized by the common (co)occurrence of *Paragloborotalia mayeri* and *Globorotalia lenguaensis* (samples 175 to 189). Characteristic faunal elements include globoquadrinids (*venezuelana, globosa, altispira*), globorotaliids (almost exclusively sinistral *menardii* "A"), *Globigerinoides* (*trilobus, obliquus*), *Sphaeroidinellopsis multiloba*, and orbulinids. The LO of *P. mayeri* (= N14/N15 boundary), occurs in sample 189, about 16 m above the base of the Main Roadcut. Zone N14 is thus about 10 m thick at Buff Bay (samples 173 to 189).

Faunas in the remaining (~60 m thick) Buff Bay Formation (samples 190 to 243) are remarkably uniform and similar to those described above. Notable events, however, include the (apparent) FO of *Neogloboquadrina acostaensis* in sample 228, just below a 2-m-thick debris-flow unit (the base of which is considered unconformable and with a hiatus of ~1 m.y.) and the presence in the interval above the debris flow (samples 230 to 243) of *Globorotalia menardii* "B" of Bolli (= *G. limbata* of Parker, Berggren inter alia), *G. plesiotumida* (sporadic, and rare) and *Globigerinoides obliquus extremus* (in addition to the continued presence of *G. lenguaensis, G. nepenthes*, and *N. acostaensis*), indicative of a late Miocene (Zone N17) age.

The Bowden Formation (about 11 to 12 m thick) is exposed in the West Roadcut and contains planktonic foraminiferal faunas that span the interval of Zone PL2 to PL3 (Berggren, 1973). Characteristic Pliocene faunal elements (*Globorotalia pseudomiocenica, G. miocenica, G. crassaformis, G. margaritae, Sphaeroidinellopsis seminulina, Globigerinoides conglobatus*) occur with forms continuing from the late Miocene. The faunal association of *Globorotalia crassaformis, G. limbata, G. margaritae, G. exilis, G. pseudomiocenica, Dentoglobigerina altispira, D. venezuelana*, and the absence of *Globoturborotalita nepenthes* and *Globigerinoides fistulosus* at the base of the section (sample 244 and ER 146-41) indicate assignment to Zone PL2. The LO of *G. margaritae* (sample 246) indicates the Zone PL2/PL3 zonal boundary and the absence of *Dentoglobigerina altispira* at the top of the section (sample 252) suggest the PL3/PL4 zonal boundary; however, faunal preservation and abundance is markedly reduced here owing to the pronounced shallowing in the upper part of the section and this determination is questionable. The stratigraphic position in the Buff Bay section of first (FO) and last (LO) occurrences of selected planktonic foraminiferal taxa is shown in Table 3.

SOME PROBLEMS IN NEOGENE PLANKTONIC FORAMINIFERAL BIOSTRATIGRAPHY

The stratigraphic section along the coast road east of the town of Buff Bay exposes rocks belonging to (and representing the type localities of) the Spring Garden Member and the Buff

TABLE 3. STRATIGRAPHIC POSITION OF FIRST AND LAST OCCURRENCES OF SELECTED PLANKTONIC FORAMINIFERA IN THE BUFF BAY SECTION, JAMAICA

	Taxon	FO*	LO*	Sample	Formation
1	Globoquadrina altispira		X	251	
2	G. miocenica	X		247	
3	G. margaritae		X	246	Bowden
4	G. menardii "B"		X	245	
5	Candeina nitida	X		241	
6	N. acostaensis	X		228	
7	Paraloborotalia mayeri		X	189	Buff Bay
8	Globorotalia lenguaensis	X		173	
9	G. nepenthes	X		168	
10	Globigerinoides subquadratus		X	SG 6	
11	G. menardii "A"	X		SG 12	
12	G. fohsi robusta		X	SG 12	
13	G. fohsi lobata		X	SG 12	
14	G. fohsi robusta	X		SG 23	Spring Garden
15	G. foshi lobata	X		SG 23	
16	G. fohsi fohsi		X	SG 23A	
17	G. fohsi fohsi	X		SG 24	
18	G. peripheroacuta		X	SG 24	
19	G. peripheroronda		X	SG 25	
20	G. peripheroacuta	X		SG 26	
21	Orbulina universa	X		SG 28A	
22	G. diminutus		X	SG 29A	
23	Globorotalia birnageae		X	SG 29	

*FO = first; LO = last.

Bay and Bowden Formations. Robinson (1969b) has described this section and the history of terminology associated with the lithostratigraphic units exposed there. A series of samples collected by Robinson in the 1960s have become the focal point of considerable stratigraphic interest because they have figured significantly in the formulation of our concepts of late Neogene planktonic foraminiferal biostratigraphy. Blow (1969) used several of these samples as "holotype" or "paratype" localities for some of his late Neogene planktonic foraminiferal zones (Table 4). Bolli (1970) and, more recently, Liska (1985) have also examined and commented on the (bio)stratigraphic affinities of these samples (Table 2).

At this point it should be pointed out that the outcrops formerly exposed along the road east of Buff Bay have been slightly modified by a new roadcut along the coast made in the 1970s, which has resulted in a slight rearrangement of the geometry of the former exposures. In the course of measuring the Buff Bay section in the spring of 1987, and with the help of Edward Robinson, we attempted to plot (project) the position of his "ER" sample series into our measured section. Subsequent analysis of splits of the original "ER" sample series by Blow, Bolli, and Liska (Table 2), our own examination of samples subsequently collected on various field trips to Jamaica by others under Robinson's supervision (Table 5), and of our own samples have yielded some minor discrepancies with earlier interpretations.

We consider the problems associated with the Buff Bay section biostratigraphy below.

1. The Zone of N13 problem

A. The *Globorotalia mayeri* Zone of Bolli (1957) represents the stratigraphic interval between the LO of *"Globorotalia fohsi robusta"* and the LO of *Globorotalia mayeri* (recte *G. siakensis* of Blow) of the Lengua Formation of southern Trinidad. *Globoturborotalita nepenthes* was found to occur in all but the lowest part of the *mayeri* Zone in Trinidad (Bolli, 1957, p. 99, 103). Blow (1959) found that the interval between the LO of *G. fohsi robusta* and the FO of *G. nepenthes* contained *G. lenguaensis* and *Globigerinoides subquadratus* over a considerable stratigraphic interval in the Pozon Formation of Venezuela.

B. In the belief that this interval was not represented in Trinidad, Blow (1959) subdivided the *"G. mayeri"* Zone into a lower *G. mayeri/G. lenguaensis* Subzone and an upper *G. mayeri/G. nepenthes* Subzone. The upper subzone was considered equivalent to all of Bolli's *G. mayeri* Zone, whereas the lower subzone was believed to be represented in Trinidad only by the basal Lengua beds and the unconformity between the Lengua and subjacent Cipero Formation.

C. The lower *G. mayeri/G. lenguaensis* Subzone was defined (Blow, 1959) as the concurrent range of the nominate taxa between the LO of *G. fohsi robusta* and the FO of *G. nepenthes*.

D. Blow (1969, p. 239–243) subsequently expressed the opinion that the LO of *G. fohsi robusta* was erratic owing to "increasingly great ecological and geographical restriction." He defined his Zone N13—the *Sphaeroidinellopsis subdehiscens subdehiscens/Globigerina druryi* Partial-range Zone—as the interval between the evolutionary FO of *S. subdehiscens* s.s. (from its ancestor *S. seminulina* s.s.) to the FO of *G. nepenthes*. The base of the zone was said to lie within the upper part of the *G. fohsi robusta* Zone of Bolli (1957, 1966) and Blow (1959).

E. Bolli (1966) defined the *Globigerinoides ruber* Zone as the biostratigraphic interval between the LO of *G. fohsi robusta* and the LO of the nominate taxon. The zone was further discussed in Bolli and Saunders (1985, p. 167) who observed that in the Caribbean *G. ruber* disappeared simultaneously with *G. fohsi lobata* or shortly afterwards, whereas in Java *G. ruber* persisted for a considerable stratigraphic interval (over 200 m) above the LO of *G. fohsi robusta*. The *G. ruber* Zone was thus introduced to subdivide the thick *G. mayeri* Zone in Java.

F. The Buff Bay section, as currently exposed at Buff Bay, is about 300 m thick (our samples SG 33 to 252). The Buff Bay Formation was interpreted by Robinson (1969b, p. 5) to span the stratigraphic interval from between samples ER146-32 and 33 to the unconformity between samples ER146-40 and 41. We have placed the boundary between the (underlying) Spring Garden Member of the Montpelier Formation and the (overlying) Buff Bay Formation at a level about 223.5 m above the base of the section, between samples 177 and 178 in the Main Roadcut section (Fig. 1).

TABLE 4. SAMPLES USED TO DENOTE HOLOTYPE AND PARATYPE LOCALITIES FOR N-ZONES OF BLOW*

Formation	Age	N-Zone†	Sample	Locality
Manchioneal	Pleistocene	22 HT	ER 143-24	Navy Island Member of Manchioneal Formation, San San Bay section
Bowden	Pliocene	20 HT	ER 156	Bowden type section
		PT	ER 193	Drivers River section (near Manchioneal Harbor)
		PT	ER 538	Folly Point section
		19 HT	ER 146-44	Buff Bay section
		PT	ER 300	Bowden section (Blow, 1969, p. 254) spot sample at Arcadia Road (personal communication to H. M. Bolli from Dr. Robinson)
Bowden	Miocene	18 HT	ER 146-41	Buff Bay section
		PT	WHB 181B	Buff Bay section
		PT	ER 143-7	San San Bay section (Buff Bay Formation of Robinson, 1969a, b)
Buff Bay	Miocene	17 HT	ER 146-40	Buff Bay-type section
		PT	ER 305	Bowden-type section (Blow, 1969, Fig. 29) spot sample of Arcadia Road (personal communication to H. M. Bolli from Dr. Robinson; see also Robinson, 1969b, p. 17)
Buff Bay	Miocene	16 PT	ER 146-37	Buff Bay section
Montpelier	Miocene	13 PT	ER 143-4	San San Bay

*Blow, 1969 (modified from Bolli, 1970).
†HT = holotype; PT = paratype.

In the Spring Garden Formation of the Buff Bay section we have observed that the LO of *G. subquadratus* (= *G. ruber* of Bolli) in sample SG6 occurs about 10 m above that of *G. fohsi robusta* in sample SG12 and about 12 m below the FO of *G. nepenthes* at the base of the Buff Bay Formation, in sample 168 and 25 m below the LO of *P. mayeri* in sample 189, corroborating the observations of Bolli and Saunders (1985, p. 167) regarding the range of *G. ruber* (= *G. subquadratus*) in the Caribbean. Thus we have estimated a relatively thin (~10 m) *G. ruber* Zone and relatively thick *G. mayeri* Zone (~25 m) at Buff Bay, but we have no evidence in the Spring Garden Formation for an unconformity within the *G. ruber–G. mayeri-siakensis* biostratigraphic interval.

G. Further we have not observed typical *G. lenguaensis* together with *P. mayeri* over the 22-m-thick stratigraphic interval between the LO of *G. fohsi robusta* (sample SG12) and the FO of *G. nepenthes* (sample 168). Small indeterminate specimens of an unkeeled globorotaliid similar, but not referable, to *G. lenguanesis* do occur sporadically in the lower part of the Spring Garden Member over the stratigraphic interval referable to Zone N10–N12. However, the first truly typical forms referable to *G. lenguaensis* occur in sample 173 (and become common from sample 175 upwards) in the upper part of the Spring Garden Member near the base of the Main Roadcut section and within the basal part of the range of *G. nepenthes* (FO in sample 168), that is, within Zone N14. Thus we view the use of a "*Globorotalia mayeri/G. lenguaensis* Subzone" for the lower part of the "*G. mayeri*" Zone as less than useful, if not, indeed, misleading. We have adopted the provisional name *Paragloborotalia mayeri* interval zone for the stratigraphic interval in the Spring Garden Member (samples 565 to 168) characterized by the partial range of the nominate taxon between the LO of *G. subquadratus* and the FO of *G. nepenthes* (Fig. 1; see discussion in following section).

2. The Zone of N14 problem

A. The *Globorotalia mayeri* Zone of Bolli (1966; see also Bolli and Saunders, 1985) was defined as the biostratigraphic interval containing the nominate taxon between the mid-Miocene LO of *G. ruber* (= *G. subquadratus*) and the last occurrence of the nominate taxon. It corresponds to Blow's (1959) *Globorotalia mayeri/Globigerina nepenthes* Subzone. Blow (1969, p. 245) subsequently defined his Zone N14: *Globigerina nepenthes/*

TABLE 5. FAUNAL DETERMINATIONS AND AGE ASSIGNMENT OF ROBINSON (EHR) SAMPLES FROM BUFF BAY SECTION COLLECTED BY CHEVRON OIL COMPANY*

Sample (EHR)	Faunal Data	Age	Zone	Remarks
43	Globorotalia crassaformis, G. crassula, G. miocenica, G. exilis, G. altispira, Sph. dehiscens s.s., N. humerosa, Sph. seminulina, G. trilobus	Late Pleiocene	PL3-4	Varying preservation, some specimens discolored, i.e., D. venezuelana (reworked).
42	Globorotalia crassaformis, Gl. crassula, Gl. margaritae, Gl. pseudomiocenica, Neoglogoquadrina humerosa, Globigerinoides conglobatus, G. trilobus, G. fistulosus, G. obliquus extremus, Sph. seminulina, S. dehiscens, G. aequilateralis	Early Pliocene	P13	
41	Globorotalia crassaformis, Gl. multicamerata, Gl. pseudomiocenica, Gl. margaritae, G. scitula, Globigerinoides conglobatus, G. trilobus, G. fistulosus, G. obliquus extremus, Dentoglobigerina altispira, Sphaeroidinellopsis seminulina, S. dehiscens, Globigerinella aequilateralis	Early Pliocene	PL1C or PL2	Weathered (oxidized, reddish color), poor preservation; displaced benthic fauna (including amphistegenids).
40	Gl. menardii "A," Gl. limbata, G. "plesiotumida," G. lenguaensis, G. scitula, Candeina nitida, Gl. trilobus, G. ruber, D. altispira, D. venezuelana, Sph. multiloba, Globigerinella aequilateralis, G. praesiphonifera, N. acostaensis	Late Miocene	N17	Preservation excellent; large, robust specimens.
39	Globorotalia menardii "A," G. "pleisiotumida," G. lenguaensis, G. scitula, D. altispira, D. venezuelana, D. globosa, Gl, trilobus, G. obliquus, G. nepenthes, G. praesiphonifera, Sph. multiloba, Gl. hexagonus, N. acostaensis	Late Miocene	N16-17	
38	Gl. menardii "A," Gl. lenguaensis, Gl. scitula, G. nepenthes, D. altispira, D. venezuelana, D. globosa, G. praesiphonifera, ?Candeina nitida, G. trilobus, G. obliquus	Middle Miocene	N15	P. mayeri and N. acostaensis not present. G. fohsi robusta (rare, reworked).
37	Gl. menardii "A," G. lenguaensis, D. altispira, D. globosa, D. venezuelana, G. hexagonus, G. obliquus, G. nepenthes, Sph. multiloba, Candeina nitida	Middle Miocene	N15	N. acostaensis not present; P. mayeri very rare (and probably reworked). Rare mid-Eocene morozovellids (reworked).
36	Globorotalia menardii "A" (rare), G. lenguaensis, D. altispira, D. venezuelana, G. trilobus, G. obliquus, Sph. multiloba, G. siphonifera	Middle Miocene	N55	P. mayeri (rare, some discolored, presumed to be reworked).
35	Globorotalia menardii "A," G. lenguaensis, D. altispira, D. venezuelana, D. globosa, G. dehiscens, G. nepenthes, G. trilobus, G. obliquus, G. praesiphonifera, G. variabilis, Sph. multiloba	Middle Miocene	N15	P. mayeri (rare, presumed reworked); displaced amphisteginids; varying preservation, some specimens discolored.
31	Globorotalia menardii "A," G. scitula, G. cf. lenguaensis, P. mayeri, D. venezuelana, D. baroemoenensis, G. dehiscens, G. trilobus, G. obliquus, G. praesiphonifera, Sph. multiloba, G. decoraperta, G. druryi, G. cf. nepenthes, ?Cassigerinella sp.	Middle Miocene	N13-14	Individuals similar, but not typical of the species, referred here to G. cf. lenguaensis and to G. cf. nepenthes.
30	Globorotalia menardii "A," G. scitula, G. fohsi lobata, G. altispira, D. venezuelana, G. praesiphonifera, G. trilobus, G. obliquus, Sph. multiloba, P. mayeri	Middle Miocene	N12	

*Samples loaned through the courtesy of B. Kohl, Chevron USA, New Orleans.

Globorotalia mayeri Concurrent-range Zone as the biostratigraphic interval between the evolutionary FO of *G. nepenthes* and the LO of *G. mayeri.*

B. Blow (1969, p. 246) observed that almost all of Bolli's *"Globorotalia mayeri"* Zone is equivalent to his Zone N14 (because of the supposed close juxtaposition of the LO of *G. ruber* in the upper part of Zone N13 and the FO of *G. nepenthes,* the definitive criterion for the base of Zone N14 which he said occurred in the expanded Pozon section of Venezuela and which Bolli, 1966, confirmed in Java). Indeed, Bolli and Saunders (1985, p. 169, Fig. 7) show these two bioevents coinciding at the *G. ruber/ G. mayeri* (= N13/N14 boundary).

C. However, I would point out that in Jamaica the LO of *G. ruber* follows very shortly that of *G. fohsi robusta* (~10 m) and occurs some 12 m below the FO of *G. nepenthes,* resulting in an approximately 25 m thick *G. mayeri* Zone (Bolli, 1966); a 22-m-thick Zone N13 and an approximately 13-m-thick Zone N14 (Blow, 1969) characterized by the brief overlap in the lower part of the Bluff Bay Formation of *G. nepenthes* and *P. mayeri.* We have no evidence of a hiatus within this stratigraphic interval in Jamaica that would explain the abbreviated upper limit of *G. ruber* vis à vis the FO of *G. nepenthes* and view the range of *G. ruber* (= *G. subquadratus*) as being normal and typical of the Caribbean region (cf. Bolli and Saunders, 1985, p. 167). Thus, in Jamaica there may be an "expanded" *G. mayeri* Zone at the expense of a "reduced" *G. subquadratus* Zone following original biostratigraphic criteria for zonal differentiation.

D. Robinson (1969b, p. 5, Fig. 1) shows sample ER 146-31 situated in the uppermost part of the Spring Garden Formation and *within* Zone N13. He draws the N13/N14 boundary between samples ER 146-31 and ER 146-32 (see also Robinson, personal communication, in Lamb and Beard, 1972, p. 33, Fig. 16). In 1987 the position of sample ER 146-31 was projected into our field section at the level of sample 165 in the uppermost Spring Garden Formation, about 9 to 10 m *below* the Spring Garden/Buff Bay Formation boundary, which is located between samples 177 and 178, about 3 to 4 m above the projected level of sample ER 146-32 (Fig. 1). Sample ER 146-32 was projected into sample level 173, approximately 5 m above sample ER 146-31. Sample ER 146-33 was projected between sample levels 184/185, about 6.5 m above sample ER 146-32 and about 12 to 13 m above sample ER 146-31 (= sample 165) at the base of the Main Roadcut section. We have examined several samples in the series 165 to 189 as well as samples ER 146-31, 32, and 33. The FO of *Globoturborotalita nepenthes* occurs in sample 168, about 6.5 m below the Spring Garden/Buff Bay contact. The LO of *P. mayeri* occurs in sample 189, about 6 m above this contact.

E. In terms of the ER sample series we record the FO of *G. nepenthes* in sample ER 146-32; we have not found it in sample ER 146-31 (our record agrees with that of Robinson in Lamb and Beard, 1972, p. 33, Fig. 16; cf. Bolli, 1970, p. 595 who recorded *G. nepenthes* in both ER 146-31 and 146-32). We attribute this difference to preservation (preservation in sample ER 146-31 is poor, in ER 146-32 it is excellent) or perhaps to slight differences in taxonomic concept between authors, that is *G. druryi* versus *G. nepenthes.* Alternatively the discrepancy may be due to the projection of sample ER 146-31 into the base of the more recently exposed section in the Main Roadcut, equivalent to our sample level BB1 (= sample 165).

F. It should be noted here that whereas Bolli (1957, 1966; see also Bolli and Saunders, 1985, p. 167) stated that both *Globorotalia lenguaensis* and *Globigerina nepenthes* appear at the base of the *G. mayeri* Zone, Blow (1969, p. 244) indicated that *G. lenguaensis* appeared in the mid-part of Zone N12 but was not observed as a common form until within Zone N13. Our observations in Jamaica agree with those of Bolli for we have found the FO of *G. lenguaensis* to occur about 3 m above the FO of *G. nepenthes,* that is, near the base of Zone N14.

3. The Zone N15 problem

A. The *Globorotalia menardii* Zone of Bolli (1966) represents the interval with the nominate taxon between the LO of *G. mayeri* and the FO of *G. acostaensis.* It is essentially equivalent to the *Globorotalia (T.) continuosa* Consecutive-range Zone of Blow (1969) characterized by the partial range of the nominate taxon between the LO of *P. mayeri* and the FO of *G. acostaensis.* Indeed, the type locality of the *G. menardii* Zone of Bolli (1966) is the same as that chosen by Blow (1969) for his *G. (T.) continuosa* (N15) Consecutive-range Zone. The only essential difference between these two zones is the use of different nominate forms, although reference to Bolli and Saunders (1985, p. 102) shows that they consider that *N. continuosa* is absent in the interval denoted by their *G. menardii* Zone. A recent examination of the type sample (KR 23425) of these two zones indicates that they may be stragraphically correlative with the *G. (T.) acostaensis acostaensis-G. (T.) merotumida* (N16) Zone of Blow (1969) inasmuch as both *G. acostaensis* and *Polyperibola christiani,* have been found there (Liska, 1991). I have verified this by my own examination of the type sample as well.

B. The close juxtaposition of the LO and FO of *P. mayeri/siakensis* and *N. acostaensis,* respectively, at North Atlantic DSDP Sites 558 and 563 led Miller et al. (1985) to suggest that Zone N15 was an extremely short zone, if in fact real at all.

C. In the Buff Bay Formation the stratigraphic interval between the LO of *P. mayeri* (sample 189) and the FO of *N. acostaensis* (sample 228) spans about 45 m (but see discussion below). This interval is wholly within Zone NN9 (Aubry, this volume; cf. Ryan et al., 1974). This "restoration" of Zone N15 = *G. menardii* Zone to proper dimensions suggests that the close juxtaposition of the LO of *P. mayeri/siakensis* and FO of *N. acostaensis* in the North Atlantic may have been due to a mid-Miocene deep sea hiatus. This point will be discussed in greater detail elsewhere.

D. The fact that the *G. menardii* Zone = *G. continuosa* (N15) Zone may, at their type locality level, be stratigraphically correlative with a level within Zone N16 notwithstanding, there is a biostratigraphic interval between the LO of *P. mayeri/sia-*

kensis and the FO of *N. acostaensis*. I have (informally) designated this biostratigraphic interval the *G. menardii* "A"–*G. lenguaensis* Partial-range Zone here. It is clear that further work is needed to clarify biostratigraphic relationships over this critical stratigraphic interval before a fundamental modification/revision to current low-latitude Neogene zones is made.

E. Liska (1991) noted that the type level/sample of calcareous nannoplankton Zone NN9 (*Discoaster hamatus*) is the same sample KR 23425 described above. Inasmuch as the type level of Zone NN9 is equivalent to Zone N16, he questions whether Zone NN9 spans the middle-late Miocene transition or is restricted to the late Miocene. Our studies show clearly (see Aubry, this volume) that Zone NN9 essentially spans the interval from the LO of *P. mayeri/siakensis* to a level within the upper part of Zone N16 and clearly spans the middle-late Miocene transition as currently delineated at the type locality of the Tortonian Stage in northern Italy.

F. Zone N15 is at least 45 m thick in the Buff Bay section of eastern Jamaica. Inasmuch as the type level in Trinidad is biostratigraphically equivalent to a level within Zone N16, I have chosen to demarcate a new type locality for this zone in the Buff Bay section. The type level of the *Globorotalia menardii* Zone of Bolli (1966) = *Globorotalia (Turborotalia) continuosa* (N15) Zone of Blow (1969) is herein designated as sample level 194 in the Buff Bay Formation at Buff Bay, Jamaica (see Fig. 2). This level is 232 m above the base of the measured section (beginning with the Pots and Pans section at the bottom), 2 m above sample 189 in which the LO of *P. siakensis* has been observed, 43 m below the base of a prominent 2 m thick cobble-boulder breccia bed just below which (sample 228) the FO of *N. acostaensis* have been recorded, and ~68 m below the (covered) unconformity between the Buff Bay Formation (below) and the Bowden Formation (above). Characteristic faunal elements in sample 194 include *Globorotalia menardii* "A," *G. lenguaensis, Dentoglobigerina venezuélana, D. globosa, D. altispira, Globoquadrina dehiscens* (rare), *Globigerinoides obliquus, G. trilobus, Sphaeroidinellopsis multiloba, Globoturborotalita nepenthes* (rare), *Hastigerina siphonifera,* and orbulinids. No individuals referable to *N. acostaensis* or *N. continuosa* were observed. Sample level 194 is within Zone NN9 (see Aubry, this volume).

4. Intra-Buff Bay Formation unconformity

Our examination of the stratigraphic succession of the Buff Bay section supports the suggestion of Liska (1985) of an intra–Buff Formation unconformity in the interval between samples ER 146-38 and ER 146-39 (Fig. 2). However, we would interpret the unconformity as spanning (i.e., separating) beds of Zone N16 to 17 age from those of N15 to N16 age, rather than N17 to N14.

Samples ER 146-38 and ER 146-39 are located, respectively, about 1.5 m below (sample 225) and within (sample 230) a 2-m-thick cobble to boulder marl breccia debris-flow bed in the upper part of the Buff Bay Formation exposed in the Slide Roadcut (Fig. 2). We would interpret this debris-flow as a late Miocene (Zone N17) event that eroded the underlying upper middle Miocene marls. We assign the underlying marls to Zone N15 to N16 and NN9 (see separate paper by Aubry, this volume).

Paragloborotalia mayeri is a consistent, and often common, component of samples in the Spring Garden and lower Buff Bay Formation up to sample 189 above which we have not found it in any sample, other than as rare, sporadic specimens, interpreted here as being reworked. Robinson in Lamb and Beard (1972, p. 33, Fig. 16) recorded a similar stratigraphic range showing the youngest occurrence of *G. mayeri* in his sample ER 146-32 (Table 2). Liska (1985, p. 374) recorded the presence of *G. mayeri* and *G. ruber* in samples ER 146-33, 33A, 35, 36, 37, and 39; and *G. plesiotumida* and *G. ruber* from sample ER 146-39 (Table 2); and noted the absence of *G. acostaensis*, nominate form of Zone N16, in the paratype locality sample of the zone ER 146-37 as did Bolli (1970, p. 595; cf. Robinson in Lamb and Beard, 1972, p. 33, Fig. 16, who recorded the initial appearance of *G. acostaensis* in the Buff Bay Formation in his sample ER 146-37).

I attribute these different interpretations to differing taxonomic concepts (*G. ruber, G. nepenthes, G. plesiotumida*) and to the influence of reworking (*G. mayeri, G. ruber*). There is clear evidence, visible in the outcrop sections, of downslope transport of material in the Buff Bay Formation. For instance, we find rare, yet well preserved, specimens of the *Globorotalia fohsi* group together with amphisteginids and other neritic elements scattered throughout the Main and Slide Roadcut sections (e.g., samples 228 and 229 in the Slide Roadcut section). I have also found rare occurrences of *P. mayeri* in samples 223, 225, 227, and 229 in the Slide Roadcut, in a fauna that otherwise requires assignment to the *Globorotalia menardii* Zone of Bolli (= N15 of Blow) and *N. acostaensis* (N16) Zone. I would regard these isolated individuals as reworked and believe that the record cited by Liska (1985, p. 374) of *G. mayeri* in samples ER 146-5 to 37 and 39 refers to reworked specimens (in the absence of definitive information on abundance). In our material *P. mayeri/siakensis* is common to abundant up to sample 189 (about 2 m above sample ER 146-33) and is present only as isolated individuals in the section above.

A single specimen of *Neogloboquadrina acostaensis* was found in sample 228, just below the debris-flow bed, in the "Slide Roadcut section" above the Main Roadcut section. The species occurs relatively commonly in all samples (231 to 240) from the Landslide section (although, admittedly, these samples may not represent a real stratigraphic interval) and continues to the top of the Buff Bay Formation exposed in Dead Goat Gully (samples 241 to 243). Robinson (during fieldwork of spring 1987) projected his samples ER 146-35 to 37 into the covered interval between the top of the Main Roadcut and the Slide Roadcut. The absence of *N. acostaensis* in sample ER 146-37 is curious inasmuch as Blow (1969, p. 51) chose it as a "paratype" locality of N16. We have been unable to verify the presence of *N. acostaensis* in the lower part of the Slide Roadcut section (samples 223 to

227). This interval contains the level of sample ER 146-38. Its (?) FO in sample 228 is only 2 m below the supposed level of sample ER 146-39 from which *N. acostaensis* is recorded (Robinson in Lamb and Beard, 1972, p. 33, Fig. 16; Liska, 1985, p. 374). Like Liska (1985), we would assign this level (i.e., ER 146-39) to Zone N17 (and NN10; Aubry, this volume).

It is not clear whether the lone occurrence of *N. acostaensis* in sample 228 represents its true FO. However, the FO of *Discoaster hamatus* in sample 184 and in Zone N14, some 50 m below the Boulder Bed and the initial occurrence of *N. acostaensis*, while agreeing with earlier records of their relative biostratigraphic sequence (Ryan et al., 1974), is in marked contrast to more recent records (Berggren et al., 1985). Calcareous nannoplankton data (Aubry, this volume) suggest that the hiatus represented by the erosional surface at the base of the Boulder Bed (between samples 228 and 229) is on the order of 1 m.y. or less, and spans the upper part of Zone NN9 to the upper part of Zone NN10.

The upper part of the Buff Bay Formation, above the debris-flow bed (samples 230 to 243) and below the unconformity separating the Buff Bay and Bowden Formations, is characterized by *Globorotalia menardii* "B" (= *G. limbata* of Parker, Berggren, *G. plesiotumida* (rare, sporadic), and *Globigerinoides obliquus extremus*, which argues for assignment of this interval to Zone N17.

5. (Bio)stratigraphic position and correlation of the type level of some late Neogene planktonic foraminiferal zones

Blow (1969) selected several of the original samples collected by E. R. Robinson from the Buff Bay section, as well as other localities on the North Coast of Jamaica as the "holotype" or "paratype" locality of several of his late Neogene tropical planktonic foraminiferal zones (Table 4). In the Buff Bay Formation at Buff Bay samples, ER 146-37 was selected as the "paratype" locality sample of Zone N16, and samples ER 146-40 and ER 146-41 as the "holotype" localities of Zones N17 and N18, respectively, whereas ER 146-44 from the top of the Bowden Formation exposed in the West Roadcut at Buff Bay was chosen as the "holotype" level for his Zone N19. The "holotype" level for Zone N20 is represented by sample ER 156 (our sample number 163) in the Arcadia Road section, southeastern Jamaica. "Paratype" samples for Zone N20 are located in the Drivers River section (ER 193) and Folly Point section (ER 538 = our sample FP-2).

"Paratype" samples for Zones N17 (ER 305 = our sample 160) and N19 (ER 300 = our sample 154) are located in the Arcadia Road section, southeastern Jamaica, whereas one of the "paratype" samples of Zone N18 (ER 143-7) is located in the San San Clay Member of the Bowden Formation exposed in the San San Bay section, northeastern coast of Jamaica. Although considered by Blow (1969) to be roughly correlative, our examination of these samples as well as a more detailed examination of the section along the road to Arcadia (see below) has revealed that there are problems with the biostratigraphic and biochronologic position of some of the "holotype" sample levels, as well as distinct spatial and temporal separation between the "holotype" and "paratype" samples of some zones (e.g., Zone N18).

Sample levels ER 146-35 to 37 are no longer visible in the Buff Bay section; they occur within what is now a covered interval of some 20 m between the main body of the Buff Bay Formation exposed in the Main Roadcut and the short exposure referred to as the "Slide Roadcut." Sample ER 146-40 is situated near the top of the Buff Bay Formation (equivalent to our sample 241) about 2.5 m below the unconformity with the Bowden Formation. Sample ER 146-41 is located just above the unconformity at the base of the exposed section in the West Roadcut section and is equivalent to our sample 244.

Samples 241 and 243 (Dead Goat Gully) are characterized by abundant *Globorotalia limbata* (= *Globorotalia menardii* "B" of Bolli), *G. nepenthes*, and *N. acostaensis* (rare). In addition, sample 243 contains specimens referable to *G. plesiotumida* Blow. Sample 242 is characterized by moderate to strong dissolution and globorotaliids are very rare. Sample ER 146-40 (which is from about the same level as sample 242) is the holotype locality of Blow's (1969) Zone N17, the *Globorotalia plesiotumida* Consecutive-range Zone. This sample contains a fauna similar to that of sample 243. Specimens that Blow probably would have considered referable to *G. plesiotumida* and that Bolli (1970, p. 595) referred to as *"G. ? tumida* cf. *plesiotumida"* intergrade perceptibly with the *G. menardii* "B" complex and we would agree with Bolli and Saunders (1985, p. 227) that the *merotumida-plesiotumida* group is related to the *G. menardii* stock rather than the *G. paralenguaensis-lenguaensis* stock. However, problems arise from assignment of this level to Zone N17 because of its firm correlation here to calcareous nannoplankton Zone NN10 (see Aubry, this volume). Previous correlations (Berggren et al., 1985) have shown that the N16/N17 zonal boundary lies well within Zone NN11, at ca. 7 Ma. The problem is exacerbated by our observation that the *G. tumida* stock is a predominantly Indo-Pacific group and it is relatively easier (although still in some instances difficult) to distinguish typical *G. plesiotumida* in late Miocene assemblages from the Indo-Pacific region. In tropical Atlantic-Caribbean assemblages, on the other hand, it is often extremely difficult to distinguish *G. plesiotumida* consistently, although *plesiotumida* morphotypes may be observed sporadically, usually associated with *menardii* "B" assemblages. It is unfortunate, then, that Blow (1969, p. 250) chose a sample from the Buff Bay Formation of Jamaica as the "holotype" level for his Zone N17. *Globorotalia plesiotumida* is anything but common in the Buff Bay Formation, or in the Caribbean region, in general for that matter, where a major hiatus generally characterizes the upper Miocene–lower Pliocene stratigraphic sequences. It may be recalled, however, that the holotype of *G. plesiotumida* was originally described by Banner and Blow (1965a) from the lower Pliocene Cubagua Formation of Venezuela. In short, it may be that we are dealing here in Jamaica with atypical morphotypes of *G. plesiotumida* as opposed to more typical representatives of the taxon that are generally found in the younger part of the late Miocene (≈ Messinian) of the Indo-Pacific region. The correlation here of the type level of Zone N17 with Zone NN10 suggests an age of about 8 Ma or older (Berggren et al., 1985).

There is a further problem with the correlation of the "holotype" samples of the Pliocene Zones N18, and N19, and N20 (samples ER 146-41, and ER 146-44 and ER 156, respectively). The former sample is stratigraphically equivalent to our sample 244, located at the base of the Pliocene exposure in the West Roadcut above the Buff Bay/Bowden unconformity (the unconformity lies within a covered interval almost 15 m thick) and contains a fauna referable to Zone PL2 (Berggren, 1973) characterized by the concurrence of *Globorotalia margaritae, G. exilis, G. crassaformis, G. pseudomiocenica,* and *Globigerinoides conglobatus* (but without the nominate taxon *G. tumida*). This level is placed in Zone NN15 by Aubry (this volume).

Sample ER 146-44 (type level of Zone N19) is located at the top of the exposure in the West Roadcut, and is equivalent to our sample 252. Faunas from these samples contain *Globorotalia exilis, G. miocenica, G. pseudomiocenica,* and *Sphaeroidinella dehiscens* (with well-developed supplementary aperture on the spiral side) but do not contain *G. margaritae* and are equivalent to Zone PL3 or PL4 of Berggren (1973). The base of Zone N19 was placed at the level of the first evolutionary appearance of *S. dehiscens* s.l. (including forma *immatura,* with small, discrete supplementary aperture) from its ancestor *Sphaeroidinellopsis subdehiscens paenedehiscens.* This event occurs very shortly after the first evolutionary appearance of *Globorotalia tumida* (Berggren, 1973). However, if the type level of N18 is biostratigraphically equivalent to Zone PL2 at <3.7 Ma and the type level of Zone N19 is equivalent to Zone PL3 to PL4 (ca. 3+ Ma), but the defining criterion of Zone N19 occurs at a level biochronologically older than the type level of Zone N18, we have a dilemma with regards to the biostratigraphic and biochronologic position of these two zones. That is, the type level of Zone N18 is biostratigraphically within Zone N19 (Fig. 3).

The type level of Blow's (1969) redefined Zone N20 is from an artificial outcrop near but not a part of the type section of the Bowden Formation at Bowden Parish of St. Thomas, southeastern Jamaica (see Blow, 1969, Fig. 29, p. 99). Bolli (1970) examined the "holotype" sample (ER 156) of this zone and noted that the association of globorotaliids (*crassaformis, miocenica, multicamerata*), *G. dutertrei* s.l., and *S. dehiscens,* coupled with the absence of *G. extremus* and *G. fistulosus,* indicated a stratigraphic position high in his *G. exilis/G. miocenica* Zone because both *extremus* and *fistulosus* are restricted to the lower part of the zone at Caribbean Deep Sea Drilling Project (DSDP) sites 29, 30, and 31. Both taxa are present, he stated, in sample ER 529 about 30 m stratigraphically higher (see Blow, 1969, Fig. 29, p. 297). Bolli indicated that the Bowden type section may be tectonically disturbed, the planktonic foraminifera may be reworked, or sample ER 156 may actually be stratigraphically above ER 529.

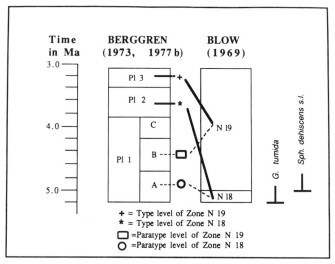

Figure 3. Comparative biostratigraphy and biochronology of the type levels of Zones N18 and N19 (Blow, 1969). Note that the type level of Zone N18 is correlative with Zone PL 2 of Berggren (1973); ca. 4 Ma) and that of Zone N19 with Zone PL 3 (ca. 3 Ma). The defining criteria for the base of these zones occur within close stratigraphic and temporal sequence with the result that the type level of Zone N18 is well within Zone N19. The paratype levels for Zones N18 and N19 are within Zones PL 1a (ca. 5.0 Ma) and PL 1b (ca. 4.4 Ma), respectively.

Tectonic disturbance is most likely as slopes between sample stations are heavily overgrown with vegetation. The stratigraphic position of sample ER 156 was obtained by extrapolating dips and is not tied positively either to the Bowden type locality (Woodring's mollusc beds) or to the relatively well exposed coastal section south of Bowden, containing samples ER 523 and 529 (E. R. Robinson, written communication, February, 1991). Saito (1985) also examined sample ER 156 and recorded, inter alia, the presence of *Globorotalia truncatulinoides, G. crassaformis, G. exilis, G. miocenica,* and *Globigerinoides fistulosus.* He correlated this sample with the uppermost part of the Matuyama Magnetochronozone, indicated that the type level of Zone N20 is stratigraphically younger than that of Zone N21, and excluded Zone N20 from his biostratigraphic subdivision of eastern equatorial Pacific upper Neogene sediments. My examination of sample ER 156 supports Bolli's (1970) assignment of the holotype level of Zone N20 to the *G. exilis/miocenica* Zone and this is consistent with the assignment of this sample to Zone NN16 correlative by Aubry (this volume). I have not found *G. truncatulinoides* in this material. A sample from the paratype locality ER 538 at Folly Point yielded a diverse and well-preserved fauna characterized by, inter alia, *Globorotalia exilis, G. miocenica, G. crassaformis, G. crassula, Neogloboquadrina acostaensis, N. dutertrei,* and *N. humerosa.* The absence of *multicamerata* and any specimens of *Sphaeroidinellopsis* as well suggests that this sample is of comparable age to the holotype sample ER 156 (Bowden type section) and in my Zone PL 4, consistent with the assignment by Aubry (this volume) of sample ER 538 to Zone NN16 and Subzone CN12a. Bolli (1970) examined samples ER 529 and ER 523, presumably well above the position of sample ER 156 (see Blow, 1969, Fig. 29, p. 297) and did not record *G. truncatulinoides* in either sample. Bolli (1970, p. 595) pointed out the fact that the Bowden Formation is a shallow-water, coastal deposit subject to faunal mixing and reworking and that "until a possible heterogeneity of its fauna can be disproved, the Bowden Formation should not be used for the designation of type localities for faunal zones." I concur wholeheartedly with this assessment. A comparison of the stratigraphic and biochronologic position of various of Blow's holotype and paratype sample levels has revealed considerable spatial and temporal separation in some cases.

I have found the Pliocene part (Zones N18 to N21) to be the least satisfactory of Blow's (1969) zonal scheme for the Neogene and for this reason have proposed an alternative six-fold scheme. The basal part of this zonal scheme (Zone PL 1a) may require revision (Hodell and Kennett, 1986). The remainder of the zonation appears applicable to low and mid-latitudes of the Atlantic-Caribbean region (Berggren, 1973, 1977b), although the FO of *G. puncticulata* (predominantly a temperate water form) remains problematic in this low-latitude zonal scheme. We have followed this zonal scheme here (Fig. 2).

THE ARCADIA ROAD SECTION

UNOCAL geologists collected a series of samples from the lower part of the Arcadia road section (see Blow, 1969, Fig. 29, p. 297) during our spring 1987 field work in Jamaica. This section represents a shallowing upward sequence of brown clays and marls (lower Bowden Formation) that terminate upward in shallow water calcarenites (shell beds). It should be pointed out here that the Arcadia road section and the type section of the Bowden Formation are two different sections and do not lie in stratigraphic continuity as shown in Blow (1969, Fig. 29, p. 297; cf. Bolli, 1970, p. 594). The distribution of some planktonic foraminifera is shown in Table 6. Salient features include the following:

1. The faunal association of samples 160 to 159 (including *G. plesiotumida*) indicates a late Miocene (Zone N17) age and this is supported by the presence of *G. lenguaensis* in sample 160 (= ER 305).

2. The presence of *G. cibaoensis* from samples 159 to 156 (= ER 302) indicates a correlation with Zone PL 1a of Berggren (1973, 1977b).

3. The presence of *G. margaritae* from sample 157b to 154 (i.e., between ER 302/304 to ER 300) supports a latest Miocene–early Pliocene age for this interval (<5.6 to >3.4 Ma = Zone PL 1 to PL 2). The LO of *G. cibaoensis* in sample 156 (= ER 302) indicates the PL 1a/b boundary (~4.7 Ma).

4. A notable change from sinistral to dextral coiling in the *Globorotalia limbata-pseudomiocenica* complex occurs between samples 155 (ER 301) and 154 (ER 302). At DSDP site 502 (Colombia Basin) sinistrally coiled populations of *G. limbata-pseudomiocenica* occur over a stratigraphic interval within the lower Gilbert Chronozone up to the lower part of the Thvera Subchronozone (~5.4 to 4.7 Ma; Keigwin, 1982; Kent and Spariosu, 1982). A distinct coiling change from sinistral to dextral occurs at about 4.7 Ma followed by a long (lower Pliocene) interval of dextral coiling in this group up to its transition to *G. miocenica* near the Mammoth Subchronozone, ~3.2 Ma.

5. Sample 154 (= ER 300), paratype locality for Zone N19 contains a characteristic late early Pliocene fauna including *G. margaritae*, dextrally coiled *G. limbata-pseudomiocenica* (see above), and *S. dehiscens immatura*. It is referable to Zone PL 1b (≤4.7 Ma) and is assigned to Zone NN13/14 (Aubry, this volume; see also Fig. 3, this chapter).

The data summarized above suggest that the Arcadia road section (samples 160 [ER 305] to 154 [ER 300]) spans a stratigraphic section corresponding to the uppermost Miocene to lower Pliocene (Zones N17, PL 1a to PL 1b) representing a time interval of ~6 Ma to ≤4.7 Ma.

THE SAN SAN BAY TYPE SECTION

The San San Clay Member of the Bowden Formation was formerly exposed along the road at San San Bay and represented by a series of samples collected by Robinson, ER 143-1 to 21 (see Robinson, 1969b, p. 6, Fig. 2 on p. 8). This section is no longer exposed along the roadside, being covered by a retaining stone wall. This is all the more unfortunate because it appears to have been the only section in Jamaica spanning the Miocene/Pliocene boundary in a deep-water facies and with excellent preservation of the microfauna. It should be noted that a fault separates samples ER 143-13 and 18 and that the interval between (samples ER 143-14 to 17) repeats the section below (Aubry, this volume). The stratigraphic distribution of planktonic foraminiferal taxa in the Coastal Group of Jamaica, including the San San Bay Member, was presented in Lamb and Beard (1972, Fig. 16, p. 33) based on data supplied by Robinson while a study of coiling direction changes in planktonic foraminifera, including the *G. menardii* group, was done by Robinson (1969a). I have examined several of these samples as part of this study (Table 7). While Robinson in Lamb and Beard (1972, Fig. 16) clearly recognized that the upper part of the section at Buff Bay is younger than the San San Clay at San San Bay (cf. Blow, 1969, Fig. 28), I draw attention to several points and differences in the interpretation made by Robinson (stratigraphic distribution of taxa) and biostratigraphic zonation and chronostratigraphic subdivision made by Lamb and Beard (1972, Fig. 16):

1. I have observed *Globoquadrina dehiscens* in sample ER 143-7; it does not appear to be reworked. It was not recorded by Robinson in Lamb and Beard (1972, Fig. 6) above sample ER 146-39 in the Buff Bay section.

2. I have observed *G. margaritae* in all samples from ER 143-6 to ER 143-21; Robinson (in Lamb and Beard, 1972) recorded it only in samples ER 143-19 to 21.

3. The FO of *Sphaeroidinella dehiscens* forma *immatura* is

TABLE 6. DISTRIBUTION OF SOME PLANKTONIC FORAMINIFERA IN SAMPLES COLLECTED FROM LOWER PART OF BOWDEN FORMATION, ARCADIA ROAD SECTION, JAMAICA*

Sample Unocal	Sample ER	Globorotalia cibaoensis	Globorotalia limbata	Globorotalia margaritae	Globorotalia plesiotumida	Globorotalia pseudomiocenica	Globorotalia lenguaensis	Globigerinoides conglobatus	Globigerinoides extremus	Globigerinoides obliquus	Globigerinoides trilobus	Sph. dehiscens immatura	Sph. seminulina	Sph. multiloba	G. nepenthes	D. altispira	D. venezuelana	N. humerosa	G. aequilateralis	Blow (1969; Fig. 29)	Berggren (this chapter)	Aubry (this volume)	Epoch/Series
154	300†	cf.	D	X		D		X	X		X	X	X			X				X	N19	B	CN10c (NN13-14)	Pliocene Early
155	301		S	X		S		X	X		X	X			X	X				X	N19			
156	302	X	S			S		X	X		X				X	X		X		X				
157A	-	X	S			S		X	X		X	X	X	X	X	X	X			X	N18	A	Not Zoned	
157B	-	X	S	X		S		X	X		X				X	X				X		PL 1		
157C	-		S	X		S		X	X		X				X	X				X				
158	304	X	S			S			X	X	X	X				X				X	N17		CN9b (=NN11b)	Miocene Late
159	-	X	S		X	S			X	X	X								X	X	N17			
160	305§		S		X	S	X			X	X					X	X			X		N17		

*See text for further discussion.
†Paratype locality of Zone N 19.
§Paratype locality of Zone N 17.

TABLE 7. DISTRIBUTION OF SOME PLANKTONIC FORAMINIFERA IN SEVERAL SAMPLES FROM THE TYPE LOCALITY OF THE SAN SAN MEMBER OF THE BOWDEN FORMATION, SAN SAN BAY, JAMAICA*

Sample	Globorotalia cibaoensis	Globorotalia crassaformis	Globorotalia limbata	Globorotalia margaritae	Globorotalia multicamerata	Globorotalia pseudomiocenica	Globorotalia puncticulata	Globorotalia menardii "A"	Globorotalia lenguaensis	Globorotalia plesiotumida	P. mayeri	S. multiloba	S. seminulina	S. dehiscens immatura	D. altispira	D. globosa	D. venezuelana	G. dehiscens	G. conglobatus	G. obliquus	G. extremus	G. trilobus	G. nepenthes	N. continuosa	N. acostaensis	N. humerosa	Planktonic Foram Zone	Epoch/Series	Estimated Biochronology (Ma)
21A		X	D	X	X	D	X						X	X	X		X		X	X	X	X	X		X	X		C	Early Pliocene	<4.3
21		X	D	X	X	D	X						X	X	X		X		X	X	X	X	X		X	X		C		
19			D	X	X	D	X						X	X	X		X		X	X	X	X	X		X	X	PL 1	B		≤4.7
17			S	X		S	X						X	X	X		X		X	X	X	X	X		X	X		B		
15	X		S	X		S							X	X	X		X		X	X	X	X	X		X	X				~4.7
14	X		S	X		S								X	X		X		X	X	X	X	X		X	X		A		
12	X		S	X		S								X	X		X		X	X	X	X	X		X	X				
8	X		S	X		S						X	X	X	X		X		X	X	X	X	X		X	X				
7	X		S	X		S						X	X	X	X		X		X	X	X	X	X	X	X	X				≥5.0
6	X		S	X		S			X			X	X	X	X		X		X	X	X	X	X		X	X				
5	†		S	†					X	X		X	X	X	X		X		X	X	X	X	X		X	X	N17			
4								X	X	X	X	X	X					X		X	X	X	X				N14	Late Miocene	
3								X	X	X	X	X	X			X	X	X		X	X	X	X	X						
2								X	X	X	X	X	X			X	X	X		X	X	X	X	X						

*See text for further discussion.
† = Aff.
S = sinistral; D = dextral.

in sample ER 143-15 (which should be equivalent to part of the section below).

4. The FO of *G. crassaformis* is recorded in sample ER 143-21, whereas Robinson (in Lamb and Beard, 1972) recorded it in sample ER 143-19. The difference may be the result of taxonomic concepts. In sample ER 143-15, I have observed a distinct association of forms representing the transition from *G. cibaoensis* to *G. puncticulata* as shown in Berggren (1977b, Fig. 19) representing the boundary between Zone PL 1a/PL 1b with an estimated age of 4.7 Ma (Berggren, 1977b). The FO of *G. crassaformis* (= Zone PL 1b/c is identified at 4.3 Ma at DSDP site 502 (Colombia Basin; Keigwin, 1982; see also Berggren et al., 1985).

5. The FO of *Globorotalia multicamerata* is recorded in sample ER 143-19 (Robinson, in Lamb and Beard, 1972, recorded this event in sample ER 143-18, not examined here).

6. A change from sinistral to dextral coiling in the *Globorotalia limbata-pseudomiocenica* complex (noted above in the Arcadia Road section) occurs between samples ER 143-14 and ER 143-19 consistent with the observation by Robinson (1969a, p. 556) that this change occurred between samples ER 143-12 and 18 and with the age estimate of ~4.7 Ma for a level between samples ER 143-12 and 18.

The biochronology derived from the above data suggests that samples ER 143-6 to 21 span a time interval of about 1 m.y. from about 5 to 4 Ma and represents Zone PL 1a (part) to PL 1c (part) of Berggren (1977b). It suggests that sample ER 143-7 (paratype locality of Zone N18) is approximately 5 Ma (equivalent to Zone PL 1a). Indeed Aubry (this volume) places this sample in the upper part of Zone NN11 (i.e., 4.8 Ma). The paratype level of Zone N18 is thus seen to be about 1.5 m.y. older than the biostratigraphic level (Zone PL 2) of the holotype sample of Zone N18 at Buff Bay (Fig. 3). It is notable that neither Robinson (in Lamb and Beard, 1972) nor I have recorded *Globorotalia tumida*, nominate form of Zone N18, in any of the type samples of that zone, let alone any of the Pliocene samples examined from the Coastal Group of Jamaica, with a notable exception in the Ecclesdown section (see below) and a rare occurrence in sample ER 143-19. Finally Lamb and Beard (1972, Fig. 16) placed the Miocene/Pliocene boundary (based on the FO of *G. margaritae*) between samples ER 143-18 and 19. I would estimate the age of this level as <4.7 Ma and within the lower Pliocene. I would suggest that the Miocene/Pliocene boundary, as currently recognized (see discussion below) at 4.9 Ma, is between samples ER 143-8 and 15.

Samples 143-1 to 4 (Robinson, 1969b) from the San San Bay section are from the Spring Garden Member of the Montpelier Formation. Samples 143-2 to 4 contain *Globoturborotalita nepenthes* and *Paragloborotalia mayeri* and are assigned here to Zone N14 (middle Miocene). The faunal association in sample ER 143-5 (and in particular the occurrence of *G. limbata*, *G. plesiotumida*, and *G. lenguaensis*) is indicative of a late Miocene (Zone N17) age.

THE ECCLESDOWN ROAD SECTION

UNOCAL geologists collected seven samples from a section on the road to Ecclesdown over a stratigraphic interval of about 150 m. I have examined samples from this section and while precise biostratigraphic determination/assignment is difficult, an approximate relationship can be made between this section and standard biostratigraphic schemes. The distribution of some of the taxa is shown in Table 8. Pertinent data relevant to this section include the following:

1. The section spans the late Miocene to mid-Pliocene (>8 Ma to >3 Ma). There is probably an unconformity between the lower two samples because of their distinctly different stratigraphic position.

2. The lowest sample (137) is of late Miocene age. It contains *G. lenguaensis*, *G. plesiotumida*, *G. scitula*, and sinistrally coiled *G. limbata* "A" and is referable to Zone N17.

3. Dextrally coiled *G. limbata-pseudomiocenica* and *S. dehiscens* forma *immatura* characterize the remainder of the section, save for the uppermost sample (137c where specimens are predominantly sinistral). This suggests that the interval between (and including samples 136 to 137b) is between 4.7 Ma and 3.2 Ma.

4. Sinistrally coiled *Globorotalia tumida* occurs in samples 135, 134, and 137A over a 50-m interval in the middle part of the section. This is the only location where *G. tumida* has been observed in the samples examined from the Pliocene of Jamaica in this study, except for an occurrence at San San Bay (sample ER 143-19; see Table 7).

5. The absence of *G. margaritae* in samples 137b and 137a at the top of this section and the continued presence of robust individuals of *D. altispira* and *S. seminulina* suggest placement of these samples in Zone PL3.

There is evidence of transport of shallow-water benthic foraminifera (amphisteginids, inter alia) in the Pliocene part of this section, preservation is variable and there is evidence of hematite staining on tests and mechanical breakage, indicating possible reworking. Careful field measurement and additional collection of this section is required before it can be correlated precisely with other upper Miocene-Pliocene sections on Jamaica.

STONY HILL ROAD SECTION

About ten samples were collected by UNOCAL geologists from exposures along the Stony Hill road to the west of Port Antonio during our spring 1987 field work in Jamacia (see Robinson, 1969b, Fig. 3). Changing strike and dips and the discontinuous nature of the exposures made it impossible to place the samples in precise stratigraphic order. Examinations of these samples indicate that the stratigraphic interval collected is:

1. of early late Miocene age at the base (equivalent to Zone N16 with *Globorotalia menardii* "A");

2. latest Miocene age in the mid-part (Zone N17 with *G. plesiotumida*, sinistrally coiled *G. limbata-pseudomiocenca*, inter

TABLE 8. DISTRIBUTION OF SOME PLANKTONIC FORAMINIFERA IN SEVERAL SAMPLES FROM EXPOSURES ALONG ROAD TO ECCLESDOWN, JAMAICA*

Epoch/Series		Planktonic Foraminiferal Zone	Approximate height above base of section (m)	Sample	Globorotalia								Globigerinoides				Sph. dehiscens immatura	Sph. seminulina	Sph. multiloba	G. nepenthes	D. altispira	D. venezuelana	N. acostaensis	G. aequilateralis
					lenguaensis	limbata	margaritae	menardii "A"	plesiotumida	pseudomiocenica	scitula	tumida	conglobatus	obliquus	extremus	trilobus								
Pliocene	Late	PL 3-4	149	137c						S	X		X	X	X	X	X	X		X		X	-	X
		Pl 3	142	137b		D				D			X	X	X	X	X	-		X			S	X
		Pl 2	117	137a		D	X			D	X	S	X	X	X	X	X	X		X			D	X
	Early	Pl 1† (c)	114	134		D	X			D		S	X	X	X	-	X	X		X	X	D		-
		Pl 1† (b)	66	135		-	X			D		S	X	X	X	X	X	-		X		-	-	X
			22	136		D	X			D			X	X	X	X	X	X	X	X		X	-	D	X
Late Miocene		N 17	0	137	X	S		X	S	S			-	X	-	X	-	X	X	X	X	X	D	X

*See text for further explanation.
†PL 1(b) and PL 1(c) = approximate correlation to Berggren (1973, 1977b) zonal scheme; nominate taxa absent.
D = dextral; S = sinistral.

alia and Zone PL 1a with *G. cibaoensis, G. margaritae,* and sinistrally coiled *G. limbata-miocencia,* inter alia) in the middle part;

3. early Pliocene (Zone PL 1b with dextrally coiled *G. limbata-pseudomiocenica, G. margaritae,* and *S. dehiscens* forma *immatura* inter alia near the top.

An unconformity is probably located in the lower part of the section separating the lower upper Miocene (Zone N16 equivalent) from the uppermost Miocene (N17). This upper Miocene unconformity is similar to that seen elsewhere on Jamaica.

THE MIOCENE/PLIOCENE BOUNDARY

I present a brief review of the current status of the Miocene-Pliocene boundary in so far as it relates to attempts to clarify stratigraphic relationships in different sections in Jamaica. Current practice in establishing global chronostratigraphic units links the hierarchical elements in a lock-step equivalency. Thus the epoch-series boundaries coincide with the limits of sequential age-stage boundaries. The Miocene/Pliocene boundary (= Messinian/Zanclean boundary) serves as a case in point, and as a point of departure for the discussion below.

Recent magnetobiostratigraphic studies have shown that the base of the Zanclean Stage in Calabria and Sicily is situated stratigraphically just below the base of the Thvera Subchronozone of the Gilbert Chronozone and has an estimated age of 4.86 to 4.93 Ma based on different techniques of extrapolation: sedimentation rates (Zijderveld et al., 1986; Hilgen and Langereis, 1988) or limestone-marl couplet duration, which correspond to orbital precession periodicity (Channell et al., 1988). The latter age estimate is accepted here. Similar results have been obtained in western Mediterranean Ocean Drilling Project (ODP) sites 652 anda 654 (Channell et al., 1990). The presently accepted Miocene/Pliocene boundary definition at Capo Rosello (Sicily; Cita, 1975) is based on the reestablishment of marine conditions in the Mediterranean Basin above the unconformity separating marine chalks (Trubi Formation) from the terminal Miocene (Messinian) lacustrine-evaporite deposits. This level is extremely difficult to characterize biostratigraphically in the Mediterranean, the first occurrence (FO) of *Ceratolithus acutus* about 6 m above the base of the Zanclean at Capo Rosella (Cita and Gartner, 1973) being essentially the sole criterion available for regional correlation/identification of this level.

However, Benson et al. (in preparation) discuss the problem of linking the Messinian/Zanclean boundary with the Miocene/Pliocene boundary and render the following observations:

1. The boundary, as currently accepted, situated close to an unconformity and based essentially upon a local-regional event (i.e., establishment of the Neo-Mediterranean following the Messinian Salinity Crisis), is not directly relevant to the designation of a base Pliocene Global Boundary Stratotype Section and Point (GSSP). The boundary location at this level does not allow for the recognition of the sequence of historical events directly prior to the boundary point and time and precludes precise biostratigraphic recognition and correlation elsewhere.

2. Strict adherence to the historical definition of the base of the Pliocene introduces an unacceptable modicum of circular reasoning into the situation: the event under investigation at Cape Rosello (reestablishment of normal marine conditions) becomes the criterion by which the event is denoted, if the definitions of the chronohorizon is accepted literally. But interpretation of this event outside the Mediterranean will vary with interpretations

and correlations. Definition and recognition are two separate concepts.

3. Nineteenth century subdivisions of the Cenozoic by Lyell and Mayer-Eymar, inter alia, were imprecise and basically biostratigraphic and/or lithostratigraphic (sedimentary cycles; Berggren, 1971) in concept, and boundaries were frequently located at unconformities. The boundary, as currently accepted, separates faunas (primarily molluscan) that historically exhibit a "Miocene" and "Pliocene" aspect, respectively. However, Pliocene molluscan faunas continue to exhibit a Miocene aspect well into the Pliocene. Calcareous plankton faunas and floras characteristic of the early Pliocene were already becoming established over 1 m.y. earlier in the Atlantic and Indo-Pacific Oceans (i.e., outside the Mediterranean).

Accordingly Benson et al. (in preparation) have argued for recognizing the Messinian and Zanclean stages for what they are, namely the local-regional expression of particular historical events in the late Neogene evolution of the Mediterranean—the Messinian Salinity Crisis followed by the Zanclean Deluge (with a chronologic boundary age of ~4.9 Ma)—and decoupling them from the question of the Miocene/Pliocene boundary. They propose linking the latter boundary with the Gilbert/5 Magnetochronozone boundary at 5.35 Ma in a newly proposed boundary stratotype section in the Bou Regreg section near Rabat, Morocco, in a continuous marine sequence of Atlantic facies. This level was shown to correspond biostratigraphically to the development of *Globorotalia margaritae* in its typical morphology, a short distance above its FO in Magnetochronozone 5 at about 5.8 Ma.

A review of the literature and our own expereince (see Aubry, this volume) reveals that a precise biochronology for late Miocene–early Pliocene time remains elusive. In addition to the evidence cited in Berggren et al. (1985), we can cite the following (Table 9):

1. At DSDP site 502 (Colombia Basin) the FO of *Ceratolithus acutus* occurs just below the Thvera Subchronozone (Prell et al., 1982, Appendix: Table 1, see also Fig. 16, p. 34) between 140.05 and 141.53 mbsf estimated age of ~4.85 Ma (Berggren et al., 1985). This is consistent with its identification (as *C.* sp. by Gartner, 1973, Fig. 3, p. 2026) in the lower Gilbert Magnetozone in RC 12-66. Preliminary data from equatorial ODP site 758 (Ninety East Ridge, Indian Ocean) suggest its FO in the lower Gilbert Magnetozone (Shipboard Scientific Party, 1989, site 758), there as well.

The FO of *C. acutus* in the lower Pliocene at about 4.8 to 4.9 Ma appears to be a reasonable estimate.

2. The range of *C. acutus* spans the interval from below the Thvera Subchronozone to the top of the Thvera Subchronozone (i.e., about 4.85 to 4.45 Ma) at site 502 (Prell et al., 1982, Appendix: Table 1, Fig. 16, p. 34). The LO of *C. acutus* is shown to be essentially contemporaneous with the FO of *Cerotolithus rugosus* (= NN12/NN13). In the Mediterranean this event has been recently correlated with the top of the Nunivak Subchronozone at about 4.13 Ma (Rio et al., 1990, p. 522).

3. Preliminary results from ODP site 758 suggest that the FO of *Globorotalia tumida* occurs in the lower Gilbert Magnetochronozone (Shipboard Scientific Party, 1989, site 758), supporting earlier reports at this level (Berggren et al., 1985).

4. The FO of *Globorotalia cibaoensis* has been linked with the Magnetochronozone 6R1 at about 6.3 Ma (essentially contemporaneous with the FO of *G. conomiozea* and the Global Carbon Shift) at Boug Regreg, Morocco (Benson et al., in preparation).

The stratigraphy and suggested correlations based on planktonic foraminifera of the upper part of the Neogene of eastern Jamaica was shown by Blow (1969, Figs. 28 to 30). Salient points include the following:

1. An unconformity was suggested between samples ER 146-31 and ER 146-32 at about the junction between our Middle and Main Roadcut sections and separating parts of Zones N13 and N14 (Blow, 1969; Fig. 30), and the Montpelier from the Buff Bay Formations.

We believe the evidence supports a gradational lithostratigraphic succession here; however, a brief hiatus or condensed section over the interval is possible (Aubry, this volume).

2. An unconformity separating upper Miocene from lower Pliocene between samples ER 146-40 and ER 146-41 was shown (Blow, 1960) and corresponds well with our results. However, Blow (1969) did not appreciate the temporal/biostratigraphic implications of this hiatus to his zonal scheme. The hiatus is between 4 to 5 m.y. long and the Miocene/Pliocene boundary is marked by a hiatus at Buff Bay.

3. The lower 50 m of the "Bowden Formation" in the Arcadia Road section appears to span a time interval between >6 Ma to <3.5 Ma, and includes the Miocene/Pliocene boundary in both its currently (~4.9 Ma; and just below the Thvera Event) and previously (~5.3 Ma; just above the Gilbert/Chron boundary) accepted sense (Blow, 1969, Fig. 29).

4. The San San Bay section (now unfortunately mostly covered by a cement retaining wall) contains a major hiatus be-

TABLE 9. BIOCHRONOLOGY OF BIOSTRATIGRAPHIC EVENTS SPANNING THE MIOCENE/PLIOCENE BOUNDARY

	Datum Event	Paleomag. Calibration	Age (Ma)	Reference
1.	FO *C. acutus*	Early Gilbert	4.85	Prell, Gardner, and others (1982)
2.	LO *C. acutus*	Top Thvera	4.45	Prell, Gardner, and others (1982)
3.	FO *G. tumida*	Basal Gilbert	5.2	Various sources; Berggren and others (1985)
4.	FO *G. cibaoensis*	C6R1	6.3	Benson and others (in preparation)

tween middle Miocene/late Miocene (Zones N14/N17), but then exhibits an apparently continuous upper Miocene–lower Pliocene sequence (Zone N17 to PL1; ~>5.0 Ma to <4.3 Ma).

5. Blow (1969, Fig. 28) correlated the section above the main unconformity at Buff Bay with the stratigraphic succession of the type section of the San San Clay Member of the San San Bay section, placing the lower sample at San San Bay (ER 146-5) in his Zone N17 (uppermost Miocene). He recognized that the lowest part of his Zone N18 is missing in the Buff Bay section. However, correlation of the Pliocene interval at the two sections reveals a significant spatial and temporal difference that Blow (1969) did not appreciate, due in great part to the fact that he did not incorporate/integrate examination of calcareous nannoplankton biostratigraphy in his studies. The main fact that emerges from a biostratigraphic comparison of the two sections is that the Pliocene part of the section exposed at the two localities are essentially mutually exclusive.

SUMMARY AND CONCLUSIONS

The Buff Bay section exposed on the north coast of Jamaica provides a well-preserved record of early middle Miocene to early Pliocene tropical planktonic foraminifera, although a brief (~1 m.y.) earliest late Miocene and larger (~4 m.y.) late Miocene–early Pliocene hiatus punctuates the stratigraphic sequence. Samples have been examined from measured sections of the Spring Garden Member of the Montpelier Formation, the Buff Bay Formation, and the Bowden Formation.

This study supports, in general, previous investigations on the late Neogene planktonic foraminiferal biostratigraphy of Jamaica and, by extension, the Caribbean region. Several anomalous results suggest, however, that parts of Blow's (1969, 1979) late Neogene Zone scheme should be used with caution, if not discarded. Results of this study include:

1. Highly variable, but apparently taxonomically low diversity populations of *Globorotalia menardii* "A" and *G. limbata* characterize the Miocene and *G. pseudomiocenica* and *G. miocenia* the Pliocene part of the section. *Dentoglobigerina venezuelana* is the most abundant form and ranges throughout the section.

2. A normal sequence of biostratigraphic events from the (approximate) FO of *Orbulina suturalis* to the FO of *Globorotalia robusta-lobata* group supports previous subdivision of this interval by Bolli, Blow, and Banner, inter alia. However, the simultaneous FO of *G. fohsi lobata* and *G. fohsi robusta* (as well as *Triquetrorhabdulus rugosus* and *Sphenolithus heteromorphus*; see Aubry, this volume) suggests the presence of an unconformity within the Spring Garden Member of the Montpelier Formation between Zones N11 and N12 with a hiatus of about 2 m.y.

3. Bolli's (1957) *Globorotalia mayeri* Zone (= LO of *G. fohsi robusta* to LO of *G. mayeri*) is supported. Blow's (1959) subdivision of this zone into a lower *G. mayeri/G. lenguaensis* Subzone (based on the concurrent range of *G. lenguaensis* and *G. "ruber"* = *subquadratus* before the FO of *G. nepenthes*) and an upper *G. mayeri/G. nepenthes* Subzone is not supported. I have not found *G. lenguaensis* below the FO of *G. nepenthes* in Jamaica, or elsewhere. Blow (1969, p. 243) subsequently defined his Zone N13: *S. subdehiscens subdehiscens–G. druryi* Partial-range Zone (essentially equivalent to his 1959 *G. mayeri/G. lenguaensis* Subzone) but this zone has proved to be inadequately recognizable owing to the questionable identify of *S. subdehiscens subdehiscens* (see Bolli and Saunders, 1981, 1985). In this paper I have substituted a provisional *Paragloborotalia mayeri* interval zone (between the LO of *G. subquadratus* and the FO of *G. nepenthes*).

4. An approximately 12-m-thick overlap of *G. nepenthes* and *P. mayeri* occurs in the lower part of the Buff Bay Formation and characterizes the concurrent range zone of these two taxa (Blow, 1969).

5. The *Globorotalia menardii* Zone of Bolli (1966) is equivalent to the *G. (T.) continuosa* Consecutive-range Zone of Blow (1969); both are defined by the LO *G. mayeri* to the FO of *G. acostaensis*. In Jamaica there is at least a 45-m interval between these two datum levels, wholly within Zone NN9 (Aubry, this volume), suggesting that Zone N15 should be restored to the late Neogene zonal hagiography and that its apparent reduction/virtual absence elsewhere may be due to a latest middle Miocene hiatus (cf. Miller et al., 1985).

6. An apparent unconformity occurs at the base of a boulder bed in the upper part of the Buff Bay Formation and appears to separate Zone N16 from N17; the estimated duration of the hiatus is ≈ 1 m.y.

7. A more obvious (although in a covered interval) unconformity separates the Buff Bay and Bowden Formations and Zones N17 and PL 2. The hiatus is estimated at about 4 m.y.

8. The Pliocene part of the section exposed at Buff Bay and San San Bay is essentially temporally mutually exclusive, rather than virtually equivalent (Blow, 1969). The section exposed at Buff Bay spans a time interval of approximately >3.5 to <3.0 Ma; that at San San Bay from >5 to >3.5 Ma.

9. The Miocene/Pliocene boundary is apparently represented in continuous stratigraphic sequence in the San San Bay, Arcadia Road, and Ecclesdown Road sections.

10. An examination of some type level samples upon which Blow (1969) based several upper Neogene zones has revealed interesting and, in some cases, anomalous results of significance to late Neogene biostratigraphy:

a. The holotype sample/level of the *Globorotalia menardii* Zone (Bolli, 1966) is the same as that of the *Globorotalia (Turborotalia) continuosa* Zone of Blow (1969) and contains *Neogloboquadrina acostaensis* and *Polyperibola christiana,* inter alia, and N15 is correlative with a level within Zone N16 of Blow (1969; Liska, 1991).

b. The (holo)type level of Zone N17 (ER 146-40) is within Zone NN10 (rather than NN11 as previously believed) and is about 8 Ma. The paratype level of Zone N17 (ER 305) is in upper Zone NN11 (= CN 9b) and is >6 Ma.

c. The type level of Zone N18 (ER 146-41) is in Zone PL 2

($\cong$3.5 to 3.6 Ma) and Zone NN15 (Aubry, this volume), whereas the paratype level (sample ER 143-7) is in Zone PL 1a (~5.0 Ma) and in the uppermost part of Zone NN11 (Aubry, this volume) indicating a temporal separation of about 1.5 m.y. between the two. *G. tumida,* nominate form of the zone, has not been observed in these, or any other Pliocene samples from Jamaica, except over a short interval spanning Zones PL 1b to PL 1c in the Ecclesdown Road section and a rare occurrence in sample ER 143-19 in the San San Bay section.

d. The type level of Zone N19 (ER 146-44) is equivalent to Zone PL 3 ($\cong$3 Ma; and is in Zone NN16, Aubry, this volume). The paratype sample, ER 300, is in Zone PL 1b ($\leq$4.7 Ma) (and NN13-14, Aubry, this volume) indicating a temporal separation of over 1.5 m.y. between the two.

e. The defining criterion of Zone N19 (FO *S. dehiscens* s.l.) occurs at a level biochronologically older ($\cong$5.1 Ma) than the type level of Zone N18 ($\cong$3.5 to 3.6 Ma). Thus the lower part of Zone N19 is older than that of Zone N18.

f. The type level of Zone N20 (ER 156) contains an anomalous fauna referable to the latest Pliocene (Bolli, 1970; Saito, 1985) and may be biostratigraphically younger than that of Zone N21.

I would suggest avoidance of Blow's (1969) Zones N18 to N21 for tropical upper Neogene biostratigraphy and use of one of several other currently available schemes. That of Berggren (1973, 1977b) is used here (but see comments by Bolli and Krasheninnikov, 1977).

SYSTEMATIC PALEONTOLOGY

Order Foraminiferida Eichwald, 1830
Suborder Globigerinina Delage and Hérouard, 1896
Superfamily Globorotaliacea Cushman, 1927
Family Globorotaliidae Cushman, 1927

Globorotalia archeomenardii Bolli, 1957
(Figure 4.1–6)

Globorotalia archeomenardii Bolli, 1957, p. 119, Plate 28, Figs. 11a–c. — Kennett and Srinivasan, 1983, p. 122, Plate 28, Figs. 3–5.

Relatively small, weakly keeled form that probably evolved from *G. praescitula* Blow. Differs from the latter in its more acute profile in edge view, keeled periphery, and in more finely perforate test. Sporadic distribution in lower part of Pots and Pan section of Buff Bay outcrop in association with *G. peripheroronda, G. peripheroacuta,* and *Orbulina universa,* inter alia.

Globorotalia menardii "A" Bolli, 1970
(Figure 5.13–15)

Globorotalia menardii "A" Bolli, 1970, p. 582, p. 5, Figs. 1–4. — Bolli and Saunders, 1985, p. 223, Figs. 34.1, 6, 10, 12.

The mid-Miocene (Zone NN11-13) flaring (C-type) chamber development in the menardine globorotaliids represents a major evolutionary innovation (see discussion in Cifelli and Scott, 1986, p. 35–39; Fig. 5) in the planktonic foraminifera. It was quickly adopted by a group of forms that formed the major late Neogene radiation, which exploited and adapted to the tropical realm.

The state of menardine taxonomy remains unsatisfactory. While most authors appear to recognize a long-ranging *Globorotalia menardii/ cultrata* stem form that has given rise to several other menardine forms (see Cifelli and Scott, 1986, Fig. 2), I believe that the modern *G. menardii/cultrata* is restricted to the late Pliocene–Pleistocene. While difficult, under some circumstances, to differentiate, I prefer to recognize a (predominantly) middle Miocene, five- to six chambered, distinctly keeled form possessing distinctly limbate intercameral sutures on the spiral side: *Globorotalia* sp. "A" of Bolli. In the late Miocene a larger, more robust, six- to seven-chambered form with attendant thicker peripheral keel and intercameral sutures is recognized; this is *Globorotalia* sp. "B" of Bolli = *Globorotalia limbata* (Fornasini) as lectotypified by Banner and Blow (1960) and used by several authors (Parker and Berggren, inter alia; see further discussion in Berggren and Amdurer, 1973, p. 361).

Globorotalia sp. "A" is the dominant consistently sinistrally coiled (menardine) globorotaliid in the Montpelier Formation and Buff Bay Formation. It overlaps with *G. limbata* in the uppermost part of the Buff Bay Formation (samples 241–243).

Globorotalia limbata (Fornasini), 1902 (ex d'Orbigny, 1826)
(Figure 5.16–30)

Rotalia limbata Fornasini, 1902, p. 56, Plate 55.
Globorotalia (Globorotalia) cultrata limbata (Fornasini). — Banner and Blow, 1960, p. 30, Plate 5, Fig. 3 (holotype).
Globorotalia sp. "B", Bolli, 1970, p. 582, Plate 5, Figs. 5–7. — Bolli and Saunders, 1985, p. 223, Figs. 34.2.
Globorotalia (Menardella) limbata (Fornasini). — Kennett and Srinivasan, 1983, p. 124, p. 29, Figs. 4–6.

This form appears in the upper part of the Buff Bay Formation (where it is sinistrally coiled) and ranges into the Bowden Formation (sample 245), where it is dextrally coiled. A change from sinistral to dextral coiling has been observed in this taxon in the Caribbean-Colombian Basin (DSDP site 502) within the Thvera Subchron (ca. 4.7 Ma) and in a deep-water section (Rivière Bois de Chêne) in Haiti, supporting an age estimate of <4.7 Ma for the younger surface of the unconformity separating the Buff Bay and Bowden Formations.

Globorotalia multicamerata Cushman and Jarvis, 1930
(Figure 6.29–31)

Globorotalia menardii (d'Orbigny) var. *multicamerata* Cushman and Jarvis, 1930, p. 367, Plate 34, Figs. 8a–c.
Globorotalia multicamerata Cushman and Jarvis.—Parker, 1967, p. 180, Plate 31, Figs. 5, 6. — Bolli, 1970, p. 582, Plate 7, Figs. 17–20. — Lamb and Beard, 1972, p. 54, 55, Plate 11, Figs. 4–6; Plate 12, Figs. 4–5; Plate 13, Figs. 6–8; Plate 14, Figs. 5–8. — Bolli and Saunders, 1985, p. 226, Figs. 32.5; 35.16–19.

Figure 4. All specimens from Buff Bay section. 1–6, *Globorotalia archeomenardii* Bolli, sample SG25; 7–12, *Globorotalia praemenardii* Bolli, sample SG25 (7–9) and sample SG16 (10–12); 13–15, *Globorotalia peripheroronda* Blow and Banner, sample SG25; 16–18, *Globorotalia peripheroacuta* Blow and Banner, sample SG25; 19–20, *Globorotalia fohsi* Cushman and Ellisor, sample SG26; 21–23, *Globorotalia lobata* Bermúdez, sample SG1; 24–26, *Globorotalia robusta* Bolli, sample SG17; 27–31, *Globorotalia lenguaensis* Bolli, sample 226 (27, 28) and sample 228 (29–31).

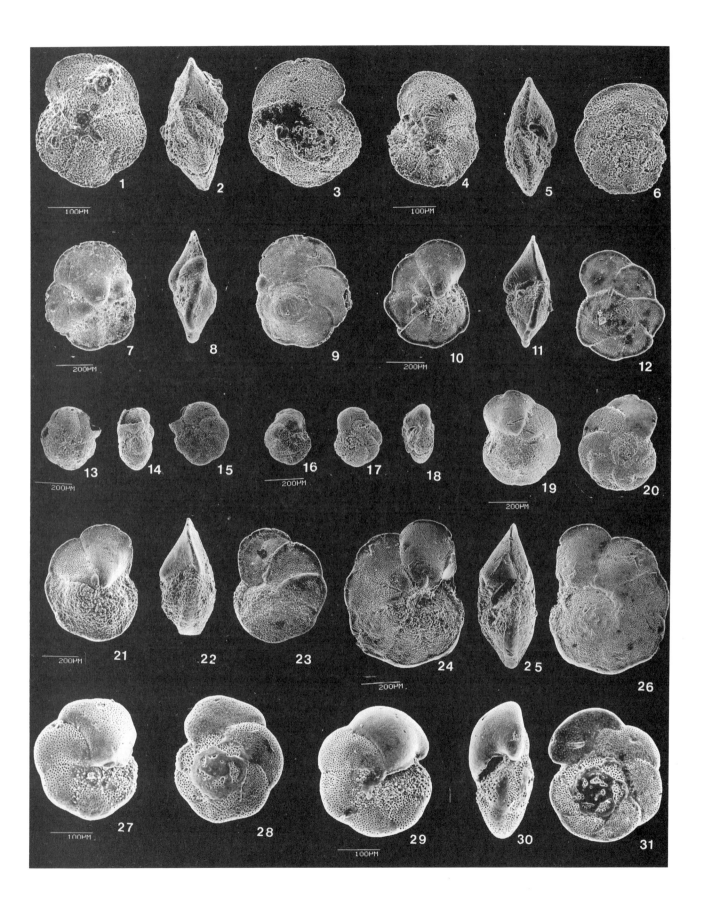

Globorotalia (Globorotalia) multicamerata Cushman and Jarvis. — Blow, 1969, p. 367–368, Plate 7, Figs. 7–9; Plate 42, Fig. 7.
Globorotalia (Menardella) multicamerata Cushman and Jarvis. — Kennett and Srinivasan, 1983, p. 126, Plate 29, Figs. 7–9.

This species is distinguished by its large, robust test, with seven to ten (rarely up to 12) chambers in last whorl, thick peripheral keel, and markedly limbate and recurved ("hockey stick") sutures on spiral side. *Globorotalia multicamerata* differs from *G. limbata,* from which it probably evolved in the early Pliocene, in its larger number of chambers and larger, thicker test.

This species occurs sporadically in the Pliocene of the Buff Bay section.

Globorotalia praemenardii Cushman and Stainforth, 1945
(Figure 4.7–12)

Globorotalia praemenardii Cushman and Stainforth, 1945, p. 70, Plate 13, Figs. 14a–c. — Bolli and Saunders, 1985, p. 223, Figs. 32–7 (holotype refigured).
Globorotalia (Menardella) praemenardii Cushman and Stainforth. — Kennett and Srinivasan, 1983, 0.122, Plate 28, Figs. 6–8. — Scott et al., 1990, p. 36, Figs. 24, 25 A–I.

This form is characterized by a five- to six-chambered, elongate-oval test and weakly to moderately developed keel. Scott et al. (1990, p. 38) suggested that tropical populations of *G. praemenardii* were ancestral to *G. menardii* (here interpreted as *G. menardii* "A" of Bolli), whereas mid-high (austral) latitude populations evolved into *G. miotumida.*

Globorotalia praemenardii occurs in the Spring Garden Member of the Montpelier Formation in the Pots and Pans section (samples SG25A to SG24).

Globorotalia pertenuis Beard, 1969
(Figure 6.22–28)

Globorotalia pertenuis Beard, 1969, p. 552, Plate 1, Fig. 1, Plate 2, Figs. 5–6. — Lamb and Beard, 1972, p. 55, Plate 14, Fig. 4; Plate 15, Figs. 1–6; Plate 16, Figs. 5–6; Plate 17, Figs. 5, 7. — Bolli and Saunders, 1985, p. 226, Figs. 35.3, 35.12.
Globorotalia (Menardella) pertenuis Beard. — Kennett and Srinivasan, 1983, p. 130, Plate 30, Figs. 7–9.

This relatively large lobulate, subcircular form is characterized by its thin delicate test, six to eight chambers in last whorl, and flaring, broadly lipped aperture.

Globorotalia pertenuis occurs in the Pliocene above the main unconformity at Buff Bay.

Globorotalia pseudomiocenica Bolli and Bermúdez, 1965
(Figure 6.13–15)

Globorotalia pseudomiocenica Bolli and Bermúdez, 1965, p. 140, Plate 1, Figs. 13–15. — Bolli and Saunders, 1985, p. 230, Figs. 33.1, 6, 10, 11.

Blow (1969, p. 359) indicated that *G. pseudomiocenica* is a junior synonym of *G. limbata,* that the holotype and lectotype of the two forms, respectively, are essentially "identical in all characters" and that *pseudomiocenica* is unrelated to *miocenica.* Similar views have been expressed by Berggren and Amdurer (1973, p. 361) and Kennett and Srinivasan (1983, p. 124). However, we have observed that in late Miocene–early Pliocene assemblages of menardine globorotaliids from the Caribbean region (DSDP site 502; Haiti) including the Buff Bay section of Jamaica, there are forms with relatively convex umbilical side, slightly elevated spiral side, and weakly lobate periphery. This morphotype can be observed to pass insensibly, by developing a flattened spiral side, into *G. miocenica* in the mid-Pliocene. The name *G. pseudomiocenica* is retained here for this morphotype, although its close relationship, including the same coiling patterns, to *G. limbata* is recognized. Bolli and Saunders (1985, p. 230) considered *G. pseudomiocenica* to have evolved from *Globorotalia* sp. "A" and to be the link between that form and *G. miocenica.* In the Buff Bay section *G. pseudomiocenica* occurs in all samples above the unconformity in the West Roadcut section.

Globorotalia miocenica Palmer, 1945
(Figure 6.19–21)

Globorotalia menardii (d'Orbigny) var. *miocenica* Palmer, 1945, p. 20, Plate 1, Fig. 10a–c. — Bolli and Saunders, 1985, p. 230, Figs. 33.2, 35.1.

This form is characterized by its perfectly flat spiral side and characteristically, but not exclusively, nearly circular and weakly lobate test outline. In the Colombian Basin (DSDP site 502) *G. miocenca* evolves from the *G. pseudomiocenica* morphotype in the early part of the Gauss Chron (ca. 3.2–3.3 Ma; personal observation).

In the Buff Bay section *G. miocenca* appears in sample 247 (ca. 4 m above the base) of the West Roadcut section.

Globorotalia plesiotumida Banner and Blow, 1965
(Figure 6.1–6)

Globoratalia (Globorotalia) plesiotumida Banner and Blow, 1965a, p. 1353, Figs. 2a–c. — Kennett and Srinivasan, 1983, p. 156, Plate 37, Figs. 7–9. — Malmgren et al., 1983, p. 378, Fig. II: 11–17. — Bolli and Saunders, 1985, p. 227, Figs. 33.5.

This form is characterized by its biconvex, six-chambered test. It is ancestral to the larger, more biconvex, thicker-walled *G. tumida* (Malmgren et al., 1983). Banner and Blow (1965a) suggested derivation of the *G. merotumida-plesiotumida-tumida* lineage from an unkeeled form *G. paralenguaensis* in the early late Miocene. I would agree with Bolli and Saunders (1985, p. 227 cf. Kennett and Srinivasan, 1983, p. 154) that this derivation is unlikely. *Globorotalia tumida* is more closely linked with *G. menardii* s.l. *Globorotalia merotumida* is fully keeled at an early ontogenetic stage, whereas *G. paralenguaensis* is unkeeled, although it possesses an imperforate peripheral band. The keel is essentially an adult character in modern globorotaliid taxa and thus the case for a non-menardine origin for the *G. tumida* stock would appear to be weak.

Globorotalia plesiotumida occurs in the upper part of the Buff Bay Formation exposed in the Dead Goat Gully section (samples 241 to 243) and in sample ER40 (see discussion in text above). Specimens are not fully typical of *G. plesiotumida,* and it was pointed out above that the problem may be in the fact that the *G. plesiotumida-tumida* group is a predominantly and characteristically Indo-Pacific group, which is less commonly found in the Atlantic-Caribbean region.

Figure 5. 1–3, *Globorotalia margaritae* Bolli and Bermúdez, sample 245; 4–6, *Globorotalia* sp., sample SG23A; compare this middle Miocene form with its early Pliocene homeomorph *G. margaritae,* Figure 2, 1–3 (above); 7–12, *Globoratalia scitula* Brady, sample SG6 (7–9) and sample 231 (10–12); 13–15, *Globorotalia menardii* "A" Bolli, sample 225; 16–30, *Globorotalia limbata* (Fornasini), sample 227 (16–18), sample 243 (19–21, unencrusted; 22–24, encrusted), sample 251 (25–27, unencrusted; 28–30, encrusted). Note coiling change from sinistral to dextral between Miocene (16–24) and Pliocene (25–30) forms.

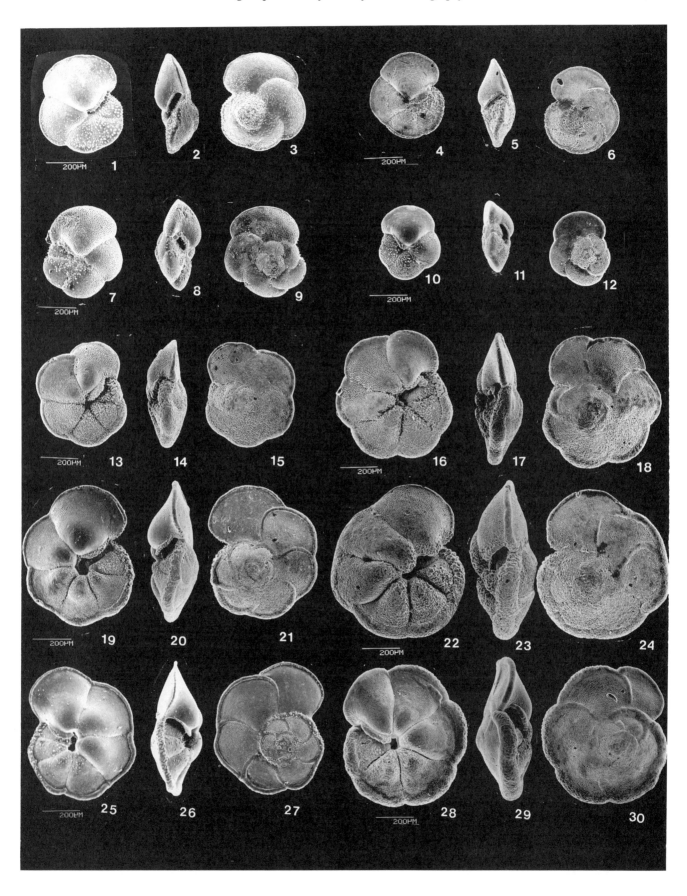

Globorotalia tumida (Brady), 1887
(Figure 6.7–12)

Pulvinulina menardii (d'Orbigny) var. *tumida* Brady, 1877, p. 535, no figures.
Pulvinulina tumida Brady, 1884, p. 692, Plate 103, Figs. 4–5.
Globorotalia tumida (Brady). — Banner and Blow, 1960, p. 26, Plate 5, Fig. 1a–c. — Bolli and Saunders, 1985, Plate 227, Figs. 33.8, 34.11–13.
Globorotalia (*Globorotalia*) *tumida* (Brady). — Kennett and Srinivasan, 1983, p. 158, Plate 36, Figs. 1, 2; Plate 38, Figs. 1–3.

This form is characterized by its large, inflated (tumid) test, evolute spire and thick keel. It evolved from *G. plesiotumida* near the Miocene/Pliocene boundary (Berggren and Poore, 1974; Malmgren et al., 1983).

This is a predominantly Indo-Pacific taxon and occurs only sporadically in the Atlantic-Caribbean region during the Pliocene. In fact I have observed this form only in a sample (ER 143-19) from the San San Clay at San San Bay and a few samples from the Ecclesdown Road section (Table 8) in an early Pliocene association with *Globorotalia margaritae*, *G. pseudomiocenica-limbata* (dextral), and *Sphaeroidinella dehiscens immatura,* inter alia. Its (apparent) absence at Buff Bay is enigmatic in as much as Blow (1969) used sample 244 from that section as the holotype sample of his *G. tumida* Zone.

Globorotalia birnageae Blow, 1959

Globorotalia birnageae Blow, 1959, p. 210, Plate 17, Fig. 108a–c.
Globorotalia (*Fohsella*) *birnageae* (Blow).—Kennett and Srinivasan, 1983, p. 94, Plate 21, Figs. 6–8.

This relatively small globorotaliid, is characterized by its nearly circular, and weakly lobate text. We would agree with Kennett and Srinivasan (1983, p. 94) that this form is closely related to *G. peripheroronda.*

In the Buff Bay section *G. birnageae* occurs sporadically in the Pots and Pans Roadcut section (sample 56 29A and 29), that is, in the Spring Garden number of the Montpelier Formation.

Globorotalia peripheroronda Blow and Banner, 1966
(Figure 4.13–15)

Globorotalia (*Turborotalia*) *peripheroronda* Blow and Banner, 1966, p. 294, Plate 1, Figs. 1a–c.
Globorotalia peripheroronda Blow and Banner. — Berggren, 1977a, p. 595, Plate 3, Figs. 14–16; cf. Plate 3, Figs. 11–13.
Globorotalia (*Fohsella*) *peripheroronda* Blow and Banner. — Kennett and Srinivasan, 1983, p. 96, Plate 22, Fig. 1–3.
Globorotalia fohsi peripheroronda Blow and Banner. — Bolli and Saunders, 1985, p. 213, Figs. 29.6, 14.
Fohsella peripheroronda (Blow and Banner). — Cifelli and Scott, 1986, p. 17, Figs. 8h–j, n, o.

Five- to six-chambered, elongate (in equatorial view) form with subangular, arched profile and low essentially flat ridged texture. The tendency to lose this ridged texture in the course of phyletic evolution (through *G. fohsi* and the *G. robusta-lobata* group) is seen also in the mid- to high latitudes lineage of *G. zealandica-G. praescitula-G. panda* (Berggren, 1992).

This taxon occurs relatively commonly in the Pots and Pans section, up to sample SG25.

Globorotalia peripheroacuta Blow and Banner, 1966
(Figure 4.16–18)

Globorotalia (*Turborotalia*) *peripheroacuta* Blow and Banner, 1966, p. 294, Plate 1, Figs. 2a–c.
Globorotalia (*Fohsella*) *peripheroacuta* Blow and Banner. — Kennett and Srinivasan, 1983, p. 96, Plate 22, Figs. 4–6.
Globorotalia fohsi peripheroacuta Blow and Banner. — Bolli and Saunders, 1985, p. 213. Figs. 29.5, 13.

The larger test size, smoother, more finely perforate and distinct peripheral compression of the terminal chamber(s) distinguishes this form from *G. peripheronda* and heralds the initiation of the relatively rapid sequential evolutionary sequence of the *G. fohsi* group.

This form has been observed over a 30-m interval (samples SG26 to SG24) in the Pots and Pans section.

Globorotalia fohsi Cushman and Ellisor, 1939
(Figure 4.19, 20)

Globorotalia fohsi Cushman and Ellisor, 1939, p. 12, p. 2, Figs. 6a–c. — Blow and Banner, 1966, Plate 1, Figs. 5–7, Plate 2, Figs. 8, 9, 12 (holotype refigured, Figs. 5–7). — Bolli and Saunders, 1985, p. 213, Figs. 29.4, 12.
Globorotalia (*Fohsella*) *fohsi fohsi* Cushman and Ellisor. — Kennett and Srinivasan, 1983, p. 100, Plate 23, Figs. 1–3.
Fohsella fohsi (Cushman and Ellisor). — Cifelli and Scott, 1986, p. 18, Figs. 8a–c.

The larger test size, increased peripheral compression and acquisition of a peripheral keel distinguish this form from *G. peripheroacuta.* The reader is referred to Bolli and Saunders (1985, p. 213, 214) and Cifelli and Scott (1986, p. 18) for discussion of problems differentiating *fohsi* and *praefohsi* (not recognized here).

This form occurs sporadically in the upper part of the Pots and Pans section; its highest occurrence is noted in sample SG23A at the top of that section within the Spring Garden Member of the Montpelier Formation.

Globorotalia lobata Bermúdez, 1949
(Figure 4. 21–23)

Globorotalia lobata Bermúdez, 1949, p. 286, p. 22, Figs. 15–17.
Globorotalia fohsi forma *lobata* Bermúdez. — Blow and Banner, 1966, text Fig. 4, Figs. 1–3 (holotype and paratype refigured).
Globorotalia fohsi lobata Bermúdez. — Bolli and Saunders, 1985, Figs. 29.2 9.
Globorotalia (*Fohsella*) *lobata* Bermúdez. — Kennett and Srinivasan, 1983, p. 100, Plate 21, Fig. 2; Plate 23, Figs. 4–6.

Distinguished by its distinctly larger sized test and typical undulating cockscomb-shaped peripheral margin. Blow and Banner (1966), Blow (1969), and Cifelli and Scott (1986) regard *lobata* and *robusta* as phenotypic variants, whereas Kennett and Srinivasan (1983) consider them valid phylogenetic species. I have retained them as separate entities because of their distinctive morphologies (see below).

Figure 6. 1–6, *Globorotalia plesiotumida* Blow and Banner, sample 229 (1–3, unencrusted), sample 228 (4–6, encrusted); 7–12, *Globorotalia tumida* (Brady); 13–15, *Globorotalia pseudomiocenica* Bolli and Bermúdez, sample 244; 16–18, *Globorotalia* sp. ex interconnec *G. pseudomiocenica-G. miocenica,* sample 245; 19–21, *Globorotalia miocenica* Palmer, sample 249; 22–28, *Globorotalia pertenuis* Beard, sample 244 (22–25) and sample 248 (26–28); 29–30, *Globorotalia multicamerata* Cushman and Jarvis, sample 244.

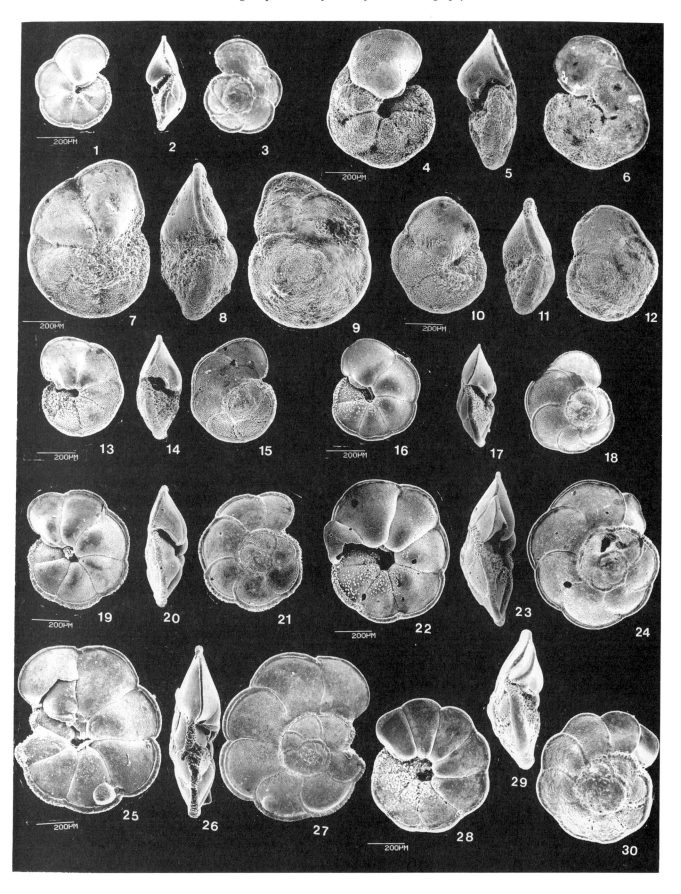

Globorotalia lobata occurs in the lower 15 m of the Middle Roadcut section (up to sample SG20).

Globorotalia robusta Bolli, 1950
(Figure 4. 24–26)

Globorotalia fohsi robusta Bolli, 1950, p. 89, Plate 15, Figs. 3a–c. — Bolli and Saunders, 1985, p. 215, Figs. 29.1, 7.
Globorotalia (Fohsella) fohsi lobata Bermúdez. — Kennett and Srinivasan, 1983, p. 102, p. 23, Figs. 7–9.
Fohsella robusta (Bolli). — Cifelli and Scott, 1986, p. 18, Figs. 8d–e.

This is a relatively large, robust form, characterized by a test with seven to eight chambers developed in a circular outline, and represents the end member of the *G. fohsi* phyletic branch. In Buff Bay outcrop it ranges as high as sample SG12, about 45 m above the base of the Middle Roadcut section.

Globorotalia lenguaensis Bolli, 1957
(Figure 4. 27–31)

Globorotalia lenguaensis Bolli, 1957, p. 120, Plate 29, Figs. 5a–c. — Berggren, 1977a, p. 595, Plate 6, Figs. 1–13. — Berggren, 1977b, p. 308, Plate 4, Figs. 22, 23. — Kennett and Srinivasan, 1983, p. 152, Plate 36, Figs. 5–7. — Bolli and Saunders, 1985, p. 215, Fig. 30.25. — Cifelli and Scott, 1986, p. 44, Figs. 18a, e.

This form is characterized by its relatively small size, circular and only slightly lobate outline, compressed test and thickened, imperforate peripheral rim. Relatively large pores occur on the spiral side. This form appears (in sample 173) a short distance (about 3 m) above the FAD of *Globoturborotalita nepenthes* in the lower part of the Main Roadcut section at Buff Bay, (i.e., within the lower part of Zone N14). Small, indeterminate forms, similar but not referable to *G. lenguaensis* occur in the lower part of the Buff Bay outcrop. It ranges to the unconformity (between samples 243 and 244) that separates the Buff Bay and Bowden Formations. It is one of the most distinctive elements of the late middle to early late Miocene faunas becoming more abundant in younger parts of the Miocene section. In the Gulf of Mexico (Eureka core holes) I have found that the LO of *G. lenguaensis* occurs close to, but above, the FO of *G. plesiotumida* at a level estimated at about 8 Ma. The LO of *G. lenguaensis* may serve, at least locally, as a proxy for the approximate recognition of a level equivalent to the N16/N17 boundary.

Globorotalia margaritae Bolli and Bermúdez, 1965
(Figure 5.1–3)

Globorotalia margaritae Bolli and Bermúdez, 1965, p. 132, Plate 1, Figs. 16–18. — Blow, 1969, p. 363, Plate 64, Figs. 1–6; Plate 45, Figs. 1–3, 5, 6. — Berggren, 1977a, p. 596, Plate 7, Figs. 17–20. — Berggren, 1977b, p. 308, Plate 4, Figs. 17–19. — Bolli and Saunders, 1985, p. 217, Plate 30.9–14. — Cifelli and Scott, 1986, p. 43, Figs. 12i, j.
Globorotalia (Hirsutella) margaritae Bolli and Bermúdez, 1965. — Kennett and Srinivasan, 1983, p. 136, p. 32, Figs. 4–6.

The arcuate spiroconvex profile and well-developed flat keel are the characteristic features of this cosmopolitan taxon widely used in late Neogene biostratigraphy although its total range remains uncertain. A recent compilation by Hodell and Kennett (1986) suggests that this form does indeed have an initial occurrence in the late Miocene (during the interval of ca. 6 to 5 Ma), as indicated initially by Hays et al. (1969) and supported by Berggren et al. (1985), that is, within magnetopolarity Chron 5. Its LO appears to occur within the late Gilbert Chron (ca. 3.4 to 3.5 Ma; see discussion in Channell et al., 1990, p. 690).
Globorotalia margaritae occurs just above the unconformity at the base of the Bowden Formation and ranges up (ca. 3 m) to sample 245, where the boundary between Zones PL 2 and PL 3 is drawn.

Globorotalia scitula Brady, 1882
(Figure 5.7–12)

Pulvinulina scitula Brady, 1882, p. 716; Brady, 1884, Plate 103, Figs. 7a–c (given as *Pulvinulina patagonica* d'Orbigny).
Globorotalia (Turborotalia) scitula (Brady). — Banner and Blow, 1960, p. 27, Plate 5, Figs. 5a–c (lectotype).
Globorotalia (Hirsutella) scitula (Brady). — Kennett and Srinivasan, 1983, p. 134, Plate 31, Figs. 1, 3–5.
Globorotalia scitula scitula (Brady). — Bolli and Saunders, 1985, p. 217, Figs. 30.26–29, 31.3–4.

This form is characterized by a subquadrate, densely perforate test with a thickened, imperforate rim.

In the Buff Bay section *G. scitula* occurs sporadically in the upper part of the Buff Bay Formation (in the Landslide and Dead Goat Gully samples).

Globorotalia crassaformis, Galloway and Wissler, 1927
(Figure 7.1–6)

Globigerina crassaformis, Galloway and Wissler, 1927, p. 41, Plate 7, Fig. 12.
Globorotalia Turborotalia crassaformis (Galloway and Wissler). — Blow, 1969, p. 347; Plate 4, Figs. 1–3; Plate 37, Figs. 104.
Globorotalia (Truncorotalia) crassaformis (Galloway and Wissler). — Kennett and Srinivasan, 1983, p. 146, p. 34, Figs. 6–8.
Globorotalia crassaformis (Galloway and Wissler). —Bolli and Saunders, 1985, p. 233, Figs. 36. 6–7; Cifelli and Scott, 1986, p. 49, Fig. 20 a–i. — Scott et al., 1990, p. 91, Figs. 61, 62A–L.

Discoidal to weakly planoconvex generally unkeeled test with subquadrate outline. Various morphotypes differentiated on the degree of compactness, axial angularity, and wall thickness (*hessi, ronda, oceanica*) are treated here as ecophenotypic variants (see also Kennett and Srinivasan, 1983, p. 146). The *G. crassaformis* lineage would appear to have evolved from a scituline ancestor of the *cibaoensis-juanai* group (Berggren, 1977a, b; Kennett and Srinivasan, 1983; Cifelli and Scott, 1986) in the early Pliocene. Blow (1969, p. 328, 390) regarded the origin of the *G. crassaformis* group to occur in the middle part of his Zone N16 with the first appearance of a form referred to as *G. crassaformis* s.l., which differentiated rapidly into *crassaformis* s.s. and *crassaformis oceanica*. In a postscript (1969, p. 220) published in the original work and a subsequent postscript (1979, p. 418) written three years after the completion of his original manuscript (but not published until a decade later). Blow indicated that these early morphotypes of *crassaformis* s.l. were actually referable to *G. (T.) subscitula*. The FO of *G. crassaformis oceanica* was considered to occur within the mid-part of Zone N17. Indeed the FO of the *G. crassaformis* s.l. group was considered to denote

---

Figure 7. 1–6, *Globorotalia crassaformis*, samples 249 (1–3) and 251 (4–6); 7–9, *Globorotalia crassula*, sample 244; 10–11, *Neogloboquadrina acostaensis*, sample 229; 12–14, *Neogloboquadrina humerosa*, sample 244; 15–18, *Neogloboquadrina continuosa*, samples SG25 (15–16) and 168 (17–18); 19–24, *Paragloborotalia mayeri*, samples SG6 (19–20), 174 (21–22), and 181 (23–24).

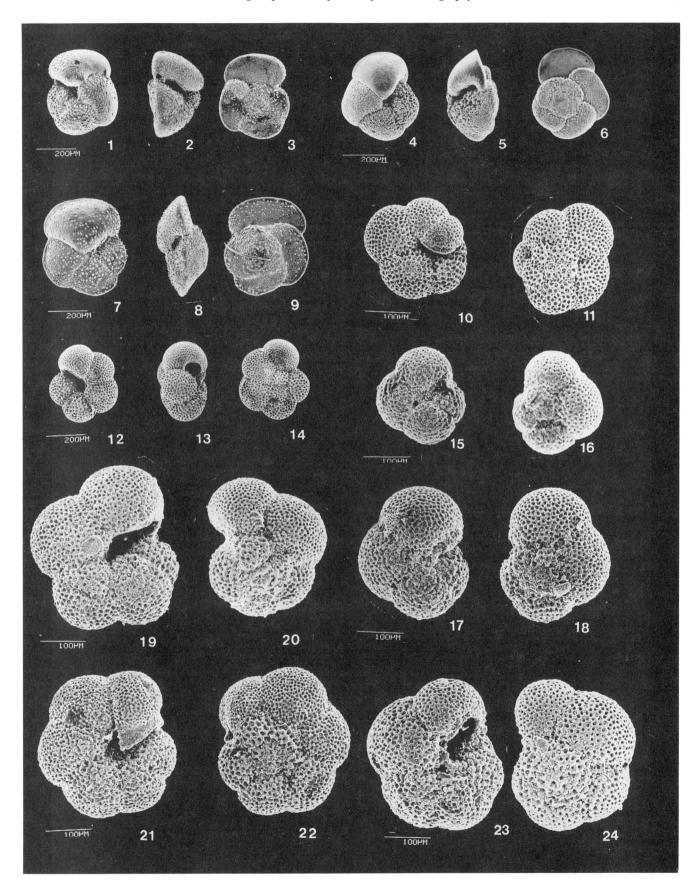

a useful guide to post–Zone N16 horizons. An indication of Blow's confusion of the stratigraphic level of the FO *G. crassaformis* group may be seen in the fact that he considered *crassaformis* to be ancestral to *G. crassula conomiozea* in Zone N17 (Blow, 1969, p. 360). In actual fact these two taxa are never seen together in late Miocene (Zone N17 and equivalent) faunas.

Subsequent studies by numerous investigators (see citations above) have shown that the initial occurrence of *G. crassaformis* is within the lower Pliocene and not in the upper Miocene. In the Buff Bay section *G. crassaformis* occurs immediately above the unconformity at the base of the West Roadcut section and ranges to the top of the exposure).

Globorotalia crassula Cushman and Stewart, 1930
(Figure 7.7–9)

Globorotalia crassula Cushman and Stewart, 1930, p. 77, Plate 7, Figs. 1 a–c. — Cifelli and Scott, 1986, p. 53, Figs. 22 a–d, f–i.
Globorotalia hirsuta (d'Orbigny) subsp. *aemiliana,* Colalongo and Sartoni, 1967, p. 267, text-Fig. 2 (holotype), p. 30, Figs. 1 (holotype), 1–5; Plate 31, Figs. 2–4.
Globorotalia crassacrotenensis Conato and Follador, 1967, p. 557, text-Fig. 2, Pl. 4, Fig. 3 a–c (holotype) (objective junior synonym of *G. aemiliana,* see Ellis and Messina, suppl. for 1972, no. 1; Berggren and Amdurer, 1973, p. 363).
Globorotalia crotonensis Conato and Follador, 1967, p. 556, text-Fig. 1, Figs. 4 (1 a–c: holotype), 2 a–c.
Globorotalia (Globorotalia) crassula Cushman and Stewart. — Blow, 1969, p. 361, Plate 9, Figs. 1–3 (holotype refigured).
Globorotalia (Truncorotalia) crassula Cushman and Stewart. — Kennett and Srinivasan, 1983, p. 144, p. 34, Figs. 3–5.

Low conical, distinctly keeled, subquadrate form with four subrectangular chambers in final whorl. The rather intricate taxonomic problems presented by the virtually simultaneous publication of the virtually identical Mediterranean forms *aemiliana, crassacrotenensis,* and *crotonensis* in 1967 are discussed more fully by Berggren and Amdurer (1973, p. 363–365) and the objective synonymy of *crassacrotenensis* with *aemiliana* pointed out in Ellis and Messina, suppl. for 1972, no. 1 (under entry for *G. crassacrotonensis* Conato and Follador). These forms are considered ecophenotypic variants of *G. crassula* which is, in turn, interpreted as closely related to, but distinct from, *G. crassaformis* (see also Cifelli and Scott, 1986, p. 54; and Scott et al., 1990, p. 104; Hornibrook et al., 1989, p. 23, consider *crassula* a synonym of *crassaformis*). In the Buff Bay section *G. crassula* occurs only at the top of the West Roadcut section (sample 252).

Neogloboquadrina Bandy, Frerichs and Vincent, 1967

Type Species. Globigerina dutertrei d'Orbigny, 1839.

Kennett and Srinivasan (1983, p. 190) argue for the separation of *Paragloborotalia* and *Neogloboquadrina* with which I would agree. However, they state that the two genera are phylogenetically unrelated. This seems unlikely, however, if, as they suggest, the earliest member of the neogloboquadrinid lineage is *N. continuosa* which, in turn, developed from *"Gr." nana* in the early Miocene. *Paragloborotalia nana* is a spinose, phenotypic (size) variant of *Paragloborotalia opima* (see discussion in Bolli and Saunders, 1985, p. 202). Blow (1969, p. 347) observed that *continuosa* and *nana* have different wall structures and textures and considered them homeomorphic rather than phylogenetically related. At the same time *G. continuosa* was considered a four-chambered variant of *G. mayeri* by Bolli and Saunders (1982, p. 49; Bolli and Saunders, 1985, p. 206). The phylogenetic ancestry/derivation of *Neogloboquadrina* remains enigmatic.

Neogloboquadrina acostaensis (Blow), 1959
(Figure 7.10–11)

Globorotalia acostaensis Blow, 1959, p. 208, Plate 17, Figs. 106 a–c. — Bolli and Saunders, 1985, p. 210, Figs. 27.10–11; 28.16–24.
Neogloboquadrina acostaensis (Blow). — Kennett and Srinivasan, 1983, p. 196, Plate 47. Figs. 1; Plate 48, Figs. 1–3.

Bolli and Saunders (1985, p. 210) have noted the trend towards increase from four to five chambers concomitant with an increase in size in this taxon in the Bodjonegoro 1 well of Java (which I have observed also in tropical to warm subtropical regions) but then make the curious observation that it seems to intergrade with *N. pachyderma* in upper Miocene to lower Pliocene sediments of temperate to polar regions. I believe that (at least part of) this intergradation may refer to *Neogloboquadrina nympha* (Jenkins), a distinct, but smaller form common in the austral regions of the Southern Hemisphere.

The species occurs in the upper part of the Buff Bay Formation, but does not occur in the Bowden Formation, above the unconformity. We have found a couple of specimens in sample 228, just *below* a 2-m-thick debris-flow bed.

Above this, *N. acostaensis* occurs in low numbers as a persistent component of the upper Miocene faunas. It is for this reason that we ascribe the unconformity at the base of the debris-flow bed to a level within Zone N16, rather than within Zone N17.

Neogloboquadrina humerosa (Takaganagi and Saito), 1962
(Figure 7.12–14)

Globorotalia humerosa Takayanagi and Saito, 1962, p. 78, p. 28, Figs. 1a–c. — Bolli and Saunders, 1985, p. 211, Figs. 27.8, 28.15.
Neogloboquadrina humerosa (Takayanagi and Saito). — Kennett and Srinivasan, 1983, p. 1961, p. 48, Figs. 4–6.

The larger test size and larger number of chambers (six to seven) serve to distinguish this form from its ancestor, *N. acostaensis.* In the Buff Bay, this form was observed in a single West Roadcut sample, 246 (Zone PL3, lower Pliocene).

Neogloboquadrina continuosa (Blow), 1959
(Figure 7.15–18)

Globorotalia opima Bolli subsp. *continuosa* Blow, 1959, p. 218, Plate 19, Figs. 125a–c.
Globorotalia (Turborotalia) continuosa Blow. — Blow, 1969, p. 347, Plate 3, Figs. 4–6 (holotype refigured).
Neogloboquadrina continuosa (Blow). — Kennett and Srinivasan, 1983, p. 192, Plate 47, Figs. 3–5.
Globorotalia continuosa Blow. — Bolli and Saunders, 1985, p. 206, Figs. 26.8–14.

This relatively small form is distinguished by its four-chambered, cancellate test and distinctly rimmed, high-arched aperture. Generally interpreted as the ancestor of *N. acostaensis* (Blow, 1959, 1969; Kennett and Srinivasan, 1983), Bolli and Saunders (1985) considered it a four-chambered variant of *"Globorotalia" mayeri.*

In Jamaica this form occurs sporadically in the Spring Garden Member and lower part of the Buff Bay Formation.

Paragloborotalia Cifelli, 1982

Type Species. Globorotalia opima opima Bolli.

The genus *Jenkinsella* Kennett and Srinivasan, 1983, is considered a junior synonym of *Paragloboratalia* Cifelli, 1982 (see also Loeblich and Tappan, 1988, p. 476, 477). The latter authors, it should be noted, included both the non-spinose *Neogloboquadrina* and the spinose *Paragloborotalia* in the non-spinose family Globorotaliidae.

Paragloborotalia mayeri (Cushman and Ellisor), 1939
(Figure 4.19–24)

Globorotalia mayeri Cushman and Ellisor, 1939, p. 11, Plate 2, Figs. 4 a–c. — Bolli and Saunders, 1982, p. 39, Plate 1, Figs. 7–15 (holotype), 16–18; plate 2, Figs. 1–47; Plate 3, Figs. 1–43.
Globorotalia mayeri LeRoy, 1939, p. 262, Plate 4, Figs. 20–22. — Bolli and Saunders, 1982, p. 39, Plate 1, Figs. 1–6 (holotype). — Kennett and Srinivasan, 1983, p. 172, Plate 42, Figs. 1, 6–8.

Bolli and Saunders (1982, 1985) have argued that differentiation between *Globorotalia siakensis* LeRoy and *G. mayeri* Cushman is not possible on either morphologic or stratigraphic grounds, whereas Kennett and Srinivasan (1983, p. 172) and Iaccarino (1985, p. 309) have argued for their separation. We have not distinguished consistently between these two forms here and have used the designation *"P. mayeri/siakensis"* in the discussion part of this paper. The name *mayeri* has preference by publication date and we use this name here in the taxonomic section. This form occurs commonly in the Buff Bay section from (at least) sample SG29A (about 25 m above the base of the Pots and Pan section) to sample 189 (Main Roadcut section).

Candeininae Cushman, 1927

Candeina d'Orbigny, 1839

Type species. Candeina nitida d'Orbigny, 1839.

Candeina nitida d'Orbigny, 1839
(Figure 8.1–4)

Candeina nitida d'Orbigny, 1839, p. 107, Plate 2, Figs. 27–28. — Bolli et al., 1957, p. 35, Plate 6, Figs. 1-a–11. — Blow, 1969, p. 335, Plate 23, Figs. 1–4. — Kennett and Srinivasan, 1983, p. 229, Plate 57, Figs. 6–8.

High trochospiral, microperforate test with numerous small, secondary sutural apertures on later chambers.

At Buff Bay this form was found only in the Dead Goat Gully section (samples 241 to 243) that is, between the two unconformities and within the Zone N17 interval.

Polyperibola christiana Liska, 1980
(Figure 8.5–6)

Polyperibola christiana Liska, 1980, p. 137, Plate 1, Figs. 1–3. — Bolli and Saunders, 1985, p. 188, Figs. 18.1–2.

This taxon is distinguished by a globular, lobulate, distinctly perforate test, a small circular bullae, and sutural apertures.

The general shape of the test and the minute sutural apertures suggests a relationship with *Candeina nitida* (see also Bolli and Saunders, 1985, p. 188, 189), although it is distinctly more perforate. This form apparently has its initial occurrence in Zone N16 (Bolli and Saunders, 1985) and it was found here only in Sample 228 from which the FO of *N. acostaensis* is also recorded.

Dentoglobigerina Blow, 1979

Type species. Globigerina galavisi Bermúdez, 1961.
Blow (1979, p. 1294) created the family Globoquadrindae for forms in which the primary aperture is either interiomarginal, intraumbilical or interiomarginal, umbilical-extraumbilical, or exhibiting a combination of both during ontogeny, in which a portical tooth is present and in which the test is spinose and cancellate with mural pores opening into pore pits. He included the following genera in this new family: *Dentoglobigerina* n. gen; *Globigerinita* (primarily the catapsydracid types), *Globoquadrina*, and *Globorotaloides*. *Dentoglobigerina* n. gen. (type species: *Globigerina galavisi* Bermúdez) was created for cancellate forms with an interiomarginal, intraumbilical primary aperture that bears a portical umbilical tooth. Banner (1982, p. 200) removed *Dentoglobigerina* from the Globoquadrinidae, placed it in the (presumably non-spinose) porticated Globigerininae, and included *Globoquadrina* in the spinose Globorotaloidinae (= Globoquadrinidae), whereas *Catapsydrax* was included in the spinose Catapsydracinae. These placements by Blow (1979) and Banner (1982) are difficult to reconcile with the generally accepted knowledge that *Globoquadrina* and *Catapsydrax* are non-spinose (Parker, 1962; Loeblich and Tappan, 1984, 1988; Hemleben et al., 1989. Blow (1979, p. 1300) included in *Dentoglobigerina* essentially all forms previously assigned to *Globoquadrina* (*altispira, baroemoenensis, venezuelana,* etc.) except for the genotype, *G. dehiscens*, which he retained for *Globoquadrina*, as emended. *Dentoglobigerina* was said to have been derived from *Subbotina triangularis* (a spinose form); it is not known at present whether *D. galavisi* is spinose, but based on similarity with other subbotinids I would expect it to be and would suggest that spinosity of the test is lost in younger members of the lineage (? *pseudovenezuelana, tripartita, venezuelana*).

Kennett and Srinivasan (1983, p. 178) have restricted *Dentoglobigerina* to the evolutionary lineage leading from *galavisi* to *globularis* to *altispira globosa* to *altispira* s.s. They pointed out that *Dentoglobigerina* is inappropriate for forms (*venezuelana, tripartita/sellii, binaiensis, praedihescens,* and *dehiscens*) which are part of a separate phyletic lineage. One could question this conclusion by Kennett and Srinivasan (1983, p. 178) in that Blow (1979, p. 1298–1316) considered *galavisi* as ancestral to both the *altispira-globosa* branch and to the *tripartita-praedihescens-dehiscens* branch, whereas *winkleri-pseudovenezuelana* and *tapuriensis-selli-binaiensis* were all considered short-ranging offshoots of *galavisi*. Thus the *selli/tripartita galavisi-globularis* and *baroemoenensis-larmeui* lineages were all considered rooted in *Dentoglobigerina galavisi* (see also Blow, 1969, p. 339, where the distinction between *Globoquadrina* and *Dentoglobigerina* was presaged) and as such I would disagree with Kennett and Srinivasan (1983, p. 178) that splitting of the *tripartita-praedihescens-dehiscens* lineage into two separate genera *Dentoglobigerina* and *Globoquadrina* is "nonphyletic and hence artificial." I have retained the name *Globoquadrina* here for forms with an umbilical-umbilical primary aperture with porticated tooth, and a non-spinose wall (*dehiscens*), which is believed to have evolved from one of several lineages rooted in *Dentoglobigerina galavisi*. All other forms discussed above are included in *Dentoglobigerina*.

Dentoglobigerina altispira (Cushman and Jarvis), 1936
(Figure 8.8–11)

Globoquadrina altispira altispira (Cushman and Jarvis). — Bolli, 1957, p. 111, Plate 24, Figs. 7a–8b. — Berggren, 1977a, p. 294, Plate 1, Figs. 25–27. — Berggren, 1977b, p. 595, Plate 2, Figs. 7–9. — Kennett and Srinivasan, 1983, p. 188, Plate 46, Figs. 4–6. — Bolli and Saunders, 1985, p. 183, Figs. 15.1–3.

Characterized by its high trochospiral test and strongly porticated umbilical teeth in the umbilical region. This is one of the most distinct late Neogene taxa in tropical assemblages.

This form ranges essentially throughout the section exposed at Buff Bay except in the lower part of the Pots and Pans section.

Dentoglobigerina baroemoenensis (LeRoy), 1939

Globigerina baroemoenensis LeRoy, 1939, p. 263, Plate 6, Figs. 1, 2.
Globoquadrina baroemoenensis (LeRoy). — Blow, 1969, p. 340, Plate 27, Figs. 4, 8, 9. — Berggren et al., 1983, p. 708, Plate 1, Fig. 12. — Kennett and Srinivasan, 1983, p. 186, Plate 46, Figs. 1–3.
Globoquadrina langhiana Cita and Gelati, 1960, p. 242, Plate 29, Figs. 1–20.

Characterized by its subquadrate test, four to four and a half chambers and distinctly porticated, interiomarginal, umbilical aperture, this form is considered to have evolved from *D. galavisi* in the mid-Oligocene (i.e., prior to *Globoquadrina dehiscens*). At Buff Bay typical individuals were observed only in the Pots and Pans Roadcut (samples SG26 to SG23A).

Dentoglobigerina globosa (Bolli), 1957
(Figure 8.12–15)

Globoquadrina altispira globosa Bolli, 1957, p. 111, Plate 24, Figs. 9a–10c. — Bolli and Saunders, 1985, p. 183, Fig. 15.3.
Dentoglobigerina altispira globosa (Bolli). — Kennett and Srinivasan, 1983, p. 189, Plate 44, Fig. 4; Plate 46, Figs. 7–9.

Distinguished from *D. altispira* primarily by its low trochospiral and less inflated chambers.

This form occurs irregularly over the interval from the upper part of the Middle Roadcut (Sample SG6) to the top of the Landslide section (sample 239). It was not observed in the Pliocene (West Roadcut section).

Dentoglobigerina venezuelana (Hedberg), 1937
(Figure 8.16–19)

Globigerina venezuelana Hedberg, 1937, p. 68, p. 92, Figs. 7a–b.
Globoquadrina venezuelana (Hedberg). — Kennett and Srinivasan, 1983, p. 180, Plate 44, Figs. 5–7.

The lobulate periphery, four to five globular chambers, and relatively wide umbilicus (which is frequently obscured by a relatively large and enveloping final chamber) characterize this form. Although often placed in *Globigerina,* its non-spinose, distinctly cancellate test indicates affinities with the *Dentoglobigerina-Globoquadrina* group. Although Blow (1979, p. 1300–1317) did not mention *venezuelana* in his discussion of *Dentoglobigerina,* he earlier (1969, p. 322, 323) indicated that *venezuelana* was probably descended from the *galavisi–altispira globularis* plexus (and not from either *pseudovenezuelana* or *prasaepis* (= *euapertura*). This would suggest that Blow would have considered *venezuelana* to be referable to *Dentoglobigerina*; its interiomarginal, umbilical aperture supports this assignment also.

This is the most common and persistent form in the Buff Bay section occurring in essentially every sample examined. It ranges to the top of the section exposed in the West Roadcut (i.e., mid-Pliocene, Zone PL 3/4, ca. 3 Ma).

Globoquadrina Finlay, 1947

Type species. Globorotalia dehiscens Chapman, Parr, and Collins, 1934.

Globoquadrina dehiscens (Chapman, Parr, and Collins), 1934
(Figure 9.1–3)

Globorotalia dehiscens Chapman, Parr, and Collins, 1934, p. 569, Plate 11, Figs. 36a–c.
Globoquadrina sp. Bolli, Loeblich, and Tappan, 1957, p. 31, Plate 5, Fig. 6.
Globoquadrina dehiscens (Chapman, Parr, and Collins). — Bolli et al., 1957, p. 31, Plate 5, Figs. 5a–e. — Blow, 1969, p. 341, Plate 29, Fig. 1. — Berggren, 1977a, p. 595, Plate 2, Figs. 4–6. — Berggren, 1977b, p. 296, Plate 1, Figs. 19–23. — Blow, 1979, p. 1353, Plate 29, Fig. 1 (pt. 1). — Kennett and Srinivasan, 1983, p. 184, Plate 44, Fig. 2; Plate 45, Figs. 7–9. — Bolli and Saunders, 1985, p. 183, Figs. 15.4–7.

Genotype of *Globoquadrina,* this taxon is characteristic of and apparently restricted to the Miocene (Hodell and Kennett, 1986). It is characterized by its robust, (sub)quadrate test, triangular-cuneiform shaped chambers on the umbilical side, and interiomarginal umbilical-extraumbilical aperture. For a detailed description of this taxon reference is made to Blow (1979, p. 1353–1355). In the Buff Bay section, *G. dehiscens* is restricted to the mid-part of the sequence (samples 5G6 to 199), that is, the middle Miocene.

Globorotaloides Bolli, 1957

Type species. Globorotaloides variabilis Bolli, 1957.

Globorotaloides hexagonus (Natland), 1935
(Figure 9.4–5)

Globigerina hexagona Natland, 1938, p. 149, Plate 7, Figs. 1a–c.
Globorotaloides hexagona (Natland). — Kennett and Srinivasan, 1983, p. 216, Plate 54, Figs. 1, 3–5.

This taxon is characterized by a compressed five-chambered, strongly cancellate test, and pronounced apertural plate or rim.

In the Buff Bay section this form occurs sporadically, but in relatively high numbers, in the lower part (Pots and Pans section) of the Spring Garden Member of the Montpelier Formation.

Globigerinacea Carpenter, Parker and Jones, 1862
Globigerinidae Carpenter, Parker and Jones, 1862
Globigerininae Carpenter, Parker and Jones, 1862

Globigerinella Cushman, 1927

Type species. Globigerina aequilateralis Brady, 1879.

Globigerinella aequlateralis (Brady), 1879
(Figure 9.9–11)

Globigerina aequilateralis Brady, 1879, p. 285; Brady, 1884, Plate 80, Figs. 18–21.
Globigerinella aequilateralis (Brady). — Kennett and Srinivasan, 1983, p. 238, Plate 59, Fig. 1; Plate 60, Figs. 4–6.
Hastigerina siphonifera (d'Orbigny). — Bolli and Saunders, 1985, p. 251, Figs. 42.1–4, 43.1–2 (?non d'Orbigny).

The taxonomic history of this group is complex. Banner and Blow (1960) selected and synonymized two virtually identical specimens to lectotypify and synonymize *G. siphonifera* d'Orbigny and *G. aequilateralis* Brady. Todd (1964) and Saito et al., (1976) have called attention to the fact that while the lectotypes are probably conspecific, the one se-

---

Figure 8. 1–4, *Candeina nitida,* sample 227; 5–7, *Polyperibola christiani,* sample 228; 5 and 6 are different views of same individual; 8–11, *Dentoglobigerina altispira,* samples 231 (8–9) and 248 (10–11); 12–15, *Dentoglobigerina globosa,* samples 561 (12–13) and 199 (14–15); 16–19, *Dentoglobigerina venezuelana,* sample 248.

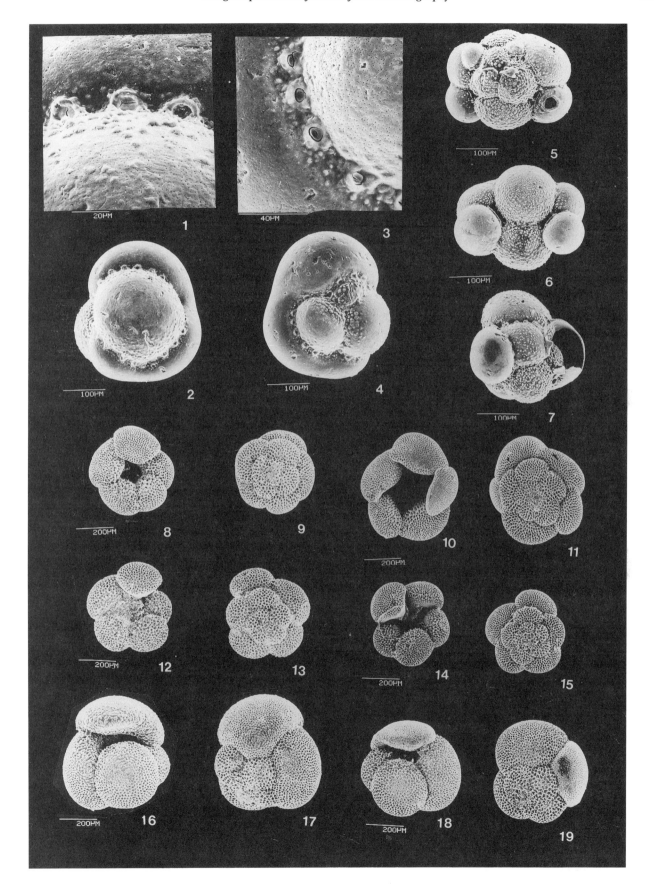

lected for *siphonifera* differs significantly from d'Orbigny's original illustration and suggested retention of the name *aequilateralis,* which was followed by Loeblich and Tappan (1988). Bolli and Saunders (1985, p. 251) distinguished between more tightly coiled forms (*siphonifera,* as lectotypified by Banner and Blow, 1960) and more evolute forms (*aequilateralis* Brady, non Banner and Blow). Banner and Blow added to the confusion by designating the more involute form *Hastigerina siphonifera* (d'Orbigny) despite the fact that they state (1960, p. 253) that the lectotypes of both *siphonifera* and *aequilateralis* differ from d'Orbigny's original. In view of the wide morphologic variation seen in living representations of this group (Parker, 1962; Hemleben et al., 1989) it would appear appropriate to recognize a single taxon *Globigerinella aequilateralis* (Brady) and to consider *Globigerina siphonifera* d'Orbigny *nomen dubium.*

At Buff Bay *G. aequilateralis* occurs in the Main Roadcut to the basal part of the West Roadcut sections (samples 173 to 241).

Globigerinoides Cushman, 1927

Type species. Globigerina rubra d'Orbigny, 1839

Globigerinoides conglobatus Brady, 1879
(Figure 9.12–13)

Globigerina conglobata Brady, 1879, p. 28B; Brady, 1884, Plate 80, Figs. 1–5.
Globigerinoides conglobatus (Brady). — Banner and Blow, 1960, p. 6, Plate 4, Figs. 4a–c (lectotype). — Kennett and Srinivasan, 1983, p. 58, Plate 12, Figs. 4–6.

The quadrate test shape and heavily cancellate surface texture are characteristic features of this taxon. In the Buff Bay section *G. conglobatus* ranges from samples 244 to 249 in the West Roadcut (i.e., lower Pliocene).

Globigerinoides diminutus Bolli, 1957

Globigerinoides diminutus Bolli, 1957, p. 114, Plate 25, Figs. 11a–c. — Blow, 1969, p. 324, Plate 20, Fig. 4. — Kennett and Srinivasan, 1983, p. 74, Figs. 4–6.

This taxon is distinguished by its small size, compact, strongly cancellate test, and low, symmetrically located aperture.

This form occurs sporadically in the lower part of the exposed section (samples 5G33 to 5G294) of the Pots and Pans Roadcut.

Globigerinoides extremus Bolli and Bermúdez, 1965
(Figure 9.14–15)

Globigerinoides obliquus extremus Bolli and Bermúdez, 1965, p. 139, Plate 1, Figs. 10–12. — Kennett and Srinivasan, 1983, p. 58, Plate 12, Figs. 1–3.

The skewed (laterally compressed) chambers and "cocked" (asymmetric) or flattened last chamber serve to distinguish this form from its ancestor *G. obliquus.* It is the taxon *G. extremus* that Berggren (1973, 1977b) had in mind in designating his late Neogene *G. obliquus* (P16) Zone (see comments by Bolli and Krasheninnikov, 1977).

This form was found only in sample 244 at the base of the West Roadcut (i.e., lower Pliocene).

Globigerinoides fistulosus Schubert, 1910

Globigerinoides fistulosus Schubert, 1910, p. 323, text-Fig. 1
(Figure 9.16–17)

Globigerinoides fistulosus (Schubert). — Kennett and Srinivasan, 1983, p. 68, Plate 14, Figs. 22.5–11.

The single to multiple digitate to "staghorn"-like chamber extensions are characteristic features of this form. It is characteristic of mid–late Pliocene tropical assemblages and disappeared in the early Pleistocene just above the Olduvai Subchron (ca. 1.5 Ma).

A lone early Pliocene occurrence of *G. fistulosus* was observed in sample 249, mid-part of the West Roadcut (Zone PL 3).

Globigerinoides obliquus Bolli, 1957
(Figure 9.24–27)

Globigerinoides obliqua Bolli, 1957, p. 113, Plate 25, 9a–10c; text-Fig. 21, no. 5.
Globigerinoides obliquus Bolli. — Kennett and Srinivasan, 1983, p. 56, Plate 11, Figs. 7–9.

The laterally oblique compression of the last chamber and the higher arched and wider primary aperture distinguishes this taxon from *G. trilobus* with which it is commonly associated.

This species occurs relatively commonly in the Miocene part of the section exposed at Buff Bay (from at least Sample 5G23A at the top of the Pots and Pans section). It does not occur in the Pliocene above the main unconformity.

Globigerinoides ruber (d'Orbigny), 1839
(Figure 9.18–19)

Globigerina rubra d'Orbigny, 1839, p. 82, Plate 4, Figs. 12–14.
Globigerinoides rubra (d'Orbigny). — Banner and Blow, 1960, p. 19–21, Plate 3, Fig. 8 (lectotype).
Globigerinoides ruber (d'Orbigny). — Bermúdez, 1960, p. 1233, Plate 11, Fig. 1. — Blow, 1969, p. 326, Plate 21, Figs. 4, 7.) Lamb and Beard, 1972, p. 49, Plate 33, Figs. 1–3, 5. — Kennett and Srinivasan, 1983, p. 78, Plate 10, Fig. 6; Plate 17, Figs. 1–3.

While exhibiting marked morphologic similarity with *G. subquadratus,* the two forms are distinguished here because of their disjunct stratigraphic range and the generally larger, higher spired test seen in the late Neogene *ruber.*

Typical specimens of *G. ruber* are recorded only from the Landslide and Dead Goat Gully sections (Samples 231 to 243; upper Miocene). It was not observed in the Pliocene in the West Roadcut section.

Globigerinoides subquadratus (Brönnimann, 1954)
(Figure 9.20–21)

Globigerinoides subquadrata Brönnimann, 1954, in Todd et al., 1954, p. 680, Plate 1, Figs. 5, 8a–c.
Globigerinoides subquadratus Brönnimann. — Bermúdez, 1961, p. 1244, Plate 12, Fig. 4. — Cordey, 1967, p. 650, Figs. 1, 2; Plate 103, Fig. 1–4. — Blow, 1969, p. 326–327, Plate 21, Figs. 5, 6. — Kennett and Srinivasan, 1983, p. 74, Plate 10, Fig. 2; Plate 16, Figs. 1–3.

Figure 9. 1–3, *Globoquadrina dehiscens,* sample SG1. 4–5, *Globorotaloides hexagonus,* sample 225; 6–11, *Globigerinella praesiphonifera,* sample SG1; 12–13, *Globigerinoides conglobatus,* sample 245; 14–15, *Globigerinoides extremus,* sample 231; 16–17, *Globigerinoides fistulosus,* sample 246; 18–19, *Globigerinoides ruber,* sample 231; 20–21, *Globigerinoides subquadratus,* sample SG17; 22–23, *Globigerinoides trilobus,* sample 224; 24–27, *Globigerinoides obliquus,* sample SG1.

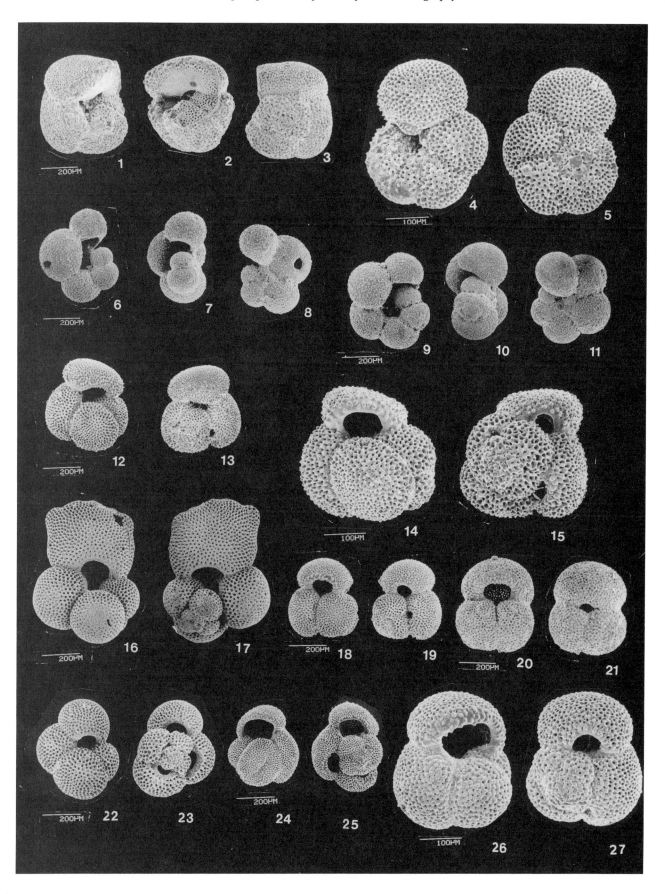

This relatively small, subquadrate form occurs in the lower part of the exposed section at Buff Bay from samples 5G 28A (Pots and Pans section) to Sample 5G 6 (Middle Roadcut). It figures prominently in various middle Miocene stratigraphies (see discussion in text).

Globigerinoides triloba (Reuss), 1850
(Figure 9.22–23)

Globigerinoides triloba (Reuss), 1850, p. 374, Plate 47, Figs. 10a–11c. — Banner and Blow, 1965c, p. 108, Fig. 2 (line drawing of cotype after Reuss, 1850).
Globigerinoides triloba (Reuss). — Cushman, 1946, p. 20, Plate 3, Fig. 8; Plate 4, Figs. 16–18. — Kennett and Srinivasan, 1983, p. 62, Plate 10, Fig. 4; Plate 13, Figs. 1–3.
Globigerinoides trilobus (Reuss). — Bermúdez, 1961, p. 1244, Plate 12, Fig. 6.
Globigerinoides triloba triloba (Reuss). — Bolli, 1957, p. 112, Plate 25, Fig. 2. — Blow, 1969, p. 367, Plate 11, Fig. 60.
Globigerinoides quadrilobatus trilobus. — Blow and Banner, 1962, p. 137. — Banner and Blow, 1965, p. 105–112, Plate 16, Fig. 4.
Globigerinoides quadrilobatus triloba (Reuss, 1850). — Stainforth et al., 1975, p. 310, Fig. 138. 1–5.

The coarsely cancellate trilobate test and low, slit-like primary and supplementary apertures are characteristic features of this long-ranging Neogene taxon. *Globigerinoides sacculifer* (Brady) is characterized by a terminal stage sac-like kummerform chamber that is related to reproduction (Brummer et al., 1986, 1987) and the two forms are considered synonymous (Hemleben et al., 1989). The name *trilobus* is retained over *sacculifer* owing to its priority and more extensive usage in stratigraphic literature.

This form ranges throughout the exposed section at Buff Bay. Forms with sac-like final chambers occur only in the Pliocene part of the section (West Roadcut).

Globoturborotalita Hofker, 1976

Type species. Globigerina rubescens Hofker, 1956.

The genus *Zeaglobigerina* (type species: *Globigerina woodi* Jenkins, 1960) was created by Kennett and Srinivasan (1983) for cancellate, (apparently) non-spinose globigerinids. The monotypic *Globoturborotalita* Hofker (type species: *Globigerina rubescens* Hofker, 1956) was considered to have questionable priority. Loeblich and Tappan (1988, p. 490, 491) placed *Zeaglobigerina* in the synonymy of *Globoturborotalita* and noted that the latter genus is characterized by "true spines of circular cross section" (Loeblich and Tappan, 1988, p. 491; see also Hemleben et al., 1989, p. 12). *Globoturborotalita rubescens* may be a late Neogene spinose end member of this group. Indeed the role of spines in a vertical (phylogenetic) as opposed to a horizontal classification remains uncertain.

Globoturborotalita druryi (Akers), 1955
(Figure 10.1–2)

Globoturborotalita druryi Akers, 1955, p. 654, Plate 65, Figs. 1a–c. — Blow, 1969, p. 318, Plate 14, Fig. 4. — Berggren, 1977a, p. 594, Plate 1, Fig. 1.
Globigerina (Zeaglobigerina) druryi Akers. — Kennett and Srinivasan, 1983, p. 46, Plate 8, Figs. 7–9.

The coarsely pitted wall and thickly rimmed, arched umbilical aperture are characteristic features of this taxon.

This form occurs in low numbers near the top of the Middle Roadcut (sample SG1) and lower part of the Main Roadcut (samples 165 to 173).

Globoturborotalita nepenthes (Todd), 1957
(Figure 10.3–8)

Globigerina nepenthes Todd, 1957, p. 301, Figs. 7a–b. — Bolli, 1957, p. 111, Plate 24, Figs. 2a–c. — Blow, 1959, p. 178, p. 8, Figs. 44, 45. — Blow, 1969, p. 320, Plate 14, Fig. 5. — Lamb and Beard, 1972, p. 47, Plate 4, Figs. 1–8. — Berggren, 1977a, p. 594, Plate 1, Figs. 2, 3. — Bolli and Saunders, 1985, p. 201, Figs. 25.1–4.
Globigerina (Zeaglobigerina) nepenthes (Akers). — Kennett and Srinivasan, 1983, p. 48, Plate 9, Figs. 1–3.

The protruding, thumb-like extension of the final chamber is a distinctive feature of this taxon. It exhibits a clear trend toward increase in test size through time; Pliocene individuals are significantly larger, on average, than Miocene forms.

Globoturborotalita nepenthes occurs persistently, but usually in low numbers, in the Buff Bay Formation. Its initial occurrence is in sample 168 (about 3 m above the base of the Main Road Cut section at a level situated about halfway between projected ER samples 31 (≅ sample 165 at the base of the Main Roadcut section) and ER 32 (≅ sample 173, about 3 m above sample 168).

Sphaeroidinella Cushman, 1927

Type species. Sphaeroidina dehiscens Parker and Jones, 1865.

Sphaeroidinella dehiscens (Parker and Jones), 1865
(Figure 10.9–10)

Sphaeroidina bulloides d'Orbigny var. *dehiscens* Parker and Jones, 1865, p. 369, Plate 19, Fig. 5. — Banner and Blow, 1960, p. 35, Plate 7, Fig. 3 (subsequent description and illustrations of lectotype designated by Bolli et al., 1957, p. 33).
Sphaeroidinella dehiscens (Parker and Jones). — Bolli et al., 1957, p. 32, Plate 6, Figs. 1–5. (lectotype designated but not described). — Parker, 1962, p. 234, Plate 5, Fig. 1. — Lamb and Beard, 1972, p. 59, Plate 1, Figs. 1–2; Plate 34, Figs. 1–2. — Kennett and Srinivasan, 1983, p. 212, Plate 51, Fig. 2; Plate 52, Figs. 7–9.

This three-chambered form is characterized by a weakly lobulate outline, thick cortex, the development of one or two supplementary apertures on the spiral side, and the deeply scalloped sutures bordered by everted crenulated margins of the cortex. In advanced forms the everted cortical margins form a distinct girdle that essentially divides the test into two hemispheres.

Forms with a small, discrete supplementary aperture have been identified as *S. dehiscens immatura* Cushman and *S. ionica ionica* Cita and Ciaranfi; the gradual increase in size in the supplementary aperture is useful in early Pliocene stratigraphy (ca. 5 to 3 Ma).

Figure 10. 1–2, *Globoturborotalita druryi,* sample 561; 3, *Globoturborotalita nepenthes,* samples 168 (3, 4: lowest occurrence of *G. nepenthes*), 168 (5, 6), and 222 (7–8). Note evolution of quadrate form and high apertural arch (*druryi*) to elongate-oval test shape and gradual test shape transformation and thumb-like extension of last chamber in nepenthes. 9–10, *Sphaeroidinella dehiscens,* sample 245; 11–12, *Sphaeroidinellopsis seminulina,* sample 250, decorticated specimen; 13–16, *Sphaeroidinellopsis multiloba,* samples SG1 (13–14) and 199 (15–16). Note difference in presentation owing to strong dissolution in older (sample SG1) levels.

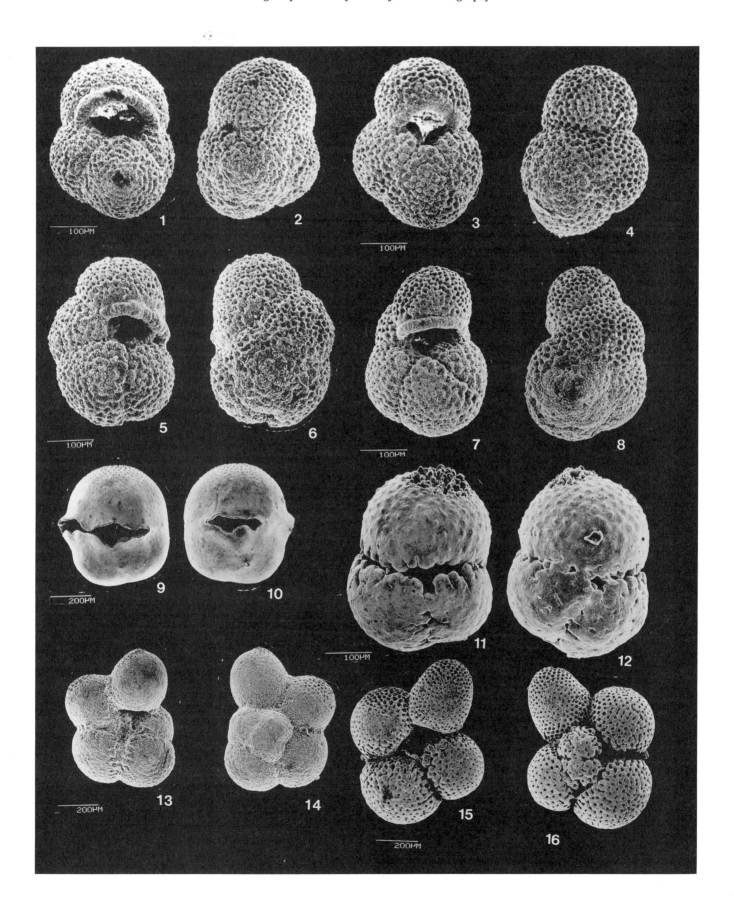

This form was recorded only in samples 245, 247, and 249, midpart of the West Roadcut.

Sphaeroidinellopsis Banner and Blow, 1959

Type species. Sphaeroidinellopsis dehiscens (Parker and Jones) subsp. subdehiscens Blow, 1959

A distinct dichotomy exists between two recent attempts to clarify the taxonomy and systematics of this group.

An analysis of the genus *Sphaeroidinellopsis* led Bolli and Saunders (1981) to conclude that (1) three- to five-chambered forms of *Sphaeroidinellopsis* with apertural flanges and thick cortex (generally referred to *seminulina*) appear only in the late Miocene; *S. paenedehiscens* is considered to be a junior synonym of *S. seminulina*; (2) forms with a rough, cancellate surface and imperforate apertural rim (rather than apertural flange) are referable to *disjuncta* (synonym: *grimsdalei*); (3) forms intermediate between *disjuncta* and *seminulina*, with a weakly developed glassy cortex and apertural structures closer to *disjuncta* than *seminulina* are referable to *multiloba, S. kochi*; and *S. subdehiscens*, type species of the genus, is considered a three-chambered juvenile variant; (4) *S. sphaeroides* is a distinct taxon, but related to the *seminulina* plexus.

Kennett and Srinivasan (1983), on the other hand, interpreted the middle Miocene *multiloba* and *rutschi* as synonyms (as did Bolli and Saunders) but consider both as junior synonyms of *S. kochi*, whereas the latter authors place *kochi* in questionable synonymy with *hancocki* (noting that the holotype of *kochi* is damaged and represented only by fragments), a form they indicate is restricted to the Pliocene and is distinguishable from *seminulina* solely on the basis of the larger number of chambers (4½ to 7). They consider *sphaeroides* a synonym of *paenedehiscens* and the progenitor of *Sphaeroidinella dehiscens* (with the minute supplementary aperture).

In this study we recognize three taxa of *Sphaeroidinellopsis: disjuncta, multiloba* (both in the Miocene below the main unconformity), and *seminulina* (occurring above the main unconformity).

Sphaeroidinellopsis disjuncta (Finlay), 1940

Sphaeroidinella disjuncta Finlay, 1940, p. 467, Plate 67, Figs. 224–228.
Sphaeroidinellopsis disjuncta (Finlay). — Bolli and Saunders, 1981, p. 18, Plate 1, Fig. 12 (holotype refigured): Plate 2, Figs. 30–35. — Kennett and Srinivasan, 1983, p. 206, Plate 51, Figs. 3–5.
Globigerina grimsdaeli Keijzer, 1945, p. 205, Figs. 33a–c.

Coarsely cancellate, predominantly three-chambered form. Forms transitional to *S. multiloba* exhibit four to five chambers concomitant with development of glassy surficial cortex over the interval of Zones N12 to "N13" (Blow, 1969). This form occurs sporadically in the Spring Garden Member in the Pots and Pans section (Samples 5628A to 5623A).

Sphaeroidinellopsis multiloba (LeRoy), 1944
(Figure 10.13–16)

Sphaeroidinella multiloba LeRoy, 1944, p. 91, Plate 4, Figs. 7–9.
Sphaeroidinellopsis multiloba (LeRoy). — Bolli and Saunders, 1981, p. 19, Plate 1, Fig. 9 (holotype, refigured); Plate 2, Figs. 18–20, 22–27.
Sphaeroidinella rutschi Cushman and Renz, 1941, p. 25, Figs. 5a–c.
Sphaeroidinellopsis kochi (Caudri). — Kennett and Srinivasan, 1983, p. 210, Plate 52, Figs. 1–3 (non Caudri).

Sphaeroidinellopsis multiloba possesses typically four to five chambers and a relatively thin cortex. Forms assigned to *S. kochi* (Caudri) by Kennett and Srinivasan (1983) are included here under *S. multiloba*. (The name *Globigerina kochi* Caudri is considered *nomen dubium* here; see Bolli and Saunders, 1981, p. 24).

This form ranges from (at least) sample SG 23A (top of Pots and Pans section) to sample 239 (top of Landslide section), which is Zones N11–12 to N17, in the Buff Bay outcrop.

Sphaeroidinellopsis seminulina (Schwager), 1866
(Figure 10.11–12)

Globigerina seminulina Schwager, 1866, p. 256, Plate 7, Fig. 112.
?*Sphaeroidinella dehiscens subdehiscens* Blow, 1959, p. 195, Plate 12, Figs. 71a–c.
Sphaeroidinellopsis seminulina (Schwager). — Banner and Blow, 1960, p. 24, Plate 7, Figs. 2a–b (neotype). — Bolli and Saunders, 1981, p. 19, Plate 1, Figs. 7–8 (holotype, neotype, refigured); Plate 2, Figs. 6–14. — Kennett and Srinivasan, 1983, p. 206, Plate 51, Figs. 1, 6–8.
Sphaeroidinellopsis subdehiscens paenedehiscens Blow, 1969, p. 386, 387, Plate 30, Figs. 4, 5, 9. — Kennett and Srinivasan, 1983, p. 210, Plate 52, Figs. 4–6.

I use the name *seminulina* in the sense given it by Bolli and Saunders (1981) for distinctly corticated, trilobate (less commonly subquadrate) forms with a narrow, slit-like scalloped apertural flange. Kennett and Srinivasan (1983) included *S. subdehiscens* in the synonymy of *S. seminulina* (thus making the latter type species of the genus), whereas Loeblich and Tappan (1988, p. 491) considered the two forms distinct. Bolli and Saunders (1981, p. 19) considered *subdehiscens* to be a three-chambered juvenile form of *S. multiloba*. The type level of *S. subdehiscens* is in the *Globorotalia mayeri* Zone of Bolli (1957; = N13–14 of Blow, 1969) of Venezuela, that of *S. seminulina* in the mid-Pliocene of Kar Nicobar, Indian Ocean. I would tentatively agree with Bolli and Saunders (1981) that *subdehiscens* Blow is best considered a variant of *multiloba* LeRoy. Pending resolution of the taxonomic status of *subdehiscens* Blow, however, I believe it most appropriate to consider it as type species of the genus by original designation (see Loeblich and Tappan, 1988, p. 491).

This form is found only in the Pliocene part of the Buff Bay outcrop occurring throughout the West Roadcut section.

ACKNOWLEDGMENTS

This study is an outgrowth of a project devoted to a study of Neogene Deep Water Benthic Foraminifera of the Gulf of Mexico and Caribbean region and is funded by a consortium of oil companies (British Petroleum, Chevron, Marathon, Texaco, and Unocal). The cooperation and assistance of the 1987 field party to Jamaica is gratefully acknowledged (M.-P Aubry, G. Jones, D. V. Kent, K. G. Miller, E. Robinson, C. Stuart, R. C. Tjalsma, and M. Van Fossen). Field work in Jamaica was funded by Unocal; sample preparation was also conducted by Unocal to whom I would like to extend my thanks. Particular thanks are extended to D. V. Kent, K. G. Miller, and M. Van Fossen for providing paleomagnetic data and discussions on the interpretation of the data; M.-P. Aubry for discussions on biostratigraphic interpretations; B. Kohl (Chevron) and J. Hardenbol (Exxon) for providing original samples collected in the 1960s from Robinson's localities; D. Grieg (Chevron) for providing the SEM micrographs; and L. Beaufort for technical assistance. Critical review of an early version of the paper by H. M. Bolli, E. R. Robinson, and R. K. Olsson is gratefully acknowledged and has contributed substantially to its improvement. This is Woods Hole Oceanographic Institution Contribution No. 7887.

REFERENCES CITED

Akers, W. H., 1955, Some planktonic foraminifera of the American Gulf Coast and suggested correlations with the Caribbean Tertiary: Journal of Paleontology, v. 29, p. 647–664, 3 figs., Plate 65.

Banner, F. T., 1982, A classification and introduction to the Globigerinacea, *in* Banner, F. T., and Lord, A. T., eds., Aspects of micropaleontology: George Allen and Unwein, p. 142–239.

Banner, F. T., and Blow, W. H., 1959, The classification and stratigraphical distribution of the Globigerinaceae: Palaeontology, v. 2, p. 1–27.

Banner, F. T., and Blow, W. H., 1960, Some primary types of species belonging to the superfamily Globigerinacea: Contributions from the Cushman Foundation, v. 11, p. 1–41.

Banner, F. T., and Blow, W. H., 1965a, Two new taxa of the Globorotaliinae (Globigerinacea, Foraminifera) assisting determination of the late/middle Miocene boundary: Nature, v. 207, p. 11351–11354

Banner, F. T., and Blow, W. H., 1965b, Progress in the planktonic foraminiferal biostratigraphy of the Neogene: Nature, v. 208, p. 1164–1166.

Banner, F. T., and Blow, W. H., 1965c, *Globigerinoides quadrilobatus* (d'Orbigny) and related forms: Their taxonomy, nomenclature and stratigraphy: Cushman Foundation Foraminiferal Research, Contributions, v. 16, p. 105–115.

Beard, J. H., 1969, Pleistocene paleotemperature record based on planktonic foraminifers, Gulf of Mexico: Transactions of the Gulf Coast Association of Geological Societies, v. 19, p. 535–553.

Benson, R. H., and 7 others, in prep., The Bou-Regreg section, Morocco: Proposed global boundary stratotype section and point of the Pliocene.

Berggren, W. A., 1971, Tertiary boundaries, *in* Funneell, B. F., and Riedel, W. R., eds., The micropalaeontology of the oceans: Cambridge, Cambridge University Press, p. 693–808.

Berggren, W. A., 1973, The Pliocene time scale: Calibration of planktonic foraminifera and calcareous nannoplankton zones: Nature, v. 241, no. 5407, p. 391–397.

Berggren, W. A., 1977a, Late Neogene planktonic foraminiferal biostratigraphy of DSDP Site 357 (Rio Grande Rise), *in* Supko, P., Perch-Nielsen, K., and others, 1977, Initial Reports of The Deep Sea Drilling Project, 39: Washington, D.C., U.S. Government Printing Office, p. 591–614.

Berggren, W. A., 1977b, Late Neogene planktonic foraminiferal biostratigraphy of the Rio Grande Rise (South Atlantic): Marine Micropaleontology, v. 2, no. 3, p. 265–313.

Berggren, W. A., 1992, Neogene planktonic foraminifer magnetobiostratigraphy of the southern Kerguelen Plateau (Sites 747, 748 and 751), *in* Wise, S. W., Jr., Schlich, R., and others, Proceedings, Ocean Drilling Program, Scientific Results, 120: College Station, Texas (Ocean Drilling Program), p. 631–647.

Berggren, W. A., and Amdurer, M., 1973, Late Paleogene (Oligocene) and Neogene planktonic foraminiferal biostratigraphy of the Atlantic Ocean (lat. 30°N to lat. 30°S): Rivista Italiana Paleontologia, v. 79, no. 3, p. 337–392.

Berggren, W. A., and Poore, R. Z., 1974, Late Miocene–early Pliocene planktonic foraminiferal biochronology: *Globorotalia tumida* and *Sphaeroidinella dehiscens* lineages: Rivista Italiana Paleontologia, v. 80, no. 4, p. 689–698.

Berggren, W. A., Aubry, M. -P., and Hamilton, N., 1983, Neogene magnetobiostratigraphy of Deep Sea Drilling Project Site 516 (Rio Grande Rise, South Atlantic), *in* Barker, P. F., Carlos, R. L., Johnson, D. A., and others, Initial Reports of the Deep Sea Drilling Project, v. 72: Washington, D.C., U.S. Government Printing Office, p. 675–713.

Berggren, W. A., Kent, D. V., and Van Couvering, J. A., 1985, Neogene geochronology and chronostratigraphy, *in* Snelling, N. J., ed., The chronology of the geological record: Geological Society Memoir 10, p. 211–260.

Bermúdez, P. J., 1949, Tertiary smaller foraminifera of the Dominican Republic: Cushman Laboratory for Foraminiferal Research Special Publications, v. 25, p. 1–322.

Bermúdez, P. J., 1961, Contribución al estudio de las Glogiberinidea de la region Caribe-Antillana (Paleoceno-Reciente), Memoria del Tercer Congreso Geológico Venezolano, Caracas, v. 3: Boletín de Geología, Publicación Especial 3, p. 1119–1393.

Blow, W. H., 1959, Age, correlation and biostratigraphy of the upper Tocuyo (San Lorenzo) and Pozon Formations, eastern Falcón, Venezuela: Bulletins of American Paleontology, v. 39, p. 67–251.

Blow, W. H., 1969, Late middle Eocene to Recent planktonic foraminiferal biostratigraphy, *in* Brönnimann, P., and Renz, H. H., eds., Proceedings, First International Conference on Planktonic Microfossils, Geneva, 1967, v. 1: Leiden, E. J. Brill, p. 199–421.

Blow, W. H., 1979, The Cainozoic Globigerinida, v. I and II: Leiden, E. J. Brill, 1413 p.

Blow, W. H., and Banner, F. T., 1962, Part 2: The Tertiary (Upper Eocene to Aquitanian) Globigerinaceae, *in* Eames, F. E., and others, eds., Fundamentals of mid-Tertiary stratigraphical correlation: Cambridge, Cambridge University Press, p. 61–151.

Blow, W. H., and Banner, F. T., 1966, The morphology, taxonomy and biostratigraphy of *Globorotalia barisanensis* LeRoy, *Globorotalia fohsi* Cushman and Ellisor, and related taxa: Micropaleontology, v. 12, p. 286–301.

Bolli, H. M., 1950, The direction of coiling in the evolution of some Globorotalidae: Cushman Foundation for Foraminiferal Research, v. 1, p. 82–89.

Bolli, H. M., 1957, Planktonic foraminifera from the Oligocene-Miocene Cipero and Lengua Formations of Trinidad, B.W.I.: U.S. National Museum Bulletin, v. 215, p. 97–123, Figs. 17–21, Plate 22–29.

Bolli, H. M., 1966, The planktonic foraminifera in well Bodjonegoro-1 of Java: Eclogae Geologiae Helvetiae, v. 59, p. 449–466.

Bolli, H. M., 1970, The foraminifera of Sites 23–31, Leg 4, *in* Bader, R. G., and others, Initial Reports of the Deep-Sea Drilling Project, v. 4: Washington, D.C., U.S. Government Printing Office, p. 577–643.

Bolli, H. M., 1972, Correlacion de las estaciones JOIDES 29, 30 y 31 del Caribe con Jamaica, Venezuala y Trinidad, *in* Memoria, Congreso Geologico Venezolano, 4th: Caracas, Venezuela, Direccion de Geologia, Boletin de Geologia, Publicacion Especial, v. 3, p. 1315–1336.

Bolli, H. M., and Bermúdez, P. J., 1965, Zonation based on planktonic foraminifera of middle Miocene to Pliocene warm-water sediments: Asociación Venezolana de Geología, Minería y Petróleo Boletín Informativo, v. 8, p. 119–149, 1 Pl.

Bolli, H. M., and Krashennikov, V. A., 1977, Problems in Paleogene and Neogene correlations based on planktonic foraminifera: Micropaleontology, v. 23, no. 4, p. 436–452.

Bolli, H. M., and Saunders, J. B., 1981, The species of *Sphaeroidinellopsis* Banner and Blow, 1959: Cahiers de Micropaleontologie, v. 4, p. 13–25.

Bolli, H. M., and Saunders, J. B., 1982, *Globorotalia mayeri* and its relationship to *Globorotalia mayeri* and *Globorotalia continuosa*: Journal of Foraminiferal Research, v. 12, no. 1, p. 39–50, Plates 1–4.

Bolli, H. M., and Saunders, J. B., 1985, Oligocene to Holocene low latitude planktonic foraminifera, *in* Bolli, H. M., Saunders, J. B., and Perch-Nielsen, K., eds., Plankton stratigraphy: Cambridge University Press, p. 155–262.

Bolli, H. M., Loeblich, A. R., and Tappan, H., 1957, Planktonic foraminiferal families Hantkeninidae, Orbulinidae, Globorotaliidae and Globotruncanidae: U.S. National Museum Bulletin, v. 215, p. 3–50.

Brady, H., 1877, Supplementary notes on the foraminifera of the Chalk (?) of the New Britain group: Geological Magazine, new series, decade 2, v. 4, p. 534–536.

Brady, H. B., 1879, Notes on some Reticularian Rhizopoda of the *Challenger* Expedition, Part 2, Additions to the knowledge of porcellanous and hyaline types: Quarterly Journal of Microscopical Science, new ser., v. 19, p. 261–299.

Brady, H., 1882, Report on the Foraminifera: Proceedings of the Royal Society Edinburgh, v. 11, p. 708–717.

Brady, 1884, Report on the Foraminifera dredged by H.M.S. *Challenger*, during the years 1873–1876: London, Reports of the Scientific Results of H.M.S. *Challenger* 1873–1876, v. 9 (Zoology), p. 1–814.

Brummer, G.J.A., Hemleben, C., and Spindler, M., 1986, Planktonic foraminiferal ontogeny and new perspectives for micropaleontology: Nature, v. 319, p. 50–52.

Brummer, G.J.A., Hemleben, C., and Spindler, M., 1987, Ontogeny of extant spinose planktonic foraminifera (Globigerinidae): A concept exemplified by *Globigerinoides sacculifer* (Brady) and *G. ruber* (d'Orbigny): Marine Micropaleontology, v. 2, p. 357–381.

Channell, J. J., Rio, D., and Thunell, R. C., 1988, Miocene/Pliocene boundary magnetostratigraphy at Capo Spartivento, Calabria, Sicily: Geology, v. 16, p. 1096–1099.

Channell, J.E.T., Rio, D., Sprovieri, R., and Glason, G., 1990, Biomanetostratigraphic correlations from Leg 107 in the Tyrrharian Sea, *in* Kastens, K. A., Masch, J., and others, eds., Proceedings, Ocean Drilling Program, Scientific Results, v. 107: College Station, Texas (Ocean Drilling Program, p. 669–693.

Chapman, F., Parr, W. J., and Collins, A. C., 1934, Tertiary foraminifera of Victoria Australia—The Balcombian deposits of Port Phillip, Part III: Journal of the Linnaean Society of London, Zoology, v. 38, p. 553–577.

Cifelli, R., 1982, Early occurrences and some phylogenetic implications of spiny, honeycomb-textured planktonic foraminifera: Journal of Foraminiferal Research, v. 12, no. 2, p. 105–115.

Cifelli, R., and Scott, G., 1986, Stratigraphic record of the Neogene globorotaliid radiation (planktonic Foraminiderida): Smithsonian Contributions to Paleobiology, v. 58, 101 p.

Cita, M. B., 1975, The Miocene-Pliocene boundary: History and definition: Micropaleontology, Special Publication no. 1, p. 1–30.

Cita, M. B., and Gartner, S., 1973, Studi sul Pliocene e sugli srati di passasgio dal Miocene al Pliocene, IV, The stratotype Zanclean, foraminiferal and nannofossil biostratigraphy: Rivista Italiana Paleontologia, v. 79, p. 503–558.

Cita, M. B., and Gelati, R., 1960, *Globoquadrina langhiana* n. sp. del Langhianotipo: Rivista Italiana Paleontologia e Stratigrafia, v. 66, p. 241–246, 1 Fig., Plate 29.

Colalongo, M. L., and Sartoni, S., 1967, *Globorotalia hirsuta aemiliana* nuova sottospecie cronologica del Pliocene in Italia: Giornale di Geologia, ser. 3, v. 34, p. 255–284, 2 Fig., 2 Plates.

Conato, V., and Follador, U., 1967, *Globorotalia crotonensis* e *Globorotalia crassacrotenensis* nuove species del Pliocene Italiano: Bollettino della Società Geologica Italiana, v. 84, p. 555–563, 6 Fig.

Cordey, W. G., 1967, The development of *Globigerinoides ruber* (d'Orbigny, 1839) from the Miocene to Recent: Paleontology, v. 10, no. 4, p. 647–659.

Cushman, J. A., 1946, The species of *Globigerina* described between 1839 and 1850: Cushman Laboratory of Foraminiferal Research Contributions, v. 22, p. 15–21.

Cushman, J. A., and Ellisor, A., 1939, New species of foraminifera from the Oligocene and Miocene: Cushman Laboratory of Foraminiferal Research, Contributions, v. 15, p. 1–14.

Cushman, J. A., and Renz, H. H., 1941, New Oligocene-Miocene Foraminifera from Venezuelana: Cushman Laboratory of Foraminiferal Research, Contributions, v. 17, p. 1–27.

Cushman, J. A., and Stainforth, R. M., 1945, The Foraminifera of the Cipero Marl Formation of Trinidad, British West Indies: Cushman Laboratory for Foraminiferal Research, Special Publication, 14, p. 1–75.

Cushman, J. A., Stewart, R. E., and Stewart, K. C., 1930, Tertiary foraminifera from Humboldt County, California: A preliminary survey of the fauna: San Diego Society of Natural History Transactions, v. 6, no. 2, p. 41–94.

Finlay, H. J., 1940, New Zealand Foraminifera: Key species in stratigraphy—No. 4: Transactions of the Royal Society of New Zealand, v. 69, p. 448–472.

Fornasini, C., 1902, Sinossi metodica dei foraminideri sin qui rinvenuti nella sabbia de Lido di Rimini: Memorie della R. Accademia della Scienze dell'Istituto de Bologna, ser. 5, v. 10, p. 1–68.

Galloway, J. J., and Wissler, S. G., 1927, Pleistocene foraminifera from the Lomita Quarry, Palos Verdes Hills, California: Journal of Paleontology, v. 1, no. 1, p. 35–87, Plates 7–12, Tables 1–2.

Gartner, S., 1973, Absolute chronology of the late Neogene calcareous nannofossil succession in the equatorial Pacific: Bulletin of the Geological Society of America, v. 84, p. 2021–2034.

Hays, J. D., Saito, T., Opdyke, N. D., and Burckle, L. H., 1969, Pliocene-Pleistocene sediments of the equatorial Pacific—Their paleomagnetic, biostratigraphic and climatic record: Bulletin of the Geological Society of America, v. 80, p. 1481–1514.

Hedberg, H. D., 1937, Foraminifera of the middle Tertiary Carapita Formation of northeastern Venezuela: Journal of Paleontology, v. 11, p. 661–697, Plates 90–92.

Hemleben, C., Spindler, M., and Anderson, O. R., 1989, Modern planktonic foraminifera: New York-Berlin, Springer Verlag, 363 p.

Hilgen, F. J., and Langereis, C. G., 1988, The age of the Miocene-Pliocene boundary in the Capo Rosselo area (Sicily): Earth and Planetary Science Newsletters, v. 91 (1988), p. 214–222.

Hill, R. T., 1989, The geology and physical geography of Jamaica: Study of a type of Antillean development: Cambridge, University Press, Harvard College, Museum of Comparative Zoology Bulletin, v. 34, 326 p.

Hodell, D., and Kennett, J. P., 1986, Late Miocene and early Pliocene stratigraphy and paleoceanography of the South Atlantic and southwest Pacific Oceans: A synthesis: Paleoceanography, v. 1, no. 3, p. 285–311.

Hornibrook, N.deB., Brazier, R. C., and Strong, C. P., 1989, Manual of New Zealand Permian to Pleistocene foraminiferal biostratigraphy: New Zealand Geological Survey Paleontological Bulletin 56, 175 p.

Iaccarino, S., 1985, Mediterranean Miocene and Pliocene planktic foraminifera, *in* Bolli, H. M., Saunders, J. B., and Perch-Nielsen, K., eds., Plankton Stratigraphy: Cambridge, Cambridge University Press, p. 283–314.

Keigwin, L. D., Jr., 1982, Neogene planktonic foraminifers from Deep Sea Drilling Project Sites 502 and 503, *in* Prell, W. S., Gardner, J. V., and others, Initial Reports of the Deep Sea Drilling Project, v. 68: Washington, D.C., U.S. Government Printing Office, p. 269–288 (see also Appendix, p. 493–495).

Keijzer, F. G., 1945, Outline of the geology of the eastern part of the Province of Oriente, Cuba (E of 76°WL) with notes on the geology of other parts of the island: Geographische en Geologische Mededelingen. Publicaties uit het Geographisch en uit het Mineralogisch–Geologisch Instituut der Rijksuniversiteit te Utrecht. Physiographisch-Geologische Reeks Service II, no. 6, p. 1–239.

Kennett, J. P., and Srinivasan, M. S., 1983, Neogene Planktonic Foraminifera: A Phylogenetic Atlas: Stroudsburg, Pennsylvania, Hutchinson and Ross Publication Co., 265 p.

Kent, D. V., and Spariosu, D. J., 1982, Magnetostratigraphy of Caribbean Site 502 hydraulic piston cores, *in* Prell, W. L., Gardner, J. V., and others, Initial Reports of the Deep Sea Drilling Project, 68: Washington, D.C., U.S. Government Printing Office, p. 419–440.

Lamb, J. L., and Beard, J. H., 1972, Late Neogene planktonic foraminifers in the Caribbean, Gulf of Mexico and Italian stratotypes: University of Kansas Paleontological Contributions, Article 57, Protozoa, v. 8, p. 1–67.

Leroy, L. W., 1939, Some small Foraminifera, Ostracoda, and otoliths from the Neogene ("Miocene") of the Rokan-Tapaoeli area, central Sumatra: Natuurkundig Tijdschrift voor Nederlandsch-Indië, v. 99, no. 6, p. 215–296.

Leroy, L. W., 1944, Miocene Foraminifera from Sumatra and Java, Netherlands East Indies: Colorado School of Mines Quarterly, v. 39, no. 3, pt. 1, p. 9–69, Plates 1–8.

Liska, R. D., 1985, The range of *Glogiberinoides ruber* (d'Orbigny) from the middle to late Miocene in Trinidad and Jamaica: Micropaleontology, v. 31, no. 4, p. 372–379.

Liska, R. D., 1980, *Polyperibola,* a new planktonic foraminiferal genus from the late Miocene of Trinidad and Tobago: Journal of Foraminiferal Research, v. 10, p. 136–142.

Liska, R. D., 1991, The history, age and significance of the *Globorotalia menardii* Zone in Trinidad and Tobago, West Indies: Micropaleontology, v. 37, no. 2, p. 173–182.

Loeblich, A. R., Jr., and Tappan, H., 1984, Suprageneric classification of the Foraminiferida (Protozoa): Micropaleontology, v. 30, no. 1, p. 1–70.

Loeblich, A. R., and Tappan, H., 1988, Foraminiferal Genera and their Classification: New York, Van Nostrand Reinhold, Co., v. 1, Text, 970 p.; v. 2, 1212 p., 847 Plates.

Malmgren, B. A., Berggren, W. A., and Lohmann, G. P., 1983, Evidence for punctuated gradualism in the late Neogene *Globorotalia tumida* lineage of planktonic foraminifera: Paleobiology, v. 9, no. 4, p. 377–389.

Miller, K. G., Kahn, M. J., Aubry, M. -P., Berggren, W. A., Kent, D. V., and Melillo, A., 1985, Oligocene to Miocene bio-, magneto-, and isotope stratigraphy of the western North Atlantic: Geology, v. 13, p. 257–261.

Natland, M. L., 1938, New species of Foraminifera from off the west coast of North America and from the later Tertiary of the Los Angeles Basin: University of California, Scripps Institute of Oceanography, Technical Series Bulletin 4, p. 137–152.

d'Orbigny, A. D., 1839, Foraminifères, *in* Ramon de la Sagra, Histoire physique, politique et naturelle de l'Ile de Cuba: 224 p., atlas, 12 plates.

Palmer, D. K., 1945, Notes on the foraminifera from Bowden, Jamaica: Bulletin of American Paleontology, v. 115, p. 3–83.

Parker, F. L., 1962, Planktonic foraminiferal species in Pacific sediments: Micropaleontology, v. 8, p. 219–254.

Parker, F. L., 1967, Late Tertiary biostratigraphy (planktonic foraminifera) of tropical Indo-Pacific deep sea cores: Bulletins of American Paleontology, v. 52, p. 115–208.

Prell, W. L., Gardner, J. V., and others, 1982, Initial reports of the Deep Sea Drilling Project, v. 68: Washington, D.C., U.S. Government Printing Office, 495 p.

Reuss, A. E., 1850, Neues Foraminiferen aus den Schichten des österreichischen Tertiärbeckens: Denkschriften der Kaiserlichen Akademie der Wissenschaften, Mathematisch-Naturwissenschaftlische Klasse, v. 1, p. 365–390.

Rio, D., Raffi, I., and Villa, G., 1990, Pliocene, Pleistocene calcareous nannofossil distribution patterns in the western Mediterranean, *in* Kastens, K., Mascle, J., and others, Proceedings, Ocean Drilling Program Science Results, v. 107: College Station, Texas, Ocean Drilling Program, p. 513–533.

Robinson, E., 1965, Tertiary rocks of the Yallahs area, Jamaica: Geological Society of Jamaica Quarterly Journal (Geonotes), v. 7, p. 18–27.

Robinson, E., 1967, Biostratigraphic position of late Cenozoic rocks in Jamaica: Geological Society of Jamaica Journal, v. 9, p. 32–41.

Robinson, E., 1969a, Coiling directions in planktonic foraminifera from the Coastal Group of Jamaica: Gulf Coast Association of Geological Societies, Transactions, v. 19, p. 555–558.

Robinson, E., 1969b, Geological field guide to Neogene sections in Jamaica: Geological Society of Jamaica Journal, v. 10, p. 1–24.

Ryan, W.B.F., Cita, M. B., Rawson, M. D., Burckle, L. H. and Saito, T., 1974, A paleomagnetic assignment of Neogene stage boundaries and the development of isochronous datum planes between the Mediterranean and the Pacific and Indian Oceans in order to investigate the response of the world ocean to the Mediterranean "Salinity Crisis": Rivisita Italiana Paleontologia, v. 80, p. 631–688.

Saito, T., 1985, Planktonic foraminiferal biostratigraphy of eastern equatorial Pacific sediments, Deep Sea Drilling Project, Leg 85, *in* Mayer, L., Theyer, F., and others, Initial Reports of the Deep Sea Drilling Project, v. 85: Washington, D.C., U.S. Government Printing Office, p. 621–653.

Saito, T., Thompson, P. R., and Berger, D., 1976, Skeletal ultramicrostructure of some elongate-chambered planktonic foraminifera and related species, *in* Takayanagi, Y., and Saito, T., eds., Progress in micropaleontology: New York, Micropaleontology Press, p. 278–304.

Schwager, C., 1866, Fossile Foraminiferen von Kar Nikobar: Novara Expedition, 1857–1859: Wein, Geologischer Theil, v. 2, p. 187–268, Plate 4–7.

Scott, G. H., Bishop, S., and Burt, B. J., 1990, Guide to some Neogene Globorotaliids from New Zealand: New Zealand Geological Survey Bulletin 61, 135 p.

Shipboard Scientific Party, 1989, Site 758, *in* Pierce, J., Weissel, J., and others, Proceedings, Ocean Drilling Program Initial Reports, v. 121: College Station, Texas, Ocean Drilling Program, p. 359–453.

Stainforth, R. M., Lamb, J. L., Luterbacher, H., Beard, J. H., and Jeffords, R. M., 1975, Cenozoic planktonic foraminiferal zonation and characteristics of index forms: Lawrence, University of Kansas Paleontological Contributions, v. 62, p. 1–425.

Steineck, P. L., 1974, Foraminiferal paleoecology of the Montpelier and Lower Coastal Groups (Eocene-Miocene), Jamaica, West Indies: Palaeogeography, Palaeoclimatology, Palaeoecology, v. 16, (1974), p. 217–242.

Steineck, P. L., 1981, Upper Eocene to middle Miocene ostracode faunas and paleoceanography of the North Coastal Belt, Jamaica, West Indies: Marine Micropaleontology, v. 6 (1981), p. 339–366.

Takayanagi, Y., and Saito, T., 1962, Planktonic foraminifera from the Nobori Formation, Shikoku, Japan: Tohoku Imperial University Science Reports, series 2 (Geology), Special Volume 5, p. 67–105.

Todd, R., Cloud, P. E., Jr., Low, D., and Schmidt, R. G., 1954, Probable occurrence of Oligocene in Saipan: American Journal of Science, v. 252, p. 673–682, 1 Plate.

Todd, R., 1964, Planktonic foraminifera from deep-sea cores off Enewetak Atoll: U.S. Geological Survey Professional Paper 260-CC, p. 1067–1100.

Todd, R., Cloud, P. E., Jr., Low, D., and Schmidt, R. G., 1954, Probable occurrence of Oligocene in Saipan: American Journal of Science, v. 252, p. 673–682, 1 Plate.

Wright, R. M., ed., 1974, Field guide to selected Jamaican geological localities: Mines and Geology Division Special Publication 1, 57 p.

Zijderveld, J.S.A., Zachariasse, J. W., Verhallen, J.J.M., and Hilgen, F., 1986, The age of the Miocene-Pliocene boundary; Newsletters in Stratigraphy, v. 16, no. 3, p. 169–181.

MANUSCRIPT ACCEPTED BY THE SOCIETY NOVEMBER 11, 1992

Miocene-Pliocene bathyal benthic foraminifera and the uplift of Buff Bay, Jamaica

Miriam E. Katz
Lamont-Doherty Earth Observatory of Columbia University, Palisades, New York 10964
Kenneth G. Miller
Department of Geological Sciences, Rutgers University, New Brunswick, New Jersey 08903, and Lamont-Doherty Earth Observatory of Columbia University, Palisades, New York 10964

ABSTRACT

We sampled lowermost middle Miocene to Pliocene sections exposed in roadcuts east of Buff Bay, Jamaica, for quantitative benthic foraminiferal studies. Our analyses document paleobathymetric changes and sediment sources. The Spring Garden Member of the Montpelier Formation and the Buff Bay Formation (middle to upper Miocene) were deposited at lower bathyal paleodepths (1,000 to 2,000 m) that probably exceeded 1,300 m, while the overlying Bowden Formation (Pliocene) was deposited within the upper bathal zone (200 to 500 m). A significant portion of the benthic foraminifera found in Buff Bay sections consists of transported shallow-water reefal elements (*Amphistegina* spp./*Asterigerina* spp.) occurring in otherwise in situ deep-water bathyal biofacies.

The lower part of the Spring Garden Member (middle Miocene) contains a typical cosmopolitan lower bathyal biofacies dominated by *Globocassidulina subglobosa, Cibicidoides mundulus, Stilostomella aculeata,* and *Oridorsalis* spp. A biofacies change that occurs within the Spring Garden Member may reflect the well-documented global benthic foraminiferal taxonomic turnover that occurred throughout the deep sea from early to middle Miocene Biochrons N8 to N11. The upper Spring Garden Member and portions of the Buff Bay Formation (middle Miocene) faunas are characterized by the apparently endemic species *Compressigerina coartata* and *Siphouvigerina porrecta*; we interpret this as an in situ lower bathyal biofacies. Much of the Buff Bay Formation (middle-upper Miocene) is dominated by *Reusella spinulosa* var. *pulchra, Uvigerina proboscidea, Rosalina* sp. 1, and *Siphouvigerina porrecta*; although this assemblage may be an in situ lower bathyal biofacies, there is evidence to suggest that it may be a transported thanatofacies. The Bowden Formation (Pliocene) is characterized by an upper bathyal (200 to 600 m) biofacies dominated by *Bulimina aculeata, Globobulimima* spp., *Planulina foveolata* (lower depth limit is 500 m), and *Uvigerina* sp. 1. The change in benthic foraminiferal biofacies between the Buff Bay and Bowden Formations reflects the tectonic uplift of the north coast of Jamaica. We estimate that at least 800 m of shallowing occurred between the late Miocene (ca. 8 Ma) and Pliocene (ca. 4 to 3 Ma) during the hiatus that separated deposition of the Buff Bay and Bowden Formations. A subsequent shallowing occurred from the Pliocene (bathyal paleodepths >200 m) to the Pleistocene (neritic depths <60 m; Robinson, 1969).

BACKGROUND

Regional and local setting

Hill (1899) described strata that are exposed east of the town of Buff Bay, Jamaica (Fig. 1), in surveys that he made for Alexander Agassiz. He designated over 30 m (100 ft) of "... bluish white, earthy, semi-indurated marl...." as the Buff Bay Beds, and reported that these beds rest directly on a "...pure, white chalky marl which in its lower part grades into the Montpelier beds...." (Hill, 1899, p. 84). The latter were believed to be lower Oligocene. Hill (1899) suggested that

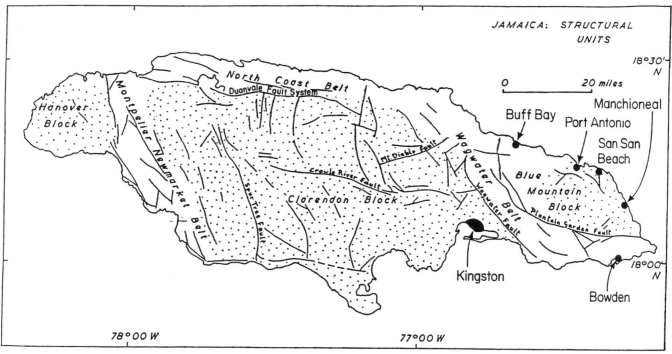

Figure 1. Structural map of Jamaica (from Wright, 1974).

the previously reported Bowden Beds correlate with the Buff Bay Beds.

Robinson (1969) proposed formal names for Jamaican sections, including those east of the town of Buff Bay. He refined Hill's (1899) designations, assigning only part of the original beds near Buff Bay to the Montpelier and Bowden Formations. He named Hill's white marl the Spring Garden Member of the uppermost Montpelier Formation and used Bolli's (1957) zones to determine that it is Miocene rather than Oligocene, as previously believed. A conformable contact separates the Spring Garden Member from the overlying semi-indurated marl of the Buff Bay Formation. As formally designated, the Buff Bay Formation includes those strata that Hill referred to as the Buff Bay Beds (Robinson, 1969). An unconformable contact separates the Buff Bay Formation from the overlying unit, which Robinson (1969) described as a series of calcareous foraminiferal clays irregularly interbedded with clayey bioclastic limestones with a rich coral fauna. Hill (1899) assigned the upper beds near Buff Bay to the Bowden Marls, although his Figure 23 and Plate 17 indicate that he included the Buff Bay Beds as part of the Bowden. Robinson (1969) correlated the strata overlying the Buff Bay Formation at its type section to the Bowden Formation, which has its type locality near Bowden on the southeast coast (Fig. 1).

Caribbean foraminiferal studies

Uplifted exposures of deep-water sediments provided the only available pre-Pleistocene deep-sea record until the advent of the Deep Sea Drilling Project in 1967. The Caribbean region contains many of these classic uplifted deep-water sections. The sedimentary sequence exposed east of the town of Buff Bay on the north coast of Jamaica (Fig. 1) contains one of the most complete middle Miocene deep-water sequences in the Caribbean and has been the focus of several micropaleontological studies.

Vaughan (1928) pioneered benthic foraminiferal research in Jamaica. Cushman and Jarvis (1930, 1936) conducted preliminary studies of species from the Buff Bay outcrops. Cushman and Todd (1945) studied the benthic foraminiferal taxonomy of the Buff Bay faunas in greater detail, although most of their samples were from the Pliocene Bowden Formation (Robinson, 1969). The diverse assemblages described by Cushman and Todd (1945) include twenty-four new species and eight new varieties. Parker (1945) addressed foraminiferal taxonomy from the Bowden Formation of Jamaica, although her samples were not from the Buff Bay section. Steineck (1974) used benthic foraminifera to estimate the paleobathymetry of the Cenozoic Jamaican sections. He later (1981) modified his benthic foraminiferal estimates based on ostracods (see "Paleobathymetry" section).

Other publications on Caribbean Cenozoic benthic foraminifera describe faunas similar to those found at Buff Bay. Coryell and Rivero (1940) focused on the Miocene microfauna of Haiti. Cuban Miocene assemblages were the subject of several papers (Palmer and Bermúdez, 1936a, b; Palmer, 1940, 1941). Renz (1948) studied the Agua Salada Group of Venezuela. Bermúdez (1949) reported on the Tertiary smaller foraminifera of the Dominican Republic. McLaughlin and Sen Gupta (1989a, b, in preparation) discussed basin evolution and the paleoenvironmental significance of Neogene benthic foraminifera from the Dominican Republic. Deep-water calcareous benthic foraminiferal faunas were described from Miocene sections in Costa Rica by Cassel and Sen Gupta (1989a, b).

Robinson (summary in 1969) supplied samples (designated ER21 through ER44) from the Buff Bay region to micropaleontologists throughout the world. These samples provided the basis for the construction of much of the Cenozoic biostratigraphic framework. The sections exposed near Buff Bay were used as type localities for many of Blow's (1969, 1979) planktonic foraminiferal zones (Berggren, this volume). Bolli (1970) studied ER samples and identified the index planktonic foraminiferal species from the upper Montpelier and Buff Bay Formations (Liska, 1985). Although these studies are critical to global biostratigraphy, the previous sample coverage of the sections exposed near Buff Bay was broad (e.g., the ER21 to ER44 samples were taken at an average of ~1.2 m [4 ft] apart, representing three samples/planktonic foraminiferal zone).

The previous foraminiferal studies of the sections exposed near Buff Bay concentrated on taxonomy (e.g., Cushman and Jarvis, 1930, 1936; Cushman and Todd, 1945) or biostratigraphy (e.g., Blow, 1969, 1979; Bolli, 1970; Liska, 1985). The foraminiferal biostratigraphy of these sections is not straightforward because of the wide sample spacing, problems in the local sections (reworking of older material and downslope contamination), and lack of independent stratigraphic control (e.g., magnetostratigraphy, stable isotope stratigraphy). These problems were discussed by Liska (1985), and are addressed in studies by our joint field party (Berggren, this volume; Aubry, this volume; Kent and van Fossen, unpublished data of 1992; Miller, unpublished data of 1992).

The sections exposed in Jamaica provide a record of a critical interval of deep-water benthic foraminiferal evolution. The early to middle Miocene was a period of change in benthic foraminiferal faunas worldwide (Berggren, 1972; Schnitker, 1979). Schnitker (1986), Thomas (1986a, b), and Miller and Katz (1987a) reported that a late early to middle Miocene taxonomic turnover occurred in the North Atlantic, and Thomas (1985, 1986b), Woodruff (1985), Boersma (1986), and Thomas and Vincent (1987) documented this event in the Pacific. Comparison of Caribbean deep-water sections with Miocene open-ocean sites shows that greater diversity is found in the Caribbean benthic foraminiferal faunas. Despite the differences in assemblages, the well-documented global late early to middle Miocene benthic foraminiferal taxonomic turnover event (Thomas, 1985, 1986a, b; Boersma, 1986; Miller and Katz, 1987a; for a different view see Boltovskoy, 1978) may be reflected in the Buff Bay faunas.

Our field party sampled the sections exposed near Buff Bay, Jamaica, in 1987. These samples provide an opportunity to: (1) revise the biostratigraphy of the section; (2) evaluate the global correlations of these strata; (3) quantitatively describe middle Miocene to Pliocene benthic foraminiferal distributions; and (4) relate paleoenvironmental changes in Jamaica to global faunal changes and regional or local tectonic changes. This contribution focuses on the paleobathymetric changes within the Spring Garden Member of the Montpelier Formation, the Buff Bay Formation, and the Bowden Formation primarily based on evidence from benthic foraminifera.

METHODS

Field sampling and lithologic description

A joint field party from Unocal (G. Jones, R. C. Tjalsma, C. Stuart), Woods Hole Oceanographic Institution (W. A. Berggren, M.-P. Aubry), and Lamont-Doherty Earth Observatory (D. V. Kent, K. G. Miller, M. van Fossen) sampled the sections exposed by roadcuts near Buff Bay (Fig. 1) in April 1987. E. Robinson (Florida International University) estimated where his ER samples were located in the new roadcut, which was modified by road construction since his (1969) samples were obtained. The roadcut exposed six sections: Lower (Pots and Pans), Middle, Main, Slide, Dead Goat Gully, and West (Fig. 2). The Pots and Pans Section is the lower Spring Garden Member of the Montpelier Formation and the Middle Section represents the upper Spring Garden Member as defined by Robinson (1969). The Main, Slide, and Dead Goat Gully Roadcuts represent the Buff Bay Formation as originally defined by Robinson (1969). The West Roadcut exposes sections correlated by Robinson (1969) to the Bowden Formation.

We collected samples from all six sections and oriented magnetostratigraphic samples from the Pots and Pans, Middle, Main, and Slide Sections. The Dead Goat Gully and West Sections were not suitable for paleomagnetic analyses because of poor exposure. Samples from the Main and Slide Roadcuts were numbered upsection BB1 through BB64, while samples from Dead Goat Gully and West Sections were numbered upsection 242 through 252 (Fig. 2). Samples from the Middle and Pots and Pans Roadcuts were numbered downsection SG1 through SG32. The sections were measured and described by C. Stuart (personal communication, 1987; Fig. 2).

The general lithologies were similar to those previously described (e.g., Hill, 1899; Robinson, 1969), although closer examination of the lithology of the Main Section resulted in the relocation of the formational boundary. The Spring Garden Member is a pure, white indurated chalk with characteristic black lichen overgrowths. In contrast, the Buff Bay Formation is a sandier, yellow-tan carbonate sediment that is only moderately indurated; lichens are not present on most of the Buff Bay Formation. Exposure of the unweathered Buff Bay Formation shows it to be slightly bluish, while the Spring Garden Member is consistently white. The contact between the underlying Spring Garden Formation and the Buff Bay Formation is gradational, with the dark lichens extending upsection into the lower one-third of the Main Section. C. Stuart (personal communication, 1987) considers the formational boundary to be at the top of this level in the Main Section; we follow this placement (Fig. 2).

There are coarse sandy layers in the Main and Slide Sec-

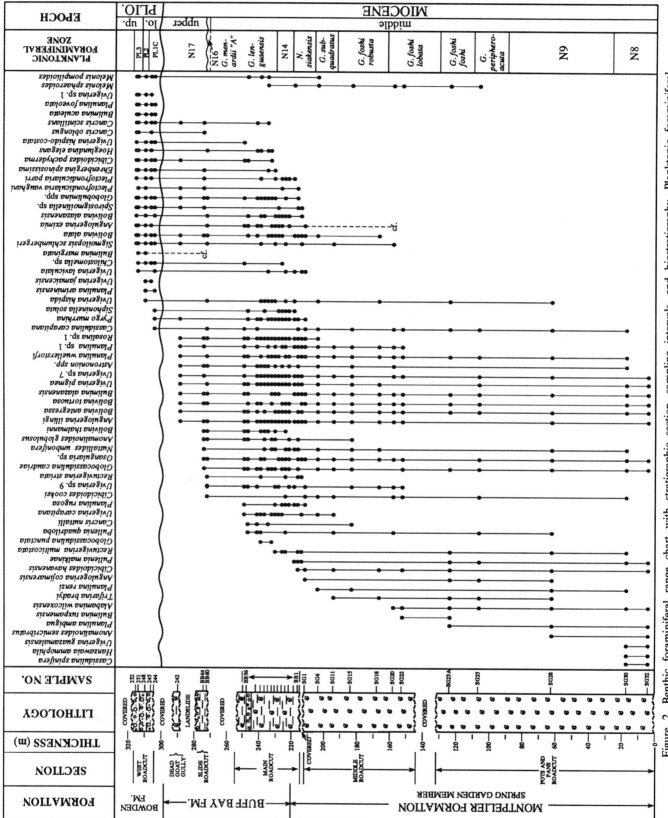

Figure 2. Benthic foraminiferal range chart with stratigraphic section, sampling intervals, and biostratigraphy. Planktonic foraminiferal biostratigraphy after Berggren (this volume).

tions. The very coarse sand and larger fraction (>1 mm) is comprised mostly of *Amphistegina* spp./*Asterigerina* spp. and reefal fragments (see "Results and Discussion"). Similar layers are not present in the Pots and Pans and Middle Sections (the Spring Garden Member sensu Robinson, 1969), but are found in the lower Main Section, assigned to the Spring Garden Member by C. Stuart (personal communication, 1987).

Biostratigraphy and magnetostratigraphy

Paleontological samples were examined for planktonic foraminiferal (W. A. Berggren) and calcareous nannoplankton (M.-P. Aubry) biostratigraphy. The planktonic foraminiferal results (Berggren, this volume) are included on Figure 2 (note that a greater number of samples were examined by Berggren for qualitative planktonic foraminiferal biostratigraphy than were examined here). Magnetostratigraphic data have been generated and will be published elsewhere (D. V. Kent and M. van Fossen, unpublished data of 1992). The integration of the biostratigraphic and magnetostratigraphic results will be published elsewhere.

We rely on a hybrid planktonic foraminiferal biostratigraphic zonation (Berggren, this volume; Fig. 2) for our correlations that uses criteria of Blow (1969, 1979) and Bolli (1957) because neither of these standard zonations is entirely suitable for correlation of the Jamaican samples (see Berggren, this volume for discussion). The magnetobiostratigraphic relationships observed in Jamaica differ somewhat from those in our previous studies (e.g., Miller, Aubry et al., 1985), and correlations to the Geomagnetic Polarity Time Scale are uncertain. Because of these difficulties, the Jamaican data are presented versus thickness of section (e.g., Figs. 2 to 12).

Benthic foraminiferal taxonomic and quantitative studies

Samples examined for benthic foraminifera were disaggregated in a rock crusher and then soaked in Alconox (an industrial detergent) for two days. Spring Garden Member samples were boiled subsequently in Quaternary-O for one hour. All samples were washed with an Alconox spray through a 63-μm sieve and air-dried. Benthic foraminifera were picked from aliquots of the greater than 150-μm-size fraction and mounted on reference slides to be used for qualitative and quantitative analyses. In general, 300 to 500 specimens were picked per sample, although several samples contained 500 to 550 specimens while others contained 240 to 300 specimens.

Our identifications of cosmopolitan deep-water taxa are based on literature and museum studies by W. A. Berggren and K. G. Miller (summarized in van Morkhoven et al., 1986). We identified taxa that were previously described from the Caribbean primarily using the species concepts of Cushman and Todd (1945) and Bermúdez (1949). Our taxonomic base for most of the uvigerinids follows Boersma (1984). Brief descriptions of some of the lesser known taxa are included in the "Taxonomic Notes" section. Not all species present were identified; for example, species of *Lagena, Fissurina,* and *Oolina* were not differentiated.

We performed Q-mode Principal Components and Varimax Factor Analyses and R-mode Principal Components Analysis on the relative abundance (percentage) data using modifications of programs provided by Lohmann (1980). The Q-mode Principal Component and Varimax Factor programs utilize a Cosine-theta matrix, standardizing each sample to unit length. R-mode Principal Components Analysis uses a correlation coefficient matrix with zero mean and unit variance. All species contribute equally in this analysis, and as a result, it explains less of the total variation in the faunal data. These programs were modified to run on a Macintosh microcomputer. These multivariate analyses objectively identify natural associations of taxa in the relative percentage data. Four Q-mode Principal Components and Factors and five R-mode Principal Components adequately describe the variance observed in the Jamaican benthic foraminiferal faunas (68 and 43% of the total faunal variance, respectively). Species that constitute greater than 1% in at least one sample were included in these multivariate analyses.

The benthic foraminiferal and percent planktonic foraminiferal (Table 1) data from the Jamaican samples were compared with faunas from three Gulf of Mexico Eureka boreholes that were drilled near the De Soto Canyon (Table 2; van Morkhoven, et al., 1986). The Eureka borehole paleodepth estimates were calculated assuming simple thermal subsidence and empirical age-subsidence curves ("backtracking"; Sclater et al., 1971; Berger and Winterer, 1974; see Miller et al., 1989, for examples). The backtracking of the Eureka boreholes will be presented elsewhere (Katz and Miller, in preparation). These De Soto Canyon sites lie on transitional or continental crust with a Jurassic rifting age; because of this, thermal subsidence has been minor since the Miocene, and sediment loading corrections make up most of the minor differences between the present water depths and paleodepths of these locations. We estimate the paleodepths at E68-136, E66-73, and E68-151A to be 600, 900, and 1,300 m, respectively.

RESULTS AND DISCUSSION

Paleobathymetry

Paleobathymetric estimates for the Jamaican sections were made by: (1) examining the Jamaican sections for depth-diagnostic taxa (depth ranges after van Morkhoven et al., 1986); and (2) calculating the percentages of planktonic foraminifera relative to total foraminifera (greater than 150-μm size fraction; Fig. 3, Table 1). We use the bathymetric terminology described by van Morkhoven et al. (1986) and Berggren and Miller (1989) in subdividing the bathyal zone into upper (200 to 600 m), middle (600 to 1,000 m), and lower (1,000 to 2,000 m) bathyal.

Benthic foraminiferal faunas indicate that both the Spring Garden Member and the Buff Bay Formation were deposited at lower bathyal depths. Comparison of the Jamaican benthic foram-

TABLE 1. PERCENT PLANKTONIC FORAMINIFERA
RELATIVE TO TOTAL FORAMINIFERA AT BUFF BAY

Thickness (m)	Sample	Planktonics (%)
315.9	BW252	68.90
315.0	BW251	69.96
310.3	BW248	55.30
306.6	BW245	58.50
304.8	BW244	69.40
290.0	BB242	97.19
274.7	BB64	95.58
273.1	BB60	94.13
249.4	BB56	94.10
247.1	BB52	94.44
242.3	BB48	94.65
240.0	BB44	93.50
237.3	BB40	97.23
235.1	BB36	95.67
233.4	BB32	94.60
230.8	BB28	97.60
228.9	BB24	97.30
226.4	BB20	98.16
224.3	BB16	97.98
222.3	BB12	94.70
219.2	BB8	96.30
216.9	BB4	97.28
214.6	BB1	91.29
212.3	SG1	96.53
204.2	SG6	95.70
194.6	SG11	97.03
183.8	SG15	95.00
167.8	SG18	95.50
158.1	SG20	94.50
153.4	SG22	94.76
124.8	SG23A	97.60
107.2	SG25	95.00
63.2	SG28	97.80
17.4	SG30	99.00
4.4	SG32	99.27

TABLE 2. DEPTH-DIAGNOSTIC BENTHIC FORAMINIFERAL TAXA
COMPARED AMONG THE BUFF BAY SECTION
AND THE EUREKA BOREHOLES*

Upper to middle bathyal taxa absent from the Spring Garden Member and the Buff Bay Formation which are present in the Eureka boreholes:
   Cibicidoides alazanensis
   Cibicidoides crebbsi (only at E68-136 and E66-73)
   Rectuvigerina nodifera (only at E68-136)
   Rectuvigerina transversa
   Planulina ariminensis (only at E68-136 and Bowden Formation)
   Planulina dohertyi

Middle to lower bathyal taxa found in the Spring Garden Member and the Buff Bay Formation which are absent from the Eureka boreholes:
   Anomalinoides globulosus
   Cibicidoides robertsonianus (also present at lower bathyal E68-151A)
   Cibicidoides havanensis
   Planulina rugosa
   Anomalinoides pseudogrosserugosus

*Paleobathymetric ranges after van Morkhoven and others, 1986.

iniferal assemblages with faunas from the three Eureka boreholes (Miller and Katz, in preparation) indicates that depths of deposition exceeded 1,300 m for the Spring Garden Member and the Buff Bay Formation (Table 2). There are a number of upper to middle bathyal marker species (van Morkhoven et al., 1986) that are absent from these units that are present in the Eureka boreholes, including *Cibicidoides alazanensis, Cibicidoides crebbsi* (present only at E68-136 and E66-73), *Rectuvigerina nodifera* (only at the shallowest site, E68-136), *Rectuvigerina transversa, Planulina ariminensis* (found in E68-136 and the Bowden Formation), and *Planulina dohertyi*. Similarly, several species characteristic of a middle to lower bathyal setting are present in the Spring Garden Member and in the Buff Bay Formation, but are absent from the Eureka boreholes, including *Anomalinoides globulosus, Cibicidoides havanensis, Planulina rugosa,* and *Anomalinoides pseudogrosserugosus. Cibicidoides robertsonianus* is a middle bathyal to abyssal taxon that is present at E68-151A (1,300 m paleodepth), in the Spring Garden Member, and in the Buff Bay Formation. The presence or absence of these depth-diagnostic species implies that the paleodepths were 1,300 to 2,000 m during the deposition of the Spring Garden Member and the Buff Bay Formation.

Several first and last occurrences of benthic foraminiferal species suggest that there was a shallowing within the lower bathyal zone from the uppermost Spring Garden Member to the lowermost Buff Bay Formation. There is a transition from the deeper-water *Melonis sphaeroides* to the shallower-water morphotype *Melonis pompilioides* in this interval. In addition, several species first occur in the uppermost Spring Garden Member or lower Buff Bay Formation that are generally rare at lower bathyal depths and increase in abundance at shallower depths (van Morkhoven et al., 1986; Katz and Miller, in preparation): *Cancris nuttali, Cibicidoides pachyderma, Plectofrondicularia parri, Plectofrondicularia vaughani,* and *Rectuvigerina striata*. Although the presence or absence of a few key benthic foraminiferal taxa suggests that a shallowing occurred, attempting to subdivide the lower bathyal zone into finer depth increments may not be warranted at this time. Nevertheless, this gradual shallowing is consistent with the gradual nature of the lithologic change from the upper Spring Garden Member to the Buff Bay Formation.

Benthic foraminiferal faunas indicate that the Bowden Formation was deposited in the upper bathyal zone. *Planulina foveolata* and *Bulimina marginata* occur only in the Bowden Formation samples at Buff Bay. These taxa are recognized as generally restricted to upper bathyal and shallower depths (van Morkhoven et al., 1986), and they are found throughout the Bowden Formation. In fact, *P. foveolata* has a lower depth limit of 500 m (van Morkhoven et al., 1986). Benthic foraminiferal faunal abundance changes indicate that there may have been a

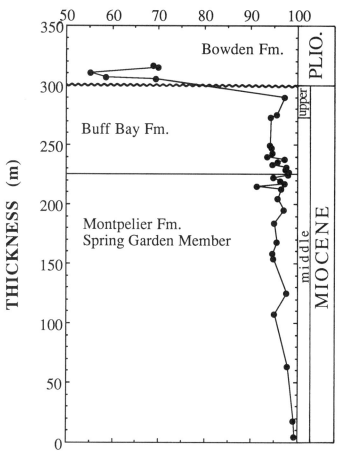

Figure 3. Percent planktonic foraminifera at Buff Bay. Refer to Figure 2 for the positions of unconformities.

the Spring Garden Member and in the Buff Bay Formation (Fig. 3, Table 1), consistent with deposition at lower bathyal paleodepths (1,000 to 2,000 m; Grimsdale and van Morkhoven, 1955). Planktonic foraminiferal percentages were lower in the Eureka samples (Katz and Miller, in preparation) than in the Spring Garden Member and the Buff Bay Formation, reflecting the shallower paleodepths at the Eureka locations.

There is a marked decrease in the planktonic foraminiferal percentages (to 55 to 70%) across the unconformity separating the Buff Bay Formation from the overlying Bowden Formation (Fig. 3). This decrease in the percentage of planktonic foraminifera is consistent with the inferred decrease in water depth. The lower planktonic foraminiferal percentages observed in the Bowden Formation samples are consistent with deposition in the upper bathyal zone (200 to 600 m; Grimsdale and van Morkhoven, 1955).

The paleodepth estimates made here compare well with those based on benthic foraminifera and ostracod faunas (Steineck, 1974, 1981). Steineck (1974) initially estimated that the Spring Garden Member of the Montpelier Formation and the basal Buff Bay Formation were deposited at paleodepths of approximately 1,500 m. He noted that there was a shallowing within the Buff Bay Formation; our paleobathymetric estimates agree with this. Steineck (1974) estimated that the middle and upper Buff Bay Formation was slightly shallower than the section below, with an upper depth limit of 1,000 m. He later (1981) recalculated these paleobathymetric estimates using ostracods. The lowermost Spring Garden Member section that we sampled (Zone N8) corresponds to the upper portion of Steineck's (1981) ostracod Biofacies 3. He interpreted this ostracod biofacies to represent depths of 1,500 to 2,000 m. Above this, the remaining section of the Spring Garden Member and the entire Buff Bay Formation is characterized by Steineck's (1981) ostracod Biofacies 4, representing paleodepths of 1,000 to 1,500 m. This corroborates our paleobathymetric estimates and our suggestion that shallowing (and inferred uplift; see also Steineck, 1974, 1981) began in the middle Miocene.

Biofacies

Downslope contamination. Q-mode Principal Components Analysis of the combined faunal datasets from the Spring Garden Member of the Montpelier Formation, the Buff Bay Formation, and the Bowden Formation reveals two primary benthic foraminiferal associations (Fig. 4, Table 3). High negative loadings on Q-mode Principal Component II represent relatively deep-water assemblages. Dominant deep-water benthic foraminifera include *Globocassidulina subglobosa, Bulimina aculeata, Gyroidinoides* spp., and *Cibicidoides mundulus.*

Approximately 5 to 45% of the benthic foraminifera identified in the Jamaican samples consists of shallow-water (neritic) species that were transported downslope and mixed with an in situ deep-water fauna (Fig. 4). These shallow-water species include *Amphistegina* spp., *Asterigerina* spp., and *Cibicides lobatu-*

slight deepening during the deposition of the Bowden Formation (see "In Situ Assemblages"). In addition, several depth-diagnostic benthic foraminiferal species found in the deeper-water Spring Garden Member and Buff Bay Formation are absent from the Bowden Formation, including *Anomalinoides globulosus, C. robertsonianus, P. rugosa,* and *P. wuellerstorfi.* The benthic foraminiferal faunas indicate that a shallowing of at least 800 m occurred at Buff Bay between the early late Miocene (ca. 8 Ma) and the Pliocene (ca. 4 to 3 Ma). Additional shallowing occurred between the deposition of the upper bathyal Pliocene Bowden Formation (this study) and the Pleistocene upper Coastal Group (Robinson, 1969). The Coastal Group contains neritic reefal foraminifera (Robinson, 1969) that lived in the photic zone (<60 m), implying that between ~140 and 500 m of shallowing occurred from the Pliocene to the Pleistocene at Buff Bay.

The relative abundances of planktonic foraminifera are consistent with the inferred paleobathymetry. Planktonic foraminifera constitute about 94 to 99% of the total foraminifera both in

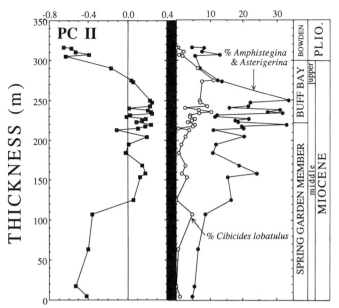

Figure 4. Q-mode Principal Component II and percentage plots of *Amphistegina* spp./*Asterigerina* spp. and *Cibicides lobatulus* at Buff Bay. Refer to Figure 2 for the positions of unconformities.

lus. Amphistegina spp. and *Asterigerina* spp. are typical of reefs, lagoons, outer shelf areas, guyots, and atolls (Todd, 1976). *Cibicides lobatulus* is typical of shallow-water and reefal environments (Brasier, 1975; Sen Gupta and Schaefer, 1973; McLaughlin, 1989). These are the three most abundant transported taxa in our samples, although other species may also have been transported downslope (see "In Situ Assemblages"). High positive loadings on Q-mode Principal Component II (Fig. 4) represents these shallow-water benthic foraminiferal assemblages that are dominated by the taxa *Amphistegina* spp./*Asterigerina* spp. and *Cibicides lobatulus*. These high loadings characterize the Buff Bay Formation and the upper Spring Garden Member, which contain the highest abundances of transported material.

Amphistegina spp. and *Asterigerina* spp. are morphologically similar genera with controversial taxonomic histories (see Todd, 1976, for discussion). Todd (1976) concludes that *Amphistegina* spp. and *Asterigerina* spp. should be maintained as distinct genera, even though the separation may be more an artificial than a natural one. These two taxa are combined in this study for the multivariate analyses because they represent similar paleoenvironmental and paleobathymetric settings.

There are numerous discrete, coarse, relatively thin, dark layers that are irregularly spaced in the Main Section (Fig. 2). Quantitative analysis reveals a difference between the benthic foraminiferal assemblages contained in these layers and the faunas yielded by the remaining sectin. These dark layers contain fragments of corals and bivalves in addition to many benthic foraminiferal specimens greater than 1,000 μm. This large size fraction primarily contains *Amphistegina* spp., *Asterigerina* spp., and *Gypsina* spp., along with a few *Lenticulina* spp. Two *Amphistegina* spp./*Asterigerina* spp.–rich layers that we examined are composed of 11 and 12% benthic foraminifera in the greater than 1,000 μm size fraction and 33 and 40% benthic foraminifera in the 355 to 1000 μm size fraction, respectively. Shallow-water benthic foraminiferal specimens (*Amphistegina* spp., *Asterigerina* spp., and *Gypsina* spp.) constitute 65 and 58% of the assemblages, respectively. The planktonic foraminiferal percentages (80 and 75%) were lower than those in the remaining Buff Bay Formation and Spring Garden Member samples.

The coarse nature of the dark layers, the high percentages of the shallow-water taxa *Amphistegina* spp., *Asterigerina* spp., and *Gypsina* spp., the fragments of corals and bivalves, and the lower percentages of planktonic foraminifera suggest that the discrete layers formed by downslope transport. High percentages of *Amphistegina* spp. and *Asterigerina* spp. were observed throughout the rest of the section (Fig. 4), although the large size fractions and large *Gypsina* spp. were not observed outside of the discrete layers. We suggest that the contamination by *Amphistegina* spp./*Asterigerina* spp. outside of the layers occurred either as a result of the steady downslope transport of reefal elements during storms or as a result of the mixing of these discrete layers after deposition.

In situ assemblages. We deleted the major shallow-water components (*Amphistegina* spp., *Asterigerina* spp., and *Cibicides lobatulus*) from the faunal dataset and conducted a Q-mode Factor Analysis on the remaining deep-water assemblages in order to examine the changes in the in situ benthic foraminiferal faunas. While there are other probable shallow-water taxa present in the Buff Bay samples (e.g., *Discorbis* spp., *Planorbulina* spp.), they are very rare and do not significantly affect the outcome of the multivariate analyses. The Q-mode Factor Analysis of the deep-water taxa delineated four major benthic foraminiferal biofacies, which will be discussed in ascending stratigraphic order.

Q-mode Factor 3 has the highest negative loadings in the lower part of the Spring Garden Member, and explains 13% of the faunal variation (Fig. 5). It is dominated by *Globocassidulina subglobosa, Cibicidoides mundulus, Stilostomella aculeata,* and *Oridorsalis* (Fig. 6). An abundance change reflected by the transition from the high negative values of Q-mode Factor 3 (the *G. subglobosa–C. mundulus–S. aculeata–Oridorsalis* biofacies) to high negative values of Factor 4 (a *Compressigerina coartata–Siphouvigerina porrecta* biofacies, see below) occurs within the Spring Garden Member (between 100 and 150 m). Steineck (1981) found a change in ostracod biofacies near this level. The relative abundance change in benthic foraminifera may reflect the well-documented global benthic foraminiferal taxonomic turnover from planktonic foraminiferal Zones N8 to N11 (e.g., Thomas, 1985, 1986a, b; Woodruff, 1985; Boersma, 1986; Miller and Katz, 1987a). One point in favor of the global nature of the faunal abundance change is that a coeval decrease occurred in the percentages of *Globocassidulina subglobosa* and *Cibicidoides mundulus* at western North Atlantic Deep Sea Drilling Project (DSDP) Site 563 (paleodepth >3 km; Miller and Katz, 1987a).

TABLE 3. BUFF BAY BENTHIC FORAMINIFERAL TAXA USED IN MULTIVARIATE ANALYSES

Alabamina wilcoxensis	Eggerella bradyi	pleurostomellids
Amphistegina spp./Asterigerina spp.	Ehrenbergina caribbea	polymorphinids
Anomalinoides globulosus	Ehrenbergina spinea	Pseudononion sp.
Anomalinoides semicribratus	Ehrenbergina spinosissima	Pullenia bulloides
Anomalinoides spp.	Elphidium spp.	Pullenia malkinae
Astrononion pusillum	Eponides spp.	Pullenia quadriloba
Bolivina aenariensiformis	Fissurina spp.	Pullenia quinqueloba
Bolivina alata	Fursenkoina spp.	Pyrgo murrhina
Bolivina alazanensis	Globobulimina spp.	Quinqueloculina spp.
Bolivina antegressa	Globocassidulina subglobosa	Rectuvigerina multicostata
Bolivina thalmanni	Globocassidulina palmerae	Rectuvigerina striata
Bolivina tortuosa	Globocassidulina caudriae	Reusella spinulosa var. pulchra
Bolivina floridana and pseudoplicata	Gyroidinoides spp.	Rosalina sp. 1
Bulimina alazanensis	Hanzawaia mantaensis	Sigmoilopsis schlumbergeri
Bulimina mexicana and macilenta	Hanzawaia ammophila	Siphoninella soluta
Bulimina marginata	Hanzawaia sp. 3	Siphotextularia spp.
Bulimina tuxpamensis	Hanzawaia spp.	Siphouvigerina porrecta
Bulimina impendens	Hoeglundina elegans	Sphaeroidina bulloides
Bulimina aculeata	Karreriella subglabra	Spirosigmoilinella sp.
Cancris nuttalli	Lagena and Oolina spp.	Stilostomella aculeata
Cancris oblongus	Laticarinina pauperata	Stilostomella curvatura
Cancris scintillans	Lenticulina spp.	Stilostomella modesta
Cancris spp.	Loxostomum isidroensis	Stilostomella subspinosa
Cassidulina crassa	Loxostomum spp.	Stilostomella spp.
Cassidulina carapitana	Melonis barleeanum	Textularia spp.
Cassidulina laevigata	Melonis pompilioides and sphaeroides	Trifarina bradyi
Cassidulina reflexa	Neoeponides spp.	Uvigerina laviculata
Cassidulina spinifera	Nonion spp.	Uvigerina carapitana
Cassidulinoides spp.	Nonionellina sp.	Uvigerina hispida
Chilostomella sp.	Nuttallides umbonifera	Uvigerina pigmea
Cibicidoides bradyi	Oridorsalis spp.	Uvigerina sp. 1
Cibicidoides cicatricosus	Osangularia sp.	Uvigerina cojimarensis
Cibicidoides havanensis	Orthomorphina spp.	Uvigerina proboscidea
Cibicidoides incrassatus	Pararotalia sp.	Uvigerina hispido-costata
Cibicidoides matanzasensis	Planorbulina sp.	Uvigerina jamaicensis
Cibicidoides mundulus	Planulina ariminensis	Uvigerina sp. 7
Cibicidoides pachyderma	Planulina ambigua	Uvigerina sp. 9
Cibicidoides coryelli	Planulina foveolata	Uvigerina sp. 53
Cibicidoides dominicus	Planulina renzi	Angulogerina illingi
Cibicidoides sp. 33	Planulina rugosa	Angulogerina eximia
Cibicidoides cookei	Planulina wuellerstorfi	Vulvulina spp.
Cibicides lobatulus	Planulina spp.	Verneuilina spp.
Compressigerina coartata	Plectofrondicularia parri	Siphonina spp.
Dentalina spp.	Plectofrondicularia vaughani	
Discorbis spp.	Pleurostomella spp.	

Alternatively, the faunal abundance change may have resulted from a local change. This is supported by the nature of the change from a cosmopolitan lower bathyal biofacies (the *G. subglobosa–C. mundulus–S. aculeata–Oridorsalis* biofacies) to an apparently endemic biofacies (*Compressigerina coartata–Siphouvigerina porrecta* biofacies; see below) that also contains cosmopolitan elements. This local change may have been caused by: (1) regional tectonic basinal isolation; (2) increased downslope transport that caused dilution of the in situ, deeper-water biofacies; or (3) progressive shallowing of this part of the Jamaican section within the lower bathyal zone. We find that the faunal changes (see "Paleobathymetry") indicate that there was a shallowing from the deeper part of the lower bathyal zone during the deposition of the Spring Garden Member to the upper part of the lower bathyal zone during the deposition of the Buff Bay Formation.

Q-mode Factor 4 explains 22% of the faunal trend and it characterizes the upper part of the Spring Garden Member and a portion of the Buff Bay Formation (Fig. 5). Dominant taxa in the Q-mode Factor 4 biofacies are *Compressigerina coartata* and *Siphouvigerina porrecta* (Fig. 7). Interpreting the paleoenvironmental significance of these two species is difficult. Neither has been reported in studies of North Atlantic, South Atlantic, and Pacific deep-water Miocene sections recovered by the Deep Sea Drilling Project (e.g., Thomas, 1985, 1986a, b; Woodruff, 1985; Boersma, 1986; Miller and Katz, 1987a), or observed in the

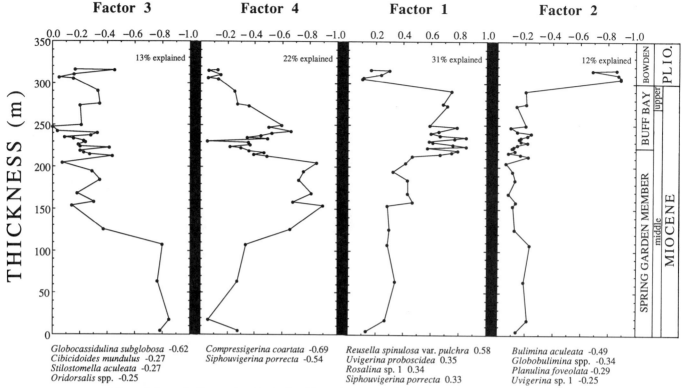

Figure 5. Q-mode Factors 1 to 4 calculated from the deep-water benthic foraminiferal dataset at Buff Bay. Refer to Figure 2 for the positions of unconformities.

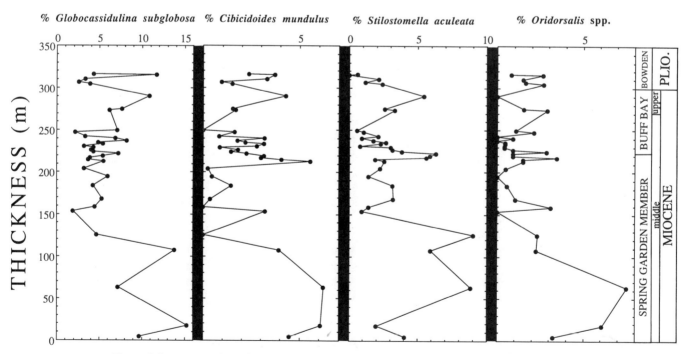

Figure 6. Percentages plots of the taxa that characterize Q-mode Factor 3. Refer to Figure 2 for the positions of unconformities.

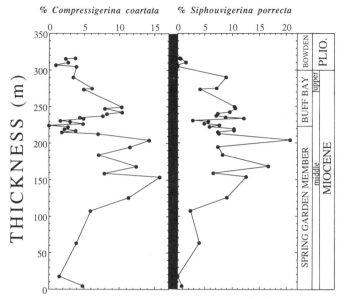

Figure 7. Percentages plots of the taxa that characterize Q-mode Factor 4. Refer to Figure 2 for the positions of unconformities.

coeval Eureka bathyal boreholes (Katz and Miller, in preparation), the Miocene of Haiti (Coryell and Rivero, 1940), the Tertiary Agua Salada group of Venezuela (Renz, 1948), or Tertiary sections from Cuba (Palmer and Bermúdez, 1936a, b; Palmer, 1940). Bermúdez (1949) reported that *Compressigerina coartata* and *Siphouvigerina porrecta* are present in samples from the Dominican Republic. Since the abundances of these two taxa do not co-vary with the abundances of the transported shallow-water taxa, they are interpreted as being in situ deep-water forms. In fact, the interval of high negative Q-mode Factor 4 scores of the deep-water dataset (the *C. coartata* and *S. porrecta* biofacies) alternates with the *Amphistegina* spp./*Asterigerina* spp. thanatofacies.

R-mode Principal Components Analysis of the complete Buff Bay section dataset (including the shallow-water taxa) supports the in situ nature of the Q-mode Factor 4 biofacies. Negative loadings on R-mode Principal Component III are characterized by *Compressigerina coartata*, *Cassidulina reflexa*, and *Siphouvigerina porrecta*, while *Amphistegina* spp./*Asterigerina* spp. does not load onto this component. Shallow-water taxa dominate the negative loadings on R-mode Principal Component II and characterize the upper Spring Garden Member and the Buff Bay Formation with *Cibicides lobatulus* and *Amphistegina* spp./*Asterigerina* spp., in addition to *Rosalina* sp. 1, *Uvigerina pigmea*, and *Reussella spinulosa* var. *pulchra* (Fig. 8). Negative loadings on R-mode Principal Component III and negative loadings on R-mode Principal Component II do not co-vary.

Q-mode Factor 1 accounts for 31% of the faunal variation and is found in the upper Spring Garden Member and most of the Buff Bay Formation (Fig. 5). It is dominated by *Reussella spinulosa* var. *pulchra*, *Uvigerina proboscidea*, *Rosalina* sp. 1, and *Siphouvigerina porrecta* (Fig. 9). *Reussella spinulosa* has been reported as being abundant in neritic biofacies (e.g., McLaughlin, 1989). In addition, *R. spinulosa* is common in faunas that contain both in situ deep-water species and transported species off the northeastern Australian margin (Katz and Miller, 1993). With the available data, we cannot say for certain if Q-mode Factor 1 represents an in situ biofacies or a transported thanatofacies. This factor does not consistently coincide with the high positive loadings on Q-mode Principal Component II that represent the highest abundances of the *Amphistegina* spp./*Asterigerina* spp. thanatofacies, suggesting that it may be an in situ biofacies. However, percentage plots comparing *Cibicides lobatulus* and *Reussella spinulosa* var. *pulchra* (Fig. 10) show three distinct common peaks. In addition, two of the dominant Q-mode Factor 1 species, *Reussella spinulosa* var. *pulchra* and *Rosalina* sp. 1, load onto R-mode Principal Component II, which is characterized by the shallow-water forms *Cibicides lobatulus* and *Amphistegina* spp./*Asterigerina* spp. (Fig. 8). This may indicate that *Reussella spinulosa* var. *pulchra* and *Rosalina* sp. 1 are shallow-water taxa that were transported downslope with *Cibicides lobatulus* and *Amphistegina* spp./*Asterigerina* spp.; alternatively, this simply may reflect high abundances of *Reussella spinulosa* var. *pulchra* and *Rosalina* sp. 1, along with *Uvigerina proboscidea*, in the Buff Bay Formation (Fig. 9). R-mode Principal Components and Factor Analyses on the Buff Bay Formation dataset without the Spring Garden Member and Bowden Formation samples help to clarify this issue. The results of the analyses show that neither *Reussella spinulosa* var. *pulchra* nor *Rosalina* sp. 1 load onto the same principal component or factor as the shallow-water taxa *Amphistegina* spp./*Asterigerina* spp. and *Cibicides lobatulus*. While the limited evidence cited above suggests that the *Reussella spinulosa* var. *pulchra*-dominated assemblage may be a transported thanatofacies, Q-mode Factor 1 may represent an in situ biofacies primarily based on results of the R-mode Principal Components and Factor Analyses on the Buff Bay Formation dataset.

Q-mode Factor 2 characterizes the Bowden Formation with 12% of the faunal variation explained (Fig. 5). It is dominated by *Bulimina aculeata*, *Globobulimina* spp., *Planulina foveolata*, and *Uvigerina* sp. 1 (Fig. 11). Separate factor analysis of the Bowden Formation faunal dataset yields two important factors (Fig. 12). Q-mode Factor 1 of the isolated Bowden Formation dataset represents 52% of the faunal variation and is most pronounced in the lower three samples. This biofacies is dominated by *Bulimina aculeata*, *Globobulimina* spp., *Lenticulina* spp., *Melonis barleeanum*, and *Planulina foveolata*. The upper two samples have higher loadings on Q-mode Factor 2, explaining 37% of the faunal variation. Q-mode Factor II is dominated by *Globocassidulina subglobosa*, *Melonis pompiloioides*, *Sphaeroidina bulloides*, *Uvigerina* sp. 1, *Gyroidinoides* spp., and *Cibicidoides mundulus*.

Both of the Bowden Formation biofacies delineated by the factor analysis were deposited in the upper bathyal zone (200 to 600 m; see "Paleobathymetry"). The presence of *Planulina foveolata* in this section indicates that the paleodepth estimate can be

R-MODE PRINCIPAL COMPONENT II

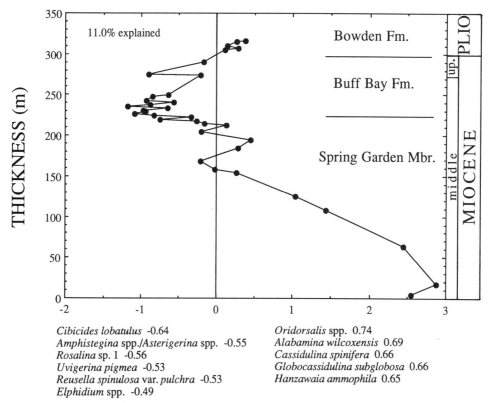

Figure 8. R-mode Principal Component II. Refer to Figure 2 for the positions of unconformities.

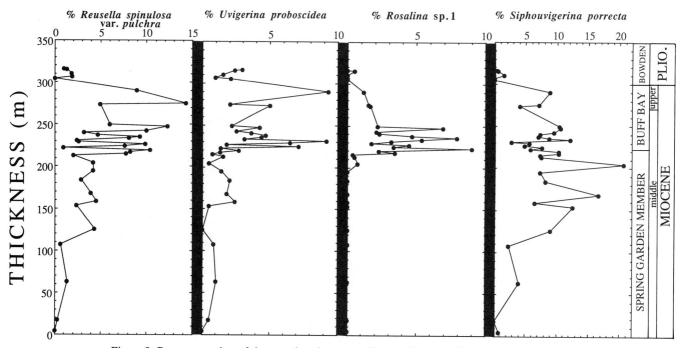

Figure 9. Percentages plots of the taxa that characterze Q-mode Factor 1. Refer to Figure 2 for the positions of unconformities.

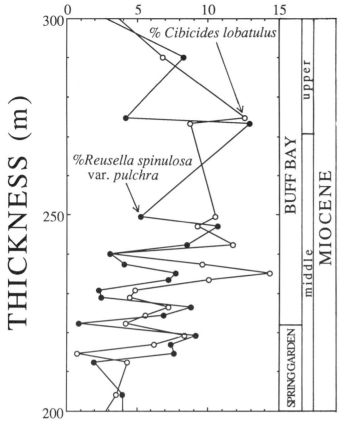

Figure 10. Percentage plots of *Reusella spinulosa* var. *pulchra* and *Cibicides lobatulus* (calculated from a dataset with *Amphistegina* spp. and *Asterigerina* spp. deleted). Refer to Figure 2 for the positions of unconformities.

further refined to 200 to 500 m. The higher abundances of *Planulina foveolata* in the lower of the two Bowden Formation biofacies and the higher abundances of typically deeper-water taxa in the upper Bowden Formation biofacies indicate that there may actually have been a slight deepening from the upper part of the upper bathyal zone to the lower part during the Pliocene. However, subdividing the upper bathyal zone with this precision may not be warranted.

Stratigraphic ranges

We compiled a benthic foraminiferal range chart for the entire Buff Bay section (Fig. 2). Because of space limitations, about 90 continuously ranging or very rare taxa were omitted from Figure 2. Thirty-six of the 64 taxa shown here last occur within this section; 19 of these last occur near the top of the Main Section within the Buff Bay Formation immediately below the unconformity. Many of these last occurrences are local; for example, *Melonis sphaeroides, Rectuvigerina striata, Anomalinoides globulosus,* and *Uvigerina carapitana* are all known to range above this interval elsewhere (van Morkhoven et al., 1986).

The calibration of the first and last occurrences at Buff Bay to planktonic foraminiferal biostratigraphy (Fig. 2) shows that a few may be biostratigraphically useful. For example, *Anomalinoides globulosus* first occurs in Zone N12 both in Jamaica and at Deep Sea Drilling Site 563, although van Morkhoven et al., 1986) reported that this species first occurred in Biochron N15 (see "Taxonomic Notes"). The first occurrence of *Planulina wuellerstorfi* is in Zone N8 in Jamaica; Thomas (1985, 1986b) also reported the first occurrence of this taxon in Zone N8 in the

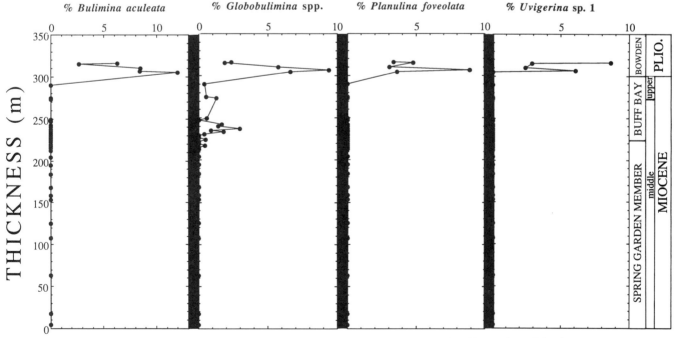

Figure 11. Percentages plots of the taxa that characterize Q-mode Factor 2. Refer to Figure 2 for the positions of unconformities.

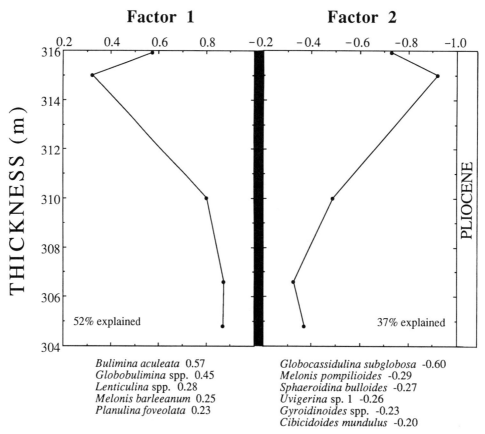

Figure 12. Q-mode Factors 1 and 2 calculated from the deep-water benthic foraminiferal dataset for the Bowden Formation samples.

Pacific. These are among the earliest reported occurrences of *P. wuellerstorfi*, although an older occurrence (Zones N6 to N7) has been reported from Costa Rica (Cassel and Sen Gupta, 1989a, b).

Some of the first and last occurrences found in this section resulted from paleobathymetric changes. The ranges of the following taxa reflect the shallowing from the late Miocene to Pliocene:

1. *Nuttallides umbonifera* has been reported as abundant at abyssal depths and as scattered through the bathyal zone in Miocene sections from the Pacific (Woodruff, 1985), and it is present only in the Spring Garden Member and the Buff Bay Formation at Jamaica.

2. *Planulina wuellerstorfi* is present only in the Spring Garden Member and the Buff Bay Formation; van Morkhoven et al. (1986) report that it is primarily a lower bathyal to abyssal taxon, while Katz and Miller (in preparation) have found that it occurs as shallow as 500 m in the Gulf of Mexico and increases in abundance with depth.

3. *Planulina ariminensis* extends from outer neritic depths to approximately 800 m (van Morkhoven et al., 1986); it is present in the Bowden Formation, yet is absent from the rest of the section.

4. *Bulimina marginata* is present only in the Bowden Formation and is generally restricted to upper bathyal and shallower depths (van Morkhoven et al., 1986).

5. *Planulina foveolata* is useful as both an age and a depth indicator in the Bowden Formation, as it is known to first occur in the early Pliocene (Zone N18) and it is regarded as a useful guide fossil for Pliocene-Pleistocene outer neritic-upper bathyal environments (van Morkhoven et al., 1986) with a lower depth limit of 500 m.

The ranges of the following taxa indicate that a shallowing may have occurred during the middle-late Miocene deposition of the Spring Garden Member and Buff Bay Formation:

1. The transition from the deeper-water morphotype *Melonis sphaeroides* to the shallower-water form *M. pompilioides* occurs in the lower part of the Main Section (Fig. 2; see van Morkhoven et al., 1986, for discussion of these morphotypes).

2. Van Morkhoven et al. (1986) report that *Plectofrondicularia parri* and *P. vaughani* are commonly associated with outer neritic to middle bathyal environments, while Katz and Miller (in preparation) found rare specimens of *P. vaughani* in the upper lower bathyal zone; these two species are absent from the Spring Garden samples, yet they are found in the Buff Bay and Bowden Formation samples.

3. *Cancris nuttalli* is reported primarily from upper and

middle bathyal deposits (van Morkhoven et al., 1986), although it may extend into the upper lower bathyal zone (Katz and Miller, in preparation); it is present only in one Spring Garden Member sample and several Buff Bay Formation samples, but is absent from most of the deeper Spring Garden section.

4. *Rectuvigerina striata* is primarily an upper and middle bathyal taxon (van Morkhoven et al., 1986) that is sometimes rare in the lower bathyal zone (Katz and Miller, in preparation); it is present only in the Buff Bay Formation and the uppermost Spring Garden Member.

5. *Cibicidoides pachyderma* is primarily an upper bathyal species, although it has been found in shelf assemblages and at depths of 3,615 m (van Morkhoven et al., 1986); it is present in the Buff Bay and Bowden Formations, but is absent from the Spring Garden Member.

Two of the taxa discussed above (*M. pompilioides* and *P. parri*) were reported previously to be middle bathyal or shallower (van Morkhoven et al., 1986). These taxa are found in the uppermost Spring Garden Member and the Buff Bay Formation, suggesting that these units may have been deposited at middle bathyal or shallower depths. However, based on the evidence presented here, we interpret the Spring Garden Member and the Buff Bay Formation as lower bathyal with estimated paleodepths of 1,300 to 2,000 m (below the depth of E68-151A). We suggest that the above two taxa have had wider depth ranges than suggested by van Morkhoven et al. (1986).

SUMMARY AND CONCLUSIONS

Sections exposed east of the town of Buff Bay, Jamaica, provide a record of earliest middle Miocene to Pliocene paleoenvironmental and paleoceanographic changes. Quantitative analyses of benthic foraminifera contained in these sections delineates a transported thanatofacies of shallow-water reefal elements and four deep-water (bathyal) biofacies. The transported elements are found in discrete layers that contain high percentages of large (>1 mm) *Amphistegina* spp., *Asterigerina* spp., and *Gypsina* spp. along with coral and bivalve fragments. In addition, high abundances (5 to 45%) of *Amphistegina* spp./*Asterigerina* spp. are scattered throughout the sections; these result either from the steady downslope transport of reefal elements by storms or from the mixing of the *Amphistegina* spp./*Asterigerina* spp.-rich layers after deposition.

The four biofacies in the Buff Bay sections record deposition in the bathyal zone and progressive upsection shallowing:

1. A typical lower bathyal biofacies (estimated paleodepth 1,300 to 2,000 m) dominates the lower part of the Spring Garden Member (lower middle Miocene).

2. An unusual *Compressigerina coartata* and *Siphouvigerina porrecta* biofacies is found in the middle Miocene upper Spring Garden Member and lower Buff Bay Formation. R-mode Principal Components Analysis supports the in situ lower bathyal nature of this biofacies, which is apparently unique to Jamaica and portions of the Dominican Republic.

3. A *Reusella spinulosa* var. *pulchra* biofacies found in the Buff Bay Formation may represent either in situ lower bathyal deposition or transport from further upslope and mixing with in situ deeper-water species.

4. A typical upper bathyal (200 to 500 m estimated paleodepth) biofacies dominates the Bowden Formation (*Bulimina aculeata*, *Globobulimina* spp., *Planulina foveolata*, and *Uvigerina* sp. 1).

Benthic foraminiferal biofacies, key bathymetric marker species, and the relative abundances of planktonic foraminifera document that there was a distinct shallowing from the lower bathyal Miocene Spring Garden Member of the Montpelier Formation and the Buff Bay Formation (1,300 to 2,000 m estimated paleodepth) to the upper bathyal Pliocene Bowden Formation (200 to 500 m estimated paleodepth). This shallowing occurred during an approximately 4 to 5 m.y. hiatus between the late Miocene (ca. 8 Ma) and the Pliocene (ca. 4 to 3 Ma; Berggren, this volume; Aubry, this volume), and reflects the local tectonic uplift of the north coast of Jamaica. There may have been a gradual shallowing within the lower bathyal zone during the middle to early late Miocene, as is indicated by the benthic foraminiferal biofacies changes and the presence/absence of key bathymetric marker species. A biofacies change that occurred in the early middle Miocene (during the deposition of the Spring Garden Member) may reflect either this local paleobathymetric change or the well-documented global benthic foraminiferal taxonomic turnover from Zones N8 to N11. Changes in benthic foraminiferal faunas suggest that a slight deepening within the upper bathyal realm may have occurred during the deposition of the Pliocene Bowden Formation. Further uplift from the upper bathyal zone to the neritic zone occurred between the Pliocene Bowden Formation and the Pleistocene Upper Coastal Group.

TAXONOMIC NOTES

This section provides references for most of the benthic foraminiferal species included in this study. Brief descriptions of some of the lesser known taxa are included. We describe "rare" taxa as those with fewer than two to three specimens in a sample. Not all species present were identified; for example, species of *Lagena*, *Fissurina*, and *Oolina* were not differentiated (Table 3). Most identifications were based on literature studies, except for those taxa also considered by van Morkhoven et al. (1986), who made comparisons with type material.

Alabamina wilcoxensis Toulmin

Pulvinulinella exigua (H. B. Brady) var. *obtusa* (Burrows and Holland).-Cushman and Ponton, 1932, p. 71, Plate 9, Figs. 9a–c.-Jennings, 1936, p. 192, Plate 31, Figs. 4a–b.-Howe, 1939, p. 81, Plate 11, Figs. 4–6.
Pulvinulinella obtusa (Burrows and Holland).-Cushman and Garrett, 1939, p. 87, Plate 15, Figs. 12–13.
Alabamina wilcoxensis Toulmin, 1941, p. 603, Plate 81, Figs. 10–14; p. 605, cf. 4A–C.

We distinguish this species from the similar form *A. mississippiensis* Todd on the basis of the plano-convex nature of the Buff Bay specimens.

Amphistegina spp.

Amphistegina spp. and *Asterigerina* spp. are morphologically similar genera with controversial taxonomic histories (see Todd, 1976, for discussion). Todd (1976) concludes that *Amphistegina* spp. and *Asterigerina* spp. should be maintained as distinct genera, even though the separation may be more an artificial than a natural one. These two taxa are combined in this study for statistical analyses because they represent the same paleoenvironmental and paleobathymetric settings.

Angulogerina cojimarensis Palmer
Figure 15, Photos 11–12

Angulogerina cojimarensis Palmer, 1941, p. 186, Plate 15, Fig. 8.-Bermúdez, 1949, p. 214, Plate 13, Figs. 49–50.

The test is irregularly triangular in cross section with blunt angles. The coiling tends to loosen in the upper portions.

Angulogerina eximia Cushman and Jarvis
Figure 18, Photos 9–10

Angulogerina eximia Cushman and Jarvis, 1936, p. 3, Plate 1, Figs. 11–12.-Coryell and Rivero, 1940, p. 342, Plate 44, Fig. 26.-Palmer, 1941, p. 186, Plate 15, Fig. 18.-Cushman and Todd, 1945, p. 52, Plate 8, Fig. 1.-Bermúdez, 1949, p. 214–215, Plate 13, Figs. 59–63.

Test is subtriangular in cross section with small, longitudinal costae.

Angulogerina illingi Cushman and Renz
Figure 18, Photos 5–6

Angulogerina illingi Cushman and Renz, 1941, p. 21, Plate 3, Figs. 19–20.-Cushman and Todd, 1945, p. 52, Plate 8, Fig. 2.-Renz, 1948, p. 114, Plate 7, Figs. 31–32.-Bermúdez, 1949, p. 216, Plate 13, Fig. 52.
Angulogerina yumuriana Palmer, 1941, p. 186, Plate 15, Fig. 8.

This form is strongly triangular in cross section. Three keels at the triangle apices run the length of the test in larger specimens. The three corners of each chamber tend to partially extend over the previous chamber.

Angulogerina selseyensis Heron-Allen and Earland

Uvigerina selseyensis Heron-Allen and Earland, 1909, p. 437, Plate 18, Figs. 1–3.-Cushman, 1913, p. 93, Plate 42, Figs. 5a–b.
Angulogerina selseyensis (Heron-Allen and Earland).-Bermúdez, 1949, p. 218, Plate 13, Fig. 53.

This rare form (generally less than 1%) resembled *Siphouvigerina porrecta* at Buff Bay and was combined with it for multivariate analyses.

Anomalinoides globulosus (Chapman and Parr)
Figure 19, Photos 11–13

Anomalina globulosa Chapman and Parr, 1937, v. 1, pt. 2, p. 117, Plate 9, Fig. 27.
Anomalinoides globulosus (Chapman and Parr).-van Morkhoven et al., 1986, p. 36–38, Plate 9, Figs. 1–3.

Anomalinoides globulosus is very similar to *Anomalinoides semicribratus*. *A. globulosus* has a distinct lip, while *A. semicribratus* is more coarsely perforate on the umbilical surface. Discrepancies in the reported stratigraphic ranges result from difficulties in differentiating *Anomalinoides globulosus* from *Anomalinoides semicribratus*. We report the first occurrence of *Anomalinoides globulosus* in Zone N12 at Jamaica and at other locations (Miller and Katz, 1987a), while van Morkhoven et al. (1986) report its first occurrence in Zone N15. We believe that forms with the distinct apertural lips of *Anomalinoides globulosus* are recognized in upper middle Miocene sediments (Zone N12), representing the global first occurrence of this taxon.

Anomalinoides semicribratus (Beckmann)

Anomalina pompilioides Galloway and Heminway var. *semicribrata* Beckmann, 1954, v. 2, p. 400, Plate 27, Fig. 3 (above).
Anomalinoides semicribratus (Beckmann).-van Morkhoven et al., 1986, p. 147–149, Plate 48, Figs. 1–3.

See *Anomalinoides globulosus* for comments.

Asterigerina spp.
Figure 13, Photos 4–6

See notes on *Amphistegina* spp.

Astrononion pusillum Hornibrook

Astrononion pusillum Hornibrook, 1961, p. 96, Plate 12, Figs. 229, 236.-Miller, 1983, p. 431, Plate 3, Fig. 3.

Bolivina aenariensiformis Subbotina
Figure 17, Photos 3–4

Bolivina aenariensiformis Subbotina, 1960, vypusk 153, sbornk 11, p. 223–224, Plate 5, Figs. 1–6.

Tests tend to be flat with an acute periphery. The medial suture forms a zig-zag pattern. Longitudinal costae may be faintly visible.

Bolivina alata (Seguenza)
Figure 17, Photo 7

Vulvulina alata Seguenza, 1862, p. 15, Plate 2, Figs. 5, 5a.
Bolivina alata (Seguenza).-Cushman and Todd, 1945, p. 42, Plate 6, Fig. 25.-Renz, 1948, p. 116, Plate 12, Figs. 12a–b.-Bermúdez, 1949, p. 187, Plate 12, Fig. 25.-Phleger and Parker, 1951, p. 12, Plate 6, Fig. 11.

A spine extends from the end of each chamber.

Bolivina alazanensis Cushman
Figure 17, Photos 5–6

Bolivina alazanensis Cushman, 1926, v. 1, pt. 4, p. 82, Plate 12, Fig. 1.-Renz, 1948, p. 117, Plate 12, Fig. 7.

A narrow keel widens at the base of the test, often terminating in several short spines. Chambers appear to be separated by clear shell material.

Bolivina antegressa Subbotina

Bolivina antegressa Subbotina, 1953, p. 226, Plate 10, Figs. 11–16.-Miller, Curry, et al., 1985, Plate 4, Fig. 11.-Miller and Katz, 1987a, p. 121, Plate 1, Figs. 4a–b.-Miller and Katz, 1987b, p. 279, Plate 2, Figs. 1, 2.
Bolivina tectiformis Cushman-Bermúdez, 1949, p. 195, Plate 12, Fig. 47.-Tjalsma, 1983, p. 739, Plate 1, Figs. 3a–b.

Cushman (1926) illustrated faint striae on his *B. tectiformis*. However, his type specimen shows a faint medial ridge with no striae. Therefore, we follow Miller and Katz (1987a) in adopting Subbotina's taxon for this striate bolivinid.

Bolivina thalmanni Renz
Figure 17, Photos 1–2

Bolivina thalmanni Renz, 1948, p. 120, Plate 12, Fig. 13.

The most distinctive feature of this taxon is the prominent ornamentation of the test. Two sharp ridges run from the initial end roughly parallel to the periphery on either side of the medial line. Two additional peripheral ridges may form on larger specimens. The ridges are connected by irregular, sharply raised lamellar ridges, forming irregular deep depressions. This ornamentation gives this distinctive test surface a strongly reticulate appearance.

Bolivina tortuosa Brady
Figure 15, Photo 5

Bolivina tortuosa Brady, 1884, v. 9, p. 420, Plate 52, Figs. 31–34.-Cushman and Todd, 1945, p. 44, Plate 7, Fig. 6.-Bermúdez, 1949, p. 195, Plate 12, Fig. 45.

This small *Bolivina* is easily recognized by the characteristic twist in the early portion of its test.

Bolivinita quadrilatera (Schwager)

Textularia quadrilatera Schwager
Bolivinita quadrilatera (Schwager).-Cushman, 1927c, p. 90.

Rare specimens were scattered through the Buff Bay exposure.

Bolivinita sp.

Rare specimens were scattered through the Buff Bay exposure.

Bulimina aculeata d'Orbigny

Bulimina aculeata d'Orbigny, 1826, p. 269, type figure not given.-Parker et al., 1871, Plate 11, Fig. 128.-Fornasini, 1902, p. 153, Fig. 4.-Bermúdez, 1949, p. 179, Plate 12, Fig. 10.-van Morkhoven et al., 1986, p. 31–33, Plate 7, Figs. 1–3.

Bulimina alazanensis Cushman

Bulimina alazanensis Cushman, 1927a, p. 161, Plate 25, Fig. 4.-Bermúdez, 1949, p. 180, Plate 12, Fig. 1.-Tjalsma and Lohmann, 1983, p. 24, Plate 14, Fig. 4.-Miller, Curry, et al., 1985, Plate 4, Fig. 6.-Miller and Katz, 1987a, p. 124, Plate 2, Fig. 7.

Bulimina impendens Parker and Bermudez

Bulimina impendens Parker and Bermúdez, 1937, p. 514, Plate 58, Figs. 7–8.-Bermúdez, 1949, p. 181, Plate 12, Fig. 9.-Proto Decima and Bolli, 1978, p. 791, Plate 2, Figs. 11–12.-Tjalsma and Lohmann, 1983, p. 25, Plate 14, Figs. 2a–b.-van Morkhoven et al., 1986, p. 236–238, Plate 79, Figs. 1–2.-Miller and Katz, 1987a, p. 125, Plate 2, Fig. 3.

Bulimina impendens can be distinguished from *B. trinitatensis* Cushman and Jarvis by its squat test and absence of ribs.

Bulimina marginata d'Orbigny

Bulimina marginata d'Orbigny, 1826, p. 269, Plate 12, Figs. 10–12.-Bermúdez, 1949, p. 182, Plate 12, Fig. 11.-van Morkhoven et al., 1986, p. 18–21, Plate 2, Fig. 1.

Bulimina mexicana Cushman
Figure 17, Photo 14–15

Bulimina inflata Seguenza var. *mexicana* Cushman, 1922a, p. 95, Plate 21, Fig. 2.
Bulimina mexicana Cushman.-Bermúdez, 1949, p. 182, Plate 12, Fig. 2.-van Morkhoven et al., 1986, p. 59–62, Plate 19, Figs. 1–4.

Bulimina tuxpamensis Cole

Bulimina tuxpamensis Cole, 1928, p. 212, Plate 32, Fig. 23.-Parker and Bermúdez, 1937, p. 513, Plate 58, Figs. 1a–c.-Tjalsma and Lohmann, 1983, p. 26, Plate 12, Figs. 1a–c.-van Morkhoven et al., 1986, p. 155–158, Plate 51A, Figs. 1–4; Plate 51B, Figs. 3–4.

Cancris nuttalli (Palmer and Bermudez)

Valvulineria nuttalli Palmer and Bermúdez, 1936a, p. 300, Plate 19, Figs. 3–5.-Bermúdez, 1949, Plate 18, Figs. 25–27.
Cancris nuttalli (Palmer and Bermudez)-van Morkhoven et al., 1986, p. 120–122, Plate 38, Figs. 1–3.
?*Cancris sagra* Nuttall (not d'Orbigny), 1932, p. 27, Plate 6, Figs. 6, 7.

The type locality of *Cancris nuttalli* is in Cuba, and it has been reported from other Caribbean localities (e.g., the Ponce Formation of Puerto Rico; Galloway and Hemingway, 1941; and the Sombrerito Formation of the Dominican Republic; Bermúdez, 1949). van Morkhoven et al. (1986) suggest that *Cancris sagra* (a species commonly identified in Caribbean faunas) is a synonym of *Cancris nuttalli*, which would establish *C. sagra* as the senior synonym.

Cancris oblongus (Williamson)

Pulvinulina oblonga (Williamson).-Barker, 1960, Plate 106, Figs. 4–5.

Cancris scintillans (Coryell and Mossman)
Figure 19, Photos 3–4

Valvulineria scintillans Coryell and Mossman, 1942, p. 236, Plate 36, Figs. 13–15.

We combined *Cancris scintillans* and *Cancris scintillans* var. *sinecarina* for multivariate analyses.

Cancris scintillans var. *sinecarina* (Coryell and Mossman)
Figure 21, Photos 1–4

Valvulineria scintillans Coryell and Mossman var. *sinecarina* Coryell and Mossman, 1942, p. 236, Plate 36, Figs. 16–18.

Cancris scintillans var. *sinecarina* is most common in the Bowden Formation. This taxon has prominent knobs surrounding the umbilical region.

Cassidulina carapitana Hedberg

Cassidulina carapitana Hedberg, 1937, p. 680, Plate 92, Fig. 6.-Renz, 1948, p. 124, Plate 9, Fig. 8.-Bermúdez, 1949, p. 267, Plate 20, Figs. 7–9.

This form has about ten chambers visible in the final whorl. These chambers are sharply curved at the inner ends, forming hooks at the center of the test. *Cassidulina carapitana* can be distinguished from *C. reflexa* by its blunt, subrounded periphery, more robust test, and consistently hooked chambers, while *C. reflexa* has an acute periphery and hooked chambers only in the first few visible chambers.

Cassidulina crassa d'Orbigny

Cassidulina crassa d'Orbigny, 1839, p. 56, Plate 7, Figs. 18–20.-Renz, 1948, p. 124, Plate 9, Figs. 13a–b; Plate 12, Fig. 23.-Miller and Katz, 1987a, p. 125, Plate 3, Figs. 1a–b.

This small *Cassidulina* has a rounded periphery and broad chambers.

Cassidulina laevigata d'Orbigny

Cassidulina laevigata d'Orbigny, 1826, p. 282, Plate 15, Figs. 4–5.-Cushman and Todd, 1945, p. 62. Plate 10, Fig. 10–11.-Renz, 1948, p. 125, Plate 9, Fig. 9.-Bermúdez, 1949, p. 268, Plate 20, Figs. 16–21.

Cassidulina laevigata has a relatively flat test with a very sharp periphery. It is distinguished from the other *Cassidulina* spp. at Buff Bay by its straighter sutures and broader chambers.

Cassidulina reflexa Galloway and Wissler

C. reflexa Galloway and Wissler, 1927, p. 80, Plate 12, Fig. 13.

See discussion of *Cassidulina carapitana*.

Cassidulina spinifera Cushman and Jarvis
Figure 14, Photos 10–11

Cassidulina spinifera Cushman and Jarvis, 1929, p. 17, Plate 3, Fig. 1

This distinctive taxon is easily recognized by the spines extending from each chamber.

Cibicides lobatulus (Walker and Jacob) d'Orbigny
Figure 13, Photos 1–3

Nautilus lobatulus Walker and Jacob, 1798, p. 642, Plate 14, Fig. 36.
Nautilus spiralis lobatus anfractibus supra rotundatis subtus depressioribus Walker and Boys, 1784, p. 20, Plate 3, Fig. 71.
Nautilus farctus Fitchell and Moll, 1798, p. 64, tab. 9, Figs. g,h,i.
Serpula lobata Montagu, 1803, p. 515.
Polyxenes cribratus Montfort, 1808, p. 139, tab. 9, Figs. g,h,i.
Truncatulina tuberculata d'Orbigny, 1826, p. 279.
Truncatulina lobata d'Orbigny, in Barker et al., 1839, p. 134, Plate 2, Figs. 22–24.
Truncatulina lobatula d'Orbigny, 1846, p. 168, Plate 9, Figs. 18–23. -Brady, 1884, Plate 95, Figs. 4–5.-Cushman, 1918, p. 16, 60, Plate 1, Fig. 10.
Truncatulina lobatula (Walker and Jacob)).-Nuttall, 1928, p. 98.
Cibicides lobatulus (Walker and Jacob).-Cole, 1931, Plate 56.-Cushman, 1931, p. 118, Plate 21, Fig. 3.
Cibicides lobatulus (d'Orbigny).-Galloway and Heminway, 1941, p. 393, Plate 24, Fig. 4.

Cibicidoides bradyi (Trauth)

Truncatulina dutemplei Brady (not d'Orbigny), 1884, p. 665, Plate 95, Fig. 5 (type figure).
Truncatulina bradyi Trauth, 1918, p. 235 (type reference).
Cibicidoides haitiensis (Coryell and Rivero).-Tjalsma and Lohmann, 1983, p. 26, Plate 17, Figs. 6a–b.-Miller, 1983, p. 433, Plate 2, Fig. 5.
Cibicidoides bradyi (Trauth).-Pflum and Frerichs, 1976, Plate 3, Figs. 6, 7.-van Morkhoven et al., 1986, p. 100–102, Plate 30, Figs. 1–2.-Miller and Katz, 1987a, p. 126, Plate 7, Figs. 2a–c.

Cibicidoides cicatricosus (Schwager)

Anomalina cicatricosa Schwager, 1866, p. 260, Plate 7, Figs. 4, 108.
Cibicidoides cicatricosus (Schwager).-van Morkhoven et al., 1986, p. 53–55, Plate 16, Figs. 1a–c.

Cibicidoides compressus (Cushman and Renz)

Cibicides floridanus (Cushman) var. *compressus* Cushman and Renz, 1941, p. 26, Plate 4, Fig. 9.-Renz, 1948, p. 127, Plate 10, Figs. 9a–c.
Cibicidoides compressus (Cushman and Renz).-van Morkhoven et al., 1986, p. 137–139, Plate 44, Figs. 1–2.

Cibicidoides cookei (Cushman and Garrett)

Cibicides cookei Cushman anda Garrett, 1938, p. 65, Plate 11, Fig. 3.

This species is reminiscent of *C. bradyi*, but the test is larger, the spiral side sutures are limbate, the chambers are not inflated, and the periphery is subacute.

Cibicidoides coryelli (Bermúdez)
Figure 14, Photos 4–6

Cibicidoides coryelli Bermúdez, 1949, p. 296, Plate 25, Figs. 7–9.

This small, strongly biconvex test has a subacute periphery. The contact between the whorls on the spiral side is very distinct. The umbilical side, while strongly convex, tends to be flattened at the end.

Cibicidoides dominicus (Bermúdez)
Figure 14, Photos 1–3

Cibicidoides dominicus Bermúdez, 1949, p. 298, Plate 25, Figs. 25–27.

This taxon is similar to *Cibicidoides coryelli* in that the contact between whorls on the spiral side is very distinct; however, it is not as strongly biconvex and the end of the umbilical side is not flattened. The early chambers tend to be obscured and are frequently perforate.

Cibicidoides havanensis (Cushman and Bermúdez)

Cibicides havanensis Cushman and Bermúdez, 1937, p. 28, Plate 3, Figs. 1–3.
Cibicidoides havanensis (Cushman and Bermúdez).-Tjalsma and Lohmann, 1983, p. 27, Plate 22, Figs. 4a–c.-Miller, 1983, p. 433, Plate 2, Figs. 9–10.-van Morkhoven et al., 1986, p. 189–193, Plate 64A, Figs. 1–4; Plate 64B, Figs. 1–2.-Miller and Katz, 1987a, p. 128, Plate 7, Figs. 5a–c.

Cibicidoides incrassatus (Fichtel and Moll)

Nautilus incrassatus Fichtel and Moll, 1798, p. 38, Plate 4, Figs. a–c.-Rogl and Hansen, 1984, p. 36, Plate 8, Figs. 4–6.
Cibicidoides incrassatus (Fichtel and Moll).-van Morkhoven et al., 1986, p. 83–89, Plate 25A, Figs. 1–2; Plate 25B, Figs. 1–4; Plate 25C, Figs. 1–4.

Cibicidoides matanzasensis (Hadley)
Figure 20, Photos 1–3

Planulina matanzasensis Hadley, 1934, p. 27, Plate 4, Figs. 1–3.
Cibicides matanzasensis (Hadley).-Renz, 1948, p. 129, Plate 11, Figs. 12a–b.-Bermúdez, 1949, p. 302, Plate 24, Figs. 4–6.
Cibicidoides matanzasensis (Hadley).-van Morkhoven et al., 1986, p. 158–161, Plate 52, Figs. 1–5.

Cibicidoides mundulus (Brady, Parker, and Jones)

Truncatulina mundula Brady, Parker, and Jones, 1888, p. 228, Plate 45, Fig. 25a–c.
Cibicidoides mundulus (Brady, Parker, and Jones).-Loeblich and Tappan, 1955, Plate 25, Figs. 4a–c.-van Morkhoven et al., 1986, p. 65–67, Plate 21, Figs. 1a–c.-Miller and Katz, 1987a, p. 130, Plate 7, Figs. 3a–c.
Cibicides kullenbergi Parker (in Phleger et al., 1953), p. 49, Plate 11, Figs. 7–8.

Cibicidoides pachyderma (Rzehak)

Truncatulina pachyderma Rzehak, 1886, p. 87, Plate 1, Figs. 5a–c.
Cibicidoides pachyderma (Rzehak).-van Morkhoven et al., 1986, p. 68–71, Plates 22, Figs. 1a–c.

Cibicidoides robertsonianus (Brady)

Planorbulina robertsoniana Brady, 1881, p. 65 (type reference).
Truncatulina robertsoniana Brady, 1884, p. 664, Plate 95, Figs. 4a–c (type figure).
Cibicides robertsonianus (Brady).-Pflum and Frerichs, 1976, Plate 3, Figs. 3–5.
Cibicidoides robertsonianus (Brady) van Morkhoven et al., 1986, p. 41–43, Plate 11, Figs. 1a–c.-Miller and Katz, 1987a, p. 132, Plate 7, Figs. 1a–c.

Cibicidoides sp. 33

This plano-convex test has ten chambers in the rapidly inflating final whorl. The initial whorl on the spiral side is obscured by a perforate covering. It has limbate sutures, a small umbilical knob, and a narrow keel.

Compressigerina coartata (Palmer)
Figure 18, Photos 3–4

Uvigerina compressa Palmer, 1941, p. 182, Plate 15, Figs. 10–11.
Uvigerina coartata Palmer (new name), 1945, p. 51.
Compressigerina coartata (Palmer).-Bermúdez, 1949, p. 220, Plate 13, Figs. 71–74.

This form is strongly compressed with only the earliest portion triserial. Most of the test appears to be biserial. There is a characteristic twist in the test. The aperture is on a short neck.

Eggerella bradyi (Cushman)

Verneuilina bradyi Cushman, 1911, pt. 2, p. 67, Figs. 107a–c.
Eggerella bradyi (Cushman).-Phleger and Parker, 1951, p. 6, Plate 3, Figs. 1–2.-Miler, 1983, p. 435, Plate 5, Fig. 5.-Miller and Katz, 1987a, p. 132, Plate 1, Figs. 9a–b.

Ehrenbergina caribbea Galloway and Heminway
Figure 17, Photos 10–11

Ehrenbergina caribbea Galloway and Heminway, 1941, p. 342, Plate 44, Fig. 22.-Renz, 1948, p. 131, Plate 9, Figs. 17a–b.

We include in our species concept forms that have two rows of knobs along with less common specimens in which the knobs appear to fuse into a single row of knobs or even disappear in rare instances.

Ehrenbergina spinea Cushman
Figure 15, Photos 6–8

Ehrenbergina spinea Cushman, 1935, p. 8, Plate 3, Figs. 10–11.-Cushman and Todd, 1945, p. 63, Plate 11, Fig. 2.

This very tightly enrolled species of *Ehrenbergina* has one spine on each of the final two chambers, one on either side of the test.

Ehrenbergina spinosissima Cushman and Jarvis
Figure 21, Photos 13–14

Ehrenbergina spinosissima Cushman and Jarvis, 1936, p. 5, Plate 1, Figs. 15–16.-Cushman and Todd, 1945, p. 63, Plate 11, Figs. 1a–b.-Bermúdez, 1949, p. 272, Plate 20, Figs. 32–34.

This distinctive species is characterized by its prominently raised plate-like chamber margins which extend into spines on the end of each chamber.

Globocassidulina caudriae (Cushman and Stainforth)

Cassidulina caudriae Cushman and Stainforth, 1945, p. 64, Plate 12, Figs. 2–3.

This species possesses chambers which tend to be more inflated than *Cassidulina*; therefore, we assign it to *Globocassidulina*.

Globocassidulina palmerae (Bermúdez and Acosta)
Figure 16, Photos 4–7

Cassidulina palmerae Bermúdez and Acosta, 1940, p. 57, Plate 9, Figs. 6–8.-Bermúdez, 1949, p. 269, Plate 20, Figs. 25–28.

This taxon similar in size and overall shape to *Globocassidulina subglobosa* and displays the same inflated test. It is characterized by irregular, sharply raised lamellar ridges, forming irregular, deep depressions. This ornamentation gives this distinctive test surface a strongly reticulate appearance.

Globocassidulina subglobosa (Brady)

Cassidulina subglobosa Brady, 1881, p. 60 (type reference).
Cassidulina subglobosa Brady.-Brady, 1884, p. 430, Plate 54, Figs. 17a–c (type figure).
Globocassidulina subglobosa (Brady).-Lohmann, 1978, p. 26, Plate 2, Figs. 8–9.-Tjalsma and Lohmann, 1983, p. 31, Plate 16, Fig. 9.-Miller and Katz, 1987a, p. 134, Plate 3, Fig. 4.

Gypsina sp.

We found this shallow-water taxon only in the *Amphistegina/Asterigerina*-rich layers. They tend to be concentrated in the >1,000 µm size fraction.

Hanzawaia ammophila (Guembel)

Rotalia ammophila Guembel, 1868, p. 652, Plate 2, Figs. 90a–b.
Cibicides cushmani Nuttall, 1930, p. 291, Plate 25, Figs. 3, 5–6.
Hanzawaia cushmani (Nuttall).-Tjalsma and Lohmann, 1983, p. 32, Plate 17, Figs. 1a–c.-Miller, 1983, p. 437, Plate 1, Fig. 12.
Hanzawaia ammophila (Guembel).-van Morkhoven et al., 1986, p. 168–171, Plate 56, Figs. 1–3.-Miller and Katz, 1987a, p. 134, Plate 6, Figs. 3a–b.

***Hanzawaia mantaensis* Galloway and Morrey**
Figure 19, Photos 8–10

Anomalina mantaensis Galloway and Morrey, 1929, p. 28, Plate 4, Figs. 5a–c.
Cibicides mantaensis (Galloway and Morrey).-Renz, 1948, p. 128, Plate 11, Figs. 8a–b.-Bermúdez, 1949, p. 302, Plate 25, Figs. 22–24.
Hanzawaia mantaensis (Galloway and Morrey).-van Morkhoven et al., 1986, p. 105–107, Plate 32, Figs. 1–2.

This plano-convex test has strongly curved sutures, a clear umbilical boss, and an acute periphery.

Hanzawaia sp. 3
Figure 19, Photos 5–7

This species is very similar to the Eocene form *Hanzawaia caribaea* (Cushman and Bermudez) with its thick, circular test that is quadrate to subquadrate in peripheral view. The spiral side is planar to slightly convex with broad, limbate sutures. There is a peripheral thickening of shell material with a keel-like appearance.

Hoeglundina elegans (d'Orbigny)

Rotalia (Turbinulina) elegans d'Orbigny, 1826, p. 276 (type reference).
Rotalia (Turbinulina) elegans d'Orbigny.-Parker et al., 1865, Plate 12, Fig. 142 (type figure).
Hoeglundina elegans (d'Orbigny).-Phleger and Parker, 1951, p. 22, Plate 12, Figs. 1a–b.-Lohmann, 1978, p. 33, Plate 4, Figs. 10–12.-van Morkhoven et al., 1986, p. 97–99, Plate 29, Figs. 1–2b.

Karreriella subglabra (Guembel)

Gaudryina subglabra Guembel, 1868, p. 602, Plate 1, Figs. 4a–b.
Gaudryina bradyi Cushman, 1911, p. 67, Fig. 107.
Karreriella bradyi (Cushman).-Cushman, 1937, p. 135, Plate 16, Figs. 6–11.-Cushman and Todd, 1945, p. 8, Plate 1, Fig. 20.-Barker, 1960, Plate. 46, Figs. 1–4.
Karreriella subglabra (Guembel).-Tjalsma and Lohmann, 1983, p. 34, Plate 9, Figs. 1a–b.-Miller and Katz, 1987a, p. 134, Plate 1, Figs. 3a–b.

Laticarinina pauperata (**Parker and Jones**)

Pulvinulina repanda Fichtel and Moll, var. *menardii* d'Orbigny, subvar. *pauperata* Parker and Jones, 1865, p. 395, Plate 16, Figs. 50–51b.
Laticarinina pauperata (Parker and Jones).-Bermúdez, 1949, p. 309, Plate 23, Figs. 43–45.-Phleger et al. 1953, p. 49, Plate 11, Figs. 5–6.-van Morkhoven et al., 1986, p. 89–91, Plate 26, Figs. 1a–c.-Miller and Katz, 1987a, p. 134, Plate 3, Figs. 7a–b.

Loxostomum isidroensis (**Cushman and Renz**)
Figure 17, Photos 8–9

Bolivina isidroensis Cushman and Renz, 1941, p. 17, Plate 3, Fig. 8.-Cushman and Todd, 1945, p. 43, Plate 6, Fig. 30.-Renz, 1948, p. 118, Plate 7, Figs. 5a–b.

Melonis barleeanum (**Williamson**)

Noniona barleeana Williamson, 1858, p. 32, Plate 3, Figs. 68–69.
Nonion barleeanum (Williamson).-Phleger et al., 1953, p. 30, Plate 6, Fig. 4.
Melonis barleeanum (Williamson).-Pflum and Frerichs, 1976, Plate 7, Figs. 5–6.-Miller and Katz, 1987a, p. 136, Plate 4, Figs. 5a–b.

Melonis pompilioides (Fichtel and Moll)
Figure 15, Photos 1–2

Nautilus pompilioides Fichtel and Moll, 1798, p. 31, Plate 2, Figs. a–c (above).
Noniona soldanii d'Orbigny, 1846, p. 109, Plate 5, Figs. 15–16.
Nonion soldanii (d'Orbigny).-Cushman and Todd, 1945, p. 36, Plate 5, Fig. 25.
Melonis pompilioides (Fichtel and Moll).-van Morkhoven et al., 1986, p. 72–77, Plate 23A, Figs. 1a–2c; Plate 23B, Figs. 1a–2b; Plate 23C, Figs. 1a–d.

Melonis pompilioides and *Melonis sphaeroides* are morphologically similar, bathymetrically segregated species. We follow van Morkhoven et al. (1986) in distinguishing between these species. *Melonis sphaeroides* is the more inflated, deeper-water form that is restricted to middle bathyal to abyssal depths. *Melonis pompilioides* is usually larger and has more chambers in the final whorl (10 to 11) and increases in width less rapidly, resulting in a different outline in apertural view. *Melonis pompilioides* has a more finely perforate test and a larger umbilicus. Its sutures are broad, imperforate bands, rather than the thinner, straight, and slightly depressed sutures of *M. sphaeroides*.

Melonis sphaeroides Voloshinova
Figure 15, Photos 3–4

Melonis sphaeroides Voloshinova, 1958, p. 153, Plate 3, Figs. 1a–b.-Miller and Katz, 1987a, p. 136, Plate 4, Figs. 3a–b.
Melonis pompilioides (Fichtel and Moll), forma *sphaeroides* Volo-

shinova.-van Morkhoven et al., 1986, p. 76–80, Plate 23D, Figs. 1a–d; Plate 23E, Figs. 1a–c.

See discussion of *Melonis pompilioides.*

Nuttallides umbonifera (Cushman)

Pulvinulinella umbonifera Cushman, 1933, p. 90, Plate 9, Figs. 9a–c.
Epistominella(?) *umbonifera* (Cushman).-Phleger et al., 1953, p. 43, Plate 9, Figs. 33–34.
Nuttallides umbonifera (Cushman).-Miller, 1983, p. 439, Plate 1, Figs. 1–3.-Miller, Curry, et al., 1985, Plate 3, Figs. 7–8.-Miller and Katz, 1987a, p. 136, Plate 5, Figs. 5a–c.

Pararotalia sp.

Our biconvex, subcircular specimens have a large umbilical knob. Only the first several chambers of the final whorl touch this knob. Test margins range from a peripheral thickening to a well-developed keel. Some specimens display a small knob on the central terminus of each chamber on the umbilical side. Occurrences are scattered throughout the Spring Garden Member, the Buff Bay Formation, and the Bowden Formation.

Planulina ambigua (Franzenau)
Figure 14, Photo 7 9

Rotalia ambigua Franzenau, 1888, p. 106, Plate 2, Figs. 9–11.
Planulina ambigua (Franzenau).-van Morkhoven et al., 1986, p. 220–235, Plate 78A, Figs. 1–8; Plate 78B, Figs. 1–4.

van Morkhoven et al. (1986) reported that *Planulina ambigua* last occurred in Zone N5. In contrast, this taxon last occurred within the *G. fohsi fohsi* Zone (approximately Zone N11) in the Spring Garden Member.

Planulina foveolata (Brady)
Figure 21, Photos 5–8

Anomalina foveolata Brady, 1884, p. 674, Plate 94, Fig. 1a–c.
Planulina foveolata (Brady).-Cushman and Todd, 1945, p. 69, Plate 12, Figs. 2a–b.-Bermúdez, 1949, p. 291, Plate 23, Figs. 22–24.-Phleger and Parker, 1951, p. 33, Plate 18, Figs. 9a–10b.-Poag, 1981, p. 76, Plate 43, Fig. 1; Plate 44, Fig. 1a–b.-van Morkhoven et al., 1986, p. 26–28, Plate 5, Figs. 1–3.

This distinctive taxon is easily recognized by its nearly flat, parallel sides and its strong surface ornamentation.

Planulina renzi Cushman and Stainforth

Planulina renzi Cushman and Stainforth, 1945, p. 72, Plate 15, Figs. 1a–c.-Bermúdez, 1949, p. 292, Plate 23, Figs. 31–33.-Tjalsma, 1983, p. 743, Plate 3, Fig. 4; Plate 6, Figs. 6–7.-van Morkhoven et al., 1986, p. 133–137, Plate 43A, Figs. 1–5; Plate 43B, Figs. 1–2.-Miller and Katz, 1987a, p. 136, Plate 6, Figs. 1a–c.

Planulina rugosa (Phleger and Parker)
Figure 20, Photos 4–6

Cibicides rugosa Phleger and Parker, 1951, p. 31, Plate 17, Figs. 5–6.
Planulina rugosa (Phleger and Parker).-van Morkhoven et al., 1986, p. 45–47, Plate 13, Figs. 1–2.

Planulina wuellerstorfi (Schwager)

Anomalina wuellerstorfi Schwager, 1866, p. 258, Plate 7, Figs. 105, 107.
Cibicides wuellerstorfi (Schwager).-Pflum and Frerichs, 1976, p. 116, Plate 4, Figs. 2–4.
Planulina wuellerstorfi (Schwager).-Bermúdez, 1949, p. 293, Plate 23, Figs. 37–39.-Phleger et al., 1953, p. 26, Plate 11, Figs. 1–2.-van Morkhoven et al., 1986, p. 48–50, Plate 14, Figs. 1–2.-Miller and Katz, 1987a, p. 136, Plate 6, Figs. 2a–c.

Planulina sp. 1
Figure 20, Photos 7–9

This small planulinid is reminiscent of *Planulina foveolata,* but it lacks surface ornamentation.

Plectofrondicularia parri Finlay

Plectofrondicularia parri Finlay, 1939, p. 516, Plate 68, Figs. 4a–b.-van Morkhoven et al, 1986, p. 128–130, Plate 41, Figs. 1–2.
Plectofrondicularia diversicostata Cushman and Todd, 1945, non *Frondicularia diversicostata* Neugeboren, 1850, p. 37, Plate 6, Fig. 3.

Cushman and Todd's (1945) illustration of *Plectofrondicularia diversicostata* is indistinguishable from our specimens of *Plectofrondicularia parri.* However, Neugeboren's (1850) type figure of *Frondicularia diversicostata* illustrates distinct, continuous longitudinal costae that cover the entire test. Cushman and Todd (1945) note that while a number of their Buff Bay specimens fit this description, they show considerable variation in the length of the intermediate costae. Cushman and Todd's (1945) illustration of *Plectofrondicularia diversicostata* lacks these pronounced, continuous longitudinal costae. van Morkhoven et al. (1986) report that *Plectofrondicularia parri* last occurred in the late Miocene (Zone N17); however, we found this species throughout the Pliocene Bowden Formation.

Plectofrondicularia vaughani Cushman

Plectofrondicularia jarvisi Cushman and Todd, 1945, p. 38, Plate 6, Fig. 5.
Plectofrondicularia vaughani Cushman 1927b, p. 112, Plate 23, Fig. 3.-van Morkhoven et al., 1986, p. 130–133, Plate 42, Figs. 1–2.

Our specimens of *Plectofrondicularia vaughani* tend to be broken. Despite this, we see little difference between *Plectofrondicularia jarvisi* illustrated by Cushman and Todd (1945) and *Plectofrondicularia vaughani.* Therefore, we tentatively consider *Plectofrondicularia jarvisi* to be a junior synonym of *Plectofrondicularia vaughani.* Comparison of type specimens would clarify this issue. As with *Plectofrondicularia parri,* van Morkhoven et al. (1986) report that *Plectofrondicularia vaughani* last occurred in the late Miocene (Zone N17); however, we found this species throughout the Pliocene Bowden Formation.

Pullenia bulloides (d'Orbigny)

Nonionina bulloides d'Orbigny, 1826, p. 293.
Pullenia bulloides (d'Orbigny).-Cushman and Todd, 1945, p. 64, Plate 11, Fig. 5.-Bermúdez, 1949, p. 276, Plate 21, Figs. 28–29.-Lohmann, 1978, p. 26, Plate 1, Figs. 10–11.-Miller and Katz, 1987a, p. 136, 138, Plate 4, Figs. 4a–b.

Pullenia malkinae Coryell and Mossman

Pullenia malkinae Coryell and Mossman, 1942, p. 234, Plate 36, Figs. 3–4.-Cushman and Todd, 1945, p. 65, Plate 11, Fig. 8.

Pullenia malkinae has a circular outline and a compressed test in axial view.

Pullenia quadriloba Reuss

Pullenia compressiuscula Reuss var. *quadriloba* Reuss, 1867, p. 87, Plate 3, Fig. 8.
Pullenia quadriloba (Reuss).-Cushman and Todd, 1945, p. 65, Plate 11, Fig. 7.

This species differs from *Pullenia quinqueloba* in having four chambers in the final whorl rather than five.

Pullenia quinqueloba (Reuss)

Nonionina quinqueloba Reuss, 1851, p. 71, Plate 5, Figs. 31a–b.
Pullenia quinqueloba (Reuss).-Bermúdez, 1949, p. 276, Plate 21, Figs. 32–33.-Tjalsma and Lohmann, 1983, p. 36, Plate 16, Fig. 2.-Miller and Katz, 1987, p. 138, Plate 4, Figs. 2a–b.

Pyrgo murrhina (Schwager)

Biloculina murrhina Schwager, 1866, p. 203, Plate 4, Figs. 15a–c.
Pyrgo murrhina (Schwager).-Cushman and Todd, 1945, p. 12, Plate 2, Fig. 5.-van Morkhoven et al., 1986, p. 50–52, Plate 15, Figs. 1–2.

Rectuvigerina multicostata (Cushman and Jarvis)
Figure 16, Photo 9–11

Siphogenerina multicostata Cushman and Jarvis, 1929, p. 14, Plate 3, Fig. 6.
Rectuvigerina multicostata (Cushman and Jarvis).-van Morkhoven et al., 1986, p. 115–117, Plate 36, Figs. 1–4.

Rectuvigerina striata (Schwager)
Figure 16, Photo 8

Dimorphina striata Schwager, 1866, p. 251, Plate 7, Figs. 2, 99 (above left).
Rectuvigerina striata (Schwager).-van Morkhoven et al., 1986, p. 110–112, Plate 34, Figs. 1–3.

Reusella spinulosa (Reuss) var. *pulchra* (Cushman)
Figure 17, Photos 12–13

Verneulina spinulosa Reuss, 1850, p. 374, Plate 47, Fig. 12a–c.
Reusella spinulosa (Reuss).-Renz, 1948, p. 156, Plate 7, Figs. 16–17.-Bermúdez, 1949, p. 198, Plate 12, Fig. 59.
Reusella pulchra Cushman, 1945, p. 34, Plate 6, Figs. 11–12.-Cushman and Todd, 1945, p. 49, Plate 7, Fig. 21.
Reusella simplex (Cushman).-Barker, 1960, p. 96, Plate 47, Fig. 1.
Reusella aculeata Cushman.-Barker, 1960, p. 96, Plate 47, Fig. 2–3.

This form has a carinate periphery with a spine at the base of each chamber. The chambers are distinct with limbate, slightly raised sutures and a finely spinose surface. *Reusella pulchra* Cushman appears to be a junior synonym of *Reusella spinulosa* (Reuss). We believe that it is separable as a subspecies on the basis of its finely spinose ornamentation and therefore we include it as a variety.

Rosalina sp. 1
Figure 19, Photos 1–2

This species has a highly perforate test and an open umbilicus.

Sigmoilopsis schlumbergeri (Silvestri)

Sigmoilina schlumbergeri Silvestri, 1904, p. 267, 269 (type reference).-Schlumberger, 1887, p. 481–482, Plate 7, Figs. 12–14 (type figure).-Cushman and Todd, 1945, p. 11, Plate 2, Fig. 3.
Sigmoilopsis schlumbergeri (Silvestri).-van Morkhoven et al., 1986, p. 57–59, Plate 18,, Figs. 1a–e.

Siphogenerina advena var. *ornata* Cushman

Siphogenerina advena Cushman, 1922b, p. 35, Plate 5, Fig. 2.-Bermúdez, 1949, p. 221, Plate 13, Fig. 76.
Siphogenerina advena Cushman var. *ornata* Palmer and Bermúdez, 1936b, p. 249, Plate 22, Figs. 4, 7.-Cushman and Todd, 1945, p. 52, Plate 7, Fig. 32.

Siphonina tenuicarinata Cushman
Figure 16, Photos 1–3

Siphonina tenuicarinata Cushman, 1927a, p. 166, Plate 26, Figs. 11–12.-van Morkhoven et al., 1986, p. 206–209, Plate 70, Figs. 1–3.

The majority of our specimens are *S. tenuicarinata*; however, several specimens may be *S.* cf. *pulchra* (see Cushman and Todd, 1945, for illustrations and discussion).

Siphoninella soluta (Brady)
Figure 15, Photos 9–10

Truncatulina soluta Brady, 1884, in Barker, 1960, p. 198, Plate 96, Figs. 4a–c.
Siphoninella soluta (Brady).-Cushman, 1927c, p. 77, Plate 16, Fig. 13.

This species tends to have a broken keel and pustular ornamentation, particularly along the sutures.

Siphouvigerina porrecta (Brady)
Figure 18, Photos 1–2

Uvigerina porrecta Brady, 1884, in Barker 1960, p. 156, Plate 74, Figs. 21–23.
Neouvigerina porrecta (Brady).-Hofker, 1951, p. 213.
Angulogerina porrecta (Brady).-Bermúdez, 1949, p. 218, Plate 13, Fig. 56.

This distinctive taxon is common in the Buff Bay samples. It has discontinuous costae on its loosely coiled chambers.

Sphaeroidina bulloides d'Orbigny

Sphaeroidina bulloides d'Orbigny, 1826, p. 267.-Parker et al., 1865, Plate 2, Fig. 58.-Bermúdez, 1949, p. 277, Plate 21, Figs. 34–38.-van Morkhoven et al. 1986, p. 80–83, Plate 24, Figs. 1–2.

Stilostomella aculeata (Cushman and Renz)

Ellipsonodosaria nuttalli Cushman and Jarvis var. *aculeata* Cushman and Renz, 1948, p. 32, Plate 6, Fig. 10.
Stilostomella aculeata (Cushman and Renz).-Miller, 1983, p. 439, Plate 4, Fig. 1.-Miller and Katz, 1987a, p. 138, Plate 1, Fig. 11.

Stilostomella curvatura (Cushman)

Ellipsonodosaria curvatura Cushman, 1939, p. 71, Plate 12, Fig. 6.
Stilostomella curvatura (Cushman).-Beckmann, 1954, p. 371, Plate 21, Figs. 26–27.-Tjalsma, 1983, p. 743, Plate 1, Figs. 7, 11.

Stilostomella modesta (Bermudez)

Ellipsonodosaria modesta Bermúdez, 1937, p. 238, Plate 20, Fig. 3.
Stilostomella modesta (Bermudez).-Beckmann, 1954, p. 371, Plate 21, Fig. 32.-Tjalsma, 1983, p. 743, Plate 1, Fig. 10.
Siphonodosaria modesta (Bermudez).-Douglas, 1973, Plate 5, Fig. 4.
Orthomorphina modesta (Bermudez).-Boltovskoy, 1978, p. 163, Plate 5, Fig. 25.

Stilostomella subspinosa (Cushman)

Ellipsonodosaria subspinosa Cushman, 1943, p. 92, Plate 16, Figs. 6–7b.
Stilostomella subspinosa (Cushman).-Tjalsma and Lohmann, 1983, p. 36, Plate 14, Figs. 16–17.-Miller and Katz, 1987a, p. 138, Plate 1, Fig. 12.

Trifarina bradyi Cushman

Trifarina bradyi Cushman, 1923, p. 99, Plate 22, Figs. 3–9.-Phleger and Parker, 1951, p. 18, Plate 8, Figs. 10–11.-Renz, 1948, p. 172, Plate 7, Fig. 33.-Bermúdez, 1949, p. 225, Plate 13, Fig. 75.

Uvigerina carapitana Hedburg

Uvigerina carapitana Hedburg, 1937, p. 677, Plate 91, Fig. 20.-Bermúdez, 1949, p. 202, Plate 13, Fig. 1.-Boersma, 1984, p. 28–30, Plate 1, Figs. 1–5.

Uvigerina hispida Schwager

Uvigerina hispida Schwager, 1866, p. 249, Plate 7, Fig. 95.-Boersma, 1984, p. 76–74, Plate 1, Figs. 1–4.-van Morkhoven et al., 1986, p. 62–64, Plate 20, Figs. 1–4.

Uvigerina hispido-costata Cushman and Todd
Figure 21, Photos 11–12

Uvigerina hispido-costata Cushman and Todd, 1945, p. 51, Plate 7, Figs. 27, 31.-Bermúdez, 1949, p. 206, Plate 13, Figs. 37–38.-Boersma, 1984, p. 77–81, Plate 1, Figs. 1–4.

Uvigerina jamaicensis (Cushman and Todd)
Figure 21, Photos 9–10

Angulogerina jamaicensis Cushman and Todd, 1945, p. 53, Plate 8, Fig. 3.

Uvigerina laviculata Coryell and Rivero

Uvigerina laviculata Coryell and Rivero, 1940, p. 343.-Cushman and Todd, 1945, p. 50, Plate 7, Figs. 25.-Bermúdez, 1949, p. 207, Plate 13, Fig. 34.-Boersma, 1984, p. 91–94, Plate 1, Figs. 1–4.

Uvigerina pigmea d'Orbigny
Figure 18, Photos 7–8

Uvigerina pigmea d'Orbigny, 1826, p. 269.-Bermúdez, 1949, p. 209, Plate 13, Fig. 44.-Boersma, 1984, p. 127–130, Plate 1, Figs. 1–6; Plate 2, Figs. 1–5.

Uvigerina proboscidea Schwager
Figure 18, Photos 13–14

Uvigerina proboscidea Schwager, 1866, p. 250, Plate 7, Fig. 96.-Cushman and Todd, 1945, p. 50, Plate 7, Figs. 28–29.-Bermúdez, 1949, p. 209, Plate 13, Fig. 45.-Boersma, 1984, p. 131–134, Plate 1, Figs. 1–5.-van Morkhoven et al., 1986, p. 28–30, Plate 6, Figs. 1–4.

Uvigerina sp. 1

Uvigerina sp. 1 is primarily a costate form, although costae may break into hisps or short ridges on a few specimens. Costae are restricted to individual chambers. Coiling is generally triserial throughout, although rare specimens are elongate with a biserial final whorl. Although this species resembles *Uvigerina parvula* Cushman (see Boersma, 1984), we did not adopt this name because of its different stratigraphic and bathymetric ranges.

Uvigerina sp. 7
Figure 18, Photos 11–12

The final chambers of this costate uvigerinid tend to become angular.

Figure 13. Shallow-water taxa. Scale bar = 100 μm. 1–3, *Cibicides lobatulus* (Walker and Jacob), sample BB64; 4–6, *Asterigerina* sp., sample BB52; 7–8, *Elphidium* sp., sample BB8; 9–10, *Planorbulina* sp., sample BB52.

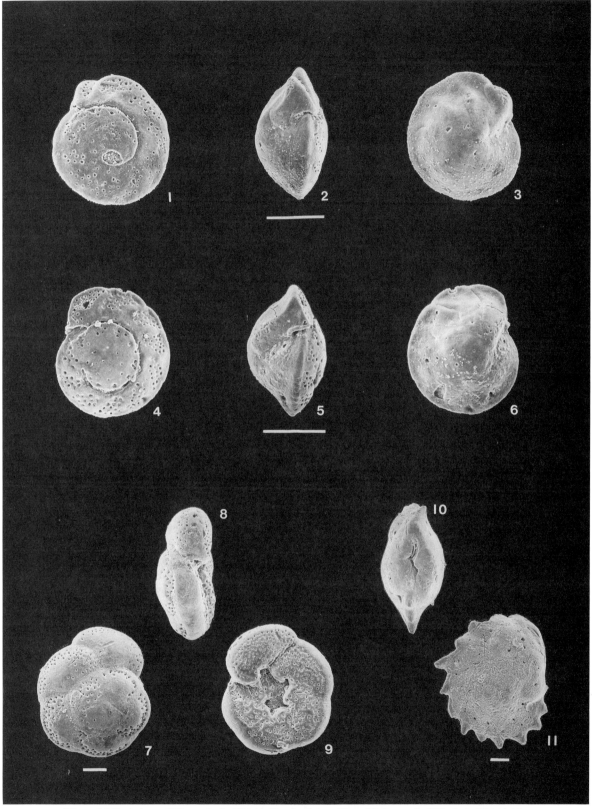

Figure 14. Scale bar = 100 μm. 1–3, *Cibicidoides dominicus* (Bermúdez), sample SG1; 4–6, *Cibicidoides coryelli* (Bermúdez), sample SG1; 7–9, *Planulina ambigua* (Franzenau), sample SG28; 10–11, *Cassidulina spinifera,* Cushman and Jarvis, sample SG30.

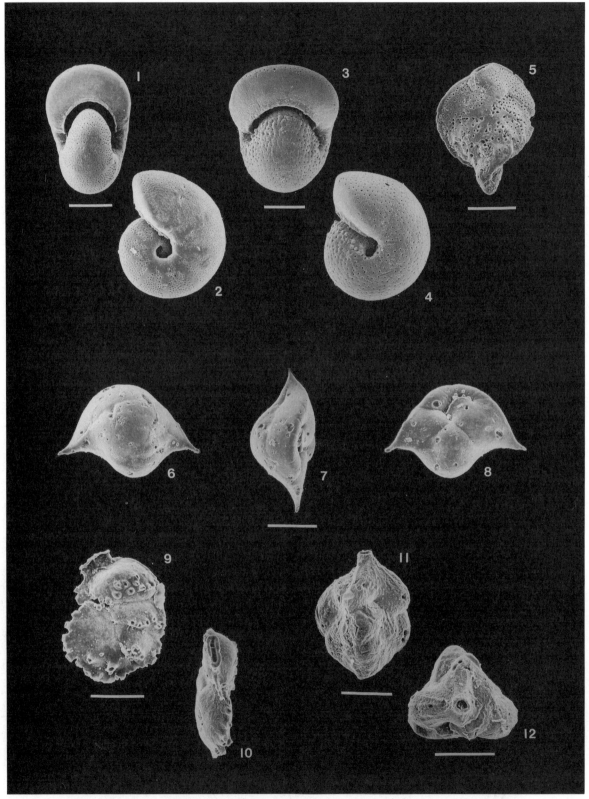

Figure 15. Scale bar = 100 μm. 1–2, *Melonis pompilioides* (Fichtel and Moll), sample BW248; 3–4, *Melonis sphaeroides* Voloshinova, sample SG23A; 5, *Bolivina tortuosa* Brady, sample SG30; 6–8, *Ehrenbergina spinea* Cushman, sample BB4; 9–10, *Siphoninella soluta* (Brady), sample BB24; 11–12. *Angulogerina cojimarensis* Palmer, sample SG25.

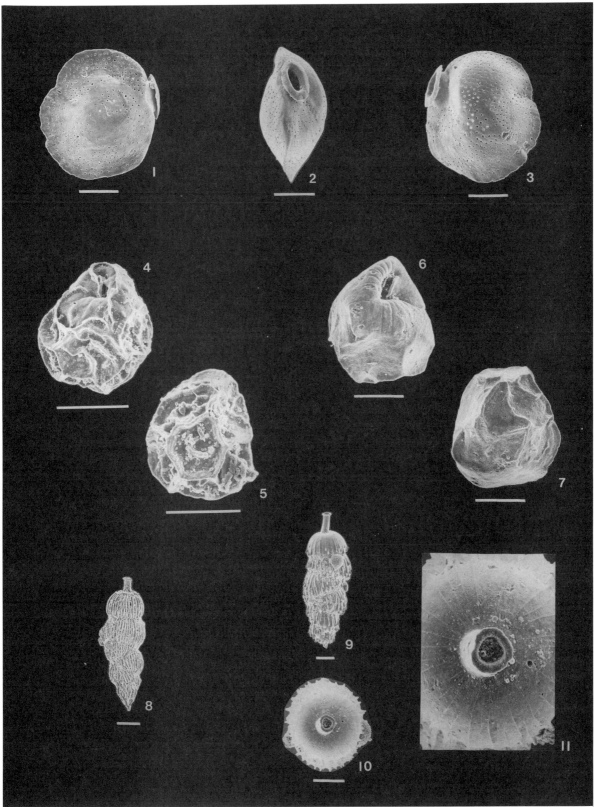

Figure 16. Scale bar = 100 μm. 1–3, *Siphonina tenuicarinata* Cushman, sample BB64. 4–7; *Globocassidulina palmerae* (Bermúdez and Acosta), sample BB8; 8, *Rectuvigerina striata* (Schwager), sample BB4; 9–11, *Rectuvigerina multicostata* (Cushman and Jarvis), sample BB4.

Figure 17. Scale bar = 100 μm. 1–2, *Bolivina thalmanni* Renz, sample BB44; 3–4, *Bolivina aenariensiformis* Subbotina, sample BB48; 5–6, *Bolivina alazanensis* Cushman, sample BB28; 7, *Bolivina alata* (Seguenza), sample BB8; 8–9, *Loxostomum isidroensis* (Cushman and Renz), sample BB12; 10–11, *Ehrenbergina caribbea* Galloway and Hemingway, sample BB36; 12–13, *Reusella spinulosa* (Reuss) var. *pulchra* (Cushman), sample BB64; 14–15, *Bulimina mexicana* Cushman, sample BB20.

Figure 18. Scale bar = 100 μm. 1–2, *Siphouvigerina porrecta* (Brady), sample BB64; 3–4, *Compressigerina coartata* (Palmer), sample BB64; 5–6, *Angulogerina illingi* Cushman and Renz, sample BB64; 7–8, *Uvigerina pigmea* d'Orbigny, sample BB44; 9–10, *Angulogerina eximia* Cushman and Jarvis, sample BB48; 11–12, *Uvigerina* sp. 7, sample BB48; 13–14, *Uvigerina proboscidea* Schwager, sample BB56.

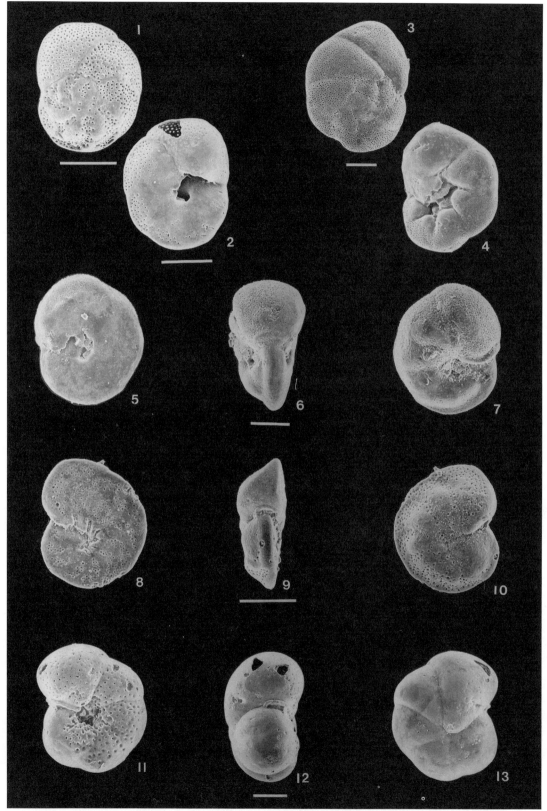

Figure 19. Scale bar = 100 μm. 1–2, *Rosalina* sp. 1, sample BB64; 3–4, *Cancris scintillans* (Coryell and Mossman), sample BB60; 5–7, *Hanzawaia* sp. 3, sample BB1; 8–10, *Hanzawaia mantaensis* Galloway and Morrey, sample SG6; 11–13, *Anomalinoides globulosus* (Chapman and Parr), sample BB64.

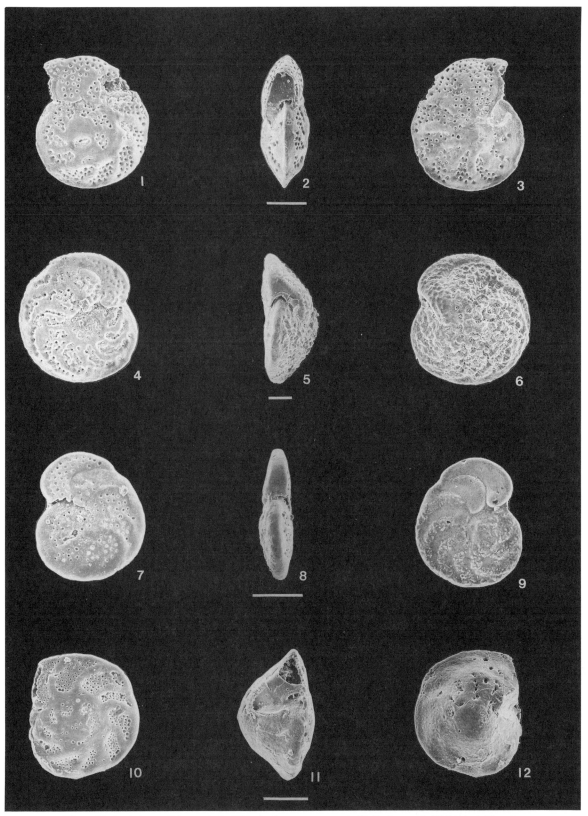

Figure 20. Scale bar = 100 μm. 1–3, *Cibicidoides matanzasensis* (Hadley), sample BB16; 4–6, *Planulina rugosa* (Phleger and Parker), sample BB20; 7–9, *Planulina* sp. 1, sample BB48; 10–12, *Cibicidoides* sp., sample BB4.

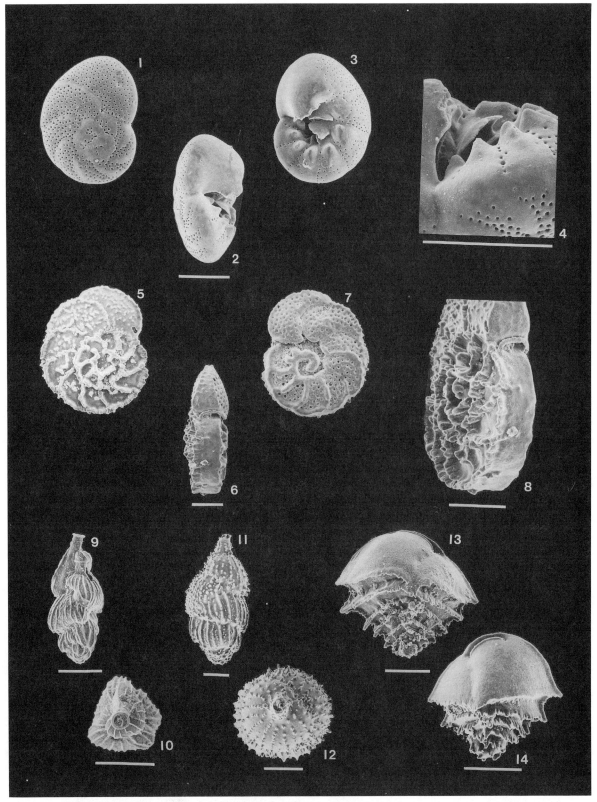

Figure 21. Scale bar = 100 μm. 1–4; *Cancris scintillans* var. *sinecarina* (Coryell and Mossman), sample BW252; 5–8, *Planulina foveolata* (Brady), sample BW 245; 9–10, *Uvigerina jamaicensis* (Cushman and Todd), sample BW245; 11–12, *Uvigerina hispido-costata* Cushman and Todd, sample BW252; 13–14, *Ehrenbergina spinosissima* Cushman and Jarvis, sample BW244.

ACKNOWLEDGMENTS

This project would not have been possible without the support and collaboration of other members of the field party (M.-P. Aubry, W. A. Berggren, G. Jones, D. V. Kent, E. Robinson, C. Stuart, R. C. Tjalsma, and M. van Fossen). We thank W. A. Berggren for organizing the field expedition, and Unocal, Chevron, and Marathon for funding the field work. We thank C. Stuart for lithological descriptions and measurements of the sections, W. A. Berggren and M.-P. Aubry for biostratigraphic data, G. Jones and R. C. Tjalsma for discussions of the work, Unocal for processing the samples, D. V. Kent and M. van Fossen for supplying magnetostratigraphic data and discussions, D. Grieg (Chevron) for producing the SEM micrographs, and R. Free and J. Zhang for technical assistance. We also thank M.-P. Aubry, A. Boersma, S. K. Donovan, G. Jones, P. P. McLaughlin, Jr., E. Robinson, B. Sen Gupta, E. Thomas, and R. C. Tjalsma for reviews. The study of Neogene deep-water benthic foraminifera of the Gulf of Mexico and Caribbean is funded by a consortium of oil companies (Unocal, Chevron, Marathon, British Petroleum, and Texaco). Studies at Lamont-Doherty Earth Observatory were funded by NSF Grant OCE85-21690 and OCE88-11834. This is Lamont-Doherty Earth Observatory Contribution No. 5106.

REFERENCES CITED

Barker, R. W., 1960, Taxonomic notes on the species figured by H. B. Brady in his "Report on the foraminifera dredged by H.M.S. *Challenger* during the years 1873–1876": Society of Economic Paleontologists and Mineralogists Special Publication 9, 238 p.

Barker-Webb, P., and Berthelot, S., 1839, Foraminifères, *in* Histoire Naturelle des Îles Canaries, v. 2, pt. 2, Zool., p. 119–146.

Beckmann, J. P., 1954, Die Foraminiferen der Oceanic Formation (Eocaen-Oligocaen) von Barbados, Kl. Antillen: Eclogae Geologicae Helvetiae, v. 46, p. 301–412.

Berger, W. H., and Winterer, E. L., 1974, Plate stratigraphy and the fluctuating carbonate line, *in* Hsü, K. J., and Jenkyns, H. C., eds., Pelagic sediments on land and under the sea: International Association of Sedimentologists, Special Publication, v. 1, p. 11–48.

Berggren, W. A., 1972, Cenozoic biostratigraphy and paleobiogeography of the North Atlantic, *in* Laughton, A. S., Berggren, W. A., et al., Initial Reports of the Deep Sea Drilling Project, v. 12: Washington, D.C., U.S. Government Printing Office, p. 965–1001.

Berggren, W. A., and Miller, K. G., 1989, Cenozoic bathyal and abyssal benthic foraminiferal zonations, Micropaleontology, v. 35, p. 308–320.

Bermúdez, P. J., 1937, Nuevas especies de foraminiferos del Eoceno de las cercanias de Guanajay, Provincia Pinar del Rio, Cuba: Sociedad Cubana de Historia Natural Memorias de la Museo Poey, p. 237–247.

Bermúdez, P. J., 1949, Tertiary smaller foraminifera of the Domincan Republic: Cushman Laboratory for Foraminiferal Research Special Publication no. 25, 322 p.

Bermúdez, P. J., and Acosta, J. T., 1940, Resultados de la primera expedicion en las Antillas del Ketch Atlantic bajo los auspicios de las Universidades de Harvard y Habana; Nuevos especies de foraminiferos recientes: Sociedad Cubana de Historia Natural Memorias de la Museo Poey, v. 14, p. 57.

Blow, W. H., 1969, Late middle Eocene to Recent planktonic foraminiferal biostratigraphy, *in* Bronnimann, P., and Renz, H. H., eds., Proceedings, International Conference on Planktonic Microfossils, 1st, Geneva: Leiden, Netherlands, E. J. Brill, p. 199–421.

Blow, W. H., 1979, The Cainozoic Globigerinida: Leiden, Netherlands, E. J. Brill, 1413 p. (not including the plates).

Boersma, A., 1984, Handbook of common Tertiary *Uvigerina:* Stony Point, New York, Microclimates Press, 207 p.

Boersma, A., 1986, Biostratigraphy and biogeography of Tertiary bathyal benthic foraminifers: Tasman Sea, Coral Sea, and on the Chatham Rise (Deep Sea Drilling Project, Leg 90, *in* Kennett, J. P., von der Borch, C. C., et al., Initial Reports of the Deep Sea Drilling Project, v. 90: Washington, D.C., U.S. Government Printing Office, p. 961–1035.

Bolli, H. M., 1957, Planktonic foraminifera from the Oligocene-Miocene Cipero and Lengua Formations of Trinidad, B.W.I.: U.S. National Museum Bulletin, v. 215, p. 97–123.

Bolli, H. M., 1970, The foraminifera of Site 23–31, *in* Bader, R. G., Gerard, R. D., et al., Initial Reports of the Deep Sea Drilling Project, v. 4: Washington, D.C., U.S. Government Printing Office, p. 577–643.

Boltovskoy, E., 1978, Late Cenozoic benthonic foraminifera of the Ninetyeast Ridge (Indian Ocean): Marine Geology, v. 26, p. 139–175.

Brady, H. B., 1881, Notes on some of the reticularian Rhisopoda of the *"Challenger"* Expedition; Part III: Quarterly Journal of Microscopical Science, v. 21, p. 31–71.

Brady, H. B., 1884, Report on the foraminifera dredged by H.M.S. *Challenger* during the years 1873–1876, *in* Murray, J., ed., Report on the scientific results of the voyage of the H.M.S. *Challenger* during the years 1873–1876: Zoology, v. 9, 814 p.

Brady, H. B., Parker, W. K., and Jones, T. R., 1888, On some foraminifera from the Abrolhos Bank: Transactions of the Zoological Society of London, v. 12, p. 40–47.

Brasier, M. D., 1975, The ecology and distribution of Recent foraminifera from the reefs and shoals around Barbuda, West Indies: Journal of Foraminiferal Research, v. 5, p. 193–210.

Cassel, D. T., and Sen Gupta, B. K., 1989a, Foraminiferal stratigraphy and paleoenvironments of the Tertiary Uscari Formation, Limon Basin, Costa Rica: Journal of Foraminiferal Research, v. 19, p. 52–71.

Cassel, D. T., and Sen Gupta, B. K., 1989b, Pliocene foraminifera and environments, Limon Basin of Costa Rica: Journal of Paleontology, v. 63, p. 146–157.

Chapman, F., and Parr, W. J., 1937, Foraminifera, *in* Johnston, T. H., ed., Australasian Antarctic Expedition, 1911–1914: Sydney, Australia, Science Reports Series C (Zoology and Botany), v. 1, p. 1–190.

Cole, W. S., 1928, A foraminiferal fauna from the Chapapote Formation in Mexico: Bulletin of American Paleontology, v. 14, p. 200–232.

Cole, W. S., 1931, The Pliocene and Pleistocene foraminifera of Florida: Florida Geological Survey, Bulletin 6, 79 p.

Coryell, H. N., and Mossman, R. W., 1942, Foraminifera from the Charco Azul Formation, Pliocene, of Panama: Journal of Paleontology, v. 16, p. 233–246.

Coryell, H. N., and Rivero, F. C., 1940, A Miocene microfauna of Haiti: Journal of Paleontology, v. 14, p. 324–344.

Cushman, J. A., 1911, A monograph of the foraminifera of the North Pacific Ocean; Part II—Textulariidae: U.S. National Museum Bulletin, no. 71, 108 p.

Cushman, J. A., 1913, A monograph of the foraminifera of the North Pacific Ocean; Part III—Lagenidae: U.S. National Museum Bulletin, no. 71, 119 p.

Cushman, J. A., 1918, Some Pliocene and Miocene foraminifera of the coastal plain of the United States: U.S. Geological Survey Bulletin, v. 676, p. 39–98.

Cushman, J. A., 1922a, The foraminifera of the Atlantic Ocean; Part III—Textulariidae: U.S. National Museum Bulletin, no. 104, 149 p.

Cushman, J. A., 1922b, Shallow-water foraminifera of the Tortugas region: Carnegie Institution of Washington, Department of Marine Biology, Publication 311, v. 17, p. 1–85.

Cushman, J. A., 1923, The foraminifera of the Atlantic Ocean; Part IV—Lagenidae: U.S. National Museum Bulletin, no. 104, 228 p.

Cushman, J. A., 1926, Some fossil *Bolivinas* from Mexico: Cushman Laboratory for Foraminiferal Research Contributions, v. 1, p. 81–85.

Cushman, J. A., 1927a, Some characteristic Mexican fossil foraminifera: Journal of Paleontology, v. 1, p. 147–172.

Cushman, J. A., 1927b, New and interesting foraminifera from Mexico and Texas: Cushman Laboratory for Foraminiferal Research Contributions, v. 3, p. 111–117.

Cushman, J. A., 1927c, An outline of a re-classification of the foraminifera: Cushman Laboratory for Foraminiferal Research Contributions, v. 3, p. 1–105.

Cushman, J. A., 1931, The foraminifera of the Atlantic Ocean; Part VIII—Rotaliidae, Amphisteginidae, Calcarinidae, Cymbaloporettidae, Globorotaliidae, Anomalinidae, Planorbulinidae, Rupertiidae and Homotremidae: United States National Museum Bulletin 104, 179 p.

Cushman, J. A., 1933, Some new Recent foraminifera from the tropical Pacific: Cushman Laboratory for Foraminiferal Research Contributions, v. 9, p. 77–94.

Cushman, J. A., 1935, Fourteen new species of foraminifera: Smithsonian Institution Miscellaneous Collection, v. 91, p. 1–9.

Cushman, J. A., 1937, A monograph of the foraminiferal family Valvulinidae: Cushman Laboratory for Foraminiferal Research, Special Publication 8, 210 p.

Cushman, J. A., 1939, Eocene foraminifera from submarine cores off the eastern coast of North America: Cushman Laboratory for Foraminiferal Research Contributions, v. 15, p. 49–76.

Cushman, J. A., 1943, Some new foraminifera from the Tertiary of the Island of St. Croix: Cushman Laboratory for Foraminiferal Research Contributions, v. 19, p. 90–93.

Cushman, J. A., 1945, The species of the subfamily Reussellinae of the foraminiferal family Buliminidae: Cushman Laboratory for Foraminiferal Research Contributions, v. 21, p. 34.

Cushman, J. A., and Bermudez, P. J., 1937, Further new species of foraminifera from the Eocene of Cuba: Cushman Laboratory for Foraminiferal Research Contributions, v. 13, p. 1–29.

Cushman, J. A., and Garrett, J. B., 1938, Three new Rotaliform foraminifera from the Lower Oligocene and upper Eocene of Alabama: Cushman Laboratory for Foraminiferal Research Contributions, v. 14, p. 62–66.

Cushman, J. A., and Garrett, J. B., 1939, Eocene foraminifera of Wilcox age from Woods Bluff, Alabama: Cushman Laboratory for Foraminiferal Research Contributions, v. 15, p. 87.

Cushman, J. A., and Jarvis, P. W., 1929, New foraminifera from Trinidad: Cushman Laboratory for Foraminiferal Research Contributions, v. 5, p. 6–17.

Cushman, J. A., and Jarvis, P. W., 1930, Miocene foraminifera from Buff Bay, Jamaica: Journal of Paleontology, v. 4, p. 353–368.

Cushman, J. A., and Jarvis, D. W., 1936, Three new foraminifera from the Miocene, Bowden marl, of Jamaica: Cushman Laboratory for Foraminiferal Research Contributions, v. 12, p. 3–5.

Cushman, J. A., and Ponton, G. M., 1932, An Eocene foraminiferal fauna of Wilcox age from Alabama: Cushman Laboratory for Foraminiferal Research Contributions, v. 8, p. 71.

Cushman, J. A., and Renz, H. H., 1941, New Oligocene-Miocene foraminifera from Venezuela: Cushman Laboratory for Foraminiferal Research Contributions, v. 17, p. 1–27.

Cushman, J. A., and Renz, H. H., 1948, Eocene foraminifera of the Navet and Hospital Hill Formations of Trinidad, B.W.I.: Cushman Laboratory for Foraminiferal Research, Special Publication 24, 42 p.

Cushman, J. A., and Stainforth, R. M., 1945, The foraminifera of the Cipero Marl Formation of Trinidad, British West Indies: Cushman Laboratory for Foraminiferal Research Contributions, v. 14, p. 3–75.

Cushman, J. A., and Todd, R., 1945, Miocene foraminifera from Buff Bay, Jamaica: Cushman Laboratory for Foraminiferal Research, Special Publication 15, 73 p.

Douglas, R. G., 1973, Benthonic foraminiferal biostratigraphy in the central North Pacific, Leg 17, Deep Sea Drilling Project, in Winterer, E. L., Ewing, J. I., et al., eds., Initial Reports of the Deep Sea Drilling Project, v. 17: Washington, D.C., U.S. Government Printing Office, p. 607–671.

Ficthel, L. V., and Moll, J.P.C.V., 1798, Testacea microscopica aliaque minuta ex generibus Argonauta et Nautilus ad naturam delineata et descripta (Microscopische und andere kleine Schalthiere aus den Geschlechtern Argonaute und Schiffer, nach der Natur gezeichnet und beschrieben.): Wien, Camesina (1803 reprint), p. 1–124, taf. 1–24.

Finlay, H. J., 1939, New Zealand foraminifera: Key species in stratigraphy—No. 1: Transactions, Royal Society of New Zealand, v. 68, p. 504–533.

Fornasini, 1902, Contributo a la conoscenza de le Bulimine adriatiche: Reale Accaedmia delle Scienze dell'Istituto di Bologna, Memorie delle Scienze Naturali, ser. 5, p. 371–381.

Franzenau, A., 1888, Adat Budapest altalajanak ismeretehez: Budapest, Foldtani Kozlony, v. 18, no. 3–4, p. 87–106, 157–174.

Galloway, J. J., and Heminway, C. E., 1941, The Tertiary foraminifera of Puerto Rico: New York Academy of Science, Scientific Survey of Puerto Rico and the Virgin Islands, v. 3, p. 275–491.

Galloway, J. J., and Morrey, M., 1929, A lower Tertiary foraminiferal fauna from Manta, Ecuador: Bulletin of American Paleontology, v. 15, p. 7–56.

Galloway, J. J., and Wissler, S. G., 1927, Pleistocene foraminifera from the Lomita Quarry, Palos Verdes Hills, California: Journal of Paleontology, v. 1, p. 35–87.

Grimsdale, T. F., and van Morkhoven, F.P.C.M., 1955, The ratio between pelagic and benthonic foraminifera as a means of estimating depth of deposition of sedimentary rocks, in Proceedings, Fourth World Petroleum Congress Section I/D: Rome, Carlo Colombo Publisher, p. 474–491.

Guembel, C. W., 1868, Beitrage zur Foraminiferenfauna der nordalpinen Eocangebilde: Abhandlungen, Munchen, Koniglich-Bayerische Akademie der Wissenschaften, Mathematisch-Physikalische Klasse, v. 10, p. 581–730.

Hadley, W. H., 1934, Some Tertiary foraminifera from the north coast of Cuba: Bulletin of American Paleontology, v. 20, p. 1–40.

Hedberg, H. D., 1937, Foraminifera of the middle Tertiary Carapitana Formation of northeastern Venezuela: Journal of Paleontology, v. 11, p. 661–697.

Heron-Allen, E., and Earland, A., 1909, On the Recent and fossil foraminifera of the shore-sands at Selsey Hill, Sussex: Part III: Journal of the Royal Microscopical Society, London, v. 49, p. 437.

Hill, R. T., 1899, The geology and physical geography of Jamaica: Study of a type of Antillean development: University Press, Harvard College, Cambridge, Bulletin of the Museum of Comparative Zoology, v. 34, 256 p.

Hofker, J., 1951, The foraminifera of the "Siboga" Expedition; Part III, Siboga Expeditie, Monographie 4a: Leiden, E. J. Brill, 513 p.

Hornibrook, N. de B., 1961, Tertiary foraminifera from Oamaru District (N.Z.): Part I—Systematics and distribution: New Zealand Geological Survey, Palaeontological Bulletin, v. 34, p. 1–192.

Howe, H. V., 1939, Louisiana Cook Mountain Eocene foraminifera: Louisiana Department of Conservation, Geological Bulletin, v. 14, p. 1–122.

Jennings, P. H., 1936, A microfauna from the Monmouth of basal Rancocas groups of New Jersey: Bulletin of American Paleontology, v. 23, p. 161–232.

Katz, M. E., and Miller, K. G., 1993, Neogene subsidence along the northeastern Australian margin: Benthic foraminiferal evidence, in Davies, P. J., MacKenzie, J. A., et al., eds., Scientific Results, 133: College Station, Texas (Ocean Drilling Program) (in press).

Liska, R. D., 1985, The range of Globigerinoides ruber (d'Orbigny) from the middle to late Miocene in Trinidad and Jamaica: Micropaleontology, v. 31, p. 372–379.

Loeblich, A. R., and Tappan, H., 1955, Revision of some Recent foraminiferal genera: Smithsonian Institution Miscellaneous Collections, v. 128, p. 1–37.

Lohmann, G. P., 1978, Abyssal benthonic foraminifera as hydrographic indicators in the western South Atlantic: Journal of Foraminiferal Research, v. 8, p. 6–34.

Lohmann, G. P., 1980, PATS-1, A package of programs for the analysis of marine micropaleontological data on the VAX 11/780 computer: Woods Hole Oceanographic Institution Technology Report, WHOI-80-27, 148 p.

McLaughlin, P. P., Jr., 1989, Neogene basin evolution in the southwestern Dominican Republic: A foraminiferal study [Ph.D. thesis]: Louisiana State University, 318 p.

McLaughlin, P. P., Jr., and Sen Gupta, B. K., 1989a, Foraminifera, paleoenvironments, and basin evolution, S.W. Dominican Republic [abs.]: Caribbean Geology Conference, 12th, St. Croix, U.S. Virgin Islands, p. 115.

McLaughlin, P. P., Jr., and Sen Gupta, B. K., 1989b, The paleoenvironmental significance of extinct benthic foraminifera in the Neogene sequence of Southwestern Dominican Republic: Geological Society of America Annual Meeting, v. 21, p. A47.

Miller, K. G., 1983, Eocene-Oligocene paleoceanography of the deep Bay of Biscay: Benthic foraminiferal evidence: Marine Micropaleontology, v. 7, p. 403–440.

Miller, K. G., and Katz, M. E., 1987a, Oligocene to Miocene benthic foraminiferal and abyssal circulation changes in the North Atlantic: Micropaleontology, v. 33, p. 97–149.

Miller, K. G., and Katz, M. E., 1987a, Oligocene to Miocene benthic foraminiferal and abyssal circulation changes in the North Atlantic: Micropaleontology, v. 33, p. 97–149.

Miller, K. G., and Katz, M. E., 1987b, Eocene benthic foraminiferal biofacies of the New Jersey transect, in Poag, C. W., Watts, A. B., et al., Initial Reports of the Deep Sea Drilling Project, v. 95: Washington, D.C., U.S. Government Printing Office), p. 267–298.

Miller, K. G., Aubry, M. -P., Khan, M. J., Melillo, A. J., Kent, D. V., and Berggren, W. A., 1985, Oligocene-Miocene biostratigraphy, magnetostratigraphy, and isotopic stratigraphy of the western North Atlantic: Geology, v. 13, p. 257–261.

Miller, K. G., Curry, W. B., and Ostermann, D. R., 1985, Late Paleogene (Eocene to Oligocene) benthic foraminiferal oceanography of the Goban Spur region, Deep Sea Drilling Project Leg 80, in de Graciansky, P. C., Poag, C. W., et al., Initial Reports of the Deep Sea Drilling Project, v. 80, Washington, D.C., U.S. Government Printing Office, p. 505–538.

Miller, K. G., Wright, J. D., and Brower, A. N., 1989, Oligocene to Miocene stable isotope stratigraphy and planktonic foraminifer biostratigraphy of the Sierra Leone Rise (DSDP Site 366 and ODP Site 667), in Ruddiman, W. F., Sarnthein, M., et al., Proceedings of the Ocean Drilling Program, Scientific Reports, v. 108: Washington, D.C., U.S. Government Printing Office, p. 279–294.

Montague, G., 1803, Testacea Britannica, or natural history of British shells, marine land and fresh-water, including the most minute: Romsey, England, J. S. Hollis, p. 515.

Montfort, P. D. de, 1808, Conchyliologie systematique et classification methodique des coquilles: Paris, France, F. Schoell, tome 1.

Neugeboren, J. L., 1850, Foraminiferen von Felso-Lapugy; zweiter Artikel: Verhandlungen und Mittheilungen des Siebenbürgischer Verein für Naturwissenschaften zu Hermannstadt, v. 1, p. 118–127.

Nuttall, W.L.F., 1928, Tertiary foraminifera from the Naparima region of Trinidad (British West Indies): Royal Geological Society of London, Quarterly Journal, v. 84, p. 57–117.

Nuttall, W.L.F., 1930, Eocene foraminifera from Mexico: Journal of Paleontology, v. 4, p. 271–293.

Nuttall, W.L.F., 1932, Lower Oligocene foraminifera from Mexico: Journal of Paleontology, v. 6, p. 3–35.

d'Orbigny, A. D., 1826, Tableau methodique de classe des Cephalopodes: Annales des Sciences Naturelles, ser. 1, p. 96–314.

d'Orbigny, A. D., 1839, Voyage dans l'Amerique meridionale; Foraminiferes: Strasbourg, France, Levrault, tome 5, p. 1–86.

d'Orbigny, A. D., 1846, Foraminiferes fossils du bassin tertiare de Vienne (Autriche): Paris, Gide et Comp., 303 p.

Palmer, D. K., 1940, Foraminifera of the upper Oligocene Cojimar Formation of Cuba; Part 3: Sociedad Cubana de Historia Natural Memorias de la Museo Poey, v. 15, p. 277–303.

Palmer, D. K., 1941, Foraminifera of the upper Oligocene Cojimar Formation of Cuba; Part 4: Sociedad Cubana de Historia Natural Memorias de la Museo Poey, v. 15, p. 181–200.

Palmer, D. K., 1945, Notes on the foraminifera from Bowden, Jamaica: Bulletin of American Paleontology, v. 115, p. 3–83.

Palmer, D. K., and Bermúdez, P. J., 1936a, An Oligocene foraminiferal fauna from Cuba; Part 2: Sociedad Cubana de Historia Natural Memorias de la Museo Poey, v. 10, p. 273–317.

Palmer, D. K., and Bermúdez, P. J., 1936b, Late Tertiary foraminifera from the Matanzas Bay region, Cuba: Sociedad Cubanade Historia Natural Memorias de la Museo Poey, v. 9, p. 237–257.

Parker, F. L., and Bermúdez, P. J., 1937, Eocene species of the genera *Bulimina* and *Buliminella* from Cuba: Journal of Paleontology, v. 11, p. 513–516.

Parker, W. K., and Jones, T. R., 1865, On some foraminifera from the North Atlantic and Arctic Oceans, including Davis Straits and Baffin's Bay: Philosophical Transactions of the Royal Society of London, v. 155, p. 325–441.

Parker, W. K., Jones, T. R., and Brady, H. B., 1865, On the nomenclature of the foraminifera; Part XII—The species enumerated by d'Orbigny in the "Annales des Sciences Naturelles", 1826, vol. 7: London, The Annals and Magazine of Natural History, ser. 3, p. 15–41.

Parker, W. K., Jones, T. R., and Brady, H. B., 1871, On the nomenclature of the foraminifera; Part XIV—The species enumerated by d'Orbigny in the "Annales des Sciences Naturelles", 1826, v. 7 (continued from Annals Natural History, ser. 3, vol. XVI, p. 41): London, The Annals and Magazine of Natural History, ser. 4, p. 145–179, 238–266.

Pflum, C. E., and Frerichs, W. E., 1976, Gulf of Mexico deep-water foraminifers: Cushman Foundation for Foraminiferal Research, Special Publication 14, 125 p.

Phleger, F. B., and Parker, F. L., 1951, Ecology of foraminifera, northwest Gulf of Mexico; Part II—Foraminifera species: Geological Society of America Memoir 46, 64 p.

Phleger, F. B., Parker, F. L., and Peirson, J. F., 1953, North Atlantic foraminifera: Reports of the Swedish Deep-Sea Expedition, 1947–1948, v. 7, p. 3–122.

Poag, C. W., 1981, Ecologic Atlas of Benthic Foraminifera of the Gulf of Mexico: Woods Hole, Massachusetts, Marine Science International, 174 p.

Proto Decima, F., and Bolli, H. M., 1978, Southeast Atlantic Leg 40 Paleogene benthic foraminifers, in Bolli, H. M., Ryan, W.B.F., et al., eds., Reports of the Deep Sea Drilling Project, v. 40: Washington, D.C., U.S. Government Printing Office, p. 783–809.

Renz, H. H., 1948, Stratigraphy and fauna of the Agua Salada group, State of Falcon, Venezuela: Geological Society of America Memoir 32, 219 p.

Reuss, A. E., 1850, Neue Foraminiferen aus den Schichten des Osterreichischen Tertiarbeckens: Wien, Osterreich, Wien, Kaiserliche Akademie der Wissenschaften, Mathematische-Naturwissenschaftlicheclasse, Denkschriften, v. 1, p. 365–390.

Reuss, A. E., 1851, Ueber die fossilen Foraminiferen und Entomostraceen der Septarienthone der Umgegend von Berlin: Deutsche Geologische Gesellschaff, v. 3, p. 49–91.

Reuss, A. E., 1867, Die fossile fauna der Steinsalzablagerung von Wieliczka in Galizien, Kaiserliche Akademie der Wissenschaften, Mathematische-Naturwissenschaftlicheclasse, Denkschriften, Wien, Osterreich, Bd. 55, Abth. 1, p. 87.

Robinson, E., 1969, Geological field guide to Neogene sections in Jamaica, West Indies: Journal of the Geological Society of Jamaica, v. 10, p. 1–24.

Rogl, F., and Hansen, J. H., 1984, Foraminifera described by Fichtel & Moll in 1798; A revision of Testacea Microscopica: Norn, Austria, Ferdinand, Berger & Sohne, 143 p.

Rzehak, A., 1886, Die Foraminiferenfauna der Neogenformation der Umgebung von Mahr: Ostrau. Naturforschender Verein Brunn, Verhandlungen, Brunn (Brno), v. 24, p. 77–126.

Schlumberger, C., 1887, Note sur le genre Planispirina: Bulletin de la Société Zoologie de France, v. 12, p. 475–488.

Schnitker, D., 1979, Cenozoic deep water foraminifers, Bay of Biscay, in Montadert, L., Roberts, D. G., et al., eds., Initial Reports of the Deep Sea Drilling Project, v. 81: U.S. Government Printing Office, Washington, D.C., p. 611–622.

Schnitker, D., 1986, North-east Atlantic Neogene benthic foraminiferal faunas-tracers of deepwater paleoceanography, *in* Summerhayes, C. P., and Shackleton, N. J., eds., North Atlantic palaeoceanography: Geological Society of London Special Publication 21, p. 191–203.

Schwager, C., 1866, Fossile Foraminiferen von Kar Nikobar: Novara Expedition 1857–1859: Wien, Osterreich, Geologische Theil, v. 2, p. 187–268.

Sclater, J. G., Anderson, R. N., and Bell, M. L., 1971, Elevation of ridges and evolutions of the central Eastern Pacific: Journal of Geophysics, v. 76, p. 7888–7915.

Seguenza, G., 1862, Prime recerche intorno ai rizopodi fossili delle argille Pleistoceniche dei dintorni di Catania: Atti Accademie Gioenia de Sci. Nat. di Catania, ser. 2, v. 18, p. 84–126.

Silvestri, A., 1904, Richerche strutturali su alcune forme dei Trube di Bonfornello (Palermo): Roma, Accademia Pontificia Romana dei Nuovi Lincei, Memorie, v. 22, p. 235–276.

Steineck, P. L., 1974, Foraminiferal paleoecology of the Montpelier and Lower Coastal Groups (Eocene-Miocene), Jamaica, West Indies: Palaeogeography, Palaeoclimatology, Palaeoecology, v. 16, p. 217–242.

Steineck, P. L., 1981, Upper Eocene to middle Miocene ostracode faunas and paleo-oceanography of the North Coastal Belt, Jamaica: Marine Micropaleontology, v. 6, p. 339–366.

Subbotina, N. N., 1953, Upper Eocene Lagenidae and Buliminidae of the southern U.S.S.R.: Trudy Vsesoyuznogo Neftyanogo Nauchno-issledovatel'skogo Geologo-razvedochrnogo Instituta (VNI-GRI), n. ser., vypisk 69 (Microfauna of the U.S.S.R., sbornik 6), p. 115–255.

Subbotina, N. N., 1960, Microfauna of the Oligocene and Miocene deposits of the Vorotyshche River (Ciscarpathinans): Leningrad, Trudy Vsesoyuznogo Neftyanogo Nauchno–Issledovatel'skogo Geologo–Razvedochnogo Instituta (VNIGRI) Microfauna of the USSR, v. 153, p. 157–263.

Thomas, E., 1985, Late Eocene to Recent deep-sea benthic foraminifers from the central equatorial Pacific Ocean, *in* Mayer, L., Theyer, F. et al., Initial Reports of the Deep Sea Drilling Project, v. 65: Washington, D.C., U.S. Government Printing Office, p. 655–694.

Thomas, E., 1986a, Late Oligocene to Recent deep-sea benthic foraminifers from DSDP Sites 608 and 610, northeastern North Atlantic, *in* Ruddiman, W. F., Kidd, R. B., et al., Initial Reports of the Deep Sea Drilling Project, v. 94: Washington, D.C., U.S. Government Printing Office, p. 997–1031.

Thomas, E., 1986b, Changes in composition of Neogene benthic foraminiferal faunas in equatorial Pacific and North Atlantic: Palaeogeography, Palaeoclimatology, Palaeoecology, v. 53, p. 47–61.

Thomas, E., and Vincent, E., 1987, Equatorial Pacific deep-sea benthic foraminifera: Faunal changes before the middle Miocene polar cooling: Geology, v. 15, p. 1035–1039.

Tjalsma, R. C., 1983, Eocene to Miocene benthic foraminifera from DSDP site 516, Rio Grande Rise, South Atlantic, *in* Barker, P. F., Carlson, R. L., Johnson, D. A., et al., Initial Reports of the Deep Sea Drilling Project, v. 72: Washington, D.C., U.S. Government Printing Office, p. 731–755.

Tjalsma, R. C., and Lohmann, G. P., 1983, Paleocene-Eocene bathyal and abyssal benthic foraminifera from the Atlantic Ocean: Micropaleontology Special Publication 4, 90 p.

Todd, R., 1976, Some observations about Amphistegina (foraminifera), *in* Takayanage, Y., and Saito, T., eds., Progress in micropaleontology: Micropaleontology Press Special Publication, p. 382–394.

Toulmin, L. D., 1941, Eocene smaller foraminifera from the Salt Mountain limestone of Alabama: Journal of Paleontology, v. 15, p. 567–611.

Trauth, F., 1918, Das Eozanvorkommen bei Radstadt im Pongou und seine Beziehungen zu den gleichalterigen Ablagerungen bei Kirchberg am Wechsel und Wimpassing am Leithagebirge: Wien, Osterreich, Kaiserliche Akademie der Wissenschaften, Mathematisch-Naturwissenschaftlicheclasse, Denkschriften, v. 95, p. 171–278.

van Morkhoven, F.P.C.M., Berggren, W. A., and Edwards, A. S., 1986, Cenozoic cosmopolitan deep-water benthic foraminifera: Bulletin des Centres de Recherches Exploration-Production, Memoir 11, 421 p.

Vaughan, T. W., 1928, Species of large arenaceous and orbitoidal foraminifera from the Tertiary deposits of Jamaica: Journal of Paleontology, v. 1, p. 277–298.

Voloshinova, N. A., 1958, On new systematics of the Nonionidae: Trudy Vsesoyuznogo Neftyanogo Nauchno-Issledovatel'skogo Geologorazvedochnogo Instituta, v. 115, p. 117–223.

Walker, G., and Boys, W., 1784, Testacea minuta rariora, nuperrime detecta in arena littoris Sandvicensis a Gul. Boys, arm S.A.S. multa addidit, et omnium figuras ope microscopii ampliatas accurate delineavit Geo. Walker: London, J. March, 25 p.

Walker, G., and Jacob, E., 1798, *in* Kanmacher, F., ed., Adams' essays on the microscope (second edition): London, Dillon and Keating, 712 p.

Williamson, W. C., 1858, On the Recent foraminifera of Great Britain: Royal Society of London, 107 p.

Woodruff, F., 1985, Changes in Miocene deep-sea benthic foraminiferal distribution in the Pacific Ocean: Relationship to paleoceanography, *in* Kennett, J. P., ed., The Miocene ocean: Paleoceanography and biogeography: Geological Society of America Memoir 163, p. 131–176.

Wright, R. M., ed., 1974, Field guide to selected Jamaican geological localities: Mines and Geology Division Special Publication 1, 57 p.

MANUSCRIPT ACCEPTED BY THE SOCIETY NOVEMBER 11, 1992

Taxonomy, biostratigraphy, and paleoecologic significance of calcareous-siliceous facies of the Neogene Montpelier Formation, northeastern Jamaica

Florentin J-M. R. Maurrasse
Department of Geology, Florida International University, Tamiami Trail and S.W. 107th Avenue, Miami, Florida 33199

ABSTRACT

Rock sequences of the Montpelier Formation exposed on the northeastern coast of Jamaica east of Buff Bay include an interval of calcareous-siliceous facies composed of abundant sponge spicules with varying amounts of Radiolaria. Hexactinellid spicules of the Hyalospongea group (60 to 99%) dominate the biogenic silica components, whereas radiolarian assemblages vary from impoverished to well diversified. *Liriospyris globosa, Calocycletta costata,* and *Dorcadospyris dentata* occur at the base of the exposed outcrop, thus indicating a minimum age of late early Miocene (*Calocycletta costata* zone) for the onset of the calcareous-siliceous facies.

This age is correlative with the *Globigerinatella insueta* zone as indicated by the co-occurrence of *Globigerinatella insueta, Globorotalia peripheroronda,* and *Globigerinoides sicanus* (without *Praeorbulina glomerosa*). Thus, the base of the silica-bearing sequence lies within Zone N7 of Banner and Blow (1965).

The upper limit of the calcareous-siliceous facies lies within the upper part of the *Dorcadospyris alata* zone, or early middle Miocene. This age is indicated by the presence of the nominate radiolarian taxon whose full range characterizes the zone, and the disappearance of distinctive taxa such as *Didymocyrtis violina* and *Calocycletta costata,* which occur up to the base of this zone. This radiolarian stratigraphic level corresponds to the *Globorotalia fohsi fohsi* zone, as indicated by the presence of the nominate taxon and *Sphaeroidinellopsis disjuncta* and the absence of *Praeorbulina glomerosa.* These foraminifera are also indicative of Zones N10/N11 of Banner and Blow (1965).

The overwhelming abundance of sponge spicules in the calcareous-siliceous facies at Buff Bay, Jamaica, implies that a significant subsurface current flowed through the area for more than four million years, between late early and middle Miocene time. Sponge spicule abundance remained continuous during that time while Radiolaria occurred with large variation in frequencies. Radiolarian recurrences are interpreted to indicate fluctuating productivity in relation to variations in climatic forcing superimposed on the long-lasting subsurface flow in the area.

Physiographic and oceanic conditions that led to the calcareous-siliceous facies at Buff Bay are compared to environmental conditions in the Hatton/Rockall Bank area, in northeastern Atlantic. Sedimentation on the bank surface is strongly influenced by variations in the subsurface transport of North Atlantic Deep Water (NADW) over the bank. The subsurface flow may result in unusually high productivity of Hyalospongea similar to the Buff Bay sediments. Such a mechanism may apply to the Jamaican Miocene Series when paleophysiographic conditions in the northern Caribbean allowed the flow of NADW into the Caribbean basin through the incipient Jamaican island.

Maurrasse, F.J-M.R., 1993, Taxonomy, biostratigraphy, and paleoecologic significance of calcareous-siliceous facies of the Neogene Montpelier Formation, northeastern Jamaica, *in* Wright, R. M., and Robinson, E., eds., Biostratigraphy of Jamaica: Boulder, Colorado, Geological Society of America Memoir 182.

INTRODUCTION

Calcareous-siliceous facies in Miocene rocks of the Caribbean area are best known in sediments recovered from the Deep Sea Drilling Project (DSDP) Legs 4 and 15 (Bader, Gerard, et al., 1970; Edgar, Saunders, et al., 1973) where radiolarian facies are recorded up to the earliest part of early Miocene (Aquitanian) at sites 31 and 149 in the Venezuela Basin (Figs. 1, 2). The level of last occurrence of significant biogenic siliceous sediments in the deep Caribbean basins (Maurrasse, 1976, 1979) lies within the *Lychnocanoma elongata* radiolarian Zone, correlative with the *Globigerinoides primordius* foraminiferal Zone (Fig. 2). Despite the occurrence of a few radiolarians, mostly orosphaerids, in medial Miocene sediments of the Aves Ridge (site 30; Figs. 1, 2), and appreciable numbers of specimens reported in some surface sediments of the present (Goll and Bjorklund, 1971), true calcareous-siliceous facies with significant Radiolaria recur intermittently only in younger Miocene deposits of some of the Caribbean islands, notably in Trinidad (Maurrasse and Keens-Dumas, 1988), and Jamaica (Fig. 2). Thus, in contrast to pre-Aquitanian widespread production of biogenic siliceous sediment in Caribbean deep-sea environments, subsequent deposits are localized in certain areas of incipient islands.

Sanfilippo and Riedel (1976) summarized the relative stratigraphic position of land-based radiolarian occurrences in the Caribbean region, including Jamaica. In the present work I will discuss radiolarian taxonomy and stratigraphy in calcareous-siliceous facies of the Neogene Montpelier Formation, which crops out east of the town of Buff Bay, Jamaica (Fig. 3). The section is of Miocene age equivalent to the Spring Garden Member of the Montpelier Formation (Robinson, 1969; Steineck, 1981). I will further comment upon the paleophysiographic and paleoecologic significance of these facies, which are unusual due to the predominance of sponge spicules among the biogenic silica components.

LOCATION AND LITHOSTRATIGRAPHIC POSITION

The section of the Montpelier Formation studied in the present work is located along the coastal road near the eastern access of the hamlet of Buff Bay in northeastern Jamaica (Fig. 4). This road cut begins 1.2 km east of the eastern limit of the hamlet of Buff Bay, which is taken as the point where there is a "bridge carrying the coast road over the railway just east of Buff Bay township" (Robinson, 1969, p. 3). This section lies farther east and stratigraphically lower relative to the section studied by Robinson (1969). The eastern limit of the study area is precisely where the old railroad track enters a tunnel that goes under the main coastal road, and immediately west of the bridge crossing White River (Fig. 4).

Samples were taken from east to west, or stratigraphically upwards, over a distance of approximately 400 m beginning below the railroad track, near the eastern entrance of the railroad tunnel (Fig. 3). These rocks correspond to the ones referred to by Hill (1899) as "Montpelier beds seen at the railway tunnel."

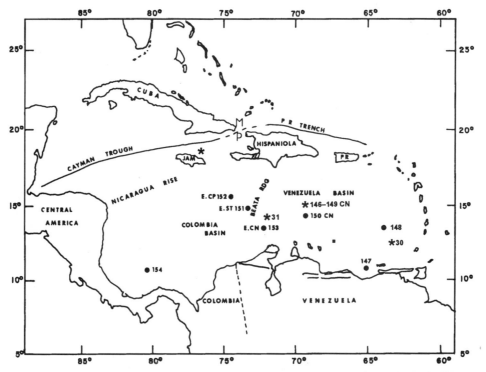

Figure 1. The Caribbean area showing site locations of Deep Sea Drilling Project (DSDP) Caribbean Legs 4 (30, 31) and 15 (146–149, 150, 151, 152, 153). *Asterisks indicate DSDP sites and Caribbean islands where calcareous siliceous facies occur in the Miocene series. MP = Mona Passage.

At the time of the initial sampling in 1976 the new road cut stood out as a magnificent white cliff of bedded chalk and partly indurated limestone, but by 1991 the cut was mostly covered by a lush vegetation (Fig. 3). In the field, the facies of pure white chalk at this location is physically identical to the sequence of "pure white chalky marl" that occurs stratigraphically higher and closer to Buff Bay. These rocks are referred to by Robinson (1969) as the Spring Garden Member of the Montpelier Formation, which is reported to span between the middle of middle Miocene and the late middle Miocene time.

The upper part of the Buff Bay sequence described in the literature (Hill, 1899; Robinson, 1969) as "pure white chalky marl" is also presently (1991) not as strikingly visible as such, because it is extensively covered by vegetation as well. The vagary of physical appearance of these cuts certainly influenced the nomenclature applied to these rocks through time (Hill, 1899; Robinson, 1969; Balkissoon, 1989). Under present (1991) conditions, for instance, unimpaired exposure of the rocks in the prominent road cuts east of Buff Bay occurs only in the sequences that correspond to the younger Buff Bay and Bowden Formations, respectively (Robinson, 1969). Rocks at these levels show fresh exposures because of continuous rock falls and slumps along the cut.

In an effort to clarify the nomenclatural complication of

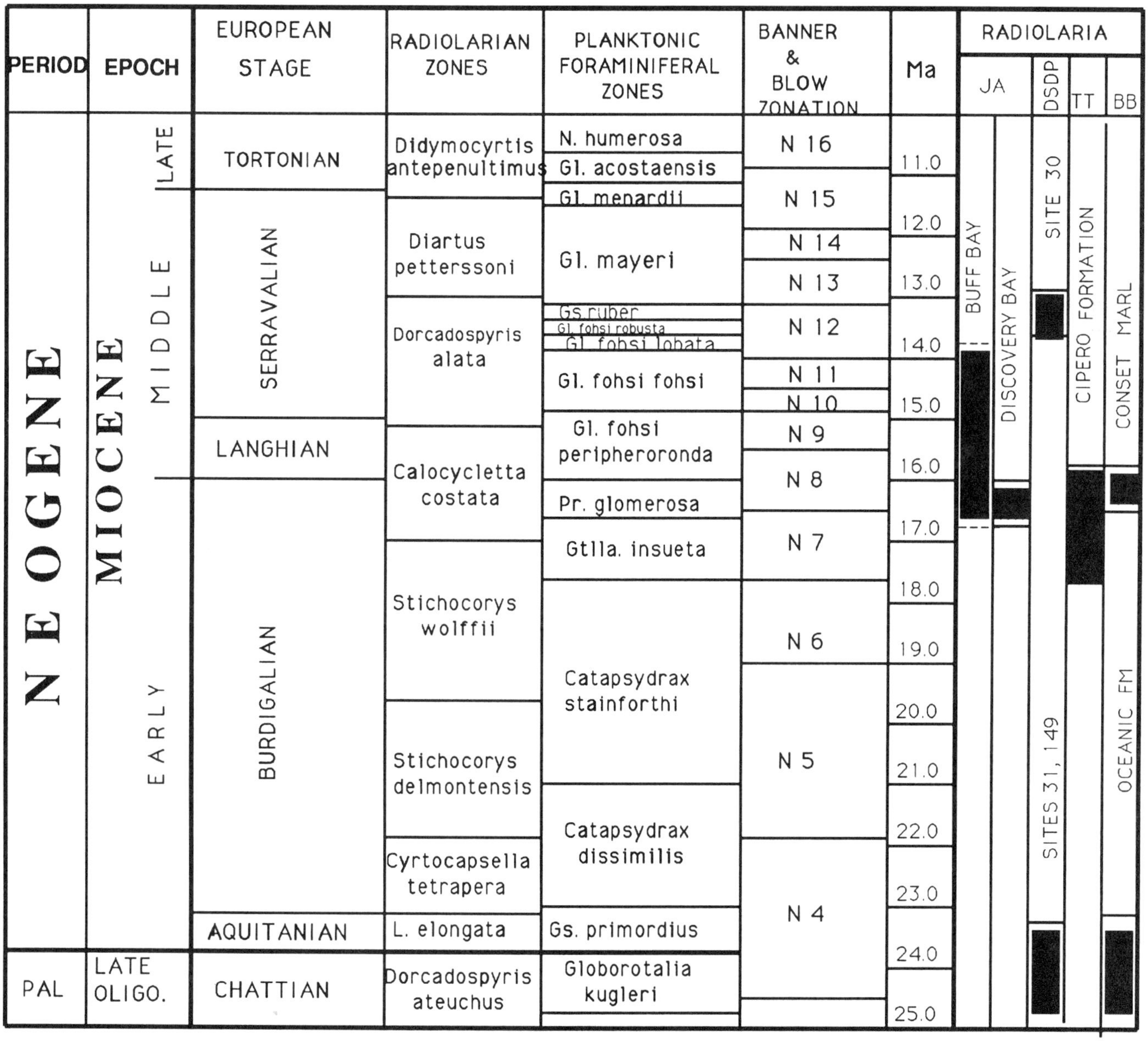

PERIOD	EPOCH		EUROPEAN STAGE	RADIOLARIAN ZONES	PLANKTONIC FORAMINIFERAL ZONES	BANNER & BLOW ZONATION	Ma	RADIOLARIA			
								JA	DSDP	TT	BB
NEOGENE	MIOCENE	LATE	TORTONIAN	Didymocyrtis antepenultimus	N. humerosa	N 16	11.0	BUFF BAY	SITE 30	CIPERO FORMATION	CONSET MARL
					Gl. acostaensis						
		MIDDLE	SERRAVALIAN		Gl. menardii	N 15	12.0				
				Diartus petterssoni	Gl. mayeri	N 14		DISCOVERY BAY			
						N 13	13.0				
				Dorcadospyris alata	Gs. ruber	N 12	14.0				
					Gl. fohsi robusta						
					Gl. fohsi lobata						
					Gl. fohsi fohsi	N 11					
						N 10	15.0				
			LANGHIAN	Calocycletta costata	Gl. fohsi peripheroronda	N 9	16.0				
					Pr. glomerosa	N 8	17.0				
		EARLY	BURDIGALIAN		Gtlla. insueta	N 7	18.0				
				Stichocorys wolffii	Catapsydrax stainforthi	N 6	19.0				OCEANIC FM
							20.0		SITES 31, 149		
				Stichocorys delmontensis		N 5	21.0				
					Catapsydrax dissimilis		22.0				
				Cyrtocapsella tetrapera			23.0				
						N 4					
			AQUITANIAN	L. elongata	Gs. primordius		24.0				
PAL	LATE OLIGO.		CHATTIAN	Dorcadospyris ateuchus	Globorotalia kugleri		25.0				

Figure 2. Radiolarian and foraminiferal biochronology applicable to the correlation of Neogene land-based sections in the northeastern coast of Jamaica. Data are compiled from various sources, including Palmer (1983), Saito (1984), Sakai (1984), and Goll and Bjorklund (1989). Radiolaria: L. = Lychnocanoma; Foraminifera: Gl = Globorotalia, Gs. = Globigerinoides, Gttla. = Globigerinatella, N. = Neogloboquadrina, Pr. = Praeorbulina; Ma = millions of years; JA = Jamaica; DSDP = Deep Sea Drilling Project Caribbean Legs 4 and 15; TT = Trinidad; BB = Barbardos.

eastern Jamaica, Balkissoon (1989) suggested that medial Miocene series of soft evenly bedded chert-free foraminiferal nannoplankton chalks exposed along the coast road east of Buff Bay be assigned to the Spring Garden Formation. Accordingly, the name Sign Member (Steineck, 1974) would designate the older series (uppermost Oligocene–lowermost Miocene) of soft, evenly bedded radiolarian- and planktonic foraminifera–rich chalks with chert.

As indicated above, however, the chalk facies at issue are in fact *physically* identical, whether or not they contain Radiolaria. Also, the part of the sequence attributable to the lower Miocene that contains Radiolaria does not have visible chert or apparent silicified horizons. Furthermore, Radiolaria are not restricted to the lower Miocene because Sanfilippo and Riedel (1976) reported numerous radiolarian species from Robinson's samples that are well constrained within middle middle Miocene, and are equivalent to the Spring Garden Member (Robinson, 1969).

Thus, both lithofacies and biofacies of the Miocene series exposed east of Buff Bay indicate that these rocks include overlapping characteristics that were proposed as criteria (Balkissoon, 1989) intended to identify a distinct Spring Garden Formation (middle Miocene) from the Sign Member (lower Miocene) of the Montpelier Formation. The consistent lithic characteristics of the sequence along the coast road near Buff Bay does not lend to easy mappable units (North American Commission on Stratigraphic Nomenclature, 1983) that would justify their separation as different entities.

Based on these arguments, the present work designates the whole sequence of bedded white chalks and partly indurated limestones without chert exposed east of Buff Bay as Spring Garden Member of the Montpelier Formation. This sequence includes intermittent levels of calcareous-siliceous facies that are physically indistinguishable in the field from the nonsiliceous levels. Its chronostratigraphic position extends between the upper part of early Miocene to the middle Miocene (Steineck, 1981).

In contrast to this area, however, similar white chalky facies that crop out farther west in road cuts south of Discovery Bay

Figure 3. A, The Spring Garden Member of the Montpelier Formation in the Buff Bay area along road cut on the northern coastal road of Jamaica. View is looking west-northwest toward hamlet of Buff Bay. Eastern entrance of tunnel is concealed by vegetation and is immediately west of bridge crossing the White River where its approximate position is marked by the arrow. B, View of the topmost part of the Montpelier Formation and the type section of the upper Miocene Buff Bay Formation. Site is farther northwest of Figure 3A (see asterisks in Fig. 4 for location) and view is also taken looking northwest toward hamlet of Buff Bay. (Pictures were taken on November 29, 1991, by the author.)

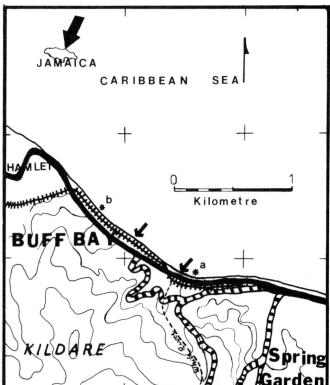

Figure 4. Location of the study area in northeastern Jamaica is indicated by tip of arrow pointed toward the island shown at the northwest corner of the map. Small arrows pointed towards coastal road delimit location of samples JA76-22 to JA76-36 of Table 1. Asterisks with letters a and b correspond to approximate positions where pictures shown in Figure 3 were taken.

(Fig. 2, on Route B3) may be compatible with the designated criteria distinctive of the Sign Member (Balkissoon, 1989). These rock sequences contain well-defined silicified horizons (chert and porcellanite), and Radiolaria are also recorded only up to the close of the early Miocene *Calocycletta costata* Zone, before the disappearance of *Dorcadosphyris dentata*. This level is correlative with the *Praeorbulina glomerosa* foraminiferal Zone (Fig. 2). Hence, the Sign Member of the Montpelier Formation defined by the presence of Radiolaria and chert would be in part chronostratigraphically equivalent to the Spring Garden Member.

MICROFACIES

The Buff Bay sequence consists predominantly of white (N9) to yellowish gray (5Y8/1) interbedded chalks, and partly indurated limestones rich in biogenic siliceous components. Most of the sequence can be described microscopically as sparse fossiliferous micrites. They contain rich assemblages of well-preserved planktonic foraminifera, and sponge spicules, which predominate in the fraction coarser than 40 μm. Radiolarians also occur in variable abundance with sponge spicules (Table 1), together with very rare diatoms. Nannoplanktons constitute the fine micritic matrix of the chalks, but more indurated levels also contain abundant microspar interspersed with micrite, and dispersive silicification affects some foraminiferal tests. Coarse fraction varies from less than 15% to greater than 30%, composed of variable amounts of calcareous foraminiferal tests and opaline sponge spicules. Secondary calcareous components include benthic foraminifera, ostracods, and echinoid spines. Hexactinellid sponge spicules of the Hyalospongea group make up 60 to 99% of the siliceous components. The sponge spicules change from very diversified assemblages with dominant large triaxonids of different morphologies (Fig. 5), to predominantly monaxonids and small sclerites toward the upper part of the section, when the siliceous components gradually vanished. The spicules show varying degrees of dissolution as axial canals may become enlarged, and holes several micrometers in diameter appear to be drilled into the spicules from the outside, as reported previously in opaline sponge spicules of the present reef environment from north Jamaica (Land, 1976). Radiolarian assemblages vary from well diversified to impoverished, and their frequency and state of preservation fluctuate intermittently irrespective of the abundance of the opaline sponge spicules (Table 1).

Although there are no distinct ash layers present in the sequence at Buff Bay, significant amount of volcanogenic feldspars (up to 10% total coarse) occurs intermittently, notably at the levels of samples 23, 25, and 32. There is also no apparent relationship between the abundance of volcanics and the abundance or preservation of the biogenic silica components (Table 1).

BIOSTRATIGRAPHY

The radiolarian taxonomy used in this study is partly after Campbell (1954), Riedel (1971) and mostly after the works of more recent investigators based upon the results of the Deep Sea Drilling Project (DSDP) reports referred to in the bibliography. The radiolarian biostratigraphic zonation is after Riedel and Sanfilippo (1978), and specific ranges are adapted from different authors as reported in the systematic section.

Biostratigraphic data shown in Table 1 include more than 90% of the taxa present in the Jamaican Miocene series. Radiolaria occur in varying frequency and degree of preservation in calcareous-siliceous facies dominated exclusively by hexactinellid sponge spicules (60 to 99%) of the Hyalospongea group (Figs. 5 to 10). Radiolarian tests always make up less than 2% of the total siliceous components, which constitute up to 10% of the total dry bulk weight of the rock. Although diversity of radiolarian assemblages varies proportionally with the absolute frequency of Radiolaria, it does not appear to be directly related to the abundance of sponge spicules. In sample 22, for instance, at the lowest level of occurrence of the calcareous-siliceous facies, sponge spicules are very abundant and well preserved, radiolarians are equally well preserved but they show very low diversity (Table 1). The same pattern is repeated at levels 32 and 33.

Carbonate components in the calcareous-siliceous facies are essentially nannoplankton and planktonic foraminifera, with less than 1% benthic foraminifera. There are also traces of diatoms (Fig. 8) at certain horizons, and preservation of the siliceous remains varies considerably (Table 1). Sponge spicules often show dissolution features similar to those observed in the modern environments (Land, 1976). Some levels contain abundant sponge spicules, but Radiolaria are virtually absent (Table 1).

The lower and upper time ranges during which abundant biogenic silica was produced and preserved in the pelagic carbonate of the sedimentary basin that is now exposed in the area east of Buff Bay can be constrained on the basis of radiolarians as follow. The numbers in parenthesis after species name corresponds to the ordinal number in the range chart of Table 1.

1. The lower limit lies within the upper part of the lower Miocene Series that correlates with the lower part of the *Calocycletta costata* Zone (Fig. 2). This biochronologic level is identified by the presence of *Liriospyris globosa* (30) and *Gorgospyris schizopodia* (26), both of which range only to the base of this zone (Goll, 1972). The lower part of this zone also includes the last occurrences of *Didymocyrtis prismatica* (10) and *Lychnocanoma elongata* (37).

In Jamaica, *Stichocorys coronata* (39) can be used to distinguish the transitional interval between late early and early middle Miocene because its range is restricted to the *C. costata* Zone (Table 1). *Liriospyris stauropora* (31), which ranges only from the upper part of the *Stichocorys wolffii* Zone to the base of the *Diartus petterssoni* Zone, is also more abundant within the *C. costata* Zone. *Dorcadospyris dentata* (28) is also typical of this zone, as found elsewhere in the tropical-equatorial realm of that time.

The age of late early Miocene based on Radiolaria from the base of the calcareous siliceous facies east of Buff Bay is in agreement with the concurrent foraminiferal assemblage comprised of *Globigerinatella insueta* Cushman and Stainforth, 1945, *Globoro-*

TABLE 1. DISTRIBUTION OF SILICEOUS MICROFOSSILS AND RANGES OF RADIOLARIAN SPECIES FROM THE MONTPELIER FORMATION, BUFF BAY, JAMAICA*

Sample Sites JA-76		22	23†	24	25†	26	27	28	29	30	31	32†	33	34	35	36
Acrosphaera spinosa echinoides	1	+					+				+			+?		
Acrosphaera transformata	2													+		
Trisolenia megalactis megalactis	3													+	+	
Trisolenia megalactis costlowi	4	+					R	+	R					R	R	
Solenosphaera omnitubus?	5														+	
Prunopyle sp.	6	R			+	F	+	+	R	F	+	F	+	F	+	+
Thecosphaera stylodendra?	7					R	+	+								
Druppatractus agostinelli	8						+			+	+			+	+	
Stylosphaera angelina	9								+		+		R	+	+	
Didymocyrtis prismatica	10					+		+								
Didymocyrtis tubaria	11					+?		+						+		
Didymocyrtis mammifera	12							+		R	+			+	+	
Didymocyrtis violina	13							+						+?		
Circodiscus microporus	14				+?	R	F							+		
Ommatocampe amphistylium	15					+										
Periphaena sp.	16					+		+						+	+	
Spongocore puer	17					+										
Rhopalodictyum malagaense	18					+	+	+		+	+			+		
Tholocubus sp.	19									+?					+	
Lithomelissa ehrenbergi	20					+										
Callimitra atavia?	21									+						
Dendrospyris bursa	22					R	+				+			+?		
Dendrospyris stabilis	23					R	R	+								
Gyraffospyris laterispina	24					F	F	+			+			+	+	
Gyraffospyris angulata	25	R				R	R	+	F	R	+	+	+	+		+?
Gorgospyris schizopodia	26							+								
Dorcadospyris forcipata	27					+	+	+	+							
Dorcadospyris dentata	28	+				+	+	R	R	+						
Dorcadospyris alata	29										+?		+	+	+	+?
Liriospyris globosa	30					+										
Liriospyris stauropora	31					F	F	F	R	F	+			+	+	?
Tholospyris anthophora	32	F				F	F	C	C	F	C	+	+	+	+	+
Tholospyris mammillaris	33	C				F	F	C	R	R	F			+		+
Bathropyramis	34					+										
Cornutella spp.	35						+			+						
Cyrtocapsella tetrapera	36					R	F	C	C	F	+			F	+	
Lychnocanoma elongata	37						+	+		?	?			+		
Eucyrtidium cienkowskii	38													+		
Stichocorys coronata	39					F	F	C	+	+						
Stichocorys delmontensis	40					R	F	?	+	+				+	+	
Stichocorys wolffii	41				+									+		
Carpocanopsis cingulata	42					F	+	C	R	R				+		
Carpocanopsis cristatum	43					R	+	+	R	R				+		
Carpocanopsis bussonii	44					F	R	+	-	F				+		
Lamprocyrtis margatensis	45					-	-									
Calocycletta costata	46					R	F	+	F	F	R		+	+?		
Calocycletta virginis	47	+				C	C	+	+	F				R	+	+
Artostrobium spp.	48						+	+		+	+			+		
Lithomitra spp.	49													+	+	
SP.SP. abundance/preservation§		D1	D3	T5	T5	A2	D2	D1	D3	D2	D2	D2	D3	D2	D2	D3
Radiolarian abundance/preservation§		C2	-	T5	T5	D1	A1	A1	A1	A1	A1	F3	T4	C1	C1	R3

*Numbers on top of chart refer to sampling stations recorded as JA76-22 to JA76-36. Sample intervals are approximately 1 m from samples 22 to 26, and 1.5 to 3 m from samples 26 to 36, over a total estimated thickness of approximately 25 m. Ordinal number associated with taxa refers to the same sequence of species presented in the taxonomic section. Bold face outlined symbols emphasize taxa of good chronostratigraphic value. Question mark indicates specimens of questionable identity. The limit of the *Calocycletta costata* Zone is taken here at the level of sample 31.
†Samples 23, 25, and 32 indicate levels with significant volcanogenic components.
§Last two rows show sponge spicule and radiolarian abundances and preservations. Symbols of relative frequency of taxa and preservation indicate the following:
Frequency: D = Dominant - more than 50 percent of siliceous components; A = Abundant - more than 5 specimens per field of view through the slide at 100X. C = Common - 3 to 5 specimens per field of view at 100X; F = Few - about 1 specimen or less per field of view at 100X; R = Rare - less than 1 specimen per ten fields of view at 100X; + = Present - three or more well-preserved specimens per slide; T = Trace - less than 2 or 3 specimens per slide, and generally in poor state of preservation. **Preservation:** 1 = Excellent, specimens are practically intact; 2 = Well preserved, specimens may show signs of partial dissolution, sponge spicules display occasional borings. 3 = Moderately preserved, specimens show advanced signs of dissolution, corrosion. Sponge spicules show extensive borings; 4 = Poorly preserved, specimens are only as fragments or recrystallized molds. Sponge spicules are fragmented due to excessive borings; 5 = Very poorly preserved, mostly as unidentifiable fragments.

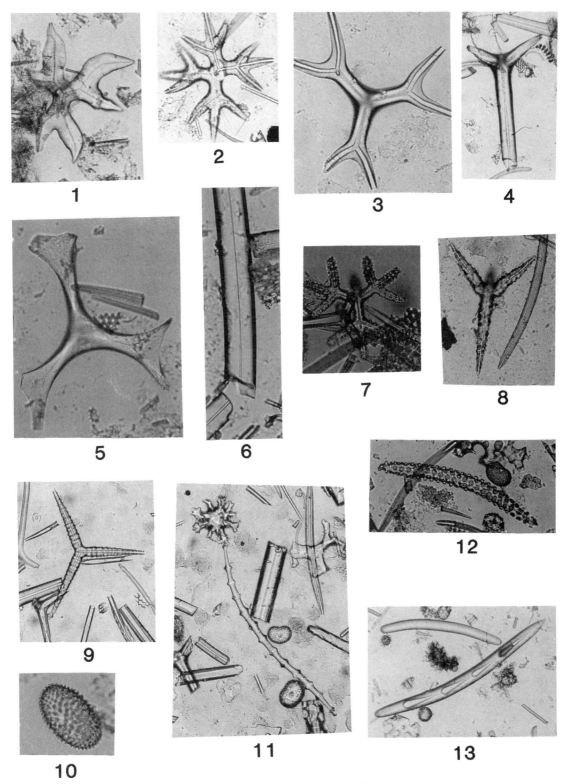

Figure 5. Magnification 100× (7–13); 200× (1–6). 1 to 13 represent different varieties of selected sponge spicules found in the Miocene series of the Montpelier Formation of Jamaica. In transmitted light sponge spicules show a typical axial canal, which differentiate them from large orosphaerid radiolarian (Friend and Riedel, 1967) spines. Many of the different shapes and sizes of hexactinellid spines shown here are also found in Quaternary sediments of the Hatton-Rockall Bank (see discussion in text).

talia peripheroronda Banner and Blow, 1965, and *Globigerinoides sicanus* De Stefani, 1952. The presence of these taxa without *Praeorbulina glomerosa* (Blow, 1969), is indicative of the late early Miocene *Globigerinatella insueta* Zone (Fig. 2), or Zone N7 (Banner and Blow, 1965; Blow, 1969).

2. The upper limit of the calcareous-siliceous facies lies entirely within the middle Miocene *D. alata* Zone. In the area investigated radiolarian assemblages include the nominate taxon and the last occurrences of *Didymocyrtis violina* and *Calocycletta costata*, which indicate the middle of the *D. alata* Zone. The latter species are not reported above this level. Worth noting is that radiolarian assemblages in the levels where siliceous components are scarce become impoverished, but remain quite well preserved, which rules out dissolution as the primary controlling factor in their disappearance. As will be discussed later, periodic upwelling conditions favorable to sustained Radiolaria productivity may have been controlled by dynamic divergence in relation to fluctuating Miocene climate. Notably absent in the Montpelier Formation at Buff Bay is *Didymocyrtis laticonus*. Elsewhere, this species has its first occurrence near the base of the *D. alata* Zone.

Concurrent foraminiferal taxa include *Globorotalia fohsi fohsi*, and *Sphaeroidinellopsis disjuncta* (Finlay, 1940). Notably absent is *Praeorbulina glomerosa* (Blow,1969), which ranges in the subjacent *Globorotalia fohsi peripheroronda* Zone. This assemblage is thus correlative with the *Globorotalia fohsi fohsi* Zone (Fig. 2), and is equivalent to Zones N10/N11 (Banner and Blow, 1965; Blow, 1969).

Both Radiolaria and sponge spicules continue to occur upward in the rock sequences of the Montpelier Formation east of Buff Bay into younger levels of the middle Miocene *Dorcadospyris alata* Zone, equivalent to Zone N12 (Robinson,1969; Sanfilippo and Riedel, 1976).

DISCUSSION

Lower to medial Miocene calcareous-siliceous facies intercalated in pelagic chalks exposed along cuts on the coastal road in northeastern Jamaica, east of Buff Bay, provide important data to further our understanding of the prevailing local and regional paleophysiographic and paleoceanographic conditions of the area at that time. As compared to other known siliceous facies in the Caribbean (Maurrasse and Keens-Dumas, 1988), these rocks are exceptional because of the predominance of sponge spicules. The significance of these facies can be evaluated on the basis of their time of occurrence and composition.

The geologic record of Jamaica available in the published literature clearly indicates that biogenic silica components are very scarce in Jamaican sediments prior to the early Miocene. Although biogenic silica productivity was widespread in the Pacific-Caribbean-Atlantic area during pre-early Miocene time (Maurrasse, 1979; Maurrasse and Keens-Dumas, 1988), it is most remarkable, however, that the primary biogenic productivity that characterizes all sedimentary rocks up to this age in Jamaica remained essentially dominated by a variety of carbonate producers. They were either benthic organisms of platform carbonates, or a combination of benthic and planktonic organisms indicative of neritic to neritopelagic and peripheral pelagic environments (Zans et al., 1963; Wright, 1974; Steineck, 1974, 1981; Robinson, 1988). These environments are usually not conducive to permanent upwelling of sufficient magnitude to maintain continuing resupply of silica to the surface waters permitting meaningful planktonic biogenic silica productivity.

Of special chronologic interest is the fact that the sedimentary basins that became part of present northern Jamaica do not show a record of biogenic silica in the Paleogene to Neogene series corresponding to the times when opaline silica accumulation was widespread in the adjacent deep basins. Instead, Radiolaria and diatoms are recorded in significant abundance (about 0.2% dry weight of total rock mass) in these environments 6 m.y. later (Fig. 2). It could be argued that the later preservation may be related to increased availability of dissolved silica in pore waters due to the presence of abundant sponge spicules or volcanogenic components. However, as pointed out before, such effects can be ruled out, because the abundance and state of preservation of Radiolaria (Table 1) vary independently of these constituents in the rocks.

Pre-Neogene siliceous sediments that accumulated intermittently in the deep basins of the Caribbean area from the Cretaceous until the earliest Miocene (Aquitanian: *Lychnocanoma elongata* Zone, equivalent to *Globigerinoides primordius* Zone) consist essentially of Radiolaria (Edgar, Saunders, et al., 1973; Maurrasse, 1973, 1976, 1979). In contrast, subsequent biogenic silica deposits that developed over isolated shallower areas of the Caribbean basin (Maurrasse and Keens-Dumas,1988) include diverse siliceous organisms, with at times predominance of diatoms and/or sponge spicules with or without Radiolaria, as found at Buff Bay. Earlier studies (Maurrasse, 1973, 1976, 1979: Maurrasse and Keens-Dumas, 1988) concluded that the production of widespread pre-Miocene biogenic silica in the deep Caribbean basins is compatible with upwelling in the area, as can be related to dynamic divergence associated with a large-scale global equatorial circulation (Sverdrup et al., 1970) flowing through the Caribbean area. The oceanographic mechanism inferred to have led to the earlier silica deposits is analogous to present circulation systems in the equatorial Pacific (Hays et al., 1969; Molina-Cruz, 1977; Leinen, 1979). It was also argued that cessation of biogenic silica accumulation in the Caribbean deep-sea basins at the onset of the early Miocene, *Lychnocanoma elongata* Zone (Fig. 2) resulted from increasing disruptive effects of the large-scale circulation system between the Atlantic and Pacific Oceans caused by the emerging Isthmus of Panama (Maurrasse, 1973, 1976, 1979).

Consequently, broad-scale upwelling practically ceased in the Caribbean area, and subsequent biogenic silica facies such as those in the Miocene rocks of Jamaica and their coeval equivalent elsewhere in the Caribbean region (DSDP Site 30; Trinidad), developed instead over physiographic areas corresponding to existing banks, submarine ridges, or emerging submarine highs (Maurrasse and Keens-Dumas, 1988). The radiolarian and diatom-rich siliceous facies of the Miocene Cipero Formation of

Trinidad, for instance, has been related to localized upwelling induced by forcing mechanisms generating eddies caused by the flow of a strong Caribbean current interfering to a certain extent with the incipient Central Range (Maurrasse and Keens-Dumas, 1988). This same mechanism appears to have induced radiolarian productivity recorded in Aves Ridge sediments at DSDP Site 30 (Figs. 1, 2).

In the case of Jamaica, however, the predominance of sponge spicules in the sediments for more than 2.5 Ma (Fig. 2) further indicates direct interactions between a subsurface current and the bottom for that length of time, because these sponges form filter-feeder communities requiring constant flow. The varying amounts of Radiolaria associated with the sponge spicules, also indicates that periodic upwelling resupplied enough silica to the surface to sustain planktonic biogenic silica productivity as well.

These conditions appear to have been unique to the northern Jamaican area in the Caribbean basin because no such combination of biogenic silica exists elsewhere.

In the present Caribbean sea topographically induced upwelling is known to occur over the Nicaragua Rise (Fig. 1) where the Caribbean western boundary current system flows past islands and banks (Hallock et al., 1988). The effect of such upwelling seems to be limited under existing conditions, because radiolarian productivity in the Caribbean Sea (Goll and Bjorklund, 1971) does not produce silica-rich sediments. Thus, it can be inferred that Radiolaria present in the Miocene Spring Garden Member of the Montpelier Formation in northern Jamaica were produced under upwelling conditions induced by either stronger flow of this current system, or that a different flow system affected the area at that time.

Upwelling attributable solely to the effects of the Caribbean Current, as argued for coeval siliceous facies in Trinidad (Maurrasse and Keens-Dumas, 1988), does not seem plausible in northern Jamaica, because that area would have been sheltered by very shallow banks to the south (Wright, 1974). Taking account of the paleogeographic reconstruction for the Miocene basins of proto-Jamaica (Wright, 1974), the alternative possibility to explain the siliceous facies in northern Jamaica requires a northerly provenance for the flow system that led to both strong subsurface flow, and simultaneously favorable upwelling conditions over the bank that constituted proto-Jamaica.

The involvement of a strong northerly subsurface flow as a causative agent for the productivity of the hyalosponges is compatible with earlier ostracod studies in the Montpelier Formation (Steineck, 1981), which suggested a depth of accumulation between 1,500 and 2,000 m, and the influence of a 3 to 4°C subsurface water-mass tentatively identified as the North Atlantic Deep Water (Steineck, 1981). Under present physiographic conditions the North Atlantic Deep Water (NADW) occurs in the Cayman Trough only (McMillen and Casey, 1978) where it enters the Caribbean basin via the Mona Passage (Fig. 1).

Development of Hyalospongea colonies in the Montpelier Formation supports the postulated flow of NADW over northern Jamaica (Steineck, 1981) at that time, because similar sponge populations develop in relatively shallow areas (between about 800 and 1,100 m) of the present Hatton-Rockall bank (Fig. 6) where the NADW overflows into the northeastern Atlantic.

The sedimentary record of spongolites associated with calcareous-siliceous facies with Radiolaria over this bank (asterisks in Fig. 6) represents the closest analog to the silica-rich facies of the Montpelier Formation. It is indeed remarkable that while the Hatton-Rockall Bank develops normal eupelagic foraminiferal-nannoplankton chalks, the shallower areas also include concomitant abundance of hyalosponge spicules and Radiolaria in proportions comparable to those found in the Jamaican Miocene facies. The calcareous-siliceous facies develop along the flow path of the North Atlantic Deep Water, and at depths where it influences the bottom. For instance, the three shallower-cores V29-196, V29-197, and V29-198 shown in Figure 6 include intermittent layers rich enough in sponge spicules to develop true spongolites. Variations in frequencies of these spongolites appear to follow the fluctuating patterns of the Pleistocene climatic cycles, which are not recorded farther north in the biogenic silica components (Ciesielski and Case, 1989; Goll and Bjorklund, 1989) where the Norwegian Sea is covered with permanent pack ice, and therefore unsuitable for generating NADW. Increased sponge spicules should thus represent periods of more open surface waters, and greater production of NADW, which is intensified over the bank permitting hyalosponges to thrive during warmer cycles. The stratigraphic record from DSDP Leg 48 (Sites 404, 405, 406) further suggests that similar conditions have been recurrent over the area since the early Eocene (Krasheninnikov, 1979).

By analogy with the Hatton-Rockall Bank in the northeast Atlantic, where hyalosponge productivity is influenced by subsurface flow, the calcareous-siliceous facies of the Montpelier Formation can be inferred to indicate upper bathyal pelagic deposits over a bank area that was affected by the flow of a subsurface water mass. The flow pattern over the Jamaican Bank was most likely affected by physiographic irregularities that caused the formation of vortices conducive to upwelling. This phenomenon would have been strong enough periodically to allow sufficient resupply of silica in the surface waters to sustain significant radiolarian productivity as well. Similar to the Hatton-Rockall Bank, the occurrence of sponge spicules in the Buff Bay section also coincides with a period of thermal high which is recorded in the Miocene (Vail et al., 1977; Ciesielski and Case, 1989; Goll and Bjorklund, 1989). It can be further postulated that the NADW overflow in the "Jamaican Passage" during Miocene time was related to greater production of this deep water mass during warmer periods when increased surface circulation reached into the Norwegian-Greenland Sea. The presence of low-latitude species such as *Cyrtocapsella tetrapera* in the Norwegian Sea sediments (Bjorklund, 1976) at that time brings supportive evidence for such transport, as the lower-latitude taxa from the Atlantic appear in the Norwegian Sea only during the thermal high.

The stratigraphic data thus suggest that the environmental

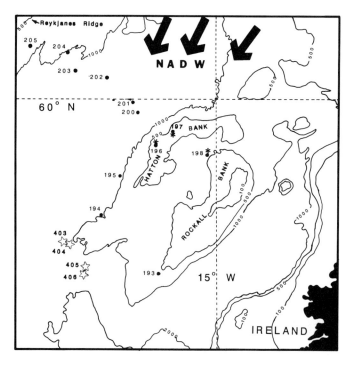

CORE NUMBER	DEPTH (FATHOMS)	GENERAL LITHOLOGY
V29-193	711	Interbedded Foram/Nanno chalks-no Sp.sp
V29-194	977	Interbedded Foram/Nanno chalks-no Sp.sp.
V29-195	1103	Interbedded Foram/Nanno marls with sandy turbidites
V29-196*	440	Interbedded Foram/Nanno chalk, marl with abundant sponge spicules at intermittent intervals
V29-197*	434	Interbedded Foram/Nanno chalk, marl with abundant sponge spicules at intermittent intervals
V29-198*	613	1,640 cm of Interbedded Foram/Nanno chalk, sandy chalk; spongolite layer only between 900 and 915 cm
V29-200	1426	Interbedded Foram/Nanno marls- no Sp.sp.
V29-201	1437	Interbedded Foram/Nanno marls, silty sand layers with cross-lamination- no Sp.sp.
V29-202	1430	Interbedded Foram/Nanno marls, basaltic ash layers no Sp.sp.
V29-203	1125	Interbedded Foram/Nanno marls and clays- no Sp.sp.
V29-204	1000	Interbedded Foram/Nanno marls and clays- no Sp.sp.
V29-205	784	Interbedded Foram/Nanno marls and clays- no Sp.sp.

Note: Asterisks denote cores with spongolites.

Figure 6. Map of the northeastern Atlantic region showing bathymetry in the area of Hatton Bank and Rockall Bank (bathymetric contour lines are in fathoms). Numbers refer to piston core locations from Vema 29 expedition in 1972 (coring sites 193 to 205), and Deep Sea Drilling Project Leg 48 in 1976 (stars show drill sites 403 to 406). NADW, North Atlantic Deep Water.

conditions conducive to NADW overflow over the site of proto-Jamaica during early and middle Miocene appear to have been enhanced in response to short- and long-term factors that affected the circulation in the Atlantic Ocean. Changing circulation in the North Atlantic with the diversion of the westward-flowing equatorial currents toward the north due to progressive disruptive effects of the emerging Isthmus of Panama (Maurrasse, 1979) caused fragmentation of the initial circulation between the Atlantic and the Pacific Oceans. Superimposed on these factors, changing global oceanic conditions (Haq et al., 1977; Vergnaud-Grazzini, 1979; Savin et al., 1985; Shackleton and Kennett, 1975) and probably cooling of deep waters occurred in conjunction with the onset of Antarctic ice sheet during early middle Miocene time (Vergnaud-Grazzini et al., 1979; Woodruff et al., 1981).

Unlike present conditions, physiographic conditions developed by the emerging Caribbean islands during Miocene time led to the flow of NADW southward into the Colombia Basin (Fig. 1) over a sill located at the site of the sedimentary basin where rocks of the Montpelier Formation accumulated. The record of fluctuating abundance of the Radiolaria (Table 1), which reflects variations in silica availability in surface waters over the Jamaican sill, was probably controlled by either interaction of periodic intensification of vortices due to increasing NADW flow, or by a combination of these factors with intermittent intensification of regional surface circulation (related to the effects of climatically induced oscillations on wind intensities). This argument is supported by the coeval sedimentary record of the Norwegian Basin (Ciesielski and Case, 1989; Goll and Bjorklund, 1989), which shows that recurring forcing mechanisms attributable to climatic factors occurred at that time (late early and the early middle Miocene).

The impoverished radiolarian assemblages observed in the Buff Bay section can thus be interpreted to reflect the recurring effects of climatically induced variations in intensity of local upwelling, while the subsurface flow of the NADW over Jamaica remained sufficiently strong to maintain continuous sponge productivity on the bottom. Perhaps the broader crisis of the nasselarians that started at the close of late early Miocene (Lombari, 1985), and apparently peaked in the *C. costata/D. alata* Zones, may have also had some effects on the radiolarian decline observed in the Montpelier Formation of Jamaica.

CONCLUSION

Bedded chalks exposed east of Buff Bay, Jamaica, include varying abundances of biogenic silica that consist mainly of sponge spicules (Figs. 5, 7, 8) and Radiolaria. These facies developed between late early and middle Miocene, subsequent to permanent interruption of widespread planktonic biogenic silica productivity in the deep basins of the Caribbean area (Maurrasse, 1979). The calcareous-siliceous facies of the Montpelier Formation in Jamaica are very different from other contemporaneous, as well as pre-Miocene, biogenic silica-rich deposits of the Caribbean region. Earlier siliceous facies of the deep-basins contain insignificant amounts or no remnants of sponge spicules. Similarly, contemporaneous calcareous-siliceous facies at other Caribbean locations contain Radiolaria, abundant diatoms, but very few sponge spicules (Maurrasse and Keens-Dumas, 1988).

Hyalosponges are benthic filter feeders, thus their presence in northern Jamaica implies the effects of a strong subsurface flow in the area during the time of accumulation of the calcareous-siliceous facies of the Montpelier Formation. The unusual characteristic of these facies in Jamaica stems from the fact that the abundance of planktonic biogenic components (foraminifers, nannoplanktons, and radiolarians) indicate a pelagic environment, whereas the predominance of hyalosponge spicules requires steady subsurface flows in contact with the sea floor for a prolonged period of time, in order to sustain continuous productivity of these sponges, which live fixed to the substrate.

The Jamaican Miocene facies are similar to sponge spicule-rich foraminiferal nannoplankton chalks that occur in Pleistocene deposits of the shallower areas of the present Hatton-Rockall Bank. These later calcareous-siliceous facies can be correlated with the effects of the North Atlantic Deep Water (NADW) overflowing into the north Atlantic (Fig. 6). Thus, siliceous facies of the Montpelier Formation are interpreted to have developed by similar mechanisms that involved overflow of the NADW through this area of Jamaica into the Caribbean Sea (Steineck, 1981). This flow caused both the unusual production of sponges, and vortices of varying intensities (depending on climate) that controlled resupply of silica to the surface waters to sustain variable Radiolaria productivity.

Conditions conducive to the deposition of the calcareous-siliceous facies of the Montpelier Formation seems to have been unique to Jamaica. In fact, Miocene calcareous-siliceous facies that developed in Trinidad (Maurrasse and Keens-Dumas, 1988) can be related essentially to the surface flow of the presumed Caribbean Current of that time crossing over the incipient Central Range. In the case of the Montpelier Formation of Jamaica, however, a steady overflow of the NADW over the bank maintained nearly continuous sponge productivity, whereas radiolarian productivity was more intermittent and is inferred to involve the interacting effects of the following factors: (1) Variations in the subsurface flow output of the NADW that caused fluctuations in the magnitude of vortices over the nascent Jamaican island, and controlled forced upwelling and recirculation of sufficient silica upward to sustain the localized surface biogenic silica productivity. (2) Variations in climatically induced surface wind stresses that further enhanced dynamic upwelling, associated with vortices resulting from NADW overflow in the northern area of the nascent Jamaican island.

Temporal variation in productivity of both benthic and planktonic biogenic silica in the Montpelier Formation resulted from a combination of local and regional physiographic factors influenced by forcing mechanisms related to climatic fluctuations of that time (Vail et al., 1977).

Independent from recurring variations in radiolarian productivity and diversity, permanent changes in the sponge spicule assemblages are recorded in the upper part of the calcareous-siliceous facies of the Buff Bay sequence. The permanent changes are probably related to declining NADW overflow over the "Jamaican Passage" as further change occurred in the Atlantic circulation system. Also, as uplift further developed in the Jamaican area, adjacent Cuba, and Hispaniola during the later part of Miocene time, the changing physiographic conditions shut off the southward flow of NADW into the Caribbean over Jamaica.

SYSTEMATIC PALEONTOLOGY

Taxa reported in this section are shown in Figures 7 to 10. Synonymies are provided only for specimens illustrated in the literature. Main emphasis is placed on radiolarian occurrences from diverse geographic areas in order to depict the overall biogeographic distribution of the different taxa relative to Jamaica and within the time frame of late early to middle middle Miocene. Number in parentheses before species name corresponds to the ordinal number in the range chart of Table 1.

Kingdom PROTISTA Haeckel, 1862
Superclass SARCODINA Hertwig and Lesser, 1874
Subclass RADIOLARIA Muller, 1858
Order POLYCYSTINA Ehrenberg, 1838, emend. Riedel, 1967b, p. 291
Suborder SPUMELLARIA Ehrenberg, 1875
Family COLLOSPHAERIDAE Haeckel, 1862

Collosphaerida **Haeckel, 1862; emend. Campbell, 1954, p. D51.**

Genus *Acrosphaera* Haeckel, 1881

Type species. Acrosphaera echinoides Haeckel, 1881, p. 471: subsequent designation by Bjorklund and Goll, 1979, p. 1307–1308.

(1) *Acrosphaera* sp. aff. *A. spinosa echinoides* Haeckel, 1887
Figure 7.11a

Acrosphaera spinosa echinoides Haeckel, 1887; Bjorklund and Goll, 1979, p. 1311–1312, Plate 4, Figs. 7–8.

Remarks. This species is reported to occur in the Pacific (Bjorklund and Goll, 1979) from the base of the *Cyrtocapsella tetrapera* Zone (lower Miocene) to the *Stichocorys peregrina* Zone (lower Pliocene). These authors also remarked that *A. spinosa echinoides* undergoes a pronounced size decrease in the upper Miocene of the equatorial Pacific region. In Jamaica very rare specimens attributable to this species occur in the *D. alata* Zone.

(2) *Acrosphaera* sp. aff. *A. transformata* Hilmers, 1906;
Bjorklund and Goll, 1979
Figure 7.12

Acrosphaera transformata Hilmers, 1906; Bjorklund and Goll, 1979, p. 1312–1315, Plate 1, Figs. 1–6.

Remarks. Specimens attributable to this taxon are reported in upper Pleistocene sediments of the central part of the equatorial Pacific region only (Bjorklund and Goll, 1979). If the specimens found in rocks of middle Miocene age in Jamaica are truly the same as this species, the present study suggests that its range is diachronous. This species appears to have originated in the Atlantic or the Caribbean, and later migrated to the eastern part of the equatorial Pacific.

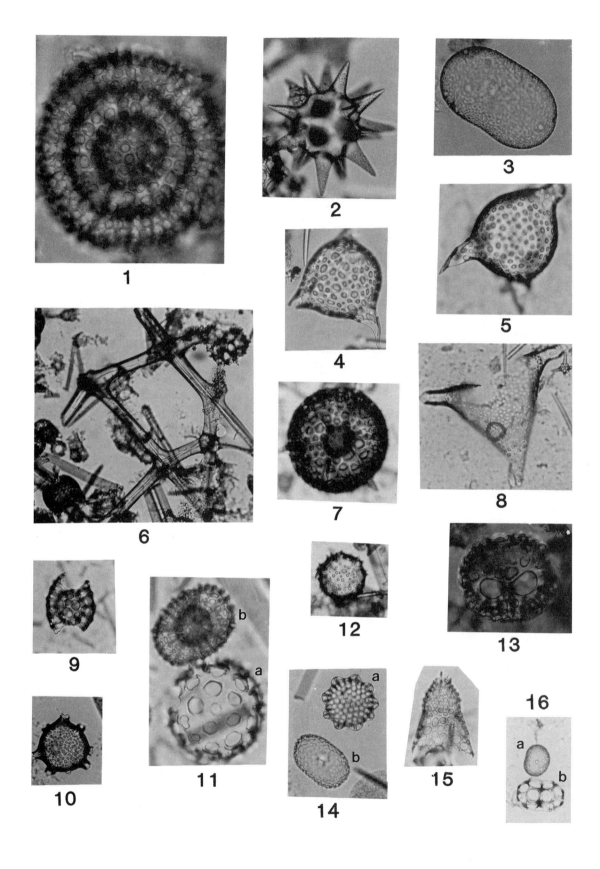

Genus *Trisolenia* Ehrenberg, 1860

Trisolenia Ehrenberg, 1860; Bjorklund and Goll, 1979, p. 1317–1318.

(3) *Trisolenia megalactis megalactis* Ehrenberg, 1872a; emend. Bjorklund and Goll, 1979
Figure 7.8

Trisolenia megalactis megalactis Ehrenberg: Bjorklund and Goll, 1979, p. 1321–1322, Plate 5, Figs. 1–11.

Remarks. *Trisolenia megalactis* has been best recorded in lower Miocene sediments of the eastern central Pacific, although it is also found in lower and middle Miocene sediments throughout the equatorial Pacific region (Bjorklund and Goll, 1979). In the Buff Bay section of Jamaica it occurs only within the middle Miocene *D. alata* Zone.

(4) *Trisolenia megalactis* (Ehrenberg, 1872) *costlowi* Bjorklund and Goll, 1979
Figure 7.4, 5

Trisolenia megalactis costlowi Bjorklund and Goll, 1979, p. 1322–1324, Plate 4, Figs. 5, 6, 9–12; Plate 6, Figs. 1–11.

Remarks. This species is reported to occur with varying frequency mostly at locations of the western equatorial Pacific during a short period of time, within the *Dorcadospyris alata* Zone (Bjorklund and Goll, 1979). These authors also suggested that *Trisolenia megalactis costlowi* may not have occupied the eastern tropical Pacific. Its penecontemporaneous occurrence in Jamaica may imply discontinuous biogeographic distributions related to existing oceanographic conditions.

(5) *Trisolenia* sp. cf. *Solenosphaera omnitubus* (Riedel and Sanfilippo, 1971)
Figure 7.10

Solenosphaera omnitubus Riedel and Sanfilippo, 1971, p. 1586, Plate 4, Figs. 1, 2.

Figure 7. Magnification 100×(*); 200×(**). 1, *Circodiscus microporus* (Stohr, 1880)**; 2, Sponge spicule aff. *Sphaeraster* in Dumitrica, 1973, p. 876, Plate 16, Fig. 3. Similar spicules are also reported as *Acanthometron (Astrolithium) astraeforme* in Miocene sediments of California (Campbell and Clark, 1944, p. 32, Plate 5, Fig. 8). This remarkable type of spicule occurs sporadically throughout the sequence, and more continually in the younger levels. ** 3, Sponge spicule, aff. *Sterraster* in Dumitrica, 1973, p. 876, Plate 16, Fig. 4**; 4–5, *Trisolenia megalactis costlowi* Bjorklund and Goll, 1979**; 6, Sponge spicules forming a solid framework similar to modern forms known as *Euplectella* spp. from the present Philippines and Japan Seas**; 7, *Thecosphaera* sp. cf. *T. stylodendra* Carnevale, 1908**; 8, *Trisolenia megalactis megalactis* Bjorklund and Goll, 1979 **; 9, *Spumellaria* gen. sp. indet.**; 10, *Solenosphaera* sp. cf. *S. omnitubus* Riedel and Sanfilippo, 1971*; 11, a, *Acrosphaera* sp., b, *Spumellaria,* gen. sp. indet.**; 12, *Acrosphaera* sp. cf. *A. transformata* Hilmers, 1906 *; 13. *Liriospyris globosa* Gol, 1968**; 14, a, *Periphaena* sp.; b, *Sponge sclerite***; 15, *Cyclampterium* (?) sp. This specimen shows a distant affinity to *C. milowi* Riedel and Sanfilippo, 1971 (p. 1593, Plate 3B, Fig. 3), but it is smaller than the latter, and the thorax (specimen from JA76-34, top of the calcareous-siliceous series) is not as thick as in typical *Cyclampterium* found in older levels (JA76-26).* 16, Sponge sclerite (a), and Trissocyclid Radiolaria (b)*.

Remarks. Specimens relating to this taxon from the Montpelier Formation in Jamaica are assigned to *Trisolenia* in concurrence with arguments presented by Bjorklund and Goll (1979, p. 1317–1318). Data presented by Riedel and Sanfilippo (1971) show a very narrow range for this species within the upper Miocene *Didymocyrtis penultima* Zone, corresponding to an interval between foraminiferal Zones 16 and 17. Specimens illustrated by Riedel and Sanfilippo (1971, Plate 1A, Figs. 17, 23, 24; Plate 4, Figs. 1, 2) are different from the Jamaican specimens, which have smaller peripheral tubes and a much larger central area than their apparent descendant in the Pacific.

Family ACTINOMMIDAE Haeckel, 1862

Actinommida Haeckel, 1862; emend. Riedel, 1967b, p. 294; Sanfilippo and Riedel, 1980, p. 1008.

Genus *Prunopyle* Dreyer, 1889, p. 3

Type species. Prunopyle pyriformis; subsequent designation by Campbell, 1954, p. D72, Fig. 30,7.

(6) *Prunopyle?* spp.
Figure 8.9

Remarks. Specimens referred to this genus-group are scarce in the Jamaican sediments. The group appears intermittently in Caribbean sediments since at least the earliest Eocene (Maurrasse, 1973), and is reported from the equatorial to the polar regions by various authors. Taxa of this group seem to be more frequent in Miocene siliceous sediments of southwestern Trinidad, and are quite common in Pliocene-Pleistocene sediments of the Antarctic region (Hays, 1965).

Genus *Thecosphaera* Haeckel, 1862

Thecosphaera Haeckel; Carnevale, 1908, p. 8, 9.

(7) *Thecosphaera?* sp. aff. *T. stylodendra* Carnevale, 1908
Figure 7.7

Thecosphaera stylodendra Carnevale, 1908, p. 8, Plate 1, Fig. 6.
Spongoplegma sp.: Chen, 1975, p. 454, Plate 22, Figs. 1, 2

Remarks. Specimens related to this species seem to occur simultaneously over widespread areas in the Miocene: Italy (Carnevale, 1908), Antarctica (Chen, 1975), and the Caribbean, where it is quite common in the Jamaican calcareous-siliceous series.

Genus *Druppatractus* Haeckel, 1887

Type species. Druppatractus hippocampus Haeckel, 1887; subsequent designation by Frizzell, 1951 (Frizzell and Middour, 1951).

(8) *Druppatractus agostinelli* Carnevale, 1908
Figure 8.2, 3, 4, 5

Druppatractus Agostinelli Carnevale, 1908, p. 20 (typographic error in p. 44, spelled as *Drappatractus*), Plate 3, Fig. 10.
Lithatractus (Lithatracta) santaennae Campbell and Clark, 1944, p. 19, Plate 2, Fig. 21.
Stylatractus santaennae (Campbell and Clark, 1944): Petrushevskaya and Kozlova, 1972, p. 520, Plate 11, Fig. 10.

Remarks. Specimens attributable to *D. agostinelli* are reported from middle and upper Miocene sediments of the subtropical regions, from the Mediterranean to the Pacific.

Genus *Stylosphaera* Ehrenberg, 1847; emend. Campbell and Clark, 1944, p. 10

Type species. Stylosphaera hispida Ehrenberg, 1854; subsequent designation by Frizzell, 1951 (Frizzell and Middour, 1951).

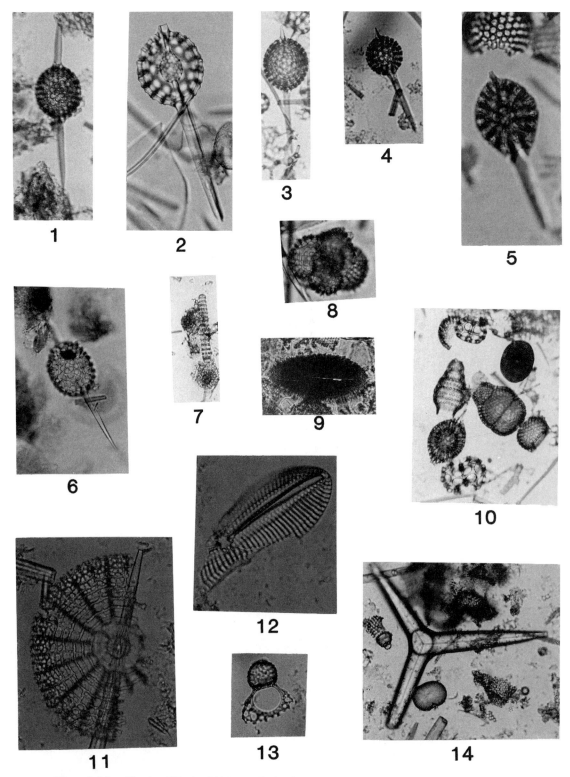

Figure 8. Magnification 100×(*); 200(**). 1, *Stylosphaera angelina* Campbell and Clark, 1944**; 2–5, *Druppatractus agostinelli* Carnevale, 1908**; 6, *Stylosphaera angelina* Campbell and Clark, 1944*; 7, Spongodiscid Radiolaria gen. and sp. indet. *; 8, *Tholocubus* sp.*; 9, *Prunopyle?* sp.**; 10, Radiolarian assemblage, including *Lithomitra, Prunopyle, Druppatractus, Cyrtocapsella tetrapera,* and Trissocyclids*; 11, Diatom, affinity *Arachnodiscus ornatus* Ehrenberg: Bukry and Foster, 1973, p. 833, Plate 8, Fig. 6, reported as rare in medial to upper Miocene sediments of the eastern equatorial Pacific.**; 12, Diatom, aff. *Nitzschia* sp.?**; 13, *Callimitra atavia* Goll, 1979*; 14, Triactine sponge spicule.*.

(9) *Stylosphaera* sp. aff. *Stylosphaera angelina* Campbell and Clark, 1944
Figure 8.1, 6

Stylosphaera (Stylosphaerella) angelina Campbell and Clark, 1944, p. 12, Plate 1, Figs. 14–20.
Axoprunum angelinum (Campbell and Clark, 1944): Kling, 1973, part, p. 634, Plate 6, Figs. 14–18; non *Stylatractus universus* Hays, 1970 (p. 215, Plate 1, Figs. 1, 2).
Amphistylus angelinus (Campbell and Clark, 1944): Chen, 1975, p. 453, Plate 21, Figs. 3, 4.

Remarks. Specimens referred to this taxon differ from *Stylosphaera laevis* Ehrenberg, 1873 (p.259; Ehrenberg, 1875, Plate 25, Fig. 6), which has bladed spines. *S. angelina* originally described from the Miocene of California was widespread, but appeared in staggered ranges at least throughout the Miocene. For instance, in Antarctica it is reported from the Oligocene to the upper Miocene, whereas in northeastern Pacific it is recorded from at least the *Calocycletta costata* Zone to the Pleistocene (Kling, 1973). In the latter case this extended range is due to the grouping by Kling of *Stylatractus universus* Hays with this species. The two taxa appear to overlap in Antarctica, with *S. universus* ranging from the upper Miocene to the Pleistocene similar to its range in the North Pacific as well (Hays, 1970). *S. angelina* may be conspecific with *Amphisphaera cristata* Carnevale, 1908 (p. 14, Plate 2, Fig. 7) from the middle Miocene of the Mediterranean region.

Family COCCODISCIDAE Haeckel, 1862

Coccodiscida Haeckel, 1862; emend. Campbell, 1954, p. D.82; Sanfilippo and Riedel, 1980, p. 1009.

Subfamily ARTISCINAE Haeckel, 1881

Artiscida Haeckel, 1881; emend. Campbell, 1954, p. D74; Sanfilippo and Riedel, 1980, p. 1010.

Genus *Didymocyrtis* Haeckel, 1860

Type species. Haliomma didymocyrtis violina Haeckel, 1862; Sanfilippo and Riedel, 1980, p. 1010.

(10) *Didymocyrtis prismatica* (Haeckel, 1887)
Figure 9.3

Pipetella prismaticus Haeckel, 1887, p. 305, Plate 39, Fig. 6.
Cannartus prismaticus (Haeckel, 1887): Riedel and Sanfilippo, 1970, p. 520, Plate 15, Fig. 1.
Cannartus prismaticus (Haeckel, 1887): Riedel and Sanfilippo, 1971, p. 1588, Plate 2C, Figs. 11–13; Plate 4, Fig. 5.
Cannartus prismaticus (Haeckel): Goll, 1972, p. 956, Plate 3, Figs. 1–3; Plate 4, Figs. 1, 2.
Cannartus sp., Petrushevskaya and Kozlova, 1972, Plate 12, Fig. 2.
Cannnartus sp. aff. *C. prismaticus* (Haeckel): Chen, 1975, p. 453, Plate 20, Fig. 7.
Cannartus prismaticus (Haeckel, 1887): Ling, 1975, p. 717, Plate 2, Figs. 7, 8.
Didymocyrtis prismatica (Haeckel, 1887): Sanfilippo and Riedel, 1980, p. 1010, text; Fig. 1c.

Remarks. D prismatica is reported to be common in the tropical/equatorial regions of the Pacific from the late Oligocene, *Theocyrtis annosa* Zone (Goll, 1972), to the lower part of the *Calocycletta costata* Zone (Riedel and Sanfilippo, 1971; Goll, 1972). It is reported as rare in lower Miocene sediments (lower part of the *Calocycletta veneris* Zone of Petrushevskaya and Kozlova, 1972) of the eastern subtropical Atlantic (*C. veneris* Zone = four radiolarian zones in the early Miocene shown below the *C. costata* Zone in Fig. 2). *D. prismatica* is not reported from the higher-latitude sediments of the northeastern Pacific (Kling, 1973), although it is recorded as rare in medial Miocene sediments of Antarctica (Chen, 1975). Its occurrence in the latter area would therefore indicate that it is diachronous, because it is recorded there at a level relatively younger than in the low-latitude regions. Specimens referred to *D. prismatica* in Jamaica appear only up to the *C. costata* Zone (Table 1).

(11) *Didymocyrtis tubaria* (Haeckel, 1887)
Figure 9.4

Pipetteria tubaria Haeckel, 1887, p. 339, Plate 39, Fig. 15.
Pipetteria tubaria Haeckel, 1887: Riedel, 1959, p. 289, Plate 1, Fig. 2.
Cannartus tubarius (Haeckel, 1887): Riedel and Sanfilippo, 1970, p. 520, Plate 15, Fig. 2.
Cannartus tubarius (Haeckel, 1887): Riedel and Sanfilippo, 1971, p. 1588, Plate 2C, Figs. 8–10.
Cannartus tubarius (Haeckel): Goll, 1972, p. 956, Plate 5, Figs. 1, 2.
Cannartus tubarius (Haeckel, 1887): Petrushevskaya and Kozlova, 1972, p. 521.
Cannartus tubarius (Haeckel, 1887): Dinkelman, 1973, p. 765, Plate 5, Figs. 3, 4.
Didymocyrtis tubaria (Haeckel, 1887): Sanfilippo and Riedel, 1980, p. 1010.

Remarks. Didymocyrtis tubaria is reported only in sediments from the tropical/equatorial region of the Pacific, Caribbean, Atlantic, and the Mediterranean. It occurred from the early Miocene *L. elongata* Zone to the base of the late middle Miocene *C. costata* Zone. Very rare specimens attributable to this taxon occur sporadically in the Buff Bay sequence.

(12) *Didymocyrtis mammifera* (Haeckel, 1887)
Figure 9.2

Cannartidium mammiferum Haeckel, 1887, p. 375, Plate 39, Fig. 16.
Cannartus mammiferus (Haeckel, 1887): Riedel, 1959, p. 291, Plate 1, Fig. 4.
Cannartus mammiferus (Haeckel, 1887): Riedel and Sanfilippo, 1970, p. 520, Plate 14. Fig. 1.
Cannartus mammiferus? (Haeckel, 1887): Petrushevskaya and Kozlova, 1972, p. 521, Plate 12, Fig. 3.
Cannartus mammiferus (Haeckel, 1887): Dinkelman, 1973, p. 765, Plate 3, Figs. 2, 3.
Cannartus mammifer (Haeckel, 1887): Sanfilippo et al., 1973, p. 216, Plate 1, Fig. 7.
Cannartus mammiferus (Haeckel): Kling, 1973, p. 634, Plate 7, Fig. 9.
Cannartus mammiferus (Haeckel): Ling, 1975, p. 717, Plate 2, Figs. 5, 6.
Didymocyrtis mammifera (Haeckel, 1887): Sanfilippo and Riedel, 1980, p. 1010.

Remarks. D. mammifera is recorded mostly in low-latitude areas of the Pacific, Caribbean, and Atlantic. Few to rare specimens are also reported in middle latitude regions of northeastern Pacific, at DSDP site 173 (off Cape Mendocino, California: Kling, 1973), and the Mediterranean (Sanfilippo et al., 1973). It ranges within the *Calocycletta costata* Zone (latest early Miocene) and the base of the *Dorcadospyris alata* Zone (early middle Miocene). This range is coincident with its occurrence in the calcareous-siliceous facies at Buff Bay (Table 1).

(13) *Didymocyrtis violina* (Haeckel, 1887)
Figure 9.1

Cannartus violina Haeckel, 1887, p. 358, Plate 39, Fig. 10
Cannartus violina Haeckel, 1887: Campbell, 1954, p. D74, Fig. 32.8
Cannartus violina Haeckel, 1887: Riedel, 1959, p. 290–291, Plate 1, Fig. 3.
Cannartus violina Haeckel, 1887: Petrushevskaya and Kozlova, 1972,

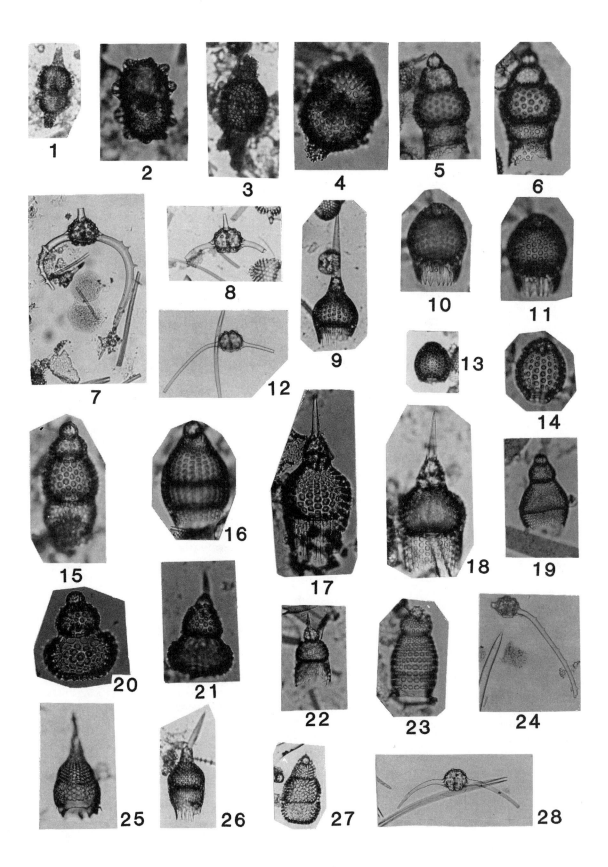

p. 522, Plate 12, Figs. 9, 10.
Cannartus violina Haeckel, 1887: Goll, 1972, p. 956, Plate 6, Figs. 1–3.
Cannartus violina Haeckel, 1887: Dinkelman, 1973, p. 765, Plate 8, Fig. 1.
Cannartus violina Haeckel, 1887: Sanfilippo et al., 1973, p. 216, Plate 1, Figs. 11, 12.
Cannarthus violina Haeckel, 1887: Kling, 1973, p. 634, Plate 7, Fig. 10.
Didymocyrtis violina (Haeckel, 1887): Sanfilippo and Riedel, 1980, p. 1010, text, Fig. 1d.

Remarks. *D. violina* shows similar paleogeographic distribution as *D. mammifera*. In Jamaica and elsewhere its ranges is mostly within the early Miocene, up to the *C. costata* Zone, but very rare specimens may occur in the *D. alata* Zone.

Family PORODISCIDAE Haeckel, 1887

Porodiscida Haeckel, 1887, p. 481; emend. Kozlova, 1967, p. 1171.

Genus *Circodiscus* Kozlova, 1972, in Petrushevskaya and Kozlova, 1972, p. 526

Type species. *Trematodiscus microporus* Stohr, 1880, p. 108, Plate 4, Fig. 17; Petrushevskaya and Kozlova, 1972, p. 526.

(14) *Circodiscus microporus* (Stohr, 1880)
Figure 7.1

Trematodiscus microporus Stohr, 1880, p. 108, Plate 4, Fig. 17.
Circodiscus microporus (Stohr): Petrushevskaya and Kozlova, 1972, p. 526, Plate 19, Figs. 7–8.

Remarks. This taxon is reported in the Atlantic from the Miocene to the Quaternary (Petrushevskaya and Kozlova, 1972). Of all the specimens illustrated by these authors, actually the specimen found in the middle Pliocene, their Figure 7, is the only one that can be equated to the specimens found in Jamaican Miocene sediments. Specimens from the Miocene of the Atlantic and Indian Oceans show a pronounced rim (Petrushevskaya and Kozlova, 1972, Plate 19, Figs. 1–6, 8), which is absent in the Jamaican ones. It is possible that the rim may have been severed in the Jamaican speciemns, although its absence seems to be constant at all levels of occurrence of this species.

Bjorklund and Goll (1979) reported the range of *C. microporus* to be concurrent with the *D. alata* Zone in the Pacific. They also suggested that it may have been restricted to the western equatorial or central water masses. In Jamaica, very rare specimens attributable to this species also occur in the upper part of the *C. costata* Zone (Table 1).

Genus *Ommatocampe* Ehrenberg, 1860

Type species. *Ommatocampe polyarthra* Ehrenberg, 1872a, Plate 6, Fig. 9.

(15) *Ommatocampe* sp. aff. *O. amphistylium* (Haeckel): Petrushevskaya and Kozlova, 1972
Figure 10.14

Ommatocampe sp. aff. *Amphymenium amphistilium* Haeckel: Petrushevskaya and Kozlova, 1972, p. 527, Fig. 2.

Remarks. Specimens attributable to *O. amphistylium* are reported in eastern Atlantic sediments from the late Eocene *Podocyrtis goetheana* Zone to the early Miocene *Cyrtocapsella tetrapera* Zone (Petrushevskaya and Kozlova, 1972). From specimens illustrated by these authors it appears that *O. amphistylium* shows a reduction in size prior to its extinction. Jamaican specimens more closely resemble Atlantic specimens from the earliest Miocene.

Family PHACODISCIDAE Haeckel, 1881

Phacodiscida Haeckel, 1881, p. 456.; emend. Campbell, 1954, p. D78.

Genus *Periphaena* Ehrenberg, 1873

Type species. By monotypy *Periphaena decora* Ehrenberg, 1873, p. 246; Ehrenberg, 1875, Plate 28, Fig. 6.
Periphaena Ehrenberg: Sanfilippo and Riedel, 1973, p. 522.

(16) *Periphaena* sp.
Figure 7.14a

Remarks. Numerous taxa attributable to periphaenids are reported in the Eocene of the Gulf of Mexico (Sanfilippo and Riedel, 1973), the Eocene and Oligocene of the equatorial Atlantic (Petrushevskaya and Kozlova, 1972), but this taxon group is not reported elsewhere or in younger sediments. In Jamaica, specimens attributed to this taxon occur intermittently at the levels of highest assemblage diversity (Table 1).

Family SPONGODISCIDAE Haeckel, 1862

Spongodiscida Haeckel, 1862, p. 452; emend. Riedel, 1967b, p. 295.

Genus *Spongocore* Haeckel, 1887

Type species. *Spongocore vellata* Haeckel, 1887: subsequent designation by Campbell, 1954, p. D74.

(17) *Spongocore puer* Campbell and Clark, 1944
Figure 10.18

Spongocore (*Spongocorisca*) *puer* Campbell and Clark, 1944, p. 22, Plate 3, Figs. 7–9.

Remarks. Specimens referred to this taxon are reported only in the Miocene of California. Nonetheless, radiolarians from the equatorial

Figure 9. Magnification 100 ×(*); 200×(**). 1, *Didymocyrtis violina* Haeckel, 1887*; 2. *Didymocyrtis mammifera* (Haeckel)**: 3, *Didymocyrtis prismatica* (Haeckel)**; 4, *Didymocyrtis tubaria* (Haeckel, 1887)**; 5–6, *Stichocorys delmontensis* (Haeckel, 1887)**; 7, *Dorcadospyris dentata* Haeckel*; 8, *Dorcadospyris forcipata* (Haeckel, 1881)*; 9, *Calocycletta costata* (Riedel, 1959); 10–11, *Carpocanopsis* sp. aff. *C. cingulata* Riedel and Sanfilippo, 1971**; 12, *Dorcadospyris alata* (Riedel, 1959)*; 13, *Carpocanopsis cristatum* (Carnevale, 1908)*; 14, *Carpocanopsis bussonii* (Carnevale, 1908)**; 15, *Stichocorys wolffii* Haeckel, 1887**; 16, *Eucyrtidium* sp. aff. *E. inflatum* ?Kling, 1973** (from JA76-26); 17, *Calocycletta virginis* (Haeckel, 1887)**; 18, *Calocycletta virginis* (Haeckel, 1887)**; 19. *Eucyrtidium cienkowskii* Haeckel, 1887*; 20–21, *Stichocorys coronata* Carnevale, 1908**; 22, Theoperid, gen. sp. indet., reminiscent of indet. Theoperid figured by Johnson (1974), Plate 2, Fig. 17, from upper medial Eocene sediments of the eastern Indian Ocean*; 23, *Artostrobium* sp.**; 24, *Dorcadospyris alata* (Riedel, 1959)*; 25, *Lamprocyrtis margatensis* Campbell and Clark, 1944 (JA76-35)*; 26, *Calocycletta virginis* (Haeckel, 1887)*; 27, *Cyrtocapsella tetrapera* (Haeckel, 1887)*; 28, *Dorcadospyris* sp. intermediate between *D. alata* and *D. simplex*.

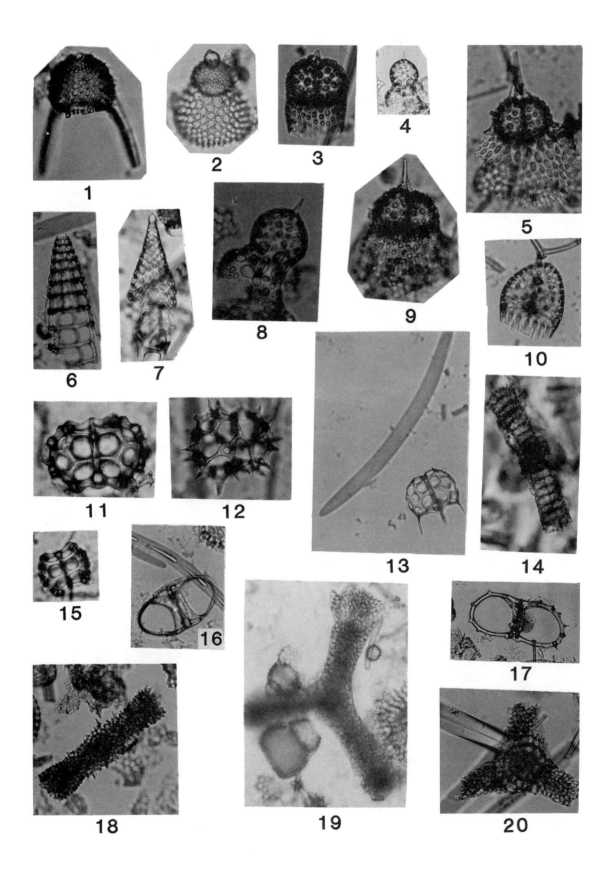

Atlantic assigned to *Amphibrachium robustum* Vinassa de Regny, 1900, p. 577, Plate 2, Fig. 11: Petrushevskaya and Kozlova, 1972, p. 528, Plate 21, Fig. 10, as well as *Ommatogramma* spp., Petrushevskaya and Koslova, 1972, Plate 21, Figs. 8, 9, 11, show close similarities with the Californian species. They are reported to occur from the Miocene to the Quaternary. In the Buff Bay sequence very rare specimens assigned to this taxon occur only within the middle part of the *C. costata* Zone.

Genus *Rhopalodictyum* Ehrenberg, 1860

Rhopalodictyum Ehrenberg, 1860, p. 830.

Type species. Rhopalodictyum abyssorum Ehrenberg, 1872b, p. 299, 392, Plate 8, Fig. 17; subsequent designation by Haeckel, 1887, p. 592; Campbell, 1954, p. D94, Fig. 46,8.

(18) *Rhopalodictyum malagaense* Campbell and Clark, 1944
Figure 10.19?, 20

Rhopalodictyum (*Rhopalodictya*) *malagaense* Campbell and Clark, 1944, p. 29, Plate 4, Figs. 4, 5.
Rhopalastrum profunda (Ehrenberg) group: Petrushevskaya and Kozlova, 1972, p. 529, Plate 20, Fig. 8, part; non *Dictyocoryne profunda* Ehrenberg, 1872b, Plate 7, Fig. 23.

Remarks. R. malagaense is reported in Neogene to Quaternary sediments of the equatorial to subtropical regions worldwide, starting at least from the early Miocene. This species is very similar to the type species which was reported from the equatorial Pacific, Philippine Seas, by Ehrenberg in 1872b, but the rare specimens observed in this study are insufficient to allow for conclusive comparison between the two species. *R. malagaense* differs from *Dictyocoryne profunda* Ehrenberg, which shows interbrachial spongy veil or patagium. This taxon is rare in the Jamaican samples and occurs intermittently throughout the interval studied.

Family THOLONIIDAE Haeckel, 1887

Tholonida Haeckel, 1887; emend. Campbell, 1954, p. D98.

Genus *Tholocubus* Haeckel, 1887

Type species. Tholocubus (*Tholocubus*) *tessellatus,* Haeckel; subsequent designation by Campbell, 1954, p. D98, Fig. 48.2.

(19) *Tholocubus* sp.
Figure 8.8

Tholocubus sp.: Benson, 1972, p. 1091, Plate 2, Figs. 3, 4.

Figure 10. Magnification 100×(*); 200×(**). 1, *Lychnocanoma elongata* (Vinassa de Regny, 1900)**; 2, *Stichocorys coronata*? Carnevale, 1908**; 3, *Dendrospyris stabilis* Goll, 1968**; 4, *Lithomelissa ehrenbergi* Butschli, 1882*; 5, *Dendrospyris bursa* Sanfilippo and Riedel, 1973**; 6, *Bathropyramis* sp.*; 7, *Cornutella* sp.**; 8, *Lithomelissa ehrenbergi* Butschli, 1882**; 9, *Dendrospyris bursa* Sanfilippo and Riedel, 1973**; 10, *Gorgospyris schizopodia* Haeckel, 1887**; 11, *Tholospyris anthopora* (Haeckel, 1887)**; 12, *Tholospyris mammillaris* (Haeckel, 1887)**; 13, Sponge spicule (monaxonid), and *Giraffospyris* sp.*; 14, *Ommatocampe amphistylum* (Haeckel, 1887)*; 15, *Tholospyris* sp.**; 16, *Liriospyris stauropora* (Haeckel, 1887)**; 17, *Giraffospyris angulata* (Haeckel, 1887)**; 18, *Spongocore puer* Campbell and Clark, 1944*; 19–20, *Rhopalodictyum malagaense* Campbell and Clark, 1944.

Remarks. Similar specimens are reported in very scant numbers in the Labrador Basin of northwestern Atlantic where they are first recorded in the *Calocycletta costata* Zone (Benson, 1972). Related taxa referred to as *Cubotholus* Haeckel, 1887, are also reported in Pleistocene sediments of the Mediterranean Sea (Dumitrica, 1973). Specimens referred to this genus are quite rare in the Jamaican Miocene section, with a more certain occurrence in the *D. alata* Zone (Table 1).

Suborder NASSELLARIA Ehrenberg, 1875
Family PLAGONIIDAE Haeckel, 1881

Plagoniida Haeckel, 1881; emend. Riedel, 1967b

Genus *Lithomelissa* Ehrenberg, 1847, p. 54

Type species. Lithomelissa tartari Ehrenberg, 1854: Campbell, 1954, p. D122, Fig. 61,1.

(20) *Lithomelissa* sp. aff. *Lithomelissa ehrenbergi* Butschli, 1882
Figure 10.4, 8

Lithomelissa(?) *ehrenbergi* Butschli, 1882: Dumitrica, 1973, p. ;837, Plate 25, Figs. 6, 7.
Lithomelissa sp. aff. *L. ehrenbergi* Butschli: Chen, 1975, p. 458, Plate 11, Figs. 1, 2.

Remarks. L ehrenbergi was originally described from specimens of the Eocene of Barbados (Butschli, 1882). Specimens referred to this species in Antarctica are reported to range from early to late Miocene (Chen, 1975), thus at a younger stratigraphic horizon than the original taxon. Likewise, Dumitrica (1973) reported specimens assigned to *L. ehrenbergi* in Quaternary sediments of the Mediterranean sea.

Jamaican specimens show similarities with their contemporaries from Antarctica, because they also do not have exposed dorsal nor lateral spines. Although these specimens may actually be a different species, their scarcity and occurrence in only one sample (Table 1) do not warrant further characterization.

Family TRISSOCYCLIDAE Haeckel, 1881

Trissocyclida Haeckel, 1881, p. 446; 1887, p. 986; emend. Goll, 1968, p. 1416.

Genus *Callimitra* Haeckel, 1881

Callimitra Haeckel, 1881, p. 431; emended Goll, 1979, p. 386.

Type species. Callimitra carolotae Haeckel, 1887; subsequent designation by Campbell, 1954, p. D122.

(21) *Callimitra* sp. aff. *Callimitra atavia* Goll, 1979?
Figure 8.13

Callimitra atavia Goll, 1979, p. 388, 390, Plate 5, Figs. 1, 5–9, 11.

Remarks. Specimens attributable to this taxon occur in lower Oligocene to upper Miocene sediments of the equatorial Pacific, as well as from the base of the Oligocene in the Oceanic Formation in Barbados (Goll, 1979). In Jamaica, it occurs in only one sample from the upper part of the *C. costata* Zone (Table 1).

Genus *Dendrospyris* Haeckel, 1881

Dendrospyris Haeckel, 1881, p. 441, emend. Goll, 1968, p. 1417.

Type species. Ceratospyris stylophora Ehrenberg, 1838 (Ehrenberg, 1875, Plate 20, Fig. 10); subsequent designation by Campbell, 1954, p. D114, Fig. 54,9.

(22) *Dendrospyris*(?) sp. aff. *Dendrospyris bursa* Sanfilippo and Riedel, 1973
Figure 10.5, 9

Dendrospyris bursa Sanfilippo and Riedel, 1973, in Sanfilippo et al., 1973, p. 217–218, Plate 2, Figs. 9–13.
Rhodospyris? spp. De 1 group Goll, 1968: Petrushevskaya and Kozlova, 1972, p. 531, Plate 38, Figs. 15(?), 16.

Remarks. Specimens referred to this taxon are reported from Oligocene and lower Miocene sediments of the equatorial/tropical regions of the Pacific, the Caribbean, and the Atlantic. In the Mediterranean region it is recorded sporadically from at least the base of the *C. tetrapera* Zone (early Miocene) to the *D. alata* zone (early middle Miocene). In the Buff Bay sequence this taxon is more frequent in the *C. costata* Zone (Table 1).

(23) *Dendrospyris stabilis* Goll, 1968
Figure 10.3

Dendrospyris stabilis Goll, 1968, p. 1422–1423, Plate 173, Figs. 16–18, 20.
Dendrospyris stabilis Goll, 1968: Chen, 1975, Plate 7, Fig. 3.

Remarks. Dendrospyris stabilis is reported as few to present in sediments of the equatorial tropical Pacific, the Caribbean, and in Antarctica (Chen, 1975, p. 455). According to Goll (1968), this taxon is found in the low-latitude areas starting from the middle Miocene, and ranges until the Quaternary. In Antarctica, however, this species is recorded only in lower Oligocene sediments (Chen, 1975). Its probable range into younger deposits in Antarctica still remains uncertain because of the existence of a hiatus at that time in the series. The remarkable time discrepancy between the occurrence of *D. stabilis* in Antarctica and the low-latitude areas may indicate that it originated from the former region, and later migrated into warmer-latitude areas during episodes of favorable climatic conditions, which occurred during changing Miocene climates (Vail et al., 1977). In the Buff Bay sequence it ranges only in the upper part of the *C. costata* Zone (Table 1).

Genus *Giraffospyris* Haeckel, 1881, emend. Goll, 1969, p. 329

Giraffospyris Haeckel, 1881, p. 442; Haeckel, 1887, p. 1056; Type species *Ceratospyris heptaceros* Ehrenberg, 1875; subsequent designation by Campbell, 1954, p. D114.

(24) *Giraffospyris* sp. aff *Giraffospyris laterispina* Goll, 1969
Figure 10.13

Giraffospyris laterispina Goll, 1969, p. 334–335, Plate 58, Figs. 15, 16, 20, 21.
Giraffospyris laterispina Goll, 1969: Goll, 1972, Plate 66, Figs. 1–4.
Acanthodesmiidae, gen. et sp. indet., Kling, 1971, Plate 8, Fig. 10.

Remarks. Specimens referred to this species are reported from upper Miocene to recent sediments of the Pacific and the Caribbean areas. Jamaican specimens attributable to this taxon are few to rare in the *C. costata* and *D. alata* Zones (middle Miocene) at Buff Bay. They are also different from the typical species, because their subcircular lattice pores appear to be wider than in Goll's species. The Jamaican specimens are very similar to an undetermined Acanthodesmiidae specimen represented by Kling (1971) shown in the *D. alata* Zone (Plate 8, Fig. 10) from northeastern Pacific.

(25) *Giraffospyris* sp. aff. *Giraffospyris angulata* Haeckel, 1887
Figure 10.17

Eucoronis angulata Haeckel, 1887, p. 978, Plate 82, Fig. 3.
Giraffospyris angulata (Haeckel): Goll, 1969, p. 331, Plate 59, Figs. 4, 6, 7, 9.

Remarks. According to Goll (1969) this taxon ranged from the late Miocene to the Quaternary in the Pacific and Caribbean regions. Specimens referred to this species in Jamaica are present from the base of the *C. costata* Zone (late early Miocene) to the *D. alata* Zone (Table 1). They show sagittal rings that are less spiny, a characteristic that may be associated with early ancestral forms, since they occur in older stratigraphic levels in Jamaica.

Genus *Gorgospyris* Haeckel, 1887, p. 1071

Gorgospyris Haeckel, 1881; Type species *Gorgospyris medusa* Haeckel, 1887; subsequent designation by Campbell, 1954, p. D114.

(26) *Gorgospyris schizopodia* Haeckel, 1887
Figure 10.10

Gorgospyris schizopodia Haeckel, 1887, p. 1071, Plate 87, Fig. 4.
Gorgospyris schizopodia Haeckel, 1887: Goll, 1972, p. 966, Plate 67, Figs. 1–3.
Thamnospyris sp. aff. *T. schizopodia* (Haeckel, 1887): Petrushevskaya and Kozlova, 1972, p. 531, Plate 38, Figs. 2.
Gorgospyris schizopodia Haeckel, 1887: Sanfilippo et al., 1973, p. 218, Plate 3, Figs. 6, 7(?), 9.

Remarks. Goll (1972) emphasized the stratigraphic significance of this taxon, which is found within a very short time interval in sediments of the equatorial Pacific and the Caribbean region. This species is recorded *only within the early Miocene,* from the *Cyrtocapsella tetrapera* Zone to the lower part of the *Calocycletta costata* Zone. Very rare specimens referred to this taxon are also reported to occur very briefly in the early Miocene of the Atlantic (Petrushevskaya and Kozlova, 1972) and the Mediterranean (Sanfilippo et al., 1973). In the Buff Bay sequence very rare specimens attributable to this taxon occur in sample 28 (Table 1), within the upper part of the *C. costata* Zone, or at a level equivalent to its uppermost known range elsewhere.

Genus *Dorcadospyris* Haeckel, 1881, emend. Goll, 1969, p. 335

Type species. Dorcadospyris dentata Haeckel, 1887; subsequent designation by Campbell, 1954, p. D112, Fig. 54,11.

(27) *Dorcadospyris forcipata* (Haeckel, 1887)
Figure 9.8

Dipodospyris forcipata Haeckel, 1887, p. 1037, Plate 85, Fig. 1.
Dipodospyris forcipata Haeckel, 1887: Riedel, 1959, p. 293, Plate 1, Fig. 9.
Dorcadospyris forcipata (Haeckel, 1887): Riedel and Sanfilippo, 1970, p. 523, Plate 15, Fig. 7.
Dipodospyris forcipata Haeckel, 1887: Goll, 1972, p. 964, Plate 54, Figs. 1, 2.

Remarks. D. forcipata is reported from the *L. elongata* Zone to the upper part of the *C. costata* Zone (base of early Miocene to the base of middle

Miocene) in low- to mid-latitude regions. Its appearance may be discontinuous or very rare in certain areas (Miocene sediments of the Indian Ocean, for instance, Johnson, 1974). This taxon shows close similarities with *D. dentata*, whose cephalis also bears a remarkable apical horn, except that the latter has developed short spinules on the outer edges of the feet. In the Jamaican Miocene series the two species are concurrent and seem to intergrade, as some specimens may display only faint development of ornamental spinules on the feet. At Buff Bay, specimens referred to *D. forcipata* occur within the *C. costata* Zone (Table 1).

(28) *Dorcadospyris dentata* Haeckel, 1887
Figure 9.7

Dorcadospyris dentata Haeckel, 1887, p. 1040, Plate 85, Fig. 6.
Dorcadospyris dentata Haeckel, 1887: Campbell, 1954, p. D112, Fig. 54, 11.
Dorcadospyris dentata Haeckel, 1887: Goll, 1969, p. 338, Plate 60, Figs. 8, 10, 11–13.
Dorcadospyris dentata Haeckel, 1887: Riedel and Sanfilippo, 1971, p. 1590, Plate 2D, Figs. 2, 3.
Dorcadospyris dentata Haeckel, 1887: Moore, 1971, p. 740, Plate 11, Figs. 1, 2.
Dorcadospyris dentata Haeckel, 1887: Goll, 1972, p. 964, Plate 56, Figs. 1, 3; Plate 57, Figs. 2, 3.
Dorcadospyris dentata Haeckel, 1887: Sanfilippo et al., 1973, p. 218, Plate 3, Figs. 2, 3.
Dorcadospyris dentata Haeckel, 1887: Johnson, 1974, p. 546, Plate 8, Fig. 10.
Dorcadospyris dentata Haeckel, 1887: Riedel and Sanfilippo, 1978, p. 68, Plate 5, Fig. 4.

Remarks. *D. dentata* is well represented in the tropical/equatorial regions of the world oceans where it ranged from the upper part of early Miocene (*S. wolffii* Zone) to the early part of middle Miocene (latest *C. costata* Zone). In the eastern Indian Ocean, specimens attributable to this species show only poorly developed spinules on the outer edges of the feet, and they ranged only within the *C. costata* Zone (Johnson, 1974). Specimens of this species found at Buff Bay show similar morphology and range (Table 1).
D. dentata is not reported in high latitude regions.

As pointed out above, *D. dentata* shows varying degrees of development of the ornamental spines on the feet, a character that does not seem to follow a distinct geographic pattern. Specimens from Majorca in the Mediterranean region, for instance, are reported to intergrade with *Dorcadospyris alata* (Sanfilippo et al., 1973), whereas in the Jamaican Miocene series *D. dentata* seems to intergrade with *D. forcipata*.

(29) *Dorcadospyris alata* (Riedel, 1959)
Figure 9.12, 24, 28?

Brachiospyris alata Riedel, 1959, p. 293–294, Plate 1, Figs. 11, 12.
Dorcadospyris alata (Riedel, 1959): Riedel and Sanfilippo, 1970, p. 523, Plate 14, Fig. 5.
Dorcadospyris alata (Riedel, 1959): Kling, 1971, p. 1087, Plate 4, Fig. 2.
Dorcadospyris sp. aff. *D. alata*: Petrushevskaya and Kozlova, 1972, p. 532, Plate 40, Fig. 13.
Dorcadospyris alata (Riedel, 1959): Sanfilippo and Riedel, 1973, *in* Sanfilippo et al., 1973, p. 218, Plate 2, Figs. 14, 15.
Dorcadospyris alata (Riedel, 1959): Riedel and Sanfilippo, 1978, p. 68, Plate 5, Fig. 2.

Remarks. *D. alata* is widespread in medial Miocene sediments of the tropical/equatorial regions where it defines a distinct biozone. Despite the fact that this species is well defined by its distinctive ornamental spines on the outer edges of the feet, variation in the development of these spines is most remarkable. In Jamaica this polymorphism leads to morphological types closely related and transitional to *D. simplex* (see, for instance, Fig. 9.12 and 28), and which show no actual ornamental short thorns on the feet. In these forms the only distinctive character from *D. Simplex* is that the latter has a larger cephalis with more pores. In the Mediterranean region, *D. alata* is reported to intergrade with *D. dentata*, because the latter may display a small cephalis that lacks the strong apical horn (see, for instance, Sanfilippo and Riedel, 1973, Plate 3, Fig. 2). The presence of *D. alata* in the siliceous facies at Buff Bay unequivocally defines its upper limits in the middle Miocene.

Genus *Liriospyris* Haeckel, 1881, emend. Goll, 1968, p. 1423–1424

Type species. Liriospyris hexapoda Haeckel, 1887; subsequent designation by Campbell, 1954, p. D.114.

(30) *Liriospyris globosa* Goll, 1968
Figure 7.13

Liriospyris globosa Goll, 1968, p. 1427–1428, Plate 176, Figs. 1–3, 5.
Liriospyris globosa Goll, 1968: Goll, 1972, p. 966, Plate 68, Figs. 1, 2.

Remarks. *L. globosa* has been reported only in the equatorial Pacific and the Caribbean regions. Goll (1972) emphasized the stratigraphic value of this taxon, which is found only in the Miocene series between the top of the *S. wolffii* Zone and the base of the *C. costata* Zone. Although this taxon is rare in the Buff Bay sediments (Table 1), its presence indicates that the onset of the calcareous siliceous facies at Buff Bay lies at least within the lower part of the *C. costata* Zone.

(31) *Liriospyris stauropora* (Haeckel, 1887)
Figure 10.16

Trissocyclus stauroporus Haeckel, 1887, p. 987, Plate 83, Fig. 5.
Liriospyris stauropora (Haeckel): Goll, 1968, p. 1431, Plate 175, Figs. 1–3, 7.
Liriospyris stauropora (Haeckel): Riedel and Sanfilippo, 1971, p. 1590–1591, Plate 2C, Figs. 16–19.
Liriospyris parkerae Riedel and Sanfilippo, 1971, p. 1590, Plate 2C, Fig. 15; Plate 5, Fig. 4.
Trissocyclus stauropora Haeckel, 1887: Petrushevskaya and Kozlova, 1972, p. 533, Plate 39, Figs. 29–31.
Liriospyris parkerae Riedel and Sanfilippo, 1971: Riedel and Sanfilippo, 1978, p. 69, Plate 5, Fig. 15.
Liriospyris stauropora (Haeckel, 1887): Riedel and Sanfilippo, 1978, p. 69, Plate 5, Fig. 16.

Remarks. Representatives of this taxon are reported in equatorial/tropical sediments of the Pacific, Caribbean, Atlantic, and Mediterranean regions, from the late Oligocene to the middle Pliocene. Riedel and Sanfilippo (1971) believed that *L. stauropora* evolved into *L. parkerae* at about the boundary between the *C. costata* and *D. alata* Zones, or early middle Miocene. In Jamaica, specimens attributable to this species occur in both the *C. costata* and *D. alata* Zones (Table 1).

Genus *Tholospyris* Haeckel, 1881, emend. Goll, 1969, p. 322

Type species. Tholospyris tripodiscus Haeckel, 1887, p. 441; Haeckel, 1887, p. 1078; subsequent designation by Campbell, 1954, p. D114.

(32) *Tholospyris anthophora* (Haeckel, 1887)
Figure 10.11, 15

Dictyospyris anthophora Haeckel, 1887, p. 1076, Plate 89, Fig. 8.

Tholospyris anthophora (Haeckel, 1887): Goll, 1969, p. 324–325, Plate 55, Figs. 1–3.
Tholospyris anthopora (Haeckel): Goll, 1972, p. 967–968, Plate 73, Fig. 1, 2; Plate 74, Figs. 1–3.
Lithotympanium tuberosum Haeckel, 1887: Petrushevskaya and Kozlova, 1972, p. 533, Plate 38, Figs. 23, 24; non *Liriospyris* sp. 2 Goll, 1968, Text, Fig. 9.

Remarks. Specimens referred to this taxon are reported from sediments of the equatorial regions of the Pacific and the Atlantic, ranging from the Oligocene to the middle Miocene *Diartus pettersonni* Zone (Goll, 1972). This taxon is well represented throughout the calcareous-siliceous facies at Buff Bay.

(33) *Tholospyris mammillaris* (Haeckel, 1887)
Figure 10.12

Dictyospyris mammillaris Haeckel, 1887, p. 1076, Plate 89, Figs. 9, 10.
Tholospyris mammillaris (Haeckel): Goll, 1969, p. 327, Plate 55, Figs. 5, 6, 8, 9, text–Fig. 1.
Tholospyris mammalaris (Haeckel): Goll, 1972, p. 968, Plate 80, Figs. 1–4 (p. 1051, also typo, spelled as *Tholospyris mannalaris*).

Remarks. This species is reported from sediments of the equatorial Pacific and the Caribbean region, where it ranged in the early Miocene from the *L. elongata* Zone to the *C. costata* Zone (Goll, 1972). This range is compatible with the biostratigraphic data at Buff Bay, although very rare specimens seem to occur also within the *D. alata* Zone (Table 1).

Family PLECTOPYRAMIDAE Haecker, 1908

Plectopyramididae Haecker, 1908; *Archiphormididae* and *Archicorythidae* (partim) Campbell, 1954, p. D118–D121.

Genus *Bathropyramis* Haeckel, 1881

Bathropyramis Haeckel, 1881, p. 428. Type species *Bathropyramis acephala* Haeckel, 1887 (p. 1159); subsequent designation by Campbell, 1954, p. D118.

(34) *Bathropyramis* sp.
Figure 10.6

Bathropyramis Haeckel: Campbell, 1954, p. D118, Figs. 59, 3a, 3b.

Remarks. Taxa of this genus are reported in different places at different times, from the Eocene to the present. Their record of occurrences indicates a cosmopolitan group with greater affinity for cold-water environments (Maurrasse, 1979). Specimens in the Jamaican section are very rare (Table 1), but they are similar to various taxa referred to this name, and illustrated by most authors from all latitudes worldwide.

Genus *Cornutella* Ehrenberg, 1838, p. 128

Type species. Cornutella clathrata Ehrenberg, 1838 (p. 129; Ehrenberg, 1844, p. 77; Ehrenberg, 1854, Plate 22, Fig. 39); subsequent designation by Campbell, 1954, p. D121.

(35) *Cornutella* sp.
Figure 10.7

Cornutella Ehrenberg: Campbell, 1954, p. D121, Fig. 60, 3a, 3b.

Remarks. Representative specimens of this genus group are very rare in the Buff Bay section (Table 1). They have affinity to the cosmopolitan and apparently long-ranged species *Cornutella profunda* Ehrenberg (= *Cornutella clathrata* [beta] *profunda* Ehrenberg, 1856, Plate 35b, Fig. 21); Petrushevskaya 1968 p. 110, Fig. 6. Maurrasse (1979) indicated that taxa of this genus occurred sporadically in Caribbean Paleogene deep-sea sediments, and were related to sporadic incoming of cooler water in the area. Similar specimens assigned to *C. profunda* reported in the northwest Pacific are restricted to the Miocene, from the *S. wolffii* Zone to the *O. antepenultimus* Zone (Ling, 1975, p. 710).

Family THEOPERIDAE Haeckel, 1881

Theoperida Haeckel, 1881, emended Riedel, 1967b, p. 296.

Genus *Cyrtocapsella* Haeckel, 1887

Cyrtocapsella Haeckel, 1887, p. 1512.

Type species. Cyrtocapsa (*Cyrtocapsella*) *tetrapera* Haeckel, 1887; subsequent designation by Campbell, 1954, p. D143, Fig. 73, 11a.

(36) *Cyrtocapsella tetrapera* (Haeckel, 1887)
Figure 9.27

Cyrtocapsa (*Cyrtocapsella*) *tetrapera* Haeckel, 1887, p. 1512, Plate 78, Fig. 5.
Cyrtocapsella tetrapera (Haeckel, 1887): Sanfilippo and Riedel, 1970, p. 453, Plate 1, Figs. 16–18.
Lithocampe tetrapera (Haeckel, 1887): Petrushevskaya and Kozlova, 1972, p. 546, Plate 25, Figs. 13, 14.
Cyrtocapsella tetrapera (Haeckel, 1887): Sanfilippo et al., 1973, p. 221, Plate 5, Figs. 4–6.
Cyrtocapsella tetrapera (Haeckel, 1887): Kling, 1973, p. 636, Plate 11, Figs. 12–15.
Cyrtocapsella tetrapera (Haeckel, 1887): Johnson, 1974, p. 547, Plate 7, Fig. 11.
Cyrtocapsella tetrapera (Haeckel, 1887): Chen, 1975, p. 460, Plate 20, Fig. 1.
Cyrtocapsella tetrapera (Haeckel, 1887): Ling, 1975, p. 728, Plate 9, Fig. 18.
Cyrtocapsella tetrapera (Haeckel, 1887): Holdsworth, 1975, p. 530, Plate 2, Figs. 9, 13–15.
Cyrtocapsella tetrapera (Haeckel, 1887): Weaver, 1976, p. 581, Plate 3, Figs. 1, 3; Plate 9, Figs. 1, 3.
Cyrtocapsella tetrapera (Haeckel, 1887): Bjorklund, 1976, p. 1124, Plate 17, Figs. 19, 20.
Cyrtocapsella tetrapera (Haeckel, 1887): Ling, 1980, p. 367, Plate 2, Fig. 5.

Remarks. Cyrtocapsellid specimens from the Montpelier Formation in Jamaica are assigned to this taxon, which is quite polymorphic and widespread in tropical-equatorial sediments of the Atlantic, Caribbean, and Pacific, from the early Miocene *L. elongata* Zone to the late middle Miocene *D. antepenultimus* Zone. Although it also appears to have a similar range in the higher-latitude areas of the north Pacific (Ling, 1980), its range seems to vary considerably in other high-latitude areas.

Benson (1972) reported the presence of *C. tetrapera* in the North Atlantic south of the Hatton-Rockall bank (Fig. 6), at DSDP Site 116 where it is also found starting within the *L. elongata* Zone. But at that site its range also extends into younger levels of middle late Miocene age. On the other hand, farther north in the Norwegian Sea, at DSDP site 338, this species is reported only in sediments assigned to the early middle Miocene (Bjorklund, 1976, p. 1111, Table 8).

In the Indian Ocean, *C. tetrapera* is reported to be restricted to the lowermost part of early Miocene (Johnson, 1974). In Antarctica it is recorded in the middle Miocene in certain areas (Weaver, 1976), and from the early Miocene to the base of late Miocene in others (Chen, 1975).

It is evident that *C. tetrapera* is diachronous in different parts of the world oceans, but its record in Jamaica (Table 1) is correlative with its recorded range in the low-latitude regions. The apparent erratic distributional range may have been controlled by migratory behavior in response to the changing oceanic circulation systems of the Miocene (Haq et al., 1977; Vergnaud-Grazzini et al., 1979).

Genus *Lychnocanoma* Haeckel 1887, emend. Foreman, 1973, p. 437

Lychnocanoma Haeckel, 1887.

Type species. Lychnocanium (Lychnocanoma) clavigerum Haeckel, 1887, p. 1230, Plate 61, Fig. 4; subsequent designation by Campbell, 1954, p. D124.

(37) *Lychnocanoma elongata* (Vinassa de Regny, 1900)
Figure 10.1

Tetrahedrina elongata Vinassa de Regny, 1900, p. 243, Plate 2, Fig. 31.
Lychnocanium bipes Riedel, 1959, p. 294, Plate 2, Figs. 5, 6.
Lychnocanium bipes Riedel, 1959: Riedel and Sanfilippo, 1970, p. 529, Plate 15, Fig. 8.
Lychnocanium bipes Riedel, 1959: Riedel and Sanfilippo, 1971, p. 1595, Plate 2F, Figs. 1, 2.
Lychnocanium bipes Riedel, 1959: Goll, 1972, p. 960, Plate 15, Fig. 1; Plate 16, Fig. 1.
Lychnocanoma elongata (Vinassa de Regny, 1900): Sanfilippo et al., 1973, p. 221–222, Plate 5, Figs. 19, 20.

Remarks. Lychnocanoma elongata is reported only from lower Miocene sediments of the tropical equatorial regions, and ranged from the base of the *L. elongata* Zone (= ex-*L. bipes* Zone), to the *C. costata* Zone. This species is rare in the Buff Bay sequence, and is recognized only within the upper part of the *C. costata* Zone (Table 1). *L. grande* Campbell and Clark, 1944, is a very close relative of this taxon, and ranged into younger levels of the Miocene (*Didymocyrtis penultima* Zone) in higher-latitude regions.

Genus *Eucyrtidium* Ehrenberg, 1847, p. 54

Type species. Lithocampe acuminata Ehrenberg, 1844, p. 84; Ehrenberg, 1854, Plate 22, Fig. 27; subsequent designation by Campbell, 1954, p. D140.

(38) *Eucyrtidium cienkowskii* Haeckel, 1887
Figure 9.19

Eucyrtidium cienkowskii Haeckel, 1887, p. 1493, Plate 80, Fig. 9.
Eucyrtidium cienkowskii Haeckel group: Sanfilippo et al., 1973, p. 221, Plate 5, Figs. 7–11.
Stichopodium cienkowskii (Haeckel, 1887): Petrushevskaya and Kozlova, 1972, p. 548, Plate 26, Figs. 18, 19.
Eucyrtidium cienkowskii Haeckel group: Chen 1975, p. 460, Plate 15, Fig. 7.
Eucyrtidium cienkowski Haeckel group: Weaver, 1976, p. 581, Plate 4, Figs. 3–5.

Remarks. E. cienkowskii is a cosmopolitan species that occurred with varying frequencies throughout the Miocene. However, in the eastern Atlantic it is reported up to the base of the *C. costata* Zone only (Petrushevskaya and Kozlova, 1972). Specimens referred to this species remain very rare and occur only within the *D. alata* Zone (Table 1), in the sequence of the Montpelier Formation at Buff Bay.

Genus *Stichocorys* Haeckel, 1881

Stichocorys Haeckel, 1881, p. 438

Type species. Stichocorys wolffii Haeckel, 1887, p. 1479, Plate 80, Fig. 10; subsequent designation by Campbell, 1954, p. D140, Fig. 72,1.
Stichocorys Haeckel: Petrushevskaya and Kozlova, 1972, p. 546.

(39) *Stichocorys coronata* (Carnevale, 1908)
Figure 9.20, 21; Figure 10.2(?)

Calocyclas coronata Carnevale, 1908, p. 33 (No. 83), Plate 4, Fig. 24.
Eucyrtidium diaphanes Sanfilippo and Riedel, 1973, *in* Sanfilippo et al., 1973, p. 221, Plate 5, Figs. 12–14.

Remarks. Specimens referred to this taxon are placed under genus *Stichocorys* because they show more closely related morphological traits with this genus than with *Eucyrtidium*. So far, *S. coronata* has been reported exclusively from lower Miocene series of the Mediterranean area. In Jamaica, this species is well represented in the upper part of the *C. costata* Zone (Table 1), which indicates a range overlapping late early and early middle Miocene.

(40) *Stichocorys delmontensis* (Campbell and Clark, 1944)
Figure 9.5, 6

Eucyrtidium delmontense Campbell and Clark, 1944, p. 56, Plate 7, Figs. 19, 20.
Stichocorys delmontensis (Campbell and Clark, 1944): Riedel and Sanfilippo, 1970, p. 530, Plate 14, Fig. 6.
Stichocorys delmontensis (Campbell and Clark, 1944): Riedel and Sanfilippo, 1971, Plate 1F, Figs. 5–7; Plate 2E, Figs. 10, 11.
Stichocorys delmontensis (Campbell and Clark, 1944): Petrushevskaya and Kozlova, 1972, p. 546, Plate 25, Figs. 11, 12.
Stichocorys delmontense (Campbell and Clark, 1944): Goll, 1972, p. 960, Plate 34, Figs. 1–3; Plate 35, Fig. 1.
Stichocorys delmontensis (Campbell and Clark, 1944): Dinkelman, 1973, p. 783, Plate 9, Fig. 1.
Stichocorys delmontensis (Campbell and Clark, 1944): Sanfilippo et al., 1973, p. 222, 224, Plate 6, Fig. 3.
Stichocorys delmontensis (Campbell and Clark, 1944): Kling, 1973, p. 638, Plate 11, Figs. 8–10.
Stichocorys delmontensis (Campbell and Clark, 1944); Ling, 1973, p. 781, Plate 2, Fig. 12.
Stichocorys delmontensis (Campbell and Clark, 1944): Johnson, 1974, p. 549, Plate 8, Fig. 13.
Stichocorys delmontenesis (Campbell and Clark, 1944): Chen, 1975, Plate 20, Fig. 10

Remarks: S. delmontensis is reported to be widespread and common in sediments of the equatorial to subtropical belt of the world oceans where it occurred from the base of the Miocene to the Pliocene. The existing record seems to indicate that its earliest and latest appearances are in the equatorial regions, and its occurrence is diachronous or apparently erratic in areas affected by cooler surface waters, such as (1) the eastern Indian Ocean where it is recorded in different locations at different times, from the base of the *S. delmontensis* Zone to the close of the Miocene (*Didymocyrtis penultima* Zone); (2) farther south in the Indian Ocean where it disappeared earlier (data from Johnson, 1974); (3) Antarctica, where

Chen (1975) reported *S. delmontensis* in DSDP Site 266 at stratigraphic levels constrained in the uppermost Miocene; (4) the North Pacific where it reached its acme in the late Miocene, but declined rapidly afterward, and is absent in the Pliocene (Ling, 1973).

Although the distribution pattern of *S. delmontensis* clearly shows a preference for low-latitude regions, its relation to surface-water temperature is unclear. For instance, it reached its southernmost limit in the Indian Ocean in the Serravalien, at a time which also correlates with a peak of a global warming trend (Vail et al., 1977). On the other hand, while it disappeared there in younger levels, it is reported in the latest Miocene of Antarctica (Chen, 1975, Table 5, p. 445) at a time when significant global cooling was in progress. In the Buff Bay section, *S. delmontensis* is best represented in the *C. costata* Zone corresponding to the Langhian (Fig. 2 and Table 1).

(41) *Stichocorys wolffii* Haeckel, 1887
Figure 9.15

Stichocorys wolffii Haeckel, 1887, p. 1479, Plate 80, Fig. 10.
Stichocorys wolffii Haeckel, 1887: Riedel, 1959, p. 300, Plate 2, Fig. 14.
Stichocorys wolffii Haeckel, 1887: Riedel and Sanfilippo, 1971, p. 1595, Plate 2E, Figs. 8, 9.
Stichocorys wolffii Haeckel, 1887: Kling, 1971, p. 1087, Plate 2, Fig. 5
Stichocorys wolffii Haeckel, 1887: Petrushevskaya and Kozlova, 1972, p. 546, Plate 25, Fig. 22.
Stichocorys wolffii Haeckel, 1887: Riedel and Sanfilippo, 1978, p. 74, Plate 1, Fig. 3; Plate 9, Fig. 12.

Remarks. Specimens assigned to this taxon are characterized by the criteria discussed in Riedel and Sanfilippo (1978). From the published literature it appears that *S. wolffii* may represent a dimorph (Holdsworth, 1975) of the more cosmopolitan *S. delmontensis.* The greatest frequency of the two forms in simultaneous occurrence is in the tropical/equatorial regions, where *S. wolffii* ranged from the *S. wolffii* Zone (middle early Miocene), to the *D. alata* Zone (Fig. 2). Elsewhere, the range and frequency of *S. wolffii* appears to vary greatly in response to local environmental factors. In the eastern Indian Ocean, for instance (Johnson, 1974), *S. wolffii* is poorly preserved and has a shorter range (*S. wolffii* Zone to the lower part of *D. alata* Zone) relative to the western Indian Ocean where this taxon is recorded up to the latest Miocene, *Stichochorys peregrina* Zone (Sanfilippo and Riedel, 1974). Similar discrepancies are recorded for assemblages between low-latitude areas of the western Pacific (Kling, 1971; Holdsworth, 1975), and higher-latitude areas (Ling, 1975). In the latter areas *S. wolffii* is recorded only in the middle Miocene (Ling, 1975, DSDP Site 296, Table 5, p. 710).

Specimens referred to this species remain very rare and of irregular occurrence in the Montpelier sequence at Buff Bay (Table 1).

Family CARPOCANIIDAE Haeckel, 1881

Carpocaniida Haeckel, 1881; emend. Riedel, 1967b.

Genus *Carpocanopsis* Riedel and Sanfilippo, 1971

Type species. Carpocanopsis cingulatum Riedel and Sanfilippo, 1971, p. 1596–1597, Plate 2G, Figs. 17–21; Plate 8, Fig. 8.

(42) *Carpocanopsis* aff. *C. cingulata* Riedel and Sanfilippo, 1971
Figure 9.10, 11

Carpocanopsis cingulatum Riedel and Sanfilippo, 1971, p. 1596–1597, Plate 2G, Figs. 17–20; Plate 8, Fig. 8.
Myllocercion sp. C: Petrushevskaya and Kozlova, 1972, p. 535, Plate 22, Figs. 13, 14.
Carpocaniidae, gen. et sp. indet.: Kling, 1973, Plate 12, Fig. 5.

Carpocanopsis cingulatum Riedel and Sanfilippo, 1971: Sanfilippo and Riedel, 1973, in Sanfilippo et al., 1973, p. 224, Plate 6, Figs. 5, 6.
Carpocanopsis cingulatum Riedel and Sanfilippo, 1971: Sanfilippo and Riedel, 1974, p. 1021, Plate 3, Fig. 7.
Carpocanopsis cingulatum Riedel and Sanfilippo, 1971: Riedel and Sanfilippo, 1978, p. 67, Plate 4, Fig. 7.

Remarks: C. cingulata is reported as a polymorphic and cosmopolitan species with geographic habitat mostly in the tropical equatorial areas of the Pacific (Riedel and Sanfilippo, 1971; Kling, 1973), the Caribbean and Atlantic (Petrushevskaya and Kozlova, 1972), as well as the Mediterranean (Sanfilippo et al., 1973). It is scarce in the Indian Ocean (Sanfilippo and Riedel, 1974). Its range is coeval in all areas and coincides with the *C. costata* Zone (late early Miocene to early middle Miocene).

Specimens from Jamaica referred to *C. cingulata* are also polymorphic and appear mostly in the *C. costata* Zone (Table 1).

(43) *Carpocanopsis cristatum* (Carnevale, 1908)
Figure 9.13

Sethocorys cristata Carnevale, 1908, p. 31, Plate 4, Fig. 18.
Sethocorys cristata var. a. Carnevale, 1908, p. 32, Plate 4, Fig. 19.
Carpocanopsis cristatum (Carnevale): Riedel and Sanfilippo, 1971, p. 1597, Plate 1G, Fig. 16; Plate 2G, Figs. 1–7.

Remarks. Specimens from Jamaica assigned to this taxon show great morphological variability with apparent intergradation with *C. cingulatum.* Worldwide data indicate that both species are cosmopolitan with overlapping range within the *Calocycletta costata* Zone, but *C. cristatum* extends into the late middle Miocene *Diartus petterssoni* Zone (Fig. 2). In Jamaica, *C. cristatum* occurs mostly in the *C. costata* Zone (Table 1).

(44) *Carpocanopsis bussonnii* (Carnevale, 1908)
Figure 9.14

Sethocorys Bussonii Carnevale, 1908, p. 31, Plate 4, Fig. 17.
Carpocanistrum spp.: Riedel and Sanfilippo, 1971, Plate 2F, Figs. 13, 16; Plate 3D, Fig. 3.

Remarks. Specimens referred to this taxon are recorded in the tropical/equatorial regions of the Pacific, Caribbean, and Mediterranean. Its total range is uncertain, but data from the published literature indicate that *C. bussonii* ranges in the *C. costata* and *D. alata* Zones. In Jamaica, it occurs mostly in the *C. costata* Zone (Table 1).

Family PTEROCORYTHIDAE Haeckel, 1881

Pterocordya Haeckel, 1881; emend Riedel, 1967b, p. 296.

Genus *Lamprocyrtis* Kling, 1973, p. 638

Type species. Lamprocyclas heteroporos Hays, 1965, p. 179, Plate 3, Fig. 1.

(45) *Lamprocyrtis margatensis* (Campbell and Clark, 1944)
Figure 9.25

Calocyclas (Calocycletta) margatensis Campbell and Clark, 1944, p. 47–48, Plate 6, Figs. 17, 18.
Lamprocyrtis hannai (Campbell and Clark, 1944): Kling, 1973, p. 638, part.

Remarks. L. margatensis is characterized by its "subequidistant, distally sharpened, *strong.** spikelike teeth" along its margin, and shows close similarities with *L. hannai* Campbell and Clark, 1944 (p. 48, Plate 6,

Figs. 21, 22). The latter is distinguished by "its margin with *discrete.** projecting, spinelike, sharp denticles" (Campbell and Clark, 1944, p. 48), and seems to be found mostly in higher-latitude regions. Both taxa are reported to range from the Miocene to the Pliocene. Specimens assigned to *L. margatensis* are very rare in the Buff Bay section, they occur at a level corresponding to the upper part of the *D. alata* Zone. (* Emphasis is of the present author.)

Genus *Calocycletta* Haeckel, 1887

Type species. Calocyclas (*Calocycletta*) *veneris* Haeckel, 1887, p. 1831, Plate 74, Fig. 5; subsequent designation by Campbell, 1954, p. D132, Fig. 68, 1c.

(46) *Calocycletta costata* (Riedel, 1959)
Figure 9.9

Calocyclas costata Riedel, 1959, p. 296, 298, Plate 2, Fig. 9.
Calocyclas costata (Riedel, 1959): Riedel and Sanfilippo, 1970, p. 535, Plate 14, Fig. 12.
Calocyclas costata (Riedel, 1959): Riedel and Sanfilippo, 1971, p. 1598, Plate 2H, Figs. 12–14.
Calocyclas costata (Riedel, 1959): Petrushevskaya and Kozlova, 1972, p. 544, Plate 35, Fig. 17.
Calocyclas costata (Riedel, 1959): Sanfilippo et al., 1973, p. 226, Plate 6, Fig. 10.
Calocyclas costata (Riedel, 1959): Riedel and Sanfilippo, 1978, p. 66, Plate 3, Fig. 9.

Remarks. *C. costata* is well represented in the tropical/equatorial regions of the world oceans, where it is reported to range from the base of the *C. costata* Zone (late early Miocene) to the lower part of the *D. alata* Zone (early middle Miocene). Only very rare specimens are reported beyond these latitudes in the North Atlantic regions. For instance, very rare specimens assigned to *C. costata* from sediments recovered in the Bay of Biscay (Benson, 1972, DSDP Site 119, Table 2) belong in the upper part of the *Calocycletta virginis* Zone (= *S. wolffii* Zone, in Fig. 2, herein) and the *C. costata* Zone. Assuming this chronology is correct, the occurrence of *C. costata* in the Bay of Biscay is earlier than its known record elsewhere.

Very rare specimens assigned to this species are also found in the *C. costata* or *D. alata* Zones (Sanfilippo et al., 1973) from Majorca, in the Mediterranean. These authors pointed out, however, that the majorcan speciments "are not so pronouncedly costate as the typical Pacific forms, and the general form of the shell resembles that of *Calocycletta virginis*" (p. 226). Specimens assigned to *C. costata* in the Buff Bay section are found mostly in the *C. costata* Zone, and disappear at the base of the *D. alata* Zone (Table 1).

(47) *Calocycletta virginis* (Haeckel, 1887)
Figure 9.17, 18, 26

Calocyclas (*Calocycletta*) *virginis* Haeckel, 1887, p. 1381, Plate 74, Fig. 4.
Calocycletta costata (Haeckel, 1887); Riedel and Sanfilippo, 1970, p. 535, Plate 14, Fig. 10.
Calocycletta virginis (Haeckel, 1887): Riedel and Sanfilippo, 1971, p. 1598, Plate 2H, Figs. 5–9, 11.
Calocycletta virginis Haeckel, 1887: Goll, 1972, p. 957–958, Plate 14, Fig. 1.
Calocycletta virginis (Haeckel, 1887) s. str.: Petrushevskaya and Kozlova, 1972, p. 544, Plate 35, Figs. 8, 9 (non Plate 35, Fig. 10).
Calocycletta virginis (Haeckel, 1887): Sanfilippo and Riedel, 1973, in Sanfilippo et al., 1973, p. 226, Plate 6, Fig. 11.
Calocycletta virginis (Haeckel, 1887): Riedel and Sanfilippo, 1978, p. 66, Plate 3, Figs. 13–14.

Remarks. *Calocycletta virginis* is well represented in the tropical/equatorial regions where it ranges from the *C. tetrapera* Zone (early Miocene) to the *C. petterssoni* Zone (middle Miocene). Petrushevskaya and Kozlova considered the range of their *C. virginis* sensu stricto to be Eocene, not Miocene (1972, p. 544), although specimens used for illustration are from sediments of the upper part of the early Miocene *C. virginis* Zone (= *S. wolffii* Zone in Fig. 2, herein). Like its close relative *C. costata,* this species seems to occur in higher-latitude areas of the North Atlantic only. It occurs in significant abundance in the Bay of Biscay sediments (DSDP Site 119, Benson, 1972), attributed to the early part of the *C. virginis* Zone (= *S. delmontensis* Zone, in Fig. 2, herein). In the Mediterranean region, however, only very rare specimens are recorded in the *C. costata* or *D. alata* Zone from Majorca (Sanfilippo et al., 1973).

Specimens assigned to *C. virginis* in the Buff Bay section occur sporadically throughout the sequence studied, but are significantly more abundant at levels equivalent to the middle part of the *C. costata* Zone (Table 1).

Family ARTOSTROBIIDAE Riedel, 1967a

Artostrobiidae Riedel, 1967a, p. 149; Riedel, 1967b, p. 296.

Subfamily *Artostrobiinae* Riedel, 1967a

Genus *Artostrobium* Haeckel, 1887

Type species. Lithocampe aurita Ehrenberg, 1844, p. 84; Ehrenberg, 1854, Plate 22, Fig. 25; subsequent designation by Campbell, 1954, p. D140; *Dictyomitra* ?Zittel, 1876: Campbell and Clark, 1944, p. 51; *Botryostrobus* Haeckel, 1887: Petrushevskaya and Kozlova, 1972 (part), p. 539.

(48) *Artostrobium* sp.
Figure 9.23

Remarks. This genus group in general is cosmopolitan, with an overall affinity for cooler-water environments in the Paleogene (Maurrasse, 1979). In the Buff Bay section, specimens referred to this genus group are rare and occur intermittently throughout the sequence studied (Table 1).

Genus *Lithomitra* Butschli, 1882

Lithomitra Butschli, 1882, p. 528.

Type species. Eucyrtidium pachyderma Ehrenberg, 1873; Ehrenberg, 1875, Plate 11, Fig. 21; subsequent designation by Campbell, 1954, p. D141.

(49) *Lithomitra* spp.
Figure 8.10

Remarks. Published data on the geographic occurrences of this genus group indicate that they are cosmopolitan with an optimum frequency in cooler-water environments.

Specimens attributable to this genus are sparsely distributed in the Jamaican Miocene series.

ACKNOWLEDGMENTS

Funds for field and laboratory works for this research were provided in part by the College of Arts and Sciences, Faculty Development Funds, the Florida International University Foundation, and private sources. Laboratory preparations benefited greatly from the help provided by Wenddell Beddoe. Cores used in this study are from the Lamont-Doherty Geological Observatory of Columbia University, Palisades, New York. Special thanks are due to W. R. Riedel, A. Sanfilippo, and R. Goll for a thorough review of the manuscript.

REFERENCES CITED

Bader, R. G., and 8 others, 1970, Initial Reports of the Deep Sea Drilling Project, v. 4: Washington, D.C., U.S. Government Printing Office, 753 p.

Balkissoon, I. G., 1989, Compilation and suggestions for a standardized nomenclature for eastern Jamaica: The Journal of the Geological Society of Jamaica, v. 25, p. 27–37.

Banner, F. T., and Blow, W. H., 1965, Progress in the planktonic foraminiferal biostratigraphy of the Neogene: Nature, v. 208, p. 1164–1166.

Benson, R. N., 1972, Radiolaria, Leg 12, Deep Sea Drilling Project, in Laughton, A. S., Berggren, W. A., and 8 others, eds., Initial Reports of the Deep Sea Drilling Project, v. 12: Washington, D.C., U.S. Government Printing Office, p. 1085–1113.

Bjorklund, K. R., 1976, Radiolaria from the Norwegian Sea, Leg 38 of the Deep Sea Drilling Project, in Talwani, M., and 11 others, eds., Initial Reports of the Deep Sea Drilling Project, v. 38: Washington, D.C., U.S. Government Printing Office), p. 1101–1168.

Bjorklund, K. R., and Goll, R. M., 1979, Internal skeletal structures of Collosphaera and Trisolenia: A case of repetitive evolution in the Collosphaeridae (Radiolaria): Journal of Paleontology, v. 53, no. 6, p. 1293–1326.

Blow, W. H., 1969, Late middle Eocene to Recent planktonic foraminiferal biostratigraphy: Proceedings, First International Conference on Planktonic Microfossils, Geneva 1967, v. 1, p. 199–422.

Bukry, D., and Foster, J. H., 1973, Silicoflagellate and diatom stratigraphy, Leg 16, Deep Sea Drilling Project, in van Andel, T. H., and 9 others, eds., Initial Reports of the Deep Sea Drilling Project, v. 16: Washington, D.C., U.S. Government Printing Office, p. 815–871.

Butschli, O., 1882, Beitrage zur Kenntnis der Radiolarienskelette, insbesondere der Cyrtida: Zeitschrift fur Wissenschaftlichte Zoologie, v. 36, p. 485–540, Plates 31–33.

Campbell, A. S., 1954, Protista 3 Radiolaria: Treatise on Invertebrate Paleontology, Part D: Geological Society of America and University of Kansas, p. D11–D163.

Campbell, A. S., and Clark, B. L., 1944, Miocene radiolarian faunas from southern California: Geological Society of America Special Papers, v. 51, p. 1–76, plates 1–7.

Carnevale, P., 1908, Radiolarie e Silicoflagellati di Bergonzano (Reggio Emilia): Memorie del Reale Istituto Veneto di Scienze, Lettere ed Arti, v. 28, no. 3, p. 1–47.

Chen, P.-H., 1975, Antarctic Radiolaria, in Hayes, D. E., and 11 others, eds., Initial Reports of the Deep Sea Drilling Project, v. 28, Washington, D.C., U.S. Government Printing Office, p. 437–513.

Ciesielski, P. F., and Case, S. M., 1989, Neogene paleoceanography of the Norwegian Sea based upon silicoflagellate assemblage changes in ODP Leg 104 sedimentary sequence, in Eldholm, O., Thiede, J., Taylor, E., and 25 others, eds., Proceedings of the Ocean Drilling Program, Scientific Results, 104: College Station, Texas (Ocean Drilling Program), p. 527–541.

Cushman, J. A., and Stainforth, R. M., 1945, The foraminifera of the Cipero marl formation of Trinidad, British West Indies: Cushman Laboratories for Foraminiferal Research Special Publication 14, p. 69.

De Stefani, T., 1952, Su alcune manifestazioni di idrocarburi in provincia di Palermo e descrizione di Foraminifer nuovi: Palermo, Italy, Plinia, v. 3, (1950–1951), nota 4, p. 9.

Dinkelman, M. G., 1973, Radiolarian stratigraphy: Leg 16, Deep Sea Drilling Project, in van Andel, T. H., Heath, G. R., and 8 others, eds., Initial Reports of the Deep Sea Drilling Project, v. 16: Washington, D.C., U.S. Government Printing Office, p. 747–813.

Dreyer, F., 1889, Morphologische Radiolarienstudien. Heft 1: Die Pylombildungen in vergleichend anatomischer und entwicklungs geschichtlicher Beziehung bei Radiolarien und bei Protisten uberhaupt, nebst System und Beschreibung neuer und der bis jetzt bekannten pylomatischen Spumellarien: Jenaische Zeitschrift fur Naturwissenschaft, v. 23 (new ser. 1, p. 1–138, Plates 1–6).

Dumitrica, P., 1973, Cretaceous and Quaternary Radiolaria in deep sea sediments from the northwest Atlantic Ocean and Mediterranean Sea, in Ryan, W.B.F., Hsu, K. J., and 8 others, eds., Initial Reports of the Deep Sea Drilling Project, v. 13: Washington, D.C., U.S. Government Printing Office, p. 829–901.

Edgar, N. T., and 13 others, eds., 1973, Initial Reports of the Deep Sea Drilling Project, v. 15: Washington, D.C., U.S. Government Printing Office, 1137 p.

Ehrenberg, C. G., 1838, Uber die bildung der Kreidefelsen und des Kreidemergels durch unsichtbare Organismen: Abhandlungen, Koniglichen Akademie der Wissenchaften zu Berlin, Jahre, 1838, p. 59–147, Plates 1–4.

Ehrenberg, C. G., 1844, Uber 2 neue Lager von Gebirgmassen aus Infusorien als Meeres-Absatz in Nord-Amerika und eine Vergleichung derselben mit den organischen Kreide-Gebilde in Europa und Afrika: Monatsberichte der Koniglichen Preussischen Akademie der Wissenchaften zu Berlin, Jahre 1844, p. 57–97.

Ehrenberg, C. G., 1847, Uber die mikroskopischen kieselschaligen Polycystinen als machtige Gebirgsmasse von Barbados und uber das Verhaltniss der aus mehr als 300 Neuen Arten bestehenden ganz eigenthumlichen formengruppe jener Felsmasse zu den jeztz lebenden Thieren und zur Kreidebildung. Eine neue Anregung zur Erforschung des Erdlebens: Monatsberichte der Koniglichen Preussischen Akademie der Wissenchaften zu Berlin, Jahre 1847, p. 40–60.

Ehrenberg, C. G., 1854, Mikrogeologie: Leipzig, Voss, xxviii + 374 p., Atlas, 31 p., 40 Plates., Fortsetzung (1856), 88 p. + 1 p. errata.

Ehrenberg, C. G., 1860, Uber den Tiefgrund des stillen Oceans zwichen Californien und den Sandwich-Inseln aus bis 15600' Tiefe nach Lieutenant Brooke: Monatsberichte der Koniglichen Preussischen Akademie der Wissenchaften zu Berlin, Jahre 1860, p. 819–833.

Ehrenberg, C. G., 1872a, Mikrogeologischen Studien als Zusammenfassung der Beobachtungen des kleinsten Lebens der Meeres Tiefgrunde aller Zonen und dessen geologischen Einfluss: Monatsberichte der Koniglichen Preussischen Akademie der Wissenchaften zu Berlin, Jahre 1872, p. 265–322.

Ehrenberg, C. G., 1872b, Mikrogeologische Studien uber das klienste Leben der Meeres-Tiefgeunde aller Zonen und dessen geologischen Einfluss: Abhandlungen Koniglichen Preussischen Akademie der Wissenchaften zu Berlin, Jahre 1872, p. 131–399, 1 chart, Plates 1–12.

Ehrenberg, C. G., 1873, Grossere Felsproben des Polycysteinen-Mergels von Barbados mit weiteren Erlauterungen: Monatsberichte der Koniglichen Preussischen Akademie des Wissenchaften zu Berlin, Jahre 1873, p. 213–263.

Ehrenberg, C. G., 1875, Fortsetzung der mikrogeologischen Studien als Gesammt-Uebersichten der mikroskopischen Palaontologie gleichartig analysirter Gebirgsarten der Erde, mit specieller Rucksicht auf den Polycystinen-Mergel von Barbados: Abhandlungen Koniglichen Preussischen Akademie der Wissenchaften zu Berlin, Jahre 1875, p. 1–225.

Finlay, H. J., 1940, New Zealand foraminifera: Key species in stratigraphy, No. 4: Transactions of the Royal Society of New Zealand, v. 69, pt. 4, p. 448–472, Plates 62–67.

Frizzell, D. L., and Middour, E. S., 1951, Paleocene Radiolaria from southeastern Missouri: University of Missouri School of Mines and Metallurgy Bulletin, Technical series 77, p. 1–41, Plates 1–3.

Foreman, H. P., 1973, Radiolaria of Leg 10 with systematics and ranges for the families Amphypyndacidae, Artostrobiidae, and Theoperidae, in Worzel, J. L., Bryant, W., and 7 others, eds., Initial Reports of the Deep Sea Drilling Project, v. 10: Washington, D.C., U.S. Government Printing Office, p. 407–474.

Friend, J. K., and Riedel, W. R., 1967, Cenozoic orosphaerid radiolarians from tropical Pacific sediments: Micropaleontology, v. 13, no. 2, p. 217–232.

Goll, R. M., 1968, Classification and phylogeny of Cenozoic Trissocyclidae (Radiolaria) in the Pacific and Caribbean basins, Part I: Journal of Paleontology, v. 42, no. 6, p. 1409–1432.

Goll, R. M., 1969, Classification and phylogeny of Cenozoic Trissocyclidae (Radiolaria) in the Pacific and Caribbean basins, Part II: Journal of Paleontology, v. 43, no. 2, p. 322–339.

Goll, R., 1972, Leg synthesis, Radiolaria, in Hays, J. D., and 7 others, eds., Initial Reports of the Deep Sea Drilling Project, v. 9: Washington, D.C., U.S. Government Printing Office, p. 947–1058.

Goll, R. M., 1979, The Neogene evolution of Zygocircus, Neosemantis and Callimitra: Their bearing on nassellarian classification: Micropaleontology, v. 25, no. 4, p. 365–396, Plates 1–5.

Goll, R. M., and Bjorklund, K.R., 1971, Radiolaria in surface sediments of the North Atlantic Ocean: Micropaleontology, v. 17, no. 4, p. 434–454.

Goll, R. M., and Bjorklund, K. R., 1989, A new radiolarian biostratigraphy for the Neogene of the Norwegian Sea: ODP Leg 104, in Eldholm, O., Thide, J., Taylor, E., and 25 others, eds., Proceedings of, Ocean Drilling Program, Scientific Results, 104: College Station, Texas, Ocean Drilling Program, p. 697–737.

Haeckel, E., 1860, Uber neue, lebende Radiolarien des Mittelmeeres und legte die dazu gehorigen: Abhandlungen Koniglichen Preussischen Akademie der Wissenschaften zu Berlin, Jahre 1860, p. 794–817.

Haeckel, E.,1862, Die Radiolarien (Rhizopoda Radiolaria): Berlin, Eine Monographie, 572 p.

Haeckel, E., 1881, Entwurf eines Radiolarien-Systems auf Grund von Studien der Challenger-Radiolarien: Jenaische Zeitschrift fur Naturwissenschaft, v. 15 (new ser., 8), no. 3, p. 418–472.

Haeckel, E., 1887, Report of the Radiolaria collected by H.M.S. *Challenger* during the years 1873–1876: Report on the scientific results of the voyage of the H.M.S. *Challenger:* Zoology, v. 18, pt. I, II, p. i–cxxxvii, 1–1803, Plates 1–140, 1 map.

Haecker, V., 1908, Tiefseeradiolarien: Wissenschaftifte Ergebnisse der Deutschen Tiefsee-Expedition auf dem Dampfer, Valdivia, v. 14, p. 477–705, Plates 86–87, 2 charts.

Hallock, P., Hine, A. C., Vargo, G. A., Elrod, J. A., and Jaap, W. C., 1988, Platforms of the Nicaraguan Rise: Examples of the sensitivity of carbonate sedimentation to excess trophic resources: Geology, v. 16, p. 1104–1107.

Haq, B. U., Lohmann, G. P., and Wise, S. W., 1977, Calcareous nannoplankton biogeography and its paleoclimatic implications: Cenozoic of the Falkland Plateau (DSDP Leg 36) and Miocene of the Atlantic Ocean, in Barker, P. F., Dalziel, I.W.D., and 10 others, eds., Initial Reports of the Deep Sea Drilling Project, v. 36: Washington, D.C., U.S. Government Printing Office, p. 745–759.

Hays, J. D., 1965, Radiolaria and late Tertiary and Quaternary history of Antarctic seas: American Geophysical Union, Antarctic Research series 5, Biology of the Antarctic Seas II, p. 125–184.

Hays, J. D., 1970, Stratigraphy and evolutionary trends of Radiolaria in North Pacific deep-sea sediments: Geological Society of America Memoir 126, p. 185–218.

Hays, J. D., Saito, T., Opdyke, N., and Burckle, L. H., 1969, Pliocene-Pleistocene sediments of the Equatorial Pacific: Their paleomagnetic, biostratigraphic, and climatic record: Geological Society of America Bulletin 80, p. 1481–1514.

Hertwig, R., and Lesser, 1874, Ueber Rhizopoden und denselben nahestenden Organismen: Archives Mikroskopische Anatomische, v. 10 (Suppl.: 35).

Hill, R. T., 1899, The geology and physical geography of Jamaica; A study of a type of Antillean development: Harvard, Bulletin Museum Comparative Zoology, v. 34, p. 1–226.

Hilmers, K., 1906, Zur Kenntnis der Collosphaeriden [Ph.D. thesis]: Kiel, Koniglichen Christian-Albrechts Universitat, 93 p.

Holdsworth, B. K., 1975, Cenozoic Radiolaria biostratigraphy: Leg 30: tropical and equatorial Pacific, in Andrews, J. E., Packham, G., and 9 others, eds., Initial Reports of the Deep Sea Drilling Project, v. 30: Washington, D.C., U.S. Government Printing Office, p. 499–537.

Johnson, D. A., 1974, Radiolaria from the eastern Indian Ocean, DSDP Leg 22, in von der Borch, C. C., Sclater, J. G., and 8 others, eds., Initial Reports of the Deep Sea Drilling Project, v. 22: Washington, D.C., U.S. Government Printing Office, p. 521–575.

Kling, S. A., 1971, Radiolaria: Leg 6 of the Deep Sea Drilling Project, in Fischer, A. G., and 10 others, eds., Initial Reports of the Deep Sea Drilling Project, v. 6: Washington, D.C., U.S. Government Printing Office, p. 1069–1117.

Kling, S. A., 1973, Radiolaria from the eastern North Pacific, Deep Sea Drilling Project, Leg 18, in Kulm, L. D., von Huene, R., and 9 others, eds., Initial Reports of the Deep Sea Drilling Project, v. 18: Washington, D.C., U.S. Government Printing Office, p. 617–671.

Kozlova, G. E., 1967, Tipy stroeniya skeleto v radiolyarii iz semeistva Porodiscidae: Zoologicheskii Zhurn, v. 46, vyp. 8.

Krasheninnikov, V. A., 1979, Stratigraphy and planktonic foraminifers of Cenozoic deposits of the Bay of Biscay and Rockall Plateau, DSDP Leg 48, in Montadert, L., Roberts, G., and 12 others, eds., Initial Reports of the Deep Sea Drilling Project, v. 48: Washington, D.C., U.S. Government Printing Office, p. 431–450.

Lamb, J. L., and Beard, J. H., 1972, Late Neogene planktonic foraminifers in the Caribbean, Gulf of Mexico, and Italian stratotypes: The University of Kansas Paleontological Contributions, Article 57 (Protozoa 8), 67 p., 36 Plates.

Land, L. S., 1976, Early dissolution of sponge spicules from reef sediments, north Jamaica: Journal of Sedimentary Petrology, 46, p. 967–969.

Leinen, M., 1979, Biogenic silica accumulation in the central equatorial Pacific and its implication for Cenozoic paleoceanography: Geological Society of America Bulletin, Part II, v. 90, p. 1310–1376.

Ling, H. Y., 1973, Radiolaria: Leg 19 of the Deep Sea Drilling Project, in Creager, J. S., Scholl, D. W., and 10 others, eds., Initial Reports of the Deep Sea Drilling Project, v. 19: Washington, D.C., U.S. Government Printing Office, p. 777–797.

Ling, H. Y., 1975, Radiolaria: Leg 31 of the Deep Sea Drilling Project, in Karig, E. D., Ingle, J. C., Jr., and 11 others, eds., Initial Reports of the Deep Sea Drilling Project, v. 31: Washington, D.C., U.S. Government Printing Office, p. 703–761.

Ling, H. Y., 1980, Radiolarians from the Emperor Seamounts of the Northwest Pacific: Leg 55 of the Deep Sea Drilling Project, in Jackson, E. D., and Koisumi, I., and 12 others, eds., Initial Reports of the Deep Sea Drilling Project, v. 55: Washington, D.C., U.S. Government Printing Office, p. 365–373.

Lombari, G., 1985, Biogeographic trends in Neogene Radiolaria from the Northern and Central Pacific: Geological Society of America Memoir 163, p. 291–303.

Maurrasse, F.J-M.R., 1973, Biostratigraphy, paleoecology, biofacies variations of middle Paleogene sediments in the Caribbean deep sea [Ph.D. thesis]: New York, Columbia University, 424 p.

Maurrasse, F.J-M.R., 1976, Paleoecologic and paleoclimatic implications of radiolarian facies in Caribbean Paleogene deep-sea sediments: Transactions, 7th Caribbean Geological Conference, Guadeloupe and Martinique, p. 185–204.

Maurrasse, F.J-M.R., 1979, Cenozoic radiolarian paleobiogeography: Implications concerning plate tectonics and climatic cycles: Palaeogeography, Palaeoclimatology, Palaeoecology, v. 26, p. 253–289.

Maurrasse, F.J-M.R., and Keens-Dumas, J. E., 1988, Calcareous-siliceous sediments in southwestern Trinidad and radiolarian lithofacies diachrony in the Caribbean Neogene, in Transactions, 11th Caribbean Geological Conference, Barbados: The Energy Division of the Ministry of Finance of the Government of Barbados, p. 7:1–7:10.

McMillen, K. J., and Casey, R. E., 1978, Distribution of living polycystine radiolarians in the Gulf of Mexico and Caribbean Sea and comparison with the sedimentary record: Marine Micropaleontology, v. 3, p. 121–145.

Molina-Cruz, A., 1977, The relation of the southern trade winds to upwelling processes during the last 75,000 years: Quaternary Research, v. 8, p. 323–339.

Moore, T., 1971, Radiolaria, in Tracey, J. I., and 8 others, eds., Initial Reports of the Deep Sea Drilling Project, v. 8, Washington, D.C., U.S. Government Printing Office, p. 727–775.

Muller, J., 1858, Uber die Thalassicollen, Polycystinen und Acanthometren des Mittelmeeres: Abhandlungen Koniglichen Preussischen Akademie der Wissenschaften zu Berlin, Jahre 1858, p. 1–62, Plates 1–11.

North American Commission on Stratigraphic Nomenclature, 1983, North American stratigraphic code: American Association of Petroleum Geologists Bulletin 67, p. 841–875.

Palmer, A. R., compiler, 1983, Decade of North American geology, 1983 geologic time scale: Geology, v. 11, p. 503–504.

Petrushevskaya, M. G., 1968, Radiolarians of orders Spumellaria and Nassellaria of the Antarctic Region (from material of the Soviet Antarctic expedition), in Studies of marine fauna IV (XII), Biological reports of the Soviet Antarctic Expedition (1955–1958), v. 3: Akademiya Nauk SSSR, Zoologicheskii Institute, 1968, p. 2–186.

Petrushevskaya, M. G., and Kozlova, G. E., 1972, Radiolaria: Leg 14, Deep Sea Drilling Project, in Hayes, D. E., and 7 others, eds., Initial Reports of the Deep Sea Drilling Project, v. 14: Washington, D.C., U.S. Government Printing Office, p. 495–648.

Riedel, W. R., 1959, Oligocene and lower Miocene Radiolaria in tropical Pacific sediments: Micropaleontology, v. 5, no. 3, p. 285–302.

Riedel, W. R., 1967a, Some new families of Radiolaria: Proceedings, Geological Society of London, v. 1640, p. 148–149.

Riedel, W. R., 1967b, Subclass Radiolaria, in Harland, W. B., and 3 others, eds., The fossil record: London, Geological Society of London, p. 291–298.

Riedel, W. R., 1971, Systematic classification of Polycystine Radiolaria, in Proceedings, "Micropaleontology of Marine Bottom Sediments" Scientific Committee on Oceanic Research, Working Group 19: Cambridge University Press, p. 64–661.

Riedel, W. R., and Sanfilippo, A., 1970, Radiolaria, Leg 4, Deep Sea Drilling Project, in Bader, R. G., and 8 others, eds., Initial Reports of the Deep Sea Drilling Project, v. 4: Washington, D.C., U.S. Government Printing Office, p. 503–575.

Riedel, W. R., and Sanfilippo, A., 1971, Cenozoic Radiolaria from the western tropical Pacific, Leg 7, in Winterer, E. L., and 9 others, eds., Initial Reports of the Deep Sea Drilling Project, v. 7: Washington, D.C., U.S. Government Printing Office, p. 1529–1672.

Riedel, W. R., and Sanfilippo, A., 1973, Cenozoic Radiolaria from the Caribbean, Deep Sea Drilling Project, Leg 15, in Edgar and 13 others, eds., Initial Report of the Deep Sea Drilling Project, v. 15, Washington, D.C., U.S. Government Printing Office, p. 705–751.

Riedel, W. R., and Sanfilippo, A., 1978, Stratigraphy and evolution of tropical Cenozoic radiolarians: Micropaleontology, v. 24, no. 1, p. 61–96.

Robinson, E., 1969, Geological Field Guide to the Neogene sections in Jamaica, West Indies, in 19th Annual Convention Gulf Coast Association of Geological Societies–Society of Economic Paleontologists and Mineralogists, Field Trip Jamaica, November, 1969: GCAGS/SEPM, 24 p.

Robinson, E., 1988, Late Cretaceous and early Tertiary sedimentary rocks of the Central Inlier, Jamaica: Journal of the Geological Society of Jamaica, 24, p. 49–67.

Saito, T., 1984, Planktonic foraminiferal datum planes for biostratigraphic correlation of Pacific Neogene sequences—1982 Status Report, in Ikebe, N., and Tsuchi, R., eds., Pacific Neogene datum planes: University of Tokyo Press, p. 3–10.

Sakai, T., 1984, Neogene radiolarian datum planes of the equatorial and northern Pacific, in Ikebe, N., and Tsuchi, R., eds., Pacific Neogene datum planes: University of Tokyo Press, p. 35–39.

Sanfilippo, A., and Riedel, W. R., 1970, Post-Eocene "closed" theoperid radiolarians: Micropaleontology, v. 16, no. 4, p. 446–462.

Sanfilippo, A., and Riedel, W. R., 1973, Cenozoic Radiolaria (exclusive of Theoperids, Artostrobids, and Amphipyndacids) from the Gulf of Mexico, Deep Sea Drilling Project Leg 10, in Worzel, J. L., Bryant, W., and 7 others, eds., Initial Reports of the Deep Sea Drilling Project, v. 10: Washington, D.C., U.S. Government Printing Office, p. 475–611.

Sanfilippo, A., and Riedel, W. R., 1974, Radiolaria from the west-central Indian Ocean and the Gulf of Aden, in Fisher, R. L., and 10 others, eds., Initial Reports of the Deep Sea Drilling project, v. 24: Washington, D.C., U.S. Government Printing Office, p. 997–1035.

Sanfilippo, A., and Riedel, W. R., 1976, Radiolarian occurrences in the Caribbean Region, in Transactions, 7th Caribbean Geological Conference, Guadeloupe and Martinique: Imprimeries Réunies de Chambéry, p. 145–168.

Sanfilippo, A., and Riedel, W. R., 1980, A revised generic and suprageneric classification of the artiscins (Radiolaria): Journal of Paleontology, v. 54, No. 5, p. 1008–1011.

Sanfilippo, A., Burckle, L. H., Martini, E., and Riedel, W. R., 1973, Radiolarians, diatoms, silicoflagellates and calcareous nannofossils in the Mediterranean Neogene: Micropaleontology, v. 19, no. 2, p. 209–234.

Savin, S. M., and 8 others, 1985, The evolution of Miocene surface and near-surface marine temperatures: Oxygen isotopic evidence: Geological Society of America Memoir 163, p. 49–82.

Shackleton, N. J., and Kennett, J. P., 1975, Paleotemperature history of the Cenozoic and the initiation of Antarctic glaciation: Oxygen and Carbon isotope analyses in DSDP sites 277, 279, and 281, in Kennett, J. P., Houtz, R. E., and 9 others, eds., Initial Reports of the Deep Sea Drilling Project, v. 29: Washington, D.C., U.S. Government Printing Office, p. 743–755.

Stainforth, R. M., Lamb, J. L., Luterbacher, H., Beard, J. H., and Jeffords, R. M., 1975, Cenozoic planktonic foraminiferal zonation and characteristics of index forms: The University of Kansas Paleontological Contributions, Article 62, 162 p.

Steineck, P. L., 1974, Foraminiferal paleoecology of the Montpelier and lower Coastal Groups (Eocene-Miocene), Jamaica, West Indies: Palaeogeography, Palaeoclimatology, and Paleoecology, v. 16, p. 217–242.

Steineck, P. L., 1981, Upper Eocene to middle Miocene ostracode faunas and paleoceanography of the north coastal belt, Jamaica, West Indies: Marine Micropaleontology, v. 6, p. 339–366.

Stohr, E., 1880, Die radiolarienfauna der Tripoli von Grotte, Provinz, Girgenti in Sicilien: Palaeontographica, v. 26 (ser. 3, v. 2), p. 69–124, Plates 17–23 (1–7).

Sverdrup, H. U., Johnson, M. W., and Fleming, R. H., 1970, The oceans: Their physics, chemistry, and general biology: New Jersey, Prentice Hall, Inc., 1087 p.

Vail, P. R., Mitchum, R. M., and Thompson, S., 1977, Seismic stratigraphy and global changes of sea level, Part 3: Relative changes of sea level from coastal onlap, in Payton, C., ed., Seismic stratigraphy—Applications to hydrocarbon exploration: American Association of Petroleum Geologists Memoir 26, p. 63–97.

Vergnaud-Grazzini, C., Muller, C., Pierre, C., Letolle, R., and Peypouquet, J. P., 1979, Stable isotopes and Tertiary paleontological paleoceanography in the northeast Atlantic, in Montadert, L., Roberts, D. G., and 12 others, eds., Initial Reports of the Deep Sea Drilling Project, v. 48: Washington, D.C., U.S. Government Printing Office, p. 475–491.

Vinassa de Regny, P. E., 1900, Radiolari Miocenici Italiani: Memorie della Reale Accademia delle Scienze dell' Instituto di Bologna, ser. 5, v. 8, p. 227–257 (565–595), Plates 1–3.

Weaver, F. M., 1976, Antarctic Radiolaria from the southeast Pacific basin, Deep Sea Drilling Project, Leg 35, in Hollister, C. D., Craddock, C., and 11 others, eds., Initial Reports of the Deep Sea Drilling Project, v. 35: Washington, D.C., U.S. Government Printing Office, p. 569–603.

Woodruff, F., Savin, S. M., and Douglas, R. G., 1981, Miocene stable isotope record: A detailed deep Pacific Ocean study and its paleoclimatic implications: Science, v. 22, p. 665–668.

Wright, R. M., ed., 1974, Field guide to selected Jamaican geological localities, Special Publication 1: Jamaica, Ministry of Mining and Natural Resources, Mines and Geology Division, 57 p.

Zans, V. A., Chubb, L. J., Versey, H. R., Williams, J. B., Robinson, E., and Cooke, D. L., 1963, Synopsis of the geology of Jamaica: Geological Survey Department of Jamaica Bulletin 4, 72 p.

Zittel, K., 1876, Uber einige fossile Radiolarien aus der nordeutschen Kreide: Zeitschrift Deutschen geologischen Gesellsch, v. 28, p. 75–86, Plate 2.

MANUSCRIPT ACCEPTED BY THE SOCIETY NOVEMBER 11, 1992

Jamaican Paleogene larger foraminifera

Edward Robinson
Department of Geology, University of the West Indies, Mona, Kingston 7, Jamaica, West Indies and Department of Geology, Florida International University, Miami, Florida 33199
Raymond M. Wright
Petroleum Corporation of Jamaica, 36 Trafalgar Road, Kingston 10, Jamaica, West Indies

ABSTRACT

Thirty-four genera and fifty species and subspecies of Paleogene larger foraminifera from Jamaican localities are reviewed and illustrated. Three species, *Athecocyclina stephensoni, Hexagonocyclina inflata,* and *Ranikothalia catenula,* are from Paleocene localities in eastern Jamaica, described for the first time. Ten species first appear in the early Eocene, sixteen in the early middle Eocene, and nine in the late middle Eocene. Four species are confined to the late Eocene. Six species first appear in the early Oligocene, including such characteristic forms as *Lepidocyclina undosa* and *L. yurnagunensis*. The first occurrence of *Miogypsinoides* spp. is used to define the base of the upper Oligocene, but in the back-reef and lagoonal carbonates of the White Limestone Group the boundary between the lower and upper Oligocene is still poorly defined. Here, peneroplid and related forms, such as *Archaias asmaricus,* not previously recorded from the Caribbean region, and *Praerhapydionina delicata* are of local biostratigraphic importance.

INTRODUCTION

Although the existence of larger foraminifera had been noted from Jamaica in publications dating from the nineteenth century (e.g., Sawkins, 1869; Hill, 1899), the first systematic treatment of these fossils was carried out by T. Wayland Vaughan (1928a, b, 1929a), based on material collected by the Jamaican government geologist, C. A. Matley, in the period 1921 to 1924. These collections were from the Yellow and White Limestone Groups (Fig. 1), mainly from central and northern Jamaica where Matley had worked on water supply problems.

Further collections made by Matley from the White Limestone Group in the Kingston district were examined by L. M. Davies (1952) who published lists of species but did not illustrate or describe the material in any detail. The collection sent by Matley to Davies is now housed in the Palaeontology Department, British Museum (Natural History). Some of Matley's Kingston district material was also examined and illustrated by Shoshiro Hanzawa (1937).

With the advent of the Jamaican Geological Survey in 1948 and the exploration for bauxite, which extended through the 1940s and 1950s over the Tertiary limestones, additional larger foraminiferal collections were made by H. R. Versey and H. R. Hose and sent to T. W. Vaughan and W. Storrs Cole for identification. The results of this work appeared in systematic papers by Cole (1956a, b). At the same time Versey and Hose compiled the first modern accounts of the biostratigraphy and microfacies associations of the larger foraminifera of the Yellow and White Limestone Groups of Jamaica (Hose and Versey, 1956; Versey, 1957; Versey, *in* Zans et al., 1963).

Examinations of larger foraminiferal assemblages from eastern Jamaica, the paleogeographic distribution of distinctive assemblages, microfacies and paleoenvironmental relationships of central Jamaican assemblages, and larger foraminiferal biostratigraphy in relation to planktic foraminiferal stratigraphy were carried out in the 1960s (Robinson, 1966, 1968; Wright, 1966). These were followed by more detailed analyses of specific groups and assemblages from the Chapelton Formation and lower part of the White Limestone Group (Eva, 1976a, b, 1980a, b; McFarlane, 1974, 1977; Robinson, 1974a, b, 1977; Wright, 1975). The composition and distribution of Paleogene shelf edge and carbonate platform assemblages, with particular reference to imperforate species, in Jamaica and other parts of the Caribbean were discussed by Robinson (1988, 1993).

Robinson, E., and Wright, R. M., 1993, Jamaican Paleogene larger foraminifera, *in* Wright, R. M., and Robinson, E., eds., Biostratigraphy of Jamaica: Boulder, Colorado, Geological Society of America Memoir 182.

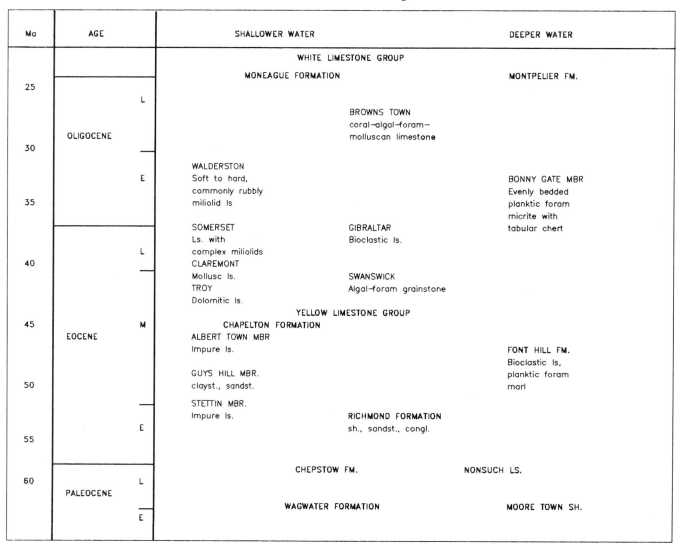

Figure 1. Diagram showing the main Paleogene lithostratigraphic units in Jamaica. Units deposited at shallow paleodepths are to the left. Boldface formational names are followed, where applicable, by names of formation subdivisions. For the Moneague Formation these are microfacies units, rather than members in the strict sense, and follow the usage of Hose and Versey (1956). The Wagwater Formation is mainly terrestrial (red beds; see Mann and Burke, 1990, for discussion).

In this paper we review available information on the main species of early Tertiary larger foraminifera and their stratigraphic distribution, and describe several species from the Paleocene, early Eocene, and Oligocene, previously recorded but not illustrated (Robinson, in Zans et al., 1963; Robinson, 1969; McFarlane, 1974, 1977; Jiang and Robinson, 1987). A range chart (Fig. 2), and keys to some of the more significant genera and species are also included. Except for two Paleocene species newly illustrated from Jamaica, the discocycliniform groups are omitted from our discussion and will be dealt with in a future paper. Classification follows that of Loeblich and Tappan (1988).

GENERAL DISTRIBUTION

Figure 1 indicates the stratigraphical distribution of the main Paleogene formations of Jamaica. Hose and Versey (1956), Eva and McFarlane (1985), Jiang and Robinson (1987), and Mann and Burke (1990) have described the nature and geographical distribution of these units. Localities of figured specimens, and of most type localities of the stratigraphic units of Figure 1, are on Figures 3, 4, 5, and 6.

Paleocene

Records of Paleocene larger foraminifera are confined to the eastern end of the island, mainly from the Parish of Portland (Jiang and Robinson, 1987; Robinson and Jiang, 1990), together with sporadic occurrences from the southwest side of the Blue Mountains, which have not been independently dated, and may be of Paleocene or early Eocene age.

Early Eocene

In eastern Jamaica shelf-edge faunas (Group I species of Robinson, 1988) are found in the Richmond Formation of the Wagwater Belt, including several discocycliniform species and *Eoconuloides*. In central Jamaica the Stettin Member of the Chapelton Formation contains a typical carbonate platform fauna (Group II species of Robinson, 1988) marking the first Tertiary marine transgression over the shelf. It includes *Verseyella jamaicensis, Fabularia colei, Coleiconus zansi, Peneroplis* spp., and *Helicostegina gyralis.*

Early middle Eocene

The middle Eocene saw a marked increase in the diversity of Group II larger foraminifera in Jamaica, with imperforate species being abundant locally and forming important, rock-building associations. Besides the forms used by Robinson (1977, 1993) for zoning these rocks, Eva (1976c) has drawn attention to the stratigraphic utility of species of *Yaberinella*. The assemblage reported from the Dump Limestone (Robinson, 1969; Hottinger, 1969) is typical.

At the shelf edge Group I forms include several discocycliniform genera: *Eoconuloides* and *Helicostegina*. Nummulitids appear to be absent, at least from the earlier part of this interval, first appearing in sediments containing *Globigerinatheka subconglobata* (P11) zone planktics.

Late middle Eocene

The late middle Eocene of central Jamaica is characterized by imperforate assemblages, including *Yaberinella jamaicensis, Fabularia, Cushmania, Coskinolina, Coleiconus, Fallotella,* and *Pseudochrysalidina*. Shelf-edge Group I faunas include nummulitids, discocycliniforms, and the first species of *Lepidocyclina.*

Late Eocene

Late Eocene Group II faunas are less diversified than those of the middle Eocene. *Fabularia verseyi* and *Cyclorbiculinoides jamaicensis* apparently are confined to Jamaica and the Niacaragua Rise (Robinson, 1988). Group I assemblages include several species of *Lepidocyclina, Asterocyclina,* and the earliest American heterosteginid, *H. ocalana.*

Oligocene

The end of the Eocene is marked conventionally in Jamaica, as elsewhere, by the extinction of discocycliniform species and *Fabiania* (Versey, *in* Zans and others, 1963). Genera continuing into the Oligocene include *Lepidocyclina, Heterostegina,* and possibly *Fallotella.*

The Indopacific *Archaias operculiniformis* (Smout and Eames, 1958) was recorded questionably from the Oligocene of Jamaica (McFarlane, 1974). Although published Oligocene records of *Archaias* are mainly from the Middle East and Indo-Pacific regions, the earliest known species, *A. columbiensis,* is from the Florida middle Eocene. Smout and Eames (1958) drew attention to the similarity between this species and the Middle Eastern *Archaias operculiniformis*. It is likely that, as with the orbitoidal genus *Lepidocyclina* (Butterlin, 1987), *Archaias* spread from the Caribbean into the Middle East/Indo-Pacific region during the earlier part of the Oligocene. The Jamaican Oligocene and Miocene limestones contain species of *Archaias* resembling Middle East forms as well as forms known from other parts of the Caribbean.

Another distinctive species in the Jamaican Oligocene is *Praerhapydionina delicata,* thought to be restricted to the Indo-Pacific-Mediterranean region until McFarlane (1974) reported it from central Jamaica.

Fallotella cookei and *F. floridana* are regarded as Eocene markers, but there are records of these species from rocks of Oligocene age in Florida and Cuba (Applin and Jordan, 1945; Beckmann, 1958). These two species are found in Jamaica at stratigraphic levels that could be regarded as early Oligocene, but independent biostratigraphic data are conflicting or lacking. In some cases reworking is probable, but in other cases no other known Eocene species are found with these "reworked" forms, which occur in carbonate platform paleoenvironments (McFarlane, 1977).

For the purposes of this review the boundary between the early and late Oligocene epochs is drawn at the first appearance of *Miogypsinoides,* which occurs within the P 21 or *Globorotalia opima opima* zone (Robinson and Persad, 1987). In the Jamaican carbonate platform sequences, which usually lack orbitoidal, miogypsinid, and planktic species, this boundary is still difficult to place with any precision.

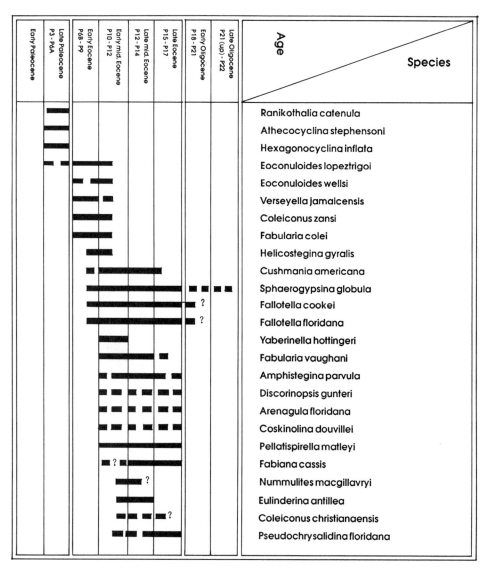

Figure 2 (on this and facing page). Ranges of Paleogene larger foraminifers in Jamaica.

SYSTEMATIC SECTION

Imperforate conical species

Species discussed in this paper have tests consisting of a uniserial series of shallow cup-shaped or saucer-shaped chambers, which increase in size to form a conical to subcylindrical test. An initial coiled series of nepionic chambers may be present or absent when the proloculus occupies the apex of the cone. Beyond the first one or two, each chamber is divisible into a central zone, with interseptal pillars, and a marginal zone without pillars, and, in simple forms, without any kind of partitioning. In more complex forms the marginal zone is divided by a system of radially arranged subepidermal partitions (major partitions), each one extending out from one of the outermost ring of pillars in the central zone to form a series of marginal chamberlets. Minor radial partitions, and short horizontal partitions parallel to the septa, may further subdivide the marginal chamberlets. The major partitions alternate in position in successive chambers. Apertures consist of circular pores scattered over the central zone of the apertural face, with the outermost ones forming a ring, inclined radially outwards towards the periphery of the test. Each radial pore is located over a major partition (if developed).

We believe the beam and rafter terminology of Hottinger and Drobne (1980) and Loeblich and Tappan (1988) is inappropriate and can lead to misconceptions regarding the position of the structural elements we call partitions.

The Jamaican imperforate conical species are divided into two families, the Coskinolinidae, with comparatively simple internal skeletal features and a keriothecal wall structure, and the Dictyoconidae, without

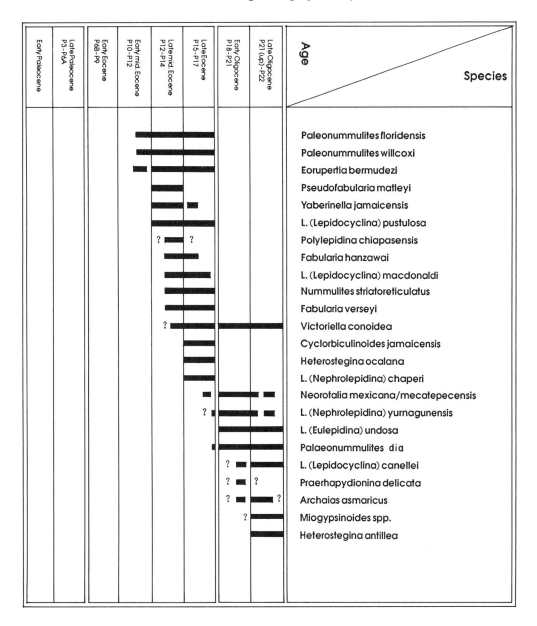

a keriothecal wall and with a generally more complex internal skeleton. Although the Dictyoconidae are closely similar in their morphology to certain Early and middle Cretaceous orbitolinid genera, no obvious Late Cretaceous connecting link has been described, and the Tertiary conical genera are presumed to have evolved from simpler arenaceous forms in the Paleocene.

Family Costknolinidae Moullade, 1965

Hottinger and Drobne (1980) suppressed the formerly widely used name *Lituonella,* and created a number of taxa of subgeneric rank within the genus *Coskinolina.* We have followed Loeblich and Tappan (1988) in elevating these to generic rank. In essence this means that most Caribbean species previously reported as *Lituonella* are now *Coskinolina* Stache, while species previously assigned to *Coskinolina* now belong either to *Coleiconus* Hottinger and Drobne or to the dictyoconid genus *Fallotella* Mangin.

Key to Jamaican genera and species:

A. Marginal zone of each chamber not subdivided .... genus *Coskinolina*
   1. Coiled early growth stage short .... *Coskinolina* sp. cf *C. douvillei*
B. Marginal zone of each chamber divided into chamberlets .... genus *Coleiconus*
   1. Coiled early growth stage short, major partitions relatively
     numerous .... *Coleiconus zansi*

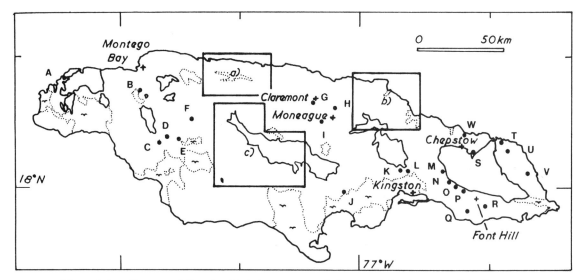

Figure 3. Locality map of Jamaica. Plus signs are named places, mainly of type localities for stratigraphic units of Figure 1; spots, localities as follows: A, ER1710; B, ER573 (Montpelier); C, WS38; D, WS83 (Ipswich); E, WS116; F, ER1654; G, WR503; H, WR508; I, WR520; J, St. Helens borehole; K, VL78; L, Red Gal Ring section; M, ER1700; N, AWK490A; O, AWK388; P, M66; Q, ER39; R, ER142; S, ER209A; T, ER2061; U, ER94 (Nonsuch); V, ER1283; W, ER126. Solid lines surround areas of Cretaceous rocks; dotted lines delineate Quaternary alluvium (with v's). Inset areas a), b), c) shown in detail as Figures 4, 5, 6, respectively.

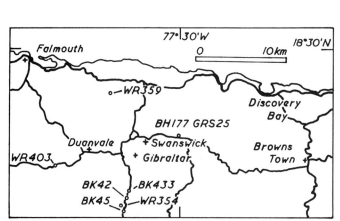

Figure 4. Enlargement of area a) on Figure 3. Plus signs, named locations; circles, localities of figured specimens.

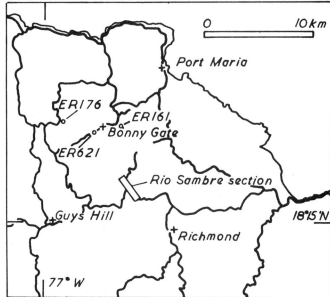

Figure 5. Enlargement of area b) on Figure 3. Explanation as for Figure 4.

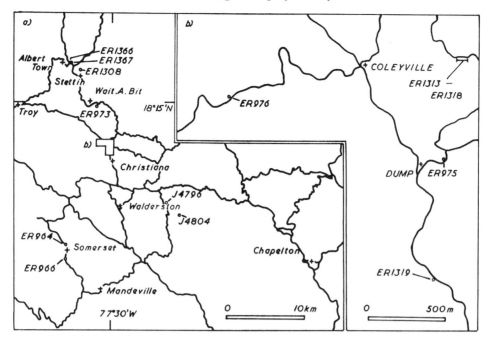

Figure 6. Enlargement of area c) on Figure 3. 6b), enlargement of area b) on Figure 6a. Explanation as for Figure 4.

2. Coiled early growth stage relatively large, major partitions less numerous ....
.... *Coleiconus christianaensis*

Genus *Coskinolina* Stache, 1875

Costkinolina sp., cf *Coskinolina douvillei* (Davies), 1930
(Figure 7.1–3)

Lituonella sp. of Vaughan, 1928a, p. 283, Plate 43, Fig. 7.
Coskinolina sp., cf. *Coskinolina douvillei* (Davies), Robinson, 1993, Plate 3, Figs. 1–8.

Although there is considerable variation in morphology, most Jamaican *Coskinolina* are best placed in the group that shows similarities to *Coskinolina douvillei* (Davies). The microspheric form has a more pronounced coiled early stage than the megalospheric form. We do not consider these, or the types of *Coskinolina inflata* Keijzer (1945), to be conspecific with *Lituonella floridana* Cole. The Jamaican specimens are distinguished by a larger test with a much broader cone than Cole's species as expressed by his types (Cole, 1942).
Distribution. Middle Eocene shelf facies units of the Yellow and White Limestone Groups.

Genus *Coleiconus* Hottinger and Drobne, 1980

Coleiconus christianaensis Robinson, 1993
(Figure 7.8, 9; 8.8)

Coleiconus christianaensis Robinson, 1993, Plate 1, Figs. 1–7; Plate 2, Fig. 1.

Remarks. This species is distinguished from *C. zansi* by possessing a much more prominent spiral stage, and a coarser internal structure, with slightly larger chamberlets in the marginal zone, with a PD index of 13.5 to 14.0. The PD index (major partition density index) is obtained by dividing the number of major partitions in any transverse section, by the mean diameter of the section in millimetres. Major partitions may be incompletely developed.
Distribution. Common in the upper part of the Chapelton Formation (Albert Town Member), central Jamaica, associated with *Yaberinella jamaicensis, Cushmania americana,* and other dictyoconids.

Coleiconus zansi Robinson, 1993
(Figures 7.4–7; 8.10)

Coleiconus zansi Robinson, 1983, Plate 2, Figs. 2–4.
Coskinolina elongata Cole, 1956a, p. 215, Plate 24, Figs. 6–11; Plate 31, Figs. 1, 2; not *Coskinolina elongata* Cole, 1942, p. 20, Plate 3, Figs. 15–17; Plate 4, Figs. 1–3; Plate 5, Figs. 2–7; Plate 16, Fig. 6.
Coskinolina (Coleiconus) elongata (Cole) of Hottinger and Drobne, 1980, p. 233, Plate 13, Figs. 7–14; text Fig. 11.

Remarks. The main features of this species are the equilateral, conical test, up to about 2.4 mm in size, a coiled early growth stage, much less pronounced than in *Coleiconus christianaensis,* and a conical stage containing thick, somewhat irregularly arranged pillars in the central zone. The major partition density (PD) index is 16.5 to 17.0.
Distribution. Stettin Limestone Member of the Chapelton Formation ("Yellow Limestone" of authors), central Jamaica, associated with *Fabularia colei* Robinson and *Verseyella jamaicensis* (Cole).

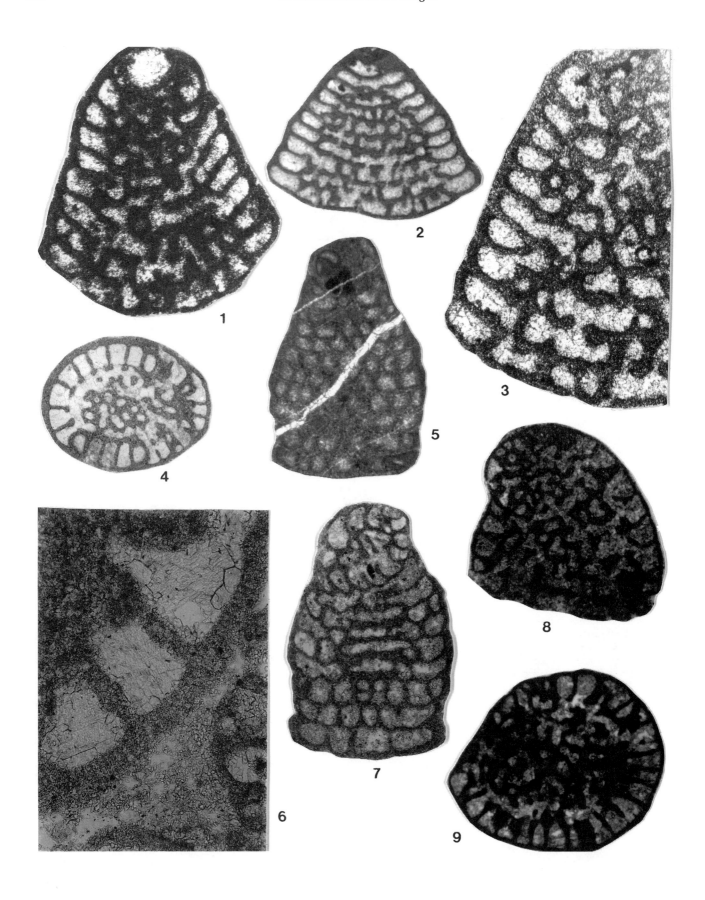

Family Dictyoconidae Moullade, 1965

Key to Jamaican genera and species.
All species have the marginal zone of each chamber divided into chamberlets by major partitions, and at least one cycle of minor radial partitions.

A. Marginal chamberlets of each chamber normally with one cycle of minor radial partitions ....
   .... genus *Fallotella*
   1. Marginal chamberlets normally lacking horizontal partitions
      .... *Fallotella floridana*
   2. Marginal chamberlets normally with one cycle of horizontal partitions ....
      .... *Fallotella cookei*
B. Marginal chamberlets of each chamber with more than one cycle of minor radial and horizontal partitions ....
   .... genus *Cushmania*
   1. with minor partitions forming a polygonal meshwork inside conical wall ....
      .... *Cushmania americana*

Genus *Fallotella* Mangin, 1954

Fallotella cookei (Moberg, 1928)
(Figures 9.1–6, 7?; 10.3, 4)

Coskinolina cookei Moberg, 1928, p. 166, Plate 3, Figs. 1–8; Plate 5, Fig. 3.
Heterodictyoconus cookei (Moberg) Butterlin and Moullade, 1968, p. 13, Plate 2, Figs. 1–11; Plate 3, Figs. 1–3.
Fallotella (Fallotella) cookei (Moberg) Hottinger and Drobne, 1980, p. 239, Plate 1, Fig. 3; Plate 16; text Figs. 9, D, E, 12, A.

Remarks. Both *Fallotella cookei* and *F. floridana* have one cycle of minor radial partitions, alternating with the major ones. But *F. cookei* is distinguished by the additional possession of a single horizontal partition in the marginal zone of each chamber. In eroded specimens, or in thin sections cut tangentially to the conical or peripheral wall these show up, with the radial partitions, as a rectangular grid. Because horizontal partitions are present in *F. cookei,* this grid consists of cells that are almost square in shape, whereas in *F. floridana,* which lacks horizontal partitions, the cells are distinctly elongate rectangular. This feature is seen in Figures 9.7 and 10.4, sections tangential to the conical wall of *F. cookei.* The short horizontal partitions, visible at the edges of the cuts, form square cells, whereas the centers of the sections, which are deeper into the cone, have cut past the inner edges of the horizontal partitions, leaving only rectangular cells, formed by the radial partitions and the chamber septa.

Distribution. In western Jamaica *F. cookei* is abundant in the Swanswick Limestone. In central Jamaica *F. cookei* ranges from the middle of the Chapelton Formation into the lowest part of the Walderston Limestone, early middle Eocene to possibly earliest Oligocene. In eastern Jamaica its lowest stratigraphic occurrence is in the upper part of the Richmond Formation, late early Eocene (P9 zone). At the lower end of its range *F. cookei* displays a coarser structure to its test, with thicker skeletal elements, than in specimens from the later part of the Eocene.

Figure 7. 1–3, *Coskinolina* sp. cf. *C. douvillei* (Davies): 1, axial section, ×45; 2, tangential section, ×24; 3, partial axial section, ×48, all from ER1319, upper Chapelton Formation. 4–7, *Coleiconus zansi* Robinson: 4, 7, transverse and axial sections, ×25, ER973, Stettin Member of the Chapelton Formation; 6, ×170, enlargement of part of 4 to show alveolar nature of the conical wall. 5, axial section, ×25, ER1700, Richmond Formation, Clydesdale. 8, 9, *Coleiconus christianaensis* Robinson, ×24, axial and transverse sections, J4804, Chapelton Formation (H. J. Mac Gillavry collection).

Fallotella floridana (Cole, 1941)
(Figures 9.8; 10.1, 2)

Coskinolina floridana Cole, 1941, p. 24, Plate 3, Figs. 1–7; Plate 4, Figs. 1–9; Plate 5, figs. 1–5, 11; Plate 18, Fig. 9.
Dictyoconus floridanus (Cole) Cole and Applin, 1964, p. 25, Plate 2, Figs. 2, 5, 8.
Heterodictyoconus floridanus (Cole) Butterlin and Moullade, 1968, p. 16, Plate 3, Fig. 7, 10–13; Plate 3, ?Figs. 8, 9.
"*Coskinolina*" *floridana* Cole, Hottinger, and Drobne, 1980, p. 243.

This species is distinguished from *F. cookei* by the normal lack of horizontal partitions in the marginal chamberlets. However, the two species are frequently associated and specimens otherwise referable to *F. floridana* occasionally develop small horizontal partitions in a few of the marginal chamberlets (see Fig. 9.8).
Distribution. In Jamaica the range of this species is similar to that of *F. cookei.*

Genus *Cushmania* Silvestri, 1925

The large, apically positioned proloculus is followed by a few, incomplete, annular chambers. Remaining chambers with pillars; marginal zone divided into chamberlets or cellules containing numerous minor partitions, which show up as a reticulate network in the conical wall.

The name *Cushmania* Silvestri, 1925, has priority over *Heterodictyoconus* Butterlin and Moullade, 1968. *Cushmania* is given generic rank because the uncoiled megalospheric embryonic apparatus, with a large proloculus, differs markedly from the coiled embryonic stage of the type species of *Dictyoconus* Blankenhorn (Butterlin and Moullade, 1968; Hottinger and Drobne, 1980). So far, the two genera have been reported from geographically separate regions.

Although Vaughan (1932) concluded that four species of the *C. americana* group were recognizable, most later workers have treated these as a single variable species, as we do here.

Cushmania americana (Cushman), 1919
(Figures 10.5–8; 11.2, 4, 5)

Conulites americana Cushman, 1919b, p. 43, text Fig. 3.
Cushmania americana (Cushman) Vaughan, 1928a, p. 281, Plate 14, Figs. 3–5.
Cushmania fontabellensis Vaughan, 1928a, p. 282, Plate 44, Fig. 3.
Dictyoconus codon Woodring, Vaughan, 1928a, p. 280, Plate 43, Figs. 1–5b; Cole, 1941, p. 28, Plate 7, Figs. 1, 7; Plate 18, Figs. 10, 11.
Dictyoconus puilboreauensis Woodring, Vaughan, 1928a, p. 281, Plate 43, Fig. 6.
Dictyoconus gunteri Moberg, Cole, 1941, p. 27–28, Plate 3, Figs. 8–10; Plate 7, Figs. 2–6, 8.
Heterodictyoconus americanus (Cushman), Butterlin and Moullade, 1968, p. 12, Plate 1, Figs. 1–10.
Dictyoconus americanus (Cushman) Cole, 1956a, p. 215, Figs. 8–11; Cole and Applin, 1964, Plate 2, Figs. 3, 6; Hofker, 1966, Plate 8, Figs. 9–16; Robinson, 1974a, p. 284, Plate 2, Figs. 1–5; Plate 3, Figs. 1, 2; text Fig. 3.
Dictyoconus (*Cushmania*) *americanus* (Cushman), Hottinger and Drobne, 1980, p. 247, Plate 1, Figs. 1, 4–9; Plate 20, Figs. 1–10; text Figs. 9, A, B, 12, C.
Dictyoconus (*Cushmania*) *puilboreauensis* (Woodring), Hottinger and Drobne, 1980, p. 248, Plate 21, Figs. 1–15.

Figure 8. 1–3, *Arenagula floridana* (Cole), ×50, spiral, apertural and side views of the same specimen, ER975, Chapelton Formation; 4, 6, 7, *Discorinopsis gunteri* Cole, ×50, side, apertural and spiral views of same specimen, ER975, Chapelton Formation; 5, 9, *Verseyella jamaicensis* (Cole), ×40, apertural face of two specimens, ER1315, Stettin Member of Chapelton Formation; 8, *Coleiconus christianaensis* Robinson, ×41, showing pronounced spiral stage, ER976, Chapelton Formation; 10, *Coleiconus zansi* Robinson, ×40, view of apertural face, ER1315, Stettin Member of Chapelton Formation.

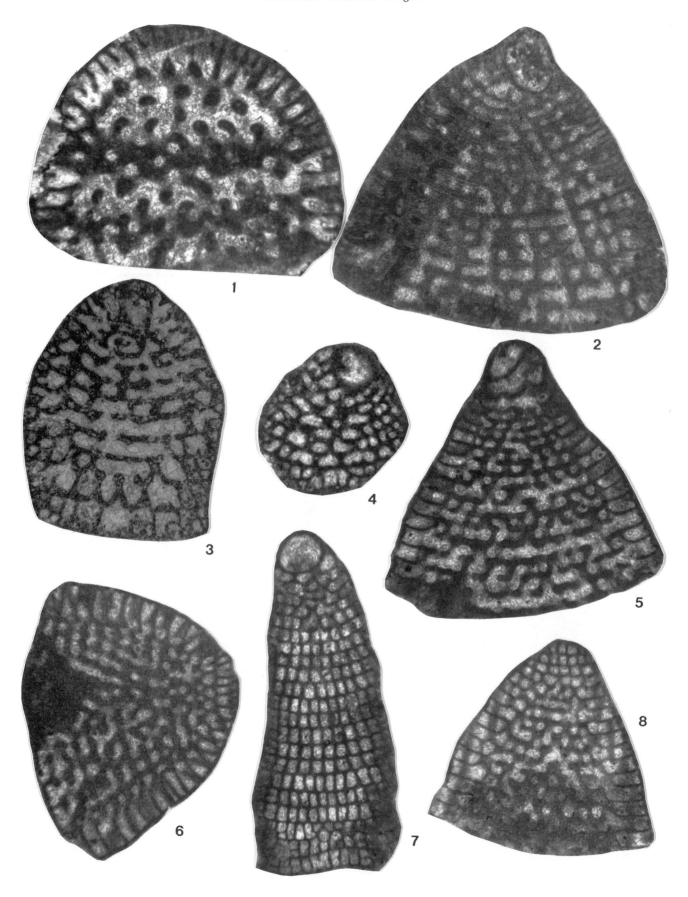

Figure 9. 1–6, *Fallotella cookei* (Moberg): 1, 3, 4, ×49, oblique sections, ER1700, Richmond Formation, Clydesdale; 2, 6, ×49, axial section, ER1319, upper Chapelton Formation, 6 has very short horizontal partitions; 5, ×49, J4796, Chapelton Formation, Ritchies (H. J. Mac Gillavry collection). 7, *Fallotella cookei?*. 8, *Fallotella floridana* (Cole), ×49, axial section (not centered), J4796. 8 shows incipient horizontal partitions in the marginal zone of one or two chambers.

Remarks. *Cushmania puilboreauensis* (Woodring) was maintained as a separate species by Hottinger and Drobne, 1980, but is provisionally included with *C. americana* in this paper, as is the Jamaican form called *C. fontabellensis* by Vaughan (1928a). Further study may indicate that these should be separated. There is wide variation in morphology of the *C. americana* group, and this appears to be correlatable with the facies associations in which the species is found.

Eva (1976a) drew attention to the problem of positioning the genus *Dictyoconus* (including *Cushmania*) in his paleoecological model for Eocene larger foraminiferal genera. In Jamaica, forms similar to the types of *C. americana* are commonly associated with Group I larger foraminiferal assemblages (Robinson, 1988) at the shelf/platform margin (as they are at the type locality in St. Bartholomew), whereas specimens with the *fontabellensis* morphology are more frequently found with Group II assemblages in platform carbonates. Specimens of the *fontabellensis* type are particularly distinctive and common in the higher part of the range of *Cushmania* in Jamaica (Versey, 1957). Forms resembling *C. puilboreauensis* occur in the Chapelton Formation of eastern Jamaica (Robinson and Jiang, 1990) as a virtually monospecific foraminiferal assemblage in micrites, as does *C. puilboreauensis* at its type locality in Haiti.

Distribution. In Jamaica *C. americana* ranges from the upper part of the Richmond Formation (Langley Member, latest early Eocene) to late middle or early late Eocene levels in the White Limestone Group. Forms with the *C. fontabellensis* morphology are found in the upper Chapelton Formation, the Claremont, and lower Somerset Limestone.

Genus *Verseyella* Robinson, 1977

Verseyella jamaicensis (Cole, 1956)
(Figures 8.5, 9; 11.1; 12.1–4)

Coskinolinoides jamaicensis Cole, 1956a, p. 216, Plate 24, Figs. 12–16; Plate 31, Figs. 3, 4.
"*Coskinolinoides*" *jamaicensis* Cole, Hottinger, and Drobne, 1980, p. 251, Plate 22, Figs. 1–7; text Fig. 14.
Verseyella jamaicensis (Cole), Robinson, 1977, p. 1414; Robinson, 1993, Plate 6, Figs. 1–10, 12.

Remarks. *Verseyella* is unusual among Tertiary larger agglutinated genera in possessing a fragile blade-shaped juvenile stage, followed by a subcylindrical later stage. The early stage is seldom preserved intact.

Verseyella is structurally distinct from both *Coskinolinoides* Keijzer and from other members of the dictyoconid and coskinolinid groups. The last two characteristically have a conical test, with internal structures alternating in position in successive chambers, and, often, an initial coiled stage, whereas in *Verseyella* internal structures are aligned and continuous from one chamber to the next (Robinson, 1993). Provisionally we have retained this genus in the Orbitolinidae, but it should probably be accommodated in a new family.

Distribution. *V. jamaicensis* is normally associated with *Coleiconus zansi* and *Fabularia colei*, less commonly with *Helicostegina gyralis* in the Stettin Limestone Member of the Chapelton Formation, central Jamaica. It also occurs in the basal part of the Yellow Limestone Group of northwestern Jamaica. The age is late early to early middle Eocene.

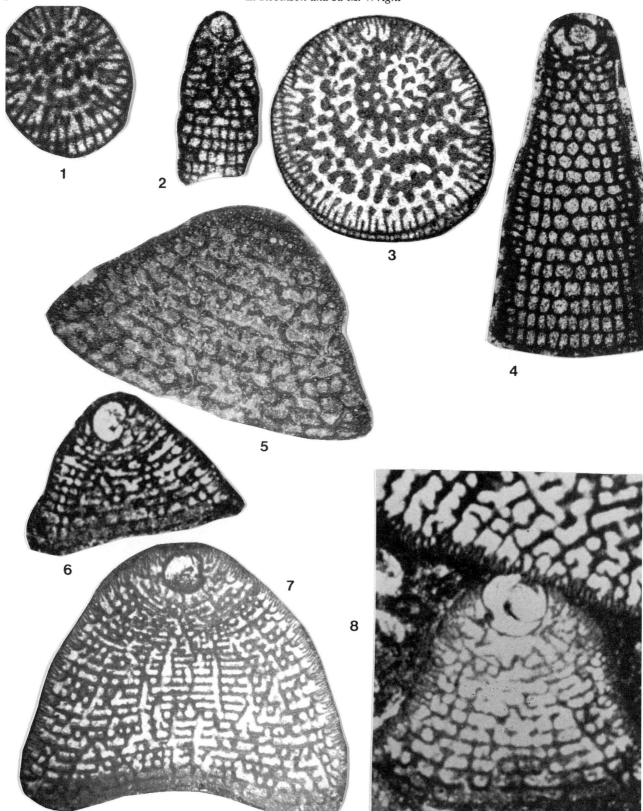

Figure 10. 1, 2, *Fallotella floridana* (Cole): 1, ×45, transverse section, 2, ×45, axial section, ER1319, upper Chapelton Formation. 3, 4, *Fallotella cookei* (Moberg). 3, ×40, ER964, Somerset Limestone; 4, ×45, Somerset Limestone. 5–8, *Cushmania americana* (Cushman): 5, ×25, oblique section, ER 1245, highest part of Richmond Formation, Rio Sambre (latest early Eocene to earliest middle Eocene); 6, ×28, axial section, 7, ×24, slightly oblique axial section, ER283, Claremont Limestone, Red Gal Ring; 8, ×28, near-axial section and partial section through marginal zone of another specimen, Troy Limestone.

Family Valvulamminidae Loeblich and Tappan, 1986

Genus *Discorinopsis* Cole, 1941

Discorinopsis spp. are a common component of foraminiferal assemblages in central Jamaica, but they have not been systematically investigated as regards number of species and total range.

Discorinopsis gunteri Cole, 1941
(Figures 8.4, 6, 7; 12.5)

Discorinopsis gunteri Cole, 1941, p. 36, Plate 1, Figs. 7–9.
Discorinopsis gunteri Cole, Robinson, 1969, p. 740, Plate II, Figs. 3, 6; not Plate III, Figs. 4, 6.

The specimens illustrated here from the late middle Eocene of central Jamaica are typical of those seen throughout the Jamaican middle Eocene.

Genus *Arenagula* Bourdon and Lys, 1955

Both *Discorinopsis* and *Arenagula* are trochospirally coiled genera. Externally *Arenagula* is distinguished from *Discorinopsis* by its possession of an areal aperture, consisting of rounded pores, scattered more or less randomly over the apertural face.

Arenagula floridana (Cole), 1942
(Figures 8.1–3)

Cribrobulimina floridana Cole, 1942, p. 17, Plate 1, Figs. 7, 8.
Discorinopsis gunteri Cole, part, Robinson, 1969, Plate III, Figs. 4–6.

The specimens illustrated here are from the late middle Eocene of central Jamaica, where *Arenagula floridana* is associated with *Cushmania, Fallotella, Coskinolina,* and *Yaberinella* spp.

Family Chrysalidinidae Neagu, 1968

Genus *Pseudochrysalidina* Cole, 1941

Pseudochrysalidina floridana Cole, 1941
(Figures 12.6–9)

Pseudochrysalidina floridana Cole, 1941, p. 36, Plate 1, Figs. 10, 11; Plate 2, Fig. 4; Cole, 1956a, p. 215, Plate 24, Figs. 1, 2; Plate 25, Figs. 1–5; Robinson, 1974a, p. 284, Plate 1, Figs. 2, 6, 8.
Chrysalidina (*Chrysalidina*) *floridana* (Cole), Hottinger and Drobne, 1980, p. 221, Plate 4, Figs. 1–3.

Remarks. Members of this family possess a well-defined canaliculate wall structure.
Distribution. This species is common in the late Eocene shelf carbonates of central Jamaica (Claremont and Somerset Limestones).

Jamaican Fabulariids and Alveolinids

Most genera in the family Alveolinidae construct an axially elongate, enrolled test with a large number of chambers per whorl, internally divided to produce one or more layers of chamberlets. Caribbean alveolinids have a sporadic stratigraphic distribution, in the Paleocene to early Eocene, the middle Eocene, and the Holocene. In Jamaica alveolinids have so far only been recognized in the middle Eocene (*Pseudofabularia,* Robinson, 1974b) and Holocene (*Borelis,* E. R. observation of Recent Jamaican sediments).

Members of the family Fabulariidae add two chambers per whorl, producing biloculine tests. Rarely, three or more chambers may be added in the later whorls. Internally the chambers are divided into one or more layers of spirally directed chamberlets. Descriptions and identifications of genera and species are based on various oriented sections. The following definitions are useful (modified from Drobne, 1974).

1. *Coiling axis,* the axis about which successive chambers are coiled; this axis passes through the *poles* of the test.
2. *Oral axis,* the axis through the zone occupied by successive apertures.
3. *Equatorial section* lies in the plane that includes the oral axis, perpendicular to the coiling axis (sometimes called the *spiral* section).
4. *Longitudinal section* lies in the plane that includes the coiling axis and the oral axis.
5. *Axial section* lies in the plane that includes the coiling axis, perpendicular to the oral axis.

The Jamaican middle Eocene alveolinids construct tests consisting of only two or three chambers per whorl. The following key includes Jamaican alveolinids and fabulariids:

A. Major diameter of test usually coincides with the coiling axis; apertural face extends to poles of test, family Alveolinidae .... genus *Pseudofabularia*
  1. Chambers about two per whorl, divided into a single row of chamberlets ....
B. Major diameter of test usually coincides with oral axis; apertural face does not reach poles of test, family Fabulariidae .... genus *Fabularia*
  1. Chambers divided into a single row of chamberlets only ......
    a) without a significant basal layer
      i) proloculus small; chamberlet cross sections subcircular ....
      .... *Fabularia colei*
      ii) proloculus medium sized; chamberlet cross sections subrectangular to wedge shaped ....
      .... *Fabularia vaughani*
    b) with significant basal layer
      i) proloculus large; chamberlet cross sections wedge shaped ....
      .... *Fabularia hanzawai*
  2. Chambers with more than one row of chamberlets ....
    .... *Fabularia verseyi*

Family Alveolinidae

Genus *Pseudofabularia* Robinson, 1974

Pseudofabularia matleyi (Vaughan, 1929)
(Figures 11.7; 13.8)

Borelis matleyi Vaughan, 1929a, Plate 40, Figs. 2, 3.
Borelis jamaicensis Vaughan, 1929a, Plate 40, Figs. 4–8.
Borelis jamaicensis var. *truncata* Vaughan, 1929a, Plate 40, Figs. 11, 12.
Pseudofabularia matleyi (Vaughan) Robinson, 1974b, Plate 1.

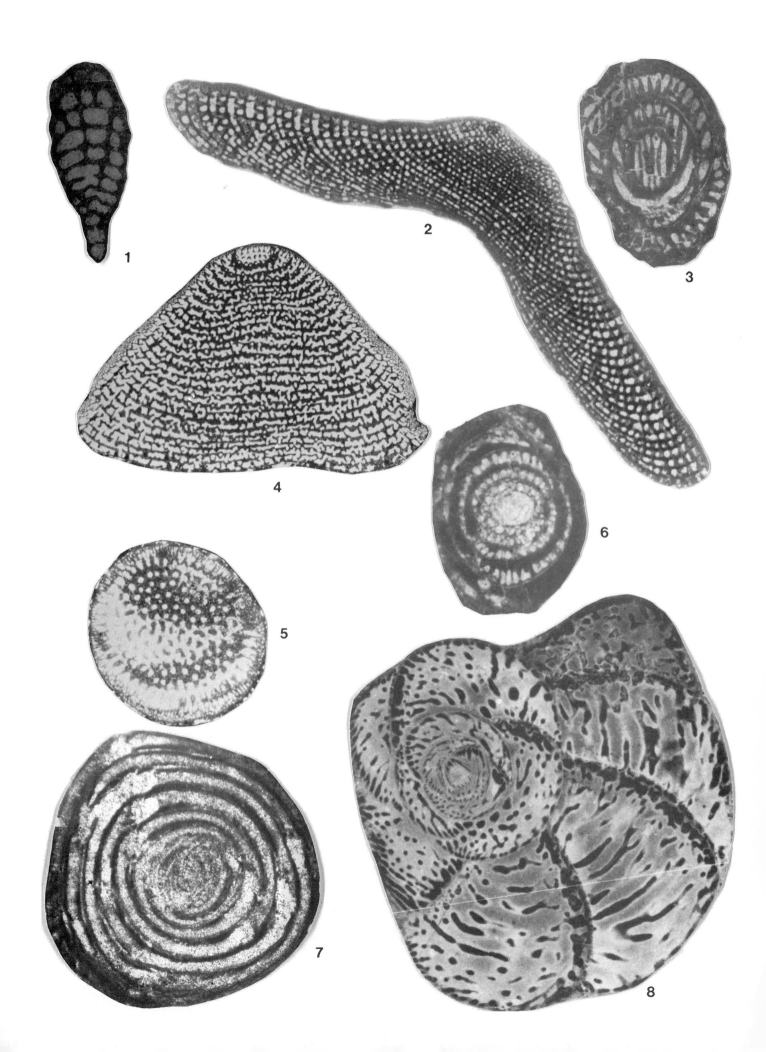

Figure 11. 1, *Verseyella jamaicensis,* ×43, off-center axial section showing blade-shaped early part of the test, ER1318, Stettin Member, Chapelton Formation. 2, 4, 5, *Cushmania americana* (Cushman): 2, ×14, axial section of a microspheric specimen (the *fontabellensis* kind) showing strongly concave apertural face; 4, ×18, slightly oblique axial section of a megalospheric specimen, both WR503, Claremont Limestone; 5, ×20, transverse section of a megalospheric specimen, showing at least three cycles of minor partitions in the marginal zone, WR508, Claremont Limestone. 3, 6, *Fabularia hanzawai* Robinson, ×40: 3, an oblique tangential section passing through an intercameral septum in the lower part of the figure; 6, sections cut obliquely between the oral and coiling axes; all ER239, Swanswick Limestone. 7, *Pseudofabularia matleyi* (Vaughan), ×40, equatorial section, Hose 18992 (Smithsonian collection). 8, *Fabularia verseyi* Cole, ×17.5, an unusual example of a specimen with seven chambers in the final whorl, V62/27, Somerset Limestone (collection of H. R. Versey).

Remarks. The generic position of this species and its relationship to American species of *Fabularia* have been discussed by Robinson (1974b) and Eva (1980b). In randomly cut sections it may be difficult to distinguish from *Fabularia.* It differs by having septula that are always parallel to one another, as in *Alveolina,* and an apertural face that extends to the poles of the test, with a preseptal passage.
Distribution. This species is locally abundant in the Chapelton Formation of central Jamaica, middle Eocene.

Family Fabulariidae Ehrenberg, 1839

Genus *Fabularia* Defrance, 1820

Fabularia colei Robinson, 1969
(Figures 13.1–4)

Fabularia colei Robinson, 1969, p. 742, Plate 1, Figs. 1–7.
?*Fabularia donatae liburnica* Drobne, 1974, p. 18, Plate 3, Figs. 4–6; Plate 4, Figs. 10–16; Plate 5, Figs. 1–10.

Remarks. A small, almost spherical species, with a streptospiral (milioline) initial stage and chambers, which are divided spirally into a single row of chamberlets that are oval to rounded in cross section. *F. colei* differs from *F. vaughani* in its smaller size, smaller proloculus (0.13 to 0.16 mm diameter), thicker chamberlet walls, and relatively fewer chamberlets per chamber at similar stages of development. *Fabularia donatae liburnica* Drobne from the Paleocene–early Eocene of Yugoslavia is very close in its features to *F. colei,* but the microspheric form appears to be at least partly lacazine in its coiling (Drobne, 1974, Plate 5, Fig. 10) as is *F. donatae* s.s., placed in the genus *Pseudolacazina* by Caus (1979). The microspheric features of *F. colei* are incompletely known.
Distribution. Often abundant in the Stettin Member of the Chapelton Formation, central Jamaica, early Eocene.

Fabularia vaughani Cole and Ponton, 1934
(Figures 13.5–7)

Fabularia vaughani Cole and Ponton, 1934, p. 130, Plate 1, Figs. 1–9.
Fabularia vaughani Cole and Ponton, Cole, 1945, p. 98, Plate 15, Fig. 6; Plate 16, Figs. 1–10.
Fabularia gunteri Applin and Jordan, 1945, p. 137, Plate 18, Figs. 12a, b.
Fabularia matleyi (Vaughan) Cole, 1956a, Plate 26, Figs. 7–10, not Figs. 11–14.

Remarks. The proloculus of *Fabularia vaughani* is larger than in *F. colei.* The chamberlets are more angular (rectangular or wedge shaped) in cross section. In "advanced" forms there may be appreciable basal thickening, although, by definition, this does not exceed 40% of the chamber height, as seen in axial section.
Fabularia gunteri Applin and Jordan is regarded here as synonymous with *F. vaughani,* but there is considerable morphological variation within the group, between *F. gunteri*-like, and *F. vaughani*-like specimens. Most Jamaican specimens resemble the types of *F. gunteri* quite closely.
Distributions. Found in the Chapelton Formation, Swanswick, and lower Claremont Limestones, middle Eocene.

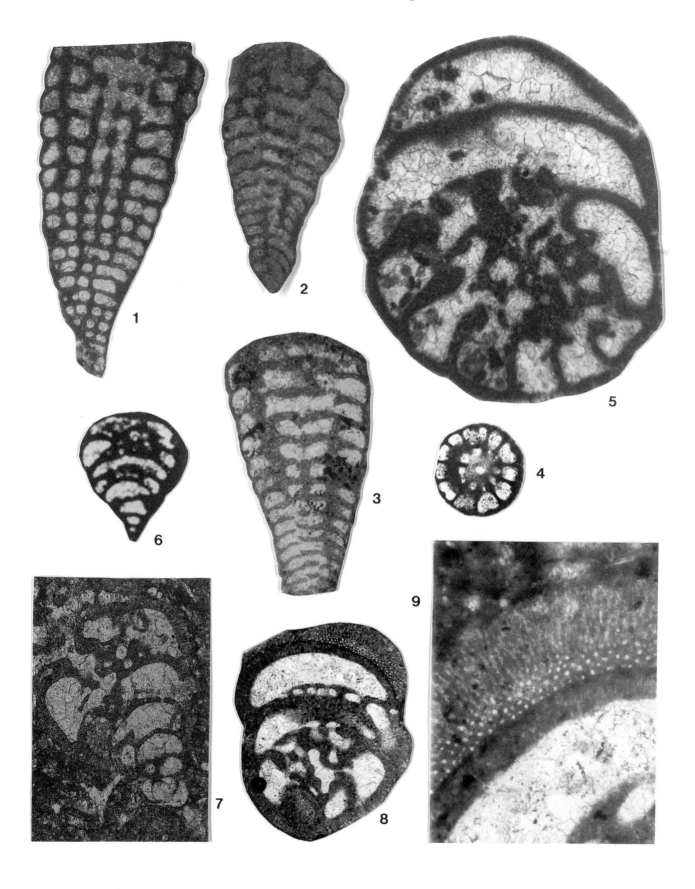

Figure 12. 1–4, *Verseyella jamaicensis* (Cole): 1, ×43, ER1317; 2, ×43, ER1308; 3, ×43, ER1318, axial sections, not centered; 2 and 3 are cut parallel to the width of the blade-shaped initial stage; 4, ×40, transverse section of mature stage, ER1315; all Stettin Member of the Chapelton Formation. 5, *Discorinopsis gunteri* Cole, ×49, spiral section, ER1319, Chapelton Formation. 6–9, *Pseudochrysalidina floridana* Cole: 6, ×28, axial section showing pillars in central area of chambers, Swanswick Limestone; 7, ×45; 8 ×55, Somerset Limestone, 9 ×150, enlargement of part of 8 to show canaliculate wall.

Fabularia hanzawai Robinson, 1993
(Figures 11.3, 6; 13.9, 10)

Fabularia hanzawai Robinson, 1993, Plate 2, Fig. 6; Plate 7, Figs. 4, 5.
Fabularia vaughani Cole and Ponton, Hanzawa, 1937, Plate 20, Figs. 1–4.
Fabularia vaughani n. var., Robinson, 1977, p. 1416, Figs. 1, 3, 4.
Fabularia vaughani var. A of Robinson, 1988, Table 1.
?*Fabularia roselli* Caus, 1979 (part?), Fig. 2b, d, i.

Remarks. This species is distinguished from *Fabularia vaughani* by possessing a larger proloculus (diameter 180 to 270 microns), and developing significant basal thickening in each chamber, the basal layer occupying 40% or more of each chamber, after about the first two whorls, as viewed in axial section. The small, more or less wedge-shaped chamberlets are crowded to the external margin of each chamber. It is distinguished from *Fabularia verseyi* by its lack of secondary chamberlets in the basal layer. Like *F. verseyi*, *F. hanzawai* may have three chambers in the final whorl. Some of the illustrations of *Fabularia roselli* Caus resemble *F. hanzawai* in the thickening of the basal layer, but on the evidence of the type description this seems to be a very variable feature in Caus' species.
Distribution. Occurs in the higher part of the Claremont and Swanswick Limestones of central Jamaica, late middle Eocene and early late Eocene, associated with *Lepidocyclina macdonaldi*.

Fabularia verseyi Cole, 1956
(Figures 11.8; 14.1, 2)

Fabularia verseyi Cole, 1956a, p. 219, Plate 26, Figs. 1–6.
Fabularia verseyi Cole, Robinson, 1974a, p. 287, Plate 4, Figs. 5, 6; Plate 5, Figs. 1–6; Plate 6, Figs. 1, 3–7; Plate 7, Fig. 1.

Remarks. *Fabularia verseyi* is similar to *F. hanzawai* but with secondary chamberlets appearing as tubes in a greatly thickened basal layer. Forms transitional between the two species lack secondary chamberlets in the first few chambers (Robinson, 1974a, Plate 5, Fig. 4). Specimens of *F. verseyi* with three chambers in the final whorl are relatively common, and specimens with up to seven such chambers have been noted (Fig. 12.8).
Distribution. This species is locally abundant in late Eocene shelf carbonates of Jamaica (Somerset Limestone). Transitional specimens occur in the upper part of the Claremont and lower Somerset.

Genus *Yaberinella* Vaughan, 1928

Species of this genus have been the subject of a detailed, unpublished study by A. N. Eva (1976c). In the present paper two distinct species are recognized, and one form with intermediate characteristics. In making specific distinctions primary emphasis is placed on proloculus size, the shape of the coil in the megalospheric test, and the regularity with which the internal chamberlets are arranged (Eva, 1976c; Robinson, 1993).

Yaberinella hottingeri Robinson, 1993
(Figure 15.11)

Yaberinella trelawniensis Vaughan, Hottinger, 1969, p. 746, Plate 5, Fig. 1–6.
Yaberinella cf. *trelawniensis* Vaughan, of Robinson, 1988, Table 1.

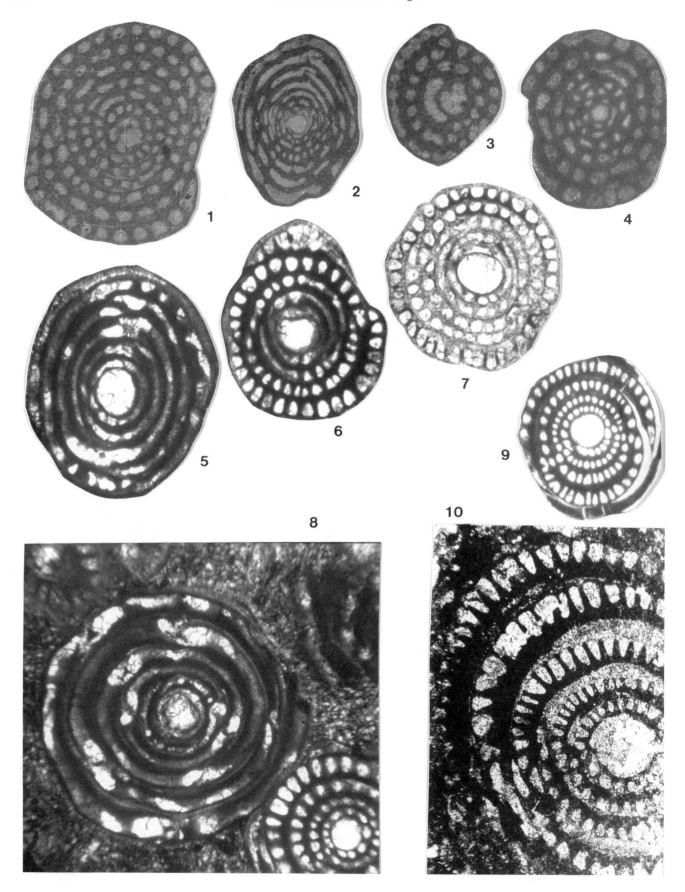

Yaberinella sp. A, Eva, 1976c, p. 63, Plate 17, Figs. 1–12.
Yaberinella hottingeri Robinson, 1993, Plate 2, Fig. 5; Plate 4, Fig. 6; Plate 5, Figs. 5–7.

Remarks. *Yaberinella hottingeri* differs from *Y. jamaicensis* Vaughan, the type for the genus, in possessing a much smaller proloculus and a test with a moderately rapidly enlarging operculine coil, but lacking the flaring final stage of *Y. jamaicensis*. Typical specimens also have a test in which the tubular chamberlets of the basal layer, although meandering, with intersecting sets, lack the well-ordered chevronlike pattern of those seen in *Y. jamaicensis*. But there is wide variation in this feature and, at the top of its stratigraphic range, *Y. hottingeri* does show quite well ordered sets of chamberlets.

Distribution. *Y. hottingeri* occurs in the middle part of the Chapelton Formation, as at the type locality of Dump, where it is associated with *Pellatispirella matleyi* (Vaughan). The age is early middle Eocene.

Yaberinella jamaicensis Vaughan, 1928
(Figure 14.3)

Yaberinella jamaicensis Vaughan, 1928b, p. 8, Plate 4, 5.
Yaberinella jamaicensis Vaughan, Lehmann, 1961, p. 656, Plate 13; Plate 14, Figs. 1–3; text Figs. 43–46.
Yaberinella jamaicensis Vaughan, Hottinger, 1969, p. 746, Plates 1–6.
Yaberinella trelawniensis Vaughan, 1929a, p. 374, Plate 39, Fig. 1.

Figure 13. 1–4, *Fabularia colei* Robinson: 1, 3, ×43, near axial sections; 2, ×25.5, oblique equatorial section showing irregular coiling of the earliest whorls, all ER1316; 4, ×43, oblique section, ER1308, all Stettin Member of Chapelton Formation. 5–7, *Fabularia vaughani* Cole and Ponton, ×49, ER1654, Chapelton Formation; 5, near-equatorial section, 6, 7, near-axial sections. 8, *Pseudofabularia matleyi* (Vaughan), ×49, equatorial section, with *F. vaughani*, ER1654, Chapelton Formation. 9, 10, *Fabularia hanzawai* Robinson: 9, ×32, axial section, ER237; 10, ×45, axial section of part of a topotype, ER235. Both from Swanswick Limestone, Dallas Mountain.

Remarks. *Yaberinella jamaicensis* is characterized by the possession of a large proloculus and well-ordered, internal chamberlets that intersect in a distinctive, regular chevronlike pattern. Proloculus size is typically over 0.5 mm diameter (*Y. trelawniensis*-like forms), and often as much as 1.0 mm. It differs from *Y. hottingeri* in possessing a more rapidly expanding operculine coil, terminated by a flaring final stage.

We regard *Y. trelawniensis* Vaughan as a variant of *Y. jamaicensis*, with a proloculus size smaller than typical *Y. jamaicensis*. The increase in proloculus size, which is one of the features we use to separate *Y. jamaicensis* from *Y. hottingeri*, occurs over a comparatively short interval up-section in both the Twara-1 and Punta Gorda-1 wells at the western end of the Nicaragua Rise (Robinson, 1993, Fig. 4). A similar transition is seen in surface sections in central Jamaica, in the Chapelton Formation.

The recent discovery of *Y. jamaicensis* in Oman (Adams and Racey, 1992) has demolished the long-accepted view that this genus is endemic to the Caribbean region.

Distribution. In the upper part of the Chapelton Formation, central Jamaica, and the higher part of the Yellow Limestone Group of the north coastal region.

Family Soritidae Ehrenberg, 1839

Genus *Praerhapydionina* van Wessem, 1943

Praerhapydionina delicata Henson, 1950
(Figures 15.1–6)

Praerhapydionina delicata Henson, 1950, Plate 2, Figs. 4, 6, 9; Plate 8, Fig. 9.
Praerhapydionina delicata Henson, Hottinger, 1963, p. 964, Plate 1, Fig. 3; Plate 2, Figs. 1–10.
Praerhapydionina delicata Henson, McFarlane, 1974, Plate 44A; 1977.
Praerhapidionina delicata Henson, Robinson, 1993, Plate 5, Figs. 1–4.

304

Remarks. Praerhapydionina delicata is characterized by a straight or slightly curved, narrow conical to subcylindrical test, 2 to 3 mm long, of uniserial chambers, divided internally by radially arranged subepidermal partitions, which project from the chamber wall inwards up to about two-thirds of the distance towards the axis, leaving the axial zone free of internal skeletal elements. Partitions are continuous from one chamber to the next. The aperture is terminal, centrally placed, and apparently stellate.

Distribution. Recorded from Oligocene levels in the shelf carbonates of central and eastern Jamaica, mainly from the Walderston Limestone, associated with peneroplid/miliolid benthic foraminiferal assemblages.

Genus *Archaias* de Montfort, 1808

Test normally planispiral and involute, often with a flaring later stage, becoming pseudevolute, sometimes cyclical. Chamber interiors with interseptal pillars but without subepidermal partitions. Aperture areal, usually consisting of two or more irregularly distributed rows of pores on the apertural face.

Archaias asmaricus Smout and Eames, 1958
(Figures 15.6–10)

Archaias asmaricus Smout and Eames, 1958, p. 220; Plate 40, Figs. 21–24; Plate 41, Figs. 6–8, 10, 20, 27.
Archaias sp. McFarlane, 1977, p. 1405.

Description. Test asymmetrically lenticular, consisting of an inflated, central region with about two whorls of involute, planispirally coiled, strongly recurved chambers, terminating in a pseudevolute, flabelliform flange. No cyclical stage seen. The internal skeleton consists of a single row of pillars in the equatorial plane of the test, more or less aligned from one chamber to the next. Details of the proloculus and earliest part of the test not seen. The maximum observed test diameter is 1.9 mm.

Remarks. Smout and Eames (1958, p. 220) mention a probable cyclical stage, at least for the microspheric form of this species. We have observed neither any undoubted microspheric individuals, nor a cyclical stage in any of the Jamaican material, which in other respects appears identical with Smout's and Eames' species.

Distribution. The figured specimens are from locality ER1283A, probably Walderston Limestone, northeastern Jamaica, associated with *Praerhapydionina delicata, Peneroplis ?evolutus, Neorotalia* sp., and *Lepidocyclina yurnagunensis*. These limestones are in faulted contact with early Miocene limestones containing *Miosorites americanus, Heterostegina (Vlerkina) antillea*, uniserial *Miogypsina* sp., and *Archaias ?floridanus*.

Age. Early to middle Oligocene.

Genus *Cyclorbiculinoides* Robinson, 1974

Cyclorbiculinoides jamaicensis Robinson, 1974
(Figures 14.4–6; 16.3)

Cyclorbiculinoides jamaicensis Robinson, 1974a, p. 289, Plate 3, Figs. 3–5; Plate 4, Figs. 1–4, 7.

Remarks. Cyclorbiculinoides jamaicensis resembles the Miocene species *Miosorites americanus* (Cushman) in many of its features, such as the large, centrally placed proloculus and radial subepidermal partitions in

Figure 14. 1, 2, *Fabularia verseyi* Cole, ×37: 1, ER955B, same as VL78, type locality of *F. verseyi*, near-equatorial section, showing irregularity of earliest whorls; 2, axial section, ER964, both Somerset Limestone. 3, *Yaberinella jamaicensis* Vaughan, ×19, equatorial section of a megalospheric individual, ER976, upper Chapelton Formation. 4–6, *Cyclorbiculinoides jamaicensis* Robinson. 4, ×37, equatorial section, refigured from Robinson, 1974a; 5, ×110, tangential section of a topotype, both ER954, upper Claremont Limestone; 6, ×95, portion of a tangential section, figured as Plate 4, Figure 7 in Robinson 1974a, where it was incorrectly referred to locality ER954, the correct locality being ER966, Somerset Limestone. Both tangential sections show apertures scattered over the entire apertural face.

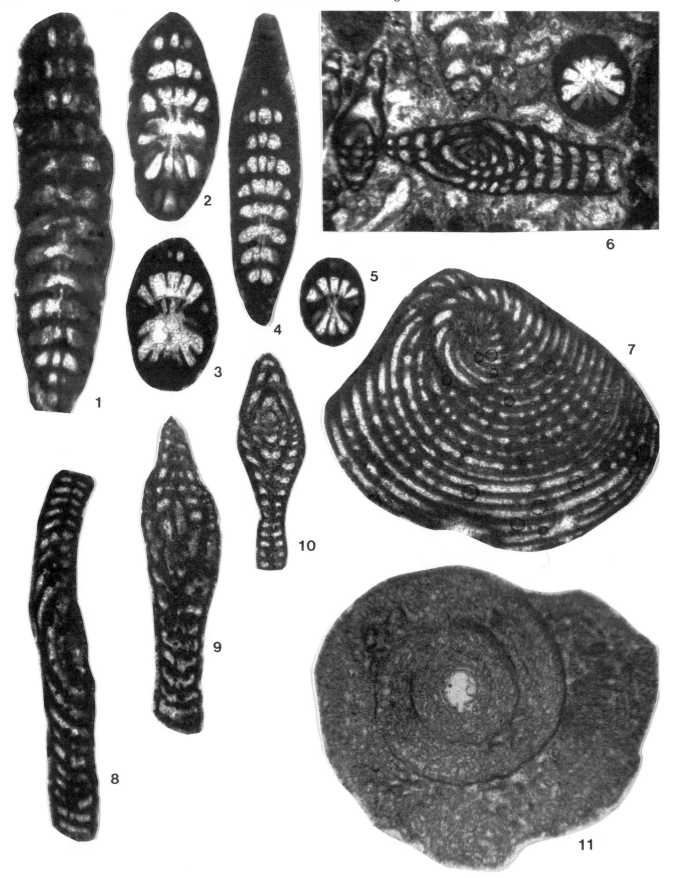

each annular chamber. But besides being a Paleogene, not a Neogene form, it may be distinguished from *M. americanus* by the arrangement of the subepidermal partitions, which are aligned from one chamber to the next in *C. jamaicensis,* and by the positions of the apertures. In *Cyclorbiculinoides* the apertures are scattered irregularly over the whole of the apertural face, whereas in *Miosorites* the apertures are restricted to an apertural band around the equator of the test, between the zones against which the subepidermal partitions terminate (e.g., Vaughan, 1929a, Plate 41, Fig. 2; Seiglie et al., 1977, p. 868, and Fig. 4a).

Distribution. This species is found commonly in the upper Claremont and Somerset Limestones of Jamaica, where it is frequently associated with species of *Fabularia, Pseudochrysalidina,* and *Fallotella.* It is also occasionally observed in well sections at the western end of the Nicaragua Rise but has not yet been reported from elsewhere in the Caribbean. It is regarded as being a late Eocene marker.

Family Cymbaloporidae Cushman, 1927

Genus *Eofabiania* Kupper, 1955

Eofabiania grahami Kupper, 1955
(Figure 17.2)

Eofabiania grahami Kupper, 1955, p. 136, Plate 19, Figs. 1–7.

Remarks. The Jamaican specimens are similar to those of *E. grahami* figured by Kupper (1955) in structure and size. The species is common in the Font Hill and Swanswick Limestones (middle–upper Eocene). Specimens from the same assemblage show a wide variation in shape ranging from low conical to high conical. This may be attributed partly to compaction of the enclosing sediment.

Genus *Fabiania* Silvestri, 1924

Fabiania cassis (Oppenheim, 1896)
(Figures 17.1, 3–6)

Eodictyoconus cubensis Cushman and Bermúdez, 1936, Plate 10, Figs. 27–30; Cole, 1941, Plate 2, Figs. 5–11; Cole, 1942, Plate 3, Fig. 4; Plate 5, Fig. 1; Cole and Bermúdez, 1944, p. 336, Plate 27, Fig. 1; Plate 28, Figs. 1–12; Plate 29, Figs. 1–5; Cole, 1945, Plate 12, Figs. 10–11.

Remarks. As pointed out by Kupper (1955) the species *Eodictyoconus cubensis* is clearly conspecific with *Fabiania cassis* (Oppenheim). It is common to abundant in the Eocene shelf carbonates of west-central Jamaica (Troy/Claremont and Swanswick Limestones) and occurs also in the Font Hill and the Bonny Gate Formations in biosparite layers.

Family Victoriellidae Chapman and Crespin, 1930
Genus *Victoriella* Chapman and Crespin, 1930
Victoriella conoidea Rutten, 1914
(Figures 16.5, 6)

Carpenteria conoidea Rutten, 1914, Plate 7, Figs. 6–9.
Victoriella conoidea, Glaessner and Wade, 1959, Plate 2, Figs. 1–5; Plate 3, Figs. 3; Reiss, 1967, Plate 1, Figs. 1–5; Plate 2, Figs. 1–14.

Remarks. All Jamaican specimens of *Victoriella* recognized in this study occur in random sections of hard rock and show features similar to *V. conoidea,* e.g., three to four chambers per whorl, and coarsely perforate,

Figure 15. 1–6, *Praerhapydionina delicata* Henson, ×49: 1, axial section, ER399, Browns Town Limestone; 2–6, oblique to transverse sections, showing the radially arranged partitions, continuous from one chamber to the next, ER1283, Walderston Limestone, northeastern Jamaica. 6–10, *Archaias asmaricus* Smout and Eames, ×49, ER1283: 6, with *P. delicata,* and *Peneroplis* sp., off-centered axial section; 7, near-equatorial section showing pillars limited to the equatorial plane of the test; 8, 9, oblique sections; 10, off-center axial section showing one row of pillars in the equatorial plane. 11, *Yaberinella hottingeri* Robinson, ×49, equatorial section of a topotype, ER975, Dump Limestone, Chapelton Formation.

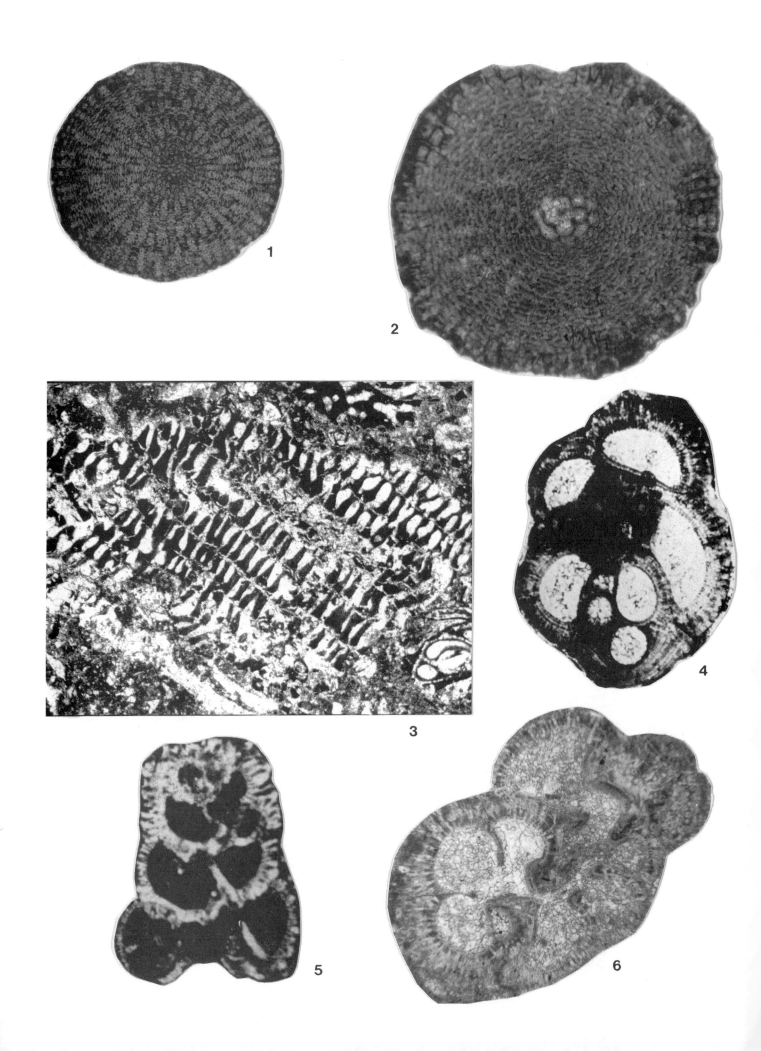

pillared walls. They occur in the Swanswick and Browns Town Limestones and thus range from upper Eocene to Oligocene, whereas *V. conoidea* has only been recorded from the Oligocene of Australia, New Guinea, and the Mediterranean area (Reiss, 1967).

Genus *Eorupertia* Yabe and Hanzawa, 1925
Eorupertia bermudezi Anisgard, 1957
(Figure 16.4)

Eorupertia bermudezi Anisgard, 1957, p. 2, Plate 1, Figs. 1–10, text Figs. 1A–C.

Remarks. Jamaican specimens resemble *E. bermudezi* Anisgard but have a particularly coarse wall structure. The coarse wall structure is more characteristic of *Victoriella* into which the species may more rightfully be placed. It occurs in the Chapelton and Swanswick, and in the bioclastic layers of the Bonny Gate Limestone.

Family Acervulinidae Schultze, 1854

Genus *Sphaerogypsina*

Sphaerogypsina globulus (Reuss), 1848
(Figures 16.1, 2)

Figure 16. 1, 2, *Sphaerogypsina globulus* (Reuss): 1, ×24, ER1234, upper Richmond Formation; 2, ×49, centered section, ER1399, upper Richmond Formation. 3, *Cyclorbiculinoides jamaicensis* Robinson, ×37, part of an equatorial section, showing interseptal pillars more or less aligned from one chamber to the next, ER966, Somerset Limestone. 4, *Eorupertia bermudezi* Anisgard, ×25, showing unilocular embryonic apparatus and thick wall with concentric lamellae. The lamellae are cut normally by fine straight tubules that extend from the exterior of the wall and terminate in the basal mantle. 5, 6, *Victoriella conoidea* (Rutten): 5, ×30, axial section, BK433, Swanswick Limestone; 6, ×40, showing coarsely perforate wall, WS38, Swanswick Limestone.

Ceriopora globulus Reuss, 1848, p. 33, Plate 5, Fig. 7.
Tinoporus pilaris Brady, 1876, p. 15.
Sphaerogypsina globula (Reuss), Cole, 1944, Plate 7, Fig. 23.
Sphaerogypsina globula (Reuss), Vavra, 1978, p. 742, Fig. 1.
Gypsina peruviana Berry, 1929, p. 240, text Figs. 1, 2.

Distribution. Specimens of *Sphaerogypsina* similar to *S. globulus* (Reuss) and *S. peruviana* (Berry) occur in the upper part of the Richmond Formation (early Eocene), in the Font Hill and Swanswick Limestone (middle to late Eocene), and, rarely, in the Browns Town Limestone (Oligocene).

Family Amphisteginidae Cushman, 1927

Genus *Amphistegina* d'Orbigny, 1826

Amphistegina parvula (Cushman, 1918)
(Figures 18.3–5)

Nummulites parvula Cushman, 1919b, p. 51, Plate 4, Figs. 3, 4, 6.
Amphistegina parvula (Cushman), Cole, 1958a, p. 201, Plate 25, Figs. 17–19.
Amphistegina parvula (Cushman), Butterlin, 1970, p. 293, Plate 3, Figs. 5–6.
Amphistegina parvula (Cushman), Caudri, 1974, p. 304, Plate 1, Figs. 8, 9; Plate 4, Figs. 7–9; Plate 5, Figs. 2–5, 10, 11.

Remarks. This species has been included in *Eoconuloides* by Cole (1969) and others, but near topotypic and other specimens we have examined lack distinguishing features of *Eoconuloides,* and, following Butterlin (1970) and Caudri (1974), we prefer to retain this species in *Amphistegina.* It is distinguished from *Eoconuloides lopeztrigoi* (Palmer) by its stout test with thick walls with a prominent radial structure, absence of

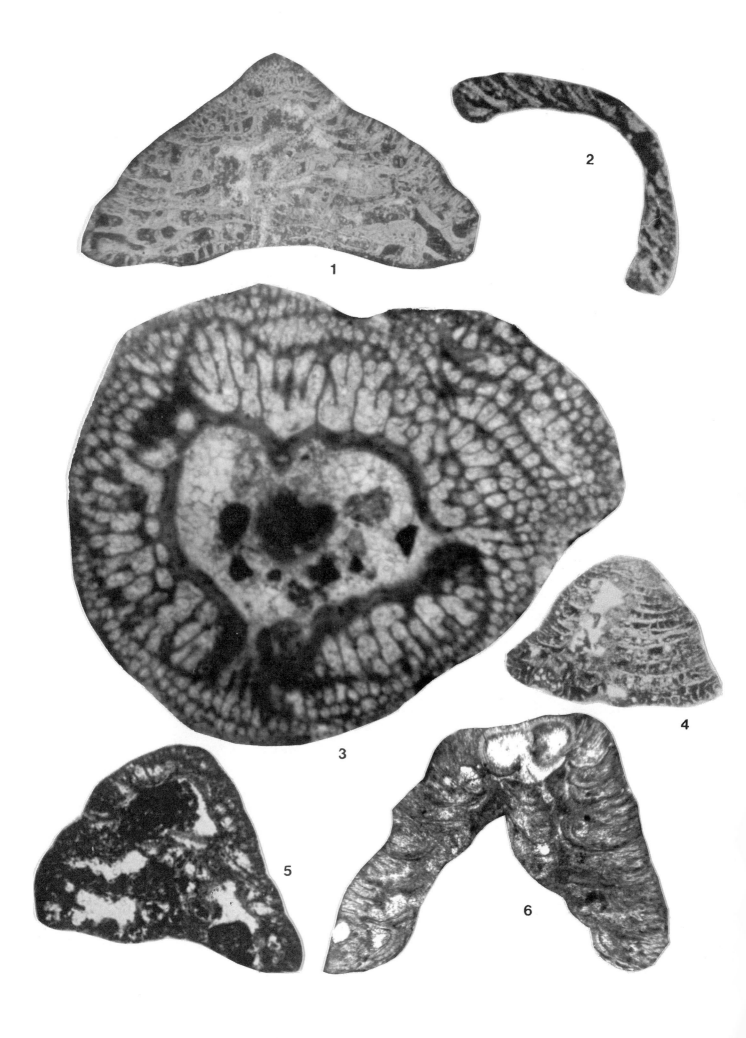

pillars, and apparent absence of the siphonate apertural features of *Eoconuloides*.

Distribution. The types are from the upper middle Eocene of St. Bartholomew, but this is a widely distributed species in the Caribbean region. In Jamaica it occurs in shelf edge carbonates of the middle and late Eocene, with *Lepidocyclina* and *Palaeonummulites,* and discocycliniform species, but is rare in the platform units of the Yellow and White Limestones.

Family Boreloididae Cole and Bermúdez, 1947

Genus *Eoconuloides* Cole and Bermúdez, 1944

Eoconuloides lopeztrigoi (Palmer) 1934
(Figures 19. 6–9)

Amphistegina lopeztrigoi Palmer, 1934, p. 255, Plate 15, Figs. 6, 8.
Amphistegina lopeztrigoi Palmer, Barker and Grimsdale, 1936, p. 233, Plate 30, Figs. 1, 2; Plate 32, Figs. 1–3; Plate 34, Fig. 1; Plate 38, Fig. 3.
Amphistegina lopeztrigoi Palmer, Cole and Gravell, 1952, p. 714, Plate 91, Figs. 6–8.

Remarks. Typified by a lenticular test with strongly developed pillars. Cuban topotypes range from 1.0 to 2.8 mm in diameter. Jamaican examples range from about 0.8 to 1.8 mm. The forms called *Amphistegina senni* Cushman (Vaughan, 1945) from Barbados are similar, except for their smaller size, to *E. lopeztrigoi.*
Distribution. Small specimens of *Eoconuloides lopeztrigoi* occur questionably in the late Paleocene Nonsuch Limestone, and are quite common in the early Eocene Richmond Formation of eastern Jamaica. Larger specimens occur up to the top of that unit (*Globigerina pentacamerata* zone age). It also appears in the Chapelton Formation (early middle Eocene) of northwest Jamaica.

Eoconuloides wellsi Cole and Bermúdez, 1944
(Figures 18. 1, 2; 19.1–5)

Eoconuloides wellsi Cole and Bermúdez, 1944, p. 341, 342; Plate 1, Figs. 4–10.
Eoconuloides wellsi Cole and Bermúdez, Cole and Gravell, 1952, p. 713, Plate 92, Figs. 1–10.

Remarks. Characterized by a medium to high conical, rather than the lenticular test of *E. lopeztrigoi.* Although *E. lopeztrigoi* and *E. wellsi* are frequently regarded as synonymous, and there is probably gradation in morphology, the acquisition of the characteristic high, conical shape appears to be of biostratigraphic significance. *Eoconuloides* with high conical tests appear at a higher stratigraphic level than those with lenticular tests, appearing first in the upper part of the Richmond Formation, near the top of the lower Eocene.
Distribution. E. wellsi is locally common in the upper part of the Richmond Formation and in the basal part of the Chapelton Formation of east-central Jamaica.

Family Lepidocyclinidae Scheffen, 1932

This group of larger foraminifera is characterized by a compressed circulate or radiate test with a distinct equatorial layer and a bilocular embryonic chamber, occurring in Jamaica in rocks ranging in age from middle Eocene to late early Miocene. The lepidocyclinids were common in the shallow seas of the warm water belts of this period, and because many

Figure 17. 1, 3–6, *Fabiania cassis* (Oppenheim): 1, ×48, near-axial section, WS116; 3, ×68, near-basal section, BK45; 4, ×36, axial section, BK42, all Swanswick Formation; 5, ×36, axial section, Swanswick Limestone; 6, ×42, axial section through proloculus, ER1710, Swanswick Limestone, Hanover. 2, *Eofabiania grahami* (Kupper), ×38, axial section showing lack of lateral chamberlets, BK42, Swanswick Limestone.

312 E. Robinson and R. M. Wright

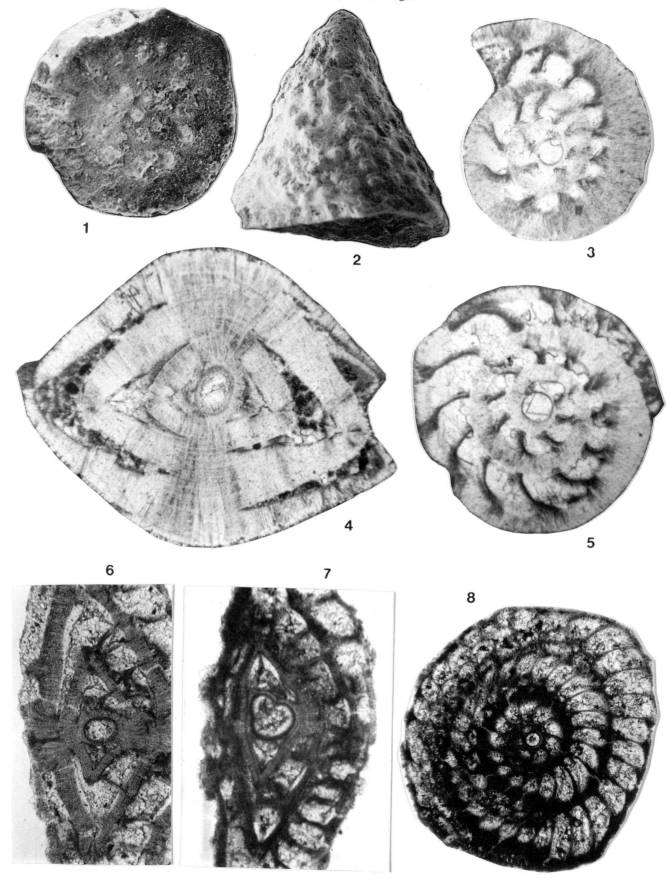

Figure 18. 1, 2, *Eoconuloides wellsi* Cole and Bermudez, ×29, spiral and axial views of the exterior of the same specimen, ER500, upper Richmond Formation. 3–5, *Amphistegina parvula* (Cushman), ×40: 3, 5, spiral sections, 4 axial section, ER39, Bonny Gate Limestone, Lloyds Quarry. 6–8, *Helicostegina gyralis* Barker and Grimsdale: 6, 7, ×82, axial sections of the central part of two specimens, showing wall structure and pores through ventral septa; 8, ×47, equatorial section, all ER1313, Stettin Member, Chapelton Formation.

species are identifiable under the low 10× magnification of a hand lens they have proved to be useful mapping and stratigraphical aids.

Gümbel (1870) first proposed the genus *Lepidocyclina* as a subgenus of *Orbitoides* based upon the type species, *Nummulites mantelli*. *Lepidocyclina* was transferred to the Orbitoididae by Lemoine and R. Douvillé (1904). H. Douvillé (1911) introduced *Nephrolepidina* and *Eulepidina* as subgenera of *Lepidocyclina,* and subsequently proposed *Isolepidina* and *Pliolepidina* (1915), and *Amphilepidina* (1922). Because of later studies relating to the use of the megalospheric embryonic chambers in subgeneric classification, *Isolepidina* was regarded as synonymous with *Lepidocyclina* (s.s.) and *Amphilepidina* was considered a synonym of *Nephrolepidina* (Vaughan, 1924; Van der Vlerk, 1928). Scheffen (1932b) erected the family Lepidocyclinidae. More recent comprehensive reviews of generic and subgeneric names proposed for the Lepidocyclinidae are given by Grimsdale (1959), Eames et al. (1962b) and Adams (1987).

The many specific descriptions of *Lepidocyclina* have been made mainly on a regional basis from localities in North America, Europe, and the western Pacific areas. Because little agreement exists as to the systematic importance of particular characteristics, comparison of external and internal features in forms from different geographic areas is difficult to achieve (Butterlin, 1987, 1990). This problem is compounded by the wide range of variability in both structure and morphology exhibited by the lepidocyclinids, resulting in the establishment of more species than were warranted. Conflicting opinions still exist regarding the overall classification of the lepidocyclinids. We have followed the classification of Scheffen (1932a), recognizing two subfamilies in the family Lepidocyclinidae, and of Adams (1987), who regarded the following morphological characters of the Lepidocyclinidae to be of taxonomic importance in ranked order:

1. The relative sizes of the first and second chambers of the embryonic apparatus.
2. Presence or absence of spiral thickening of the primary coil.
3. Number of principal auxiliary chambers.
4. Number and relative lengths of the periembryonic spires and the presence of alar prolongations.
5. Presence or absence of adauxiliary chambers.
6. Shape and mode of growth of the equatorial chambers.
7. Presence or absence of lateral chambers.

Subfamily Helicolepidininae Tan, 1936

Forms placed in this subfamily have lenticular tests, characterized by the possession of one or more trochospiral or planispiral nepionic spires of chambers. The later part of the equatorial zone is occupied by spirally to cyclically developed arcuate chambers. Characteristically the outer margin of the primary spiral sheet is thickened for some part of its length. As emphasized by Adams (1987), the genera *Helicostegina* and *Eulinderina* have a proloculus and deuteroloculus, the first two chambers of a spiral series, while in *Polylepidina* and other members of the subfamily Lepidocyclinae the first and second chambers are the protoconch and deuteroconch, forming a specialized embryonic apparatus, or nucleoconch, succeeded by one or two primary auxiliary chambers, the first chambers with a retrovert aperture.

Genus *Helicostegina* Barker and Grimsdale, 1936

Helicostegina gyralis Barker and Grimsdale, 1936
(Figures 18.6–8)

Helicostegina gyralis Barker and Grimsdale, 1936, p. 236, 237; Plate 30, Figs. 3–5; Plate 32, Figs. 4, 5; Plate 34, Figs. 2–6; Plate 37, Fig. 6; Cole and Gravell, 1952, p. 713, Plate 92, Figs. 11–21.

Remarks. The test is trochospiral, the proloculus being followed by 26 or more spiral chambers. The spire reaches the periphery of the test. Lateral chambers are absent. In random thin sections this species may be confused with *Eoconuloides lopeztrigoi* as the arcuate equatorial chambers are inserted near the periphery and may be damaged or missing in poorly preserved and worn specimens.
Distribution. H. gyralis occurs in the lower part of the Chapelton Formation (late early Eocene to earliest middle Eocene).

Genus *Eulinderina* Barker and Grimsdale, 1937

Eulinderina antillea (Cushman) 1919

Lepidocyclina antillea Cushman, 1919b, p. 63, Plate 3, Fig. 3.
?*Planorbulina* (*Planorbulinella*) *guayabalensis* Nuttall, 1930, p. 276, Plate 25, Figs. 15–17.
Lepidocyclina (*Polylepidina*) *antillea* Cushman, Cole, 1960a, Plate 10, Fig. 1; Plate 12, Figs. 1, 3, 4, 6, 7; Plate 13, Figs. 1, 2, 5.
Lepidocyclina (*Polylepidina*) *antillea* Cushman, Cole, 1956a, Plate 27, Fig. 9; Cole, 1963a, Plate 7, Fig. 5.
Eulinderina antillea (Cushman), Adams, 1987, Plate 1, Figs. 22, 23.

Remarks. Cole (1944, 1960a, 1963a) and others have usually considered *Polylepidina chiapasensis* and *Eulinderina guayabalensis* as being synonymous with the inadequately illustrated *Lepidocyclina antillea* Cushman, thus making *L. antillea* the type for *Polylepidina*, with *Eulinderina* being a junion synonym. Adams (1987) regarded this species as being the same as *E. guayabalensis* and distinct from *Polylepidina chiapasensis*, thus making it the type of *Eulinderina*.

Our examination of near-type material from St. Bartholomew confirms Adam's contention that *L. antillea* displays the generic features of *Eulinderina*. It is distinguished from *Polylepidina chiapasensis* (Vaughan) by the presence of a spiral coil with a thickened outer wall, no true primary auxiliary chambers, and possession of countersepta and apertures with anteriorly directed lips (Cole, 1960a, 1963a).
Distribution. The St. Bartholomew limestones have been assigned to the P11 or P12 planktic foraminiferal zone by Westercamp and Andreieff (1983). The example of *Eulinderina antillea* figured by Cole (1956a) is from the Chapelton Formation of north-central Jamaica.

Subfamily Lepidocyclininae Scheffen, 1932

As defined here this subfamily includes forms characterized by the possession of a true bilocular nucleoconch, followed by one or more primary auxiliary chambers, producing two to four spirals from which cyclically arranged equatorial chambers are generated. We recognize four genera and subgenera.

Key to Jamaican genera and subgenera:

A. Nucleoconch more or less isolepidine
  1. With protoconch larger than deuteroconch, with two to four nepionic spirals ....
     .... *Polylepidina,* middle Eocene
  2. With protoconch slightly smaller to slightly larger than the deuteronch; with four nepionic spirals ....
     .... *Lepidocyclina* (*Lepidocyclina*), middle Eocene to early Miocene ....
     (*Neolepidina* of Butterlin (1987) is similar) ....

Figure 19. 1–5, *Eoconuloides wellsi* Cole and Bermúdez: 1, ×46, Font Hill Formation; 2–5, ×49, ER1700 upper Richmond Formation; 1–3, 5, axial sections; 4, spiral section. 6–9, *Eoconuloides lopeztrigoi* (Palmer): 6, 8, 9, axial sections, ×43; 7, spiral section, ×122; 6, AWK490A, Richmond Formation, Green River (coll. A. W. Kemp); 7, ER1315, lower part of Stettin Member, associated with specimens we identify as *Helicostegina gyralis*; 8, ER500, upper Richmond Formation; 9, AWK388, Richmond Formation, Hagley Gap (collection of A. W. Kemp).

B. Nucleus variably nephrolepidine
  1. with relatively thick wall to nucleoconch; adauxiliary chambers relatively numerous ....
      equatorial chambers rhombic or spatulate to hexagonal ....
      ..... *Lepidocyclina (Nephrolepidina)*, early Oligocene to early Miocene. ....
  2. with relatively thin wall to nucleoconch; few adauxiliary chambers ....
      equatorial chambers arcuate to spatulate ....
      .... *Lepidocyclina (Nephrolepidina) chaperi,* late Eocene.
C. Nucleoconch eulepidine
      ... *Lepidocyclina (Eulepidina),* early Oligocene to early Miocene.

Over the last three decades an increasing awareness has developed as to the morphological variations inherent in, and between, populations of *Lepidocyclina*. This has resulted in a decline in the number of species recognized.

The evolutionary trends shown by Caribbean lepidocyclinids, based on interpretations of Eames and others (1962b), Hanzawa (1962), Barker and Grimsdale (1936, 1937), Cole (1963a, 1964b, 1969), Frost and Langenheim (1974), Sirotti (1983), Adams (1987), and our own observations are suggested on Figure 20. Although Butterlin (1987, 1990) also indicates possible lineages within the family, he does not consider that the subgenera *Lepidocyclina* s.s., *Nephrolepidina,* and *Eulepidina* can be precisely defined.

Genus *Polylepidina* Vaughan, 1924

Polylepidina chiapasensis Vaughan, 1924
(Figures 21.1–6)

Lepidocyclina (Polylepidina) chiapasensis Vaughan, 1924, p. 808, Plate 30, Figs. 1–3.

?*Lepidocyclina (Pliolepidina) kinlossensis* Vaughan, 1928a, p. 286, Plate 47, Figs. 1–6.
Lepidocyclina (Polylepidina) gardnerae Cole, 1938, Plate 9, Figs. 1–10.
Lepidocyclina (Polylepidina) antillea (Cushman), Cole, 1963a, Plate 1, Fig. 3.
Lepidocyclina (Polylepidina) antillea (Cushman), Eva, 1980a, Plate 1, Figs. 1–8.

Remarks. Probably the majority of specimens recorded from Jamaica would be regarded as belonging to *Polylepidina chiapasensis* rather than to *Eulinderina antillea*. But these two species can be difficult to separate when only random thin sections are available for study.

The specimen illustrated by Cole (1956a, Plate 27, Fig. 9) is regarded here as a *Eulinderina,* as it is unispiral sensu Eva (1980a) with a displaced third chamber without a retrovert aperture, but with countersepta and apertures with anteriorly directed lips (Cole, 1963a, Plate 7, Fig. 5, attributed by Cole to a St. Bartholomew locality; but see Cole, 1956a, Plate 27, Fig. 9, of which it is an enlargement). A bispiral example (i.e., a true *Polylepidina*) has also been illustrated from the same locality (Cole 1963a, Plate 1, Fig. 3). The vertical sections of material from the same locality (Cole 1956a, Plate 30, Figs. 7, 8) with their vacuolar-like lateral chambers also reflect the intermediate nature of these specimens.

Distribution. *Polylepidina chiapasensis* is found in the late middle Eocene (possibly also late Eocene) of Jamaica, upper part of the Chapelton Formation, Swanswick, and middle Bonny Gate Limestones.

Genus *Lepidocyclina* Gümbel, 1870

Subgenus *Lepidocyclina* Gümbel, 1870

Lepidocyclina (Lepidocyclina) canellei Lemoine and R. Douvillé, 1904
(Figure 22.6, 7)

Lepidocyclina (Lepidocyclina) canellei Lemoine and R. Douvillé, 1904,

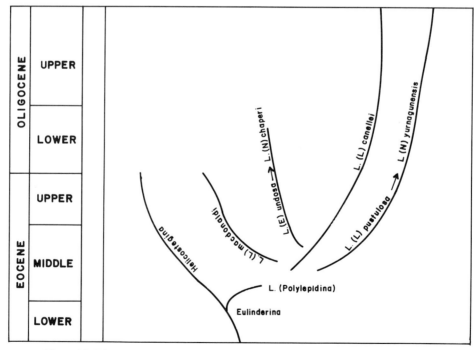

Figure 20. Diagram showing suggested evolutionary trends of Jamaican *Lepidocyclina.*

p. 20, Plate 1, Fig. 1; Plate 3, Fig. 5; Cole, 1952, p. 18–20, Plate 16, Figs. 1–22; Plate 17, Figs. 1–3 (references); Cole, 1961, p. 383–389, Plate 30, Figs. 1–13; Plate 31, Figs. 1–5; Plate 32, Figs. 1–4; Plate 33, Figs. 1–4; Plate 34, Figs. 1–8; Plate 35, Figs. 1–2, 4–5; Plate 36, Figs. 1–5; Plate 37, Figs. 1–5; Plate 38, Figs. 1–7; Plate 39, Figs. 1–9; Eames and others, 1968, p. 297, Plate 59, Fig. 7.
Lepidocyclina matleyi Vaighan, 1928a, Plate 46, Figs. 1–3.
Lepidocyclina (Lepidocyclina) giraudi Douvillé, 1907, Figs. 9–10, 15, 16. Cole, 1957a, p. 41–42, Plate 1, Fig. 6; Plate 2, Figs. 2–5, 7–9; Plate 4, Figs. 2, 5; Plate 5, Fig. 4; Plate 6, Fig. 8; Cole, 1958b, Plate 26, Fig. 10; Plate 28, Fig. 4; Butterlin, 1961, p. 16–17, Plate 5, Figs. 1–3; Sachs and Gordon, 1962, p. 14–15, Plate 1, Figs. 6, 8, 11; Plate 2, Fig. 9; Plate 3, Figs. 2, 4, 9.
Lepidocyclina (Lepidocyclina) asterodisca Nuttall, 1932, p. 34–35, Plate 7, Figs. 5, 8; Plate 9, Fig. 10; Cole, 1952, p. 17–18, Plate 17, Fig. 14; Cole, 1958a, p. 201–202, Plate 22, Fig. 6; Plate 23, Figs. 1–12; Plate 24, Figs. 6–7; Sachs, 1959, p. 406, Plate 35, Figs. 7–9.
Lepidocyclina (Lepidocyclina) waylandvaughani Cole, 1938, p. 21–22, Plate 4, Figs. 1–8; Cole, 1957a, p. 42, Plate 1, Fig. 6; Plate 2, Figs. 2–5, 7–9; Plate 4, Figs. 2, 5; Plate 5, Fig. 4; Plate 6, Fig. 8; Cole, 1958b, Plate 27, Fig. 4; Plate 28, Figs. 2–3; Sachs and Gordon, 1962, p. 15, Plate 1, Figs. 5, 10.
Lepidocyclina (Lepidocyclina) parvula Cushman, Eames and others, 1962b, p. 306, Plate 8, Figs. 2–3.

Remarks. Lepidocyclina canellei occurs in the highest part of the Bonny Gate Formation, the upper part of the Browns Town Formation, and the lower part of the Montpelier Formation. Jamaican specimens support Cole's (1961) thesis that the lateral layers show a larger degree of morphologic variation, correlatable with differences in environmental conditions, than does the equatorial layer. The *L. giraudi* type is readily distinguishable by the pillaring in the central region of the test and by low lateral chambers. The ratio of *canellei* to *giraudi* types is higher in biomicrites than in biosparites suggesting that the morphology of the lateral chambers and the development of pillars is related to environment and sea-bottom lithology.
Distribution. Early? Oligocene to early Miocene of the White Limestone Group. The illustrated specimen is from the earliest Miocene (N4 planktic zone).

Lepidocyclina (Lepidocyclina) macdonaldi Cushman, 1919
(Figures 23.5–7)

Lepidocyclina (Lepidocyclina) macdonaldi Cushman, 1919a, p. 94, Plate 40, Figs. 1–6; Cole, 1952, p. 16, Plate 7, Figs. 1–19; Plate 8, Figs. 1–4; Plate 14, Fig. 11; Plate 20, Fig. 16; Cole, 1945, p. 117–120, Plate 19, Figs. 1–13; Cole, 1956a, p. 214, Plate 27, Figs. 3–5, 8; Plate 28, Figs. 1–5; Cole, 1958b, Plate 26, Fig. 3; Hanzawa, 1962, Plate 2, Figs. 11–13.
L. (Polylepidina) proteiformis Vaughan, Barker and Grimsdale, 1936, p. 241, Plate 33, Fig. 3; Plate 36, Fig. 5; Renz and Kupper, 1947; Caudri, 1948, Plate 73, Fig. 7.
L. (Pliolepidina) proteiformis Vaughan, Cole, 1956a, p. 221, Plate 27, Figs. 6, 7; Plate 28, Figs. 7–10.
L. (Lepidocyclina) proteiformis Vaughan, Cole, 1963a, Plate 6, Fig. 3.
Lepidocyclina ariana Cole and Ponton, 1934.

Remarks. L. macdonaldi, has been distinguished from *L. proteiformis* Vaughan only by differences in vertical section, *L. macdonaldi* having pillars and thick roofs and floors to the lateral chambers whereas *L. proteiformis* has much thinner lateral chamber partitions (e.g., Cole 1956a, using Jamaican examples). *L. ariana* Cole and Ponton differs in having straight roofs and floors. We accept the conclusion of Frost and Langenheim (1974) that these species are synonymous with *L. macdonaldi.* They appear to have similar ranges in Jamaica.
The morphological variation seen can be correlated with variations in depositional environments of the rocks in which they are found. Specimens of *L. (L.) macdonaldi* with low lateral chambers and thick roofs and floors (*L. macdonaldi* s.s. and *L. ariana*) are dominant in shelf carbonates, particularly algal-foraminiferal biosparites such as the Swanswick Limestone. Specimens of the *L. proteiformis* variety with open lateral chambers and thin roof and floors are most common in biomicrite basin slope deposits such as the Bonny Gate Limestone. The specimens in shelf carbonates are invariably more thickly pillared.
Distribution. Late middle and late Eocene; abundant in the Bonny Gate, Swanswick, and Somerset Limestones.

Lepidocyclina (Lepidocyclina) pustulosa H. Douvillé, 1917
(Figures 23.1–4; 24.1., ?5)

Lepidocyclina (Lepidocyclina) pustulosa H. Douvillé 1917, p. 844, text Figs. 1–4; Vaughan and Cole, 1941, p. 65–67, Plate 25, Figs. 1–9; Plate 26, Figs. 1–9; Plate 27, Figs. 1–3; Plate 28, Figs. 1–9; Plate 30, Figs. 1–3; Cole, 1952, p. 16–17, Plate 13, Figs. 1–20; Plate 14, Figs. 1–10; Plate 15, Figs. 14–16; Cole, 1956a, Plate 27, Figs. 1–2; Plate 28, Fig. 6; Plate 30, Figs. 1–3; Butterlin, 1961, p. 19, Plate 9, Figs. 1–4; Eames and others, 1962b, p. 312, 314 (references and synonymy); Cole, 1962, Plate 6, Figs. 2–4; Plate 7, Figs. 1–6; Plate 8, Figs. 1, 2, 4–8; Cole, 1963a, p. 21–35, Plate 1, Fig. 5; Plate 2, Figs. 1–6; Plate 3, Figs. 1–6; Plate 5, Figs. 1–4; Plate 10, Figs. 1–4,9–12; Plate 14, Figs. 1–5; Cole and Applin, 1964, Plate 5, Figs. 1,3,5,6,8,9; Plate 6, Figs. 1,2,4–8; Plate 7, Figs. 1–3.
Lepidocyclina (Lepidocyclina) sherwoodensis Vaughan, 1928a, p. 287, Plate 48, Figs. 4–8.
Lepidocyclina (Lepidocyclina) trinitatis H. Douvillé, Vaughan, 1928a, p. 289, Plate 49, Figs. 10–13.

Remarks. This is, perhaps, the most widespread lepidocyclinid species in the Caribbean region. Cole's (1963a) synonyms are accepted provisionally, as is his thesis that morphological distinctions among some forms of *L. pustulosa* are attributable to differences in the depositional environment. Specimens in limestones generally possess more pillars, lower average values for the height of the lateral chamber, and higher values for thickness of chamber roofs and floors, than specimens derived from clastic depositional environments.
Forms apparently intermediate between *L. pustulosa* and *L. yurnagunensis* occur in west-central Jamaica. *L. peruviana* Cushman is probably a distinct species but has not been treated as such in this paper, because of lack of suitable material for analysis. The specimen illustrated at Figure 24.1 is one of the *L. peruviana* sort.
Distribution. Late Eocene; typically in the Swanswick Limestone, occurring with *L. (L.) macdonaldi, Fabiania cassis, Palaeonummulites willcoxi,* and *L. (E.) chaperi.*

Subgenus *Nephrolepidina* H. Douvillé, 1911

In the nephrolepidine *Lepidocyclina* the protoconch is enclosed to a variable extent by a reniform deuteroconch. Eames and others (1962b) regarded *Nephrolepidina* as a subgenus of *Lepidocyclina,* Hanzawa

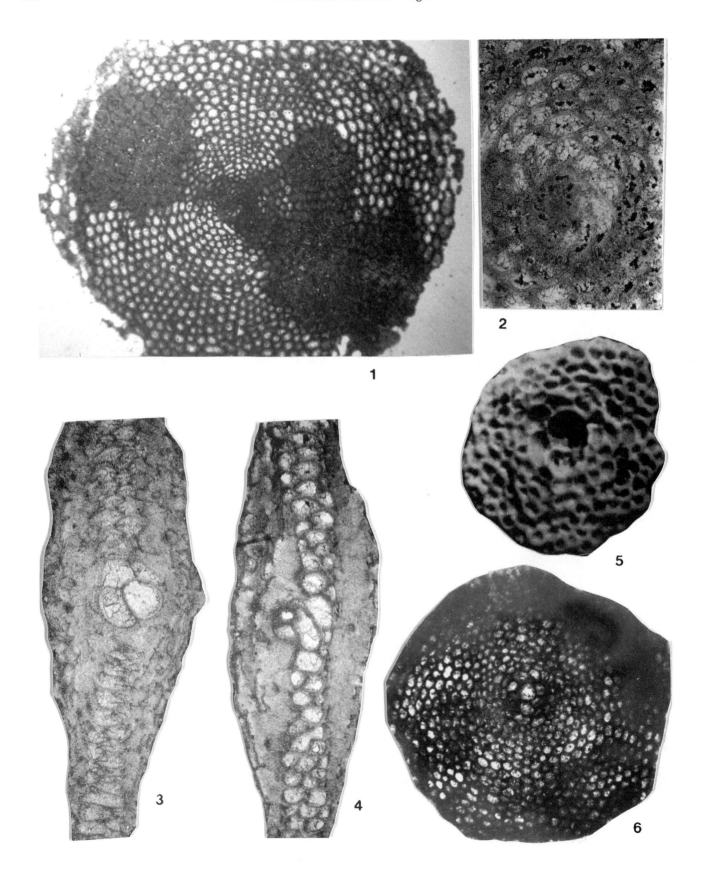

(1962) elevated it to generic rank, whereas Cole (1960a) included it in the subgenus *Eulepidina*. In this study we regard *Nephrolepidina* as a valid subgenus of *Lepidocyclina*.

Lepidocyclina (*Nephrolepidina*) *yurnagunensis* Cushman, 1919
(Figures 23.3; 24.2–4)

Lepidocyclina canellei Lemoine and Douvillé, variety *yurnagunensis*, Cushman, 1919b, p. 54, Plate 12, Figs. 7–8.
Lepidocyclina morgani Lemoine and R. Douvillé, Cushman, 1919b, p. 59, text Fig. 7, Plate 11, Figs. 1–3.
Lepidocyclina (*L.*) *yurnagunensis* Cushman, Vaughan, 1926, p. 391–393, Plate 25, Figs. 2–6; Cole, 1952, p. 22–23, Plate 15, Fig. 3; Plate 17, Figs. 5–18; Plate 20, Figs. 11–12; Cole, 1968, Plate 23, Fig. 8; Plate 24, Figs. 5, 8; Eames and others, 1968, p. 284, Plate 49, Figs. 1–5.
Lepidocyclina yurnagunensis Cushman, 1919, subsp. *morganopsis* Vaughan, 1933, Cole, 1952, p. 23, Plate 15, Figs. 1–2, 4–5; Plate 23, Figs. 1–5, 9; Eames and others, 1968, p. 293–295, Plate 55, Figs. 5–9; Plate 56, Figs. 1–2, 6–9.

Description. The test is markedly biconvex, reaching as much as 4.0 mm in diameter with diameter/thickness ratios of around 2:1. Embryonic apparatus is bilocular, varying from nearly isolepidine to nephrolepidine with a hemispherical protoconch some 0.16 to 0.27 mm in axial diameter and 0.2 to 0.3 mm wide joined to a deuteroconch 0.07 to 0.13 mm in axial diameter and 0.2 to 0.3 mm in width, reniform in shape. Both protoconch and deuteroconch walls are bilamellar with a radial microstructure with thicknesses in the range 0.03 to 0.04 mm, the protoconch wall being slightly thicker. The periembryonic chambers consist of two elongate initial chambers, one developed at each end of the protoconch/deuteroconch wall, and variably developed coils of small periembryonic chambers. Equatorial chambers are rhombic to spatulate, averaging 0.035 to 0.04 mm in diameter, with trilamellar walls and four stolons, as in *L.* (*L.*) *pustulosa*. Vertical sections have a thin equatorial layer and open lateral chambers arranged in tiers. The lateral chambers directly above the equatorial layer are low and siltlike.

Remarks. Although *L.* (*Nephrolepidina*) *yurnagunensis* has frequently been included in the subgenus *Lepidocyclina*, examination of illustrations of type material (e.g., Vaughan, 1933, Plate 11, Fig. 3; Vaughan and Cole, 1941, Plate 38, Fig. 3) indicate that the nucleoconch is distinctly nephrolepidine, with a slightly reniform deuteroconch, and with adauxiliary chambers on the deuteroconch. On the basis of embryonic characteristics Cole (1960a) regarded this species as a *Eulepidina*. However, all the early Jamaican examples are nearly isolepidine, becoming nephrolepidine at stratigraphically higher horizons.

As Butterlin (1987) pointed out, *L. yurnagunensis* is virtually indistinguishable from the Indo-Pacific species *L. isolepidinoides* van der Vlerk. Forms with a near-isolepidine nucleoconch, and characterized by a relatively thin wall to the nucleoconch are similar to those called *Lepidocyclina yurnagunensis morganopsis* Vaughan by Eames and others (1968). *L. yurnagunensis morganopsis* was originally distinguished by the presence of pillars, a feature which, in other species of *Lepidocyclina*, appears to be related to differences in paleoenvironment. Examples of *L. yurnagunensis morganopsis* occur in Jamaica in high-energy biosparites of the reef facies whilst forms similar to *L. yurnagunensis* (s.s.) occur in deep-water biomicrites of the fore-reef slope. A similar relationship between structure and ecology occurs in the pillar development in *L.* (*L.*) *canellei* and *L.* (*E.*) *undosa* where they are associated with *L. yurnagunensis* (s.l.).

Figure 21. 1, 2, equatorial sections of specimens which we identify as the microspheric form of *Polylepidina chiapasensis* Vaughan: 1, ×25.5; 2, ×95, ER161, Font Hill Formation. 3–6, *Polylepidina chiapasensis* Vaughan: 3, 4, ×43, oblique sections of megalospheric individuals, ER1367, Chapelton Formation; 5, ×18, natural equatorial section of a megalospheric individual, Borehole 177, GRS25, Font Hill Formation, Trelawny; 6, ×28, equatorial section, ER1366, Chapelton Formation.

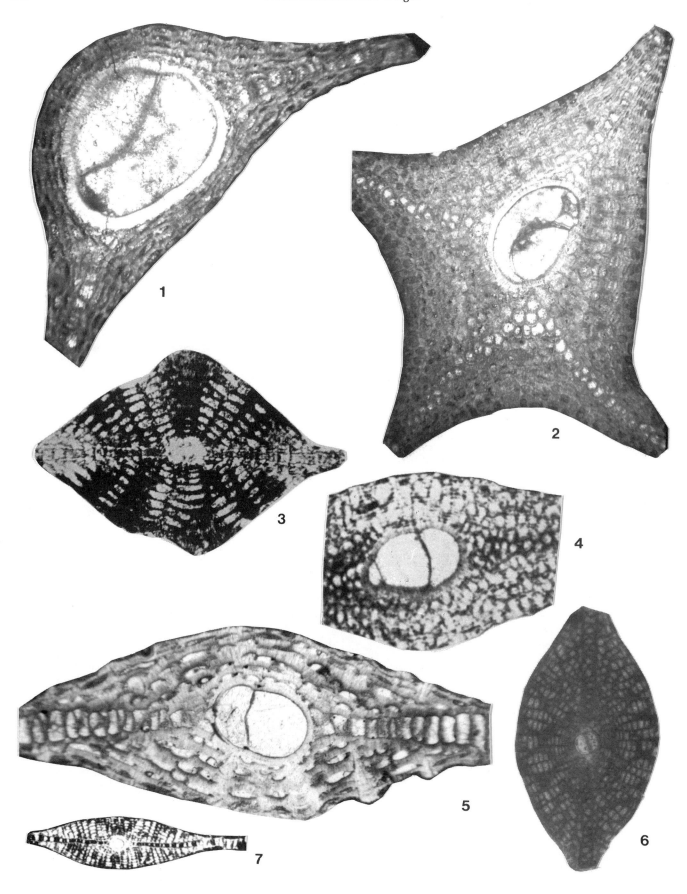

It is likely that the series *L. pustulosa–L. yurnagunensis* forms an evolutionary sequence leading to the typical *Nephrolepidina* spp. of the Indo-Pacific Miocene and to such Caribbean species as *L. (Nephrolepidina) vaughani* (Butterlin, 1987). The example of *L. yurnagunensis* figured from the Oligocene Browns Town Formation at Red Gal Ring has measured parameters (Ai = 39, C = 2; van der Vlerk, 1964; Drooger and Socin, 1959), which would place it in the *L. praetournoueri* European group (de Mulder, 1975) or the *L. isolepidinoides* Indo-Pacific group (van Vessem, 1978). It is morphologically more advanced than specimens placed in *L. yurnagunensis* by van der Vlerk and Postuma (1967) from the early Oligocene, *G. ampliapertura* zone of Trinidad, and by Wong (1976) from the early and middle Oligocene of the Guiana Basin. Future work may show Jamaica to be one place in the Caribbean where a relatively complete stratigraphic sequence of the *L. pustulosa–L. yurnagunensis* lineage is recognizable. On Figure 23.4 we illustrate a specimen that is possibly referable to the more highly evolved *L. (N.) vaughani*, from the late Oligocene Browns Town Formation.

Distribution. Latest Eocene?–late Oligocene of the White Limestone Group.

Lepidocyclina (Nephrolepidina) chaperi Lemoine and R. Douvillé, 1904
(Figures 22.5; 25.1–5)

Lepidocyclina chaperi Lemoine and R. Douvillé, 1904, p. 14–15, Plate 2, Fig. 5.
Lepidocyclina (Nephrolepidina) chaperi Lemoine and R. Douvillé, Cole, 1952, p. 23–27, Plate 8, Figs. 5–8; Plate 9, Figs. 3–19; Plate 10, Figs. 1–10; Plate 11, Figs. 1–8; Plate 12, Figs. 1–15; Plate 20, Figs. 8–10; Plate 23, Figs. 8, 11–12 (references and synonymy); Butterlin, 1961, p. 17–18, Plate 4, Fig. 3; Plate 6, Figs. 1–4.
Lepidocyclina (Eulepidina) chaperi Lemoine and R. Douvillé, Cole, 1963a, p. 10, Plate 8, Fig. 3; Plate 9, Figs. 1–3; Plate 10, Fig. 13.
Lepidocyclina (Nephrolepidina) haddingtonensis Vaughan, 1928a, p. 292, Plate 50, Figs. 1–2.

Remarks. Provisionally we retain this species in the subgenus *Nephrolepidina*, because of the shape of its nucleus, but we regard it as being part of a species group quite distinct from *L. (N.) yurnagunensis*. Adams (1987, p. 306, 307) regarded *L. chaperi* as a subgenerically problematic species, showing apparent nuclear characteristics of *Nephrolepidina* although being too early stratigraphically, to form a part of the bioseries leading to *Nephrolepidina*. Figures 17.4 and 17.5 show stolons leading from the deuteroconch of *L. (N.) chaperi* to adauxiliary chambers that are not in the equatorial plane of the test. We agree with a number of other authors (e.g., Hose and Versey, 1956, p. 24; Wright, 1966, 1975; Frost and Langenheim, 1974; Butterlin, 1987, 1990) that *L. (N.) chaperi* is a probable ancestor for the Oligocene eulepidinids, the late Eocene *L. (N.) chaperi* evolving into *L. (Eulepidina) undosa*. In the Burnt Hill–Kinloss section in Jamaica this evolution can be traced through a succession of carbonates where the evidence as to the late Eocene and Oligocene ages is provided by other foraminifera, such as *L. (L.) macdonaldi, L. (L.) canellei, L. (N.) yurnagunensis, Heterostegina ocalana,* and *Heterostegina antillea.*

Distribution. Late Eocene of the White Limestone Group.

Subgenus *Eulepidina* H. Douvillé, 1911

Among the lepidocyclines the taxonomic rank of *Eulepidina* has probably been the subject of most controversy. Eames and others

Figure 22. 1, 2, *Lepidocyclina (Eulepidina) undosa* Cushman: 1, ×30; 2, ×18, both sections through selliform individuals, Browns Town Limestone. 3, *Lepidocyclina (Nephrolepidina) yurnagunensis* Cushman, ×25, the form called *L. (N.) yurnagunensis morganopsis* Vaughan, Browns Town Limestone. 4, *Lepidocyclina (Nephrolepidina)* sp., possibly *L. (N.) vaughani*, ×20, oblique section, Browns Town Limestone. 5, *Lepidocyclina (Nephrolepidina) chaperi* Lemoine and R. Douvillé, ×17, axial section, Swanswick Limestone. 6, 7, *Lepidocyclina canellei* Lemoine and R. Douvillé: 6, ×32, oblique section, ER126, Montpelier Formation; 7, ×14, axial section, well GRS 13 at 18 m (60 ft), Browns Town Limestone.

Figure 23. 1–4, *Lepidocyclina (Lepidocyclina) pustulosa* (H. Douvillé): 1, 3, ×25, axial sections, WR403, Swanswick Limestone; 2, ×25, Swanswick Limestone, Ipswich, St. Elizabeth; 4, ×49, equatorial section, ER39, Bonny Gate Limestone. 5–7, *Lepidocyclina (Lepidocyclina) macdonaldi* Cushman, axial sections: 5, 6, ×35, ER239; 7, ×24, WR354, Swanswick Limestone.

(1962b) regarded *Eulepidina* as being of generic rank but kept *Nephrolepidina* as a subgenus of *Lepidocyclina* contending (p. 300) that *Eulepidina* can be distinguished from advanced species of *Nephrolepidina* by its much larger embryonic chambers, complete or almost complete embrace of the protoconch and deuteroconch, thicker embryonic chamber walls, and the large number of intercameral foramina ("stolons") and their associated ridges typical of the chambers in the equatorial layer. Eames and others (1968) used the same scheme, although in 1962b they had admitted that there are imperfections in the parameters used to distinguish *Nephrolepidina* and *Eulepidina*. Hanzawa (1962) recognized seven genera of lepidocyclinids by raising both *Nephrolepidina* and *Eulepidina* to generic rank. In several papers Cole (1952, 1957a, 1958a, 1960b, 1961, 1964b) has regarded *Eulepidina* as a subgenus of *Lepidocyclina* and considered *Nephrolepidina* to be a junior synonym of *Eulepidina*.

Except for the geometry of the embryonic apparatus, and the size of the equatorial chambers, no significant morphological distinction can be made between species of *Eulepidina* and *Nephrolepidina* on the one hand, and species of *Lepidocyclina (Lepidocyclina)* on the other. Many of the diagnostic parameters noted by Eames and others (1962b) can be attributed to differences in depositional environment rather than genetic difference of taxonomic rank. Size of the test, size of embryonic and equatorial chambers, thickness of equatorial layer, and number of intercameral foramina are directly relatable to environmental factors such as energy conditions generated by waves and currents, associated animals and plants, and light intensity.

However, the geometry of the protoconch and deuterconch seems to be a constant parameter in lepidocycline populations within different depositional environments. Consequently, a case can be made for considering *Nephrolepidina* and *Eulepidina* as taxa of the same rank. Hence we regard *Eulepidina* and *Nephrolepidina* as subgenera of *Lepidocyclina*. We distinguish *Eulepidina* from the subgenera *Nephrolepidina* and *Lepidocyclina* principally by the degree to which the protoconch is embraced by the deuteroconch (eulepidine in *Eulepidina*), and recognize a single species, *L. (Eulepidina) undosa*.

Lepidocyclina (Eulepidina) undosa Cushman, 1919
(Figures 22.1, 2)

Lepidocyclina undosa Cushman, 1919b, p. 65, Plate 2, Fig. 1a.
Lepidocyclina (Nephrolepidina) undosa Cushman, Vaughan, 1928a, Plate 48, Fig. 3.
L. (N.) crassata Cushman, Vaughan, 1928a, Plate 45, Figs. 4,5.
Lepidocyclina (Eulepidina) undosa Cushman, Cole, 1945, p. 43–44, Plate 1, Figs. 14–15; Plate 2, Fig. 8; Plate 8, Fig. 7; Plate 11, Fig. 8 (references and synonymy); Cole, 1952, p. 30, Plate 22, Figs. 6–8; Sachs, 1959, p. 401, Plate 34, Figs. 2,5; Sachs and Gordon, 1962, p. 15–16, Plate 1, Fig. 9; Cole, 1963b, p. 158–159.
Eulepidina undosa Cushman, Hanzawa, 1962, Plate 2, Fig. 40; Eames and others, 1968, p. 296–297, Plate 49, Figs. 6–7; Plate 57, Figs. 7–8.
Lepidocyclina gigas Cushman, 1919b, p. 64, Plate 1, Figs. 3–5; Plate 5, Fig. 4; Cole, 1945, p. 44–45, Plate 8, Figs. 5–6 (references and synonymy); Cole, 1952, p. 30, Plate 22, Fig. 9; Eames and others, 1968, p. 299.
Lepidocyclina favosa Cushman, 1919b, p. 66, Plate 3, Fig. 1b; Plate 15, Fig. 4.
Lepidocyclina (Eulepidina) favosa Cushman, Cole, 1945, p. 41–43, Plate 4, Figs. 3,4,7,11; Plate 8, Figs. 1–2; Plate 9, Figs. 1–7; Plate 10, Figs.

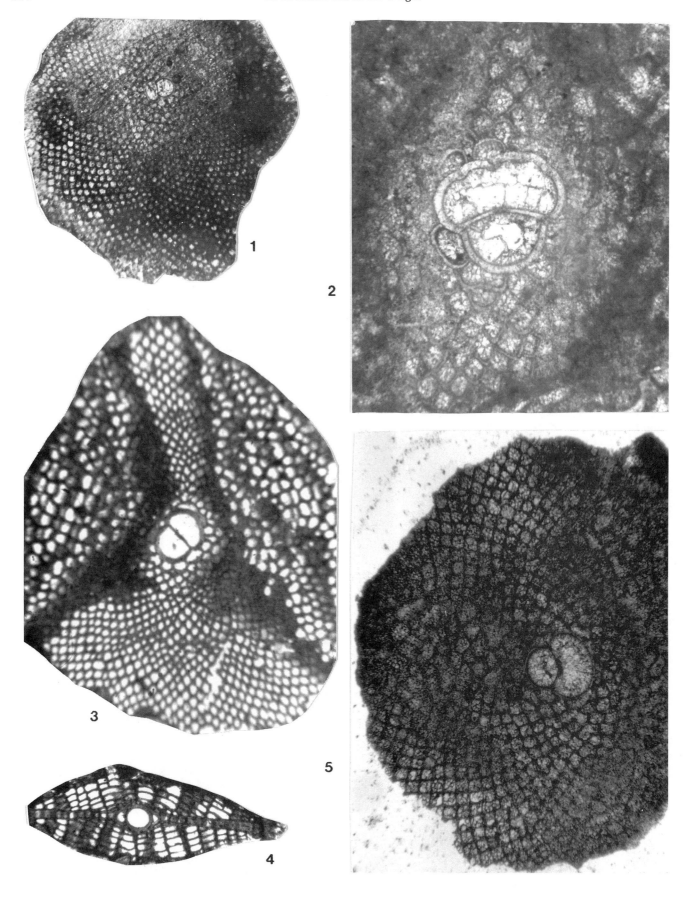

1–9; Plate 11, Fig. 9 (synonymy); Cole, 1952, p. 30, Plate 22, Figs. 1–5.
Eulepidina favosa Cushman, Hanzawa, 1962, Plate 2, Fig. 39; Eames and others, 1968, p. 296, Plate 57, Fig. 6.

Remarks. This species is a common index foraminifer of the Oligocene of Jamaica. The species shows a large degree of variability, particularly in size of the embryonic chambers, and size and degree of inflation of the test. In the Burnt Hill–Kinloss section the larger microspheric forms are usually flattened, whereas the megalospheric forms are generally markedly selliform. Traditionally the "group" has been divided into two form species, a relatively flat *L. (E.) undosa* and an inflated *L. (E.) favosa,* with a greater number of lateral layers of chambers. These original distinctions were proposed by Cushman (1919b) for specimens from the Antigua Limestone, Antigua. Cole and Applin (1961) demonstrated that both forms had similar embryonic chambers as well as equatorial chambers and formally suggested *L. (E.) favosa* as a junior synonym of *L. (E.) undosa.* Cushman (1919a, b) also described *L. gigas* at the same time as the above forms from Hodges Bluff, Antigua. *Lepidocyclina gigas* is associated with *L. favosa* and *L. undosa* as recognized by Sachs and Gordon (1962). However, some subsequent authors (e.g., Eames and others, 1968) regard the three forms as distinct species.

Distribution. The species is abundant and widespread in the Oligocene of Jamaica. It disappears at about the *Globorotalia kugleri* (N4) zone horizon (Bolli and Saunders, 1985) after a narrow overlap with the uniserial *Miogypsina panamensis,* so that its range extends into the basal part of the Miocene.

Figure 24. 1, *Lepidocyclina (Lepidocyclina) pustulosa* (H. Douvillé), ×29, a variety close to *L. peruviana* Cushman, ER1366, Chapelton Formation. 2–4, *Lepidocyclina (Nephrolepidina) yurnagunensis* Cushman: 2, ×120, equatorial section of a morphologically advanced specimen with two prominent adauxiliary chambers, ER379, Browns Town Limestone, middle to late Oliogocene, Red Gal Ring section; 3, 4, ×42, equatorial and axial sections of two specimens, ER621, from an early Oligocene horizon in the Bonny Gate Limestone. 5, *Lepidocyclina* sp. cf. *L. pustulosa*?, ×52, St. Helens borehole at 256 m (840 ft), Clarendon.

Family Elphidiidae Galloway, 1933

Genus *Pellatispirella* Hanzawa, 1937

Pellatispirella matleyi (Vaughan, 1929)
(Figures 26.1–5)

Camerina matleyi Vaughan, 1929a, p. 376, Plate 39, Figs. 2–7.
Pellatispirella matleyi (Vaughan), Cole, 1956b, p. 241 (synonymy), Plate 32, Figs. 1–8; Plate 33, Figs. 1,4,6,8,9; Plate 34, Figs. 2, 4–10. Hanzawa, 1937, Plate 21, Figs. 4–7; Butterlin, 1960, p. 86–87, Plate 1, Figs. 1–5; Robinson, 1974a, p. 290, Plate 6, Fig. 2; Plate 7, Fig. 6.
Miscellanea antillea Hanzawa, Cizancourt, 1948, p. 667–668, Plate 23, Figs. 4,7,12.
Miscellanea hedbergi Cizancourt, 1948, p. 669, Plate 23, Figs. 1–3.
Miscellanea matleyi Vaughan, Vaughan and Cole, 1941, p. 32–33, Plate 6, Fig. 1; Caudri, 1944, p. 369–371.
Miscellanea nicaraguana Cizancourt, 1948, p. 669, Plate 23, Figs. 5,6,8–11,13.

Remarks. This species was redescribed in detail by Cole (1956b). Additional illustrations to show the structure of the test wall are given here. Reexamination of Cizancourt's (1948) types of *Miscellanea hedbergi* and *M. nicaraguana* from the Twara-1 well, and many other specimens seen from this and other wells on the Nicaragua Rise, presently supports Cole's (1956b) contention that only one species of *Pellatispirella* is recognizable. The first appearance of this species in central Jamaica was used to define the base of Robinson's (1977) Zone 2.

Distribution. In Jamaica this species ranges from the base of the middle Eocene (Chapelton Formation) to the top of the Eocene.

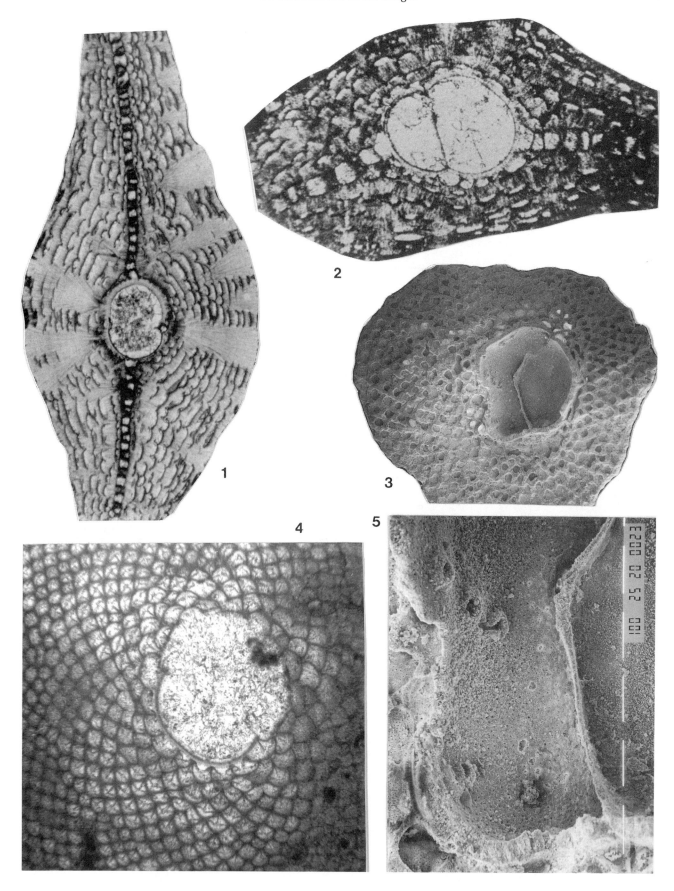

Family Rotaliidae Ehrenberg, 1839

Genus *Neorotalia* Bermúdez, 1952

Neorotalia mexicana (Nuttall) 1928
(Figure 27.7)

Rotalia mexicana Nuttall, 1928, p. 374, Plate 50, Figs. 6–8.
Rotalia mexicana Nuttall var. *mecatepecensis* Nuttall, 1932, p. 26, Plate 4, Figs. 11–12.
Neorotalia mexicana (Nuttall), Bermúdez, 1952, p. 75, Plate 12, Fig. 4.
Pararotalia mexicana (Nuttall), Poag, 1966, p. 414, Plate 6, Figs. 11–19.

Remarks. Because most specimens are observed in random thin sections, the distinction between *N. mexicana* and *N. mecatepecensis* may be arbitrary and the two forms are here considered together. *N. mecatepecensis* is normally larger, and may show small subsidiary chambers as discussed by Barker and Grimsdale (1937, Plate 9, Fig. 8). Hottinger and others (1991) have recently given reasons for retaining *Neorotalia*, of which *Rotalia mexicana* is the type, as a genus distinct from *Pararotalia* Le Calvez.

Distribution. Locally common in the lower part of the Browns Town and upper part of the Bonny Gate Limestone, ?upper Eocene and Oligocene. Although there is overlap in the ranges of this species and *Miogypsinoides*, we have not seen undoubted in situ occurrences of *N. mexicana* in sediments younger than earliest Miocene.

Family Miogypsinidae Vaughan, 1928

Genus *Miogypsinoides* Yabe and Hanzawa, 1928

Miogypsinoides sp. cf. *M. bermudezi* (Drooger) 1951
(Figures 27.1–4)

?*Miogypsina* (*Miogypsinella*) *bermudezi* Drooger, 1951, p. 357, Figs. 1–6.
Miogypsinoides cf. *M. bermudezi* Drooger, Akers and Drooger, 1957, p. 670, 674.
Miogypsina (*Miogypsinoides*) cf. *bermudezi* Drooger, Barker, 1965, p. 316, Plate 1, Figs. 9, 10.

Description. Test of average size, flattened lenticular, with excentric umbonal region. In horizontal sections the equatorial layer consists of a peripheral proloculus, followed by a spire of 9 to 14 undivided nepionic chambers, succeeded in turn by a flaring set of rhombic chambers forming the rest of the equatorial layer. In vertical sections the chambers are moderately high and the lateral walls are thick, so that the chamber cavity occupies about one-third of the total cross-sectional distance. Lateral chambers absent.

Critical measurements made on 23 near-equatorial thin sections in limestone indicate proloculus diameter is 0.38 mm, average number of spiral chambers (Mx of Drooger, 1963) is 12.3 (extremes of 9 and 14), and position of largest nepionic chamber (Mz) is 8.

Remarks. The description and dimensions of our specimens are close to those of *M.* cf. *M. bermudezi* of Akers and Drooger (1957). They differ from *M. complanata* Schlumberger (Drooger, 1951; Barker, 1965) and *M. butterlinus* Salmeron (1972), which both have 16 or more spiral chambers. Excluding the rest of the test morphology, the Mx value would place our specimens in Drooger's *M. bantamensis* group

Figure 25. 1–5, *Lepidocyclina* (*Nephrolepidina*) *chaperi* Lemoine and R. Douvillé: 1, ×40, axial section; 2, ×42, slightly oblique axial section; 3, ×31, natural equatorial section of a megalospheric individual, ER176, Bonny Gate Limestone; 5, ×136, enlargement of 3 showing stolons leading from the deuteroconch into adauxiliary chambers. Note also the pores in both deuteroconch and the floors of the chambers of the equatorial layer. 4, ×49, equatorial section, WR354.

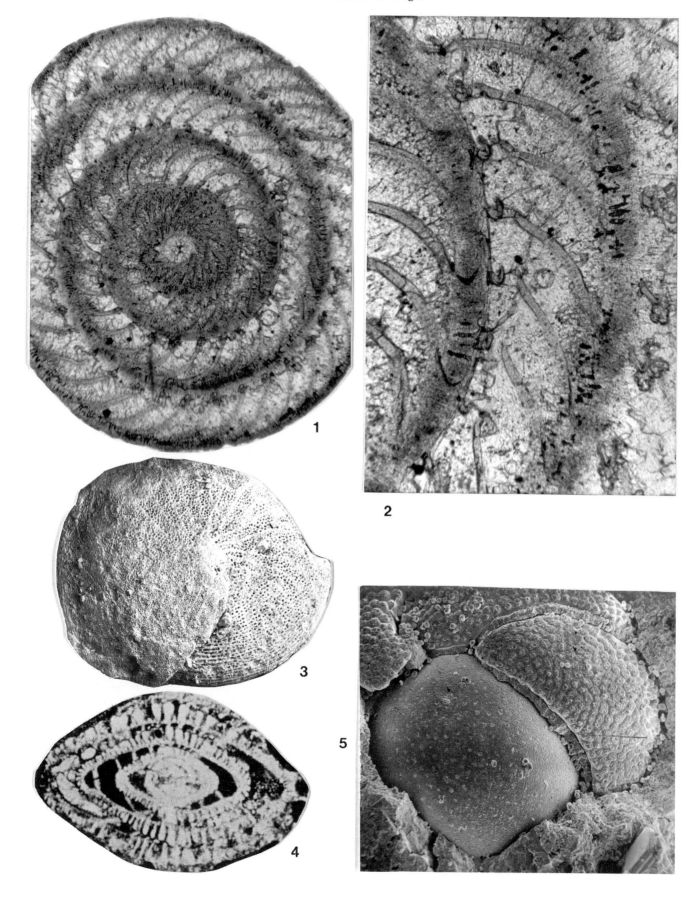

(Drooger, 1963, p. 317). Our Mx value is identical with that obtained by Akers and Drooger (1957, p. 674) for *M.* cf. *M. bermudezi* from their well number 4.

Distribution. Local occurrence in the upper Browns Town, and in the Montpelier Formation, northern Jamaica. Described and figured specimens are from locality ER573, Montpelier Formation, northwest Jamaica, late Oligocene.

Family Nummulitidae de Blainville, 1827

The Nummulitidae are a Cenozoic family of calcareous perforate foraminifera, with a planispiral coil of chambers, which have well-developed spiral and sutural septal canals, and a marginal cord. They can be divided into forms with simple, planispiral coils, comprised of undivided chambers (Nummulitinae), and forms with median chambers divided into chamberlets (Heterostegininae). Forms with divided chambers probably evolved independently several times from forms with undivided chambers. Lateral chambers (cubicula of Banner and Hodgkinson, 1991) are developed in some genera.

Key to Jamaican genera:

1. With undivided median chambers (Nummulitinae)
   a) Marginal cord prominent, with coarse canal system .... *Ranikothalia*
   b) Marginal cord moderate to fine
      i) involute, relatively tightly coiled, with numerous whorls, at least in the microspheric (B) generation, which is larger than the A generation .... *Nummulites*
      ii) involute, coiling relatively lax and variable .... *Palaeonummulites*
2. With median chambers divided into chamberlets (Heterostegininae)
   a) without lateral chambers .... *Heterostegina* (sensu lato)
   b) with well-developed lateral chambers .... *Spiroclypeus*

Subfamily Nummulitinae de Blainville, 1827

Genus *Ranikothalia* Caudri, 1944

Ranikothalia catenula (Cushman and Jarvis), 1932
(Figures 28.1–7)

Operculina catenula Cushman and Jarvis, 1932, p. 42, Plate 12, Figs. 13a, b.
Pellatispirella antillea Hanzawa, 1937, p. 116, Plate 20, Figs. 8–10; Plate 21, Fig. 1.
Camerina pellatispiroides Barker, 1939, p. 325, Plate 20, Fig. 10; Plate 22, Fig. 4.
Miscellanea tobleri Vaughan and Cole, 1941, p. 35,36, Plate 4, Figs. 5–7; Plate 7, Fig. 1.
Camerina catenula (Cushman and Jarvis), Cole, 1969, p. 78,79, Plate 17, Figs. 1–4, 6,8.
Ranikothalia antillea (Hanzawa), Caudri, 1975, Plate 6, Figs. 4–6.

Remarks. Caudri (1975) considered *R. antillea, R. tobleri,* and *R. soldadensis* to be varieties of one species. *R. antillea* represents robust speci-

Figure 26. 1–5, *Pellatispirella matleyi* (Vaughan): 1, ×53, median section; 2, ×156, enlargement of 1 to show "hook-shaped lip" on countersepta; 3, ×38, external view of an individual; all from St. Helens borehole at 259 m (850 ft), Chapelton Formation; 4, ×40, axial section showing well-developed pectination of the spiral sheet, WS 83, Swanswick Limestone; 5, ×400, natural internal casts of the proloculus, deuteroloculus and first spiral chamber of an individual from the Touche-1 well, Nicaragua (collection of de Cizancourt, Natural History Museum, Basel).

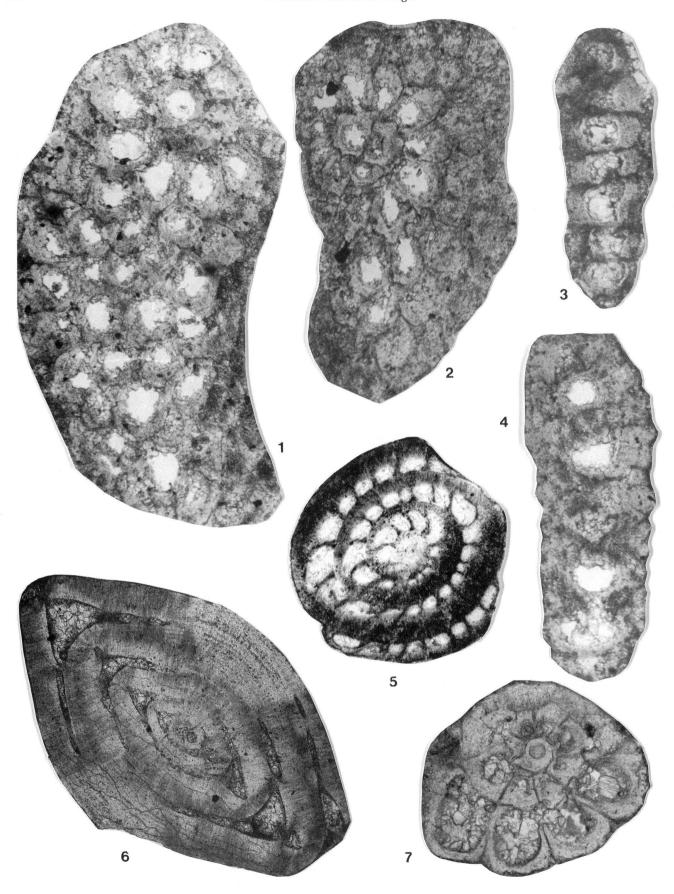

Figure 27. 1–4, *Miogypsinoides* sp. cf. *M. bermudezi* (Drooger), ×95, accidental equatorial and off-center axial sections, ER573, Montpelier Formation, Trelawny, 5, 6, *Palaeonummulites dia* (Cole and Ponton): 5, ×40, near-equatorial section; ER394; 6, ×62, axial section, ER379, both from Browns Town Limestone, Red Gal Ring. 7, *Neorotalia mexicana* (Nuttall), ×122, oblique section, ER393, Browns Town Limestone, Red Gal Ring, St. Andrew.

mens of the A form, *R. tobleri* a more flaring variety, of both A and B forms, and *R. soldadensis* a large, flat B form with a thickened rim. *Operculina catenula* Cushman and Jarvis has nomenclatural priority over the other names. Cole (1969) illustrated near-topotypes of *R. catenula,* which resemble the *R. antillea* and *R. tobleri* forms. Externally *R. catenula* is similar to weathered specimens on limestone from the Chepstow Formation, eastern Jamaica, while our sectioned material resembles *R. antillea, R. tobleri,* and *R. catenula.* We regard *R. antillea, R. pellatispiroides,* and *R. tobleri,* at least, as junior synonyms of *R. catenula.*

Distribution. Locally common in the Nonsuch and Chepstow Limestones, northeastern Jamaica; rare in limestone lenses in the base of the Richmond Formation, near Arntully, southern end of the Wagwater Belt. In all cases where independent dating is possible, *Ranikothalia*-bearing rocks have proved to be of late Paleocene age (NP5-NP7 zones, Jiang and Robinson, 1987; Robinson and Jiang, 1990).

Genus *Nummulites* Lamarck, 1801

We use the generic name *Nummulites* for Eocene forms characterized by involute tests in which the spiral sheet increases in size at a low, more or less constant rate, producing numerous whorls in which the chambers are about equidimensional in spiral section, with a more or less constant height in the later whorls. The B (microspheric) generation is larger than the A (megalospheric) generation.

Nummulites macgillavryi (M. G. Rutten), 1935
(Figures 29.1–3)

Camerina macgillavryi M. G. Rutten, 1935, p. 530, Plate 59, Figs. 6–10.

Remarks. In this species the B form is very much larger than the A form, which has a proloculus much larger than any other American nummulitid. Our specimens from the Font Hill Formation have proloculus sizes ranging from 0.65 to 0.9 mm. The range of proloculus diameters for the types is 0.7 to 1.03 mm. Blondeau (1982) considered this species to be possibly conspecific with *Nummulites gizehensis* (Forskal) from the Tethys province.

Distribution. In blocks from turbidites in the Font Hill Formation, St. Thomas Parish, Jamaica, and in northern Jamaica north of the Duanvale fault, middle Eocene.

Nummulites striatoreticulatus L. Rutten, 1928
(Figures 29.5; 30.6)

Nummulites striatoreticularus L. Rutten, 1928, p. 1068.
Camerina striatoreticulata (L. Rutten), Cole, 1958c, Plate 32, Figs. 6–8.

Remarks. This species is distinguished from *N. macgillavryi* by its smaller size, much smaller size difference between A and B forms, and by the much smaller proloculus diameter (0.05 to 0.2 mm). In random sections it may be difficult to distinguish from *Palaeonummulites willcoxi* (Heilprin). *R. willcoxi* has a coil that expands at a slightly more rapid rate, containing more numerous chambers, which are relatively higher and narrower than in *N. striatoreticulatus.* The spire of *N. striatoreticulatus* expands at a very low rate, with the ratio chamber height in last whorl/ chamber height in first whorl being relatively constant at about 2 (Wright and Switzer, 1971, Table 1). In the final whorl the chambers remain at about the same height as in the penultimate whorl, or may even become lower (Fig. 29.5).

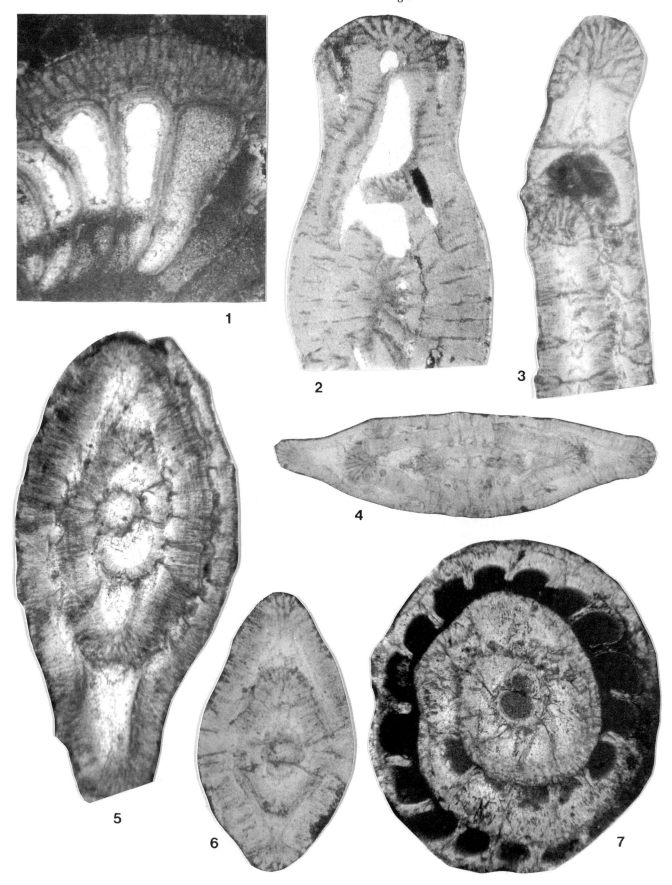

Figure 28. 1–7, *Ranikothalia catenula* (Cushman and Jarvis): 1, partial equatorial section; 2, partial oblique section, ×43, ER94, Nonsuch Limestone; 3, ×43, partial axial section; 5, ×49, oblique section; 6, ×43, axial section; 7, ×43, equatorial section, ER209A, Chepstow Limestone; 4, ×25.5, M66, limestone at base of Richmond Formation, Cedar Valley, St. Thomas (collection of M. Nath).

Distribution. This species is found in the Swanswick, Font Hill, and lower part of the Bonny Gate Limestones (particularly the Gibraltar facies), late middle to late Eocene.

Genus *Palaeonummulites* Schubert, 1908

Palaeonnumulites Schubert, 1908, is accepted as a senior synonym of the widely used name *Operculinoides* Hanzawa, 1935 (Loeblich and Tappan, 1988, p. 686; Eames and others, 1962a, p. 50, Plate 1). In a statistical analysis of three species of nummulitid in Jamaica, *Palaeonummulites willcoxi, P. floridensis,* and *N. striatoreticulatus,* Wright and Switzer (1971) concluded that tightness of coiling and total number of whorls proved the most useful of eight measured parameters in discriminating between any two groups at the specific level. They also concluded that there was little basis for distinguishing between *Nummulites* and *Operculinoides* (=*Palaeonummulites*) at the generic level.

Nevertheless, we use *Palaeonummulites,* as Cole (1958c) used the name *Operculinoides,* for small, involute nummulitids with comparatively few whorls of chambers, which have a variable rate of increase in height, but increase more rapidly than in *Nummulites* sensu stricto. Cole (1958c, p. 263) gave a key for distinguishing the various American species, which we have followed here.

Palaeonummulites floridensis (Heilprin), 1885
(Figures 29.4; 30.1–3)

Nummulites floridensis Heilprin, 1885, p. 321.
Operculinoides cushmani (Cole), Cole, 1956a, Plate 30, Figs. 11–13; Plate 31, Figs. 5,6; Cole, 1958a, Plate 18, Figs. 1,2,7–16; Plate 19, Figs. 1,2,4,,5,7–14; Plate 20, Figs. 5–9, 13–15, 17–20; Plate 22, Figs. 4,5; Cole, 1958c, Plate 33, Fig. 2; Wright and Switzer, 1971, Plate 1.
Operculinoides jennyi Barker, Cole, 1956a, Plate 31, Figs. 7–9.
Operculinoides floridensis (Heilprin), Cole, 1958c, Plate 33, Fig. 2.

Remarks. Previously recorded from Jamaica by Cole (1956a) as *Operculinoides cushmani* and *O. jennyi,* this species is characterized by having about three (more than 2.25 by Cole's definition, 1958c) rapidly enlarging whorls of high, narrow chambers. The ratio chamber height in last whorl/chamber height in first whorl is variable, but typically as much as 10 (Wright and Switzer, 1971). The species varies from forms that have a very thin, fragile test, with a prominent flange, to others that are much more robust, with a less pronounced flange to the test. Fragile specimens occur in shaly layers; thick-walled forms occur in turbidite sediments rich in shallow-water coralline algal material.
Distribution. The species is common in the Font Hill Formation, Bonny Gate, Troy/Claremont, and Swanswick Limestones; late middle to late Eocene.

Palaeonummulites willcoxi (Heilprin), 1882
(Figures 29.6; 30.4, 5)

Nummulites willcoxi Heilprin, 1882, p. 190, text Figs. 1, 2; Cole, 1958c, Plate 33, Figs. 1,3–12; Plate 20, Fig. 11; Plate 22, Fig. 3.
Operculinoides moodybranchensis Gravell and Hanna, 1935, p. 332, Cole, 1952, Plate 1, Figs. 10–19.

Remarks. Differences between this species and *Nummulites striatoreticulatus* have been noted under the latter species. In the analysis by Wright and Switzer (1971, Table 1) the ratio chamber height in last whorl/

chamber height in first whorl varies between about 2.5 and 4. Proloculus sizes are similar on average to those of *N. striatoreticulatus* but show a greater range of variation.

Distribution. This species occurs in the Font Hill, Swanswick, and Bonny Gate Limestones, middle to late Eocene.

Palaeonummulites dia (Cole and Ponton), 1930
(Figures 27.5, 6)

Operculinella dia Cole and Ponton, 1930, p. 37.
Operculinoides dia (Cole and Ponton), Cole, 1958c, Plate 34, Figs. 2–4,6,9.

Remarks. Cole (1964a) placed his *Operculinoides dia* in synonymy with *Nummulites panamensis* Cushman, 1918, a proposal that would require further study. K. M. Persad (1969) was able to distinguish two species of *Palaeonummulites* in samples from the late Oligocene Antigua Formation of Antigua, eastern Caribbean, using biometrical analysis of coiling and chamber development. Our material, in hard limestones, is not amenable to statistical treatment.

Distribution. Inflated robust specimens are common in the Browns Town Limestone, late Oligocene.

Subfamily Heterostegininae Galloway, 1933

Genus *Heterostegina* d'Orbigny, 1826

The middle Tertiary species considered here are involute forms that probably evolved from a middle Eocene nummulitid such as *Operculina gomezi* (sensu Hottinger, 1977; Herb, 1978). We are of the opinion that *Heterostegina* d'Orbigny, as used by most workers, is polyphyletic, and that d'Orbigny's genus, sensu stricto, is of late Neogene (possibly Quaternary) origin, showing primitive nepionic features (small proloculus, large number of undivided nepionic chambers we have observed in Recent *H. suborbicularis* and *H. antillarum*), in contrast to such species as the middle Tertiary *Heterostegina antillea.*

Three main heterostiginiform groups are evident. The middle Tertiary involute forms like *Grzybowskia* (Bieda, 1950) and *Vlerkina* (Eames and others, 1968), the later Tertiary, mainly Miocene, compressed, evolute forms, originating from an *Operculina complanata*–like ancestor (*Planostegina* of Banner and Hodgkinson, 1991), and the true *Heterostegina* of relatively recent origin. Jamaican species are confined to the middle Tertiary (late Eocene and late Oligocene to early Miocene) and are referable to *Heterostegina* (*Vlerkina*) of Banner and Hodgkinson. We have not seen *Heterostegina* in the Jamaican early Oligocene.

Heterostegina (*Vlerkina*) *ocalana* Cushman, 1921
(Figure 31.4)

Heterostegina ocalana Cushman, 1921, p. 130–131, Plate 21, Figs. 15–18.
Heterostegina ocalana Cushman, Cole, 1941, Plate 11, Figs. 3–6; Cole, 1952, Plate 4, Figs. 2–18.
?*Heterostegina* (*Vlerkinella*) *kugleri* Eames and others, 1968, Plate 51, Figs. 1, 2.

Remarks. This species is characterized by a small proloculus (0.05 to 0.12 mm) and, as described by Cole (1952, p. 13) a relatively large number of undivided operculine chambers (4 to 15). Using the parameters of Chaproniere (1980), our specimens have N_0 values of 4 to 6, and

Figure 29. 1–3, *Nummulites macgillavryi* (M. G. Rutten): 1, 2, ×24, axial and near-axial sections of megalospheric individuals; 3, ×24, partial axial section of a microspheric individual, ER142, Font Hill Formation. 4, *Palaeonummulites floridensis* (Heilprin), ×38, axial section, ER39, Bonny Gate Limestone; 5, *Nummulites striatoreticulatus* L. Rutten, ×, axial section not quite centered, ER39, Bonny Gate Limestone; 6, *Palaeonummulites willcoxi* (Heilprin), ×62, axial section, ER357, Gibraltar Limestone, Red Gal Ring section.

Figure 30. 1–3, *Palaeonummulites floridensis* (Heilprin): 1, 2, ×20, equatorial section and equatorial exterior view of two individuals; 3, ×25, exterior of a specimen with spiral wall eroded away. 4, 5, *Palaeonummulites willcoxi* (Heilprin), ×35, external views of complete individuals. 6, *Nummulites striatoreticulatus* L. Rutten, ×20, equatorial section. All from BH177 GRS 25, Font Hill Formation, Trelawny.

S_{4+5} values of 2 to 3. Proloculus diameters on eight specimens average 0.054 mm, range 0.045 to 0.068 mm. Including the one illustrated here, they appear to be completely involute, not maturo-evolute as suggested by Banner and Hodgkinson (1991, p. 260). There is the possibility that more than one species, such as the incompletely illustrated *H. (Vlerinella) kugleri*, is involved in the published material, but until more material is examined we accept that only one species, *H. ocalana*, is present in the Caribbean Eocene.

Age and Distributions. This species occurs sporadically in late Eocene shelf-edge or turbidite carbonate sequences, such as the Swanswick Limestone, in association with *Lepidocyclina pustulosa, Palaeonummulites willcoxi,* and *Fabiania cassis*, together with abundant coralline algae.

Heterostegina (Vlerkina) antillea Cushman, 1919
(Figures 31.3, 5)

Heterostegina antillea Cushman, 1919b, p. 49, Plate 2, Fig. 1b; Plate 5, Figs. 1, 2; Vaughan and Cole, 1941, p. 54, Plate 15, Figs. 10–15; Sachs, 1959, Plate 34, Figs. 1, 10; Plate 35, Figs. 3–6, 10–12; Plate 36, Fig. 8; Butterlin, 1961, Plate 2, Figs. 1–3; Sachs and Gordon, 1962, Plate 3, Fig. 1; Cole, 1964a, Plate 13, Figs. 1–3, 6, 8, 11–12.
Heterostegina panamensis Gravell, 1933, p. 17–18, Plate 1, Figs. 10–11; Cole, 1957b, p. 328, Plate 25, Figs. 6–7; Butterlin, 1961, p. 14, Plate 2, Figs. 4–5.
Heterostegina texana Gravell and Hanna, 1937, p. 525, Plate 63, Figs. 1–4.
Heterostegina israelskyi Gravell and Hanna, 1937, p. 524, Plate 62, Figs. 1–4; Cole, 1957b, p. 327, Plate 25, Figs. 8, 9.

Study of mainly random sections of *Heterostegina* in Jamaica presently supports Cole's (1964a) conclusion that only one Oligo-Miocene species of *Heterostegina* occurs in the Caribbean. It is characterized by an involute test, with a proloculus diameter of 0.09 to 0.23 mm. This is normally followed by the deuteroloculus and a single, crescent-shaped, operculine chamber that extends about one-quarter of the way around the perimeter of the proloculus when viewed in equatorial section. The differences between *H. israelskyi, H. texana,* and *H. antillea* relate to test shape and size, features that are governed primarily by ecological factors, not regarded as of specific significance (Reiss and Hottinger, 1984).

Distribution. This species is common in shelf-edge limestones of late Oligocene to earliest Miocene age (Browns Town Limestone and Montpelier Formation).

Discocycliniform Species

Our review of Jamaican discocycliniform species is limited here to two species from late Paleocene localities, which have not been illustrated from Jamaica previously. Classification follows Brönnimann (1945, 1951) and Less (1987).

Family Discocyclinidae Galloway, 1928

Genus *Athecocyclina* Vaughan and Cole, 1940

Athecocyclina stephensoni (Vaughan) 1929
(Figures 31.1, 2, 6)

Discocyclina stephensoni Vaughan, 1929b, p. 16, Plate 6, Figs. 1–4.
Pseudophragmina (Athecocyclina) stephensoni (Vaughan), Vaughan, 1945, Plate 45, Figs. 3, 4.

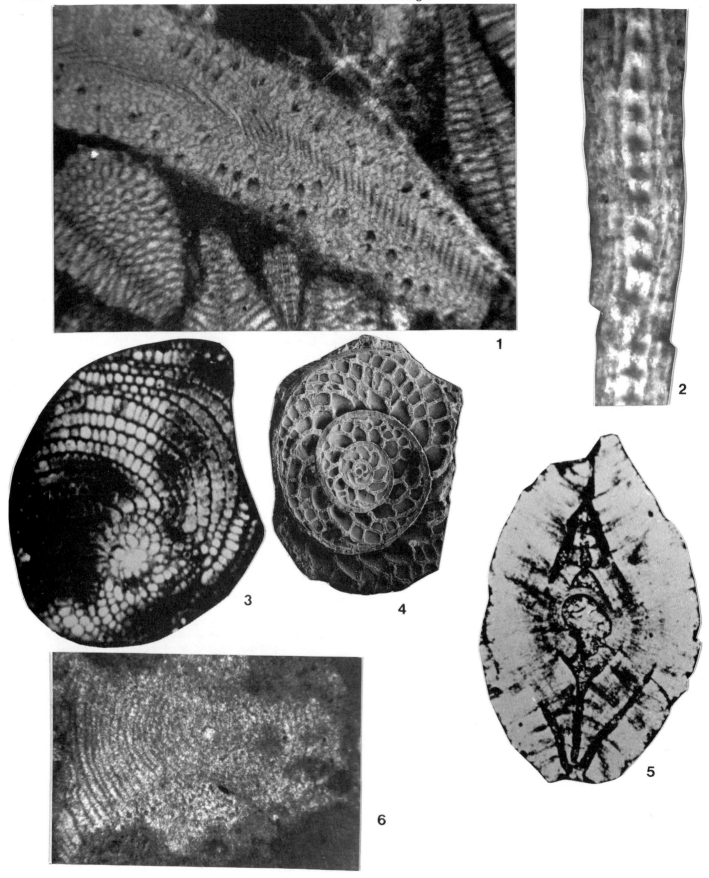

Remarks. In the Paleocene of northeastern Jamaica, a species of *Athecocyclina* occurs sporadically with other larger foraminifera. No sections passing through the nucleoconch are available, so a specific designation is not possible. In horizontal partial sections the width of the annuli average 0.05 mm. Vertical sections show the height of the equatorial layer as about 0.065 mm, with up to six layers of lateral chambers with floors that are thicker than the chamber cavities.

In a review of Paleocene larger foraminifera Cole (1959) placed all American species of *Athecocyclina* in synonymy with *A. stephensoni* (Vaughan). Although we do not necessarily agree with Cole's interpretation of this group, the observable characters in our specimens are similar to those described for the types of *A. stephensoni.*

Traces of incomplete radial septa, more or less aligned from one annulus to the next, and similar to those visible on the types of *A. stephensoni* (Vaughan, 1929b, Plate 6, Fig. 1) are present in our material. They appear to be structurally analogous to the pilintradermal partitions of the porcelaneous genus *Miosorites* (Seiglie and others, 1977).

Distribution. Athecocyclina has been collected from several localities in the Nonsuch Limestone, northeastern Jamaica, from levels assignable to the NP5 to NP7 calcareous nannofossil zones of Martini (1971), late Paleocene (Jiang and Robinson, 1987).

Family Asterocyclinidae Brönnimann, 1951

Genus *Hexagonocyclina* Caudri, 1944

Caudri's genus has been regarded as a synonym of *Discocyclina* by most workers, but we consider that it aptly distinguishes a group of early discocycliniform species characterized by subequal protoconch and deuteroconch and equatorial chambers that differ from those of other American discocycliniforms in being irregularly polygonal to hexagonal.

Whether these forms can be considered as ancestral to American *Neodiscocyclina* (of Caudri, 1972) or *Orbitoclypeus* (Less, 1987) depends on more biostratigraphic data being collected. Published occurrences from the southern Caribbean (Caudri, 1944, 1948, 1975; Cizancourt, 1951) indicate that *Hexagonocyclina* appears at a slightly higher stratigraphic level than the first species of *Neodiscocyclina* and *Orbitoclypeus.* But the stratigraphy of these occurrences is complicated by problems of reworking in turbidite and olistostrome sequences. In the northern Caribbean the stratigraphic relationships of this group, published usually as *Discocyclina cristensis,* suggest that it is among the first, if not the first to appear in Paleocene sequences (Sachs, 1957; Cole, 1959), but specific stratigraphic studies on the group are lacking. Our examples of *Hexagonocyclina* are accompanied by very rare specimens of ?*Neodiscocyclina.*

***Hexagonocyclina inflata* (Caudri, 1948)**
(Figure 32.1–6)

Bontourina inflata, Caudri, 1948, p. 477, Plate 73, Fig. 6; Plate 74, Fig. 5.
Hexagonocyclina inflata, Caudri, 1975, p. 544, Plate 3, Figs. 2,3,4,11,13; Plate 9, Figs. 2–7; Plate 10, Fig. 2.

Description. A small, inflated lenticular species of the genus *Hexagonocyclina,* characterized in equatorial sections by a figure-of-eight shaped nucleoconch, the rounded protoconch being attached to the similar or slightly smaller sized deuteroconch along about 20% of their circumferences, the common wall being slightly convex towards the deuteroconch. It has a periembryonic ring consisting of two prominent ovoid principal auxiliary chambers (p.a.c.), one at each end of the common wall between

Figure 31. 1, 2, 6, *Athecocyclina stephensoni* (Vaughan): 1, 6, ×49, oblique and partial equatorial sections; 2, ×122, partial axial section, ER2061, Nonsuch Limestone; 1, is associated with *Hexagonocyclina inflata* Caudri. 3, 5, *Heterostegina (Vlerkina) antillea* Cushman, ×40: 3, median section, WR359, Browns Town Limestone; 5, axial section, WR520, Brown Town Limestone. 4, *Heterostegina (Vlerkina) ocalana* Cushman, ×33, natural median section showing five undivided spiral chambers, ER176, Bonny Gate Limestone.

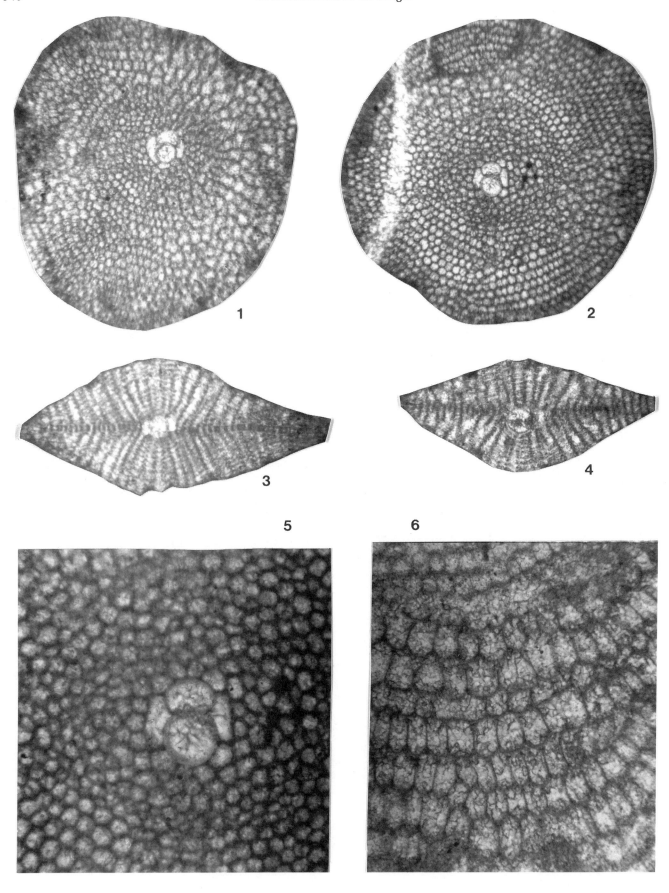

the protoconch and deuteroconch. From each p.a.c. arise two spirals, enclosing the nucleoconch with four additional chambers on the protoconch and four on the deuteroconch. Measurements of 18 centered equatorial sections gave a mean protoconch diameter (D1) of 0.118 mm and a mean deuteroconch diameter (D2) of 0.112 mm, measured along the common axis of these two chambers.

Equatorial chambers initially irregularly arranged, subarcuate to polygonal, later arranged in more or less regular annuli and hexagonal in shape, with radial dimension about equal to, or slightly more than the tangential dimension in the outermost annuli.

In vertical sections the thin-walled chambers of the lateral layers (lateral chamberlets of Ferrandez-Canadell and Serra-Kiel, 1992) are arranged in more or less regular tiers, up to as many as 18 layers, but more commonly about 12 layers, over the polar regions of the test. The range of the ratio thickness/diameter is 0.36 to 0.41.

Remarks. The specimens from the Jamaican Paleocene are identical in their features with the types described by Caudri from Trinidad, as reworked specimens in the late Eocene. The species is distinguished from *Hexagonocyclina cristensis* by a nucleus in which the deuteroconch is subcircular and normally slightly smaller than the protoconch, by the larger number of lateral chambers in an inflated test, and by the lack of rectangular chambers in the equatorial layer.

Localities. Figured specimens from locality ER2061 at side of main coast road on north side of Alligator Crawle Harbour, Portland (Fig. 2), in hard, graded, bioclastic limestones interbedded with shaly marls; abundant, associated with rare *Athecycyclina stephensoni* and very rare ?*Neodiscocyclina* sp.

Distribution. Nonsuch Limestone, late Paleocene (NP5 to NP7 zones, Jiang and Robinson, 1987).

Figure 32. 1–6, *Hexagonocyclina inflata* Caudri: 1, 2, ×49, equatorial sections; 3, 4, ×49, axial sections; 5, ×122, enlargement of 2 to show nucleus and irregular equatorial chambers of the central area; 6, ×122 another individual, showing spatulate to hexagonal chambers of the more distal part of the equatorial layer. All ER2061, Nonsuch Limestone.

ACKNOWLEDGMENTS

We have benefitted over the years from many fruitful discussions with former colleagues in Jamaica, particularly A. N. Eva, J. Krijnen, N. A. McFarlane, and H. R. Versey. Illustrations are of the authors' sample material, except where noted. Figured specimens are deposited as indicated in the figure captions. Where no depository is given they remain in the authors' collections. Photography was carried out at various times extending back to the 1960s, with assistance from facilities at University College, London (T. Barnard), Stanford University (J. C. Ingle), University of the West Indies (R. Desouza, H. L. Dixon, S. K. Donovan), Naturhistorisches Museum Basel (P. Jung), ETH Zurich (H. M. Bolli), and Florida International University (G. Draper). We thank C. G. Adams and J. B. Saunders for their very helpful reviews of the first draft of this paper.

REFERENCES CITED

Adams, C. G., 1987, On the classification of the Lepidocyclinidae (Foraminiferida) with redescriptions of the unrelated Paleocene genera *Actinosiphon* and *Orbitosiphon:* Micropaleontology, v. 33, p. 289–317.

Adams, C. G., and Racey, A., 1992, The occurrence and paleobiogeographical significance of the foraminiferid *Yaberinella* from the Eocene of Oman: Palaeontology, v. 35, p. 237–245.

Akers, W. H., and Drooger, C. W., 1957, Miogypsinids, planktonic foraminifera and Gulf Coast Oligocene-Miocene correlations: American Association of Petroleum Geologists Bulletin, v. 41, no. 4, p. 656–678.

Anisgard, H. W., 1957, *Eorupertia bermudezi,* a new foraminifera from the middle Eocene of Cuba: Cushman Foundation for Foraminiferal Research Contributions, v. 8, pt. 1, p. 1–8.

Applin, E. R., and Jordan, L., 1945, Diagnostic foraminifera from subsurface formations in Florida: Journal of Paleontology, v. 19, no. 2, p. 129–148, Plates 18–21.

Banner, F. T., and Hodgkinson, R. L., 1991, A revision of the foraminiferal subfamily Heterostegininae: Revista Española de Micropaleontología, v. 23, p. 101–140.

Barker, R. W., 1939, Species of the foraminiferal family Camerinidae in the Tertiary and Cretaceous of Mexico: U.S. National Museum Proceedings, v. 86, no. 3052, p. 305–330.

Barker, R. W., 1965, Notes on Miogypsinidae in the Gulf of Mexico region: Dr. D. N. Wadia Commemorative Volume: Mining and Metallurgical Institute of India, p. 306–342.

Barker, R. W., and Grimsdale, T. F., 1936, A contribution to the phylogeny of the orbitoidal foraminifera with descriptions of new forms from the Eocene of Mexico: Journal of Paleontology, v. 10, p. 231–247.

Barker, R. W., and Grimsdale, T. F., 1937, Studies of Mexican fossil foraminifera: Annals and Magazine of Natural History, Ser. 10, v. 19, p. 161–178, Plates 5–9.

Beckmann, J. P., 1958. Correlation of pelagic and reefal faunas from the Eocene and Paleocene of Cuba: Eclogae geologicae Helvetiae, v. 51, p. 416–421.

Bermúdez, P. J., 1952, Estudio sistematico de los foraminiferos rotaliformes: Venezuela, Boletin de Geología, v. 2, p. 1–230.

Berry, E. W., 1929, Shorter contributions to the paleontology of the Eocene of northwestern Peru: I, Solitary corals. II, Brachiopods. III, Foraminifer Gypsina: Baltimore, Maryland, Washington Academy of Science Journal, v. 19, p. 240.

Bieda, F., 1950, Sur quelques foraminifères nouveaux ou peu connus du flysch des Karpates Polonaises: Rocznik Polskiego Towarzystwa Geologicznego, v. 18, p. 167–179.

Blondeau, A., 1982, On American nummulites: Bulletin de la Société Géologique de France, v. 24, no. 2, p. 397–401.

Bolli, H. M., and Saunders, J. B., 1985, Oligocene to Holocene low latitude planktic foraminifera, in Bolli, H. M., et al., eds., Plankton stratigraphy: Cambridge Earth Science Series Cambridge University Press, p. 155–262.

Bourdon, M., and Lys, M., 1955. Foraminifères du Stampien de la carrière de la Souys-Floirac (Gironde): Compte Rendu des Séances de la Société Géologique de France 1955, p. 336–338.

Brady, H. B., 1876, Description d'une nouvelle espèce de foraminifère des couches Miocènes de la Jamaïque: Société de Malacologie Belgique, Annales (Memoires), Bruxelles, Belgique, tome 11 (ser. 2, tome 1), p. 1–103.

Brönnimann, P., 1945, Zur Frage der verwandschaftli hen Beziehungen zwischen *Discocylina* s.s. und *Asterocyclina*: Ecologae geologicae Helvetiae, v. 38, p. 579–615.

Brönnimann, P., 1951, A model of the internal structure of *Discocyclina* s.s.: Journal of Paleontology, v. 25, p. 208–211.

Butterlin, J., 1960, Presencia de *Pellatispirella* Hanzawa 1937 en Mexico: Boletin del Asociacion Mexicana Geologos Petroleros, v. 12, nos. 3, 4, p. 85–88.

Butterlin, J., 1961, Grandes foraminiferos del Pozo Palizada Numero 2, Municipio de Palizada, Estado de Campeche: Paleontología Mexicana, no. 10, p. 1–27.

Butterlin, J., 1970, Macroforaminiferos y edad de la formacion Punta Mosquito (grupo Punta Carnero) de la isla de Margarita, Venezuela: Boletin Informativo del Asociacion Venezolana de Geología, Minería y Petroleo, v. 13, no. 10, p. 273–315, 8 Plates.

Butterlin, J., 1987, Origine et evolution des lépidocyclines de la region des Caraïbes. Comparaisons et relations avec les lépidocyclines des autres régions du monde: Revue de Micropaléontologie, v. 29, no. 4, p. 203–219.

Butterlin, J., 1990, Problèmes poses par la systématique de la famille Lepidocyclinidae (Foraminiferida): Revista Española de Micropaleontología, v. 22, no. 1, p. 101–126.

Butterlin, J., and Moullade, M., 1968, Les Orbitolinidae de l'Eocène de la région des Caraïbes: Archives des Sciences, Genève, v. 21, p. 5–20, 3 Plates.

Caudri, C.M.B., 1944, The larger foraminifera from San Juan de los Morros, State of Guárico, Venezuela: Bulletins of American Paleontology, v. 28, no. 114, p. 355–404.

Caudri, C.M.B., 1948, Note on the stratigraphic distribution of *Lepidorbitoides*: Journal of Paleontology, v. 22, pt. 4, p. 473–481.

Caudri, C.M.B., 1972, Systematics of the American Discocyclinas: Ecologae geologicae Helvetiae, v. 65, pt. 1, p. 211–219.

Caudri, C.M.B., 1974, The larger foraminifera of Punta Mosquito, Margarita Island, Venezuela: Verhandlungen der Naturforschenden Gesellschaft in Basel, v. 84, p. 293–320.

Caudri, C.M.B., 1975, Geology and paleontology of Soldado Rock, Trinidad (West Indies), Part 2: The larger foraminifera: Ecologae geologicae Helvetiae, v. 68, no. 3, p. 533–589.

Caus, E., 1979, *Fabularia roselli* n. sp. et *Pseudolacazina* n. gen., Foraminifères de l'Eocène moyen du nord-est de l'Espagne: Géobios, v. 12, no. 1, p. 29–45.

Chapman, F., and Crespin, I., 1930, Rare foraminifera from deep borings in the Victorian Tertiaries—*Victoriella* gen. nov., *Cycloclypeus communis* Martin, and *Lepidocyclina borneensis* Provale: Royal Society of Victoria, Proceedings, v. 42, p. 110–115, Plates 7–8.

Chaproniere, G.C.H., 1980, Biometrical studies of early Neogene larger Foraminiferida from Australia and New Zealand: Alcheringa, v. 4, p. 153–181.

Cizancourt, M., de, 1948, Matèriaux pour la paléontologie et la statigraphie des regions Caraïbes: Bulletin de la Société Géologique de France, ser. 5, v. 18, p. 663–674.

Cizancourt, M., de, 1951, Grands foraminifères du Paleocène, de l'Eocène inférieur et de l'Eocène moyen du Venezuela: Memoires de la Société Géologique de France, nouvelle serie, tome (see Hottinger 1977) 30, memoir no. 64, p. 1–68, Plates 1–6.

Cole, W. S., 1938, Stratigraphy and micropaleontology studies of two deep wells in Florida: Florida Geological Survey Bulletin, no. 16, p. 1–168.

Cole, W. S., 1941, Stratigraphic and paleontologic studies of wells in Florida: Florida Geological Survey Bulletin 19, p. 1–53, Plates 1–16.

Cole, W. S., 1942, Stratigraphic and paleontologic studies of wells in Florida—No. 2: Florida Geological Survey Bulletin, no. 20, p. 1–89, 16 Plates.

Cole, W. S., 1944, Stratigraphic and paleontologic studies of wells in Florida, pt. 3: Florida Geological Survey Bulletin, no. 26, p. 11–168, Plates 1–29.

Cole, W. S., 1945, Stratigraphic and paleontologic studies of wells in Florida, no. 4: Florida Geological Survey Bulletin, no. 28, p. 1–160, Plates 1–22.

Cole, W. S., 1952, Eocene and Oligocene larger foraminifera from the Panama Canal Zone and vicinity: U.S. Geological Survey Professional Paper no. 244, p. 1–41, 28 Plates.

Cole, W. S., 1956a, Jamaican larger foraminifera: Bulletins of American Paleontology, v. 36, no. 158, p. 205–233, Plates 24–31.

Cole, W. S., 1956b, The genera *Miscellanea* and *Pellatispirella*: Bulletins of American Paleontology, v. 36, no. 159, p. 239–254, Plates 32–34.

Cole, W. S., 1957a, Variation in American Oligocene species of *Lepidocyclina*: Bulletins of American Paleontology, v. 38, no. 166, p. 31–43, 6 Plates.

Cole, W. S., 1957b, Late Oligocene larger foraminifera from Barro Colorado Island, Panama Canal Zone: Bulletins of American Paleontology, v. 37, p. 313–338.

Cole, W. S., 1958a, Names of and variation in certain American larger foraminifera—no. 1: Bulletins of American Paleontology, v. 38, no. 170, p. 179–213.

Cole, W. S., 1958b, Larger foraminifera from Carriacou, British West Indies: Bulletins of American Paleontology, v. 38, no. 171, p. 219–233.

Cole, W. S., 1958c, Names and variation in certain American larger foraminifera, particularly the camerinids — No. 2: Bulletins of American Paleontology, v. 38, no. 173, p. 261–284.

Cole, W. S., 1959, Faunal associations and the stratigraphic position of certain American Paleocene and Eocene larger foraminifera: Bulletins of American Paleontology, v. 39, no. 182, p. 377–393.

Cole, W. S., 1960a, Revision of *Helicostegina, Helicolepidina* and *Lepidocyclina* (*Polylepidina*): Cushman Foundation for Foraminiferal Research Contributions, v. 11, p. 57–63.

Cole, W. S., 1960b, Variability in embryonic chambers of *Lepidocyclina*: Micropaleontology, v. 6, no. 2, p. 133–140.

Cole, W. S., 1961, An analysis of certain taxonomic problems in the larger foraminifera: Bulletins of American Paleontology, v. 43, no. 197, p. 373-392, 12 Plates.

Cole, W. S., 1962, Embryonic chambers and the subgenera of *Lepidocyclina*: Bulletins of American Paleontology, v. 44, p. 28-60.

Cole, W. S., 1963a, Illustrations of conflicting interpretations of the biology and classification of certain larger foraminifera: Bulletins of American Paleontology, v. 46, no. 205, p. 5-44, 14 Plates.

Cole, W. S., 1963b, Analysis of *Lepidocyclina radiata* (Martin): Bulletins of American Paleontology, v. 46, no. 208, p. 157-176, 6 Plates.

Cole, W. S., 1964a, American mid-Tertiary miogypsinid foraminifera. Classification and zonation: Cushman Foundation for Foraminiferal Research Contributions, v. 15, no. 4, p. 138-153.

Cole, W. S., 1964b, Lepidocyclinidae, *in* Moore, R. C., ed., Treatise on invertebrate paleontology, Part C, *in* Loeblich, A. R., and Tappan, H., eds., Protista 2: Geological Society of America and University of Kansas Press, p. C717-C724.

Cole, W. S., 1968, More on variation in the genus *Lepidocyclina* (larger foraminifera): Bulletins of American Paleontology, v. 54, no. 243, p. 291-327.

Cole, W. S., 1969, Internal structure, stratigraphic range and phylogenetic relationships of certain American Eocene foraminifera: Cushman Foundation for Foraminiferal Research Contributions, v. 20, pt. 3, p. 77-86.

Cole, W. S., and Applin, E. R., 1961, Stratigraphic and geographic distribution of larger foraminifera occurring in a well in Coffee County, Georgia: Cushman Foundation for Foraminiferal Research Contributions, v. 12, p. 4, p. 127-135, 2 Plates.

Cole, W. S., and Applin, E. R., 1964, Problems of the geographic and stratigraphic distribution of American middle Eocene larger foraminifera: Bulletins of American Paleontology, v. 47, no. 212, p. 5-48, Plates 1-11.

Cole, W. S., and Bermúdez, P. J., 1944, New foraminiferal genera from the Cuban middle Eocene: Bulletins of American Paleontology, v. 28, no. 113, p. 333-350.

Cole, W. S., and Gravell, D. W., 1952, Middle Eocene foraminifera from Peñon seep, Matanzas Province, Cuba: Journal of Paleontology, v. 26, no. 5, p. 708-727.

Cole, W. S., and Ponton, G. M., 1930, The foraminifera of the Marianna limestone of Florida: Florida State Geological Survey Bulletin, no. 5, p. 1-37.

Cole, W. S., and Ponton, G. M., 1934, New species of *Fabularia, Asterocyclina,* and *Lepidocyclina* from the Florida Eocene: American Midland Naturalist, v. 15, no. 2, p. 138-147, Plates 1, 2.

Cushman, J. A. (for 1918), 1919a, The larger fossil foraminifera of the Panama Canal Zone: U.S. National Museum Bulletin, no. 103, p. 89-102, Plates 34-45.

Cushman, J. A., 1919b, Fossil foraminifera from the West Indies: Washington, D.C., Carnegie Institution, Publication 291, p. 21-71, Plates 1-15.

Cushman, J. A., 1921, A new species of *Orthophragmina* from Louisiana: U.S. Geological Survey Professional Paper 128-E, p. 139, Plate 22.

Cushman, J. A., and Bermúdez, P. J., 1936, Additional new species of foraminifera and a new genus from the Eocene of Cuba: Cushman Laboratory for Foraminiferal Research Contributions. v. 12, no. 3, p. 55-63.

Cushman, J. A., and Jarvis, P. W., 1932, Upper Cretaceous foraminifera from Trinidad: U.S. National Museum Proceedings, v. 80, Art. 14, p. 1-60.

Davies, L. M., 1930, The genus *Dictyoconus* and its allies: a review of the group, together with a description of three new species from the lower Eocene beds of northern Baluchistan: Royal Society of Edinburgh Transactions, v. 56, no. 20, p. 485-505.

Davies, L. M., 1952, Foraminifera of the White Limestone of the Kingston District, Jamaica: Edinburgh Geological Society Transactions, v. 15, p. 121-132.

Defrance, J.L.M., 1820, Dictionnaire des Sciences Naturelles, v. 16: Paris, eup–fik, F. G. Levrault, p. 1-567.

Douvillé, H., 1911, Les foraminifères dans le Tertiaire des Philippines: Philippine Journal of Science, v. 6, no. 2, p. 53-80.

Douvillé, H., 1915, Les Orbitoïdes du Danien et du Tertiaire: *Orthophragmina* et *Lepidocyclina*: Comptes Rendus Hebdomadaires des Séances de l'Académie des Sciences, Paris, v. 161, p. 721-728.

Douvillé, H., 1917, Les orbitoïdes de l'Ile de la Trinité: Comptes Rendus de l'Académie des Sciences, Paris, v. 164, p. 841-847, 6 figs.

Douvillé, H., 1922, Les lépidoclines et leur evolution: un genre nouveau *"Amphilepidina"*: Comptes Rendus Hebdomadaires des Séances de l'Académie des Sciences, Paris, v. 175, p. 550-555.

Drobne, K., 1974, Les grandes Miliolides des couches paléocènes de la Yougoslavie du Nord-Ouest (*Idalina, Fabularia, Lacazina, Periloculina*): Razprave Slovenska Akademija Znanosti in Umetnosti 17, p. 129-184, Plates 1-15.

Drooger, C. W., 1951, Notes on some representatives of *Miogypsinella*: Proceedings, Koninklijke Nederlandse Akademie van Wetenschappen, Amsterdam, ser. B, v. 4, p. 357-365, 7 figs.

Drooger, C. W., 1963, Evolutionary trends in Miogypsinidae, *in* Evolutionary trends in Foraminifera: Amsterdam, Elsevier, p. 315-349, 25 figs.

Drooger, C. W., and Socin, C., 1959, Miocene foraminifera from Rosignano, northern Italy: Micropaleontology, v. 5, p. 415-426.

Eames, F. E., Banner, F. T., Blow, W. H., and Clarke, W. J., 1962a, Fundamentals of mid-Tertiary stratigraphical correlation: Cambridge University Press, 163 p.

Eames, F. E., Banner, F. T., Blow, W. H., Clarke, W. J., and Smout, A. H., 1962b, Morphology, taxonomy, and stratigraphic occurrence of the Lepidocyclinae: Micropaleontology, v. 8, pt. 3, p. 289-322, Plates 1-8.

Eames, F. E., Clarke, W. J., Banner, F. T., Smout, A. H., and Blow, W. H., 1968, Some larger foraminifera from the Tertiary of Central America: Palaeontology, v. 11, pt. 2, p. 283-305.

Eva, A. N., 1976a, The paleoecology and sedimentology of middle Eocene larger foraminifera in Jamaica: Maritime Sediments Special Publication 1, p. 467-475.

Eva, A. N., 1976b, New data on the middle Eocene Yellow Limestone Group in western Jamaica: Transactions, 7th Caribbean Geological Conference, Guadeloupe, p. 249-254.

Eva, A. N., 1976c, The biostratigraphy and paleoecology of larger foraminifera from the Yellow Limestone Group in western Jamaica [Ph.D. thesis]: Jamaica, University of the West Indies, 166 p.

Eva, A. N., 1980a, Pre-cyclical chamber arrangement in the foraminiferal genus *Polylepidina* Vaughan, 1924: Micropaleontology, v. 26, no. 1, p. 90-94.

Eva, A. N., 1980b, Structure and systematic position of *Fabularia* DeFrance, 1820: Journal of Foraminiferal Research, v. 10, no. 4, p. 261-265.

Eva, A. N., and McFarlane, N. A., 1985, Tertiary to early Quaternary carbonate facies relationships in Jamaica: Transactions, 4th Latin American Geological Conference, Port of Spain, Trinidad and Tobago, 1979, p. 210-219.

Ferrandez-Canadell, C., and Serra-Kiel, J., 1992, Morphostructure and paleobiology of *Discocyclina* Gümbel, 1870: Journal of Foraminiferal Research, v. 22, p. 147-165.

Frost, S. H., and Langenheim, R. L., 1974, Cenozoic reef biofacies. Tertiary larger foraminifera and scleractinian corals from Mexico: Northern Illinois University Press, 388 p., 123 Plates.

Galloway, J. J., 1933, A manual of the Foraminifera: Bloomington, Indiana, Principia Press, 483 p.

Glaessner, M. F., and Wade, M., 1959, Revision of the foraminiferal family Victoriellidae: Micropaleontology, v. 5, no. 2, p. 193-212, 3 Plates, 6 figs.

Gravell, D. W., 1933, Tertiary larger foraminifera of Venezuela: Smithsonian Miscellaneous Collections, v. 89, no. 11, p. 1-44.

Gravell, D. W., and Hanna, M. A., 1935, Larger foraminifera from the Moody's Branch marl, Jackson Eocene, of Texas, Louisiana, and Mississippi: Journal of Paleontology, v. 9, no. 4, p. 327-340, Plates 29-32.

Gravell, D. W., and Hanna, M. A., 1937, The *Lepidocyclina texana* horizon in the *Heterostegina* zone, upper Oligocene of Texas and Louisiana: Journal of Paleontology, v. 11, p. 517-529.

Grimsdale, T. F., 1959, Evolution in the American Lepidocyclinidae (Cainozoic Foraminifera), an interim review 1 and 2: Proceedings, Koninklijke Nederlandse Akedemie van Wetenschappen, Ser. B, v. 62, no. 1, p. 8-33.

Gümbel, C. W., von, 1870, Beiträge zur Foraminiferenfauna der Nordalpinen Eocängebilde: Abhandlungen der Mathematisch Physikalischen Classe der Koniglich Bayerischen Akademie der Wissenschaften, v. 10, no. 2, p. 1–152.

Hanzawa, S., 1935, Some fossil *Operculina* and *Miogypsina* from Japan and their stratigraphical significance: Science Reports of the Tôhoku Imperial University, 2nd scr. (Geology), v. 18, no. 1, p. 1–29, 3 Plates.

Hanzawa, S., 1937, Notes on some interesting Cretaceous and Tertiary Foraminifera from the West Indies: Journal of Paleontology, v. 11, p. 110–117.

Hanzawa, S., 1962, Upper Cretaceous and Tertiary three-layered larger foraminifera and their allied forms: Micropaleontology, v. 8, p. 129–173.

Heilprin, A., 1882, On the recurrence of Nummulitic deposits in Florida and the association of *Nummulites* with a freshwater fauna: Proceedings, Academy of Natural Science, Philadelphia, v. 34, p. 189–193.

Heilprin, A., 1885, Notes on some new Foraminifera from the nummulitic formation of Florida: National Academy of Sciences, Proceedings, Philadelphia, v. 36, p. 321–322.

Henson, F.R.S., 1950, Middle Eastern Tertiary Peneroplidae (foraminifera) with remarks on the phylogeny and taxonomy of the family [Ph.D. thesis]: Wakefield, England, Yorkshire Printing Co., Ltd., Leiden University, p. 1–70, Plates 1–10.

Herb, R., 1978, Some species of *Operculina* and *Heterostegina* from the Eocene of the Helvetic nappes of Switzerland and from Italy: Ecologae geologicae Helvetiae, v. 71, pt. 3, p. 745–767.

Hill, R. T., 1899, The geology and physical geography of Jamaica: A study of a type of Antillean development: Harvard, Museum of Comparative Zoology Bulletin, v. 34, p. 1–226.

Hofker, J., Jr., 1966, Studies on the family Orbitolinidae: Paleontographica, v. 126, p. 1–34.

Hose, H. R., and Versey, H. R., 1956, Palaeontological and lithological divisions of the lower Tertiary limestones of Jamaica: Colonial Geology and Mineral Resources, v. 6, no. 1, p. 19–39.

Hottinger, L., 1963, Quelques foraminifères porcelanés oligocènes dans la série sédimentaire prébetique de Moratalla (Espagne meridionale): Ecologae geologicae Helvetiae, v. 56, p. 963–972, Plates 1–5.

Hottinger, L., 1969, The foraminiferal genus *Yaberinella* Vaughan, 1928, with remarks on its species and its systematic position: Ecologae geologicae Helvetiae, v. 62, no. 2, p. 745–749, Plates 1–6.

Hottinger, L., 1977, Foraminiferès operculiniformes: Mémoires du Muséum national d'Histoire Naturelle (Paris), nouvelle série C, Tome 40, pl. 159 C40, p. 1–159.

Hottinger, L., and Drobne, K., 1980, Early Tertiary conical imperforate foraminifera: Razprave Slovenska Akademija Znanosti in Umetnosti, v. 23, no. 3, p. 189–276, 22 Plates.

Hottinger, L., Halicz, E., and Reiss, Z., 1991, The foraminiferal genera *Pararotalia, Neorotalia,* and *Calcarina*: Taxonomic revision: Journal of Paleontology, v. 65, p. 18–33.

Jiang, M-J., and Robinson, E., 1987, Calcareous nannofossils and larger foraminifera in Jamaican rocks of Cretaceous to early Eocene age, *in* Ahmad, R., ed., Proceedings, Workshop on the Status of Jamaican Geology: Geological Society of Jamaica Special Publication, p. 24–51.

Keijzer, F. G., 1945, Outline of the geology of the eastern part of the province of Oriente, Cuba (E. of 76° WL) with notes on the geology of other parts of the island [Ph.D. thesis]: Utrecht, University of Utrecht, and De Vliegende Hollander, and p. 1–239, Plates 1–11.

Kupper, K., 1955, Eocene larger foraminifera near Guadelupe, Santa Clara County, California: Cushman Foundation for Foraminiferal Research Contributions, v. 6, pt. 4, p. 133–139.

Lamarck, J. B., 1801, Système des animaux sans vertébres: Paris, p. 1–432.

Lehmann, R., 1961, Strukturanalyse einiger Gattungen der Subfamilie Orbitolitinae: Ecologae geologicae Helvetiae, v. 54, no. 2, p. 597–667, Plates 1–14.

Lemoine, P., and Douvillé, R., 1904, Sur le genre *Lepidocyclina* Gümbel: Société Géologique de France, Mémoir Paléontologie, v. 12, fasc. 2, Memoir no. 32, p. 1–41.

Less, G., 1987, Paleontology and stratigraphy of the European Orthophragminae: Geologica Hungarica Ser. Paleontologica, fasc. 51, 373 p.

Loeblich, A. R., Jr., and Tappan, H., 1988, Foraminiferal genera and their classification: Van Nostrand Reinhold, New York, v. 1, p. i–x, 1–970, v. 2, p. i–viii, 1–212; Plates 1–847.

Mangin, J. P., 1954, Description d'un nouveau genre de Foraminifère: *Fallotella alavensis*: Dijon Bulletin Scientifique de Bourgognc, (1952–1953), v. 14, p. 209–219.

Mann, P., and Burke, K., 1990, Transverse intra-arc rifting: Paleogene Wagwater Belt, Jamaica: Marine and Petroleum Geology, v. 7, p. 410–427.

Martini, E., 1971, Standard Tertiary and Quaternary calcareous nannoplankton zonation, *in* Farinacci, A., ed., Proceedings, Planktonic Conference, II, Roma, 1970, v. 2, p. 739–785.

McFarlane, N. A., 1974, Geology of the Dry Harbour Mountains, St. Ann [M.Sc. thesis]: University of the West Indies, 94 p.

McFarlane, N. A., 1977, Some Eocene and Oligocene faunas from central Jamaica: Memorías Segundo Congreso Latinoamericano de Geología, v. 3, p. 1393–1411.

Moberg, M. W., 1928, New species of *Coskinolina* and *Dictyoconus* (?) from Florida: Tallahassee, Florida State Geological Survey, 19th Annual Report, p. 166–175, Plates 3–5.

Montfort, P. D., de, 1808, Conchyliologie Systématique et Classification Méthodiques des Coquilles: F. Schoell, Paris, v. 1, lxxxvii + 409 p.

Mulder, E.F.J., de, 1975, Microfauna and sedimentary–tectonic history of the Oligo-Miocene of the Ionian islands and western Epirus (Greece): Utrecht Micropaleontological Bulletin 13, p. 1–139.

Nuttall, W.L.F., 1928, Notes on the Tertiary Foraminifera of southern Mexico: Journal of Paleontology, v. 2, p. 372–376.

Nuttall, W.L.F., 1930, Eocene foraminifera from Mexico: Journal of Paleontology, v. 4, p. 271–293.

Nuttall, W.L.F., 1932, Lower Oligocene Foraminifera from Mexico: Journal of Paleontology, v. 6, p. 3–35, pls. 1–9.

Oppenheim, P., von, 1896, Das Alttertiär der Colli Berici in Venetien, die Stellung der Schichten von Priabona, und die Oligocäne Transgression in alpinen Europa: Zeitschrift der Deutschen Geologischen Gesellschaft, v. 48, p. 27–152.

Orbigny, A. d', 1826, Tableau méthodique de la classe des Céphalopodes: Annales des Sciences Naturelles, v. 7, p. 245–314.

Palmer, D. K., 1934, Some large fossil foraminifera from Cuba: Memorías de la Sociedad Cubana de Historia Natural, v. 8, p. 235–264.

Persad, K. M., 1969, Stratigraphy, paleontology and paleocology of the Antigua Formation [Ph.D. thesis]: Kingston, University of the West Indies, 222 p.

Poag, C. W., 1966, Paynes Hammock (lower Miocene?) foraminifnera of Alabama and Mississippi: Micropaleontology, v. 12, p. 393–440.

Reiss, Z., 1967, *Victoriella* (Foraminiferida) from Israel: Israel Journal of Earth Science, v. 259, p. 481–492.

Reiss, Z., and Hottinger, L., 1984, The Gulf of Aqaba: Springer Verlag, 354 p.

Renz, O., and Kupper, H., 1947, Über morphogenetische untersuchungen an Grossforaminiferen: Ecologae geologicae Helvetiae, v. 39, p. 317–342.

Reuss, A. E., 1848, Die fossilen Polyparien des Wiener Tertiärbeckens: Naturwissenschaftliche Abhandlungen, Wien, Osterreich, v. 2, p. 33.

Robinson, E., 1966, Eocene limestones in eastern Jamaica: Transactions, 3rd Caribbean Geological Conference, Kingston, Jamaica, 1962, p. 71–74.

Robinson, E., 1968, Stratigraphic ranges of some larger foraminifera in Jamaica: Transactions, 4th Caribbean Geological Conference, Port of Spain, Trinidad, 1965, p. 189–194.

Robinson, E., 1969, Stratigraphy and age of the Dump limestone lenticle, central Jamaica: Ecologae geologicae Helvetiae, v. 62, p. 737–744, 3 Plates.

Robinson, E., 1974a, Some larger Foraminifera from the Eocene limestones at Red Gal Ring, Jamaica: Verhandlung der Naturforschenden Gesellschaft in Basel, v. 84, no. 1, p. 281–292, 7 Plates.

Robinson, E., 1974b, *Pseudofabularia* n. gen., an alveolinid foraminifer from the middle Eocene Yellow Limestone Group, Jamaica: Journal of Foraminiferal Research, v. 4, p. 29–32, 1 Plate.

Robinson, E., 1977, Larger imperforate foraminiferal zones of the Eocene of central Jamaica: Memorías Segundo Congreso Latinoamericano de Geología, v. 3, p. 1413–1421.

Robinson, E., 1988, Early Tertiary larger foraminifera and platform carbonates of the northern Caribbean: Transactions, 11th Caribbean Geological Conference, Bridgetown, Barbados, July 1986, p. 5:1–5:12.

Robinson, E., 1993, Some Imperforate larger foraminifera from the Paleogene of Jamaica and the Nicaragua Rise: Journal of Foraminiferal Research, v. 23, p. 47–65.

Robinson, E., and Jiang, M. M-J., 1990, Paleogene calcareous nannofossils from western Portland, and the ages and significance of the Richmond and Mooretown Formations of Jamaica: Journal of the Geological Society of Jamaica, v. 27, p. 17–25.

Robinson, E., and Persad, K. M., 1987, The occurrence of *Miogypsinoides* in Antigua: Transactions, 11th Caribbean Geological Conference, Cartagena, Colombia, 1984, p. 250–254.

Rutten, L.M.R., 1914, Foraminiferen-führende Gesteine von Niederlandisch New-Guinea, in Uitkomsten der Nederlandsch Nieuw-Guinea Expeditie in 1903: Leiden, v. 6 (Geologie), p. 21–51.

Rutten, L.M.R., 1928, On Tertiary Foraminifera from Curacao: Proceedings, Koninklijke Nederlandse Akademie van Wetenschappen, Ser. B, v. 31, p. 1061–1070.

Rutten, M. G., 1935, Larger foraminifera of northern Santa Clara Province, Cuba: Journal of Paleontology, v. 9, p. 527–545.

Sachs, K. N., 1957, Restudy of some Cuban larger foraminifera: Cushman Foundation for Foraminiferal Research Contributions, v. 8, p. 106–120.

Sachs, K. N., 1959, Puerto Rican Oligocene larger foraminifera: Bulletins of American Paleontology, v. 39, p. 399–410.

Sachs, K. N., and Gordon, W. A., 1962, Stratigraphic distribution of middle Tertiary larger foraminifera from southern Puerto Rico: Bulletins of American Paleontology, v. 44, no. 199, p. 5–19.

Salmeron, U. P., 1972, Mutacion entre los generos *Pararotalia* y *Miogypsinoides*: Revista del Instituto Mexicano del Petroleo, v. 4, p. 5–27.

Sawkins, J. G.,1869, Reports on the geology of Jamaica: Geological Survey Memoir, London, United Kingdom, 399 p.

Scheffen, W., 1932a, Ostindische Lepidocyclinen 1: Wetenschappelijke Mededeelingen, Dienst van de Mijnbouw in Nederlandsch Oost-Indie, no. 2, p. 1–76.

Scheffen, W., 1932b, Zur morphologie und morphogenese der "Lepidocyclinen": Paläontologische Zeitschrift, v. 14, p. 233–256.

Schubert, R. J., 1908, Zur Geologie des Osterreichischen Velebit: Wien, Jahrbuch der Geologischen Reichsanstalt, v. 58, p. 345–386.

Seiglie, G. A., Grove, K., and Rivera, J. A., 1977, Revision of some Caribbean Archaiasinae, new genera, species and subspecies: Ecologiae geologicae Helvetiae, v. 70, no. 3, p. 855–883.

Silvestri, A., 1924, Revisione de fossili della Venezia e Venezia Giulia: Atti dell' Accademia Scientifica Veneto-Trentino-Istriana, Padova (1923), ser. 3, v. 14, p. 7–12.

Silvestri, A., 1925, Sulla diffusione stratigraphica del generi *"Chapmania"* Silv. e Prev.: Memorie della Accademia Pontificia della Scienze, Nuovi Lincei, ser. 2, v. 8, p. 31–60.

Sirotti, A., 1983 (for 1982), Phylogenetic classification of Lepidocyclinidae; a proposal: Bollettino della Societa Paleontologica Italiana, v. 21, no. 1, p. 99–112.

Smout, A. H., and Eames, F. E., 1958, The genus *Archaias* (Foraminifera) and its stratigraphical distribution: Palaeontology, v. 1, part 3, p. 207–225.

Stache, G., 1875, Neue Beobachtungen in den Schichten der liburnischen Stufe: Verhandlungen der Geologischen Reichsanstalt 1875, p. 334–338.

Van Wessem, A., 1943, Geology and paleontology of central Camaguey, Cuba [Ph.D. thesis]: Utrecht, Universiteit te Utrecht.

Vaughan, T. W., 1924, American and European Tertiary larger foraminifera: Geological Society of America Bulletin, v. 35, p. 785–822.

Vaughan, T. W., 1926, Species of *Lepidocyclina* and *Carpenteria* from the Cayman Islands and their geological significance: Geological Society of London Quarterly Journal, v. 82, p. 388–400.

Vaughan, T. W., 1928a, Species of large arenaceous and orbitoidal foraminifera from the Tertiary deposits of Jamaica: Journal of Paleontology, v. 1, no. 4, p. 277–298.

Vaughan, T. W., 1928b, *Yaberinella jamaicensis*, a new genus and species of arenaceous foraminifera: Journal of Paleontology, v. 2, no. 1, p. 7–12, Plates 4, 5.

Vaughan, T. W., 1929a, Additional new species of Tertiary larger foraminifera from Jamaica: Journal of Paleontology, v. 3, no. 4, p. 373–382, Plates 39–41.

Vaughan, T. W., 1929b, Descriptions of new species of foraminifera of the genus *Discocyclina* from the Eocene of Mexico: U.S. National Museum Proceedings, v. 76, Art. 3, p. 1–18, 7 Plates.

Vaughan, T. W., 1932, American species of the genus *Dictyoconus*: Journal of Paleontology, v. 6, no. 1, p. 94–99.

Vaughan, T. W., 1933, Studies of American species of foraminifera of the genus *Lepidocyclina*: Smithsonian Miscellaneous Collections, v. 89, no. 10, p. 1–53, 32 Plates.

Vaughan, T. W., 1945, American Paleocene and Eocene larger foraminifera: Geological Society of America Memoir 9, p. 1–67, 46 Plates.

Vaughan, T. W., and Cole, W. S., 1940, Discocyclinidae, in Cushman, J. A., eds., Foraminifera, their classification and economic use: Harvard University Press, p. 327–330, 424.

Vaughan, T. W., and Cole, W. S., 1941, Preliminary report on the Cretaceous and Tertiary larger foraminifera of Trinidad, British West Indies: Geological Society of America Special Paper 30, p. 1–137, 46 Plates.

Vavra, N., 1978, *Sphaerogypsina* Galloway, 1933 (Foraminfera)—von Reuss (1848) als Bryozoe (*Ceriopora globulus*) und als Koralle (*Chaetetes pygmaeus*) beschrieben: Neues Jahrbuch für Geologie und Paläontologie, Monatshefte, v. 12, p. 741–746.

Versey, H. R., 1957, The White Limestone of Jamaica and the paleogeography governing its deposition [M.Sc. thesis]: Leeds University, 56 p.

Vessem, E. J., van, 1978, Study of Lepidocyclinidae from southeast Asia, particularly from Java and Borneo: Utrecht Micropaleontological Bulletins, no. 19, 142 p., 10 Plates, 84 figs., 10 Tables.

Vlerk, I. M., van der, 1928, The genus *Lepidocyclina* in the Far East: Ecologae geologicae Helvetiae, v. 21, no. 1, p. 182–211.

Vlerk, I. M., van der, 1964, Biometric research on European Lepidocyclinas: Proceedings, Koninklijke Nederlandse Akademie van Wetenschappen, Ser. B, v. 67, p. 1–10.

Vlerk, I. M., van der, and Postuma, J. A., 1967, Oligo-Miocene lepidocyclinas and planktonic foraminifera from East Java and Madura, Indonesia: Proceedings, Koninklijke Nederlandse Akademie van Wetenschappen, Ser. B, v. 70, p. 391–398, no. 4, p. 392–399.

Westerkamp, D., and Andreieff, P., 1983, St. Barthelémy et ses îlets, Antilles françaises. Stratigraphie et évolution magmato-structurale: Bulletin de la Société Géologique de France, v. 25, pt. 6, p. 873–883.

Wong, T. E., 1976, Tertiary stratigraphy and micropalaeontology of the Guiana Basin: Geologie en Mijnbouw Dienst Surinam, Mededeling 25, p. 13–107.

Wright, R. M., 1966, Biostratigraphical studies on the Tertiary White Limestone in parts of Trelawny and St. Ann, Jamaica [M. Phil. thesis]: London University, 94 p.

Wright, R. M., 1975, Aspects of the geology of Tertiary limestones in west-central Jamaica, West Indies [Ph.D. thesis]: Stanford University, 283 p.

Wright, R. M., and Switzer, P., 1971, Numerical classification applied to certain Jamaican Eocene nummulitids: Mathematical Geology, v. 3, no. 3, p. 297–311.

Yabe, H., and Hanzawa, S., 1925, Nummulitic rocks of the islands of Amakusa (Kyushu, Japan): Sendai, Science Reports of the Tôhoku University, ser. 2, Geology, v. 7, p. 73–82.

Yabe, H., and Hanzawa, S., 1928, Tertiary foraminiferous rocks of Taiwan (Formosa): Proceedings, Imperial Academy of Japan, v. 4, p. 533–536.

Zans, V. A., Chubb, L. J., Versey, H. R., Williams, J. B., Robinson, E., and Cooke, D. L., 1963, Synopsis of Jamaica: An explanation of the 1958 provisional geological maps of Jamaica: Bulletin of the Geological Survey of Jamaica, no. 4, 72 p.

MANUSCRIPT ACCEPTED BY THE SOCIETY NOVEMBER 11, 1992

Printed in U.S.A.

Geological Society of America
Memoir 182
1993

Tertiary cephalopods from Jamaica

Winfried Schmidt
Fundação Estadual de Proteção Ambiental (FEPAM), Avenida A. J. Renner 10, 90250 Porto Alegre, RS, Brazil
Peter Jung
Naturhistorisches Museum, Augustinergasse 2, CH-4001 Basel, Switzerland

ABSTRACT

Few specimens of cephalopods are known from Jamaica. Their internal molds can only be determined at the generic level. *Hercoglossa* sp., *Hercoglossa*? sp., and *Aturia* sp., occur in bioclastic limestone sequences in Jamaica and sporadically throughout the Caribbean and Central America, suggesting a widespread distribution and open passageways to North America during middle Eocene times.

INTRODUCTION

Tertiary cephalopods, although sporadically found, occur throughout the West Indian Islands and northern South America. Only three specimens are recorded from Jamaica, all from Eocene strata.

Nautiloids of the Western Hemisphere were abundantly distributed during the Paleocene and are known from numerous localities in the Gulf Coast and Atlantic regions and in northern South America. Eocene times saw a decline in specimen diversity compared to the preceding epoch, and by the end of the Miocene only one out of the eight Tertiary genera survived.

LOCATIONS AND STRATIGRAPHY

The Jamaican specimens were found in the Chapelton Formation, Yellow Limestone Group in western Jamaica and in the Claremont Formation, White Limestone Group on a parochial road outside of Kingston. The ages are given as lower to middle Eocene and middle Eocene, respectively (Robinson, 1968, 1969, 1974). Location maps are shown in Figures 1 and 2.

The Eocene Yellow Limestone Group in western Jamaica consists of thin, rubbly limestone beds and intervening marly layers. The argillaceous component appears as clay matrix surrounding limestone nodules or embedding thin layers of carbonates. The limestones are predominantly bioclastic, at various levels dolomitic and contain abundant macro- and microfossils.

Chapelton Formation

The base of the Yellow Limestone Group in western Jamaica is marked by a 0.3- to 0.5-m-thick layer of conglomerates overlaying a thick pile of red sandstones and conglomerates of unknown age. These so-called Upper Red Beds (Fig. 3) are barren and can neither be assigned to the middle Maastrichtian oyster limestone, nor to the highly fossiliferous marly limestones of lower to middle Eocene age. The Red Beds do not yield any microfossils or pollen and spores. They are incised into the calcareous section and could represent a remnant of Paleocene deposition, because both base and top are marked by angular unconformities.

The overlying Yellow Limestone conglomerate contains well-rounded carbonate fragments. Locally andesite and sandstone pebbles appear in the marly matrix. At other localities the basal conglomerate is missing and the Eocene section commences with approximately 8 to 10 m of orange-yellow marls, thinly bedded limestones, and streaks of silt- and sandstones. The marly horizons contain a rich benthic marine fauna. Present are internal moulds of bivalves and gastropods as well as poorly preserved corals, echinoids, and abundant ostracodes and benthic foraminifera. Several bedding planes clustered with in situ and detached single valved bivalves, most notably *Lucina* sp., can be observed in the sandy calcareous layers of this part of the section. *Operculinoides* sp., *Operculinoides cushmani,* and more specifically a 0.2-m-thick grainstone marker bed entirely made up of accumulated and densely compacted *Nummulites floridanus* suggest a lower to middle Eocene age.

Towards the top, the Yellow Limestone Group consists of limestone nodules set in a marl or clay matrix, or alternatively, of thin layers of bioclastic limestone. The sequence is 20 to 30 m thick and contains abundant macro- and microfossils. Most spectacular are gigantic gastropods like *Campanile* sp. (Jung, 1987) and *Campanile* cf. *giganteum,* the former reaching lengths of 40 cm and more. The limestone layers are composed of rounded and

---

Schmidt, W., and Jung, P., 1993, Tertiary cephalopods from Jamaica, *in* Wright, R. M., and Robinson, E., eds., Biostratigraphy of Jamaica: Boulder, Colorado, Geological Society of America Memoir 182.

ellipsoidal, very hard crystalline limestone nodules that show a crude stratification. Large internal moulds of *Lucina* sp. in life position or scattered on bedding planes can be found in the softer marly beds. The very top of the Yellow Limestone contains less argillaceous material, appears more massive, building into dense crystalline carbonate layers without the spectacular amount of fossils encountered in the lower section. *Hercoglossa* sp. and a juvenile specimen identified as *Hercoglossa*? sp. were found in the upper part of the Yellow Limestone.

Claremont Formation

The Claremont Formation of the White Limestone Group had been originally named the Troy and Claremont Limestones

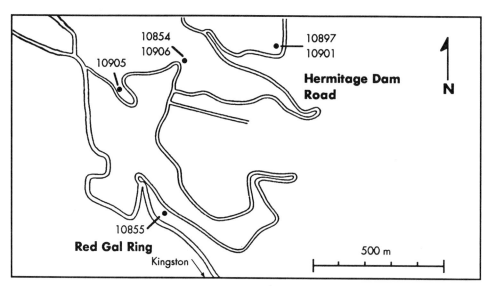

Figure 1. Map of the Red Gal Ring area, St. Andrew, showing the Natural History Museum Basel (NMB) localities.

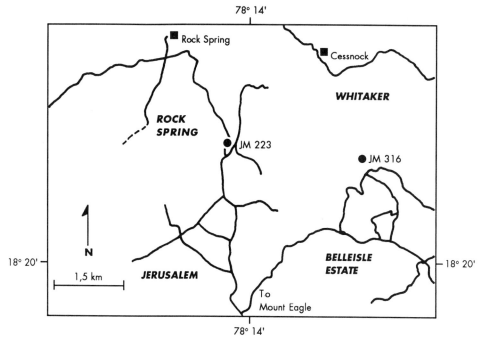

Figure 2. Location map of the Jerusalem Mountain, western Jamaica.

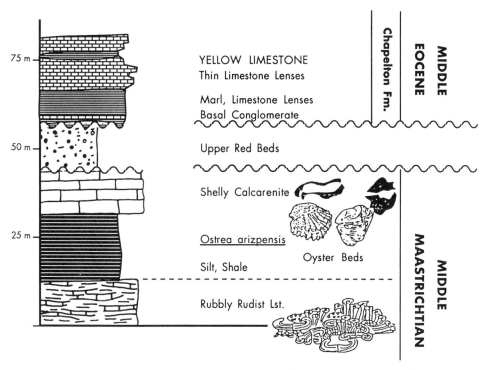

Figure 3. Generalized stratigraphic column from Belleisle Estate, western Jamaica.

(Hose and Versey, 1956). Deposited on to pre-Yellow Limestone strata as an overlapping sequence, it represents a transgressing and widening event of the Eocene sea. The two limestones are distinguishable only on the basis of the recrystallization and dolomitization of the former. The typical Claremont facies, on the other hand, is a shallow-water, bioclastic, molluscan-rich limestone that is evenly bedded and alternates with foraminiferal micrites (Wright, 1968). The foraminiferal micrites yield a fauna of abundant *Dictyoconus americanus* (Cushman), miliolids, and other small benthonic foraminifera. According to Robinson (1968, 1969), the age of the Claremont Formation is middle Eocene.

REMARKS ON THE FOSSILS

Only three specimens of Tertiary cephalopods are known from Jamaica. Recently R. Portell, Florida Museum of Natural History, has found an additional *Aturia* specimen in the Montpellier Limestone. The state of preservation of all three specimens is unsatisfactory, yet due to their scarcity it seems worth putting them on record. The fossils are catalogued at the Natural History Museum Basel (NMB).

Hercoglossa sp.
(Fig. 4.1, 4.2)

Material. One specimen; collected by W. Schmidt, 1986, NMB J 31309.
Measurements. Longest dimension: 119.2 mm; width: 57.1 mm.
Locality. Field Reg. No. JM 223 WS: Road cut 850 m north of Rock Spring Elementary School, Jerusalem Mountain, western Jamaica. Chapelton Formation (middle Eocene).

The specimen is an incomplete internal mould lacking its inner volutions. Phragmocone with seven camerae and part of the body chamber are preserved. The position of the siphuncle is obscured. Sinuous suture lines with broad and rounded lobes and saddles identify the specimen as an advanced *Hercoglossa* (Miller, 1947, p. 40, Fig. 7C).

Hercoglossa? sp.
(Fig. 4.3)

Material. One specimen; collected by W. Schmidt, 1986, NMB 31310.
Measurements. Largest dimension: 46.2 mm; width: 14.5 mm.
Locality. Field Reg. No. JM 316 WS: Trail 800 m north of road from Jerusalem Mountain to Grange, 1.4 km east of Jerusalem Mountain Road Junction, western Jamaica. Chapelton Formation (middle Eocene).

Illustrated is the internal mould of a reworked juvenile phragmocone. The generic assignment is questionable due to the poor state of preservation. Although a few saddles and lobes are visible, the entire shape of the suture cannot be traced. Eight camerae and the body chamber of a little more than half a volution are preserved. The specimen is extraordinarily narrow. A species of similar proportions is *Hercoglossa waltheri* Miller (1947, p. 62, Plate 46, Figs. 1, 2) from the Paleocene of Alabama. *H. waltheri*, however, is much larger and an identification down to species level is not attempted here.

Aturia sp.
(Fig. 4.4)

Material. One specimen; collected by P. Jung, 1968, NMB J 31311.
Measurement. Largest dimension: 142 mm.

Figure 4. (All natural size). 1, 2, *Hercoglossa* sp., adult phragmocone, lateral and apertural views. Specimen NMB J 31309. Locality JM 223 WS, road cut 2.7 km south of Rock Spring, Jerusalem Mountain, western Jamaica. Chapelton Formation (middle Eocene). 3, *Hercoglossa*? sp., juvenile phagmocone, lateral view. Specimen NMB J 31310. Locality JM 316 WS, trail 300 m north from Jerusalem Mountain to Grange, 1.4 km east of Jerusalem Mountain Road junction, western Jamaica. Chapelton Formation (middle Eocene). 4, *Aturia* sp., fragment of adult pragmocone, lateral view. Specimen NMB J 31311. NMB locality 10897, road cut on Hermitage Dam Road, Stony Hill above Kingston. Claremont Formation (middle Eocene).

Locality. NMB locality 10897: Road cut on Hermitage Dam Road, Stony Hill above Kingston. Elevation 662 m. Claremont Formation (middle Eocene). For exact location see Jung (1987, Fig. 1).

Internal mould of an adult partial phragmocone consisting of eight chambers. The position of the siphuncle is not recognizable. Septal suture with deep and narrow lobes along the ventral shoulders with hook-like constrictions on the two anterior suture lines like those of *Aturia alabamensis* (Morton) illustrated in Miller (1947, Plate 58, Fig. 2). The pinched-in chamber area between the otherwise sinous lobes and saddles is missing on all other sutures due to abrasive and weathering effects. The *Aturia* cannot, however, be positively identified to species level because of the fragmental state of the test.

GEOGRAPHIC DISTRIBUTION

Various nautiloids have been recorded and described from the West Indies, Venezuela, Colombia, and Panama (Table 1). Aturiids are known from Cuba (*A. cubaensis* Lea), Puerto Rico (*Aturia* sp.), Trinidad (*A. curvilineata* Miller and Thompson), Colombia (*A. peruviana* Olson; *A. panamensis* Miller), and Venezuela (*A. curvilineata* Miller and Thompson).

Several of these Eocene aturiids are advocated to be synonymous to other species. According to Miller (1947), *A. peruviana* Olson and *A. panamensis* Miller may be the same as *A. alabamensis,* first described by Morton (1834, p. 33, Plate 18, Fig. 3) and illustrated by Miller (1947) and Kummel (1956). *A. curvilineata* is a synonym of *A. cubaensis,* according to Jung (1966, p. 489, Plate 1, 2, Figs. 2–6).

Hercoglossidae are described from Trinidad (*Hercoglossa harrisi* Miller and Thompson; *H. waringi* Miller). Both species are of Paleocene and late Eocene age, respectively. The poor state of preservation of the Jamaican specimens does not permit of a detailed taxonomic description and discussion.

The Eocene was a warm period climatically with tropical conditions prevailing in Europe. Cooler temperatures commenced to be a factor in the beginning of the Oligocene. The presence of *Hercoglossa* and *Aturia* in the Eocene of Jamaica adds to the evidence of their widespread distribution during this epoch. Passageways to northwestern South America (Woodring, 1966) and North America (Squires, 1988) existed and an unrestricted circulation of surface waters, coupled with a high sea stand, favoured the worldwide spread of nautiloids during Eocene times.

ACKNOWLEDGMENTS

The photographs were taken by Wolfgang Suter, Natural History Museum Basel.

TABLE 1. DISTRIBUTION OF TERTIARY NAUTILOIDS IN THE CARIBBEAN*

Species	Paleocene	Eocene	Oligocene	Miocene	Pliocene
Cimomia kugleri	Trinidad				
Hercoglossa spp.	Bonaire				
H. sp.	Trinidad	Jamaica			
H. waringi		Trinidad			
Aturia peruviana		Colombia Venezuela Panama			
A. panamensis		Panama			
A. spp.		Jamaica	Puerto Rico	Martinique	
A. cubaensis			Cuba	Cuba Venezuela Trinidad	
A. sp.				Jamaica	

*Modified after Miller, 1947.

REFERENCES CITED

Hose, H. R., and Versey, H. R., 1956, Paleontological and lithological divisions of the lower Tertiary limestones of Jamaica: Colonial Geology and Mining Resources, v. 6, no. 1, p. 19–39.

Jung, P., 1966, Zwei miozäne Arten von Aturia (Nautilaceae): Eclogae Geologicae Helvetiae, v. 59, no. 1, p. 485–492.

Jung, P., 1987, Giant gastropods of the genus Campanile from the Caribbean Eocene: Eclogae Geologicae Helvetiae, v. 80, no. 3, p. 889–896.

Kummel, B., 1956, Post-Triassic nautiloid genera: Harvard College Museum of Comparative Zoology Bulletin, Harvard College, v. 114, p. 324–494.

Miller, A. K., 1947, Tertiary nautiloids of the Americas: Geological Society of America Memoir 23, 234 p.

Morton, S. G., 1834, Synopsis of the organic remains of the Cretaceous Group of the United States: Philadelphia, 88 p.

Robinson, E., 1968, Stratigraphic ranges of some larger foraminifera in Jamaica: Transactions, 4th Caribbean Geological Conference, Trinidad 1965, Port of Spain, Trinidad, p. 189–194.

Robinson, E., 1969, Stratigraphy and age of the Dump Limestone Lenticle, central Jamaica: Eclogae Geologicae Helvetiae, v. 62, no. 2, p. 737–744.

Robinson, E., 1974, Some larger foraminifera from the Eocene limestones at Red Gal Ring, Jamaica, in Contributions to the geology and paleobiology of the Caribbean and adjacent areas: Verhandlungen, Naturforschende Gesellschaft in Basel, v. 84, p. 281–292.

Squires, R. L., 1988, Cephalopods from the late Eocene Hoko River Formation, northwestern Washington: Journal of Paleontology, v. 62, no. 1, p. 76–82.

Woodring, W. P., 1966, The Panama land bridge as a sea barrier: American Philosophical Society Proceedings, v. 110, p. 425–433.

Wright, R. M., 1968, Biostratigraphical studies on the Tertiary White Limestone in parts of Trelawny and St. Ann, Jamaica [M.Phil. thesis]: London, London University, 94 p.

MANUSCRIPT ACCEPTED BY THE SOCIETY NOVEMBER 11, 1992

The fossil record of terrestrial mollusks in Jamaica

Glenn A. Goodfriend
Geophysical Laboratory, Carnegie Institution of Washington, 5251 Broad Branch Road, N.W., Washington, D.C. 20015

ABSTRACT

The Cenozoic fossil record of land snails in Jamaica consists of numerous deposits of Quaternary age and a single pre-Quaternary deposit—the Bowden shell bed, a marine deposit of early Pliocene age. Analysis of the amounts and D/L ratios of amino acids in Bowden land snails indicates that some of the material is actually of late Holocene age. Only four species can confidently be assigned stratigraphically to the Bowden shell bed; all are extinct, but most belong to extant species groups endemic to the island. The Bowden fauna provides a glimpse of the Jamaican land snail fauna 60 to 80% of the way along its time course of evolution (the island became emergent only in the middle Miocene) and shows that the endemic character of the fauna was already well advanced by this time. However, significant faunal change has occurred since then, similar to the amount of change seen in European land snail faunas over the same time period but greater than that seen in North American faunas. The Quaternary land snail faunas are quite rich in species, nearly all of which are still extant. Significant changes in species distributions during the Quaternary have occurred in some taxa, but stability of distributions is the predominant pattern.

INTRODUCTION

Jamaica is renowned for its highly diverse land snail fauna. There are about 400 to 450 species of land snails living in Jamaica, over 95% of which are endemic. There is also high regional endemism within the island—most species have ranges restricted to small portions of the island (Jarvis, 1902a, b, 1903; Paul, 1982; Goodfriend, 1986a, 1989; Goodfriend and Mitterer, 1988, 1993). Since the island was apparently entirely submerged prior to the middle Miocene (Eva and McFarlane, 1985), the land snail fauna must have evolved since that time. All of the numerous families and many of the genera of the Jamaican land snail fauna occur also on other West Indian islands or on the mainland. This indicates that the fauna is derived from a large number of colonization events, rather than just a few events followed by major evolutionary radiations.

The terrestrial fossil record of mollusks in Jamaica is rather less well known than the marine fossil record, which has been extensively studied. Yet the terrestrial deposits of Jamaica contain one of the most extensive Quaternary fossil records of terrestrial organisms in the New World tropics. This excellent Quaternary record is the result of two factors. First, there is a great abundance of land snails on the island, as is typical of many tropical islands.

Secondly, the extensive limestone deposits of the island provide ideal conditions for preservation of land snail shells. Well-developed karst features (caves, solution holes) exist over much of the island. Land snail shells get transported into these passages and tend to be preserved either because of dry conditions inside these solution features or because the water they are exposed to is already saturated with respect to carbonate. In the latter case, deposition of carbonates on the shells, rather than dissolution of shell carbonate, tends to occur. In some cases, large cave pearls may form around a land snail shell nucleus (McFarlane, 1987). However, these same conditions producing an excellent Quaternary record are also the cause of the poor pre-Quaternary record. Due to the rapid development of karst, more ancient features have long since disappeared. Dissolution is the predominant geomorphological process in the limestone regions. The insoluble residues of the limestone are deposited in the valleys between the karst hills but here they are exposed to extensive leaching by rainwater, so no fossil record of land snails is preserved.

Only one pre-Quaternary occurrence of terrestrial mollusks is known from Jamaica: that of the early Pliocene marine Bowden Beds. The Bowden fauna is reevaluated here and an overview of the Quaternary land snail faunas of the island is presented.

THE BOWDEN FAUNA

The Bowden shell bed

Toward the base of the outcrop of the Bowden Formation at Bowden, St. Thomas Parish, is a unit of gravelly marly sand, some 70 cm in thickness, which contains an extensive marine fauna (Hill, 1899; Chubb, 1958). The marine mollusks of this bed were studied by Woodring (1925, 1928), who identified about 600 species. The species assemblage of marine mollusks is peculiar in that it contains species of very diverse ecologies—both brackish water and truly marine species, and both intertidal and deep-water (200 m) species (Woodring, 1928). Woodring (1929) hypothesized that the bed may have formed in deep water and that the shallower-water species may have been washed down to this depth. However, the coarseness of the sediments and the concentration of shells suggest deposition in a higher-energy environment. The bed may have been deposited in shallow water adjacent to an eroding shoreline deposit containing mollusks. As the deposit collapsed or eroded away, the waves would have removed finer sediments and concentrated the gravels and shells. The shells of terrestrial mollusks living on top of the deposit would also have been incorporated as the deposit eroded. Under this scenario, the faunal assemblage would be of mixed age, containing older redeposited shells (perhaps predominantly from deeper-water environments) and shells (predominantly shallow water species) that were modern at the time of deposition. The age difference between these may be significant or negligible.

The age of the Bowden shell bed has been a matter of some controversy. Earlier thought to be of Oligocene age, the Bowden beds were assigned by Woodring (1928) to the middle Miocene. An age as young as late Pliocene to early Pleistocene was suggested by Lamb and Beard (1972) on the basis of its Foraminifera. However, more recent work shows that the foraminiferal fauna of the Bowden shell bed belongs to the *Globorotalia margaritae evoluta* subzone of the *Globorotalia margaritae* zone, which is of early Pliocene age (Bolli and Premoli Silva, 1973). Recent studies on Bowden mollusks (Jung, 1989) have accepted this age assignment. However, the possibility that some material may be younger or older than this must be considered, in light of the taphonomic scenario proposed above. The land snails would be among the younger elements of the deposits.

Which land snail species really belong to the Bowden fauna?

A number of collections of material from the shell bed have been made over the years and land snail shells have been found in several of these. A list of land snail species collected in the Bowden shell bed is presented in Table 1. There has been some question about the origin of some of these shells. Woodring (1928, p. 109) suggested that some of the Bowden land snails may be "the remains of living snails that fell into openings in the ground and thus were collected with the fossil material." Two of

TABLE 1. LIST OF LAND SNAIL SPECIES COLLECTED FROM BOWDEN AND CONCLUSIONS CONCERNING THEIR AGES

Family / Species	Collector*	Ref.†	Age
Helicinidae			
Lucidella costata Simpson	1	4	Early Pliocene
Poteriidae			
Poteria bowdenensis (Bartsch)§	2	5	?Early Pliocene
Incerticyclus bakeri (Simpson)	1	4	Early Pliocene
I. schermoi (Bartsch)§,**	2	5	Late Holocene
Annulariidae			
Parachondria augustae (C. B. Ads.)	1	6 (2580)	?Late Holocene
Truncatellidea			
Geomelania ?elegans C. B. Ads.	1	6 (135430)	?Late Holocene
Geomelania cf. *exilis* C. B. Ads.	?	6	?Quaternary
Urocoptidae			
Urocoptis brevis (Pfeiffer)	?	6	?Quaternary
Subulinidae			
Lamellaxis gracilis (Hutton)	1	6 (135471)	Recent
Sagdidae			
Hyalosagda (Stauroglypta) brevis (C. B. Ads.)	?	6	Recent
H. (Stauroglypta) cf. *brevior* (C. B. Ads.)	?	6	Recent
H. (Strialuna) sincera (C. B. Ads.)	?	6	Recent
Camaenidae			
Pleurodonte bowdeniana Simpson	1	4	Early Pliocene
P. bernardi Kimball	3	7	Early Pliocene
P. lucerna (Müller)	2	8	Late Holocene
	?	6	Recent
P. valida (C. B. Ads.)	2	8	Late Holocene
P. ?valida (shell fragments)	?	6	Recent
Helminthoglyptidae			
Hemitrochus bowdenensis Goodfriend	2	9	Late Holocene

*1 = Henderson; 2 = Smith and Schumo; 3 = Bernard Lewis.
†4 = Simpson, 1895; 5 = Bartsch, 1942; 6 = in collection of the National Museum of Natural History (catalogue numbers given for specimens that have been catalogued); 7 = Kimball, 1947; 8 = in collection of the Academy of Natural Sciences of Philadelphia; 9 = Goodfriend, 1992.
§Considerable confusion exists over the naming of the two poteriid species by Bartsch (1942). The species *bowdenensis* and *schermoi* were placed in the "pseudogenus" *Incerticyclus*. Morrison (1955a, p. 154) pointed out that this genus name had "no nomenclatorial standing," and consequently renamed these two species using the same specific names, with *bowdenensis* being placed in *Poteria* and *schermoi* in *Cyclochittya*; *schermoi* was later amended to *schumoi* (Morrison, 1955b). However, the non-validity of Bartsch's genus name does not invalidate the species names he introduced. Morrison's later names are therefore considered synonyms of Bartsch's names.
**This is considered a probable synonym of *Poteria adamsi*, as discussed in the text.

the land snail shells collected by Simpson and Henderson belong to modern species, which makes them suspect, as noted by Simpson (1895).

It is difficult to judge the age of shell material on the basis of its appearance. Very fresh material may have periostracum (the outer organic layer present in many species) present, but this may be lost within years or decades. Shells tend to change from slightly translucent to an opaque white as they age, but this may occur within decades or centuries. The presence of secondary carbonate deposits on a shell can be used to indicate that it is not modern. But such deposits may occur even on shells as young as late Holocene (Goodfriend and Mitterer, 1988) and are not necessarily present on older material.

To evaluate the possibility of the inclusion of more recent material along with genuine Bowden shell bed material, amino acid epimerization/racemization analysis was employed. The basis of this method is that the D/L amino acid ratio measured in modern shells is very low (1 to 5%, depending on the amino acid; Goodfriend, 1991) and increases progressively with increasing age. In ancient material, the ratio of the epimers D-alloisoleucine/L-isoleucine (A/I) reaches an equilibrium value of ca. 1.3 (Hare and Mitterer, 1969), whereas the D/L enantiomer ratios of other amino acids reach a theoretical equilibrium value of 1.0. However, in older material various processes, such as leaching of amino acids out of the shell or contamination by amino acids from the soil, can cause these equilibrium ratios not to be reached. It was possible to analyze only selected Bowden shells for D/L amino acid ratios, since many of the specimens are too small (analysis requires ca. 20 mg of shell) or too valuable scientifically to allow destructive analysis of even a part of the shell. Marine shells from the Bowden shell bed were also analyzed for comparison.

The specimen of *Pleurodonte bernardi* collected by B. Lewis in 1947 (Kimball, 1947) shows very low levels of amino acids (Table 2), too low for measurement of D/L ratios. These low levels of amino acids are not normally observed in Quaternary material (cf. *P. lucerna* from the Smith and Schumo collection, Table 2) and suggest a considerable age for this specimen. Marine taxa from the Bowden shell bed show similarly low levels of amino acids. *P. bernardi* is therefore accepted as representing Bowden-age material. The shell has light secondary carbonate deposits on it (Fig. 1A).

It was not possible to analyze any of the material from the collection of Henderson for amino acid D/L ratios. However, the collection includes several extinct and rather unusual species, including *Pleurodonte bowdeniana, Incerticyclus bakeri,*[1] and *Lucidella costata,* and these are considered to represent genuine members of the Bowden fauna. These also have light secondary carbonate deposits, as seen in *P. bernardi.*

Each of the four species analyzed from the collection of Smith and Schumo (not Schermo, as misspelled by Bartsch, 1942, p. 138 and 139, and Morrison, 1955a, and later corrected by Morrison, 1955b) shows A/I ratios of ca. 0.06 (Table 3). Based on the rate of epimerization determined at a site on the north coast of Jamaica, a late Holocene age of ca. 1,000 to 2,000 yr. B.P. is indicated (Goodfriend, 1992). The most likely source of this material is a colluvial deposit that had slumped down over the Bowden shell bed. The hill above the bed is quite steep and the bed is now covered by colluvium over much of its extent (Goodfriend, personal observation, 1991). Because of the uniformity of the A/I values, it seems probable that this collection came from a single deposit. The small differences between species may be the result of differences in epimerization rates (*Poteria* epimerizes about 15% slower than *Pleurodonte,* for example). It is surprising that one of the species in this collection, *Hemitrochus bowdenensis,* is apparently extinct, but amino acid analyses indicate that it is definitely of late Holocene age. Amino acid results on a paratype specimen of *Incerticyclus schermoi* Bartsch (Fig. 2C, D), considered by Bartsch (1942) and Morrison (1955a) to be part of the Bowden fauna, show that this species is also of late Holocene age. These specimens closely resemble *Poteria adamsi* (Bartsch) (*Ptychocochlis adamsi* Bartsch), which inhabits southeastern Jamaica. They are also very similar to *Cyclochittya dentistigmata* (Chitty), another southeastern endemic. The genera *Poteria* and *Cyclochittya* can be readily distinguished on the basis of their opercula (Morrison, 1955a). However, no operculum of *C. schermoi* is known. This species is tentatively synonymized with *P. adamsi.* Besides *Hemitrochus bowdenensis,* there is one other extinct species present in the Smith and Schumo collection: *Poteria bowdenensis* (Fig. 2E, F). This species was accepted by both Bartsch (1942) and Morrison (1955a) as being a genuine Bowden fossil, as was the late Holocene *C. schermoi.* There is no extant poteriid resembling this species. The single specimen is tentatively assigned a late Holocene age on the basis of the amino

TABLE 2. CONCENTRATIONS OF THE AMINO ACIDS ALLOISOLEUCINE + ISOLEUCINE (A+I), ASPARTIC ACID (ASP), AND GLUTAMIC ACID (GLU) IN TERRESTRIAL AND MARINE MOLLUSKS FROM BOWDEN*

Species	Amino Acid Concentration (pM/mg of shell)		
	A+I	Asp	Glu
Marine taxa			
Anadara halidonata (Dall)	2.1	1.8	1.7
Laevicardium serratum (Linne)	10.9	9.3	7.3
Terrestrial taxa			
Pleurodonte lucerna[†]	52.7	105	171
Pleurodonte bernardi	0.7	4.6	2.3

*Note: Amounts were calculated from gas chromatographic data, with the peak areas calibrated against standards of known amounts. Errors are about ±30 percent.
[†]From the Smith and Schumo collection.

---
[1]The generic nomenclature of Morrison (1955a) is used for poteriids throughout this paper.

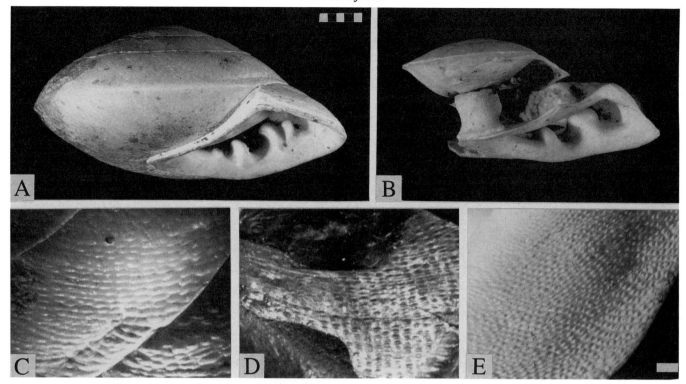

Figure 1. *Pleurodonte* from Bowden, Jamaica. A, *P. bernardi*, type (Institute of Jamaica). B, *P. bowdeniana*, type material of juvenile (above) and adult fragment (below) (NMNH), placed together to give an indication of what a more complete adult specimen would look like. The scale bar for A and B (in A) is marked in millimeters. C, microsculpture of *P. bernardi*. The growth lines are approximately horizontal in this photograph. D, microsculpture of *P. bowdeniana* (clay impression of parietal callus, where it contacted the now-missing penultimate whorl). E, microsculpture of *P. lucerna* from Bowden from the Smith and Schumo collection (ANSP). The scale bar for C, D, and E (in E) is 1 mm.

acid results on other species in this collection; but there is a possibility that this specimen could be a different age than the other Smith and Schumo shells and actually belong to the Bowden fauna.

On the basis of their very fresh appearance, several lots of shells (Table 1) can be assumed to be of recent age: *Hyalosagda* cf. *brevior*, *H. brevis*, *H. sincera*, *Pleurodonte lucerna* (NMNH collection), *P. valida* (NMNH collection), and *Lamellaxis gracilis*. This last species has been introduced into the West Indies in historical times.

TABLE 3. D/L RATIOS OF ALLOISOLEUCINE/ISOLEUCINE (A/I), ASPARTIC ACID (ASP), AND GLUTAMIC ACID (GLU) IN BOWDEN LAND SNAILS FROM THE SMITH AND SCHUMO COLLECTION

Species	A/I	D/L Asp	D/L Glu
Pleurodonte lucerna (spec. A)	0.074	0.34	0.098
Pleurodonte lucerna (spec. B)	0.072	0.34	0.099
Pleurodonte valida	0.069	0.33	0.095
*Incerticyclus schermoi**	0.056	0.33	0.079
*Hemitrochus bowdenensis**	0.053	0.27	0.062

*Analysis of paratype.

More problematic are several extant species represented by fossil material from Bowden. The specimens of *Geomelania* cf. *exilis* and *Urocoptis brevis* both have secondary carbonate deposits, indicating that they are not recent. They are tentatively assigned a Quaternary age but their being genuine Bowden faunal elements cannot be ruled out. Specimens of *Parachondria augustae* and *Geomelania* ?*elegans* have opaque white shells, indicating that they are not recent, but no secondary carbonate deposits are present on the shells. They may perhaps be recently dead or late Holocene material but are unlikely to belong to the Bowden fauna.

Relationships of the Bowden land snail species to the modern fauna

Pleurodonte bernardi was assigned by Kimball (1947) to the *P. sinuata* group on the basis of the presence of four apertural denticles. The other group of Jamaican *P.* (*Pleurodonte*) of the section *Dentellaria* (taxonomy follows Wurtz, 1955), the *P. lucerna* group (called the *P. acuta* group by Jarvis, 1902a, and others), is characterized by the presence of two apertural denticles, although additional minute denticles occur in some species. The structure of the denticles in *P. bernardi* (Fig. 1A) is like that

Figure 2. Poteriid land snails from Bowden. A, B, *Incerticyclus bakeri*, type (USNM); C, D, *Incerticyclus schermoi*, holotype (ANSP); E, F, *Poteria bowdenensis*, holotype (ANSP). The scale bar is marked in millimeters.

of the *P. lucerna* group and not like the *P. sinuata* group. In the *P. sinuata* group, the denticles are formed by a folding inward of the shell, such that each elongate denticle extending from the inside of the lip corresponds to an elongate groove on the base of the shell behind the lip. The groove corresponding to the innermost denticle can be seen on the wall of the umbilicus in adult specimens of those species in which the umbilicus is not covered over by an umbilical callus, or in subadult stages of those species that do cover over the umbilicus as adults; the umbilicus is open in juveniles and partly open in subadults. In the *P. lucerna* group, as well as in *P. bernardi,* the denticles are not as elongate and the exterior manifestation of the denticles appears as a depressed area on the base of the shell behind the lip, not an elongate groove. Examination of the unique type of *P. bernardi* shows that the innermost denticle is not homologous to that in the *P. sinuata* group, since it occupies a position on the basal lip rather than on the wall of the columella. In two species of the *P. lucerna* group, *P. chemnitziana* (Pfeiffer) and *P. ingens* (C. B. Adams), an additional one to several very small (ca. 1 to 2 mm) denticles are present on the lower lip on the inner (umbilical) side of the two major denticles. These thus occupy the same position as the innermost denticle of *P. bernardi* and are probably homologous. The second-to-innermost denticle of *P. bernardi* occupies the same position as the inner denticle in the *P. lucerna* group. It is possible that the outer pair of denticles in *P. bernardi* is homologous to the single outer denticle of the *P. lucerna* group. In some species of this group, such as *P. bowdeniana* and north coast populations of *P. lucerna,* this outer denticle is significantly wider than the inner denticle. This wide denticle may have evolved from the fusion of a pair of outer denticles (such as those seen in *P. bernardi*); or the outer pair of denticles in *P. bernardi* may have evolved from the splitting of a single wide outer denticle (such as that seen in *P. bowdeniana*). The direction of evolution is not clear, since outgroups (such as Lesser Antillean *P. (Pleurodonte)* of the section *Pleurodonte* (the only other section of this subgenus besides the Jamaican *Dentellaria*) or the other Jamaican subgenera of *Pleurodonte, Eurycratera,* and *Thelidomus*) lack well-developed denticles. These various features show that *P. bernardi* clearly belongs in the *P. lucerna* group, rather than in the *P. sinuata* group. *P. bernardi* is unusual for a member of this group in that it possesses two well-developed outer denticles. Another unusual feature concerns the microsculpture of the shell. The elongate pustules, which are characteristic of most *P. (Pleurodonte)* species, are arranged in *P. bernardi* into regular rows, perpendicular to the growth lines (Fig. 1C). In the *P. lucerna* group, these pustules are generally more irregularly arranged (Fig. 1E); only in certain areas of the shell in certain specimens will they be aligned. This alignment is more common in the *P. sinuata* group, but occurs also in *P. bowdeniana.*

Pleurodonte bowdeniana is another member of the *P. lucerna* group present in the Bowden fauna. It differs from modern species of *Pleurodonte* in two features. Although only the last whorl of the adult specimen is preserved (there is also a second specimen consisting of an apical fragment of 3.3 whorls), the high angle of the acute periphery indicates that the shell was strongly domed. An approximate idea of what the whole shell would have looked like can be obtained from placing together the pieces of the two different shells in their relative positions (Fig. 1B). Other species of the *P. lucerna* group are more flattened. Although the microsculpture of both specimens has been worn off the outside of the shells (the surfaces are abraded), the adult fragment retains a very clear impression of the sculpture of the base of the missing penultimate whorl on the parietal callus that was deposited on it. From a clay impression of this, the original microsculpture could be reconstructed (Fig. 1D). This shows the same regular alignment of pustules as seen in *P. bernardi,* discussed above. The core of the outer denticle is hollow, forming a slightly elongated depression on the base of the shell behind the lip, as noted by Simpson (1895). However, this appears to be the result of decomposition of the shell in this region, the pit being lined with soft, chalky shell material. Consequently this feature cannot be considered homologous to the basal grooves that are associated with the denticles in the *P. sinuata* group.

Thus both species of *Pleurodonte* of the Bowden fauna belong to the *P. lucerna* group. This group is endemic to Jamaica, as is the *P. sinuata* group; another section of *P. (Pleurodonte)* occurs in the Lesser Antilles (Wurtz, 1955).

Incerticyclus bakeri (Fig. 2A, B) is perhaps the most unusual species of the Bowden fauna. It cannot be placed in any modern poteriid genus. The unique type measures 30 mm in diameter, making it as large as the largest known species in the family in Jamaica; but the specimen is a *juvenile* (a fact noted by neither Bartsch, 1942, who originally described the shell, nor Morrison, 1955a, who redescribed it), as is indicated by its thin lip, the strong convexity of the basal lip, and the lack of descent of the suture behind the lip. The last ca. one-half whorl in adult Jamaican poteriids has a more flattened base than in the juvenile shell. This would imply that the specimen of *I. bakeri* had at least another one-half whorl to grow, so the adult would have been monstrous: if the aperture width/shell width ratio remained constant with growth, another half whorl of growth would have brought the shell to 53 mm in diameter! The sculpture of *I. bakeri* is also unique. The coarse ribs are strongly curved on the dorsum (the convexity points backwards). They are unusually coarse in the umbilicus (5 ribs/2 mm at the end of the last whorl), and bend strongly forward as they emerge from the umbilicus. In adults of most other Jamaican poteriids, there is a strong to weak angle (sometimes a distinct cord) between the umbilicus and the base of the shell. In *I. bakeri,* this is lacking, but the bending of the ribs occurs in the position where the angle occurs in other poteriids. It may be that this angularity would have developed later in the growth of the shell, as it is lacking in juvenile poteriids. *I. bakeri* possesses no features that relate it to any particular one of the extant Jamaican poteriid groups. Morrison (1955a, p. 156) placed it in a new genus *Incerticyclus,* together with a "probable member of this genus," the poorly known *Cyclostoma perpallidum* C. B. Adams. However, there seems to be very little to relate these two species. Unfortunately, the operculum of *I. bakeri,* which would assist in determining these relationships, is not known. *I. bakeri* shares with many other Jamaican poteriid species a well-developed coarse sculpture, a feature lacking in poteriids outside the island.

Lucidella costata (illustrated by Simpson, 1895) is similar to other (extant) Jamaican *Lucidella* (*Poenia*), such as *L. persculpta* and *L. lineata,* but is a distinct species. This subgenus occurs also in northernmost South America and in Central America (Boss and Jacobson, 1974).

In summary, only *Lucidella costata, Incerticyclus bakeri, Pleurodonte bernardi,* and *P. bowdeniana* are accepted as belonging to the Bowden fauna. Because of the paucity of species, characterization of this fauna is necessarily tentative. All of the species are extinct. Three of the species (*L. costata* and the two *Pleurodonte* species) are closely related to modern Jamaican forms—they can be placed securely within modern species groups within modern subgenera. The fourth species, *I. bakeri,* is not closely related to any modern Jamaican species and probably belongs in its own monotypic genus.

Jamaica is considered to have become emergent in the middle Miocene (Eva and McFarlane, 1985), or between 16 and 10 Ma. The Bowden fauna is about 3 to 4 m.y. old, according to the dating of associated Foraminifera (Bolli and Premoli Silva, 1973), so it provides us with a glimpse of the Jamaican land snail fauna about 60 to 80% along the time course of its evolution to the present fauna. By that time, the Jamaican fauna had already attained its distinctive endemic character; even modern species groups existed by that time.

Comparison of the Bowden fauna to fossil faunas from other regions

No other Pliocene land snail faunas are known from the tropics. However, Miocene land snails are known from other West Indian islands and from Kenya. From the lignite beds of Camp Perrin, on the southwest peninsula of Haiti, two Miocene land snail species were listed by Woodring et al. (1924) but were not described. One is a *Crocidopoma* sp. (a poteriid); the genus is endemic to Hispaniola and Cuba. The other ("*Thysanophora* sp.") is presumably some species of sagdid. From the northern Dominican Republic, Pilsbry and Olsson (1954) described two extinct species of *Cepolis* (*Coryda*) (Helminthoglyptidae) and an extinct *Urocoptis* (?*Autocoptis*) species. *Coryda* is endemic to Hispaniola and Cuba, whereas *Autocoptis,* which the fossil may or may not belong to, is endemic to Hispaniola. From the early middle Miocene of Carriacou in the Lesser Antilles, an extinct species of *Pleurodonte* is known (Jung, 1971). This closely resembles other Lesser Antillean *Pleurodonte*. A second species of *Pleurodonte* (Trechmann, 1935) from the same deposit was tentatively synonymized with the first species by Jung (1971). These various Miocene faunas, although very limited and not studied in detail, suggest that differentiation of faunas among the West Indian islands had already occurred by that time at approximately the subgeneric level. This is consistent with the occurrence of endemic species groups in Jamaica by the early Pliocene. A relatively rich lower Miocene fauna is known from Kenya (Verdcourt, 1963). All of these species can readily be placed in genera of modern species, and many of the species are similar to modern species. It thus has a more modern character than the much younger Bowden fauna of Jamaica.

Several rich Pliocene land snail faunas are known from Europe: a fauna 3.5 to 3 m.y. old from the lower Rhine Valley (Schlickum and Strauch, 1979), a fauna ca. 3 m.y. old from Alsace (Schlickum and Geissert, 1980), and a fauna 2.5 to 3 m.y. old from eastern France (Schlickum, 1975; dating according to Nordsieck, 1982). Each of these faunas consists of 47 to 56 species and 35 to 40 genera. Within each fauna, there are four to five extinct genera present; but several of the species in each fauna are still extant. In North America, late Pliocene faunas are known from Kansas. The Rexroad local fauna contains 24 land snail species, the great majority of which are still extant (Taylor, 1960). The Bender local fauna contains 18 species and here also the great majority are still extant (Miller, 1964). A lower Pliocene fauna from Oklahoma contains five species, all but one of which are extant (Leonard and Franzen, 1944); and from New Mexico a Pliocene fauna contains fifteen terrestrial species, of which all but three are extant (Leonard and Frye, 1978).

Comparison of these various faunas indicates a slow rate of faunal change in the Great Plains and southwestern North America since the Pliocene. Changes since the lower Miocene in Kenya have also been relatively small, given the considerably greater age of this fauna. A greater rate of change is seen in the Jamaican fauna since the early Pliocene, comparable to the changes seen in the European fauna over the same time period.

QUATERNARY LAND SNAIL FAUNAS

The Quaternary land snails of Jamaica and their biogeography were reviewed recently (Goodfriend, 1989), so only an overview of these faunas is presented here. Although numerous Quaternary deposits exist in Jamaica, only a small number of these have been collected, analyzed, and dated. A list of these is presented in Table 4. Faunal studies have been completed for only a few of these.

Of the many species known from the Quaternary fossil record, all but a few are still extant. The oldest known deposit, Sheep Pen Cave, of probable middle Pleistocene age or older, contains 48 species, three of which are apparently extinct (Goodfriend, 1986b, 1989). The rich late Pleistocene and Holocene faunas at Green Grotto (Goodfriend and Mitterer, 1988), Coco Ree (Goodfriend and Mitterer, 1993), Bonafide (Goodfriend, 1989), and a cave on Portland Ridge (Goodfriend, unpublished data) all consist entirely of extant species. However, the Bowden collection of Smith and Schumo contains two extinct species, one definitely, and the other probably, of late Holocene age. The reasons for these recent extinctions in the Bowden area may be the combination of the very locally endemic character of the fauna of eastern St. Thomas Parish, combined with the extensive habitat distur-

TABLE 4. DATED JAMAICAN QUATERNARY DEPOSITS CONTAINING LAND SNAILS

Deposit	Age	References*	Comments
Sheep Pen	?Middle-early Pleistocene	1	Breccia
Wallingford	Late or middle Pleistocene[†]	2	Breccia, unstudied
Coco Ree	Late Pleistocene to Holocene	3	1 m sequence in cave
Green Grotto	Late Pleistocene to Holocene	4	Series of solution hole deposits
Portland Ridge	Late Pleistocene to Holocene	2	Under study
Bonafide Cave	Latest Pleistocene to Holocene	5	Under study
Wallywash Pond	Holocene and ?last interglacial	6	Unstudied sediment core
Bonavista	Middle or late Holocene	7	Unstudied
Long Mile Cave	Late Holocene[§]	8	Unstudied
Bowden	Late Holocene	9	Smith and Schumo

*1 = Goodfriend (1986b); 2 = Florida State Museum collection; 3 = Goodfriend and Mitterer (1993); 4 = Goodfriend and Mitterer (1988); 5 = Goodfriend (1989); 6 = Paul and others (this volume); 7 = Porter and Anderson (1983); 8 = MacPhee (1984); 9 = this chapter.
[†]MacPhee et al. (1989).
[§]MacPhee (1984).

bance of this area (Goodfriend, 1992). The fauna of the Cockpit Country (which contains Sheep Pen and Bonafide Caves) also has a very locally endemic character, but forest still exists in much of this area.

Although few species extinctions have occurred during the Quaternary, there have been some significant changes in the ranges of some of the species during this period. The various species of *Sagda,* for example, have particularly labile distributions (Goodfriend, 1989). At Green Grotto Cave, a succession of four species of *Sagda* is seen from late Pleistocene to late Holocene (Goodfriend and Mitterer, 1988). Other taxa, such as chrondopomids, show much more stable distributions. There are also regional differences in faunal stability: late Quaternary climatic changes have caused significant faunal changes on the north coast of Jamaica (Goodfriend and Mitterer, 1988), whereas the interior and south coast faunas show very little change.

CONCLUSIONS

Because of the incomplete nature of the pre-Quaternary Cenozoic record of land snails in Jamaica, they would seem to have little potential as biostratigraphic markers. Only one pre-Quaternary fauna—the Bowden fauna—is known. The Quaternary record is relatively extensive but is composed predominantly of species that are still extant. The times of extinction of those few species that are no longer extant are not known with good precision and, in any case, these species are likely very local in their distributions, thus reducing their potential for stratigraphic correlation. However, the fossil record of land snails in Jamaica does contain a uniquely detailed record of Quaternary faunal change in the New World tropics which provides insight into the processes that have led to remarkable diversity observed today.

ACKNOWLEDGMENTS

Loans of specimens for study were provided by G. Rosenberg and G. M. Davis of the Academy of Natural Sciences of Philadelphia, W. Blow and T. Waller of the U.S. National Museum of Natural History, and T. Farr of the Institute of Jamaica. G. Rosenberg provided helpful suggestions on the nomenclatorial problems of the fossil poteriids. A visit to Bowden was arranged by R. M. Wright and F. Gillings-Grant. Identification of the marine mollusks from Bowden used for amino acid analysis was provided by T. Waller. Most of the amino acid analyses were carried out by E. Negreanu at the Weizmann Institute of Science. Photographic equipment was provided by P. E. Hare. Helpful comments on the manuscript were provided by P. Jung, C.R.C. Paul, and G. Rosenberg.

REFERENCES CITED

Bartsch, P., 1942, The cyclophorid mollusks of the West Indies, exclusive of Cuba, *in* de la Torre, C., Bartsch, P., and Morrison, J.P.E., The cyclophorid operculate land mollusks of America: Bulletin of the United States National Museum, v. 181, p. 43–141, Plates 9–18, 41–42.

Bolli, H. M., and Premoli Silva, I., 1973, Oligocene to Recent planktonic Foraminifera and stratigraphy of the Leg 15 sites in the Caribbean Sea, *in* Edgar, N. T., et al., eds., Initial reports of the deep sea drilling project, v. 15: Washington, D.C., U.S. Government Printing Office, p. 475–497.

Boss, K. J., and Jacobson, M. K., 1974, Monograph of the genus *Lucidella* in Cuba (Prosobranchia: Helicinidae): Harvard University, Museum of Comparative Zoology, Occasional Papers on Mollusks, v. 4, p. 1–27.

Chubb, L. J., 1958, Higher Miocene rocks of southeast Jamaica: Genotes, v. 1, p. 26–31.

Eva, A., and McFarlane, N., 1985, Tertiary to early Quaternary carbonate facies relationships in Jamaica, *in* Transactions, Fourth Latin American Geological Congress, Trinidad and Tobago, 1979, Volume 1: p. 210–219.

Goodfriend, G. A., 1986a, Radiation of the land snail genus *Sagda* (Pulmonata:

Sagdidae): Comparative morphology, biogeography and ecology of the species of north-central Jamaica: Zoological Journal of the Linnean Society, v. 87, p. 367–398.

Goodfriend, G. A., 1986b, Pleistocene land snails from Sheep Pen Cave in the Cockpit Country of Jamaica, *in* Proceedings, Eighth International Malacological Congress: Budapest, Hungarian Natural History Museum, p. 87–90.

Goodfriend, G. A., 1989, Quaternary biogeographical history of land snails in Jamaica, *in* Woods, C. A., ed., Biogeography of the West Indies: Past, present, and future: Gainesville, Florida, Sandhill Crane Press, p. 201–216.

Goodfriend, G. A., 1991, Patterns of racemization and epimerization of amino acids in land snail shells over the course of the Holocene: Geochimica et Cosmochimica Acta, v. 55, p. 293–302.

Goodfriend, G. A., 1992, A new fossil land snail of the genus *Hemitrochus* from Bowden, Jamaica: Nautilus, v. 106, p. 55–59.

Goodfriend, G. A., and Mitterer, R. M., 1988, Late Quaternary land snails from the north coast of Jamaica: Local extinctions and climatic change: Palaeogeography, Palaeoclimatology, Palaeoecology, v. 63, p. 293–311.

Goodfriend, G. A., and Mitterer, R. M., 1993, A 45,000-yr record of a tropical lowland biota: The land snail fauna from cave sediments at Coco Ree, Jamaica: Geological Society of America Bulletin, v. 105, p. 18–29.

Hare, P. E., and Mitterer, R. M., 1969, Laboratory simulation of amino-acid diagenesis in fossils: Carnegie Institution of Washington Yearbook, v. 67, p. 205–208.

Hill, R. T., 1899, The geology and physical geography of Jamaica: Study of a type of Antillean development: Harvard College, Museum of Comparative Zoology Bulletin, v. 34, p. 1–256.

Jarvis, P. W., 1902a, Notes on the distribution of the Pleurodonte acuta group: Nautilus, v. 15, p. 137–141.

Jarvis, P. W., 1902b, Notes on the distribution of the Pleurodonte sinuata group: Nautilus, v. 16, p. 1–4.

Jarvis, P. W., 1903, Distribution of Jamaican species of Colobostylus: Nautilus, v. 17, p. 62–65.

Jung, P., 1971, Fossil mollusks from Carriacou, West Indies: Bulletins of American Paleontology, v. 61, p. 147–262.

Jung, P., 1989, Revision of the Strombina-group (Gastropoda: Columbellidae), fossil and living: Schweizerische Paläontologische Abhandlungen, v. 111, p. 1–298.

Kimball, D., 1947, A new Pleurodonte from the Miocene, Bowden, Jamaica: Nautilus, v. 61, p. 37–39.

Lamb, J. L., and Beard, J. H., 1972, Late Neogene planktonic foraminifers in the Caribbean, Gulf of Mexico, and Italian stratotypes: University of Kansas Paleontological Contributions, article 57, p. 1–67.

Leonard, A. B., and Franzen, D. S., 1944, Mollusca of the Laverne (lower Pliocene) of Beaver County, Oklahoma: University of Kansas Science Bulletin, v. 30, p. 15–39.

Leonard, A. B., and Frye, J. C., 1978, Paleontology of Ogallala Formation, northeastern New Mexico: New Mexico Bureau of Mines and Mineral Resources, Circular no. 161, p. 1–21.

MacPhee, R.D.E., 1984, Quaternary mammal localities and Heptaxodontid rodents of Jamaica: American Museum Novitates, no. 2803, p. 1–34.

MacPhee, R.D.E., Ford, D. C., and McFarlane, D. A., 1989, Pre-Wisconsinan mammals from Jamaica and models of late Quaternary extinction in the Greater Antilles: Quaternary Research, v. 31, p. 94–106.

McFarlane, D. A., 1987, Radiant darkness: The many facets of the caves of Jamaica: Terra, v. 25, p. 24–26.

Miller, B. B., 1964, Additional mollusks from the late Pliocene Bender local fauna, Meade County, Kansas: Journal of Paleontology, v. 38, p. 113–117.

Morrison, J.P.E., 1955a, Notes on American cyclophorid land snails, with two new names, eight new species, three new genera, and the family Amphicyclotidae, separated on animal characters: Journal of the Washington Academy of Sciences, v. 45, p. 149–162.

Morrison, J.P.E., 1955b, A correction: Nautilus, v. 69, p. 72.

Nordsieck, H., 1982, Zur Stratigraphie der neogenen Fundstellen der Clausiliidae und Triptychiidae Mittel- und Westeuropas (Stylommatophora, Gastropoda): Mitteilungen der Bayerischen Staatssammlung für Paläontologie und Historische Geologie, v. 22, p. 137–155.

Paul, C.R.C., 1982, The Jamaican land snail genera *Geoscala* and *Simplicervix* (Pulmonata: Urocoptidae): Journal of Conchology, v. 31, p. 101–127.

Pilsbry, H. A., and Olsson, A. A., 1954, Miocene land shell fossils from the Dominican Republic: Notulae Naturae, no. 266, p. 1–4.

Porter, A.R.D., and Anderson, R. E., 1983, Age of the Bonavista gravels, St. Elizabeth, Jamaica: Journal of the Geological Society of Jamaica, v. 13, p. 90–93.

Schlickum, W. R., 1975, Die oberpliozäne Molluskenfauna von Cessey-sur-Tille (Départment Côte d'Or): Archiv für Molluskenkunde, v. 106, p. 47–79.

Schlickum, W. R., and Geissert, F., 1980, Die pliozäne Land- und Süsswassermolluskenfauna von Sessenheim/Krs. Hagenau (Unterelsass): Archiv für Molluskenkunde, v. 110, p. 225–259.

Schlickum, W. R., and Strauch, R., 1979, Die Land- und Süsswassermollusken der pliozänen Deckschichten der rheinischen Braunkohle: Abhandlungen der Senckenbergischen Naturforschenden Gesellschaft, v. 536, p. 1–144.

Simpson, C. T., 1895, Distribution of the land and fresh-water mollusks of the West Indian region, and their evidence with regard to past changes of land and sea: Proceedings, U.S. National Museum, v. 17 (for 1894), p. 423–450, Plate 16.

Taylor, D. W., 1960, Late Cenozoic molluscan faunas from the High Plains: U.S. Geological Survey Professional Paper, no. 337, p. I–IV, 1–94, Plates 1–4.

Trechmann, C. T., 1935, The geology of Carriacou, West Indies: Geological Magazine, v. 72, p. 529–555.

Verdcourt, B.,1963, The Miocene non-marine Mollusca of Rusinga Island, Lake Victoria and other localities in Kenya: Palaeontographica, Abteilung A, Band 121, p. 1–37, +7 Plates.

Woodring, W. P., 1925, Miocene mollusks from Bowden, Jamaica. Pelecypods and scaphopods: Washington, D.C., Carnegie Institution of Washington, no. 366, p. i–vii, 1–222, and 28 Plates.

Woodring, W. P., 1928, Miocene mollusks from Bowden, Jamaica; Part II. Gastropods and discussion of results: Washington, D.C., Carnegie Institution of Washington, no. 385, p. i–vii, 1–564, and 40 Plates.

Woodring, W. P., 1929, Ecology of the mollusks of the Bowden Formation, Jamaica: Geological Society of America Bulletin, v. 40, p. 259–260.

Woodring, W. P., Brown, J. S., and Burbank, W. S., 1924, Geology of the Republic of Haiti: Port-au-Prince, Republic of Haiti, Department of Public Works, 631 p.

Wurtz, C. B., 1955, The American Camaenidae (Mollusca: Pulmonata): Proceedings of the Academy of Natural Sciences of Philadelphia, v. 107, p. 99–143, Plates 1–19.

MANUSCRIPT ACCEPTED BY THE SOCIETY NOVEMBER 11, 1992

The freshwater Mollusca of Jamaica

C.R.C. Paul
University of Liverpool, Department of Earth Sciences, P.O. Box 147, Liverpool L69 3BX, United Kingdom
Philip Hales, R. A. Perrott, and F. A. Street-Perrott
University of Oxford, School of Geography, Mansfield Road, Oxford OX1 3TB, United Kingdom

ABSTRACT

Few freshwater molluscs are endemic to Jamaica, unlike most land snails, and hence they are potentially more useful in broad regional correlation. However, deposits yielding them are rare. This paper records the stratigraphic distribution of sixteen species of gastropods and two bivalves from Pleistocene and Holocene deposits of a core taken in Wallywash Great Pond, St. Elizabeth. The fauna includes the majority of the Recent freshwater species of Jamaica and one possible new taxon, *"Hydrobia"* cf. *rivularis*, which became extinct at the beginning of the Holocene. The fauna shows repeated phases of high and low diversity, which probably reflect lake levels. Nevertheless, *Drepanotrema anatima, Armigerus (A.) albicans*, and *A. (Tropicorbis) edentata* are apparently confined to the Holocene.

INTRODUCTION

Freshwater molluscs are geographically more widespread both within Jamaica and in the Caribbean generally, than most Jamaican land snails (Goodfriend, this volume), but deposits yielding freshwater fossils are less commonly encountered. However, a core from Wallywash Great Pond (Fig. 1), which goes back more than 120,000 years, gives some indication of the biostratigraphic potential of freshwater molluscs. The full limnology of Wallywash Great Pond will be recorded elsewhere (Street-Perrott et al., 1993). The core, just over 9 m deep (Fig. 2), penetrated lake marls and peats, which recorded a complex history of varying lake levels, with fluctuating diverse and restricted molluscan faunas. Radiocarbon and U/Th dates (Fig. 2) suggest that the top 4.75 m or so represent the Holocene, –6 to –4.75 m below datum may represent the last "glacial" epoch from which very few molluscs were recovered. There is then a dramatic increase in age (29,790 at –6 m, 93,000 at –6.4 m) and the lowest 3 m appear to represent the last "interglacial." For the purposes of this chapter, molluscs recorded below –4.75 m are regarded as Pleistocene; those above as Holocene (Fig. 2). The stratigraphic distribution of the species found is shown in Figure 3. The core is important, not only as the first fossiliferous freshwater deposit recorded from Jamaica, but also because the preserved mollusc fauna covers almost the entire known Jamaican fauna (Adams, 1851b; Vendryes, 1899) and includes one possible new taxon. Figure 2 gives a summary of the sedimentology and stratigraphy of the core. Apart from the Wallywash core, the only other known occurrence of fossil freshwater molluscs from Jamaica is an unconfirmed report of some external moulds in dark gray clays within a lignite-bearing part of the Eocene Chapelton Formation (S. K. Donovan, written communication, May 2, 1990).

The taxonomy and nomenclature of Jamaican freshwater molluscs are far from stable. The anatomy of most species remains totally unknown, although Baker (1945) did publish details of some planorbids. Here we use the best available name, but also quote the name under which C. B. Adams first recorded any species, as it is under these names that Johnson and Boss (1972) figured types of Adams' freshwater Jamaican taxa. In a few cases these names cannot be improved on. We are primarily concerned to record the stratigraphic occurrence of species, but one or two taxa for which there are as yet no published figures are illustrated. Representative specimens have been deposited in Liverpool Museum, National Museums on Merseyside (LIVCM), and Oxford University Museum (OUM). The systematic order follows that of Wenz (1938–1944) and Wenz and Zilch (1959–1960).

THE FAUNA

1. *Pomacea fasciata* (Lamarck, 1822)
(=*Ampullaria fasciata* Lamarck of Adams, 1851b)

The common, large apple snail of Jamaica. Rare fragments and juvenile shells occur in the Wallywash core between –8.30 and –7.75 m.

Paul, C.R.C., Hales, P., Perrott, R. A., and Street-Perrott, F. A., 1993, The freshwater Mollusca of Jamaica, *in* Wright, R. M., and Robinson, E., eds., Biostratigraphy of Jamaica: Boulder, Colorado, Geological Society of America Memoir 182.

364   C.R.C. Paul and others

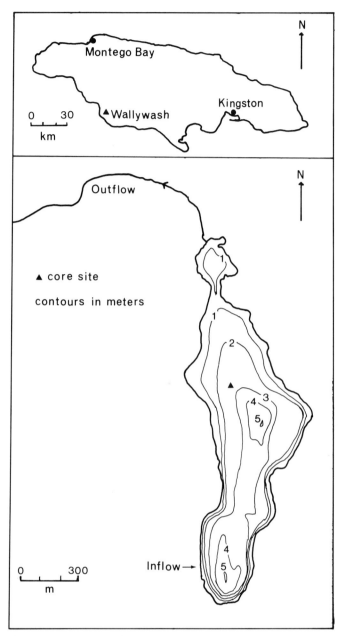

Figure 1. Location map for Wallywash Great Pond and site of the core. The inflow is from a spring and the outflow goes to the Lower Morass. After Street-Perrott et al. (1993, Figs. 2 and 4).

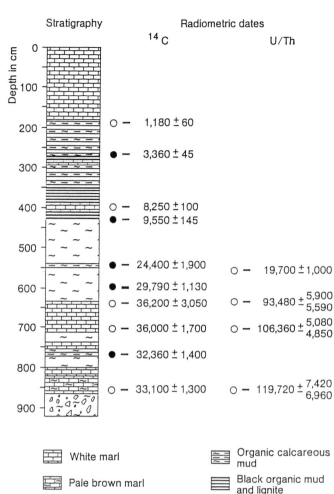

Figure 2. Summary of the stratigraphy of the Wallywash core. Radiocarbon dates are in uncorrected radiocarbon years. Those less than 30,000 years may overestimate age; those over 30,000 years may be discounted as beyond the limits of resolution. Molluscs from below –4.75 m are regarded as Pleistocene, those above as Holocene. (No molluscs are known from between –4.54 and –4.97 m.) After Street-Perrott et al. (1993, Fig. 1).

Age. Pleistocene; extant. Apparently not present at Wallywash during the Holocene.

2. *"Valvata" pygmaea* Adams, 1849.
(= *"V." inconspicua* Adams, 1850)

Minute (1.5 mm high by 1.7 mm wide) *Valvata*-like snails of uncertain systematic affinities. Adams thought two species were present in Jamaica, a more conical form (*"V." pygmaea*) and another with a much lower spire (Wallywash shells reach 1.05 mm high by 1.8 mm wide). For illustrations see Johnson and Boss (1972, Plate 41, Figs. 6 and 7).

However, the abundant material from the Wallywash core shows all gradations between these extremes. It also includes some forms with a marked external rib just behind the aperture, which are much more common in the lower part of the core. Again all gradations seem to occur between forms with unthickened shells and those with a definite and pronounced rib, so no taxonomic significance is attached to this variation. *"V." pygmaea* is probably the most abundant snail in the core. It is present in almost every sample from –9.24 m (the lowest sample available to us) to –1.40 m, and a solitary juvenile shell at –0.2 m, which could possibly be a contaminant.

Age. Pleistocene, Holocene; extant. Living examples were collected from the southern end of Wallywash Great Pond by J. Holmes in August 1990. This species was not found in any other pond investigated. Labiate forms occur in the last "interglacial" and just at the Pleistocene/Holocene boundary.

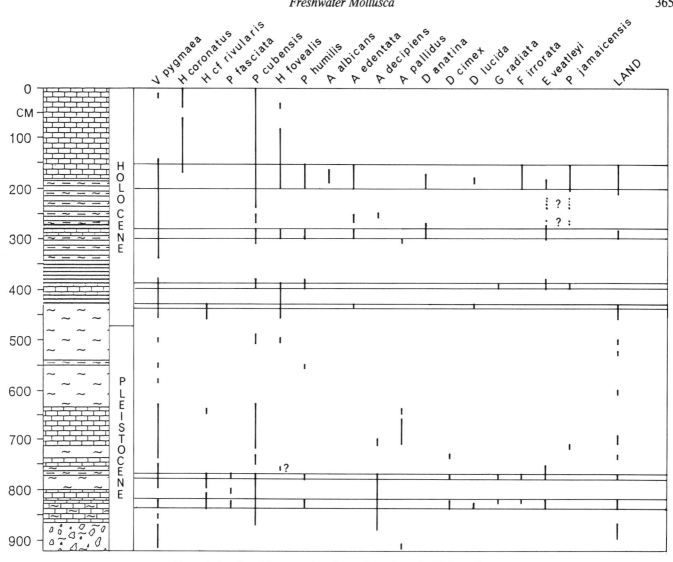

Figure 3. Stratigraphic range chart for molluscs from the Wallywash core.

3. *"Hydrobia" coronatus* (Pfeiffer, 1839a)
(=*Paludina jamaicensis* Adams, 1849;
=*Melania spinifera* Adams, 1845)

A small, high-spired hydrobiid with shells up to 3.8 mm high by 2.1 mm maximum diameter. Some examples develop a spiny keel on the periphery of the last whorl, which has led to the suggestion that they belong in the genus *Potamopyrgus* (e.g., Robart et al., 1977). As far as we are aware no anatomical evidence has been presented to support this assignment. Adams (1845, 1849) named both forms apparently, although his *"Melania" spinifera* has a very strongly developed keel and resembles the Tertiary fossil *Sellia pulchra* (see Johnson and Boss, 1972, Plate 42, Fig. 5). Recent shells from Wallywash collected by J. Holmes in August 1990 vary considerably in their apical angle and one is as strongly ornamented as the type of *"H." spinifera,* although no shells from the Wallywash core are as strongly ornamented. Thompson (1984, p. 35, Figs. 62–64) illustrates shells of *Pyrgophorus platyrachis* Thompson, that include all variations found in *"H." coronatus,* but we have no anatomical information to compare with the Floridan species. Here we refer these shells to *"H." coronatus* (but see comments on the next species). Whatever its real identity, this species appears suddenly at −1.61 m and is present in every sample to the top of the core. It effectively replaces *"V." pygmaea.*

Age. Holocene; extant. Living snails were collected at several locations in Wallywash Great Pond by J. Holmes in August 1990. Apparently this species arrived at Wallywash only about 1,000 years ago.

4. *"Hydrobia"* sp. cf. *Paludina rivularis* Adams, 1845
(?=*Paludina parvulus* Guilding, 1828)
(Fig. 4a, b)

An even smaller, typically high-spired, hydrobiid (Fig. 4a, b) with shells that strongly resemble the illustration of *P. rivularis* Adams (see Johnson and Boss, 1972, Plate 41, Fig. 4) except that they have a marked external rib on the outer lip at the aperture (Fig. 4b). The apex is smaller than that of *"H." coronatus* from Wallywash, as is the overall shell (specimens reach 3.15 mm high but most are only 2.55 mm, by 1.40 mm maximum diameter including the thickened outer lip). Robart et al., (1977, p. 161–162) recorded two species of hydrobiid from Haiti, which they referred to *Potamopyrgus coronatus* (Pfeiffer) and *P. parvulus* (Guild-

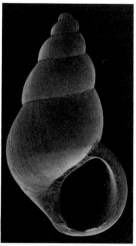

Figure 4. *"Hydrobia"* sp. cf. *Paludina rivularis* Adams, 1845 (?= *Paludina parvula* Guilding, 1828), Pleistocene, Wallywash. a, LIVCM 1991.123.a, apertural view, ×23; b, LIVCM 1991.123.b, lateral view to show prominent external rib characteristic of Wallywash shells, ×23.

ing). *P. coronatus* was said to be larger (5 to 6 mm high) and sometimes to develop a spiny keel; *P. parvulus* is smaller (not more than 4 mm high) and has a thinner shell. However, Guilding's original description and figures give no indication of size and do not match the Wallywash shells well. The illustrations (Guilding, 1828, Plate 28, Figs. 1–3) resemble unkeeled *P. coronatus* and show both spiral and longitudinal striations. The description specifically mentions longitudinal ornament (*"longitudinaliter plicatula,"* Guilding, 1828, p. 537). Guilding's original specimens came from St. Vincent. In contrast, Adams' type of *Paludina rivularis* appears smooth (see Johnson and Boss, 1972, Plate 41, Fig. 4), as are all Wallywash shells. *"P."* rivularis appears to be the best name for this species, but, even so, Wallywash shells are typically smaller than Adams' type and without exception possess a thickened outer lip. No intermediates between this and the preceding taxon have been seen and the two taxa are widely separated stratigraphically. The present form occurs commonly at −8.35 to −7.75 m and at −4.55 to −4.35 m, near the Pleistocene/Holocene boundary.

Age. Pleistocene just into the Holocene; extinct as far as is known.

5. *Physella cubensis* (Pfeiffer, 1838b)
(=*Physa jamaicensis* Adams, 1851a)

A typical, sinistral physid snail (see Johnson and Boss, 1972, Plate 38, Fig. 4) The most widely occurring snail in the Wallywash core. Physid shells are notoriously indistinguishable and it is assumed that all belong to a single species for which *P. cubensis* appears to be the oldest available name. It is possible that some fatter specimens belong to *P. marmoratus* Guilding, 1828. Thompson (1984) and Burch (1989) refer *P. cubensis* to the genus *Physella*. *Physella* occurs in almost every sample from −8.63 to −6.35 m, at about −5.0 m, and then almost continuously from −3.95 m to the top of the core.

Age. Pleistocene, Holocene; extant. Living examples were collected from Wallywash by J. Holmes in August 1990.

6. *Drepanotrema anatina* (d'Orbigny, 1835)
(=*Planorbis haldemani* Adams, 1849)

A small, distinctive planorbid with rounded outer whorl, which overlaps the preceding one asymmetrically (see Johnson and Boss, 1972, Plate 37, Fig. 2). As with most other planorbids, *D. anatina* is restricted to diverse faunas in the Wallywash core. It is known from about −2.9 m and from −1.91 to −1.79 m.

Age. Holocene; extant. Apparently reached Wallywash 3,500 years ago. A single fresh shell was collected from Wallywash by J. Holmes in 1990.

7. *Drepanotrema cimex* (Moricand, 1839)
(*Planorbis macnabianus* Adams, 1849)

A small, thin, multiwhorled planorbid, very like the European *Anisus vortex* (see Johnson and Boss, 1972, Plate 36, Fig. 2). It occurs in diverse faunas at −8.41 to −8.24 m and at about −7.75 m, also in a less diverse fauna at about −7.40 m.

Age. Pleistocene; extant. Not yet known in the Holocene. Apparently *D. cimex* became locally extinct at Wallywash around 110,000 years ago.

8. *Drepanotrema lucida* (Pfeiffer, 1838b)
(=*Planorbis redfieldi* Adams, 1849)

Similar to *D. cimex*, but larger and with fewer, broader whorls (see Johnson and Boss, 1972, Plate 39, Fig. 2). Known from −8.30 m, −4.38 m, and −1.86 m, all in diverse faunas.

Age. Pleistocene, Holocene; extant. Fresh shells were collected from Wallywash by J. Holmes in August 1990.

9. *Armigerus (Armigerus) albicans* (Pfeiffer, 1939a)
(=*Planorbis dentiferus* Adams, 1845)

A small, evolute planorbid with rounded whorls and teeth set internally well back from the aperture, which has the lip deflected on one side (see Johnson and Boss, 1972, Plate 37, Fig. 3). Difficult to separate from the next taxon when juvenile, but undoubted *A. (A.) albicans* occurs in a diverse fauna at −1.89 to −1.71 m, and possibly also at −1.65 m.

Age. Holocene; extant. A few fresh shells were collected from Wallywash by J. Holmes in August 1990. Apparently reached Wallywash about 1,200 years ago.

10. *Armigerus (Tropicorbis) edentatus* (Adams, 1850)
(=*Planorbis dentiferus* var. *edentatus* Adams, 1850)

Almost identical to the above except that it lacks the internal denticles and the outer lip of the aperture is undeflected all round (see Johnson and Boss, 1972, Plate 38, Fig. 1). It occurs in diverse faunas at −4.38 m, from −2.92 to −2.59 m, and again from −1.91 to −1.59 m.

Age. ?Latest Pleistocene, Holocene; extant. Fresh shells were collected from Wallywash by J. Holmes in August 1990. Apparently first arrived at Wallywash about 10,000 years ago, but see comments on the next species.

11. *Armigerus (Tropicorbis) decipiens* (Adams, 1849)
(=*Planorbis decipiens* Adams, 1849)

Similar to the last but with a slightly larger shell, with more weakly striated whorls (see Johnson and Boss, 1972, Plate 36, Fig. 4). Occurs in the Wallywash core in diverse faunas at −8.64 to −7.75 m, at −7.03 m, and at −2.59 m.

Age. Pleistocene, Holocene; extant. Apparently existed intermittently at Wallywash until about 3,300 years ago. However, shells of this and the last species are difficult to distinguish. In general, the Pleistocene examples more closely resemble *A. (T.) decipiens* and the Holocene examples *A. (T.) edentata*, but both do seem to occur together at −2.59 m.

12. *Armigerus (Lateorbis) pallidus* (Adams, 1846)
(=*Planorbis pallidus* Adams, 1846)

A yet larger, distinctly striate, planorbid with very broad whorl section (see Johnson and Boss, 1972, Plate 36, Fig. 3). Occurs in restricted faunas at −9.20 m, −7.08 to −6.46 m, and −3.06 m.

Age. Pleistocene, Holocene; extant. Only one certain Holocene record. *A. (L.) pallidus* appears to have become locally extinct about 4,000 years ago.

13. ?*Helisoma (Pierosoma) fovealis* (Menke, 1830)
(=*Planorbis caribaeus* d'Orbigny, 1841; =*P. affinis* Adams, 1849)

The largest Jamaican planorbid, with a flat-sided, keeled, apparently sinistral, juvenile shell that is unmistakable, and a reticulate ornament that allows even tiny shell fragments to be identified (see Johnson and Boss, 1972, Plate 36, Fig. 1). It is the third most common snail in the Wallywash core and occurs in both diverse and restricted faunas. There are single records based on fragments at −7.37 m and −5.05 m, then it occurs more or less regularly from −4.50 to −3.87 m, from −2.92 to −2.79 m, and from −1.92 to −0.82 m, with final occurrences at −0.67 and −0.38 m.

Age. Pleistocene, Holocene; extant. Living snails were collected from Wallywash by J. Holmes in August 1990.

14. "*Planorbis*" *humilis* Adams, 1850
(Fig. 5a–c)

Johnson and Boss (1972, p. 205) could not locate any specimens of this species in Adams' collection at Harvard. However, from Adams' original description we are fairly confident that the Wallywash shells belong to this species. Since there has been no illustration of it ever published, we include photos and a brief description here. The shell most closely resembles those of the genus *Menetus* (*Micromenetus*) Baker, 1945, which is commonly found in lakes.

Description. Shell minute, planispiral, with rapidly increasing whorls (Fig. 5a), peripheral keel and angular umbilical margin on the left side (Fig. 5b). Protoconch pitted, grading into juvenile whorls that have growth striae and impressed spiral lines both increasing in strength progressively so that the shell eventually appears weakly cancellate (Fig. 5a, c). Whorl-section trapezoidal, flattened on the right side with a peripheral keel, flattened outer margin with sharp, angular umbilical margin, and short columellar and parietal margins. Aperture not flared nor thickened internally (Fig. 5b), but columellar margin very slightly reflected. Plane of the aperture at about 10° to the spiral axis. Wallywash shells have about 2¼ to 2½ whorls and reach 2.2 mm diameter by 0.9 mm high.

Remarks. Adams' original description (1850, p. 131, repeated verbatim in Johnson and Boss, 1972, p. 205) mentions 2½ whorls "with a very acute periphery nearly in the plane of the spire" and "umbilical margin abruptly excavated," and cites the greatest breadth as 0.06 inch (about 1.5 mm). The colour was also unknown suggesting Adams was at least dealing with a long dead shell if not a fossil one. Thompson (1983, 1984) records *Micromenetus dilatatus avus* (Pilsbry) from Jamaica. The subspecies was originally described from the Pliocene of Florida, and differs from the nominate subspecies in being flatter above and more sharply keeled. In this respect it closely resembles the Wallywash shells. If the subspecies is valid, *humilis* is the older name.

"*P.*" *humilis* occurs in diverse faunas at −8.31 to −8.27 m, at −7.74 m, in a restricted fauna at −5.50 m, and in diverse faunas at −3.95 to −3.87 m, −2.92 to −2.89 m, and −1.92 to −1.59 m.

Age. Pleistocene, Holocene; ?extant. A single fresh shell was collected from Wallywash by J. Holmes in August 1990. It was not encountered in any other pond investigated. "*P.*" *humilis* appears to have lived in the Wallywash Great Pond during the last diverse faunal episode. It is possibly also known from the Pliocene of Florida.

15. *Gundlachia radiata* (Guilding, 1828)
(=*G. ancyliformis* Pfeiffer, 1849; *Ancylus obliquus* Adams, 1850 *non* Broderip, 1832; =*A. chittyi* Adams, 1851c [new name for *A. obliquus* Adams])
(Fig. 6a–d)

Guilding (1828, p. 535–536, Plate 26, Figs. 1–9) described and illustrated two freshwater limpets from St. Vincent under the names *Ancylus irroratus* and *A. radiatus*. Both had oval shells and the shell of *A. radiatus* was indicated to be about 7.5 mm long. As the trivial name implies, *A. radiatus* has striations radiating over the shell. Later, Pfeiffer was sent some septate freshwater limpets from Cuba for which he erected the new genus *Gundlachia* (Pfeiffer, 1849, p. 97), with the single species *G. ancyliformis*.

Johnson and Boss (1972) could not locate a type specimen for Adams' *Ancylus obliquus*, and the original description (Adams, 1850, p. 132; repeated verbatim in Johnson and Boss, 1972, p. 210) included the qualifying remarks that Edward Chitty had at that time only one specimen in his collection. Adams' original description mentions "microscopic radiating raised lines" and "apex very prominent, elevated, extending far to one side and posteriorly, and projecting nearly over the margin." Adams also remarked that the shell might belong to a new genus. All evidence leads us to conclude that Adams' species, later renamed *Ancylus chittyi* (Adams, 1851c, p. 204) was, in all probability, a *Gundlachia*.

The status of *Gundlachia* has been debated (see Basch, 1959), but it was originally erected for some freshwater limpets that develop a septum across the majority of the aperture and then grow forward to form a new, usually highly flared aperture (Fig. 6b). Since septate and nonseptate individuals occur, it has been suggested that the formation of a septum is more likely to be a response to adverse environmental conditions than a genetically controlled generic (or even specific) character. Nevertheless, two ancylids occur in the Wallywash core and only one ever develops a

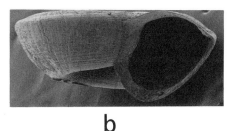

Figure 5. "*Planorbis*" *humilis* Adams, 1851a, Holocene, Wallywash. a, LIVCM 1991.123.c, right-lateral view; b, LIVCM 1991.123.d apertural view; c, LIVCM.1991.123.e, left-lateral view. All ×32.

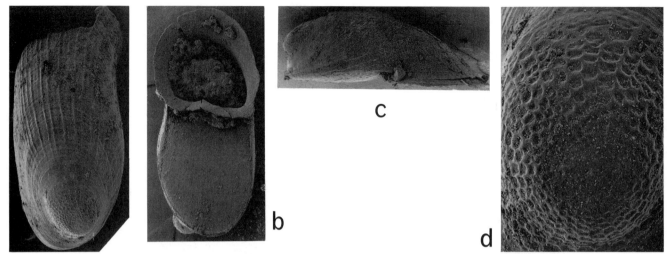

Figure 6. *Gundlachia radiata* (Guilding, 1828), Pleistocene, Wallywash. a, LIVCM 1991.123.f, dorsal view to show parallel-sided outline, striations and pitted protoconch, ×45; b, LIVCM 1991.123.g, ventral view to show septum and post-septate aperture, ×37; c, LIVCM 1991.123.g, lateral view, ×37; d, LIVCM 1991.123.f, detail of protoconch, ×170.

septate shell (Fig. 6b, c). That one is characterized by a parallel-sided shell (Fig. 6b) with conspicuous, raised, radiating lines (Fig. 6a) and a pitted protoconch (Fig. 6d). Despite the Wallywash species having a smaller, parallel-sided shell and all but one specimen being septate, whereas Guilding's original specimens from St. Vincent were larger, oval and nonseptate, we tentatively assign them to the same species. The oldest available name is, therefore, *Gundlachia radiata* (Guilding, 1828).

Septate *Gundlachia* occur not uncommonly in a sample at −8.27 m, a single septate example was found at −7.75 m and another, nonseptate individual occurred in a third sample at −3.94 m. Thus *Gundlachia* appears to have survived into the Holocene, but became locally extinct about 8,250 years ago.

Age. Pleistocene, Holocene; ?extinct. Adams' original description suggests a fresh shell, but as there have been no known recordings of the species since, it may well be extinct in Jamaica as a whole.

16. *Ferrissia (Ferrissia) irrorata* (Guilding, 1828)
(Fig. 7a–c)

This is a larger freshwater limpet, with an oval outline with rounded, not parallel, sides (Fig. 7a), a much lower apex to the posterior right (Fig. 7a, b), and ornamented with a central, dimplelike smooth area surrounded by an area of extremely fine radiating striae that are confined to the protoconch (Fig. 7c). Wallywash shells agree fairly well with Guilding's original description and figures (1828, p. 535, Plate 26).

Although Adams (1851b) reported two species of freshwater limpet from Jamaica, he regarded the first one discovered as synonymous with *Ancylus obscurus* Haldeman. Basch (1963, p. 419) regarded "*A. obscurus* Haldeman" as a junior synonym of *Laevapex fuscus* (Adams, 1840), which Johnson and Boss (1972, Plate 39, Fig. 3) illustrate and which is undoubtedly not conspecific with the Wallywash specimens. The finely

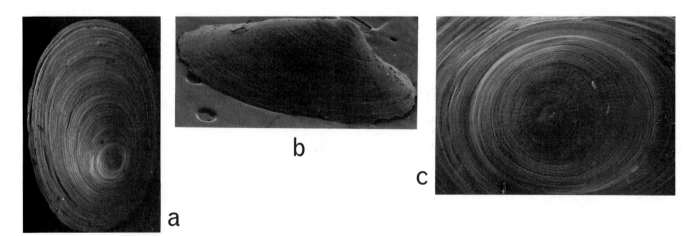

Figure 7. *Ferrissia (Ferrissia) irrorata* (Guilding, 1828), Pleistocene, Wallywash. a, LIVCM 1991.123.h, dorsal view to show oval outline and surface ornament; b, LIVCM 1991.123.i, oblique lateral view; c, LIVCM 1991.123.h, detail of protoconch of shell. a, b, ×18.5, c, ×85.

striate apex is characteristic of *Ferrissia*, which Hubendick (1964, p. 61) regarded as a valid genus and the New World species of which were all assigned to the nominate subgenus.

Ferrissia irrorata occurs rarely with septate *Gundlachia* at −8.27 and −7.75 m, and alone in a diverse fauna between −1.82 and −1.59 m. It thus appears to have become locally extinct about 1,000 years ago.

Age. Pleistocene, Holocene; ?extant.

17. *Eupera veatleyi* (Adams, 1849)
(=*Cyclas veatleyi* Adams, 1849)

A small freshwater bivalve, with a moderately inequivalve shell, prominent umbones and a straight dorsal margin (see Johnson and Boss, 1972, Plate 40, Fig. 4). It occurs in diverse faunas at −8.27 and −7.75 m, in a restricted fauna at −7.55 m, and in diverse faunas at −3.96 to −3.86 m, −2.93 to −2.79 m, and in a sample at −1.90 m.

Age. Pleistocene, Holocene. *Eupera* apparently became locally extinct during the last diverse faunal episode, about 1,200 years ago.

18. *Pisidium* sp. ?*P. jamaicensis* (Prime, 1865);
Cyclas pygmaea Adams, 1849)

An extremely small, inequivalve, bivalve with a subtriangular outline (see Johnson and Boss, 1972, Plate 39, Fig. 4). Possibly more than one species is present, but examples occur in a restricted fauna at −7.10 m and in a diverse fauna between −2.02 and −1.59 m. Fragments of *Pisidium* and/or *Eupera* also occur at −3.95 m, −2.60 m, and −2.30 m.

Age. Pleistocene, Holocene; extant. Live *Pisidium* were collected from near the inlet to Wallywash Great Pond by J. Holmes in August 1990.

ACKNOWLEDGMENTS

We are grateful to J. Holmes, Kingston University, for providing specimens of Recent molluscs collected from Wallywash and other Jamaican ponds in August 1990, and to the reviewers for improving the original draft of this manuscript.

REFERENCES CITED

Adams, C. B., 1840, Descriptions of thirteen new species of New England shells: Boston Journal of Natural History, v. 3, p. 318–332.

Adams, C. B., 1845, Specierum novarum conchyliorum, in Jamaica repetorum, Synopsis: Boston Society of Natural History Proceedings, v. 2, p. 1–17.

Adams, C. B., 1846, Descriptions of undescribed species of shells from the island of Jamaica: Boston Society of Natural History Proceedings, v. 2, p. 102–3.

Adams, C. B., 1849, Descriptions of new species of fresh water shells which inhabit Jamaica: Contributions to Conchology, no. 3, p. 42–44.

Adams, C. B., 1850, Descriptions of new species and varieties of shells, which inhabit Jamaica: Contributions to Conchology, no. 8, p. 129–140.

Adams, C. B., 1851a, Descriptions of new fresh-water shells which inhabit Jamaica: Contributions to Conchology, no. 9, p. 174–175.

Adams, C. B., 1851b, Catalogue of fresh water shells which inhabit Jamaica: Contributions to Conchology, no. 9, p. 187.

Adams, C. B., 1851c, Errata: Contributions to Conchology, no. 10, p. 204.

Baker, F. C., 1945, The molluscan family Planorbidae (with collation, revision, and additions by Harley Jones van Cleave): Urbana, Illinois, University of Illinois Press, xxxvi + 530 p., 141 Plates.

Basch, P. F., 1959, Status of the genus *Gundlachia* (Pulmonata, Ancylidae): University of Michigan Museum of Zoology Occasional Papers, no. 602, p. 1–9.

Basch, P. F., 1963, A review of Recent freshwater limpet snails of North America: Harvard, Museum of Comparative Zoology Bulletin, v. 129, p. 399–461.

Broderip, W. J., 1832, "Minutes of the meeting of Dec. 11, 1832": Zoological Society of London Proceedings, v. 2, p. 194–202.

Burch, J. B., 1989, North American freshwater snails: Hamburg, Michigan, Malacological Publications, 365 p.

Guilding, L., 1828, Observations on the zoology of the Cariboean islands. Mollusca Caribaeana: London Zoological Journal, v. 3, p. 527–544, Plates 26–28.

Hubendick, B., 1964, Studies on Ancylidae: Göteborgs Musei Zoologiska Avdelning Meddelanden, v. 137, p. 1–72.

Johnson, R. I., and Boss, K. J., 1972, The freshwater, brackish, and non-Jamaican land mollusks described by C. B. Adams: Harvard, Museum of Comparative Zoology Occasional Papers on Mollusks, v. 3, p. 193–233, Plates 36–42.

Lamarck, J.B.P.A. de M., de, 1815–1822, Histoire naturelle des animaux sans vertébrés . . .: Paris, v. 6 and 7 (Mollusques), v. 6, Part 1, p. 1–345 (1815); v. 6, Part 2, p. 1–252 (1822); v. 7, p. 1–711 (1822).

Menke, C. T., 1830, Synopsis methodica Molluscorum generum omnium et specierum earum, quae in Museo Menkeano adservantur cum synonymia critica et novarum specierum diagnosibus. Edito altero: Pyrmonti, Georgi Uslar, xvi + 169 p.

Moricand, S., 1839, Premier supplement au memoire sur les coquilles terrestres et fluviatiles de la Province de Bahia: Genève Société de Physique et d'Histoire Naturelle Mémoires, v. 8, p. 139–148, Plate 3.

d'Orbigny, A.C.V.D., 1835, Synopsis terrestrium et fluviatilium molluscorum in suo per Americam meridionalem itinere ab A. d'Orbigny collectorum: Magasin de Zoologie, Classe V, v. 5, nos. 61 and 62, 44 p.

d'Orbigny, A.C.V.D., 1841, Mollusques, *in* Sagra, R., de la, ed., 1839–1857, Histoire physique, politique et naturelle de l'Ile de Cuba: Paris, 12 vols. and atlases.

Pfeiffer, L., 1838a, Ubersicht der im Januar, Februar und Marz 1839 auf Cuba gesammelten Mollusken: Archiv für Naturgeschichte, v. 5, p. 250–261.

Pfeiffer, L., 1839b, Bericht uber die Ergebnisse meiner Reise nach Cuba im Winter 1838–39: Archiv für Naturgeschichte, v. 5, p. 346–358.

Pfeiffer, L., 1849, Neue Molluskengattungen: Zeitschrift für Malakozoologie, v. 6, p. 97–105.

Prime, T., 1865, Monograph of American Corbiculidae (Recent and fossil); Smithsonian Miscellaneous Collections, no. 145, xi + 80 p.

Robart, G., Mandahl-Barth, G., and Ripert, C., 1977, Inventaire, repartition geographique et ecologie des mollusques dulcaquicoles d'Haiti (Caraïbes): Haliotis, v. 8, p. 159–171.

Street-Perrott, F. A., Hales, P. E., and Perrott, R. A., 1993, Limnology and palaeolimnology of a tropical karstic lake: Wallywash Great Pond, Jamaica: Journal of Paleolimnology (in press).

Thompson, F. G., 1983, The planorbid snail *Micromenetus dilatatus avus* in the West Indies and Central America: Nautilus, v. 97, p. 68–69.

Thompson, F. G., 1984, Freshwater snails of Florida. A manual for identification: Gainesville, Florida, University of Florida Press, 94 p.

Vendryes, H., 1899, Systematic catalogue of the land and freshwater shells of Jamaica: Institute of Jamaica Journal, v. 2, p. 590–607.

Wenz, W., 1938–1944, Band 6, Gastropoda, Teil 1: Allgemeiner Teil und Prosobranchia, *in* Schindewolf, O. H., ed., Handbuch der Paläozoologie: Frankfurt, Borntraeger, v. 1, p. i–xii, 1–948, Figs. 1–2704; v. 2, p. 949–1649, Figs. 2705–4211.

Wenz, W., and Zilch, A., 1959–1960, Band 6, Gastropoda Teil 2: Euthyneura, *in* Schindewolf, O. H., ed., Handbuch der Paläozoologie: Frankfurt, Borntraeger, p. i–xii, 1–834, 2513 figs.

MANUSCRIPT ACCEPTED BY THE SOCIETY NOVEMBER 11, 1992

Geological Society of America
Memoir 182
1993

Jamaican Cenozoic Echinoidea

Stephen K. Donovan
Department of Geology, University of the West Indies, Mona, Kingston 7, Jamaica, West Indies

ABSTRACT

The Jamaican Cenozoic echinoid fauna is large and diverse, comprising about 95 nominal species (with further taxa presently classified under open nomenclature) divided between 43 genera. The Eocene fauna is particularly large and includes about 85% of the Jamaican Cenozoic species. In contrast, the echinoids of the Jamaican Paleocene, Oligocene, and Miocene are very poorly known. The Eocene fauna is dominated by oligopygoids, cassiduloids, and spatangoids, while the Neogene and Quaternary fauna is best known from cidaroids and clypeasteroids. Comparatively few regular echinoids are known from the Jamaican Cenozoic. The following species are described and/or reported from the Jamaican fossil record for the first time herein: cf. *Neolaganum dalli* (Twitchell); *Cubanaster* sp. cf. *c. acunai* (Lambert and Sánchez Roig); *Wythella* sp.; *Encope* sp. cf. *E. sverdrupi* Durham; and *Eurhodia* sp. cf. *E. rugosa* (Ravenel).

INTRODUCTION

The Jamaican Cenozoic echinoid fauna is particularly diverse, including about 95 nominal species distributed between 43 genera. However, the stratigraphic distribution of this fauna is very uneven, most taxa being known from the Eocene or (to a lesser extent) the Plio-Pleistocene. Between the moderately diverse fauna of the Jamaican Campanian and Maastrichtian (see Donovan, this volume) and the Eocene Yellow Limestone Group, only two occurrences of fossil echinoids have been reported. These records are not even based on tests, but on cidaroid radioles from the Paleocene (Donovan and Carby, 1989; see cidaroid radioles indet. below) and the identification of the trace fossil *Scolicia* sp. cf. *S. plana* Ksiazkiewicz from the Paleocene of the Richmond Formation (=Mooretown Formation; Pickerill and Donovan, 1991), most probably produced by spatangoid echinoids (Smith and Crimes, 1983). However, most of Jamaica's Paleogene echinoid species have been described from the slightly younger Yellow Limestone Group. The majority of the Cenozoic echinoids described from Jamaica by Arnold and Clark (1927, 1934) and Hawkins (1924, 1927) came from this unit. As Donovan et al. (1991) observed, the Yellow Limestone Group (midearly to mid-middle Eocene: Robinson, 1988, Figs. 3, 4) represents a time span of about 6.5 m.y. (Harland et al., 1982) and has so far produced about 70 nominal species of echinoids, yet the younger Eocene, including eight formations within the White Limestone Supergroup spanning about 8 m.y., has so far yielded only about 20 echinoid species. Donovan (in McKinney et al., 1992) has suggested that this apparent decline after the mid-Eocene may be at least partly due to a lack of collected material and differing diagenetic histories affecting the modes of preservation between the White and Yellow Limestones, although the contribution of the four Eocene extinction events, from the mid-middle Eocene onwards (Prothero, 1989), is difficult to assess. The succeeding Oligocene fauna is even more sparse, including less than five nominal species. The Neogene+ Quaternary have yielded a moderately diverse echinoid fauna (Donovan anad Lewis, research in progress), with a number of horizons producing monospecific assemblages of clypeasteroids (e.g., see Donovan et al., 1989b).

Other problems recognized in reviewing the Jamaican Cenozoic echinoids are that some taxa, particularly *Haimea, Clypeaster, Echinolampas,* and *Macropneustes,* would benefit from a more detailed taxonomic reexamination and revision, and that stratigraphic control is usually poor, leading to stretching of true ranges (Paul, 1985, p. 13–14, Text-fig. 5). Indeed, Arnold and Clark (1927, 1934) were not geologists and showed little interest in the relative ages of the specimens or (1934) even in the localities from which tests were collected. (I attribute the numerous specimens in the 1934 monograph from unknown localities and horizons to Arnold having made no notes about where they were collected before he died in 1932.) It was not until recently (Donovan, 1988b) that any attempt was made to determine the detailed biostratigraphy of the Jamaican echinoids. Nevertheless, we still do not know the localities and horizons that yielded 14 of the specimens described by Arnold and Clark (Table 1, herein).

Donovan, S. K., 1993, Jamaican Cenozoic Echinoidea, *in* Wright, R. M., and Robinson, E., eds., Biostratigraphy of Jamaica: Boulder, Colorado, Geological Society of America Memoir 182.

TABLE 1. JAMAICAN FOSSIL ECHINOIDS FROM UNKNOWN HORIZONS*, WITH DETAILS OF THE STRATIGRAPHIC DISTRIBUTION OF THE GENUS, WHERE KNOWN†

Taxon	Range of Genus
Clypeaster eurychorus	Late Eocene to Recent (late Oligocene to Recent in the New World: Poddubiuk, 1985).
Encope homala	Pliocene (or late Miocene?) to Recent (Mooi, 1989).
Cassidulus? platypetalus *Cassidulus? sphaeroides*	Eocene to Recent (Kier, 1962).
Lambertona jamaicensis	Eocene to Miocene (Kier, 1984)
Linthia obesa	Late Cretaceous to Pliocene
Cyclaster sterea	Late Cretaceous to Recent
Eupatagus clevei	Eocene to Recent (*E. clevei* ranges from the Eocene to early Miocene elsewhere in the Caribbean: Kier, 1984)
Macropneustes? dyscritus *Macropneustes? sinuosus* *Macropneustes? stenopetalus*	Eocene to Recent
Metalia dubia *Metalia jamaicensis*	?Eocene to Recent
Homeopetalus axiologus	Unknown (?Cenozoic)

*Derived largely from Donovan, 1988b, Table 1, with some modifications.
†Moore, 1966, and others.

The present paper discusses every nominal echinoid species known from the Jamaican Cenozoic (excluding those taxa which Gordon, 1989, 1990, has identified from the last interglacial Falmouth Formation, on the basis of disarticulated ossicles). The distribution of these species between formations is summarized in Table 2. One species from each genus is figured and described in detail. Where more than one species is known from any given genus, reference is made to published descriptions, and the differences between figured and unfigured species within the genus are discussed. For diagnoses of genera, the reader is referred to Moore (1966). All of the figured species are Jamaican, although some of the figured specimens are from Cuba and Florida. Most tests were painted with a dark water-soluble dye and whitened with ammonium chloride sublimate before photography. The classification used herein follows Moore (1966) unless otherwise stated. The morphological terminology used in the systematic descriptions follows Moore (1966) and Smith (1984). Museum abbreviations used herein are: MCZ, Museum of Comparative Zoology at Harvard; USNM, U.S. National Museum, National Museum of Natural History, Smithsonian Institution; BMNH, British Museum (Natural History); PRI, Paleontological Research Institution, Ithaca, NY; UWIGM, geological museum of the University of the West Indies at Mona. I have followed the recommendations of Bengtson (1988, p. 226) regarding open nomenclature where taxonomic assignment at the generic and/or specific level has required caution.

The geographic distribution of the Cenozoic echinoid localities of Jamaica and the stratigraphic distribution of the echinoid species are summarized in Figures 1 to 3.

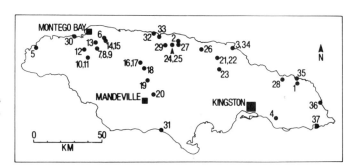

Figure 1. Geographic distribution of the principal Cenozoic echinoid localities of Jamaica (large dots). The geology has been omitted (see McFarlane, 1977a), but the locations of major towns and cities are indicated where convenient (squares). Key to localities. Chepstow Formation (Paleocene): 1, Near Port Antonio, Portland. Font Hill Formation (mid-early to mid-middle Eocene): 2, South of Mt. Zion, St. Ann; 3, Ceran Hill, Port Maria, St. Mary. "Font Hill Formation"*: 4, Easington, St. Thomas. Chapelton Formation (mid-early to mid-middle Eocene): 5, Abingdon, southwest of Green Island harbour, Hanover; 6, Glasgow, St. James; 7, Spring Mount, St. James; 8, Springfield, St. James; 9, Point, St. James; 10, Seven Rivers, St. James; 11, Cambridge, St. James; 12, Montpelier, St. James; 13, Johns Hall, St. James; 14, Ginger Valley, St. James; 15, Somerton, St. James; 16, Albert Town, Trelawny; 17, Freemans Hall, Trelawny; 18, Wait-a-Bit Cave, Trelawny; 19, South of Christiana, Manchester; 20, Peace River, Clarendon; 21, Lucky Hill, St. Mary; 22, Gayle, St. Mary; 23, Guys Hill, St. Catherine. Swanswick Formation (late middle Eocene): 24, Knapdale, St. Ann; 25, West of Bamboo, St. Ann; 26, Beecher Town, St. Ann. Claremont Formation (mid-middle to mid-late Eocene); 27, Between Brown's Town and Bamboo, St. Ann. *Non* Bonny Gate Formation** (late middle Eocene): 28, Content, near Hope Bay, Portland. Brown's Town Formation (Oligocene): 29, Brown's Town, St. Ann. Coastal Group (late Miocene to late Quaternary): 30, Round Hill, Hanover; 31, Farquhar's Beach, Clarendon; 32, East Rio Bueno harbour, St. Ann; 33, Discovery Bay, St. Ann; 34, Port Maria, St. Mary; 35, Navy Island, Portland; 36, Manchioneal, Portland; 37, Bowden, St. Thomas. * Not Font Hill Formation sensu stricto, but a facies of the Yellow Limestone Group intermediate between the shallow-water Chapelton Formation and the deep-water Font Hill Formation, possibly indicative of a shelf-edge environment (E. Robinson, personal communication, 1989). **Not Bonny Gate Formation (although mapped as such: Geological Survey Division, 1987), but an unnamed shallow water unit conformable on the underlying Font Hill Formation (E. Robinson, personal communication, 1989).

TABLE 2. THE STRATIGRAPHIC DISTRIBUTION OF JAMAICAN CENOZOIC ECHINOIDS BY FORMATION AND/OR GROUP

Chepstow Formation (Paleocene)
Cidaroid radioles indet.

YELLOW LIMESTONE GROUP (mid-early to mid-middle Eocene)

Chapelton Formation
Fellius foveatus (Jackson)*
Prionocidaris loveni (Cotteau)
Stenechinus regularis Arnold and Clark
Stenechinus perplexus Arnold and Clark
Phymosoma peloria Arnold and Clark
Trochalosoma? chondra (Arnold and Clark)
Echinopsis simplex (Hawkins)
Oligopygus jamaicensis Arnold and Clark
Haimea ovumserpentis (Guppy)
Haimea alta (Arnold and Clark)
Haimea convexa (Arnold and Clark)
Haimea cylindrica (Arnold and Clark)
Haimea elevata (Arnold and Clark)
Haimea parvipetula (Arnold and Clark)
Haimea rotunda (Arnold and Clark)
Haimea stenopetula (Arnold and Clark)
Fibularia jacksoni Hawkins
Tarphypygus notabilis Arnold and Clark
Tarphypygus ellipticus Arnold and Clark
cf. *Neolaganum dalli* Twitchell
Echinolampas clevei Cotteau
Echinolampas alta (Arnold and Clark)
Echinolampas brachytona Arnold and Clark
Echinolampas lycopersicus? Guppy
Echinolampas paragoga Arnold and Clark
Echinolampas plateia (Arnold and Clark)
Echinolampas strongyla Arnold and Clark
Rhyncholampas? matleyi (Hawkins)
Rhyncholampas? antillarum (Cotteau)
Rhyncholampas? ellipticus (Arnold and Clark)
Schizaster subcylindricus Cotteau
Schizaster bathypetalus Arnold and Clark
Schizaster dumblei? Israelsky
Schizaster hexagonalis Arnold and Clark
Caribbaster dyscritus (Arnold and Clark)
Caribbaster loveni (Cotteau)
Linthia trechmanni Hawkins
Brissus sp. cf. *B. unicolor* (Leske)
Euptagus alatus Arnold and Clark
Eupatagus sp. cf. *E. antillarum* (Cotteau)
Eupatagus defectus Arnold and Clark
Macropneustes? angustus Arnold and Clark
Macropneustes? parvus Arnold and Clark
Meoma antiqua Arnold and Clark
Plagiobrissus loveni (Cotteau)
Plagiobrissus abruptus Arnold and Clark
Plagiobrissus elevatus Arnold and Clark
Plagiobrissus latus Arnold and Clark
Plagiobrissus robustus Arnold and Clark
Asterostoma pawsoni Kier

Asterostoma excentricum L. Agassiz
Antillaster arnoldi Clark

Font Hill Formation
Fellius foveatus (Jackson)
Haimea ovumserpentis (Guppy)

"Font Hill Formation"†
Pedina? sp.
Stenechinus regularis Arnold and Clark
Stenechinus perplexus Arnold and Clark
Triadechinus multiporus Arnold and Clark
Amblypygus americanus Michelin
Haimea elevata (Arnold and Clark)
Haimea pyramoides (Arnold and Clark)
Haimea rugosa (Arnold and Clark)
Echinolampas clevei Cotteau
Echinolampas alta (Arnold and Clark)
Echinolampas altissima Arnold and Clark
Echinolampas brachytona Arnold and Clark
Rhyncholampas? punctatus (Arnold and Clark)
Schizaster altissimus Arnold and Clark
Agassizia inflata Jackson
Macropneustes? altus Arnold and Clark
Macropneustes? angustus Arnold and Clark
Plagiobrissus perplexus Arnold and Clark
Asterostoma excentricum L. Agassiz
Antillaster longipetalus (Arnold and Clark)

YELLOW LIMESTONE GROUP (Formation unknown)
Rhyncholampas? parallelus (Cotteau)

WHITE LIMESTONE SUPERGROUP

Claremont Formation (mid-middle to mid-late Eocene)
Fibularia jacksoni Hawkins
Cubanaster sp. cf. *C. acunai* (Lambert)
Wythella sp.
Eurhodia sp. cf. *E. rugosa* (Ravenel)
Rhyncholampas? matleyi (Hawkins)

Swanswick Formation (late middle Eocene)
Fellius? foveatus? (Jackson)
Prionocidaris loveni (Cotteau)
Cidaroid radioles indet.
Oligopygus wetherbyi de Loriol
Haimea alta (Arnold and Clark)
Haimea sp. or spp.
Tarphypygus sp. cf. *T. ellipticus* Arnold and Clark
Echinolampas clevei Cotteau
Rhyncholampas? alabamensis (Twitchell)
Cassiduloid sp. nov.§
Schizaster subcylindricus Cotteau

Agassizia inflata Jackson
Eupatagus alatus Arnold and Clark

Non Bonny Gate Formation† (late middle Eocene)
Eucidaris sp.
cf. *Fibularia* sp.

Somerset Formation (mid-late Eocene)
Rhyncholampas? matleyi (Hawkins)

Brown's Town Formation (Oligocene)
Cidaroid radioles indet.
Clypeaster cotteaui? Egozcue
Clypeaster lanceolatus? Cotteau
Eupatagus hildae Hawkins

Newport Formation (early to early late Miocene)
Clypeaster concavus? Cotteau
Clypeaster spp.

Sign or Spring Garden Formations (mid-Oligocene to late middle Miocene)
Brissus sp. indet.

COASTAL GROUP
Bowden Shell Bed (Pliocene)
Echinometra lucunter (Linnaeus)

August Town Formation, Round Hill Beds (Plio-Pleistocene?)
Clypeaster sp. cf. *C. rosaceus* (Linnaeus)
Encope sp. cf. *E. sverdrupi* Durham

Manchioneal Formation (Early Pleistocene)
Cidaroid radioles indet.
Echinoneus cyclostomus Leske
Echinolampas(?) sp. nov.(?)
Paleopneustes sp. or *Pericosmus* sp.

Falmouth Formation** (late Pleistocene)
Eucidaris tribuloides (Lamarck)
Diadematoid sp. or spp. (*Diadema antillarum* Philippi and/or *Astropyga magnifica* Clark)
Echinometra sp. or spp. (*E. lucunter* [Linnaeus] and/or *E. viridis* A. Agassiz)
Lytechinus? sp.
Tripneustes? sp.
Spatangoid sp. indet.

COASTAL GROUP (Formation unknown)
Prionocidaris sp.
Echinometra lucunter (Linnaeus)
Mellita quinquesperforata (Leske)

---

*Presence in this formation probable, but uncertain.
†See explanation in Figure 1.
§Awaiting description (see Donovan et al., 1989a).
**After Gordon, 1990.

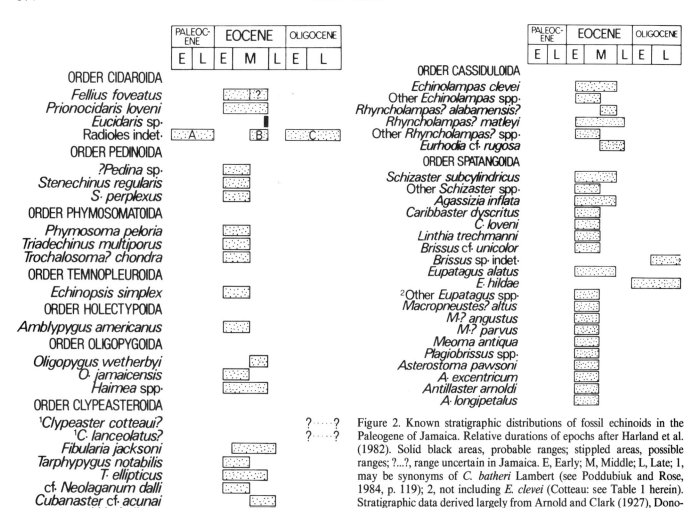

Figure 2. Known stratigraphic distributions of fossil echinoids in the Paleogene of Jamaica. Relative durations of epochs after Harland et al. (1982). Solid black areas, probable ranges; stippled areas, possible ranges; ?...?, range uncertain in Jamaica. E, Early; M, Middle; L, Late; 1, may be synonyms of *C. batheri* Lambert (see Poddubiuk and Rose, 1984, p. 119); 2, not including *E. clevei* (Cotteau: see Table 1 herein). Stratigraphic data derived largely from Arnold and Clark (1927), Donovan (1988b), Hawkins (1924, 1927), McFarlane (1977a, b), and Robinson (1988).

SYSTEMATIC PALEONTOLOGY

Class ECHINOIDEA Leske
Order CIDAROIDA Claus

Remarks. The classification of cidaroids used herein follows Smith and Wright (1989).

Family RHABDOCIDARIDAE Lambert
Subfamily RHABDOCIDARINAE Lambert

Genus *Fellius* Cutress
Fellius foveatus (Jackson, 1922)
(Figure 4.7, 4.8)

Description. (See also Jackson, 1922, p. 17, 18, Plate 1, Figs. 6, 7; Cutress, 1980, p. 116–120, Plate 11). Test moderately large, globular and angular in outline. Ambitus just below midheight of test.

Apical system unknown, but rounded in outline and slightly larger than peristome.

Ambulacra straight adapically, becoming slightly sinuous adorally. The poriferous zones occur in shallow depressions. Pores are oval and conjugate. Interporiferous zones are broad with a medium groove. The tubercles of each ambulacrum are arranged in two parallel the smaller tubercles positioned perradially. Ambulacral plates numerous and small.

Interambulacra broad, about four times the width of the ambulacra at the ambitus, where the interambulacral plates are eleven times higher than the ambulacral plates. Primary tubercles are large, conical, crenulate and perforate, with a broad, elliptical areole, and surrounded by a regularly spaced scrobicular ring. Scrobicular tubercles are larger than ambulacral tubercles and are sometimes associated with other closely packed secondary tubercles. Tertiary tubercles are small, arrayed in rows radiating interradially away from primary tubercles, the rows being separated by shallow grooves. Triangular pits occur at the triple junctions of interambulacral plates.

Peristome rounded and slightly smaller than the apical system.

Radioles are stout, elongate, with a sculpture of low longitudinal ribs. Spinules are present or absent. Where present, spinules are separate, conical to rounded, and angled away from the base of the radiole.

Material. Holotype, MCZ 3234 (Fig. 4.7–4.8), a complete test. Further specimens from Jamaica (Cutress, 1980, p. 120) include MCZ 4158a-c (three tests and six radioles) and PRI 29752 (two tests and one radiole). Two radioles and one test fragment identified as *F. foveatus*? by Dono-

Figure 3. Known stratigraphic distributions of fossil echinoids in the Neogene and Quaternary of Jamaica. Relative durations of epochs after Harland et al. (1982). Key as in Figure 2: P, Pleistocene. Stratigraphic data derived largely from Donovan (1988b), Gordon (1989, 1990, in preparation), and Robinson (1967a, 1969).

van et al. (1989a, p. 7) are to be deposited in the BMNH.
Occurrence. All occurrences appear to be Eocene. The holotype comes ". . . from yellowish limestone, probably the Cambridge formation . . ." (Jackson, 1922, p. 18), that is, the Yellow Limestone Group (mid-early to mid-middle Eocene). PRI 29752 is from Ceran Hill, Port Maria, parish of St. Mary (Cutress, 1980, p. 14), presumably from the Font Hill Formation, Yellow Limestone Group (McFarlane, 1977a). Donovan et al. (1989a) collected *F. foveatus*? from Beecher Town, parish of St. Ann, from the Swanswick Formation, Moneague Group, White Limestone Supergroup, of late middle Eocene age (Robinson, 1988, Fig. 4).
Remarks. The monospecific genus *Fellius* is only known from Jamaica and Cuba. The median row of triangular pits in the interradial suture distinguishes this species from other Caribbean fossil cidaroids.

Family CIDARIDAE Gray
Subfamily CIDARINAE Gray
Tribe CIDARINI Gray
Subtribe CIDARINA Gray

Genus *Cidaris* Leske
Subgenus *Cidaris* (*Tretocidaris*) Mortensen
Cidaris? (*Tretocidaris*?) *anguillensis*? Cutress, 1980

Description. See Cutress (1980, p. 59–63, Plate 2, Figs. 9–12).
Remarks. Arnold and Clark (1927, p. 11, 12) recognized, but did not figure or describe, a small test that they considered to be *Cidaris melitensis* Wright. The locality and horizon of this specimen are unknown, as it ". . . was picked up by a very small boy who was too bewilderd [sic] to remember where he found it . . ."! Further, Cutress (1980, p. 61) could not locate Arnold and Clark's specimen in the MCZ or elsewhere. This is particularly unfortunate, because *C. melitensis* is a Maltese species and Cutress (1980, p. 59–63) reclassified all of the Caribbean tests that had been referred to this taxon as *Tretocidaris anguillensis* Cutress, 1980. This species is otherwise limited to the upper lower Miocene of the Anguilla Formation of Anguilla.

Subtribe PHYLLACANTHINA Smith and Wright

Genus *Prionocidaris* A. Agassiz
Prionocidaris loveni (Cotteau, 1875)
(Figure 4.3, 4.4)

Description. (See also Cutress, 1980, p. 106–116, Plate 10). Test moderately large, rounded in ambital outline and varying from moderately high to globular.
The apical system is unknown, but it is about the same size as the peristome and rounded in outline (Fig. 4.3; note that the peristome in Fig. 4.4 appears unusually large due to breakage).
Ambulacra moderately broad and sinuous. Poriferous zones occur in shallow depressions. Adradial pores elliptical in outline, perradial pores more circular. Pore pairs conjugate. Interporiferous zones narrower than combined width of poriferous zones in any ambulacrum. About 11 to 12 ambulacral plates per interambulacral plate in the ambital region. Columns of larger ambulacral tubercles situated adjacent to the poriferous zones and smaller tubercles perradially.
Interambulacra with about seven to eight plates per plate column. Interambulacra three to four and a half times wider than ambulacra in the ambital region. Interambulacral plates broad and high. Primary tubercles conical, noncrenulate and perforate, with a slightly depressed platform, and broad circular to elliptical areolae. Scrobicular tubercles are circular to elliptical in outline, and larger than the ambulacral tubercles. Tertiary tubercles are small and densely packed, arrayed in rows radiating interradially from the primary tubercles, with rows of tubercles separated by grooves.
Peristome large, rounded in outline and about one-third of the test diameter.
Radioles either cylindrical or tapering gently distally, and highly variable. Sculptured by small and/or large spinules, which are either arranged in rows or are staggered. Large spinules may form a branching distal tip to the radiole.
Material. Jamaican specimens of this species include BMNH E17207, MCZ 4159, and USNM 358873 (four tests, referred to as USNM 301378, the accession number, by Cutress, 1980, p. 113: Fig. 4.3, 4.4 herein).
Occurrence. This species is known from St. Bartholomew (middle Eocene), Cuba (middle Eocene to ?Oligocene), and Jamaica (Cutress, 1980). Most of the specimens available from Jamaica are from the Yellow Limestone Group (mid-early to mid-middle Eocene). The precise locality of USNM (Donovan, 1988a, p. 36) and MCZ specimens is unknown. The BMNH test is from the Chapelton Formation at Spring Mount, parish of St. James (*Cidaris* sp. indet. in Hawkins, 1924, p. 312, 317). Donovan et al. (1989a, p. 7), reported radioles of *P. loveni* from the Swanswick Formation, Moneague Group, White Limestone Supergroup (late middle Eocene); these specimens are to be deposited in the BMNH.
Remarks. P. loveni is the commonest and most widespread of the Caribbean Eocene cidaroids. Donovan (1988b, p. 126) also reported *Prionocidaris* sp. cf. *P. spinidentatus* (Palmer in Sánchez Roig, 1949) from the Pleistocene of Jamaica (elsewhere in the Caribbean it is known from the Oligocene[?], Miocene and Pliocene; Cutress, 1980, p. 94). This identification was made on the basis of disarticulated test plates from the Coastal Group north of Port Maria, parish of St. Mary. Until more complete specimens are available, these plates are best regarded as *Prionocidaris* sp. The available specimens are deposited in the BMNH.
Cutress (1980, p. 111–112) considered *Cidaris gymnozona* Arnold

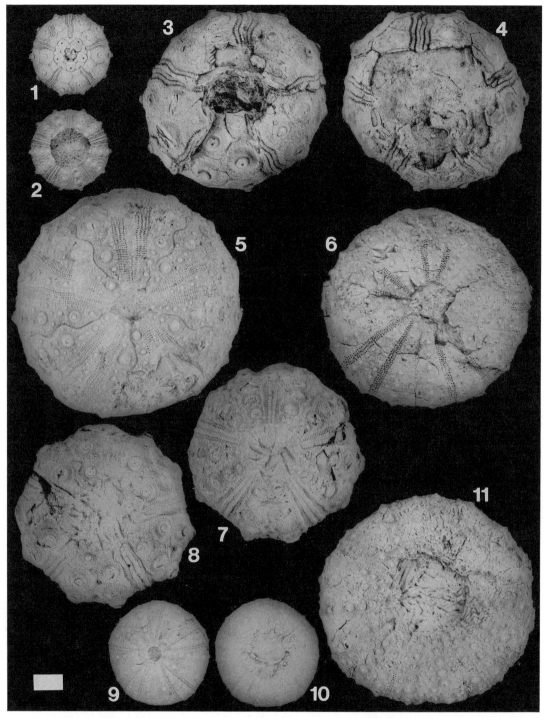

Figure 4. 1, 2, *Eucidaris tribuloides* (Lamarck), Recent, south coast of Palisadoes, parish of St. Andrew: 1, USNM E40377, apical view; 2, USNM E40374, oral view. 3, 4, *Prionocidaris loveni* (Cotteau), USNM 358873, Eocene, Yellow Limestone Group (precise locality and horizon unknown: Donovan, 1988a, p. 36): 3, apical view; 4, oral view. 5, *Triadechinus multiporus* Arnold and Clark, Eocene, "Font Hill Formation," Yallahs River valley, parish of St. Thomas, apical view. 6, 11, *Phymosoma peloria* Arnold and Clark, paratype, USNM 444305, Eocene, Chapelton Formation, probably from the Spring Mount area, parish of St. James: 6, apical view; 11, oral view. 7, 8, *Fellius foveatus* (Jackson), holotype, MCZ 3234, Eocene, Yellow Limestone Group (precise locality and horizon unknown: Cutress, 1980, p. 120): 7, apical view; 8, oral view. 9, 10, *Stenechinus regularis* Arnold and Clark, Eocene, Chapelton Formation, Spring Mount area, parish of St. James: 9, apical view; 10, oral view. Scale bar represents 8 mm for 9 and 10, 10 mm for all other figures.

and Clark, 1927 (holotype, MCZ 3261, from the Chapelton Formation at Spring Mount) to be synonymous with *P. leoni*.

Genus *Eucidaris* Döderlein
Eucidaris tribuloides (Lamarck, 1816)
(Figures 4.1, 4.2, 5)

Description. (See also Weisbord, 1969, p. 287–294, Plates 14, 15, Figs. 1–5; Phelan, 1970, p. 12, 13, Plates 16, 17, 18, Figs. 1–3; Cutress, 1980, p. 66–70). Test moderately high, but flattened adapically and adorally. Test convex in the ambital region, ambitus at about half the test height. Test outline circular, subpentagonal, or pentagonal.

Apical system dicyclic and rounded pentagonal in outline. Genital plates subtrapezoid in outline and slightly elongated, narrowing adorally (Fig. 4.1). The large genital pores occur close to, but not at, the circumference of the apical system. Madreporite slightly larger than other genital plates. Some or all genital plates separated by periproctal plates (Melville and Durham, 1966, Fig. 186.5). Ocular plates small and triangular, with the longest side at the circumference. Periproctal plates arranged in irregular, small circlets around the central, small periproct (Phelan, 1970, Plate 17, Fig. 6).

Ambulacra narrow, sinuous, expanding adambitally, but then narrowing slightly adorally. Poriferous zones occur in slightly sunken grooves. Pores round, separated by a low interporal partition and with the perradial pore adjacent to the shallow groove of the plate suture. Columns of larger tubercles prominent and arranged adjacent to the poriferous zones. Smaller tubercles offset from larger tubercles and crowded perradially. About eight to nine ambulacral plates per interambulacral plate adambitally.

Interambulacra three to four times wider than ambulacra adambitally. Interambulacral plates about one and a half times to twice as wide as high adambitally. Seven to ten plates per interambulacral column. Primary tubercles circular to elliptical in outline, becoming most elongate adorally. Primary tubercles have a low conical boss, narrow areole, are noncrenulate and perforate. Scrobicular tubercles are variable in size, often with a D-shaped boss, and are generally larger than ambulacral tubercles. The surrounding tertiary tubercles are densely packed and variable in size, arrayed in rows radiating from primary tubercles in the interradial region and separated by shallow grooves.

Peristome slightly larger than apical system and rounded pentagonal in outline (Fig. 4.2). Composed of imbricating plates.

Primary radioles (Fig. 5) have a low, smooth, gently tapered base, with an acetabulum about half the diameter of the base. The milled ring is slightly convex, with a coarser pattern of ridges and grooves than the striated, gently conical collar. Neck low and smooth. Shaft moderately long, slightly inflated in the median region. Spinules rounded to elongate, separate, arranged in longitudinal rows. Spinules larger and more closely spaced distally. Areas between rows of spinules smooth, with a coarse, reticulate stereom surface. Shaft truncated by a low, ribbed crown, with a central raised prominence. Scrobicular radioles broad, blunt, and short.

Material. Specimens from the fossil record of Jamaica include a unique test (BMNH collections) and a radiole, BMNH E83644b (Figs. 5.2, 5.3). Gordon (1989, 1990) has deposited her large collection of disarticulated *E. tribuloides* ossicles in the BMNH.

Occurrence. *E. tribuloides* is an extant, common shallow-water species (0 to 450 m) in the seas of the West Indies, Gulf of Mexico, Bermuda, and the Atlantic coast of South America to Brazil (Weisbord, 1969, p. 293, 294; Phelan, 1970, p. 13). Disarticulated ossicles of this species are locally abundant in the last interglacial Falmouth Formation at east Rio Bueno harbour, parish of St. Ann (Gordon, 1989, 1990; locality 3 of Donovan and Gordon, 1989). A test and further radioles were collected from a Pleistocene raised reef (?Falmouth Formation) exposed in a road cutting on the northeast side of Round Hill, parish of Hanover. Cutress (1980, p. 70) reported a personal communication from the late Dr. Thomas Goreau Senior, who found this species in the Pliocene(?) of the Discovery Bay area, parish of St. Ann, presumably from the Hopegate Formation (Plio-Pleistocene), but this report is unconfirmed.

Remarks. Radioles attributed to *Eucidaris* sp. have been identified from the late middle Eocene at Content, near Hope Bay, parish of Portland (Donovan et al., 1991). This is the oldest known *Eucidaris*. *E. strobilata* Fell, 1954, from New Zealand is late Eocene or early Oligocene in age and is not dissimilar to the Jamaican specimens (four radioles, BMNH EE983a-c, EE988b).

Incertae familiae

Cidaroid ossicles indet.
(Figure 6)

Remarks. McFarlane (1977b, p. 1397, 1403) recorded "Spines of *Cidaris* sp." from the Brown's Town Formation (Oligocene) and spine fragments (presumably cidaroid; see Donovan et al., 1989a) from the Swanswick Formation (late middle Eocene). The record from the Brown's Town Formation is particularly important, as very few Jamaican Oligocene echinoids are known. However, the present author has been unable to locate these specimens in the UWIGM.

The only record of a Jamaican Paleocene echinoid is based on a collection of cidaroid radioles and interambulacral plates collected from near Port Antonio, parish of Portland (Donovan and Carby, 1989; Fig. 6 herein). These ossicles may represent *Palmerius roberti* Cutress, 1980, or a species of *Temnocidaris* (*Stereocidaris*).

Hawkins (in Trechmann, 1930, p. 215, 216) recognized "*Cidaris* spines" from the Navy Island Member of the Manchioneal Formation (early Pleistocene: Robinson, 1967a, 1969), which he tentatively attributed to *Stylocidaris affinis* (Philippi).

Order PEDINOIDA Mortensen
Family PEDINIDAE Pomel

Genus *Pedina* L. Agassiz

Pedina? sp.

Description. (See also, Arnold and Clark, 1927, p. 17). Test moderately large, domed adapically and flattened adorally, with a low, rounded ambitus. The ambulacral and interambulacral regions are difficult to differentiate, except for a small area of one ambulacrum near the ambitus. Ambulacral plates appear to be trigeminate, but the precise arrangement of pore pairs and component plate boundaries is uncertain. Primary tubercles are low, domed, and ?imperforate.

Occurrence. A single specimen, MCZ 3382, from the Yallahs River valley, parish of St. Thomas, presumably from the "Font Hill Formation," Yellow Limestone Group (mid-early to mid-middle Eocene) exposed to the north of Easington Bridge (Robinson, 1965).

Remarks. This is a very poorly preserved specimen. The apex has been broken off, presumably during collection; the adoral surface is largely obscured by sediment; and the rest of the test is poorly visible or damaged. Arnold and Clark (1927, p. 17) considered that this specimen has perforate primary tubercles and trigeminate ambulacral plating.

Genus *Stenechinus* Arnold and Clark

Remarks. The genus *Stenechinus* is only known from the Eocene of Jamaica. Arnold and Clark (1927) described two species, *S. perplexus* and the type species *S. regularis*.

Stenechinus regularis Arnold and Clark, 1927
(Figure 4.9, 4.10)

Description. (See also Arnold and Clark, 1927, p. 15–17, Plate 1, Figs.

Figure 5. *Eucidaris tribuloides* (Lamarck), radioles. 1, 4, BMNH E83646b, Recent, Discovery Bay, parish of St. Ann: 1, lateral view of shaft (total length about 14 mm); 4, oblique view of distal tip. 2, 3, BMNH E83644b, Pleistocene, Round Hill, parish of Hanover: 2, oblique view of distal tip; 3, lateral view. All scanning electron micrographs of specimens coated with 60% gold-palladium. Scale bars represent 1 mm.

7–9). Test moderately high, subglobular, pentagonal to well-rounded pentagonal in outline, gently domed adapically and more flattened adorally. Ambitus slightly adoral to the midheight of the test.

Apical system dicyclic. Genital plates equal, madreporite not distinguished. Genital plates broad, with genital pores near plate centers. Oculars small and pentagonal. Periproct moderately large and rounded pentagonal in outline.

Ambulacra moderately broad, expanding adambitally and tapering more adapically than adorally. Ambulacral plates trigeminate, the median component-plate bearing a primary tubercle adjacent to the pore pairs. About 21 ambulacral plates per column. Poriferous zones straight and narrow. Pores rounded, large, closely spaced in each pore pair, with a low interporal partition. Interporiferous zones broad. Secondary tubercles sparse.

Interambulacra about twice as wide as the ambulacra are at the ambitus. About 14 interambulacral plates per column. Ambital interambulacral plates about twice as wide as high. Primary tubercles low, probably noncrenulate (but preservation of specimens is too poor to determine this with confidence) and perforate, arranged in columns towards the center of interambulacral plates. Secondary tubercles irregularly spaced adradically, few interradially.

Peristome large, rounded pentagonal in outline, with five pairs of triangular to elongate buccal notches.

Material. Holotype, MCZ 3264, plus ten paratype tests in the B. W. Arnold collection of the MCZ.

Occurrence. All known occurrences of this species are from the Yellow Limestone Group (mid-early to mid-middle Eocene). The type area is the Spring Mount region, parish of St. James (Arnold and Clark, 1927, p. 15), from the Chapelton Formation. Other specimens are recorded (Arnold and Clark, 1927, p. 17) from the "Font Hill Formation" in the Yallahs River valley, north of Easington Bridge, parish of St. Thomas; the Chapelton Formation from near Welcome Hall (=Springfield), parish of St. James; and the Chapelton Formation of the Guys Hill district, parish of St. Catherine.

Stenechinus perplexus Arnold and Clark, 1927

Description. See Arnold and Clark (1927, p. 14–15, Plate 1, Figs. 4–6).
Material. Holotype, MCZ 3263, plus two paratype tests in the B. W. Arnold collection of the MCZ.
Occurrence. All known occurrences of this species are from the Yellow Limestone Group (mid-early to mid-middle Eocene). The type area is the Spring Mount region, parish of St. James (Arnold and Clark, 1927, p. 14), from the Chapelton Formation. The paratype specimens are from the "Font Hill Formation" in the Yallahs River valley north of Easington Bridge, parish of St. Thomas.
Remarks. S. *perplexus* differs from the type species in the following characters: genital plates are subequal, with the madreporite probably distinguished; ambulacra taper almost as much adorally as adapically; and interporal partitions are more prominent (although they may be abraded in the holotype; Arnold and Clark, 1927). Whether these features are sufficiently distinct to recognize two species is debatable. A thorough reassessment of this genus is planned (Donovan, research in progress).

Order PHYMOSOMATOIDA Mortensen
Family PHYMOSOMATIDAE Pomel

Genus *Phymosoma* Haime

Phymosoma peloria Arnold and Clark, 1927
(Figure 4.6, 4.11)

Description. (See also Arnold and Clark, 1927, p. 18–20, Plate 2, Figs. 1–3). Test large, low, highly rounded pentagonal in outline. Adoral surface broad and flattened, adapical surface gently domed. Ambitus low, just above the adoral surface.

Apical system unknown, but much smaller than peristome and probably pentagonal in outline.

Ambulacra broad, widest adambitally, tapering more adapically than adorally. Ambulacral plates show diadematoid compounding. Poriferous zones straight and flush with the test surface. Pore pairs arrayed in

Figure 6. Cidaroid sp. indet., Paleocene, Chepstow Formation, between Port Antonio and Fellowship, parish of Portland. 1, BMNH E82817, oblique view of interambulacral plate; 2, BMNH E82818, shaft of radiole. Both scanning electron micrographs of specimens coated with 60% gold-palladium. Scale bars represent 1 mm.

slightly offset and closely spaced double rows adapically (Fig. 4.6), but as single rows adorally (Fig. 4.11). Pores rounded and closely spaced. Interporiferous zones broad. Ambulacral and interambulacral tubercles forming vertical columns. Primary ambulacral tubercles large, crenulate, imperforate, and situated close to the poriferous zones. Secondary and tertiary tubercles are distributed throughout the interporiferous zones.

Interambulacra are about 1.6 times broader than ambulacra adambitally. Interambulacral plates are broad, with 18 plates per ambulacral column. Primary tubercles crenulate, imperforate, present centrally on the interambulacral plates. Secondary and tertiary tubercles numerous.

Peristome sunken, large, rounded, with broad, triangular buccal notches.

Material. Holotype, MCZ 3266, plus 58 paratypes in the B. W. Arnold collection of the MCZ and a single paratype in the collection of the USNM, 444305 (ex-MCZ 3378: Fig. 4.6, 4.11 herein). USNM acc. 301378 contains four further tests (Donovan, 1988a, p. 37).

Occurrence. This species is only known from the Chapelton Formation of the Yellow Limestone Group (mid-early to mid-middle Eocene). The majority of the type specimens are from the Spring Mount area, parish of St. James, although *P. peloria* is also recorded from near Cambridge, St. James (Arnold and Clark, 1927, p. 18, 20).

Remarks. P. peloria is the commonest of the noncidaroid regular echinoids known from the Jamaican Cenozoic. It is easily distinguished by its size, shape, and the presence of adapical double columns of pore pairs.

Family STOMECHINIDAE Pomel

Genus *Triadechinus* Arnold and Clark

Remarks. The monospecific genus *Triadechinus* is based on a unique specimen from Jamaica.

Triadechinus multiporus Arnold and Clark, 1927
(Figure 4.5)

Description. (See also Arnold and Clark, 1927, p. 20–22, Plate 1, Figs. 10, 11). Test large, subglobular, circular in outline, with a low ambitus. Test flattened adapically. Adoral surface apparently flattened, but largely obscured by sediment.

Apical system unknown, but small and angular pentagonal in outline.

Ambulacra broad and slightly sinuous. Ambulacral plates show polygeminous diadematoid compounding. Poriferous zones arranged as triple columns of pore pairs (at least adapically) that are flush with the test surface, with over 300 pore pairs per poriferous zone (Arnold and Clark, 1927, p. 22). Adjacent pore pairs of each row are slightly offset. Pore pairs of adjacent columns interdigitate (Fig. 4.5) near the apex in some poriferous zones. Pores small, rounded, and closely spaced in each pore pair. Interporiferous zones narrow adapically, expanding to about as wide as the combined width of the associated poriferous zones adambitally. Ambulacral primary tubercles not present adjacent to the apical system, but becoming large adambitally and distributed in an irregularly alternating pattern on either side of the sinuous perradial suture. Smaller tubercles also irregularly distributed.

Interambulacra about twice as wide as ambulacra adambitally. Median suture sinuous and forming a prominent groove. Interambulacral plates high. Primary tubercles large, noncrenulate and imperforate, occurring adradially of the concave regions of the interradial suture (Fig. 4.5). Primary tubercles are thus strongly offset from each other. Primary tubercles surrounded by a scrobicular ring of very small tubercles and an outer ring of larger tubercles of varying sizes. At least four different orders of smaller tubercle are present on interambulacra.

Material. A unique holotype, MCZ 3267.

Occurrence. The type locality is Mt. Sinai, north of Easington Bridge, west bank of the Yallahs River, parish of St. Thomas (Arnold and Clark, 1927, p. 21), from the "Font Hill Formation," Yellow Limestone Group (mid-early to mid-middle Eocene).

Remarks. Mortensen (1935, p. 513) and Fell and Pawson (1966, p. U408) suggested that *Triadechinus* may be of Cretaceous antiquity, but there are no rocks of this age in the Yallahs River valley. The only echinoids that are known to occur in this region are from the lateral equivalent of the Font Hill Formation (Robinson, 1965, p. 21).

Arnold and Clark referred to this species as "extraordinary" and rightly so. The large size, distinctive sinuous median grooves to plate series (particularly the interradial suture), broad poriferous zones, and prominent interambulacral primary tubercles make this one of the most distinctive of Caribbean fossil echinoids.

Genus *Trochalosoma* Lambert

Trochalosoma? chondra (Arnold and Clark, 1927)
(Figure 9.3, 9.4)

Description. (See also Arnold and Clark, 1927, p. 17, 18, Plate 2, Figs. 4–6). Test flattened adapically and more adorally, rounded pentagonal in outline, elongate and bilaterally symmetrical, with the greatest test width about 65% of the length from the blunt "posterior."

Apical system unknown, but small and apparently elongate.

Ambulacra broad adambitally, tapering adorally and adapically. Ambulacral compounding polygeminous, probably diadematoid. Poriferous zones flush with test surface and composed of double columns of pore pairs. Pores rounded and closely spaced. Interporiferous zones about 65% of the ambulacral width adambitally. Primary tubercles of ambulacra arranged as columns adjacent to poriferous zones. Primary tubercles large, noncrenulate and imperforate. Smaller tubercles occur perradially.

Interambulacra slightly broader than ambulacra at the ambitus. About 16 low, broad, interambulacral plates per column. Primary tubercles imperforate, noncrenulate, arranged in columns and occurring centrally on interambulacral plates. Smaller tubercles form columns adjacent to primary tubercles, but the interradial region appears bare.

Peristome obscured by sediment, but presumably moderately large. Arnold and Clark (1927, p. 18) considered broad, shallow buccal notches to be present.

Material. A unique holotype, MCZ 3265.

Occurrence. The holotype comes from the Seven Rivers region, parish of St. James, either from the Maastrichtian *Veniella* Shales of the Seven Rivers Inlier (see Donovan, Chapter 6, this volume, Fig. 1F) or (most probably) from the Chapelton Formation, Yellow Limestone Group (mid-early to mid-middle Eocene).

Remarks. Some uncertainty exists concerning the age of this specimen. Mortensen (1935, p. 515) stated that "The age of the Jamaican species . . . is uncertain." Fell and Pawson (1966, p. U408) considered that all *Trochalosoma* species were of Late Cretaceous age (but see discussion of the affinities of *T. chondra* below). Donovan (1988b, p. 126) placed this species in the Eocene; Donovan and Bowen (1989) considered it to be Maastrichtian; and Donovan (1990) restored it to the Eocene once more, arguing that all other echinoids from the Seven Rivers district have undoubted Eocene affinities and that *T. chondra* was probably derived from the same horizon.

Further, it is debatable if the Jamaican species really is a true *Trochalosoma.* Arnold and Clark (1927, p. 18) considered that "There seems little reason to doubt that this is a typical *Leiosoma* [=*Trochalosoma*; see below], although the tuberculation is different in detail from that of any other known species." Further, the test shape and size of the apical system are atypical of the genus ("Test . . . somewhat wheel-shaped . . . apical system large"; Mortensen, 1935, p. 515). Although Arnold and Clark considered that the cracked test was probably deformed from an original radial symmetry, the outline is too regular to support this supposition. It was certainly elongate and bilaterally symmetrical in life. Also, the apical system is small, whereas in the type, *T. rugosum* (Cotteau; Mortensen, 1935, Fig. 307a, c; Fell and Pawson, 1966, Fig. 304.3b, c), the apical system and peristome are both large. It is probable that *T. chondra* represents a new genus, although further and superior material will be required before it can be diagnosed with accuracy.

Arnold and Clark (1927, p. 17) placed this species in the genus *Leiosoma* Cotteau, an objective synonym of *Trochalosoma* (Fell and Pawson, 1966, p. U408).

Order TEMNOPLEUROIDA Mortensen
Family GLYPHOCYPHIDAE Duncan

Genus *Echinopsis* L. Agassiz

Echinopsis simplex (Hawkins, 1924)
(Figure 7)

Description. (See also Hawkins, 1924, p. 317, 318, Plate 18, Figs. 4, 5). Test small, moderately high, pentagonal in outline with rounded angles, and flattened adorally and adapically.

Apical system small, dicyclic (D. N. Lewis, written communication, 1990), occurring in a shallow depression. Madreporite relatively large. Periproct hexagonal, excentric towards I.

Ambulacra moderately broad adambitally, tapering adorally and more so adapically. Ambulacral plates show trigeminate diadematoid compounding. Poriferous zones single and straight. Pores moderately large, rounded to D-shaped, separated by raised interporiferous partitions. Interporiferous zones about 60% of the width of ambulacra adambitally. Tuberculation sparse adapically. Primary tubercles about central on ambulacral plates. Tubercles perforate and noncrenulate. Smaller tubercles moderately sparse.

Interambulacra moderately broad, about one and a half times wider than ambulacra at the ambitus, tapering adapically and adorally. Primary tubercles perforate, noncrenulate, situated adradially on interambulacral plates. Secondary and tertiary tubercles sparse, particularly adapically.

Peristome moderately large, rounded pentagonal in outline, with well-developed, slightly elongate buccal notches with raised margins.
Material. A unique holotype, BMNH E17208.
Occurrence. The type area is the Spring Mount region, parish of St. James, from the Chapelton Formation, Yellow Limestone Group (mid-early to mid-middle Eocene).

Remarks. Hawkins (1924) originally classified this species within the genus *Herbertia* Lambert, an objective synonym (Fell, 1966, p. U357) of *Echinopedina* Cotteau (Donovan, 1988b, p. 126). Lewis (1986, p. 26) correctly recognized that this echinoid is an *Echinopsis*.

Order ECHINOIDEA Claus
Family ECHINOMETRIDAE Gray

Genus *Echinometra* Gray

Echinometra lucunter (Linnaeus, 1758)
(Figure 8)

Description. (See also Jackson, 1914, p. 154–157; Mortensen, 1943, p. 357–368, plus numerous plate figures; Weisbord, 1969, p. 302–310, Plates 16, 17, Figs. 1–5). Test outline varies from rounded to elliptical to elongate pentagonal, elongation occurring in the direction 3-I. Test moderately high, flattened adapically, and with a gently concave adoral surface, with the long axis of the concavity being perpendicular to the long axis of the test.

Apical system small, dicyclic apart from ocular V, which is usually insert (Mortensen, 1943, Fig. 174). Genital plates slightly elongated, hexagonal to heptagonal in outline, with the genital pores situated near the periproctal margin. Madreporite conspicuously larger than other genital plates. Oculars small, pentagonal or hexagonal. Periproct small, displaced towards genital 5, surrounded by irregular circlets of periproctal plates.

Ambulacra broadest adambitally, tapering adapically but only slightly adorally. Ambulacral plates show polygeminate echinoid compounding, with at least 21 compound plates per ambulacral plate column in larger specimens. Poriferous zones flush with test surface. Pore pairs arranged in short arcs around ambulacral tubercles, each arc being closest to the adjacent tubercle adapically. Six pore pairs per arc adambitally, varying between five and seven pore pairs per arc adapically. Poriferous zones expand to form moderately broad phyllodes adjacent to the peristome. Pores rounded, closely spaced and separated by a low interporal partition. Interporiferous zones taper adapically and adorally. Primary ambulacral tubercles moderately large, conical, imperforate and noncrenulate, about as large or larger than secondary tubercles of interambulacra. Smaller ambulacral tubercles numerous.

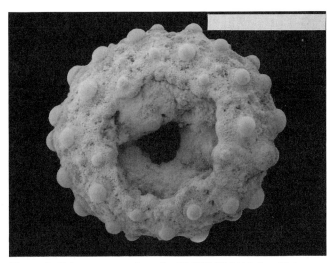

Figure 8. *Echinometra* sp. juvenile, BMNH E83640, late Pleistocene, Falmouth Formation, east Rio Bueno harbour, parish of St. Ann, oral(?) view. Scanning electron micrograph of specimen coated with 60% gold-palladium. Scale bar represents 0.5 mm.

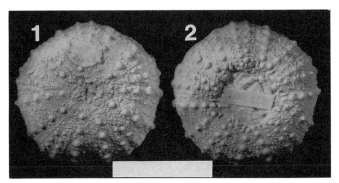

Figure 7. *Echinopsis simplex* (Hawkins), holotype, BMNH E17208, Eocene, Chapelton Formation, Spring Mount, parish of St. James. 1, apical view; 2, oral view. Scale bar represents 10 mm.

Interambulacra about 1.5 to 1.7 times broader adambitally than ambulacra in larger specimens, which have 13 or more interambulacral plates per plate column. Primary tubercles conical, imperforate, noncrenulate, with broad areole, particularly adapically. Interambulacra with a dense packing of smaller tubercles of at least three orders. Secondary tubercles are particularly prominent, both as a zigzag interradial series and adradially. Adjacent columns of primary tubercles converge adorally, where the adradial plate surfaces are bare (apart from secondary tubercles adjacent to adradial suture) just adapically of the buccal notches.

Peristome large, rounded, central, with broad, shallow buccal notches. Auricles are greatly extended by a tag produced at their symphysis (Mortensen, 1943, p. 362).

Material. A single test, MCZ 3499, plus numerous disarticulated ossicles of *Echinometra*, from both *E. lucunter* and *E. viridis* A. Agassiz, in the collection of Miss C. M. Gordon (1989, 1990; deposited in the BMNH). A juvenile test (Fig. 8), BMNH E83640, referred to *Echinometra* sp., was unfortunately severely damaged subsequent to photography.

Occurrence. E. lucunter and the superficially similar *E. viridis* are common shallow-water species in Jamaica, the Caribbean, and western Atlantic. Fossil *E. lucunter* in Jamaica is known from the Pliocene and Pleistocene of the Coastal Group. The MCZ test is from the parish of St. Elizabeth, horizon and locality uncertain (Arnold and Clark, 1934, p. 140), although the matrix is reminiscent of that found in raised reefs. Weisbord (1969, p. 309) mentioned *E. lucunter* from the late Miocene of Bowden, St. Thomas. The present author has been unable to locate this specimen. If it came from the Bowden shell bed, then it is more correctly regarded as Pliocene in age (Robinson, 1967a, 1969; Stanley and Campbell, 1981). Gordon (1989, 1990) has collected abundant *Echinometra* ossicles from the last interglacial Falmouth Formation between Discovery Bay, parish of St. Ann, and Rio Bueno, parish of Trelawny, particularly at east Rio Bueno Harbour (Donovan and Gordon, 1989, locality 3).

Remarks E. viridis differs from *E. lucunter* in having pore pairs arrayed in arcs of five rather than arcs of six; in having no insert oculars; in having fewer coronal plates; and in lacking elongate tags on the auricles (Mortensen, 1943, p. 362, 372). However, complete tests of *E. viridis* have as yet to be reported from the Jamaican fossil record.

Order HOLECTYPOIDA Duncan
Suborder ECHINONEINA H. L. Clark
Family ECHINONEIDAE L. Agassiz and Desor

Genus *Echinoneus* Leske

Echinoneus cyclostomus Leske, 1778
(Figure 9.5, 9.6)

Description. (See also Mortensen, 1948, p. 71–80, Plate 1, Figs. 14, 26). Test oval, small to moderate in size. Width about 75% of length and height about 50% of length in figured specimen (Fig. 9.5, 9.6).

Apical system central, elongate pentagonal in outline and tetrabasal, with four genital pores (genital 5 missing), and a prominent madreporite.

Ambulacra straight and widest adambitally. Pore pairs uniserial adapically and arrayed in arcs of three adorally. Poriferous zones are narrow and slightly sunken. Interporiferous zones broad.

Interambulacra about two to two and a half times wider than ambulacra adambitally. Test surface densely covered with imperforate, noncrenulate primary tubercles, which are sunken into shallow pits that are deepest adorally. Glassy tubercles and small secondary tubercles are densely packed adapically.

Peristome central, oblique, irregularly rounded triangular in outline, with the long axis orientated 2-4 and the longest side of the triangle anterior (Fig. 9.6). Periproct inframarginal, just posterior to, and slightly longer than, the peristome. Periproct teardrop-shaped, rounded anteriorly, and pointed posteriorly (Fig. 9.6).

Occurrence. Recent *E. cyclostomus* has a circumtropical distribution (Mortensen, 1948, p. 79), but is unknown from western Africa and the Pacific coast of the Americas. This led Mortensen to suggest that the Indo-Pacific and Caribbean populations of *Echinoneus* probably represent different species, although they are morphologically inseparable.

Hawkins in Trechmann (1930, p. 216) identified *E. cyclostomus* from the early Pleistocene (Robinson, 1967a) Manchioneal Formation of the type area, parish of Portland. The whereabouts of this specimen is unknown. This species is also recorded from the lower Miocene of Cuba (Sánchez Roig, 1949; Brodermann, 1949) and Anguilla (Jackson, 1922, p. 54, 55; Poddubiuk and Rose, 1984).

Remarks. Confusion appears to exist concerning the plating arrangement of the apical system in *Echinoneus cyclostomus.* Mortensen (1948, p. 72) makes the ambiguous statement that "... the genital plates are not separately limited...." This may have been the authority that influenced Wagner and Durham (1966, p. U445) to state that the apical system is monobasal. However, the apical system of the figured specimen (Fig. 9.5) is clearly tetrabasal, with a distinct madreporite, a feature that *Echinoneus* shares with other members of the Family Echinoneidae.

Incerti subordinis
Incertae familiae

Genus *Amblypygus* L. Agassiz

Amblypygus americanus Michelin, 1856
(Figure 9.1, 9.2)

Description. (See also Jackson, 1922, p. 55, 56; Arnold and Clark, 1927, p. 25, Plate 3, Figs. 1–3; Cooke, 1959, p. 27, Plate 7, Figs. 8, 9). Test large, rounded, rounded pentagonal, or oval in outline. Test moderately high with a subangular apex and a gently convex adoral surface, with a deep depression around the peristome. The ambitus is low and rounded.

Apical system central, tetrabasal, with four genital pores (genital 5 missing), the madreporite central and much larger than other genital plates, and small oculars.

Ambulacra narrow, composed of numerous ambulacral plates. Open petals developed adapically (Fig. 9.1). Pore pairs broad, conjugate, with adradial pores elongate and perradial pores rounded to weakly elliptical, occurring in a straight line. Adorally all pores are rounded in outline. Interporiferous zone widest adambitally, tapering adapically and adorally.

Interambulacra broad, about four and a half times wider adambitally than ambulacra. Interambulacral plates numerous and low. Test surface covered by very numerous tubercles that are inset in shallow pits, with closely packed secondary tubercles covering the remainder of the test surface. Primary tubercles perforate and crenulate (Wagner and Durham, 1966, p. U450).

Peristome oblique, small, situated in a central adoral concavity (Fig. 9.2). Peristome subtriangular in shape, with the longest side convex and orientated 2-4. Periproct inframarginal, slightly closer to the peristome than to the posterior margin. Periproct larger than the peristome, teardrop-shaped, and broadest posteriorly.

Material. Jamaican specimens of *A. americanus* include the holotype in the Michelin collection (Jackson, 1922, p. 55), 28 specimens in the B. W. Arnold collection of the MCZ (Arnold and Clark, 1927, p. 25) and two further tests collected by Arnold in USNM acc. 301378 (Donovan, 1988a, p. 37).

Occurrence. Where the locality is known, Jamaican specimens all come from the "Font Hill Formation," Yellow Limestone Group

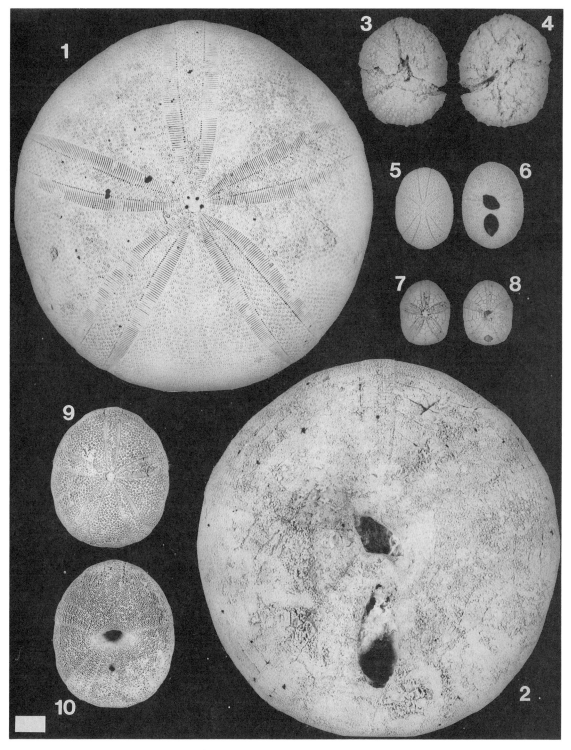

Figure 9. 1, 2, *Amblypgus americanus* Michelin, USNM 444300, Eocene, Crystal River Formation, Florida: 1, apical view; 2, oral view. 3, 4, *Trochalosoma? chondra* (Arnold and Clark), holotype, MCZ 3265, probably Eocene, Chapelton Formation, Seven Rivers area, parish of St. James: 3, apical view; 4, oral view. 5, 6, *Echinoneus cyclostomus* Leske, USNM E40375, Recent, south side of Palisadoes, parish of St. Andrew: 5, apical view; 6, oral view. 7, 8, *Haimea ovumserpentis* (Guppy), USNM 444303, Eocene, Yellow Limestone Group (precise locality and horizon unknown: Donovan, 1988a, p. 38): 7, apical view; 8, oral view. 9, 10, *Oligopygus wetherbyi* de Loriol, USNM 649833, Eocene, Crystal River Formation, Florida: 9, apical view; 10, oral view. Scale bar represents 5 mm for 5 and 6, 10 mm for all other figures.

(mid-early to mid-middle Eocene) of the Mt. Sinai area, north of Easington Bridge, west bank of the Yallahs River valley, parish of St. Thomas (Arnold and Clark, 1927). Also known from the southeastern United States (Cooke, 1959).

Remarks. This species was first described by Michelin (1856), although it is usually attributed to Desor (1858). Michelin considered the age of this species to be "supracrétacées" and not Upper Cretaceous as was misreported by Donovan (1988a). Jamaican tests are often smaller and more elongated than the large, rounded tests that are best known from Florida and Georgia. It is perhaps significant that *A. americanus* is typically a Jacksonian (=Priabonian=late Eocene) species in the southeastern United States (Cooke, 1959; McKinney and Zachos, 1986; Carter, 1987a, 1989; Carter and Hammack, 1989; Carter et al., 1989) and is thus somewhat younger than the Jamaican occurrence of this taxon. This stratigraphic distribution may be because *A. americanus* has a specific substrate preference (Carter et al., 1989) or could indicate that the Jamaican and North American specimens are not conspecific.

Order OLIGOPYGOIDA Kier

Remarks. The classification of the oligopygoids used herein follows Kier (1967). Members of this order are only known from the middle and late Eocene (Kier, 1967, Table 2) and have a distribution that is centered on, and almost restricted to, the Caribbean region.

Family OLIGOPYGIDAE Duncan
Genus *Oligopygus* de Loriol

Remarks. Three species of *Oligopygus* have been reported from Jamaica. Arnold and Clark (1927, p. 28–30) named two taxa, *O. hypselus* and *O. jamaicensis*, while McFarlane (1977b, p. 1397) recorded *O. floridanus* Twitchell. Kier regarded *O. floridanus* to be a junior synonym of the type species, *O. wetherbyi* de Loriol (Kier, 1967, p. 58, 59; see also Cooke, 1959, p. 28), and *O. hypselus* to be a junior subjective synonym of *O. jamaicensis* (Kier, 1967, p. 65–67).

Oligopygus wetherbyi de Loriol, 1887
(Figure 9.9, 9.10)

Description. (See also Cooke, 1959, p. 28, Plate 8, Figs. 9–12; Kier, 1967, p. 54–59, with numerous illustrations). Test anteroposteriorly elongated, but broad, either oval or highly rounded pentagonal in outline. Test widest just posterior of the peristome, moderately high, gently domed adapically, with a well-rounded ambitus on the lower half. Adoral surface gently concave, particularly adjacent to the peristome.

Apical system approximately central, pentagonal in outline, and monobasal with four genital pores.

Ambulacra narrow, open petaloid adapically, petals not extending to the ambitus. Petals expanding gently and attaining their greatest width adambitally. Pore pairs of petals conjugate, with adradial pores elongate and perradial pores circular. Pore pairs angled slightly away from apical system. Adoral to the petals the poriferous zones comprise single pores. Interporiferous zones widest adambitally, tapering adorally and slightly more adapically.

Interambulacra about two and a half times wider than ambulacra adambitally. Surface of test irregularly covered by numerous perforate, crenulate, tubercles inset in round, shallow pits. Remaining test surface covered by very numerous, densely packed small tubercles.

Peristome broad, low, diamond-shaped, with the long axis corresponding to the widest part of the test. Peristome occurs in a deep, broad central concavity on the adoral surface. Periproct inframarginal, small, circular, positioned slightly closer to the peristome than to the posterior margin.

Material. Although very common in the southeastern United States, only four tests of this species are known from Jamaica, UWIGM 2784a–d.

Occurrence. Jamaican specimens are only known from the Swanswick Formation, Moneague Group, White Limestone Supergroup (late middle Eocene), from the minor road from Thatchfield to Knapdale, east-northeast of Brown's Town, parish of St. Ann, GR 4392 5478 (McFarlane, 1974).

Remarks. The Jamaican occurrence of this species is somewhat older than in the United States, where it is used as a zone fossil for the late late Eocene (McKinney and Jones, 1983; McKinney and Zachos, 1986).

Oligopygus jamaicensis Arnold and Clark, 1927

Description. See Kier (1967, p. 63–67, Plate 16, Figs. 1–7; Plate 17, Figs. 1, 2; Text Fig. 29).

Material. Holotype of *O. jamaicensis*, MCZ 3270, plus 245 paratype specimens in the B. W. Arnold collection of the MCZ (Arnold and Clark, 1927, p. 29); holotype of *O. hypselus*, MCZ 3269, plus 16 paratype specimens in the B. W. Arnold collection of the MCZ; and a further 24 tests of *O. jamaicensis* in the USNM, acc. 301378 (Donovan, 1988a, p. 38).

Occurrence. All specimens came from the area around Spring Mount, parish of St. James (Arnold and Clark, 1927), from the Chapelton Formation, Yellow Limestone Group (mid-early to mid-middle Eocene).

Remarks. *O. jamaicensis* differs from *O. wetherbyi* in having a slightly broader, higher test, a more inflated adoral surface, and a relatively broader peristomal trough with a straight anterior margin (Kier, 1967, p. 67).

Genus *Haimea* Michelin

Remarks. Arnold and Clark (1927) described 12 nominal species of *Haimea* (as *Pauropygus* Arnold and Clark; see Arnold and Clark, 1934, p. 143) from the Eocene of Jamaica: *H. alta* (Arnold and Clark), *H. convexa* (Arnold and Clark), *H. cylindrica* (Arnold and Clark), *H. elevata* (Arnold and Clark), *H. lata* (Arnold and Clark), *H. ovumserpentis* (Guppy), *H. parvipetula* (Arnold and Clark), *H. platypetula* (Arnold and Clark), *H. pyramoides* (Arnold and Clark), *H. rotunda* (Arnold and Clark), *H. rugosa* (Arnold and Clark), and *H. stenopetula* (Arnold and Clark). Arnold and Clark (1934, p. 143) later suggested that *H. elevata* is a junior synonym of *H. caillaudi* Michelin, but Kier (1967, p. 101, 102) considered these two species to be morphologically distinct. However, *H. caillaudi* was reported from Jamaica by Lambert (1925). Cooke (1961, p. 15) suggested that *H. elevata* may be a junior synonym of *H. ovumserpentis*, although this was rejected by Kier (1967, p. 48).

Kier (1967) was unable to revise fully the genus *Haimea*, because most species are based on few specimens with poor stratigraphic control. Kier did, however, suggest that two Jamaican species, *H. lata* and *H. platypetula*, may be junior synonyms of *H. ovumserpentis*. Nevertheless, the Jamaican *Haimea* fauna is in need of further revision.

A key to the known species of *Haimea* appears in Kier (1967, p. 94).

Haimea ovumserpentis (Guppy, 1866)
(Figure 9.7, 9.8)

Description. (See also Kier, 1967, p. 121–126, Plate 34, text Figs. 3, 10). Test rounded to elongate, moderately high, flattened adorally and gently domed adapically. Ambitus well-rounded and about one-third of the height of the test.

Apical system central, moderately large, rounded to pentagonal in outline, monobasal, with four genital pores.

Ambulacra petaloid adapically. Petals broadly open, expanding gently adambitally and extending almost to the margin of the adapical surface (Fig. 9.7). Petals in shallow, elongate grooves adradially, with the interporiferous zones slightly swollen. Perradial pores rounded, arranged in an approximately straight column. Adradial pores elongate, arranged in a gently curved column. Pores well separated within pore pairs, which are slightly oblique to the axes of the ambulacra. Interporiferous zones about one-third the width of the petals, except at their tips, where the poriferous zones narrow. Ambulacra of approximately constant width between tips of petals and ambitus, tapering gently adorally. Pores single adorally of the petals, with the development of demiplates (Kier, 1967, Fig. 3).

Interambulacra broad at the ambitus, about three and a half to four times broader than ambulacra, and tapering adapically and adorally. Primary tubercles common over the entire test, inset in circular pits, with very numerous small secondary tubercles grouped over the intervening test surface.

Peristome moderately large, pentagonal, situated in a central concavity, with bourrelets more or less developed. Periproct inframarginal, rounded polygonal in outline, smaller than the peristome and situated close to the posterior margin (Fig. 9.8).
Material. Arnold and Clark (1927, p. 38) included 1,111 specimens in the B. W. Arnold collection of the MCZ within this species and there are 41 other tests (USNM acc. 301378) in the Arnold collection of the USNM (Donovan, 1988a, p. 38), including USNM 444303 (Fig. 9.7, 9.8). MCZ 3277 is the unique holotype of *H. platypetula.* Arnold and Clark (1927, p. 35, 36) recorded nine specimens of *H. lata,* of which MCZ 3275 is the holotype. The tests recorded by Donovan et al. (1989a, p. 8) as *Haimea* sp. or spp. include many that are probably referable to *H. ovumserpentis* (specimens to be deposited in the BMNH).
Occurrence. Arnold and Clark (1927, p. 38) recorded this species as ". . . the most common fossil sea-urchin in Jamaica, occurring . . . particularly in the Cambridge-Catadupa section and the hills between Spring Mount and Montpelier." The named localities are all within the parish of St. James and all lie within the outcrop area of the Chapelton Formation, Yellow Limestone Group (mid-early to mid-middle Eocene). McFarlane (1977b, p. 1395) also identified this species from the Font Hill Formation of the Yellow Limestone Group at GR 45105512, south of Mount Zion, parish of St. Ann (McFarlane, 1974). Donovan et al. (1989a) recorded *Haimea* sp. or spp. from the Swanswick Formation, Moneague Group, White Limestone Supergroup (late middle Eocene) at Beecher Town, parish of St. Ann. The locality of *H. platypetula* is given as "western Jamaica." The type area for *H. lata* is Spring Mount and other specimens come from the Seven Rivers district, parish of St. James, and south of Lucky Hill, parish of St. Mary (all Chapelton Formation). The species also occurs in Trinidad, St. Bartholomew, Cuba, Peru, and Bonaire (Kier, 1967, p. 125, 126).
Remarks. *H. ovumserpentis* has tended to be used as a "dustbin" taxon, including all of those *Haimea* tests not distinct enough for the erection of a new species. Thus, Arnold and Clark comment that ". . . specimens referred to Guppy's species . . . are not definitely placed in some other species" (1927, p. 38). The true limits to variation within *H. ovumserpentis* are unknown; certainly, some of Arnold and Clark's "species" may just be extreme members of a highly variable continuum (e.g., *H. alta, H. pyramoides*). However, variation within *Haimea* species has not been described from single populations, so any attempt at reclassification would be highly subjective, a problem recognized by Kier (1967).

Haimea alta (Arnold and Clark, 1927)

Description. See Arnold and Clark (1927, p. 33, Plate 4, Figs. 15–17) and Kier (1967, p. 102–107, various illustrations, particularly Plate 26).
Material. Holotype, MCZ 3271, plus 174 specimens in the Arnold collection of the MCZ including 25 paratypes, MCZ 3395, 3396 (Kier, 1967, p. 104). Donovan (1988a, p. 37) recorded 22 tests in USNM acc. 301378. A single test, to be deposited in the BMNH, was identified by Donovan et al. (1989a, p. 8).
Occurrence. The distribution of this species is similar to that of *H. ovumserpentis.* The type area is Spring Mount, parish of St. James, and it is also known from near Lucky Hill, parish of St. Mary, both from the Chapelton Formation, Yellow Limestone Group (mid-early to mid-middle Eocene), and from Beecher Town, parish of St. Ann, from the Swanswick Formation, Moneague Group, White Limestone Group (late middle Eocene). The species is also known from Cuba.
Remarks. *H. alta* is distinctive in having a highly inflated test of well-rounded elliptical outline (see also Kier, 1967, p. 105). This species is very similar to *H. cylindrica* and *H. elevata,* but the test is broader in *H. alta.* For further discussion of how these species differ, see Kier (1967).

Haimea convexa (Arnold and Clark, 1927)

Description. See Arnold and Clark (1927, p. 33, 34, Plate 4, Figs. 18–20) and Kier (1967, p. 111, 112, Plate 29, Figs. 3–6).
Material. A unique holotype, MCZ 3272.
Occurrence. The type area is Lucky Hill, parish of St. Mary, from the Chapelton Formation, Yellow Limestone Group (mid-early to mid-middle Eocene).
Remarks. Kier (1967, p. 112) suggested that *H. convexa* can be differentiated from other species of *Haimea* ". . . by the combination of its strongly inflated petals with deeply depressed poriferous zones, adapical interambulacral plates with well-developed nodes, and very high peristome with its height twice its width." What this does not say is that the specimen appears to have suffered some growth deformity, so it is not symmetrical about a plane III-5. This is particularly shown by the peristome, which is elongated V-2 (Kier, 1967, Plate 29, Figs. 5, 6). *H. convexa* may be a growth abnormality of another species, comparable with those tests figured by Kier (1967, Plate 12).

Haimea cylindrica (Arnold and Clark, 1927)

Description. See Arnold and Clark (1927, p. 34, 35, Plate 4, Figs. 12–14) and Kier (1967, p. 108–110, Plate 27, Fig. 6; Plate 28; Plate 29, Figs. 1, 2; Text fig. 45).
Material. Holotype, MCZ 3273, plus nine paratypes, MCZ 3394 (Kier, 1967, p. 108; Arnold and Clark, 1927, p. 34, referred 20 specimens to this species). Four tests in USNM acc. 301378 are referred to *H. ?cylindrica* (see Donovan, 1988a, p. 37, 38).
Occurrence. The type specimens were collected from between Seven Rivers and Welcome Hall (=Springfield), parish of St. James, from the Chapelton Formation, Yellow Limestone Group (mid-early to mid-middle Eocene).
Remarks. The similarity of this species to *H. alta* (see above) is suggested by Phelan on a label included with the USNM specimens.

Haimea elevata (Arnold and Clark, 1927)

Description. See Arnold and Clark (1927, p. 35, Plate 5, Figs. 1–3) and Kier (1967, p. 99–102, Plate 25, Figs. 1–4; Text fig. 42).
Material. Holotype, MCZ 3274, plus six paratypes, MCZ 3397 (Kier, 1967, p. 99: Arnold and Clark based their description on 32 specimens).
Occurrence. The type area for this species is the west side of the Yallahs River valley, parish of St. Thomas, presumably from the "Font Hill Formation," Yellow Limestone Group (mid-early to mid-middle Eocene) from north of Easington Bridge. Arnold and Clark also reported this species from the parish of St. James, presumably from the Chapelton Formation.
Remarks. See above for a discussion of the similarities and differences of *H. elevata, H. alta,* and *H. cylindrica* (also see Kier, 1967, p. 102).

Haimea parvipetula (Arnold and Clark, 1927)

Description. See Arnold and Clark (1927, p. 38, 39, Plate 5, Figs. 13–15) and Kier (1967, p. 119–121, Plate 32, Fig. 6; Plate 33, Figs. 1–3; Text figs. 1, 48).
Material. Holotype, MCZ 3276, plus five paratypes, MCZ 3405 (Kier,

1967, p. 121: Arnold and Clark described this species on the basis of 21 specimens).
Occurrence. The type area is the Spring Mount region, parish of St. James, from the Chapelton Formation, Yellow Limestone Group (mid-early to mid-middle Eocene).
Remarks. This species is particularly distinctive, having short, broad petals and a submarginal periproct.

Haimea pyramoides (Arnold and Clark, 1927)

Description. See Arnold and Clark (1927, p. 39, 40, Plate 5, Figs. 19–21) and Kier (1967, p. 107, 108, Plate 27, Figs. 1–5).
Material. Holotype, MCZ 3278, plus two paratypes, MCZ 3406 (Kier, 1967, p. 108: Arnold and Clark described this species on the basis of six specimens).
Occurrence. The type area is the Yallahs River valley, parish of St. Thomas, presumably from the "Font Hill Formation," Yellow Limestone Group (mid-early to mid-middle Eocene), north of Easington Bridge.
Remarks. The holotype of this species has a distinct high and conical adapical surface quite unlike any other *Haimea.* Kier (1967, p. 107) did not consider the paratypes of this species to be conspecific with the holotype, which may be an aberrant specimen of another species.

Haimea rotunda (Arnold and Clark, 1927)

Description. See Arnold and Clark (1927, p. 40, 41, Plate 6, Figs. 1–3) and Kier (1967, p. 126–128, Plate 32, Figs. 1–5).
Material. Holotype, MCZ 3279, plus two paratypes, MCZ 3407 (Kier, 1967, p. 128: Arnold and Clark described this species on the basis of six specimens). Kier (1967, p. 126–127) considered that only one of the paratypes is conspecific with the holotype.
Occurrence. The type area is the Spring Mount region, parish of St. James, from the Chapelton Formation, Yellow Limestone Group (mid-early to mid-middle Eocene).
Remarks. The test of *H. rotunda* is low and highly rounded, with a large peristome. Kier (1967, p. 128) considered that this species and *H. rugosa* are probably conspecific, slight differences between the two taxa being explainable by differences of preservation. *H. rotunda* has a lower test and the petals are flush with the test surface, unlike *H. rugosa.*

Haimea rugosa (Arnold and Clark, 1927)

Description. See Arnold and Clark (1927, p. 41, Plate 6, Figs. 4–6) and Kier (1967, p. 118, 119, Plate 31, Figs. 1–3).
Material. Holotype, MCZ 3280. Kier (1967, p. 118) was uncertain whether the six paratypes, MCZ 3408, are conspecific with the holotype (Arnold and Clark described this species from 23 or 24 specimens). Twenty other tests in USNM acc. 301378 are attributed to *H. rugosa* (Donovan, 1988a, p. 38).
Occurrence. The type area is the Yallahs River valley, parish of St. Thomas, presumably from above Easington Bridge, in the "Font Hill Formation," Yellow Limestone Group (mid-early to mid-middle Eocene).
Remarks. This taxon may be conspecific with *H. rotunda* (see above and Kier, 1967, p. 119).

Haimea stenopetula (Arnold and Clark, 1927)

Description. See Arnold and Clark (1927, p. 41–42, Plate 6, Figs. 7–9) and Kier (1967, p. 128, 129, Plate 35, Figs. 1–3; Text fig. 6).
Material. Holotype, MCZ 3281. Kier (1967, p. 128) did not consider the three paratypes in the collections of the MCZ to be conspecific with the holotype (Arnold and Clark described this species from 12 tests).
Occurrence. The type area is Spring Mount, parish of St. James, from the Chapelton Formation, Yellow Limestone Group (mid-early to mid-middle Eocene).
Remarks. Kier (1967, p. 129) suggested that the holotype of this particularly low, elongate species may be a highly weathered specimen of *H. ovumserpentis* (Guppy).

Order CLYPEASTEROIDA A. Agassiz

Remarks. The classification of clypeasteroids used herein follows Durham (1966) and Mooi (1989).

Suborder CLYPEASTERINA A. Agassiz
Family CLYPEASTERIDAE L. Agassiz

Genus *Clypeaster* Lamarck

Remarks. Poddubiuk (1985) recognized that fossil *Clypeaster* in the Caribbean has been grossly oversplit. For example, over 60 nominal species have been recorded from the interval late Oligocene to early Miocene in the West Indies, which should actually be grouped into no more than seven taxa. *Clypeaster* in Jamaica is not so well known, but Donovan (1988b, p. 127) listed five named species: *C. concavus* Cotteau, 1875 (synonymous with *C. antillarum* Cotteau, 1875: Poddubiuk, 1985, p. 76); *C. cotteaui* Egozcue in Cotteau, 1897; *C. eurychorus* Arnold and Clark, 1934; *C. lanceolatus* Cotteau, 1897; and *C.* sp. cf. *C. rosaceus* (Linnaeus, 1758). Further, Chubb (1958) and McFarlane (1977b) both recognized *Clypeaster* spp. from different horizons. Additionally, Donovan and Lewis (research in progress) have recognized two horizons in the Miocene Newport Formation, which include *Clypeaster* referable to the *"concavus"* and *"altus"* morphotypes *sensu* Rose and Poddubiuk (1987).

Of the five named *Clypeaster* species from Jamaica, *C. concavus* and *C. lanceolatus* are referred to by Arnold and Clark (1927, p. 26, 27) as being recorded on a list, provided to them by H. L. Hawkins, of fossil echinoids from the island. However, no formal description of the Jamaican specimens referred to these species has been published, although Arnold and Clark (1927, p. 26) did describe *C. antillarum* (=*C. concavus*) from the island.

The following key is to the nominal species of *Clypeaster* from Jamaica:

1. Petals open 2
   Petals closed or nearly closed 3
2. Test low, petals broadly open .... *C. cotteaui*
   Test moderately high, tips of petals incurved .... *C. rosaceus*
3. Test high 4
   Test low .... *C. concavus*
4. Test moderately high, petals short and narrow ....
   .... *C. lanceolatus*
   Test high, petals long and broad .... *C. eurychorus*

A more complete key to West Indian fossil *Clypeaster* species is given in Jackson (1922, p. 32, 33).

Clypeaster rosaceus (Linnaeus, 1758)
(Figure 10.1, 10.2)

Description. (See also Jackson, 1922, p. 33, 34). Test large, elongate pentagonal in outline, but well-rounded. Widest about two-thirds of the length from the posterior margin. Test inflated and broadly conical, the adapical surface sloping away from the apical system. Ambitus low, rounded and close to the flattened adoral surface. Test wall double, with the development of internal pillars.

Apical system central, pentagonal, monobasal, with five genital pores.

Figure 10. 1, 2, *Clypeaster rosaceus* (Linnaeus), USNM E40376, Recent, Jamaica: 1, apical view; 2, oral view. 3, 4, *Cubanaster* sp. cf. *C. acunai* (Lambert and Sánchez Roig), BMNH E83648, Eocene, Claremont Formation, between Brown's Town and Bamboo, parish of St. Ann: 3, apical view; 4, oral view, 5, 6, *Wythella* sp., Eocene, Claremont Formation, between Brown's Town and Bamboo, parish of St. Ann: 5, BMNH E83649, apical view; 6, BMNH E83650, oral view. 7, 8, *Tarphypygus notabilis* Arnold and Clark, holotype, MCZ 3283, Eocene, Chapelton Formation, Seven Rivers area, parish of St. James (apical and oral surfaces abraded): 7, apical view; 8, oral view. 9, *Cassidulus*? *sphaeroides* Arnold and Clark, holotype, MCZ 3477 (locality and horizon unknown), apical view. Scale bar represents 5 mm for 3 and 4; 8 mm for 9; and 10 mm for all other figures.

Ambulacra petaloid adapically (Fig. 10.1). Petals open, broadest near tip, and composed of alternating primary plates and demiplates (Durham, 1966, p. U461). Pore pairs conjugate. Adradial pores elongate, perradial pores circular. Poriferous zones occur in shallow depressions. Interporiferous zones inflated and broad. Ambulacra widest adambitally and broad adorally. Ambulacra in contact close to the peristome (Durham, 1966, Fig. 357.1h), with numerous small accessory pore pairs surrounding the peristome and concentrated laterally to the five simple, radial food grooves (Fig. 10.2).

Interambulacra narrow between petals, about a fifth to a third as wide as ambulacra at the ambitus, tapering adorally and discontinuous close to the margin of the peristome. Test covered by very numerous, inset, perforate primary tubercles that occur in rounded, shallow to deep pits, particularly adorally. Minute, close-packed secondary tubercles occur over the entire test surface between the sunken primary tubercles.

Peristome central, pentagonal, small, situated in a deep conical depression. Periproct smaller than peristome, rounded, posterior, inframarginal, and angled down (Fig. 10.2).

Occurrence. *C. rosaceus* is a common shallow-water echinoid in the Recent of the Greater Antilles, Florida, and South Carolina (Clark, 1919). *C.* sp. cf. *C. rosaceus* was reported from the Round Hill Beds, August Town Formation, Coastal Group (probably Plio-Pleistocene) at Farquhar's Beach, near Milk River Bath, parish of Clarendon, by Donovan et al. (1989b, p. 46; Donovan and Lewis, research in progress). Fossil *C. rosaceus* is also reported from Puerto Rico and Cuba (Jackson, 1922).

Clypeaster concavus? Cotteau, 1875

Description. See Jackson (1922, p. 34–36, Plate 2, Figs. 10–12).
Occurrence. Poddubiuk (1985, p. 76) considered *C. concavus* to be late early to early middle Miocene in age. The specimen determined by Hawkins (in Arnold and Clark, 1927, p. 26) was from the "Oligocene (white limestone)." This may conceivably have been early Miocene, by analogy with the confusion that has existed between the Oligocene and Miocene in Cuba (Kier, 1984, p. 6). Two tests, determined to be the synonymous (Poddubiuk, 1985, p. 76) *C. antillarum* Cotteau by Arnold and Clark (1927, p. 26; collection of MCZ), are "probably from . . . near Port Maria," parish of St. Mary, although there are no early or middle Miocene deposits in this area (McFarlane, 1977a).

Clypeaster cotteaui? Egozcue in Cotteau, 1897

Description. See Jackson (1922, p. 41, 42, Plate 6, Figs. 6–8) and Arnold and Clark (1927, p. 26, 27).
Occurrence. Jackson (1922, p. 42: MCZ 3255) recognized this species from Jamaica, albeit from an unknown locality and horizon. Arnold and Clark (1927, p. 27) recorded a test collected loose from a river near Port Maria, parish of St. Mary. Elsewhere in the Caribbean (Cuba and Antigua) this species is late Oligocene in age (Jackson, 1922; Poddubiuk and Rose, 1984). The latter authors consider *C. cotteaui* to be a junior synonym of *C. batheri* Lambert.

Clypeaster eurychorus Arnold and Clark, 1934

Description. See Arnold and Clark (1934, p. 141, Plate 1, Figs. 1, 2).
Material. A unique holotype, MCZ 3474.
Occurrence. Locality and horizon unknown.

Clypeaster lanceolatus? Cotteau, 1897

Description. See Jackson (1922, p. 38, 39, Plate 4, Fig. 2).
Occurrence. The specimen determined by Hawkins (in Arnold and Clark, 1927, p. 27) was from "Oligocene (white limestone)." Poddubiuk and Rose (1984, p. 119) considered this taxon to be conspecific with *C. batheri* Lambert, from the late Oligocene of Antigua (see also *C. cotteaui*, above).

Suborder LAGANINA Mortensen
Family FIBULARIIDAE Gray

Genus *Fibularia* Lamarck

Fibularia jacksoni Hawkins, 1927
(Figure 11)

Description. (See also Hawkins, 1927, p. 76, 77, Plate 22, Figs. 1–3). Test small, high, elongated anterioposteriorly with an oval to highly rounded pentagonal outline. Test more rounded anteriorly. Test domed adapically and gently convex adorally. Ambitus at about midheight of test. No internal buttressing (unlike the similar *Echinocyamus* van Phelsum: Mooi, 1989, Fig. 21b).

Apical system small, monobasal with four genital pores and positioned slightly posterior of center (Fig. 11.2).

Five ambulacral petals adapically, not extending to the ambitus and shortest posteriorly. Petals widen adambitally and broadly open. Poriferous zones straight apart from a gentle adapical curvature. Ambulacral pores rounded and close together within pore pairs. Interporiferous zones broad, about 70% of the petal width adambitally. Ambulacra widening to the ambitus, where they are about three times wider than interambulacra (Hawkins, 1927, p. 77), and tapering adorally.

Interambulacra wider than ambulacra adapically, but narrowing adorally of the petals. Test bearing numerous small tubercles inset in shallow pits.

Peristome small, central, rounded, and apparently slightly elongated anterioposteriorly (Fig. 11.1). Periproct rounded in outline, inframarginal, smaller than and slightly posterior to the peristome.
Material. Holotype, BMNH E17665, plus 12 paratypes recorded by Hawkins (1927, p. 76) and five additional specimens in the BMNH, E75860 (Trechmann Bequest) and E78498 (four tests in the Matley collection).
Occurrence. The type area is the "Main road, northeast of Albert Town" (Hawkins, 1927, p. 76), that is, on the B5 road to Ulster Spring and Rio Bueno, parish of Trelawny, in the type area of the Albert Town Member, Chapelton Formation, Yellow Limestone Group (middle Eocene: Robinson, 1988). The species is also known from the B11 road between Brown's Town and Bamboo, parish of St. Ann, at GR 44565418, from

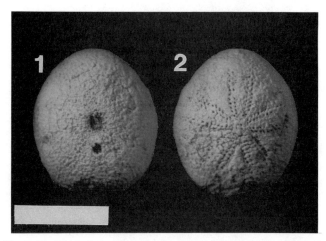

Figure 11. *Fibularia jacksoni* Hawkins, holotype, BMNH E17665, Eocene, Chapelton Formation, near Albert Town, parish of Trelawny. 1, oral view; 2, apical view. Scale bar represents 0.5 mm.

the Claremont Formation, Moneague Group, White Limestone Supergroup (mid-middle to mid-late Eocene: McFarlane, 1974, 1977b).
Remarks. *F. jacksoni* is the only common, small irregular echinoid in the Jamaican fossil record. Donovan et al. (1991) have described and figured cf. *Fibularia* sp. from an undescribed lithostratigraphic unit of late middle Eocene age from Content, near Hope Bay, parish of Portland.

Genus *Tarphypygus* Arnold and Clark

Remarks. Arnold and Clark (1927) classified two Jamaican species within this genus, *T. notabilis* and the type species *T. ellipticus.* McFarlane (1977b, p. 1397) recognized *Tarphypygus* sp. from the Swanswick Formation, which Donovan (1988b, p. 127) misread as *Taphropygus* Pomel, a synonym of *Nucleolites* Lamarck. *Tarphypygus* is also known from Cuba.

Tarphypygus notabilis Arnold and Clark, 1927
(Figure 10.7, 10.8)

Description. (See also Arnold and Clark, 1927, p. 43, 44, Plate 6, Figs. 13–15). Test oval in outline, moderately high, domed adapically, and more flattened adorally. Ambitus slightly lower than the midheight of the test. Apical system obscured by abrasion (Fig. 10.7).
Ambulacra petaloid adapically. Petals open and extending almost to the ambitus. Pore series within petals straight, except adjacent to the apical system, and slightly divergent. Pores circular, occurring at the adoral margins of plates. Interporiferous zones broad, about 50% of the petal width close to the ambitus. Ambulacra expand slightly to the ambitus and then taper to the peristome. Ambulacral plates low and hexagonal adorally, very low adapically.
Interambulacra continuous, about as wide as ambulacra at the ambitus, tapering adorally, but expanding slightly adapically, so that interambulacra are about one and a half times wider than the extremities of the petals. Tuberculation not preserved.
Peristome small, central, and apparently rounded. Periproct inframarginal, smaller than, and situated just posterior to, the peristome (Fig. 10.8; see also Durham, 1966, Fig. 362.3d).
Material. A unique holotype, MCZ 3283.
Occurrence. The type area is the Seven Rivers district, parish of St. James (Arnold and Clark, 1927, p. 43), from the Chapelton Formation, Yellow Limestone Group (mid-early to mid-middle Eocene).
Remarks. It is highly unlikely that this specimen came from the Maastrichtian of the Seven Rivers Inlier (Donovan, 1990; see remarks under ?*Trochalosoma chondra* above). Mooi (1989, p. 38) considered all species of *Tarphypygus* to be Eocene and the oldest known clypeasteroid sensu lato is Paleocene in age (Kier, 1982).

Tarphypygus ellipticus Arnold and Clark, 1927

Description. See Arnold and Clark (1927, p. 43, Plate 6, Figs. 10–12).
Material. A unique holotype, MCZ, 3282.
Occurrence. The type area is the Spring Mount region, parish of St. James (Arnold and Clark, 1927, p. 43), from the Chapelton Formation, Yellow Limestone Group (mid-early to mid-middle Eocene).
Remarks. This species differs from *T. notabilis* in having a more inflated test and an anterior petal that is gently incurved adambitally. Another specimen (UWIGM 2775), collected by McFarlane (1974, Plate 34, Figs. A, B; 1977b, p. 1397), is regarded herein as *Tarphypygus* sp. cf. *T. ellipticus.* This specimen is poorly preserved, but is undoubtedly *Tarphypygus,* as indicated by the sizes and relative positions of the peristome and periproct, and from the geometry of the adoral plating, where apparent. This specimen came from GR 43845458, on the minor road from Knapdale to Thatchfield, east-northeast of Brown's Town, parish of St. Ann, from the Swanswick Formation, Moneague Group, White Limestone Supergroup (late middle Eocene; Robinson, 1988, Fig. 4).

Family NEOLAGANIDAE Durham

Genus *Neolaganum* Durham
cf. *Neolaganum dalli* (Twitchell in Clark and Twitchell, 1915)
(Figure 12)

Description. (See also Hawkins, 1927, p. 78, 79, Plate 22, Figs. 4, 5; Cooke, 1959, p. 51, 52, Plate 21, Figs. 1–4). Test rounded, slightly elongated, and very low, with a flattened adoral surface. Test also flattened adapically, apart from the area around the apical system, which is weakly elevated, and the slightly raised margin.
Apical system central, slightly elevated, moderately large, with four genital pores.
Ambulacra petaloid adapically, petals not extending to the ambitus (Fig. 12.1). Petals closed (Hawkins, 1927, p. 78), broadest at their midpoint, where the interporiferous zone is about 65% of the petal width. Pore pairs conjugate, with pores elongate adradially and rounded perradially. Other details of test plating not apparent. Test covered by small inset primary tubercles in circular pits, with numerous granular secondary tubercles in the intervening areas.
Peristome central, rounded, and small. Periproct inframarginal, rounded, small, and situated close to the posterior margin of the test.
Material. Syntypes of *Sismondia crustula* Hawkins, 1927, BMNH E17667, 17668, plus 33 additional, but unlocated, tests (Hawkins, 1927, p. 78).
Occurrence. The type area of *S. crustula* is around Peace River, parish of Clarendon, from the Chapelton Formation, Yellow Limestone Group (mid-early to mid-middle Eocene). Hawkins (1927, p. 78) also recorded specimens from south of Christiana, parish of Manchester, presumably from the Chapelton Formation.
Remarks. Hawkins (1927) classified these specimens as a new species, *Sismondia crustula.* The particularly low test and closed petals of this species indicate that it is not a true *Sismondia.* Open petals are diagnostic features of the laganids (Durham, 1966, p. U472) and the type species of *Sismondia, S. occitana* (Defrance), has a more elevated and rounded profile (compare with Durham, 1966, Fig. 364.1a–d; see also Kier, 1982, Plate 2, Figs. 1–3, for *S. logotheti* Frass). Drs. M. J. McKinney and B. D. Carter (written communications, 1990) have independently suggested that these specimens are closest to *Neolaganum dalli* and may be conspecific, an interpretation with which I tentatively concur.

Genus *Cubanaster* Sánchez Roig
Cubanaster sp. cf. *C. acunai* (Lambert and Sanchez Roig in Sánchez Roig, 1926)
(Figure 10.3, 10.4)

Description. (See also Sánchez Roig, 1926, p. 59, Plate 13, Figs. 6, 7, 9; Sánchez Roig, 1949, p. 102, 103; Durham, 1966, Fig. 365.3d: the de-

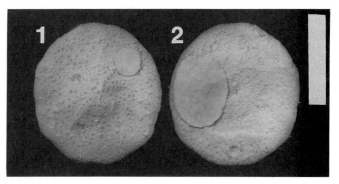

Figure 12. Cf. *Neolaganum dalli* Twitchell, BMNH E17668, Eocene, Chapelton Formation, Peace River, parish of Clarendon. 1, apical view; 2, oral view. Scale bar represents 10 mm.

scription below is based on the Jamaican specimen only). Test elongated pentagonal in outline, with well-rounded angles and almost parallel sides posteriorly, widest anterior of the apical system. Test very low, flattened adapically, with the circumference and apical system slightly inflated. Adoral surface gently concave. Ambitus above midheight of test and well rounded.

Apical system small, approximately central, monobasal, with four genital pores.

Ambulacra petaloid adapically. Petals long, open, reaching to the inner margin of the raised circumference. Petals longest posteriorly. Rounded perradial pores occur in straight columns, with elongate adradial pores in a gentle arc. Pore pairs are apparently conjugate. Interporiferous zones moderately broad. Ambulacra expand adambitally and taper adorally.

Interambulacra continuous, broadest between petals, tapering adambitally but of approximately constant width adorally except adjacent to the peristome. Tuberculation not preserved.

Peristome small, central, ?pentagonal, situated at the center of the adoral concavity. Periproct not preserved, but it must have occurred close to the posterior margin of the test.

Material. A single specimen, BMNH E83648.

Occurrence. McFarlane (1974, locality MU 392) collected this specimen from the B11 road between Brown's Town and Bamboo, parish of St. Ann, GR 44565418, from the Claremont Formation, Moneague Group, White Limestone Supergroup (mid-middle to mid-late Eocene). Brodermann (1949) considered *C. acunai* to occur in both the middle and late Eocene of Cuba.

Remarks. McFarlane (1974, 1977b, p. 1398) originally reported this specimen as *Jacksonaster* cf. *remediensis* Sánchez Roig, but it has been reidentified by Dr. Rich Mooi (personal communication, 1989).

Genus *Wythella* Durham

Wythella sp.
(Figure 10.5, 10.6)

Description. Test rounded in outline, either irregularly circular or slightly elongated and tapering anteriorly, with the test widest poseriorly of the apical system. Test very low, disc-shaped, and flattened adorally. Flattened marginally adapically, but elevated and gently domed centrally. Ambitus moderately sharp. Internal buttresses well developed.

Apical system central, moderately large, pentagonal, and monobasal. Genital pores not apparent due to poor preservation.

Ambulacra pelatoid adapically (Fig. 10.5). Petals short, only extending about two-thirds of the distance to the ambitus and situated on the raised central portion of the test. Petals closed, widest slightly adapically of their midpoint. Poriferous zone curved. Perradial pores rounded, adradial pores elongate, pore pairs conjugate and angled away from the apical system. Interporiferous zone broad, about 65% of the petal width at the widest point. Ambulacra widening to the ambitus and tapering adorally.

Interambulacra continuous, broadest between petals, tapering adambitally and adorally. Interambulacra narrower than ambulacra at the ambitus. Test covered by very numerous, small, primary tubercles that are inset in circular pits, with very numerous granular secondary tubercles in the intervening space.

Peristome small, central, and pentagonal: E83650 preserves three teeth (Fig. 10.6). Periproct inframarginal, smaller than peristome, rounded and situated close to the posterior margin of the test.

Material. Two tests, BMNH E83649, E83650, plus six more poorly preserved specimens, UWIGM 2764.

Occurrence. All specimens come from the B11 road between Brown's Town and Bamboo, parish of St. Ann, GR 44565418 (McFarlane, 1974), from the Claremont Formation, Moneague Group, White Limestone Supergroup (mid-middle to mid-late Eocene).

Remarks. McFarlane (1977b, p. 1398) originally reported this species as *Scutella cubae*? Weisbord, but it has been reidentified by Dr. Rich Mooi (personal communication, 1989). Mooi (1989, p. 38) considered *Wythella eldridgii* (Twitchell, in Clark and Twitchell, 1915) to be the only known species within this genus. The Jamaican specimens attributed to *Wythella* are much more rounded than the somewhat elongate type species and probably represents a new species (compare Figs. 10.5 and 10.6 herein with Cooke, 1959, Plate 21, Figs. 8–10; Durham, 1966, Figs. 365.5a, b).

Suborder SCUTELLINA Haeckel
Family MELLITIDAE Stefanini

Genus *Mellita* L. Agassiz

Mellita quinquesperforata (Leske, 1778)
(Figure 13)

Description. See Cooke (1959, p. 46, Plate 19, Figs. 6, 7).

Remarks. Arnold and Clark (1927, p. 28) reported that "Hawkins, in his manuscript list gives this well-known species as from the Pleistocene deposits of Jamaica." However, the present author is unaware of the whereabouts of any fossil specimens of this species from the island. It is conceivable that a fragment of *Encope* (see below) could have been mistaken for *M. quinquesperforata*, though unlikely. However, *M. quinquesperforata* is a common shallow-water echinoid in the Caribbean at the present day (Clark, 1919) and it may well be a locally abundant Pleistocene fossil in terrigenous deposits (Mooi, 1989, p. 41). Neither Vaughan (1922) nor Brodermann (1949) include this species in their tables of fossil echinoids from the Caribbean and Cuba.

Genus *Encope* L. Agassiz

Remarks. Two Jamaican fossil species have previously been assigned to the genus *Encope*, *E. homala* Arnold and Clark and *E.* sp. cf. *E. sverdrupi* Durham.

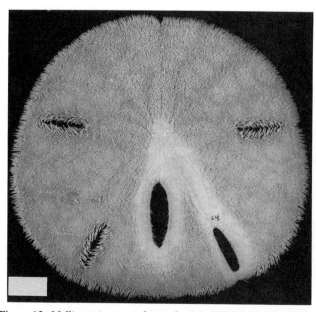

Figure 13. *Mellita quinquesperforata* (Leske), USNM E36726, Recent, Fort Pierce, Florida, apical view. Specimen uncoated and retaining radioles, except where these have been removed in ambulacrum I and interambulacrum 5. Scale bar represents 10 mm.

Encope sp. cf. *E. sverdrupi* Durham, 1950
(Figure 14)

Description. (See also Durham, 1950, p. 48, 49, Plate 37, Fig. 6, Plate 39, Figs. 4, 6). Test highly rounded but tapering anteriorly, with a rounded anterior and a straight posterior margin. Test widest in the region of the anus and the anal lunule. Test low, flattened adorally, and low domed adapically, with the apex just anterior of the apical system. Five ambulacral lunules, three of which are closed and two open. Anal lunule closed and closer to the center of the test than the ambulacral lunules. The three posterior lunules are particularly elongate, while the anterior ambulacral lunule (III) is shortest.

Apical system monobasal and pentastellate in outline. Poorly preserved, but at least three genital pores are apparent.

Ambulacra simple, biserial, petaloid in the central part of the aboral surface. Petals II and IV are shortest. Petals open, broadest adambitally and elliptical (II and IV) to pear-shaped in outline. Pore pairs conjugate within petals, with rounded perradial and elongate adradial pores. Interporiferous zone of petals moderately broad, particularly at the widest parts of the petals, especially in II and IV. Ambulacra greatly expanded between the petals and the ambitus and on the adoral surface. Channels with numerous small tubercles link the ambulacral lunules to the peristome. Tuberculation within the petals not apparent, probably due to poor preservation. Adorally of the petals ambulacral tuberculation is similar to that of interambulacra.

Interambulacra narrower than ambulacra, except adjacent to the petals. Each plate bears very numerous primary tubercles adapically and fewer larger primaries adorally.

Peristome small, pentagonal, and situated slightly anterior of center. Ten radiating, branching channels are grouped in adradial pairs about the peristome, extending to near the ambitus and enclosing the ambulacral lunules. Periproct inframarginal, elliptical, small, situated about 10 mm posterior of the peristome and close to the anterior lip of the anal lunule.

Material. A single test, BMNH E83647, plus numerous test fragments (to be deposited in the BMNH).

Occurrence. The figured specimen was collected from the northwestern end of Farquhar's Beach, Round Hill, parish of Clarendon, GR 413347, from the Round Hill Beds, August Town Formation, Coastal Group (probably Plio-Pleistocene). This is locality 2a of Donovan et al. (1989b). Fragments of test are commonest in chalky beds with a vertical orientation in the middle part of the Farquhar's Beach section.

Remarks. This specimen is a typical *Encope,* with an apical system and a peristome that are both anterior of center, elongate posterior petals, a continuous posterior interambulacrum and the anal lunule positioned anteriorly to the two posterior ambulacral petals. *E. sverdrupi* sensu stricto is from the early Pliocene of the Gulf of California (Durham, 1950, p. 12, 49), so it is suspected that the Jamaican species is not conspecific, although the two are very similar. The Round Hill species does not closely resemble either of the two extant Caribbean species of *Encope, E. emarginata* (Leske) and *E. michelini* L. Agassiz (Phelan, 1972). *Scutella* sp., which was mentioned by Chubb (1958) from the type section of the August Town Formation, may be *E.* sp. cf. *E. sverdrupi.*

Encope homala Arnold and Clark, 1934

Description. See Arnold and Clark (1934, p. 142, 143, Plate 2, Fig. 1).
Material. A unique holotype, MCZ 3475.
Occurrence. From the parish of Manchester, precise locality unknown.
Remarks. This specimen is poorly preserved, but it obviously differs from the Round Hill species in having more slender, relatively more elongate petals (particularly anteriorly and posteriorly), possibly lacking posterior ambulacral lunules and in having a rounded posterior margin. It may also have a more circular test outline.

Order CASSIDULOIDA Claus

Remarks. Specimens of at least some of the cassiduloid species discussed herein are too poorly preserved to permit determination of the details of the apical plating and of the phyllodes. Determination of some taxa to generic level is, therefore, extremely tentative in many instances.

Family ECHINOLAMPADIDAE Gray

Genus *Echinolampas* Gray

Remarks. Nine nominal species of *Echinolampas* were identified from Jamaica by Arnold and Clark: *E. altissima* Arnold and Clark, *E. anguillae* Cotteau, *E. brachytona* Arnold and Clark, *E. clevei* Cotteau, *E. lycopersicus* Guppy, *E. paragoga* Arnold and Clark, *E. strongyla* Arnold and Clark, *E. alta* (Arnold and Clark), and *E. plateia* (Arnold and Clark). The last two species were originally placed in the genus *Palaeolampas* Bell, which Kier (1962, p. 106, 1966, p. U506) considered to be a junior synonym of *Echinolampas*. The reference to *E. anguillae* is based on a list supplied by Hawkins to Arnold and Clark. Hawkins (1924, p. 319) further mentioned a poorly preserved test from Spring Mount, parish of St. James (Chapelton Formation), which he referred to as *Echinolampas* sp. indet. (BMNH E17212) and (in Trechmann, 1930, p. 216) identified a specimen from the Manchioneal Formation as *Echinolampas*(?) sp. nov.(?). The whereabouts of the latter specimen is unknown to the present author.

Echinolampas clevei Cotteau, 1875
(Figure 15.4, 15.5)

Description. (See also Jackson, 1922, p. 63, Plate 10, Figs. 6, 7; Plate 11, Figs. 1, 2; Arnold and Clark, 1927, p. 49, 50, Plate 9, Figs. 7, 8). Test elongate pentagonal in outline, rounded anteriorly, and more blunt posteriorly. Test widest posterior of the apical system. Test high, rounded adapically and flat adorally. Ambitus at or slightly less than half the test height.

Apical system small, well anterior of center, monobasal, with four genital pores (genital 5 missing).

Five elongate, open petals (Fig. 15.4), extending almost to the ambitus. Petals widest medially, comprising two gently curved columns of conjugate pore pairs. Pores rounded perradially and more elongate adradially. Interporiferous zone broad, about 60% of the petal width at the widest point. In all ambulacra except III one column of pores within each petal is longer. Ambulacra expand slightly adambitally and taper gently adorally. Ambulacral pores adjacent to the peristome appear single.

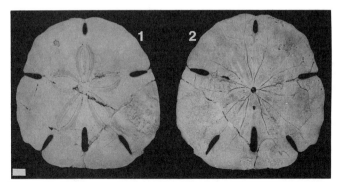

Figure 14. *Encope* sp. cf. *E. sverdrupi* Durham, BMNH E83647, Plio-Pleistocene, Farquhar's Beach, parish of Clarendon. 1, apical view; 2, oral view. Scale bar represents 10 mm.

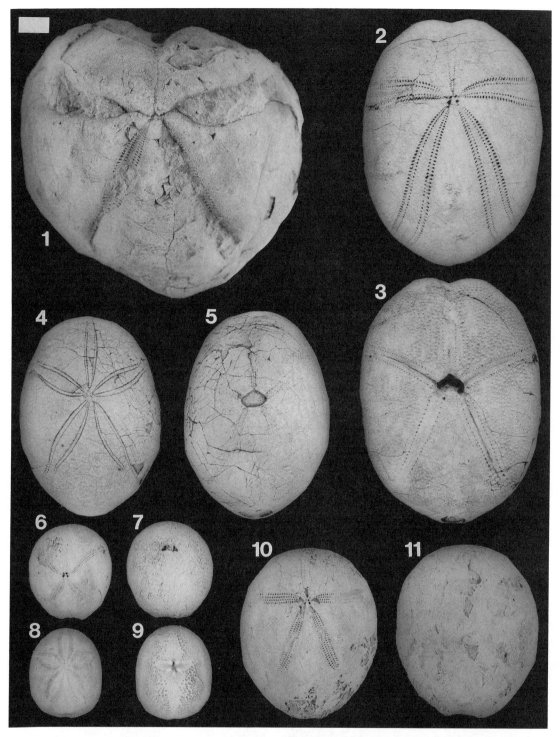

Figure 15. 1, *Cyclaster sterea* Arnold and Clark, holotype, MCZ 3483 (locality and horizon unknown), apical view. 2, 3, *Asterostoma pawsoni* Kier, holotype, USNM 358866, Eocene, Chapelton Formation (precise locality uncertain): 2, apical view; 3, oral view. 4, 5, *Echinolampas clevei* Cotteau, USNM 444304, Eocene, Yellow Limestone Group (precise locality unknown: Donovan, 1988a, p. 39, 40): 4, apical view; 5, oral view. 6, 7, *Agassizia inflata* Jackson, BMNH E83643, Eocene, Swanswick Formation, Beecher Town, parish of St. Ann: 6, apical view; 7, oral view. 8, 9, *Ryncholampas*? *matleyi* (Hawkins), USNM 444301, Eocene, Chapelton Formation (precise locality unknown: Donovan, 1988a, p. 38, 39): 8, apical view; 9, oral view. 10, 11, *Brissus* sp. cf. *B. unicolor* (Leske), MCZ 3469, Eocene, Chapelton Formation, Lucky Hill, parish of St. Mary: 10, apical view; 11, oral view. Scale bar represents 10 mm.

Interambulacra wide, composed of broad, moderately high plates. Interambulacra about seven times wider than ambulacra at the ambitus and tapering to the peristome, where they are still much broader than ambulacra. Test surface covered by numerous primary tubercles in shallow pits, which are most densely packed adorally. Any secondary tubercles of the available specimens have been lost through abrasion.

Peristome wider than high, pentagonal in outline, with rounded angles corresponding to the ambulacra (Fig. 15.5). Periproct posterior, wide, submarginal, larger than the peristome, and rounded triangular in outline.

Material. Jamaican specimens of *E. clevei* include 116 tests identified by Arnold and Clark (1927, p. 49) and presumably in the MCZ; 12 tests in USNM acc. 301378 (Donovan, 1988a, p. 39, 40), including USNM 444304 (Fig. 15.4–15.5); and a test identified by Donovan et al. (1989a; to be deposited in the BMNH).

Occurrence. Arnold and Clark (1927, p. 49) found this species to be abundant in the Spring Mount area, parish of St. James (Chapelton Formation) and in the Yallahs River valley, parish of St. Thomas ("Font Hill Formation"), both in the Yellow Limestone Group (mid-early to mid-middle Eocene). Donovan et al. (1989a) collected their test from the Swanswick Limestone, Moneague Group, White Limestone Supergroup (late-middle Eocene). This species is also known from the Eocene of St. Bartholomew and Cuba (Jackson, 1922, p. 63).

Remarks. The above description is based largely on USNM 444304 (Fig. 15.4, 5), which agrees well with the figures of syntypes in Jackson (1922, Plate 10, Figs. 6, 7), but poorly with the atypical specimen illustrated by Arnold and Clark (1927, Plate 9, Figs. 7, 8). The following comments on the specimens of *E. clevei* in USNM acc. 301378 were made by Donovan (1988a, p. 40).

These specimens are all more or less well preserved and, taken as a group, together show most of the features of the adoral and aboral [=adapical] surfaces. The aboral surface is 'flatter' than in *E. altissima*. A note in the specimen tray suggests, 'Should be referred to *E. nuevitasensis'*. *E. nuevitasensis* Weisbord (1934, pp. 229–231, Plate 7, Figs. 4–6) from Cuba differs from *E. clevei* principally in having a larger periproct and a more rostrate posterior test. The periproct of *E. clevei* is 'little developed, oval, subtriangular' (Cotteau, 1875; *loc. cit.* Weisbord, 1934), whereas in *E. nuevitasensis* it is 'at the posterior margin, subtriangular, large, seemingly somewhat larger than the peristome' (*ibid.*). Arnold and Clark (1927, p. 48) do not discuss the shape of the periproct in *E. clevei* [sic: this should read *E. altissima*], while stating 'submarginal, under the conspicuously overhanging end of the keel . . . about 6 mm wide by 3.5 mm high'. This indicates a probably oval or triangular outline. The largest of the HC [ex-Hamilton College; USNM acc. 301378] specimens of *E. clevei* (about 69 mm long by 51 mm wide by 37 mm high) has a rounded triangular periproct approximately 10.5 mm wide by 7.5 mm high. Other tests preserve a more rounded, pentagonal periproct, yet in all other characters these tests appear similar within the range of morphology that might be expected within a single species. It is possible that *E. clevei*, *E. nuevitasensis* and *E. altissima* are all variants within a single species but . . . large collections from single horizons and comparison with the type material are required.

E. altissima Arnold and Clark is relatively higher and narrower than *E. clevei,* with a sharp anteroposterior "crest" adapically (Donovan, 1988a, p. 39). Comparison of Figure 15.4 and 15.5 herein with Arnold and Clark (1927, Plate 9, Figs. 1, 2) shows how similar are these two species and it is suggested, at least tentatively, that *E. altissima* is probably a junior synonym of *E. clevei.*

Echinolampas alta (Arnold and Clark, 1927)

Description. See Arnold and Clark (1927, p. 52, 53, Plate 10, Figs. 1–3).
Material. Holotype, MCZ 3289, plus a further 40 specimens mentioned by Arnold and Clark. USNM acc. 301378 includes five tests (Donovan, 1988a, p. 39). An additional test (BMNH E75960) is in the Trechmann Collection.

Occurrence. The type area is Easington, Yallahs River valley, parish of St. Thomas, from the "Font Hill Formation," Yellow Limestone Group (mid-early to mid-middle Eocene). The BMNH test probably comes from Spring Mount, parish of St. James, from the Chapelton Formation, Yellow Limestone Group.

Remarks. *E. alta* is a particularly distinctive species of *Echinolampas,* being moderately large, with a flattened adoral surface, domed adapical surface and elongate outline that tapers posteriorly. *E. alta* is more inflated than *E. plateia* and less inflated than both *E. clevei* and *E. strongyla.*

Echinolampas altissima Arnold and Clark, 1927

Description. See Arnold and Clark (1927, p. 47, 48, Plate 9, Figs. 1–4).
Material. Holotype, MCZ 3285, plus three other specimens examined by Arnold and Clark (1927, p. 48). Two tests in USNM acc. 301378 are assigned to this species (Donovan, 1988a, p. 39).
Occurrence. All of Arnold and Clark's specimens were collected from the west side of the Yallahs River valley, presumably from the Easington district, parish of St. Thomas, from the "Font Hill Formation," Yellow Limestone Group (mid-early to mid-middle Eocene).
Remarks. *E. altissima* is probably a junior synonym of *E. clevei* Cotteau (see above). It is perhaps significant to note that the type area of *E. altissima* has also produced abundant specimens of *E. clevei* (Arnold and Clark, 1927, p. 49).

Echinolampas? anguillae (Cotteau, 1875)

Remarks. Arnold and Clark (1927, p. 48), quoting Hawkins, stated that this species was known from the "Oligocene (white limestone)" of Jamaica. Poddubiuk and Rose (1984, p. 121) considered *E. anguillae* to be based on a deformed specimen of *E. lycopersicus* Guppy (see below).

Echinolampas brachytona Arnold and Clark, 1927

Description. See Arnold and Clark (1927, p. 49, Plate 9, Figs. 5, 6).
Material. Holotype, MCZ 3286, plus one other specimen recognized by Arnold and Clark.
Occurrence. The type area is Guys Hill, parish of St. Mary, from the Chapelton Formation, Yellow Limestone Group (mid-early to mid-middle Eocene). The second specimen is from the Yallahs River valley, parish of St. Thomas, from the "Font Hill Formation," Yellow Limestone Group.
Remarks. The poriferous zones in each posterior petal of *E. brachytona* are extremely unequal in length and the test is low, with a particularly blunt posterior.

Echinolampas lycopersicus? Guppy, 1866

Description. See Jackson (1922, p. 64–66, Plate 11, Figs. 3–6).
Material. Arnold and Clark included 13 specimens in this species (1927, p. 50).
Occurrence. Jamaican specimens attributed to this species all came from the Spring Mount area, parish of St. James, from the Chapelton Formation, Yellow Limestone Group (mid-early to mid-middle Eocene). Otherwise known from the early Miocene of Anguilla (Poddubiuk and Rose, 1984) and the ?early Miocene of Cuba (reported as Oligocene by Jackson, 1922, p. 65, 66, but see comments on the "Oligocene" echinoids of Cuba by Kier, 1984, p. 6).
Remarks. The presence of *E. lycopersicus* in the Eocene of Jamaica is considered doubtful, as all other reports of this species refer it to the Miocene or (doubtfully) to the Oligocene. The highly rounded *E. strongyla* Arnold and Clark is superficially similar to the more oval *E. lycopersicus*. It may be that the specimens attributed to *E. lycopersicus*

by Arnold and Clark are merely more oval *E. strongyla*. The Jamaican specimen of *E. anguillae* (=*E. lycopersicus*?: see above) may have been genuinely Miocene in age, as the White Limestone Supergroup, from which it was derived, ranges from mid-middle Eocene to early-late Miocene in age (Robinson, 1988, Fig. 4), but the whereabouts of this specimen is unknown.

Echinolampas paragoga Arnold and Clark, 1927

Description. See Arnold and Clark (1927, p. 50, 51, Plate 9, Figs. 9–13).
Material. Holotype, MCZ 3287, plus 14 additional specimens identified by Arnold and Clark.
Occurrence. The type area is the Montpelier–Spring Mount region, parish of St. James, from the Chapelton Formation, Yellow Limestone Group (mid-early to mid-middle Eocene).
Remarks. *E. paragoga* differs from *E. clevei* in having a small, moderately high, oval test and a rounded peristome just anterior of center.

Echinolampas plateia (Arnold and Clark, 1927)

Description. See Arnold and Clark (1927, p. 53, 54, Plate 10, Figs. 4–6).
Material. Holotype, MCZ 3290, plus 27 other tests identified by Arnold and Clark. Four tests in USNM acc. 301378 belong to this species (Donovan, 1988a, p. 40).
Occurrence. The type area is the Montpelier to Spring Mount region, parish of St. James, from the Chapelton Formation, Yellow Limestone Group (mid-early to mid-middle Eocene).
Remarks. *E. plateia* is a large, oval *Echinolampas*, more elongate than *E. alta*, with a higher ambitus and relatively more slender poriferous zones within petals. The adoral surface is planar in *E. plateia* and the adapical surface is gently domed.

Echinolampas strongyla Arnold and Clark, 1927

Description. See Arnold and Clark (1927, p. 51, Plate 9, Figs. 14–17).
Material. Holotype, MCZ 3288, plus a further 13 tests identified by Arnold and Clark. Three tests in USNM acc. 301378 that are referred to this species do not agree well with Arnold and Clark's figured holotype (Donovan, 1988a, p. 40).
Occurrence. The type area is the Seven Rivers district, parish of St. James, from the Chapelton Formation, Yellow Limestone Group (mid-early to mid-middle Eocene).
Remarks. This species is superficially similar to *E. lycopersicus* (see above). Adult *E. strongyla* has a nearly circular outline, is flattened adorally and gently domed adapically, with a high, blunt posterior margin.

Family CASSIDULIDAE L. Agassiz and Desor

Genus *Cassidulus* Lamarck

Remarks. Two species of *Cassidulus* have been described from Jamaica, *C.?* *platypetalus* Arnold and Clark and *C.?* *sphaeroides* Arnold and Clark.

Cassidulus? sphaeroides Arnold and Clark, 1934
(Figure 10.9)

Description. (See also Arnold and Clark, 1934, p. 144, 145, Plate 1, Figs. 6–8). Adoral surface largely obscured by sediment. Test rounded, broadest posterior of the apical system, rounded anteriorly, but more blunt posteriorly. Test high, flattened adorally, gently domed adapically, with steep sides. Ambitus low.

Apical system not preserved, but anterior of center and probably small.

Five ambulacral petals developed adapically (Fig. 10.9). Petals long, open, extending to the margin of the adapical surface. Petal III narrowest. Petals broadest medially. Pore pairs composed of elongate, well-spaced pores. Interporiferous zones broad, between 43 and 57% of median petal width. Ambulacra expanding only slightly adorally of the petals.

Interambulacra broad, tapering adapically and adorally, but of approximately constant width on the high sides of the test, where the interambulacra are three to four times wider than the ambulacra. Primary tubercles only preserved posteriorly on the adoral surface.

Peristome obscured. Periproct large, posterior, supramarginal (Fig. 10.9) and higher than wide.
Material. A unique holotype, MCZ 3477.
Occurrence. Locality and horizon unknown.
Remarks. The apical system of *Cassidulus* is monobasal with four genital pores and the peristome is anterior, pentagonal in outline and wider than high, with well-developed bourrelets and phyllodes with few pores (Kier, 1962, p. 174, 1966, p. U514). These features are not preserved in *C.? sphaeroides*.

Cassidulus? platypetalus Arnold and Clark, 1934

Description. See Arnold and Clark (1934, p. 144, Plate 1, Figs. 3–5).
Material. A unique holotype, MCZ 3476.
Occurrence. Locality and horizon unknown.
Remarks. *C.? platypetalus* differs from *C.? sphaeroides* in having a more elongate, relatively lower test that is rounded both anteriorly and posteriorly. The peristomal region of *C.? platypetalus* is gently convex.

Genus *Eurhodia* Haime

Eurhodia sp. cf. *E. rugosa* (Ravenel, 1848)

Description. See Cooke (1959, p. 63, 64, Plate 22, Figs. 1–4) and Kier (1962, p. 214, 216, Plate 41, Figs. 1–5, Text fig. 177).
Material. A single test from Jamaica, UWIGM 2764.
Occurrence. On the B11 road between Brown's Town and Bamboo, near Burts Run, parish of St. Ann, GR 44565418, from the Claremont Formation, Moneague Group, White Limestone Supergroup (mid-middle to mid-late Eocene: Robinson, 1988, Fig. 4). *E. rugosa* occurs in the middle, and possibly late, Eocene of South and North Carolina (Cooke, 1959, p. 64; Zullo and Harris, 1986).
Remarks. This specimen requires considerable preparation before it can be figured and adequately described. The adoral surface is obscured by sediment and a clypeasteroid is cemented adapically. However, what is apparent shows this specimen to be very close to *E. rugosa*, although the adoral surface is flat, the periproct is lower and the test is less inflated.

Genus *Rhyncholampas* A. Agassiz sensu lato

Remarks. Six other species of Cenozoic cassiduloids have been described or mentioned from Jamaica. These are *Anisopetalus ellipticus* Arnold and Clark, *Parapygus antillarum* (Cotteau), *Parapygus parallelus* (Cotteau), *Rhynchopygus matleyi* Hawkins, *Rhynchopygus punctatus* Arnold and Clark, and *Rhyncholampas* sp. cf. *R. alabamensis* (Twitchell). Hawkins (1924, p. 319) also mentioned a *Rhynchopygus* sp. indet. (BMNH E17213, E17214, probably *R. matleyi*). Kier (1962, p. 183, 184) showed that *A. ellipticus* is really a *Rhyncholampas*. Kier further demonstrated that *Parapygus* is limited to the Upper Cretaceous of Europe and Africa (1962, p. 100) and that *Rhynchopygus* is only known from the Upper Cretaceous of Belgium, Holland, and France (1962, p. 160). It is thus apparent that most of the Jamaican taxa listed above area in need of revision. Where uncertainty exists, species are included within the genus *Rhyncholampas* sensu lato herein, following Donovan (1988a), although it is recognized that further revision is still required.

Rhyncholampas? matleyi (Hawkins, 1927)
(Figure 15.8, 15.9)

Description. (See also Hawkins, 1927, p. 79, 80, Plate 22, Figs. 6–8; Arnold and Clark, 1927, p. 54, 55, Plate 11, Figs. 3, 4). Test small, oval in outline, rounded anteriorly and more blunt posteriorly, almost straight sided. Test moderately high, with a slightly anterior apex. Test domed adapically and concave adorally, with the posterior adoral surface slightly raised and angled away from the peristome. Ambitus low and moderately angular.

Apical system slightly anterior of center, small, monobasal, with four genital pores.

Ambulacra forming five petals adapically (Fig. 15.8). Petals open (but closing) and widest medially. Petals II and IV shortest. Pore pairs conjugate, with pores rounded perradially and elliptical adradially. Interporiferous zones broad, about 55 to 60% of the petal width at the broadest point. Ambulacra do not appreciably expand adorally of the petals. Phyllodes petaloid and sunken.

Interambulacra broad, composed of wide, moderately high plates adapically. Bourrelets well developed, bearing abundant secondary tubercles. Adapically primary tubercles are small and closely spaced, inset, occurring in circular shallow pits. Adorally primary tubercles are larger and more prominent (Fig. 15.9). Primary tubercles excluded from an anteroposteriorly elongated, lens-shaped naked zone (sensu Mooi, 1990a) in ambulacrum III and interambulacrum 5. Naked zone bears a pitted sculpture.

Peristome slightly anterior, small, pentagonal, and elongated anteroposteriorly. Periproct posterior, supramarginal, small, elliptical or circular, with a slight adapical overhang.

Material. Holotype, BMNH E17666. Arnold and Clark (1927, p. 54) based their assessment of this species on a collection of 137 tests. Other specimens in the BMNH include E75873 to E75879, E75945 to E75947, E75949, E75950, E78496, E78497. A further 17 specimens in USNM acc. 301378 (Donovan, 1988a, p. 38) belong to this species, including USNM 444301 (Figs. 15.8, 15.9).

Occurrence. The type area is the Spring Mount region, parish of St. James, from the Chapelton Formation, Yellow Limestone Group (mid-early to mid-middle Eocene). Arnold and Clark (1927, p. 54) also recorded this species from the region around Lucky Hill and Gayle, parish of St. Mary (Chapelton Formation) and from the Yallahs River valley, parish of St. Thomas, presumably from the Easington district ("Font Hill Formation," Yellow Limestone Group). This species is also known from Wait-a-Bit Cave (Robinson, 1988, p. 60) and Freemans Hall, parish of Trelawny (Stettin Member, Chapelton Formation, late early Eocene). BMNH E75873 to E75879 are from Guys Hill, parish of St. Mary (Chapelton Formation). McFarlane (1974, 1977b) recorded this species from the Claremont (mid-middle to mid-late Eocene) and Somerset (early to mid-late Eocene) Formations of the Dry Harbour Mountains, parish of St. Ann.

Remarks. *R.? matley* is one of the commonest fossil echinoids known from Jamaica. Donovan (1988a, p. 38) reclassified this taxon as a *Rhyncholampas*, although it is considered possible that it is really a *Eurhodia* (Mooi, 1990b; Mooi, Suter, and Donovan, research in progress).

Rhyncholampas? alabamensis? (Twitchell in Clark and Twitchell, 1915)

Description. See Twitchell in Clark and Twitchell (1915, p. 172, Plate 80, Figs. 3a–d; Cooke, 1959, p. 57).
Material. A single specimen from Jamaica, UWIGM 2771 (figured McFarlane, 1974, Plate 33, Figs. C–E; mentioned McFarlane, 1977b, p. 1397).
Occurrence. From the B11 road west of Bamboo, parish of St. Ann, GR 45585428, from the Swanswick Formation, Moneague Group, White Limestone Supergroup (late middle Eocene).

Remarks. This unique specimen is poorly preserved, but is more inflated than *R.? matleyi*, with the periproct situated at about half the test height on the high posterior margin and the pentagonal peristome wider than high. *R. alabamensis* sensu stricto (Kier, 1962, p. 180) is found in the Oligocene of Florida, Georgia and Mississippi (Cooke, 1959, p. 58).

Rhyncholampas? antillarum (Cotteau, 1875)

Description. See Jackson (1922, p. 56, 57, Plate 9, Figs. 6–9). The Jamaican specimens are discussed in Arnold and Clark (1927, p. 54, Plate 11, Figs. 1–2).
Material. Arnold and Clark based their description on a collection of 20 tests. Four tests in USNM acc. 301378 are doubtfully assigned to this species (Donovan, 1988a, p. 39).
Occurrence. Jamaican specimens come from the Spring Mount area, parish of St. James, from the Chapelton Formation, Yellow Limestone Group (mid-early to mid-middle Eocene). This species is also known from the Eocene of St. Bartholomew and Cuba (Jackson, 1922, p. 57).
Remarks. *R.? antillarum* differs from *R.? matleyi* in being larger and more inflated, with a rounded ambitus at about the midheight of the test and a pentagonal peristome that is wider than high.

Rhyncholampas ellipticus (Arnold and Clark, 1927)

Description. See Arnold and Clark (1927, p. 45, 46, Plate 6, Figs. 16–20) and Kier (1962, p. 183, 184, Plate 29, Figs. 1–3, Text fig. 153).
Material. Holotype, MCZ 3284, plus six paratypes in the B. W. Arnold collection of the MCZ.
Occurrence. The type area is on the western side of the hills east of Montpelier, parish of St. James, presumably from the Chapelton Formation, Yellow Limestone Group (mid-early to mid-middle Eocene).
Remarks. *R. ellipticus* differs from *R.? matleyi* in having a relatively lower, more oval test with a more anterior apex, a more flattened adoral surface, a lower, wider periproct, and a more evenly pentagonal peristome.

Rhyncholampas? parallelus (Cotteau, 1897)

Remarks. Arnold and Clark (1927, p. 54) recorded this taxon from Jamaica only on the basis of "Hawkin's manuscript list." The horizon given is "Eocene (yellow limestone)." I am unaware of any specimens from Jamaica that have been assigned to this taxon. This species is otherwise recorded from the Eocene of Cuba (see Jackson, 1922, p. 57, for description).

Rhyncholampas? punctatus (Arnold and Clark, 1927)

Description. See Arnold and Clark (1927, p. 55, 56, Plate 11, Figs. 5–7).
Material. Holotype, MCZ 3291, plus a paratype specimen (MCZ) and a third test, MCZ 3478 (Arnold and Clark, 1934, p. 146).
Occurrence. The holotype comes from the west side of the Yallahs River, south of Easington Bridge, parish of St. Thomas (Arnold and Clark, 1927, p. 55). The only echinoid-bearing deposits in this region (Robinson, 1965; Kier, 1967) outcrop to the north of Easington Bridge, in the "Font Hill Formation," Yellow Limestone Group. The holotype may have been collected from river alluvium or alternately from an olistostromic block (cf., Robinson, 1967b). The other two specimens are from unknown localities and horizons.
Remarks. Donovan (1988a, p. 38) considered *Rhynchopygus punctatus* more probably to be a *Cassidulus* or a *Rhyncholampas* and (1988b, p. 127) referred the species to *Cassidulus*. *R.? punctatus* has a small, distinctive test that is highest anteriorly and slopes posteriorly in the holotype. MCZ 3478 (Arnold and Clark, 1934, Plate 1, Figs. 9–11) is very different from the holotype, being low, but highest centrally; it may not be conspecific.

Order SPATANGOIDA Claus
Family SCHIZASTERIDAE Lambert

Genus *Schizaster* L. Agassiz

Remarks. Arnold and Clark (1927) identified six species of fossil *Schizaster* from Jamaica: *S. altissimus* Arnold and Clark, *S. bathypetalus* Arnold and Clark, *S. brachypetalus* Arnold and Clark, *S. dumblei* Israelsky, *S. dyscritus* Arnold and Clark, and *S. hexagonalis* Arnold and Clark. *S. brachypetalus* has subsequently been shown to be a junior synonym of *S. subcylindricus* Cotteau by Kier (1984, p. 56) and *S. dyscritus* has been placed in the genus *Caribbaster* Kier (1984, p. 68; see below). Hawkins (1924, p. 319) has earlier identified *S.* sp. cf. *S. subcylindricus* Cotteau from the island and suggested that *Paraster* (=*Schizaster*) sp. a of Jackson (1922, p. 79) was probably conspecific. Hawkins (1924, p. 321) identified another specimen (BMNH E17221) as ?*Schizaster* sp. juvenile.

Schizaster subcylindricus Cotteau, 1875
(Figure 16.5, 16.6)

Description. (See also Kier, 1984, p. 56–59, Plates 27, 28, Figs. 1–4, Text fig. 21). Test rounded hexagonal in outline, blunt anteriorly, and more rounded posteriorly. Test widest at or anterior to the apical system. Test inflated, highest posterior to the apical system, with a high posterior margin and the adapical surface sloping anteriorly. Adoral surface convex with an anteroposterior keel. Ambitus below the midheight of the test.

Apical system posterior, small, with two genital pores only (1 and 4), ethmolytic, with the madreporite elongated posteriorly and genital 5 missing (Kier, 1984, Fig. 21B).

Four ambulacral petals adapically (not III: Fig. 16.5). Petals situated in deep, elongate grooves. Posterior petals very short and straight, petals II and IV longer and slightly sinuous. Pore pairs conjugate, composed of elliptical pores. Interporiferous zones poorly developed. Ambulacrum III straight and moderately broad, in an elongate groove adapically, either flush with the test surface or in an extremely shallow sulcus anteriorly. Ambulacrum III has a distinctive sinuous adradial suture adapically. Pore pairs are widely spaced adapically, composed of rounded pores. Adorally of the petals ambulacra are narrow and single pored. Posterior ambulacra are elongate adorally, flanking the plastron and separated by the labrum. A pentastellate pattern of phyllodes is developed (Fig. 16.6).

Interambulacra broad, composed of high, wide plates. Interambulacra 1 and 4 may not reach the peristome. Plastron broad and meridosternous. Primary tubercles numerous, perforate, crenulate, inset in shallow, rounded pits, and particularly abundant on the plastron. Secondary tubercles granular and very numerous. Peripetalous and latero-anal fascioles developed (Kier, 1984, p. 57).

Peristome anterior (Fig. 16.6) and directed anteriorly, moderately large, kidney-shaped, with a broad labrum overhanging the posterior third. Periproct high on the posterior margin, elliptical to teardrop-shaped, and higher than wide.

Material. Jamaican specimens include the holotype of *S. brachypetalus,* MCZ 3295, plus a further 126 specimens identified by Arnold and Clark (1927, p. 59, 60, Plate 11, Figs. 14–16); five tests in USNM acc. 301378 (Donovan, 1988a, p. 41); and six tests in the collection of Donovan et al. (1989a, p. 7), to be deposited in the BMNH. Specimens referred to *S.* cf. *subcylindricus* by Hawkins (1924) are BMNH E17217 and E17218. *Paraster* sp. A of Jackson (1922, p. 79) is MCZ 3236.

Occurrence. The type area for *S. brachypetalus* is near Abingdon, southwest of Green Island Harbour, parish of Hanover, presumably from the Chapelton Formation, Yellow Limestone Group (mid-early to mid-middle Eocene). Arnold and Clark (1927, p. 60) also noted that this species is common in the parish of St. James, presumably from the same horizon. The specimens referred to *S.* cf. *subcylindricus* by Hawkins (1924, p. 319) came from Spring Mount, parish of St. James. *S. subcylindricus* is also known from Beecher Town, parish of St. Ann, from the Swanswick Formation, Moneague Group, White Limestone Supergroup, of late middle Eocene age (Donovan et al., 1989a). This species is also known from the Eocene of St. Bartholomew and Cuba (Kier, 1984, p. 57). The figured specimen is Cuban.

Schizaster altissimus Arnold and Clark, 1927

Description. See Arnold and Clark (1927, p. 58, Plate 11, Figs. 11–33).
Material. Holotype, MCZ 3293, plus two further (but poorly preserved) specimens referred to by Arnold and Clark.
Occurrence. The type area is the Yallahs River valley above Easington Bridge, parish of St. Thomas, from the "Font Hill Formation," Yellow Limestone Group (mid-early to mid-middle Eocene).
Remarks. This species is larger, rounder and more highly inflated than *S. subcylindricus,* being almost as high as long, with a central apical system, very high posterior margin, and abrupt anterior margin with a shallow anterior sulcus.

Schizaster bathypetalus Arnold and Clark, 1927

Description. See Arnold and Clark (1927, p. 58, 59, Plate 12, Figs. 1–4) and Kier (1984, p. 36–39, Plates 14, 15, Text figs. 12, 13).
Material. Jamaican specimens of this species include the holotype, MCZ 3294, plus a further 23 specimens identified by Arnold and Clark. USNM acc. 301378 includes 12 tests of this species (Donovan, 1988a, p. 40, 41).
Occurrence. The type area is the Spring Mount region towards Seven Rivers, parish of St. James, from the Chapelton Formation, Yellow Limestone Group (mid-early to mid-middle Eocene). This species is also known from the Eocene of Cuba (Kier, 1984).
Remarks. Kier (1984, p. 36) stated that "This species is easily distinguished from *Schizaster sybcylindricus* Cotteau . . . by its far larger test, more central apical system with 4 genital pores, longer posterior petals and anterior petals that curve in an opposite direction." The anterior sulcus is markedly more pronounced in *S. bathypetalus.*

Schizaster dumblei? Israelsky, 1924

Description. See Israelsky (1924, p. 141) and Arnold and Clark (1927, p. 60, Plate 12, Figs. 5, 6).
Material. Arnold and Clark attributed 17 specimens to this species. USNM acc. 301378 includes five tests of this taxon (Donovan, 1988a, p. 41).
Occurrence. Jamaican tests attributed to this species come from the Spring Mount district, parish of St. James, from the Chapelton Formation, Yellow Limestone Group (mid-early to mid-middle Eocene). The type area for *S. dumblei* is the Tampico Region of Mexico, from the Rafael Beds (Israelsky, 1924).
Remarks. Kier (1984, p. 7) considered the Jamaican specimens attributed to *S. dumblei* by Arnold and Clark to be very similar to *S. subcylindricus.* Arnold and Clark's comments (1927, p. 60) certainly suggest that the specimens of *S. dumblei* described in their monograph may be "lumped" together. Further, Donovan (1988a, p. 41) regarded the three smallest specimens of *S. dumblei* in USNM acc. 301378 to resemble the holotype of *S. brachypetalus* (=*S. subcylindricus*: Kier, 1984, Plate 28, Figs. 1–4). However, the specimens figured by Arnold and Clark (1927, Plate 12, Figs. 5, 6) are distinctive from *S. subcylindricus* in having a more pronounced anterior sulcus. These two morphologies are tentatively considered to represent two separate species herein.

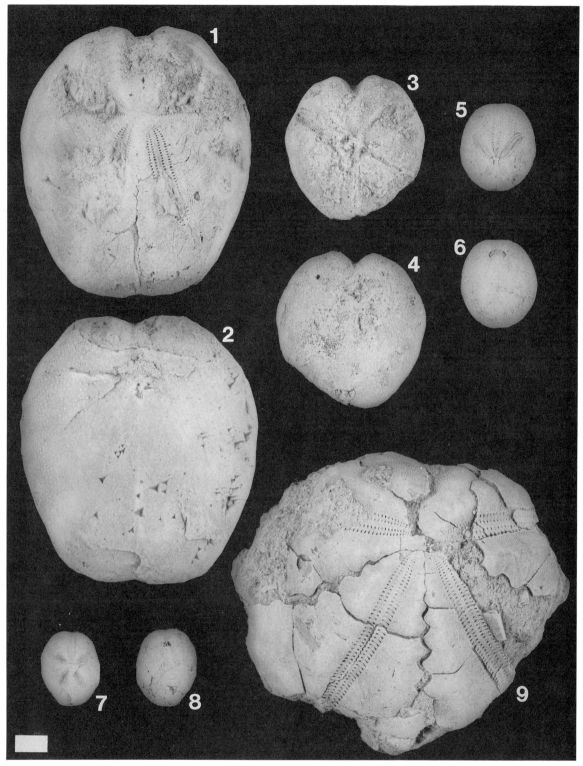

Figure 16. 1, 2, *Metalia jamaicensis* Arnold and Clark, holotype, MCZ 3489 (locality and horizon unknown): 1, apical view; 2, oral view. 3, 4, *Linthia trechmanni* Hawkins, BMNH E83642, Eocene, Chapelton Formation, Freeman's Hall, parish of Trelawny: 3, apical view; 4, oral view. 5, 6, *Schizaster subcylindricus* Cotteau, USNM 563308a, Eocene, Loma Caoba, Pinar del Rio Province, Cuba: 5, apical view; 6, oral view. 7, 8, *Caribbaster dyscritus* (Arnold and Clark), holotype, MCZ 3296, Eocene, Chapelton Formation, Seven Rivers district, parish of St. James. 7, apical view; 8, oral view. 9, *Macropneustes*? *sinuosus* Arnold and Clark, holotype, MCZ 3486 (locality and horizon unknown), apical view. Scale bar represents 10 mm.

Schizaster hexagonalis Arnold and Clark, 1927

Description. See Arnold and Clark (1927, p. 61, 62, Plate 12, Figs. 10, 11, Plate 13, Figs. 1–3).
Material. Holotype, MCZ 3297, plus a further 12 specimens identified by Arnold and Clark. USNM acc. 301378 includes five tests of this species (Donovan, 1988a, p. 41).
Occurrence. The type area is west of Springfield (=Welcome Hall), parish of St. James, from the Chapelton Formation, Yellow Limestone Group (mid-early to mid-middle Eocene).
Remarks. *S. hexagonalis* differs from *S. subcylindricus* in being larger, having a pronounced anterior sulcus, an apical system that is central or anterior of center, in being broadest anteriorly, in being more rounded anteriorly and more flattened posteriorly.

Genus *Agassizia* L. Agassiz and Desor

Agassizia inflata Jackson, 1922
(Figure 15.6, 15.7)

Description. (See also Jackson, 1922, p. 70, 71, Plate 12, Figs. 2–4; Kier, 1984, p. 65, Plate 32). Test oval in outline, rounded anteriorly, blunt posteriorly, widest just anterior of the apical system. Test inflated, adoral surface gently convex, sloping away from an anteroposterior axis. Ambitus about one-third height of test.

Apical system small, central to slightly posterior, tetrabasal, ethmolytic, with a prominent madreporite that is slightly elongated posteriorly and with four genital pores (genital 5 missing). Apical area flattened, with immediately adapical areas of interambulacra 1 to 4 slightly inflated.

Four petaloid ambulacra (I, II, IV, V) slightly sunken (Fig. 15.6). Petals closed. Posterior petals short, approximately straight, with plates arranged in two equal columns. Interporiferous zone very narrow. Pores elliptical perradially, elliptical to teardrop-shaped adradially, pores circular adapically and adambitally. Petals in ambulacra II and IV long, almost reaching ambitus, slightly sinuous, composed of unequal columns of pores. Posterior columns similar to those of posterior petals, anterior columns narrower and comprising rounded pores. Ambulacrum III straight, narrow but expanding slightly adambitally, comprising two columns of keyhole-shaped plates, each plate with a single, central pore. Adambitally of the petals ambulacra are slightly expanded, with plates apparently bearing single pores. Adorally ambulacra I+II and IV+V are in contact immediately adjacent to the peristome. Ambulacra I and V are long and crescentic adorally, composed of elongate plates and flanking the plastron, although separated by the labrum adjacent to the peristome. Other ambulacra are short adorally. Ambulacral plates are double-pored adjacent to the peristome.

Interambulacra broad adapically, composed of broad, high plates. Adorally interambulacrum 5 forms a broad, elongate, shield-shaped meridosternous plastron. Interambulacra 2 and 3 in contact with the peristome by only one plate each. Interambulacra 1 and 4 do not extend to the peristome. Test covered by numerous small primary tubercles with broad areolae, surrounded by closely packed, granular secondary tubercles. Primary tubercles are particularly well developed on the plastron, where they form columns radiating from the posterior margin. Tuberculation is generally poorly preserved, but part of the right lateral fasciole is apparent in one specimen (see Kier, 1984, Figs. 22–24, for diagrams of the arrangement of peripetalous and latero-anal fascioles on other Caribbean fossil species of *Agassizia*).

Peristome anterior, small, wide, kidney-shaped, directed anteriorly, with a small, curved labrum overhanging the posterior one-quarter of the peristome (Fig. 15.7). Periproct high on the posterior margin of the test, slightly larger than the peristome, oval in outline and wider than high.
Material. Arnold and Clark (1927, p. 56) recorded 29 specimens of this species from Jamaica, although Kier (1984, p. 65) mentioned no specimens of *A. inflata* from the MCZ collections. Kier did examine four tests from Jamaica in the USNM, although these were not located by the present author. Donovan et al. (1989a, p. 7) identified 15 specimens of this species from Jamaica (including BMNH E83643; Fig. 15.6, 15.7), which are to be deposited in the BMNH.
Occurrence. The specimens collected by Arnold and Clark are from Easington, parish of St. Thomas, from the "Font Hill Formation," Yellow Limestone Group (mid-early to mid-middle Eocene). Those collected by Donovan et al. (1989a) are from Beecher Town, parish of St. Ann, from the Swanswick Formation, White Limestone Supergroup (late middle Eocene). This species is also known from St. Bartholomew, Cuba, and North and South Carolina (Jackson, 1922, p. 71; Kier, 1984, p. 65).
Remarks. The above description is based mainly on the specimens from Beecher Town. *A. inflata* is easily distinguished from other Jamaican heart urchins by the distinctive arrangement of the petals in ambulacra II and IV.

Genus *Caribbaster* Kier

Remarks. Kier (1984, p. 68) considered that two Jamaican species should be assigned to the genus *Caribbaster*, *C. dyscritus* (Arnold and Clark) and the type species *C. loveni* (Cotteau).

Caribbaster dyscritus (Arnold and Clark, 1927)
(Figure 16.7, 16.8)

Description. (See also Arnold and Clark, 1927, p. 61, Plate 12, Figs. 7–9). Test small, oval in outline, blunt posteriorly and anteriorly, with a very shallow anterior sulcus. Test widest just anterior to the apical system. Test moderately high, steep-sided, gently domed adapically and more flattened adorally, with an anterioposterior keel. Ambitus at about midheight of the test.

Apical system small, slightly anterior of center, ethmolytic (Kier, 1984, p. 67), with two genital pores.

Ambulacra form four petals adapically (not III: Fig. 16.7). Petals straight and situated in shallow grooves. Posterior petals short, petals II and IV longer and extending close to the ambitus. Each petal composed of two equal columns of pore pairs. Pore pairs conjugate within petals, all pores elliptical in outline. Interporiferous zones narrow. Ambulacrum III broad and straight, in a shallow anterior sulcus, with two columns of pore pairs separated by a very broad interporiferous zone. Pore pairs of ambulacrum III comprise narrow D-shaped pores adapically, separated by a thick bar (cf., Kier, 1974, Text fig. 12A; aboral ambulacrum III isopore of Smith, 1980, Text fig. 14D). Adorally the posterior ambulacra are elongate and flank the plastron. Ambulacra form a pentastellate phyllodal region adjacent to the peristome (Fig. 16.8).

Interambulacra are broad adapically, two to three times broader than the ambulacra at the ambitus. Plastron broad and meridosternous. Primary tubercles numerous and occurring in shallow pits. Fascioles not apparent (Kier, 1984, p. 67, 69, noted peripetalous and latero-anal fascioles in the type species, *C. loveni*).

Peristome anterior, small, kidney-shaped and directly anteriorly, with a very weakly developed labrum overhang. Periproct higher than wide, occurring on the posterior margin.
Material. A unique holotype, MCZ 3296 (Fig. 16.7–16.8).
Occurrence. The type area is the Seven Rivers district, parish of St. James, from the Chapelton Formation, Yellow Limestone Group (mid-early to mid-middle Eocene).
Remarks. Arnold and Clark (1927, p. 61) originally placed this species in the genus *Schizaster*, while noting that "It is even uncertain whether it should be considered a Schizaster [sic]."

Caribbaster loveni (Cotteau, 1875)

Description. See Arnold and Clark (1927, p. 56, 57, Plate 11, Figs. 8–10) and Kier (1984, p. 67–69, Plate 34, Text fig. 25).

Material. The holotype of *Hypselaster perplexus* is MCZ 3292, plus 85 other tests noted by Arnold and Clark.
Occurrence. Jamaican specimens come from the Spring Mount area, parish of St. James, from the Chapelton Formation, Yellow Limestone Group (mid-early to mid-middle Eocene). The species is also known from the Eocene of St. Bartholomew and Cuba (Kier,1984, p. 69).
Remarks. Kier (1984, p. 68) showed *H. perplexus* Arnold and Clark to be a junior synonym of *C. loveni* (Cotteau). *C. loveni* is much more inflated than *C. dyscritus,* is proportionally wider and has no anterior sulcus.

Genus *Lambertona* Sánchez Roig

Lambertona jamaicensis (Arnold and Clark, 1934)
(Figure 17.1, 17.2)

Description. (See also Arnold and Clark, 1934, p. 150, Plate 3, Figs. 1–3). Only known from an internal mould. Test large, heart-shaped, rounded anteriorly with a deep anterior sulcus and more blunt posteriorly. Test low, conical adapically and flattened adorally. Ambitus low.

Apical system not preserved, but situated posterior of center, with the widest part of the test anterior to the apex.

Four ambulacral petals developed adapically (not III), occurring in deep, straight, elongate depressions (Fig. 17.1). Petals open. Posterior petals short, petals in ambulacra II and IV long and extending to close to the ambitus. Ambulacral pores in petals rounded and closely spaced. Interporiferous zone barely developed. Ambulacra becoming broader between petals and the ambitus, composed of high plates that apparently bear single pores. Ambulacrum III broader adapically than the petals, in a deeply sunken, straight anterior sulcus that expands gently adambitally. Ambulacra I and V are long adorally, flanking the plastron. Ambulacral plates with single pores adorally.

Interambulacra broad adapically, composed of broad, high plates. The most anterior plate columns of interambulacra 2 and 3 are composed of geniculate plates that are partially folded into the anterior sulcus. Adorally interambulacrum 5 forms a broad, shield-like plastron (Fig. 17.2). Interambulacra 1 and 4 do not reach the peristome.

Peristome anterior, kidney shaped, apparently directed anteriorly. Periproct presumably posterior and marginal.
Material. A unique holotype, MCZ 3484.
Occurrence. Locality and horizon unknown.
Remarks. Arnold and Clark (1934) originally placed this species in the genus *Victoriaster,* a classification subsequently followed by Donovan (1988b, p. 128), who referred this taxon to *Pericosmus* (*Victoriaster*) *jamaicensis,* following the classification of Fischer (1966, p. U569). However, Kier (1984, p. 69) referred *V. jamaicensis* to the genus *Lambertona* and noted that it was very similar to the holotype of *Lambertona lamberti* (Sánchez Roig), although the Jamaican specimen is too poorly preserved to permit a complete comparison to be made.

Genus *Linthia* Desor

Remarks. Two Jamaican spatangoids have been classified as *Linthia, L. trechmanni* Hawkins and *L. obesa* Arnold and Clark.

Linthia trechmanni Hawkins, 1924
(Figure 16.3, 16.4)

Description. (See also Hawkins, 1924, p. 319–321, Plate 18, Figs. 6, 7). Test moderately large, heart-shaped, with a deep anterior sulcus and a blunt posterior margin. Test broadest just anterior of the apical system. Test moderately high, steep sided, with a gently domed adapical surface. An anteriopoterior keel is developed adorally. Ambitus just below the midheight of the test.

Apical system approximately central, but not seen in the available specimens (the apical system of *Linthia* is ethmolytic with four genital pores: Kier, 1984, p. 71).

Four ambulacral petals developed adapically (not III; Fig. 16.3). Petals occur in moderately deep, straight grooves. Posterior petals shorter than those of ambulacra II and IV, which reach the ambitus. Petals open, comprising two equal columns of ambulacral pores, with elliptical pores perradially and teardrop-shaped pores adradially. Pore pairs conjugate. Interporiferous zones narrow. Ambulacrum III occurs in a broad, deep anterior sulcus. Ambulacra not readily apparent adambitally or adorally, except adjacent to the peristome, where a pentastellate arrangement of phyllodes (Fig. 16.4) is defined by the occurrence of single pores. Ambulacra I+II and IV+V are so close as to appear to be in contact adjacent to the peristome, with I and V separated by the labrum.

Interambulacra broad adapically, but plate sutures only poorly apparent. Interambulacral plates wide and moderately high adapically. Most anterior plate columns of interambulacra 2 and 3 composed of geniculate plates that fold into the anterior sulcus. Adorally, interambulacrum 5 forms a broad, raised plastron and interambulacra 1 and 4 do not appear to reach the peristome. Numerous low primary tubercles cover the test surface, appearing to be particularly dense adjacent to the petals. Very numerous, granular secondary tubercles occur in the intervening areas. Peripetalous and anal fascioles are well developed (Hawkins, 1924, Plate 18, Fig. 6).

Peristome small, anterior, kidney-shaped, wider than high and directed anteriorly. A labrum extends over the posterior ?half of the peristome. Periproct elliptical, higher than wide and occurring high on the posterior margin.
Material. Holotype, BMNH E17219, plus another specimen, BMNH E83642 (Fig. 16.3, 16.4). Arnold and Clark (1927, p. 57) had three tests in their collection.
Occurrence. The type area is around Spring Mount, parish of St. James, from the Chapelton Formation, Yellow Limestone Group (mid-early to mid-middle Eocene). The Arnold and Clark specimens came from the same formation at Springfield (=Welcome Hall), parish of St. James. BMNH E82642 is from Freemans Hall, parish of Trelawny, presumably from the Stettin Member of the Chapelton Formation, from the same horizon at Wait-a-Bit Cave (Robinson, 1988, Fig. 3, horizon of locality 10; late early Eocene).
Remarks. Kier (1984, p. 7) considered *L. trechmanni* to be very different from any Cuban species of spatangoid.

Linthia obesa Arnold and Clark, 1934

Description. See Arnold and Clark (1934, p. 148, 149, Plate 2, Figs. 4–6).
Material. A unique holotype, MCZ 3482.
Occurrence. Locality and horizon unknown.
Remarks. This species differs from *L. trechmanni* in being more conical adapically and more convex adorally. Kier (1984, p. 7, 8) considered *L. obesa* to be based on a specimen too poorly preserved for comparison with Cuban spatangoids.

Genus *Periaster* d'Orbigny

Remarks. Hawkins (1924, p. 312, 319) considered two specimens from the Chapelton Formation collected at Spring Mount, parish of St. James, to belong to the genus *Periaster.* The better preserved of these tests (BMNH E17216) was classified as *Periaster* cf. *elongatus* Cotteau. This specimen is, in fact, *Eupatagus alatus* Arnold and Clark (see below). The Hawkins specimen is similar in size and morphology to the test figured by Arnold and Clark (1927, Plate 13, Figs. 5–7). Further, it is of incidental significance that Kier (1984, p. 8) considered that "*Periaster elongatus* Cotteau from St. Bartholomew is based on a holotype too poorly preserved for comparison."

The second specimen of Hawkins (BMNH E17215) was referred to

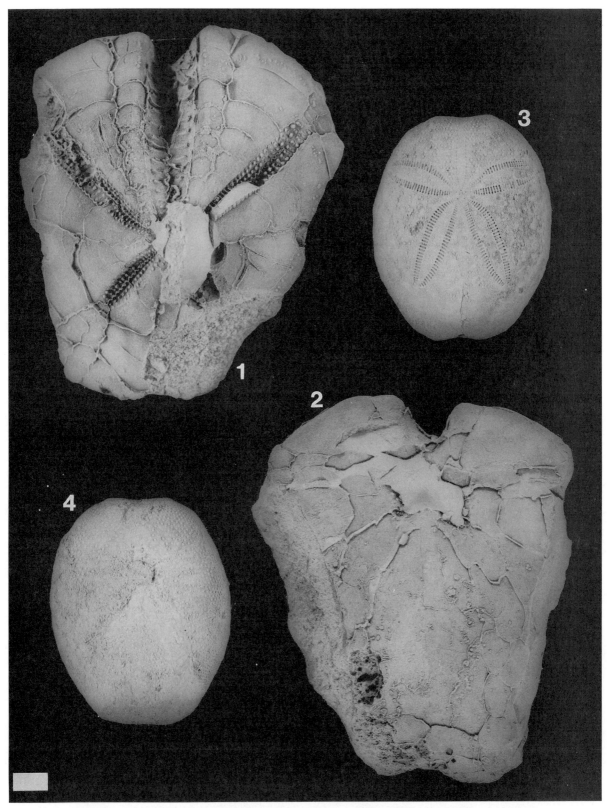

Figure 17. 1, 2, *Lambertona jamaicensis* (Arnold and Clark), holotype, MCZ 3484 (locality and horizon unknown), internal mould: 1, apical view; 2, oral view. 3, 4, *Eupatagus alatus* Arnold and Clark, USNM 341257, Eocene, Loma Caoba, Pinar del Rio Province, Cuba: 3, apical view; 4, oral view. Scale bar represents 10 mm.

Periaster sp. nov. It is a very battered test of a spatangoid, not comparable with the type species (*P. elatus* (Desmoulins): Fischer, 1966, Fig. 460.2a–e). Of the Jamaican Eocene echinoids, BMNH E17215 is closest to *Schizaster hexagonalis* Arnold and Clark, 1927, but only on the basis of test outline. It is perhaps best to regard the BMNH specimen as spatangoid sp. indet.

Family BRISSIDAE Gray

Genus *Brissus* Gray

Brissus sp. cf. *B. unicolor* (Leske, 1778)
(Figure 15.10, 15.11)

Description. (See also Cooke, 1959, p. 81, 82, Plate 36, Figs. 1–4). Test oval in outline, more rounded anteriorly than posteriorly, and moderately high. Adapical surface highest near the posterior margin, sloping gently anteriorly. Anteroposterior keel developed adorally. Ambitus below mid-height of test.

Apical system small, well anterior of center, with four large genital pores (genital 5 missing) and apparently ethmolytic (Cooke, 1959, p. 81; Fischer, 1966, p. U582, U583).

Four ambulacral petals developed adapically (not III; Fig. 15.10). Petals not extending to the margin of the adapical surface, particularly posteriorly. Petals approximately straight, situated in shallow, elongate grooves. Petals II and IV slightly longer than I and V. Pore pairs within petals conjugate, composed of elliptical pores. Interporiferous zones narrow. Ambulacrum III about as wide as petals adapically, but flush with the test surface (no anterior sulcus developed), with single, well-separated ambulacral pores. Other ambulacra expanding adambitally of the petals and tapering adorally to the peristome. Posterior ambulacra elongate adorally, flanking the plastron. Ambulacra single pored adorally, forming a pentastellate pattern of phyllodes (Fig. 15.11).

Interambulacra broad adapically, although interambulacrum 5 is markedly the narrowest. Interambulacra broadest adambitally, tapering adapically and adorally. Moderately broad plastron developed adorally in interambulacrum 5. Low, domed primary tubercles numerous, particularly on the plastron, with very numerous secondary tubercles occurring in the intervening space. Fascioles not preserved (but see Fischer, 1966, Fig. 466).

Peristome moderately large, kidney-shaped, anterior in position and directed anteriorly, with a very weak labral lip (Fig. 15.11). Periproct elliptical, higher than wide, posterior and marginal.
Material. A single test, MCZ 3469.
Occurrence. Collected at Lucky Hill, parish of St. Mary, from the Chapelton Formation, Yellow Limestone Group (mid-early to mid-middle Eocene).
Remarks. Arnold and Clark (1927, p. 63) classified this specimen as *B. brissus* (=*unicolor*) and it is certainly extremely similar to Recent tests of this species (compare Fig. 15.10–15.11 with Cooke, 1959, Plate 36, Figs. 1–4). However, Cooke (1959, p. 82) considered this species to be limited to the Quaternary and Jackson (1922, p. 87) could not confirm that *B. unicolor* was otherwise known as a fossil in the West Indies. The Jamaican specimen does not resemble any of the Cuban species of *Brissus* figured by Kier (1984), with the possible exception of *B. camagueyensis* Weisbord (Kier, 1984, Plate 43, Figs. 4–6). Until further specimens are found from Lucky Hill, the identification of the Jamaican test should be regarded as tentative.

Dr. M. L. McKinney (written communication, 1990) considers that the length of the posterior petals in the Jamaican specimens is reminiscent of *B. bridgeboroensis* Carter, 1987b. However, the width-to-length ratio appears to be higher and the petals appear to be narrower in the Jamaican test than in either *B. unicolor* or *B. bridgeboroensis*.

Hawkins (1924, p. 322) briefly described an internal mould (BMNH E17226) that he regarded as *Brissus* sp. indet. This specimen is younger than the Lucky Hill test, coming from the "White Limestone, Montego Bay," parish of St. James, presumably from the Montpelier "Formation," that is, from the Sign+Spring Garden Formations, Montpelier Group, White Limestone Supergroup (mid-Oligocene to mid-middle Miocene: Robinson, 1988, Fig. 4).

Genus *Cyclaster* Cotteau

Cyclaster sterea Arnold and Clark, 1934
(Figure 15.1)

Description. (See also Arnold and Clark, 1934, p. 149, 150, Plate 3, Figs. 4, 5). Test has been compressed anterioposteriorly, the anterior margin is damaged and the adoral surface is obscured by limestone. Test large, inflated, rising to a steeply conical apex. Test rounded in outline, blunt anteriorly, more pointed posteriorly and apparently broadest anterior of the apical system.

Apical system not preserved, but small and apparently situated well anterior of center.

Four straight, elongate petals developed adapically (not III; Fig. 15.1). Petals developed in narrow, elongate grooves that extend to close to the ambitus and taper adapically. Petals in ambulacra II and IV appear as grooves adapically (due to the test being crushed anteroposteiorly) and are largely obscured by sediment more adorally. Posterior petals are long, ?closed, with interporiferous zones narrower than a single pore pair. Pores elongate. Ambulacrum III straight, sloping steeply adambitally, with a broad, deep anterior sulcus and sparse pores.

Interambulacra broad adapically. Tubercles abraded, but where preserved primary tubercles are small and close packed. Fascioles not preserved.

Peristome and periproct obscured.
Material. A unique holotype, MCZ 3483.
Occurrence. Locality and horizon unknown.
Remarks. This specimen is too poorly preserved for Kier (1984, p. 88) to compare it with the Cuban species of *Cyclaster*.

Genus *Eupatagus* L. Agassiz

Remarks. Arnold and Clark (1927, p. 63–67, 1934, p. 155, 156; see also Hawkins, 1927, p. 81, 82) recognized seven species of *Eupatagus* from Jamaica: *E. alatus* Arnold and Clark, *E. antillarum* (Cotteau), *E. attenuatus* Arnold and Clark, *E. defectus* Arnold and Clark, *E. grandiflorus* (Cotteau), *E. hildae* Hawkins, and *E. longipetalus* Arnold and Clark. Kier (1984, p. 7) considered *E. attenuatus* to be a synonym of *E. alatus*, *E. grandiflorus* to be synonymous with *E. clevei* (1984, p. 7, 8, 98), and *E. longipetalus* to be more correctly classified as an *Antillaster* (1984, p. 7; see below). *Eupatagus* sp. a of Jackson (1922, p. 90), from the Yellow Limestone Group, may be an *E. alatus,* based on Jackson's description. Hawkins (1924, p. 322) referred a further test to *E.* cf. *antillarum.*

Eupatagus alatus Arnold and Clark, 1927
(Figure 17.3, 17.4)

Description. (See also Arnold and Clark, 1927, p. 63, 64, Plate 13, Figs. 4–7; Kier, 1984, p. 95–98, Plate 54; Plate 55, Figs. 1, 2; Text fig. 34). Test moderately large, oval in outline, blunt anteriorly and posteriorly, with or without a very shallow anterior sulcus. Test broadest posterior of the apical system, moderately high, highest posteriorly, flattened to gently domed adapically and convex adorally, with an anteroposterior keel developed on the plastron. Ambitus rounded, at about the mid-height of the test.

Apical system anterior, small, with four genital pores (genital 5 missing), ethmolytic, with the elongate madreporite extending posteriorly (Kier, 1984, Fig. 34B).

Four ambulacral petals adapically (not III; Fig. 17.3). Petals closed, widest medially, extending to near the adapical margin (particularly II and IV). Posterior petals longest. Petals flush with test surface, although pore columns may be slightly sunken. Pore pairs conjugate, composed of rounded pores. Interporiferous zones broad, about 50 to 55% of the petal width medially. Ambulacrum III narrow, expanding slightly adambitally, where a very shallow anterior sulcus may be developed. Ambulacrum III has nonconjugate anisopores (Kier, 1984, p. 97; Smith, 1980) arranged in adradial columns. Ambulacra expanding adorally of the petals. Ambulacra II and IV widest adambitally and tapering adorally. Ambulacrum III expanding adorally. Ambulacra I and V widest just posterior of the plastron, being separated by the labrum adorally. Ambulacra all single pored adorally, with a pentastellate development of phyllodes (Fig. 17.4).

Interambulacra broadest adambitally, where they are two to three and a half times wider than ambulacra, and tapering adorally and adapically. Interambulacra composed of broad, moderately high plates. Keeled, moderately broad, meridosternous plastron adorally in interambulacrum 5. Primary tubercles small, abundant, absent adorally on the phyllodes and labrum (Fig. 17.4). Secondary tubercles granular and very numerous. Peripetalous and anal fascioles present (Kier, 1984, p. 97).

Peristome anterior, moderately large, oval in outline, wider than high and not overhung by the elongate labrum (Kier, 1984, Fig. 34A). Periproct posterior, elliptical, higher than wide and situated high on the posterior margin.

Material. Holotype, MCZ 3298, plus a further 243 specimens referred to this species by Arnold and Clark (1927, 1934). Eighteen other tests in USNM acc. 301378 (Donovan, 1988a, p. 41) were mentioned by Kier (1984, p. 96). Four tests of this species were identified by Donovan et al. (1989a; to be deposited in the BMNH), as well as a fifth classified as *E.* sp. cf. *E. attenuatus*. The holotype of *E. attenuatus* is MCZ 3299.

Occurrence. The type area is west of Springfield (= Welcome Hall) towards Seven Rivers, parish of St. James, from the Chapelton Formation, Yellow Limestone Group (mid-early to mid-middle Eocene). All of Arnold and Clark's specimens were collected in St. James, presumably from the same horizon. The type area for *E. attenuatus* is the Seven Rivers region, parish of St. James, from the Chapelton Formation. The specimens collected by Donovan et al. (1989a) are from Beecher Town, parish of St. Ann, from the Swanswick Formation, Moneague Group, White Limestone Supergroup (late middle Eocene). This species is also known from the Eocene of Cuba (Kier, 1984, p. 97, 98: Fig. 17.3, 17.4 herein).

Remarks. *E. alatus* is the commonest species of *Eupatagus* in Jamaica.

Eupatagus sp. cf. *E. antillarum* (Cotteau, 1875)

Description. See Jackson (1922, p. 91, 92, Plate 16, Figs. 5, 6).
Material. Jamaican specimens include BMNH E17225, plus a second test in the B. W. Arnold collection of the MCZ.
Occurrence. The BMNH specimen is from Glasgow, presumably Glasgow, parish of St. James (all other localities mentioned in Hawkins, 1924, p. 312, are in St. James), from the Chapelton Formation, Yellow Limestone Group (mid-early to mid-middle Eocene). The MCZ specimen is from near Point, parish of St. James, from the same horizon. The type area of *E. antillarum* is St. Bartholomew (Jackson, 1922, p. 92). This species is widespread in the Caribbean and southeastern United States (Cooke, 1959, p. 90).
Remarks. Hawkins (1924, p. 322) reported *E.* cf. *antillarum* from Jamaica on the basis of a "much worn specimen" (BMNH E17255). Arnold and Clark (1927, p. 64, 65) identified their specimen as *E. antillarum* and compared it with Jackson's figured specimen (1922, Plate 16, Figs. 5, 6). Neither of the Jamaican tests have been figured.

Kier (1984, p. 98) stated that "*Eupatagus alatus* differs from the lectotype . . . of *E. antillarum* (Cotteau) . . . in having a broader test with wider petals and a much higher peristome."

Eupatagus clevei (Cotteau, 1875)

Description. See Kier (1984, p. 98–103, Pl. 61, Figs. 3–6, Plates 62–66, Text fig. 35).
Material. A single specimen (originally described as the synonymous *E. grandiflorus* [Cotteau]) in the B. W. Arnold collection of the MCZ.
Occurrence. Locality and horizon of the Jamaican specimen unknown. Otherwise known from the Eocene of St. Bartholomew, Cuba, Panama, and Florida, the Oligocene to Miocene of Cuba and also from Curaçao (Kier, 1984, p. 99).
Remarks. Kier (1984, p. 98) stated that "*Eupatagus clevei* (Cotteau) . . . differs from *E. alatus* in its much narrower test, wider petals and lower peristome."

Eupatagus defectus Arnold and Clark, 1927

Description. See Arnold and Clark (1927, p. 65, 66, Plate 14, Figs. 1–3).
Material. Holotype, MCZ 3300, plus 21 other tests in the B. W. Arnold collection of the MCZ.
Occurrence. The type area is the Spring Mount region, parish of St. James, from the Chapelton Formation, Yellow Limestone Group (mid-early to mid-middle Eocene).
Remarks. *E. defectus* differs from *E. alatus* in having a relatively shorter, broader and lower test, with posterior paired petals shorter than anterior paired petals.

Eupatagus hildae Hawkins, 1927

Description. See Hawkins (1927, p. 81, 82, Plate 22, Figs. 9, 10; Text fig. 1).
Material. A unique holotype, BMNH E17664.
Occurrence. From St. Hilda's School, Brown's Town, parish of St. Ann, from the Brown's Town Formation, Moneague Group, White Limestone Supergroup (Oligocene: Robinson, 1988, Fig. 4).
Remarks. Although still to be confirmed, Kier (1984, p. 99) considered *E. hildae* to be a junior synonym of *E. clevei* (see above). This is the only species of echinoid from the Oligocene of Jamaica to have so far received adequate description.

Genus *Macropneustes* L. Agassiz sensu lato

Remarks. Arnold and Clark (1927, 1934) described six species from Jamaica which they assigned to *Macropneustes*: *M. altus* Arnold and Clark, *M. angustus* Arnold and Clark, *M. dyscritus* Arnold and Clark, *M. parvus* Arnold and Clark, *M. sinuosus* Arnold and Clark, and *M. stenopetalus* Arnold and Clark. Of these species, Kier (1984, p. 7, 8) considered four to be too poorly preserved to compare with the Cuban spatangoids (*M. angustus*, *M. dyscritus*, *M. parvus*, *M. sinuosus*) and *M. stenopetalus* to be very different from any Cuban spatangoid (*M. altus* was not mentioned). Of the Jamaican Cenozoic echinoids, those assigned to *Macropneustes* are the least well known and most species would greatly benefit from redescription based on more, and superior, specimens. Arnold and Clark (1927, p. 67) certainly gave the impression that they may have forced various loosely related taxa into *Macropneustes*.

In recognition that Jamaican *Macropneustes* are generally poorly known, it is considered important to reproduce the following diagnosis of the genus (Kier, 1984, p. 115) as an aid to recognition:

Test large, low, broad, with anterior groove; paired petals long; large tubercles confined within peripetalous fasciole; peristome large; subanal fasciole present.

Macropneustes? sinuosus Arnold and Clark, 1934
(Figure 16.9)

Description. (See also Arnold and Clark, 1934, p. 152, Plate 4, Fig. 3). Adoral surface and posterior of test not preserved. Test large, rounded, blunt anteriorly with a broad, shallow anterior sulcus and widest posterior of the apical system. Test apparently wider than long (but incomplete and slightly crushed). Adapical surface domed, sloping gently posteriorly and more steeply anteriorly. Ambitus low.

Apical system anterior and small, but not visible.

Ambulacra I, II, IV, and V petaloid adapically (Fig. 16.9). Petals long, straight, sunken slightly. Posterior petals apparently longest, petals probably extending to close to the ambitus. Pore pairs conjugate, perradial pores rounded to elliptical and adradial pores elliptical to teardrop-shaped. Interporiferous zones narrow. Ambulacrum III wide, flush with the test surface adapically and occurring in a broad, shallow sulcus.

Interambulacra very broad adambitally, tapering adorally. Interradial suture conspicuously sinuous. Interambulacra composed of broad, moderately high plates. Primary tubercles small and numerous adapically, largest anteriorly, and separated by very numerous, granular secondary tubercles. Fascioles not visible.

Material. A unique holotype, MCZ 3486.
Occurrence. Locality and horizon unknown.
Remarks. Despite its very poor condition, this specimen, with its large test, elongate petals, extreme width, and distinctive sinuous interradial sutures, is distinct from all other Jamaican fossil echinoids. It is relevant to note that the figure of the holotype in Arnold and Clark (1934, Plate 4, Fig. 3) is incorrectly mounted, with ambulacrum V (instead of III) orientated towards the top of the page. The illumination of the specimen from the top left (that is, interambulacrum 3) and the description show this to be an error of mounting only, rather than of interpretation.

Macropneustes? *altus* Arnold and Clark, 1927

Description. See Arnold and Clark (1927, p. 68, Plate 14, Figs. 7–9).
Material. A unique holotype, MCZ 3302.
Occurrence. The type area is the Yallahs River valley, presumably north of Easington Bridge, parish of St. Thomas, from the "Font Hill Formation." Yellow Limestone Group (mid-early to mid-middle Eocene).
Remarks. The holotype of this species is very abraded and poorly preserved. It differs from *M.? sinuosus* in being smaller, oval in outline and moderately elongate, widest in the region of the apical system, rounded anteriorly and blunt posteriorly, with a high apex situated posterior of the apical system. The appearance of this test is reminiscent of a small asterostomatid.

Macropneustes? *angustus* Arnold and Clark, 1927

Description. See Arnold and Clark (1927, p. 68, 69, Plate 14, Figs. 10, 11, Plate 15, Fig. 1).
Material. Holotype, MCZ 3303, plus three paratype specimens in the B. W. Arnold collection of the MCZ.
Occurrence. The type area is the Spring Mount region, parish of St. James, from the Chapelton Formation, Yellow Limestone Group (mid-early to mid-middle Eocene). This species is also reported from the western hillsides of the Yallahs River valley, parish of St. Thomas, from the "Font Hill Formation," Yellow Limestone Group.
Remarks. These tests are poorly preserved and it is uncertain whether they are *Macropneustes* or even if they are conspecific (Arnold and Clark, 1927, p. 69). The "young paratype" figured by Arnold and Clark (1927, Plate 14, Figs. 10, 11) is highly deformed and has a narrow, elongate, inflated outline. It could easily be some other taxon of spatangoid apart from *Macropneustes*. The holotype (Arnold and Clark, 1927, Plate 15, Fig. 1) is highly battered, but is obviously narrower and more elongate than *M.? sinuosus*.

Macropneustes? *dyscritus* Arnold and Clark, 1934

Description. See Arnold and Clark (1934, p. 151, Plate 4, Figs. 1, 2).
Material. A unique holotype, MCZ 3485.
Occurrence. Locality and horizon unknown.
Remarks. *M.? dyscritus* differs from *M.? sinuosus* in being slightly larger, more elongate, and very oval in outline. The apex is very anterior, sloping steeply anteriorly and gently posteriorly.

Macropneustes? *parvus* Arnold and Clark, 1927

Description. See Arnold and Clark (1927, p. 69, 70, Plate 18, Figs. 1–3).
Material. Holotype, MCZ 3306, plus two paratype specimens in the B. W. Arnold collection of the MCZ.
Occurrence. The type area is near Johns Hall, Springfield road, parish of St. James, presumably from the Chapelton Formation, Yellow Limestone Group (mid-early to mid-middle Eocene).
Remarks. *M.? parvus* differs from *M.? sinuosus* in being much smaller, oval in outline, blunt anteriorly and posteriorly, with a moderately deep anterior sulcus and in being relatively more inflated, with a more flattened adapical surface.

Macropneustes? *stenopetalus* Arnold and Clark, 1934

Description. See Arnold and Clark (1934, p. 152, 153, Plate 4, Figs. 4, 5).
Material. A unique holotype, MCZ 3487.
Occurrence. Locality and horizon unknown.
Remarks. This is perhaps the best preserved of the Jamaican *Macropneustes* sensu lato. The test is smaller than *M.? sinuosus*, with a more oval outline (although still broad), a more gently domed adapical surface, and an ambitus at about the midheight of the test.

Genus *Meoma* Gray

Meoma antiqua Arnold and Clark, 1927
(Figure 18.3)

Description. (See also Arnold and Clark, 1927, p. 70, Plate 15, Fig. 2). Test large, elongate, angular hexagonal in outline and with a deep anterior sulcus. Test moderately high, widest at about midlength. Adapical surface gently convex, with interambulacra 2 and 3 inflated adapically. Sides of test steep. Adoral surface not preserved.

Apical system small, anterior, poorly preserved, but with three large genital pores apparent.

Four ambulacral petals developed adapically (not III; Fig. 18.3). Petals straight and probably open, expanding adambitally, long and occurring in deep grooves. Posterior petals longest. Pore pairs conjugate, composed of elliptical pores. Interporiferous zones weakly developed. Ambulacrum III moderately broad, flush with test surface adapically, but in a deep sulcus anteriorly, with single pores forming two adradial columns.

Interambulacra broad, composed of wide, moderately high plates. Primary tubercles fairly common, separated by very numerous, granular secondary tubercles. Part of the peripetalous fasciole is preserved (Arnold and Clark, 1927, p. 70).

Periproct large, posterior, and marginal.
Material. The holotype, MCZ 3304, is the only specimen known from Jamaica.
Occurrence. The type area is near Lucky Hill, parish of St. Mary, from the Chapelton Formation, Yellow Limestone Group (mid-early to mid-middle Eocene). This species is also known from the middle Eocene of Cuba (Kier, 1984, p. 120, 121).
Remarks. This specimen is poorly preserved, but Arnold and Clark (1927, p. 70) considered that the preserved part of the fasciole, "Around tip of IV and across interambulacrum 4," was typical of *Meoma*. Kier's diagnosis of the genus (1984, p. 118, 119) included the following relevant comments: ". . . apical system ethmolytic with 4 genital pores . . . peripetalous fasciole deeply indented between petals, subanal fasciole

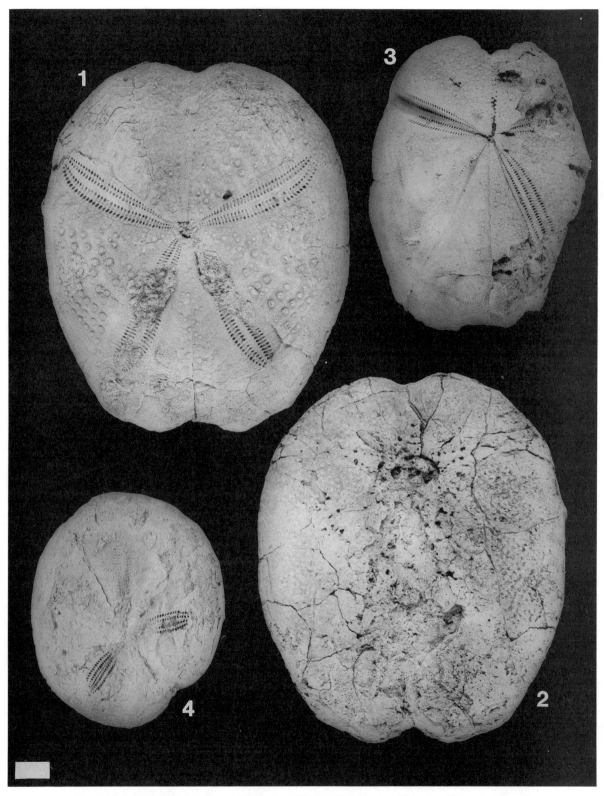

Figure 18. 1, 2, *Plagiobrissus loveni* (Cotteau), USNM 444302, Eocene, Chapelton Formation (locality unknown: Donovan, 1988a, p. 41, 42): 1, apical view; 2, oral view. 3, *Meoma antiqua* Arnold and Clark, holotype, MCZ 3304, Eocene, Chapelton Formation, near Lucky Hill, parish of St. Mary, apical view. 4, *Homeopetalus axiologus* Arnold and Clark, holotype, MCZ 3479 (locality and horizon unknown), apical view. Scale bar represents 1 mm.

bilobed. . . ." The apical system of the holotype is poorly preserved, so it is probably best to say that at least three genital pores are present in *M. antiqua.*

Genus *Metalia* Gray

Remarks. Arnold and Clark (1934) described two species of *Metalia* from Jamaica, *M. dubia* and *M. jamaicensis.*

Metalia jamaicensis Arnold and Clark, 1934
(Figure 16.1, 16.2)

Description. (See also Arnold and Clark, 1934, p. 154, 155, Plate 5, figs. 1–3). Test large, steep-sided, oval in outline and moderately elongate, rounded anteriorly and blunt posteriorly, with a deep anterior sulcus. Test moderately high, highest just anterior to the apical system, sloping posteriorly gently and anteriorly steeply. Adoral surface gently convex, due to the presence of a low anteroposterior keel. Ambitus rounded and at about midheight of the test.

Apical system anterior, small, but poorly preserved (like *Brissus*, the apical system of *Metalia* is ethmolytic with four gonopores: Fischer, 1966, p. U582, U583, U597).

Four ambulacral petals developed adapically (not III; Fig. 16.1). Petals occur in deep grooves. Petals II and IV largely obscured by sediment, but short and apparently curving anteriorly. Posterior petals longer, straight and open, although closing adambitally. Pore pairs conjugate, composed of elliptical pores. Interporiferous zones about one-third of the petal width. Ambulacrum III largely obscured by sediment adapically, but in a deep anterior sulcus. Ambulacra expanding adambitally of petals, then tapering adorally, forming a pentastellate pattern of single-pored phyllodes adjacent to the peristome (Fig. 16.2). The posterior ambulacra are broad adorally and flank the large, triangular plastron.

Interambulacra much broader than ambulacra adambitally, tapering adapically and adorally. Interambulacral plates broad. Interambulacrum 5 forms a broad, triangular plastron adorally. Primary tubercles less common adapically than adorally, separated by very numerous, granular secondary tubercles. Fascioles not preserved (peripetalous and subanal fascioles usually occur in brissids: Kier, 1984, p. 81).

Peristome anterior and directed anteriorly, wide and low, kidney-shaped, only slightly overhung by the labrum posteriorly. Periproct higher than wide, posterior and occurring on the posterior margin.
Material. A unique holotype, MCZ 3489.
Occurrence. Locality and horizon unknown.

Metalia dubia Arnold and Clark, 1934

Description. See Arnold and Clark (1934, p. 153, 154, Plate 5, Figs. 4–6).
Material. A unique holotype, MCZ 3488.
Occurrence. Locality and horizon unknown.
Remarks. M. dubia is quite unlike any other fossil echinoid from Jamaica, the two posterior petals (I and V) being depressed and confluent adapically. *M. dubia* further differs from *M. jamaicensis* in being shorter and relatively broader, without an anterior sulcus.

Genus *Plagiobrissus* Pomel

Remarks. Arnold and Clark (1927) recognized six species of *Plagiobrissus* from Jamaica: *P. abruptus* Arnold and Clark, *P. elevatus* Arnold and Clark, *P. latus* Arnold and Clark, *P. loveni* (Cotteau), *P. perplexus* Arnold and Clark, and *P. robustus* Arnold and Clark. Kier (1984, p. 7) considered that all of these taxa were very different from any Cuban species; indeed, Kier did not recognize any species of *Plagiobrissus* in the Cuban fossil record. Other reports of *Plagiobrissus* in Jamaica include the recognition of a poorly preserved test as *P.* cf. *loveni* by Hawkins (1924, p. 321). Hawkins (1927) later identified two further tests of this species.

This genus is similar to *Eupatagus.* Fischer (1966, p. U602) discerned *Plagiobrissus* ". . . from *Eupatagus* chiefly in having anal branches on subanal fasciole, long plastron, short labrum, and long, narrow, flexed petals."

Plagiobrissus loveni (Cotteau, 1875)
(Figure 18.1, 18.2)

Description. (See also Jackson, 1922, p. 83, 84, Plate 14, Fig. 5; Arnold and Clark, 1927, p. 73, 74, Plate 20, Figs. 1–5; Hawkins, 1927, p. 82–84, Plate 22, Figs. 11–13, Text figs. 2, 3). Test large, broad, with a moderately deep anterior sulcus and an equally deep anal sulcus. Test widest just anterior of apical system. Test flattened adapically and adorally, with a gently convex anterioposterior adoral keel. Ambitus lower than the midheight of the test and rounded.

Apical system small, anterior of center and in a slight depression, ethmolytic, probably with four genital pores.

Four petaloid ambulacra developed adapically (not III; Fig. 18.1). Petals elongate, open, situated in shallow grooves. Petals II and IV slender, curving anteriorly, broadest close to their adambital tips. Petals I and V straight. Ambulacra wider than the petals, particularly adambitally. Pore pairs conjugate, composed of elliptical to teardrop-shaped pores. Interporiferous zones narrow. Ambulacrum III straight, expanding adambitally, occurring in a shallow groove that forms the anterior sulcus and with two adradial columns of small, sunken pore pairs adapically. Ambulacra expand adambitally from petals and then taper adorally. Ambulacra single pored adorally, forming a pentastellate pattern of phyllodes (Fig. 18.2). Ambulacra I and V broad and elongate adorally, flanking a relatively narrow plastron.

Interambulacra broad adambitally, tapering adorally and adapically, composed of broad, moderately high, V-shaped plates. Interambulacra slightly inflated adapically, rising above the level of the apical system. Interambulacrum 5 forms a relatively narrow, triangular plastron adorally. Primary tubercles prominent, perforate, crenulate, with broad, circular areolae, confined to interambulacral areas within the peripetalous fascioles adapically and mainly to the interambulacra adorally. Secondary tubercles very small and granular, very common adapically, but apparently less common adorally. Peripetalous fasciole approximately parallel to the margin of the adapical surface. Subanal fasciole present.

Peristome anterior, large, wider than high, kidney-shaped, with the labrum overhanging the posterior quarter. Periproct posterior, marginal, large and rounded, slightly more elevated than the ambitus and within the anal sulcus.
Material. Arnold and Clark based their description on a collection of 116 tests, now in the collection of the MCZ. USNM acc. 301378 includes a further seven Arnold tests (Donovan, 1988a, p. 41, 42). The specimens studied by Hawkins are BMNH E17220 (*P.* sp. cf. *P. loveni*), E17663, plus five other specimens.
Occurrence. Reported localities for *P. loveni* in Jamaica include along the Springfield Road near Point, at Ginger Valley, and from near Somerton, all in the parish of St. James and all within the outcrop of the Chapelton Formation, Yellow Limestone Group (mid-early to mid-middle Eocene). This species is also known from the Eocene of St. Bartholomew (Jackson, 1922, p. 84).
Remarks. The large, flattened test of *P. loveni* make it one of the most distinctive of Jamaican fossil spatangoids. Lewis (1986, p. 8) incorrectly placed E17663 in the genus *Caribbaster*, confusing *Plagiobrissus loveni* with *Prenaster* (=*Caribbaster*) *loveni* (Kier, 1984, p. 68).

Plagiobrissus abruptus Arnold and Clark, 1927

Description. See Arnold and Clark (1927, p. 71, Plate 18, Figs. 4, 5).
Material. Holotype, MCZ 3307, plus a further 17 paratypes in the B. W.

Arnold collection of the MCZ. USNM acc. 301378 includes three tests of this species (Donovan, 1988a, p. 41).
Occurrence. The type area is the Spring Mount region, parish of St. James, from the Chapelton Formation, Yellow Limestone Group (mid-early to mid-middle Eocene).
Remarks. *P. abruptus* differs from *P. loveni* in being more inflated, having a gently domed adapical surface with a slightly more posterior apical system and apparently lacking an anal sulcus, giving the posterior margin of the test a truncated appearance. The petals of *P. abruptus* extend to or near to the ambitus.

Plagiobrissus elevatus Arnold and Clark, 1927

Description. See Arnold and Clark (1927, p. 71, 72, Plate 19, Figs. 2–4).
Material. Holotype, MCZ 3308, plus a second, poorly preserved specimen mentioned by Arnold and Clark that may be conspecific.
Occurrence. The type area is the Spring Mount region, parish of St. James, from the Chapelton Formation, Yellow Limestone Group (mid-early to mid-middle Eocene).
Remarks. *P. elevatus* differs from *P. loveni* in having a much more inflated, domed test, with a relatively larger peristome, lower ambitus, and no anal sulcus. The holotype of *P. elevatus* is about as large as the largest *P. loveni* (Arnold and Clark, 1927, p. 71, 73). However, the largest *P. loveni* is only 37 mm high, compared with 48 mm in *P. elevatus*. This difference in height is accentuated by *P. elevatus* having a low ambitus, so that the adapical surface is domed rather than flattened. The petals of *P. elevatus* are elongate and extend to close to the ambitus (cf. *P. perplexus*, below).

Plagiobrissus latus Arnold and Clark, 1927

Description. See Arnold and Clark (1927, p. 72, 73, Plate 18, Figs. 6, 7; Plate 19, Fig. 1).
Material. A unique holotype, MCZ 3309.
Occurrence. The type area is on the Springfield road near Point, parish of St. James, from the Chapelton Formation, Yellow Limestone Group (mid-early to mid-middle Eocene).
Remarks. *P. latus* differs from *P. loveni* in being much more rounded in outline and relatively more inflated, with a more anterior peristome that is more overhung by the labrum. Petals I and V are shorter than II and IV, which extend to close to the ambitus. The superficial appearance of this test is reminiscent of a larger, more rounded *Linthia trechmanni* (see above).

Plagiobrissus perplexus Arnold and Clark, 1927

Description. See Arnold and Clark (1927, p. 74, 75, Plate 20, Fig. 6; Plate 21, Figs. 1, 2).
Material. A unique holotype, MCZ 3310.
Occurrence. The type area is the west bank of the Yallahs River valley, above Easington Bridge, parish of St. Thomas, from the "Font Hill Formation," Yellow Limestone Group (mid-early to mid-middle Eocene).
Remarks. *P. perplexus* differs from *P. loveni* in being narrower and more inflated, with a low ambitus and elongate petals that extend to the ambital region. Although relatively as inflated as *P. elevatus*, *P. perplexus* slopes more steeply to the ambitus and has longer petals. In both *P. elevatus* and *P. perplexus* the apical system is approximately central, although it is slightly more posterior in *P. perplexus*.

Plagiobrissus robustus Arnold and Clark, 1927

Description. See Arnold and Clark (1927, p. 75, Plate 20, Fig. 7; Plate 21, Figs. 3, 4).
Material. Holotype, MCZ 3311, plus four paratype specimens. USNM acc. 301378 includes two tests of this species (Donovan, 1988a, p. 42).
Occurrence. The type area is near Johns Hall, on the Springfield road, parish of St. James, from the Chapelton Formation, Yellow Limestone Group (mid-early to mid-middle Eocene).
Remarks. *P. robustus* differs from *P. loveni* in being higher and more narrow. Of the other Jamaican species of *Plagiobrissus*, *P. robustus* is morphologically closest to *P. perplexus,* but is lower, with a more anterior apical system and a very shallow anterior sulcus. In *P. perplexus* petals II and IV curve anteriorly (cf. *P. loveni*; Fig. 18.1), but these petals are straight in *P. robustus*.

Family ASTEROSTOMATIDAE Pictet

Genus *Asterostoma* L. Agassiz

Remarks. Two species of *Asterostoma* have been described from Jamaica, *A. excentricum* L. Agassiz and *A. pawsoni* Kier. This genus is only known from Jamaica and Cuba.

Asterostoma pawsoni Kier, 1984
(Figure 15.2, 15.3)

Description. (See also Kier, 1984, p. 128–130, Plate 78, Text fig. 40). Test moderately large, oval in outline, rounded anteriorly with a shallow anterior sulcus, but more angular posteriorly. Test moderately high, with the apex posterior to the apical system, sloping more steeply anteriorly than posteriorly. Adoral surface flattened apart from a very low keel in interambulacrum 5 and the five ambulacral grooves. Ambitus lower than the midheight of the test and rounded.

Apical system anterior, small, with four genital pores (genital 5 missing), ethmolytic, with the madreporite alongate, extending and expanding posteriorly.

Four ambulacral petals developed adapically (not III; Fig. 15.2). Petals open, either flush with the test surface or slightly depressed. Petals slightly sinuous, particularly in ambulacra II and IV. Petals long, particularly posteriorly, almost extending to the ambitus. Pore pairs conjugate, with pores rounded to elliptical perradially and elliptical to teardrop-shaped adradially. Interporiferous zones broad, about 55 to 60% of petal width at broadest point. Poriferous zones taper slightly adambitally. Ambulacrum III flush with the test surface adapically, extending into a shallow anterior sulcus adambitally. Pore pairs small, forming two narrow columns at the adradial margins of the ambulacrum. Ambulacra expand slightly between the petals and the ambitus. Adorally ambulacra are straight, with shallow ambulacral grooves forming elongate phyllodes which extend to near the margin of the adoral surface (Fig. 15.3). Phyllodes expand slightly towards the peristome and are single pored, with columns of pores occurring near the adradial margins of the ambulacra.

Interambulacra broad, particularly adambitally, tapering adapically and adorally. Interambulacral plates wide and moderately high. Plastron triangular and meridosternous. Primary tubercles fairly numerous, small, perforate, crenulate, with broad areolae. Secondary tubercles very numerous and granular. No fascioles (Kier, 1984, p. 131).

Peristome slightly anterior of center, wider than high, V-shaped and pointing anteriorly, with an angular labrum extending about one-third of the peristome length. Periproct posterior, submarginal (Fig. 15.3), oval, and higher than wide.
Material. Holotype, USNM 358866, plus two numbered paratypes, USNM 358867 and 358868, and three unnumbered paratypes, all in USNM acc. 301378 (Donovan, 1988a, p. 42).
Occurrence. The type locality is uncertain, as these specimens were originally considered to be small tests of *A. excentricum* (Donovan, 1988a, p. 42), which Arnold and Clark (1927, p. 46) recorded from a

number of Yellow Limestone Group (mid-early to mid-middle Eocene) localities (see below). *A. pawsoni* definitely occurs in the Chapelton Formation at Guys Hill, parish of St. Mary.

Remarks. Asterostomatid spatangoids lack fascioles. *A. pawsoni* is the smaller of the two Jamaican species of *Asterostoma,* the genus easily being recognized by its extraordinarily long phyllodes.

Asterostoma excentricum L. Agassiz in Agassiz and Desor, 1847

Description. See Kier (1984, p. 125–128, Plates 70–75; Plate 76, Fig. 1; Text fig. 39).

Material. Arnold and Clark (1927, p. 47) noted that there were only a few large tests in a collection of 79 specimens that they attributed to this species (the small tests are *A. pawsoni* Kier; see above). Two other tests form part of USNM acc. 301378 (Donovan, 1988a, p. 42).

Occurrence. Arnold and Clark listed the following localities (all Yellow Limestone Group; mid-early to mid-middle Eocene) as producing specimens of *Asterostoma* (both *A. excentricum* and *A. pawsoni*): Yallahs River, presumably north of Easington Bridge, parish of St. Thomas ("Font Hill Formation"); Lucky Hill region, parish of St. Mary; and Spring Mount region, parish of St. James (both Chapelton Formation). This species also occurs in Cuba (Kier, 1984).

Remarks. Kier (1984, p. 130) stated that *A. pawsoni* ". . . differs from *A. excentricum* in having a smaller, more elongate test, its first petaloid pore pairs in ambulacrum II in plates 35–38 versus 40–45 in *A. excentricum,* and in plates 30–34 in ambulacrum I versus 40 . . . the labrum in *A. pawsoni* . . . is [relatively] longer . . . and it extends posteriorly to the third adjacent ambulacral plate instead of the second."

A. excentricum is the largest of the Jamaican fossil echinoids (see Arnold and Clark, 1927, Plates 7, 8). The largest Jamaican specimen recorded is 187 mm long by 147 mm wide by 50 mm high (but crushed; Arnold and Clark, 1927, p. 46), and most tests are about 150 mm in length. The only species of comparable size, *Antillaster arnoldi* (see below), lacks the distinctive elongate phyllodes typical of *Asterostoma.*

A. excentricum is the type species of this genus.

Genus *Antillaster* Lambert

Remarks. Two species of *Antillaster* are recognized from Jamaica, *A. arnoldi* Clark and *A. longipetalus* (Arnold and Clark).

Antillaster arnoldi Clark in Arnold and Clark, 1927
(Figure 19)

Description. (See also Arnold and Clark, 1927, p. 62, 63, Plate 15, Fig. 3, Plates 16, 17; Kier, 1984, p. 135, 136, Plate 82, Text fig. 43). Test very large (holotype 165 mm long; Fig. 19.1) and oval in outline. Test rounded anteriorly with a shallow anterior sulcus, more angular posteriorly, widest posterior of the apical system. Test high, with apex just anterior of center, but posterior of apical system. Adapical surface domed, adoral surface flattened, but with a low anterioposterior keel. Ambitus low and rounded.

Apical system anterior of center, but not preserved on the holotype. (The holotype of the synonymous *A. rojasi* Sánchez Roig has an ethmolytic apical system with four genital pores: Kier, 1984, p. 136.)

Four ambulacral petals developed adapically (not III; Fig. 19.1). Petals open, long, flush with test surface, extending to the ambital region (Fig. 19.2). Petals widening adambitally, with columns of pore pairs slightly incurved at their tips. Posterior petals straight, petals II and IV curving gently anteriorly. Pore pairs conjugate. Pores rounded perradially and elongate adradially. Interporiferous zones broad. Ambulacrum III straight, narrower than petals, forming a shallow anterior sulcus. Ambulacra single pored adorally of petals and narrower than interambulacra. Posterior ambulacra flank the narrow plastron adorally. Phyllodes single pored, arrayed in a pentastellate pattern adjacent to the peristome.

Interambulacra broad adapically (Fig. 19.1), composed of broad, moderately high plates. Plastron relatively narrow and elongate. Holotype preserving only a few low primary tubercles adorally. No fascioles.

Peristome moderately large, wide, kidney-shaped, anterior of center, directed anteriorly, with a well-developed labrum lip extending about half the length. Periproct large, posterior and supramarginal.

Material. Holotype, MCZ 3305, plus a second Jamaican specimen (Arnold and Clark, 1934, p. 151).

Occurrence. The type area is Spring Mount, parish of St. James, from the Chapelton Formation, Yellow Limestone Group (mid-early to mid-middle Eocene). This species is also known from the Oligo-Miocene of Cuba (Kier, 1984, p. 136).

Remarks. A. arnoldi and *Asterostoma excentricum* are the two largest echinoid species known from the Jamaican fossil record. *A. arnoldi* lacks deep adoral ambulacral grooves (=phyllodes), a diagnostic feature of *Asterostoma* (Fig. 15.3).

Antillaster longipetalus (Arnold and Clark, 1927)

Description. See Arnold and Clark (1927, p. 67, Plate 14, Figs. 4–6).

Material. A unique holotype, MCZ 3301.

Occurrence. From near the "short cut" road to Kingston (presumably the road from Llandewey to Eleven Mile), Yallahs River valley, parish of St. Thomas, presumably from the "Font Hill Formation," Yellow Limestone Group (mid-early to mid-middle Eocene).

Remarks. This specimen was originally classified as *Eupatagus* by Arnold and Clark, but Kier (1984, p. 7) considered that it ". . . should be referred to *Antilaster* and is similar to the Cuban *Antillaster vaughani* (Jackson)" (for a description of *A. vaughani* see Kier, 1984, p. 140–144, Plates 84–86; Plate 87, Figs. 1–3). *A. longipetalus* differs from *A. arnoldi* in being smaller (only 53 mm in length), relatively broader, more rounded, lower, with the apex posterior and the test sloping markedly anteriorly and less so posteriorly, where there is a blunt posterior margin, and with a prominent keel formed by the plastron.

Incerti subordinis
Incertae familiae

Genus *Homeopetalus* Arnold and Clark

Homeopetalus axiologus Arnold and Clark, 1934
(Figure 18.4)

Description. (See also Arnold and Clark, 1934, p. 146–148, Plate 2, Figs. 2, 3). Test oval in outline, low and flattened adapically. Adoral surface not preserved. Ambitus about midheight of test, with the margin of the corona well rounded.

Apical system not preserved, but posterior in position.

Ambulacra petaloid adapically, except in ambulacrum III. Petals short (particularly posteriorly), closed and slightly sunken. Pores rounded, close together in each pore pair. Interporiferous zones extremely narrow. Ambulacrum III narrow and elongate adapically, expanding towards ambitus.

Interambulacra at most twice as wide as ambulacra adambitally. Fascioles not preserved. At least two orders of small tubercles present, forming a dense covering adapically.

Material. A unique holotype, MCZ 3479.

Occurrence. Locality and horizon unknown.

Remarks. Fischer (1966, p. U625) noted that the ". . . shape and narrowness of interporiferous zones in petals make [*H. axiologus*] very distinctive." However, the specimen is very poorly preserved (Fig. 18.4), lacking an oral surface, fascioles and an apical system, and is a dreadful type for the monospecific genus *Homeopetalus.*

Figure 19. *Antillaster arnoldi* Clark, holotype, MCZ 3305, Eocene, Chapelton Formation, Spring Mount area, parish of St. James. 1, apical view; 2, lateral view (anterior to right). Scale bars represent 10 mm. Copyright President and Fellows of Harvard College. Photograph by R. C. Eng, courtesy of Museum of Comparative Zoology, Harvard University.

OTHER TAXA

Donovan (1988b, p. 128) referred to "*Paleopneustes* sp. or *Pericosmus* sp." from the Pleistocene of Jamaica. These specimens come from the early Pleistocene Manchioneal Formation near Port Maria, parish of St. Mary, and are now in the collection of the BMNH awaiting adequate description (Donovan and Lewis, research in progress).

ACKNOWLEDGMENTS

I am most grateful to David N. Lewis and Andrew B. Smith (British Museum [Natural History]), Felicita d'Escrivan (Museum of Comparative Zoology, Harvard), Frederick J. Collier and Jan Thomson (U.S. National Museum, Department of Paleobiology), and David L. Pawson (U.S. National Museum, Department of Invertebrate Zoology) for loaning and allowing me access to specimens in their care. Figures 5 and 6 were taken by the author in the EM Unit, Department of Botany, University of the West Indies, Mona. I thank Cornelis J. Veltkamp (University of Liverpool, England) and D. George Soloman (University of the West Indies, Mona) for taking Figures 8 and 14, respectively. Andrew Smith kindly made his photomicroscope available for the author to photograph the specimens in Figures 7, 11, and 12. I particularly thank Felicita d'Escrivan for arranging for the photography of *Antillaster arnoldi* at the Museum of Comparative Zoology, Harvard. Other photographs were taken by the author during the period of a Smithsonian Short Term Visitor Grant in 1989, using equipment kindly made available by David Pawson. Carla M. Gordon (University of the West Indies, Mona) kindly gave permission for reference to be made to her unpublished research results. Constructive review comments by Michael L. McKinney (University of Tennessee), Burchard D. Carter (Georgia Southwestern College), and David Lewis are gratefully acknowledged.

REFERENCES CITED

Agassiz, L., and Desor, P., 1847, Catalogue raisonné des espèces, des genres et des familles d'echinides: Paris, Annales des Sciences Naturelles, Zoologie, series 3, v. 7, p. 129–168; v. 8, p. 5–35, p. 355–380.

Arnold, B. W., and Clark, H. L., 1927, Jamaican fossil echini: Harvard, Museum of Comparative Zoology Memoirs, v. 50, p. 1–75.

Arnold, B. W., and Clark, H. L., 1934, Some additional fossil echini from Jamaica: Harvard, Museum of Comparative Zoology Memoirs, v. 54, p. 139–156.

Bengtson, P., 1988, Open nomenclature: Palaeontology, v. 31, p. 223–227.

Brodermann, J., 1949, Significacion estratigrafica de los equinodermos fosiles de Cuba: Paleontologia Cubana, v. 1, p. 305–330.

Carter, B. D., 1987a, Paleogene echinoid distributions in the Atlantic and Gulf Coastal Plains: Palaios, v. 2, p. 390–404.

Carter, B. D., 1987b, *Brissus bridgeboroensis*, a new spatangoid echinoid from the lower Oligocene of southwestern Georgia: Journal of Paleontology, v. 61, p. 1043–1046.

Carter, B. D., 1989, Echinoid biofacies and lithofacies distributions in the upper Eocene of the Dougherty Plain, southwestern Georgia: Southeastern Geology, v. 30, p. 175–191.

Carter, B. D., and Hammack, R. E., 1989, Stratigraphic distribution of Jacksonian (Priabonian) echinoids in Georgia: Comparison and suggested correlations with Florida and the Carolinas: Palaoios, v. 4, p. 86–91.

Carter, B. D., Beisel, T. H., Branch, W. B., and Mashburn, C. M., 1989, Substrate preferences of late Eocene (Priabonian/Jacksonian) echinoids of the eastern Gulf Coast: Journal of Paleontology, v. 63, p. 495–503.

Chubb, L. J., 1958, Higher Miocene rocks of southeast Jamaica: Geonotes, v. 1, p. 26–31.

Clark, H. L., 1919, The distribution of littoral echinoderms of the West Indies: Carnegie Institution of Washington, Papers from the Department of Marine Biology, v. 13, p. 51–73.

Clark, W. B., and Twitchell, M. W., 1915, The Mesozoic and Cenozoic Echinodermata of the United States: U.S. Geological Survey Monograph, v. 54, 341 p.

Cooke, C. W., 1959, Cenozoic echinoids of eastern United States: U.S. Geological Survey Professional Paper, v. 321, 106 p.

Cooke, C. W., 1961, Cenozoic and Cretaceous echinoids from Trinidad and Venezuela: Smithsonian Miscellaneous Collections, v. 142, no. 4, 35 p.

Cotteau, G. H., 1875, Description des échinides Tertiaires des îsles St. Barthélemy et Anguilla: Kungliga Svenska Vetenskaps-Akademiens Handlingar, v. 13, no. 6, 48 p.

Cotteau, G. H., 1897, Descripción de los equinoides fósiles de la Isla de Cuba por M. G. Cotteau, adicionada por D. Justo Egozcue y Cia: Boletin de la Comisión del Mapa Geológico España, v. 22, 99 p. [Not seen].

Cutress, B. M., 1980, Cretaceous and Tertiary Cidaroida (Echinodermata: Echinoidea) of the Caribbean area: Bulletins of American Paleontology, v. 77, no. 309, 221 p.

Desor, E., 1858, Synopsis des échinides fossiles: Paris, Reinwald, and Weisbaden, Kriedel und Niedner, 490 p.

Donovan, S. K., 1988a, A second B. W. Arnold collection of Jamaican fossil echinoids: Geological Society of Jamaica Journal, v. 24 (for 1987), p. 36–43.

Donovan, S. K., 1988b, A preliminary biostratigraphy of the Jamaican fossil Echinoidea, *in* Burke, R. D., Mladenov, P. V., Lambert, P., and Parsley, R. L., eds., Echinoderm biology: Proceedings, Sixth International Echinoderm Conference, Victoria, British Columbia, August 23–28, 1987: Rotterdam, A.A. Balkema, p. 125–131.

Donovan, S. K., 1990, Jamaican Cretaceous Echinoidea. 2. *Goniopygus supremus* Hawkins, 1924, *Heterosalenia occidentalis* Hawkins, 1923, and a comment on *Trochalosoma chondra* (Arnold and Clark, 1927): Mesozoic Research, v. 2, 205–217.

Donovan, S. K., and Bowen, J. F., 1989, Jamaican Cretaceous Echinoidea. 1. Introduction and reassessment of ?*Pygopistes rudistarum* (Hawkins, 1923) n. comb.: Mesozoic Research, v. 2, p. 57–65.

Donovan, S. K., and Carby, B. E., 1989, Disarticulated echinoid plates from the Paleocene of Jamaica: Geological Society of Jamaica Journal, v. 26, p. 1–4.

Donovan, S. K., and Gordon, C. M., 1989, Report of a field meeting to selected localities in St. Andrew and St. Ann, 25 February 1989: Geological Society of Jamaica Journal, v. 26, p. 51–54.

Donovan, S. K., Gordon, C. M., Schickler, W. F., and Dixon, H. L., 1989a, An Eocene age for an outcrop of the "Montpelier Formation" at Beecher Town, St. Ann, Jamaica, using echinoids for correlation: Geological Society of Jamaica Journal, v. 26, p. 5–9.

Donovan, S. K., Jackson, T. A., and Littlewood, D.T.J., 1989b, Report of a field meeting to the Round Hill region of southern Clarendon, 9 April 1988: Geological Society of Jamaica Journal, v. 25 (for 1988), p. 44–47.

Donovan, S. K., Scott, A. D., and Veltkamp, C. J., 1991, A late middle Eocene echinoid fauna from Portland, northeastern Jamaica: Geological Society of Jamaica Journal, v. 28, p. 1–8.

Durham, J. W., 1950, The 1940 *E. W. Scripps* cruise to the Gulf of California. Part 2. Megascopic paleontology and marine stratigraphy: Geological Society of America Memoir, v. 43, 216 p.

Durham, J. W., 1966, Clypeasteroids, *in* Moore, R. C., ed., Treatise on Invertebrate Paleontology, Part U, Echinodermata 3(2): New York, Geological Society of America, and Lawrence, University of Kansas Press, p. U450–U491.

Fell, H. B., 1954, Tertiary and Recent Echinoidea of New Zealand: Cidaridae: New Zealand Geological Survey, Paleontological Bulletin, v. 23, 62 p.

Fell, H. B., 1966, Diadematacea, *in* Moore, R. C., ed., Treatise on Invertebrate Paleontology, Part U, Echinodermata 3(1): New York, Geological Society of America, and Lawrence, University of Kansas Press, p. U340–U366a.

Fell, H. B., and Pawson, D. L., 1966, Echinacea, *in* Moore, R. C., ed., Treatise on invertebrate paleontology, Part U, Echinodermata 3(2): New York, Geological Society of America, and Lawrence, University of Kansas Press, p. U367–U440.

Fischer, A. G., 1966, Spatangoida, *in* Moore, R. C., ed., Treatise on invertebrate paleontology, Part U, Echinodermata 3(2): New York, Geological Society of America, and Lawrence, University of Kansas Press, p. U543–U628.

Geological Survey Division, 1987, Geological sheet 26, Swift River: Kingston, Government of Jamaica, scale 1:50,000.

Gordon, C. M., 1989, Fossil echinoids of the late Pleistocene Falmouth Formation (reef terrace 1) of Jamaica's north coast: Twelfth Caribbean Geological Conference, St. Croix, August 7th–11th, Abstracts, p. 60.

Gordon, C. M., 1990, Taxonomy and palaeoecology of the echinoids of the late Pleistocene Falmouth Formation of Jamaica [M. Phil. thesis]: Mona, University of the West Indies, 297 p.

Guppy, R.J.L., 1866, On Tertiary echinoderms from the West Indies: Quarterly Journal of the Geological Society, London, v. 22, p. 297–301.

Harland, W. B., Cox, A. V., Llewellyn, P. G., Pickton, C.A.G., Smith, A. G., and Walters, R., 1982, A geologic time scale: Cambridge, Cambridge University Press, 131 p.

Hawkins, H. L., 1924, Notes on a new collection of fossil Echinoidea from Jamaica: Geological Magazine, v. 61, p. 312–324.

Hawkins, H. L., 1927, Descriptions of new species of Cainozoic Echinoidea from Jamaica: Harvard, Museum of Comparative Zoology Memoirs, v. 50, p. 76–84.

Israelsky, M. C., 1924, Notes on some new echinoids from the San Rafael and Tuxpam Beds of the Tampico region, Mexico: Proceedings, California Academy of Sciences, ser. 4, v. 13, p. 137–145.

Jackson, R. T., 1914, Studies on Jamaican echini: Carnegie Institution, Washington, Publication no. 182, p. 141–162.

Jackson, R. T., 1922, Fossil echini of the West Indies: Carnegie Institution, Washington, Publication no. 306, p. 1–103.

Kier, P. M., 1962, Revision of the cassiduloid echinoids: Smithsonian Miscellaneous Collections, v. 144, no. 3, 262 p.

Kier, P. M., 1966, Cassiduloids, in Moore, R. C., ed., Treatise on Invertebrate Paleontology, Part U, Echinodermata 3(2): New York, Geological Society of America, and Lawrence, University of Kansas Press, p. U492–U523.

Kier, P. M., 1967, Revision of the oligopygoid echinoids: Smithsonian Miscellaneous Collections, v. 152, no. 2, 149 p.

Kier, P. M., 1974, Evolutionary trends and their functional significance in the post-Paleozoic echinoids: Paleontological Society Memoir, v. 5, 95 p.

Kier, P. M., 1982, Rapid evolution in echinoids: Palaeontology, v. 25, p. 1–9.

Kier, P. M., 1984, Fossil spatangoid echinoids of Cuba: Smithsonian Contributions to Paleobiology, v. 55, 336 p.

Lamarck, J.B.P.A. de M., de, 1816, Histoire naturelle des animaux sans vertèbres, présentant les caractères, généralaux et particuliers de ces animaux, leur distribution, leurs classes, leurs familles, leurs genres et la citation synonymique des principales espèces qui s'y rapportent, 1st edition, v. 3: Paris, 586 p. [Not seen].

Lambert, J., 1925, Sur la récente découverte d'*Haimea caillaudi* dans l'Eocène supérieur de la Jamaïque: Compte Rendu Sommaire des Séances de la Société Géologique, no. 17 for 1925, p. 232.

Leske, N. G., 1778, Iacobi Theodori Klein Naturalis disposito Echinodermatum, edita et aucta a N. G. Leske: Lipsiae, 278 p. [Not seen].

Lewis, D. N., 1986, Catalogue of the type and figured specimens of fossil Echinoidea in the British Museum (Natural History): London, British Museum (Natural History), 86 p.

Linnaeus, C., 1758, Systema naturae per regna tria naturae, secundum classes, ordines, genera, species, cum characteribus differentis, synonymis, locis: Holmiae, 824 p. [Not seen].

Loriol, P., de, 1887, Notes pour servir à l'étude des échinodermes: Recueil Zoologie Suisse, sér. 7, no. 2, v. 4, p. 365–407. [Not seen].

McFarlane, N., 1974, The geology of the Dry Harbour Mountains, St. Ann, Jamaica [M.Sc. thesis]: University of the West Indies, Mona, 87 p.

McFarlane, N., 1977a, Jamaica geological sheet: Kingston, Ministry of Mining and Natural Resources, 1:250,000.

McFarlane, N., 1977b, Some Eocene and Oligocene faunas from central Jamaica: Memoria Segundo Congreso Latinoamericano de Geologia, Caracas, Venezuela, 11 al 16 de Noviembre de 1973, v. 3, p. 1393–1411.

McKinney, M. L., and Jones, D. S., 1983, Oligopygoid echinoids and the biostratigraphy of the Ocala Limestone of peninsular Florida: Southeastern Geology, v. 24, p. 21–30.

McKinney, M. L., and Zachos, L. G., 1986, Echinoids in biostratigraphy and paleoenvironmental reconstruction: A cluster analysis from the Eocene Gulf Coast (Ocala Limestone): Palaios, v. 1, p. 420–423.

McKinney, M. L., McNamara, K. J., Carter, B. D., and Donovan, S. K., 1992, Evolution of Paleogene echinoids: A global and regional view, in Prothero, D. R., and Berggren, W. A., eds., Eocene-Oligocene climatic and biotic evolution: Princeton, New Jersey, Princeton University Press, p. 349–367.

Melville, R. V., and Durham, J. W., 1966, Skeletal morphology, in Moore, R. C., ed., Treatise on Invertebrate Paleontology, Part U, Echinodermata 3(1): New York, Geological Society of America, and Lawrence, University of Kansas Press, p. U220–U252.

Michelin, H., 1856, [No title]: Bulletin de la Société Géologique de France, sér. 2, v. 13, p. 222.

Mooi, R., 1989, Living and fossil genera of the Clypeasteroida (Echinoidea: Echinodermata): An illustrated key and annotated checklist: Smithsonian Contributions to Zoology, v. 488, 51 p.

Mooi, R., 1990a, Living cassiduloids (Echinodermata: Echinoidea): a key and annotated list: Proceedings, Biological Society of Washington, v. 103, p. 63–85.

Mooi, R., 1990b, A new "living fossil" echinoid (Echinodermata) and the ecology and paleobiology of Caribbean cassiduloids: Bulletin of Marine Science, v. 46, p. 688–700.

Moore, R. C., ed., 1966, Treatise on Invertebrate Paleontology, Part U, Echinodermata 3 (in 2 volumes): New York, Geological Society of America, and Lawrence, University of Kansas Press, 695 p.

Mortensen, T., 1935, A monograph of the Echinoidea. II. Bothriocidaroida, Melonechinoida, Lepidocenteroida and Stirodonta: Copenhagen, Reitzel, 647 p.

Mortensen, T., 1943, A monograph of the Echinoidea. III (3). Camarodonta II: Copenhagen, Reitzel, 446 p.

Mortensen, T., 1948, A monograph of the Echinoidea. IV(1). Holectypoida, Cassiduloida: Copenhagen, Reitzel, 371 p.

Paul, C.R.C., 1985, The adequacy of the fossil record reconsidered: Special Papers in Palaeontology, v. 33, p. 7–15.

Phelan, T., 1970, A field guide to the cidaroid echinoids of the northwestern Atlantic Ocean, Gulf of Mexico, and the Caribbean Sea: Smithsonian Contributions to Zoology, v. 40, 67 p.

Phelan, T., 1972, Comments on the echinoid genus *Encope,* and a new subgenus: Proceedings, Biological Society of Washington, v. 85, p. 109–129.

Pickerill, R. K., and Donovan, S. K., 1991, Observations of the ichnology of the Richmond Formation of eastern Jamaica: Geological Society of Jamaica Journal, v. 28, p. 19–35.

Poddubiuk, R. H., 1985, Evolution and adaptation in some Caribbean Oligo-Miocene *Clypeasters, in* Keegan, B. F., and O'Connor, B.D.S., eds., Echinodermata: Proceedings, Fifth International Echinoderm Conference, Galway, 24–29 September, 1984: Rotterdam, A.A. Balkema, p. 75–80.

Poddubiuk, R. H., and Rose, E.P.F., 1984, Relationships between mid-Tertiary echinoid faunas from the central Mediterranean and eastern Caribbean and their palaeobiogeographic significance: Annales Géologiques des Pays Hélleniques, v. 32, p. 115–127.

Prothero, D. R., 1989, Stepwise extinctions and climatic decline during the later Eocene and Oligocene, in Donovan, S. K., ed., Mass extinctions: Processes and evidence: London, Belhaven Press, p. 217–234.

Ravenel, E., 1848, Echinidae, Recent and fossil, of South Carolina: Charleston, South Carolina, Burgeo and James, 4 p.

Robinson, E., 1965, Tertiary rocks of the Yallahs area, Jamaica: Geological Society of Jamaica Journal, v. 7, p. 18–27.

Robinson, E., 1967a, Biostratigraphic position of late Cainozoic rocks in Jamaica: Geological Society of Jamaica Journal, v. 9, p. 32–41.

Robinson, E., 1967b, Submarine slides in White Limestone Group, Jamaica: American Association of Petroleum Geologists Bulletin, v. 51, p. 569–578.

Robinson, E., 1969, Geological field guide to Neogene sections in Jamaica West Indies: Geological Society of Jamaica Journal, v. 10, p. 1–24.

Robinson, E., 1988, Late Cretaceous and early Tertiary sedimentary rocks of the Central Inlier, Jamaica: Geological Society of Jamaica Journal, v. 24 (for 1987), p. 49–67.

Rose, E.P.F., and Poddubiuk, R. H., 1987, Morphological variation in the Cenozoic echinoid *Clypeaster* and its ecological and stratigraphical significance: Hungarian Institute of Geology Annals, v. 70, p. 463–469.

Sánchez Roig, M., 1926, Contribucion a la paleontologia Cubana: Los equinodermos fósiles de Cuba: Boletin de Minas, v. 10, 179 p. [Not seen].

Sánchez Roig, M., 1949, Los equinodermos fosiles de Cuba: Paleontologia Cubana, v. 1, p. 1–302.

Smith, A. B., 1980, The structure, function and evolution of tube feet and ambulacral pores in irregular echinoids: Palaeontology, v. 23, p. 39–83.

Smith, A. B., 1984, Echinoid palaeobiology: London, George Allen and Unwin, 191 p.

Smith, A.B., and Crimes, T. P., 1983, Trace fossils by heart-urchins (Echinoidea): A study of *Scolicia* and related traces: Lethaia, v. 16, p. 79–92.

Smith, A. B., and Wright, C. W., 1989, British Cretaceous echinoids. Part 1, general introduction and Cidaroida: London, Palaeontographical Society Monograph, v. 141 (for 1987), no. 578, p. 1–101.

Stanley, S. M., and Campbell, L. D., 1981, Neogene mass extinction of western Atlantic molluscs: Nature, v. 293, p. 457–459.

Trechmann, C. T., 1930, The Manchioneal Beds of Jamaica: Geological Magazine, v. 67, p. 199–218.

Vaughan, T. W., 1922, Stratigraphic significance of the species of West Indian fossil echini: Carnegie Institution, Washington, Publication no. 306, p. 105–122.

Wagner, C. D., and Durham, J. W., 1966, Holectypoids, *in* Moore, R. C., ed., Treatise on Invertebrate Paleontology, Part U, Echinodermata 3(2): New York, Geological Society of America, and Lawrence, University of Kansas Press, p. U440–U450.

Weisbord, N., 1934, Some Cretaceous and Tertiary echinoids from Cuba: Bulletins of American Paleontology, v. 20, no. 70C, p. 165–270.

Weisbord, N., 1969, Some late Cenozoic Echinoidea from Cabo Blanco, Venezuela: Bulletins of American Paleontology, v. 56, no. 252, p. 275–371.

Zullo, V. A., and Harris, W. B., 1986, Introduction: Sequence stratigraphy, lithostratigraphy, and biostratigraphy of the North Carolina Eocene carbonates, *in* Textoris, D. A., ed., Field Guidebook, Third Annual Midyear Meeting: Society of Economic Paleontologists and Mineralogists, p. 257–263.

MANUSCRIPT ACCEPTED BY THE SOCIETY NOVEMBER 11, 1992

Jamaican Tertiary marine Vertebrata

Daryl P. Domning
Laboratory of Paleobiology, Department of Anatomy, Howard University, Washington, D.C. 20059
James M. Clark
Department of Vertebrate Paleontology, American Museum of Natural History, New York, New York 10024

ABSTRACT

The record of Tertiary marine vertebrates from Jamaica is fragmentary in the extreme, but nonetheless includes some significant specimens. The Eocene section has produced the type specimens of a crocodile and of the most primitive known sea cow, together with unidentifiable fragments of a turtle, whereas the Pliocene has yielded a needlefish. It is likely that a much richer fauna remains to be discovered, once exposures of the appropriate facies are located and collected.

SYSTEMATIC PALEONTOLOGY

CLASS OSTEICHTHYES
SUBCLASS ACTINOPTERYGII
DIVISION TELEOSTEI
ORDER ATHERINIFORMES
Family Belonidae

cf. *Platybelone argalus* (LeSueur, 1821)

The only fossil fish specimen reported from the Tertiary of Jamaica comprises two associated fragments of a mandible of a needlefish (Family Belonidae). The specimen was collected from the Bowden Formation near Bowden in St. Thomas Parish (Caldwell, 1965), and was deposited in the Bowden collection at the Institute of Jamaica in Kingston. The Bowden Formation has a rich fauna of marine invertebrates, but this specimen is the only fossil vertebrate known from it. The age of the Bowden Formation was said by Caldwell to be Miocene, but it is now considered Pliocene (see Berggren, this volume; Aubry, this volume).

Caldwell noted that the dentition of the specimen is very similar to that of *Platybelone argalus* (the keeltail needlefish) and differs from that of the other five extant belonid species known to occur near Jamaica. He therefore identified the specimen as "*Platybelone* cf. *argalus*." As he noted, however, this genus is monotypic (Collette and Berry, 1965), and this identification is more correctly expressed as cf. *Platybelone argalus*. This species is commonly found today in nearshore environments of Jamaica.

CLASS REPTILIA
ORDER CROCODYLIA
SUBORDER EUSUCHIA
Family Thoracosauridae

?*Characlosuchus kugleri* Berg, 1969

The Tertiary Crocodylia of Jamaica are known from a mandibular symphysis and isolated bones and teeth found near the village of Dump, near Christiana in northern Manchester Parish. The beds exposed at this locality form the type section of the Dump Limestone Lenticle, Guys Hill Sandstone Member, Chapelton Formation, and are of basal middle Eocene age (Robinson, 1969). The crocodylian fossils are from a poorly cemented, muddy limestone (bed 3 of Donovan et al., 1990, Fig. 2).

The mandibular symphysis (now in the Naturhistorisches Museum, Basel) was designated the holotype of *Characlosuchus kugleri* Berg, 1969; the genus is otherwise known only from the Miocene of Colombia (Langston, 1965) and, possibly, the latest Miocene of coastal South Carolina and Florida (Webb and Tessman, 1968). The mandible of *Characlosuchus* shows that the skull was very long and narrow, more so than in any alligator or true crocodile (*Crocodylus*) but similar to some very long-snouted crocodylians (*Gavialis gangeticus* and *Tomistoma schlegelii*) that today live only in Asia. *Characlosuchus* is distinguished by the raised or everted margins of its tooth sockets and by its moderately long symphysis (including about ten teeth) in which the splenial bone forms only a short part.

As noted by Langston (1965), *Characlosuchus* is most similar to the thoracosaurs, an extinct group of long-snouted crocodylians that may be related to the extant *Tomistoma schlegelii*. Among thoracosaurs, *C. kugleri* is especially similar to *Dollosuchus dixoni* from the middle Eocene of England and Belgium (Swinton, 1937). Although we have not examined any of the relevant specimens, the figures in the literature indicate that the mandibular symphyses of the two species are extremely similar. In his paper describing *C. kugleri*, Berg (1969) did not compare it with *D. dixoni*, and the diagnosis of *C. kugleri* he gave (p. 733) differentiates it from the Colombian species but does not differentiate it from the nearly contemporaneous *D. dixoni*. It is therefore possible that *C. kugleri* is synonymous with *D. dixoni*, although direct comparisons are needed to determine this.

Prevailing opinion has been that thoracosaurs are related to the living false gharial of Southeast Asia, *Tomistoma schlegelii*. However, most points of resemblance between the false gharial and thoracosaurs (Troedsson, 1924) are associated with the great length of the rostrum, a feature that has evolved repeatedly in unrelated groups of crocodylians (Benton and Clark, 1988), and the relationships of thoracosaurs have not been addressed using modern methods of phylogenetic analysis. The relationships of *T. schlegelii* are themselves currently the focus of a debate, with molecular evidence indicating

---

Abbreviations: BMNH, British Museum (Natural History), London; USNM, U.S. National Museum of Natural History, Washington, D.C.

Domning, D. P., and Clark, J. M., 1993, Jamaican Tertiary marine Vertebrata, *in* Wright, R. M., and Robinson, E., eds., Biostratigraphy of Jamaica: Boulder, Colorado, Geological Society of America Memoir 182.

a close relationship to true gharials (Densmore and Owen, 1989) and morphological evidence suggesting that it is related to true crocodiles (Frey et al., 1989; Norell, 1990). Thoracosaurs are clearly members of the group of advanced crocodylians to which all living forms belong (the Eusuchia), and they are quite possibly related to *T. schlegelii,* but that is all that can be said at present.

Recent collecting efforts at the *Characto suchus kugleri* type locality have located several isolated crocodylian bones (Donovan et al., 1990). These include a tooth, an isolated osteoderm (bony scute), and a phalanx, all probably representing one individual (USNM 437770). The proximal end of a right femur (USNM 456635) has also been collected from a site at Coleyville, about a kilometer northwest of Dump. The tooth differs somewhat from typical thoracosaur teeth; none of the other specimens is diagnostic at the generic or specific level. All could belong to *C. kugleri.*

The tooth is slightly flattened and has anterior and posterior keels, a slight lingual curvature, and a short crown coming to a blunt tip. The enamel is finely crenulated on both sides, more so lingually than labially, but lacks the longitudinal grooves and ridges present on many thoracosaur teeth. It is difficult to determine from the published descriptions, but this tooth may be similar to those of *Characto suchus fieldsi* from South America and *Dollosuchus dixoni:* the *C. fieldsi* holotype is described as having a tooth with "delicate vertical ridges" (Langston, 1965, p. 45) that are more pronounced lingually than labially, and the lower teeth of *D. dixoni* are described as having "longitudinal lines" (Swinton, 1937, p. 21). In general, isolated crocodylian teeth cannot be identified with certainty because there is considerable variation in tooth shape within a tooth row, and similar tooth shapes have evolved repeatedly in unrelated groups of crocodylians.

The square osteoderm is from the dorsal side of the animal, as indicated by the presence of a low median keel. Its lateral edges demonstrate that it was sutured on both sides to neighboring osteoderms. The indication that there were more than two osteoderms in a transverse row and the square shape together demonstrate that this is from a eusuchian (Norell and Clark, 1990). The well-developed smooth anterior part of the dorsal surface, where the adjacent anterior osteoderm overlay it, is a primitive feature of eusuchians; in extant crocodylians this area is not so distinctly set off from the rest of the dorsal surface.

The phalanx is a proximal element from the second or third toe. It is probably from the hindfoot of a small animal (less than 2 m in total length), but it may be from the forefoot of a very large animal. Femora are very similar among crocodylians, so the femoral fragment cannot be identified beyond Crocodylia.

CLASS MAMMALIA
ORDER SIRENIA
Family Prorastomidae

Prorastomus sirenoides Owen, 1855

The Sirenia (manatees and dugongs) are a cosmopolitan order of herbivorous marine mammals, represented in Jamaica today by the Antillean manatee (*Trichechus manatus manatus*). The extinct species *Prorastomus sirenoides* is assigned to a monotypic genus and family, and is known only from Jamaica. It was described by Owen (1855) on the basis of a single skull, now in the British Museum (BMNH 44897), which was found in a loose boulder of Yellow Limestone in the bed of the Quashies River near Freemans Hall in southern Trelawny Parish. The source unit was evidently the Stettin Member of the Chapelton Formation, Yellow Limestone Group, and is of probable late early Eocene age (Robinson, 1988) rather than middle Eocene as stated in most of the literature. If this early Eocene age estimate is correct, *Prorastomus* is the only named sirenian taxon known to have existed prior to the middle Eocene. As it is also the most primitive known member of the order, it is in most respects a plausible structural ancestor for all other sirenians.

No further specimens of *Prorastomus* were discovered until 1989, when fragments of a skeleton (USNM 437769) were collected near the village of Dump in northern Manchester Parish. The locality is the same as the type locality of the crocodylian *Characto suchus kugleri* (see above), but the sirenian came from a slightly higher bed than the latter (Donovan et al., 1990). A detailed description of the new *Prorastomus* material, together with a redescription of the holotype, is being prepared by R.J.G. Savage, D. P. Domning, and J.G.M. Thewissen.

Prorastomus, like other sirenians, is characterized by extremely dense (osteosclerotic) bone in many parts of the skeleton, especially the ribs, which also have a swollen (pachyostotic) appearance. *Prorastomus,* however, was very small compared to most sirenians; its ribs are only about 1.0 to 1.5 cm in diameter. It resembles other Eocene sirenians in having an extremely primitive dental formula (3.1.5.3) including five premolars, but differs from them in having a less downturned rostrum and a more primitive ear region. Its appendicular skeleton is unknown, but it is probable that it still retained functional hind limbs and an ability to walk on land. Like modern manatees, it probably fed on freshwater aquatic plants in rivers and on seagrasses in nearshore marine and estuarine waters. The depositional environment of the Dump Limestone Lenticle was possibly a brackish lagoon (Donovan et al., 1990).

In the middle and late Eocene, sirenians were distributed throughout the Tethyan Realm, including the southeastern United States (Domning et al., 1982). They are believed to have originated in the Old World, together with their nearest living relatives, the Proboscidea. Therefore, the unique occurrence of their most primitive member in Jamaica is and has always been a fascinating anomaly. It must be taken as evidence that they were able to disperse widely in, or along the borders of, the tropical marine waters of the world. This is interesting in the light of debates over the paleogeography of the Tethys Sea in the early Eocene, and the isolation of Jamaica from other Eocene landmasses.

ACKNOWLEDGMENTS

We thank Hal Dixon, Steve Donovan, and Frank Garcia for assistance in the field, and Bruce Collette for discussion of the needlefish. Domning's work in Jamaica was supported by National Science Foundation grant BSR 84-16540.

REFERENCES CITED

Benton, M. J., and Clark, J. M., 1988, Archosaur phylogeny and the relationships of the Crocodylia, in Benton, M. J., ed., The phylogeny and classification of the tetrapods: Oxford, Clarendon Press, p. 295–338.

Berg, D. E., 1969, *Characto suchus kugleri,* eine neue Krokodilart aus dem Eozän von Jamaica: Eclogae Geologicae Helvetiae, v. 62, p. 731–735.

Caldwell, D. K., 1965, A Miocene needlefish from Bowden, Jamaica: Florida Academy of Sciences Quarterly Journal, v. 28, no. 4, (1966), p. 339–344.

Collette, B. B., and Berry, F. H., 1965, Recent studies on the needlefishes (Belonidae): An evaluation: Copeia, p. 386–392.

Densmore, L. D., III, and Owen, R. D., 1989, Molecular systematics of the order Crocodilia: American Zoologist, v. 29, p. 831–841.

Domning, D. P., Morgan, G. S., and Ray, C. E., 1982, North American Eocene sea cows (Mammalia: Sirenia): Smithsonian Contributions to Paleobiology, v. 52, 69 p.

Donovan, S. K., Domning, D. P., Garcia, F. A., and Dixon, H. L., 1990, A bone bed in the Eocene of Jamaica: Journal of Paleontology, v. 64, p. 660–662.

Frey, E., Riess, J., and Tarsitano, S. F., 1989, The axial tail musculature of Recent crocodiles and its phyletic implications: American Zoologist, v. 29, p. 857–862.

Langston, W., Jr., 1965, Fossil crocodilians from Colombia and the Cenozoic history of the Crocodilia in South America: University of California Publications in Geological Science, v. 52, 169 p.

LeSueur, C. A., 1821, Observations on several genera and species of fish, belonging to the natural family of the Esoces: Journal of the Academy of Natural Sciences of Philadelphia, v. 2, no. 1, p. 124–138.

Norell, M. A., 1990, The higher relationships of the living Crocodylia: Journal of Herpetology, v. 23, no. 4, p. 325–334.

Norell, M. A., and Clark, J. M., 1990, A reanalysis of *Bernissartia fagesii,* with comments on its phylogenetic position and its bearing on the origin and diagnosis of the Eusuchia: Bulletin de l'Institut Royal des Sciences Naturelles de Belgique, v. 60, p. 115–128.

Owen, R., 1855, On the fossil skull of a mammal (*Prorastomus sirenoides,* Owen), from the island of Jamaica: Geological Society of London Quarterly Journal, v. 11, p. 541–543.

Robinson, E., 1969, Stratigraphy and age of the Dump Limestone Lenticle, central Jamaica: Eclogae Geologicae Helvetiae, v. 62, p. 737–744.

Robinson, E., 1988, Late Cretaceous and early Tertiary sedimentary rocks of the Central Inlier, Jamaica: Geological Society of Jamaica Journal, v. 24, p. 49–67.

Swinton, W. E., 1937, The crocodile of Maransart (*Dollosuchus dixoni* [Owen]): Mémoires du Musée Royal d'Histoire Naturelle de Belgique, v. 80, 46 p.

Troeddson, G. T., 1924, On crocodilian remains from the Danian of Sweden: Acta Universitatis Lundensis, n.s., v. 20, no. 2, p. 1–76.

Webb, S. D., and Tessman, N., 1968, A Pliocene vertebrate fauna from low elevation in Manatee County, Florida: American Journal of Science, v. 266, p. 777–811.

MANUSCRIPT ACCEPTED BY THE SOCIETY NOVEMBER 11, 1992

Quaternary land vertebrates of Jamaica

Gary S. Morgan
Florida Museum of Natural History, University of Florida, Gainesville, Florida 32611

ABSTRACT

Quaternary land vertebrates have been reported from 21 sites in Jamaica, almost all of which are located in caves. These cave fossil deposits are widely distributed throughout the island in regions of limestone karst. Each cave deposit consists of autochthonous sediments that cannot be biostratigraphically correlated with other sites containing sediments of similar origin. The age of these deposits has been determined primarily through absolute dating and faunal comparisons. An informal temporal sequence previously proposed for Jamaican vertebrate-bearing strata was based on relative stratigraphic position and faunal content and included four layers (from youngest to oldest): (1) *Rattus* layer, (2) *Oryzomys* layer, (3) lizard/bat layers, (4) hard breccias. These strata range in age from post-Columbian (<500 yr B.P.) for the *Rattus* layer to middle Pleistocene for the hard breccias. Absolute ages have been obtained for six Jamaican cave deposits. The oldest dated Quaternary terrestrial vertebrate faunas are preserved in indurated bone breccias and conglomerates from Wallingford Roadside Cave in St. Elizabeth Parish that range in age from 100 to 250 ka. Undated breccias from Sheep Pen Cave in Trelawny Parish, Molton Fissure in Manchester Parish, and Lluidas Vale Cave in St. Catherine Parish, are probably similar in age. Most of the remaining Jamaican cave faunas are derived from unconsolidated sediments of late Pleistocene and Holocene age.

The published Quaternary vertebrate fauna of Jamaica consists of 49 species, including 3 amphibians, 18 reptiles, 9 birds, and 19 mammals. The amphibians include three extant frogs, while the reptile fauna is composed of a turtle, a crocodile, 12 species of lizards, and 4 snakes. Three of the lizards, the large gecko *Aristelliger titan,* the giant anguid *Celestus* cf. *C. occiduus,* and the iguanid *Leiocephalus jamaicensis,* are extinct. Lizard bones are so abundant in Dairy Cave in St. Ann Parish and several other Jamaican caves that the strata from which they were derived have been called the "lizard layers." Three species of birds from Quaternary sites are now extinct in Jamaica, although the burrowing owl *Athene cunicularia* still survives elsewhere in the West Indies and the endemic nightjar *Siphonorhis americana* disappeared very recently. The extinct flightless ibis *Xenicibis xympithecus* is an endemic genus and species known only from Jamaica. The Quaternary mammalian fauna of Jamaica includes 12 species of bats, 3 primates, and 4 rodents. None of the bats are extinct, but three of the species have disappeared from Jamaica. *Brachyphylla nana* still survives on Cuba and Hispaniola, whereas *Tonatia bidens* and *Mormoops megalophylla* are now restricted to the continental Neotropics. Jamaica has three extinct species of monkeys, which constitutes the largest primate fauna of any Antillean island. *Xenothrix mcgregori,* an endemic Jamaican genus and species, is so unlike other genera of Neotropical ceboid monkeys that it has been placed in a separate family, the Xenotrichidae. The two other taxa of Jamaican fossil primates are undescribed. Both are represented only by isolated postcranial elements that suggest generic-level distinction from one another and from *Xenothrix.* The Quaternary rodent fauna of Jamaica consists of three extinct and one living species. The

Morgan, G. S., 1993, Quaternary land vertebrates of Jamaica, *in* Wright, R. M., and Robinson, E., eds., Biostratigraphy of Jamaica: Boulder, Colorado, Geological Society of America Memoir 182.

two described species of the large, endemic Jamaican heptaxodontid rodent genus *Clidomys* apparently went extinct prior to the late Pleistocene. The rice rat *Oryzomys antillarum* is the only sigmodontine rodent known from the Greater Antilles. *Oryzomys* is common in late Pleistocene and Holocene cave deposits throughout Jamaica, but has gone extinct within the last century. The capromyid rodent *Geocapromys brownii* is the sole surviving rodent in Jamaica. *G. brownii* is the most abundant and widespread species in Jamaican Quaternary deposits.

Jamaica was almost completely submerged in the Oligocene, indicating that most species of land vertebrates in the modern and late Quaternary faunas colonized Jamaica from the Miocene onward. Endemic Jamaican genera, including the primate *Xenothrix*, the rodent *Clidomys*, the bat *Ariteus*, and the ibis *Xenicibis*, among others, probably have inhabited Jamaica since the Miocene or Pliocene. The large number of Jamaican vertebrates that are either conspecific with or closely related to mainland species suggests periodic overwater dispersal from Middle or South America, probably during periods of low sea level (e.g., late Miocene, ca. 10 Ma; latest Miocene, ca. 6 Ma; latest Pleistocene, 10 to 20 ka) when the Nicaraguan Rise was partially emergent. Fifteen of the 49 species (31%) of fossil vertebrates recorded from Jamaican Quaternary deposits are now extinct on the island. Of the 15 species no longer found in Jamaica, 11 are totally extinct, and 4 are locally extinct in Jamaica but still survive elsewhere in the West Indies or on the continent. Most of the extinct and extirpated species of Jamaican vertebrates survived into the late Quaternary. The localized extinction in Jamaica of the burrowing owl and several species of bats probably was related to environmental changes during the late Pleistocene and early Holocene. Many of the vertebrate extinctions in Jamaica and elsewhere in the West Indies were human-caused, resulting from predation, habitat destruction, and the introduction of exotic species such as rats, dogs, cats, and mongoose.

INTRODUCTION

Large regions of Jamaica are underlain by Cenozoic limestones (Zans et al., 1963) that have been eroded to form caves, sinks, fissures, and other karst geomorphic features. More than 950 caves and sinkholes have been recorded from Jamaica (Fincham, 1977). The extensive tropical karst terrain in Jamaica has had a significant effect on the distribution of Quaternary terrestrial vertebrate sites since the vast majority of these fossil deposits occur in caves. Pleistocene and Holocene vertebrate sites have been reported from nine of Jamaica's thirteen parishes. The published fossil sites are concentrated in three areas: the Cockpit Country in Trelawny Parish; the vicinity of Worthy Park and Lluidas Vale in northwestern St. Catherine Parish; and the Portland Ridge in southernmost Clarendon Parish. These apparent concentrations of cave fossil deposits probably are at least partially a result of collecting bias, and other regions of the island may eventually prove to have as many or more vertebrate fossil sites.

H. E. Anthony was the first paleontologist to collect Quaternary vertebrate fossils in Jamaica. In 1919 and 1920 he explored more than 70 caves throughout Jamaica and found significant concentrations of terrestrial vertebrate fossils in 10 of those caves (Anthony, 1920a, b; Koopman and Williams, 1951). Anthony (1920b) published only one major paper on his paleontological discoveries, the description of several new genera and species of large heptaxodontid rodents from Wallingford Roadside Cave near Balaclava in St. Elizabeth Parish. However, a number of subsequent papers were based primarily on his collections. Koopman and Williams (1951) reported on the fossil bats collected by Anthony from Wallingford Roadside Cave and Dairy Cave near Discovery Bay in St. Ann Parish and Hecht (1951) described the extinct gecko, *Aristelliger titan*, and other lizards from these same two localities. Williams and Koopman (1952) described a new genus and species of ceboid monkey, *Xenothrix mcgregori*, and Olson and Steadman (1977) described an extinct genus of flightless ibis, *Xenicibis xympithecus*, both from Anthony's excavations in Long Mile Cave near Windsor, Trelawny Parish. MacPhee (1984) and MacPhee et al. (1989) reported on specimens of heptaxodontid rodents collected by Anthony from Wallingford Roadside Cave and Sheep Pen Cave near Windsor, Trelawny Parish. MacPhee and Fleagle (1991) discussed postcranial remains of *Xenothrix mcgregori* and several other species of mammals in Anthony's collection from Long Mile Cave.

The second expedition to collect vertebrate fossils in Jamaica was during the summer of 1950, and consisted of B. M. Hecht, M. K. Hecht, K. F. Koopman, and E. E. Williams. The members of the 1950 expedition excavated two caves originally discovered by Anthony, Wallingford Roadside Cave and Dairy Cave, and one new site, Portland Cave on the Portland Ridge in Clarendon Parish. Hecht (1951) discussed the fossil lizards obtained by the 1950 expedition, Williams (1952) reported on the fossil bats, and Etheridge (1966) described specimens of the extinct lizard *Leiocephalus jamaicensis*. W. Auffenberg of the Florida State Museum (now the Florida Museum of Natural History) visited Jamaica in 1957 and 1958 and carried out additional excavations in Dairy Cave and Portland Cave. He also discovered several new localities, including Lluidas Vale Cave in St. Catherine Parish and Montego Bay Airport Cave in St. James Parish. Etheridge (1966) reported specimens of *L. jamaicensis* from Montego Bay Airport Cave and MacPhee (1984) recorded teeth of the heptaxodontid rodent *Clidomys* collected by Auffenberg in Lluidas Vale Cave.

The next major vertebrate fossil collections from Jamaica were obtained by T. H. Patton and Florida State Museum field crews between 1966 and 1970. Patton conducted new excavations in Wallingford Roadside Cave, the Portland Cave system, Sheep Pen Cave, and Lluidas Vale Cave. He also discovered many new fossiliferous cave deposits, primarily in the vicinity of Worthy Park in St. Catherine Parish including Braham Cave, Coco Ree Cave, Mt. Diablo Cave, Swansea Cave, and Tydixon Cave. Several papers have been published on Patton's extensive collections, but much of this material remains unstudied. Olson and Steadman (1979) described a complete humerus of *Xenicibis xympithecus* from Swansea Cave, MacPhee (1984) figured an articulated hind foot of *Clidomys osborni* from Wallingford Roadside Cave, Ford and Morgan (1986) recorded a ceboid monkey femur from Coco Ree Cave, and Humphrey et al. (1993) reported fossils of the extinct rice rat *Oryzomys antillarum* from all five of Patton's caves in the Worthy Park area.

In 1981, R.D.E. MacPhee revisited and made additional excavations in three of Anthony's most important localities: Long Mile Cave, Sheep Pen Cave, and Wallingford Roadside Cave (MacPhee, 1984). G. K. Pregill, D. W. Steadman, and a field crew from the Smithsonian Institution carried out excavations in Marta Tick Cave in the Cockpit Country in Trelawny Parish in 1983 (Pregill et al., 1991). F. V. Grady excavated Bonafide Cave in the Cockpit Country in Trelawny Parish in 1985 and 1986. Humphrey et al. (1993) recorded *Oryzomys antillarum* from Bonafide Cave based on material collected by Grady. Savage (1990) reported on a small sample of vertebrate fossils, mostly consisting of the hutia *Geocapromys brownii*, collected by S. K. Donovan in 1988 from Red Hills Road Cave in St. Andrew Parish.

Despite the efforts of the many paleontologists mentioned above, the Quaternary vertebrate fauna of Jamaica is incompletely known compared to some other Antillean islands. Amphibians are represented by fossils of three species of frogs (Pregill et al., 1991): *Eleutherodactylus* sp., *Hyla* sp., and *Osteopilus brunneus*. Two extinct lizards have been described from fossil deposits on the island: the large gecko *Aristelliger titan* (Hecht, 1951), and the curly-tailed lizard *Leiocephalus jamaicensis* (Etheridge, 1966). The recently extinct giant anguid *Celestus* cf. *C. occiduus* occurs in many fossil cave deposits in Jamaica (Hecht, 1951; Williams, 1952; Pregill et al., 1991). Several other genera of lizards and snakes also have been recorded from Jamaica (Hecht, 1951; Etheridge, 1966; MacPhee, 1984), including the lizards *Ameiva*, *Anolis*, *Cyclura*, and *Sphaerodactylus*; and the snakes *Alsophis*, *Arrhyton*, *Tropidophis*, and *Typhlops*. Other reptiles reported from fossil sites in Jamaica include the freshwater turtle, *Trachemys terrapen*, and a crocodile, *Crocodylus* sp. (Anthony, 1920a; Williams, 1950; Koopman and Williams, 1951; MacPhee, 1984). Only one extinct species of fossil bird has been described from Jamaica, the flightless ibis, *Xenicibis xympithecus* (Olson and Steadman, 1977, 1979). Other fossil birds reported from Jamaica include the recently extinct Jamaican pauraque, *Siphonorhis americana*, and the burrowing owl, *Athene cunicularia*, which is locally extinct in Jamaica but survives on several other West Indian islands and in North and South America.

As with the other Greater Antillean islands, the best-known fossil vertebrates from Jamaica are mammals. The extant terrestrial or nonvolant mammalian fauna of Jamaica consists of a single species, the large capromyid rodent, *Geocapromys brownii*. A second species of rodent, the Jamaican rice rat, *Oryzomys antillarum*, has gone extinct within the past 100 years. These two rodents are also known from Quaternary deposits in Jamaica. Extinct mammals recorded from the island consist of two large heptaxodontid rodents, *Clidomys osborni* and *C. parvus*, and at least three primates, *Xenothrix mcgregori*, and two undescribed species represented only by fragmentary postcranial remains (Ford and Morgan, 1986, 1988; Ford, 1990a, b; MacPhee and Fleagle, 1991). The entire Quaternary terrestrial mammalian fauna of Jamaica is composed of only seven species: three primates and four rodents. The Quaternary land mammals of Puerto Rico consist of six species: one species of insectivore, one edentate, and four rodents. Cuba and Hispaniola have much richer nonvolant mammal faunas, composed of approximately 25 species each.

The chiropteran fauna of Jamaica is rather diverse, with 21 living species and 3 locally extinct species that still survive either in Middle America (*Mormoops megalophylla* and *Tonatia bidens*) or elsewhere in the Greater Antilles (*Brachyphylla nana*). Cuba, with 26 living and 7 extinct species, is the only West Indian island that has a richer chiropteran fauna than Jamaica. Hispaniola has a Recent bat fauna of 18 species as well as 1 locally extinct species from Quaternary deposits. Puerto Rico supports only 13 living species of bats along with 3 additional species known from paleontological sites.

In addition to bats, only one other species of mammal survives in Jamaica at the present time, the rodent *Geocapromys brownii*. Six other species of terrestrial mammals went extinct during the late Pleistocene and Holocene. This reflects a trend observed throughout the West Indies in which almost 90% of the land mammals recorded from Quaternary fossil deposits are now extinct (Morgan and Woods, 1986). These widespread extinctions have been attributed to one of two general causes: (1) climatic change since the end of the Pleistocene; (2) the arrival of humans after 4.5 ka. Human-caused extinctions are well documented in the West Indies and resulted from predation, habitat destruction, and introduction of exotic mammals. The timing of these extinctions, whether human-caused or a result of natural factors such as climatic change, provides the best prospect for constructing a biochronology of Jamaican land vertebrates.

BIOCHRONOLOGY

The great majority of Quaternary vertebrate fossils presently known from Jamaica have been recovered from sediments deposited in caves or fissures. Each cave deposit or fissure filling represents an autochthonous sedimentary system that is very local in origin. Consequently, the sediments in the various fossiliferous caves found throughout the island have no obvious stratigraphic relationships with one another. Because these cave and fissure deposits are so localized they do not meet the criteria for formal lithostratigraphic or biostratigraphic units as proposed in the North American Stratigraphic Code (1983). The lack of stratigraphic continuity between cave fossil sites in Jamaica, and elsewhere in the West Indies, makes it very difficult to conduct biostratigraphic analyses using Quaternary land vertebrates. Furthermore, the terrestrial vertebrate faunas of the individual islands in the Greater Antilles are so highly endemic at the specific and even generic levels that comparisons between islands are often of limited utility.

Many previous attempts to determine the age of Jamaican fossil vertebrates were subjective and primarily based on faunal associations. Other factors that have been used in determining the relative age of fossil deposits in Jamaica are the stratigraphic position within a sequence of cave sediments, degree of fossilization of the bones, and color of the sediments. In proposing a tentative stratigraphy for the vertebrate-bearing cave deposits of Jamaica, Koopman and Williams (1951, p. 2) noted, "Within caves . . . a number of different layers can be distinguished, the correlation of which in different caves and their real temporal sequence are in considerable part conjectural." The temporal sequence of Jamaican vertebrate-bearing strata proposed by Koopman and Williams (1951) and Williams (1952) is as follows (from youngest to oldest): (1) *Rattus* layers, (2) *Oryzomys* strata, (3) lizard/bat layers, and (4) hard breccias.

The *Rattus* layers are characterized by their surficial position, gray or black color, totally unfossilized bones, and presence of post-Columbian (<500 yr B.P.) introduced mammals such as *Rattus, Mus,* etc. The next oldest or subsurface layers, the *Oryzomys* strata, grade into, and are difficult to separate from, the *Rattus* layers. A more brownish color of the sediments and a less fresh appearance of the bones characterize the *Oryzomys* strata (Koopman and Williams, 1951).

Below the surface and subsurface strata are the lizard and bat layers. These two strata are characterized by the great abundance of bones; yellow, orange, or reddish brown colored sediments; and the absence of introduced mammals, the latter suggesting a pre-Columbian age (Koopman and Williams, 1951; Williams, 1952). The bones from these two layers are not heavily mineralized, although they may be encrusted with a carbonate matrix. Williams (1952) separated the lizard and bat layers primarily on the relative abundance of lizards and bats in the two layers. However, he did not find a cave in which these two layers were deposited in stratigraphic superposition. There is as yet no strong evidence that the lizard and bat layers from the various cave deposits are temporally equivalent. It seems more likely that the abundance of lizards and bats in these deposits is related to paleoecological and/or taphonomic factors, rather than the age of the sediments. In addition to the large number of bones of lizards and bats found in these layers, frogs, snakes, birds, and the rodent *Geocapromys* are also relatively common.

The oldest fossil-bearing strata in Jamaican caves are the indurated bone breccias and conglomerates, typified by the deposits at Wallingford Roadside Cave and Sheep Pen Cave (Anthony, 1920b; MacPhee, 1984; MacPhee et al., 1989; Goodfriend, 1986, 1989). The bones from the breccias are highly mineralized and the matrix is often extremely hard and dense. The most characteristic vertebrate of the hard breccias is the large extinct heptaxodontid rodent *Clidomys*. These breccias do not underlie the lizard and bat layers, but instead are attached to the ceiling or walls above the present floor of the caves in which they have been found. Several of the breccia deposits were not found in caves at all, but either were plastered to the limestone walls of overhangs, such as Sheep Pen, or deposited in fissures like Molton Fissure (MacPhee, 1984).

Quaternary fossil sites in the West Indies were placed in three general time periods by Morgan and Woods (1986): (1) late Pleistocene and early Holocene (approximately 40 ka to 4.5 ka); (2) Amerindian (4.5 ka to 500 yr B.P.), corresponding to the arrival of Amerindian peoples; (3) post-Columbian (500 yr B.P. to present), corresponding to the arrival of Europeans. In the absence of absolute dates these periods are useful for placing sites in a broad time framework. However, as more absolute dates become available, a refinement and subdivision of these time periods will be necessary. When Morgan and Woods (1986) wrote their paper, no Quaternary fossil sites from the West Indies were known to be older than 40 ka; however, two fossil sites from Jamaica recently have been dated at greater than 100 ka (Goodfriend, 1989; MacPhee et al., 1989). It would be useful if vertebrate faunas from the late Pleistocene and early Holocene could be separated, considering the major climatic changes that occurred at the termination of the Pleistocene about 10 ka. At present, there is no reliable faunal basis for distinguishing between late Pleistocene and early Holocene terrestrial vertebrate faunas in Jamaica.

Vertebrate fossil sites in Jamaica and elsewhere in the West Indies usually have been regarded as Pleistocene in age (i.e., >10 ka) if they lacked evidence of human occupation or introduced mammals. The assumption of a Pleistocene age for pre-human faunas in the West Indies is not entirely accurate since Amerindians did not colonize the West Indies until the middle Holocene at approximately 4.5 ka, and apparently the Tainos did not reach Jamaica until after 1.5 ka (Rouse, 1989). Consequently, both late Pleistocene and early Holocene sites in the West Indies lack evidence of humans. There are also many Antillean vertebrate fossil sites that lack evidence of humans, but are late Holocene in age based on radiocarbon dates. Therefore, using the absence of humans to infer the age of Antillean fossil sites may lead to erroneous conclusions. If the age of a site is not obvious from either radiocarbon dates (or other types of dates) or evidence from the vertebrate fauna itself, the site is here regarded as late Quaternary in age. The use of the term late Quaternary in this paper is specifically intended to reflect the uncertainty regarding the age of a particular deposit.

The presence in Jamaican fossil deposits of the introduced Old World murid rodents *Rattus* and *Mus* and most domesticated mammals, with the exception of the dog *Canis familiaris,* is indicative of a post-Columbian age (<500 yr B.P.), as these species were introduced into the West Indies by Europeans. The dog was brought to the Caribbean by Amerindian peoples sometime after 4.5 ka. The recently extinct rice rat, *Oryzomys antillarum,* does not occur in the indurated bone breccias in association with *Clidomys* and thus may not have reached Jamaica until later in the Pleistocene (Humphrey et al., 1993). The late arrival of *O. antillarum* in Jamaica, probably during the last glacial between 20 and 10 ka, is supported by the fact that this species is only weakly differentiated from *O. couesi* of Central America. *O. antillarum* is also known from Amerindian archaeological sites and from the late Holocene deposits containing *Rattus* and *Mus,* and did not go extinct until the late nineteenth century.

Giant rodents of the genus *Clidomys* seem to have gone extinct in Jamaica before the end of the Pleistocene (MacPhee, 1984; Morgan and Woods, 1986; MacPhee et al., 1989). The *Clidomys*-bearing bone breccias and conglomerates are clearly older than the more ubiquitous unconsolidated sediments deposited on the floor of caves. The comparatively great age of these breccias is confirmed by several absolute and relative dates that indicate a pre–late Pleistocene age (e.g., Goodfriend, 1989; MacPhee et al., 1989). Other factors that tend to substantiate the great age for the breccias are: (1) they are highly indurated and the bones contained within them are usually heavily mineralized; (2) they are most commonly found attached to the ceiling or walls of caves suggesting they may represent previous cave infillings that have been mostly eroded away.

Only within the past ten years have fossil vertebrate faunas from Jamaica been subjected to absolute dating techniques (MacPhee, 1984; Goodfriend and Mitterer, 1987; Goodfriend, 1989; MacPhee et al., 1989). See Table 1 for a summary of all published dates from Jamaican vertebrate fossil deposits. Radiocarbon (^{14}C) dating is the most promising method for developing a biochronology of late Quaternary vertebrate fossil sites in Jamaica, and on other islands in the West Indies as well, since most of these sites are within the range of radiocarbon (<40 ka). Radiocarbon dates have been published from a limited number of fossil sites in Puerto Rico (Pregill, 1981), Hispaniola (Woods, 1989), Jamaica (MacPhee, 1984; Goodfriend, 1989; Pregill et al., 1991), the Cayman Islands (Steadman and Morgan, 1985; Morgan, 1993), and Antigua (Steadman et al., 1984; Pregill et al., 1988). Wood and charcoal are generally regarded as the preferred organic materials for radiocarbon dating; however, both

TABLE 1. ABSOLUTE AND RELATIVE DATES FROM JAMAICAN QUATERNARY VERTEBRATE LOCALITIES*

Locality	Age (yr B.P.)	Type of Date[†]	Material Dated	Reference[§]
Bonafide Cave	c. 13,000	^{14}C	Land snail shell	1
Coco Ree Cave**	13,400 (10 to 20 cm layer)	^{14}C	Land snail shell	2
	30,100 (55 to 65 cm layer)	^{14}C	Land snail shell	2
	36,300 (80 to 90 cm layer)	^{14}C	Land snail shell	2
Green Grotto Cave	30,000	^{14}C	Land snail shell	1
	16,000	AA	Land snail shell	1
	19,000	AA	Land snail shell	1
Long Mile Cave	2,145	^{14}C	Rodent and lizard bones	3
Marta Tick Cave	770	^{14}C	Charcoal	4
Wallingford Roadside Cave conglomerate	33,250	^{14}C	Bone apatite from turtle shell	3
	120,000 to 250,000	U/Th	Speleothem	5
	65,000 to 150,000	ESR	Speleothem	5
	136,000 to 253,000	ESR	*Clidomys* long bones	5
Cave fill	270	^{14}C-AMS	Collagen from *Oryzomys* bone	5

*The original publications should be consulted for error ranges and laboratory numbers.
[†]Dating methods: ^{14}C = radiocarbon; ^{14}C-AMS = radiocarbon using accelerator mass spectrometry; U/Th = uranium series; AA = amino acid epimer analysis; ESR = electron spin resonance.
[§]1 = Goodfriend, 1989; 2 = Goodfriend and Mitterer, 1987; 3 = MacPhee, 1984; 4 = Pregill and others, 1991; 5 = MacPhee and others, 1989.
**Amino acid epimer analysis of land snail shells from Coco Ree Cave reveals that each of the three layers dated contains a mixture of shells of variable ages, and thus the ^{14}C dates represent only average ages for the shells in each layer (Goodfriend and Mitterer, 1987).

are notably rare in Antillean cave deposits. Most radiocarbon dates from these sites have been obtained from either bones or land snail shells. There are certain problems associated with dates from both of these organic materials.

Goodfriend and Stipp (1983) summarized the problems encountered when dating land snail shell carbonate. They demonstrated that anomalies may occur when dating land snails as a result of ingestion of "old" or "dead" carbon by the living animal, thus giving erroneously old dates in some instances. Goodfriend and Stipp (1983) obtained a maximum dating error of 3,000 yr using modern land snail shells from Jamaica. This anomaly may present a major problem for Holocene sites, but its significance is diminished in Pleistocene sites. Furthermore, they note that this dating error may be minimized by selecting ecologically appropriate species of land snails, specifically species that do not ingest ^{14}C-free limestone and incorporate it into their shells. All published radiocarbon dates from Jamaica that are based on land snail shells have been corrected for the age anomaly from old limestone (Goodfriend, 1989).

There are also problems associated with radiocarbon dates on bones. It is now generally accepted that accurate bone dates can be obtained only from the organic or collagen fraction of the bone, which comprises about 20% of the bone, and not the inorganic carbon or apatite fraction (Taylor, 1980; Stafford, 1990). Many previously published radiocarbon dates from the West Indies obtained by "whole bone" analysis are very likely to be inaccurate since they were primarily based on the inorganic carbon component of the bones. The dominant sources of error in radiocarbon dates using bone collagen are diagenesis and contamination by exogenous carbon, particularly humic and fulvic acids (Stafford, 1990). According to Stafford (1990) the best method for obtaining accurate radiocarbon dates from bone is to analyze the collagen fraction using accelerator mass spectrometry (AMS), which can be performed on very small samples. Very few AMS analyses have been conducted on fossil bones from the West Indies, but future work in this region will almost certainly include more accurate bone dates of this type.

Several Jamaican Quaternary fossil sites that are beyond the range of radiocarbon dating have been dated using other techniques. The pre-Wisconsinan age (between 100 and 200 ka) of the *Clidomys*-bearing breccia deposits at Wallingford Roadside Cave has been established by uranium series (U/Th) dating and electron spin resonance of cave speleothems (MacPhee et al., 1989). By using both radiocarbon dating and amino acid epimer analysis of land snail shells and geological evidence, Goodfriend (1986, 1989) proposed that the indurated bone breccia from Sheep Pen Cave is pre-late Pleistocene in age as well. These studies indicate that the fossil deposits in both Wallingford Roadside Cave and Sheep Pen Cave are older than 100 ka, making them the oldest documented Quaternary sites in the West Indies. By using uranium series dating, amino acid racemization, and other dating methods for pre-late Pleistocene sites and radiocarbon dating for sites younger than 40 ka, paleontologists should soon be able to develop a reliable biochronology for Quaternary vertebrate faunas in Jamaica and elsewhere in the West Indies.

BIOGEOGRAPHY

The availability of Jamaica for colonization by terrestrial organisms depends heavily upon theories pertaining to the geological evolution of the island, and in particular the sea-level history. Jamaica was mostly submerged in the Oligocene and probably has been continuously above sea level only since the early Miocene (Buskirk, 1985). The middle Cenozoic inundation of Jamaica probably eliminated much of the earlier biota that inhabited the island in the Late Cretaceous or early Cenozoic (Perfit and Williams, 1989), although a limited number of terrestrial vertebrates could have survived as relicts provided that some small islands existed in the region.

Next to Cuba, Jamaica is the closest of the large West Indian islands to a continental land mass. Jamaica and the closest point in Middle America are currently separated by about 600 km. However, during the late Pleistocene (Wisconsinan) glacial interval when sea levels were as much as 100 m lower than present, the eastern coast of Honduras and Nicaragua would have extended almost 300 km northeastward into the Caribbean Sea as shallow banks on the Nicaraguan Rise emerged above sea level. Furthermore, two large shallow banks nearer to Jamaica, the Rosalind and Pedro banks, probably were emergent as well and would have served as steppingstones for land vertebrates colonizing Jamaica from Middle America. Although the late Pleistocene marked one of the most extreme, and certainly the most recent, eustatic sea level lows, there were other periods of low sea level during the Neogene. The most significant low sea level stands during the Neogene occurred in the late Miocene (late Serravallian) at about 10 Ma, latest Miocene (Messinian) at about 6 Ma, and episodically throughout the late Pliocene and Pleistocene from about 2.5 Ma onward (Haq et al., 1987). During all three of these periods the distance between Middle America and Jamaica would have been significantly reduced, thereby facilitating overwater dispersal.

Overwater dispersal of terrestrial organisms has been criticized by many workers studying Caribbean biogeography, particularly those favoring vicariance (e.g., Rosen, 1975). The biota of Jamaica has a complex origin, and probably includes some components that colonized the island by dispersal and others that originated through vicariance of ancestral proto-Antillean stocks. Among mammals, which are the best-known group of fossil vertebrates in the West Indies, the majority of taxa appear to have colonized Jamaica and the other Greater Antilles from the Miocene onward by overwater dispersal from either South America or Middle America. Evidence for more recent dispersal comes from Jamaican species that are either conspecific with or very closely related to species from the Middle American mainland. The close proximity of Jamaica to Middle America is reflected in the presence of *Oryzomys antillarum*, the only sigmodontine rodent known from the Greater Antilles, and by the richer species diversity of bats compared to all other Antillean islands except Cuba. The large number of bats from Jamaica that are conspecific with Middle American species suggests they colonized Jamaica by overwater dispersal in the very recent geological past, probably during the late Pleistocene low sea level stand.

The antiquity of the Jamaican land mammal fauna can be roughly approximated by analyzing the levels of endemism observed and by making comparisons of Jamaican taxa with fossils of known age from North and South America. Among the seven species of nonvolant mammals known from Jamaica, only *Oryzomys antillarum* belongs to a genus that occurs outside of the West Indies. The close similarity between *O. antillarum* and the Middle American species *O. couesi* indicates a probable late Pleistocene colonization of Jamaica by *Oryzomys* (Humphrey et al., 1993). The Jamaican hutia, *Geocapromys brownii*, belongs to the endemic Antillean family Capromyidae. The genus *Geocapromys* is widespread in the West Indies, with endemic species in Jamaica, the Swan Islands, Cayman Islands, Cuba, and Bahamas (Morgan, 1985; Morgan and Woods, 1986). The Capromyidae belong to the Superfamily Octodontoidea, a predominantly South American group of caviomorph rodents first known from the late Oligocene (Deseadan) of Argentina and Bolivia. The large rodents of the endemic Jamaican genus *Clidomys* belong to the Antillean family Heptaxodontidae, a group of rodents of unknown phylogenetic position (Patterson and Wood, 1982). Like the capromyids, the heptaxodontids are clearly of South American origin. The fossil record of caviomorph rodents in South America suggests that the progenitors of the endemic Antillean groups reached the islands in the Miocene or thereafter. Although three genera of fossil primates are known from Jamaica, only the endemic genus *Xenothrix* is represented by sufficient material to allow phylogenetic comparisons with other New World ceboid monkeys. *Xenothrix* almost certainly was derived from a South American form more advanced than *Branisella*, the only known Deseadan primate, and must stem from Miocene or younger stock.

Twelve of the 24 species of bats (50%) recorded from Jamaica are conspecific with living mainland forms suggesting relatively recent overwater dispersal, probably during the Pleistocene. The fossil record indicates that most extant species of mammals evolved during the Pleistocene. Various levels of endemism are evident among the remaining 12 species of Jamaican bats: there is one species in the endemic Jamaican genus *Ariteus*; one species in each of the four endemic Antillean genera *Brachyphylla, Erophylla, Phyllonycteris,* and *Monophyllus*; and seven endemic Antillean species belonging to genera also found on the mainland. Although pre-late Pleistocene records of the endemic Antillean species of bats are lacking, most of these taxa probably reached the West Indies in the Miocene or Pliocene, primarily from South America.

Jamaica lacks three of the endemic West Indian mammal groups characteristic of the other three Greater Antilles, including the primitive solenodontoid insectivores *Solenodon* and *Nesophontes*, small ground sloths of the family Megalonychidae, and caviomorph rodents of the family Echimyidae (Simpson, 1956; MacPhee et al., 1983; Morgan and Woods, 1986). Morgan and Woods (1986) analyzed the living and extinct mammalian fauna of the West Indies in light of possible times of origin based on the fossil record of closely related taxa from both North and South America. Among terrestrial mammals, which have the best fossil record both on the mainland and in the West Indies, most of the endemic Antillean taxa seem to have their closest relationships with taxa from the Miocene and Pliocene of South America. The only obvious exceptions are *Solenodon* and *Nesophontes*, which are undoubtedly of North American origin. The solenodontoid insectivores probably were derived from some North American geolabidid insectivore during the Eocene or Oligocene (Lillegraven et al., 1981). The megalonychid ground sloths, primates, and three Antillean families of caviomorph rodents (Capromyidae, Echimyidae, and Heptaxodontidae) are almost certainly of South American origin. The fossil record indicates that the closest mainland sister taxa of these groups evolved sometime after the Oligocene (Simpson, 1956; Paula Couto, 1967; Morgan and Woods, 1986; Pascual et al., 1990).

Changes in climate, vegetation, and sea level during the late Pliocene and Pleistocene certainly had an effect on the Jamaican land vertebrate fauna. Previous authors have used evidence from the distribution of fossil lizards, birds, and bats (Pregill and Olson, 1981; Morgan and Woods, 1986; Morgan, 1989a) to show that changes in vegetation and sea level strongly affected patterns of extinction in the West Indies, particularly during the latest Pleistocene and early Holocene. Between about 12 and 8 ka, many species of land vertebrates disappeared from the West Indies and many more species underwent significant range contractions or localized extinctions. The restriction in the geographic range of certain Antillean birds and reptiles has been attributed to the change from predominantly xeric habitats in the late Pleistocene and early Holocene to more mesic habitats in the late Holocene (Pregill and Olson, 1981). The localized extinction of many bats in the West Indies seems to be related to the post-glacial rise in sea level and the subsequent flooding of large cave systems, particularly on smaller islands such as the Bahamas and Cayman Islands (Morgan and Woods, 1986; Morgan, 1989a, 1993). A change in cave microenvironments related to overall climatic change may have resulted in the extinction of selected species of cave-dwelling bats in the Greater Antilles, including *Brachyphylla nana* and *Mormoops megalophylla* in Jamaica.

Human-caused extinctions began shortly after the arrival of Amerindian peoples in the West Indies in the late Holocene about 4.5 ka (Rouse, 1989). Extinctions of native Antillean vertebrates in the late Holocene resulted from direct exploitation for food, habitat destruction, or through competition with or predation by introduced mammals, including the Old World murid rodent *Rattus*, the mongoose *Herpestes*, dogs, cats, and other domesticated animals. As discussed above, these extinctions of

424   G. S. Morgan

native vertebrates also may provide important biochronological information since they were concentrated during certain time periods.

QUATERNARY FOSSIL LOCALITIES IN JAMAICA

Brief descriptions are presented of the published Quaternary fossil localities from Jamaica that have produced remains of land vertebrates. No attempt is made here to place the large number of unpublished Jamaican fossil localities on record. The published vertebrate fossil localities are plotted on an outline map of Jamaica in Figure 1. The original publications should be consulted to obtain detailed geologic and geographical data for these localities. All published absolute dates are listed. Only fossil deposits containing indigenous Jamaican land vertebrates are included here. With a few minor exceptions, archaeological sites and post-Columbian deposits are not discussed.

The vertebrate fossils from Jamaica are housed primarily in three museums. The collections from Anthony's 1919 to 1920 expedition are in the American Museum of Natural History (AMNH), as are the fossils obtained during the Hecht, Koopman, and Williams expedition of 1950. The vertebrate fossils collected by Auffenberg in 1957 and 1958 and by Patton and field crews between 1966 and 1970 are housed in the Florida Museum of Natural History (formerly the Florida State Museum), University of Florida (UF). Fossils from Marta Tick Cave collected by Pregill, Steadman, and others in 1983 and from Bonafide Cave collected by Grady in 1985 and 1986 are housed in the U.S. National Museum of Natural History, Smithsonian Institution (USNM).

Bonafide Cave

This cave is in the Cockpit Country, about 10 km southwest of Windsor, Trelawny Parish (18°18′N, 77°43′W). Bonafide Cave was excavated by F. V. Grady in 1985 and 1986. A description of this cave is provided by Baker et al. (1986). Humphrey et al. (1993) record the rice rat, *Oryzomys antillarum,* from the upper layer of Bonafide Cave. The uppermost layer (0 to 15 cm) of Bonafide Cave is late Holocene in age based on radiocarbon dates from land snail shells, while the deeper layers (15 to 40 cm) have been ^{14}C dated at about 13 ka (Goodfriend, 1989).

Braham Cave

Located alongside the Rio Cobre in Thetford, just southeast of Worthy Park in St. Catherine Parish (18°08′N, 77°09′W). Braham Cave was discovered and excavated by T. H. Patton in the late 1960s. This cave has produced a rich sample of *Oryzomys antillarum,* including several nearly complete skulls (Humphrey et al., 1993). The fossils are highly mineralized and some are preserved in a semi-indurated breccia, perhaps indicating a greater age than most other Jamaican deposits from which *O. antillarum* has been recovered.

Cambridge Cave

Located near Cambridge, St. James Parish (18°19′N, 77°54′W). The fossiliferous deposit in Cambridge Cave was ex-

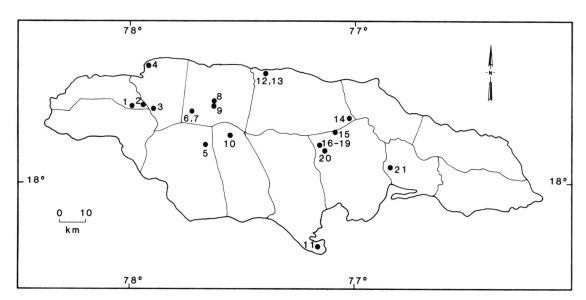

Figure 1. Map of Jamaica showing location of Quaternary terrestrial vertebrate fossil sites discussed in text. *Hanover Parish:* 1, Knockalva Cave; 2, Seven Rivers Cave; *St. James Parish:* 3, Cambridge Cave; 4, Montego Bay Airport Cave; *St. Elizabeth Parish:* 5, Wallingford Roadside Cave; *Trelawny Parish:* 6, Bonafide Cave; 7, Marta Tick Cave; 8, Long Mile Cave; 9, Sheep Pen Cave; *Manchester Parish:* 10, Molton Fissure; *Clarendon Parish:* 11, Portland Ridge Caves; *St. Ann Parish:* 12, Dairy Cave; 13, Green Grotto Cave; 14, Moseley Hall Cave; *St. Catherine Parish:* 15, Mt. Diablo Cave; 16, Braham Cave; 17, Coco Ree Cave; 18, Swansea Cave; 19, Tydixon Cave; 20, Lluidas Vale Cave; *St. Andrew Parish:* 21, Red Hills Road Cave.

cavated by H. E. Anthony in 1919 or 1920 and consists of a poorly consolidated breccia containing a single bat mandible and abundant remains of the Jamaican hutia, *Geocapromys brownii*. The bones are white in color and unfossilized suggesting an age similar to that of the lizard layers. Neither introduced mammals nor *Oryzomys* are present in this deposit (Koopman and Williams, 1951).

Coco Ree Cave

This cave is 4 km west-northwest of Worthy Park, St. Catherine Parish (18°09′N, 77°11′W). Coco Ree Cave was excavated by Patton and a Florida State Museum field party in June 1970. Ford and Morgan (1986) described the cave and its associated fauna, including the proximal femur of a primate of unknown affinities, *Geocapromys brownii*, and *Oryzomys antillarum*. Anderson et al. (1983) reported *G. brownii* from Coco Ree Cave and Humphrey et al. (1993) analyzed the fossil *O. antillarum*. Goodfriend and Mitterer (1987) provided radiocarbon dates for three layers from Coco Ree Cave based on bulk shell samples of the land snail *Poteria varians*: 13,400 ± 180 yr B.P. (10 to 20 cm layer); 30,100 ± 1100 yr B.P. (55 to 65 cm layer); 36,300 ± 4300 yr B.P. (80 to 90 cm layer). The monkey femur was found in the 80 to 90 cm layer.

Dairy Cave

Dairy Cave (part of Runaway Bay Caves according to Fincham, 1977) is located about 3 km east of Discovery Bay (formerly called Dry Harbor) on the north coast of Jamaica in St. Ann Parish (18°28′N, 77°23′W). Anthony first collected vertebrate fossils from Dairy Cave in 1919 and 1920. M. K. Hecht, K. F. Koopman, and E. E. Williams revisited this cave in 1950 and W. Auffenberg excavated additional material there in 1958. Koopman and Williams (1951) and Hecht (1951) give general descriptions of Dairy Cave, and Etheridge (1966) provided a map showing the location of the various fossil-producing sites within the cave. Koopman and Williams (1951) described the fossil bats from Dairy Cave; Hecht (1951) and Etheridge (1966) reported on fossil lizards; and Anderson et al. (1983) noted the presence of *Geocapromys brownii*. No radiocarbon dates are available from Dairy Cave. The concept of the "lizard layers" and "bat layers" in Jamaican fossil cave deposits was derived primarily from fossiliferous strata in Dairy Cave (Koopman and Williams, 1951; Williams, 1952).

Green Grotto Cave

Green Grotto Cave (part of Runaway Bay Caves according to Fincham, 1977) is 3 km east of Discovery Bay and 4 km west of Runaway Bay, St. Ann Parish (18°28′N, 77°23′W). Goodfriend (1989) provided data on the location and stratigraphy of this cave, which is best known for its rich fauna of terrestrial snails. Anderson et al. (1983) listed *Geocapromys brownii* from Green Grotto Cave.

Knockalva Cave

Located near the village of Knockalva, Hanover Parish, in western Jamaica (18°19′N, 77°59′W). Knockalva Cave was discovered by Patton in 1968. Only a small sample of fossils was recovered from this cave including specimens of *Geocapromys brownii* reported by Anderson et al (1983).

Lluidas Vale Cave

Located 0.5 km southwest of Lluidas Vale, St. Catherine Parish (18°07′N, 77°09′N). Lluidas Vale Cave (= River Sink Cave) is very near the Lluidas Sinkhole (Fincham, 1977), but is about 20 m above it in elevation. Auffenberg first recovered fossils from Lluidas Vale Cave in the summer of 1957 and it was visited again in 1966 by Patton. MacPhee (1984) noted the presence of the giant heptaxodontid rodent *Clidomys* from Lluidas Vale Cave, suggesting that this deposit is older than late Pleistocene.

Long Mile Cave

Located about 2 km north of Windsor, Trelawny Parish (18°23′N, 77°39′W). Long Mile Cave was first excavated in 1920 by Anthony. This cave is the type locality of the extinct monkey, *Xenothrix mcgregori* (Williams and Koopman, 1952), as well as the extinct flightless ibis, *Xenicibis xympithecus* (Olson and Steadman, 1977). R.D.E. MacPhee carried out further excavations at Long Mile Cave in 1981. He provided a description of the cave, discussion of the excavations conducted there, and a faunal list (MacPhee, 1984). MacPhee and Fleagle (1991) described additional primate fossils from Long Mile Cave. A sample of *Geocapromys* and lizard bones from the same layer that produced *Xenothrix* were radiocarbon dated at 2145 ± 220 yr B.P. (MacPhee, 1984).

Marta Tick Cave

This cave is very near Bonafide Cave in the Cockpit Country, Trelawny Parish, about 8 km west-northwest of Quickstep and 10 km southwest of Windsor (18°18′N, 77°44′W). Marta Tick Cave was excavated in 1983 by G. K. Pregill, D. W. Steadman, and a field party from the Smithsonian Institution. Anderson et al. (1983) reported *Geocapromys brownii* from this cave (called Quickstep Cave in their paper). Pregill et al. (1991) recorded 26 species of vertebrates from Marta Tick Cave, including the extinct lizards, *Celestus* cf. *C. occiduus* and *Leiocephalus* cf. *L. jamaicensis,* and the rodents, *G. brownii* and *Oryzomys antillarum*. Two samples of wood charcoal from Marta Tick Cave were radiocarbon dated. One sample was modern and a sample from "pocket 1" dated at 770 ± 70 yr B.P. (Pregill et al., 1991).

Molton Fissure

Located near the town of Molton (Malton on some maps), about 7 km west-northwest of Christiana, Manchester Parish (18°12′N, 77°33′W). A small sample of fossils was recovered from Molton Fissure by Patton in the late 1960s consisting entirely of teeth and limb fragments of *Clidomys* preserved in an extremely hard breccia. MacPhee (1984) reported the fossils of *Clidomys* from Molton Fissure.

Montego Bay Airport Cave

Located near the western end of the runway at the Montego Bay airport, 2.5 km north of Montego Bay, St. James Parish (18°30′N, 77°54′W). This site was excavated by Auffenberg in 1957 and 1958. Etheridge (1966) provided a general description of the cave and reported specimens of the extinct iguanid lizard *Leiocephalus jamaicensis*.

Moseley Hall Cave

Located near Guys Hill, 0.5 km north of the main road from Blackstonedge to the Mosely Hall District, St. Ann Parish (18°15′N, 77°02′W). Patton and Florida State Museum field crews excavated this cave in the late 1960s. Anderson et al. (1983) reported fossils of *Geocapromys brownii* from Moseley Hall Cave.

Mt. Diablo Cave

This cave is on Mt. Diablo about 3 to 4 km north of Ewarton on the border between St. Catherine and St. Ann Parishes (18°12′N, 77°06′W). Patton and Florida State Museum field crews excavated this cave in the late 1960s. Humphrey et al. (1993) reported fossils of *Oryzomys antillarum* from Mt. Diablo Cave.

Portland Ridge caves

A number of caves (including Lower Portland Cave, Upper Portland Cave, and Portland Caves 1, 2, and 3) located on the Portland Ridge in southernmost Clarendon Parish (17°45′N, 77°09′W) have produced vertebrate fossils. Anthony first collected fossils from the Portland Ridge caves in 1919 or 1920, followed by the Hecht, Koopman, and Williams expedition in 1950, Williams in 1953, Auffenberg in 1957, and Patton in 1966. It is unclear how these caves were named or numbered by the paleontologists who discovered and excavated them. It is quite possible, indeed probable, that several caves in the Portland Ridge system have been given more than one name. Koopman and Williams (1951) briefly mentioned Anthony's collections from Portland Cave and Williams (1952) reported *Geocapromys brownii* and the bats from this locality. Etheridge (1966) provided a description of Portland Cave 1 and Cave 3 and reported fossils of *Leiocephalus jamaicensis*. No absolute dates are available for any of the Portland Ridge caves.

Red Hills Road Cave

Located on the south side of Red Hills Road, about 2 km northeast of the town of Red Hills, St. Andrew Parish (18°04′N, 76°52′W). A. G. Godwin and M. M. Britton discovered Red Hills Road Cave in the summer of 1988. The occurrence of this cave was first noted by Donovan and Gordon (1989). Savage (1990) reported amphibians, reptiles, birds, and *Geocapromys brownii* from Red Hills Road Cave.

Seven Rivers Cave

Seven Rivers Cave (called Hazelymph Cave in Fincham, 1977) is located near the village of Hazelymph, Hanover Parish (18°20′N, 77°55′W). There is some confusion regarding the location of this cave. Koopman and Williams (1951) placed Seven Rivers Cave in St. James Parish, whereas Fincham (1977) stated that Seven Rivers Cave is a synonym of Hazelymph Cave, which is located in Hanover Parish. The town of Seven Rivers is less than 1 km east of the St. James/Hanover Parish boundary and Hazelymph is situated virtually on the boundary in Hanover Parish. Since Koopman and Williams do not provide more detailed locality information for Seven Rivers Cave, I am following Fincham's (1977) locality data. Anthony excavated Seven Rivers Cave in 1919 or 1920. He found *Rattus* associated with *Geocapromys*, giant *Celestus*, and several species of extant bats, suggesting a post-Columbian age for the deposit.

Sheep Pen Cave

Sheep Pen is actually not a cave, but rather is an overhang located on a hill 0.5 km south of Windsor, near Windsor Great House, Trelawny Parish (18°21′N, 77°39′W). Sheep Pen Cave originally was discovered by Anthony in 1920 and was later visited by Patton in the late 1960s. The deposit at Sheep Pen consists of an indurated bone breccia plastered on the face of an overhang. The rodents *Clidomys* and *Geocapromys* have been reported from Sheep Pen Cave (Koopman and Williams, 1951; MacPhee, 1984), as has the proximal femur of a ceboid primate (Ford and Morgan, 1988; Ford, 1990a, b). Lithological evidence from Sheep Pen Cave indicates several dissolution-precipitation cycles, with phases of deposition presumably representing drier climates and dissolution under wetter conditions (Goodfriend, 1986). Goodfriend suggested that these supposed wet and dry phases may correspond with interglacial and glacial cycles, respectively, and therefore that this deposit probably represents more than one glacial cycle and may be several hundred thousand years old. Bones and enamel from Sheep Pen lack detectable levels of amino acids suggesting a long period of leaching, and further indicating a considerable age for this deposit (Goodfriend, 1986, 1989). The presence of *Clidomys* from Sheep Pen Cave tends to substantiate the hypothesis of a pre-late Pleistocene age.

Swansea Cave

Located 4 km north of Worthy Park in northwestern St. Catherine Parish (18°11′N, 77°09′W). This cave was excavated by Patton in 1966. Olson and Steadman (1979) described a complete humerus of the ibis *Xenicibis xympithecus* from Swansea Cave; Anderson et al. (1983) noted the presence of *Geocapromys brownii*; and Humphrey et al. (1993) reported remains of *Oryzomys antillarum*.

Tydixon Cave

This cave is 3 km northwest of Worthy Park, St. Catherine Parish (18°09'N, 77°11'W). Tydixon Cave was excavated by Patton and UF field crews in the late 1960s. Humphrey et al. (1993) reported cranial remains of *Oryzomys antillarum* from Tydixon Cave.

Wallingford Roadside Cave

This cave is located approximately 1 km north of Balaclava along the main road between Balaclava and Oxford in St. Elizabeth Parish (18°11'W, 77°39'W). Wallingford Roadside Cave originally was discovered by Anthony (1920a, b), who described four genera and five species of large heptaxodontid rodents from this site (later reduced to two species of *Clidomys* by MacPhee et al., 1983; and MacPhee, 1984). Anthony collected fossils from a highly indurated calcareous conglomerate attached to the ceiling and rear wall of the cave. Much younger unconsolidated sediments from the floor of this cave were excavated in 1950 by Hecht, Koopman, and Williams (Koopman and Williams, 1951). Patton visited Wallingford Roadside Cave in 1966 and removed samples of both the hard conglomerate and the unconsolidated deposit. Additional fossils from Wallingford Roadside Cave were collected by MacPhee in 1981 and by D. A. McFarlane in 1985 (MacPhee, 1984; McFarlane and Gledhill, 1985; MacPhee et al., 1989). MacPhee (1984) provided a map, vertebrate faunal list, and discussion of the discovery and excavation of Wallingford Roadside Cave. The indurated conglomerate has produced the heptaxodontid rodents, *Clidomys osborni* and *C. parvus*; the capromyid rodent, *Geocapromys brownii*; the crocodile, *Crocodylus*; the turtle, *Trachemys terrapen*; and the anguid lizard, *Celestus*. A sample of turtle shell fragments from the upper calcareous conglomerate at Wallingford Roadside Cave was radiocarbon dated at 33,650 ± 2550 yr B.P. (MacPhee, 1984). Uranium series and electron spin resonance dates from the calcite flowstone in Wallingford Roadside Cave suggest that this fauna is older than 100 ka, but younger than 200 ka (MacPhee et al., 1989). *Oryzomys* teeth and bones from 20 to 30 cm below the surface in the unconsolidated deposit on the floor of this cave have yielded a ^{14}C date of 270 ± 67 yr B.P. (MacPhee et al., 1989).

SYSTEMATIC PALEONTOLOGY

This section consists of concise systematic accounts of all vertebrate species recorded from Pleistocene and early Holocene fossil deposits in Jamaica. Extant taxa known only from late Holocene localities and archaeological sites are not discussed individually, but are listed in Table 2, which is a complete faunal list of all species of vertebrates recorded from the Quaternary of Jamaica. In several instances, new records are presented (indicated by UF—denoting specimens in the vertebrate paleontology collection of Florida Museum of Natural History, University of Florida). Each account is divided into four sections: synonymy, fossil localities, age and stratigraphic occurrence, and remarks. The synonymy provides the original citation for the description of the species, other taxonomic names used for the species, and references to the best-available morphological descriptions and figures. The locality section lists all published fossil sites in Jamaica from which the species has been identified. The age and stratigraphic occurrence provides radiocarbon dates or other absolute dates that have been obtained from bones of the species, or from other materials in stratigraphic association with the species in question. If no absolute dates are available for a species, this section contains a brief statement on its stratigraphic range in Jamaica. The remarks includes any additional information on the species that is not covered in the other sections, especially data on systematics. Several of the diagnostic species of Jamaican fossil vertebrates are figured. For most of the species not figured here, a literature citation is provided for the best available illustration.

Class REPTILIA Laurenti, 1768
Order TESTUDINES Batsch, 1788
Family EMYDIDAE Lydekker, 1889

Genus *TRACHEMYS* Agassiz, 1857

Trachemys terrapen (Lacépède, 1788)

[*Testudo*] *terrapen* Lacépède, 1788, p. 129.
Pseudemys terrapen (Lacépède). Barbour and Carr, 1940, p. 391–394, Plate 4.
Pseudemys terrapen (Lacépède). Williams, 1950, p. 7.
Pseudemys floridana (Le Conte, 1830). MacPhee, 1984, p. 10.
Trachemys terrapen (Lacépède). Seidel, 1988, p. 23–27, Figs. 8, 11.

Fossil localities. Wallingford Roadside Cave, St. Elizabeth Parish (Anthony, 1920a; Williams, 1950; MacPhee, 1984; MacPhee et al., 1989); Lluidas Vale Cave, St. Catherine Parish (UF).
Age and Stratigraphic Occurrence. A sample of turtle shell fragments from the upper calcareous conglomerate at Wallingford Roadside Cave was dated at 33,250 ± 2950 yr B.P. (MacPhee, 1984). This deposit is now thought to be between 100 and 200 ka in age based on uranium series dates and electron spin resonance dates obtained from the calcareous conglomerate (MacPhee et al., 1989). The stratigraphic range of *Trachemys terrapen* in Jamaica extends from at least 100 ka in the indurated breccia at Wallingford Roadside Cave to the present.
Remarks. Anthony (1920b, p. 161) briefly mentioned remains of a fossil turtle discovered in December 1919 in Wallingford Roadside Cave, "Associated with this large rodent [*Clidomys*] were found a terrapin, larger considerably than the one living on the island today, probably a large tortoise..." Williams (1950) identified as *Pseudemys terrapen* a partial turtle shell collected by Anthony from the conglomerate in Wallingford Roadside Cave in association with remains of heptaxodontid rodents. It seems clear that the fossil Williams (1950) referred to *P. terrapen* was part of the same material originally identified as tortoise by Anthony (1920b), as Williams does not list Jamaica among the West Indian islands possessing remains of fossil tortoises. Koopman and Williams (1951) noted that abundant remains of *Pseudemys* occurred in the breccia in Wallingford Roadside Cave. MacPhee (1984) erroneously referred the Wallingford turtle fossils to *Pseudemys floridana*, a species confined to the southeastern United States. MacPhee (1984, p. 10) stated that, "The *Pseudemys* from Roadside Cave is large and very thick-shelled compared with extant *P. floridana* [presumably he meant the living Jamaican form]; this may have led Anthony (1920a) to infer that another chelonian was represented in the Wallingford material..." In 1957 Auffenberg obtained shell fragments of *T. terrapen* from a breccia

TABLE 2. FAUNAL LIST OF FOSSIL VERTEBRATES IDENTIFIED FROM PUBLISHED LATE QUATERNARY SITES IN JAMAICA

Class Amphibia
    Order Anura
        Family Hylidae
            Hyla sp.*
            Osteopilus brunneus
        Family Leptodactylidae
            Eleutherodactylus sp.*

Class Reptilia
    Order Testudines
        Family Emydidae
            Trachemys terrapen
    Order Squamata
        Suborder Sauria
        Family Anguidae
            Celestus cf. *C. occiduus*†
            Celestus - 2 species
        Family Gekkonidae
            Aristelliger praesignis
            Aristelliger titan†
            Sphaerodactylus sp.
        Family Iguanidae
            Anolis - 3 species
            Cylura collei
            Leiocephalus jamaicensis†
        Family Teiidae
            Ameiva cf. *A. dorsalis*
        Suborder Serpentes
        Family Boidae
            Tropidophis cf. *T. haitanus*
        Family Typhlopidae
            Typhlops sp.
        Family Colubridae
            Alsophis sp.
            Arrhyton sp.
    Order Crocodylia
        Family Crocodylidae
            Crocodylus sp.

Class Aves
    Order Ciconiiformes
        Family Plateleidae
            Xenicibis xympithecus†
    Order Strigiformes
        Family Strigidae
            Athene cf. *A. cunicularia*§
    Order Caprimulgiformes
        Family Caprimulgidae
            Siphonorhis americana†
        Family Nyctibiidae
            Nyctibius griseus

Class Aves (continued)
    Order Columbiformes
        Family Columbidae
            Columba inornata
            Geotrygon versicolor
    Order Passeriformes
        Family Turdidae
            Turdus aurantius
            Turdus jamaicensis
        Family Fringillidae
            Loxigilla violacea

Class Mammalia
    Order Chiroptera
        Family Mormoopidae
            Mormoops blainvillii
            Mormoops megalophylla§
            Pteronotus parnellii
        Family Phyllostomidae
            Ariteus flavescens
            Brachyphylla nana pumila§
            Erophylla sezekorni
            Macrotus waterhousii
            Monophyllus redmani
            Phyllonycteris aphylla
            Tonatia bidens saurophila§
        Family Natalidae
            Natalus major
        Family Vespertilionidae
            Eptesicus lynni
    Order Primates
        Family Xenotrichidae
            Xenothrix mcgregori†
        Family indeterminate
            Genus and species indeterminate A†
            Genus and species indeterminate B†
    Order Rodentia
        Family Capromyidae
            Geocapromys brownii
        Family Heptaxodontidae
            Clidomys osborni†
            Clidomys parvus†
        Family Muridae
            Subfamily Sigmodontinae
                Oryzomys antillarum†

*Undetermined number of species present.
†Extinct species.
§Extinct in Jamaica, but extant elsewhere in West Indies and/or continental Neotropics.

deposit in Lluidas Vale Cave in St. Catherine Parish. The presence of turtles in this second deposit has not been noted before in the literature. Seidel and Smith (1986) transferred all West Indian emydid turtles from the genus *Pseudemys* to *Trachemys*. The Jamaican slider, *Trachemys terrapen*, is the only species of freshwater turtle presently inhabiting Jamaica (Seidel, 1988). No fossil specimens of *T. terrapen* from Jamaica have been figured.

Order SQUAMATA Oppell, 1811
Family ANGUIDAE Gray, 1825

Genus *CELESTUS* Gray, 1839

Celestus cf. *C. occiduus* (Shaw, 1802)

Lacerta occidua Shaw, 1802, p. 288.
Celestus sp. Hecht, 1951, p. 2
Celestus sp. Koopman and Williams, 1951, p. 4.
Celestus sp. Williams, 1952, p. 174.
Celestus cf. *C. occiduus* (Shaw). Pregill et al., 1991, p. 4, 10.

Fossil localities. Dairy Cave, St. Ann Parish (Hecht, 1951); Long Mile Cave, Trelawny Parish (Williams, 1952); Seven Rivers Cave, St. James Parish (Koopman and Williams, 1951); Marta Tick Cave, Trelawny Parish (Pregill et al., 1991); Wallingford Roadside Cave (both conglomerate and cave fill deposits), St. Elizabeth Parish (Hecht, 1951; MacPhee, 1984). Remains of a giant *Celestus* are known from many additional Jamaican cave faunas in the UF collection.
Age and Stratigraphic Occurrence. MacPhee (1984) reported a single dentary of *Celestus* sp. from the indurated conglomerate deposit in Wallingford Roadside Cave that has since been dated at between 100 and 200 ka (MacPhee et al., 1989). However, he does not indicate if this dentary was from the giant *Celestus* or one of the smaller species of this genus known from Jamaica. The next oldest records of the giant *Celestus* are in the "lizard layers" in Dairy Cave and in the unconsolidated floor sediments of Wallingford Roadside Cave (Hecht, 1951). Pregill et al. (1991) obtained a radiocarbon date of 770 yr B.P. on charcoal from strata in Marta Tick Cave containing fossils referred to *Celestus* cf. *C. occiduus.* The giant *Celestus* was found with *Rattus* in Seven Rivers Cave, although their association was thought to be a result of faunal mixing (Koopman and Williams, 1951). However, if the giant *Celestus* is conspecific with the recently extinct *C. occiduus,* then there is no need to explain its association with *Rattus* by reworking (Williams, 1952; Pregill et al., 1991).
Remarks. Hecht (1951) mentioned abundant remains of a giant, extinct species of the anguid lizard *Celestus* from the unconsolidated floor sediments in Wallingford Roadside Cave and the lizard layers in Dairy Cave. In their description of the "lizard layers" from Jamaican cave deposits, Koopman and Williams (1951, p. 4) noted that these layers are ". . . distinguished by the presence of a large number of bones, including the osteoscutes of a giant member of the lizard genus *Celestus*." Williams (1952) first suggested that the giant fossil *Celestus* might be conspecific with *C. occiduus,* a view shared by Pregill et al. (1991). *C. occiduus* is a large, recently extinct species of galliwasp last collected in Jamaica over 100 years ago. Williams (1952) hypothesized that some individuals of the giant *Celestus* were over two feet (600 mm) in length and Pregill et al. (1991) cited a preserved specimen of *C. occiduus* that had a snout-vent length of over 300 mm. There are no published illustrations of the giant Jamaican *Celestus*.

Family GEKKONIDAE Gray, 1825

Genus *ARISTELLIGER* Cope, 1861

Aristelliger titan Hecht, 1951

Aristelliger titan Hecht, 1951, p. 5–18, Figs. 1–6.

Fossil Localities. Dairy Cave, St. Ann Parish (Hecht, 1951); Lower Portland Cave, Clarendon Parish (Hecht, 1951); seaside cave in Healthshire (= Hellshire) Hills, St. Catherine Parish (Hecht, 1951).
Age and Stratigraphic Occurrence. No absolute dates are available for localities containing the extinct gecko *Aristelliger titan*. The oldest strata in which this species occurs are the "lizard layers" in Dairy Cave. The youngest record of *A. titan* appears to be a single frontal from a cave in the Hellshire Hills (Hecht, 1951).
Remarks. Aristelliger titan is a large extinct species of gecko currently known from three cave fossil deposits in Jamaica (Hecht, 1951). The only deposit where *A. titan* and the living species *A. praesignis* were found in association was in the upper strata in Dairy Cave (Hecht, 1951). The older lizard layers in Dairy Cave contain only *A. titan*. There appears to be no obvious reason for the extinction of *A. titan.* However, it is intriguing that large species of both *Aristelliger* and *Celestus* disappeared from Jamaica during the late Quaternary, apparently in the late Holocene. Illustrations of the frontal and dentary bones of *A. titan* are reproduced in Figures 2A and 2B (from Hecht, 1951, Fig. 1).

Aristelliger praesignis (Hallowell, 1857)

Hemidactylus praesignis Hallowell, 1857, p. 222.
Aristelliger praesignis (Hallowell). Hecht, 1951, p. 18–20, Fig. 6.

Fossil Locality. Dairy Cave, St. Ann Parish (Hecht, 1951).
Age and Stratigraphic Occurrence. The only published fossil record of *Aristelliger praesignis* is from strata above the "lizard layers" in Dairy Cave (Hecht, 1951). This species still occurs in Jamaica.
Remarks. Aristelliger praesignis is a living species of gecko that is much smaller than the extinct *A. titan. A. praesignis* occurs throughout Jamaica, as well as the Cayman Islands and Swan Islands (Schwartz and Thomas, 1975). Hecht (1951, Fig. 6) provided comparative photographs of the dentaries of *A. praesignis* and *A. titan.*

Family IGUANDIAE Gray, 1827

Genus *CYCLURA* Harlan, 1829

Cyclura collei Gray, 1845

Cyclura Collei Gray, 1845, p. 190.
Cyclura collei Gray. Grant, 1940, p. 97–100.
Cyclura collei Gray. Wing, 1972, p. 20, 23.
Cyclura collei Gray. Schwartz and Carey, 1977, p. 56–59, Fig. 14.
Cyclura sp. MacPhee, 1984, p. 17.

Fossil Locality. Long Mile Cave, Trelawny Parish (MacPhee, 1984). *Cyclura* also has been reported from one archaeological site in Jamaica, the White Marl Site, located about 5 km east of Spanish Town in St. Catherine Parish (Wing, 1972).
Age and Stratigraphic Occurrence. A radiocarbon date of 2145 yr B.P.

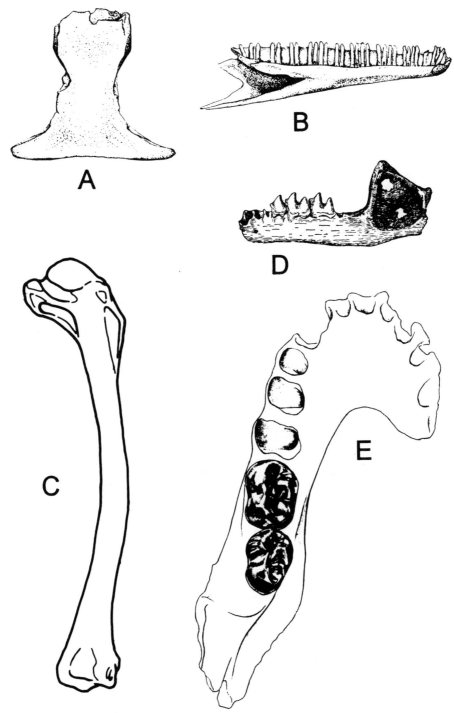

Figure 2. A, Frontal (AMNH 7504) and B, left dentary (AMNH 7503-holotype) of *Aristelliger titan* from Dairy Cave, St. Ann Parish, Jamaica, 2.5× natural size (from Hecht, 1951, Figs. 1B, 1C). C, Right humerus (UF 23768) of *Xenicibis xympithecus* from Swansea Cave, St. Catherine Parish, Jamaica, natural size. D, Left mandible (AMNH 147207-paratype) of *Tonatia bidens saurophila* from Wallingford Roadside Cave, St. Elizabeth Parish, Jamaica, 3× natural size (from Koopman and Williams, 1951, Fig. 1A). E, Left mandible of *Xenothrix mcgregori* (AMNH 148198-holotype) from Long Mile Cave, Trelawny Parish, Jamaica, 3× natural size (from Rosenberger et al., 1990, Fig. 1).

on a combined sample of rodent and lizard bones was obtained by MacPhee (1984) for the stratigraphic layer (Unit 2) in Long Mile Cave in which *Cyclura* was found. Radiocarbon dates for strata containing *Cyclura* in the White Marl Site range from 1,100 to 500 yr B.P. (Wing, 1989). The rock iguana, *C. collei,* still occurs in Jamaica, but is nearly extinct (Grant, 1940; Schwartz and Thomas, 1975).

Remarks. MacPhee (1984) identified *Cyclura* sp. from Long Mile Cave in the Cockpit Country in the north-central region of Jamaica. This record is the only fossil evidence of the rock iguana from Jamaica. Wing (1972) identified a large sample (minimum number of 38 individuals) of *Cyclura collei* from the White Marl Site located between Kingston and Spanish Town on the south coast of Jamaica about 6 km inland from Kingston Harbour. Grant (1940) noted that the living population of *C. collei* in Jamaica was restricted to the Goat Islands located just offshore from the Hellshire Hills in St. Catherine Parish along the southern coast. The fossil and archaeological record indicates that iguanas were more widespread in Jamaica during the Holocene, and presumably the Pleistocene as well.

Genus *LEIOCEPHALUS* Gray, 1827

Leiocephalus jamaicensis Etheridge, 1966

Leiocephalus jamaicensis Etheridge, 1966, p. 49–57.

Fossil Localities. Dairy Cave, St. Ann Parish (Hecht, 1951; Etheridge, 1966); Montego Bay Airport Cave, St. James Parish (Etheridge, 1966); Portland Ridge caves (including Portland Cave I and Portland Cave III), Clarendon Parish (Etheridge, 1966); Marta Tick Cave, Trelawny Parish (Pregill et al., 1991).

Age and Stratigraphic Occurrence. Pregill et al. (1991) published a radiocarbon date of 770 yr B.P. for strata containing *Leiocephalus* cf. *L. jamaicensis* in Marta Tick Cave, confirming that this extinct species survived into the late Holocene. Based on the associated vertebrate faunas and stratigraphic relationships, the three other cave deposits in which *L. jamaicensis* occurs are older than the surface (*Rattus*) and subsurface (*Oryzomys*) layers of Koopman and Williams (1951). The type specimens of *L. jamaicensis* were recovered from the "lizard layers" in Dairy Cave, while the fossils from Montego Bay Airport Cave and the two Portland Caves were found well below the surface in strata containing only indigenous species of Jamaican vertebrates. Etheridge (1966) gave the age of *L. jamaicensis* from Dairy Cave as "probably Late Pleistocene." However, as noted earlier, late Pleistocene and early Holocene faunas are difficult to distinguish in Antillean vertebrate fossil sites.

Remarks. Etheridge (1966) described the extinct lizard *Leiocephalus jamaicensis* based on small samples from three cave fossil deposits in Jamaica. Although the genus *Leiocephalus* no longer occurs on Jamaica, *L. jamaicensis* was rather widespread on the island having been recorded from the northwest coast at Montego Bay, about midway along the northern coast near Discovery Bay, from the Cockpit Country near Quickstep, and on the Portland Ridge about midway along the southern coast. Study of the extensive microvertebrate faunas present in the caves excavated by T. H. Patton and UF field crews undoubtedly would yield more records of *L. jamaicensis.* The endemic Antillean genus *Leiocephalus* was much more widespread in the late Quaternary than at present, with extinct species known from Barbuda and Martinique in the Lesser Antilles, Puerto Rico, and Hispaniola, as well as Jamaica (Pregill, 1981). Among these islands, only Hispaniola still supports extant species in this genus. The reasons for the widespread extinction of *Leiocephalus* are not well understood, but Pregill (1986) suggested that some insular extinctions of lizards in the Antilles may have been caused by humans during the late Holocene.

Order CROCODYLIA Gmelin, 1789
Family CROCODYLIDAE Cuvier, 1807

Genus *Crocodylus* Laurenti, 1768

Crocodylus sp.

Crocodylus sp. Anthony, 1920a, p. 161.
Crocodylus sp. Koopman and Williams, 1951, p. 5.
Crocodylus sp. MacPhee, 1984, p. 10.

Fossil Localities. Wallingford Roadside Cave, St. Elizabeth Parish (Anthony, 1920a; Koopman and Williams, 1951; Williams, 1952; MacPhee, 1984); Dairy Cave, St. Ann Parish (UF).

Age and Stratigraphic Occurrence. A crocodile vertebra from Wallingford Roadside Cave was collected from the hard breccia or calcareous conglomerate dated at between 100 and 200 ka (MacPhee et al., 1989). Several crocodile teeth from Dairy Cave were recovered from the as yet undated "lizard layers." If these specimens belong to the living Jamaican species *Crocodylus acutus,* then crocodiles have inhabited Jamaica for at least the past 100,000 years or more, and continue to live along the southern coast of the island.

Remarks. Late Quaternary fossils of crocodiles (*Crocodylus*) are rare in Jamaica, and until now have been reported from only one locality, the hard breccias of Wallingford Roadside Cave. Anthony (1920a) first mentioned the presence of crocodiles in the bone breccia in Wallingford Roadside Cave. Koopman and Williams (1951) noted that Anthony obtained a single crocodile vertebra from this cave. Wallingford Roadside Cave is located approximately 25 km inland at about 230 m in elevation near the One Eye River, a tributary of the Black River (MacPhee, 1984). Crocodiles presently occur at the mouth of the Black River. During the last interglacial when sea level was somewhat higher than present, crocodiles apparently inhabited the One Eye River in the vicinity of Balaclava.

I recently have identified three fossil crocodile teeth (UF 22740) from Dairy Cave, which is situated about midway along the northern coast of Jamaica about 3 km east of Discovery Bay. Dairy Cave is about 1 km inland from the coast at an elevation of 15 m. This appears to be the first report of crocodiles from the north coast of Jamaica (Grant, 1940). The crocodile vertebra from Wallingford Roadside Cave and the teeth from Dairy Cave are identifiable only as *Crocodylus* sp. Further identification is not warranted without more diagnostic material, preferably including at least a portion of the skull.

The rarity of crocodile fossils in Jamaica is not surprising considering that most Quaternary vertebrates from the island are found in caves. The American crocodile, *Crocodylus acutus,* inhabits fluvial, estuarine, and nearshore marine habitats along the southern coast of Jamaica. Specimens have been recorded from the Cabarita River and Savanna-la-Mar in Westmorland Parish in western Jamaica to Holland Bay in St. Thomas Parish at the eastern end of the island (Grant, 1940). Both localities where fossil crocodiles have been found in Jamaica are outside the modern range of *C. acutus.*

Class AVES Linnaeus, 1758
Order CICONIIFORMES Bonaparte, 1854
Family PLATELEIDAE Bonaparte, 1838

Genus *XENICIBIS* Olson and Steadman, 1977

Xenicibis xympithecus Olson and Steadman, 1977

Xenicibis xympithecus Olson and Steadman, 1977, p. 449–454, Figs. 1–3.

Xenicibis xympithecus Olson and Steadman. Olson and Steadman, 1979, p. 23–26, Figs. 1–2.

Fossil Localities. Long Mile Cave, Trelawny Parish (Olson and Steadman, 1977); Swansea Cave, St. Catherine Parish (Olson and Steadman, 1979).

Age and Stratigraphic Occurrence. Olson and Steadman (1977) stated that the type series of *Xenicibis xympithecus* from Long Mile Cave had no stratigraphic provenience. Anthony's original field notes and further excavations by MacPhee in 1981 indicated that most of the vertebrate remains from Long Mile Cave, including the ibis and the monkey, *Xenothrix mcgregori*, were removed from the "yellow limestone detritus." This layer has been ^{14}C dated at 2,145 yr B.P. based on a combined sample of hutia and lizard bones (MacPhee, 1984). The fossil humerus of *Xenicibis xympithecus* from Swansea Cave occurs in undated strata containing only indigenous Jamaican vertebrates, presumably indicating a pre-Columbian age (Olson and Steadman, 1979).

Remarks. The extinct flightless ibis, *Xenicibis xympithecus*, is an endemic genus and species known only from Jamaica. *X. xympithecus* was described from Long Mile Cave near Windsor in the Cockpit Country, the same fossil site that produced the extinct ceboid monkey, *Xenothrix mcgregori*. This large ibis has no known close relatives elsewhere in the West Indies (Olson and Steadman, 1977), and thus was probably isolated in Jamaica for a considerable period of time. This species apparently survived into the late Holocene based on the radiocarbon date from Long Mile Cave (MacPhee, 1984). It is possible that *X. xympithecus* was exterminated by Amerindian peoples, although no specimens have yet been recovered from archaeological sites. A number of partial postcranial elements of *X. xympithecus,* including the type proximal tarsometatarsus, are figured by Olson and Steadman (1977, Figs. 1–3), and Olson and Steadman (1979, Figs. 1–2) provided photographs of a complete humerus of this species from Swansea Cave.

The humerus of *X. xympithecus* from Swansea Cave is illustrated in Figure 2C.

Order STRIGIFORMES Wagler, 1830
Family STRIGIDAE Vigors, 1825

Genus *ATHENE* Boie, 1822

Athene cf. *A. cunicularia* (Molina, 1782)

Strix Cunicularia Molina, 1782, p. 263.
Speotyto cf. *cunicularia* (Molina). Olson and Steadman, 1977, p. 454–455.
Athene cunicularia (Molina). Olson and Hilgartner, 1982, p. 37–42, Figs. 4–7.

Fossil Locality. Dairy Cave, St. Ann Parish (Olson and Steadman, 1977).

Age and Stratigraphic Occurrence. No radiocarbon dates are available for Dairy Cave. Olson and Steadman (1977) did not indicate which stratigraphic layer in Dairy Cave produced the fossils of *Athene cunicularia*. Koopman and Williams (1951) noted that bird bones were relatively common in the "lizard layers."

Remarks. The burrowing owl, *Athene cunicularia*, is no longer found in Jamaica, although this small owl still occurs in the West Indies on Hispaniola and certain of the Bahamas. The fossil record documents that *A. cunicularia* has suffered localized extinctions on a number of Caribbean islands besides Jamaica, including Cayman Brac (Morgan, 1977, 1993), Mona Island, Puerto Rico, and Barbuda (Pregill and Olson, 1981; Olson and Hilgartner, 1982). Olson and Hilgartner (1982) pointed out that the burrowing owl is an excellent ecological indicator of xeric habitats such as open grasslands, savannas, and scrub forests. Pregill and Olson (1981) and Olson and Hilgartner (1982) proposed that the wider distribution of *A. cunicularia* in the West Indies in the late Pleistocene was a result of the more arid climate and predominance of open, xeric habitats during the terminal Pleistocene glaciation.

Order CAPRIMULGIFORMES Ridgway, 1881
Family CAPRIMULGIDAE Vigors, 1825

Genus *SIPHONORHIS* Sclater, 1861

Siphonorhis americana (Linnaeus, 1758)

Caprimulgus americanus Linnaeus, 1758, p. 193.
Siphonorhis americana (Linnaeus). Olson and Steadman, 1977, p. 456, Fig. 4.
Siphonorhis americana (Linnaeus). Olson, 1985, p. 528–531, Figs. 2–3.

Fossil Localities. Dairy Cave, St. Ann Parish (Olson and Steadman, 1977); Wallingford Roadside Cave, St. Elizabeth Parish (Olson, 1985).

Age and Stratigraphic Occurrence. Neither Jamaican fossil site containing *Siphonorhis americana* has been precisely dated. The stratigraphic provenience for these fossils is unknown as well. It seems most likely that the *S. americana* humeri from Dairy Cave reported by Olson and Steadman (1977) were derived from the "lizard layers" of Koopman and Williams (1951). The preservation of the single bone of this species from Wallingford Roadside Cave suggests it came from the unconsolidated cave floor sediments rather than the considerably older breccia deposit. The only age that can be confidently assigned to the fossil material of *S. americana* is late Quaternary. The Jamaican pauraque is now extinct in Jamaica, although it survived into the nineteenth century (Olson, 1985).

Remarks. *Siphonorhis americana* is known as a fossil from four specimens collected from two cave fossil sites in Jamaica (Olson and Steadman, 1977; Olson, 1985). The endemic Antillean genus *Siphonorhis* is represented by three species, a fossil species from eastern Cuba, *S. daiquiri*; the recently extinct species from Jamaica, *S. americana*; and a living species from Hispaniola, *S. brewsteri* (Olson, 1985). Although this genus has not been recorded from pre-late Quaternary deposits, its highly distinctive morphology suggests a long period of isolation and independent evolution in the Caribbean (Olson, 1978). The recent extinction of *S. americana* in Jamaica has been attributed to the introduction of the mongoose in the late nineteenth century (Olson, 1985). Olson and Steadman (1977, Fig. 4) illustrated a humerus of *S. americana,* while Olson (1985, Fig. 2) figured a distal tarsometatarsus.

Family NYCTIBIIDAE Bonaparte, 1853

Genus *NYCTIBIUS* Vieillot, 1816

Nyctibius griseus (Gmelin, 1789)

Caprimulgus griseus Gmelin, 1789, p. 1029.
Nyctibius griseus (Gmelin). Olson and Steadman, 1977, p. 455–456.

Fossil Locality. Long Mile Cave, Windsor, Trelawny Parish (Olson and Steadman, 1977).

Age and Stratigraphic Occurrence. The single fossil specimen of *Nyctibius griseus* reported by Olson and Steadman (1977) from Long Mile Cave was probably derived from the same stratum that produced the majority of the vertebrate remains, the "yellow limestone detritus." This layer has been ^{14}C dated at 2,145 yr B.P. (MacPhee, 1984). *N. griseus* still lives in Jamaica.

Remarks. The common potoo, *Nyctibius griseus,* occurs only in Jamaica and Hispaniola in the West Indies, as well in the mainland Neotropics from Mexico south to Argentina.

Class **MAMMALIA** Linnaeus, 1758
Order **CHIROPTERA** Blumenbach, 1779
Family **MORMOOPIDAE** Saussure, 1860

Genus *MORMOOPS*, Leach, 1821

Mormoops blainvillii Leach, 1821

Mormoops Blainvillii Leach, 1821, p. 77.
Aello Cuvieri Leach, 1821, p. 71.
Mormoops blainvillei Leach. Koopman and Williams, 1951, p. 7.
Mormoops blainvillii Leach. Smith, 1972, p. 108–113, Figs. 4–8, 37.
Aello cuvieri Leach. Hall, 1981, p. 97.
Mormoops blainvilli Leach. MacPhee, 1984, p. 9.

Fossil Localities. Dairy Cave, St. Ann Parish (Koopman and Williams, 1951; Williams, 1952); Wallingford Roadside Cave (cave fill deposit), St. Elizabeth Parish (MacPhee, 1984).
Age and Stratigraphic Occurrence. No fossil deposit from Jamaica containing *Mormoops blainvillii* has been dated by absolute methods. This species occurs in both the bat and lizard layers in Dairy Cave, as well as the unconsolidated cave floor sediments in Wallingford Cave (Koopman and Williams, 1951; Williams, 1952). *M. blainvillii* is still a rather common and widespread inhabitant of Jamaican caves.
Remarks. *Mormoops blainvillii* is endemic to the West Indies where it is now restricted to the four Greater Antilles. This species had a considerably wider distribution during the late Quaternary. *M. blainvillii* has been reported from cave fossil deposits on five islands where it no longer occurs, including Barbuda (Morgan and Woods, 1986) and Antigua (Pregill et al., 1988) in the northern Lesser Antilles, La Gonave off the west coast of Haiti (Koopman, 1955), and Little Exuma and New Providence in the Bahamas (Koopman, 1951; Morgan, 1989a). *M. blainvillii* is an obligate cave-dwelling bat that roosts primarily in extensive cave systems having a stable microenvironment characterized by high temperature and humidity (Goodwin, 1970; Silva Taboada, 1979; Morgan, 1989a).

Mormoops megalophylla Peters, 1864

Mormops megalophylla Peters, 1864, p. 381.
Mormoops megalophylla Peters. Smith, 1972, p. 113–117, Figs. 4–8, 37.
Aello megalophylla (Peters). Hall, 1981, p. 97, Fig. 57.
Mormoops megalophylla Peters. Morgan, 1989a, p. 694, Table 3.

Fossil Locality. Swansea Cave, St. Catherine Parish (Morgan, 1989a).
Age and Stratigraphic Occurrence. The single distal humerus of *Mormoops megalophylla* from Swansea Cave occurs in undated strata containing only indigenous Jamaican vertebrates, including the extinct ibis, *Xenicibis xympithecus,* presumably indicating a pre-Columbian age (Morgan, 1989a).
Remarks. *Mormoops megalophylla* is unknown in the modern fauna of the West Indies. The current distribution of this species is restricted to the mainland from southern Arizona and Texas south through Mexico and Central America to northern South America and Trinidad (Smith, 1972). Fossils of *M. megalophylla* have been recorded from late Quaternary cave deposits on four islands in the West Indies: Cuba (Silva Taboada, 1974), Hispaniola, Jamaica, and Andros in the Bahamas (Morgan and Woods, 1986; Morgan, 1989a). The presence of *M. megalophylla* in Jamaica in the late Quaternary, as well as the more diverse mormoopid faunas in Cuba and the Bahamas during this same time period, provides strong evidence that environmental changes contributed to the localized extinction of certain cave-dwelling bats in the West Indies. Other fossil occurrences of *M. megalophylla* from outside the modern range of the species include late Pleistocene records from Tobago (Eshelman and Morgan, 1985) and Florida (Ray et al., 1963; Morgan, 1991).

Genus *PTERONOTUS* Gray, 1838

Pteronotus parnellii (Gray, 1843)

Phyllodia Parnellii Gray, 1843a, p. 50.
Chilonycteris parnelli (Gray). Williams, 1952, p. 172.
Pteronotus parnellii (Gray). Smith, 1972, p. 57–67, Figs. 4–8, 23.
Pteronotus parnellii (Gray). Hall, 1981, p. 90–91, Figs. 52–53.

Fossil Localities. Dairy Cave, St. Ann Parish; Portland Cave, Clarendon Parish (Williams, 1952).
Age and Stratigraphic Occurrence. There are no dates available for the two Jamaican cave deposits from which *Pteronotus parnellii* has been identified. Williams (1952) reported this species from the bat layers in both Dairy Cave and Portland Cave. *P. parnellii* is currently a common and widespread inhabitant of Jamaican caves.
Remarks. *Pteronotus parnellii* now occurs on the four Greater Antilles, as well as on the mainland from Mexico south to northern South America (Smith, 1972). As with the two previous species of Mormoopidac, *P. parnellii* has undergone a significant reduction in its Antillean range during the late Quaternary due to localized extinctions. Extinct Antillean populations of *P. parnellii* are known from La Gonave (Koopman, 1955), Isla de Pinos (Silva Taboada, 1979), Antigua (Pregill et al., 1988), Grand Cayman (Morgan and Woods, 1986; Morgan, 1993), and New Providence, Bahamas (Morgan, 1989a).

Family **PHYLLOSTOMIDAE** Gray, 1825

Genus *MACROTUS* Gray, 1843

Macrotus waterhousii Gray, 1843

Macrotus waterhousii Gray, 1843b, p. 21.
Macrotus waterhousei Gray. Williams, 1952, p. 172.
Macrotus waterhousii Gray. Anderson and Nelson, 1965, p. 13–22, Figs. 3–4.
Macrotus waterhousii Gray. Hall, 1981, p. 104–106, Fig. 66.

Fossil Localities. Dairy Cave, St. Ann Parish; Portland Cave, Clarendon Parish (Williams, 1952).
Age and Stratigraphic Occurrence. There are no dates from Jamaican fossils deposits containing *Macrotus waterhousii*. Williams (1952) reported this species from the bat layers in both Dairy Cave and Portland Cave. *M. waterhousii* is one of the more abundant and widespread bats in Jamaica.
Remarks. *Macrotus waterhousii* is currently widespread in the West Indies where it occurs in Cuba, Jamaica, Hispaniola, Bahamas, and the Cayman Islands. Extinct populations of *M. waterhousii* are known from Puerto Rico and Barbuda, both of which are east of the species' present range (Morgan and Woods, 1986).

Genus *TONATIA* Gray, 1827

Tonatia bidens saurophila Koopman and Williams, 1951

Tonatia saurophila Koopman and Williams, 1951, p. 11–12, Figs. 1–2.
Tonatia bidens saurophila Koopman and Williams. Koopman, 1976, p. 45.

Fossil Localities. Dairy Cave, St. Ann Parish; Wallingford Roadside Cave, St. Elizabeth Parish (Koopman and Williams, 1951).

Age and Stratigraphic Occurrence. The fossils of *Tonatia saurophila* (= *T. bidens saurophila*) from Dairy Cave and Wallingford Cave were derived from the lizard layers (Koopman and Williams, 1951; Williams, 1952). No radiocarbon dates are available for these strata. This species no longer occurs in Jamaica.

Remarks. Koopman and Williams (1951) originally described *Tonatia saurophila* as an endemic species from Jamaica. Subsequent taxonomic study indicates that the Jamaican form is best regarded as subspecies of the widespread Neotropical species, *Tonatia bidens* (Koopman, 1976). *T. bidens* is no longer found in Jamaica, nor has it been reported from fossil sites elsewhere in the West Indies. This species is still extant on the continent from Guatemala to South America. Koopman and Williams (1951) and Williams (1952) noted that *Tonatia* occurred only in the stratigraphically deeper and presumably older lizard layers in Dairy Cave and Wallingford Cave. Its absence from the younger bat layers, as well as the surface and subsurface strata, indicates that this bat probably went extinct in Jamaica in pre-Columbian times, probably in the late Pleistocene or early Holocene. A partial skull and several mandibles of *T. bidens saurophila* from Dairy Cave and Wallingford Roadside Cave are figured by Koopman and Williams (1951, Figs. 1–2). The paratype mandible of *T. bidens saurophila* is illustrated in Figure 2D (herein; from Koopman and Williams, 1951, Fig. 1A).

Genus BRACHYPHYLLA Gray, 1834

Brachyphylla nana pumila Miller, 1918

Brachyphylla nana Miller, 1902, p. 409.
Brachyphylla pumila Miller, 1918, p. 39.
Brachyphylla pumila Miller. Koopman and Williams, 1951, p. 12–14, Figs. 3–4.
Brachyphylla cavernarum pumila Miller. Varona, 1974, p. 27.
Brachyphylla nana Miller. Swanepoel and Genoways, 1978, p. 49–51.
Brachyphylla nana Miller. Silva Taboada, 1979, p. 139–147, lam. 2J, 3F, 4F, 5F, 6B, 7B.
Brachyphylla cavernarum pumila Miller. Hall, 1981, p. 170.

Fossil Localities. Dairy Cave, St. Ann Parish (Koopman and Williams, 1951; Williams, 1952; Swanepoel and Genoways, 1978); Portland Cave, Clarendon Parish (Williams, 1952).

Age and Stratigraphic Occurrence. *Brachyphylla nana* occurs in the lizard layers in Dairy Cave and in the bat layers in Portland Cave (Koopman and Williams, 1951; Williams, 1952). There are no radiocarbon dates associated with *B. nana*, although the species seems to have gone extinct in Jamaica in pre-Columbian times.

Remarks. Koopman and Williams (1951) referred a small sample of Jamaican bat fossils to *Brachyphylla pumila*, a species originally described from Hispaniola. In a review of the endemic Antillean bat genus *Brachyphylla*, Swanepoel and Genoways (1978) synonymized *B. pumila* with *B. nana*. However, more recent study indicates that these two taxa may indeed be distinct species (Morgan, 1989a). Until more detailed taxonomic analyses can be undertaken, *B. nana* and *B. pumila* are tentatively regarded as separate subspecies of *B. nana*. This species still survives in Cuba, Hispaniola, Grand Cayman, and Grand Caicos. *B. nana* is one of three species of bats recorded from fossil deposits in Jamaica that is now extinct on the island. Other locally extinct populations of *B. nana* have been reported from Cayman Brac and New Providence and Andros in the Bahamas (Morgan, 1989a). Fossil skulls and mandibles of *B. nana* from Jamaica are illustrated by Koopman and Williams (1951, Figs. 3–4).

Genus EROPHYLLA Miller, 1906

Erophylla sezekorni (Gundlach, 1861 *in* Peters)

Phyllonycteris sezekorni Gundlach, 1861, p. 818.
Erophylla sp. Williams, 1952, p. 172, tab. 1.
Phyllonycteris sezekorni Gundlach. Varona, 1974, p. 29–30.
Erophylla sezekorni (Gundlach). Hall, 1981, p. 171, Fig. 137.

Fossil Localities. Dairy Cave, St. Ann Parish; Portland Cave, Clarendon Parish (Williams, 1952).

Age and Stratigraphic Occurrence. This species is found in the bat layers in Dairy Cave and Portland Cave (Williams, 1952).

Remarks. *Erophylla sezekorni* is an Antillean endemic that occurs in Jamaica, Cuba, the Bahamas, and the Cayman Islands.

Genus PHYLLONYCTERIS Gundlach, 1861

Phyllonycteris aphylla (Miller, 1898)

Reithronycteris aphylla Miller, 1898, p. 334.
Reithronycteris aphylla Miller. Koopman and Williams, 1951, p. 14–17, Fig. 5.
Phyllonycteris (Reithronycteris) aphylla (Miller). Koopman, 1952, p. 255–257.
Phyllonycteris aphylla (Miller). Hall, 1981, p. 173, Fig. 139.

Fossil Localities. Dairy Cave, St. Ann Parish (Koopman and Williams, 1951); Portland Cave, Clarendon Parish (Williams, 1952); Wallingford Roadside Cave, St. Elizabeth Parish (Koopman and Williams, 1951).

Age and Stratigraphic Occurrence. Fossils of *Phyllonycteris aphylla* have been recovered from the lizard layers of Dairy Cave and Wallingford Cave and the bat layers of Portland Cave (Koopman and Williams, 1951; Williams, 1952). None of these specimens are associated with radiocarbon dates.

Remarks. *Phyllonycteris aphylla* is endemic to Jamaica. This species has in the past been referred to the monotypic genus *Reithronycteris* (see discussion in Koopman, 1952). *Reithronycteris* is now generally considered a subgenus of the Antillean genus *Phyllonycteris*. Koopman and Williams (1951) and Williams (1952) suggested that *P. aphylla* was extremely rare if not extinct in Jamaica. However, mammalogists have since captured live individuals of *P. aphylla* in several regions of Jamaica where this bat prefers hot, humid chambers deep within extensive cave systems (Goodwin, 1970). Koopman and Williams (1951, Fig. 5) illustrated a partial skull of *P. aphylla* from Dairy Cave and a mandible from Wallingford Roadside Cave.

Genus MONOPHYLLUS Leach, 1821

Monophyllus redmani Leach, 1821

Monophyllus redmani Leach, 1821, p. 76.
Monophyllus sp. Williams, 1952, p. 172.
Monophyllus redmani Leach. Schwartz and Jones, 1967, p. 4–6.
Monophyllus redmani Leach. Hall, 1981, p. 125–126, Fig. 89.

Fossil Locality. Portland Cave, Clarendon Parish (Williams, 1952).

Age and Stratigraphic Occurrence. Williams (1952) recorded *Monophyllus redmani* from the bat layers in Portland Cave.

Remarks. *Monophyllus redmani* is endemic to the Greater Antilles, Bahamas, and Caicos Islands. This bat is common and widespread in Jamaica where it generally roosts in large, deep caves with high temperature and humidity (Goodwin, 1970).

Genus *ARITEUS* Gray, 1838

Ariteus flavescens (Gray, 1831)

Istiophorus flavescens Gray, 1831, p. 37.
Ariteus flavescens (Gray). Koopman and Williams, 1951, p. 7.
Ariteus sp. Williams, 1952, p. 177.
Ariteus flavescens (Gray). MacPhee, 1984, p. 9.

Fossil Localities. Dairy Cave, St. Ann Parish (Koopman and Williams, 1951; Williams, 1952); Wallingford Roadside Cave, St. Elizabeth Parish (MacPhee, 1984).
Age and Stratigraphic Occurrence. Ariteus flavescens occurs in both the lizard and bat layers in Dairy Cave (Koopman and Williams, 1951; Williams, 1952) and in the unconsolidated cave fill sediments in Wallingford Roadside Cave (MacPhee, 1984).
Remarks. Ariteus flavescens is an endemic genus and species restricted to Jamaica. Koopman and Williams (1951) considered *A. flavescens* to be rare and local in Jamaica based on the fact that in 1919 and 1920 Anthony recovered large samples of skulls and mandibles of this species in surficial cave deposits, but failed to obtain live specimens. With the advent of mist netting for collecting bats, many tree-roosting species such as *A. flavescens* have proven to be more common than was previously thought. *A. flavescens* is now known to occur more widely throughout Jamaica.

Family NATALIDAE Miller, 1899

Genus *NATALUS* Gray, 1838

Natalus major Miller, 1902

Natalus major Miller, 1902, p. 389.
Natalus major Miller. Koopman and Williams, 1951, p. 17–18, Fig. 6.
Natalus major jamaicensis Goodwin, 1959, p. 9–10.
Natalus stramineus jamaicensis Goodwin. Varona, 1974, p. 32.
Natalus stramineus Gray. Silva Taboada, 1979, p. 225–227, lam. 8.
Natalus major Miller. Morgan, 1989a, p. 701–703, Table 6.

Fossil Locality. Wallingford Roadside Cave, St. Elizabeth Parish (Koopman and Williams, 1951).
Age and Stratigraphic Occurrence. A single mandible of *Natalus major* was recovered from the lizard layers in Wallingford Roadside Cave (Koopman and Williams, 1951).
Remarks. Varona (1974) and many subsequent authors have regarded the endemic Antillean species *Natalus major* as a subspecies of the mainland and Lesser Antillean species, *N. stramineus*. However, Morgan (1989a) discussed a number of morphological characters that clearly distinguish these two species. When first reported as a fossil from Jamaica (Koopman and Williams, 1951), *N. major* was thought to be extinct on the island. Shortly thereafter, Goodwin (1959) described an extant subspecies, *N. major jamaicensis*, based on a specimen from St. Clair Cave in St. Catherine Parish. Extinct populations of *N. major* are known from Cuba, Isla de Pinos, Andros and New Providence in the Bahamas, Grand Caicos, and Grand Cayman (Morgan, 1989a; 1993). Like *Phyllonycteris aphylla*, which was also thought to be extinct in Jamaica, *N. major* inhabits chambers deep within extensive cave systems (Goodwin, 1970). Koopman and Williams (1951, Fig. 6) illustrated the single fossil mandible of *N. major* from Wallingford Roadside Cave.

Family VESPERTILIONIDAE Gray, 1821

Genus *EPTESICUS* Rafinesque, 1820

Eptesicus lynni Shamel, 1945

Eptesicus lynni Shamel, 1945, p. 107.
Eptesicus sp. Koopman and Williams, 1951, p. 18–19.
Eptesicus lynni Shamel. Baker and Genoways, 1978, p. 64.
Eptesicus lynni Shamel. Hall, 1981, p. 217.

Fossil Localities. Cambridge Cave, St. James Parish; Wallingford Roadside Cave, St. Elizabeth Parish (Koopman and Williams, 1951).
Age and Stratigraphic Occurrence. Eptesicus sp. was reported from the lizard layers in Wallingford Cave and the "loose breccia" in Cambridge Cave, neither of which have been precisely dated (Koopman and Williams, 1951).
Remarks. Koopman and Williams (1951) did not refer their two fragmentary fossil mandibles of *Eptesicus* from Jamaica to a species because at that time both *E. fuscus* and *E. lynni* had been reported from the island. They noted that their fossils seemed to be closer to *E. lynni* than to the larger *E. fuscus*. The small endemic Jamaican species *E. lynni* is now generally regarded as the only species of *Eptesicus* extant on Jamaica (Baker and Genoways, 1978).

Order PRIMATES Linnaeus, 1758
Superfamily CEBOIDEA Simpson, 1931
Family XENOTRICHIDAE Hershkovitz, 1970

Genus *XENOTHRIX* Williams and Koopman, 1952

Xenothrix mcgregori Williams and Koopman, 1952

Xenothrix mcgregori Williams and Koopman, 1952, p. 12–14, Figs. 1–4.
Xenothrix mcgregori Williams and Koopman. Hershkovitz, 1970, p. 2–3, Plates 2, 3, 12.
Xenothrix mcgregori Williams and Koopman. Rosenberger, 1977, p. 470–473, Plates 1–2.
Xenothrix mcgregori Williams and Koopman. Ford, 1990b, p. 242–244, Fig. 6.
Xenothrix mcgregori Williams and Koopman. Rosenberger et al., 1990, p. 213–216, Fig. 1.
Xenothrix mcgregori. Williams and Koopman. MacPhee and Fleagle, 1991, p. 291–308, Figs. 2–10.

Fossil Locality. Long Mile Cave, Trelawny Parish (Williams and Koopman, 1952; MacPhee, 1984; Ford, 1990b; MacPhee and Fleagle, 1991).
Age and Stratigraphic Occurrence. Geocapromys and lizard bones from the same layer in Long Mile Cave that produced the type mandible of *Xenothrix mcgregori* have been radiocarbon dated at 2145 ± 220 yr B.P. (MacPhee, 1984).
Remarks. The type specimen of the endemic Jamaican monkey *Xenothrix mcgregori*, a left mandible with two teeth, is one of the most frequently discussed and figured vertebrate fossils from the West Indies (Williams and Koopman, 1952; Hershkovitz, 1970; Rosenberger, 1977; MacPhee and Woods, 1982; Ford, 1990b; Rosenberger et al., 1990). MacPhee and Fleagle (1991) also attributed a femur, innominate, and two proximal tibiae from Long Mile Cave to *X. mcgregori.* No two authors seem to completely agree on the phylogenetic position of *Xenothrix* within the New World platyrrhine monkeys. Hypotheses range from a possible relationship with the marmoset family,

Callitrichidae (Ford, 1986), to placement in the family Atelidae (MacPhee and Woods, 1982), to recognition of a separate monotypic family for this genus, the Xenotrichidae (Hershkovitz, 1970; MacPhee and Fleagle, 1991). The latter interpretation is followed here. There is little argument that *Xenothrix* is a very distinct endemic genus of Antillean primate that was isolated in Jamaica for a considerable period of time. It seems reasonably certain that *Xenothrix* was ultimately derived from some ancestral South American primate, probably during the Miocene or Pliocene (Simpson, 1956; Morgan and Woods, 1986). The holotype mandible of *Xenothrix mcgregori* is shown in Figure 2E (from Rosenberger et al., 1990, Fig. 1).

Superfamily CEBOIDEA Simpson, 1931

genus and species indeterminate A

Ceboidea, gen. et sp. indet. Ford and Morgan, 1986, p. 281–287, Fig. 2.
Cebidae, gen. et sp. indet. Hershkovitz, 1988, p. 366–377, Figs. 1, 3, 4, 6.
Ceboidea, gen. et sp. indet. Ford, 1990a, p. 164, Fig. 9.
Ceboidea, gen. et sp. indet. Ford, 1990b, p. 245–247, Fig. 9.

Fossil Locality. Coco Ree Cave, St. Catherine Parish (Ford and Morgan, 1986; Ford, 1990a).

Age and Stratigraphic Occurrence. The proximal femur of a monkey was recovered from a layer 80 to 90 cm below the surface of Coco Ree Cave. Land snails from this layer have been ^{14}C dated at 36,300 ± 4,300 yr B.P. (Goodfriend and Mitterer, 1987).

Remarks. The proximal femur of a primate from Coco Ree Cave has been discussed and/or figured by numerous authors (Ford and Morgan, 1986; Goodfriend and Mitterer, 1987; Hershkovitz, 1988; Ford, 1990a, b; MacPhee and Fleagle, 1991). This specimen is clearly an indigenous Jamaican monkey as it predates any human occupation of the island. It was originally suspected that this femur might pertain to *Xenothrix*, a primate of similar size described from Jamaica some 30 years previously. However, a primate femur recently identified from Long Mile Cave, presumably belonging to *Xenothrix*, is quite distinct from the femur of the Coco Ree monkey (Ford, 1990a; MacPhee and Fleagle, 1991). A third type of primate femur described below provides strong evidence that there were actually three distinct lineages of primates in Jamaica during the late Pleistocene (Ford, 1990a, b). Ford and Morgan (1986) suggested that despite its much larger body size the Coco Ree femur had derived affinities with the Callitrichidae, whereas Hershkovitz (1988) thought this specimen more closely resembled the cebid, *Cebus*.

genus and species indeterminate B

Ceboidea, gen. et sp. indet. Ford, 1990a, p. 163–164, Fig. 8.
Ceboidea, gen. et sp. indet. Ford, 1990b, p. 245, Fig. 8.

Fossil Locality. Sheep Pen Cave, Trelawny Parish (Ford and Morgan, 1988; Ford, 1990a, b).

Age and Stratigraphic Occurrence. The proximal femur of a primate was recovered from highly indurated sediments plastered to an overhang known as Sheep Pen Cave (MacPhee, 1984; Goodfriend, 1986, 1989). Sheep Pen Cave is beyond the limits of radiocarbon dating, thus indicating an age greater than 40 ka. Bones and enamel from Sheep Pen lacked detectable levels of amino acids suggesting a long period of leaching, and further indicating a considerable age for this deposit (Goodfriend, 1986, 1989). Goodfriend thought this deposit probably represented more than one glacial cycle and was older than 100 ka. The presence of the extinct rodent genus *Clidomys* from Sheep Pen Cave tends to substantiate the hypothesis of a pre-late Pleistocene age.

Remarks. The proximal primate femur from Sheep Pen Cave has only recently been described and figured (Ford and Morgan, 1988; Ford, 1990a, b; MacPhee and Fleagle, 1991). The limited sample of Jamaican primate fossils includes three partial femora: one each from Coco Ree Cave, Sheep Pen Cave, and Long Mile Cave. Although comparative study of these femora has not yet been completed, they clearly represent three distinct taxa, most likely three different genera. Jamaica has the most diverse primate fauna of any Antillean island with three genera and three species. Cuba has two genera and two species of primates and Hispaniola has one genus and species (Ford, 1990a). The proximal monkey femur from Sheep Pen Cave is the oldest primate presently known from the West Indies, and along with material from Wallingford Roadside Cave, is one of the most ancient terrestrial vertebrate fossils known from Jamaica (Goodfriend, 1989; MacPhee et al., 1989).

Order RODENTIA Bowdich, 1821
Family CAPROMYIDAE Smith, 1842

Genus *GEOCAPROMYS* Chapman, 1901

Geocapromys brownii (Fischer, 1830)

Capromys brownii Fischer, 1830, p. 389.
Capromys (Geocapromys) brownii Fischer. Chapman, 1901, p. 320–321, Plate 40.
Capromys (Geocapromys) brownii Fischer. Varona, 1974, p. 67.
Geocapromys brownii (Fischer). Anderson et al., 1983, p. 1–3, Figs. 2–7.
Geocapromys brownii (Fischer). Morgan, 1985, p. 31–41, Figs. 1–3.
Geocapromys brownii (Fischer). Savage, 1990, p. 33–34, Fig. 1.

Fossil Localities. St. Andrew Parish: Red Hills Road Cave (Savage, 1990); *St. Ann Parish:* Dairy Cave (Koopman and Williams, 1951; Anderson et al., 1983); Green Grotto Cave (Anderson et al., 1983); Mosely Hall Cave (Anderson et al., 1983); *St. Catherine Parish:* Coco Ree Cave (Anderson et al., 1983; Ford and Morgan, 1986); Lluidas Vale Cave (UF); Mt. Diablo Cave (UF); Swansea Cave (Anderson et al., 1983); Tydixon Cave (UF); *Clarendon Parish:* Portland Ridge Caves (Williams, 1952; UF); *Trelawny Parish:* Long Mile Cave (Williams, 1952; MacPhee, 1984); Marta Tick Cave (also known as Quickstep Cave; Anderson et al., 1983; Pregill et al., 1991); Sheep Pen Cave (MacPhee, 1984); *St. James Parish:* Cambridge Cave (Koopman and Williams, 1951); Montego Bay Airport Cave (UF); *St. Elizabeth Parish:* Wallingford Roadside Cave (both conglomerate and cave fill deposits; Koopman and Williams, 1951; Anderson et al., 1983; MacPhee, 1984); *Hanover Parish:* Knockalva Cave (Anderson et al., 1983); Seven Rivers Cave (Koopman and Williams, 1951).

Age and Stratigraphic Occurrence. Geocapromys brownii occurs in the two oldest dated Jamaican fossils deposits, the breccias in Wallingford Roadside Cave and Sheep Pen Cave (MacPhee, 1984). The conglomerate in Wallingford Roadside Cave has been dated at between 100 and 200 ka (MacPhee et al., 1989), while Sheep Pen Cave supposedly predates the last glacial interval and thus is also older than 100 ka (Goodfriend, 1986). *G. brownii* also has been identified from a third undated breccia deposit, Lluidas Vale Cave in St. Catherine Parish. Unconsolidated cave sediments containing *G. brownii* from Coco Ree Cave have been radiocarbon dated at between 13,400 and 36,300 yr B.P. (Goodfriend and Mitterer, 1987). The remaining cave deposits from which the Jamaican hutia has been reported have not been precisely dated. Unfossilized bones of *G. brownii* are commonly encountered in surficial deposits in association with *Rattus*, oftentimes in regions of Jamaica where this species is no longer known to occur. Although uncommon, this species still survives in several widely scattered localities in Jamaica (Oliver, 1982).

Remarks. The Jamaican hutia, *Geocapromys brownii*, is the most common and widespread vertebrate in Jamaican Quaternary fossil deposits. *G. brownii* is an endemic Jamaican species of the Antillean rodent family Capromyidae. *Geocapromys* has been regarded as a subgenus of *Capromys* by some authors (e.g. Varona, 1974), but Morgan (1985) argued that several important morphological characters, including the much shorter tail, reduced first digit on the front foot, and the inclination of the occlusal surface of the cheekteeth, were sufficient to separate *Geocapromys* as a distinct genus. The only other extant species in this genus, *G. ingrahami*, inhabits a tiny island in the southeastern Bahamas. A third species, *G. thoracatus*, survived until the 1950s on Little Swan Island located several hundred kilometers west of Jamaica (Morgan, 1985, 1989b). Miller (1916) reported *G. thoracatus* from an archaeological site near Salt River in Clarendon Parish, but Morgan (1989b) has shown that these specimens are actually juveniles of *G. brownii*. Morgan (1985) proposed that *G. brownii* is the most highly derived species in the genus indicating a long period of isolation in Jamaica. This hypothesis is strengthened by data indicating that *G. brownii* has inhabited Jamaica for at least the past 100 ka, but appears to have undergone very little evolutionary change during that time. *G. brownii* is the only member of the Capromyidae known from Jamaica, whereas substantial adaptive radiations of capromyids, including ten or more species, occurred on both Cuba and Hispaniola (Morgan and Woods, 1986). The tiny Cayman Islands located between Jamaica and Cuba supported a fauna of three capromyids during the late Quaternary (Morgan, 1993).

Specimens from paleontological and archaeological sites establish that *Geocapromys brownii* was widespread in Jamaica until quite recently. The 18 fossil localities listed above do not include nearly all of the Jamaican cave deposits from which this species is known, only those sites that have been published. The Jamaican hutia is also common in Amerindian kitchen midden deposits suggesting that this large rodent was an important food item for aboriginal peoples (Wing, 1972, 1989). The White Marl archaeological site located between Kingston and Spanish Town in St. Catherine Parish contained more than 300 individuals of *G. brownii* (Wing, 1972). Oliver (1982) provided a map of Jamaica showing the widespread occurrence of *G. brownii* in Amerindian midden deposits, including records from St. Thomas Parish and St. Mary Parish. Fossils of *G. brownii* are unknown from these two parishes. The range of the Jamaican hutia has been greatly restricted over the past several hundred years due to hunting pressure and habitat destruction. *G. brownii* is currently found in the vicinity of Worthy Park and the Hellshire Hills in St. Catherine Parish and the John Crow Mountains in Portland Parish (Oliver, 1982; Wilkins, personal communication, 1993). Photographs of the skull and mandible and illustrations of the upper and lower dentition of a modern specimen of *G. brownii* are presented in Figures 3A–E.

Family HEPTAXODONTIDAE Anthony, 1917

Genus CLIDOMYS Anthony, 1920

Clidomys osborni **Anthony, 1920**

Clidomys osborni Anthony, 1920b, p. 469–472, text Figs. 1–3, Plate 33 (Figs. 1–5).
Speoxenus cundalli Anthony, 1920b, p. 473–474, text Fig. 3, Plate 33 (Fig. 12).
Alterodon major Anthony, 1920b, p. 474–475, text Fig. 4, Plate 33 (Fig. 11).
Alterodon major Anthony. MacPhee et al., 1983, p. 831–834, Figs. 1–2.
Clidomys osborni Anthony. MacPhee, 1984, p. 19–30, Figs. 6, 8–12, 13.

Fossil Localities. Wallingford Roadside Cave (bone breccia/conglomerate), St. Elizabeth Parish (Anthony, 1920b; Koopman and Williams, 1951; MacPhee et al., 1983; MacPhee, 1984; MacPhee et al., 1989); Lluidas Vale Cave, St. Catherine Parish (MacPhee, 1984); Sheep Pen Cave, Trelawny Parish (MacPhee, 1984); Molton Fissure, Manchester Parish (MacPhee, 1984).
Age and Stratigraphic Occurrence. The large extinct heptaxodontid rodent, *Clidomys osborni*, is one of only two species of Jamaican vertebrates that appears to have gone extinct prior to the end of the Pleistocene (MacPhee, 1984; Morgan and Woods, 1986; MacPhee et al., 1989). One of the four localities from which *C. osborni* is known, Wallingford Roadside Cave, has been dated. MacPhee et al. (1989) used both uranium series dates and electron spin resonance to come up with an age of 100 to 200 ka for the calcareous conglomerate ("hard breccias" of Koopman and Williams, 1951) in Wallingford Roadside Cave containing bones of *Clidomys*. Goodfriend was not able to obtain reliable amino acid racemization dates from a second locality, Sheep Pen Cave, but suggested an age of greater than 100 ka for this deposit based on geological and climatological considerations. Lluidas Vale Cave and Molton Fissure both consist of highly indurated breccia deposits similar to those from Wallingford and Sheep Pen. The absence of *C. osborni* from the large number of unconsolidated cave fill deposits of late Quaternary age found throughout Jamaica strongly suggests that this large rodent went extinct before the end of the Pleistocene.
Remarks. *Clidomys osborni* is a very large extinct genus and species of caviomorph rodent endemic to Jamaica. *Clidomys* and several related genera belong to the extinct Antillean rodent family Heptaxodontidae (Simpson, 1945; MacPhee, 1984). *C. osborni* was originally described based on a mandible and isolated teeth from Wallingford Roadside Cave near Balaclava (Anthony, 1920b). Anthony (1920b) recognized that *Clidomys* was related to several other genera of large Antillean rodents, including *Elasmodontomys* from Puerto Rico and *Amblyrhiza* from Anguilla and St. Martin in the Lesser Antilles. Two other Antillean genera have been referred to the Heptaxodontidae, *Heptaxodon* (now considered a junior synonym of *Elasmodontomys*, after Ray, 1964) from Puerto Rico (Anthony, 1926), and *Quemisia* from Hispaniola (Ray, 1965).

Anthony (1920b) described four genera and five species of large caviomorph rodents from Wallingford Roadside Cave: *Clidomys osborni*, *C. parvus*, *Spirodontomys jamaicensis*, *Speoxenus cundalli*, and *Alterodon major*. Except for *C. osborni*, all of these taxa were based on very small samples of isolated teeth. MacPhee et al. (1983) first suggested that *Alterodon* might be an aberrant specimen of *Clidomys*, rather than a member of the otherwise strictly South American family Octodontidae as proposed by other authors (Anthony, 1926; Simpson, 1945). In a revision of the Jamaican heptaxodontids, MacPhee (1984) documented the range of variation in the dental pattern of these large rodents. Based on this analysis he proposed that all isolated teeth previously referred to *Alterodon*, *Speoxenus*, and *Spirodontomys* by Anthony (1920b) could be placed in two species of the genus *Clidomys*. The larger and more common of these two species, *C. osborni*, includes *Speoxenus cundalli* and probably *Alterodon major* as synonyms, whereas the smaller *C. parvus* (see next account) includes *Spirodontomys jamaicensis* as a synonym (MacPhee, 1984).

Both species of *Clidomys* are known from the type locality, Wallingford Roadside Cave, whereas *C. osborni* occurs in three other sites: Molton Fissure, Sheep Pen Cave in the Cockpit Country, and Lluidas Vale Cave in the Worthy Park area. Although fossils of *Clidomys* are known from only four localities, these sites are located in four different parishes in the western and central part of the island suggesting that this rodent was widespread in Jamaica. The rarity of *Clidomys* probably is related to its occurrence only in pre–late Pleistocene indurated breccia deposits that are much less common than unconsolidated cave fill deposits of latest Pleistocene and Holocene age. The holotype mandible of *C. osborni* is illustrated in Figures 3F and 3G (from Anthony, 1920b, Fig. 1A, 1B).

Clidomys parvus **Anthony, 1920**

Clidomys parvus Anthony, 1920b, p. 472–473, text Fig. 2, Plate 33 (Figs. 6, 7).
Spirodontomys jamaicensis Anthony, 1920b, p. 473, text Fig. 3, Plate 33 (Figs. 8–10).
Clidomys parvus Anthony. MacPhee, 1984, p. 19–30, Figs. 7–9, 11, 14.

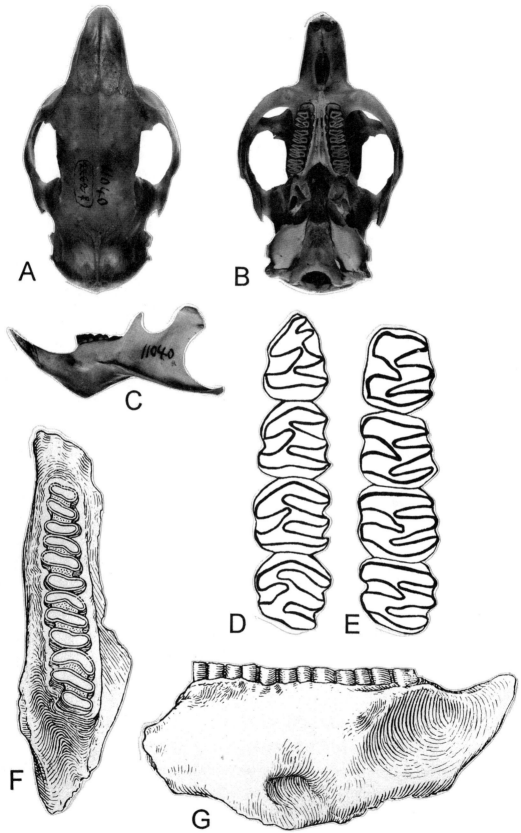

Figure 3. A, dorsal and B, ventral view of skull; C, lateral view of left mandible; D, occlusal view of lower dentition; E, occlusal view of upper dentition of modern *Geocapromys brownii* from Jamaica (MCZ 11040). A to C are natural size, D and E are 4× natural size. F, occlusal view and G, lateral view of left mandible of *Clidomys osborni* (AMNH 17364-holotype) from Wallingford Roadside Cave, St. Elizabeth Parish, Jamaica, 1.2× natural size (from Anthony 1920b, Figs. 1A, 1B).

Fossil Locality. Wallingford Roadside Cave, St. Elizabeth Parish (Anthony, 1920b; Koopman and Williams, 1951; MacPhee et al., 1983; MacPhee, 1984; MacPhee et al., 1989).

Age and Stratigraphic Occurrence. Clidomys parvus, along with the more common *C. osborni,* seems to have gone extinct prior to the end of the Pleistocene (MacPhee, 1984; Morgan and Woods, 1986; MacPhee et al., 1989). MacPhee et al. (1989) dated the calcareous conglomerate containing the *Clidomys* fossils in Wallingford Roadside Cave at between 100 to 200 ka.

Remarks. Clidomys parvus is the smaller and rarer of the two species of *Clidomys* recognized by MacPhee (1984). As noted in the more detailed discussion above (under *C. osborni*), *C. parvus* includes *Spirodontomys jamaicensis* Anthony (1920b) as a junior synonym. MacPhee (1984) investigated the possibility that there was actually only one species of Jamaican heptaxodontid, namely *C. osborni,* and that *C. parvus* represented small female individuals of *C. osborni.* For several reasons he rejected this hypothesis and recognized *C. parvus* as a valid species that was approximately 25% smaller than *C. osborni. C. parvus* is known only from the type locality (MacPhee, 1984).

Family MURIDAE Gray, 1821
Subfamily SIGMODONTINAE Thomas, 1897

Genus ORYZOMYS Baird, 1858

Oryzomys antillarum **Thomas, 1898**

Oryzomys antillarum Thomas, 1898, p. 177–178.
Oryzomys palustris antillarum Thomas. Hershkovitz, 1966, p. 736.
Oryzomys palustris antillarum Thomas. Hall, 1981, p. 609.
Oryzomys couesi antillarum Thomas. Honacki et al., 1982, p. 438–439.

Fossil Localities. St. Catherine Parish: Braham Cave (Humphrey et al., 1993); Coco Ree Cave (Ford and Morgan, 1986; Humphrey et al., 1993); Mt. Diablo Cave (site is actually on border between St. Catherine Parish and St. Ann Parish; Humphrey et al., 1993); Swansea Cave (Humphrey et al., 1993); Tydixon Cave (Humphrey et al., 1993); *Trelawny Parish:* Long Mile Cave (MacPhee, 1984; Humphrey et al., 1993); Bonafide Cave (Goodfriend, 1989; Humphrey et al., 1993); Marta Tick Cave (Pregill et al., 1991); *St. Elizabeth Parish:* Wallingford Roadside Cave (cave fill deposit only; MacPhee, 1984; MacPhee et al., 1989).

Age and Stratigraphic Occurrence. Oryzomys antillarum has not been identified from any of the older breccia and conglomeratic deposits in Jamaica. *O. antillarum* occurs in seven undated fossil cave deposits in Jamaica. These sites contain no evidence of humans or introduced mammals indicating that they are older than 1,400 yr B.P., the oldest date for the appearance of Amerindians in Jamaica (Rouse, 1989). This species probably arrived in Jamaica in the late Pleistocene (<100 ka). *O. antillarum* is common in a number of surface and subsurface deposits in Jamaican caves, the so-called *Rattus* and *Oryzomys* strata of Koopman and Williams (1951). In most of these deposits *O. antillarum* occurs in association with *Rattus* and *Mus* indicating a post-Columbian age (<500 yr B.P.). All radiocarbon-dated sites in Jamaica containing *O. antillarum* are late Holocene in age. Pregill et al. (1991) obtained a date of 770 yr B.P. on charcoal derived from strata in Marta Tick Cave containing *O. antillarum.* An accelerator radiocarbon date of 270 ± 67 yr B.P. was obtained from teeth and bones of *Oryzomys* collected 20 to 30 cm below the surface from the unconsolidated cave fill sediments in Wallingford Roadside Cave (MacPhee et al., 1989). *O. antillarum* has been reported from two radiocarbon-dated archaeological sites in Jamaica: the Bellevue Site outside of Kingston in St. Andrew Parish dated at 1,000–1,100 yr B.P. and the White Marl Site in St. Catherine Parish dated at 500 to 1,000 yr B.P. (Wing and Reitz, 1982; Wing, 1989). *O. antillarum* is almost certainly extinct in Jamaica, as living specimens were last captured in 1877 (Humphrey et al., 1993).

Remarks. The recently extinct Jamaican rice rat, *Oryzomys antillarum,* originally was described as an endemic Jamaican species based on three skins and skulls collected between 1845 and 1877 (Thomas, 1898). Recent authors have synonymized *O. antillarum* with either *O. couesi* from Middle America (Honacki et al., 1982; Corbet and Hill, 1986) or *O. palustris* from the southeastern United States (Hershkovitz, 1966; Hall, 1981). The latter two authors considered *O. couesi* to be a synonym of *O. palustris* as well. However, Humphrey et al. (1993) demonstrated that *O. antillarum* is a distinct species based on an analysis of samples from the late Pleistocene and Holocene fossil deposits listed above. They also concluded that *O. antillarum* was derived from Central America where the closely related species *O. couesi* occurs. Jamaica is the only island in the Greater Antilles known to have been inhabited by an indigenous species of oryzomyine rodent. An impressive radiation of endemic oryzomyines, including several species of large size, is known from the Lesser Antilles (Morgan and Woods, 1986).

The fossil sites listed above for *Oryzomys antillarum* include only those localities that are pre-Columbian in age. Humphrey et al. (1993) discussed additional samples of *O. antillarum* from post-Columbian surficial deposits in eight other cave sites in Jamaica. These bone deposits are primarily decomposed owl pellets consisting mainly of bones of the introduced Old World murid rodents, *Rattus rattus* and *Mus musculus. O. antillarum* has been reported from three Amerindian archaeological sites in Jamaica (Wing, 1972, 1989; Wing and Reitz, 1982; Humphrey et al., 1993). The Jamaican rice rat disappeared shortly after the Indian mongoose, *Herpestes auropunctatus,* was introduced into Jamaica in 1872 to control rats. Competition with *Rattus* and predation by feral dogs and cats also may have contributed to the extinction of *O. antillarum* (Morgan and Woods, 1986).

ACKNOWLEDGMENTS

I am grateful to Raymond M. Wright for inviting me to participate in the Biostratigraphy of Jamaica project. Charles A. Woods and Robert J. G. Savage provided helpful comments on the manuscript. Linda Chandler drew Figure 2C. The editors of the Bulletin of the American Museum of Natural History, the American Museum Novitates, and the Journal of Human Evolution kindly permitted me to reproduce figures originally published in these journals. This is University of Florida Contribution to Paleontology Number 422.

REFERENCES CITED

Anderson, S., and Nelson, C. E., 1965, A systematic revision of *Macrotus* (Chiroptera): American Museum Novitates, no. 2212, p. 1–39.

Anderson, S., Woods, C. A., Morgan, G. S., and Oliver, W.L.R., 1983, *Geocapromys brownii:* Mammalian Species, no. 201, p. 1–5.

Anthony, H. E., 1920a, A zoologist in Jamaica: Natural History, v. 20, p. 156–168.

Anthony, H. E., 1920b, New mammals from Jamaica: American Museum of Natural History Bulletin, v. 42, p. 469–475.

Anthony, H. E., 1926, Mammals of Porto Rico, living and extinct–Rodentia and Edentata: New York Academy of Sciences, Scientific Survey of Porto Rico and the Virgin Islands, v. 9, no. 2, p. 97–238.

Baker, L. L., Devine, E. A., and DiTonto, M. A., 1986, Jamaica: The 1985 expedition of the NSS Jamaica Cockpits project: National Speleological

Society News, v. 44, p. 4–15.

Baker, R. J., and Genoways, H. H., 1978, Zoogeography of Antillean bats: *in* Gill, F. B., ed., Zoogeography in the Caribbean: Academy of Natural Sciences of Philadelphia, Special Publication, no. 13, p. 53–97.

Barbour, T., and Carr, A. F., Jr., 1940, Antillean terrapins: Memoirs of the Museum of Comparative Zoology, no. 54(5), p. 381–415.

Buskirk, R. E., 1985, Zoogeographic patterns and tectonic history of Jamaica and the northern Caribbean: Journal of Biogeography, v. 12, p. 445–461.

Chapman, F. M., 1991, A revision of the genus *Capromys:* American Museum of Natural History Bulletin, v. 14, p. 313–323.

Corbet, G. B., and Hill, J. E., 1986. A world list of mammalian species: London, British Museum (Natural History), 254 p.

Donovan, S. K., and Gordon, C. M., 1989, Report on a field meeting to selected localities in St. Andrews and St. Ann, 25 February, 1989: Geological Society of Jamaica Journal, v. 26, p. 51–54.

Eshelman, R. E., and Morgan, G. S., 1985, Tobagan Recent mammals, fossil vertebrates, and their zoogeographical implications: National Geographic Society, Research Reports, v. 21, p. 137–143.

Etheridge, R., 1966, An extinct lizard of the genus *Leiocephalus* from Jamaica: Florida Academy of Sciences Quarterly Journal, v. 29, p. 47–59.

Fincham, A. G., 1977, Jamaica underground: The Geological Society of Jamaica, Kingston, Jamaica, 247 p.

Fischer, 1830, Synopsis Mammalium, addenda, p. 389.

Ford, S. M., 1986, Subfossil platyrrhine tibia (Primates: Callitrichidae) from Hispaniola: A possible further example of island gigantism: American Journal of Physical Anthropology, v. 70, p. 47–62.

Ford, S. M., 1990a, Locomotor adaptations of fossil platyrrhines: Journal of Human Evolution, v. 19, p. 141–173.

Ford, S. M., 1990b, Platyrrhine evolution in the West Indies: Journal of Human Evolution, v. 19, p. 237–254.

Ford, S. M., and Morgan, G. S., 1986. A new ceboid femur from the late Pleistocene of Jamaica: Journal of Vertebrate Paleontology, v. 6, p. 281–289.

Ford, S. M., and Morgan, G. S., 1988, Earliest primate fossil from the West Indies: American Journal of Physical Anthropology, v. 75, p. 209.

Gmelin, J. F., 1789, Systema Naturae, pt. 2, p. 1029.

Goodfriend, G. A., 1986, Pleistocene land snails from Sheep Pen Cave in the Cockpit Country of Jamaica: Proceedings, 8th International Malacological Congress, Budapest, 1983, p. 87–90.

Goodfriend, G. A., 1989, Quaternary biogeographical history of land snails in Jamaica, *in* Woods, C. A., ed., Biogeography of the West Indies: Past, present, and future: Gainesville, Florida, Sandhill Crane Press, p. 201–216.

Goodfriend, G. A., and Mitterer, R. M., 1987, Age of the ceboid femur from Coco Ree, Jamaica: Journal of Vertebrate Paleontology, v. 7, p. 344–345.

Goodfriend, G. A., and Stipp, J. J., 1983, Limestone and the problem of radiocarbon dating of land-snail shell carbonate: Geology, v. 11, p. 575–577.

Goodwin, G. G., 1959, Bats of the subgenus *Natalus:* American Museum Novitates, no. 1977, p. 1–22.

Goodwin, R. E., 1970, The ecology of Jamaican bats: Journal of Mammalogy, v. 51, p. 571–579.

Grant, C., 1940, The herpetology of Jamaica, II. The reptiles: Bulletin of the Institute of Jamaica, Science Series no. 1, p. 61–148.

Gray, J. E., 1831, Zool. Misc., no. 1, p. 37.

Gray, J. E., 1843a, [Letter addressed to the Curator]: Proceedings of the Zoological Society of London, p. 50.

Gray, J. E., 1843b, List of the specimens of Mammalia in the collection of the British Museum (Natural History), 216 p.

Gray, J. E., 1845, Catalogue of the lizards in the British Museum, p. 190.

Gundlach, J. (in Peters, W.), 1861, Eine neue von Hrn. Dr. Gundlach beschriebene Gattung von Flederthieren aus Cuba: Berlin, Monatsber. K. Preuss Akad. Wiss., p. 817–818.

Hall, E. R., 1981, The mammals of North America (second edition): New York, John Wiley and Sons, 1175 p.

Haq, B. U., Hardenbol, J., and Vail, P. R., 1987, Chronology of fluctuating sea levels since the Triassic: Science, v. 235, p. 1156–1167.

Hallowell, E., 1857, Notes on the reptiles in the collection of the Academy of Natural Sciences: Proceedings, Academy of Natural Sciences, Philadelphia, v. 8, p. 221–238.

Hecht, M. K., 1951, Fossil lizards of the West Indian genus *Aristelliger* (Gekkonidae): American Museum Novitates, no. 1538, p. 1–33.

Hershkovitz, P., 1966, Mice, land bridges, and Latin American faunal interchange, *in* Wenzel, R. L., and Tipton, V. J., eds., Ectoparasites of Panama: Chicago, Illinois, Field Museum of Natural History, p. 725–751.

Hershkovitz, P., 1970, Notes on Tertiary platyrrhine monkeys and description of a new genus from the late Miocene of Colombia: Folia Primatologica, v. 12, p. 1–37.

Hershkovitz, P., 1988, The subfossil monkey femur and subfossil monkey tibia of the Antilles: A review: International Journal of Primatology, v. 9, p. 365–384.

Honacki, J. H., Kinman, K. E., and Koeppl, J. W., 1982, Mammal species of the world: Lawrence, Kansas, Allen Press, Inc. and Association of Systematics Collections, 694 p.

Humphrey, S. R., Setzer, H. W., and Morgan, G. S., 1993, Description and taxonomic status of the extinct Jamaican rice rat (*Oryzomys antillarum*): Journal of Mammalogy (in press).

Koopman, K. F., 1951, Fossil bats from the Bahamas: Journal of Mammalogy, v. 32, p. 229.

Koopman, K. F., 1952, The status of the bat genus *Reithronycteris:* Journal of Mammalogy, v. 33, p. 255–258.

Koopman, K. F., 1955, A new subspecies of *Chilonycteris* from the West Indies and a discussion of the mammals of La Gonave: Journal of Mammalogy, v. 36, p. 109–113.

Koopman, K. F., 1976, Zoogeography, *in* Baker, R. J., Jones, J. K., Jr., and Carter, D. C., eds., Biology of bats of the New World family Phyllostomatidae, Part 1: Texas Tech University, The Museum, Special Publications, v. 10, p. 39–47.

Koopman, K. F., and Williams, E. E., 1951, Fossil Chiroptera collected by H. E. Anthony in Jamaica, 1919–1920: American Museum Novitates, no. 1519, p. 1–29.

Lacépède, B.G.E., 1788, Histoire naturelle des quadrupèdes ovipares et des serpens: Syn. Meth. 1, p. 443–462.

Leach, W. E., 1821, The characters of seven genera of bats with foliaceous appendages to the nose: Transactions of the Linnean Society, v. 13, p. 73–82.

Lillegraven, J. A., McKenna, M. C., and Krishtalka, L., 1981, Evolutionary relationships of middle Eocene and younger species of *Centetodon* (Mammalia, Insectivora, Geolabididae) with a description of the dentition of *Ankylodon* (Adapisoricidae): University of Wyoming Publications, v. 45, p. 1–115.

Linnaeus, C., 1758, Systema Naturae, 10th ed., p. 193.

MacPhee, R.D.E., 1984, Quaternary mammal localities and heptaxodontid rodents of Jamaica: American Museum Novitates, no. 2803, p. 1–34.

MacPhee, R.D.E., and Fleagle, J. G., 1991, Postcranial remains of *Xenothrix mcgregori* (Primates, Xenotrichidae) and other late Quaternary mammals from Long Mile Cave, Jamaica: American Museum of Natural History Bulletin, no. 206, p. 287–321.

MacPhee, R.D.E., and Woods, C. A., 1982, A new fossil cebine from Hispaniola: American Journal of Physical Anthropology, v. 58, p. 419–436.

MacPhee, R.D.E., Woods, C. A., and Morgan, G. S., 1983, The Pleistocene rodent *Alterodon major* and the mammalian biogeography of Jamaica: Palaeontology, v. 26, p. 831–837.

MacPhee, R.D.E., Ford, D. C., and McFarlane, D. A., 1989, Pre-Wisconsinan mammals from Jamaica and models of late Quaternary extinction in the Greater Antilles: Quaternary Research, v. 31, p. 94–106.

McFarlane, D. A., and Gledhill, R. E., 1985, The Quaternary bone caves at Wallingford, Jamaica: Cave Science, v. 12, p. 127–128.

Miller, G. S., Jr., 1898, Descriptions of five new phyllostome bats: Proceedings, Academy of Natural Sciences, Philadelphia, v. 50, p. 326–337.

Miller, G. S., Jr., 1902, Twenty new American bats: Proceedings, Academy of

Natural Sciences, Philadelphia, v. 54, p. 389–412.

Miller, G. S., Jr., 1916, Remains of two species of *Capromys* from ancient burial sites in Jamaica: Proceedings, Biological Society of Washington, v. 24, p. 48.

Miller, G. S., Jr., 1918, Three new bats from Haiti and Santo Domingo: Proceedings, Biological Society of Washington, v. 31, p. 39–40.

Molina, 1782, Saggio Stor. Nat. Chili, p. 263.

Morgan, G. S., 1977, Late Pleistocene fossil vertebrates from the Cayman Islands, British West Indies [M.S. thesis]: Gainesville, Florida, University of Florida, 273 p.

Morgan, G. S., 1985, Taxonomic status and relationships of the Swan Island hutia, *Geocapromys thoracatus* (Mammalia: Rodentia: Capromyidae), and the zoogeography of the Swan Islands vertebrate fauna: Biological Society of Washington Proceedings, v. 98, p. 29–46.

Morgan, G. S., 1989a, Fossil Chiroptera and Rodentia from the Bahamas, and the historical biogeography of Bahamian mammal fauna, *in* Woods, C. A., ed., Biogeography of the West Indies: Past, present, and future: Gainesville, Florida, Sandhill Crane Press, p. 685–740.

Morgan, G. S., 1989b, *Geocapromys thoracatus:* Mammalian Species, no. 341, p. 1–5.

Morgan, G. S., 1991, Neotropical Chiroptera from the Pliocene and Pleistocene of Florida: American Museum of Natural History Bulletin, no. 206, p. 176–213.

Morgan, G. S., 1993, Late Quaternary fossil vertebrates from the Cayman Islands, *in* Davies, J. E., and Brunt, M., eds., Biogeography and ecology of the Cayman Islands: Dordrecht, Netherlands, Monographiae Biologicae, Kluwer Academic Publishers (in press).

Morgan, G. S., and Woods, C. A., 1986, Extinction and the zoogeography of West Indian land mammals: Linnean Society Biological Journal, v. 28, p. 167–203.

North American Stratigraphic Code, 1983, American Association of Petroleum Geologists Bulletin, v. 67, p. 841–875.

Oliver, W.L.R., 1982, The coney and the yellow snake: The distribution and status of the Jamaican Hutia *Geocapromys brownii* and the Jamaican Boa *Epicrates subflavus:* Dodo, Journal of the Jersey Wildlife Preservation Trust, v. 19, p. 6–33.

Olson, S. L., 1978, A paleontological perspective of West Indian birds and mammals, *in* Gill, F. B., ed., Zoogeography in the Caribbean: Academy of Natural Sciences of Philadelphia, Special Publication 13, p. 99–117.

Olson, S. L., 1985, A new species of *Siphonorhis* from Quaternary cave deposits in Cuba (Aves: Caprimulgidae): Proceedings, Biological Society of Washington, v. 98, p. 526–532.

Olson, S. L., and Hilgartner, W. B., 1982, Fossil and subfossil birds from the Bahamas, *in* Olson, S. L., ed., Fossil vertebrates from the Bahamas: Smithsonian Contributions to Paleobiology, no. 48, p. 22–56.

Olson, S. L., and Steadman, D. W., 1977, A new genus of flightless ibis (Threskiornithidae) and other fossil birds from cave deposits in Jamaica: Biological Society of Washington Proceedings, v. 90, p. 447–457.

Olson, S. L., and Steadman, D. W., 1979, The humerus of *Xenicibis,* the extinct flightless ibis of Jamaica: Biological Society of Washington Proceedings, v. 92, p. 23–27.

Pascual, R., Vucetich, M. G., and Scillato-Yané, G. J., 1990, Extinct and Recent South American and Caribbean Megalonychidae edentates and Hystricognathi rodents: Outstanding examples of isolation, *in* Biogeographical aspects of insularity: Accademia Nazionale dei Lincei, Roma, Atti dei Convegni Lincei, v. 85, p. 627–640.

Patterson, B., and Wood, A. E., 1982, Rodents from the Deseadan Oligocene of Bolivia and the relationships of the Caviomorpha: Museum of Comparative Zoology Bulletin, v. 149, no. 7, p. 371–543.

Paula Couto, C., de, 1967, Pleistocene edentates of the West Indies: American Museum Novitates, no. 2304, p. 1–55.

Perfit, M. R., and Williams, E. E., 1989, Geological constraints and biological retrodictions in the evolution of the Caribbean Sea and its islands, *in* Woods, C. A., ed., Biogeography of the West Indies: Past, present, and future: Gainesville, Florida, Sandhill Crane Press, p. 47–102.

Peters, W., 1864, Über einige neue säugethiere: Monatsber. K. Preuss. Akad. Wiss., Berlin, 1865, p. 381–399.

Pregill, G. K., 1981, Late Pleistocene herpetofaunas from Puerto Rico: University of Kansas Museum of Natural History, Miscellaneous Publication no. 71, p. 1–72.

Pregill, G. K., 1986, Body size of insular lizards: A pattern of Holocene dwarfism: Evolution, v. 40, p. 997–1008.

Pregill, G. K., and Olson, S. L., 1981, Zoogeography of West Indian vertebrates in relation to Pleistocene climatic cycles: Annual Review of Ecology and Systematics, v. 12, p. 75–98.

Pregill, G. K., Steadman, D. W., Olson, S. L., and Grady, F. V., 1988, Late Holocene fossil vertebrates from Burma Quarry, Antigua, Lesser Antilles: Smithsonian Contributions to Zoology, no. 463, p. 1–27.

Pregill, G. K., Crombie, R. I., Steadman, D. W., Gordon, L. K., Davis, F. W., and Hilgartner, W. B., 1991, Living and late Holocene fossil vertebrates, and the vegetation of the Cockpit Country, Jamaica: Atoll Research Bulletin, no. 353, p. 1–19.

Ray, C. E., 1964, The taxonomic status of *Heptaxodon* and dental ontogeny in *Elasmodontomys* and *Amblyrhiza* (Rodentia: Caviomorpha): Museum of Comparative Zoology Bulletin, v. 131, no. 5, p. 107–127.

Ray, C. E., 1965, The relationships of *Quemisia gravis* (Rodentia:? Heptaxodontidae): Smithsonian Miscellaneous Collections, v. 149, no. 3, p. 1–12.

Ray, C. E., Olsen, S. J., and Gut, H. J., 1963, Three mammals new to the Pleistocene fauna of Florida, and a reconsideration of five earlier records: Journal of Mammalogy, v. 44, p. 373–395.

Rosen, D. E., 1975, A vicariance model of Caribbean biogeography: Systematic Zoology, v. 24, p. 431–461.

Rosenberger, A. L., 1977, *Xenothrix* and ceboid phylogeny: Journal of Human Evolution, v. 6, p. 461–481.

Rosenberger, A. L., Setoguchi, T., and Shigehara, N., 1990, The fossil record of callitrichine primates: Journal of Human Evolution, v. 19, p. 209–236.

Rouse, I., 1989, Peopling and repeopling of the West Indies, *in* Woods, C. A., ed., Biogeography of the West Indies: Past, present, and future: Gainesville, Florida, Sandhill Crane Press, p. 119–136.

Savage, R.J.G., 1990, Preliminary report on the vertebrate fauna of the Red Hills Road Cave, St. Andrew, Jamaica: Geological Society of Jamaica Journal, v. 27, p. 33–35.

Schwartz, A., and Carey, M., 1977, Systematics and evolution in the West Indian iguanid genus *Cyclura:* Studies on the Fauna of Curacao and other Caribbean Islands, v. 53, p. 15–97.

Schwartz, A., and Jones, J. K., Jr., 1967, Review of bats of the endemic Antillean genus *Monophyllus:* Proceedings of the United States National Museum, v. 124, p. 1–20.

Schwartz, A., and Thomas, R., 1975, A check-list of West Indian amphibians and reptiles: Pittsburgh, Carnegie Museum of Natural History, 216 p.

Seidel, M., 1988, Revision of the West Indian emydid turtles (Testudines): American Museum Novitates, no. 2918, p. 1–41.

Seidel, M. E., and Smith, H. M., 1986, *Chrysemys, Pseudemys, Trachemys* (Testudines: Emydidae): Did Agassiz have it right?: Herpetologica, v. 42, p. 242–248.

Shaw, G., 1802, General Zoology, 3 (1), p. 288.

Silva Taboada, G., 1974, Fossil Chiroptera from cave deposits in central Cuba, with description of two new species (genera *Pteronotus* and *Mormoops*) and the first West Indian record of *Mormoops megalophylla:* Acta Zoologica Cracoviensia, v. 19, p. 33–73.

Silva Taboada, G., 1979, Los murciélagos de Cuba: La Habana, Cuba, Editorial Academia, 423 p.

Simpson, G. G., 1945, The principles of classification and a classification of mammals: American Museum of Natural History Bulletin, no. 85, p. 1–350.

Simpson, G. G., 1956, Zoogeography of West Indian land mammals: American Museum Novitates, no. 1759, p. 1–28.

Smith, J. D., 1972, Systematics of the chiropteran family Mormoopidae: University of Kansas Museum of Natural History, Miscellaneous Publication no. 56, p. 1–132.

Stafford, T. W., Jr., 1990, Late Pleistocene megafauna extinctions and the Clovis culture: Absolute ages based on accelerator ^{14}C dating of skeletal remains, *in* Agenbroad, L. D., Mead, J. I., and Nelson, L. W., eds., Megafauna and man: Discovery of America's heartland: The Mammoth Site of Hot Springs, South Dakota, Inc., Scientific Papers, v. 1, p. 118–122.

Steadman, D. W., and Morgan, G. S., 1985, A new species of bullfinch (Aves: Emberizinae) from a late Quaternary cave deposit on Cayman Brac, West Indies: Biological Society of Washington Proceedings, v. 98, p. 544–553.

Steadman, D. W., Pregill, G. K., and Olson, S. L., 1984, Fossil vertebrates from Antigua, Lesser Antilles: Evidence for late Holocene human-caused extinctions in the West Indies: National Academy of Sciences of the U.S.A. Proceedings, v. 83, p. 4448–4451.

Swanepoel, P., and Genoways, H. H., 1978, Revision of the Antillean bats of the genus *Brachyphylla* (Mammalia: Phyllostomatidae): Carnegie Museum of Natural History Bulletin, no. 12, p. 1–53.

Taylor, R. E., 1980, Radiocarbon dating of Pleistocene bone: Toward criteria for the selection of samples: Radiocarbon, v. 22, p. 969–979.

Thomas, O., 1898, On indigenous Muridae in the West Indies; with the description of a new Mexican *Oryzomys:* Annals and Magazine of Natural History, ser. 7, v. i, p. 176–180.

Varona, L. S., 1974, Catálogo de los mamíferos vivientes y extinguidos de las Antillas: La Habana, Cuba, Academia de Ciencias de Cuba, 139 p.

Williams, E. E., 1950, *Testudo cubensis* and the evolution of Western Hemisphere tortoises: American Museum of Natural History Bulletin, v. 95, p. 1–36.

Williams, E. E., 1952, Additional notes on fossil and subfossil bats from Jamaica: Journal of Mammalogy, v. 33, p. 171–179.

Williams, E. E., and Koopman, K. F., 1952, West Indian fossil monkeys: American Museum Novitates, no. 1546, p. 1–16.

Wing, E. S., 1972, Identification and interpretation of faunal remains, *in* Silverberg, J., ed., The White Marl Site in Jamaica: Report of the 1964 Robert R. Howard Excavation: University of Wisconsin-Milwaukee, Department of Anthropology, p. 18–35.

Wing, E. S., 1989, Human exploitation of animal resources in the Caribbean, *in* Woods, C. A., ed., Biogeography of the West Indies: Past, present, and future: Gainesville, Florida, Sandhill Crane Press, p. 137–152.

Wing, E. S., and Reitz, E. J., 1982, Prehistoric fishing economies of the Caribbean: New World Archaeology, v. 5, p. 13–32.

Woods, C. A., 1989, A new capromyid rodent from Haiti, the origin, evolution, and extinction of West Indian rodents and their bearing on the origin of New World hystricognaths, *in* Black, C.C., and Dawson, M. R., eds., Papers on fossil rodents in honor of Albert Elmer Wood: Natural History Museum of Los Angeles County, Science Series, no. 33, p. 59–89.

Zans, V. A., Chubb, L. J., Versey, H. R., Williams, J. B., Robinson, E., and Cooke, D. L., 1963, Synopsis of the geology of Jamaica: Geological Survey Department, Jamaica, Bulletin 4, p. 1–72.

MANUSCRIPT ACCEPTED BY THE SOCIETY NOVEMBER 11, 1992

Geological Society of America
Memoir 182
1993

Contribution toward a Tertiary palynostratigraphy for Jamaica: The status of Tertiary paleobotanical studies in northern Latin America and preliminary analysis of the Guys Hill Member (Chapelton Formation, middle Eocene) of Jamaica

Alan Graham
Department of Biological Sciences, Kent State University, Kent, Ohio 44242

ABSTRACT

Approximately 185 types of pollen grains and spores have been identified from 13 Tertiary formations in the Gulf/Caribbean region (northern Latin America), providing a revised data base for a published palynostratigraphy for the area. Such forms as *Magnastriatites howardii* (= the modern *Ceratopteris*; middle(?) to late Eocene to Recent), *Echiperiporites estelae* (= *Hampea/Hibiscus*; late Eocene to Recent), *Echitricolporites spinosus* (= Compositae-Heliantheae type; late Oligocene to Recent), *Fenestrites spinosus* (= Compositae-Vernonieae type; late Miocene to Recent), *Psilatricolporites crassus* (= *Pelliceria*; middle Eocene to Recent), *Monoporites annuloides* (= Gramineae [grass]; late Eocene [in our area] to Recent), *Zonocostites ramonae* (= *Rhizophora*; late Eocene to Recent), and *Psilatricolporites operculatus* (= *Alchornea*; middle Eocene to Recent) are proving of stratigraphic value in Tertiary deposits of northern Latin America. These ranges will likely be refined and augmented by current studies on other Gulf/Caribbean Tertiary formations, including the Chapelton Formation (middle Eocene) of Jamaica. Palynological evidence from the Guys Hill Member of the Chapelton Formation suggests deposition in near-shore marine habitats under warm-temperate to tropical conditions. The estimated age, based on a preliminary study of the plant microfossils, is middle Eocene, consistent with previous assignments based on other lines of evidence. In the meantime, data from northern South America (isolated from northern Latin America for most of the Tertiary), and from the southeastern United States (both geographically removed and likely in a somewhat different climatic zone from our area) can be applied, to some extent, to the correlation of terrestrial strata in the Antilles and the surrounding region. Older data from megafossil paleobotanical studies are proving unreliable, both as a basis for correlation and for biogeographic considerations.

INTRODUCTION

The published paleobotanical and paleopalynological record for Jamaica is meager. No plant megafossils have been studied, and none are presently known to occur on the island. The plant microfossil record is limited to six papers by Germeraad (1978a, b, c, 1979a, b, 1980; four mimeographed) on various algae (including dinoflagellates), fungi, scolecodonts, and palynomorphs from Cretaceous and Tertiary strata; a report by Graham (1977) on pollen of the mangrove *Pelliceria* (*Psilatricolporites crassus*) from the middle Eocene Guys Hill Member of the Chapelton Formation; and a record of two species of the calcareous alga *Neomeris* from the Pliocene Bowden Beds (Racz, 1971). Consequently, a broader region must be encompassed to establish Tertiary terrestrial paleoenvironments and a general palynostratigraphy for the Caribbean. By extrapolation, this information may

Graham, A., 1993, Contribution toward a Tertiary palynostratigraphy for Jamaica: The status of Tertiary paleobotanical studies in northern Latin America and preliminary analysis of the Guys Hill Member (Chapelton Formation, middle Eocene) of Jamaica, *in* Wright, R. M., and Robinson, E., eds., Biostratigraphy of Jamaica: Boulder, Colorado, Geological Society of America Memoir 182.

be applied to Jamaica until studies on the Chapelton Formation, and assemblages from other localities in the Caribbean region, now underway are completed.

A meaningful delimitation of the relevant areas to be included is difficult because of the complex tectonic history of the Caribbean Plate. Among the various models available (e.g., Mann and Burke, 1984; Pindell and Dewey, 1982; Sykes et al., 1982; see review by Buskirk, 1985), some place Jamaica off the southwestern coast of Mexico during the early Tertiary, and subsequently moving eastward 1,000 km or more through the portal between North and South America until reaching its present position in the late Miocene (Wadge and Burke, 1983); since the early Pliocene, movement has been mainly vertical through local uplift and subsidence. In other models (Donnelly, 1985), horizontal movement along the Motagua-Polichic fault system is viewed as considerably less, with the islands of the Antilles originating closer to their present positions. In both models, however, South America constituted a separate physiographic province throughout most of the Tertiary, becoming connected to North America through uplift of the Panamanian isthmus only about 3 to 2.5 Ma (Stehli and Webb, 1985) to possibly as late as 1.8 Ma (Keller et al., 1989). Therefore, the zonation and palynostratigraphy recently summarized for northern South America (Muller et al., 1987) is not directly applicable even to adjacent southern Central America. The region most relevant to the vegetational history, terrestrial paleoenvironments, and palynostratigraphy of Jamaica for early and middle Tertiary times is Mexico, Central America, and the adjacent islands of the Greater Antilles. Palynofloras from the southeastern United States are similar mostly with regards to stratigraphically and geographically widespread or generalized forms, and South American assemblages become important only in the late Tertiary and especially in the Quaternary.

With the paleophysiographic province so defined, a summary of the fossil plant record involves compilation of the literature on paleobotany and paleopalynology for the region; compilation of the plant taxa represented in Tertiary strata of the Gulf/Caribbean area; assessment of the reliability of the megafossil and microfossil records, recognizing that most megafossil studies were completed 50 to 60 years ago (e.g., Berry, 1921, 1923; Hollick, 1928); and a summary of the stratigraphic range of the palynomorphs, recognizing that taxa, ranges, and zones established for northern South America may not be precisely the same for adjacent Central America, Mexico, and the Antilles (northern Latin America) because they belonged to different physiographic provinces.

THE LITERATURE ON VEGETATIONAL HISTORY FOR NORTHERN LATIN AMERICA

Approximately 400 papers have been published on the paleobotany and paleopalynology of Mexico (252 papers), Central America (84), and the Antilles (59; Graham, 1973, 1979, 1982, 1986, and in preparation). The earliest reference to plant fossils in the Antilles is by an anonymous author (1818) describing petrified wood from the Antigua, as listed in LaMotte's (1952) catalog. All studies on Jamacia are quite recent, as previously noted. There are other unpublished, and generally unavailable, technical and progress reports resulting from studies by industry, government, agencies, and consulting firms.

Although 400 papers would seem to provide an adequate data base for reconstructing terrestrial paleoenvironments and for establishing a workable palynostratigraphy, the list includes unpublished theses and dissertations, abstracts, general review papers, and those briefly mentioning fossil plants in connection with other types of studies. Only 43 papers have been published that actually list identified plant fossils from Tertiary strata in northern Latin America (Appendix 1). Of these, 14 were published more than 40 years ago, and only 13 (all dealing with microfossil assemblages) treat relatively large floras studied within the last 20 years. Thus the real data base is 13 papers for all of northern Latin America, for all of Tertiary time.

THE MEGAFOSSIL RECORD

In an earlier paper (Graham, 1988a), a preliminary summary was provided for the literature, megafossil, and microfossil plant record for the Gulf/Caribbean Tertiary through about 1985. At that time study of four microfossil assemblages in our project had been completed. Now data is available for nine floras, in addition to three more by other authors (Figs. 1, 2), and the new information is incorporated into this revised summary. Preliminary results from the Chapelton Formation of Jamaica are also discussed.

Tertiary plant megafossils described for northern Latin America are listed by age, country, and biological affinity in Appendix 1. A total of 123 genera are included. Almost all of these are from the works of Arthur Hollick and especially E. W. Berry in the early 1900s (Appendix 3). In the absence of extensive comparative herbarium material, and modern methodologies and approaches (cleared leaf collections, detailed cuticular studies, SEM, leaf physiognomy), it is probable that many identifications and interpretations based on fossil leaf material from these highly diverse tropical assemblages are incorrect. Dilcher (1973) has estimated that 60% of the taxa listed for the tropical to warm-temperate Eocene floras of the Mississippi Embayment by Berry are misidentified. In addition, Berry was the principal North America spokesman for paleobotany arguing for permanence of continents and ocean basins. As a consequence, the early geological, paleoenvironmental, and biogeographical conclusions derived from this data have also proven inconsistent with more recent information. None of the material from northern Latin America studied by Berry and Hollick has been revised. Although plant megafossils are very important in the study of vegetational history, in this particular instance the information is so out of date that it should not be incorporated into developing models of Caribbean biostratigraphy.

THE MICROFOSSIL RECORD

Through the mid-1960s only a single study had been published on a relatively extensive Tertiary microfossil flora from northern Latin America (Langenheim et al., 1967; Oligo-Miocene Simojovel Group, Chiapas, Mexico). Subsequently, a series of studies was initiated in my laboratory on pollen and spore assemblages derived from Tertiary lignites and associated sediments (Figs. 1, 2). These include the middle(?) to upper Eocene Gatuncillo Formation of Panama (Graham, 1985), the middle to upper Oligocene San Sebastian Formation of Puerto Rico (Graham and Jarzen, 1969), the lower Miocene Culebra, Cucaracha, and La Boca Formations of Panama (Graham, 1988b, c, 1989), the lower Miocene Uscari Formation of Costa Rica (Gra-

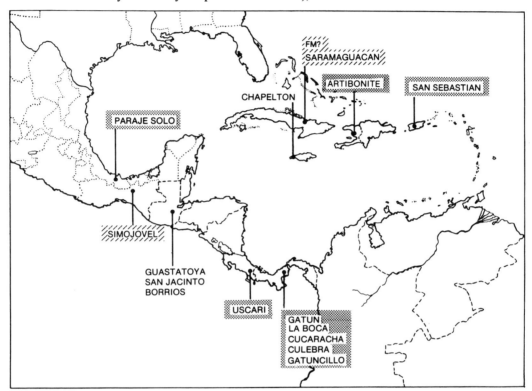

Figure 1. Distribution of Tertiary plant microfossil assemblages in the Gulf/Caribbean region. Stippling indicates studies by the author that are complete and the results published; no marking indicates studies currently underway; hatching indicates published studies by other authors.

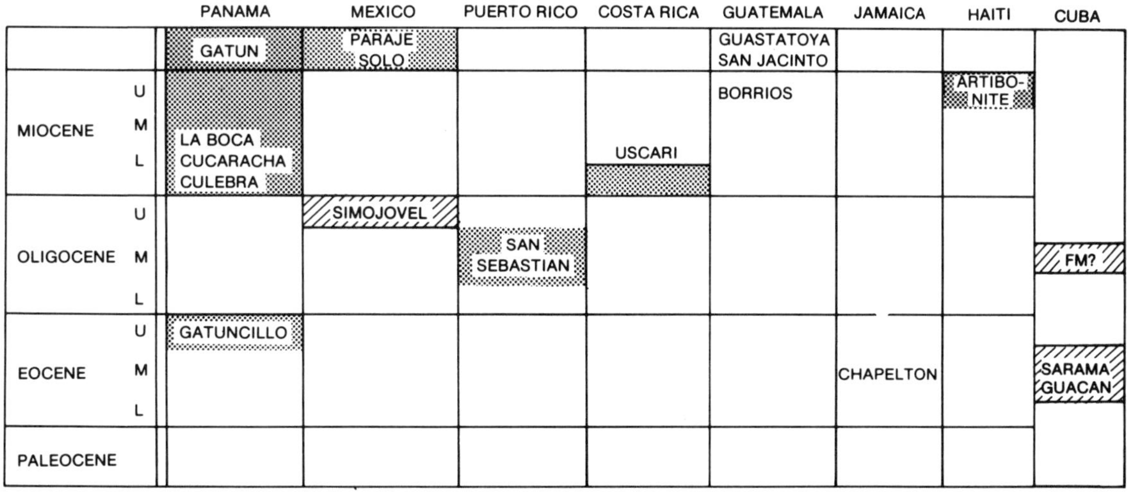

Figure 2. Age assignments of Tertiary plant microfossil assemblages in the Gulf/Caribbean region. Markings as in Figure 1. Position on the chart does not imply formal correlation.

ham, 1987), the late Tertiary (Mio-Pliocene?) Artibonite Group of Haiti (Graham, 1990), the Pliocene Paraje Solo Formation of Mexico (Graham, 1976), and the Pliocene Gatun Formation of Panama (Graham, 1991a, b, c). A new project will include the middle Eocene Guys Hill Member of the Chapelton Formation (Yellow Limestone Group) of Jamaica, new localities in Costa Rica, and several upper Tertiary formations in Guatemala (Figs. 1, 2). Areces-Mallea (1985, 1987, 1988) has recently initiated some Tertiary palynological studies in Cuba (the middle Eocene Saramaguacan Formation, and an unnamed middle Oligocene unit in the Ciego de Avila Province of central Cuba). To date approximately 185 types of pollen grains and spores, excluding unknowns, have been recognized from 13 Tertiary formations, and these are listed in Appendix 2.

SOME PALYNOMORPHS OF STRATIGRAPHIC SIGNIFICANCE FOR THE GULF/CARIBBEAN TERTIARY

It is recognized that age assignments and stratigraphic correlation among terrestrial strata in Latin America are not dependent exclusively on palynomorph assemblages. Several formations have marine and nonmarine phases (e.g., the Gatun Formation of Panama), and can be correlated with Deep Sea Drilling Project (DSDP) zones based on foraminifera, dinoflagellates, calcareous nannofossils, and other microfossils. The general trend has been toward younger age estimates than given in the older literature (e.g., Cooke et al., 1943). For example, the Paraje Solo and Gatun Formations, both earlier considered Miocene, are now assigned to the Pliocene (Akers, 1979, 1981, 1984; Machain-Castillo, 1985). In other instances, however, the strata do not interfinger with marine deposits and the palynomorphs become more prominent in stratigraphic considerations, as well as in paleoenvironmental reconstructions.

Nine fossil pollen and spore types are selected to illustrate ranges emerging as both applicable to Tertiary deposits in northern Latin America and consistent between at least several formations. *Ceratopteris,* known in the stratigraphic literature as *Magnastriatites howardii* (Fig. 3.1), ranges from middle(?) to late Eocene to Recent. It is a fresh-water, floating, aquatic fern, and is known from the Gatuncillo, Cucaracha, Paraje Solo, and Gatun Formations. By contrast, its range in northern South America is from the base of the Oligocene to Recent. *Echiperiporites estelae* (Fig. 3.2) is similar to pollen of the modern genera *Hampea* and *Hibiscus,* but most specimens likely represent the latter, which includes *H. tiliaceus,* often associated with the mangrove *Rhizophora* in tropical environments. Its presence, particularly in lignites containing *Rhizophora,* is consistent with warm-temperate to tropical, coastal, brackish-water environments. Its range is from late Eocene to Recent. Pollen of the Compositae is especially useful because the various types appear at different times in the late Tertiary (Elsik and Tomb, 1988). The most simple forms (*Echitricolporites spinosus*; small, echinate; Fig. 3.8) appear in low numbers late in the Oligocene, but do not become abundant until the Miocene. More complex forms (*Fenestrites spinosus*; large, reticulate, echinate; Fig. 3.3) range from late Miocene to Recent. Because of the ecological diversity within the family, precise paleoecological conditions usually cannot be inferred from the presence of pollen. *Psilatricolporites crassus* (Fig. 3.4) is produced by the mangrove *Pelliceria,* a tree now restricted to lowland coastal areas from Puntarenas Province in Costa Rica through Panama and northwestern Colombia, to Esmeraldas Province in Ecuador (Graham, 1977). It ranges from middle Eocene to Recent and is known from the Chapelton, Gatuncillo, San Sebastian, Simojovel, and La Boca Formations, as well as from several localities in northern South America. *Striatricolpites catatumbus* pollen (Fig.3.5) is produced by the modern legume genus *Crudia.* Its paleodistribution, and hence its stratigraphic range in different parts of Latin America, are only now becoming clear. It presently occurs in Amazonian Brazil, but in the Tertiary it extended into Panama (Gatuncillo, Cucaracha, La Boca, and Gatun Formations), and pods similar to *Crudia* have recently been reported from the middle Eocene Claiborne Formation of western Tennessee (Herendeen and Dilcher, 1990). Its present stratigraphic range is from early Eocene to Recent. Grass pollen (*Monoporites annuloides*; Fig. 3.6) appears in South America in the Paleocene, but not until the latest Eocene (Muller, 1981), middle Eocene (Frederiksen, 1988), or early Eocene (Wilcox) in North America. Remains from beds previously thought to be Paleogene (Crepet and Feldman, 1991) are now assigned to the early Eocene (Frederiksen, Personal Communication, 1992). In our area it has been recovered in small amounts from the Culebra, La Boca, Paraje Solo, and Gatun Formations. The most common fossil pollen in the Gulf/Caribbean Tertiary, from the late Eocene to the Recent, especially in lignites, is the mangrove *Rhizophora* (*Zonocostites ramonae*; Fig. 3.7). The genus is presently restricted to coastal, brackish-water habitats. Prior to the late Eocene its presumed ecological equivalent is *Brevitricolpites variabilis* of unknown biological affinities. By the middle Eocene *Psilatricolporites crassus* (*Pelliceria*) had also become established as a prominent and widespread Caribbean mangrove. The geographic range of *Pelliceria* became progressively restricted during the middle and late Tertiary to its present locales in southern Central America and northern South America. *Psilatricolporites operculatus* (Fig. 3.9) is produced by the modern genus *Alchornea,* distinguished by conspicuous copal opercula and evident in polar view. The pollen ranges from middle Eocene to Recent.

A useful palynostratigraphy is emerging from published studies on Tertiary formations in northern Latin America, but establishing accurate and complete ranges is presently hindered by three factors. First, there are too many gaps in the record. There are no data on the Paleocene, lower Eocene, and lower Oligocene, which makes the presently known range bases of the palynomorphs tentative, and provides no information on the range tops of taxa becoming extinct in the lower and mid-Tertiary. Many of the Tertiary epochs are represented by a single assemblage for the entire region. Second, core material is not available for many sites, or, if available has not been published,

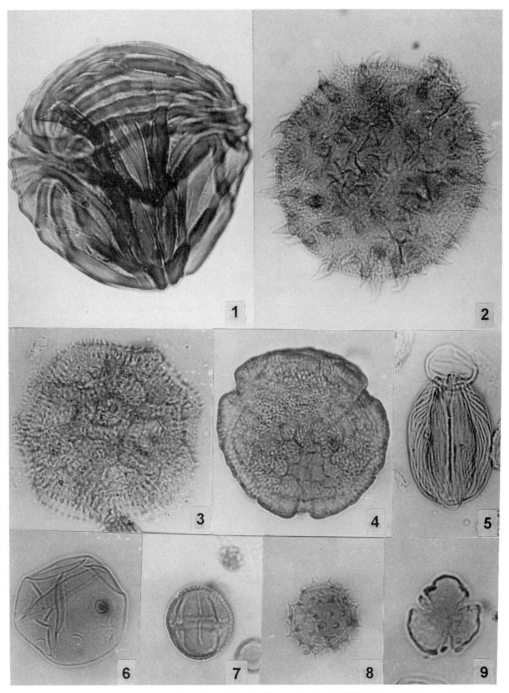

Figure 3. Selected palynomorphs of stratigraphic significance in the Gulf/Caribbean Tertiary. 1, *Magnastriatites howardii,* Cucaracha Formation, Panama, 90 μm; 2, *Echiperiporites estelae,* Gatun Formation, Panama, 80 μm; 3, *Fenestrites spinosus,* Paraje Solo Formation, Mexico, 63 μm; 4, *Psilatricolporites crassus,* San Sebastian Formation, Puerto Rico, 60 μm; 5, *Striatricolpites catatumbus,* La Boca Formation, Panama, 47 μm; 6, *Monoporites annuloides,* Paraje Solo Formation, Mexico, 37 μm; 7, *Zonocostites ramonae,* Paraje Solo Formation, Mexico 26 μm; 8, *Echitricolporites spinosus,* La Boca Formation, Panama, 28 μm; 9, *Psilatricolporites operculatus,* Culebra Formation, Panama, 20 μm.

precluding better resolution through zonation. Third, the published studies on large Tertiary plant microfossil assemblages from the Caribbean region have mostly emphasized biogeography, paleoecology, and the history of modern taxa and vegetation types. Consequently, the nomenclature used is that of modern genera and families, with unidentified taxa referred to as a sequence of numbered unknowns. In contrast, stratigraphically oriented studies for the Caribbean utilize an artificial nomenclature, with biological affinities cited secondarily. Therefore, it will eventually be necessary to better equate the modern generic and family names, and the numbered unknowns, with the artificial names to standardize nomenclature and facilitate comparison between the various assemblages. Each of these problems is currently being addressed, and a more useful palynostratigraphy is forthcoming. Until then, the existing microfossil data, augmented by studies in northern South America (e.g., Germeraad et al., 1968; González Guzmán, 1967; Lorente, 1986; Muller et al., 1987) and southern North America (e.g., Elsik, 1968; Frederiksen, 1980, 1988), provide the most recent and reliable source of paleobotanical information on Tertiary deposits for the Caribbean.

PALYNOMORPHS IN THE GUYS HILL MEMBER OF THE CHAPELTON FORMATION

The Chapelton Formation was named for exposures of interbedded, impure limestones, calcareous siltstones, and sandstones along the road through Chapelton, central Jamaica (Robinson, 1987). It is divided into three members—the Stettin, Guys Hill, and Albert Town. The Guys Hill Member consists of channeled, cross-bedded sandstones with larger foraminifera, oyster beds, and dispersed carbonized material and lignitic lenses 1 to 2.5 m thick. The organic-rich lenses are not true lignites because considerable mineral matter is present (Robinson, 1976). The Guys Hill Member is dated as middle Eocene on the basis of planktonic foraminifera and other invertebrate fossils. In 1974, samples of the lignitic material were provided by E. Robinson from a road section at Wait-a-Bit, Trelawny Parish; in 1984, J. Bujak sent a slide from the Central Inlier region; and in 1990, collections were made from five additional exposures of the Guys Hill Member. One of these (our locality 1, Christiana-Bryce; = Cascade, locality 10 in Robinson, 1976), along with the Bujak and Wait-a-Bit samples, yielded palynomorphs of fair preservation and diversity (Figs. 4.1–4.14). The assemblage is important for palynostratigraphy in Jamaica and northern Latin America because it is the oldest Tertiary plant microfossil-bearing material published for the region. The next oldest is from the middle(?) to upper Eocene Gatuncillo Formation of Panama.

Descriptions

Eight types of palynomorphs have been recovered from the Guys Hill Member sediments, and three of these (Figs. 4.1, 4.2–4.4, 4.7) constitute the most abundant forms present. The preliminary identifications and ranges are based on Germeraad et al. (1968), González Guzmán (1967), Lorente (1986), Muller et al. (1987), and references associated with localities shown in Figures 1 and 2. Location of specimens on the slides is given by England Slide Finder (ESF) coordinates. All materials are deposited in the palynology collections at Kent State University.

Deltoidospora [cf. *D. adriennis* (Potonié and Gelletich) Frederiksen]
(Fig. 4.1)

Amb triangular to oval-triangular, apices rounded; trilete, laesurae straight, 18 to 20 μm long, small in relation to size of spore; laevigate, wall 2 to 3 μm thick; 65 to 70 μm. Locality 1 (Christiana-Bryce), slide 1, ESF coordinates E-34, 1.

This spore is a generalized type that ranges throughout the Cenozoic section in the Caribbean region. It is one of the three most abundant palynomorphs in the Guys Hill assemblage (along with *Psilatricolporites crassus* and *Bombacacidites* sp. discussed below). *Psilatricolporites* represents pollen of the modern mangrove *Pelliceria,* and the lithology of the Guys Hill Member clearly indicates the sediments were deposited in near-shore marine to brackish-water environments. Considering the number of these spores recovered, the depositional environment, and the morphological features of the spores, they likely represent a plant similar to the modern fern *Acrostichum* (e.g., *A. aureum*), which is commonly associated with mangroves in the neotropics.

Psilatricolporites crassus van der Hammen and Wijmstra
(Figs. 4.2–4.4)

Oblate, amb oval-triangular to nearly circular; tricolporate, colpi straight, 15 μm long (equator to apex), equatorially arranged, meridionally elongated, equidistant, colpi transversalis 1 μm × 8 to 12 μm, situated at midpoint of colpus; tectate to tectate-perforate, wall 4 to 7 μm thick; sculpture variable from large (ca. 4 to 5 μm diameter) verrucae to reticulate; size variable, 40 to 65 μm. Figure 4.2, Locality 8 (Wait-a-Bit), slide 8, ESF S-26, 1; Figure 4.3, Locality 1 (Christiana-Bryce), slide 11, ESF D-47, 2-4; Figure 4.4, Locality 1 (Christiana-Bryce), slide 11, X-33.

These pollen grains are similar to those of the modern *Pelliceria,* which also include the variation in sculpture and size exhibited by the microfossils. They are abundant in the samples. In the absence of the common modern mangrove *Rhizophora* (until the late Eocene), *Pelliceria* occupied a similar ecological niche in the northern Caribbean region. During the middle Tertiary its geographic range was increasingly restricted by expanding populations of *Rhizophora*, and *Pelliceria* now occupies a zone along the Pacific between Costa Rica and Ecuador. The stratigraphic range of the microfossil is middle Eocene to Recent.

Retimonocolpites sp.
(Fig. 4.5)

Prolate; monosulcate, sulcus straight, 40 to 42 μm long, extending nearly to apies of the grain; tectate-perforate, wall 2 μm thick; microreticulate; 60 × 30 μm. Locality 8 (Wait-a-Bit), slide 8, ESF F-29.

This pollen represents a general palm type produced by many members of the family, and is widespread in the Cenozoic of the Caribbean region. Figure 4.6 (Locality 1, Christiana-Bryce, slide 11, ESF L-36, size 36 by 30 μm) is another type of *Retimonocolpites* and is also a member of the Palmae.

Figure 4. Fossil palynomorphs from the Guys Hill Member of the Chapelton Formation, middle Eocene, Jamaica (pollen/spore size given in text). 1, *Deltoidospora* (cf. *D. adriennis*); 2–4, *Psilatricolporites crassus*; 5, 6, *Retimonocolpites*; 7, *Bombacacidites*; 8, *Cupanieidites*; 9, *Corsinipollenites*; 10, *Mauritiidites*; 11–14, Dinoflagellates (11, Body 27 μm, processes to 32 μm; 12, 42 μm, to 15 μm; 13, 63 × 42 μm, to 6 μm; 14, 57 × 45 μm, to 5 μm).

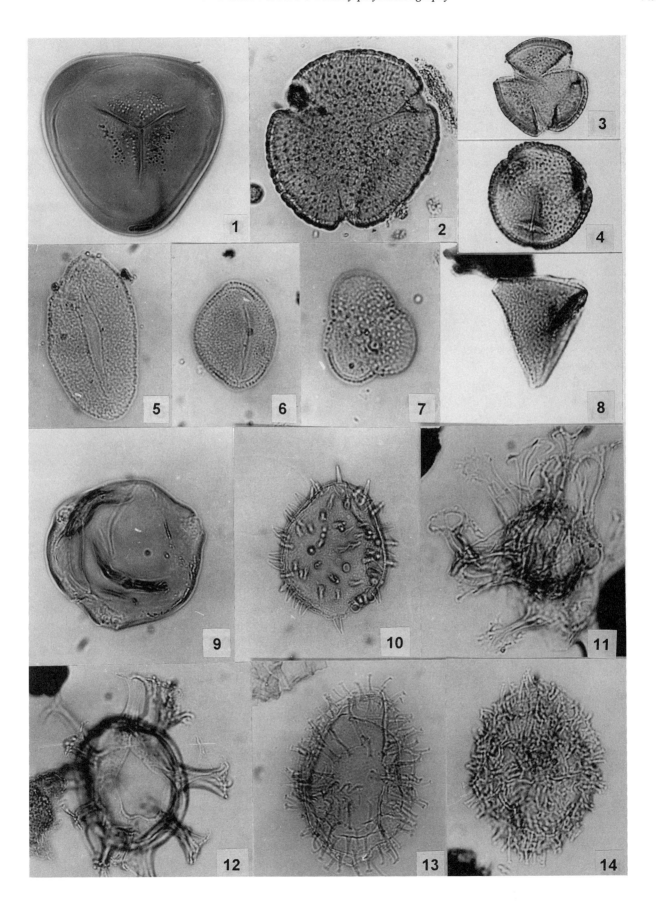

Bombacacidites sp.
(Fig. 4.7)

Oblate, amb triangular, apices rounded; tricolpate, colpi short (5 to 6 µm, apex to equator), equatorially arranged, meridionally elongated, equidistant; tectate-perforate, wall 2 to 3 µm thick; finely reticulate (diameter of lumen ca. 1 to 2 µm), becoming finer toward the apices; 26 to 32 µm. Locality 8 (Wait-a-Bit), slide 1, ESF H-31, 2-4.

These grains are small for the Bombacaceae and their affinities are not established beyond the family. They are abundant in the Guys Hill Member, but I have not found the exact type illustrated elsewhere in the literature. Other types of Bombacaceae pollen are widespread in the Cenozoic of the Caribbean region, and the family is a prominent member of the modern tropical vegetation.

Cupanieidites sp.
(Fig. 4.8)

Oblate, amb distinctly triangular; tricolporate, colpi faint, straight, narrow, equatorially arranged, equidistant, syncolpate(?), pores situated at midpoint of colpus on equator (apices of grain), 3 to 4 µm diameter; tectate-perforate, wall 3 µm thick; finely reticulate; 30 to 32 µm. Slide from J. Bujak, Central Inlier, ESF R-46, 2-4.

The specimen is similar to pollen of the family Sapindaceae, but an exact match could not be found among pollen of extant genera in our reference collection. The family is widespread in the neotropics, and pollen of the Sapindaceae ranges throughout the Cenozoic section in the region.

Corsinipollenites sp.
(Fig. 4.9)

Oblate, amb oval-triangular; triporate, pores equatorially arranged, equidistant, 6 µm diameter, protruding, surrounded by granular costae colpi; tectate, wall 3 to 4 µm thick; psilate to faintly scabrate; 55 µm. Locality 1 (Christiana-Bryce), slide 1, ESF H-39, 2.

The specimen is a corroded pollen of the Onagraceae and cannot be referred to a modern genus. The pollen of the family is frequent but never abundant in Cenozoic deposits of northern Latin America.

Mauritiidites sp.
(Fig. 4.10)

Prolate-spheroidal, amb oval; monosulcate(?)-aperture obscure); tectate, wall 3 µm thick; scabrate, echinate, spines 5 to 8 µm long, gradually tapering from base to apex (no collar evident at base of spines); 50 to 56 × 42 to 45 µm (excluding spines). Locality 1 (Christiana-Bryce), slide 1, ESF K-46, 3.

This pollen type is similar to that of the modern palm *Mauritia*, widespread in northern South America. The specimens are frequent but not abundant in the Cenozoic of northern Latin America.

Depositional Environments and Age

Lignitic materials containing mangrove pollen are deposited under coastal, brackish-water conditions in warm-temperate to tropical climates. The sediments accumulate in lagoons, bays, estuaries, and other low-energy environments where coastal currents do not disperse the organic material. The Guys Hill carbonaceous material contains fine-grained sand and other mineral particles, suggesting the organic material was transported a short distance into near-shore marine waters (Robinson, 1976). This view is supported by two facets of the palynomorph assemblage. There is a relatively diverse dinoflagellate component (Figs. 4.11–4.14) mixed with the pollen and spores. These are marine organisms and have been recovered from other lignitic sediments accumulating in near off-shore environments (e.g., the La Boca Formation of Panama; Graham, 1989, Figs. 2–4). Also, the preservation of the palynomorphs is only fair, and most are thick-walled forms suggesting differential preservation commonly associated with transport. The plant microfossils recovered to date from these sediments are useful for purposes of stratigraphy and the reconstruction of depositional environments, but provide a limited and biased picture of the vegetation.

The most useful aspect of the Guys Hill assemblage for making an age assignment is the absence of five pollen types, common in other Cenozoic deposits in northern Latin America, with ranges presently extending from the Recent back only into the late Eocene—*Ceratopteris* (*Magnastriatites howardii*, Fig. 3.1); *Hampea/Hibiscus* (probably akin to the mangrove-associated *Hibiscus tiliaceus*; *Echiperiporites estelae*, Fig. 3.2); *Crudia* (*Striatricolpites catatumbus*, Fig. 3.5); Gramineae (grasses, *Monoporites annuloides*, Fig. 3.6); and especially *Rhizophora* (*Zonocostites ramonae*, Fig. 3.7). The latter is abundant in lignitic sediments of Cenozoic age in northern Latin America, reaching values of 60% in samples from the Cucarcha Formation (Graham, 1988c, Table 1), 88% in the La Boca Formation (Graham, 1989, Table 1), and present virtually to the exclusion of other palynomorphs in additional miscellaneous material from the region (e.g., in core material from the Cucaracha Formation). The lithology of the Guys Hill sediments, and the presence of *Pelliceria* (and possibly *Acrostichum*), along with an abundance of independent geological and paleontological evidence (Robinson, 1987), document deposition in coastal habitats. Thus, conditions were ideal for *Rhizophora* had it been present in the region. The absence of *Rhizophora,* and the other late Eocene palynomorphs, together with the generally modern aspect of all the specimens, argues for an age just older than late Eocene, an estimate consistent with the middle Eocene assignment based on other stratigraphic and paleontological evidence (Robinson, 1987).

ACKNOWLEDGMENTS

Field work in Jamaica was facilitated by Stephen Donovan, and valuable information on the Tertiary geology of Jamaica and the Chapelton Formation was gained through discussions with Edward Robinson. His cooperation in the project is sincerely appreciated. The author gratefully acknowledges review of the manuscript by Norman Frederiksen, David Jarzen, and Edward Robinson. Research supported by NSF grant BSR-8819771.

APPENDIX I. PLANT MEGAFOSSILS REPORTED FROM THE GULF/CARIBBEAN TERTIARY

Many of the identifications made in the early 1900s are likely incorrect, and none of the floras have been revised. Spelling of generic names and family assignments follow that in the original citations listed in parentheses. Asterisk (*) designates taxa of unknown generic affinity. References refer to papers cited in Appendix 3.

Arranged by Age/Country

Tertiary undifferentiated (most mid- to late Tertiary, probably Oligo-Miocene to Miocene)
Mexico
*Palmoxylon, *Acacioxylon (Felix and Nathorst, 1893).
Juglans (Mullerried, 1938).
Costa Rica
Anona, Buttneria?, Ficus, Goeppertia, Heliconia, Hieronymia, Inga, Nectandra, *Phyllites, *Piperites (Berry, 1921a).
*Palmacites (Gomez-P., 1971).
*Karatophyllum (Gomez-P., 1972).
Panama
Banisteria, Calyptranthes, Cassia, Diospyros, Ficus, Guatteria, Hieronymia, Hiraea, Inga, *Melastomites, Mespilodaphne, *Myristicophyllum, *Palmoxylon, Rondeletia, *Rubiacites, Schmidelia, *Taenioxylon (Berry, 1918).
Antigua
*Palmoxylon (Stenzel, 1897).
Dominican Republic
Bucida, Bumelia, Calyptranthes, Guettarda, Inga, *Melastomites, Pisonia, Pithecolobium, *Poacites, Sapindus, Sophora (Berry, 1921).
Haiti
Chara, Gymnogramme, Bumelia, Chrysophyllum, Guettarda, Mespilodaphne, Mimusops, Pisonia, Simaruba (Berry, 1923).
Miocene
Mexico
Acrostichum, Gymnogramme, Allamanda, *Anacardites, Anona, *Apocynophyllum, *Bignonoides, Cedrela, Connarus, Coussapoa, Crescentia, Dioclea?, Drypetes, Fagara, Ficus?, Goeppertia, Gouania, Guettarda, Gymnocladus, Inga, *Lecythidophyllum, *Leguminosites, Liquidambar, *Melastomites, Mespilodaphne, Moquillea, Myrcia, Nectandra, Rondeletia?, Simaruba (Berry, 1923).
Panama
*Iriartites (Berry, 1921b), *Palmocarpon (Berry, 1928).
Antigua
Phytelephas (Kaul, 1943).
Cuba
Gleichenia, *Antholithus, Bignonia, Bumelia, Caesalpinia, *Caesalpinites, Calophyllum, Capparis, Cassia, Celastrus, Dalbergia, Dodonaea, Drypetes, Eugenia, Exostema, Fagara, Heliconia, Inga, Laguncularia, Metopium, Mimusops, Myrcia, Pisonia, Pithecolobium, Pseudolmedia, Reynosia, Rheedia, Sapindus, Simaruba, Sophora, Swietenia, Trichilia, Zizyphus (Berry, 1939).
Dominican Republic
Acacia (Dilcher et al., 1990).
Oligo-Miocene
Mexico
Acacia, Tapirira (Miranda, 1963).
Costa Rica
*Mixoneura?, *Pecopteris?, Thelypteris? (Gomez-P., 1970; tentative identifications, especially for first two compared to genera of Paleozoic ferns; material from private collection).

Oligocene
Dominican Republic
Grammitis (Gomez-P., 1982).
Puerto Rico
Isoetes?, Hemitelia, Zamia, Acrodiclidium, Aniba, Annona, *Apocynophyllum, Aspidosperma, Bactris, Cassia, *Chondrites, Chrysophyllum, Combretum, Copaiva, Cynometra, Dipholis, Echites, Echitonium?, Eugenia, Ficus, Guarea, Guettarda, Hancornia, Hufelandia, Icacorea, Inga, Iriartea, Lonchocarpus, *Malvocarpon, Manicaria, Melicocca, Misanteca, *Mussophyllum, Myrcia, Myrsine, Oreodaphne, *Palmocarpon, *Palmacites, *Palmophyllum, Pithecellobium, Plumiera, Psidium, *Ramulus, Rhizophora?, Sapota, Sapindus, Sideroxylon, Sophora?, Trichilia, Zizyphus (Hollick, 1928).

Arranged Taxonomically

Incertae Sedis
*Antholithus (Miocene, Cuba, Berry, 1939).
*Chondrites (Oligocene, Puerto Rico, Hollick, 1928).
*Phyllites (Tertiary undifferentiated, Costa Rica, Berry, 1921a).
*Ramulus (Oligocene, Puerto Rico, Hollick, 1928).
Algae
Chara (Characeae; Tertiary undifferentiated, Haiti, Berry, 1923).
Lycopsida
Isoetes? (Isoetaceae; Oligocene, Puerto Rico, Hollick, 1928).
Filicineae (ferns)
Acrostichum (Pteridaceae; Miocene, Mexico, Berry, 1923).
Gleichenia (Gleicheniaceae; Miocene, Cuba, Berry, 1939).
Gymnogramme (Gymnogrammaceae; Tertiary undifferentiated, Haiti, Berry, 1923; Miocene, Mexico, Berry, 1923).
Grammitis (Polypodiaceae; Oligocene, Dominican Republic, Gomez-P., 1982).
Hemitelia (Cyatheaceae; Oligocene, Puerto Rico, Hollick, 1928).
*Mixoneura? (tentative identification of specimen in private collection, compared to genus of Paleozoic ferns; Oligo-Miocene(?), Costa Rica, Gomez-P., 1970).
*Pecopteris? (tentative identification of specimen in private collection, compared to genus of Paleozoic ferns; Oligo-Miocene(?), Costa Rica, Gomez-P., 1970).
Thelypteris? (Thelypteridaceae; tentative identification of specimen in private collection; Oligo-Miocene(?), Costa Rica, Gomez-P., 1970).
Gymnospermae
Zamia (Cycadaceae; Oligocene, Puerto Rico, Hollick, 1928).
Angiospermae
Acacia (Leguminosae, Subfam. Mimosoideae; Oligo-Miocene, Mexico, Miranda, 1963; Miocene, Dominican Republic, Dilcher et al., 1990).
*Acacioxylon (Leguminosae, Subfam. Mimosoideae; Tertiary undifferentiated, Mexico, Felix and Nathorst, 1893).
Acrodiclidium (Lauraceae; Oligocene, Puerto Rico, Hollick, 1928).
Allamanda (Apocynaceae; Miocene, Mexico, Berry, 1923).
*Anacardites (Anacardiaceae; Miocene, Mexico, Berry, 1923).
Aniba (Lauraceae; Oligocene, Puerto Rico, Hollick, 1928).
Annona/Anona (Annonaceae; Tertiary undifferentiated, Costa Rica, Berry, 1921a; Miocene, Mexico, Berry, 1923; Oligocene, Puerto Rico, Hollick, 1928).
*Apocynophyllum (Apocynaceae; Miocene, Mexico, Berry, 1923; Oligocene, Puerto Rico, Hollick, 1928).
Aspidosperma (Apocynaceae; Oligocene, Puerto Rico, Hollick, 1928).
Bactris (Palmae; Oligocene, Puerto Rico, Hollick, 1928).
Banisteria (Malpighiaceae; Tertiary undifferentiated, Panama, Berry, 1918).
Bignonia (Bignoniaceae; Miocene, Cuba, Berry, 1939).

Bignonoides (Bignoniaceae; Miocene, Mexico, Berry, 1923).
Bucida (Combretaceae; Tertiary undifferentiated, Dominican Republic, Berry, 1921).
Bumelia (Sapotaceae; Tertiary undifferentiated, Dominican Republic, Berry, 1921; Tertiary undifferentiated, Haiti, Berry, 1923; Miocene, Cuba, Berry, 1939).
Buttneria? (Sterculiaceae; Tertiary undifferentiated, Costa Rica, Berry, 1921a).
Caesalpinia (Leguminosae, Subfam. Caesalpinioideae; Miocene, Cuba, Berry, 1939).
Caesalpinites (Leguminosae, Subfam. Caesalpinioideae; Miocene, Cuba, Berry, 1939).
Calophyllum (Guttiferae; Miocene, Cuba, Berry, 1939).
Calyptranthes (Myrtaceae; Tertiary undifferentiated, Panama, Berry, 1918; Tertiary undifferentiated, Dominican Republic, Berry, 1921).
Capparis (Capparidaceae; Miocene, Cuba, Berry, 1939).
Cassia (Leguminosae, Subfam. Caesalpinioideae; Tertiary undifferentiated, Panama, Berry, 1918; Miocene, Cuba, Berry, 1918; Miocene, Cuba, Berry, 1939; Oligocene, Puerto Rico, Hollick, 1928).
Cedrela (Meliaceae; Miocene, Mexico, Berry, 1923).
Celastrus (Celastraceae; Miocene, Cuba, Berry, 1939).
Chrysophyllum (Sapotaceae; Tertiary undifferentiated, Haiti, Berry, 1923; Oligocene, Puerto Rico, Hollick, 1928).
Combretum (Combretaceae; Oligocene, Puerto Rico, Hollick, 1928).
Connarus (Connaraceae; Miocene, Mexico, Berry, 1923).
Copaiva (=*Copaifera*; Leguminosae, Subfam. Caesalpinioideae; Oligocene, Puerto Rico, Hollick, 1928).
Coussapoa (Urticaceae; Miocene, Mexico, Berry, 1923).
Crescentia (Bignoniaceae; Miocene, Mexico, Berry, 1923).
Cynometra (Leguminosae, Subfam. Caesalpinioideae; Oligocene, Puerto Rico, Hollick, 1928).
Dalbergia (Leguminosae, Subfam. Papilionoideae; Miocene, Cuba, Berry, 1939).
Dioclea? (Leguminosae, Subfam. Papilionoideae; Miocene, Mexico, Berry, 1923).
Diospyros (Ebenaceae; Tertiary undifferentiated, Panama, Berry, 1918).
Dipholis (Sapotaceae; Oligocene, Puerto Rico, Hollick, 1928).
Dodonaea (Sapindaceae; Miocene, Cuba, Berry, 1939).
Drypetes (Euphorbiaceae; Miocene, Mexico, Berry, 1923; Miocene, Cuba, Berry, 1939).
Echites (Apocynaceae; Oligocene, Puerto Rico, Hollick, 1928).
Echitonium? (Apocynaceae; Oligocene, Puerto Rico, Hollick, 1928).
Eugenia (Myrtaceae; Miocene, Cuba, Berry, 1939; Oligocene, Puerto Rico, Hollick, 1928).
Exostema (Rubiaceae; Miocene, Cuba, Berry, 1939).
Fagara (fossils subsequently referred to *Xanthoxylum* = *Zanthoxylum*, Rutaceae; Miocene, Mexico, Berry, 1923; Miocene, Cuba, Berry, 1939).
Ficus (Moraceae; Tertiary undifferentiated, Costa Rica, Berry, 1921a; Tertiary undifferentiated, Panama, Berry, 1918; Oligocene, Puerto Rico, Hollick, 1928; *Ficus*?, Miocene, Mexico, Berry, 1923).
Goeppertia (Lauraceae; Tertiary undifferentiated, Costa Rica, Berry, 1921a; Miocene, Mexico, Berry, 1923).
Gouania (Rhamnaceae; Miocene, Mexico, Berry, 1923).
Guarea (Meliaceae; Oligocene, Puerto Rico, Hollick, 1928).
Guatteria (Annonaceae; Tertiary undifferentiated, Panama, Berry, 1918).
Guettarda (Rubiaceae; Tertiary undifferentiated, Dominican Republic, Berry, 1921; Tertiary undifferentiated, Haiti, Berry, 1923; Miocene, Mexico, Berry, 1923; Oligocene, Puerto Rico, Hollick, 1928).
Gymnocladus (Leguminosae, Subfam. Caesalpinioideae; Miocene, Mexico, Berry, 1923).
Hancornia (Apocynaceae; Oligocene, Puerto Rico, Hollick, 1928).
Heliconia (Mussaceae; Tertiary undifferentiated, Costa Rica, Berry, 1921a; Miocene, Cuba, Berry, 1939).
Hieronymia (Euphorbiaceae; Tertiary undifferentiated, Costa Rica, Berry, 1921a).
Hiraea (Malpighiaceae; Tertiary undifferentiated, Panama, Berry, 1918).
Hufelandia (Lauraceae; Oligocene, Puerto Rico, Hollick, 1928).
Icacorea (Myrsinaceae; Oligocene, Puerto Rico, Hollick, 1928).
Inga (Leguminosae, Subfam. Mimosoideae; Tertiary undifferentiated, Costa Rica, Berry, 1921a; Tertiary undifferentiated Panama, Berry, 1918; Tertiary undifferentiated, Dominican Republic, Berry, 1921; Miocene, Mexico, Berry, 1923; Miocene, Cuba, Berry, 1939; Oligocene, Puerto Rico, Hollick, 1928).
Iriartea (Palmae; Oligocene, Puerto Rico, Hollick, 1928).
Iriarites (Palmae; Miocene, Panama, Berry, 1921b).
Juglans (Juglandaceae; Tertiary undifferentiated, Mexico, Mullerried, 1938).
Karatophyllum (Bromeliaceae; Tertiary undifferentiated, Costa Rica, Gomez-P., 1972).
Laguncularia (Combretaceae; Miocene, Cuba, Berry, 1939).
Lecythidophyllum (Lecythidaceae; Miocene, Mexico, Berry, 1923).
Leguminosites (Leguminosae; Miocene, Mexico, Berry, 1923).
Liquidambar (Hamamelidaceae; Miocene, Mexico, Berry, 1923).
Lonchocarpus (Leguminosae, Subfam. Papilionoideae; Oligocene, Puerto Rico, Hollick, 1928).
Malvocarpon (Malvaceae; Oligocene, Puerto Rico, Hollick, 1928).
Manicaria (Palmae; Oligocene, Puerto Rico, Hollick, 1928).
Melastomites (Melastomataceae; Tertiary undifferentiated, Panama, Berry, 1918; Tertiary undifferentiated, Dominican Republic, Berry, 1921; Miocene, Mexico, Berry, 1923).
Melicocca (=*Melicoccus*, Sapindaceae; Oligocene, Puerto Rico, Hollick, 1928).
Mespilodaphne (Lauraceae; Tertiary undifferentiated, Panama, Berry, 1918; Tertiary undifferentiated, Haiti, Berry, 1923; Miocene, Mexico, Berry, 1923).
Metopium (Anacardiaceae; Miocene, Cuba, Berry, 1939).
Mimusops (Sapotaceae; Tertiary undifferentiated, Haiti, Berry, 1923; Miocene, Cuba, Berry, 1939).
Misanteca (Lauraceae; Oligocene, Puerto Rico, Hollick, 1928).
Moquillea (Chrysobalanaceae; Miocene, Mexico, Berry, 1923).
Mussophyllum (Mussaceae; Oligocene, Puerto Rico, Hollick, 1928).
Myrcia (Myrtaceae; Miocene, Mexico, Berry, 1923; Miocene, Cuba, Berry, 1939; Oligocene, Puerto Rico, Hollick, 1928).
Myristicophyllum (Myristicaceae; Tertiary undifferentiated, Panama, Berry, 1918).
Myrsine (Myrsinaceae; Oligocene, Puerto Rico, Hollick, 1928).
Nectandra (Lauraceae; Tertiary undifferentiated, Costa Rica, Berry, 1921a; Miocene, Mexico, Berry, 1923).
Oreodaphne (Lauraceae; Oligocene, Puerto Rico, Hollick, 1928).
Palmacites (Palmae; Tertiary undifferentiated, Costa Rica, Gomez-P., 1971; Oligocene, Puerto Rico, Hollick, 1928).
Palmocarpon (Palmae; Miocene, Panama, Berry, 1928; Oligocene, Puerto Rico, Hollick, 1928).
Palmophyllum (Palmae; Oligocene, Puerto Rico, Hollick, 1928).
Palmoxylon (Palmae; Tertiary undifferentiated, Mexico, Felix and Nathorst, 1893; Tertiary undifferentiated, Panama, Berry, 1918; Tertiary undifferentiated, Antigua, Stenzel, 1897).
Phytelephas (Palmae; Miocene, Antigua, Kaul, 1943).
Piperites (Piperaceae; Tertiary undifferentiated, Costa Rica, Berry, 1921a).
Pisonia (Nyctaginaceae; Tertiary undifferentiated, Dominican Republic, Berry, 1921; Tertiary undifferentiated, Haiti, Berry, 1923; Miocene, Cuba, Berry, 1939).
Pithecolobium (=*Pithecellobium*; Leguminosae, Subfam. Mimosoideae; Tertiary undifferentiated, Dominican Republic, Berry, 1921;

Miocene, Cuba, Berry, 1939; Oligocene, Puerto Rico, Hollick, 1928).
Plumiera (Apocynaceae; Oligocene, Puerto Rico, Hollick, 1928).
**Poacites* (Gramineae; Tertiary undifferentiated, Dominican Republic, Berry, 1921).
Pseudolmedia (Moraceae; Miocene, Cuba, Berry, 1939).
Psidium (Myrtaceae; Oligocene, Puerto Rico, Hollick, 1928).
Reynosia (Rhamnaceae; Miocene, Cuba, Berry, 1939).
Rheedia (Guttiferae; Miocene, Cuba, Berry, 1939).
Rhizophora? (Rhizophoraceae; Oligocene, Puerto Rico, Hollick, 1928).
Rondeletia (Rubiaceae; Tertiary undifferentiated, Panama, Berry, 1918; *Rondeletia*?, Miocene, Mexico, Berry, 1923).
**Rubiacites* (Rubiaceae; Tertiary undifferentiated, Panama, Berry, 1918).
Sapindus (Sapindaceae; Tertiary undifferentiated, Dominican Republic, Berry, 1921; Miocene, Cuba, Berry, 1939; Oligocene, Puerto Rico, Hollick, 1928).
Sapota (Sapotaceae; Oligocene, Puerto Rico, Hollick, 1928).
Schmidelia (Sapindaceae; Tertiary undifferentiated, Panama, Berry, 1918).
Sideroxylon (Sapotaceae; Oligocene, Puerto Rico, Hollick, 1928).
Simaruba (Simarubaceae = Simaroubaceae; Tertiary undifferentiated, Haiti, Berry, 1923; Miocene, Mexico, Berry, 1923; Miocene, Cuba, Berry, 1939).
Sophora (Leguminosae, Subfam. Papilionoideae; Tertiary undifferentiated, Dominican Republic, Berry, 1921; Miocene, Cuba, Berry, 1939; *Sophora*?, Oligocene, Puerto Rico, Hollick, 1928).
Swietenia (Meliaceae; Miocene, Cuba, Berry, 1939).
**Taenioxylon* (Leguminosae, Subfam. unassigned, Papilionoideae?; Tertiary undifferentiated, Panama, Berry, 1918).
Tapirira (Anacardiaceae; Oligo-Miocene, Mexico, Miranda, 1963).
Trichilia (Meliaceae; Miocene, Cuba, Berry, 1939; Oligocene, Puerto Rico, Hollick, 1928).
Zizyphus (Rhamnaceae; Miocene, Cuba, Berry, 1939; Oligocene, Puerto Rico, Hollick, 1928).

Family Index

Incertae Sedis
  Antholithus
  Chondrites
  Phyllites
  Ramulus
Charophyta (algae)
  Characeae
    Chara
Lycopsida
  Isoetaceae
    Isoetes?
Filicineae (ferns)
  Cyatheaceae
    Hemitelia
  Gleicheniaceae
    Gleichenia
  Gymnogrammaceae
    Gymnogramme
  Pteridaceae
    Acrostichum
  Polypodiaceae
    Grammitis
  Thelypteridaceae
    Thelypteris?
  Uncertain
    **Mixoneura*?
    **Pecopteris*?
Gymnospermae
  Zamia
Angiospermae
  Anacardiaceae
    **Anacardites*
    Metopium
    Tapirira
  Annonaceae
    Annona/Anona
    Guatteria
  Apocynaceae
    Allamanda
    **Apocynophyllum*
    Aspidosperma
    Echites
    Echitonium?

Angiospermae (continued)
  Hancornia
  Plumiera
Bignoniaceae
  Bignonia
  **Bignonoides*
  Crescentia
Bromeliaceae
  **Karatophyllum*
Capparidaceae
  Capparis
Celastraceae
  Celastrus
Chrysobalanaceae
  Moquillea
Combretaceae
  Bucida
  Combretum
  Laguncularia
Connaraceae
  Connarus
Ebenaceae
  Diospyros
Euphorbiaceae
  Drypetes
  Hieronymia
Gramineae
  **Poacites*
Guttiferae
  Calophyllum
  Rheedia
Hamamelidaceae
  Liquidambar
Juglandaceae
  Juglans
Lauraceae
  Acrodiclidium
  Goeppertia
  Hufelandia
  Mespilodaphne
  Misanteca
  Nectandra
  Oreodaphne

Angiospermae (continued)
  Lecythidaceae
    **Lecythidophyllum*
  Leguminosae
    Caesalpinioideae
      Caesalpinia
      **Caesalpinites*
      Cassia
      Cynometra
      Gymnocladus
    Mimosoideae
      Acacia
      **Acacioxylon*
      Inga
      Pithecolobium
    Papilionoideae
      Dalbergia
      Dioclea?
      Lonchocarpus
      Sophora
    Subfam. unassigned
      **Leguminosites*
      **Taenioxylon*
  Malpighiaceae
    Banisteria
    Hiraea
  Malvaceae
    **Malvocarpon*
  Melastomataceae
    **Melastomites*
  Meliaceae
    Cedrela
    Guarea
    Swietenia
    Trichilia
  Moraceae
    Ficus
    Pseudolmedia
  Mussaceae
    **Mussophyllum*
  Myristicaceae
    **Myristicophyllum*
  Myrsinaceae

Angiospermae (continued)
  Icacorea
  Myrsine
Myrtaceae
  Calyptranthes
  Eugenia
  Myrcia
  Psidium
Nyctaginaceae
  Pisonia
Palmae
  Bactris
  Iriartia
  **Iriartites*
  Manicaria
  **Palmacites*
  **Palmocarpon*
  **Palmophyllum*
  **Palmoxylon*
  Phytelephas
Piperaceae
  **Piperites*
Rhamnaceae
  Gouania
  Reynosia
  Zizyphus
Rhizophoraceae
  Rhizophora?
Rubiaceae
  Exostema
  Guettarda
  Rondeletia
  **Rubiacites*
Rutaceae
  Fagara
Sapindaceae
  Dodonaea
  Melicocca (=*Melicoccus*)
  Sapindus
  Schmidelia
Sapotaceae
  Bumelia
  Chrysophyllum

454 A. Graham

Angiospermae (continued)
Dipholis
Mimusops
Sapota
Sideroxylon
Simarubaceae
Simaruba
Sterculiaceae
Buttneria?
Urticaceae
Coussapoa

APPENDIX II. PLANT MICROFOSSILS REPORTED FROM THE GULF/CARIBBEAN TERTIARY

Asterisk (*) designates genera of exclusively fossil plants. References refer to papers cited in Appendix 3.

Arranged by Age/Country

Eocene
  Panama
    Selaginella, Ceratopteris, Pteris, Alfaroa/Engelhardia, cf. Araliaceae, Arrabidaea (as cf. Paragonia/Arrabidaea), cf. Campnosperma, Cardiospermum, Casearia, cf. Chrysophyllum, Coccoloba, Combretum/Terminalia, Crudia, Engelhardia (as Alfaroa/Engelhardia), Eugenia/Myrcia, Faramea, cf. Ficus, Ilex, Lisianthius, Malpighiaceae, Mortoniodendron, Palmae, cf. Paragonia/Arrabicaea, Paullinia, Pelliceria, cf. Protium, Rhizophora, Serjania, Terminalia (as Combretum/Terminalia), cf. Tetragrastis, cf. Tillandsia, cf. Tontalea (Graham, 1985; Lisianthius also Graham, 1984; Pelliceria also Graham, 1977).
  Cuba
    *Bombacacidites (Areces-Mallea, 1985); *Ornatisporites, Lygodiumsporites, *Concavisporites, *Laevigatosporites, *Polypodiisporites, *Arecipites, *Monocolpopollenites, *Liliacidites, *Graminidites (Areces-Mallea, 1988).
  Jamaica
    Pelliceria (Graham, 1977).
Oligocene
  Puerto Rico
    Lycopodium, Selaginella, Cyathea, Hemitelia (Cnemidaria), Jamesonia (Eriosorus), Pteris, Podocarpus, Abutilon, Acacia, Aetanthus, Alchornea, Bernoullia, Bombax, Brunellia, Bursera, Casearia, Catostemma, Chrysophyllum, Corynostylis, Dendropanax, Engelhardia, Eugenia, Fagus, Faramea, Guarea, Hauya, Ilex, Jacaranda, Liquidambar, Marcgravia, Merremia, Myrcia, Norantea, Nyssa, Oxalis, Palmae, Pelliceria, Pleodendron, Rauwolfia, Rhizophora, Salix, Tecoma, Tetrorchidium, Tournefortia, Zanthophyllum (Graham and Jarzen, 1969).
  Cuba
    Pinus (Areces-Mallea, 1987).
Oligo-Miocene
  Mexico
    Podocarpus, Engelhardia, Pachira-type, Pelliceria, Rhizophora (Langenheim et al., 1967).
  Dominican Republic
    *Geotrichites (fungi; Stubblefield et al., 1985).
Lower Miocene
  Costa Rica
    Microthyrium-type (fungi), Phaeoceros (Bryophyta), Muscae, Lycopodium, Selaginella, Cnemidaria (Hemitelia), cf. Hymenophyllum, Lophosoria, Pityrogramma, Pteris, Podocarpus, Alchornea, cf. Banisteriopsis, Bombacaceae, Compositae, Ericaceae, Eugenia/Myrcia, cf. Glycydendrum, cf. Hiraea, Ilex, Lisianthus,

Melastomataceae, Rhizophora (Graham, 1987a).
  Panama
    Micractinium (algae-Chlorophyta, Graham, 1981), *Operculodinium, *Spiniferites [algae-Pyrrophpyta (dinoflagellates)], Ascomycete cleistothecium (fungi), Lycopodium, Selaginella, cf. Anthophyum, Ceratopteris, Cyathea, Danaea, Lygodium, Pteris, Acacia, cf. Aguiaria, Alchornea, Alfaroa/Engelhardia, Allophylus, Casearia, cf. Ceiba, Chenopodiaceae/Amaranthaceae, Combretum/Terminalia, Compositae, Crudia, Cupania, Dioscorea/Rajania, cf. Doliocarpus, Eugenia/Myrcia, cf. Gramineae, Guazuma, Hampea/Hibiscus, Ilex, Malpighiaceae, Matayba, Mortoniodendron (Graham, 1979). Pelliciera (Graham, 1977), cf. Pouteria, Pseudobombax, Rhizophora, cf. Rourea, Sabicea (Graham, 1987b), cf. Sapium, Tetrorchidium, Utricularia; (Graham, 1987a, 1988a, b, 1989).
Mio-Pliocene
  Mexico
    Psilotum, Lycopodium, Selaginella, Alsophila, Ceratopteris, Cnemidaria [as Hemitelia, =Cnemidaria], Cyathea, Dicranopteris, Hemitelia (Cnemidaria), Lomariopsis (Stenochlaena), Pityrogramma, Pteris, Sphaeropteris/Trichipteris, Stenochleana [as Lomariopsis (Stenochlaena)], Trichipteris (as Sphaeropteris/Trichipteris, Abies, Picea, Pinus, Podocarpus, cf. Acacia, Alchornea, cf. Alibertia, Allophylus, Alnus, Amaranthaceae (as Amaranthaceae-Chenopodiaceae) cf. Astrocaryum, cf. Attalea, cf. Bernardia, Borreria, cf. Brahea, Bravaisia, cf. Bredemeyera, Buettneria, Bursera, Casearia, Cedrela, Celtis, cf. Chamaedorea, Chenopodiaceae-Amaranthaceae, Cleyera, Coccoloba, Combretum/Terminalia, Comocladia, Compositae, Cupania, Cuphea, Cyperaceae. Daphnopsis, Desmanthus, Dichapetalum, Engelhardia, Eugenia/Myrcia, Faramea, Gramineae, Guarea, Gustavia, Hampea/Hibiscus, Hedyosmum, cf. Hiraea, Ilex, Iresine, Juglans, Justicia, Laetia, Laguncularia, Liquidambar, Ludwigia, cf. Malpighia, Matayba, cf. Maximiliana type, Meliosma, cf. Mezia(?)-type, Mimosa, Mortoniodendron, Myrica, Passiflora, cf. Paullinia, Populus, Protium, Quercus, Rajania, Rhizophora, cf. Sapium, cf. Securidaca, Serjania, Smilax, Spathiphyllum, cf. Stillingia, Struthanthus, Symphonia, Terebrania, Terminalia (as Combretum/Terminalia), cf. Tetrorchidium, Thalictrum, cf. Tithymalus, Tournefortia, Ulmus, Utricularia (Graham, 1976); Mortoniodendron, Sphaeropteris/Trichipteris (also Graham, 1979).
  Panama
    Lycopodium, Selaginella, Alsophila, Ceratopteris, Cnemidaria, Ctenitis, Cyathea, Grammitis, Ophioglossum, Pteris, Podocarpus, Acacia, Aegiphila, cf. Aguiaria, Alchornea, Alfaroa/Oreomunnea, Allophylus, Bernoullia, cf. Bucida, cf. Bumelia, Bursera, Cabomba, Casimiroa, Cedrela, Ceiba, Chomelia type, cf. Cionosicys, Combretum/Terminalia, Cosmibuena, Crudia, Cupania, Cymbopetalum, Ericaceae, Erythrina, Eugenia/Myrcia, Faramea, cf. Glycydendrum, Gramineae, Guarea, Hauya, Hampea/Hibiscus, Ilex, cf. Jatropha, Loranthaceae, Malpighiaceae, Melastomataceae, Mortoniodendron, Mutisieae type, Palmae, Paullinia, Petraea, Posqueria, Pseudobombax, Quercus, Rhizophora, Serjania, cf. Stillingia, Symplocos, Utricularia (Graham, 1991a, b, c).
  Haiti
    Alsophila, cf. Anthophyum, Pteris, Pinus, Alchornea, Alfaroa/Oreomunnea, Chenopodiaceae/Amaranthaceae, Compositae, Hygrophila, Malpighiaceae, Oryctanthus, Palmae (Graham, 1990).

Arranged Taxonomically

Algae
  Micractinium (Chlorophyta; lower Miocene, Panama, Graham, 1989; see also Graham, 1981).
  *Operculodinium, *Spiniferites [Pyrrophyta (dinoflagellates); lower

Miocene, Panama, Graham, 1989).

Fungi
Ascomycete cleistothecium (Ascomycetes; lower Miocene, Panama, Graham, 1989).
Geotrichites (Deuteromycotina, fungi imperfecti; Oligo-Miocene, Dominican Republic, Stubblefield et al., 1985).
Microthyrium-type (Ascomycetes, Microthyriaceae; lower Miocene, Costa Rica, Graham, 1987a).

Bryophyta
Phaeoceros (Anthocerotaceae; lower Miocene, Costa Rica, Graham, 1987a).
Muscae (lower Miocene, Costa Rica, Graham, 1987a).

Psilopsida
Psilotum (Psilotaceae; Mio-Pliocene, Mexico, Graham, 1976).

Lycopsida
Lycopodium (Lycopodiaceae; Oligocene, Puerto Rico, Graham and Jarzen, 1969; lower Miocene, Costa Rica, Graham, 1987a; lower Miocene, Panama, Graham, 1988a, 1989; Mio-Pliocene, Mexico, Graham, 1976; Mio-Pliocene, Panama, Graham, 1991a).
Selaginella (Selaginellaceae; Eocene, Panama, Graham, 1985; Oligocene, Puerto Rico, Graham and Jarzen, 1969; lower Miocene, Costa Rica, Graham, 1987a; lower Miocene, Panama, Graham, 1988a, b, 1989; Mio-Pliocene, Mexico, Graham, 1976; Mio-Pliocene, Panama, Graham, 1991a).

Filicineae (ferns)
Alsophila (Cyatheaceae; Mio-Pliocene, Mexico, Graham, 1976; Mio-Pliocene, Panama, Graham, 1991a; Mio-Pliocene, Haiti, Graham, 1990).
cf. *Antrophyum* (Vittariaceae; lower Miocene, Panama, Graham, 1988a, b, 1989; Mio-Pliocene, Haiti, Graham, 1990).
Ceratopteris (Pteridaceae; Eocene, Panama, Graham, 1985; lower Miocene, Panama, Graham, 1988b; Mio-Pliocene, Mexico, Graham, 1976; Mio-Pliocene, Panama, Graham, 1991a).
Cnemidaria [also as *Hemitelia* (*Cnemidaria*), Cyatheaceae; Oligocene, Puerto Rico, Graham and Jarzen, 1969; lower Miocene, Costa Rica, Graham, 1987a; Mio-Pliocene, Mexico, Graham, 1976; Mio-Pliocene, Panama, Graham, 1991a].
Concavisporites (Cyatheacea; Eocene, Cuba, Areces-Mallea, 1988).
Ctenitis (Dryopteridaceae; Mio-Pliocene, Panama, Graham, 1991a).
Cyathea (Cyatheaceae; Oligocene, Puerto Rico, Graham and Jarzen, 1969; lower Miocene, Panama, Graham, 1988a, b, 1989; Mio-Pliocene, Mexico, Graham, 1976; Mio-Pliocene, Panama, Graham, 1991a).
Danaea (Marattiaceae; lower Miocene, Panama, Graham, 1988a).
Dicranopteris (Gleicheniaceae; Mio-Pliocene, Mexico, Graham, 1976).
Eriosorus (see *Jamesonia*).
Grammitis (Polypodiaceae; Mio-Pliocene, Panama, Graham, 1991a).
Hemitelia (see *Cnemidaria*).
cf. *Hymenophyllum* (Hymenophyllaceae; lower Miocene, Costa Rica, Graham, 1987a).
Jamesonia [as *Jamesonia* (*Eriosorus*), Pteridaceae; Oligocene, Puerto Rico, Graham and Jarzen, 1969].
Laevigatosporites (Polypodiaceae?; Eocene, Cuba, Areces-Mallea, 1988).
Lomariopsis [as *Lomariopsis* (*Stenochlaena*), Dryopteridaceae; Mio-Pliocene, Mexico, Graham, 1976].
Lophosoria (Lophosoriaceae; lower Miocene, Costa Rica, Graham, 1987a).
Lygodium (Schizaeaceae; lower Miocene, Panama, Graham, 1988a).
Lygodiumsporites (Schizaeaceae, Cyatheaceae?; Eocene, Cuba, Areces-Mallea, 1988).
Ophioglossum (Ophioglossaceae; Mio-Pliocene, Panama, Graham, 1991a).
Ornatisporites (Pteridaceae?; Eocene, Cuba, Areces-Mallea, 1988).

Pityrogramma (Pteridaceae; lower Miocene, Costa Rica, Graham 1987a; Mio-Pliocene, Mexico, Graham, 1976).
Polypodiisporites (Polypodiaceae, Pteridaceae?; Eocene, Cuba, Areces-Mallea, 1988).
Pteris (Pteridaceae; Eocene, Panama, Graham, 1985; Oligocene, Puerto Rico, Graham and Jarzen, 1969; lower Miocene, Costa Rica, Graham, 1987a; lower Miocene, Panama, Graham, 1988a, b, 1989; Mio-Pliocene, Mexico, Graham, 1976; Mio-Pliocene, Panama, Graham, 1991a; Mio-Pliocene, Haiti, Graham, 1990).
Sphaeropteris (as *Sphaeropteris/Trichipteris*, Cyatheaceae; Mio-Pliocene, Mexico, Graham, 1976).
Stenochlaena (see *Lomariopsis*).
Trichipteris (see *Sphaeropteris*).

Gymnospermae
Abies (Pinaceae; Mio-Pliocene, Mexico, Graham, 1976).
Picea (Pinaceae; Mio-Pliocene, Mexico, Graham, 1976).
Pinus (Pinaceae; Oligocene, Cuba, Areces-Mallea, 1987; Mio-Pliocene, Mexico, Graham, 1976; Mio-Pliocene, Haiti, Graham, 1990).
Podocarpus (Podocarpaceae; Oligocene, Puerto Rico, Graham and Jarzen, 1969; Oligo-Miocene, Mexico, Langenheim et al., 1967; lower Miocene, Costa Rica, Graham, 1987a; Mio-Pliocene, Mexico, Graham, 1976; Mio-Pliocene, Panama, Graham, 1991a).

Angiospermae
Abutilon (Malvaceae; Oligocene, Puerto Rico, Graham and Jarzen, 1969).
Acacia (Leguminosae, Subfam. Mimosoideae; Oligocene, Puerto Rico, Graham and Jarzen, 1969; lower Miocene, Panama, Graham, 1988a; Mio-Pliocene, Panama, Graham, 1991b; cf. *Acacia*, Mio-Pliocene, Mexico, Graham, 1976).
Aegiphila (Verbenaceae; Mio-Pliocene, Panama, Graham, 1991b).
Aetanthus (Loranthaceae; Oligocene, Puerto Rico, Graham and Jarzen, 1969).
cf. *Aguiaria* (Bombacaceae; lower Miocene, Panama, Graham, 1989; Mio-Pliocene, Panama, Graham, 1991b).
Alchornea (Euphorbiaceae; Oligocene, Puerto Rico, Graham and Jarzen, 1969; lower Miocene, Costa Rica, Graham, 1987a; lower Miocene, Panama, Graham, 1988a, b, 1989; Mio-Pliocene, Mexico, Graham, 1976; Mio-Pliocene, Panama, Graham, 1991b; Mio-Pliocene, Haiti, Graham, 1990).
Alfaroa (as *Alfaroa/Engelhardia*, Juglandaceae; Eocene, Panama, Graham, 1985; lower Miocene, Panama, Graham, 1988b, 1989; as *Alfaroa/Oreomunnea*- Mio-Pliocene, Panama, Graham, 1991b; Mio-Pliocene, Haiti, Graham, 1990; as *Engelhardia*- Oligocene, Puerto Rico, Graham and Jarzen, 1969; Oligo-Miocene, Mexico, Langenheim et al., 1967; Mio-Pliocene, Mexico, Graham, 1976).
cf. *Alibertia* (Rubiaceae; Mio-Pliocene, Mexico, Graham, 1976).
Allophylus (Sapindaceae; lower Miocene, Panama, Graham, 1988a; Mio-Pliocene, Mexico, Graham, 1976; Mio-Pliocene, Panama, Graham, 1991b; Mio-Pliocene, Haiti, Graham, 1990).
Alnus (Betulaceae; Mio-Pliocene, Mexico, Graham, 1976).
Amaranthaceae (see Chenopodiaceae/Amaranthaceae).
cf. Araliaceae (Eocene, Panama, Graham, 1985).
Arecipites (Araceae?; Eocene, Cuba, Areces-Mallea, 1988).
Arrabidaea (see *Paragonia/Arrabidaea*).
cf. *Astrocaryum* (Palmae; Mio-Pliocene, Mexico, Graham, 1976).
cf. *Attalea* (Palmae; Mio-Pliocene, Mexico, Graham, 1976).
cf. *Banisteriopsis* (Malpighiaceae; lower Miocene, Costa Rica, Graham, 1987a).
cf. *Bernardia* (Euphorbiaceae; Mio-Pliocene, Mexico, Graham, 1976).
Bernoullia (Bombacaceae; Oligocene, Puerto Rico, Graham and Jarzen, 1969; Mio-Pliocene, Panama, Graham, 1991b).
Bombacaceae (lower Miocene, Costa Rica, Graham, 1987a).
Bombacacidites (Bombacaceae; Eocene, Cuba, Areces-Mallea, 1985).

Bombax (Bombacaceae; Oligocene, Puerto Rico, Graham and Jarzen, 1969.
Borreria (Rubiaceae; Mio-Pliocene, Mexico, Graham, 1976).
cf. *Brahea* (Palmae; Mio-Pliocene, Mexico, Graham, 1976).
Bravaisia (Acanthaceae; Mio-Pliocene, Mexico, Graham, 1976).
cf. *Bredemeyera* (Polygalaceae; Mio-Pliocene, Mexico, Graham, 1976).
Brunellia (Brunelliaceae; Oligocene, Puerto Rico, Graham and Jarzen, 1969).
cf. *Bucida* (Combretaceae; Mio-Pliocene, Panama, Graham, 1991b).
Buettneria (Sterculiaceae; Mio-Pliocene, Mexico, Graham, 1976).
cf. *Bumelia* (Sapotaceae; Mio-Pliocene, Panama, Graham, 1991b).
Bursera (Burseraceae; Oligocene, Puerto Rico, Graham and Jarzen, 1969; Mio-Pliocene, Mexico, Graham, 1976; Mio-Pliocene, Panama, Graham, 1991b).
Cabomba (Nymphaeaceae; Mio-Pliocene, Panama, Graham, 1991b).
cf. *Campnosperma* (Anacardiaceae; Eocene, Panama, Graham, 1985).
Cardiospermum (Sapindaceae; Eocene, Panama, Graham, 1985).
Casearia (Flacourtiaceae; Eocene, Panama, Graham, 1985; Oligocene, Puerto Rico, Graham and Jarzen, 1969; lower Miocene, Panama, Graham, 1988a; Mio-Pliocene, Mexico, Graham, 1976).
Casimiroa (Rutaceae; Mio-Pliocene, Panama, Graham, 1991b).
Catostemma (Bombacaceae; Oligocene, Puerto Rico, Graham and Jarzen, 1969).
Cedrela (Meliaceae; Mio-Pliocene, Mexico, Graham, 1976; Mio-Pliocene, Panama, Graham, 1991b).
Ceiba (Bombaceae; Mio-Pliocene, Panama, Graham, 1991b; as cf. *Ceiba*, lower Miocene, Panama, Graham, 1989).
Celtis (Ulmaceae; Mio-Pliocene, Mexico, Graham, 1976).
cf. *Chamaedorea* (Palmae; Mio-Pliocene, Mexico, Graham, 1976).
Chenopodiaceae (as Chenopodiaceae/Amaranthaceae; lower Miocene, Panama, Graham, 1988a; Mio-Pliocene, Mexico, Graham, 1976; Mio-Pliocene, Haiti, Graham, 1990).
Chomelia type (Rubiaceae; Mio-Pliocene, Panama, Graham, 1991b).
Chrysophyllum (Sapotaceae; Oligocene, Puerto Rico, Graham and Jarzen, 1969; cf. *Chrysophyllum*, Eocene, Panama, Graham, 1985).
cf. *Cionosicys* (Cucurbitaceae; Mio-Pliocene, Panama, Graham, 1991b).
Cleyera (Theaceae; Mio-Pliocene, Mexico, Graham, 1976).
Coccoloba (Polygonaceae; Eocene, Panama, Graham, 1985; Mio-Pliocene, Mexico, Graham, 1976).
Combretum/Terminalia (Combretaceae; Eocene, Panama, Graham, 1985; lower Miocene, Panama, Graham, 1988a; Mio-Pliocene, Mexico, Graham, 1976; Mio-Pliocene, Panama, Graham, 1991b).
Comocladia (Anacardiaceae; Mio-Pliocene, Mexico, Graham, 1976).
Compositae (lower Miocene, Costa Rica, Graham, 1987a; lower Miocene, Panama, Graham, 1988a,b; Mio-Pliocene, Mexico, Graham, 1976; Mio-Pliocene, Haiti, Graham, 1990).
Corynostylis (Violaceae; Oligocene, Puerto Rico, Graham and Jarzen, 1969).
Cosmibuena (Rubiaceae; Mio-Pliocene, Panama, Graham, 1991b).
Crudia (Leguminosae, Subfam. Caesalpinioideae; Eocene, Panama, Graham, 1985; lower Miocene, Panama, Graham, 1988b, 1989; Mio-Pliocene, Panama, Graham, 1991b).
Cupania (Sapindaceae; lower Miocene, Panama, Graham, 1988a; Mio-Pliocene, Mexico, Graham, 1976; Mio-Pliocene, Panama, Graham, 1991b).
Cuphea (Lythraceae; Mio-Pliocene, Mexico, Graham, 1976).
Cymbopetalum (Annonaceae; Mio-Pliocene, Panama, Graham, 1991b).
Cyperaceae (Mio-Pliocene, Mexico, Graham, 1976).
Daphnopsis (Thymeliaceae; Mio-Pliocene, Mexico, Graham, 1976).
Dendropanax (Araliaceae; Oligocene, Puerto Rico, Graham and Jarzen, 1969).
Desmanthus (Leguminosae, Subfam. Papilionoideae; Mio-Pliocene, Mexico, Graham, 1976).
Dichapetalum (Dichapetalaceae; Mio-Pliocene, Mexico, Graham, 1976).
Dioscorea/Rajania (Dioscoreaceae; lower Miocene, Panama, Graham, 1988a).
cf. *Doliocarpus* (Dilleniaceae; lower Miocene, Panama, Graham, 1988a).
Engelhardia (see *Alfaroa/Engelhardia*).
Ericaceae (lower Miocene, Costa Rica, Graham, 1987a; Mio-Pliocene, Panama, Graham, 1991b).
Erythrina (Leguminosae, Subfam. Papilionoideae; Mio-Pliocene, Panama, Graham, 1991b).
Eugenia (see *Eugenia/Myrcia*).
Eugenia/Myrcia (Myrtaceae; Eocene, Panama, Graham, 1985; lower Miocene, Costa Rica, Graham, 1987a; lower Miocene, Panama, Graham, 1988a, b; Mio-Pliocene, Mexico, Graham, 1976; Mio-Pliocene, Panama, Graham, 1991b; as *Eugenia*, Oligocene, Puerto Rico, Graham and Jarzen, 1969).
Fagus (Fagaceae; Oligocene, Puerto Rico, Graham and Jarzen, 1969).
Faramea (Rubiaceae; Eocene, Panama, Graham, 1985; Oligocene, Puerto Rico, Graham and Jarzen, 1969; Mio-Pliocene, Panama, Graham, 1991b; Mio-Pliocene, Mexico, Graham, 1976).
cf. *Ficus* (Moraceae; Eocene, Panama, Graham, 1985).
cf. *Glycydendrum* (Euphorbiaceae; lower Miocene, Costa Rica, Graham, 1987a; Mio-Pliocene, Panama, Graham, 1991b).
Gramineae (Lower Miocene, Panama, Graham, 1988a, 1989; Mio-Pliocene, Mexico, Graham, 1976; Mio-Pliocene, Panama, Graham, 1991b).
*Graminidites (Gramineae; Eocene, Cuba, Areces-Mallea, 1988).
Guarea (Meliaceae; Oligocene, Puerto Rico, Graham and Jarzen, 1969; Mio-Pliocene, Mexico, Graham, 1976; Mio-Pliocene, Panama, Graham, 1991b).
cf. *Guazuma* (Sterculiaceae; lower Miocene, Panama, Graham, 1988a).
Gustavia (Lecythidaceae; Mio-Pliocene, Mexico, Graham, 1976).
Hampea/Hibiscus (Malvaceae; lower Miocene, Panama, Graham, 1988a; Mio-Pliocene, Mexico, Graham, 1976; Mio-Pliocene, Panama, Graham, 1991b).
Hauya (Onagraceae; Oligocene, Puerto Rico, Graham and Jarzen, 1969; Mio-Pliocene, Panama, Graham, 1991b).
Hedyosmum (Chloranthaceae; Mio-Pliocene, Mexico, Graham, 1976).
Hibiscus (see *Hampea/Hibiscus*).
cf. *Hiraea* (Malpighiaceae; lower Miocene, Costa Rica, Graham, 1987a; Mio-Pliocene, Mexico, Graham, 1976).
Hygrophila (Acanthaceae; Mio-Pliocene, Haiti, Graham, 1990).
Ilex (Aquifoliaceae; Eocene, Panama, Graham, 1985; Oligocene, Puerto Rico, Graham and Jarzen, 1969; lower Miocene, Costa Rica, Graham, 1987a; lower Miocene, Panama, Graham, 1988a, b, 1989; Mio-Pliocene, Mexico, Graham, 1976; Mio-Pliocene, Panama, Graham, 1991b).
Iresine (Amaranthaceae; Mio-Pliocene, Mexico, Graham, 1976).
Jacaranda (Bignoniaceae; Oligocene, Puerto Rico, Graham and Jarzen, 1969).
cf. *Jatropha* (Euphorbiaceae; Mio-Pliocene, Panama, Graham, 1991b).
Juglans (Juglandaceae; Mio-Pliocene, Mexico, Graham, 1976).

Justicia (Acanthaceae; Mio-Pliocene, Mexico, Graham, 1976).
Laetia (Flacourtiaceae; Mio-Pliocene, Mexico, Graham, 1976).
Laguncularia (Combretaceae; Mio-Pliocene, Mexico, Graham, 1976).
Liliacidites (Liliaceae? Eocene, Cuba, Areces-Mallea, 1988).
Liquidambar (Hamamelidaceae; Oligocene, Puerto Rico, Graham and Jarzen, 1969; Mio-Pliocene, Mexico, Graham, 1976).
Lisianthius (Gentianaceae; Eocene, Panama, Graham, 1985; lower Miocene, Costa Rica, Graham, 1987a).
Loranthaceae (Mio-Pliocene, Panama, Graham, 1991b).
Ludwigia (Onagraceae; Mio-Pliocene, Mexico, Graham, 1976).
cf. *Malpighia* (Malpighiaceae; Mio-Pliocene, Mexico, Graham, 1976).
Malpighiaceae (Eocene, Panama, Graham, 1985; lower Miocene, Panama, Graham, 1988a, 1989; Mio-Pliocene, Panama, Graham, 1991b; Mio-Pliocene, Haiti, Graham, 1990).
Marcgravia (Marcgraviaceae; Oligocene, Puerto Rico, Graham and Jarzen, 1969).
Matayba (Sapindaceae; lower Miocene, Panama, Graham, 1988a; Mio-Pliocene, Mexico, Graham, 1976).
cf. *Maximiliana* type (Palmae; Mio-Pliocene, Mexico, Graham, 1976).
Melastomataceae (lower Miocene, Costa Rica, Graham, 1987a; Mio-Pliocene, Panama, Graham, 1991b).
Meliosma (Sapindaceae; Mio-Pliocene, Mexico; Graham, 1976).
Merremia (Convolvulaceae; Oligocene, Puerto Rico, Graham and Jarzen, 1969).
cf. *Mezia*(?)-type (Malpighiaceae; Mio-Pliocene, Mexico, Graham, 1976).
Mimosa (Leguminosae, Subfam. Mimosoideae; Mio-Pliocene, Mexico, Graham, 1976).
Monocolpopollenites (Palmae; Eocene, Cuba, Areces-Mallea, 1988).
Mortoniodendron (Tiliaceae; Eocene, Panama, Graham, 1985; Mio-Pliocene, Mexico, Graham, 1976; Mio-Pliocene, Panama, Graham, 1991b).
Mutisieae type (Compositae; Mio-Pliocene, Panama, Graham, 1991b).
Myrcia (see *Eugenia/Myrcia*).
Myrica (Myricaceae; Mio-Pliocene, Mexico, Graham, 1976).
Norantea (Marcgraviaceae; Oligocene, Puerto Rico, Graham and Jarzen, 1969).
Nyssa (Nyssaceae; Oligocene, Puerto Rico, Graham and Jarzen, 1969).
Oreomunnea (see *Alfaroa*).
Oryctanthus (Loranthaceae; Mio-Pliocene, Haiti, Graham, 1990).
Oxalis (Oxalidaceae; Oligocene, Puerto Rico, Graham and Jarzen, 1969).
Pachira-type (Bombacaceae; Oligo-Miocene, Mexico, Langenheim et al., 1967).
Palmae (Eocene, Panama, Graham, 1985; Oligocene, Puerto Rico, Graham and Jarzen, 1969; lower Miocene, Panama, Graham, 1988a, b, 1989; Mio-Pliocene, Panama, Graham, 1991b; Mio-Pliocene, Haiti, Graham, 1990).
cf. *Paragonia/Arrabicaea* (Bignoniaceae; Eocene, Panama, Graham, 1985).
Passiflora (Passifloraceae; Mio-Pliocene, Mexico, Graham, 1976).
Paullinia (Sapindaceae; Eocene, Panama, Graham, 1985; Mio-Pliocene, Mexico, Graham, 1976; Mio-Pliocene, Panama, Graham, 1991b).
Pelliceria (Theaceae; Eocene, Panama, Graham, 1985; Eocene, Jamaica, Graham, 1977; Oligocene, Puerto Rico, Graham and Jarzen, 1969; Oligo-Miocene, Mexico, Langenheim et al., 1967; lower Miocene, Panama, Graham, 1989).
Petraea (Verbenaceae; Mio-Pliocene, Panama, Graham, 1991b).

Pleodendron (Canellaceae; Oligocene, Puerto Rico, Graham and Jarzen, 1969).
Populus (Salicaceae; Mio-Pliocene, Mexico, Graham, 1976).
Posqueria (Rubiaceae; Mio-Pliocene, Panama, Graham, 1991b).
cf. *Pouteria* (Sapotaceae; lower Miocene, Panama, Graham, 1988a).
Protium (Burseraceae; Mio-Pliocene, Mexico, Graham, 1976; cf. *Protium*, Eocene, Panama, Graham, 1985).
Pseudobombax (Bombaceaeae; lower Miocene, Panama, Graham, 1989; Mio-Pliocene, Panama, Graham, 1991b).
Quercus (Fagaceae; Mio-Pliocene, Mexico, Graham, 1976; Mio-Pliocene, Panama, Graham, 1991b).
Rajania (Dioscoreaceae; Mio-Pliocene, Mexico, Graham, 1976).
Rauwolfia (Apocynaceae; Oligocene, Puerto Rico, Graham and Jarzen, 1969).
Rhizophora (Rhizophoraceae; Eocene, Panama, Graham, 1985; Oligocene, Puerto Rico, Graham and Jarzen, 1969; Oligo-Miocene, Mexico, Langenheim et al., 1967; lower Miocene, Costa Rica, Graham, 1987a, lower Miocene, Panama, Graham, 1988a, b, 1989; Mio-Pliocene, Mexico, Graham, 1976; Mio-Pliocene, Panama, Graham, 1991b).
cf. *Rourea* (Connaraceae; lower Miocene, Panama, Graham, 1988a).
Rubiaceae (lower Miocene, Panama, Graham, 1989).
Sabicea (Rubiaceae; lower Miocene, Panama, Graham, 1988a; see also Graham, 1987b).
Salix (Salicaceae; Oligocene, Puerto Rico, Graham and Jarzen, 1969).
cf. *Sapium* (Euphorbiaceae; lower Miocene, Panama, Graham, 1988a [as *Sapium*]; Mio-Pliocene, Mexico, Graham, 1976).
cf. *Securidaca* (Polygalaceae; Mio-Pliocene, Mexico, Graham, 1976).
Serjania (Sapindaceae; Eocene, Panama, Graham, 1985; Mio-Pliocene, Mexico, Graham, 1976; Mio-Pliocene, Panama, Graham, 1991b).
Smilax (Liliaceae; Mio-Pliocene, Mexico, Graham, 1976).
Spathiphyllum (Araceae; Mio-Pliocene, Mexico, Graham, 1976).
cf. *Stillingia* (Euphorbiaceae; Mio-Pliocene, Mexico, Graham, 1976; Mio-Pliocene, Panama, Graham, 1991b).
Struthanthus (Loranthaceae; Mio-Pliocene, Mexico, Graham, 1976).
Symphonia (Guttiferae; Mio-Pliocene, Mexico, Graham, 1976).
Symplocos (Symplocaceae; Mio-Pliocene, Panama, Graham, 1991b).
Tecoma (Bignoniaceae; Oligocene, Puerto Rico, Graham and Jarzen, 1969).
Terebrania (Rubiaceae; Mio-Pliocene, Mexico, Graham, 1976).
Terminalia (see *Combretum/Terminalia*).
cf. *Tetragrastis* (Burseraceae; Eocene, Panama, Graham, 1985).
Tetrorchidium (Euphorbiaceae; Oligocene, Puerto Rico, Graham and Jarzen, 1969; lower Miocene, Panama, Graham, 1988a; cf. *Tetrorchidium*, Mio-Pliocene, Mexico, Graham, 1976).
Thalictrum (Ranunculaceae; Mio-Pliocene, Mexico, Graham, 1976).
cf. *Tillandsia* (Bromeliaceae; Eocene, Panama, Graham, 1985).
cf. *Tithymalus* (Euphorbiaceae; Mio-Pliocene, Mexico, Graham, 1976).
cf. *Tontalea* (Hippocrateaceae; Eocene, Panama, Graham, 1985).
Tournefortia (Boraginaceae; Oligocene, Puerto Rico, Graham and Jarzen, 1969; Mio-Pliocene, Mexico, Graham, 1976).
Ulmus (Ulmaceae; Mio-Pliocene, Mexico, Graham, 1976).
Utricularia (Lentibulariaceae; lower Miocene, Panama, Graham, 1989; Mio-Pliocene, Mexico, Graham, 1976; Mio-Pliocene, Panama, Graham, 1991b).
Zanthophyllum (Rutaceae; Oligocene, Puerto Rico, Graham and Jarzen, 1969).

Family Index

Algae
  Chlorophyta
    Micractinium
Fungi
  Deuteromycotina
    *Geotrichites
  Ascomycetes
    Microthyriaceae
      Microthyrium-type
Bryophyta
  Anthocerotaceae
    Phaeoceros
  Muscae
Psilopsida
  Psilotaceae
    Psilotum
Lycopsida
  Lycopodiaceae
    Lycopodium
  Selaginellaceae
    Selaginella
Filicineae (ferns)
  Cyatheaceae
    Alsophila
    Cnemidaria
    Cyathea
    Hemitelia (=*Cnemidaria*)
    Sphaeropteris/Trichipteris
  Dryopteridaceae
    Ctenitis
    Lomariopsis (*Stenochlaena*)
  Gleicheniaceae
    Dicranopteris
  Hymenophyllaceae
    cf. *Hymenophyllum*
  Lophosoriaceae
    Lophosoria
  Ophioglossaceae
    Ophioglossum
  Polypodiaceae
    Grammitis
  Pteridaceae
    Ceratopteris
    Jamesonia (*Eriosorus*)
    Pityrogramma
    Pteris
  Vittariaceae
    cf. *Antrophyum*
Gymnospermae
  Pinaceae
    Abies
    Picea
    Pinus
  Podocarpaceae
    Podocarpus
Angiospermae
  Acanthaceae
    Bravaisia
    Hygrophila
    Justicia
  Amaranthaceae (as Chenopodiaceae/Amaranthaceae)
  Amaranthaceae

Angiospermae (continued)
    Iresine
  Anacardiaceae
    cf. *Campnosperma*
    Comocladia
  Annonaceae
    Cymbopetalum
  Apocynaceae
    Rauwolfia
  Aquifoliaceae
    Ilex
  Araceae
    Spathiphyllum
  Araliaceae
    Dendropanax
  cf. **Araliaceae**
  Betulaceae
    Alnus
  Bignoniaceae
    Jacaranda
    Paragonia/Arrabidaea
    Tecoma
  Bombacaceae
    cf. *Aguiaria*
    Bernoullia
    *Bombacacidites
    Bombax
    Catostemma
    Ceiba
    Pachira-type
    Pseudobombax
  Boraginaceae
    Tournefortia
  Bromeliaceae
    cf. *Tillandsia*
  Brunelliaceae
    Brunellia
  Burseraceae
    Bursera
    Protium
    cf. *Protium*
    cf. *Tetragrastis*
  Chenopodiaceae/Amaranthaceae
  Combretaceae
    cf. *Bucida*
    Combretum/Terminalia
    Laguncularia
  Compositae
    Mutisieae type
  Cucurbitaceae
    cf. *Cionosicys*
  Cyperaceae
  Dichapetalaceae
    Dichapetalum
  Dioscoreaceae
    Rajania
  Ericaceae
  Euphorbiaceae
    cf. *Bernardia*
    cf. *Glycydendrum*
    cf. *Jatropha*
    cf. *Sapium*
    cf. *Stillingia*

Angiospermae (continued)
    Tetrorchidium
    cf. *Tithymalus*
  Fagaceae
    Fagus
    Quercus
  Flacourtiaceae
    Casearia
    Laetia
  Gentianaceae
    Lisianthius
  Gramineae
  Guttiferae
    Symphonia
  Hamamelidaceae
    Liquidambar
  Hippocrateaceae
    cf. *Tontalea*
  Juglandaceae
    Alfaroa/Engelhardia (= *Alfaroa/Oreomunnea*)
    Alfaroa/Oreomunnea
    Engelhardia (= *Alfaroa/Oreomunnea*)
    Juglans
  Lecythidaceae
    Gustavia
  Leguminosae
    Caesalpinioideae
      Crudia
    Mimosoideae
      Acacia
      cf. *Acacia*
      Mimosa
    Papilionoideae
      Desmanthus
      Erythrina
  Lentibulariaceae
    Utricularia
  Liliaceae
    Smilax
  Loranthaceae
    Aetanthus
    Oryctanthus
    Struthanthus
  Lythraceae
    Cuphea
  Malpighiaceae
    cf. *Banisteriopsis*
    cf. *Hiraea*
    cf. *Malpighia*
    cf. *Mezia*(?)-type
  Malvaceae
    Abutilon
    Hampea/Hibiscus
  Marcgraviaceae
    Marcgravia
    Norantea
  Melastomataceae
  Meliaceae
    Cedrela
    Guarea
  Moraceae

Angiospermae (continued)
    cf. *Ficus*
  Myricaceae
    Myrica
  Myrtaceae
    Eugenia (= *Eugenia/Myrcia*)
    Eugenia/Myrcia
  Nymphaeaceae
    Cabomba
  Nyssaceae
    Nyssa
  Onagraceae
    Hauya
    Ludwigia
  Oxalidaceae
    Oxalis
  Palmae
    cf. *Astrocaryum*
    cf. *Attalea*
    cf. *Brahea*
    cf. *Chamaedorea*
    cf. *Maximiliana*
  Passifloraceae
    Passiflora
  Polygalaceae
    cf. *Bredemeyera*
    cf. *Securidaca*
  Polygonaceae
    Coccoloba
  Ranunculaceae
    Thalictrum
  Rhizophoraceae
    Rhizophora
  Rubiaceae
    cf. *Alibertia*
    Borreria
    Chomelia type
    Cosmibuena
    Faramea
    Posqueria
    Terebrania
  Rutaceae
    Casimiroa
    Zanthophyllum
  Salicaceae
    Populus
    Salix
  Sapindaceae
    Allophylus
    Cardiospermum
    Cupania
    Matayba
    Meliosma
    Paullinia
    Serjania
  Sapotaceae
    cf. *Bumelia*
    Chrysophyllum
    cf. *Chrysophyllum*
  Sterculiaceae
    Buettneria
  Symplocaceae
    Symplocos

Angiospermae (continued)
Theaceae
Cleyera
Pelliceria
Thymeliaceae
Daphnopsis
Tiliaceae
Mortoniodendron
Ulmaceae
Celtis
Ulmus
Verbenaceae
Aegiphila
Petraea
Violaceae
Corynostylis

APPENDIX III. SELECTED LITERATURE ON TERTIARY PALEOBOTANY AND PALEOPALYNOLOGY FOR NORTHERN LATIN AMERICA

References include only papers citing identified plant fossils; abstracts are not included. For more general studies and reviews see bibliographies by Graham (1973, 1979, 1982, 1986).

Mexico

Berry, E. W., 1923, Miocene plants from southern Mexico: Proceedings, United States National Museum, v. 62, p. 1–27.

Felix, J., and Nathorst, A. G., 1893, Versteinerungen aus dem Mexicanischen Staat Oaxaca, *in* Felix, J. and Lenk, H., eds., Beitrage Geologie und Paleontologie der Republik Mexico, Teil 2: Leipzig, Felix Verlag, p. 39–54.

Graham, A., 1976, Studies in neotropical paleobotany. II. The Miocene communities of Veracruz, Mexico: Annals of the Missouri Botanical Garden, v. 63, p. 787–842.

Graham, A., 1979, *Mortoniodendron* (Tiliaceae) and *Sphaeropteris/Trichipteris* (Cyatheaceae) in Cenozoic deposits of the Gulf-Caribbean region: Annals of the Missouri Botanical Garden, v. 66, p. 572–576.

Langenheim, J. H., Hackner, B. L., and Bartlett, A. H., 1967, Mangrove pollen at the depositional site of Oligo-Miocene amber from Chiapas, Mexico: Harvard University, Botanical Museum Leaflets, v. 21, p. 289–324.

Miranda, F., 1963, Two plants from the amber of the Simojovel, Chiapas, Mexico, area: Journal of Paleontology, v. 37, p. 611–614.

Mullerried, F.K.G., 1938, Informe der Sr. F. K. G. Mullerried, paleontologo del Instituto de Geologia, acerca del material colectado en el municipio de Tlacolulan, estado de Veracruz: Boletim Sociadad Geologia de Mexico, v. 10, p. 203–206.

Central America

Berry, E. W., 1918, The fossil higher plants from the Canal Zone, *in* Contributions to the geology and paleontology of the Canal Zone, Panama, and geologically related areas in Central America and the West Indies: Bulletin of the United States Museum, v. 103, p. 15–44.

Berry, E. W., 1921a, Tertiary fossil plants from Costa Rica: Proceedings, United States National Museum, v. 59, p. 169–185.

Berry, E. W., 1921b, A palm nut from the Miocene of the Canal Zone: Proceedings, United States National Museum, v. 59, p. 21–22.

Berry, E. W., 1928, A palm fruit from the Miocene of western Panama: Washington Academy of Science Journal, v. 18, p. 455–457.

Gomez-P., L. D., 1970, A first report of fossil fern-like Pteropsida from Costa Rica: Revista Biologia Tropical, v. 16, p. 255–258.

Gomez-P., L. D., 1971, *Palmacites berryanum,* a new palm fossil from the Costa Rican Tertiary: Revista Biologia Tropical, v. 19, p. 121–132.

Gomez-P., L. D., 1972, *Karatophyllum bromelioides* L. D. Gomez (Bromeliaceae), nov. gen. et sp., del Terciario Medio de Costa Rica: Revista Biologia Tropical, v. 20, p. 221–229.

Graham, A., 1977, New records of *Pelliceria* (Theaceae/Pelliceriaceae) in the Tertiary of the Caribbean: Biotropica, v. 9, p. 48–52.

Graham, A., 1979, *Mortoniodendron* (Tiliaceae) and *Sphaeropteris/Trichipteris* (Cyatheaceae) in Cenozoic deposits of the Gulf/Caribbean region: Annals of the Missouri Botanical Garden, v. 66, p. 572–576.

Graham, A., 1981, Un alga fosil Micractiniaceae de la Zona del Canal de Panama: Biotica, v. 6, p. 229–232.

Graham, A., 1984, *Lisianthus* pollen from the Eocene of Panama: Annals of the Missouri Botanical Garden, v. 71, p. 987–993.

Graham, A., 1985, Studies in neotropical paleobotany. IV. The Eocene communities of Panama: Annals of the Missouri Botanical Garden, v. 72, p. 504–534.

Graham, A., 1987a, Miocene communities and paleoenvironments of southern Costa Rica: American Journal of Botany, v. 74, p. 1501–1518.

Graham, A., 1987b, Fossil pollen of *Sabicea* (Rubiaceae) from the lower Miocene Culebra Formation of Panama: Annals of the Missouri Botanical Garden, v. 74, p. 868–870.

Graham, A., 1988a, Studies in neotropical paleobotany. V. The lower Miocene communities of Panama—The Culebra Formation: Annals of the Missouri Botanical Garden, v. 75, p. 1440–1466.

Graham, A., 1988b, Studies in neotropical paleobotany. VI. The lower Miocene communities of Panama—The Cucaracha Formation: Annals of the Missouri Botanical Garden, v. 75, p. 1467–1479.

Graham, A., 1989, Studies in neotropical paleobotany. VII. The lower Miocene communities of Panama—The La Boca Formation: Annals of the Missouri Botanical Garden, v. 76, p. 50–66.

Graham, A., 1991a, Studies in neotropical paleobotany. VIII. The Pliocene communities of Panama—Introduction and ferns, gymnosperms, angiosperms (monocots): Annals of the Missouri Botanical Garden, v. 78, p. 190–200.

Graham, A., 1991b, Studies in neotropical paleobotany. IX. The Pliocene communities of Panama—Angiosperms (dicots): Annals of the Missouri Botanical Garden, v. 78, p. 201–223.

Graham, A., 1991c, Studies in neotropical paleobotany. X. The Pliocene communities of Panama—Composition, numerical representations, and paleocommunity paleoenvironmental reconstructions: Annals of the Missouri Botanical Garden, v. 78, p. 465–475.

Antilles

Areces-Mallea, A. E., 1985, Una nueva especies de *Bombacacidites* Couper emend. Krutzsch del Eocéno Medio de Cuba: Revista Tecnologíca, v. 15, Series Geologia, no. 1, p. 3–7.

Areces-Mallea, A. E., 1987, Consideraciónes sobre la supuesta preséncia de *Pinus sylvestris* L. en el Oligocene de Cuba: Publicació del Centro de Investigaciones y Desarrollo del Petroleo, Serie Geológica, no. 2, p. 27–40.

Areces-Mallea, A. E., 1988, Palinomorfos de la costa del Golfo de Norteamerica en el Eoceno medio de Cuba: Revista Tecnologica (Cuba), v. 17, p. 15–25.

Berry, E. W., 1921, Tertiary fossil plants from the Dominican Republic: Proceedings, United States National Museum, v. 59, p. 117–127.

Berry, E. W., 1923, Tertiary fossil plants from the Republic of Haiti: Proceedings, United States National Museum, v. 62, 10 p.

Berry, E. W., 1939, A Miocene flora from the gorge of the Yumari River, Matanzas, Cuba: John Hopkins University Studies in Geology, v. 13, p. 95–135.

Dilcher, D. L., Herendeen, P. S., and Huber, F., 1990, Fossil *Acacia* flowers with attached glands from Dominican Republic amber [abs.]: American Journal of Botany, v. 77 (supplement), p. 128.

Gomez-P., L. D., 1982, *Grammitis succinea,* the first New World fern found in amber: American Fern Journal, v. 72, p. 49–52.

Graham, A., 1977, New records of *Pelliceria* (Theaceae/Pelliceriaceae) in the Tertiary of the Caribbean: Biotropica, v. 9, p. 48–52.

Graham, A., 1990, Late Tertiary microfossil flora from the Republic of Haiti: American Journal of Botany, v. 77, p. 911–926.
Graham, A., and Jarzen, D. M., 1969, Studies in neotropical paleobotany. 1. The Oligocene communities of Puerto Rico: Annals of the Missouri Botanical Garden, v. 56, p. 308–357.
Hollick, A., 1928, Paleobotany of Porto Rico: New York Academy of Sciences, Scientific Survey of Porto Rico and the Virgin Islands, v. 7, p. 177–393.
Kaul, K. N., 1943, A palm stem from the Miocene of Antigua, W. I.—*Phytelephas sewardii* sp. nov.: Proceedings, Linnean Society of London, v. 155, p. 3–4.
Stenzel, K.G.W., 1897, *Palmoxylon iriateum* n. sp., ein fossiles Palmholz aus Antigua: Königlichen Svenska Vetenskaps Akademiens Handlingar, v. 22, p. 1–17.
Stubblefield, S. P., Miller, C. E., Taylor, T. N., and Cole, G. T., 1985, *Geotrichites glaesarius* a conidial fungus from Tertiary Dominican amber: Mycologia, v. 77, p. 11–16.

REFERENCES CITED

Anonymous, 1818, Petrified wood from Antigua: American Journal of Science, v. 1, p. 56–57.
Akers, W. H., 1979 (v. 15), 1981 (v. 16), 1984 (v. 18), Planktic foraminifera and calcareous nannoplankton biostratigraphy of the Neogene of Mexico: Tulane Studies in Geology and Paleontology, v. 15, p. 1–32; v. 16, p. 145–148; v. 18, p. 21–36.
Areces-Mallea, A., 1985, Una nueva especie de *Bombacacidites* Couper emend. Krutzsch del Eocéno medio de Cuba: Revista Tecnologíca, v. 15, Series Geológia, no. 1, p. 3–7.
Areces-Mallea, A., 1987, Consideraciónes sobra la supuesta preséncia de *Pinus sylvestris* L. en el Oligoceno de Cuba: Publicacion del Centra de Investigaciones y Desarrollo del Petroleo, Serie Geológica No. 2, p. 27–40.
Areces-Mallea, A., 1988, Palinomorfos de la costa del Golfo de Norteamerica en el Eocéno medio de Cuba: Revista Tecnologíca, v. 17, p. 15–26.
Berry, E. W., 1921, Tertiary fossil plants from the Dominican Republic: Proceedings, United States National Museum, v. 59, p. 117–127.
Berry, E. W., 1923, Tertiary fossil plants from the Republic of Haiti: Proceedings, United States National Museum, v. 62, 10 p.
Buskirk, R. E., 1985, Zoogeographic patterns and tectonic history of Jamaica and the northern Caribbean: Journal of Biogeography, v. 12, p. 445–461.
Cooke, C. W., Gardner, J., and Woodring, W. P., 1943, Correlation of the Cenozoic formations of the Atlantic and Gulf Coastal Plain and the Caribbean region: Geological Society of America Bulletin, v. 54, p. 1713–1723.
Crepet, W. L., and Feldman, G. D., 1991, The earliest remains of grasses in the fossil record: American Journal of Botany, v. 78, p. 1010–1014.
Dilcher, D. L., 1973, A paleoclimatic interpretation of the Eocene floras of southeastern North America, *in* Graham, A., ed., Vegetation and vegetational history of northern Latin America: Amsterdam, Elsevier Scientific Publishing Company, p. 39–59.
Donnelly, T. W., 1985, Mesozoic and Cenozoic plate evolution of the Caribbean region, *in* Stehli, F. G., and Webb, S. D., eds., The Great American Biotic Interchange: New York, Plenum Publishing Company, p. 89–121.
Elsik, W. C., 1968, Palynology of a Paleocene Rockdale lignite, Milam County, Texas: Pollen et Spores, v. 10, p. 263–314, 599–664.
Elsik, W. C., and Tomb, A. S., 1988, Compositae pollen morphophytes in the Gulf Coast Neogene: Program and Abstracts, 21st Annual Meeting, The American Association of Stratigraphic Palynologists, Inc. (Houston, 1988), without pagination.
Frederiksen, N. O., 1980, Sporomorphs from the Jackson Group (upper Eocene) and adjacent strata of Mississippi and western Alabama: U.S. Geological Survey Professional Paper 1084, 75 p.
Frederiksen, N. O., 1988, Sporomorph biostratigraphy, floral changes, and paleoclimatology, Eocene and earliest Oligocene of the eastern Gulf Coast: U.S. Geological Survey Professional Paper 1448, 68 p.
Germeraad, J., 1978a, Contribution to the palynology of Jamaica (B.W.I.)—A progress report: 10 p. (mimeographed; distributed by the author).
Germeraad, J., 1978b, Contribution to the palynology of the Cretaceous of Jamaica (B.W.I.)—A progress report: 7 p. (mimeographed; distributed by the author).
Germeraad, J., 1978c, Displaced sporomorphs and dinoflagellates in Jamaican Cainophytic strata—a progress report: 2 p. (mimeographed; distributed by the author).
Germeraad, J., 1979a, Literature on fossil and recent fungi, algae, etc., related *to* "Fossil remains of fungi, algae and other organisms from Jamaica": 39 p. (mimeographed; distributed by the author).
Germeraad, J., 1979b, Fossil remains of fungi, algae and other organisms from Jamaica: Scripta Geologica, v. 52, 41 p.
Germeraad, J., 1980, Dispersed scolecodonts from Cainozoic strata of Jamaica: Scripta Geologica, v. 54, 24 p.
Germeraad, J., Hopping, C. A., and Muller, J., 1968, Palynology of Tertiary sediments from tropical areas: Review of Palaeobotany and Palynology, v. 6, p. 189–348.
González Guzmán, A. E., 1967, A palynological study on the upper Los Cuervos and Mirador Formations (lower and middle Eocene; Tibú area, Colombia): Leiden, Brill, 68 p.
Graham, A., 1973, Literature on vegetational history in Latin America, *in* Graham, A., ed., Vegetation and vegetational history of northern Latin America: Amsterdam, Elsevier Scientific Publishing Company, p. 315–350.
Graham, A., 1976, Studies in neotropical paleobotany. II. The Miocene communities of Veracruz, Mexico: Annals of the Missouri Botanical Garden, v. 63, p. 787–842.
Graham, A., 1977, New records of *Pelliceria* (Theaceae/Pelliceriaceae) in the Tertiary of the Caribbean: Biotropica, v. 9, p. 48–52.
Graham, A., 1979, Literature on vegetational history in Latin America. Supplement I: Review of Palaeobotany and Palynology, v. 27, p. 29–52.
Graham, A., 1982, Literature on vegetational history in Latin America. Supplement II: Review of Palaeobotany and Palynology, v. 37, p. 185–223.
Graham, A., 1985, Studies in neotropical paleobotany. IV. The Eocene communities of Panama: Annals of the Missouri Botanical Garden, v. 72, p. 504–534.
Graham, A., 1986, Literature on vegetational history in Latin America. Supplement III: Review of Palaeobotany and Palynology, v. 48, p. 199–239.
Graham, A., 1987, Miocene communities and paleoenvironments of southern Costa Rica: American Journal of Botany, v. 74, p. 1501–1518.
Graham, A., 1988a, Some aspects of Tertiary vegetational history in the Gulf/Caribbean region: Transactions, 11th Caribbean Geological Conference (Barbados,1986), p. 3:1–3:18.
Graham, A., 1988b, Studies in neotropical paleobotany. V. The lower Miocene communities of Panama—The Culebra Formation: Annals of the Missouri Botanical Garden, v. 75, p. 1440–1466.
Graham, A., 1988c, Studies in neotropical paleobotany. VI. The lower Miocene communities of Panama—The Cucaracha Formation: Annals of the Missouri Botanical Garden, v. 75, p. 1467–1479.
Graham, A., 1989, Studies in neotropical paleobotany. VII. The lower Miocene communities of Panama—The La Boca Formation: Annals of the Missouri Botanical Garden, v. 76, p. 50–66.
Graham, A., 1990, Late Tertiary microfossil flora from the Republic of Haiti: American Journal of Botany, v. 77, p. 911–926.
Graham, A., 1991a, Studies in neotropical paleobotany. VIII. The Pliocene communities of Panama—Introduction and ferns, gymnosperms, angiosperms

(monocots): Annals of the Missouri Botanical Garden, v. 78, p. 190–200.
Graham, A., 1991b, Studies in neotropical paleobotany. IX. The Pliocene communities of Panama—Angiosperms (dicots): Annals of the Missouri Botanical Garden, v. 78, p. 201–223.
Graham, A., 1991c, Studies in neotropical paleobotany. X. The Pliocene communities of Panama—Composition, numerical representations, and paleocommunity paleoenvironmental reconstructions: Annals of the Missouri Botanical Garden, v. 78, p. 465–475.
Graham, A., and Jarzen, D. M., 1969, Studies in neotropical paleobotany. I. The Oligocene communities of Puerto Rico: Annals of the Missouri Botanical Garden, v. 56, p. 308–357.
Herendeen, P. S. and Dilcher, D. L., 1990, Reproductive and vegetative evidence for the occurrence of *Crudia* (Leguminosae, Caesalpinioideae) in the Eocene of southeastern North America: Botanical Gazette, v. 151, p. 402–413.
Hollick, A., 1928, Paleobotany of Porto Rico: New York Academy of Sciences, Scientific Survey of Porto Rico and the Virgin Islands, v. 7, p. 177–393.
Keller, G., Zenker, C. E., and Stone, S. M., 1989, Late Neogene history of the Pacific-Caribbean gateway: Journal of South American Earth Science, v. 2, p. 73–108.
LaMotte, R. S., 1952, Catalogue of the Cenozoic plants of North America through 1950: Geological Society of America Memoir 51, 381 p.
Langenheim, J. H., Hackner, B. L., and Bartlett, A., 1967, Mangrove pollen at the depositional site of Oligo-Miocene amber from Chiapas, Mexico: Botanical Museum Leaflets of Harvard University, v. 21, p. 289–324.
Lorente, M. A., 1986, Palynology and palynofacies of the upper Tertiary in Venezuela: Berlin, J. Cramer, Dissertationes Botanicae, v. 99, 222 p.
Machain-Castillo, M., 1985, Ostracode biostratigraphy and paleoecology of the Pliocene of the Isthmian salt basin, Veracruz, Mexico: Tulane Studies in Geology and Paleontology, v. 19, p. 123–139.
Mann, P., and Burke, K., 1984, Neotectonics of the Caribbean: Review of Geophysics and Space Physics, v. 22, p. 309–362.
Muller, J., 1981, Fossil pollen records of extant angiosperms: Botanical Review, v. 47, p. 1–142.
Muller, J., Di Giacomo, E., and van Erve, A. W., 1987, A palynological zonation for the Cretaceous, Tertiary, and Quaternary of northern South America: American Association of Stratigraphic Palynologists Contribution Series Number 19, p. 7–76.
Pindell, J., and Dewey, J. F., 1982, Permo-Triassic reconstruction of western Pangea and the evolution of the Gulf of Mexico/Caribbean region: Tectonics, v. 1, p. 179–211.
Racz, L., 1971, Two new Pliocene species of *Neomeris* (calcareous algae) from the Bowden Beds, Jamaica: Paleontology, v. 14, p. 623–628.
Robinson, E., 1976, Lignite in Jamaica, with additional remarks on peat: A report of the Capital Development Fund: Kingston, Jamaica, Ministry of Mining and Natural Resources, 27 p. (mimeographed; distributed by the author).
Robinson, E., 1987 (1988), Late Cretaceous and early Tertiary sedimentary rocks of the Central Inlier, Jamaica: Journal of the Geological Society of Jamaica, v. 24, p. 49–67.
Stehli, F. G., and Webb, S. D., eds., 1985, The Great American Biotic Interchange: New York, Plenum, 532 p.
Sykes, L. E., McCann, W. R., and Kafka, A. L., 1982, Motion of Caribbean Plate during last 7 million years and implications for earlier Cenozoic movements: Journal of Geophysical Research, v. 87, p. 10656–10676.
Wadge, G., and Burke, K., 1983, Neogene Caribbean Plate rotation and associated Central American tectonic evolution: Tectonics, v. 2, p. 633–643.

MANUSCRIPT ACCEPTED BY THE SOCIETY NOVEMBER 11, 1992

Index

[Italic page numbers indicate major references]

A

abies, Sphenolithus, 137, 144, 145, 150, 153, 162, 177
Abingdon area, 396
abruptus, Plagiobrissus, 373, 405, 406
abyssorum, Rhopalodictyum, 273
Acanthodesmiidae sp., *274*
Acanthometron (Astrolithium) astraeforme, 266
acanthus, Zygodiscus, 23, 27
acephala, Bathropyramis, 276
acostaensis, Globorotalia, 148, 188, 189, 197, 206, 257
  Neogloboquadrina, 143, 144, 145, 146, 147, 182, 183, 184, 185, 187, 188, 189, 190, 191, 195, 197, 205, *206*
Acrosphaera, 265
  echinoides, 265
  sponosa echinoides, 260, 265
  transformata, 160, 265, 266
  sp., 266
Acrostichum, 448, 450
  aureum, 448
aculeata, Bulimina, 222, 225, 227, 228, 229, 231, 233, *235*
  Reusella, 240
  Stilostomella, 226, 227, 228, *241*
aculeus, Ceratolithoides, 8, 9, 14, 16, 22, 24, 25, 26
acuminata, Lithocampe, 277
acunai, Cubanaster, 373, 374, 387, 389, 390
acuta, Pleurodonte, 356
Acuturris scotus, 9, 14
acutus, Ceratolithus, 139, 140, 150, 152, 155, 195, 196
  Crocodylus, 431
  Lithraphidites, 13
adamsi, Poteria, 354
  Ptychocochlis, 355
adriennis, Deltoidospora, 448, 449
advena, Siphogenerina, 240
  ornata, Siphogenerina, 240
Aello cuvieri, 433
  megalophylla, 433
aemiliana, Globorotalia, 206
  hirsuta, 206
aenariensiformis, Bolivina, 234, 246
Aequacytheridea, 71
  sp., 71, 74, 75
aequilateralis, Globigerina, 208
  Globigerinella, 187, 193, 195, 208, 210
aequiscutum, Cyclococcolithus, 131, *166*
  Umbilicosphaera, 137, *166*, 175
affinis, Stylocidaris, 377
Africa, 394

africana, Braarudosphaera, 12, 14
Agaricia sp., 118, 121
Agassizia, 398
  inflata, 373, 374, 392, 398
agostinelli, Druppatractus, 260, 267, 268
Agua Salada Group, 220, 229
Ahmuellerella octoradiata, 8, 9, 12, 14, 22, 26
  regularis, 22, 26
alabamensis, Aturia, 350, 351
  Rhyncholampas, 373, 374, 395
Alabamina mississippiensis, 233
  wilcoxensis, 222, 227, 230, *233*
Alaska, 88
alata, Bolivina, 222, 227, *234*, 246
  Brachiospyris, 275
  Dorcadospyris, 257, 260, 267, 270, 275
  Vulvulina, 234
Alatacythere sp., 70, 71, 75
alatus, Eupatagus, 373, 374, 399, 400, 401, 402
alazanensis, Bolivina, 222, 227, *234*, 246
  Bulimina, 222, 227, *235*
  Cibicidoides, 224
Albert Town area, 388
Albert Town Member, 129, 289, 388, 448
albicans, Amerigus, 365
Albunea, 120, 122
  sp., 119, 122
Alchornea, 446
Alevolina, 299
Alligator Crawle Harbour, 341
Alligator Pond, 118
alluvium, 395
Alsace, 359
Alsophis, 419
  sp., 428
Alston-Baillieston traverse, 35, 39
Alston-Borobridge traverse, 39
Alston sections, *35*, 45
Alston traverse, 54, 59
alta, Echinolampas, 373, 391, 393, 394
  Haimea, 373, 384, 385
Alterodon, 437
  major, Alterodon, 437
(Althecocyclina) stephensoni, Pseudophragmina, 337
altispira, Dentoglobigerina, 184, 185, 189, 193, 194, 195, *207*, 208, 209
  Globigerina, 144, 145, 182, 187
  Globoquadrina, 207
  altispira, 207
  altispira, Globoquadrina, 207
  globosa, Dentoglobigerina, 208
  Globoquadrina, 207

altissima, Echinolampas, 373, 391, 393
altissimus, Schizaster, 373, 396
altus, Macropneustes, 373, 374, 402, 403
Amaurolithus, 154, 163
  delicatus, 149, 150, 155, 162
  ninae, 149
  primus, 139, 149, 150, 152, 153, 155, 158, 162, 173
  tricorniculatus, 131, 139, 149, 150, 152, 154, 155, 158, 163
  spp., 146, 150
ambigua, Planulina, 222, 227, *239*, 243
  Rotalia, 239
Amblypygus, 382
  americanus, 373, 374, 382, 383, 384
Amblyrhiza, 437
Ameiva, 419, 428
  dorsalis, 428
American Coastal Plain, 69
americana, Conulites, 293
  Cushmania, 286, 289, 291, 293, 295, 296, 298
  Siphonorhis, 419, 428, 432
americanus, Amblypygus, 373, 374, 382, 383, 384
  Caprimulgus, 432
  Dictyoconus, 293, 349
  (Cushmania), 293
  Heterodictyoconus, 293
  Miosorites, 305, 307
Amerigus albicans, 365
  decipiens, 365
  edentata, 365
  pallidus, 365
  (Lateorbis) pallidus, 366, 367
  (Tropicorbis) decepiens, 366
  edentatus, 366
ammonite record, levels, *84*
ammonites, 13, 82
  Upper Cretaceous, *77*
Ammonites galicianus, 89
  garuda, 86
  oldhami, 89, 90
ammophila, Rotalia, 238
ammophilus, Hanzawaia, 222, 227, 230, *238*
Amphibrachium robustum, 273
Amphicytherura, 71
Amphilepidina, 313
amphipons, Chiastozygus, 22, 26
Amphisphaera cristata, 269
Amphistegina, 238, *309*
  lopeztrigoi, 311
  parvula, 286, 309, 312
  senni, 311
  spp., 182, 223, 225, 226, 227, 229, 230, 233, *234*

amphistilium, Amphymenium, 271
amphistylium, Ommatocampe, 260, 271, 272
Amphistylus angelinus, 269
Amphizygus brooksii, 22, 26
Amphymenium amphistilium, 271
ampliaperta, Helicosphaera, 137, 138, 139, 140, 144, 145, 146
Ampullaria fasciata, 363
Anadara halidonata, 355
Anaklinoceras reflexum, 66
anatina, Drepanotrema, 365, 366
anceps, Eiffellithus, 22, 27
Ancycloceras retrorsum, 86
ancyliformis, Gundlachia, 367
Ancylus cittyi, 367
  irroratus, 367
  obliquus, 367
  obscurus, 368
  radiatus, 367
andesite, 347
Andros, 433
angelina, Stylosphaera, 260, 268, 269
  (Stylosphaerella), 269
angelinum, Axoprunum, 269
angelinus, Amphistylus, 269
Anguilla, 375, 393, 437
Anguilla Formation, 375
anguillae, Echinolampas, 391, 393, 394
anguillensis, Cidaris (Tretocidaris), 375
  Tretocidaris, 375
angularis, Liliasterites, 4, 5, 8, 15
angulata, Eucoronis, 274
  Giraffospyris, 272, 274
  Gyraffospyris, 260
Angulogerina cojimarensis, 222, *234,* 244
  eximia, 222, 227, *234,* 247
  illingi, 222, 227, *234,* 247
  porrecta, 240
  selseyensis, 234
  yumuriana, 234
angusta, Vekshinella, 8, 15
angustus, Macropneustes, 373, 374, 402, 403
  Rhagodiscus, 8, 9, 11, 12, 15, 23, 27
Anisopetalus ellipticus, 394
Anisus vortex, 366
annuloides, Monoporites, 446, 447, 450
Anolis, 419, 428
anomala, Argyrotheca, 110
Anomalina foveolata, 239
  globulosa, 234
  mantaensis, 238
  pompilioides, 234
  wuellerstorfi, 239
Anomalinoides globulosus, 222, 224, 225, 227, 231, *234,* 248
  pseudogrosserugosus, 224
  semicribratus, 222, 227, *234*
  spp., 227
Anomia vitrea, 108
Antarctic ice sheet, 264

Antarctica, 269, 274, 227
antegressa, Bolivina, 222, 227, *234*
antepenultimus, Didymocyrtis, 257, 276
anthophora, Dictyospyris, 275
  Tholospyris, 260, 272, 275, 276
anthophorus, Reinhardtites, 4, 8, 9, 11, 12, 13, 15, 16, 23, 24, 25, 26, 27
anthropods, inventory (Jamaican), *120*
anticlines, 35, 37, 82
Antigua, 325, 335, 388, 421, 433
Antigua Formation, 325, 335
antillarum, Argiope, 111
  Clypeaster, 386, 388
  Dictyococcites, 131, *166*
  Eupatagus, 373, 401, 402
  Heterostegina, 335
  Oryzomys, 419, 421, 423, 424, 425, 426, 427, 428, 439
    couesi, 439
    palustris, 439
  Parapygus, 394
  Rhyncholampas, 373, 395
Antillaster, 401, *407*
  arnoldi, 373, 374, 407, 408, 409
  longipetalus, 373, 374, 407
  rojasi, 407
  vaughani, 407
antillea, Eulinderina, 286, 315, 316
  Heterostegina, 287, 321, 335, 337
  (Vlerkina), 305, 337, 338
  Lepidocyclina, 315
  (Polylepidina), 315, 316
  Miscellanea, 325
  Pellatispirella, 329
  Ranikothalia, 329, 331
Antillean islands, 127
Antilles, 444
Antillocaprina, 32, 34, 35, 37, 38, 39, 40, 41, 44, 45, 46, 67
  sp., 34
antiqua, Meoma, 373, 374, 403, 404
antiquus, Eprolithus, 7, 9, 15
aphylla, Phyllonycteris, 428, 434, 435
  (Reithronycteris), 434
  Reithronycteris, 434
apiculatus, Orbitoides, 57
Applinocrinus, 127
  cretacea, 127
ara, Vekshinella, 23, 27
Arachnodiscus ornatus, 268
arca, Globotruncana, 34
Arcadia Road section, *153,* 155, *156, 163,* 164, *190, 192, 196*
Archaias, 285, *305*
  asmaricus, 287, 305, 306
  columbiensis, 285
  floridanus, 305
  operculiniformia, 285
  sp., 305
archeomenardii, Globorotalia, 183, *198,* 199
arcuatus, Ceratolithoides, 4, 10, 11, 12, 13, 22, 24, 25, 26
  Lucianorhabdus, 23, 27

Arenagula, 297
  floridana, 286, 292, 297
argalus, Platybelone, 413
Argentina, 423, 432
Argiope antillarum, 111
  barrettiana, 111
Argyrotheca, 105, 107, *110*
  anomala, 110
  barrettiana, 107, 111
  magnicostata, 107, 110, 111
  spp., 107, *110, 111*
ariana, Lepidocyclina, 317
ariminensis, Planulina, 222, 224, 227, 232
Aristelliger, 429
  praesignis, 428, 429
  titan, 418, 419, 428, 429, 430
Ariteus, 423, *435*
  flavescens, 428, 435
  sp., 435
Arizona, 433
arizpensis, Lopha, 73
Arkhangelskiella cymbiformis, 9, 14, 22, 24, 26
  zone, 2
  specillata, 22, 24, 26
Armatobalanus duvergieri, 118
  (Hexacreusia) durhami, 118
armilla, Loxolithus, 23, 27
arnoldi, Antillaster, 373, 374, 407, 408, 409
Arntully area, 331
Arrhyton, 419
  sp., 428
arthropods, Cenozoic, *116*
  Cretaceous, *116*
  fossil, *115*
Arthur's Seat area, 84
Artibonite Group, 446
Artostrobium, 279
  spp., 260, 270, 279
Ascetoleberis, 69
Ascioythere sp., 74, 75
asmaricus, Archaias, 287, 305, 306
asper, Rhagodiscus, 11, 12, 15
Aspidolithus sp., 8
  parcus, 4, 22, 24, 25, 26
  constrictus, 8, 9, 10, 12, 13, 14, 16
assemblages
  carbonate-platform, 283
  deep-water, 226
  in situ, 226
  megafossil, *444, 451*
  microfossil, *445,* 448, *454*
  mollusc, 82
  nannofossil, 2, 132, *135*
  nannoplankton, 4, 6, 13
  palynomorph, *446,* 448, 450
  radiolarian, 264
  rudist, *29*
  shelf-edge, 283
  spore, 445
Assipetra infracretacea, 10, 11, 14
Asterigerina, 238
  spp., 182, 223, 225, 226, 227, 229, 230, 233, *234,* 242
asterius, Cenocrinus, 125, 126, 128

Asterocyclina, 285
asterodisca, Lepidocyclina (Lepidocyclina), 317
asteroids, *125*
　marginal ossicles, 129
Asterorbis, 37, 38, 40, 41, 45, 47, *59*
　havanensis, 35, 37, 45, 46, *59*, 62
　rooki, 59
Asterostoma, *406*
　excentricum, 373, 374, 406, 407
　pawsoni, 373, 374, 406, 407
astraeforme, Acanthometron (Astrolithium), 266
(Astrolithium) astraeforme, Acanthometron, 266
Astrononion pusillum, 227, *234*
　spp., 222
Astropyga magnifica, 373
asymmetricus, Cylindralithus, 22, 27
　Discoaster, 135, 137, 139, 142, 150, 153, 158, 162, 165, 168
atavia, Callimitra, 260, 268, 273
Athecocyclina, 337
　stephensoni, 286, 337, 338, 339, 341
Athene, 428, *432*
　cunicularia, 419, 428, 432
Atlantic coast, South America, 377
Atlantic region, 347
attenuatus, Eupatagus, 401, 402
Aturia, 349, 350, 351
　alabamensis, 350, 351
　cubaensis, 351
　curvilineata, 351
　panamensis, 351
　peruviana, 351
　sp., *349*, 350, 351
　spp., 351
aturiids, 351
August Town Formation, 118, 123, 155, 388, 391
augustae, Parachondria, 354, 356
aulakas, Discoaster, 137, 149
aurantius, Turdus, 428
aureum, Acrostichum, 448
aurita, Lithocampe, 279
auropunctatus, Herpestes, 439
Australia, 118, 229, 309
australis, Okkolithus, 23, 27
Austria, 88
Autocoptis, 359
(Autocoptis), Urocoptis, 359
avus, Menetus dilatatus, 367
axiologus, Homeopetalus, 404, 407
　Scoliechinus, 93, 94, 95, 98, 99
Axopodorhabdus dietzmannii, 22, 26
Axoprunum angelinum, 269
Ayalaina, 32, 34, 37, 40, 41, 67, 69

B

Back Rio Grande, 30
Back Rio Grande Formation, *41*, 45
Back Rio Grande Limestone Member, 41, 44
Back Rio Grande streambed, 44
Back Rio Grande–Stony River confluence, 44

Baculite Bed, 84
Baculites, *88*
　capensis, 88
　　tenuetuberculata, 88
　incurvatus, 88
　reesideri, 67
　　zone, 67
　tenuetuberculatus, 84, 87, 88
　yokoyamai, 80, 82
　sp., 88
Bahamas, 423, 432, 433, 434
bajuvaricum, Gauthiericeras, 80, 82, 84
bakeri, Incerticyclus, 354, 355, 357, 358, 359
Balaclava area, 418, 427
Balanocrinus haitiensis, 127
Balanus, 123
　eburneus, 118, 120, 121, 122
　improvisus, 123
　　assimilis, 118, 121
balticus, Inoceramus, 80, 82, 85
　balticus, 84
　balticus, Inoceramus, 84
　kunimiensis, Inoceramus, 79, 80, 84
　toyajoanus, Inoceramus, 84
Bamboo area, 395
bantamensis, Miogypsinoides, 327
barabini, Inoceramus, 84, 85
barbadensis, Ceratoconcha, 118, 121, 123
　Discoaster, 137, 147
Barbados, 273
Barbuda, 431, 432, 433
barleeana, Noniona, 238
barleeanum, Melonis, 227, 229, *238*
　Nonion, 238
barnesae, Watznaueria, 23, 24, 27
baroemoenensis, Globigerina, 208
　Globoquadrina, 207, 208
Barrettia, 21, 32, 34
　gigas, 10, 30, 32, 34, 45, 46, 81, 82
　monilifera, 41, 45, 46
　multilirata, 32, 45, 46, 80, 82
　rusea, 79, 80
Barrettia gigas Limestone, 13
Barrettia Limestone, 10, 30, 32, 34, 80, 82, 84
　age, *85*, 98, 99
Barrettia Limestone Member, 32
barrettiana, Argiope, 111
　Argyrotheca, 107, 111
　Cistella, 111
bartholomaeensis, Hepatiscus, 118, 119, 122
bartletti, Terebratula, 108
basin evolution, 220
Bath Fountain Road, 45
batheri, Clypeaster, 388
Bathropyramis, 260, 272, *276*
　acephala, 276
　sp., 276
bathypetalus, Schizaster, 373, 396
bats, fossil, 418
Beecher Town, 106, 113, 123, 129, 375, 385, 396, 398, 402

belemnos, Sphenolithus, 139, 146, 147
belgicus, Microrhabdulus, 8, 15, 23, 27
Belgium, 394, 413
Bellevue Formation, 30
Bellevue Site, 439
bellus, Discoaster, 137
Benbow–Guy Hills Inlier, 98
Benbow Inlier, *13*, 19
Bender local fauna, 359
berggrenii, Discoaster, 137, 138, 139, 149, 153, 162, 169
Bermuda, 377
bermudezi, Eorupertia, 287, 308, 309
　Miogypsina (Miogypsinella), 327
　Miogypsinoides, 327, 329, 330
bernardi, Pleurodonte, 354, 355, 356, 358, 359
biarcus, Stoverius, 11, 15
bidens, Tonatia, 419, 434
　saurophila, Tonatia, 428, 419, 430, 433, 434
Bidiscus ignotus, 22, 24, 26
bifarius, Chiastozygus, 22, 26
bifax, Discoaster, 147
bigelowii, Braarudosphaera, 8, 14, 22, 26, 137
bijugatus, Zygrhablithus, 137, 147
Biloculina murrhina, 240
biochronology, *420*
biofacies, *225*, *233*
biogeography, *422*
bioherms, 30
biostratigraphy, *1*, *69*, *155*, *190*, 221, *223*, *255*, *259*
　ostracode, *65*
　planktonic foraminiferal, *179*, *182*, *184*, 231
biozonal scheme, *134*
biozones
　lower Pliocene, *156*
　nannoplankton, *13*
　upper Miocene, *156*
bipes, Lynchocanium, 277
biporta, Watznaueria, 8, 9, 11, 12, 15, 23, 27
Biradiolites, 35, 37, 38, 39, 40, 41, 45, 46, 67
birnageae, Globorotalia, 183, 184, 185, *202*
　(Fohsella), 202
Biscutum constans, 22, 26
　patella, 22, 24, 26
bisecta, Reticulofenestra, 137, 146, 147
bitraversus, Parhabdolithus, 27
　Tranolithus, 23, 27
bivalves, 226, 233, 347
Black River, 431
blackstockae, Discoaster, 137
blainvilli, Mormoops, 428, 433
Blue Mountain Formation, 30
Blue Mountain Inlier, 19, *41*, 45, 50, 52, 54, 57, 77, 93, 102
Blue Mountain Peak area, 44
Blue Mountain Range, 30, 67

Blue Mountain Shale, 84
Blue Mountains, 285
Bodionegoro 1 well, 206
Bohemia, 115
Bolivia, 423
Bolivina, 235
  aenariensiformis, *234*, 246
  alata, 222, 227, *234*, 246
  alazanensis, 222, 227, *234*, 246
  antegressa, 222, 227, *234*
  floridana, 227
  isidroensis, 238
  pseudoplicata, 227
  tectiformis, 234
  thalmanni, 222, 227, *235*, 246
  tortuosa, 222, 227, *235*, 244
bollii, *Discoaster*, 137, 138, 140, 142, 144, 145, 146, 162, 167, *168*
Bolvinita quadrilatera, 235
  sp., *235*
Bombacacidites, 449
  sp., 448, 450
Bonafide Cave, 419, *424*, 439
Bonaire, 385
Bonnie View Andesite, 84
Bonny Gate Formation, 307, 309, 316, 317, 327, 333, 335
Bontourina inflata, 339
boreholes, 223, 229
Borelis, 297
  jamaicensis, 297
    truncata, 297
  matleyi, 297
Borobridge-Alston traverse, 35
borroi, *Hercothyris*, 113
Botriopygus, 99
  rudistarum, 93, 99
Botryostrobus, 279
Boug Regreg section, 196
Boulder Bed, 142, 144, 146, 190
Bournonia, 32, 40, 41, 45, 46
Bowden, 147, 354, 382, 413
Bowden Beds, 156, 179, 220, 353, 443
Bowden District, 142
Bowden fauna, *354*, *359*, 360
Bowden Formation, 131, 132, 134, 140, 142, 143, *147*, 150, 152, *155*, *156*, *158*, 164, 180, 182, 184, 189, 190, 196, 197, 204, 220, *224*, *229*, 232, 233, 239, 257, 354, 413
  Arcadia Road section, *153*, 155, 156, 163, 164, 190, *192*, 196
  correlative, *142*
  Drivers River section, 150, *153*, 156, 190
  Ecclesdown Road section, *155*, *158*, *194*, 198, 202
  Folly Point section, *152*, 156, 190
  Innes Bay section, 111, *153*, 156
  San San Bay type section, *192*
  Stony Hill Road section, *155*, *156*, 159, *194*
  type section, *152*, 191
Bowden Pen Formation, 44, 84

Bowden section, 150, *156*
Bowden Shell, 153, *354*
bowdenensis, *Hemitrochus*, 354, 355, 356
bowdeniana, *Pleurodonte*, 354, 355, 356, 358, 359
bowdensis, *Poteria*, 354, 357
braarudii, *Discoaster*, 137, 149
Braarudosphaera africana, 12, 14
  bigelowii, 8, 14, 22, 26, 137
brachiopods, fossil, *105*
Brachiospyris alata, 275
Brachycythere sp., 71, 72, 75
brachypetalus, *Schizaster*, 396
Brachyphylla, 423, *434*
  cavernarum pumila, 434
  nana, 419, 423, 434
    pumila, 423, 434
  pumila, 434
brachytona, *Echinolampas*, 373, 391, 393
bradyi, *Cibicidoides*, 227, *236*
  Eggerella, 227, *237*
  Gaudryina, 238
  Karreriella, 238
  Trifarina, 222, 227, *241*
  Truncatulina, 236
  Verneuilina, 237
Braham Cave, 419, *424*
Branisella, 423
breccias, 420, 421, 422, 427, 431, 435, 436, 439
brevior, *Hyalosagda*, 356
  (*Stauroglypta*), 354
brevis, *Hyalosagda*, 356
  (*Stauroglypta*), 354
  Urocoptis, 354, 356
Brevitricolpites variabilis, 446
brewsteri, *Siphonorhis*, 432
bridgeboroensis, *Brissus*, 401
Brissus, *401*
  bridgeboroensis, 401
  brissus, 401
  camagueyensis, 401
  unicolor, 373, 374, 392, 401
  sp., 373, 374, 401
brissus, *Brissus*, 401
britannica, *Ellipsagelosphaera*, 8, 9, 15
British Honduras, 65
Broinsonia lacunosa zone, 4
  matalosa, 22, 26
  parca zone, 4
  sp., 8
brooksii, *Amphizygus*, 22, 26
brouweri, *Discoaster*, 137, 138, 142, 149, 150, 153, 154, 158, 162, 169
brownii, *Capromys*, 436
  (*Geocapromys*), 436
  Geocapromys, 419, 423, 425, 426, 427, 428, 436, 437, 438
Browns Town, 129, 384, 389, 402
Browns Town Formation, 129, 309, 317, 321, 327, 329, 335, 337, 377, 402
brunneus, *Osteopilus*, 419, 428
bryozoans, 68

Buff Bay, 132, 158, 179, 184, 185, 189, 190, 197, 200, 204, 206, 207, 208, 210, 212, *219*, 233, 234, 256, 264, 276, 277
  upper, 142
Buff Bay Beds, 179, 219, 220
Buff Bay Bowden unconformity, 191
Buff Bay Formation, 123, 131, 132, 134, 140, *141*, 143, 145, 146, 150, *152*, *155*, *156*, *158*, 163, 165, 180, 182, 184, 186, 188, 189, 190, 197, 198, 204, 206, 220, *221*, *223*, 225, 226, 229, 231, 232, 233, 257
Buff Bay outcrop, 198, 204, 214
Buff Bay region, 221
Buff Bay roadcuts, 221
Buff Bay section, *132*, *140*, *142*, *146*, 163, 174, 165, 180, *182*, *185*, 186, 189, 190, *197*, 200, 202, 204, 207, 221, 229, 231, 263, 276
  biostratigraphy, *185*
Buff Bay sequence, 259, 274
Bukryaster hayi, 4, 6, 8, 10, 14, 16, 22, 24, 25, 26
bukryi, *Cretarhabdus*, 22, 26
  striatus, 26
  Russellia, 23, 27
Bulgaria, 89
Bulimina aculeata, 222, 225, 227, 228, 229, 231, 233, *235*
  alazanensis, 222, 227, *235*
  impendens, 227, *235*
  inflata mexicana, 235
  macilenta, 227
  marginata, 222, 224, 227, 232, *235*
  mexicana, 227, *235*, 246
  trinitatensis, 235
  tuxpamensis, 222, 227, *235*
bulloides, *Globotruncana*, 34
  Nonionina, 239
  Pullenia, 227, *239*
  Sphaeroidina, 212, 227, 229, *241*
Buntonia, 69
  sp., 70, 75
burkei, *Helicosphaera*, 131, 139, *165*, 177
Burnt Hill–Kinloss section, 321, 325
bursa, *Dendrospyris*, 260, 272, 274
Burts Run area, 394
bussonii, *Carpocanopsis*, 260, 270, 278
  Sethocorys, 278
butterlinus, *Miogypsinoides*, 327
Bythoceratina sp., 74

C

Cabarita River, 431
Caicos Islands, 434
caillaudi, *Haimea*, 384
Calabria, 195
Calappa, 120, 121, 122
  gallus, 120, 121, 122
calcarata, *Globotruncana*, 66
calcarenites, 30, 192

calcaris, Discoaster, 137, 138, 140, 169
Calcidiscus leptoporus, 137
  macintyrei, 137, 153, 173, 175, 177
  zone, 158
calcite, 69
Calculites, 2
  obscurus, 2, 8, 9, 10, 11, 12, 14
  zone, 2, 7, 10
  ovalis, 8, 9, 10, 14
  sp., 11
Callianassa, 118, 120, 120
  gigantea, 119, 120
  subplana, 118, 119, 120
  trechmanni, 118, 119, 122
  sp., 118, 121
  spp., 118, 122
Callimitra, 273
  atavia, 260, 268, 273
  carolotae, 273
Callinectes, 122
  jamaicensis, 118, 119, 122
Calocyclas coronata, 277
  (Calocycletta) margatensis, 278
    veneris, 279
    virginia, 279
Calocycletta, 279
  costata, 257, 260, 262, 270, 279
    zone, 259, 264, 269, 271, 273, 274, 275, 276, 277, 278, 279
  virginis, 260, 270, 279
    zone, 279
  (Calocycletta) margatensis, Calocyclas, 278
    veneris, Calocyclas, 279
    virginia, Calocyclas, 279
Calton Hill Inlier, 67
calyculus, Catinaster, 137, 138, 139, 140, 141, 142, 143, 144, 145, 162, 171
Camaguey Province, 113
camagueyensis, Brissus, 401
Cambridge area, 424
Cambridge Cave, 435, 436
Cambridge-to-Catadupa railroad locality, 67, 71, 73
Camerina catenula, 331
  macgillavryi, 325
  matleyi, 325
  pellatispiroides, 329
  striatoreticulata, 331
Camp Perrin, 359
Campanian, *19*
  Early, 4, *44*
  early Late, 25
  Late, 3, 15, *45*
  late Early, 24
  late Late, 25
  Middle, *45*
Campanile giganteum, 116, 347
  sp., 347
campanula, Scyphosphaera, 137
Cancris nuttalli, 222, 224, 227, 232, *235*
  oblongus, 222, 227, *235*
  sagra, 235

Cancris (continued)
  scintillans, 222, 227, *235*, 248
  sinecarina, 235, 250
  spp., 227
Candeina, 207
  nitida, 185, 187, *207*, 209
  praenitida, 182
canellei, Lepidocyclina, 317, 319, 320
  (Lepidocyclina), 287, 316, 319, 321
Canis familiaris, 421
Cannartidium mammiferum, 269
Cannartus mammiferus, 269
  prismaticus, 269
  tubarius, 269
  violina, 269, 271
  sp., 269
Cape Matteras, 115
capensis, Baculites, 88
  tenuetuberculata, Baculites, 88
Capo Rosello, 195
Caprimulgus americanus, 432
  griseus, 432
Capromys, 436
  brownii, 436
  (Geocapromys) brownii, 436
caput-serpentis, Terebratulina, 113
carapitana, Cassidulina, 222, 227, *235*, 236
  Uvigerina, 222, 227, 231, *241*
carbonate platform, 132
carbonates, 132, 259, 262, 297, 305, 307, 317, 337, 353, 355
Carcineretes, 116, 120
  woolacotti, 116, 117, 120
cardenasensis, Chubbina, 35
caribaea, Hanzawaia, 238
caribaeus, Planorbis, 367
caribbaeus, Mithrax, 119, 120, 122
Caribbaster, 396, *398*, 405
  dyscritus, 373, 374, *397*, 398, 399
  loveni, 373, 374, 398, 399
caribbea, Ehrenbergina, 227, *237*, 246
Caribbean-Colombian Basin, 198
Caribbean Current, 263, 265
caribbeanensis, Lacazella, 106, 107, 108
carinatus, Triquetrorhabdulus, 146, 147
carniolensis, Lithraphidites, 23, 24, 27
Carolinas, 115
carolotae, Callimitra, 273
Carpenteria conoidea, 307
Carpocaniidae sp., 278
Carpocanistrum spp., 278
Carpocanopsis, 278
  bussonii, 260, 270, 278
  cingulata, 260, 270, 278
  cristatum, 260, 270, 278
  sp., 270
carteri, Helicosphaera, 137
Cascade Formation, 95
Cassidulina, 236
  carapitana, 222, 227, *235*, 236
  crassa, 227, *236*

Cassidulina (continued)
  laevigata, 227, *236*
  palmerae, 237
  reflexa, 227, 229, *236*
  spinifera, 222, 227, 230, *236*, 243
  subglobosa, 238
  spp., 236
Cassidulinoides spp., 227
Cassiduloid sp., *100*, 101
Cassidulus, 394, 395
  platypetalus, 394
  sphaeroides, 387, 394
cassis, Fabiania, 286, 307, 310, 317, 337
Castle Hayne Formation, 106
Catadupa area, 102, 120
Catapsydrax, 207
  dissimilis, 257
  stainforthi, 257
catatumbus, Striatricolpites, 446, 447, 450
catenula, Camerina, 331
  Operculina, 329, 331
  Ranikothalia, 286, 329, 331, 332
Cathalina River confluence area, 41
Catinaster calyculus, 137, 138, 139, 140, 141, 142, 143, 144, 145, 162, 171
  coalitus, 137, 138, 139, 140, 141, 142, 144, 145, 146, 152, 162
  sp., 137, 171
caudriae, Globocassidulina, 222, 227, *237*
Cave River, 37
cavernarum pumila, Brachyphylla, 434
caves, 353, *418*
  Bonafide, 359, 360
  Coco Ree, 359
  deposits, 420
  fossil deposits, *418*, 420
  Green Grotto, 359, 360
  pearls, 353
  Portland Ridge, 359
  Sheep Pen, 359, 360
cayeuxi, Pachydiscus, 79, 80, 90
cayeuxii, Lucianorhabdus, 9, 15, 23, 27
Cayman Brac, 432, 434
Cayman Islands, 421, 423, 429, 437
Cayman Trough, 132, 263
Celestus, 426, 427, 428, *429*
  occiduus, 425, 428, 429
  sp., 429
Cenocrinus, 125
  asterius, 125, 126, 128
Cenozoic, *116*, *120*, *137*
Central America, 433, 444
Central Inlier, 19, 29, *35*, 44, 54, 57, 59, *67*, 71, 73, 96, 98, 99, 100, 102, 116, 450
Central Range, 263
cephalopods, Tertiary, *347*
Cepolis (Cordya), 359
Ceran Hill, 375
Ceratoconcha, 118, 121, 123
  barbadensis, 118, 121, 123
  creusioides, 118, 123
  jungi, 118, 123

Ceratolithoides aculeus, 8, 9, 14, 16, 22, 24, 25, 26
  arcuatus, 4, 10, 11, 12, 13, 22, 24, 25, 26
  verbeekii, 7, 8, 9, 12, 14, 22, 24, 25, 26
ceratoliths, 154
Ceratolithus acutus, 139, 140, 150, 152, 155, 195, 196
  cristatus, 137
  rugosus, 137, 138, 139, 142, 144, 145, 150, 153, 154, 155, 158, 162, 164, 196
    zone, 164
  sp., 196
Ceratopteris, 446
  (Magnastriatites) howardii, 450
Ceratospyris heptaceros, 274
  stylophora, 274
Ceriopora globulus, 309
Cerotolithoides aculeus zone, 3, *24*
chalk, 184, 257, 258, 259, 262, 264
Chalk of England, 115
challengeri, Discoaster, 131, 137, 138, 140, 141, 149, 150, 154, 162, 165, 167
Chapelton, 120
Chapelton Formation, 99, 116, 118, 120, 122, 123, 129, 283, 285, 289, 291, 295, 299, 303, 309, 311, 315, 316, 325, *347*, 349, 363, 375, 377, 379, 380, 381, 384, 385, 386, 388, 389, 390, 391, 393, 394, 395, 396, 398, 399, 402, 403, 405, 406, 407, 413, 414, *443*, 446, *448*
chaperi, Lepidocyclina, 317
  (Eulepidina), 317, 321
  (Nephrolepidina), 287, 316, 320, 321, 326
Charactosuchus fieldsi, 414
  kugleri, 413, 414
Chatsworth School area, 73
chemnitziana, Pleurodonte, 358
Chepstow Formation, 331
chert, 259
Chiapasella, 35, 37, 38, 45, 46
chiapasensis, Lepidocyclina (Polylepidina), 316
  Polylepidina, 287, 315, 316, 318
Chiasmolithus consuetus, 137, 146, 147
  grandis, 137, 147
  solitus, 147
Chiastozygus amphipons, 22, 26
  bifarius, 22, 26
  dennisonii, 22, 26
  garrisonii, 22, 26
  litterarius, 8, 9, 11, 12, 14, 22, 26
  mediaquadratus, 26
  propagulis, 22, 26
  synquadriperforatus, 22, 26
  sp., 22, 26
chiastus, Neochiastazygus, 137, 147
Chilonycteris parnelli, 433
Chilostomella sp., 222, 227
chondra, Trochalosoma, 93, 373, 374, 380, 383, 389

Christiana area, 389, 413, 425
christianaensis, Coleiconus, 286, 289, 290, 292
christiani, Polyperibola, 188, 197, *207*, 209
Chrysalidina (Chrysalidina) floridana, 297
(Chrysalidina) floridana, Chrysalidina, 297
chubbi, Pseudorbitoides, 34, 44, 46, *48*
Chubbina, 37, 38, 40, 41, 69
  cardenasensis, 35
cibaoensis, Globorotalia, 150, 192, 193, 194, 195, 196
Cibicides cushmani, 238
  floridanus, 236
  kullenbergi, 237
  lobatulus, 225, 226, 229, 230, 231, *236*, 242
  mantaensis, 238
  rugosa, 239
  wuellerstorfi, 239
Cibicidoides alazanensis, 224
  bradyi, 227, *236*
  cicatricosus, 227, *236*
  compressus, *236*
  cookei, 222, 227, *236*
  coryelli, 227, *236*, 243
  crebbsi, 224
  dominicus, 227, *236*, 243
  havanensis, 222, 224, 227, *236*
  incrassatus, 227, *237*
  lobatulus, 227
  matanzasensis, 227, *237*, 249
  mundulus, 225, 226, 227, 228, 229, *237*
  pachyderma, 222, 224, 227, 233, *237*
  robertsonianus, 224, 225, *237*
  sp., 227, *237*, 249
cicatricosus, Cibicidoides, 227, *236*
Cidaris, 375
  gymnozona, 375
  melitensis, 375
  (Tretocidaris), 375
    anguillensis, 375
  sp., 375, 377, 379
Cidaroid sp., 96
Ciego de Avila Province, 446
cienkowskii, Eucyrtidium, 260, 270, 277
  Stichopodium, 277
cimex, Drepanotrema, 365, 366
Cimomia kugleri, 351
cingulata, Carpocanopsis, 260, 270, 278
Cipero Formation, 185, 262
Circodiscus, 271
  microporus, 260, 266, 271
circula, Cribrosphaera, 22
  Cribrosphaerella, 27
circumradiatus, Haqius, 9, 15
cirripedes, 118
Cistella barrettiana, 111
cittyi, Ancylus, 367
Claiborne Formation, 446

Claremont Formation, 295, 297, 299, 301, 307, 333, 347, 351, 389, 394
Clarendon Group, 132, 146
Clarendon Parish, 67, 73, 110, 120, 123, 388, 389, 391, 418, 426, 429, 431, 433, 434, 436
clathrata, Cornutella, 276
  profunda, Cornutella, 276
claviger, Rhabdosphaera, 137
clavigerum, Lynchocanium (Lynchnocanoma), 277
clay, 67, 73, *149*, 180, 192, 220, 347, 363
claystone, 35
Cletocytheseis sp., 70, 75
clevei, Echinolampas, 373, 374, 391, 392, 393, 394
  Eupatagus, 402
Clidomys, 419, 420, 421, 422, 423, 425, 426, 427, 436, *437*
  osborni, 419, 427, 428, 437, 438, 439
  parvus, 419, 427, 428, 437, 439
Clifton Limestone, 79, 80, 85, 89, 90, 102
Clypeaster, 371, *386*
  antillarum, 386, 388
  batheri, 388
  concavus, 373, 386, 388
  cotteaui, 373, 374, 386, 388
  eurychorus, 386, 388
  lanceolatus, 373, 374, 386, 388
  rosaceus, 373, 386, 387, 388
  sp., 129, 373, 388
  spp., 373, 386
coalitus, Catinaster, 137, 138, 139, 140, 141, 142, 144, 145, 146, 152, 162
coartata, Compressigerina, 226, 227, 228, 229, 233, *237*, 247
  Uvigerina, 237
Coastal Belt, North, 180
Coastal Group, 123, 131, *132*, 194, 225, 375, 391
  Lower, *147*, *156*, *158*, 382, 388
Cobre Formation, 165
Cobre Member, 165
coccolith, 15
Coccolithus miopelagicus, 137, 146, 147, 152, 162, 173
  pataecus, 137
  pelogicus, 137
Cockpit Country, 360, 418, 419, 424, 431, 432, 437
Coco Ree, 359
Coco Ree Cave, 419, *425*, 436, 439
codon, Dictyoconus, 293
cojimarensis, Angulogerina, 222, 234, 244
  Uvigerina, 227
colei, Fabularia, 285, 286, 289, 295, 297, 299, 302
Coleiconus, 285, 287, *289*
  christianaensis, 286, 289, 290, 292
  zansi, 286, 287, 289, 290, 292, 295
(Coleiconus) elongata, Coskinolina, 289

Coleyville, 414
collections
  AMNH, 424
  Anthony, 418, 424
  Arnold, 379, 380, 382, 384, 385, 403, 406
  BMNH, 377, 409, 414
  Bowden, 413
  central Inlier, *71*
  Donovan, 396
  Florida Museum of Natural History, 125, 349, 424, 427
  Geological Survey of Jamaica, 82
  Institute of Jamaica, 413
  Jerusalem Mountain Inlier, *73*
  Liverpool Museum, 363
  Lucas Barrett, 106, 108, 113, 116, 120
  Maldon Inlier, *73*
  Marchmont, *73*
  MCZ, 386, 388, 398, 405
  Michelin, 382
  Museum of Comparative Zoology, 94
  National Museum of Natural History, 125
  National Museum (Netherlands), 29, 59
  Natural History London Museum, 77, 94, 108, 116, 125, 283
  Natural History Museum Basel, 349
  Oxford University Museum, 363
  Pattons, 419
  Smithsonian, 65, 94, 113, 125
  Trechmann, 393
  USNM, 380, 385, 424
collei, Cyclura, 428, 429, 431
colluvium, 355
Colombia, 351, 446
Colombia Basin, 192, 196, 200, 264
Columbia inomata, 428
columbiensis, Archaias, 285
columnata, Prediscosphaera, 12, 15
Comatulid sp., *127*
compacta, Helicosphaera, 137, 146, 147
compactus, Isocrystallithus, 12, 15
complanata, Miogypsinoides, 327
  Operculina, 335
compressa, Uvigerina, 237
Compressigerina coartata, 226, 227, 228, 229, 233, *237*, 247
compressiuscula quadriloba, Pullenia, 240
compressus, Cibicidoides, 236
concava, Micula, 8, 9, 11, 12, 15
concavata, Dicarinella, 82, 84
concavus, Clypeaster, 373, 386, 388
concinnus, Neochiastazygus, 147
conglobatus, Globigerinoides, 184, 187, 191, 193, *210*, 211
conglomerates, 20, 21, 30, 34, 44, 67, 79, 80, 84, *347*, 420, 421, 427, 439
Coniacian, 4, 15
  Latest, 4
conicus, Cretarhabdus, 8, 14, 22, 24, 26

conoidea, Carpenteria, 307
  Victoriella, 287, 307, 308, 309
conomiozea, Globorotalia, 196
  crassula, 206
Conorbitoides, 48
constans, Biscutum, 22, 26
constrictus, Aspidolithus parcus, 8, 9, 10, 12, 13, 14, 16
consuetus, Chiasmolithus, 137, 146, 147
continuosa, Globorotalia, 188, 206
  opima, 206
  (*Turborotalia*), 206
  Neogloboquadrina, 188, 189, 193, 205, *206*
contusa, Contusotruncana, 82
Contusotruncana contusa, 82
  fornicata, 82
Conulites americana, 293
Conusphaera mexicana, 4, 5, 6, 8, 9, 14
convallis, Minylitha, 137, 138, 140, 146, 162
  Mynilitha, 142, 177
convexa, Haimea, 373, 384, 385
cookei, Cibicidoides, 222, 227, *236*
  Coskinolina, 291
  Fallotella, 285, 286, 291, 293, 294, 296
  (*Fallotella*), 291
  Heterodictyoconus, 291
Coopers Hill area, 30
corals, 68, 105, 180, 220, 226, 233
(*Cordya*), *Cepolis*, 359
cores, *159*, 164
  wallywash, *363, 366*
Corn Husk River, 30, 41, 45
Cornutella, 276
  clathrata, 276
    profunda, 276
  profunda, 276
  spp., 260, 272, 276
Cornwall-Middlesex Platform, 132
Corollithion exiguum, 11, 14, 22, 26
  kennedyi, 4, 12, 14
  signum, 8, 11, 14, 22, 26
  sp. 9, 11
coronata, Calocyclas, 277
  Globotruncana, 34
  Stichocorys, 259, 260, 270, 272, 277
coronatus, Hydrobia, 365
  Potamopyrgus, 365
Coronocyclus nitescens, 137
Corsinipollenites, 448
  sp., 450
Coryda, 359
coryelli, Cibicidoides, 227, *236*, 243
Coskinolina, 285, 287, *289*, 297
  cookei, 291
  douvillei, 286, 287, 289, 290
  elongata, 287
  floridanus, 293
  inflata, 289
  (*Coleiconus*) *elongata*, 289
  sp., 287, 289, 290
Coskinolinoides, 295
  jamaicensis, 295, 307

Costa Rica, 221, 232, 445, 446
costata, Calocycletta, 257, 260, 262, 270, 279
  Lucidella, 354, 355, 359
costlowi, Trisolenia megalactis, 260, 266, 267
cotteaui, Clypeaster, 373, 374, 386, 388
couesi, Oryzomys, 421, 423, 439
  antillarum, Oryzomys, 439
Cphthalmoplax stephensoni, 116
crabs, *116*
crassa, Cassidulina, 227, *236*
crassacrotenensis, Globorotalia, 206
crassaformis, Globorotalia, 150, 152, 183, 184, 187, 191, 193, 194, *204*, 205
  (*Turborotalia*), 204
  ronda, Globorotalia (*Globorotalia*), 182
  viola, Globorotalia (*Globorotalia*), 182
crassata, Lepidocyclina (*Nephrolepidina*), 323
Crassostrea virginica, 118, 121, 123
crassula, Globorotalia, 187, 191, 205, 206
  (*Globorotalia*), 206
  (*Turborotalia*), 206
  conomiozea, Globorotalia, 206
crassus, Cylindralithus, 8, 9, 11, 14, 22, 27
  Pelliceria (*Psilatricolporites*), 443
  Psilatricolporites, 446, 447, 448, 449
crebbsi, Cibicidoides, 224
creber, Rhabdolithus, 137
crebra, Rhabdosphaera, 147
crenulata, Retecapsa, 8, 9, 11, 12, 15
cretacea, Applinocrinus, 127
  Prediscosphaera, 8, 9, 11, 12, 15, 23, 24, 27
  Saccocoma, 127
Cretaceous, *93*, *116*, *120*
  Late, 1
Cretacoranina, *120*
  (*Cretacoranina*), *Notopocorystes*, 116
  trechmanni, Notopocorystes, 116, 117, 120
Cretarhabdus bukryi, 22, 26
  conicus, 8, 14, 22, 24, 26
  schizobrachiatus, 22, 26
  sinuosus, 22, 24, 26
  striatus bukryi, 26
  surirellus, 22, 24, 27
creusioides, Ceratoconcha, 118, 123
cribatus, Polyxenes, 236
Cribrobulimina floridana, 297
Cribrocorona, sp., 22, 27
Cribrosphaera circula, 22
  ehrenbergii, 22, 24
Cribrosphaerella circula, 27
  ehrenbergii, 8, 9, 12, 14, 27
cricota, Umbilicosphaera, 137, *166*, 175
crinoid, *125*
crippsi, Inoceramus, 82
crisotus, Cyclococcolithus, 166

cristata, Amphisphaera, 269
　Sethocorys, 278
cristatum, Carpocanopsis, 260, 270, 278
cristatus, Ceratolithus, 137
cristensis, Discocyclina, 337
　Hexagonocyclina, 341
Crocidopoma, sp., 359
Crocodylus, 413, 427, *431*
　acutus, 431
　sp., 419, 428, 431
Cross Pass Formation, 84
crotonensis, Globorotalia, 206
Cruciplacolithus delus, 137, 147
　staurion, 147
Crudia, 446
crust, continental, 223
crustula, Sismonda, 389
Ctenorbitoides, 48
Cuba, 65, 69, 96, 108, 110, 113, 229, 235, 285, 351, 359, 372, 375, 382, 385, 388, 393, 395, 396, 398, 399, 402, 419, 423, 433, 434, 437, 446
cubaensis, Aturia, 351
Cubagua Formation, 190
Cubanaster, *389*
　acunai, 373, 374, 387, 389, 390
　sp., 373, 387, 389
cubensis, Dyscritothyris, 110
　Eodictyoconus, 307
　Physella, 365, 366
　Terebratula, 108
　Vaughanina, 35, 37, 45, 46, *54*, 55, 56
Cubotholus, 273
Cucaracha Formation, 445, 446
Culebra Formation, 445, 446
cultrata exilis, Globorotalia, 182
　limbata, Globorotalia (Globorotalia), 198
cundalli, Speoxenus, 437
cuneata, Terebratula, 110
cunicularia, Athene, 419, 428, 432
　Speotyto, 432
　Strix, 432
Cupaniedites, 449
　sp., 450
Curacao, 402
curvatura, Ellipsonodosaria, 241
　Stilostomella, 227, *241*
curvilineata, Aturia, 351
cushmani, Cibicides, 238
　Hanzawaia, 238
　Operculinoides, 333, 347
Cushmania, 285, 291, *293*, 295, 297
　americana, 286, 289, 291, 293, 295, 296, 298
　fontabellensis, 293, 295
　puilboreauensis, 295
　(Cushmania) americanus, Dictyoconus, 293
　puilboreauensis, Dictyoconus, 293
cuvieri, Aello, 433
Cyclagelosphaera margerelii, 9, 14
　sp., 22, 27
Cyclampterium, 266
　milowi, 267

Cyclas veatleyi, 369
Cyclaster, *401*
　sterea, 392, 401
Cyclochittya dentistigmata, 355
　schermoi, 355
Cyclococcolithina macintyrei zone, 153
Cyclococcolithus aequiscutum, 131, *166*
　crisotus, 166
　leptoporus, 131
　macintyrei, 153
Cyclorbiculinoides, *305*, 307
　jamaicensis, 285, 287, 304, 305, 309
Cyclostoma perpallidum, 358
cyclostomus, Echinoneus, 373, 382, 383
Cyclura, 419, *429*
　collei, 428, 429, 431
Cyclus pygmaea, 369
Cylindralithus asymmetricus, 22, 27
　crassus, 8, 9, 11, 14, 22, 27
　serratus, 14, 22, 27
　sp., 12
　spp., 22, 27
cylindrica, Haimea, 373, 384, 385
cymbiformis, Arkhangelskiella, 9, 14, 22, 24, 26
Cyntheseis segurai, 69
Cyrtocapsa (Cyrtocapsella) tetrapera, 276
Cyrtocapsella, 276
　tetrapera, 257, 260, 262, 270, 276, 277
　zone, 265, 271, 274, 279
　(Cyrtocapsella) tetrapera, Cyrtocapsa, 276
Cytherella tuberculifera, 69, 74, 75
Cytherelloidea sp., 74, 75
Cytherura sp., 71, 74, 75

D

Dairy Cave, 418, *425*, 429, 431, 432, 433, 434, 435, 436
dalli, Neolaganum, 373, 374, 389
Dead Goat Gully Roadcut outcrop, 134, 141, 142, 263, 180, 189, 207, 210, 221
debris, coralline, 182
debris flow, 141, 163, 190
decapods, 115, 118
decipiens, Amerigus, 365
　(Tropicorbis), 366
　Planorbis, 366
decora, Periphaena, 271
decoratus, Microrhabdulus, 8, 9, 11, 12, 15, 16, 23, 27
decorus, Tetrapodorhabdus, 12, 15, 23, 27
decussata, Micula, 4, 8, 9, 11, 12, 15, 23, 24, 27, 80
defectus, Eupatagus, 373, 401, 402
deflandrei, Discoaster, 137, 139, 140
deformis, Inoceramus, 82

dehiscens, Globoquadrina, 150, 152, 187, 189, 192, 207, 208, 211
　Globorotalia, 208
　Sphaeroidina, 212
　Sphaeroidinella, 213, 214
　Sphaeroidinellopsis, 187, 191, 195, 198, 214
　immatura, Sphaeroidinella, 192, 193, 194, 195, 202, 212
　subdihiscens, Sphaeroidinella, 214
delicata, Praerhapydionina, 285, 287, 303, 305, 306
delicatus, Amaurolithus, 149, 150, 155, 162
delmontense, Eucyrtidium, 277
　Stichocorys, 277
delmontensis, Stichocorys, 257, 260, 270, 277, 278, 279
Deltoidospora, 448, 449
　adriennis, 448, 449
delus, Cruciplacolithus, 137, 147
Dendrospyris, 273
　bursa, 260, 272, 274
　stabilis, 260, 272, 274
　sp., 274
dennisonii, Chiastozygus, 22, 26
　Eiffellithus, 26
Dentalina spp., 227
dentata, Dorcadospyris, 259, 260, 270, 274, 275
Dentellaria, 356, 358
dentiferus, Planorbis, 366
　edentatus, Planorbis, 366
dentistigmata, Cyclochittya, 355
Dentoglobigerina, 207, 208
　altispira, 184, 185, 189, 193, 194, 195, *207*, 208, 209
　globosa, 208
　galavisi, 207, 208
　globosa, 185, 187, 189, *208*, 209
　naroemoenensis, *208*
　venezuelana, 184, 187, 189, 193, 195, 197, *208*, 209
deposition, *13*, *142*
Desmophyllites, *88*
　phyllimorphus, 84, 87, 88
DeSota Canyon, 223
Devils River, 41, 45
dia, Operculinella, 335
　Operculinoides, 335
　Palaeonummulites, 287, 330, 335
Diablo Cave, 436
diagenesis, 2
diaphanes, Eucyrtidium, 277
Diartus petterssoni, 257
　zone, 259, 276, 278, 279
diastypus, Discoaster, 147
diatoms, 262
Dicarinella concavata, 82, 84
Dictyococcites antillarum, 131, *166*
Dictyoconus, 293, 295
　americanus, 293, 349
　codon, 293
　gunteri, 293
　puilboreauensis, 293
　(Cushmania) americanus, 293
　　puilboreauensis, 293
Dictyocoryne profunda, 273

Dictyomietra, 279
Dictyospyris anthophora, 275
　mammillaris, 276
Didymoceras nebrascense zone, 67
didymocyrti violina, Haliomma, 269
Didymocyrtis, 269
　antepenultimus, 257, 276
　laticonus, 262
　mammifera, 260, 269, 270, 271
　penultima zone, 267, 277
　prismatica, 259, 260, 269
　tubaria, 260, 269, 270
　violina, 260, 262, 269, 270, 271
dietzmannii, Axopodorhabdus, 22, 26
dihiscens, Sphaeroidinella dehiscens, 214
dilatatus avus, Menetus, 367
diminutus, Globigerinoides, 183, 184, 185, *210*
Dimorphina striata, 240
dimorphism, 57
Diozoptyxiz, 84
Diplocrinus, *126*
　maclearanus, 127
　sp., *126*, 129
diplogrammus, Zygodiscus, 23, 27
discapora, Pontosphaera, 137, 173
Discoaster asymmetricus, 135, 137, 139, 142, 150, 153, 158, 162, 165, 168
　aulakas, 137, 149
　barbadiensis, 137, 147
　bellus, 137
　berggrenii, 137, 138, 139, 149, 153, 162, 169
　bifax, 147
　blackstockae, 137
　bollii, 137, 138, 140, 142, 144, 145, 146, 162, 167, *168*
　braarudii., 137, 149
　brouweri, 137, 138, 142, 149, 150, 153, 154, 158, 162, 169
　calcaris, 137, 138, 140, 169
　challengeri, 131, 137, 138, 140, 141, 149, 150, 154, 162, 165, 167
　deflandrei, 137, 139, 140
　diastypus, 147
　exilis, 137, 140, 141, 149, 162, 169
　hamatus, 137, 138, 139, 140, 141, 142, 143, 144, 145, 150, 152, 155, 169, 189, 190
　　zone, 141
　kuepperi, 137, 146, 147
　kugleri, 137, 138, 139, 141, 145, 147, 149, 162, 165, 167
　lodoensis, 137, 146, 147
　misconceptus, 137, 142, 150, 153, 154, 162
　mohleri, 137, 146, 147
　moorei, 137, 149, 162, 167
　multiradiatus, 147
　neohamatus, 137, 138, 140, 142, 152, 155, 162
　neorectus, 139
　obtusus, 131

Discoaster (continued)
　pentaradiatus, 137, 138, 139, 140, 142, 149, 150, 153, 154, 162, 169
　perclarus, 137, 167
　prepentaradiatus, 137, 139, 146, 149
　quinqueramus, 132, 139, 144, 145, 149, 150, 151, 152, 153, 154, 155, 158, 162, 163, 164, 169
　　zone, 132, 163
　saipanensis, 137, 146, 147
　sanmiguelensis, 141
　signus, 137, 149
　sublodoensis, 146, 147
　subsurculus, 169
　surculus, 137, 138, 139, 142, 146, 149, 150, 153, 154, 158, 162, 165, 169
　tamalis, 135, 137, 138, 139, 142, 144, 145, 146, 150, 153, 162, 165, *168*
　triradiatus, 137
　variabilis, 137, 150, 154, 169
Discocyclina, 339
　cristensis, 337
　stephensoni, 337
Discolithina millapunctata, 131, *166*
Discorbis spp., 226, 227
Discorinopsis, 297
　gunteri, 286, 292, 297, 300
　spp., 297
Discovery Bay, 118, 123, 126, 377, 418, 425, 431
disjuncta, Sphaeroidinellopsis, *214*, 262
dissimilis, Catapsydrax, 257
dissolution, 353
Distefanella, 44
　mooretownensis, 44, 45, 46
diversicostata, Frondicularia, 239
　Plectofrondicularia, 239
dixoni, Dollosuchus, 413, 414
Dodekapodorhabdus noeliae, 22, 27
dohertyi, Planulina, 224
Dollosuchus dixoni, 413, 414
Dominican Republic, 105, 107, 108, 220, 229, 233, 235, 359
dominicus, Cibicidoides, 227, *236*, 243
donatae, Fabularia, 299
　liburnica, Fabularia, 299
donnatensis, Nannoconus, 4, 12, 15
Dorcadospyris, 274
　alata, 257, 260, 267, 270, 275
　　zone, 262, 264, 265, 267, 269, 271, 273, 274, 275, 276, 277, 278, 279
　dentata, 259, 260, 270, 274, 275
　forcipata, 260, 270, 274, 275
　simplex, 270, 275
　sp., 270, 275
dorsalis, Ameiva, 428
douvillei, Coskinolina, 286, 287, 289, 290
downslope contamination, *225*
Drepanotrema anatina, 365, 366
　cimex, 365, 366

Drepanotrema (continued)
　lucida, 365, 366
Drivers River bank, 153
Drivers River section, 145, 147, 150, *153*, 156, 190
Druppatractus, 267, 268
　agostinelli, 260, 267, 268
　hippocampus, 267
druryi, Globigerina, 184
　(*Zeaglobigerina*), 212
　Globoturborotalita, *212*, 213
Dry Harbour, 425
Dry Harbour Mountains, 395
DSDP legs, 256
DSDP sites, 163, 164, 191, 192, 196, 198, 200, 226, 231, 256, 262, 276, 279
Duanvale fault area, 331
Duanvale-Wagwater escarpment, 132
dubia, Metalia, 405
dubius, Neococcolithes, 137, 147
Ducketts Land Settlement, 67, 71, 73
dumblei, Schizaster, 373, 396
Dump Limestone, 123, 285, 413
Dump type locality, 303
Dump Village, 413, 414
Duncans police station area, 127
Duncans Quarry, 127
durhami, Armatobalanus (Hexacreusia), 118
dutemplei, Truncatulina, 236
dutertrei, Globigerina, 206
　Globorotalia, 148, 191
　Neogloboquadrina, 191
duvergieri, Armatobalanus, 118
Dyscritothyris, 105, 109, *110*
　sp., 110
dyscritus, Caribbaster, 373, 374, 397, 398, 399
　Macropneustes, 402, 403
　Schizaster, 396

E

Easington Bridge area, 377, 379, 380, 384, 386, 393, 395, 396, 398, 403, 406, 407
Eastern Gulf Region, 115
eburneus, Balanus, 118, 120, 121, 122
Ecclesdown Road section, *155*, *158*, *194*, 198, 202
echinoid spines, 259
echinoides, Acrosphaera, 265
　sponosa, 260, 265
echinoids, 79, 93, 105, *371*, 403
　Cenozoic, *371*
　Cretaceous, *93*
Echinolampas, 371, *391*
　alta, 373, 391, 393, 394
　altissima, 373, 391, 393
　anguillae, 391, 393, 394
　brachytona, 373, 391, 393
　clevei, 373, 374, 391, 392, 393, 394
　lycopersicus, 373, 391, 393, 394
　nuevitasensis, 393
　paragoga, 373, 391, 394
　plateia, 373, 391, 393, 394

Echinolampas (continued)
　strongyla, 373, 391, 393, 394
　sp., 373, 391
　spp., 374
Echinometra, *381*
　lucunter, 373, 381, 382
　viridis, 373, 382
　sp., 373, 381, 382
　spp., 373
Echinoneus, *382*
　cyclostomus, 373, 382, 383
Echinopedina, 381
Echiopsis, *381*
　simplex, 373, 374, 381
Echiperiporites estelae, 446, 447, 450
Echitricolporites spinosus, 446, 447
Ecuador, 446
edentata, *Amerigus*, 365
edentatus, *Amerigus (Tropicorbis)*, 366
　Planorbis dentiferus, 366
Eggerella bradyi, 227, *237*
ehrenbergii, *Cribrosphaera*, 22, 24
　Cribrosphaerella, 8, 9, 12, 14, 27
　Lithomelissa, 260, 272, 273
Ehrenbergina, 237
　caribbea, 227, *237*, 246
　spinea, 227, *237*, 244
　spinosissima, 222, 227, *237*, 250
Eiffellithus anceps, 22, 27
　dennisonii, 26
　eximius, *3*, 8, 9, 10, 11, 12, 13, 15, 16, 22, 24, 25, 26, 27, 80
　　zone, 3, 5, 6, 7, 10, 13
　gorkae, 3, *4*, 8, 9, 11, 12, 13, 15, 16
　　zone, *4*, 5, 6, 7, 10, 13
　parallelus, 22, 24, 25, 27
　primus, 15
　turriseiffelii, 8, 9, 11, 12, 15, 22, 27
　sp., 8, 11, 12, *15*, 16, 24
Elasmodontomys, 437
elatus, *Periaster*, 401
Elderslie corehole, *10*
eldridgii, *Wythella*, 390
elegans, *Geomelania*, 354, 356
　Hoeglundina, 222, 227, *238*
　Rotalia (Turbinulina), 238
Eleutherodactylus sp., 419, 428
elevata, *Globotruncana*, 34
　Haimea, 373, 384, 385
elevatus, *Plagiobrissus*, 373, 405, 406
Ellipsagelosphaera britannica, 8, 9, *15*
Ellipsodiscoaster lidzii, 175
Ellipsonodosaria curvatura, 241
　modesta, 241
　nittalli, 241
　subspinosa, 241
ellipticus, *Anisopetalus*, 394
　Rhyncholampas, 373, 395
　Tarphypygus, 373, 374, 389
Elm Creek, Kinney County, Texas, 50
elongata, *Coskinolina*, 287
　(Coleiconus), 289
　Lychnocanoma, 257, 259, 260, 272, 277
　Tetrahedrina, 277

elongatus, *Nannoconus*, 8, 12, 15
　Periaster, 399
Elphidium spp., 227, 230, 242
emarginata, *Encope*, 391
embergeri, *Zeugrhabdotus*, 8, 9, 11, 12, 15, 23, 27
Encope, 390
　emarginata, 391
　homala, 390, 391
　michelini, 391
　sverdrupi, 373, 390, 391
　sp., 373, 390, 391
endemism, 353
Endocostea sp., 82, 85
(Endocostea) typicus, *Inoceramus*, 84, 85
England, 90, 413
Eocene, *285*, *443*
Eoceratoconcha, 118, 123
　renzi, 118, 123
　sp., 118, 123
Eoconuloides, 285, 309, *311*
　lopeztrigoi, 286, 309, 311, 314, 315
　wellsi, 286, 311, 312, 314
Eocytheropteron sp., 70, 71, 75
Eodictyoconus cubensis, 307
Eofabiana, 307
　grahami, 307, 310
Eorupertia, 309
　bermudezi, 287, 308, 309
Epiglyptoxoceras, 86
Epigoniceras, *86*
　sp., 86
epigonus, *Tetragonites*, 86
Epistominella umbonifera, 239
Eponides spp., 227
Eprolithus antiquus, 7, 9, 15
　floralis, 4, 9, 11, 12, 15
Eptesicus, *435*
　fuscus, 435
　lynni, 428, 435
　sp., 435
Ericsonia formosa, 137, 147
Erophylla, 423, *434*
　sezekorni, 428, 434
　sp., 434
eruption centers, 30
Esmeraldas Province, 446
estelae, *Echiperiporites*, 446, 447, 450
Eucidaris, 377
　strobilata, 377
　tribuloides, 373, 376, 377, 378
　sp., 374, 377
Eucoronis angulata, 274
Eucyrtidium, 277
　cienkowskii, 260, 270, 277
　delmontense, 277
　diaphanes, 277
　inflatum, 270
　pachyderma, 279
　sp., 270
Eulepidina, 313, 316, 319, 321, 323
　favosa, 325
　undosa, 323
(Eulepidina), *Lepidocyclina*, 316
　chaperi, *Lepidocyclina*, 317, 321

(Eulepidina) (continued)
　favosa, *Lepidocyclina*, 323, 325
　undosa, *Lepidocyclina*, 287, 319, 320, 321, 323, 325
Eulinderina, 313, *315*, 316
　antillea, 286, 315, 316
　guayabalensis, 315
Eupatagus, *401*, 405, 407
　alatus, 373, 374, 399, 400, 401, 402
　antillarum, 373, 401, 402
　attenuatus, 401, 402
　clevei, 402
　defectus, 373, 401, 402
　grandiflorus, 401, 402
　hildae, 373, 374, 401, 402
　longipetalus, 401
　sp., 373, 402
　spp., 374
euphratis, *Helicosphaera*, 137
Euplectella spp., 266
Eureka boreholes, 223, 229
Eurepa, 369
　veatleyi, 365, 369
Eurhodia, *394*, 395
　rugosa, 373, 374, 394
　sp., 373, 394
Europe, 359, 394
eurychorus, *Clypeaster*, 386, 388
Eurycratera, 358
Eurypanopeus, 122
　abbreviatus, 120, 121, 122
evolution, benthic foraminiferal, 221
evolutus, *Peneroplis*, 305
Ewarton area, 426
excentricum, *Asterostoma*, 373, 374, 406, 407
exigua obtusa, *Pulvinulinella*, 233
exiguum, *Corollithion*, 11, 14, 22, 26
exilis, *Discoaster*, 137, 140, 141, 149, 162, 169
　Geomelania, 354, 356
　Globorotalia, 184, 187, 191
　cultrata, 182
eximia, *Angulogerina*, 222, 227, *234*, 247
eximius, *Eiffellithus*, *3*, 8, 9, 10, 11, 12, 13, 15, 16, 22, 24, 25, 26, 27, 80
extremus, *Globigerinoides*, 191, 193, *210*, 211
　obliquus, 184, 187, 189, 210
eyrei, *Vaccinites*, 79, 80

F

Fabiania, 285, *307*
　cassis, 286, 307, 310, 317, 337
Fabularia, 285, 297, 299, 307
　colei, 285, 286, 289, 295, 297, 299, 302
　donatae, 299
　　liburnica, 299
　gunteri, 299
　hanzawai, 287, 297, 298, 301, 302
　matleyi, 299
　roselli, 301
　vaughani, 286, 297, 299, 301, 302

Fabularia (continued)
 verseyi, 285, 287, 297, 298, 301, 304
facies
 calcareous-siliceous, *255*, 259, 262
 oolitic, 102
Fallotella, 285, 291, 297, 307
 cookei, 285, 286, 291, 293, 294, 296
 floridana, 285, 286, 291, 293, 294, 296
 (*Fallotella*) *cookei*, 291
 (*Fallotella*) *cookei*, *Fallotella*, 291
Falmouth Formation, 105, 113, 118, 122, 123, 130, 377, 382
falsostuarti, *Gansserina*, 66
 Globotruncana, 34
familiaris, *Canis*, 421
farctus, *Nautilus*, 236
farinacciae, *Nannoconus*, 27
Farquhar's Beach, 123, 388, 391
fasciata, *Ampullaria*, 363
 Pomacea, 363, 365
Fasciculithus tympaniformis, 146, 147
fauna
 fossil, *359*
 land mammal, 423
 land snail, *353*, *359*
 modern, *356*
 ostracode, *68*
 rudist, 13
favealis, *Helisoma* (*Piersona*), 367
favosa, *Eulepidina*, 325
 Lepidocyclina, 323
 (*Eulepidina*), 323, 325
feldspars, 259
Fellius, *374*
 foveatus, 373, 374, 376
Fenestrites spinosus, 446, 447
Ferrissia, 369
 irrorata, 365, 369
 (*Ferrissia*) *irrorata*, 368
 (*Ferrissia*) *irrorata*, *Ferrissia*, 368
Fibularia, 373, *388*
 jacksoni, 373, 374, 388, 389
 sp., 389
fibuliformis, *Placozygus*, 4, 8, 9, 11, 15
fieldsi, *Charactosuchus*, 414
Fiji, 163, 164
fissure deposits, 420
Fissurina, 223, 233
 spp., 227
fistulosus, *Globigerinoides*, 184, 187, 210, 211
 Globorotalia, 191
Flamstead area, 73
flavescens, *Ariteus*, 428, 435
 Istiophorus, 435
floralis, *Eprolithus*, 4, 9, 11, 12, 15
Florida, 285, 372, 384, 388, 395, 402, 413
floridana, *Arenagula*, 286, 292, 297
 Bolivina, 227
 Chrysalidina (*Chrysalidina*), 297
 Cribrobulimina 297
 Fallotella, 285, 286, 291, 293, 294, 296

floridana (continued)
 Lituonella, 289
 Pseudemys, 427
 Pseudochrysalidina, 286, 297, 300
 Reticulofenestra, 137, 140, 146, 147, 168
floridanus, *Cibicides*, 236
 Coskinolina, 293
 Heterodictyoconus, 293
 Nummulites, 347
 Oligopygus, 384
floridensis, *Nummulites*, 333
 Operculinoides, 333
 Palaeonummulites, 287, 333, 334, 336
 Terebratula, 108
flos, *Micrantholithus*, 137, 147
Fohsella fohsi, 202
 peripheroronda, 202
 robusta, 204
(*Fohsella*) *birnageae*, *Globorotalia*, 202
 fohsi fohsi, *Globorotalia*, 202
 lobata, *Globorotalia*, 202
 lobata, *Globorotalia*, 202
 peripheroacuta, *Globorotalia*, 202
 peripheroronda, *Globorotalia*, 202
fohsi, *Fohsella*, 202
 Globorotalia, 184, 189, 199, *202*, 204
 fohsi, 148, 183, 184, 185, 239, 257, 262
 fohsi, *Globorotalia*, 148, 183, 184, 185, 239, 257, 262
 (*Fohsella*), 202
 lobata, *Globorotalia*, 144, 145, 148, 183, 184, 185, 197, 202, 257
 (*Fohsella*), 202
 peripheroacuta, *Globorotalia*, 202
 peripheroronda, *Globorotalia*, 202, 257
 robusta, *Globorotalia*, 143, 144, 145, 146, 148, 183, 185, 186, 188, 197, 204, 257
Folly Peninsula, 123
Folly Point, 111, 165
Folly Point section, *152*, 156, 190
Font Hill Formation, 307, 309, 331, 333, 335, 375, 377, 379, 380, 382, 385, 386, 393, 395, 396, 398, 403, 406, 407
fontabellensis, *Cushmania*, 293, 295
foraminifera, 77, 105
 bathyal benthic, *219*
 benthic, 131, 194, *223*, 226, 233, 259
 larger, *283*
 orbitoidal larger, *29*
 Paleogene, *283*
 planktonic, 13, 82, 131, 142, 143, 150, *156*, *179*, *180*, *182*, *184*, *190*, 192, 196, 197, 223, 225, 259
 reefal, 225
forceps, *Mithraculus*, 119, 120, 122
forcipata, *Dorcadospyris*, 260, 270, 274, 275

formosa, *Ericsonia*, 137, 147
fornicata, *Contusotruncana*, 82
 Globotruncana, 34
fossilis, *Scapholithus*, 23, 27, 139, 177
fossils
 freshwater, *363*
 Quaternary sites, *420*
 record, 115, *125*, *353*
 vertebrate, *418*, *420*
fovealis, *Helisoma*, 365, 367
foveatus, *Fellius*, 373, 374, 376
foveolata, *Anomalina*, 239
 Planulina, 222, 224, 227, 228, 229, 231, 232, 233, *239*, 250
fragilis, *Nannoconus*, 4, 12, 15
 Oolithotus, 152, 153, 162
France, 88, 89, 90, 394
Frankfield, Clarendon Parish, 73
Freemans Hall, 118, 123, 395, 399, 414
Frondicularia diversicostata, 239
furcatolithoides, *Sphenolithus*, 147
furcatus, *Marthasterites*, 4, 5, 8, 9, 10, 11, 12, 15, 24, 25
Fursenkoina spp., 227
fuscus, *Eptesicus*, 435
 Laevapex, 368

G

gabalus, *Tranolithus*, 23, 27
galavisi, *Dentoglobigerina*, 207, 208
 Globigerina, 207
galicianus, *Ammonites*, 89
gallus, *Calappa*, 120, 121, 122
gammation, *Toweius*, 137, 147
gangeticus, *Gavialis*, 413
gansseri, *Gansserina*, 67
Gansserina falsostuarti, 66
 gansseri, 67
 wiedenmayeri, 66, 67
Garcia Pond, 155
gardetae, *Hexalithus*, 8, 15
gardnerae, *Lepidocyclina* (*Polylepidina*), 316
Garlands Formation, 67
garrisonii, *Chiastozygus*, 22, 26
Gartnerago obliquum, 22, 27
 sp., 23, 27
gartneri, *Quadrum*, 4, 7, 8, 9, 10, 12, 15
garuda, *Ammonites* 86
 Tetragonites, 84, 86, 87
gastropods, 347
Gatun Formation, 446
Gatuncillo Formation, 445, 446, 448
Gaudryina bradyi, 238
 subglabra, 238
Gauthiericeras bajuvaricum, 80, 82, 84
Gavialis gangeticus, 413
Gayle area, 395
Geocapromys, 420, 423, 425, 426, 435, *436*
 brownii, 419, 423, 425, 426, 427, 428, 436, 437, 438
 ingrahami, 437
 thoracatus, 437

(Geocapromys) brownii, Capromys, 436
Geomelania elegans, 354, 356
  exilis, 354, 356
Georgia, 384, 395
Geotrygon versicolor, 428
Germany, 88, 89
Ghareb Formation, 66
gigantea, Callianassa, 119, 120
giganteum, Campanile, 116, 347
gigas, Barrettia, 10, 30, 32, 34, 45, 46, 81, 82
  Lepidocyclina, 323, 325
Ginger Valley, 405
Giraffospyris, 274
  angulata, 272, 274
  laterispina, 274
  sp., 272, 274
giraudi, Lepidocyclina, 317
  (Lepidocyclina), 317
gizehensis, Nummulites, 331
glacial cycle, 426
glacial epoch, 363
glacial interval, 422
glaciation, 155
Glasgow, 118, 122, 402
Global Boundary Stratotype section and point, 195
Globigerina, 208
  aequilateralis, 208
  altispira, 144, 145, 182, 187
  baroemoenensis, 208
  druryi, 184
  dutertrei, 206
  galavisi, 207
  grimsdaeli, 214
  hexagona, 208
  kochi, 214
  nepenthes, 143, 144, 145, 146, 184, 185, 186, 187, 188, 189, 193, 195, 204, 212
  nepenthes/Globoratialia mayeri zone, 186, 188
  pentacamerata, 311
  praesiphonifera, 187
  rubescens, 212
  rubra, 210
  seminulina, 214
  siphonifera, 208, 210
  venezuelana, 208
  woodi, 212
  (Zeaglobigerina) druryi, 212
    nepenthes, 212
Globigerinatella insueta, 257, 259
  zone, 262
Globigerinatheka subconglobata, 285
Globigerinella, 208
  aequilateralis, 187, 193, 195, 208, 210
  praesiphonifera, 211
Globigerinita, 207
Globigerinoides, 210
  conglobatus, 184, 187, 191, 193, 210, 211
  diminutus, 183, 184, 185, 210
  extremus, 191, 193, 210, 211
  fistulosus, 184, 187, 210, 211
  margaritae, 184

Globigerinoides (continued)
  obliquus, 184, 187, 189, 193, 210, 211
    extremus, 184, 187, 189, 210
  parkerae, 182
  primoridius, 257
    zone, 256, 262
  quadrilobatus trilobatus, 212
  ruber, 185, 186, 187, 188, 189, 197, 210, 211, 257
    zone, 185, 186
  rubra, 210
  sacculifer, 184, 212
  sicanus, 262
  subquadratus, 184, 185, 186, 188, 197, 210, 211
  triloba triloba, 212
  trilobus, 184, 187, 189, 193, 210, 211, 212
  sp., 222, 227
  spp., 228, 229, 231, 233
Globocassidulina caudriae, 222, 227, 237
  palmerae, 227, 237, 245
  punctata, 222
  subglobosa, 225, 226, 227, 228, 229, 230, 237, 238
Globoquadrina, 207, 208
  altispira, 207
    altispira, 207
    globosa, 207
  baroemoenensis, 207, 208
  dehiscens, 150, 152, 187, 189, 192, 207, 208, 211
  langhiana, 208
  venezuelana, 207, 208
  sp., 208
globoquadrinids, 184
Globorotalia Turborotalia Continuoso zone, 188, 189, 197
  acostaensis, 148, 188, 189, 197, 206, 257
    acostaensis–Globoratalia Turborotalia merotumida zone, 188
  aemiliana, 206
  ampliapertura zone, 321
  archeomenardii, 183, 198, 199
  birnageae, 183, 184, 185, 202
  cibaoensis, 150, 192, 193, 194, 195, 196
  conomiozea, 196
  continuosa, 188, 206
  crassacrotenensis, 206
  crassaformis, 150, 152, 183, 184, 187, 191, 193, 194, 204, 205
  crassula, 187, 191, 205, 206
    conomiozea, 206
  crotonensis, 206
  cultrata exilis, 182
  dehiscens, 208
  dutertrei, 148, 191
  exilis, 184, 187, 191
  exilis/miocenica zone, 191
  fistulosus, 191
  fohsi, 184, 189, 199, 202, 204
    fohsi, 148, 183, 184, 185, 239, 257, 262

Globorotalia fohsi (continued)
    fohsi zone, 239, 262
    lobata, 144, 145, 148, 183, 184, 185, 197, 202, 257
    peripheroacuta, 202
    peripheroronda, 202, 257, 262
    robusta, 143, 144, 145, 146, 148, 183, 185, 186, 188, 197, 204, 257
  hexagonus, 187
  hirsuta aemiliana, 206
  humerosa, 206
  kugleri, 257, 325
  lenguaensis, 148, 184, 185, 186, 187, 188, 189, 192, 193, 194, 197, 199, 204
  limbata, 184, 187, 190, 193, 194, 197, 198, 200, 201
  limbata-miocencia, 195
  limbata-pseudomiocenica, 154, 192, 194, 195
  lobata, 199, 202
  margaritae, 144, 145, 148, 150, 184, 185, 187, 191, 192, 193, 194, 195, 196, 201, 202, 204
    zone, 354
  mayeri, 148, 185, 188, 189, 197, 206, 207, 214, 257
    zone, 185, 186, 188, 197, 214
  menardii, 148, 184, 185, 187, 188, 189, 190, 192, 193, 194, 197, 198, 200, 257
  miocenica, 182
  multicamerata, 182
    zone, 188, 189, 197
  menardii/cultrata, 198
  menardii/Globorotalia lenguaensis zone, 189
  merotumida, 200
  merotumida-pleisotumida-tumida, 200
  miocenica, 144, 145, 148, 184, 185, 187, 191, 197, 200, 203
  multicamerata, 150, 152, 157, 191, 193, 194, 198, 203
  opima continuosa, 206
    opima, 206, 285
  panda, 202
  paralenguaensis, 200
  pentacameratea zone, 110
  peripheroacuta, 144, 145, 148, 185, 198, 199, 202
  peripheroronda, 144, 145, 148, 184, 185, 198, 199, 202, 261
  pertenuis, 200, 203
  plesiotumida, 184, 187, 189, 190, 192, 193, 194, 200, 202, 203, 204
    zone, 190
  praemenardii, 184, 199, 200
  praescitula, 198, 202
  pseudomiocenica, 184, 187, 191, 193, 197, 200, 203
  pseudomiocenica-limbata, 202
  puncticulata, 150, 192, 193, 194
  robusta, 199, 204
  robusta-lobata, 197, 202

Globorotalia (continued)
  scitula, 187, 201, 204
    scitula, 204
  siakensis, 207
  truncatulinoides, 191, 192
  tumida, 191, 194, 196, 198, 200, 202, 203
    zone, 202
  zealandica, 202
  (Fohsella) birnageae, 202
    fohsi fohsi, 202
    fohsi lobata, 202
    lobata, 202
    peripheroacuta, 202
    peripheroronda, 202
  (Globorotalia) crassaformia ronda, 182
    crassaformia viola, 182
    crassula, 206
    cultrata limbata, 198
    multicamerata, 200
    plesiotumida, 200
    truncatulinoides, 182
    tumida, 202
  (Hirsutella) margaritae, 204
    scitula, 204
  (Menardella) limbata, 198
    multicamerata, 200
    pertenuis, 200
    praemendardii, 200
  (Turborotalia) continuosa, 206
    crassaformis, 204
    crassula, 206
    peripheroacuta, 202
    peripheroronda, 202
    scitula, 204
    subscitula, 204
  sp., 198, 200, 201, 203
(Globorotalia) crassaformia ronda, Globorotalia, 182
  viola, Globorotalia, 182
  crassula, Globorotalia, 206
  cultrata limbata, Globorotalia, 198
  multicamerata, Globorotalia, 200
  plesiotumida, Globorotalia, 200
  truncatulinoides, Globorotalia, 182
  tumida, Globorotalia, 202
Globorotaloides, 207, *208*
  hexagonus, *208*, 211
  variabilis, 208
globosa, Dentoglobigerina, 185, 187, 189, *208*, 209
  altispira, 208
  Globoquadrina altispira, 207
  Liriospyris, 259, 260, 266, 275
Globotruncana arca, 34
  bulloides, 34
  calcarata, 66
  coronata, 34
  elevata, 34
  falsostuarti, 34
  fornicata, 34
  lapparenti, 34
  linneiana, 34
  rosetta, 34
  stuarti, 34
  ventricosa, 34

Globoturborotalita, *212*
  druryi, *212*, 213
  nepenthes, 189, 194, *212*, 213
globula, Sphaerogypsina, 286, 308, 309
globulosa, Anomalina, 234
globulosus, Anomalinoides, 222, 224, 225, 227, 231, *234*, 248
globulus, Ceriopora, 309
glomerosa, Praeorbulina, 148, 257, 262
Glyptoxoceras, 86
  retrorsum, 84, 86, 87, 88
  rugatum, 86, 88
gollevillensis, Parapachydiscus, 89
gomezi, Operculina, 335
Goniopygus, 95, *97*
  supremus, 93, 94, 95, 96, 97, 98
  (Tetragoniopygus), 98
  sp., *98*
Gorgospyris, 274
  medusa, 274
  schizopodia, 259, 260, 272, 274
gorkae, Eiffellithus, 3, *4*, 8, 9, 11, 12, 13, 15, 16
gothicum, Quadrum, 8, 9, 10, 15, 16
gothicus, Uniplanarius, 27
grahami, Eofabiana, 307, 310
grainstone, 79
Grand Caicos, 434, 435
Grand Cayman, 433, 434, 435
grande, Lychnocanoma, 277
grandiflorus, Eupatagus, 401, 402
grandis, Chiasmolithus, 137, 147
Grange Inlier, *30*, 32, 45
Grange section, *30*
granulata, Helicosphaera, 137
gravels, 354
Great Plains, 359
Great River valley, 98, 99
Greater Antilles, 388, 419, 420, 423, 433, 434, 439, 444
Green Grotto Cave, 359, 360, *425*, 436
Green Island Formation, 30, 45
Green Island Harbour, 396
Green Island Inlier, 29, *30*, 35, 45, 50, 54, 99
Green Island section, *30*
grillii, Lithastrinus, 5, 8, 9, 10, 11, 12, 13, 15, 16, 23, 24, 25, 26, 27, 80
grimsdaeli, Globigerina, 214
griseus, Caprimulgus, 432
  Nyctibius, 428, 432
Gryphus, 105, *108*, 109
  vitreus, 108
  sp., 108, 109
Grzybowskia, 335
Guatemala, 434, 446
Guava River, 30
guayabalensis, Eulinderina, 315
  Planorbulina (Planorbulinella), 315
  Uvigerina, 222
Guembelina, 57
Guiana Basin, 321

Guinea Corn Formation, 10, 35, 44, 57, 67, 71, 73, 93, 97, 98, 99, 100, 102, 105, 110, 116, 120
Guinea Corn type section, *10*, 14
Gulf Coast region, 347
Gulf of California, 141, 391
Gulf of Mexico, 166, 223, 271, 377
Gundlachia, 367, 368, 369
  ancyliformis, 367
  radiata, 365, 367, 368
gunteri, Dictyoconus, 293
  Discorinopsis, 286, 292, 297, 300
  Fabularia, 299
Gurabo Formation, 108
Guys Hill district, 379, 393, 395, 407
Guys Hill Member, 123, *443*, 446, *448*
Guys Hill Sandstone Member, 413
gymnozona, Cidaris, 375
Gypsina peruviana, 309
  sp., *238*
  spp., 226, 233
Gyraffospyris angulata, 260
  laterispina, 260
gyralis, Helicostegina, 285, 286, 295, 312, 313, 314, 315
Gyroidinoides spp., 225, 227, 229

H

haddingtonensis, Lepidocyclina (Nephrolepidina), 321
Haimea, 371, *384*
  alta, 373, 384, 385
  caillaudi, 384
  convexa, 373, 384, 385
  cylindrica, 373, 384, 385
  elevata, 373, 384, 385
  lata, 384, 385
  ovumserpentis, 373, 383, 384, 385, 386
  parvipetula, 373, 384, 385
  platypetula, 384, 385
  pyramoides, 373, 384, 385, 386
  rotunda, 373, 384, 386
  rugosa, 373, 384, 386
  stenopetula, 373, 384, 386
  sp., 373, 385
  spp., 373, 374, 385
haitanus, Tropidiphis, 428
Haiti, 118, 198, 220, 229, 295, 359, 433, 446
haitiensis, Balanocrinus, 127
haldemani, Planorbis, 366
halidonata, Anadara, 355
Haliomma didymocyrti violina, 269
hamatus, Discoaster, 137, 138, 139, 140, 141, 142, 143, 144, 145, 150, 152, 155, 169, 189, 190
Hampea, 446, 450
Hamulus, 68
hancocki, Sphaeroidinellopsis, 214
hannai, Lamprocyrtis, 278
hannovrensis, Inoceramus waltersdorfensis, 80, 82, 84, 85
Hanover, 80, 113
Hanover Formation, 5, 80, 85
Hanover Group, 78, 80

Hanover Parish, 79, 80, 118, 120, 122, 123, 129, 377, 426, 435, 436
hanzawai, Fabularia, 287, 297, 298, 301, 302
Hanzawaia ammophilus, 222, 227, 230, *238*
   caribaea, 238
   cushmani, 238
   mantaensis, 227, *238*, 248
   spp., 227, *238*, 248
Haqius circumradiatus, 9, 15
Harbour View, 118, 123
harrisi, Hercoglossa, 351
Harvey River Formation, 5
Hastegerina siphonifera, 189, 208, 210
Hatton-Rockall Bank, 263, 265
Haughton Hall area, 32
havanensis, Asterorbis, 35, 37, 45, 46, *59*, 62
   Cibicidoides, 222, 224, 227, *236*
Hayaster perplexus, 137, 153, 174
hayi, Bukryaster, 4, 6, 8, 10, 14, 16, 22, 24, 25, 26
   Rucinolithus, 5, 8, 11, 15
Hazelymph area, 426
Hazelymph Cave, 426
Heathshire Hills, 429
hedbergi, Miscellanea, 325
heezenii, Helicosphaera, 147
heldemsis, Pachydiscus, 89
helicoideus, Microrhabdulus, 23, 27
Helicolithus trabeculatus, 8, 11, 12, 15
Helicosphaera ampliaperta, 137, 138, 139, 140, 144, 145, 146
   burkei, 131, 139, *165*, 177
   carteri, 137
   compacta, 137, 146, 147
   euphratis, 137
   granulata, 137
   heezenii, 147
   intermedia, 137
   lophota, 147
   paleocarteri, 165
   recta, 137, 146, 147
   sellii, 152, 153, 162
   seminulum, 137, 147
   stalis, 139
   vedderi, 177
Helicostegina, 285, 313
   gyralis, 285, 286, 295, 312, 313, 314, 315
   ocalana, 285
Heliolithus kleinpellii, 137, 146, 147
Helisoma fovealis, 365, 367
   (Piersona) favealis, 367
Hellshire Hills, 429, 431, 437
helvetica, Praeglobotruncana, 82
Hemiaster, 101
   sp., 93, 94, 95, 101
Hemidactylus praesignis, 429
Hemitrochus bowdenensis, 354, 355, 356
Hepatiscus, 122
   bartholomaeensis, 118, 119, 122
heptaceros, Ceratospyris, 274
Heptaxodon, 437

Herbertia, 381
herbstii, Panopeus, 120, 122
Hercoglossa, 351
   harrisi, 351
   waringi, 351
   sp., *349*, 350, 351
   spp., 351
Hercothyris, 105, 112, *113*
   borroi, 113
   semiradiata, 112, 113
   sp., 112
Hermitage Dam Road, 351
Herpestes, 423
   auropunctatus, 439
Heterodictyoconus, 293
   americanus, 293
   cookei, 291
   floridanus, 293
heteromorphus, Sphenolithus, 136, 137, 138, 139, 140, 144, 145, 146, 147, 152, 177, 184, 197
heteroporos, Lamprocyclas, 278
Heterorhabdus sinosus, 26
Heterosalenia, 97
   occidentalis, 93, 94, 95, 97
Heterostegina, 285, 329, *335*
   antillarum, 335
   antillea, 287, 321, 335, 337
   israelskyi, 337
   ocalana, 287, 321, 335, 337
   panamensis, 337
   suborbicularis, 335
   texana, 337
   (Vlerkina) antillea, 305, 337, 338
      ocalana, 338
   (Vlerkinella) kugleri, 335, 337
(Hexacreusia) durhami, Armatobalanus, 118
hexagona, Globigerina, 208
hexagonalis, Schizaster, 373, 396, 398, 401
Hexagonocyclina, 339
   cristensis, 341
   inflata, 286, 338, 339, 340
hexagonus, Globorotalia, 187
   Globorotaloides, 208, 211
Hexalithus gardetae, 8, 15
hexapoda, Liriospyris, 275
Hibiscus, 446, 450
   tiliaceus, 446, 450
hildae, Eupatagus, 373, 374, 401, 402
hillae, Reticulofenestra, 137, 146, 147
hippocampus, Druppatractus, 267
Hippurites, 39, 40, 41
hirsuta aemilliana, Globorotalia, 206
(Hirsutella) margaritae, Globorotalia, 204
   scitula, Globorotalia, 204
Hispaniola, 359, 419, 421, 431, 432, 433, 434, 436, 437
hispida, Stylosphaera, 267
   Uvigerina, 222, 227, *241*
hispido-costata, Uvigerina, 222, 227, *241*, 250
Historbitoides, 50, 52
histrica, Syracosphaera, 137
Hodges Bluff, 325
Hoeglundina elegans, 222, 227, 238

Holland, 115, 394
Holland Bay, 431
holmesii, Rhynchonella, 106
Holodiscolithus solidus, 137, 177
homala, Encope, 390, 391
Homeopetalus, 407
   axiologus, 404, 407
Honduras, 422
Hope Bay area, 127, 377, 389
Hope River Gorge, 118, 123
Hopegate Formation, 377
horticus, Sollasites, 23, 27
hottingeri, Yaberinella, 286, 301, 303, 306
howardii, Ceratopteris (Magnastriatites), 450
   Magnastriatites, 446, 447
hulburtiana, Umbilicosphaera, 166
humerosa, Globorotalia, 206
   Neogloboquadrina, 187, 191, 193, 205, *206*, 257
humilis, Planorbis, 365, 367
Hyalosagda brevior, 356
   brevis, 356
   sincera, 356
   (Stauroglypta) brevior, 354
      brevis, 354
      sincera, 354
hyalosponges, 263, 265
Hydrobia coronatus, 365
   rivularis, 365
   spinifera, 365
   sp., 365, 366
Hyla sp., 428
Hypselaster perplexus, 399
hypselus, Oligopygus, 384

I

ignotus, Bidiscus, 22, 24, 26
illingi, Angulogerina, 222, 227, *234*, 247
immatura, Sphaeroidina, 180, 182
   Sphaeroidinella dehiscens, 192, 193, 194, 195, 202, 212
   Sphaeroidinellopsis, 180
impendens, Bulimina, 227, *235*
improvisus, Balanus, 123
   assimilis, Balanus, 118, 121
Incerticyclus, 358
   bakeri, 354, 355, 357, 358, 359
   schermoi, 354, 355, 356, 357
inconspicua, Valvata, 364
inconstans, Inoceramus, 82
incrassatus, Cibicidoides, 227, *237*
   Nautilus, 237
incurvatus, Baculites, 88
India, 69, 86, 88
Indian Ocean, 275, 277
Indo-Pacific-Mediterranean region, 285
Indo-Pacific region, 191, 285, 319, 321, 382
indra, Pseudophyllites, 86
inflata, Agassizia, 373, 374, 392, 398
   Bontourina, 339
   Coskinolina, 289

inflata (continued)
   Hexagonocyclina, 286, 338, 339, 340
   Rhabdosphaera, 146, 147
  mexicana, Bulimina, 235
inflatum, Eucyrtidium, 270
infracretacea, Assipetra, 10, 11, 14
ingens, Pleurodonte, 358
ingrahami, Geocapromys, 437
Innes Bay section, 111, *153,* 156
inoceramids, 77, 84, 86
Inoceramus, 83, 95
  balticus, 80, 82, 85
    balticus, 84
    kunimiensis, 79, 80, 84
    toyajoanus, 84
    spp., 85
  barabini, 84, 85
  crippsi, 82
  deformis, 82
  inconstans, 82
  muelleri, 79, 80, 85
  proximus subcircularis, 84, 85
  waltersdorfensis hannovrensis, 80, 82, 84, 85
  (Endocostea) typicus, 84, 85
Inoceramus Shales, 80, *82,* 84
insueta, Globigerinatella, 257, 259
interglacial, 363, 426
intermedia, Helicosphaera, 137
  Scyphosphaera, 137, 173
intra-Buff Bay Formation unconformity, *189*
ionica, Sphaeroidinella ionica, 212
ionica, Sphaeroidinella, 212
Ireland, 90
irrorata, Ferrissia (Ferrissia), 368
irroratus, Ancylus, 367
isidroensis, Bolivina, 238
  Loxostomum, 227, *238,* 246
Isla de Pinos, 433
island arc, volcanic, 30
Isocrinid columnals, 125
Isocrinid sp., *127*
Isocrystallithus compactus, 12, 15
  sp., 8
Isolepidina, 313
isolepidinoides, Lepidocyclina, 301, 317, 320
israelskyi, Heterostegina, 337
Isthmolithus recurvus, 146, 147
Isthmus of Panama, 262, 264
Istiophorus flavescens, 435
Italy, 135

J

Jacksonaster remediensis, 390
Jamaica
  central, 50, 85, 289, 291, 297, 299, 303, 325, 448
  eastern, *131,* 142, *155, 179,* 291, 331
  north-central, 431
  north coast, 190, 355
  northeastern, 126, *255,* 295, 305, 331, 339
  northern, 329, 331

Jamaica (continued)
  northwestern, 311
  southeastern, 355
  waters, 107
  western, *19,* 291, 347, 349, 425, 431
Jamaican Bank, 263
Jamaican Passage, 265
jamaicensis, Borelis, 297
  Callinectes, 118, 119, 122
  Coskinolinoides, 295, 307
  Cyclorbiculinoides, 285, 287, 304, 305, 309
  Lambertona, 399, 400
  Leiocephalus, 419, 425, 426, 428, 431
  Metalia, 397, 405
  Natalus major, 435
    stramineus, 435
  Oligopygus, 373, 374, 384
  Pachydiscus, 79, 80, 85, 89
    (Pachydiscus), 83, 85, 90
  Paludina, 365
  Pericosmus (Victoriaster), 399
  Pisidium, 369
  Spirodontomys, 437, 439
  Turdus, 428
  Uvigerina, 222, 227, *241,* 250
  Verseyella, 285, 286, 289, 292, 295, 298, 300
  Victoriaster, 399
  Yaberinella, 285, 287, 289, 303, 304
  truncata, Borelis, 297
japonica, Pontosphaera, 137, 166
jarvisi, Plectofrondicularia, 239
Java, 188, 206
Jenkinsella, 206
jennyi, Operculinoides, 333
Jericho Formation, 5
Jerusalem Limestone, 68
Jerusalem Limestone Member, *73*
Jerusalem Mountain area, *5,* 349
Jerusalem Mountain Formation, 5, 71, 127
Jerusalem Mountain Inlier, 19, 56, 66, 67, *68,* 71, 73, 102, 127, 129
John Crow Mountains, 437
Johns Hall area, 403, 406
Johns Hall Formation, 21
jungi, Ceratoconcha, 118, 123

K

Kansas, 359
Karreriella bradyi, 238
  subglabra, 227, *238*
karst, 353, *418*
Kathina, 35, 37, 38, 40, 41, 68, 69
kennedyi, Corollithion, 4, 12, 14
Kenya, 359
Kingston, Jamaica, 118, 123, 165, 283, 347, 439
kinlossensis, Lepidocyclina (Polylepidina), 316
kleinpellii, Heliolithus, 137, 146, 147
Knockalva, Jamaica, 425
Knockalva Cave, *425,* 436

kochi, Globigerina, 214
  Sphaeroidinellopsis, 214
koeneni, Pachydiscus, 89
  (Pachydiscus), 83, 84, 85, 86, 87, 88, *89*
kuepperi, Discoaster, 137, 146, 147
kugleri, Charactosuchus, 413, 414
  Cimomia, 351
  Discoaster, 137, 138, 139, 141, 145, 147, 149, 162, 165, 167
  Globorotalia, 257, 325
  Heterostegina (Vlerkinella), 335, 337
kullenbergi, Cibicides, 237
kunimiensis, Inoceramus balticus, 79, 80, 84

L

La Boca Formation, 445, 446
La Gonave, 433
Lacazella, 105, *106,* 107
  caribbeanensis, 106, 107, 108
  mediterranea, 106
  sp., 106
lacunosa, Pseudoemiliania, 137, 138, 142, 145, 150, 152, 153, 162, 166
Laevapex fuscus, 368
Laevicardium serratum, 355
laevigata, Cassidulina, 227, *236*
laffittei, Rotelapillus, 8, 15, 23, 27
Lagena, 223, 227, 233
lamberti, Lambertona, 399
Lambertona, 399
  jamaicensis, 399, 400
  lamberti, 399
Lambs River road, 71
Lamellaxis gracilis, 354, 356
Lamprocyclas heteroporos, 278
  hannai, 278
  margatensis, 260, 270, 278, 279
Lamprocyrtis, 278
lanceolatus, Clypeaster, 373, 374, 386, 388
Landslide section, 208, 214
langhiana, Globoquadrina, 208
Lapideacassis tricornus, 23, 27
lapparenti, Globotruncana, 34
Las Villas Province, Cuba, 52
lata, Haimea, 384, 385
(Laterorbis) pallidus, Amerigus, 366, 367
laterispina, Giraffospyris, 274
  Gyraffospyris, 260
Laticarinina pauperata, 227, *238*
laticonus, Didymocyrtis, 262
Latin America, northern, *443, 444,* 450
latus, Plagiobrissus, 373, 405, 406
laviculata, Uvigerina, 222, 227, *241*
leaching, 2, 353
lecta, Terebratula, 108
Leiocephalus, 431
  jamaicensis, 419, 425, 426, 428, 431
Leiocidaris sp., 93
Leiosoma, 380, 381

lemarchandi, Nowakites, 80, 82, 83, 84, 85, 88, 89
  Puzosia, 88
Lengua Formation, 185
lenguaensis, Globorotalia, 148, 184, 185, 186, 187, 188, 189, 192, 193, 194, 197, 199, *204*
Lenticulina, 116
  sp., 116
  spp., 226, 227, 229
lenuis, Rhabdolithus, 137
leoni, Phyllacanthus, 93, 94, 95, 96, 98
  Prionocidaris, 373, 374, 375, 376
Lepidocyclina, 285, 311, 313, *316*, 319, 323
  antillea, 315
  ariana, 317
  canellei, 317, 319, 320
  chaperi, 317
  favosa, 323
  gigas, 323, 325
  giraudi, 317
  isolepidinoides, 301, 317, 320
  macdonaldi, 301, 317
  matleyi, 317
  morgani, 319
  peruviana, 317, 324
  praetournoueri, 321
  proteiformis, 317
  pustulosa, 317, 321, 324, 337
  undosa, 323
  yurnagunensis, 305, 317, 319, 321
  (Eulepidina), 316
    chaperi, 317, 321
    favosa, 323, 325
    undosa, 287, 319, 320, 321, 323, 325
  (Lepidocyclina), 315, 323
    asterodisca, 317
    canellei, 287, 316, 319, 321
    giraudi, 317
    macdonaldi, 287, 317, 321, 322
    parvula, 317
    proteiformis, 317
    pustulosa, 287, 317, 319, 322, 324
    sherwoodensis, 317
    trinitatis, 317
    waylandvaughani, 317
    yurnagunensis, 319
  (Nephrolepidina), 316, 320
    chaperi, 287, 316, 320, 321, 326
    crassata, 323
    haddingtonensis, 321
    undosa, 323
    vaughani, 320, 321
    yurnagunensis, 287, 319, 320, 321, 324
  (Pliolepidina) proteiformis, 317
  (Polylepidina) antillea, 315, 316
    chiapensis, 316
    gardnerae, 316
    kinlossensis, 316
    proteiformis, 317
    sp., 324

(Lepidocyclina), Lepidocyclina, 315, 323
  asterodisca, Lepidocyclina, 317
  canellei, Lepidocyclina, 287, 316, 319, 321
  giraudi, Lepidocyclina, 317
  macdonaldi, Lepidocyclina, 287, 317, 321, 322
  parvula, Lepidocyclina, 317
  proteiformis, Lepidocyclina, 317
  pustulosa, Lepidocyclina, 287, 317, 319, 322, 324
  sherwoodensis, Lepidocyclina, 317
  trinitatis, Lepidocyclina, 317
  waylandvaughani, Lepidocyclina, 317
  yurnagunensis, Lepidocyclina, 319
leptoporus, Calcidiscus, 137
  Cyclococcolithus, 131
Lesser Antilles, 358, 431, 433, 437, 439
levis, Reinhardtites, 8, 15, 25
Leyden region, 99
liburnica, Fabularia donatae, 299
lichens, 141, 221
lidzii, Ellipsodiscoaster, 175
Liliasterites angularis, 4, 5, 8, 15
limbata, Globorotalia, 184, 187, 190, 193, 194, 197, *198*, 200, 201
  (Menardella), 198
  Rotalia, 198
limbata-miocencia, Globorotalia, 195
limbata-pseudomiocenica, Globorotalia, 154, 192, 194, 195
limbatus, Uhlias, 119, 120, 122
Limburgina sp., 70, 75
limestone, 5, 13, 30, 32, 34, 35, 41, 44, 52, 79, 105, 125, 132, 148, 165, 180, 220, 257, 258, 259, 283, 305, 331, 347, 353, 413, 418, 448
  rudist, 21
  Titanosarcolites-bearing, *65*
lineata, Lucidella, 359
linneiana, Globotruncana, 34
  Marginotruncana, 82
Linthia, 399
  obesa, 399
  trechmanni, 373, 374, 397, 399, 406
Liriospyris, 275
  globosa, 259, 260, 266, 275
  hexapoda, 275
  parkerae, 274
  stauropora, 259, 260, 272, 275
  sp., 276
Lithastrinus grillii, 5, 8, 9, 10, 11, 12, 13, 15, 16, 23, 24, 25, 26, 27, 80
  moratus, 4, 8, 9, 11, 12, 13, 15, 16
  zone, 4, 5, 7, 10, 13
  septanarius, 4, 8, 9, 11, 12, 15, 16
  zone, 4, 5, 7, 10, 13
  (Lithatractus) santaennae, 267
(Lithatractus) santaennae, Lithatractus, 267

Lithocampe acuminata, 277
  aurita, 279
Lithomelissa, *273*
  ehrenbergii, 260, 272, 273
  tartari, 273
  sp., 273
Lithomitra, 268, *279*
  spp., 260, 279
lithostratigraphy, *132*, *155*, *256*
Lithostromation perdurum, 137
Lithotympanium tuberosum, 276
Lithraphidites acutus, 13
  carniolensis, 23, 24, 27
  praequadratus, 8, 15, 16, 23, 24, 26, 27
  sp., 9
litterarius, Chiastozygus, 8, 9, 11, 12, 14, 22, 26
Little Exuma, 433
Little Swan Island, 437
Lituonella floridana, 289
  sp., 289
Lluidas Sinkhole, 425
Lluidas Vale, 425
Lluidas Vale Cave, 418, 419, *425*, 427, 429, 436, 437
lobata, Globorotalia, 199, *202*
  fohsi, 144, 145, 148, 183, 184, 185, 197, 202, 257
  (Fohsella), 202
  Serpula, 236
  Truncatulina, 236
lobatulus, Cibicides, 225, 226, 229, 230, 231, *236*, 242
  Cibicidoides, 227
  Nautilus, 236
lobatus, Nautilus spiralis, 236
localities
  holotype, 163, 165, 190, 191, 194, 197
  paratype, 163, 164, 165, 190, 194, 197
lodoensis, Discoaster, 137, 146, 147
Logie Green, 120
Logie Green section, 10, *35*, 39, 45
Logie Green traverse, 54
logotheti, Sismonda, 389
Long Mile Cave 418, 419, *425*, 429, 431, 432, 435, 436, 439
longipetalus, Antillaster, 373, 374, 407
  Eupatagus, 401
lopeztrigoi, Amphistegina, 311
  Eoconuloides, 286, 309, 311, 314, 315
Lopha arizpensis, 73
Lophodolithus nascens, 137, 147
lophota, Helicosphaera, 147
loridanus, Archaias, 305
Lorrimers road, 129
loveni, Caribbaster, 373, 374, 398, 399
  Plagiobrissus, 373, 404, 405, 406
  Prenaster, 405
  Prionocidaris, 373, 374, 375, 376
Lower Roadcut outcrop, 134, 180, 221
Loxigilla violacea, 428
Loxolithus armilla, 23, 27

Loxostomum isidroensis, 227, *238*, 246
  spp., 227
Lucea area, 120, 122
Lucea Inlier, *5*, 19, 77, *78*, 89, 102
lucema, Pleurodonte, 354, 355, 356, 358
Lucianorhabdus arcuatus, 23, 27
  cayeuxii, 9, 15, 23, 27
  maleformis, 12, 15, 23, 27
  quadrifidus, 8, 15, 23, 27
lucida, Drepanotrema, 365, 366
Lucidella costata, 354, 355, 359
  lineata, 359
  persculpta, 359
  (*Poenia*), 359
Lucina sp., 347
Lucky Hill area, 385, 395, 403, 407
lucunter, Echinometra, 373, 381, 382
Lychnocanoma, 277
  bipes zone, 277
  elongata, 257, 259, 260, 272, 277
    zone, 256, 262, 269, 274, 276, 277
  grande, 277
    zone, 277
lycopersicus, Echinolampas, 373, 391, 393, 394
(*Lynchnocanoma*) *clavigerum, Lynchocanium*, 277
Lynchocanium bipes, 277
  (*Lynchnocanoma*) *clavigerum*, 277
lynni, Eptesicus, 428, 435
Lytechinus sp., 373

M

Maastricht, the Netherlands, 57
Maastrichtian, Early, 2, 3, *45*
macdonaldi, Lepidocyclina, 301, 317
  (*Lepidocyclina*), 287, 317, 321, 322
macgillavryi, Camerina, 325
  Nummulites, 286, 331, 334
macilenta, Bulimina, 227
macintyrei, Calcidiscus, 137, 153, 173, 175, 177
  Cyclococcolithus, 153
maclearanus, Diplocrinus, 127
macleodae, Zygodiscus, 23, 27
macnabianus, Planorbis, 366
Macropneustes, 373, *402*
  altus, 373, 374, 402, 403
  angustus, 373, 374, 402, 403
  dyscritus, 402, 403
  parvus, 373, 374, 402, 403
  sinosus, 397, 402, 403
  stenopetalus, 402, 403
Macrotus, *433*
  waterhousii, 428, 433
maculatus, Phymodius, 120, 121, 123
Madagascar, 86, 88, 90
Magnastriatites howardii, 446, 447
  (*Magnastriatites*) *howardii, Ceratopteris*, 450
magnetostratigraphy, *223*
magnicostata, Argyrotheca, 107, 110, 111

magnicrassus, Toweius, 137, 147
magnifica, Astropyga, 373
magnificum, Neoglyptoxoceras, 88
Main Roadcut outcrop, 134, 141, 146, 180, 182, 184, 185, 186, 189, 190, 204, 210, 212, 221, 226, 231, 232
major, Natalus, 428, 435
  jamaicensis, Natalus, 435
Majorca, 279
malagaense, Rhopalodictyum, 260, 272, 273
  (*Rhopalodictya*), 273
Maldon Formation, 67, 71, 73
Maldon Inlier, 19, *67*, 71
Maldon School, 73
maleformis, Lucianorhabdus, 12, 15, 23, 27
malkinae, Pullenia, 222, 227, *240*
mammifera, Didymocyrtis, 260, 269, 270, 271
mammiferum, Cannartidium, 269
mammiferus, Cannartus, 269
mammillaris, Dictyospyris, 276
  Tholospyris, 260, 272, 276
manatus, Trichechus manatus, 414
  manatus, Trichechus, 414
Manchester, 118, 123
Manchester Parish, 118, 129, 389, 391, 413, 414, 425
Manchioneal, 108, 113, 123
Manchioneal district, 132, 153
Manchioneal Formation, 105, 108, 111, 113, 118, 123, 126, 132, 149, 152, *155*, 158, 197, 377, 382, 391, 409
Mandeville area, 122
Manivitella pemmatoidea, 8, 9, 11, 12, 15, 23, 27
mantaensis, Anomalina, 238
  Cibicides, 238
  Hanzawaia, 227, *238*, 248
mantelli, Nummulites, 313
Marchmont Inlier, 10, 19, *67*, 71, 96, 97, 98, 99, 102, 116, 129
margaritae, Globigerinoides, 184
  Globorotalia, 144, 145, 148, 150, 184, 185, 187, 191, 192, 193, 194, 195, 196, 201, 202, *204*
  (*Hirsutella*), 204
margatensis, Calocyclas (*Calocycletta*), 278
  Lamprocyrtis, 260, 270, 278, 279
margerelii, Cyclagelosphaera, 9, 14
marginata, Bulimina, 222, 224, 227, 32, *235*
Marginotruncana linneiana, 82
Marhasterites furcatus zone, 4
marl, 21, 34, 35, 142, 148, 152, 153, 179, 189, 192, 220, 257, 347, 363
marmoratus, Physella, 366
Maroon Town area, 73
Marta Tick Cave, 419, *425*, 429, 436, 439
Marthasterites furcatus, 4, 5, 8, 9, 10, 11, 12, 15, 24, 25
Martinique, 431

Maryland Road Junction area, 79
Maryland Waterfall, 79, 89, 90
Masemure Formation, 68
matalosa, Broinsonia, 22, 26
  Vagalapilla, 9, 11, 15
Matanzas Province, 108
matanzasensis, Cibicidoides, 227, *237*, 249
  Planulina, 237
matleyi, Borelis, 297
  Camerina, 325
  Fabularia, 299
  Lepidocyclina, 317
  Miscellanea, 325
  Pellatispirella, 286, 303, 325, 328
  Pseudofabularia, 287, 297, 298, 302
  Rhyncholampas, 373, 374, 392, 395
  Rhynchopygus, 394
Mauritia, 450
Mauritiidites, 449
  sp., 450
mayeri, Globorotalia, 148, 185, 188, 189, 197, 206, 207, 214, 257
  Neogloboquadrina, 144, 145
  Paragloborotalia, 184, 185, 186, 187, 188, 189, 194, 197, 205, 207
mayeri-mayeri, Paragloborotalia, 184
mayeri/siakensis, Paragloborotalia, 207
  Polyperibola, 188, 189
mcgregori, Xenothrix, 418, 419, 425, 428, 430, 432, 435, 436
mecatepecensis, Neorotalia, 327
media, Orbitoides, 57
  megaloformis, Orbitoides, 29, 35, 37, 45, 46, *57*, 58, 60
mediaquadratus, Chiastozygus, 26
mediterranea, Lacazella, 106
  Thecidae, 106
Mediterranean area, 309
Mediterranean Basin, 195
Mediterranean Sea, 106, 273
mediterraneum, Thecidium, 106
medusa, Gorgospyris, 274
megafossils, *444*, *451*
megalactis, Trisolenia, 267
  megalactis, 260, 267
  costlowi, Trisolenia, 260, 266, 267
  megalactis, Trisolenia, 260, 267
megaloformis, Orbitoides media, 29, 35, 37, 45, 46, *57*, 58, 60
megalophylla, Aello, 433
  Mormoops, 419, 423, 428, 433
Melania spinifera, 365
melitensis, Cidaris, 375
Mellita, 390
  quinquesperforata, 373, 390
Melonis barleeanum, 227, 229, *238*
  pompilioides, 222, 224, 227, 229, 233, *238*, 239, 244
  sphaeroides, 222, 224, 227, 231, 232, *238*, 244
(*Menardella*) *limbata, Globorotalia*, 198
  multicamerata, Globorotalia, 200
  pertenuis, Globorotalia, 200

(Menardella) (continued)
　praemenardii, Globorotalia, 200
menardii, Globorotalia, 148, 184, 185, 187, 188, 189, 190, 192, 193, 194, 197, 198, 200, 257
　Pulvinulina, 202, 238
　miocenica, Globorotalia, 182
　multicamerata, Globorotalia, 182
menardii/cultrata, Globorotalia, 198
Menetus dilatatus avus, 367
　(Micromenetus), 367
Meoma, 403
　antiqua, 373, 374, 403, 404
merotumida, Globorotalia, 200
merotumida-pleisotumida-tumida, Globorotalia, 200
Messinian/Zanclean boundary, 195
Metalia, 405
　dubia, 405
　jamaicensis, 397, 405
Metholectypus, 99
　trechmanni, 93, 94, 95, 99, 100
mexicana, Bulimina, 227, *235,* 246
　inflata, 235
　Conusphaera, 4, 5, 6, 8, 9, 14
　Neorotalia, 327, 330
mexicana/mecatepecensis, Neorotalia, 287
Mexico, 396, 432, 433, 444, 446
michelini, Encope, 391
Micrantholithus flos, 137, 147
micrites, 259
microfacies, *259*
microfossils, *445, 454*
　calcareous, 131
(Micromenetus), Menetus, 367
Micropanope, 120, 121, 122
　polita, 120, 122
microporus, Circodiscus, 260, 266, 271
　Trematodiscus, 271
Microrhabdulus belgicus, 8, 15, 23, 27
　decoratus, 8, 9, 11, 12, 15, 16, 23, 27
　helicoideus, 23, 27
　stradneri, 23, 27
Micula concava, 8, 9, 11, 12, 15
　decussata, 4, 8, 9, 11, 12, 15, 23, 24, 27, 80
　murus, 15
　praemurus, 8, 15
Middle East, 285
Middle Roadcut outcrop, 134, 140, 141, 144, 180, 184, 204, 212, 221
miliolids, 349
Milk River Bath, 388
millapunctata, Discolithina, 131, *166*
milowi, Cyclampterium, 267
Mincopanope spinipes, 120, 121, 122
minimus, Zygodiscus, 27
Minylitha convallis, 137, 138, 140, 146, 162
Miocene, *195, 219*
　upper, *156*

Miocene/Pliocene boundary, *195*
miocenica, Globorotalia, 144, 145, 148, 184, 185, 187, 191, 197, *200,* 203
　menardii, 182
Miogypsina panamensis, 325
　(Miogypsinella) bermudezi, 327
　sp., 305
(Miogypsinella) bermudezi, Miogypsina, 327
Miogypsinoides, 285, *327*
　bantamensis, 327
　bermudezi, 327, 329, 330
　butterlinus, 327
　complanata, 327
　spp., 287, 327, 330
miopelagicus, Coccolithus, 137, 146, 147, 152, 162, 173
Miosorites, 307
　americanus, 305, 307
Miscellanea antillea, 325
　hedbergi, 325
　matleyi, 325
　nicaraguana, 325
　tobleri, 329
misconceptus, Discoaster, 137, 142, 150, 153, 154, 162
Misorites, 339
Mississippi, 116, 395
Mississippi Embayment, 444
mississippiensis, Alabamina, 233
Mithraculus, 119, 120, 122
　forceps, 119, 120, 122
Mithrax, 119, 120, 122
　caribbaeus, 119, 120, 122
　spinosissimus, 120, 122
　spp., 120
Mitrocaprina, 32, 45, 46
Mocho Toms River, 30, 44
models
　Buff Bay section, 144
　sedimentation rate curve, 144
modesta, Ellipsonodosaria, 241
　Orthomorphina, 241
　Siphonodosaria, 240
　Stilostomella, 227, *241*
mohleri, Discoaster, 137, 146, 147
mollusks, 68, 105
　freshwater, *363*
　terrestrial, *353*
Molton area, 425
Molton Fissure, 420, *425,* 437
Mona Island, 432
Moneague Group, 375, 384, 385, 389, 390, 393, 394, 395, 396, 402
monilifera, Barrettia, 41, 45, 46
Monophyllus, 423, *434*
　redmani, 428, 434
　sp., 434
Monoporites annuloides, 446, 447, 450
Montego Bay, 118, 122, 401, 426
Montego Bay Airport, 426
Montego Bay Airport Cave, 419, *426,* 431, 436
Montpelier Beds, 179, 219

Montpelier Formation, 118, 123, 131, *132,* 134, 140, 142, 148, 165, 180, 184, 194, 198, 200, 202, 208, 220, 221, 225, 233, *255,* 263, 276, 317, 329, 337, 349, 401
Montpelier Group, 127, 132, 146, 401
Montpelier–Spring Mount region, 394
moodybranchensis, Operculinoides, 333
moorei, Discoaster, 137, 149, 162, 167
Mooretown Formation, 371
mooretownensis, Distefanella, 44, 45, 46
moratus, Lithastrinus, 4, 8, 9, 11, 12, 13, 15, 16
morgani, Lepidocyclina, 319
morionum, Rhabdosphaera, 147
Mormoops, 433
　blainvilli, 428, 433
　megalophylla, 419, 423, 428, 433
Morocco, 196
Moseley Hall Cave, *426,* 436
Motagua-Polichic fault system, 444
Mt. Diablo, 426
Mt. Diablo Cave, 419, *426,* 439
Mt. Friendship, 118, 123
Mount Peace Formation, 5
Mt. Sinai area, 380, 384
Mount Zion area, 385
moureti, Peroniceras, 80, 82, 84
mudstone, 20, 34, 79
muelleri, Inoceramus, 79, 80, 85
multicamerata, Globorotalia, 150, 152, 157, 191, 193, 194, *198,* 203
　menardii, 182
　(Globorotalia), 200
　(Menardella), 200
multicarinata, Prolatipatella, 23, 27
multicostata, Rectuvigerina, 222, 227, *240,* 245
　Siphogenerina, 240
multilirata, Barrettia, 32, 45, 46, 80, 82
multiloba, Sphaeroidinella, 214
　Sphaeroidinellopsis, 184, 187, 189, 193, 195, 213, *214*
multipora, Pontosphaera, 137
multiporus, Triadechinus, 373, 374, 376, 380
multiradiatus, Discoaster, 147
mundula, Truncatulina, 237
mundulus, Cibicidoides, 225, 226, 227, 228, 229, *237*
murrhina, Biloculina, 240
　Pyrgo, 222, 227, *240*
murus, Micula, 15
Mus, 420, 421, 439
　musculus, 439
musculus, Mus, 439
mutiplus, Octolithus, 23, 27
Myllocercion sp., 278
Mynilitha convallis, 142, 177

N

nana, Brachyphylla, 419, 423, 434
   Paragloborotalia, 206
   pumila, Brachyphylla, 423, 434
Nannoconus, 4
   donnatensis, 4, 12, 15
   elongatus, 8, 12, 15
   farinacciae, 27
   fragilis, 4, 12, 15
   quadriangulus, 4, 12, 15
   regularis, 8, 11, 12, 15
   truitti, 8, 11, 12, 15
   sp., 8, 9
   spp., 8, 11, 12, 15, 23, 27
nannofossils, calcareous, 2, *19, 21,* 24, *131,* 133, *135, 140,* 142, 143, 149, *159*
nannoplankton, *1,* 259
naroemoenensis, Dentoglobigerina, 208
nascens, Lophodolithus, 137, 147
Natalus, 435
   major, 428, 435
      jamaicensis, 435
   stramineus, 435
      jamaicensis, 435
nautiloids, 347, 351
Nautilus farctus, 236
   incrassatus, 237
   lobatulus, 236
   pompilioides, 238
   spiralis lobatus, 236
Navy Island Member, 108, 111, 113, 126, 149, 152, 156, 158, 377
Necrocarcinus, 116, 120
   sp., 116, 117, 120
needlefish, 413
Neger Desert, 66
neoabies, Sphenolithus, 137, 140, 142, 144, 145, 150, 152, 153, 162
Neochiastazygus chiastus, 137, 147
   concinnus, 147
Neococcolithes dubius, 137, 147
Neocrioceras sp., 80, 82
Neodiscocyclina sp., 341
Neoeponides spp., 227
Neogene, *131, 155, 179, 184, 190,* 255
Neogloboquadrina, 143, *206*
   acostaensis, 143, 144, 145, 146, 147, 182, 183, 184, 185, 187, 188, 189, 190, 191, 195, 197, 205, *206*
   continuosa, 188, 189, 193, 205, *206*
   dutertrei, 191
   humerosa, 187, 191, 193, 205, *206,* 257
   mayeri, 144, 145, 148
   nympha, 206
   pachyderma, 206
Neoglyptoxoceras, 86, 88
   magnificum, 88
   retrorsum, 88
neohamatus, Discoaster, 137, 138, 140, 142, 152, 155, 162

Neolaganum, 389
   dalli, 373, 374, 389
Neolepidina, 315
Neomeris, 443
neorectus, Discoaster, 139
Neorotalia, 327
   mecatepecensis, 327
   mexicana, 327, 330
   mexicana/mecatepecensis, 287
   sp., 305
Neotropics, 432
Neouvigerina porrecta, 240
Nepenthes acostaensis zone, 189
nepenthes, Globigerina, 143, 144, 145, 146, 184, 185, 186, 187, 188, 189, 193, 195, 204, 212
   (Zeaglobigerina), 212
   Globoturborotalita, 189, 194, *212,* 213
Nephrolepidina, 313, 316, 317, 319, 321, 323
   spp., 321
   (Nephrolepidina), Lepidocyclina, 316, 320
   chaperi, Lepidocyclina, 287, 316, 320, 321, 326
   crassata, Lepidocyclina, 323
   haddingtonensis, Lepidocyclina, 321
   undosa, Lepidocyclina, 323
   vaughani, Lepidocyclina, 320, 321
   yurnagunensis, Lepidocyclina, 287, 319, 320, 321, 324
neritic deposit, 132
Nesophontes, 423
Netherlands, 57
New Ground, 84, 122
New Guinea, 309
New Jersey, 116
New Mexico, 359
New Providence, 433, 434
New Zealand, 377
Newman Hall Formation, 10, 21, *24,* 26, 30, 32, 34, 45, 84, 85, 99
Newport Formation, 118, 123, 386
Newton Farm, 71, 73
Nicaragua, 422
Nicaragua Rise, 263, 285, 303, 307, 325, 422
nicaraguana, Miscellanea, 325
nicholasi, Durania, 32, 35, 37, 45, 46
   Radiolites, 32, 38, 40, 41
Niedeösterreich, Austria, 57
ninae, Amaurolithus, 149
Ninety East Ridge, Indian Ocean, 196
nitescens, Coronocyclus, 137
nitida, Candeina, 185, 187, *207,* 209
   praenitida, Candeina, 182
nittalli, Ellipsonodosaria, 241
Nitzschia sp., 268
nodifera, Rectuvigerina, 224
noeliae, Dodekapodorhabdus, 22, 27
Nonion barleeanum, 238
   soldanii, 238
   spp., 227
Noniona barleeana, 238
   soldanii, 238
Nonionellina sp., 227

Nonionina bulloides, 239
   quinqueloba, 240
Nonsuch Limestone, 146, 311, 331, 339
North American Coastal Plain, 69
North Atlantic Basin, 163
North Atlantic Deep Water, 263, 265
North Carolina, 106, 398
Norway, 15
Norwegian Sea sediments, 263
Norwegian waters, 111
Nostoceras stantoni zone, 67
notabilis, Tarphypygus, 373, 374, 387, 389
Notocorystes, 116, *120*
Notopocorystes (Cretacoranina), 116
   trechmanni, 116, 117, 120
Nowakites, 82, 85, *88*
   lemarchandi, 80, 82, 83, 84, 85, 88, 89
   pailletteanus, 88
   paillettei, 82, 88
Nucleolites, 389
nucleolitid, 102
nuevitasensis, Echinolampas, 393
Nummulites, 329, *331,* 333
   floridanus, 347
   floridensis, 333
   gizehensis, 331
   macgillavryi, 286, 331, 334
   mantelli, 313
   panamensis, 335
   parvula, 309
   striatoreticulatus, 287, 331, 333, 334, 336
   willcoxi, 333
nuttalli, Cancris, 222, 224, 227, 232, *235*
   Valvulineria, 235
Nyctibius, 432
   griseus, 428, 432
nympha, Neogloboquadrina, 206

O

obliquum, Gartnerago, 22, 27
obliquus, Ancylus, 367
   Globigerinoides, 184, 187, 189, 193, *210,* 211
   extremus, Globigerinoides, 184, 187, 189, 210
oblongus, Cancris, 222, 227, *235*
obscurus, Ancylus, 368
   Calcidiscus, 2, 8, 9, 10, 11, 12, 14
   Calculites, 2, 8, 9, 10, 11, 12, 14
   Phanulithus, 23, 27
obtusa, Pulvinulinella, 233
   exigua, 233
obtusus, Discoaster, 131
ocalana, Helicostegina, 285
   (Vlerkina), 338
   Heterostegina, 287, 321, 335, 337
occidentalis, Heterosalenia, 93, 94, 95, 97
occiduus, Celestus, 425, 428, 429
occitana, Sismonda, 389
Ocho Rios area, 123
Octolithus mutiplus, 23, 27

octoradiata, Ahmuellerella, 8, 9, 12, 14, 22, 26
ODP sites, 195, 196
Okkolithus australis, 23, 27
Oklahoma, 359
oldhami, Ammonites, 89, 90
  Pachydiscus, 89
    (Pachydiscus), 83, 84, 85, 90
Oligocene, 285
Oligopygus, 384
  floridanus, 384
  hypselus, 384
  jamaicensis, 373, 374, 384
  wetherbyi, 373, 374, 383, 384
Oman, 303
Ommatocampe, 271
  amphistylium, 260, 271, 272
  polyarthra, 271
Ommatogramma spp., 273
omnitubus, Solenosphaera, 260, 266, 267
ompilioides, Anomalina, 234
One Eye River, 431
Oolina, 223, 233
  sp., 227
Oolithotus fragilis, 152, 153, 162
operculatus, Psilatricolporites, 446, 447
Operculina catenula, 329, 331
  complanata, 335
  gomezi, 335
Operculinella dia, 335
operculiniformis Archaias, 285
Operculinoides, 333
  cushmani, 333, 347
  dia, 335
  floridensis, 333
  jennyi, 333
  moodybranchensis, 333
  willcoxi, 317
  sp., 347
Ophiuroid vertebral ossicles, 130
ophiuroids, *125*
opima, Globorotalia opima, 206
  Paragloborotalia, 206
  continuosa, Globorotalia, 206
  opima, Globorotalia, 206
Oppel zones, *69*
Orange River locality, 21
Orbitoclypeus, 339
Orbitocyclina, 59
Orbitoides, 35, 37, 38, 40, 41, 44, 47, 57, 313
  apiculatus, 57
  media, 57
    megaloformis, 29, 35, 37, 45, 46, 57, 58, 60
Orbulina, 184
  suturalis, 184, 197
  universa, 144, 145, 185, 198
  spp., 183
organisms, terrestrial, 422
Oridorsalis, 227
  spp., 227, 228, 229
orionatus, Tranolithus, 23, 27
ornata, Siphogenerina advena, 240
ornatus, Arachnodiscus, 268
orosphaerids, 256

Orthomorphina modesta, 241
  spp., 227
Orthopsis, 102
  sp., 94, 95
orthostylus, Tribrachiatus, 137, 146, 147
Oryzomys, 420, 423, 425, 427, 431, *439*
  antillarum, 419, 421, 423, 424, 425, 426, 427, 428, 439
  couesi, 421, 423, 439
    antillarum, 439
  palustris antillarum, 439
Osangularia sp., 222, 227
osborni, Clidomys, 419, 427, 428, 437, 438, 439
Osteopilus brunneus, 419, 428
ostracodes, 67, 68, 105, 127, 131, 259
  assemblages, *69*
  zones, *69, 71*
Ottavianus terrazetus, 23, 27
ovalis, Calcidiscus, 8, 9, 10, 14
  Calculites, 8, 9, 10, 14
  Phanulithus, 23, 24, 27
ovata, Watznaueria, 23, 27
Ovocytheridea, 69
  sp., 72, 75
ovumserpentis, Haimea, 373, 383, 384, 385, 386
oyster limestones, 71
oysters, 68

P

pachyderma, Cibicidoides, 222, 224, 227, 233, *237*
  Eucyrtidium, 279
  Neogloboquadrina, 206
  Truncatulina, 237
Pachydiscus cayeuxi, 79, 80, 90
  heldemsis, 89
  jamaicensis, 79, 80, 85, 89
  koeneni, 89
  oldhami, 89
  spissus, 80, *89*
  (Pachydiscus) jamaicensis, 83, 85, 90
    koeneni, 83, 84, 85, 86, 87, 88, *89*
    oldhami, 83, 84, 85, 90
    spissus, 83, 84
  (Pachydiscus) jamaicensis, Pachydiscus, 83, 85, 90
    koeneni, Pachydiscus, 83, 84, 85, 86, 87, 88, *89*
    oldhami, Pachydiscus, 83, 84, 85, 90
    spissus, Pachydiscus, 83, 84
Pachygrapsus, 120, 122
  sp., 119, 122
paenedehiscens, Sphaeroidinellopsis, 214
  subdehiscens, 182, 191, 214
pailletteanus, Nowakites, 88
paillettei, Nowakites, 82, 88
Pakistan, 69
Palaeolampas, 391

Palaeonummulites, 311, 329, *333*
  dia, 287, 330, 335
  floridensis, 287, 333, 334, 336
  willcoxi, 287, 331, 333, 334, 336, 337
paleobathymetry, *223*
paleobiogeography, *85*
paleobotany, *443, 459*
paleocarteri, Helicosphaera, 165
paleoecology, *255*
paleoenvironments, 5
Paleogene, *283*
paleogeography, *77*
paleontology, *47*
  systematic, *105, 374, 413*
paleopalynology, *459*
Paleopneustes sp., 373, 409
pallidus, Amerigus, 365
  (Lateorbis), 366, 367
  Planorbis, 366
palmerae, Cassidulina, 237
  Globocassidulina, 227, *237*, 245
palmeri, Terebratulina, 113
Palmerius roberti, 377
Paludina jamaicensis, 365
  parvulus, 365
  rivularis, 365, 366
palustris antillarum, Oryzomys, 439
palynomorphs, *446, 448*
palynostratigraphy, Tertiary, *443, 446*
Panama, 351, 402, 445, 446
Panamanian isthmus, 444
panamensis, Aturia, 351
  Heterostegina, 337
  Miogypsina, 325
  Nummulites, 335
panda, Globorotalia, 202
Panopeus, 122
  herbstii, 120, 122
Papua, 163
Parachondria augustae, 354, 356
Paragloborotalia, 206
  mayeri, 184, 185, 186, 187, 188, 189, 194, 197, 205, *207*
    zone, 197
  mayeri-mayeri, 184
  mayeri/siakensis, 207
  nana, 206
  opima, 206
paragoga, Echinolampas, 373, 391, 394
Paraje Solo Formation, 446
paralenguaensis, Globorotalia, 200
parallelus, Eiffellithus, 22, 24, 25, 27
  Parapygus, 394
  Rhyncholampas, 373, 395
Paranecrocarcinus, 116, 120
  sp., 116, 117, 120
Parapachydiscus gollevillensis, 89
  stallauensis, 89
Parapygus, 394
  antillarum, 394
  parallelus, 394
Pararotalia, 327
  sp., 227, *239*
Paraster, 396
Parastroma trechmanni, 32, 45, 46

parcus, Aspidolithus, 4, 22, 24, 25, 26
   constrictus, Aspidolithus, 8, 9, 10, 12, 13, 14, 16
Parhabdolithus bitraversus, 27
Parish of Hanover, 30
Parish of St. James, 30
Parish of Westmoreland, 30
parkerae, Globigerinoides, 182
   Liriospyris, 274
parnelli, Chilonycteris, 433
   Phyllodia, 433
   Pteronotus, 428, 433
parri, Plectofrondicularia, 222, 224, 227, 232, 233, *239*
parvidentatum, Repagulum, 23, 27
parvipetula, Haimea, 373, 384, 385
parvula, Amphistegina, 286, 309, 312
   Lepidocyclina (Lepidocyclina), 317
   Nummulites, 309
   Uvigerina, 241
parvulus, Paludina, 365
parvus, Clidomys, 419, 427, 428, 437, 439
   Macropneustes, 373, 374, 402, 403
pataecus, Coccolithus, 137
patagonica, Pulvinulina, 204
patella, Biscutum, 22, 24, 26
pauperata, Laticarinina, 227, *238*
Pauropygus, 384
pawsoni, Asterostoma, 373, 374, 406, 407
Peace River, 389
peat, 363
pecinatus, Radiorbitoides, 32, 34, 45, 46, *52*
Pecten Bed, 123
pectinata, Pseudorbitoides trechmanni, 48, 52
Pedina, 377
   sp., 373, 374, 377
Pedro bank, 422
Pellatispirella, *325*
   antillea, 329
   matleyi, 286, 303, 325, 328
pellatispiroides, Camerina, 329
   Ranikothalia, 331
Pelliceria, 446, 448, 450
   (Psilatricolporites) crassus, 443
pelogicus, Coccolithus, 137
peloria, Phymosoma, 373, 374, 376, 379, 380
Pembroke Hall area, 98
pemmatoidea, Manivitella, 8, 9, 11, 12, 15, 23, 27
Peneroplis evolutus, 305
   sp., 306
   spp., 285
pentacamerata, Globigerina, 311
pentaradiatus, Discoaster, 137, 138, 139, 140, 142, 149, 150, 153, 154, 162, 169
Percivalia pontilithus, 27
perclarus, Discoaster, 137, 167
perculiniformia, Archaias, 285
perdurum, Lithostromation, 137
Periaster, *399*
   elatus, 401
   elongatus, 399

Pericosmus (Victoriaster) jamaicensis, 399
   sp., 373, 409
periostracum, 355
Periphaena, 271
   decora, 271
   sp., 260, 271
peripheroacuta, Globorotalia, 144, 145, 148, 185, 198, 199, *202*
   fohsi, 202
   (Fohsella), 202
   (Turborotalia), 202
peripheroronda, Fohsella, 202
   Globorotalia, 144, 145, 148, 184, 185, 198, 199, *202*, 261
   fohsi, 202, 257
   (Fohsella), 202
   (Turborotalia), 202
Peroniceras moureti, 80, 82, 84
perpallidum, Cyclostoma, 358
perplexus, Hayaster, 137, 153, 174
   Hypselaster, 399
   Plagiobrissus, 373, 405, 406
   Stenechinus, 373, 374, 377, 379
persculpta, Lucidella, 359
pertenuis, Globorotalia, *200*, 203
   (Menardella), 200
Peru, 385
peruviana, Aturia, 351
   Gypsina, 309
   Lepidocyclina, 317, 324
   Sphaerogypsina, 309
Petrochirus, 120, 122
   sp., 121, 122
Petrolisthes, 120, 122
   sp., 121, 122
petterssoni, Diartus, 257
phacelosus, Tranolithus, 8, 9, 11, 12, 13, 15
Phanulithus obscurus, 23, 27
   ovalis, 23, 24, 27
   zone, *24*
   sp., 23, 27
Phyllacanthus, 95, 96
   leoni, 93, 94, 95, 96, 98
phyllimorphus, Desmophyllites, 84, 87, 88
Phyllodia parnelli, 433
Phyllonycteris, 423, *434*
   aphylla, 428, 434, 435
   sezekorni, 434
   (Reithronycteris) aphylla, 434
Phymodius, 120, 121, 123
   maculatus, 120, 121, 123
Phymosoma, 93, *379*
   peloria, 373, 374, 376, 379, 380
Physa jamaicensis, 366
Physella cubensis, 365, 366
   marmoratus, 366
(Piersona) favealis, Helisoma, 367
pigmea, Uvigerina, 222, 227, 229, 230, *241*, 247
pilaris, Tinoporus, 309
Pimento Mill, 118, 13, 129
Pindars, River bed, 84
Pipetella prismaticus, 269
Pipetteria tubaria, 269

Pisidium jamaicensis, 369
   sp., 369
placoliths, 155, 166
Placozygus fibuliformis, 4, 8, 9, 11, 15
Plagiobrissus, *405*
   abruptus, 373, 405, 406
   elevatus, 373, 405, 406
   latus, 373, 405, 406
   loveni, 373, 404, 405, 406
   perplexus, 373, 405, 406
   robustus, 373, 405, 406
   sp., 405
   spp., 373
Plagioptychus, 34, 35, 37, 38, 39, 40, 41, 45, 46
Plaisance Formation, 118
planktonic foraminiferal zones, *159*, *190*
Planorbis caribaeus, 367
   decipiens, 366
   dentiferus, 366
   edentatus, 366
   haldemani, 366
   humilis, 365, 367
   macnabianus, 366
   pallidus, 366
   redfieldi, 366
Planorbulina robertsoniana, 237
   (Planorbulinella) guayabalensis, 315
   sp., 227, 242
   spp., 226
(Planorbulinella) guayabalensis, Planorbulina, 315
Planostegnia, 335
Planulina ambigua, 222, 227, *239*, 243
   ariminensis, 222, 224, 227, 232
   dohertyi, 224
   foveolata, 222, 224, 227, 228, 229, 231, 232, 233, *239*, 250
   matanzasensis, 237
   renzi, 222, 227, *239*
   rugosa, 222, 224, 225, 227, *239*, 249
   wuellerstorfi, 222, 225, 227, 231, 232, *239*
   sp., 222, *239*, 249
   spp., 227
plateia, Echinolampas, 373, 391, 393, 394
Platybelone argalus, 413
platycopids, 69
Platycosta sp., 70, 75
platypetalus, Cassidulus, 394
platypetula, Haimea, 384, 385
platyrachis, Pyrogophorus, 365
Plectofrondicularia diversicostata, 239
   jarvisi, 239
   parri, 222, 224, 227, 232, 233, *239*
   vaughani, 222, 224, 227, 232, *239*
plesiotumida, Globorotalia, 184, 187, 189, 190, 192, 193, 194, *200*, 202, 203, 204
   (Globorotalia), 200

Pleurodonte, 355, 356, 358, 359
  acuta, 356
  bernardi, 354, 355, 356, 358, 359
  bowdeniana, 354, 355, 356, 358, 359
  chemnitziana, 358
  ingens, 358
  lucema, 354, 355, 356, 358
  sinuata, 356, 358
  valida, 354, 356
  (Pleurodonte), 356, 358
  (Pleurodonte), Pleurodonte, 356, 358
Pleurostomella spp., 227
Pleurostomellids, 227
Pliocene, *156, 193, 219, 285*
  lower, *156*
Pliolepidina, 313
  (Pliolepidina) proteiformis, Lepidocyclina, 317
plugs, 30
plutons, 30
Podocyrtis geotheana zone, 271
Podorhabdus sp., 11
(Poenia), Lucidella, 359
Point area, 405, 406
Poland, 88, 89, 90
polita, Micropanope, 120, 122
pollen, 445, 450
polyarthra, Ommatocampe, 271
Polylepidina, 313, 315, *316*
  chiapasensis, 287, 315, 316, 318
  (Polylepidina) antillea, Lepidocyclina, 315, 316
  chiapensis, Lepidocyclina, 316
  gardnerae, Lepidocyclina, 316
  kinlossensis, Lepidocyclina, 316
  proteiformis, Lepidocyclina, 317
Polymorphinids, 227
Polyperibola christiani, 188, 197, *207,* 209
  mayeri/siakensis, 188, 189
  siakensis, 189, 193
Polyxenes cribatus, 236
Pomacea fasciata, 363, 365
pompilioides, Melonis, 222, 224, 227, 229, 233, *238,* 239, 244
  Nautilus, 238
Ponce Formation, 235
pontilithus, Percivalia, 27
  Zygodiscus, 23, 27
Pontosphaera discapora, 137, 173
  japonica, 137, 166
  multipora, 137
  pulchra, 137, 147
  scutellum, 137
Popkin Formation, 67, *73*
porcellanite, 259
Porites zone, 118
porrecta, Angulogerina, 240
  Neouvigerina, 240
  Siphouvigerina, 226, 227, 228, 229, 230, 233, 234, *240,* 247
  Uvigerina, 240
Port Antonio, 77, *84,* 86, 90, 111, 113, 123, 132, 152, 155, 377
Port Maria area, 110, 375, 388, 409
Portland, 341

Portland caves, 418, 426, 429, 431, 433, 434
Portland Parish, 67, 108, 111, 113, 123, 126, 127, 285, 377, 382, 389, 437
Portland Ridge, 418
  caves, *426,* 431, 436
Portland system, 419
Potamopyrgus, 365
  coronatus, 365
Poteria, 355
  adamsi, 354
  bowdensis, 354, 357
  varians, 425
Pots and Pans Roadcut outcrop, 134, 140, 143, 144, 180, 182, 184, 198, 200, 202, 206, 208, 210, 212, 214, 221
Pozon Formation, 185
Pozon section, 188
Praeglobotruncana helvetica, 82
praemenardii, Globorotalia, 184, 199, 200
  (Menardella), 200
praemurus, Micula, 8, 15
praenitida, Candeina nitida, 182
Praeorbulina, 184
  glomerosa, 148, 257, 262
  zone, 259
praequadratus, Lithraphidites, 8, 15, 16, 23, 24, 26, 27
Praeradiolites sp., 80, 84
Praerhapydionina, *303*
  delicata, 285, 287, 303, 305, 306
praescitula, Globorotalia, 198, 202
praesignis, Aristelliger, 428, 429
  Hemidactylus, 429
praesiphonifera, Globigerina, 187
  Globigerinella, 211
praetournoueri, Lepidocyclina, 321
Prebarrettia sparcilirata, 35, 37, 38, 39, 40, 41, 45, 46, 57
Prediscosphaera columnata, 12, 15
  cretacea, 8, 9, 11, 12, 15, 23, 24, 27
  quadripunctata, 23, 27
  spinosa, 23, 27
predistentus, Sphenolithus, 137, 146, 147
Prenaster loveni, 405
prepentaradiatus, Discoaster, 137, 139, 146, 149
primordius, Globigerinoides, 257
primus, Amaurolithus, 139, 149, 150, 152, 153, 155, 158, 162, 173
  Eiffellithus, 15
Prionocidaris, 375
  leoni, 373, 374, 375, 376
  loveni, 373, 374, 375, 376
  spinidentatus, 375
  sp., 373, 375
prismatica, Didymocyrtis, 259, 260, 269
prismaticus, Cannartus, 269
  Pipetella, 269
Probolarina, 105, *106,* 112
  transversa, 106
  sp., 112

proboscidea, Uvigerina, 227, 228, 230, *241,* 247
procera, Rhabdosphaera, 137, 177
profunda, Cornutella, 276
  clathrata, 276
  Dictyocoryne, 273
  Rhopapastrum, 273
Prolatipatella multicarinata, 23, 27
propagulis, Chiastozygus, 22, 26
Prophyllacantus, 96
Prorastomus, 414
  sirenoides, 414
proteiformis, Lepidocyclina, 317
  (Lepidocyclina), 317
  (Pliolepidina), 317
  (Polylepidina), 317
Providence, 111
Providence Shale, 13, *84,* 86, 88, 90, 105, 111
proximus subcircularis, Inoceramus, 84, 85
Prunopyle, 267, 268
  pyriformis, 267
  sp., 260, 268
pseudanthophorus, Zeugrhabdotus, 8, 9, 11, 12, 15
Pseudemys, 427
  floridana, 427
  terrapen, 427
Pseudochrysalidina, 285, *297,* 307
  floridana, 286, 297, 300
Pseudodiadema sp., 93
Pseudoemiliania lacunosa, 137, 138, 142, 145, 150, 152, 153, 162, 166
Pseudofabularia, 297
  matleyi, 287, 297, 298, 302
pseudogrosserugosus, Anomalinoides, 224
Pseudolacazina, 299
pseudomiocenica, Globorotalia, 184, 187, 191, 193, 197, *200,* 203
pseudomiocenica-limbata, Globorotalia, 202
Pseudononion sp., 227
Pseudophragmina (Althecocyclina) stephensoni, 337
Pseudophyllites indra, 86
pseudoplicata, Bolivina, 227
Pseudorbitoides, 34, 41, 47, 48, 50, 52, 54
  chubbi, 34, 44, 46, *48*
  rutteni, 44, 45, 46, *52,* 53
  trechmanni, 30, 32, 34, 35, 41, 45, 46, 47, 48, *50,* 51, 52, 53, 54
  pectinata, 48, 52
  trechmanni, 50
  spp., 53
pseudorbitoides family, *47*
pseudoumbilicus, Reticulofenestra, 137, 139, 140, 142, 144, 145, 146, 150, 152, 153, 158, 162, 165, 166, 175
Psilatricolporites, 448
  crassus, 446, 447, 448, 449
  operculatus, 446, 447
  (Psilatricolporites) crassus, Pelliceria, 443

Pteronotus, 433
  parnelli, 428, 433
Ptychocochlis adamsi, 355
puer, Spongocore, 260, 271, 272
  (Spongocorisca), 271
Puerto Rico, 69, 85, 235, 351, 388, 419, 421, 431, 432, 433, 437, 445
puilboreauensis, Cushmania, 295
  Dictyoconus, 293
  (Cushmania), 293
pulcherrima, Scyphosphaera, 137
pulchra, Pontosphaera, 137, 147
  Reusella, 240
    spinolosa, 227, 228, 229, 230, 231, 233, *240*, 246
  Sellia, 365
Pullenia bulloides, 227, *239*
  compressiuscula quadriloba, 240
  malkinae, 222, 227, *240*
  quadriloba, 222, 227, *240*
  quinqueloba, 227, 240
Pulvinulina menardii, 202, 238
  patagonica, 204
  repanda, 238
  scitula, 204
  tumida, 202
Pulvinulinella exigua obtusa, 233
  obtusa, 233
  umbonifera, 239
pumila, Brachyphylla, 434
  cavernarum, 434
  nana, 423, 434
punctata, Globocassidulina, 222
punctatus, Rhyncholampas, 373, 395
  Rhynchopygus, 394
puncticulata, Globorotalia, 150, 192, 193, 194
Puntarenas Province, 446
pusillum, Astrononion, 227, *234*
pustulosa, Lepidocyclina, 317, 321, 324, 337
  (Lepidocyclina), 287, 317, 319, 322, 324
Puzosia le marchandi, 88
pygmaea, Cyclus, 369
  Valvata, 364, 365
Pygopistes, 99
  rudistarum, 94, 95, 99, 100
Pygorhynchus, 99
pyramoides, Haimea, 373, 384, 385, 386
Pyrgo murrhina, 222, 227, *240*
pyriformis, Prunopyle, 267
Pyrogophorus platyrachis, 365

Q

Q-mode Factor Analysis, 226, 227, 229
quadriangulus, Nannoconus, 4, 12, 15
quadrifidus, Lucianorhabdus, 8, 15, 23, 27
quadrilatera, Bolvinita, 235
quadriloba, Pullenia, 222, 227, *240*
  compressiuscula, 240
quadrilobatus trilobatus,
  Globigerinoides, 212
quadriperforata, Watznaueria, 23, 27
quadripunctata, Prediscosphaera, 23, 27
Quadrum gartneri, 4, 7, 8, 9, 10, 12, 15
  gothicum, 8, 9, 10, 15, 16
  sissinghii, 8, 9, 11, 12, 15, 16, 27
  trifidum, 2, *3*, 4, 5, 8, 9, 10, 12, 13, 15, 16, 66, 67
  zone, *3*, 4, 5, 6, 10, 13
Quashies River, 414
Quaternary, *359*, *417*, *424*
Quemisia, 437
Quickstep Cave, 425, 436
quinqueloba, Nonionina, 240
  Pullenia, 227, 240
Quinqueloculina spp., 227
quinqueramus, Discoaster, 132, 139, 144, 145, 149, 150, 151, 152, 153, 154, 155, 158, 162, 163, 164, 169
quinquesperforata, Mellita, 373, 390

R

Rabat, Morocco, 196
radians, Sphenolithus, 137, 147
radiata, Gundlachia, 365, 367, 368
radiatus, Ancylus, 367
radiocarbon dates, 422
radiolarians, 256, 259
radioles, 93, 94, 95, 96, 377
Radiolites, 32, 44, 45, 46
  nicholasi, 32, 38, 40, 41
Radiorbitoides, 34, 41, 47, 48, *52*
  pecinatus, 32, 34, 45, 46, *52*
Rafael Beds, 396
ramonae, Rhizophora (Zonocostites), 446, 450
  Zonocostites, 447
Ranikothalia, *329*
  antillea, 329, 331
  catenula, 286, *329*, 331, 332
  pellatispiroides, 331
  soldadensis, 329, 331
  tobleri, 331
  willcoxi, 331
rathbunae, Xanthilites, 117, 118, 122
Rattus, 420, 421, 423, 426, 429, 431, 436, 439
  rattus, 439
rattus, Rattus, 439
recrystallization, 69
recta, Helicosphaera, 137, 146, 147
Rectuvigerina multicostata, 222, 227, *240*, 245
  nodifera, 224
  striata, 222, 224, 227, 231, 233, *241*, 245
  transversa, 224
recurvus, Isthmolithus, 146, 147
Red Beds, 347
Red Gal Ring, 321
Red Hills, 426
Red Hills Road Cave, *426*, 436
redfieldi, Planorbis, 366
redmani, Monophyllus, 428, 434
reefs, 30, 34
  coral-associated, 120

reefs (continued)
  fragments, 223
  patch, 118
  raised, 382
reflexa, Cassidulina, 227, 229, *236*
reflexum, Anaklinoceras, 66
regularis, Ahmuellerella, 22, 26
  Nannoconus, 8, 11, 12, 15
  Stenechinus, 373, 374, 376, 377
Reinhardtites anthophorus, 4, 8, 9, 11, 12, 13, 15, 16, 23, 24, 25, 26, 27
  levis, 8, 15, 25
Reithronycteris, 434
  aphylla, 434
  (Reithronycteris) aphylla,
    Phyllonycteris, 434
remediensis, Jacksonaster, 390
reniformis, Rhagodiscus, 23, 27
renzi, Eoceratoconcha, 118, 123
  Planulina, 222, 227, *239*
Repagulum parvidentatum, 23, 27
repanda, Pulvinulina, 238
residei, Baculites, 67
Retecapsa crenulata, 8, 9, 11, 12, 15
reticulata, Reticulofenestra, 137, 146, 147
Reticulofenestra, 155
  bisecta, 137, 146, 147
  floridana, 137, 140, 146, 147, 168
  hillae, 137, 146, 147
  pseudoumbilicus, 137, 139, 140, 142, 144, 145, 146, 150, 152, 153, 158, 162, 165, 166, 175
  reticulata, 137, 146, 147
  umbilica, 137, 146, 147
Retimonocolpites, *449*
  sp., 448
retrorsum, Ancycloceras, 86
  Glyptoxoceras, 84, 86, 87, 88
  Neoglyptoxoceras, 88
Reusella aculeata, 240
  pulchra, 240
  simplex, 240
  spinolosa, 229, 240
    pulchra, 227, 228, 229, 230, 231, 233, *240*, 246
Rexroad local fauna, 359
Rhabdolithus creber, 137
  lenuis, 137
  sp., 177
Rhabdosphaera claviger, 137
  crebra, 147
  inflata, 146, 147
  morionum, 147
  procera, 137, 177
  tenuis, 147
  sp., 173
Rhagodiscus angustus, 8, 9, 11, 12, 15, 23, 27
  asper, 11, 12, 15
  reniformis, 23, 27
  splendens, 8, 11, 12, 15, 23, 27
  sp., 8, 23, 27
Rhine Valley, lower, 359
Rhizophora, 446, 448, 450
  (Zonocostites) ramonae, 446, 450
Rhodospyris spp., 274

rhombicum, Rhombolithion, 23, 27
Rhombolithion rhombicum, 23, 27
*(Rhopalodictya) malagaense,
    Rhopalodictyum*, 273
Rhopalodictyum, 273
    abyssorum, 273
    malagaense, 260, 272, 273
    (Rhopalodictya) malagaense, 273
Rhopapastrum profunda, 273
Rhyncholampas, 394
    alabamensis, 373, 374, 395
    antillarum, 373, 395
    ellipticus, 373, 395
    matleyi, 373, 374, 392, 395
    parallelus, 373, 395
    punctatus, 373, 395
    sp., 394
    spp., 374
Rhynchonella holmesii, 106
Rhynchopygus matleyi, 394
    punctatus, 394
    sp., 394
Richmond Formation, 385, 291, 295, 309, 311, 331, 371
Richmond Group, 146
Rio Bueno, 118, 120, 122
Rio Bueno Harbour, 122, 123, 130, 377, 382
Rio Cobre Canyon, 165, 424
Rio Grande, 3, *13*, 30, 44
Rio Grande Limestone, 30, 41, *44*, 45, 54, 57
Rio Grande Valley, 44
Rio Gurabo section, 108
Rio Minho, 71, 110
Rio Minho Bed, 120
Rio Minho section, 71
Rio Minho valley, 67
Rio Nuevo, 98
Rio Nuevo Formation, 13
Rio Sambre, 110
Rivière Bois de Chêne, 198
rivularis, Hydrobia, 365
    Paludina, 365, 366
roberti, Palmerius, 377
robertsoniana, Planorbulina, 237
    Truncatulina, 237
robertsonianus, Cibicidoides, 224, 225, *237*
robusta, Fohsella, 204
    Globorotalia, 199, *204*
        fohsi, 143, 144, 145, 146, 148, 183, 185, 186, 188, 197, 204, 257
robusta-lobata, Globorotalia, 197, 202
robustum, Amphibrachium 273
robustus, Plagiobrissus, 373, 405, 406
Rock Spring, 5, 15, 351
Rock Spring Elementary School, 349
rocks
    carbonate, 1
    clastic, 78
    igneous, 1, 84
    sedimentary, 1
    volcaniclastic, 79
rodents, heptaxodontid, 418, 419

rojasi, Antillaster, 407
rooki, Asterorbis, 59
rosaceus, Clypeaster, 373, 386, 387, 388
Rosalina sp., 222, 227, 228, 229, 230, *240*, 248
Rosalind bank, 422
roselli, Fabularia, 301
rosetta, Globotruncana, 34
Rotalia ambigua, 239
    ammophila, 238
    limbata, 198
    (Turbinulina) elegans, 238
Rotelapillus laffittei, 8, 15, 23, 27
    sp., 9
rotunda, Haimea, 373, 384, 386
Round Hill, 113, 123, 377, 391
Round Hill Beds, 118, 123, 388, 391
Round Hill Bluff, 118, 123
ruber, Globigerinoides, 148, 185, 186, 187, 188, 189, 197, *210*, 211, 257
rubescens, Globigerina, 212
rubra, Globigerina, 210
    Globigerinoides, 210
Rucinolithus hayi zone, 4
    hayi, 5, 8, 11, 15
    sp., 5, 11, 80
rudist, 66, 67, 68, 77
    bivalves, 79, 105
rudistarum, Botriopygus, 93, 99
    Pygopistes, 94, 95, 99, 100
rugatum, Glyptoxoceras, 86, 88
rugosa, Cibicides, 239
    Eurhodia, 373, 374, 394
    Haimea, 373, 384, 386
    Planulina, 222, 224, 225, 227, *239*, 249
rugosum, Trochalosoma, 381
rugosus, Ceratolithus, 137, 138, 139, 142, 144, 145, 150, 153, 154, 155, 158, 162, 164, 196
    Triquetrorhabdulus, 137, 138, 139, 144, 145, 146, 149, 162, 184, 187
Runaway Bay, 425
Runaway Bay Caves, *425*
rusea, Barrettia, 79, 80
Russellia bukryi, 23, 27
Russia, 88, 89, 90
rutschi, Sphaeroidinella, 214
rutteni, Pseudorbitoides, 44, 45, 46, *52*, 53

S

Saccocoma, 68, 69, 127
    cretacea, 127
sacculifer, Globigerinoides, 184, 212
Sagda, 360
sagra, Cancris, 235
St. Andrew Parish, 419, 426, 436, 439
St. Ann, 113
St. Ann Group, 80, 84
St. Ann Inlier, 77, *80*, 84, 95, 96, 98, 127

St. Ann Parish, 106, 118, 122, 123, 126, 127, 129, 130, 375, 377, 382, 384, 385, 388, 389, 390, 394, 395, 396, 398, 402, 418, 425, 426, 429, 431, 432, 433, 434, 435, 436, 439
St. Ann's Bay, 118, 122
St. Ann's Great River, 2, 4, 5, *10*, 13, 15, 80, 84, 89, 98
St. Bartholomew, 311, 315, 375, 385, 395, 396, 398, 399, 402, 405
St. Bartholomew locality, 316
St. Catherine Parish, 118, 123, 379, 418, 419, 424, 425, 426, 427, 429, 431, 432, 433, 436, 437, 439
St. Clair Cave, 435
St. Elizabeth Parish, 118, 382, 418, 427, 429, 431, 433, 434, 435, 436, 437, 439
St. Hilda's School, 402
St. James Inlier, 84, 85
St. James Parish, 67, 73, 97, 98, 99, 118, 120, 375, 379, 380, 381, 384, 385, 386, 389, 391, 393, 394, 395, 396, 398, 399, 401, 402, 403, 405, 406, 407, 419, 424, 426, 429, 431, 435, 436
St. Martin, 437
St. Mary Parish, 375, 385, 388, 393, 395, 403, 407, 409
St. Thomas Parish, 354, 359, 377, 379, 380, 382, 384, 385, 386, 393, 395, 396, 398, 403, 406, 407, 413, 431
St. Vincent, 368
saipanensis, Discoaster, 137, 146, 147
Salenia, 93
Salinity Crisis, messinian, 196
Salt River area, 437
San San Bay, 108, 118, 123, 126, 132, 142, *148*, 154, 199, 202
San San Bay section, *147, 148, 156,* 159, 164, 165, 190, *192*, 196, 197, 198
    stratigraphy, *152*
San San Clay Member, 190, 192, 197
San Sebastian Formation, 445, 446
sanchezi, Torreites, 32, 34, 45, 46
sandstone, 20, 35, 78, 80, 84, 142, 347, 448
Sanguinetti, 37, 57
Sanguinetti section, *35*, 39
Sanguinetti traverse, 54
sanmiguelensis, Discoaster, 141
Santa Cruz, 118
santaennae, Lithatractus (Lithatractus), 267
    Stylatractus, 267
Santee Formation, 106
Santonian, 4
    Latest, 4
Saramaguacan Formation, 446
Saudi Arabia, 69
saurophila, Tonatia, 434
    bidens, 428, 419, 430, 433, 434

Sauvagesia, 32, 35, 37, 38, 45, 46
  sp., 80
Savanna-la-Mar, 431
Scampanella sp., 12
Scapholithus fossilis, 23, 27, 139, 177
Schaw Castle, 73
schermoi, Cyclochittya, 355
  Incerticyclus, 354, 355, 356, 357
Schizaster, 396, 398
  altissimus, 373, 396
  bathypetalus, 373, 396
  brachypetalus, 396
  dumblei, 373, 396
  dyscrtius, 396
  hexagonalis, 373, 396, 398, 401
  subcylindricus, 373, 374, 396, 397, 398
  sp., 396
  spp., 374
schizobrachiatus, Cretarhabdus, 22, 26
schizopodia, Gorgospyris, 259, 260, 272, 274
  Thamnospyris, 274
Schizoptocythere, 69
  segurai, 69, 70, 75
  sp., 70, 75
schlegelii, Tomistoma, 413
schlumbergeri, Sigmoilopsis, 222, 227, 240
scintillans, Cancris, 222, 227, 235, 248
  Valvulineria, 235
  sinecarina, Cancris, 235, 250
scitula, Globorotalia, 187, 201, 204
  scitula, 204
    (*Hirsutella*), 204
    (*Turborotalia*), 204
  Pulvinulina, 204
  scitula, Globorotalia, 204
sclerite, Sponge, 266
Scolicia plana, 371
  sp., 371
Scoliechinus, 98
  axiologus, 93, 94, 95, 98, 99
Scutella cubae, 390
  sp., 391
scutellum, Pontosphaera, 137
Scyphosphaera, 139
  campanula, 137
  intermedia, 137, 173
  pulcherrima, 137
  tubifera, 137
sedimentation, 30, 158
  rates, *146*, 150, 195
sediments, 115, 140, 146, 263, 420, 450
  deep-water, 220
  siliceous, 256
segurai, Cyntheseis, 69
  Schizoptocythere, 69, 70, 75
Sellia pulchra, 365
sellii, Helicosphaera, 152, 153, 162
selseyensis, Angulogerina, 234
  Uvigerina, 234
semicribratus, Anomalinoides, 222, 227, *234*
seminulina, Globigerina, 214

seminulina (continued)
  Sphaeroidinellopsis, 184, 185, 187, 193, 194, 195, 213, *214*
seminulum, Helicosphaera, 137, 147
semiradiata, Hercothyris, 112, 113
senni, Amphistegina, 311
septanarius, Lithastrinus, *4*, 8, 9, 11, 12, 15, 16
Serpula lobata, 236
serratum, Laevicardium, 355
serratus, Cylindralithus, 14, 22, 27
Sethocorys bussonii, 278
  cristata, 278
Seven Rivers Cave, *426*, 429, 436
Seven Rivers Inlier, 380, 389
Seven Rivers region, 380, 385, 394, 398, 402
sezekorni, Erophylla, 428, 434
  Phyllonycteris, 434
shales, 21, 29, 30, 34, 35, 67, 68, 69, 71, 79, 80, 82, 84, 95, 127
Sheep Pen Cave, 418, 419, 420, 422, *426*, 436, 437
shells, 354, 363
  beds, 192, 382
  debris, 79
  land snail, 353
  snail, 422
Shepherd's Hall Formation, 10, 20, 21, *24*, 26, 30, 32, *34*, 45, 52
sherwoodensis, Lepidocyclina (*Lepidocyclina*), 317
siakensis, Globorotalia, 207
  Polyperibola, 189, 193
sicanus, Globigerinoides, 262
Sicily, 195
Sigmoilopsis schlumbergeri, 222, 227, *240*
Sign Beds, 180
Sign Formation, 127
Sign Member, 258, 259
signum, Corollithion, 8, 11, 14, 22, 26
signus, Discoaster, 137, 149
Sigsbee Knolls, 166
silica, 259, 262, 264
silt, 347
siltstones, 79, 84
Simojovel Formation, 446
simplex, Dorcadospyris, 270, 275
  Echiopsis, 373, 374, 381
  Reusella, 240
sincera, Hyalosagda, 356
  (*Stauroglypta*), 354
sinecarina, Cancris scintillans, 235, 250
sinkholes, 418
sinosus, Heterorhabdus, 26
  Macropneustes, 397, 402, 403
sinuata, Pleurodonte, 356, 358
sinuosus, Cretarhabdus, 22, 24, 26
Siphogenerina advena, 240
  ornata, 240
  multicostata, 240
siphonifera, Globigerina, 208, 210
  Hastegerina, 189, 208, 210
Siphonina tenuicarinata, 240, 245
  spp., 227

Siphoninella soluta, 222, 227, *240*, 244
Siphonodosaria modesta, 240
Siphonorhis, *432*
  americana, 419, 428, 432
  brewsteri, 432
Siphotextularia spp., 227
Siphouvigerina porrecta, 226, 227, 228, 229, 230, 233, 234, *240*, 247
sirenoides, Prorastomus, 414
Sismonda, 389
  crustula, 389
  logotheti, 389
  occitana, 389
sissinghii, Quadrum, 8, 9, 11, 12, 15, 16, 27
  Uniplanarius, 23, 24, 25, 27
site, deep sea, 163
Slide Roadcut outcrop, 134, 141, 42, 144, 146, 163, 180, 182, 189, 190, 221
Slippery Rock Formation, 10
slumping, 146
smaricus, Archaias, 287, 305, 306
snails, land, *353*, *359*, 360, 422, 425
soldadensis, Ranikothalia, 329, 331
soldanii, Nonion, 238
  Noniona, 238
Solenodon, 423
Solenosphaera omnitubus, 260, 266, *267*
  sp., 266
solidus, Holodiscolithus, 137, 177
solitus, Chiasmolithus, 147
Sollasites horticus, 23, 27
soluta, Siphoninella, 222, 227, *240*, 244
  Truncatulina, 240
Sombrenito Formation, 235
Somerset Limestone, 295, 297, 301, 307, 317, 395
Somerton area, 405
South America, 414, 423, 433
  northern, 347
South Atlantic, 159
South Atlantic Basin, 163
South Carolina, 106, 388, 398, 413
Southampton, 123
Spanish River drainage system, 30
Spanish Town area, 429
sparcilirata, Prebarrettia, 35, 37, 38, 39, 40, 41, 45, 46, 57
species
  calcareous nannofossil, *26*
  imperforate conical, *286*
  land snail, *354*, *356*
  See also specific species
specillata, Arkhangelskiella, 22, 24, 26
Speotyto cunicularia, 432
Speoxenus, 437
  cundalli, 437
Sphaeraster, 266, 267
Sphaerodactylus, 419
  sp., 423

Sphaerogypsina, 309
　globula, 286, 308, 309
　peruviana, 309
sphaeroides, Cassidulus, 387, 394
　　Melonis, 222, 224, 227, 231, 232, 238, 244
　　Sphaeroidinellopsis, 214
Sphaeroidina bulloides, 212, 227, 229, *241*
　dehiscens, 212
　immatura, 180, 182
Sphaeroidinella, 212
　dehiscens, 213
　　dihiscens, 214
　　immatura, 192, 193, 194, 195, 202, 212
　ionica ionica, 212
　multiloba, 214
　rutschi, 214
Sphaeroidinellopsis, 191, *214*
　dehiscens, 187, 191, 195, 198, 214
　disjuncta, 214, 262
　hancocki, 214
　immatura, 180
　kochi, 214
　multiloba, 184, 187, 189, 193, 195, 213, *214*
　paenedehiscens, 214
　seminulina, 184, 185, 187, 193, 194, 195, 213, *214*
　sphaeroides, 214
　subdehiscens, 185, 214
　　paenedehiscens, 182, 191, 214
　　subdehiscens, 197
Sphenolithus abies, 137, 144, 145, 150, 153, 162, 177
　belemnos, 139, 146, 147
　furcatolithoides, 147
　heteromorphus, 136, 137, 138, 139, 140, 144, 145, 146, 147, 152, 177, 184, 197
　neoabies, 137, 140, 142, 144, 145, 150, 152, 153, 162
　predistentus, 137, 146, 147
　radians, 137, 147
　subdehiscens subdehiscens-Globorotalia druryi zone, 197
　spp., 139
spicules, sponge, 259, 262, 263
spinea, Ehrenbergina, 227, 237, 244
spinidentatus, Prionocidaris, 375
spinifera, Cassidulina, 222, 227, 230, 236, 243
　Hydrobia, 365
　Melania, 365
spinipes, Mincopanope, 120, 121, 122
Spinoleberia sp., 70, 75
spinolosa, Reusella, 229, 240
　pulchra, Reusella, 227, 228, 229, 230, 231, 233, *240*, 246
spinosa, Prediscosphaera, 23, 27
spinosissima, Ehrenbergina, 222, 227, 237, 250
spinosissimus, Mithrax, 120, 122
spinosus, Echitricolporites, 446, 447
　Fenestrites, 446, 447

spiralis, Zygodiscus, 4, 23, 27
　lobatus, Nautilus, 236
Spiroclypeus, 329
Spirodontomys, 437
　jamaicensis, 437, 439
Spirosigmoilinella sp., 222, 227
spissus, Pachydiscus, 80, *89*
　(*Pachydiscus*), 83, 84
splendens, Rhagodiscus, 8, 11, 12, 15, 23, 27
Sponge sclerite, 266
Spongocore, 271
　puer, 260, 271, 272
　vellata, 271
　(*Spongocorisca*) *puer*, 271
(*Spongocorisca*) *puer, Spongocore*, 271
spongolites, 263
Spongoplegma sp., 267
sponosa echinoides, Acrosphaera, 260, 265
Spring Garden Formation, 159, 165, 186, 188, 258, 401
Spring Garden Member, 132, 134, 136, *140*, 142, 143, 146, 148, 152, *155*, 180, 184, 189, 194, 197, 200, 202, 206, 208, 214, 220, *221*, *223*, 225, 226, 229, 232, 233, 257, 263
Spring Garden section, 233
Spring Mount, 99, 116
Spring Mount region, 375, 377, 379, 380, 381, 384, 385, 386, 391, 393, 395, 396, 399, 402, 403, 406, 407
Springfield area, 398, 399, 402, 405, 406
Spumellaria sp., 266
stabilis, Dendrospyris, 260, 272, 274
stainforthi, Catapsydrax, 257
stalis, Helicosphaera, 139
stallauensis, Parapachydiscus, 89
Stapleton Formation, 10, 20, 30, 32, 34, 45, 52, 102
Stapleton Limestone Member, 34
staurion, Cruciplacolithus, 147
(*Stauroglypta*) *brevior, Hyalosagda*, 354
　brevis, Hyalosagda, 354
　sincera, Hyalosagda, 354
stauropora, Liriospyris, 259, 260, 272, 275
　Trissocyclus, 275
Stenechinus, 377
　perplexus, 373, 374, 377, 379
　regularis, 373, 374, 376, 377
stenopetalus, Macropneustes, 402, 403
stenopetula, Haimea, 373, 384, 386
stephensoni, Athecocyclina, 286, 337, 338, 339,
　Cphthalmoplax, 116
　Discocyclina, 337
　Pseudophragmina (*Althecocyclina*) *stephensoni*, 337
sterea, Cyclaster, 392, 401
(*Stereocidaris*), *Temnocidaris*, 94, 95, 377

Sterraster, 266
Stettin Member, 285, 289, 295, 299, 395, 399, 414, 448
Stichocorys, 277
　coronata, 259, 260, 270, 272, 277
　delmontense, 277
　delmontensis, 257, 260, 270, 277, 278, 279
　peregrina zone, 265, 278
　wolffii, 257, 260, 270, 277, 278
　zone, 259, 275, 276, 278, 279
Stichopodium cienkowskii, 277
Stilostomella aculeata, 226, 227, 228, *241*
　curvatura, 227, *241*
　modesta, 227, *241*
　subspinosa, 227, *241*
　spp., 227
Stony Hill, 351
Stony Hill Road section, *155*, *156*, 159, *194*
Stony River, 30
Stoverius biarcus, 11, 15
stradneri, Microrhabdulus, 23, 27
　Tegumentum, 9, 11, 12, 15
　Vekshinella, 12, 15
stramineus, Natalus, 435
　jamaicensis, Natalus, 435
stratigraphy, *5, 20, 77, 84, 105, 231, 446*
　calcareous nannofossil, *131, 132, 140, 147*, 151, *156*
　planktonic foraminiferal, *156*, 164
striata, Dimorphina, 240
　Rectuvigerina, 222, 224, 227, 231, 233, *241*, 245
striatoreticulata, Camerina, 331
striatoreticulatus, Nummulites, 287, 331, 333, 334, 336
Striatricolpites catatumbus, 446, 447, 450
striatus, Triquetrorhabdulus, 137
　bukryi, Cretarhabdus, 26
Strix cunicularis, 432
strobilata, Eucidaris, 377
strongyla, Echinolampas, 373, 391, 393, 394
stuarti, Globotruncana, 34
Stylatractus santaennae, 267
　universus, 269
Stylocidaris affinis, 377
stylodendra, Thecosphaera, 260, 266, 267
stylophora, Ceratospyris, 274
Stylosphaera, 267
　angelina, 260, 268, 269
　hispida, 267
　(*Stylosphaerella*) *angelina*, 269
　sp., 269
(*Stylosphaerella*) *angelina, Stylosphaera*, 269
Subbotina triangularia, 207
subcircularis, Inoceramus proximus, 84, 85
subconglobata, Globigerinatheka, 285
subcylindricus, Schizaster, 373, 374, 396, 397, 398

subdehiscens, Sphaeroidinellopsis, 185, 214
  subdehiscens, 197
  paenedehiscens, Sphaeroidinellopsis, 182, 191, 214
  subdehiscens, Sphaeroidinellopsis, 197
subglabra, Gaudryina, 238
  Karreriella, 227, *238*
subglobosa, Cassidulina, 238
  Globocassidulina, 225, 226, 227, 228, 229, 230, 237, *238*
sublittoral littoral zones, 115
sublodoensis, Discoaster, 146, 147
suborbicularis, Heterostegina, 335
subplana, Callianassa, 118, 119, 120
subquadratus, Globigerinoides, 148, 184, 185, 186, 188, 197, *210,* 211
  zone, 188
subscitula, Globorotalia (Turborotalia), 204
subsidence, 30, 223, 444
subspinosa, Ellipsonodosaria, 241
  Stilostomella, 227, *241*
subsurculus, Discoaster, 169
Sulcoperculina, 32, 34, 35, 37, 38, 40, 41, 44, 45, 47
Sulcorbitoides, 47, 48
Sunderland Formation, 10, 21, *24,* 26, 30, 32, 44, 99
Sunderland Inlier, *10,* 13, *19,* 29, 30, *32, 34,* 44, 45, 48, 50, 52, 54, 67, 102
Sunderland–Newman Hall formational contact, 24
Sunderland section, *32*
supracretacea, Watznaueria, 23, 27
supremus, Goniopygus, 93, 94, 95, 96, 97, 98
surculus, Discoaster, 137, 138, 139, 142, 146, 149, 150, 153, 154, 158, 162, 165, 169
surirellus, Cretarhabdus, 22, 24, 27
suturalis, Orbulina, 184, 197
sverdrupi, Encope, 373, 390, 391
Swan Islands, 423, 429
Swansea Cave, 419, *426,* 432, 433, 436, 439
Swanswick Formation, 105, 106, 113, 123, 129, 375, 377, 384, 385, 389, 395, 396, 398, 402
Swanswick Limestone, 291, 299, 301, 307, 309, 316, 317, 333, 335, 337, 393
Sweden, 88, 89
Swift River drainage system, 30
Swift River valley, 44
synclines, 35
synquadriperforatus, Chiastozygus, 22, 26
Syracosphaera, 139
  histrica, 137
  sp., 177
systematic descriptions, *286*
  ammonites, *86*
  echinoidea, *94*

systematic paleontology, *125, 198,* 265, *427*

T

talus deposit, 180
tamalis, Discoaster, 135, 137, 138, 139, 142, 144, 145, 146, 150, 153, 162, 165, *168*
Tampico Region, Mexico, 396
Tangle River area, 73
tarboulensis, Zygodiscus, 27
Tarphypygus, 373, *389*
  ellipticus, 373, 374, 389
  notabilis, 373, 374, 387, 389
  sp., 373, 389
tartari, Lithomelissa, 273
taxonomy, *165, 233, 255*
  benthic foraminiferal, 223
tectiformis, Bolivina, 234
Tegumentum stradneri, 9, 11, 12, 15
Temnocidaris, 95
  (Stereocidaris), 94, 95, 377
Tennessee, 116, 446
tenuetuberculata, Baculites capensis, 88
tenuetuberculatus, Baculites, 84, 87, 88
tenuicarinata, Siphonina, 240, 245
tenuis, Rhabdosphaera, 147
Terebratella sp., 111
Terebratula bartletti, 108
  cubensis, 108
  cuneata, 110
  floridensis, 108
  lecta, 108
Terebratulina, 105, *111,* 112, 113
  caput-serpentis, 113
  palmeri, 113
  sp., *111,* 112, 113
terrace, raised-reef, 118
terrapen, Pseudemys, 427
  Testudo, 427
  Trachemys, 419, 427, 428
terrazetus, Ottavianus, 23, 27
Tertiary, *347, 413, 443, 446, 451, 454, 459*
tessellatus, Tholocubus (Tholocubus), 273
Testudo terrapen, 427
Tethys Sea, 414
(Tetragoniopygus), Goniopygus, 98
Tetragonites, 86
  epigonus, 86
  garuda, 84, 86, 87
Tetrahedrina elongata, 277
tetrapera, Cyrtocapsa (Cyrtocapsella), 276
  Cyrtocapsella, 257, 260, 262, 270, 276, 277
Tetrapodorhabdus decorus, 12, 15, 23, 27
texana, Heterostegina, 337
Texas, 433
Textularia spp., 227
thalmanni, Bolivina, 222, 227, *235,* 246

Thamnospyris schizopodia, 274
  sp., 274
Thecidae mediterranea, 106
Thecidium mediterraneum, 106
Thecosphaera, 267
  stylodendra, 260, 266, 267
Thelidomus, 354
Theocyrtis annosa zone, 269
  veneris zone, 269
Thetford area, 424
Thicket River Member, 68, 71, *73*
Tholocubus, 273
  (Tholocubus) tessellatus, 273
  sp., 260, 266, 268, 273
  (Tholocubus) tessellatus, Tholocubus, 273
Tholospyris, 275
  anthophora, 260, 272, 275, 276
  mammillaris, 260, 272, 276
  sp., 272
thoracatus, Geocapromys, 437
Thoracosphaera sp., 12
Thyrastylon, 35, 37, 38, 45, 46, 67
Thysanophora sp., 359
Tiber River Formation, 13
Tichosina, 105, 108, 109
  sp., *108,* 109
tiliaceus, Hibiscus, 446, 450
Tinoporus pilaris, 309
titan, Aristelliger, 418, 419, 428, 429, 430
Titanosarcolites, 10, 35, 37, 38, 39, 40, 41, 44, 45, 46, *65, 66,* 67, 68, 69, 71
Titanosarcolites Limestone, 97, *98,* 99, 102, 129
tobleri, Miscellanea, 329
  Ranikothalia, 331
Tomistoma schlegelii, 413
Tonatia, 433
  bidens, 419, 434
    saurophila, 428, 419, 430, 433, 434
  saurophila, 434
Torreites sanchezi, 32, 34, 45, 46
tortuosa, Bolivina, 222, 227, *235,* 244
Toweius gammation, 137, 147
  magnicrassus, 137, 147
toyajoanus, Inoceramus balticus, 84
trabeculatus, Helicolithus, 8, 11, 12, 15
Trachemys, 427
  terrapen, 419, 427, 428
Tranolithus bitraversus, 23, 27
  gabalus, 23, 27
  orionatus, 23, 27
  zone, 26
  phacelosus, 8, 9, 11, 12, 13, 15
transformata, Acrosphaera, 160, 265, 266
transport, 227
transversa, Probolarina, 106
  Rectuvigerina, 224
trechmanni, Callianassa, 118, 119, 122
  Linthia, 373, 374, 397, 399, 406
  Metholectypus, 93, 94, 95, 99, 100

trechmanni (continued)
   Notopocorystes (Cretacoranina), 116, 117, 120
   Parastroma, 32, 45, 46
   Pseudorbitoides, 30, 32, 34, 35, 41, 45, 46, 47, 48, *50*, 51, 52, 53, 54
      trechmanni, 50
   pectinata, Pseudorbitoides, 48, 52
   trechmanni, Pseudorbitoides, 50
trelawniensis, Yaberinella, 301, 303
Trelawny Parish, 118, 122, 123, 127, 382, 388, 395, 399, 414, 418, 419, 424, 425, 426, 429, 431, 432, 435, 436, 437, 439, 448
Trematodiscus microporus, 271
Tretocidaris anguillensis, 375
(Tretocidaris), Cidaris, 375
   anguillensis, Cidaris, 375
Triadechinus, 380
   multiporus, 373, 374, 376, 380
triangularia, Subbotina, 207
Tribrachiatus orthostylus, 137, 146, 147
tribuloides, Eucidaris, 373, 376, 377, 378
Trichechus manatus manatus, 414
tricorniculatus, Amaurolithus, 131, 139, 149, 150, 152, 154, 155, 158, 163
tricornus, Lapideacassis, 23, 27
Trifarina bradyi, 222, 227, *241*
trifidum, Quadrum, 2, *3*, 4, 5, 8, 9, 10, 12, 13, 15, 16, 66, 67
trifidus, Uniplanarius, 23, 24, 25, 26, 27
triloba, Globigerinoides triloba, 212
   triloba, Globigerinoides, 212
trilobatus, Globigerinoides quadrilobatus, 212
trilobus, Globigerinoides, 184, 187, 189, 193, 210, 211, *212*
Trinidad, 108, 189, 256, 262, 263, 265, 321, 341, 351, 385, 433
   southern, 185
trinitatensis, Bulimina, 235
trinitatis, Lepidocyclina (Lepidocyclina), 317
Tripneustes sp., 373
tripodiscus, Tholospyris, 275
Triquetrorhabdulus carinatus, 146, 147
   rugosus, 137, 138, 139, 144, 145, 146, 149, 162, 184, 187
   striatus, 137
triradiatus, Discoaster, 137
Trisolenia, 267
   megalactis, 267
      costlowi, 260, 266, 267
      megalactis, 260, 267
Trissocyclus stauropora, 275
Trochalosoma, 380
   chondra, 93, 373, 374, 380, 383, 389
   rugosum, 381
Trondheim, Norway, 15
(Tropicorbis) decepiens, Amerigus, 366
   edentatus, Amerigus, 366

Tropidiphis, 419, 428
   haitanus, 428
Trout Hall area, 120
Troy Limestone, 307, 333
Trubi Formation, 195
truitti, Nannoconus, 8, 11, 12, 15
truncata, Borelis jamaicensis, 297
Truncatulina bradyi, 236
   dutemplei, 236
   lobata, 236
   mundula, 237
   pachyderma, 237
   robertsoniana, 237
   soluta, 240
   tuberculata, 236
truncatulinoides, Globorotalia, 191, 192
   (Globorotalia), 182
tubaria, Didymocyrtis, 260, 269, 270
   Pipetteria, 269
tubarius, Cannartus, 269
tuberculata, Truncatulina, 236
tuberculifera, Cytherella, 69, 74, 75
tuberosum, Lithotympanium, 276
tubifera, Scyphosphaera, 137
tuffs, 30, 34
tumida, Globorotalia, 191, 194, 196, 198, 200, *202*, 203
   (Globorotalia), 202
   Pulvinulina, 202
turbidite, 337, 339
(Turbinulina) elegans, Rotalia, 238
(Turborotalia) continuosa, Globorotalia, 206
   crassaformis, Globorotalia, 204
   crassula, Globorotalia, 206
   peripheroacuta, Globorotalia, 202
   peripheroronda, Globorotalia, 202
   scitula, Globorotalia, 204
   subscitula, Globorotalia, 204
Turdus aurantius, 428
   jamaicensis, 428
turriseiffelii, Eiffellithus, 8, 9, 11, 12, 15, 22, 27
tuxpamensis, Bulimina, 222, 227, *235*
Two Meeting section, *10*, 14
Tydixon Cave, 419, *427*, 436, 439
tympaniformis, Fasciculithus, 146, 147
Typhlops, 419
   sp., 428
typicus, Inoceramus (Endocostea), 84, 85

U

Uca, 120, 122
   sp., 122
Uhlias, 119, 120, 122
   limbatus, 119, 120, 122
Ulster Spring, 118, 122
umbilica, Reticulofenestra, 137, 146, 147
Umbilicosphaera, 166
   aequiscutum, 137, *166*, 175
   cricota, 137, *166*, 175
   hulburtiana, 166

umbonifera, Epistominella, 239
   Nuttallides, 222, 227, 232, *239*
   Pulvinulinella, 239
unconformity, 140, 146, 158, 159, 163, 168, 180, 182, 184, 185, 190, 194, 196, 197, 200, 204, 207, 214, 225
undosa, Eulepidina, 323
   Lepidocyclina, 323
      (Eulepidina), 287, 319, 320, 321, 323, 325
      (Nephrolepidina), 323
unicolor, Brissus, 373, 374, 392, 401
Uniplanarius gothicus, 27
   sissinghii, 23, 24, 25, 27
   zone, *25*
   trifidus, 23, 24, 25, 26, 27
   zone, *25*
universa, Orbulina, 144, 145, 185, 198
universus, Stylatractus, 269
uplift, *219*, 444
Upper Coastal Group, 126
Upper Cretaceous, *29*, 77
Upper Red Beds, 347
Upper White Limestone, 132
Upper White Limestone Group, 131
Urocoptis brevis, 354, 356
   (Autocoptis), 359
Uscarl Formation, 445
Uvigerina carapitana, 222, 227, 231, *241*
   coartata, 237
   cojimarensis, 227
   compressa, 237
   guazamalensis, 222
   hispida, 222, 227, *241*
   hispido-costata, 222, 227, *241*, 250
   jamaicensis, 222, 227, *241*, 250
   laviculata, 222, 227, *241*
   parvula, 241
   pigmea, 222, 227, 229, 230, *241*, 247
   porrecta, 240
   proboscidea, 227, 228, 230, *241*, 247
   selseyensis, 234
   spp., 222, 227, 228, 229, 231, 233, *241*, 247

V

Vaccinites, 32, 45, 46
   eyrei, 79, 80
Vagalapilla matalosa, 9, 11, 15
valida, Pleurodonte, 354, 356
Valudayur Beds, 86
Valvata inconspicua, 364
   pygmaea, 364, 365
Valvulineria nuttalli, 235
   scintillans, 235
variabilis, Brevitricolpites, 446
   Discoaster, 137, 150, 154, 169
   Globorotaloides, 208
varians, Poteria, 425
Varuna, 122
   sp., 116, 118, 122

vaughani, Antillaster, 407
　Fabularia, 286, 297, 299, 301, 302
　Lepidocyclina (Nephrolepidina), 320, 321
　Plectofrondicularia, 222, 224, 227, 232, *239*
Vaughanina, 29, 37, 38, 40, 41, 44, 47, 48, 52, *54*
　cubensis, 35, 37, 45, 46, *54*, 55, 56
Vaughnsfield Formation, 67, 68, 71, *73*
veatleyi, Cyclas, 369
　Eurepa, 365, 369
vedderi, Helicosphaera, 177
Vekshinella angusta, 8, 15
　ara, 23, 27
　stradneri, 12, 15
　sp., 23, 27
vellata, Spongocore, 271
veneris, Calocyclas (Calocycletta), 279
Venezuela, 185, 188, 190, 220, 229, 351
Venezuela Basin, 256
venezuelana, Dentoglobigerina, 184, 187, 189, 193, 195, 197, *208*, 209
　Globigerina, 208
　Globoquadrina, 207, 208
Veniella, 68
Veniella Shales, 102, 116, 120, 380
ventricosa, Globotruncana, 34
verbeekii, Ceratolithoides, 7, 8, 9, 12, 14, 22, 24, 25, 26
Verneuilina bradyi, 237
　spp., 227
Verseyella, 295
　jamaicensis, 285, 286, 289, 292, 295, 298, 300
verseyi, Fabularia, 285, 287, 297, 298, 301, 304
versicolor, Geotrygon, 428
Vertebrata, *413*, *417*
vertebrate strata, sequences, *420*
Victoriaster, 399
　jamaicensis, 399
(Victoriaster) jamaicensis, Pericosmus, 399
Victoriella, *307*, 309
　conoidea, 287, 307, 308, 309
violacea, Loxigilla, 428
violina, Cannartus, 269, 271
　Didymocyrtis, 260, 262, 269, 270, 271
　Haliomma didymocyrti, 269
virginia, Calocyclas (Calocycletta), 279
virginica, Crassostrea, 118, 121, 123
virginis, Calocycletta, 260, 270, 279
viridis, Echinometra, 373, 382
vitrea, Anomia, 108
vitreus, Gryphus, 108
Vlerkina, 335
　(Vlerkina) antillea, Heterostegina, 305, 337, 338
　　ocalana, Heterostegina, 338
　(Vlerkinella) kugleri, Heterostegina, 335, 337

vortex, Anisus, 366
Vulvulina alata, 234
　spp., 227

W

Wagwater Belt, 285, 331
Wait-a-Bit Cave, 395, 399, 448
Walderston Limestone, 291, 305
Wallingford Roadside Cave, 418, 419, 420, 422, *427*, 429, 431, 432, 433, 434, 435, 436, 437, 439
Wallywash Great Pond, *363*, 364, 366, 367, 369
waltersdorfensis hannovrensis, Inoceramus, 80, 82, 84, 85
waringi, Hercoglossa, 351
waterhousii, Macrotus, 428, 433
waters, marine, 450
Watznaueria barnesae, 23, 24, 27
　biporta, 8, 9, 11, 12, 15, 23, 27
　ovata, 23, 27
　quadriperforata, 23, 27
　supracretacea, 23, 27
waylandvaughani, Lepidocyclina (Lepidocyclina), 317
Welcome Hall, 379, 398, 399, 402
wells
　corehole, 5
　Hertford, 1, 2. 3, 4, *5*, 13, 15
　Punta Gorda-1, 303
　Retrieve, 1, 2, 3, 4, *6*, 13
　Twara-1, 303, 325
　Windsor, 1, 2, 4, *10*, 15
wellsi, Eoconuloides, 286, 311, 312, 314
West Africa, 69
West Indian Islands, 347
West Indies, *29*, 351, 353, 356, 359, 377, 386, 401, 419, *420*, 432, 433, 435
West Roadcut outcrop, 134, 142, 180, 184, 190, 191, 200, 206, 208, 210, 212, 214, 221
West Town River bank, 88
Westmoreland Parish, 66, 67, 73, 99, 127, 129, 431
wetherbyi, Oligopygus, 373, 374, 383, 384
White Limestone, 311
　upper, 165, 166
White Limestone Formation, 122, 132
White Limestone Group, 123, 283, 285, 295, 317, 321, 347
White Limestone Supergroup, 127, 129, 371, 375, 384, 385, 389, 390, 393, 394, 395, 396, 398, 401, 402
White Marl site, 431, 437, 439
White River, 256
wiedenmayeri, Gansserina, 66, 67
wilcoxensis, Alabamina, 222, 227, 230, *233*
willcoxi, Nummulites, 333
　Operculinoides, 317
　Palaeonummulites, 287, 331, 333, 334, 336, 337
　Ranikothalia, 331

windows, erosional, 1
Windsor area, 418, 425, 426, 432
Windsor Shale, 80, *82*, 84, 95
wolffii, Stichocorys, 257, 260, 270, 277, 278
woodi, Globigerina, 212
Woodland Shale, 67, 73
woolacotti, Carcineretes, 116, 117, 120
Worthy Park, 418, 419, 424, 425, 426, 427, 437
wuellerstorfi, Anomalina, 239
　Cibicides, 239
　Planulina, 222, 225, 227, 231, 232, *239*
Wythella, 390
　eldridgii, 390
　sp., 373, 374, 387, 390

X

Xanthilites, 118, 122
　rathbunae, 117, 118, 122
Xanthopsis sp., 118
Xenicibis, *431*
　xymphithecus, 418, 419, 425, 426, 428, 430, 431, 432, 433
Xenothrix, 425, *435*
　mcgregori, 418, 419, 425, 428, 430, 432, 435, 436
xymphithecus, Xenicibis, 418, 419, 425, 426, 428, 430, 431, 432, 433

Y

Yaberinella, 285, *301*
　hottingeri, 286, 301, 303, 306
　jamaicensis, 285, 287, 289, 303, 304
　trelawniensis, 301, 303
　spp., 297, 303
Yallahs River Valley, 377, 379, 380, 384, 385, 386, 393, 395, 396, 403, 406, 407
Yellow Limestone Group, 20, 110, 116, 118, 123, 129, 283, 289, 295, 303, 311, *347*, 371, 375, 377, 379, 380, 381, 382, 384, 385, 386, 388, 389, 393, 394, 395, 396, 398, 399, 402, 403, 405, 406, 407, 414, 446
yokoyamai, Baculites, 80, 82
Yugoslavia, 299
yumuriana, Angulogerina, 234
yurnagunensis, Lepidocyclina, 305, 317, 319, 321
　(Lepidocyclina), 319
　(Nephrolepidina), 287, 319, 320, 321, 324

Z

Zanclean Deluge, 196
Zanclean Stage, 195
zansi, Coleiconus, 286, 287, 289, 290, 292, 295

(Zeaglobigerina) druryi, Globigerina, 212
  nepenthes, Globigerina, 212
zealandica, Globorotalia, 202
Zeugrhabdotus embergeri, 8, 9, 11, 12, 15, 23, 27
  pseudanthophorus, 8, 9, 11, 12, 15
  sp., 11, 23, 27
zones
  calcareous nannofossil, 2, 24, *146*

zones (continued)
  calcareous nannoplankton, *1*, 189
  planktonic foraminiferal *146, 159, 190*
  sublittoral littoral, 115
Zonocostites ramonae, 447
(Zonocostites) ramonae, Rhizophora, 446, 450
Zygodiscus acanthus, 23, 27

Zygodiscus (continued)
  diplogrammus, 23, 27
  macleodae, 23, 27
  minimus, 27
  pontilithus, 23, 27
  spiralis, 4, 23, 27
    zone, 4
  tarboulensis, 27
  sp., 23, 27
Zygrhablithus bijugatus, 137, 147

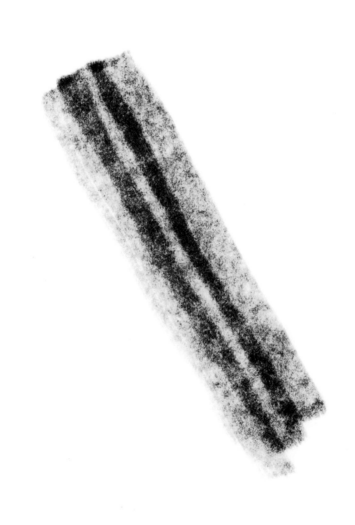